COMBUSTION EMISSIONS

COMBUSTION EMISSIONS

Formation, Reaction, and Removal of Trace Metals in Combustion Products

KEITH SCHOFIELD

Materials Research Laboratory, University of California Santa Barbara, Santa Barbara, CA, United States

ELSEVIER

ACADEMIC PRESS

An imprint of Elsevier

Academic Press is an imprint of Elsevier
125 London Wall, London EC2Y 5AS, United Kingdom
525 B Street, Suite 1650, San Diego, CA 92101, United States
50 Hampshire Street, 5th Floor, Cambridge, MA 02139, United States
The Boulevard, Langford Lane, Kidlington, Oxford OX5 1GB, United Kingdom

Notices

Knowledge and best practice in this field are constantly changing. As new research and experience broaden our understanding, changes in research methods, professional practices, or medical treatment may become necessary.

Practitioners and researchers must always rely on their own experience and knowledge in evaluating and using any information, methods, compounds, or experiments described herein. In using such information or methods they should be mindful of their own safety and the safety of others, including parties for whom they have a professional responsibility.

To the fullest extent of the law, neither the Publisher nor the authors, contributors, or editors, assume any liability for any injury and/or damage to persons or property as a matter of products liability, negligence or otherwise, or from any use or operation of any methods, products, instructions, or ideas contained in the material herein.

British Library Cataloguing-in-Publication Data
A catalogue record for this book is available from the British Library

Library of Congress Cataloging-in-Publication Data
A catalog record for this book is available from the Library of Congress

ISBN: 978-0-12-819126-2

For Information on all Academic Press publications
visit our website at https://www.elsevier.com/books-and-journals

Publisher: Joe Hayton
Acquisition Editor: Maria Convey
Editorial Project Manager: Naomi Robertson
Production Project Manager: Nirmala Arumugam
Cover Designer: Greg Harris

Typeset by MPS Limited, Chennai, India

Contents

About the author ix
Preface xi

1. Introduction

2. Recent research concepts

3. The generalities that now are apparent

4. The importance of the temperature, interaction time, and heterogeneous collision frequency constraints: fuel-rich or -lean differences

5. The nature of flame gases free of nonmetallic content

6. The role of particles, ash, or surfaces

7. Trace species reported combustion behavior and their chemistry

7.1 Alkalis (Li, Na, K, Rb, Cs)

Health concerns with the alkalis 25
Sodium and potassium's basic chemistry 25
Gas phase stabilities 28
Laboratory combustion studies: homogeneous conditions 31
Laboratory combustion studies: heterogeneous conditions 37
Basic fundamental discoveries 38
Initial implications 40
Manipulations 43
Alkali gaseous transport 44
Combustion power plant data 47

7.2 Aluminum

Health concerns with aluminum 55
Aluminum's basic chemistry 56
Gas phase stabilities and transport 58
Laboratory combustion related studies 63
Combustion power plant data 65

7.3 Antimony

Health concerns with antimony 68
Antimony's basic chemistry 69
Gas phase stabilities and transport 72
Laboratory scale combustion studies 74
Combustion power plant data 74

7.4 Arsenic

Health concerns with arsenic 78
Arsenic's basic chemistry 80
Gas phase stabilities and transport 82
Laboratory scale combustion studies 86
Combustion power plant data 87

7.5 Barium

Health concerns with barium 94
Barium's basic chemistry 95
The dissociation energy of BaO and its meaning 97
Gas phase stabilities and transport 100
Laboratory combustion studies 101
Combustion power plant data 105

7.6 Beryllium

Health concerns with beryllium 108
Beryllium's basic chemistry 109
The dissociation energy of BeO and its meaning 111
Gas phase stabilities and transport 113
Combustion power plant data 116

7.7 Bismuth

Health concerns with bismuth 118
Bismuth's basic chemistry 118
Gas phase stabilities and transport 120
Combustion power plant data 122

7.8 Boron

Health concerns with boron 125
Boron's basic chemistry 126
Gas phase stabilities and transport 128
Combustion power plant data 132

7.9 Cadmium

Health concerns with cadmium 135
Cadmium's basic chemistry 137
Gas phase stabilities and transport 139
Chemical kinetic constraints 141
Laboratory combustion studies 142
Combustion power plant data 143

7.10 Calcium

Health concerns with calcium 148
Calcium's basic chemistry 149
The dissociation energy of CaO and its meaning 151
Gas phase stabilities and transport 153
Laboratory combustion systems 156
Combustion power plant data 158

7.11 Chromium

Health concerns with chromium 162
Chromium's basic chemistry 163
Gas phase stabilities and transport 166
Laboratory combustion studies and gas phase stabilities 171
Chemical kinetic constraints 178
Combustion power plant data 180

7.12 Cobalt

Health concerns with cobalt 187
Cobalt's basic chemistry 188
Gas phase stabilities and transport 191
Laboratory combustion studies 193
Combustion power plant data 197

7.13 Copper

Health concerns with copper 200
Copper's basic chemistry 201
Gas phase stabilities and transport 203
Laboratory combustion studies 208
Combustion power plant data 209

7.14 Iron

Health concerns with iron 214
Iron's basic chemistry 216
Gas phase stabilities and transport 219
Laboratory combustion studies 222
Combustion power plant data 225

7.15 Lead

Health concerns with lead 231
Lead's basic chemistry 233
Gas phase stabilities and transport 235
Laboratory combustion studies 238
Combustion power plant data 238

7.16 Magnesium

Health concerns with magnesium 246
Magnesium's basic chemistry 247
The dissociation energy of MgO and its meaning 248
Gas phase stabilities and transport 252
Laboratory combustion studies 255
Combustion power plant data 256

7.17 Manganese

Health concerns with manganese 260
Manganese's basic chemistry 261
Gas phase stabilities and transport 264
Laboratory combustion studies 267
Combustion power plant data 268

7.18 Mercury

Health concerns with mercury 275
Mercury's basic chemistry 277
Gas phase stabilities and transport 279
Chemical kinetic constraints 281
Laboratory combustion studies 282
Combustion power plant studies 286

7.19 Molybdenum

Health concerns with molybdenum 294
Molybdenum's basic chemistry 294
Gas phase stabilities and transport 297
Potential role for molybdates 301
Laboratory combustion studies 303
Combustion power plant data 308

7.20 Nickel

Health concerns with nickel 310
Nickel's basic chemistry 311
Gas phase stabilities and transport 312
Laboratory combustion studies 318
Combustion power plant data 318

7.21 Phosphorus

Health concerns with phosphorus 324
Phosphorus's basic chemistry 325
Gas phase stabilities 327
Laboratory combustion studies 331
Molecular gas phase transport in combustion 334
Combustion power plant data 335

7.22 Scandium

Health concerns with scandium 337
Scandium's basic chemistry 337
Gas phase stabilities and transport 338
Combustion power plant data 342

7.23 Selenium

The important health aspects of selenium 344
Selenium's basic chemistry 348
Gas phase stabilities and transport 349
Combustion power plant data 353

7.24 Silver

Health concerns with silver 359
Silver's basic chemistry 360
Gas phase stabilities and transport 362
Combustion power plant data 365

7.25 Strontium

Health concerns with strontium 367
Strontium's basic chemistry 368
The dissociation energy of SrO and its meaning 370
Gas phase stabilities and transport 372
Laboratory combustion studies 375
Combustion power plant data 376

7.26 Thallium

Health concerns with thallium 379
Thallium's basic chemistry 379
Gas phase stabilities and transport 380
Combustion power plant data 383

7.27 Tin (Germanium)

Health concerns with tin 385
Tin's basic chemistry 386
Gas phase stabilities and transport 388
Combustion power plant data 391
Supplemental comments on Germanium (relative of Tin) 392

7.28 Titanium

Health concerns with titanium 394
Titanium's basic chemistry 395
Gas phase stabilities and transport 398
Laboratory combustion studies 400
Combustion power plant data 401

7.29 Tungsten

Health concerns with tungsten 403
Tungsten's basic chemistry 404
Gas phase stabilities and transport 405
Potential role for tungstates 408
Laboratory combustion studies 411
Combustion power plant data 415

7.30 Uranium

Health concerns with uranium 418
Uranium's basic chemistry 419
Gas phase stabilities and transport 420
Combustion power plant data 425

7.31 Vanadium

Health concerns with vanadium 428
Vanadium's basic chemistry 429
Laboratory combustion studies 435
Combustion pilot plant data 436

7.32 Yttrium

Health concerns with yttrium 440
Yttrium's basic chemistry 440
Gas phase stabilities and transport 442
Combustion power plant data 446

7.33 Zinc

Health concerns with zinc 448
Zinc's basic chemistry 449
Gas phase stabilities 451
Chemical kinetic constraints 454
Gaseous transport of zinc 454
Combustion power plant data 456

7.34 Zirconium

Health concerns with zirconium 461
Zirconium's basic chemistry 461
Gas phase stabilities and transport 463
Zirconium molecular spectroscopy 464
Laboratory combustion studies 465
Combustion power plant data 466

Bibliography 469
Index 591

About the author

The author was honored to be a Major Scholar, educated at the University of Cambridge where he had a very broad education in the sciences, taught by several Nobel laureates and tutors. His postgraduate thesis work was in the then new Department of Chemistry with Sir Morris Sugden, studying combustion chemistry. Moving to the United States, he joined Prof. Broida freshly located at the University of California in Santa Barbara, where the studies involved optical spectroscopy relating to chemical lasers. From there, he spent 2 years at the National Center for Atmospheric Research in Boulder, CO, where he embraced atmospheric chemistry. A brief stay of 18 months, with an emphasis on electronically excited species, at the Cornell Aeronautical Laboratory in Buffalo, NY, was interesting, but the climate returned him to the General Motors Defense Research Laboratories in Santa Barbara, CA, to become immersed in missile defense technology and also later auto emission concerns, before transferring back into the University of California in its Physics, Chemistry, and finally School of Engineering Departments. He has been a Consultant for the General Motors Defense Research Laboratories in Santa Barbara, CA, the US Department of the Air Force in missile defense, and for the British Coal Research Establishment concerning coal combustion emissions control before it closed. For 20 years, until displaced by current computer databases, he published for the Department of Energy Basic Energy Sciences a distributed bimonthly current awareness "*The Combustion Research Bulletin*." Final issues contained about 800 references every two months. The bulletin annotated all the current published references connected to combustion appearing in the various latest issues of the science and engineering journals. For the last 15 years he has been self-employed publishing papers on combustion emissions and flame deposition mechanisms, and now medical analyses concerning neurotoxicity and current neural epidemics affecting people. He holds three major patents. One involves a method to minimize alkali-induced corrosion in gas turbines, one a passive control method for mercury emissions from coal power plants, and now a new preventive medical blood test panel to monitor a small blood sample simultaneously for the six neurotoxin levels in people.

Preface

This book has survived along with the author since 2010. It started as a submitted paper for the 33rd Symposium on Combustion to be held in Beijing. It illustrated the new findings that were emerging concerning trace metal depositions in combustion. This was particularly for the alkalis, mercury, selenium, and arsenic that were all showing similar traits. Fortunately or not, the reviewers mistook it to be only a review and rejected it. On attending the symposium, I mentioned this to Prof. Norman Chigier, then editor of the Journal, *Progress in Energy and Combustion Science.* He was kind enough then to commission it for publication in an enlarged version. The contract was signed with Elsevier for a 50 typed page manuscript. As I began to modify it, I realized it had implications for the whole periodic table and moreover there were thousands of appropriate publications, all fortunately available through the excellent facilities of the University's library. It proved to be a very active field, especially now that there are significant concerns about the metals and their effects on health. As a result of it never being critically assessed before, the manuscript grew over the years, and was constantly being rewritten to remain current. In the meantime, Dr. Chigier retired, and it was decided by the new editor, Prof. Hai Wang that it was getting rather large for the Journal, especially with its thousands of references. Fortunately its value was realized, and publication as this book accepted. After initial reviews and suggested changes from a journal to book format, it has been rewritten again and is very current. Through this period, my own interests have been redirected toward the medical connections and specifically neurotoxicology, and the potential chemical metal roles in current neural epidemics. As a result, the book does stress, more than previously considered, the health risks of many metals and why it is important to control their potential combustion emissions to the environment. Since I seriously started a study of chemistry and the sciences now over 60 years ago, the volume of publications has grown astronomically. At present, the *Web of Science* database lists about 400 publications per month simply under the keyword *"mercury."* I have been privileged to read the progress of science over the years and witnessed the struggles and ingenuity of many individuals to continue to push forward our knowledge even though not often correct at times. There have been many real gems found in the older literature that were overlooked and never appreciated. I have tried to make amends for that fact. Science climbs over many mistakes constantly, but often there is some value and that is science's strength.

The book is meant to be a reference source of current awareness. I have tried to include all the important steps forward but the volume of literature now is so huge that innumerable publications could not be included and more recent work has tended to be favored in some cases. Nevertheless, it is noted that older works have formed the foundation for our understanding, although we have moved on leaving holes in the previous databases. What has become clearly apparent are several things, one is the meaning of "dew point." This term now has to be more carefully applied, it is simply the temperature when a deposit is seen. It can convey no chemical implications without further information. Another is the care that should be taken with thermochemical equilibrium calculations. They are useful only to the extent of the quality of their database and also are only a hypothetical assessment as to what might be possible. Without care they can introduce wild speculations of little value. Kinetic and temperature constraints can often produce very different realities. Another unfortunate trait is to introduce models that are speculative and are really hypotheses. They can gain the stamp of approval by publication and although never validated, if adequate, they become entrenched in usage without acceptance being questioned. This has happened with the chemistry of two elements included in this book and remains of concern to the author. Researchers should not delve into speculation unless there is at least some meaningful support. The greatest risk to science now appears to be the overwhelming volume of publications. This has to be forcing researchers into ever narrower fields of study simply to keep up with advances. The resulting outcome is that fewer and fewer can observe the interdisciplineral consequences and value of discoveries that may be going unnoticed. I fear many great ideas can easily be overlooked. This book hopes to bring some of these now to light.

Fortunately born a Brit with perseverance
Keith Schofield
June 2019

CHAPTER

1

Introduction

Combustion remains a major pillar of global economies by providing about 80% of the energy needs. Consequently, as may not be fully realized by many, this subject necessarily affects everyone in one way or another. Until the mid-20th century, major concerns centered mainly on optimizing the efficiencies of energy production. The physics and chemistry of the combustion processes were still largely academic but a foundation of understanding was increasingly being established. However, it was only when environmental concerns became more noticeably apparent in the form of acid rain and smog that a realization was accepted that the scientific and engineering aspects of combustion were important and had to be studied in much greater detail. This initiated major efforts aimed at understanding photochemical smog, NO_x and SO_2 chemistries, and resolving issues of control methods. Such programs have been extremely successful and by the beginning of this century it can be said that such polluting emissions can now be significantly controlled at an economically acceptable cost. However, during this period, the era of environmental concerns flourished covering all anthropogenic aspects, especially with many major classic examples such as Lake Erie in the United States being named a dead lake in 1970, particularly when one of its river tributaries caught fire. Also there was the 1978 evacuation of 36 square residential blocks in Love Canal near Niagara Falls in New York State due to extensive chemical pollution dumping. Such cases emphasized that other uncontrolled sources of pollution clearly were evident in society that could have significant medical and environmental consequences.

It was a result of such events that the United States Environmental Protection Agency (EPA) and corresponding agencies around the world turned their attention to examining the nature of many potential chemical pollutants and identify their sources. Around 1970 EPA commissioned many reports on trace species on which atmospheric sampling and analysis data were emerging. These included pollens and natural biological organisms, materials such as asbestos, radioactive elements, various metallic and nonmetallic elements, together with pesticides and other organic carcinogens (Miner, 1969). These studies formed initial reviews and were part of the foundation that became the 1990 Clean Air Act in the United States. Due to particular concerns with metals (Nriagu, 1988), carcinogenic and toxic metals such as As, Be, Cd, Co, Cr, Hg, Mn, Ni, Pb, Sb, and Se were included. Barium, thallium, and others, although toxic, were not included in this Act but were later present in the Governmental Safe Drinking Water Acts. All of these metals have a variety

Combustion Emissions
DOI: https://doi.org/10.1016/B978-0-12-819126-2.00001-7

of sources but as will be seen also can be present at various trace levels in fuels or be potential emitters in other high-temperature processes and so are relevant to this combustion review.

Early studies concerning the chemistries of these metals, particularly in the gas phase, were basically constrained by a lack of available experimental techniques. Their spectroscopy was the initial dominant monitor with the major emphasis concentrating on their presence in the burned gases. Such studies resolved the basic chemistry of flame gases and introduced quantitative flame photometric techniques and spectroscopic identifications (Herrmann and Alkemade, 1963; Mavrodineanu and Boiteux, 1965; Gaydon, 1974). However, at that time the lack of understanding of the overall chemical behavior of the trace metals in particular is clearly evidenced by the book entitled *Metal Vapors in Flames* that appeared in 1982. This significant contribution contained information, but predominantly only on the alkali and alkaline earths (Alkemade et al., 1982). Other metal behaviors remained unknown until kinetic studies involving metals began to appear a decade later (Fontijn, 1992). Moreover, chemical thermodynamic data also remained sparse for many metals as emphasis had been largely placed only on those elements relating more to defense and missile technologies (Chase, 1998). Without a reasonable chemical foundation or quantitative monitoring ability, it is difficult to understand combustion emission processes or establish control standards. Moreover, in order to enforce any such type of regulation, an ease of monitoring and cost-effective control is necessary. Although this is obvious, it has remained the reason why trace metals such as mercury and selenium, classified as pollutants in the 1990 Clean Air Act, still remain unregulated in coal burning facilities. It is only in recent years that the monitoring of mercury at the parts per billion level in the coal flue gases can now be measured with a certain reliability, and moreover that commercially viable control methods have finally begun to appear, yet remaining too costly in some cases. As a result, mercury's emission levels may soon be regulated for combustion power systems and the cement and other industries. The same cannot be said for selenium emissions in that it remains difficult to measure them in real time at such levels and there remain no acceptable control methods in place.

The occurrence of many trace metals or other sensitive elements such as boron is encountered with coals, and although relative concentrations can vary, typical average values are well documented as approximately indicated in Table 1.1 (Raask, 1985; Rowe and Steinnes, 1977; Germani et al., 1980; Kronberg et al., 1981; Xu et al., 2003; Hu et al., 2006; Wang et al., 2008c; Vejaati et al., 2010). They are present also in biomass fuels (Kaufmann et al., 2000; Lind et al., 2003) and other solid wastes (Sable, 1990; Becidan et al., 2010). Sewage sludges similarly can be a significant source (Reed et al., 2005; Han et al., 2006). Although raw diesel, heavy and bio-oils, also contain traces, removal methods are available for their separation (Maryutina and Soin, 2009; Bakkali et al., 2012). Petroleum coke and residual fuel oil are particularly rich in Ni and V (Lee et al., 1999a). Petroleum, natural gas or hydrogen generally are cleaned fuels. Generally, it remains common to consume the solid fuels in their raw state or now quite commonly in a cofired mode. Pretreatment has never been a particularly acceptable mode for solid fuels unless based on liquefaction or gasification. This stems from the fact that such pretreated fuels, particularly for coal, become very flammable and then introduce dangerous handling or storage problems. The gaseous or liquid fuels can be pretreated and become less of a concern

TABLE 1.1 Approximate concentration ranges of the major elements expressed as their oxide content in the fly ash and of representative trace metals in various fuel sources.

Element	US coals			Biomass	Municipal waste	Sewage Sludge
	Bituminous	Subbituminous	Lignite			
Ash major elements (wt.%)						
SiO_2	20–60	40–60	15–45	0.1–40	5–40	10–40
Al_2O_3	5–35	20–30	20–25	0.01–10	0.3–20	1–25
Fe_2O_3	10–40	4–10	4–15	0.01–0.5	0.5–5	1–30
CaO	1–12	5–30	15–40	0.1–10	1–20	1–40
P_2O_5	0.1–1.5	2	0.1–2	0.01–4	0.5–1.5	2–25
MgO	0.1–4	0.1–7	1–15	0.1–10	0.1–3	0.5–4
SO_3	0.1–12	0.1–2	0.3–15	0.01–3	0.2–0.4	0.2–12
Na_2O	0.3–3	0–2	0.1–5	0.01–0.6	0.1–0.7	0.1–3.5
K_2O	0.2–2.5	0–4	0.1–2.0	0.1–2.0	0.1–0.5	0.4–3
TiO_2	0.1–3.0	0.1–3.0	0.1–3.0	0.01–2.0	0.1–2.0	0.1–2.0
Cl,ppmw	500–8000	50–500	100–200	50–10,000	1000–14,000	300–1800
Trace metals (ppmw)						
Ag	0.02–2	0.02–2	0.2–2	0–2	5	2–10
As	0.5–30	1–30	1–30	0.2–10	1–40	1–300
B	5–400	50–250	5–200	1–30	5–50	10–100
Ba	20–2000	10–1000	20–1000	10–1000	20–500	50–500
Be	0.1–5	0.2	0.1–15	0–1.0	0.01–0.3	0.1–1.5
Bi	0.01–2.5	0.01–2.5	0.01–2.5	0.1–0.4	1–50	1–50
Cd	0.1–4	0.06–3	0.05–2	0.01–10	0.5–20	1–300
Co	1–70	0.5–20	0.5–30	0.4–10	1–50	3–200
Cr	0.5–60	2–30	6–40	0.2–100	8–300	20–500
Cu	5–100	10–100	10–200	1–100	10–1000	100–1000
Hg	0.02–1	0.05–1	0.1–1	0.01–0.4	0.01–2	1–20
Mn	5–500	5–300	15–300	20–1000	25–500	100–1000
Mo	2–40	5–40	5–40	1–6	10–40	1–100
Ni[a]	0.5–70	0.5–30	3–40	0.2–80	0.1–200	15–300
Pb	1–250	3–50	3–50	0.5–20	50–1000	20–500
Sb	0.05–10	0.1–0.2	0.2–2	0.1–1.0	10–60	0.4–7

(Continued)

TABLE 1.1 (Continued)

Element	US coals			Biomass	Municipal waste	Sewage Sludge
	Bituminous	Subbituminous	Lignite			
Sc	1–10	1–10	1–10	0–2	Low	1–10
Se	0.2–10	1–5	0.5–5	0.2–1.5	5–10	1–15
Sn	0.3–100	1–100	1–250	1–10	250	3–300
Sr	15–1000	20–200	100–600	10–500	50–500	100–500
Tl	0.01–1	0.5	0.5	0–0.02	0–0.05	0.05–3
U	0.5–3	0.5–3	0.5–3	0–0.5	0–2	2–3
V[a]	10–100	10–100	10–100	1–25	10–50	15–80
W	1–10	1–10	1–10	0–2	1–20	1–20
Y	3–25	3–25	3–25	–	–	5–25
Zn	5–200	5–100	5–100	5–100	40–4000	100–1000
Zr	5–200	5–100	5–100	2–10	5–100	20–200

[a] *Petroleum coke and residual oil fuels are rich in nickel and vanadium.*
Values based on summarizing approximately 100 published analyses.

regarding their trace metal content. Nevertheless, there are also those cases where metal additives are invoked with such fuels to initiate desirable combustion features. One analysis of fine particulate matter from US atmospheric data indicated various source categories as being major contributors for the various metals. This indicated notable contributions from the metal industry (Pb, Zn), the steel industry (Fe, Mn), Coal (As, Se), Oil (V, Ni), and from biomass (K) (Thurston et al., 2011).

Reading the older literature one becomes aware of this need for understanding, otherwise false concepts can easily become established. This was illustrated by publications that appeared in the mid-1990s and even earlier. These assessed the potential emissions of radioactive uranium, thorium, potassium, and radium from large coal-fired power plants. The implication was that these could be emitting more radioactivity than a nuclear power plant (Miner, 1969; Aly et al., 1994). Additionally, it was even suggested that such emissions of uranium to the atmosphere might breed the formation of plutonium at higher altitudes (Gabbard, 1995). It was never emphasized that these were not in fact potential gaseous emissions. As a result it is necessary to establish the chemical histories of any such trace metal in various combustion modes and be able to predict its nature and provide a full mass balance between the quantity in the fuel source and its redistribution over combustion products. Moreover, as combustion databases and regulations grow, there is an ever-increasing need by the engineering community for practical models that readily display their mass balance over a combustor system (Senior, 2000). Consequently, due to recent developments in a better understanding of the chemistry of numerous trace metals it would seem timely to illustrate that this potential goal now is plausible and that behavior is becoming predictable knowing the nature of a specific fuel, the basic chemistry that is available to an element, and the combustion and system parameters. As a result, analyses for those trace metals mentioned, together with various other metals that are not especially of environmental concern have been reviewed in depth up to the present time. There is now a wealth of information available and it is hoped this will reduce unnecessary duplication of studies and help to reduce the level of speculation in many publications. Additionally this is very timely as medical and environmental concerns over the trace or "heavy metals" being connected with major health problems are correctly identified (Pacyna et al., 2016; Anyanwu et al., 2018; Basu and Lanphear, 2019). Additionally, there remain concerns in some areas at the United Nations level, for example, with mercury still remaining about 4.5 times above natural levels in the 15th century, that it may never be reducible for millennia (Outridge et al., 2018). An important consequence of combustion is that it reduces the volume of the fuel producing end products that now are regarded as having value than being simply landfill. As the world runs low on fertilizers, ash has value for agriculture as also being a source of elements. In connection with toxicants various new conclusions are emerging. An obvious one is that all people are different and respond differently in many ways (Jager, 2013). Another is that neurotoxins not only are being more recognized as health hazards (Zahra et al., 2017), they may be in the human body for a while (Grandjean and Herz, 2015), and can be detrimental without necessarily indicating symptoms (Bellinger, 2018). Additionally the concept of single cause effects is now being replaced with synergistic interaction effects.

CHAPTER

2

Recent research concepts

The alkali metals are by no means trace elements in combustion, especially for sodium and potassium. Coal is rich in sodium, whereas biomass is rich in potassium. This seems strange especially as indications are that coal deposits originated from vegetation. It may have resulted over time by leaching and dilution in the earth of their ionic salts. Regardless of this quandary of nature, on combustion their concentrations that can be on a percentage scale in the solid fuels are evident only on parts per million scales in the burned gases. There have been many publications relating to this relatively small release of gaseous alkalis from fuels (Thompson and Argent, 1999; Bojer et al., 2008; Novakovic et al., 2009; Saw et al., 2010a; Jordan and Akay, 2012). The majority of the alkali is either retained in the fuel particles or is rapidly removed from the gas phase by heterogeneous processes. As a result the alkalis behave in a sense as trace elements. Although they are not a hazard to health or an atmospheric emissions problem, their relevance in combustion stems more from the numerous engineering difficulties they can induce even at low concentration levels. This includes their corrosive nature, and other potential slagging, fouling, and agglomeration problems. Due to its ubiquitous nature, sodium even occurring in the ambient air in some marine regions can be a problem for gas turbines and has made the alkalis the most studied of all metals. Consequently, although not really a trace constituent, it was serendipitous that they have been one key in resolving the chemical behavior of others. It was only recently, when the combustion chemistries of mercury, arsenic, and selenium became more closely examined and compared to the alkalis that noteworthy patterns of behavior became evident (Schofield, 2005a). A pronounced underlying commonality of behavior was clearly apparent for these otherwise totally unrelated metals lying in different regions of the periodic table. This will be briefly summarized here, and presented more fully and individually later, because it lays the foundation for understanding the observations of other trace metals in combustion environments as they are individually reviewed and resolves their behaviors and provides the predictive tools for their modeling.

The five alkalis, Li, Na, K, Rb, and Cs, all have correspondingly similar chemistries in combustion. They form the same type of molecules and differ only slightly due to their variations in the chemical stabilities of their bonding. Ionization of their atoms is solely a thermal process and is a kinetic process controlled by temperature, pressure, and the level of free neutral atoms and electrons. It is independent of the fuel and only coupled to the flame chemistry through its dependence on these parameters (Schofield and Sugden, 1965;

Combustion Emissions
DOI: https://doi.org/10.1016/B978-0-12-819126-2.00002-9

Goodings, 1992). It is well established that their gaseous flame composition is a distribution exemplified for sodium as being Na, NaOH, and NaCl if chlorine is present (Hynes et al., 1984; Schofield, 2012a; Wei et al., 2005). Other halogens can produce the corresponding halide. In comparison, levels of ionization are negligible. Other gaseous alkali molecules such as NaO, NaO_2 are not sufficiently stable, or Na_2SO_4 formation that is too kinetically constrained in the gas phase, remain minor. Heterogeneous chemistry on and within ash particles (Anderson et al., 2000; Nielsen et al., 2000) or on any surface (Steinberg and Schofield, 1996, 2002; Schofield, 2012a) has been shown to be very efficient over a wide range of temperatures at all equivalence ratios. This can form, for example, deposits of Na_2SO_4, NaCl, or Na_2CO_3 depending on whether sulfur or chlorine are present and even complex K-Al silicates or other condensed phase molecules within the ash with its components at appropriate temperatures. This is a major reason that gas turbines utilize pure fuels to alleviate this possibility of high-temperature gaseous alkali corrosion. Moreover, the important aspect that emerges is that deposited molecules are seen to have no relationship with the gas phase speciation other than it contains the sodium element. On an ash particle it has the preference for producing the most thermodynamically stable sodium molecule from the availability of elements (Schofield, 2003).

Mercury in its general chemistry has few similarities to the alkalis. Due to its lack of thermally stable molecules, combustion produces flame gases that contain only elemental mercury. Its only stable gaseous molecules are the dihalides such as $HgCl_2$ but their formation is kinetically constrained in the gas phase and mercury remains in elemental form until flue gas temperatures become much reduced. As a result, when the flame chemical behavior of the alkali metals and mercury were compared it was a great surprise to find that they were essentially alike (Schofield, 2005a). The gaseous mercury, although very volatile, also showed a significant, previously not suspected, propensity for heterogeneous chemistry. With an oxygen-rich flame, containing low levels of sulfur dioxide, deposits could be readily collected on a cooled metal probe that is inserted into the burned gases.

This is seen in Fig. 2.1 for a deposit of $HgSO_4$ forming on both the cooled probe and the cooled burner top. Moreover, if equal atomic quantities of the individual alkali elements

FIGURE 2.1 Atomic mercury efficiently depositing as pure $HgSO_4$ from a 2000°C fuel-lean C_3H_8/air flame containing traces of mercury and sulfur (parts per million by volume). It occurs on either a Pt sheathed or bare steel probe here cooled to about 250°C. Deposition is noted also on the water-cooled burner top from the inner experimental core flame. Flame coloration is due to a parts per billion addition of Na used as a monitor of flame stability. This is part of the basic proof for heterogeneous noncatalytic chemideposition. Source: *From Schofield, K., 2004. Let them eat fish: hold the mercury. Chem. Phys. Lett. 386, 65–69.*

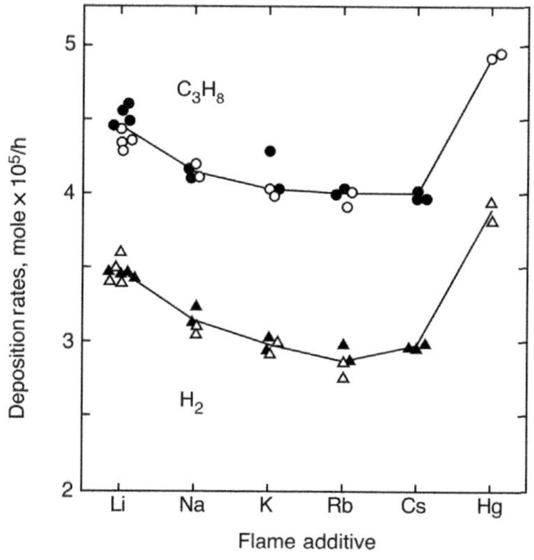

FIGURE 2.2 Mass independent chemideposition. Quantitative separate deposition rates for each alkali and mercury sulfate from either a $C_3H_8/O_2/N_2$ (0.9/5/20) or $H_2/O_2/N_2$ (1.4/1/5) flame. These contained identical (atomic) concentrations of each alkali or mercury salt aerosol together with an excess of 75 ppmv SO_2. Collected on a steel probe at 270°C under constant conditions. Both an inductively coupled emission plasma analyzer (ICP-AES) (*open symbols*) was used for all, and a simple flame photometer (*filled symbols*) for the alkalis only. The curves are displaced due to differing burned gas temperatures and flame volumes. Source: *From Schofield, K., 2005a. Mercury emission chemistry: The similarities or are they generalities of mercury and alkali combustion deposition processes? Proc Combust Inst 2005a;30:1263–1271.*

and mercury are added as a trace nebulized aerosol of their aqueous salt solutions to either a hydrogen flame or a propane flame, each case produces a pure deposit of its thermodynamically most stable entity. In the case illustrated in Fig. 2.2, this shows the rates of deposition of the individual alkali metals sulfates compared to that for mercury.

Surprising the rates of deposition, measured as the molecular quantity of the metal element were essentially almost the same for all six metals: for lightweight lithium through to heavy-element mercury. Also, each of the metal sulfate deposits were originating from differing flame gaseous speciations. Moreover, the alkalis that reflect a single valency, Na_2SO_4, and mercury a divalent nature, $HgSO_4$, each display a rate that displays a first-order concentration dependence on the total metal mole addition to the flame. These data also were unaffected by the flame temperature and whether the sulfur was present in the flame or not. In such a case, the same deposition rates were obtained and contained the same quantity of the metal, but were now white alkali carbonates and black mercuric oxide, respectively. Later, when a similar understanding explained the corresponding behavior of selenium and arsenic in flames it seemed plausible, because of the diverse nature of all these elements that a general pattern of behavior was being observed. As a result a review of other trace metals was initiated and lays the foundation for this current overview of them all.

CHAPTER

3

The generalities that now are apparent

This book aims at examining the behavior of trace elements in all combustion modes. It is concluded that from a textbook understanding of their basic inorganic chemistries, the nature of a particular coal, and the combustion parameters, the observed mass balance distributions become understandable and predictable. What is now being realized and accepted is that the gaseous species in any exhaust or flue are necessarily structurally simple. They are generally either bi- or triatomic oxides, halides, or hydroxides, and their distributions controlled by fast gas phase kinetics that adjust as flame species concentrations change. The nature of practical combustor boilers is not only the rapid generation of energy but also its rapid removal as it is converted from thermal energy to a work mode. As a result, invariably it is the limited time, the temperature decay profile, and kinetic constraints that prevent the formation of more complex molecules in the gas phase and determine the nature of any potential environmentally sensitive emissions. Therefore the gaseous components in an exhaust constitute the flowing transport mechanism in a combustion system controlled by these constraints. Additionally, parameters specific to a combustor and the available chemistry of any one individual element contribute to behavior. As a result, time, temperatures, concentrations, kinetics, and gas/surface interactions become the major controlling constraints, coupled to the chemical limitations of each metal. Although potentially complex, it is these constraints that simplify their individual natures and raise the possibility of being able to reliably model such systems with some reasonable degree of approximation.

It has now become clear that the nature of a specific coal or any fuel will control the subsequent overall scenario that is observed. Coals in particular are extremely variable and are a rich source of trace metals. A coal not only defines the temperature but together with the combustor design defines the time/temperature history of each trace metal and its potential interactions. The variations between coals now are better understood and classification systems have been suggested (Vassilev and Vassilev, 2009; Vassilev et al., 2009). More recently, the same group have repeated the study but examining and categorizing the mineral content and utilization of biomass ash (Vassilev et al., 2013, 2013a). Most solid fuels will be affected by their Si, Al, Fe, and Ca element content, which can form molecules with high melting and boiling points that largely remain as ash particles or slag and have

Combustion Emissions
DOI: https://doi.org/10.1016/B978-0-12-819126-2.00003-0

difficulty leaving the fuel particle (Van Dyk et al., 2009; Hu et al., 2018a). An additional important factor concerns the integrity of a particle. The theory and modeling of particle combustion is still very active (Jaisankar et al., 2017; Tufano et al., 2018,2018a; Sankar et al., 2019; Senneca et al., 2019). The extensive degree of fragmentation that can occur down to nanometric sizes helps to understand the elemental distributions reported. The general pathways of the major elements in coal or biomass through power plants have been known for many years and have identified this strong retention of the four basic coal elements together with a large proportion of any Mg, Na, or K (Klein et al., 1975). Moreover, it is realized that by mixing fuels such as coal with biomass, that the ratios of these condensed phase species can be used to manipulate the desired composition of slag or ash.

In recent years the importance of the concentration levels of chlorine and sulfur in the fuel also has been realized more fully. For example, in numerous older studies, it is noticeable that the chlorine content of a fuel is not even mentioned. As will be seen, these two significant and quite variable components of a fuel can in fact have a major impact on the trace elements, which occasionally even compete for them. The total lack of ignoring bromine particularly in some Chinese coals may also become an oversight that we regret.

The major recent change in understanding has arisen from an acceptance of the limitations imposed by the various constraints on gas phase kinetics for low concentration species. Under such conditions, the role of heterogeneous mechanisms becomes dominant. For example, gaseous Hg atoms in a flue gas at a ppbv (parts per billion by volume) concentration rarely sees any reactant in the three-dimensional space of the gas phase at atmospheric pressure. However, on colliding with a surface, the dynamics drastically change to two dimensions and involve efficient condensed phase kinetics and thermodynamics, also lengthening available interaction times. As a result, for the trace elements examined, their chemistries are in fact dominated by heterogeneous mechanisms if surfaces are available at the right conditions for reaction. Moreover, on surfaces, very complex molecules can be formed that can never conceivably exist or be produced in the gas phase (Schofield, 2005a).

The concepts of chemical kinetics and chemical thermodynamics are fundamental to combustion. Unfortunately, there are still misconceptions that thermodynamic predictions can provide evidence for supporting kinetic mechanisms. Also, the quality of a thermodynamic calculation has to be remembered to be only as good as the reliability of its database. If a species is not in the database it will not be included in the consideration. Nevertheless, noting these prerequisites such calculations can be useful even if only qualitative or approximate in defining the possible relative stabilities of various molecules under a specific set of conditions. However, it should remain a guide and is only an indicator of possibilities and may not relate to reality. In combustion, the gas phase components also can have nonequilibrium or partial equilibrium distributions and take a relatively long time to approach equilibration. Fortunately, in the often complex condensed phase transformations, it appears that their distributions do approximate closely to thermodynamic equilibrium expectations, especially if temperatures are not too low. Condensed phase chemistry can be extremely rapid. This certainly has facilitated the understanding of ash and slag chemistries, and acceptable models, as indicated in a recent review, have matured over the past 40 years (Frandsen, 2009).

Another very intriguing aspect is how the trace metals effectively reach a surface from flowing flue gases. For example, the high and similarly efficient rates of chemical deposition (30%) illustrated in Fig. 2.2 resulted from gaseous components that differed in mass by at least eightfold (LiOH or Hg). As a result, they remain at odds with current diffusion based theories (Schofield 2005a). Moreover, Na_2CO_3 deposition differences were seen to be slight in either very fuel-rich H_2 or D_2 flames (fuel/O_2/N_2, 6/1/5), or propane flames having the major inert diluent gas changed from N_2, to either He, Ne, or Ar (Steinberg and Schofield, 2002). Such hydrogen flames nebulized a sodium formate solution into the gas flows to ensure sufficient flame carbon. In all cases, even though flame diffusion rates through the boundary layer were significantly modified, similar rates of deposition were clearly evident. The deposition process can no longer be regarded as a simple diffusional process or be related to any form of condensation. It is better described as chemideposition, and appears to be a strong surface chemical–thermodynamic-driven interaction.

Previous aspects of condensation onto a surface under a label of dew point considerations still apply but now require a modified interpretation. The pure definition of dew point remains, strictly as that temperature at which a liquid or solid begins to condense and display a product as the temperature of the gas mixture is lowered. It remains solely an indicator of deposit formation and conveys no information about the mechanism or the species involved. What now has been proven is that the elemental or molecular forms of a metal arriving at the surface generally will differ significantly from the product that is formed on the surface. The case of mercury is undisputable. It is so volatile it can have no condensation dew point at normal temperatures. Nevertheless, its deposition as either $HgSO_4$ or HgO from flames can be readily obtained (Schofield, 2004, 2008). This is chemideposition, when the influx, deposition, and chemical change of gaseous atomic mercury on the surface simply becomes greater than the loss rate of the mercury or the condensed phase mercuric sulfate or oxide. The mechanism undoubtedly has to involve gaseous SO_2 and O_2 species on the surface but certainly not in the gas phase. Also, with insufficient sulfur in the flame gases, the rate of mercury deposition was seen to remain totally unaffected except that an equivalent mixture of sulfate and oxide was produced (Schofield, 2008). In all such chemideposition processes it is noted that within its functioning temperature window, the controlling rate is that at which the element is getting to the surface either in its elemental or molecular form. Once there, the surface chemically bonds it rapidly as its most favorable choice based on the composition of the burned gases. Consequently, in these deposition studies, dew point rarely implies any balance between the gaseous and condensed phases of the same molecule. The ingress and egress rates can be very different. The only exceptions are cases where pure homogeneous nucleation and condensation might occur. In combustion this has to be rare, possible for example with gaseous KCl (vapor pressures 1000, 100 ppmv; at 790°C, 720°C), in a sulfur-free environment, and KOH (1000, 100 ppmv; at 700°C, 600°C) if neither chlorine, sulfur, nor carbon are present. Also, practical combustion is a heterogeneous environment in which homogeneous chemistry is rare except under special laboratory conditions.

As a result, it has now been possible to reanalyze the behavior of trace metals in combustion systems to examine the nature of deposition generally. This has been undertaken for most trace metals found in coal and if conforming can readily be extrapolated to any fuel and combustion arrangement. The factors involved in each individual case depends on being able to quantify the specific answers to various chemical questions. These briefly can be summarized as:

- What is the full elemental composition of the fuel?
- What is the combustion equivalence ratio, fuel-rich or lean?
- What are the gas and surface temperatures, and velocity/time map through the combustion train?
- Which molecules are thermodynamically feasible at a specific temperature considering the trace metal concentrations at that point?
- From the particle size distributions can the ash growth illustrate the preferred deposition channels of the trace metal?
- What is the extent of fly ash/slag and its speciation and the degree of elemental removal?
- Can time constraints occur due to an inadequate availability of sulfur and/or chlorine to certain elements?
- What are the physical and chemical properties of each molecule with respect to its thermal stability and chemical reactivity in relationship to the other elements present?
- As temperatures decrease, which heterogeneous chemical options become possible for the gaseous trace metal components?
- What is the nature of the gas phase minerals and how are they distributing the available chlorine?
- With time and temperature how is the list of potential interactions changing?
- What is the efficiency of particle control: fabric filter or electrostatic precipitator and operating temperature effects?
- Do the scrubbers remove escaping gaseous trace elements or the submicron particles?
- Are fine particles and volatile emissions still to be expected in the flue stack?
- Is the understanding consistent with mass balance measurements?

These are not actually demanding questions, and answers are readily available for many practical combustors with now much improved monitoring capabilities. Also detailed mass balances provide additional guidance and indicate the major channels. Much information has already emerged on the trace metal content of ash and its sizing. However, the ash experiences the full combustion train and care has to be taken in interpreting its data. Obviously, it will depend on the location point of any sample, the basic nature of the ash, its interaction times with the flows, temperatures, and the size distribution at that point. Nevertheless, parameters such as the extent of the gas/surface interactions and their efficiencies will be controlling. Fortunately, although many trace elements are involved and their concentrations are small, they still largely retain their individual independent nature. Mutual interactions are negligible outside of the ash and interferences only come if there are deficiencies in species such as sulfur or chlorine. Even so, these are always on ppmv scales, which generally will ensure a sufficient excess. Consequently, modeling is not out of the question but will always remain an approximation.

CHAPTER

4

The importance of the temperature, interaction time, and heterogeneous collision frequency constraints: fuel-rich or -lean differences

The four systems, the alkalis, arsenic, selenium, and mercury, span the types of behavior seen across the spectrum of elements. Although these are each discussed in detail later, to facilitate the understanding of the analysis of other elements and picture what is happening and why in each specific situation, it is worthwhile to briefly summarize these four examples to show how their constraints can be readily seen as controlling the overall observations.

As will be apparent, temperature plays a pivotal role in establishing what can exist mainly due to thermal stability constraints. It controls the extent of reaction within the fuel particle and the degree of species release. Consequently, the mode of combustion and its operational temperatures will define the chemistry. Tillman (1994) presents a useful summary of combustion processes from grate burning to fluidized bed methods with their typical operating temperatures. Incorporating such information into the chemical scenarios that are possible provides a satisfactory understanding of what should be expected for specific cases. The valency displayed by a species in the gas phase generally is irrelevant. With arsenic, for example, as temperatures are raised it is released forming the trivalent toxic gaseous As_2O_3 that above about 650°C becomes gaseous AsO(g), a divalent radical. However, on cooling this will generally have sufficient time to react with some component of a fly ash, possibly calcium, and form the thermally stable arsenate or deposit as As_2O_5(s) that both contain pentavalent As^V that is nontoxic. Also, arsenic has a lengthy interaction time that can remove all the arsenic depending on the nature of the fly ash.

The alkalis have a rich inventory of mixed metal compounds that are thermally very stable. As a result these can form immediately at the high boiler temperatures on heat exchangers and fly ash and then throughout the flue gas train. They have the longest interaction time that is possible. The available time begins to shorten for selenium, which favors selenite formation but requires temperatures below about 800°C and so begins to

Combustion Emissions
DOI: https://doi.org/10.1016/B978-0-12-819126-2.00004-2

encounter a time constraint and a greater dependence on the nature of the fly ash. The situation finally becomes critical for mercury that has no stable mixed metal chemical compounds and few thermally stable options. Consequently, its potential chemistry is limited to below about 400°C, the region in a combustor where waste heat recovery is being invoked and surface collision rates are low. It is the most constrained element with respect to time and so is least modified.

As a result, an analysis of these various constraints convey a satisfactory explanation of why certain elements can pose problems but moreover indicates the steps that can be invoked to resolve such situations. Coal combustion encompasses significant chemistries that occur naturally. It is the engineer's task to manipulate these to achieve desirable features without invoking further more sophisticated technologies.

CHAPTER

5

The nature of flame gases free of nonmetallic content

The most commonly encountered combustion involves the energetic oxidation of carbon and hydrogen. Their flame chemistries are now well described and complexities tend to remain only in the soot forming or pyrolysis regimes of fuel-rich organic fuels. As a result their flame gases involve species that are mainly CO, CO_2, H_2O, OH, O, and O_2 for hydrocarbons and H_2O, H_2, H, OH, O, and O_2 with hydrogen. The nitrogen in the air can introduce low levels of NO, with potential traces of NO_2 and N_2O, generally of no consequence in the flame gases. As will be seen, the major concern for trace metal elements often centers more on the influence of the common impurities found in most fuels, namely sulfur and chlorine (Tillman et al., 2009), as well as bromine, now quite evident in Chinese coals (Peng et al., 2012; Peng and Wu, 2015). Sulfur introduces the potential availability for the formation of sulfates if fuel lean or sulfides if fuel rich, and the halogens the various halides or oxyhalides. Any trace metal that vaporizes and leaves the fuel particle experiences the flame gas temperature. It transports the metal through the flow train. Any particle surface or other surface encountered solely regards the gas flow as a provenience of enthalpy or ingredients. Due to their generally low concentration, the trace metals tend to remain somewhat independent of one another in the gas phase. As a result, unless other additives are introduced, such as boron, phosphorus, or silicon, for example, the major chemistries encountered involve considerations of oxides, hydroxides, carbonates, sulfates or sulfides, and halides. Nitrates generally are less thermally stable as are hydrides and any organometallics. As will be seen, the gas phase generally consists of simple molecular structures while the condensed phase on fly ash or other surfaces can be quite complex.

Although the nature of combustion technology is forever changing, from fluidized bed, to cofiring of fuels (Zue et al., 2019), chemical looping (Adanez and Abad, 2019), and now oxy-burn fueling (Li et al., 2018f), the trace metals remain present in the fuels and although temperatures and interaction times can significantly differ, the same chemistry exists, with emissions control and disposal of final products (Gingerich et al., 2018), that need addressing in each case. The ultralow emissions from the latest power plants appear to be achieving this extremely well (Wu et al., 2018c).

Combustion Emissions
DOI: https://doi.org/10.1016/B978-0-12-819126-2.00005-4

CHAPTER

6

The role of particles, ash, or surfaces

As in atmospheric chemistry, the processes enabling particle formation and growth in combustion still remain poorly understood. Combustion involves a very rapid release of energy and temperature rise. The fuel particle will modify rapidly, fragmenting and possibly melting. Retention will either involve stability or rapid chemical interactions. Enhanced particle formation or reactive surface area obviously will become important as temperatures begin to cool. As a result, recent detailed reviews of the very extensive research concerning nanoparticle nucleation and growth in the atmosphere are a valuable cross-discipline resource for anyone studying combustion generated particles (Zhang et al., 2012a; Wyslouzil and Wolk, 2016). These confirm, like other studies, that homogeneous nucleation is difficult (Toxvaerd, 2015). It requires a supersaturation of the gas. Apparently, water vapor in the atmosphere requires a several-fold saturation factor and invariably prefers a heterogeneous channel by utilizing condensation nuclei. Without such nuclei in the atmosphere it is indicated that Earth would have no clouds and human societies may never have existed. As a result, one unanswered question is how many molecules are needed for a critical nucleus. What is its nature and does charging of particles play any role? As will be clear in the present book, it seems apparent like the atmosphere that any practical combustor with its multitude of parameters will never be able to function as a homogeneous medium. More than two decades ago it was shown that a vaporizing fuel ash particle could readily spawn submicron particles within its boundary layers and that from 200 to 500 of these could be generated from a single coal particle (Helble et al., 1986; Baxter, 1990). Some particles fragment but some coals do not (Friedemann et al., 2016). The higher temperatures of coal combustion can in fact produce cenospheric ash that will readily fragment into many small particles (Li et al., 2013c,d). Recent studies have modeled biomass combustion (Lu et al., 2008) and explored the tailoring of fly ash by cofiring biomass and coal to modify the ash size distribution (Vassilev et al., 2014; Al-Naiema et al., 2015). In fact condensation nuclei that consist of less than 1 nm (0.001 μm) are all that is needed for nucleation and growth, but these are sizes that have been difficult to monitor. This is changing, and ultrafine and nanosize particles are being reported from coal combustion (Carbone et al., 2010), and from waste incineration (Buonanno et al., 2012; Cernuschi et al., 2012; Buha et al., 2014). Their control utilizing fabric filter bags has been shown to be very effective (Jones and Harrison, 2016). Solid particle fragmentation models also are in current development (Senneca et al., 2016, 2019).

Combustion Emissions
DOI: https://doi.org/10.1016/B978-0-12-819126-2.00006-6

Moreover, in combustion, great care is still needed in probe sampling to ensure any such sample does reflect the true particle size distribution and composition. Like the atmospheric nuclei that condense water vapor and give us clouds, it is fortunate that fly ash in all its sizes and slag are present to absorb or retain many of the trace metallic elements, greatly lessening the problems of environmental concern. A recent review summarizes the extensive Russian work concerning their mass spectrometric studies of the vaporization and thermodynamic properties of innumerable metal oxide systems (Stolyarova, 2016). This includes many systems and especially metallurgical slags containing Na/Al/Si and glass melts of Na/Si/B very appropriate to fuel combustion. An additional paper of significant insight came from a study of the behavior of titanium in coal combustion. This element has a ceramic nature with no possibility of gasification and release and the authors must be credited with having the tenacity to examine it (Baxter et al., 1997). A normal response would have been that there was little point in such a study. The remarkable discovery was that titanium was seen to be lost from the particles. It was clear evidence that on combustion, nonvolatile components can be transported out of a fuel particle mechanically as it gasifies and burns. This is why analysis of the very smallest particles in combustion can still display many elements that reflect the core ash such as Ti or Fe and do not involve their volatization or nucleation but originate from particle disruption initiated by other processes (Wang et al., 2008).

Although coal ash still is not labeled as toxic waste it has been involved in costly environmental disasters. Additionally it is the second largest waste material in the United States (Ritter, 2016). Currently, environmental aquatic wild life concerns may force the US EPA to introduce more stringent rules concerning its disposal (Lemly, 2014). There is no doubt that any combustion bottom or fly ash contains the accumulated element ingredients of most of that found in the initial coal, biomass, or waste fuels (Vejahati et al., 2010, Vassilev et al., 2014a; Saqib and Backstrom, 2016). Nevertheless, for several years now, as concerns have been raised over the sustainability of certain elements in the economy, it has been realized that fuel ash is a rich source of minerals and can have potential value (Ahmaruzzaman, 2010; Blissett and Rowson, 2012). This will be addressed further under each element but includes such aspects as soil improvement (Shaheen et al., 2014), fertilizer value (Yunusa et al., 2012, Mupambwa et al., 2015), possible animal mineral feed supplement (Sheppard and Sanipelli, 2012), and as a valuable metal source for their recovery (Brett et al., 2015, 2015a).

For many years there has been a controversial debate as to the various roles of homogeneous nucleation or thermodynamically driven chemical deposition on surfaces. In reality, whether one or the other occurs becomes indistinguishable in combustion. However, homogeneous nucleation will become less likely as concentrations decrease to low levels. Also, no combustion system can yet be monitored that is free of any form of nucleating centers. As seen in atmospheric studies (Leppa et al., 2011; Zhang et al., 2012a), homogeneous nucleation is a difficult process and certainly requires a number of like molecules to assemble. As a consequence, if the possibility exists of utilizing some other condensation nucleus to initiate the process it would appear favored. Similarly, chemideposition only needs some finite-sized nanoparticle on which to initiate deposition and growth.

Both combustion and atmospheric studies have shared several unresolved chemical difficulties. A major one is the mechanism for H_2SO_4 formation with subsequent $(NH_4)_2SO_4$

particles in the atmosphere (Kulmala et al., 2014), and also for its formation in the cool end of a power plant where it can produce problem deposits of ammonium bisulfate, $(NH_4)HSO_4$ (ABS), with the NH_3 slip from an SCR. Although SO_2 is abundant in flame gases its oxidation mechanism to the necessary hexavalent sulfur in SO_3 is a problem as the gas phase poses a kinetic constraint. The recent observation that fly ash can heterogeneously catalyze this conversion would resolve this (Belo et al., 2014). This also seems compatible with the efficient low-temperature oxidation of NO to NO_2 now being reported on surfaces (Minissale et al., 2013). Additionally, sulfate formation with many metals has been noted in atmospheric particles. This has been explained as a heterogeneous mechanism whereby atmospheric SO_2 is oxidized on a particle containing a specific metal compound (Clements et al., 2013). Sulfates are invariably the most thermodynamically stable molecule for many metals for most general fuel combustion systems.

CHAPTER

7

Trace species reported combustion behavior and their chemistry

The combustion behavior of each metallic element is outlined individually in this chapter, each in a self-contained subchapter and can be referred to separately. The format is approximately consistent for each. It entails the sources of the element and its natural and anthropogenic occurrence in the environment. The question as to whether it is now rated as a pollutant is addressed, the role of combustion, and whether there are health concerns about toxicity that might be appropriate for stricter controls on its possible emissions release or safe disposal of wastes. The basic condensed phase chemical properties are summarized together with gas phase stabilities and questions of likely transport mechanisms. With some elements chemical kinetic constraints are apparent that eradicate any chemical thermodynamic predictions. Laboratory studies are then considered although these actually can at times not be easily correlated to practical full-scale combustors. After assessing the known chemistry, it is compared to the many observations reported in power plants, the nature of the final distribution of the element, and its role in subsequent disposal.

7.1 Alkalis (Li, Na, K, Rb, Cs)

Sodium and potassium's important role in combustion partly arises as a result of their significant presence in nature. They are abundant elements. They rank sixth and eighth in abundance with concentrations of 2.4% and 2.1%, respectively, in the Earth's crust. Sodium also is the fourth most important element in the oceans at a 1% concentration after H, O, and Cl. Probably due to its less soluble mineral sources, potassium is at a 27-fold lower level. It has been mainly due to the roles of sodium and potassium in combustion system performance relating to slag formation, fouling, agglomeration, and corrosion problems that has made them by far the most studied metal elements of all (Raask, 1985; Bryers, 1996; Niu et al., 2016). Their volume of research literature tends to dwarf that of any other metal. In reality, three of the alkali metals, lithium, rubidium, and cesium are

Combustion Emissions
DOI: https://doi.org/10.1016/B978-0-12-819126-2.00007-8

only trace metals in all fuels and prove to be of little interest. Although variable, lithium concentrations can approximate to a range of about 10–80 ppmw in some coals (Zhang et al., 2003; Wang et al., 2008c; Font et al., 2010). Corresponding values for rubidium are about 10–60 ppmw (Rowe and Steinnes, 1977; Xu et al., 2003; Font et al., 2010) and cesium 1–30 ppmw (Xu et al., 2003; Hu et al., 2006; Font et al., 2010). All three are much lower in other fuels. As a result, these three mainly have been of academic interest in combustion by illustrating the overall similarity of behavior for all five. All five are commonly encountered elements. Lithium, although found in numerous minerals and abundant, is now extracted from salt brines economically. Worldwide production is about 84,000 tons $year^{-1}$ and demand will probably increase with Li-battery technology growth. The elemental pure metal is about \$1.50 g^{-1}. Sodium and potassium are the major commercial alkalis, both elements isolated by Davy in 1807 by electrolysis of their respective hydroxides. Sodium is found mainly as the chloride in well-defined deposits such as salt caves. About, 300 million tons $year^{-1}$ are mined with about 100,000 tons of pure metal produced. Liquid sodium has high heat transfer properties and finds use in some nuclear reactors. In the United States almost half is used for road deicing. At a cost of about \$0.20 kg^{-1} it is the lowest cost of all metals. Potassium is not as extensively produced, about 42 million tons $year^{-1}$, but has adequate sources. It has major use as a fertilizer. It is known as the first metal isolated by electrolysis. It has one naturally radioactive isotope ^{40}K to the compositional extent of 0.012%, but with a half-life of 1.3 billion years and so is regarded as totally safe. The pure metal can have a range of prices from \$1–5 depending on purity. Rubidium and cesium are less encountered in nature. They were both spectrally discovered in 1860/1861 by Bunsen and Kirchoff. They are mainly mined as byproducts of Li production. The United States imports 100% of its Cs needs. CsOH is the strongest molecular base molecule and Cs is now famous in defining the unit of time by one of its spectral transitions. Either metal element is expensive at \$6–35 g^{-1} and \$3–20, respectively, depending on purity. The chemical differences between the five stems from slight changes in stabilities of their molecular bonding strengths (Schofield, 1992). Of the lesser three, cesium appears to have found some application as an additive to a fluidized bed combustor as a tracer for monitoring the removal rate of potassium from the bed when utilizing biomass fuels (Svane et al., 2006). Radioactive ^{137}Cs is also an important monitor of pollution, particularly in aquatic environments (Ashraf et al., 2014). It is not a natural isotope, but arises from nuclear accidents and has a half-life of about 30 years. Nevertheless, it is true that all five alkalis have always been of great interest with regards to their different extents of ionization in flames, which is thermal in origin, cesium being important for having the lowest ionization potential of all the elements (Goodings, 1992).

In spite of these potentially high quantities in the fuels, alkali gas phase measurements in combustion systems have always reflected only low concentrations, and speciation monitoring a dilemma. This was an initial enigmatic question as to what was their distribution because the only thing measurable was the free atomic Na or K concentrations. Also if low gaseous levels remained, could these be further controlled and mitigated in the hot gases to the ppbv concentrations required, for example, for direct high temperature flue gas interfacing with gas turbines (Lee et al., 1992). Over numerous decades, multiple studies in full scale and in the laboratory have constantly addressed such questions. This major problem that still remains to a large degree is that the gaseous molecular alkali compounds are difficult to monitor directly, nevertheless a reasonably clear picture of the basic controlling chemistry now has emerged with a few minor remaining controversies. As a result, this review emphasizes the major conclusions that have now become reasonably well established in recent years and provides references that relate back to earlier studies.

Health concerns with the alkalis

The salts of the two alkali metals, sodium and potassium, are the basis for human and biological life and as elements in the body rank in magnitude only after calcium. The average human body contains slightly more potassium (140 g) than sodium (100 g). They are essential for cellular function and neural communication in the body. An average diet supplies about 3 g day^{-1} Na and 3.5–7 g day^{-1} K and any excess is excreted via the kidneys. The two are maintained in a necessary balance. The potassium is almost entirely located in the cells where it is needed for normal functioning, and is about 30-fold lower in concentration if exterior. On the other hand, sodium is mainly in the blood and 10-fold lower within the cells (Crichton, 2017). If the body cannot process the elimination of any excess, diseases affecting blood pressure arise such as hypertension result. Due to the observed scale of magnitudes, a change in blood monitoring units has arisen, and the normal range for Na now is quoted as 135–145 millimole L^{-1} (3.1–3.3 g L^{-1}) and for K, 3.5–5.1 millimole L^{-1} (0.14–0.20 g L^{-1}). Measuring the sodium level in blood is not entirely accurate though as a total body content as the volume of blood can automatically adjust if necessary, for balancing levels. The other alkali elements in the body are at much lower levels, normally in blood ranges of 0.3–10 μg L^{-1} for Li (Cesbron et al., 2013), 1.4–2.5 mg L^{-1} for Rb (Goulle et al., 2005; Cesbron et al., 2013), and <11 μg L^{-1} for urine levels of Cs (CDC fourth Rept., 2018). Available toxicology reports are available only for Cs and this is concerned mainly with radioactive ^{137}Cs in the atmosphere following any nuclear power plant accidents (Tox Rpt Cs, 2004). Blood, serum, urine, and hair sampling are all used as biomarkers by the differing test centers for the alkali elements. Levels of Li, Rb, and Cs are very low in hair samples. Lithium is noteworthy as having emerged as a medical medication for psychiatric disorders such as bipolar illness. Discovered in 1948 and having a narrow therapeutic/toxic dosage range, its mechanism remains unknown but it is an effective and inexpensive treatment. As a result, concerns with the alkali metals in combustion center totally on Na and K from practical engineering considerations rather than environmental health.

Although alkali combustion emissions are of less concern with regards to pollution or health, the alkali metals have been the basis for major research studies due to difficulties arising in the combustion process itself. By chance, this research has generated invaluable insights into their combustion deposition behavior that now has broad implications for all the trace elements.

Sodium and potassium's basic chemistry

Alkali metals have a very rich chemistry that exhibits possibly the broadest range of chemicals in the periodic table. Their salts are often white, ionic with monovalency and nearly always soluble in water. They produce salts with almost every element of the periodic table with the exception of the alkaline-earths and rare gas elements. Those most appropriate to combustion are listed in Tables 7.1 and 7.2. They produce organic-coupled salts (such as sodium acetate) that do not involve direct alkali-carbon bonding. A rare exception is the ionic sodium acetylide, NaCCH that does not occur in nature. Although many alkali compounds may be present in a fuel, on combustion the list of those of interest becomes narrowed due to their generally low thermal stability.

TABLE 7.1 Properties of sodium and potassium condensed phase molecules relevant to combustion, °C

Sodium m.pt. b.pt.	Potassium m.pt. b.pt.
Na 98 881	K 63 763
$NaHCO_3$ 50, dissoc 270	$KHCO_3$ 292, dissoc >150
Na_2O 1132, sublimes >1275, reacts w wat	K_2O 740, decomp >350, reacts w wat
Na_2O_2 460, decomp >675, reacts w wat	K_2O_2 490 dissoc, reacts w wat
NaO_2 552, dissoc >180, reacts w wat	KO_2 509 dissoc, reacts w wat
$NaNO_3$ 307 380, decomp	KNO_3 335 400, decomp
$NaNO_2$ 280, decomp >320	KNO_2 441 537, decomp
Na_2S 1172, dissoc 2050, reacts w wat, air	K_2S 948, dissoc >912, reacts w wat, air
NaH >425 dissoc, reacts w wat	KH >417 dissoc, reacts w wat, air
$NaAlO_2$ 1650	$KAlO_2$ 1900
$NaBO_2$ 967 1475	KBO_2 947 1400
$Na_2B_4O_7$ 743	$K_2B_4O_7$ 816
Na_2CrO_4 797	K_2CrO_4 977
$NaFeO_2$ 1347, hygroscopic	$KFeO_2$ unstable, reacts w air
$NaPO_3$ 630	KPO_3 804
Na_3PO_4 1583	K_3PO_4 1340
Na_2SiO_3 1089, reacts w hot wat	K_2SiO_3 976
Na_2SiO_4 884	K_2SiO_4 1069
$Na_2Si_2O_5$ 874	$K_2Si_2O_5$ 1045
Na_4SiO_4 1120	K_4SiO_4 912
Na_2TiO_3 1030, dissoc >965	K_2TiO_3 1370, reacts w wat
$NaVO_3$ 630	KVO_3 520
Na_3VO_4 900	K_3VO_4 >1000

All water soluble: *decomp*, decomposes: *dissoc*, dissociates; *reacts w wat*, reacts with water.

TABLE 7.2 Melting and boiling points, gas phase bond strengths and equilibrium molecular or dissociative vapor pressures as a function of temperature (°C) for the major gaseous sodium and potassium molecules, and also bond strengths for their known weaker bound molecules.*

	Vap.	press.	(ppmv)	Vapor (%)				
	1.0	10	100	Molecular	m.pt.	b.pt.	D_0	References
Na	228	280	344	–	98	882	–	Barin and Platzki (1995)
NaOH	621	702	800	100	323	1388	Na-OH 333 ± 8	Gurvich et al. (1996)
NaCl	592	662	746	100	801	1484	Na-Cl 408 ± 1	Barin and Platzki (1995)
Na_2CO_3	814	910	1021	$\approx 10^{-2}$	851	1600	–	Motzfeldt (1955), Hildenbrand and Lau (1991)
Na_2SO_4	936	1036	1154	≈ 30	884	1429	–	Halle and Stern (1980), Lau et al. (1985)
K	154	199	256	–	64	764	–	Barin and Platzki (1995)
KOH	467	537	620	100	406	1321	K-OH 357 ± 3	Gurvich et al. (1997)
KCl	565	634	716	100	771	1437	K-Cl 422 ± 0.5	Barin and Platzki (1995)
K_2CO_3	644	725	821	0	901	1500	–	Kosugi (1972)
K_2SO_4	910	1015	1135	60	1069	1689	–	Kosugi (1972), Lau et al. (1979)

*Additional D_0 values (kJ mol^{-1}) recommended in the text.
Na-O 266 ± 4, Na-O_2 155 ± 10, Na-ONa 224 ± 10, Na-H 182.2 ± 0.01, Na-S 257 ± 10, K-O 276 ± 4, K-O_2 177 ± 5, K-OK 204 ± 10, K-H 170.9 ± 0.5, K-S 264 ± 10 All these sodium/potassium salts are water soluble.

Also those that are stable can remain in the ash and not vaporize. The noteworthy observation is that their hydroxides and chlorides sublime to very stable gas species that are thermally stable to high temperatures (Table 7.2).

In comparison the sodium and potassium carbonates differ. As seen in Table 7.2, their condensed phases have high boiling points of about 1600°C and 1500°C and have reasonably high vapor pressures below 1000°C, but their vaporization is a dissociative process to the oxide and then element at elevated temperatures. Thermogravimetric analysis (TGA) of these carbonates confirmed the pronounced decomposition starting even in the liquid phase around the melting points (Lehman et al., 1998). Alkali carbonates cannot exist in the gas phase in flame gases. The two sulfates are similar with boiling points of about 1430°C and 1690°C, their gaseous molecules can exist but are partially stable on vaporizing, maintaining about 30% and 60% in molecular form, respectively. However, in flame gases they will be rapidly broken down to the hydroxide at elevated temperatures having a short half-life. Of the other appropriate sodium or potassium chemicals that might be encountered in fuels and combustion, as seen in Table 7.1, these are thermally unstable dissociating to the oxide (bicarbonate, nitrate, nitrite), others either decompose or react vigorously with water producing the hydroxide (oxide, peroxide, superoxide, sulfide, hydride). Sulfur also has a rich collection of alkali compounds labeled bisulfate, thio-, dithio-, pyro-, per-, and also sulfites, all of which readily dissociate with low melting points. Options of more thermally stable mixed element oxides that may arise with fuels are formation of the aluminate, ferrite, phosphate, silicate, or titanate. For sodium, these all have melting points above 1000°C. However, $NaAlO_2$ was found to dissociatively vaporize above 900°C (Popkov and Semenov, 1971). They have not been as well characterized for potassium but the $KFeO_2$ is reported to be unstable in moist air. Also alkali phosphates appear to dissociate on vaporization above 1000°C forming stable gaseous $NaPO_2$ and $NaPO_3$ molecules and similarly for potassium (Rudnyi et al., 1986; Lopatin and Shugurov, 2013). In this respect, coals or biofuels are a mix of such ingredients and it is evident that any alkali molecule present will have an affinity for the major ash components such as aluminum, iron, phosphorus, silicon, or titanium and be retained (Wibberley and Wall, 1982; Mobin and Malik, 1990; Gallagher et al., 1996; Wee et al., 2005; Zheng et al., 2007; Yu et al., 2009; Wu et al., 2009). Consequently, such complex interactions are plausible in the ash but introduce a parameter that can be difficult to model. The extensive mass spectrometric work on such alkali/mixed element systems has been recently reviewed (Stolyarova, 2016). This considers the vaporization processes and thermodynamic properties of slags such as Na/Al/Si and glass melts of Na/Si/B. The effects of additives on $K_2O/SiO_2/Al_2O_3$ melts relating to biomass combustion have also been addressed recently (Niu et al., 2016a). Particularly with coal, the melting point of such mixtures is very dependent on the composition and can be modified by the presence of Fe, Ca, and P, as well as the alkali concentration (Wang et al., 2016h).

Gas phase stabilities

The relative importance of these molecules in the gas phase is controlled by their chemical thermodynamic stabilities in both the fuel and in the flame gases, coupled to their vapor pressures and boiling points. In this regard, the examination of their chemistry immediately indicates that the atomic element, its hydroxide, chloride, carbonate, and sulfate will be of major importance. Because of this, they are listed separately in Table 7.2 with melting and boiling points and the nature and extent of their vaporization. This illustrates the significant similarities between Na and K, with their chemistries heavily reflecting each other. Their gas phase bond strengths also have been listed. The value now recommended for NaOH, $D_0(\text{Na-OH}) = 333 \pm 8\ \text{kJ mol}^{-1}$ is $7\ \text{kJ mol}^{-1}$ weaker than the value listed in the two databases (Knacke et al., 1991; Chase, 1998) but is in agreement with these sources for $D_0(\text{K-OH}) = 357 \pm 3\ \text{kJ mol}^{-1}$. The thermochemistry of the chlorides is well established from their condensed phase heat of formation coupled to that of their vaporization. These indicate very stable molecules with $D_0(\text{Na-Cl}) = 408 \pm 1$ and $D_0(\text{K-Cl}) = 422 \pm 0.5\ \text{kJ mol}^{-1}$ (Kubaschewski et al., 1993; Chase, 1998). This is of a stability range equal to HCl, OH, and H_2. As outlined already the gaseous carbonates and sulfates are not expected to be present.

Although gaseous oxides such as NaO, NaO_2, and Na_2O are known, in addition to their weak bonding, they are all easily hydrolyzed by water to the hydroxide. In normally encountered flame gases their concentrations are minor (Hildenbrand and Murad, 1970;

Ehlert, 1977; Partridge et al., 1992a; Hildenbrand and Lau, 1993), but undoubtedly, they can play important kinetic roles in facilitating a balance between the major gaseous Na and NaOH species (Hynes et al., 1984). Their bond strengths, some of which have been uncertain for a significant period are now finally becoming better established. For the monoxides, theory has confirmed their differing electronic nature with an $NaO(X^2\Pi)$ ground state and low-lying $A^2\Sigma^+$ state at 24.5 kJ mol^{-1} (2050 cm^{-1}), whereas it is KO $(X^2\Sigma^+)$ and a very close lying $A^2\Pi$ at 2.4 kJ mol^{-1} (200 cm^{-1}) (Lee et al., 2002; Mintz et al., 2009). For sodium, a recommended value D_0(Na-O) = 266 ± 4 kJ mol^{-1} now is larger than the earlier mass spectrometric result but is solidly supported by a reassessment and several reliable theoretical estimates (Steinberg and Schofield, 1991; Soldan et al., 2000; Mintz et al., 2009; Vasiliu et al., 2010a). That for potassium, D_0(K-O) = 276 ± 4 kJ mol^{-1} remains consistent with an earlier evaluation (Lamoreaux and Hildenbrand, 1984) that is now additionally supported by two recent theoretical calculations (Mintz et al., 2009; Vasiliu et al., 2010a). Extensive conflict reigned for a while over the possible importance of the MO_2(g) species. The oxides were considered as participants in the alkali flame inhibition and reburn effects for NOx control and playing catalytic roles and this sparked extensive interest in their chemistries. However, one review of alkali oxide vaporization noted that MO_2 species were not even seen in mass spectrometric studies and were regarded as negligible in the vapor (Lamoreaux and Hildenbrand, 1984). However, it had been noted that atomic Na, K, and several other metal atoms could insert into O_2 even at lower temperatures producing these dioxides. Measurements of the rate constants for this three-body type reaction displayed a very high efficiency, their rates of formation aided at atmospheric pressure by the large concentrations of O_2 and the third body such as N_2 (Marshall et al., 1990; Plane et al., 1990; Vinckier et al., 1992).

$$Na + O_2 + N_2 = NaO_2 + N_2$$

Additionally, flame studies with the alkalis were implying a much greater stability and importance for NaO_2 (Hynes et al., 1984; Steinberg and Schofield, 1987). With this three-body measured reaction rate constant, coupled to the concentrations of O_2 and N_2, the flux rate of this reaction in flames was implying a half-life for Na depletion on a <100 μs timescale, and could be 20-fold faster than the alternate direct endothermic depletion reaction of Na with H_2O. In nonequilibrated flames it was also found that NaOH concentrations could overshoot their equilibrium values because of an alternate kinetic route using this intermediate. Several metals including Na and K have this ability to react with O_2 even at temperatures slightly above room temperature to form this ionic molecule. It is regarded as a cation bonded to a superoxide anion, O_2^- and its bonding is best indicated by a D(M-O_2) value rather than individual bonds. It has been difficult to derive this strength for the alkalis in that systems are lacking where the formation/dissociation reaction above is equilibrated. As indicated, initial mass spectrometric evaporation cell studies could not detect it and suggested a very low bond strength (Lamoreaux and Hildenbrand, 1984). However, it was later questioned whether the NaO_2^+ ion formed and monitored by mass spectrometry was actually stable, which turned out to be the correct resolution of the disagreement (Steinberg and Schofield, 1991; Partridge et al., 1992). As a result, the vaporization work was repeated measuring instead torsion-effusion pressures that now indicated an upper limit bound of D_0(Na-O_2) <180 kJ mol^{-1} (Hildenbrand and Lau, 1993). Various theoretical calculations over a span of 25 years now have consistently agreed to a value of D_0(Na-O_2) = 155 ± 10 kJ mol^{-1} that would appear to be the most reliable value (Horner et al., 1991; Partridge et al., 1992a; Lee et al., 1999; Vasiliu et al., 2010a). Fewer studies emerged for potassium. However, the most recent is important in lending credence to the theoretical results. It measured the rate of depletion of the atomic K with O_2 in a 475°C–1050°C range (Sorvajarvi et al., 2015). Changes were slight for the rate constant through this temperature range suggesting a consistent mechanism with no difference in the nature of the KO_2 product. At the lower temperature range a termolecular rate constant was determined, but at higher temperatures an onset of equilibration was beginning to be observed allowing a better estimate of the dissociation rate constant and hence the equilibrium constant. A value of D_0(K-O_2) = 181.5 ± 4 kJ mol^{-1} resulted. This compared to three theoretical estimates of 170, 168, and 173, respectively (Partridge et al., 1992a; Lee and Wright, 2002; Vasiliu et al., 2010a). KO_2 also had not been observed in the earlier mass spectrometric vaporization study, possibly also indicating its ion instability (Lamoreaux and Hildenbrand, 1984). However, an earlier mass spectrometric KOH vaporization experiment did note its presence leading to a value of about D_0(K-O_2) = 153, about 20 kJ mol^{-1} slightly below the currently suggested value, possibly implying only a partial fragmentation loss of the ion seen in the mass spectrometer data in potassium's case (Hastie et al., 1984). It would be unexpected for potassium to significantly differ in behavior from sodium in this respect. Using unimolecular dissociation theory and Troe's formalism, approximate values of 189 and >202 kJ mol^{-1} were suggested for K-O_2 utilizing the three-body recombination rate data (Patrick and Golden, 1984; Plane et al., 1990). The higher temperature flame kinetic model for potassium was found to be compatible with the D_0(K-O_2) = 170 kJ mol^{-1} value (Steinberg and Schofield, 1987). As a result of these very extensive research efforts, it appears that a current recommended value with potassium would be D_0(K-O_2) = 177 ± 5 kJ mol^{-1}.

The M_2O(g) species generally are less important and have been but briefly examined. The heats of formation in two assessments indicate $D_{0Atomiz}$(Na_2O) = 494 and 490 kJ mol^{-1} (Wagman et al., 1882; Lamoreaux and Hildenbrand, 1984). A more recent theoretical calculation indicates 482 (Soldan et al., 2000). The value of 490 might be recommended for the present, which indicates a second bond strength of D_0(Na-ONa) = 224 ± 10 kJ mol^{-1}, 84% of the monoxide Na-O bond. Less reliable data are available for potassium but the same two assessments suggest at 298K, atomization energies of 490 and 486 kJ mol^{-1}, respectively. An earlier mass spectrometric measure gave a much higher value, 569 kJ mol^{-1} (Ehlert, 1977) that is not supported by the approximate theoretical value of 473 (Schulz et al., 1999). Until resolved an atomization value of D_0 = 480 kJ mol^{-1} is proposed implying D_0(K-OK) = 204 ± 10 kJ mol^{-1}, 74% of the monoxide bond.

The alkali hydrides also react with water and oxygen and are not thermally stable. They have been of interest mainly in astronomy for their spectroscopy and also as a great testing ground for theory. However, other than the extensive analyses of their $A^1\Sigma^+$ and $X^1\Sigma^+$ electronic states and their potential energy curves, no other determinations of their bond strengths are available. However, these data now appear to be accurate and reliable indicating D_0(Na-H) = 182.2 ± 0.01 and D_0(K-H) = 170.9 ± 0.5 kJ mol^{-1} supported by theoretical estimates (Walji et al., 2015; Vasiliu et al., 2010a; Stwalley et al., 1991; Fan et al., 2009; Chu et al., 2018). There has been little interest in the gaseous sulfides. An early theoretical estimate remains acceptable indicating the only available values of their bond strengths. This predicted D_0(Na-S) = 257 and D_0(K-S) = 264 kJ mol^{-1} (Partridge et al., 1988). One experimental study was attempted using an effusion mass spectrometric method examining the equilibrated dissociating vaporization above K_2S(s) that suggested a value of D_0(K-S) = 251 (Farber et al., 1983). Now it has been established that the ground electronic state for both NaS and KS is $X^2\Pi$, the same as NaO. The electronic structure, low-lying states, and potential energy curves have been calculated but possible spectral transitions not yet characterized (Lee and Wright, 2004, 2005; Khadri and Hochlaf, 2012). At present, the theoretical bond strengths are preferred. These values are slightly below those for their diatomic oxides as is expected. These bond strengths have been collectively listed in Table 7.2. The chemistry clearly conveys that the gas phase in combustion will predominantly indicate a possible balance between M, MCl, and MOH, with only minor roles played by MO and MO_2. For oxygen-free gasification or pyrolysis conditions, condensed phase sulfides or carbonates can begin to play roles of alkali retention with mixed metal oxides.

Laboratory combustion studies: homogeneous conditions

One dilemma that arises in combustion is that it is really impossible to derive fundamental understanding from a full-scale power plant. The range of conditions, variations of parameters, and too limited monitoring capability make for too much complexity. It can involve severe temperature, kinetic, and time constraints on the chemistry within the flame gases. For correctness, interpreting any available data has to be based on simpler experimental lab-based or even pilot-plant measurements that may provide a more fundamental foundation that is less speculative. Then care has to be taken in extrapolating laboratory results to larger scale applications. However, quite often there is little choice if progress is to be made. Nevertheless, the volume of studies with the alkalis in combustion is so huge that laboratory studies have significantly led to a close understanding of its gas phase reactivity.

With the alkali metals, several aspects have always been known. Initial conclusions were evident from their basic chemical analyses. It was evident that care had to be taken in any analysis that required a sample to be initially reduced to ash. In such cases, elevated temperatures had to be avoided or else alkali would be lost. Also in the very foundation of flame photometric analysis methods it was the simplest method for spectrally analyzing alkalis. For this, with an unknown sample in solution and aspirated into a flame, the intensity of the alkali atomic flame emission or absorption related directly and linearly to its total concentration indicated by a calibrating alkali solution. It was widely accepted that all alkali salts in the unknown sample matrix were fully dissociated in the flame and could be readily calibrated against standard solutions for the same flame condition. This has always implied a simplicity of the gaseous alkali speciation in the flame gases that is invariant of the nature of the sample solution's molecular contents. Few interferences were noted and it was soon proven that the speciation in the flame would be solely gaseous Na, NaOH, and or NaCl with a few Na^+ ions. Most importantly it was fortunately realized, although the alkali salts are highly ionic, that when such salts dissociate in the flame gases, they became neutral components. This results from efficient crossing of the potential energy curves associated with dissociating ionic states that transfer to the lower energy curve correlating to neutral products. It is clearly illustrated in a theoretical paper displaying the potential energy curves for LiCl (Kurosaki and Yokoyama, 2012). Consequently, this was extremely important in establishing that atomic alkali ions that may be present in flames result from thermal ionization that has originated from their flame injection time and starting with a zero-initial value. Because of their strong atomic transition probabilities, Na and K atoms have always been easy to monitor spectrally with high sensitivity. Measuring the atomic concentrations alone was sufficient for analytical analysis, and still can provide data that examines biomass release, in one example, of at least some potassium atoms (Mason et al., 2016, 2017). Due to the pronounced ionic nature of the gaseous molecular alkali compounds of NaOH, NaCl, NaO, or NaS, their required spectroscopic transitions are not available. Because of this, the specific speciation of flame alkalis has not been satisfactorily measured. Dissociative methods coupled to detection have been developed, but not fully embraced concerning accuracy. These have attempted to track the chloride/hydroxide speciation and examine for other species such as the sulfates. This monitoring has been extensively reviewed (Monkhouse, 2011). On using laser-induced fluorescence (LIF) to monitor directly elemental Na from a coal particle at 930°C–1330°C, maximum concentrations of 60 ppbv were noted (van Eyk et al., 2009). The magnitude of this increased significantly using laser-induced fragmentation spectroscopy that measured the resulting atom emission from both atomic and some molecular forms of the gaseous alkali species (Greger et al., 1996; Gottwald et al., 2001, 2002; Schurmann et al., 2001). This did manage to confirm that elevated chlorine levels in the fuel facilitated a greater release of Na and K as their chlorides. From such studies a complimentary method of laser-induced breakdown spectroscopy has since been developed but is also limited to obtaining total sodium or potassium temporal release data (Hsu et al., 2011; Yuan et al., 2015; Fatehi et al., 2015, 2017; Zhang et al., 2015c). Nevertheless, with such scant data this has not prevented the presentation of proposed full mechanisms for the release of sodium from coal or biomass particles (van Eyk et al., 2011, 2011a; Fatehi et al., 2015, 2017). This illustrates the dilemma and frustration with practical systems if adequate monitoring methods are not available even at the laboratory level. The speciation data (NaOH and NaCl) has always been difficult to measure but has been shown to be locally in balance in the gas phase with the basic flame species and controlled by thermochemical considerations. This has been adequate for understanding the relative concentrations of the gaseous components in the vapor, but the extent of retention and deposition in ash is more complex involving other chemistries, such as formation of more refractory type molecules with the other fuel components at all stages throughout the combustion temperature range. Still rather approximate, methods to monitor gaseous NaCl or NaOH by laser-fragmentation fluorescence now have been developed to resolve the vapor phase (Helble et al., 1992; Chadwick et al., 1995; Sorvajarvi et al., 2014). These have confirmed that the presence of gaseous NaCl is strongly dependent on coal type in combustion and increases with coal chlorine content. Also the rather demanding mass spectrometric studies with coal- and biomass-fueled gases now have illustrated the dominance of gaseous KCl and NaCl while finding no evidence for gaseous Na_2SO_4 or K_2SO_4 (Dayton et al., 1995; Oleschko and Muller, 2007; Blasing and Muller, 2010, 2010a). These methods have been implemented further by these and others and practical results will be indicated below. The main advance has been this possibility to examine both the extent of the release and its final speciation in the flue gas and on the fly ash.

In research laboratory flames that inject a known concentration of alkali into the flame gases it is well known that the kinetics are rapid producing a steady state distribution controlled by the concentrations of the flame radicals. For example, in a hydrogen flame the species are in steady state with all the gaseous species H, OH, H_2, O, O_2, H_2O connected by fast reactions and controlled by known chemical thermodynamics. Knowing the concentration solely of one trace flame species such as H-atom, or OH provides them all. Similarly, with sodium in the flame, the gaseous Na, NaOH, and NaCl are similarly connected and their distribution resolved, being controlled by the larger concentrations of the hydrogen/oxygen species. This is defined by the partial equilibrium balance that can be represented by the simplified reactions shown here, their equilibrium constants and the levels of the nonalkali gas component concentrations.

$$Na + H_2O = NaOH + H$$

$$Na + HCl = NaCl + H$$

Other chemical reactions facilitate and can enhance the apparent kinetic rates implied by these two reactions such as NaOH being formed also by a sequence $Na \rightarrow NaO \rightarrow NaOH$ (Hynes et al., 1984; Schofield, 2012a). As a result, the overall kinetics are rapid and display steady-state balances that can occur at atmospheric pressure on a microsecond timescale at elevated combustion temperatures. Reaction rates favor formation of gaseous NaOH and NaCl and when temperatures decrease and kinetic freezing occurs it will leave these as dominant. The main form of chlorine in combustion gases is HCl, whether fuel-rich or lean, with Cl only relevant at much higher temperatures and Cl_2 is negligible (Schofield, 2012a). The high temperature chemistry of HCl and Cl_2 have recently been reviewed (Pelucchi et al., 2015). An example of this is shown in a study resolving the deposition chemistry of NaCl (Schofield, 2012a). The alkali studies have managed to resolve the flame chemistry in laboratory flames solely by monitoring atomic sodium in well characterized flames.

It had been realized for quite a while that Na and K compounds were quite good combustion inhibitors (Babushok and Tsang, 2000; Linteris et al., 2008), and had a meaningful munitions application of reducing muzzle flash from weapons firing particularly at night. They have the necessary atoms and hydroxides to provide the necessary catalytic kinetic cycles for recombining and reducing the flame radical concentrations. Also it was also noted that addition of these alkali compounds in the reburn control method for reducing NO_x emissions improved its efficiency (Zamansky et al., 1996; Maly et al., 1999; Yang et al., 2008; Li and Wei, 2011; Lu et al., 2016). As a result, a new interest in alkali flame speciation began questioning the roles played by NaO and NaO_2 in such systems. This then introduced a controversy concerning the stability and importance of NaO_2. Atomic sodium concentrations were measured in a number of well-defined nonequilibrated hydrogen/oxygen flames. All the data then was then fit to a chemical kinetic model. The concentrations of all the basic flame gases were known. This involved the four gaseous alkali species, Na, NaO, NaO_2, NaOH in these flames interconnected by 17 possible chemical reactions and the monitored Na decay profiles in 10 flames of differing temperatures and compositions. This appeared to be a simple model with not too many variable parameters that could be adjusted to fit this reasonably large flame data base. It appeared to be a reducible concept. However, the faith in the computer was misplaced. As a result, even with this small model, although an apparently adequate unique fit was obtained, the level of accuracy in the data apparently was insufficient for exactness. It was a classic example of accepting modeling as reliable, after exhaustedly testing, over several years, 500,000 parametric variations of rate constants and bond strengths. It was a remarkable example of the risks involved in computer kinetic modeling! A reassessment of the flame work obtained a too high value of $D_0(Na\text{-}O_2) = 243 \pm 8$ kJ mol^{-1} (Steinberg and Schofield, 1991). As seen in Table 7.2, the accepted value is now 155 kJ mol^{-1}. However, the studies did imply that the oxides provided a more efficient conversion of Na to NaOH than with catalytic recombination of flame radicals explaining its inhibition properties (Hynes et al., 1984). In this period another approximate approach used to resolve this dilemma was introduced. This used a forward exothermic termolecular rate constant to try and extract the reverse rate constant and enthalpy. From unimolecular theory, a method known as the Troe formalism was suggested using only the measured values for this forward rate constant and the generic information from termolecular dissociations (Silver et al., 1984; Husain et al., 1985; Vinckier et al., 1992). However, because this is endeavoring to extract a bond strength that is controlling both the reverse rate constant and the connecting equilibrium constant from an exothermically driven forward rate constant, it also is a too demanding and insensitive method that failed. As a result, some values in a range 160–240 kJ mol^{-1} emerged questioning its value (Patrick and Golden, 1984; Husain et al., 1985; Plane and Rajasekhar, 1989; Marshall et al., 1990; Vinckier et al., 1992). It was finally resolved, but the nature of the bonding of these MO_2 dioxides still remains unclear. A later study has now modeled the flame inhibition of potassium compounds. This is a very ambitious kinetic model considering 12 species and 85 reactions (Babushok et al., 2017a). It implies quite minor roles for KO and KO_2 that may be correct. However, it does invoke such gaseous species as K_2CO_3, $KHCO_3$, and KCO_3, which are incompatible with flame gases. It is complex and must remain speculative.

Nevertheless, it was felt that the homogeneous gas phase properties of the alkali metals now were becoming well understood including the substantial ionization studies. It had always been noted that cesium ionized thermally at lower temperatures, due to its low ionization potential. Moreover, the degree of ionization could be readily calculated from the ionization potential that was 520 kJ mol^{-1} for Li, changing to 376 kJ mol^{-1} for Cs. Any level of speciation affecting the overall atomic concentration would lower the level of ionization accordingly. Nevertheless, with cesium and high temperatures the ionic levels can hardly raise to perturb the levels of the neutral species. Thermal ionization also is the only channel for the alkalis. Bond strengths are not strong enough to invoke chemiionization. As a result, when it was noted that particularly Na, K, and Cs were involved in soot suppression this was coupled to their lower ionization levels and a precursor role (Howard and Kausch, 1980). Only recently has this aspect been examined further. Soot formation has been measured in ethylene flames with added NaCl and KCl. Soot was measured by laser-induced incandescence and elastic light scattering (Simonsson et al., 2018). They noted that both the soot particle size is reduced and the soot volume fraction. An additional study with a laminar coflow methane diffusion flame studied the addition of 100 nm NaCl particles. A large number of 10 nm particles appeared where soot inception is typically not expected. It was concluded that NaCl reduces the coagulation between primary soot particles (Moallemi et al., 2018). However, any role of ions in such behavior still remains unproven.

A remaining controversy, fundamental to numerous similar situations, centers on the possibility of Na_2SO_4 or K_2SO_4 formation in the gas phase with its subsequent homogeneous nucleation and particle growth. There is still no conclusive proof of this and much evidence against it. Nevertheless, one or two observations do require resolution. It is known that such gaseous molecules can be detected in the gas phase above vaporizing condensed phase sulfate but it is also clear that these appear unstable, are partially fragmented and are short-lived (Table 7.2) (Kosugi, 1972; Lau et al., 1979, 1985; Halle and Stern, 1980). This is true of many Knudsen cell-mass spectrometric studies of mixed metal oxides when larger molecules have been detected (Lopatin and Shugurov, 2013). Mechanisms are easy to speculate by purely chemical algebraic equations such as

$$2Na + SO_2 + O_2 = Na_2SO_4$$

$$2NaOH + SO_2 + 0.5O_2 = Na_2SO_4 + H_2O$$

$$2NaCl + SO_3 + H_2O = Na_2SO_4 + 2HCl$$

that reformulation of the sulfate might occur in flame gases, but as seen by any such suggestions, this is meaningless and acceptable kinetic mechanisms are necessary. To some degree with the alkalis two major kinetic problems exist. One is that two Na atoms are involved for Na_2SO_4 but only one for NaCl or NaOH formation. Also the sulfur has to oxidize its valency from SO_2 to SO_3. It is true that dissociations of simple molecules can often be reversible in the gas phase. We live as a result of the O_3/O_2 dissociative/recombination reactions that occur in the upper atmosphere. However, few if any dissociations of multiatom molecules are known that are easily reversed. The basis for this suggestion of gaseous alkali sulfate formation resulted from observations of enhanced nucleation and growth of ultrafine particles in the appropriate nucleation temperature range (Table 7.2) that might correspond to these sulfates (Jensen et al., 2000; Jimenez and Ballester, 2005, 2005a). They were not specifically analyzed though. As a result, a gas phase kinetic mechanism was presented invoking theoretically speculated unknown intermediates of $KHSO_4$ and/or KSO_3Cl formed from either KOH or KCl with SO_3 that are then further converted to K_2SO_4 by four-center reactions with KOH or KCl (Glarborg and Marshall, 2005; Hindiyarti et al., 2008). A major concern with the mechanism centers on its need for $SO_3(g)$ as the oxidant, which restricts the mechanism to lower temperatures and introduces time constraints critical to gas kinetics, and invokes KOH or KCl as equivalent alternate participants. Moreover, the process is required to be very efficient kinetically. The requirement of SO_3 reduces the functioning temperature to a range where any such sulfate formed will be at its condensing temperature if it can nucleate (Table 7.2). As mentioned earlier, it is an interesting dilemma that is seen to parallel similar current mechanistic difficulties concerning the analogous formation of H_2SO_4 both in the atmosphere (Du and Yu, 2009; Zhang et al., 2012a) and also in cooling combustion flue gases (Srivastava et al., 2004). In the atmosphere, mechanisms similarly remain very uncertain. It is known that a homogeneous SO_2/SO_3 mechanism does exist (Fleig et al., 2013), but in combustion a heterogeneous mechanism on fly ash has been seen to be significantly faster (Belo et al., 2014). A faster mechanism also appears needed in the atmosphere and one now has suggested atmospheric organic carbonyl oxides as the gas phase catalyst (Percival et al., 2013). More recently, the O_2^- superoxide ion that can be produced in the atmosphere from galactic cosmic rays has been invoked as the agent for this oxidation (Tsona et al., 2018). This now extends further. With the use of NH_3 in selective catalytic converters (SCRs), this has introduced problems of ammonium bisulfate, NH_4HSO_4, similarly concerning its formation and deposition in the cool end of air preheaters at about 190°C–240°C (Ellison et al., 2002; Menasha et al., 2011; Ma et al., 2014; Muzio et al., 2017). Again, it is intuitively expected that this results from the NH_3 slip, SO_3, and H_2O, and whether this can happen in the gas phase. Whether the SO_2 needs to be catalyzed to the trioxide or whether it centers on heterogeneous chemideposition remains to be seen. The necessary formation of all these sulfate compounds has to invoke some mechanism of oxidation of SO_2 to SO_3, but none have a universally accepted mechanism. Homogeneous nucleation normally is not an easy process and generally needs an excess concentration (Zhang et al., 2012a; Wyslouzil and Wolk, 2016). To suppose that only one molecule can be its own nucleation center as posed by one theory is rather questionable (Christensen and Livbjerg, 2000). Also to kinetically form in the gas phase and then instantly condense would appear to be difficult to distinguish from heterogeneous chemideposition on some nondescript nanoparticle. The rate constants for SO_3's three-body formation from SO_2 and O have been established (Yilmaz et al., 2006; Hwang et al., 2010), but it has long been known that surfaces also can readily catalyze this reaction (Fenimore, 1972; Alemozafar and Madix, 2005). As a result, although attempts have been reported to monitor SO_3 through coal-fired utility boilers especially fit with an SCR and other air pollution control devices it remains difficult to interpret such data (Cao et al., 2010). It actually remains very difficult to reliably monitor SO_3 (Fleig et al., 2012; Vainio et al., 2013). As a result, these must all remain speculative. It does appear that once formed, atmospheric nucleation of SO_3 with water vapor can be readily triggered and for example the presence of 2–45 pptv of NH_3 is reported to result in up to millions of times more particles (Zollner et al., 2012). Although still without proof, this sulfate forming gas phase mechanism is widely accepted in many publications yet the question remains whether it plays any role of significance or is in fact even needed. A major test of a kinetic model is that it invokes time, concentrations, and rates. In the present case, models fail as they require a certain level of concentration to satisfy the two potassiums required in the product. Moreover, as displayed in Fig. 2.1, efficient heterogeneous deposition is seen from high temperature flames that contain no SO_3 and can have very low ppmv additions onto cooler surfaces. The consideration of a homogeneous kinetic mechanism is mainly based on one experimental study (Jensen et al., 2000). This utilized a laboratory tubular reactor at about 750°C with 200 ppmv of KCl released from a packed bed of inert porous pellets impregnated with the salt. This was in a flow of N_2, O_2, and H_2O. The unexplained result concerned the abundance of new nucleating smaller diameter particles when 200 ppmv of SO_2 was added. This requires further study and either is valid evidence for some possible homonucleation of K_2SO_4 or has to be explained by chemideposition on contaminating submicron particles or even charged dust. A recent study attempting to resolve this issue similarly has observed small particles (less than 100 nm diameter) on mixing gaseous KCl with SO_2 (Li et al., 2013). However, again an $\Phi = 0.9$ CH_4 flame was used containing 10–50 ppmv of KCl. On addition of SO_2, weak laser-scattering from particles on a scale comparable to that observed from dust in the ambient air was monitored at temperatures slightly above those for KCl deposition. In a period when the heterogeneous mechanism has been proven quantitatively beyond a doubt as being very efficient, it is disturbing that yet another experiment has emerged in an attempt to show some role for the homogeneous contribution (Weng et al., 2018). This had a counterflow of a premixed methane/air flame containing either 215 ppm of nebulized aqueous KOH or KCl against a nitrogen flow containing 245 or 980 ppmv SO_2. The mixing monitored region was below 1000K but not well defined as to temperature. A small signal of laser scattering was observed and labeled K_2SO_4. There was no method used to try and quantify this by introducing known amounts of a nonvolatile oxide or other particles. Moreover, there was a suggestion with KOH seeding that K_2CO_3 was forming. This really implies available nucleating particles in the flame that could be the centers for efficient heterogeneous formation as absolutely needed for carbonate that cannot exist in the gas phase. If there for carbonate, they were there also for sulfate. The high levels of seeding required in the experiment, and the lack of any positive product or quantitative analysis flaws the experiment. The attempt to verify any of this with added tailoring of the kinetic model with hypertheoretical molecules also is totally unacceptable.

Laboratory combustion studies: heterogeneous conditions

The decision to start a program examining the heterogeneous aspects of the alkali element's chemistry resulted from the availability of new analytical instrumentation. Namely, a Nicolet Fourier transform Raman spectrometer (Nd:YAG laser) and a high-resolution inductively coupled plasma (ICP)-atomic emission spectrometer. These proved excellent in identifying the chemical nature of the deposits, with the ICP accurately measuring their sodium and sulfur content ratio. In several chlorine experiments an Orion Model 94-17B Combination Chloride Electrode also was available to measure the chlorine content of a deposit. In addition to this newfound availability, an important paper had been published using Raman spectroscopy for the first time to monitor Na_2SO_4 deposition (McCarty and Anderson, 1987). Using seeded methane premixed fuel-lean flames, they collected a pure deposit of sodium sulfate. Moreover, this grew uniformly, the rate independent of thickness. However, for fuel-rich flames the deposit changed from sulfate to pure carbonate. This result was in itself intriguing. Because of their experimental monitoring requirements, their flame burned in a fused-cylinder chimney to isolate the flame gases from the exterior air. As a result, the collection probe was necessarily significantly downstream.

In the new program, a cylindrical Inconel-600 stainless steel was fabricated with a thermocouple just below the collection area skin with a central hole through its core for air or even water cooling. This now could be located horizontally only milliseconds downstream across a vertical flame, above the flat faced burner. Any deposit collection was on a uniform area of the probe. One convenience of the alkali salts is that they are white and readily water soluble making capture and analysis trivial. Condensed phase deposits were removed directly from the probe surface in flowing dry nitrogen and immediately sealed in test capillary tubes for analysis. Rate measurements were obtained by carefully dissolving deposits off the probe for analysis with the ICP. Alkali salts are water soluble. This placed the studies on a much more reliable quantitative basis displaying significant accuracy especially for combustion measurements. The studies started in 1995 and continued for 10 years involving 775 separate experiments. The experiments were run during the day and analyzed at night to get the results. As a result, the program directed itself from day by day. Even after the first 2 weeks it was realized that the findings were highly significant. Results are summarized below, many for the first time. All parameter dependencies were exhaustively studied.

Basic fundamental discoveries

It was immediately realized that deposition rates of an alkali (measured as atomic weight content, grams of sodium), whether forming the halide, carbonate, or sulfate deposit, were all essentially the same under similar circumstances in temperature regions where their surface vaporization was negligible. Gas phase speciation was irrelevant, the surface having total control and accepting equally any alkali compound. The curves in Fig. 7.1 were obtained. These are all on the same scale and have not been scaled together.

The corresponding curves for potassium mirror those seen for sodium solely being moved slightly to lower temperatures due to potassium's higher surface vapor pressure losses for these molecules. Levels of several ppmv of alkali in the flame would produce deposits measured as several milligrams of alkali/hour, sufficient for accurate analyses. Overall rates of deposition were controlled solely by the total alkali influx reaching the probe surface irrespective of the speciation in the flame, allowing for any vaporization or ablative losses from the surface. Moreover, whether NaCl, Na_2CO_3, or Na_2SO_4 is produced

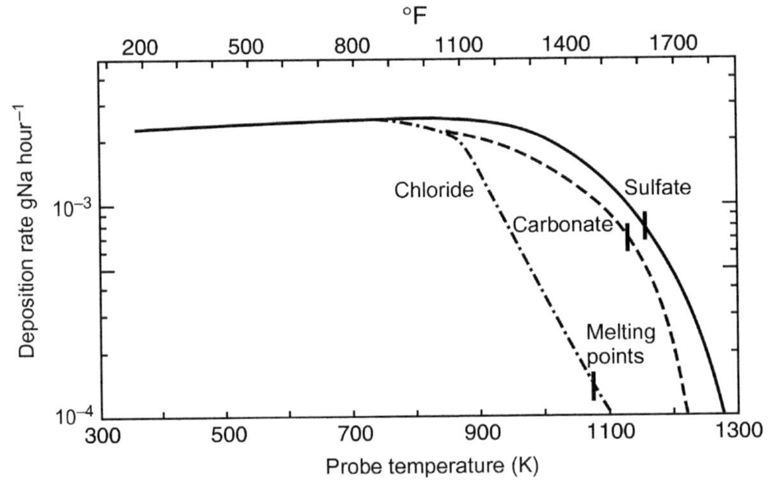

FIGURE 7.1 Rates of deposition for NaCl, Na_2CO_3, and Na_2SO_4 several milliseconds downstream in a fuel-lean hydrogen or propane/air flame under identical conditions as a function of the steel probe surface temperature. The concentration of flame sodium remained constant for the three individual cases of either no addition to the fuel gases (carbonate formed), with a slight excess addition of chlorine (chloride formation), or with sulfur (SO_2) addition (sulfate deposition). Source: *From Schofield, K., 2012a. The chemical nature of combustion deposition and corrosion: the case of alkali chlorides. Combust. Flame 159, 1987–1996.*

depended on the available elements in the flame gases. Whether it was NaCl with one atom of sodium or the sulfate with two made no difference. Deposition rates measured as atomic alkali content are always first order in the total sodium flame concentration. Also provided that they are present in excess of the chemical formula requirements, formation rates are zero order in chlorine, carbon or sulfur concentrations (Steinberg and Schofield, 1996, 2002; Schofield, 2012a). The zero order for the sulfur dependence implies an alkali sulfate deposit will always form to its full extent no matter how small the flame sulfur concentration may be. The nature of the alkali speciation in the flame gases is fully disconnected to the heterogeneous alkali molecular formation that occurs on a surface. The process is totally controlled by the preferential thermodynamic stabilities of the components on the surface at its conditions. Flames that are predominant in either atomic sodium, NaOH, or NaCl will produce exactly equivalent quantities (measured as the same atomic quantity of the alkali). This will be a Na_2CO_3 deposit if the flame is free of chlorine and sulfur, NaCl if only chlorine is present, or Na_2SO_4 if sulfur is present and chlorine is not in excess. Moreover the quantity of deposit is independent of the flame temperature or equivalence ratio for these not fully equilibrated flames. Such findings were totally unexpected. These deposition data in Fig. 7.1 were initially obtained about 20 years ago but the full significance and implications for the alkali metals have only fully emerged in recent years when coupled to other metal deposition data (Schofield, 2005a, 2012a). The most important parameter in chemical deposition was seen to be the total quantity of gaseous alkali in the flame, and not even its concentration, but its total flux rate to a surface, namely gas flow rate *and* concentration. This became obvious from the earliest quantitative measures of rates of sodium deposition from differing flame equivalence ratios but was not sufficiently stressed therein and went unnoticed (Steinberg and Schofield, 1996). However, on burning various propane or hydrogen flames on a flat flame burner, because of their differing burning velocities, gas flows differ, producing different volumes of burned gases. It was noticed with Na_2SO_4 deposits that if the total addition of sodium to the flame was held constant the same amount of deposit was noted in all cases. More recently, this has been clearly illustrated with NaCl deposition when raw deposition data and its scatter were collected for a rich and lean propane flame, and a fuel lean hydrogen flame. The total sodium rate of addition in these three differing flames was held constant resulting in burned flame gas concentrations of 16, 28, and 38 ppmv yet indicating the same magnitude of deposit in each case (Schofield, 2012a). The kinetic rate of deposit formation is first order in Na concentration, but the quantity produced depends on the total flux reaching the surface. Consequently, chlorine-induced corrosion rates will depend on flow rates and total flux (Balan et al., 2013). Most importantly product formation reflects thermodynamic preferences and simply depends on the elements that are available in the gas phase. In many flames this preference is $Na_2SO_4 > NaCl > Na_2CO_3 > NaOH$.

The alkalis are insensitive to the nature or material of the surface on which they are depositing. This explains why no material has yet been fabricated for gas turbine blades that alleviates high temperature corrosion by NaCl, Na_2SO_4, or even the carbonate (Lehmusto et al., 2012). Also as noted in Fig. 7.1, deposition is not affected whether the product is forming either on top of a solid or a liquid phase deposit. However, data do indicate that with these alkali molecules the liquid phase is far more corrosive (Raask, 1985; Nielsen et al., 2000a). Additionally, with parts per million alkali flame concentrations it can be calculated that after a few seconds a monolayer of deposit has formed on the collection probe. As collection continues, this was confirmed to grow linearly with time. After collecting an analyzable quantity of a few milligrams it is obvious that the deposit is building on itself, and a still intriguing question is whether the surface is still engaged in any manner?

The falloff of deposit at higher temperatures illustrated in Fig. 7.1 simply indicates that the surface loss rates are becoming larger than the ingress flux at those temperatures. For the halide this closely approximates to the expected evaporation curve at higher temperature. However, when the measured vapor pressure data is used to construct the falloff for the sulfate or carbonate it is clearly evident that the falloff is happening too soon. Further data in different flames and at different flame locations confirms that deposit loss for these is not solely evaporative but also chemical ablation by interaction with flame radicals (Schofield, 2005a).

Initial implications

The most significant finding made is in Fig. 7.1, that the deposition of the sulfates closely mirrors that of the carbonates implying similar formation mechanisms. As the carbonates do not exist in the gas phase in flames, this would argue against any complex gas kinetic mechanism for the sulfates that is tailored only for that. It became clear that care should now be used with terms such as condensation and "dew point." Combustion deposition can no longer be considered as a simple condensation process. It is the surface and molecular thermodynamics that controls the resulting product. The flame gases are solely a provenience for the surface, and can be in any molecular form, together with any other preferred flame species. A "dew point" has to retain only its original meaning: that is the temperature at which a deposit is observed. From this it may be possible in some cases to guess what the deposit may be, but in the case of mercury even this is not so. Most importantly, "dew point" can convey no chemical information about the mechanistic formation process of the deposit, what it is, or how it relates to the gas phase species. Moreover, deposition is always occurring on surfaces at all temperatures, at all times, but is only noticeable when the influx to the surface is greater than the outflow. Deposition now is regarded as a heterogeneous process whereby a steady state is established between an ingress flux of ingredients to a surface with the possible egress of materials. It is not an equilibrated system and can only be so if no deposit is obtained. Moreover, the flux in or out may be totally different species.

Alkali chloride deposits do not require gaseous NaCl in the flame gases. Equal rates of NaCl formation have been observed from flame gases in which alkali flame compositions are either $>90\%$ Na(g) or $>90\%$ NaCl(g). As a result, there is a very fundamental difference apparent between condensation and chemideposition processes. Both are processed on the surface producing the same end product at the same rate with no kinetic differences observable. With chlorine in combustion gases, depending on flame conditions there will generally always be a distribution between gaseous NaCl and NaOH with smaller amounts of atomic Na that decrease to lower temperatures. The decreasing falloff rate of NaCl deposition to higher temperatures, indicated in Fig. 7.1, simply reflects the onset of vaporization with some slight chemical erosion also noted (Schofield, 2012a). The extent of deposit results from this difference between the surface total alkali ingress and egress. In this case, the observed deposition data appear to conform reasonably well to published vapor pressure data. The exact temperature location of the falloff curve in fact can very roughly calibrate the total gaseous alkali at the deposit surface in the flame. As alkali flame concentrations increase, the position of the falloff will gradually change to be at higher temperatures, moving about 22°C if the alkali concentration is doubled on a low ppm scale.

In flames with a high content of gaseous sodium or potassium molecules or even Hg, it is possible to obtain the preferred heterogeneous molecular deposit at temperatures slightly above the boiling points of Na (881°C), K (764°C), and Hg (357°C). It is also possible to see the consequences of temporary deposition above such temperatures when a deposit cannot remain on the surface but for a brief instant before fully vaporizing. This illustrates the efficiency and high rates of heterogeneous surface chemical reactions (Schofield, 2008, 2012a). The chemistry is faster than the time needed for the metal to simply rebound off the surface and escape. It also implies that deposition is always occurring on surfaces at all times and temperatures, but only is retained if it can form a compound that is thermodynamically stable at that condition and does not vaporize. In cases of not being retained, it still forms, but is too short-lived to be observed. The surface chemistry is predominantly controlled by thermodynamic stabilities. Kinetic limitations appear rare and do not affect rates of deposition.

Whereas the nature of the alkali in the gas phase generally is no more complex than a triatomic molecule, surface deposition appears to have no limits to its extent of complexity of molecular formation. Molecules such as $Na_2W_4O_{13}$ have been deposited that cannot be explained by gas phase kinetics (Schofield, 2005a). Such deposits become possible due to the two-dimensional nature of surfaces and the change from gaseous to condensed phase kinetics and thermodynamics. It is now probable that early mass spectrometric studies of flame gases that reported monitoring gaseous species such as $KHVO_3$, K_2VO_3, $KHCrO_4$, and K_2CrO_4 (Farber and Srivastava, 1973), KH_2AsO_4, KH_2SbO_4 (Farber and Srivastava, 1975), and Na_2SO_4, K_2SO_4 (Stearns et al., 1977; Fryburg et al., 1978) were a reflection of probe sampling difficulties and heterogeneous formation that have since been confirmed as possible (Jiminez and Ballester, 2005).

Fossil-fueled flames contain large concentrations of CO_2 and H_2O, with concentrations of HCl and SO_2 generally orders of magnitude smaller. As a result, because we now know that the alkali is chemically compelled to react the instant it reaches a surface it is plausible that deposition is a rather chaotic process. Obviously the alkali will most likely react with the first adequate molecule it encounters to ensure attachment to the surface. If this is not its thermodynamic primary ranked preference then subsequent conversion will immediately occur. Such conversions are easily observed in the laboratory when one layer of deposit can be converted to another or even be removed by inducing vaporization. For example, a carbonate deposit formed on a surface at 800°C will disappear on adding chlorine to the flame, due to such conversion and vaporization. In fact, conversion now is considered as a probable major mechanism contributing to corrosive behavior. An alkali chloride molecule that is initially formed cannot only corrode, but also continues to do so if converted to the more thermodynamically stable sulfate due to the release of highly corrosive chlorine on the surface.

Another point of interest is that with laboratory flames it is possible to accurately calibrate the total addition of alkali to flame gases by several methods (Schofield, 2003a). Then by measuring the extent of deposit collected on a probe it is possible to estimate quite well the fraction that is being deposited. Results repeated many times of such an exercise for the alkalis and also for mercury gave a surprisingly high value of about 30% probe collection efficiency (Steinberg and Schofield, 1996; Schofield, 2005a). Because this is a large fraction, it not only implies that 30% is reaching the surface but is passing through the boundary level. As an already large quantity, this would seem to be controlled mainly by the boundary level transit and that the surface itself is possibly at close to 100% retention efficiency.

These high deposition rate efficiencies, currently being analyzed theoretically, do not conform to present transport theories through a boundary layer based on diffusion, thermophoretic, Soret, or Fick mechanisms (Mulholland et al., 1999; Pyykonen and Jokiniemi, 2003; Akbar et al., 2010). As reproduced in Fig. 2.2, with the same molecular trace additions of any of the five alkalis and even mercury to flames, their rates of sulfate deposition on a surface at 270°C in either a propane or a hydrogen flame were all of a very similar magnitude (Schofield, 2005a). Depositing either Li(g) of atomic weight 7 or LiOH(g) molecular weight 24, or the other alkalis, behaved very like mercury, which is totally atomic in the flame gases and has an atomic weight of 200. It cannot simply be a diffusional mechanism and remains an unexpected quandary beyond current understanding. This was further apparent in two auxiliary sets of experiments (Steinberg and Schofield, 2002). $C_3H_8/O_2/N_2$ flames, rich and lean were burned with high percentages of the nitrogen being replaced in turn by He, Ne, or Ar. In each case the same quantity of sodium carbonate was deposited in spite of the transport differences. Also in further test flames of $H_2/O_2/N_2$ and $D_2/O_2/N_2$ with an unburned gas ratio of 6/1/5, the flames similarly deposited the same quantity of carbonate in spite of a gas phase containing different isotopic masses of species. Irrespective of surface or flame temperature, as seen in Fig. 7.1, the alkali has a constant influx to the surface. Only upon reaching the surface can the preferred chemistry be invoked. Consequently, the alkali is colliding with the surface whether a chloride, carbonate or sulfate is to result. The choice is controlled by the thermodynamic preference but occurs irrespective of any possible subsequent consequences. In other words, in a flame free of sulfur, but containing chlorine, the thermodynamic preference is always to deposit the alkali chloride whether it can remain on the surface or be instantly vaporized. Intuitively it would seem more sensible for the alkali at elevated temperatures to produce the thermally stable carbonate that could exist to higher temperatures but the thermodynamic driving force is controlling and functioning in real time for those exact conditions. Consequently, if chlorine is present with the alkali in an equal atomic quantity, a deposit of the chloride always forms even to high temperatures where it vaporizes. It can only be retained by the surface at those temperatures when its vapor pressure falls below the total alkali ingress. Carbonate deposits can never be expected unless chlorine insufficiencies ($Cl < Na$) are present. Rigorously, it cannot be totally ruled out that a carbonate deposit might possibly be formed temporarily as an oasis on a surface where no HCl is present at that exact instant. Nevertheless, if so, this would then be instantly converted to the preferred chloride on encountering HCl and if too high a temperature instantly evaporate.

This is not expected to be the case but is an indistinguishable option. On the same basis, hydroxides can never be permanently formed in carbon-based flames. The preferred ranking of chloride>carbonate>hydroxide has been rigorously proven for the alkalis in flames at all temperatures and equivalence ratios.

Manipulations

Sulfur has a very strong affinity for the alkalis and most metal sulfates are thermodynamically very stable. From an experimental laboratory flame containing SO_2 that is exactly half the concentration of the alkali a pure deposit of Na_2SO_4 can be produced. Any less than this and the alkali will use the sulfur to the fullest extent possible, and then fill in the deficit by producing Na_2CO_3 (Schofield, 2003a). In other words it is an essentially exact titration. Any proposed method that endeavors to control sulfate corrosion or deposition by reducing sulfur in the fuel will not be effective unless every trace of sulfur is removed. It might be noted that although this dominance of sulfates is apparent it is only under oxygen-rich conditions that is the normal combustion mode in practice. This is because such flame sulfur is SO_2. However, for fuel-rich conditions that relate to pyrolysis or gasification processes with negligible O_2, the sulfur is mainly present as H_2S at equilibrium and not compatible with sulfate formation. This was why in the early experiments of McCarty and Anderson (1987) they observed carbonate with fuel-rich flames even with sulfur present. They were so far downstream that the flame gases were fully equilibrated and their sulfur was H_2S on the fuel-rich equivalence. This exact titration effect of SO_2 on alkalis is quite remarkable and was developed into a new method for calibrating alkali concentrations in flames. Using a certified SO_2 gas mixture concentration and measuring the produced deposit composition (exact Na to S ratio) an absolute calibration of the sodium concentration in a flame could be made. The method was further validated by comparing this method to other standard approaches (Schofield, 2003a).

Chlorine also has a strong affinity for alkalis at all equivalence ratios but less so than that of sulfur under oxygen-rich conditions. When not in competition with sulfur it is similarly dominant but does not titrate the alkali quite as exactly. There has to be a slight Cl excess over that of the alkali but only by several-fold in concentration to obtain pure chloride deposits (Schofield, 2012a). In other words, by adding SO_2 and HCl to a flame, each at a concentration of one-third the sodium, a deposit would result being one third each of sulfate, chloride and carbonate. It can be fully manipulated.

It is readily noted that deposition can also be modified by conversions. This is easily witnessed by initially having a flame with no additions and collecting a carbonate deposit. Even a hydrogen flame will produce this if normal ambient air is used, due to its CO_2 content. Then by adding chlorine to the flame the initial deposit will fully convert to the chloride and by further adding sulfur fully convert to the sulfate. Such conversions to more stable molecules are irreversible other than by raising temperatures to dissociate or vaporize the deposit. As was finally discovered by further adding a tungsten or molybdenum salt to the flame gases, even an alkali sulfate deposit could be converted further or even prevented in the first place from forming. Such an addition produced instead the more thermodynamically stable polytungstates or polymolybdates (Schofield, 2003, 2005). Sequences of thermodynamic preferences for alkalis that have been proven are, for example:

$$Na_2SO_4 > NaCl > Na_2CO_3 > NaOH$$

$$K_2SO_4 > KCl > K_2CO_3 > KOH$$

$$Na_2W_4O_{13} > Na_2W_2O_7 > Na_2WO_4 > Na_2SO_4 > NaCl > Na_2CO_3 > NaOH \text{ (Schofield, 2003)}$$

$$K_2Mo_4O_{13} > K_2Mo_3O_{10} > K_2Mo_2O_7 > K_2SO_4 > K_2MoO_4 > KCl > K_2CO_3 > KOH \text{ (Schofield, 2005)}$$

This concept now is invoked in one major mitigation method to minimize chlorine deposition and corrosion. By adding large quantities of sulfur to burning biomass, rich in potassium and chlorine but low in sulfur content, sulfate deposits are preferentially formed. This generally utilizes ammonium sulfate additions, for example, that prevent or rapidly convert any chloride deposits to sulfate (Brostrom et al., 2007; Kassman et al., 2011). Although alkali sulfates are also corrosive it is to a much lesser degree (Raask, 1985; Mobin et al., 1996; Nielsen et al., 2000a). Such conversions of layers of KCl to K_2SO_4 have been reported in straw-fired boilers (Hansen et al., 2000).

It does appear, if chlorine is in large excess, and an NaCl deposit is formed in significant amounts, then, depending on the temperature, there is some indication simply because deposits are condensed and multilayered that subsequent conversion by sulfur to the preferred sulfate may be rate controlled. This slowness of the chloride to sulfate conversion is in fact well documented (Henriksson and Warnqvist, 1979; Fielder et al., 1984; Boonsongsup et al., 1997; Ozawa et al., 2003), but with suggestions now of enhanced rates possible if SO_3 is present (Fielder et al., 1984). The conversion may also be kinetically limited by a required surface formation rate of SO_3. It is noteworthy that a commercial production method for Na_2SO_4 has been suggested utilizing conversion of NaCl with SO_3 and water in a fluidized bed at 500°C−600°C (Satrio et al., 2002). In atmospheric chemistry research, it has also been reported that SO_2 oxidizes on alkali halide surfaces and forms sulfates above 175°C (Rechtsteiner and Ganske, 1997). Also one method already mentioned to mitigate corrosion by alkali halides is to add $(NH_4)_2SO_4$ downstream where this initially dissociates to SO_3 (Becidan et al., 2009; Kassman et al., 2010).

An interesting aspect of such conversions is that condensed phase molecules have specific crystal structures. As a result, when being forced to change from one deposit to another this will lead to crystal defects in the initial preformed deposit as it is being converted, while continuing to have the new deposit layering on top at the same time.

Alkali gaseous transport

It is now obvious that alkali transport in the gas phase is solely by Na/NaOH/NaCl and similarly for potassium. There is no evidence for gas phase sulfate formation. Consequently, observed results depend very heavily on what was actually measured and what parameters were changed. A most recent study for example was comparing the effects of oxy-burn on alkali release from coal (Li et al., 2015). The measurement of alkali release was the difference in alkali mass between the coal sample and the collected char. This confirmed the ready release of alkalis but did not correlate in any way to the fuller history of the alkali in practical coal combustion where it experiences a full range of temperature change within a certain time period in an everchanging flue gas environment. Measuring alkali speciation remains difficult. A very ambitious recent study endeavored to resolve the degree of sulfation of potassium with injected KCl into a cofired coal/biomass system. The measurements were solely HCl(g) concentrations, obtained by probe sampling and then Fourier-transform infrared spectroscopy (FTIR) absorption. KCl(g) was monitored directly by ultraviolet absorption, which remains spectrally very difficult. The assumed K_2SO_4(g) was calculated by difference between the input and measured KCl levels. This is fraught with questionable experimental uncertainties and unfortunately the study necessarily has unknown accuracies and remains unacceptable (Ekvall et al., 2017).

With any solid fuel containing sodium or potassium, a fraction of these will be released on combustion into the gas phase. The extent of this is important due to the adverse practical consequences. Consequently, this has been very extensively studied with all fuel types in spite of the experimental difficulties. As a result, coupling to the broad-based knowledge of the possible chemistry, various conclusions now are clearly evident. Condensed phase chemistry is largely controlled by temperature and thermodynamic stabilities of the compositional components. Moreover, if formed in the condensed phase by a thermodynamic preference, the temperature then will control any subsequent vaporization. Also heterogeneous reactions if favored by the thermodynamics can be very rapid and this is why large quantities of the alkalis can be retained directly by the fuel or ash particle as thermally stable refractory glass-like formulations of silicates or aluminosilicates that have low vapor pressures. Coal or biofuels contain a myriad of alkali chemicals. However, few of these are thermally stable and on combustion all the weakly bound organic forms, and the inorganic nitrates and most sulfur compounds will rapidly break down, become chemically modified and either react or begin to be released at temperatures below about 600°C. Existing or formulated molecules such as NaCl, KCl, NaOH, and KOH are stable to high temperatures but are quite volatile as seen in Table 7.2. As a result, these are released in increasing amounts as temperatures increase.

For coal or biofuels, equilibrium calculations or models of alkali behavior are difficult due to the incomplete nature of the thermochemical database, the strong dependence on the fuel composition, and the temperature regimes. Nevertheless, they may be of limited use for qualitatively indicating the potential relative ash distributions that might be expected. Numerous examples of such equilibrium calculations are available (Yan et al., 1999; Wei et al., 2002, 2005; Poole et al., 2008). One considers the interactions between Na and Al/silicon minerals for low-rank coals during gasification (Kosminski et al., 2006). The actual degree of alkali release can be qualitatively understood if the appropriate Al/Si/Fe compounds are included in the database together with sulfur, chlorine, and their interactions considered with the other major players such as Ca, Mg, and even P. A recent effort has considered the alkali behavior for coal/biomass cofiring considering Ca and Si effects and the degree of alkali aluminosilicate ash retention (Wu et al., 2016). Modeling examples also help to illustrate the difficulty and what information is required (Pyykonen et al., 2003; Joller et al., 2007; Akbar et al., 2010; Garba et al., 2012). This has to include a formulation of release physical processes coupled to gas phase chemical distributions, interactions in the condensed phase, and the nature of deposition mechanisms on cooling. A recent detailed study of biomass combustion and deposition in a vertical refractory tube endeavored to analyze and model the observed alkali deposition rates (Capablo, 2016). It concluded that the data could be consistent with an earlier suggested thermophoresis driven mechanism (Castillo and Rosner, 1989). However, closer examination of such studies now may indicate a level of adequacy, but they all remain inconsistent with the recent major state of the art heterogeneous mechanistic conclusions already summarized.

The difficulties with alkali characterization arise from the higher temperatures of coal combustion that introduce retention due to well-known alkali aluminosilicate formation within the ash (Kosminski et al., 2006b; Gornostayev et al., 2007). Direct retention is controlled by the temperature and the Si/Al/Ca composition of the fuel (Novakovic et al., 2009). Coupled to the alkali there are glass-forming components having elevated thermal stabilities but melting points very much controlled by the alkali content (Raask, 1985; Bryers, 1996; Thy et al., 2006). Most importantly, there is chemistry available that can initiate at higher temperatures and so this has a much longer interaction period in the cooling flow gas train than for most other metal elements. Biomass is quite different from coal in that K>>Na with low levels of chlorine and sulfur. The atomic weights of K and Cl are almost the same so examining the analyses of biomass mineral content based on mass weight is directly apparent. Levels of Si can be comparable to potassium and Al can be very low. Invariably, although KCl is implied to be important there is often insufficient chlorine for this and the predominant species will be gaseous KOH and K (Dayton et al., 1995; Knudsen et al., 2004; van Lith et al., 2008; Wu et al., 2011). Additionally retention as a silicate is possible with some biomass forms but generally is much reduced (Valmari et al., 1999; Wu et al., 2009). As a move to reduce overall "old carbon" CO_2 emissions by using "new" carbon in the form of biomass, the partnering of biomass with coal has proved advantageous. They are complementary fuels, the coal supplying the higher levels of Si, Al and Cl to balance the high level K in biomass. The same is found in coal/refuse-derived fuel co-combustion where the high Cl in the waste fuel requires the coal ingredients for balance (Wei et al., 2009). As a result, by such tailoring, it is possible to burn blends where 70%–80% of the alkali is retained as the aluminosilicate, and possibly ferrite or titanate, and the free alkali levels and HCl are reduced to acceptable levels (Glazer et al., 2005; Zheng et al., 2007; Doshi et al., 2009; Fryda et al., 2010; Piotrowska et al., 2010; Li et al., 2014).

Similar patterns occur with biomass pyrolysis and municipal waste incineration when only a fraction of the alkali is released but gradually increases from 500°C to 1000°C (Olsson et al., 1997; Jensen et al., 2000a; Bojer et al., 2008; Pedersen et al., 2010). Sampling hot flue gases under coal gasification conditions at temperatures up to 1500°C and 1700°C the sodium and potassium were monitored mass spectrometrically as NaCl, NaOH, KCl, and KOH the only gaseous species stable at these temperatures (Blasing et al., 2010, 2014a). Such observations confirmed the nature of alkali molecular thermal instabilities and the simplified gas phase chemistry.

As flue gases cool and lower temperature ash or metal surfaces are encountered, deposition from the gas phase occurs. Until recent decades this has always been considered as a condensation process when the concentration of the molecule present in the gas phase becomes saturated (its so-called dew point) and simply condenses as a liquid or solid on the surface. However, this now has been conclusively shown not to be the case, either for the alkalis or any of the other trace metal elements. Heterogeneous chemical thermodynamics becomes the major controlling mechanism and the old "dew point" term loses any chemical implication it may have acquired and simply returns to its original definition of being the temperature when a deposit is seen. In recent years, advances in the laboratory relating to the controlling details of chemical deposition have been resolved with the availability of more advanced analytical ability. These quantitative findings now severely question previous concepts, speculations, and the theoretical models.

Generally, as a result, the information being sought relates to several components in the fuel combustion cycle. That is, what is the total alkali being released from the fuel, how much deposits on the ash or other surfaces, how much remains in the gas phase and what is the nature of its molecular speciation in the gas and the ash? The transport of chlorides through the gas phase generally poses the undesired repercussions for industry and is the one aspect that needs to be controlled.

Combustion power plant data

Of the two more major alkali metals, sodium tends to relate more to coal combustion and potassium to biofuels although both generally are always present to their differing extents. As indicated in Table 1.1, their concentrations can vary over a wide range in coals and generally refer to the low percentage scale of 0.1–3.0 wt.% for sodium and are either comparable for potassium or lower (Coles et al., 1979; Hu et al., 2006). Municipal wastes can exhibit similar values (Becidan et al., 2010; Pedersen et al., 2010), while sewage sludge has the potential for levels up to 2–3.5 wt.% (Han et al., 2006). Biofuels are consistently much richer in potassium than sodium and can have a very wide range of concentrations dependent on the exact nature of the biomass ranging from 0.1 wt.% with sawdust/sugarcane (Okuna et al., 2005), to 1.0 wt.% for wheat (Johansen et al., 2011). Straw and wood, rich in potassium, still can contain a wide range of levels of Si, yet for some, low levels of Cl and S (Bashir et al., 2012; Chandrasekaran et al., 2012). An extreme case, listed in Table 7.3, is that of black liquor waste fuels that are generated by the pulp and paper industry and can contain 20% sodium and up to 3.0% potassium (Saw et al., 2010a). In comparison to these fuels, alkali presence in pet-coke is negligible but can be present in limestone on a ppmw scale.

TABLE 7.3 A typical dry black liquor composition by weight.

Major elements	%	Minor elements	ppm
Carbon	30–37	Aluminum	50
Oxygen	35–39	Calcium	600
Sodium	18–23	Silicon	0–1000
Hydrogen	3.1–3.7	Iron	150
Potassium	1.0–3.0		
Sulfur	3.0–6.0		
Chlorine	0.3–4.5		

Significant attention currently concentrates on the huge Zhundong coal reserves in China. This is a 390 billion ton source sufficient for China's energy needs for more than 100 years. The only problem is that it is a high sodium, calcium coal that is difficult to burn without causing severe problems. Extensive studies are underway to either pretreat the coal (Yang et al., 2018), or enhance retainment in the combustion process. This is mainly by invoking aluminosilicate formation in the ash or absorption by adding kaolin or a similar clay (Wang et al., 2017b, 2018a; Guo et al., 2017; Liu et al., 2018). Otherwise, a solution is to cofire this coal with either biomass or sludge to better balance the ingredients. What is evident from all these studies is that monitoring remains very difficult and often meaningless and moreover every study is unique depending on the composition of the coal or fuel mix, the temperatures, and interaction time. In other words it is difficult to predict the alkali release, without an individual measurement.

Alkali release in coal combustors tends to show final distributions between any bottom ash or slag and the fly ash (Querol et al., 1995; Martinez-Tarazona and Spears, 1996; Nielsen et al., 2002). In laboratory scale measurements it is clear that silica and Al in coal can retain alkali and also be useful as scavengers (Bencze et al., 2016). High calcium levels can interfere by directly reacting with these. It is evident that release is very much dependent on coal type with chlorine aiding release and sulfur inhibiting it (Gallagher et al., 1996; Wee et al., 2005; Schurmann et al., 2007; Zhang et al., 2011).

Coal, municipal wastes, and biomass can all have chlorine contents over a wide range of 100 ppmw to several thousand (Tillman et al., 2009). As a result, the ratio of gaseous NaCl to NaOH can vary quite considerably depending on temperature and chlorine content. Nevertheless, because of this efficient chemistry, as to the exact nature of how the alkali leaves a fuel particle still remains unknown but becomes somewhat irrelevant. Once released, the alkali will be instantly controlled by this kinetically efficient flame-gas imposed distribution.

If necessary, the total amount in the gas phase can always be measured simply by collecting flue gas samples using wet chemistry methods followed by a total analyzer such as an ICP (Lee et al., 1992; Bojer et al., 2008). This appears rarely performed though. More recent studies used an ICP/MS to examine in real time the total gaseous Na and K content from a single particle of biomass combustion (Sommersacher et al., 2016), and laser-induced breakdown spectroscopy for K, but remains speculative and not persuasive (Liu et al., 2019). Many studies have examined both the size distributions and composition of fly ash and provide some basis of understanding. However, it is to be remembered that the data relate to where samples were obtained in the combustion train and how they were collected. Within any combustor, the ash particle is in constant change with time as are the flue gases. Consequently, both the structural features and the chemistry are each under temperature, time, and concentration constraints that together will produce a fly ash that is unique and can only be referred to in general terms. Recent interest in oxygen-enriched combustion has added new interest in better characterizing the fly ash by introducing this further additional parameter.

It is hardly addressed, but because particles remain burning as a function of time there is always a question of whether particles establish their own temperature or can reflect any equilibration with the burned gases in which they coexist. Depending on the mode of combustion and the extent of fragmentation, fly ash will exist in all diameter sizes below that of the fuel (Nielsen et al., 2002). Useful attempts at modeling particle size distributions and the combustion phases and history of coal and biomass particles have been extensively developed (Baxter, 1990; Baxter et al., 1997; Lu et al., 2008, 2010). Over a range from <2.5 to >65 μm the alkalis can reflect a similar wt.% with size fraction (Martinez-Tarazona and Spears, 1996). However, further more refined examinations of small-size particles clearly do display a bimodal nature with peaks in the 0.1–0.5 and 1–10 μm ranges that indicate alkali enrichment (Nielsen et al., 2002; Lind et al., 2003, 2006; Takuwa et al., 2006; Wiinikka et al., 2006, 2007). Although the very small sizes <1.0 μm constitute only a 1%–2% weight fraction of the total ash load they have importance with respect to health concerns and also ESP control (Linak et al., 2007). Modern electrostatic precipitators have efficiencies for fly ash removal of above 99%, but their leakage lies mainly in this low <0.5 μm size range (Lind et al., 2003). Using a lignite-fired plant, Wang et al. (2008) size-segregated the fly ash at the ESP outlet at 129°C into five categories: nano 0.03–0.06 μm, ultrafine 0.06–0.1 μm, submicron 0.7–1.1 μm, fine 1.1–1.8 μm, and fly ash <10 μm fractions. The content of 20 elements was analyzed in each fly ash fraction as well as the major components of the coal. One interesting aspect was that each fraction, even the nanosize (30–60 nm) particles contained all 20 elements indicating more than anything that with coal there always remains a level of chemical complexity at any size. Consequently, to imagine that certain chemistries occur in isolation without possible perturbation by other elements in the system is always an oversimplification. In this particular study, these particles in one form or another had all traversed the full temperature and time of the combustor with each element vying for some share of the available chemicals. As a result, such data that has experienced all the parametric variables of a combustor has to be unique to that system. Normally, only broad generalities will be possible with no finer details from this type of full or pilot-plant scale operation utilizing raw fuels. Consequently, even though the data indicated that Na and S had the highest element masses in the nanosize particles confirming their chemideposition, other gaseous metallic elements also deposit on the particles or exist from the original ash particle due to fragmentation transport. There also is coalescence at some point with other very small primal fragmentations to account for all the elements some of which, like iron, V, and Ti, are not volatile. The nanosize fraction had diameters of only 30–60 nm and since one molecule of Na_2SO_4 has a diameter of 0.5–1.0 nm they do have a limited molecular content. Four decades ago, coal-fired plume sampling clearly showed an abundance of submicron particles as small as 0.1 μm that still reflected a core of the coal matrix coated with deposited vapor species (Pueschel, 1976). Also in one recent study of these ultrafine coal fly ash particles (<0.5 μm), although the analysis raised environmental inhalation concerns by indicating that carbon, iron, or chromium were present, it also introduced the question of whether these elements were actually the formation nuclei (Linak et al., 2007). A more extensive analysis examined the variation of behavior of 15 coals of differing compositions (Teramae and Takarada, 2009). This clearly indicated the carryover of Si as still an important component of the smallest <0.06 μm particles and the major role of alkali deposition in the growth of particles less than 0.76 μm. With some biomass fuels, as indicated in Table 1.1; Si levels can at times be much reduced below that of coals, resulting in a dominant deposition of KCl, K_2SO_4, and $CaSO_4$ components (Li et al., 2012). Biomass particle emission can be dominated by submicron size particles in number and mass of composition mainly K,

S, Cl, and O (Johansson et al., 2003). One study burning straw discusses the problems of sampling particles. It is additionally interesting in that the compositional analyses of the small particles were reported (Christensen and Livbjerg, 1996). The habit of listing weight % in analysis is not a particularly useful unit for automatically visualizing associations for the chemist. However, conversion of such data to molecular quantities indicates that for the seven size ranges collected between 0.024 and 1.95 μm, each was alike concerning the relative molar ratios of K/Cl/S, being approximately 54:34:10. Such a ratio conveys that each particle size had exactly the same relative composition in the ratio of 10 moles of K_2SO_4 to 34 moles of KCl and apparently contained no carbonate. The straw K fuel content used did appear to be in excess of its total chloride and sulfate formation needs. As a result, the release of potassium did appear totally released from the straw and could deposit on the cooling ash. Absolute masses of the differing sizes was not measured but it would be expected that enrichment had occurred on the smaller particles solely from the surface area to volume dependence on particle diameter. It provided no mechanistic information. It did confirm this uniformity of chemistry onto particle surfaces when the gas phase contained the K/Cl/S on a constant ratio. Also thermodynamic expectations appear fully satisfied with sulfate dominant and utilizing all the sulfur with chloride making up the difference. Moreover, another intriguing question not usually addressed is the nascency of these different sized particles; it being likely that they originate at different locations in the flow train. The combustion temperatures were about 1000°C–1200°C in that study so in their history they could start only as sulfates at the higher temperatures as seen in Table 7.2 and Fig. 7.1 but then continue retaining first the sulfate and then also chloride as temperatures cooled. Consequently it is curious that both the 0.024 and 1.95 μm particles that differ in diameter by 80-fold and may have had differing times of transit through the flue gas still reflected the same relative proportions of K/Cl/S in each. Whether this is simply connected to a tracking equilibrium of the surface needs study.

Other current interest with ash particles now concerns the effects on such particle formation under oxygen-enriched conditions. Two such studies, one burning sewage-sludge (Zhang et al., 2007) and the other coal/biomass blends (Fryda et al., 2010) size-segregated the particles and provide a 30-element analysis of their contents. Again they have a broad distribution of elements in all size ranges and although there are minor differences between air and oxygen enrichment, there are equally only slight differences between whether a particle is 0.06 or 0.76 μm. With fly ash, major concerns still tend to center on where the sampling occurred and whether the data do reflect the flue gas environment or are perturbed to some degree by the probe sampling. Without experience and great care, too detailed an interpretation of some published data still has to remain in doubt (Jiminez and Ballester, 2005). The cofiring of biomass and sludge has been reported as having the benefit of reducing corrosion and decreasing the submicron particles (Amand et al., 2006). This had also been noted for coal-biomass cofiring. In that case, it was reported if the sulfur to Na ratio was at least five times the stoichiometric ratio, that no chloride was deposited and fouling difficulties were absent (Robinson et al., 2002).

It is well documented that the alkali release from fuels does appear to be facilitated by fuel-chlorine content. In some cases this may simply result from the chlorides significant vapor pressures. Nevertheless, the ratio of alkali to chlorine in the fuel is important and if chlorine is in excess it has been noted that there can be a correlation between the two, implying this KCl vaporization (Johansen et al., 2011). However, contrary to this in other biomass studies it is seen that chlorine is readily released at low temperatures (500°C) ahead of the release of K or Na (>800°C) (Knudsen et al., 2004; van Lith et al., 2008).

As a result, released alkali in the form of gaseous Na, NaOH, and NaCl if chlorine is present, become the species that transport the element through the combustion train. As flue gases and ash particles cool, controlling temperature ranges are reached when appropriate efficient chemical deposition can occur on, and interact with, the ash or any other encountered surface depleting the concentration.

As noted, to endeavor to portray a general behavior for the alkalis is difficult because of the many parameters that describe the life of a particle in a combustor. Ash is often sampled near the exit of combustors and so relates to an end product. Its creation and history can only be conjectured in a statistical manner. The fuel type alone determines many aspects as this contributes the chemicals and the enthalpy content. Combustion temperature and equivalence ratio control the extent of metallic release or retention and the course of particle fragmentation and resizing. A recent method may begin to shed better light on the process. This sampled the ultrafine particles from the 1530°C, 1230°C, and 930°C regions of a pulverized lignite combustor with in situ analysis of sodium content indicating that particle size distributions and composition can change with residence time and conditions (Gao et al., 2015, 2016; Xiao et al., 2016).

A fuel very rich in sodium, and of a unique nature is black liquor. This is a major byproduct of the paper-pulp industry that in its dry state is twice the quantity of the pulp production. In spite of the alkali level, which originates from soaps used in the process, its volume and energy content necessitates its recycling of the chemicals and the energy in a recovery boiler. Although the composition depends on the specific biomass, the approximate components of black liquid are as listed already in Table 7.3. It is interesting from a chemical viewpoint due to its very high sodium level yet low levels of sulfur, chlorine, and silicon. Clearly, combustion is going to produce not only the sulfate but some chloride and extensive carbonate deposits. Measurements of atomic sodium in the gas phase with laser induced fluorescence (LIF) indicate again very low ppmv levels (Saw et al., 2010a,b). In the gas phase, sodium will be mainly NaOH while the extent of NaCl will depend on whether there is sufficient chlorine. Nevertheless, in combustion the temperatures are such that the alkali will be released from the fuel and as temperatures cool toward 1000°C will rapidly deposit as the molten carbonate and sulfate. These form a smelt and a fume of small particles that are submicron and have been suggested to originate through thermophoretic chemideposition onto nucleating centers possibly produced from the traces of silicon and calcium (Cameron and Goerg-Wood, 1999). This was consistent with earlier molecular-beam mass spectrometer monitoring of the combustion at 1100°C. This noted the gaseous products of NaCl, NaOH, and even Na_2SO_4, but had illustrated the experimental difficulties of such measurements in practical systems (Dayton and Frederick, 1996). The system is complex but an approximate model for the inorganic emissions has been published (Wag et al., 1997). Black liquor droplet combustion has been examined in detail to characterize the individual physical stages (Ip et al., 2008). This illustrates the complexities of particle swelling and mass loss and observed that temperature differences between the surface and interior could be in excess of 300°C indicating that isothermal models were inadequate. Additions of boron can be utilized to retain some of the sodium as sodium borate (Saw et al., 2010). It also is involved in further reactions with the carbonate and permits removal as NaOH important to the process (Lindberg et al., 2007). If the black liquor is pyrolyzed or gasified, the reduced oxygen content prevents any sulfur from forming SO_2 but instead H_2S. Alkali behavior will be very similar to its combustion except Na_2SO_4 formation is not compatible with H_2S, but NaCl and the carbonate still can be formed. Although thermodynamic calculations predict the stability of Na_2S(s), its competition with the predominant carbonate formation either appears to be rate controlled or significant uncertainties remain concerning its thermodynamic parameters. Discussions involving Na_2S(s) formation all seem to be based on thermodynamic expectations and no clear analysis appears to be evident for its actual presence in the literature. Gaseous Na_2S or K_2S are not stable in combustion. The bond strengths for gaseous NaS and KS are respectively 257 and 264 kJ mol^{-1} (Table 7.2) and for K_2S(g) a tentative D_0(K-KS) ≈ 200 kJ mol^{-1} was measured, slightly less than the K-KO (Farber et al., 1983). In the gasifier, the fume, particles, and any H_2S or possibly OCS generally are quenched by particle scrubbing methods in these processes and removed to produce a usable synthetic gas (Larson et al., 2000; Ohrman et al., 2012).

In fuel pyrolysis/gasification processes that are oxygen-lean, as indicated above, without oxygen the alkali sulfates have to abdicate their normally dominant role to the halides and carbonate. As the equivalence ratio changes from oxygen-rich to lean, this change in deposition nature, is well established (McCarty and Anderson, 1987). In fact, if sulfate deposits are observed under fuel-rich conditions it implies either the presence of oxygen or a region of nonequilibration of SO_2 and H_2S, any SO_2 formed in the reaction zone still requiring further time to finally equilibrate to the latter (Muller et al., 1979). Pyrolysis/gasification processes strive to remove as much of the alkali as condensable aluminosilicates as possible. Deposition and gas phase studies confirm this behavior and also note the absence of Na_2S(s) even for high sulfur coal (Kosminski et al., 2006c; Blasing et al., 2010; Jordan and Akay, 2012). More recently, low-temperature gasification of biomass in a circulating fluidized bed shows increased retention of the alkali in the ash (Narayan et al., 2016). Release of K, Na, and Cl from rice and corn straw below 1000°C also was recently reported. Most of the alkalis were converted and in the ash (Yang et al., 2016d).

Possible mitigation or gaseous alkali cleanup methods have been studied for decades. Invariably they are tailored to specific problems such as boiler corrosion, bed agglomeration, or high-temperature gas cleanup. Methods center either on absorbing the alkali or minimizing alkali halide formation. The magnitude of the various problems center not only on the alkali/Cl/S fuel content but also on its Si/Al/Ca quantities and the interaction time and mode of combustion. Some biomass fuels with their high potassium levels pose the greatest problems and although now utilized directly (Bashir et al., 2012), it is still considered better to use co-combustion with coal or gas to moderate the levels (Glazer et al., 2005; Wu et al., 2011). One study proposed adding aluminosilicates to the biomass fuel for ash retention (Clery et al., 2018). Studies of co-combustion of coal and biomass in a fluidized bed reactor have indicated a required index of (Al + Si)/Fuel Cl mole fraction >8–10 was needed to prevent chloride deposition (Aho and Silvennoinen, 2004; Silvennoinen et al., 2005). One described this as Al/Si/S being a buffer to the alkali release (Doshi et al., 2009). Another study has noted that although coal was blended with bone meal, rich in chlorine, the combination of a coal rich in Al/Si and Ca controlled any chlorine deposition

(Aho and Ferrer, 2005). However, as Ca can reduce the sulfur content by forming $CaSO_4$, further assessments of the required $S/(Ca + 2K + 2Na)$ mole fractions to prevent Cl deposition also have been illustrated (Yrjas et al., 2005). An alternate method mentioned earlier is to add $(NH_4)_2SO_4$ or some other sulfate that dissociates directly to SO_3 to ensure only dominant sulfate deposition (Brostrom et al., 2007; Kassman et al., 2011, 2013), with a suggestion of S/Cl molar ratio >4–6 being required (Khalil et al., 2011a). Addition of solid sorbents to the fuel, center mainly on kaolin clay-type materials rich in Si/Al or also including Ca. Of these kaolinite, bauxite, boehmite, emathlite, and alumina were suggested even two decades ago (Uberoi et al., 1990; Spiro et al., 1990). At high temperatures, it has been seen that silica alone, SiO_2, has a great ability to react with Na, K, and Al (Gornostayev et al., 2007). The kaolin clays are regarded the most cost effective and are equally effective for Na or K in coal combustion (Kyi et al., 1999; Schurmann et al., 2001; Takuwa and Naruse, 2007; Zheng et al., 2008; Zhou et al., 2010), and also in gasification (Kosminski et al., 2006a). Kaolin use with biomass is not detrimental to the ash with regards to possible changes in leaching or agricultural use (Steenari and Fedje, 2010). Kaolin capture of potassium molecules has been reported recently but implied the presence again of gaseous carbonate and sulfate (Wang et al., 2018d,e). The past several years has seen numerous well defined studies involving biomass combustion that clearly illustrate this importance of Al/Si and a role of Ca whereby particulate mass concentrations can also play a major role in controlling the alkali levels (Zeuthen et al., 2007; Xiong et al., 2008; Bartels et al., 2008; Davidsson et al., 2008; Becidan et al., 2011; Wu et al., 2011). For a high alkali coal, adding 6 wt.% of 3 μm kaolin to 30 μm coal particles produced 80% sodium retention at 1100°C (Xu et al., 2015). In both the addition of kaolin and also TiO_2, an additional factor is the mitigation of small $PM_{0.5-2.5}$ particle sizes favoring particle growth (Si et al., 2014; Wiinikka et al., 2009). In a laboratory fixed-bed reactor, such absorbents were shown to be capable also of removing gaseous HCl (Dou et al., 2007). For biomass gasifiers, numerous sorbents have been examined at length showing the role of Al and indicating a chemisorption mechanism (de Oliveira et al., 2015). Gasification processes with high temperature cleaning of the syngas have always been a goal due to the resulting overall gain in efficiencies from direct gas turbine firing. Until recently, due to the need for cleaning of alkalis down to ppbv concentrations this has been too demanding. Normally synthetic gases are cooled, then scrubbed and cleaned before burning. However, now there is a greater range of high temperature filters (Sharma et al., 2010; Lupion et al., 2010), sorbents (Li et al., 2005a) and aluminosilicate absorbers used as drip-off getters even at 1400°C (Willenborg et al., 2006; Escobar and Muller, 2007; Waindich and Muller, 2014). These studies now report a possible cleaning down to 80 ppbv for potassium and 100 ppbv for NaCl. Additionally, a polishing technique based on adding very low ppmv levels of either ammonium molybdate or tungstate to the cleaned hot gases has shown that any remaining trace of alkali can be removed by its preferential deposition as a benign alkali polymolybdate or tungstate bronze that in fact is protective of the turbine blades (Schofield, 2003b, 2005, 2009). Such developments now can lead to more highly efficient combined cycle approaches using coal or biomass fuels, or operate in marine environments when even the salt in the air can be a problem.

As a result, due to the combination of extensive retention and reaction with the ash to form aluminosilicates and the efficient chemical deposition on particles and surfaces, the alkali metals can be effectively contained in combustion systems. Its multifaceted chemistry can initiate in the very early stages of combustion at elevated temperatures, so removing the oft-encountered time and temperature constraints. Consequently, now in power plants equipped with particle controls coupled to scrubbers (Cordoba et al., 2012), it would seem that any alkali aerosol or gaseous component emitted will be negligible.

In summary, the combustion behavior of sodium and potassium remains actively studied. It now illustrates the importance of heterogeneous chemistry. A reliable speciation monitor in the gas phase remains lacking but only the chloride content has any importance. This is the means of "nascent" chlorine transport through the flue gas. On deposition its conversion to sulfate releases active chlorine directly on the surface. The alkali fuel release and capture are of major importance to minimize corrosion, fouling and agglomeration problems, or catalyst deactivation. Because the selective catalytic reduction (SCR) and air preheater are at lower temperatures and are primarily surface components it is obvious these will be major regions of deposition and concern. Cofiring of higher content coals with biomass or sludge to modify the balance of fuel compositions can be beneficial. It is realized that outcomes are dependent on specific fuel composition, equivalence ratios, and temperature/ash particle histories. Capture/retention can be particularly on fly ash (in the form of aluminosilicate formation) or on additives such as kaolin. High chlorine fuels will facilitate the release of alkalis as their gaseous chlorides that do require mitigation.

7.2 Aluminum

Aluminum is the most common metal (8.8 wt.%) and the third most abundant element (after oxygen and silicon) in the Earth's crust so it is not surprising it is a significant component together with silicon in coals and its resulting ash (Table 1.1). It is present less in biomass (Chandrasekaran et al., 2012) and can show variable concentrations in wastes and sewage sludge. Its concentration is very low in pet coke but can be more significant in limestone (Lind et al., 1994; Cordoba et al., 2012). Recent work suggests its levels in synthetic biogas may be very high compared to natural gas and require processing (Cacchia et al., 2018). Although removed from Environmental Protection Agency (EPA's) lists of environmentally sensitive chemicals there is now a growing concern again about its potential neurotoxicity (Namiesnik and Rabajczyk, 2010). For combustion, this relates to the escape rate of fine particles containing traces of the element. Also it is well to remember that the large-scale production of aluminum has only been known for a relatively short historical time and in the last 80 years its mining production has increased 375-fold to a current scale of about 48 million tons $year^{-1}$ (Nriagu, 1988; US Geological Survey, 2015). It is an easily recycled metal and in the United States this is now at about 40%. It has become inexpensive with an ingot cost in the United States of about \$1 lb^{-1}. As a result, its presence in the environment has grown and it is of interest to examine its combustion chemistry in detail to assess its possible health hazards. Bauxite, the oxyhydroxy mineral is the major ore and is processed to the oxide, Al_2O_3, before being electrolyzed to the metal. The electrolysis consists of dissolving the oxide in cryolite, Na_3AlF_6, and utilizing carbon electrodes. This in itself is not without problems. The first aspect is the production of "Red Mud" a by-product in the initial processing of the oxide. Rather like black liquor in the pulp and paper industry it poses a severe disposal problem. It is produced in twice the volume of the alumina and is very strongly basic with a pH of 10–13 and comprised mainly of iron, titanium oxides and silica wastes. One method of disposal is to blend it into the production of cement (Vangelatos et al., 2009). The aluminum industry also consumes about 3% of the total world's electricity emphasizing the value in recycling that only requires one-tenth the energy in such secondary processing. Additionally, it was indicated 25 years ago that gaseous fluorocarbons were emitted from the carbon electrodes in the electrolysis and were a nonnegligible contributor to greenhouse gases (Abrahamson, 1992). The industry remains the largest source of atmospheric CF_4/C_2F_6 fluorocarbons but the United States has reduced the 2700 tons $year^{-1}$ emitted in 1990 by 80% and globally there has been a 40% reduction that continues (Scheehle, 2006). The global natural atmospheric burden of Al in dust appears to remain unassessed as does the anthropogenic contribution. Its presence in the Earth's crust and the large quantity in coal and the size of its industrial metallurgy would suggest that atmospheric particles probably are quite rich in its oxide. However, from a health perspective, the particle and its other ingredients remain more of a concern at present than the actual Al content.

Although not found in nature in the elemental state, aluminum is widely used due to its lightweight and for metal usage generally is slightly alloyed. It has strength, passive resistance to oxidation and corrosion with good electrical and thermal conductivity properties. Other than the metal it has its widely used ceramic oxide, Al_2O_3 corundum. Its sulfate is found in water purification and waste water treatment use, in baking powder and as an antiperspirant ingredient. The hydroxide is used medically as an antacid and is now the major adjuvant in most US vaccines.

Health concerns with aluminum

Aluminum is not known in any life form but is tolerated by animals and plants. Although previously considered nontoxic, and noncarcinogenic, medical concerns that centered mainly on its ability to replace other metals in the body, such as calcium in bone, now are reassessing its health aspects. Long-term effects remain controversial and elevated brain levels are noted in autopsies. It is now labeled a neurotoxin and known capable of crossing the blood/brain barrier. In certain cases it is seen that plant growth can be reduced by aluminum. As a result, like mercury, aluminum is alien to the human body but is ingested daily in diets or by medication. Whether it has long-term medical effects remains unknown (Schofield, 2016). Recently, it has been noted as a major toxic element in edible seaweed (Paz et al., 2019). Detailed toxicological reports are available on the Web (World Health Organization Trace elements, 1996; World Health Organization Al, 1997; ToxReport Al, 2008) but now are outdated having only emphasized oral intakes. These indicated that dietary exposure was on average about 0.12 mg Al/kg body wt. day^{-1}. A minimal risk level was suggested at a 1 mg level in these same units. Drinking water in the United States is now $\leq 0.2\ \mu g\ L^{-1}$ in most regions and negligible. This is regarded of minor importance because of the very low intestinal absorption for dietary aluminum compounds (0.1%–0.4%), or (0.5%–5%) for bioorganic forms of aluminum, with ready excretion from the body. There is no evidence of neurological effects from the use of antacid medications, antiperspirants and other oral sources and based on such data it currently has been accepted as harmless by the medical profession. Nevertheless, in regions of the world with areas of high aluminum deposits and content in the drinking water, it is quite well documented as a neurotoxin. However, this also can be hard to validate because such water often contains other neurotoxins such as Pb and Mn (Bakar et al., 2010). The body burden of aluminum is mainly present in the skeleton and lungs. Normal oral rejection is on a timescale of several days. Nevertheless, more recent experiments with mice, rats, and human autopsies clearly indicate that although inorganic, low levels of aluminum can cross into the brain and induce damage to the nervous system (Crepeaux et al., 2015; Cabus et al., 2015; Jiang, 2017). Effects of the combination of Al and Se on the brains of mice has indicated protection by the Se from its neural toxicity (Sadauskiene et al., 2016). As a result, the now increased medical use of aluminum hydroxide to replace mercury (thimerosal) in vaccines has raised grave concerns. When injected, the aluminum now may become $\approx 100\%$ absorbed by the body and introduces an increased body burden for its elimination. This differs significantly from any orally digested aluminum salts that are minimally absorbed by the intestines (1%). Analyses of dose levels have suggested that when injected from vaccines this may have severe potential medical implications and can correlate as a potential cause for autism (Tomljenovic and Shaw, 2011; Schofield, 2016) and Alzheimer's disease (Tomljenovic, 2011) and need its safe use to be reassessed (Lyons-Weiler and Ricketson, 2018; Masson et al., 2018). As a result, monitoring for aluminum by such body markers as blood, plasma, urine, hair or maternal milk sampling now are finally emerging (Goulle et al., 2005; Schofield, 2016; Nisse et al., 2017; Drobyshev et al., 2017; Bastos et al., 2018). This will determine the preferred monitoring method and establish recommended safe levels that still remain uncertain. The earliest of these studies examined 100 healthy volunteers in France for 27 elements and included aluminum. From this, the blood samples illustrated the generally observed broad unimodal distribution for aluminum spreading over a severalfold range and a reflection of people's differences and their genetic susceptibilities to either retain or reject a toxic material from the body. The 5th–95th statistical range for blood samples indicated values of 1.3 to 6.4 $\mu g\ L^{-1}$ similar to the values of Nisse et al. (2017) obtained later but from a similar group sample. How vaccines may bias these values remains for new surveys coupled to a better assessment of the actual toxicity curve. The Mayo Clinic in Rochester, New York is currently recommending a serum test that should be $<6\ \mu g\ L^{-1}$ (Mayo Clinic, 2018). Recent analyses of various groups indicate that blood, hair or urine samples may be valuable monitors (Oggiano et al., 2018). Aluminum plant workers did exhibit a normal doubling or larger content of Al in their hair (Skainy et al., 2018). Comparisons of these measures would appear to suggest that serum, whole blood, urine, or hair are all adequate as biomarkers of toxicity.

Aluminum's basic chemistry

In combustion, aluminum's major interest has always centered on the metal's very high heat of oxidation that justified its use as a propellant in rocket solid fuels. The metal (m.pt.660, b.pt.2519°C) displays mainly trivalency in a limited range of cationic, covalent compounds and mixed metal aluminates. The physical properties of those compounds relevant to combustion are summarized in Table 7.4.

TABLE 7.4 Properties of aluminum's condensed phase molecules relevant to combustion, °C.

m.pt. b.pt. vapor pressure
Al 660 2519, 10 ppmv/1209°C, ins
Al_2O_3 α 2054 $\approx$3000, 1 ppmv/$\approx$2000°C, ins
δ 2035 transitional phases
γ 2016
k 2039
$Al(OH)_3$ 300, dissoc to oxide 280°C–400°C, ins
AlO(OH) ins
$Al_2(CO_3)_3$ decomp, not pure
$Al(NO_3)_3$ >200 dissoc, unstable, s
$Al_2(SO_4)_3$ 1040, >500 decomp to oxide, s
$AlCl_3$ 193, >180 sublimes, 100 ppmv/77°C, s
AlH_3 decomp 100–170, reacts with water
$Al(CH_3)_3$ 15 125, volatile, reacts with water
Al_2S_3 $\approx$1230, hydrolyzed by water
AlP 2550, reacts with water
$AlPO_4$ 2000, ins
$Al(PO_3)_3$ (meta), $\approx$1525, ins
Al_2SiO_5 >1730
$Ba(AlO_2)_2$ 1827
$Ca(AlO_2)_2$ 1605, reacts with water
$Fe^{II}(AlO_2)_2$ 1800
$Mg(AlO_2)_2$ 2135
$NaAlO_2$ 1650, dissoc as vapor,10 ppmv/890°C, s
$KAlO_2$ s
$Ni(AlO_2)_2$ 2110

decomp, decomposes; *dissoc*, dissociates; *s*, water soluble; *ins*, water insoluble.

In combustion, like boron its fellow periodic table member, it is largely controlled by its Al/H/O and glass-forming chemistries. With aluminum, its main oxide is the stable ceramic αAl_2O_3, Corundum (m.pt.2054, b.pt.$\approx$3000°C) that unlike boron is not hydrolyzed. In its formation it can have differing Al_2O_3 transitional structures that are characterized by a Greek symbol. Although insoluble and generally unaffected by water it can react with acids and bases. At high temperatures it dissociates on vaporization to gaseous Al, AlO, and Al_2O with a minimal contribution of AlO_2 but has a total vapor pressure of 1 ppmv only well above 2000°C (Ho and Burns, 1980; Kashireninov et al., 1982; Lamoreaux et al., 1987). Its water soluble trichloride (m.pt.193°C) is stable but volatile, subliming below its melting at about 180°C directly as the gaseous dimer Al_2Cl_6. However, it is hydrolyzed by water. The water insoluble trihydroxide, $Al(OH)_3$ (m.pt.300°C), has found numerous uses. It is formed as one stage in the refining of bauxite when it precipitates from an intermediate sodium aluminum hydroxide solution. When heated above its melting point in the 280°C–400°C range it dissociates to the oxide by losing water and does not vaporize. Also on heating above 500°C, the sulfate $Al_2(SO_4)_3$ similarly dissociates directly to the oxide (Stern and Weise, 1966; Tagawa, 1984). The nitrate is water soluble but thermally unstable. Its dissociation kinetics above 200°C have been studied mass spectrometrically (Mahadevan et al., 2002). A pure carbonate, $Al_2(CO_3)_3$, has not been characterized, but many combinations with ammonia, sodium, and other elements do form double carbonates. The trihydride, AlH_3, is a hazardous material, decomposing above 150°C and is readily oxidized and hydrolyzed by water. It is regarded as a source of H_2 and its ignition and propulsion properties are still studied (Young et al., 2015). Aluminum also is known as the component of kaolin clays and zeolites and can form complex silicates in combination with other metals such as sodium, potassium, calcium, or magnesium. Similarly the water insoluble orthophosphate, $AlPO_4$ (m.pt.2000°C), and metaphosphate, $Al(PO_3)_3$ (m.pt.1525°C), exist along with other more complex phosphorus formulations. These are hard quartz-like materials, thermally stable, and are used in optical fibers and as ceramic bonding agents or molecular sieves. Under reducing conditions, molecules such as Al_2S_3 (m.pt. ≈ 1230°C) (Ferrante et al., 1981), or AlP (m.pt.2550°C), can be encountered. These are decomposed by water and are hazardous, the phosphide is a fumigant, pesticide, and is very toxic (Moghadamnia, 2012). Like boron, it does have a very volatile aluminum trimethyl that forms a dimer gas $Al_2(CH_3)_6$ (m.pt.15, b.pt.125°C). This finds use as a volatile source along with the trichloride in CVD (chemical vapor deposition) and the nanoparticle formation industry. It is hazardous being very flammable and reacts with water. As with many metals, aluminum oxide with other oxides can combine to form aluminum double salts or aluminates. The silicate Al_2SiO_5 (m.pt. > 1730°C) exists in various structural forms and other refractories are $BeAl_2O_4$ (m.pt.1867°C), $MgAl_2O_4$ (m.pt.2135°C), $CaAl_2O_4$ (m.pt.1605°C), $SrAl_2O_4$ (long-lived phosphor), and $BaAl_2O_4$ (m.pt.1827°C). Water soluble alkali aluminates also are known: $Li_2Al_2O_4$ (m.pt.1625°C, vapor pressure 1 ppm, 950°C; 10 ppm, 1140°C), $Na_2Al_2O_4$ (m.pt.1650°C, vapor pressure 1 ppm, 730°C; 10 ppm, 890°C) (Popkov and Semenov, 1971). Both of these alkali aluminates dissociate on vaporizing.

Gas phase stabilities and transport

As a result of its use as Al metal particles added to solid fuel propellants, there have been many studies characterizing its chemistry. The basic goal was always to understand this sufficiently to produce models of value to engineers. As a result many Knudsen cell and mass spectrometric studies have examined its vaporization chemistry supported by numerous theoretical calculations identifying its possible species. With an Al/O system, that is hydrogen-free, the gaseous species were initially reported to be Al, AlO, Al_2O, and

AlO_2. The Al, AlO, and Al_2O clearly were dominant but the role of the dioxide was controversial for many years. As a result mainly from extensive work by the Russian groups it was finally proven that there was no evidence for the presence of AlO_2 (Chervonnyl et al., 1977; Kashireninov et al., 1982). More recent work now has further confirmed this (Chervonnyl, 2010). Matrix isolation studies at low temperatures did identify an AlOO bent structure that suggested covalent bonding and recorded its basic vibrational frequencies (Snochik et al., 1983). Otherwise the other gaseous oxides are thermally stable and strongly bound. Gurvich et al. (1994) have reviewed their thermochemical values. Seventeen experimental studies up to 1977 suggested D_0(Al-O) = 507 ± 5 kJ mol^{-1}. This was confirmed by a similar analysis by Pedley and Marshall (1983). To these can now be added a value of 505 ± 10 and 510.7 ± 3.3 (Ho and Burns, 1980; Chervonnyl, 2010) from Knudsen cell-mass spectrometry, and a value of 508.5 ± 1 kJ mol^{-1} (Naulin and Costes, 1999) from assessing the reaction threshold of Al/O_2 in a crossed beam study. Consequently, this most accurate value appears reliable for recommending D_0(Al-O) = 508.5 ± 1 kJ mol^{-1}. Its electronic structure with an $X^2\Sigma^+$ ground state has been treated theoretically (Patrascu et al., 2014). In many of the same studies, the atomization energy of Al_2O also was determined. This was also characterized spectroscopically by matrix isolation and shown to be a linear $X^1\Sigma^+$ with a symmetric AlOAl structure (Douglas et al., 1983), confirmed by monitoring its fluorescence spectrum (Cai et al., 1991) and theory (Bencivenni et al., 1992). Gurvich et al. (1994) included 15 studies up to1987 in their analysis. From these, nine were selected and averaged to give $D_{0Atomiz}(Al_2O)$ = 1050 ± 20 kJ mol^{-1}. Since that appraisal a more recent value of 1067.2 ± 6.9 kJ mol^{-1}has been published (Chervonnyl, 2010). Again, it appears the value has been reasonably established. The NIST-JANAF thermochemical table dated 1979 had in fact suggested 1046 ± 18 kJ mol^{-1} (Chase, 1998). Consequently, until refined further, a value of $D_{0Atomiz}(Al_2O)$ = 1060 ± 10 kJ mol^{-1} would appear reasonable. From the value for AlO above, this indicates D_0(AlO-Al) = 552 ± 10 kJ mol^{-1}, more stable than AlO. Although many theoretical papers were published concerning the structure and stability of AlO_2 it has never been positively observed and appears to be negligible compared to the other two oxides. Theoretically its linear structure was predicted to be much more stable than a cyclic shape (Bu and Song, 2000; Pak and Gordon, 2001). As indicated above, the Russian studies concluded that if it is a product of vaporization it is in very small quantities, not detected, and they have not even considered estimating its bond strength. In their evaluation Gurvich et al. (1994) took the mass spectrometric data of Chervonnyl et al. (1977) and Ho and Burns (1980) and estimated an upper limit value of $D_{0Atomiz}(AlO_2)$ < 860 ± 20 kJ mol^{-1}. A theoretical estimate suggested 869 kJ mol^{-1} (Shirley et al., 2009) supported by an acceptable calculation also of 499 kJ mol^{-1} for AlO. This would indicate the second bond strength D_0(OAl-O) <352 ± 20 kJ mol^{-1}, considerably weaker. It is to be noted as well that this is consistent with the periodic table family trend. No dioxide has been established for the other members, Ga, In and Tl (Table 7.5). A study showing the actual vaporization speciation of Al assumed thermochemical values for AlO (506) and Al_2O(1045) but used a $D_{atomiz}(AlO_2)$ value of 907 rather than the <860 kJ mol^{-1} now accepted (Lamoreaux et al., 1987). A current study also used a value of D_0(OAl-O) = 572, even stronger than Al-O and consequently predicted similar unacceptable very significant vapor pressure values for AlO_2 (Baijot et al., 2016).

TABLE 7.5 Stability trends for relevant gaseous molecules of the B/Al/Ga/In/Tl group, kJ mol^{-1}

Bond	B[a,b]	Al	Ga[a]	In[a]	Tl[a]
D_0(M-O)	800 ± 10	508.5 ± 1[c]	370 ± 20	340 ± 25	≈260 ± 30
D_0(OM-O)	564 ± 22	≤352 ± 20[b]	–	–	–
D_0(M-OM)	≈400 ± 100?	552 ± 10[b]	515 ± 20	419 ± 30	≈324 ± 30
D_0(M-OH)	652 ± 25[d]	550 ± 10[b]	450 ± 10	400 ± 15	≈310
D_0(MO-H)	274 ± 25[d]	466 ± 10[b]	504 ± 20	484 ± 25	≈490 ± 30
D_0(H-MO)	460 ± 25[d]	284 ± 10[b]	–	–	–
D_0(OH-MOH)	462 ± 30	319 ± 15	–	–	–
D_0(M-H)	338.1 ± 0.4	294 ± 2	271.6 ± 2.5	240 ± 0.4	190.5 ± 0.2
D_0(HM-H)	331 ± 5	196 ± 5	–	–	–
D_0(M-Cl)	508 ± 5[e]	498 ± 4	460 ± 5	431.4 ± 5.4	367 ± 2
D_0(ClM-Cl)	353 ± 12[e]	335 ± 20	270 ± 50	249 ± 50	≈213[f]
D_0(M-S)	575 ± 30	370 ± 6	–	–	–
D_0(SM-S)	485 ± 25	260 ± 50	–	–	–
D_0(M-SM)	≈225 ± 100?	340 ± 15	–	–	–
D_0(M-M)	290 ± 15	155 ± 20	110 ± 10[g,h]	75 ± 10[h]	40 ± 5[i,j]
Bond ratios					
D_0 OM-O/M-O	0.71	≤0.69	–	–	–
M-OH/HOM-OH	1.41	1.72	–	–	–
D_0 M-Cl/ClM-Cl	1.44	1.49	1.70	1.73	≈1.72
D_0 M-OH/M-Cl	1.28	1.10	0.98	0.93	0.84
D_0 M-S/M-O	0.72	0.73	–	–	–
D_0 SM-S/M-S	0.84	0.70	–	–	–

[a]*Gurvich et al. (1994).*
[b]*See text.*
[c]*Naulin and Costes (1999), see text.*
[d]*Chin et al. (2004).*
[e]*Hildenbrand (1996).*
[f]*estimated from trends.*
[g]*Shim et al. (1991).*
[h]*Balducci et al. (1998).*
[i]*Froben et al. (1983).*
[j]*Han and Hirao (2000).*

As a result, due to these 47 and 220 kJ mol^{-1} respective differences in the thermochemical bond strengths, their vapor pressure predictions are heavily biased to favor the dioxide in oxygen-rich conditions. These are good examples of the importance of accurate heats of formation in chemical thermodynamic equilibrium calculations. Significant errors can make such exercises meaningless.

Like boron, the presence of hydrogen can drastically modify the role of these gaseous oxides. It had been noted that water vapor can vaporize Al_2O_3(s) but slowly (Opila and Myers, 2004). It formed gaseous $Al(OH)_3$ and a potentially complex range of Al/O/H gaseous species such as AlOH, HAlO, HAl(OH), AlO(OH), $Al(OH)_2$, $Al(OH)_3$, and others that have been predicted by theory with their heats of formation estimated (Swihart and Catoire, 2002; Cobos, 2002; Allendorf et al., 2002; Guo et al., 2007; Sebbar et al., 2012). However, it should be noted that with the exception of AlOH(g) (Milushin et al., 1986; Farber et al., 1974a), these remain largely uncharacterized experimentally but are regarded as potential intermediates in flame gases and may be involved in numerous kinetic channels that interconnect the aluminum oxides and OH, H and water (Sharipov et al., 2012). Gas phase equilibrium calculations are only a rough indication of relative stabilities and importance, but do suggest in this case that along with AlO from the Al/O/H list of potential species, major roles mainly for AlOH(g), OAlOH(g) and $Al(OH)_3$ in H_2 fuel-rich or fuel-lean flame conditions at 2500°C (Guo et al., 2007). It should be noted that Al_2O_3(g) has never been monitored and theoretically is predicted to be unimportant as seem to also be other Al_2 containing molecules in the presence of hydrogen (Guo et al., 2007). The monohydroxide does have a pronounced stability, D_0(Al-OH) = 550 ± 10 kJ mol^{-1} based on two mass spectrometric and five theoretical studies (Milushin et al., 1986; Allendorf et al., 2002; Sebbar et al., 2012). Although the other gaseous hydroxides, $Al(OH)_2$ and $Al(OH)_3$, have been invoked (Jensen and Jones, 1972; Guo et al., 2007), kinetic constraints suggest their roles would appear minor. However, it is true that their infrared spectra have been characterized in matrix isolation studies condensing laser-ablated Al in H_2/O_2mixtures (Wang and Andrews, 2007). Nevertheless, in combustion they are expected to shed water very easily on heating, modifying to attain the very stable AlOH, AlO structures. Such volatile metal hydroxides have importance for corrosion, and thermochemical data are now well known for water's corrosion of Al_2O_3 to the trihydroxide (Myers et al., 2019). However, it should be noted that rates of corrosion are generally small and more important to combustion for practical maintenance concerns rather than flame propagation. Even though the thermodynamic values have been estimated for the di- and trihydroxide molecules they were not included in kinetic analyses (Sharipov et al., 2012), nor in Al particle combustion or $AlCl_3/H_2$ chemical vapor deposition (CVD) kinetic models (Washburn et al., 2008; Nitodas and Sotirchos, 2002). Calculations of various studies do suggest D_0(HOAl-OH) = 319 ± 15 kJ mol^{-1} and D_0($Al(OH)_2$-OH) = 540 ± 25 kJ mol^{-1}, the latter strongly bound but undoubtedly kinetically constrained from formation in flames. It is to be noted that mass spectrometric analysis of such complex Al/H/O mixtures is very difficult partly due to ion formation of several species having the same mass (Crovisier et al., 1992). Additionally, it has been reported that probe sampling from flame systems in which the aluminum is present at only 1 ppmv can have problems with Al_2O_3 deposition (Goodings, 1992). Although AlOH(g) can isomerize to its HAlO structure, calculations have confirmed that this is about 170 kJ mol^{-1} less stable (Politzer et al., 2001; Allendorf et al., 2002; Handy et al., 2006; Sikorska and Skurski, 2009). Also the isomerization transition between the two has an additional energy barrier of about 180 kJ mol^{-1}, which could make the HAlO a metastable state if it is formed (Sikorska and Skurski, 2009). Although Al particles have always been a component of propellants, new interest has turned in recent years to Al nanoparticle/water mixtures as a new H_2 source for fuel cells with distant space travel. Al can insert into the water molecule forming short lived HAlOH that fragments to AlOH and hydrogen. This mechanism is thought to explain the chemiluminescence seen from this system (Douglas et al., 1982; Alvaeez-Barcia et al., 2013). It has now been examined further theoretically invoking this greater reactivity of active Al nanoparticles (Ho et al., 2016).

The volatile aluminum molecule is the trihalide, $AlCl_3$. This sublimes at about 180°C as the dimer with a vapor pressure of about 10 ppmv even at 100°C. It is thermally stable in the gas phase but gradually dissociates being about 60% monomer at 600°C and more than 90% above 700°C (Gesenhues and Wendt, 1986; Gurvich et al., 1994). However, in combustion systems where water vapor is present it is readily hydrolyzed to one of the numerous gaseous oxyhydroxy or hydroxide molecules. Five such molecules are known to exist, namely ClAlO, Cl_2AlO, ClAlOH, $ClAl(OH)_2$, and Cl_2AlOH that have been appraised by Gurvich et al. (1994) and treated theoretically in various studies. These are unlikely to have much of a lifetime in combustion where oxygen, water, and radicals will oxidize them to Al_2O_3(s), forming on surfaces or nucleating into particle growth. AlCl(g) is well characterized spectroscopically, has no lying electronic states and a $X^1\Sigma^+$ ground state. Its thermochemistry has been reviewed by Gurvich et al. (1994) recommending D_0(Al-Cl) = 498 ± 4 kJ mol^{-1} derived from transpiration experiments of the Al(*l*)/$AlCl_3$(g)/AlCl(g) at elevated temperatures. This value was discussed and accepted by Hildenbrand and Theard (1969) as a reference value in the determination of other bond strengths. Theoretical calculations are satisfactory suggesting 506, 497, 504, and 503 kJ mol^{-1} (Petrie, 1998; Allendorf et al., 2002; Shirley et al., 2009; Sebbar et al., 2012). The dichloride also exists but has been very difficult to characterize accurately (Gurvich et al., 1994). They reanalyzed the data of four vaporization experiments and found values of D_0(ClAl-Cl) = 345, 387, 362, 445 kJ mol^{-1}, which they discounted as unacceptable being inconsistent or too large. An additional more recent study suggested 364 kJ mol^{-1} using both their AlCl and $AlCl_2$ data (Numata et al., 1993). The databases had suggested 348 and 309 (Chase, 1998; Gurvich et al., 1994). On the other hand, theoretical calculations have predicted 292, 308 and 287 kJ mol^{-1}, producing a full range of values from 287 to 445 (Petrie, 1998; Politzer et al., 2001; Allendorf et al., 2002). Fortunately, the thermochemistry for $AlCl_3$ is better fixed in that an accurate measure can be obtained from the heats of formation and vaporization of Al_2Cl_6(s). This indicates a quite accurate atomization energy, $D_{0Atomiz}(AlCl_3)$ = 1269 ± 6 kJ mol^{-1} (Gurvich et al., 1994). This implies that by subtracting the energy of the first bond (498), the sum of the second and third bonds can only be 771 kJ mol^{-1} and both weaker. In such a dilemma, a useful method is to examine trends. In this case aluminum's chemistry mirrors that of boron. The three bonds in BCl_3 are better characterized and indicate values of 510, 351, and 461, also indicating a similar scaled overall atomization energy of 1322 ± 5 kJ mol^{-1} (Gurvich et al., 1994). In $AlCl_3$, the three bonds together will need to be 53 kJ mol^{-1} weaker. That for AlCl is seen to be 12 kJ $mol^{-1}$and from the apparent trend with boron the value for ClAl-Cl would appear to require a value of 330–340 kJ mol^{-1} and cannot be as low as the 300 range of theory. This leaves Cl_2Al-Cl = 446 kJ mol^{-1} for acceptable consistency. Consequently, the first bond, AlCl, is the most stable, a trend seen also in B, Ga, and In. However, it is unlikely to play any effective role in most combustion systems except possibly in facilitating initial release.

Similarly, potential roles for gaseous hydrides are very unlikely in general combustion due to their affinity for oxygen or water. Its first bond again is much stronger being determined accurately by spectral predissociation, D_0(Al-H) = 294 ± 2 kJ mol^{-1}. An early mass spectrometric study also obtained 284 ± 12 lending support for their other derived values (Kant and Moon, 1981). The ground state is $X^1\Sigma^+$ with no lying states (Gurvich et al., 1994). Thermochemical data for the dihydride, X^2A_1, has only been obtained theoretically but indicate atomization energies of 491,489 and 492 supporting an average value of 490 and a much weaker bonding of D_0(HAl-H) = 196 ± 5 kJ mol^{-1} (Swihart and Catoire, 2000; Politzer et al., 2001; Allendorf et al., 2002). Their infrared vibrational frequencies were noted in low-temperature matrix isolation experiments (Chertihin and Andrews, 1993).

Reflecting the oxides, the three sulfides have been characterized AlS($X^2\Sigma^+$), linear AlS_2 with ground state suggestions of either $X^2\Pi$ like the oxide or $X^2\Sigma$ (Zhong et al., 2013), and a linear Al_2S($X^1\Sigma^+$). The thermochemical stabilities of these show the normal weakening from the oxides. The most recent assessment is that of Gurvich et al. (1994), who have reanalyzed the two main mass spectrometric studies leading to D_0(Al-S) = 370 ± 6 and D_0(AlS-Al) = 340 ± 16 kJ mol^{-1} (Ficalora et al., 1968; Uy and Drowart, 1971). As with the oxide, no data are available for the disulfide due to its lower stability and a value of D_0(SAl-S) = 260 ± 50 kJ mol^{-1} was estimated.

These bond strengths for gaseous aluminum compounds along with the trends in values for B/Ga/In and Tl are summarized in Table 7.5.

Laboratory combustion related studies

Other aspects can be considered that have implications concerning its combustion chemistry. These relate to flame generation of Al_2O_3 particles or nanoparticles and its surface film deposition by CVD methods. Flame synthesis now is an industrialized method to tailor ceramic type materials into specific nanoparticle sizes or to coat substrate materials (Roth, 2007). High-rate production methods for such nanopowders as SiO_2, TiO_2, and Al_2O_3 now are well established (Chen et al., 1998b). Formation generally consists of adding aluminum to the fuel in the form of an aerosol water solution of its trichloride or an organoform such as an alkoxide or acetylacetonate dissolved in an organic solvent such as alcohol. This general use of $AlCl_3$ as the flame precursor for oxide nanoparticles clearly indicates no flame role for the chlorides (Fredriksson and Carlsson, 1993; Fennell et al., 2007a). An alternate method can be to add the vapor of a volatile aluminum compound such as $Al(CH_3)_3$ to the fuel that is spontaneously inflammable (Pratsinis, 1998). In this manner it is found that the aluminum is very rapidly released and Al_2O_3 particles almost instantly formed with a primary size of <25 nm and an average aggregate secondary size of about 100 nm (Singhal et al., 1999). Rapid formation of 100 nm nanoparticles of alpha-Al_2O_3 can also result from adding micron-sized Al powder into a C_2H_2/O_2 flame (Kathirvel et al., 2014). A recent study examined the combustion of a 200 μm Al particle in O_2/Ar. The rapid formation of Al_2O_3 and clustering was modeled (Savel'ev and Starik, 2018). These examples are especially noteworthy as they imply that the formation is from a gaseous Al species initiating rapid nucleation, deposition and growth (Kilian and Morse, 2001). Gaseous Al_2O_3(g) does not exist so formation is from two molecules containing Al, probably shedding water and then instantly nucleating. Otherwise it may be induced by deposition on any condensation nucleus. It remains unexplained and initial attempts to try and formulate a mechanism are gradually emerging. A recent theoretical approach examined these difficulties in terms of cluster growth. However, this was not effective and an adequate theory remains to be found (Lam et al., 2015). As a result, the very rapid mechanism of particle formation in flames remains unexplained. From the industrial formation of Al_2O_3(s) by heating the trihydroxide, it does seem that any oxide-hydroxy molecule or pair can readily shed water. In one case the particle formation was observed to occur in flames in less than 0.1 ms at 2000K as monitored by particle laser scattering, thermophoretic and molecular beam sampling with aerosol mass spectrometry (Xing et al., 1996; Lindackers et al., 1998). No other aluminum gaseous compounds were detectable in these studies. Although illustrated mainly for TiO_2(s) particles, the use of Raman scattering now appears to be a useful in situ diagnostic to detect and characterize the particle crystallinity. In this way it was seen that if 100 nm diameter γ-phase Al_2O_3 nanoparticles were injected into a coflow jet diffusion flame at about 1440°C their rapid crystalline phase change to the α-phase could be observed (Liu et al., 2010a). The rapid kinetics also was illustrated by showing that the ignition delays of H_2/O_2 mixtures could be significantly shortened by the presence of Al particles at only parts per billion concentrations (Catoire et al., 2003).

An extremely remarkable phenomenon is the formation of cenospheres in coal combustion. These spherical hollow ash particles can be obtained from coals rich in Si and Al and are now being considered more seriously as a valuable ceramic powder product. Their composition of Al/Si is >90% and the combustion conditions can be tailored to maximize their production (Fomenko et al., 2015, 2016).

It would appear that if aluminum is in the gas phase its chemistry is kinetically controlled but is very rapid. Consequently, the recent suggestion that Al_2O_3 particle formation in flames containing about 10 ppmv of aluminum at about 1750°C is a slow kinetic process requiring AlO and AlO_2 participation now is inconsistent and may be a reflection of certain sampling difficulties (Fennell et al., 2007a). It is clear that nanoparticle flame formation suffers the same dilemma as soot formation that, after more than 40 years, still has no satisfactory mechanism for initiation and growth. Soot appears to grow from C_2H_2 fragments forming hydrocarbon clusters that then stabilize by losing hydrogen. Nanoparticle oxides probably follow a similar trend of rapidly clustering hydroxyl-oxide molecules that shed water to stabilize as the oxide.

Flame CVD is well established not only for producing Al_2O_3(s) deposits but also mixed oxides known as spinels. This is possible by adding an aerosol that contains a mixture of two compounds. An example of this used organocompounds of Al and either Mg or Ni dissolved in toluene. Using the toluene as its own fuel it was then possible to deposit films of $MgAl_2O_4$ or $NiAl_2O_4$ on a substrate cooled to about 1000°C (Stollberg et al., 1997). Aluminates with about 13 other metallic elements also are known (Kurihara and Suib, 1993). That such mixed metal oxide particles can also form and possibly nucleate in the gas phase has been evident from flame emission and atomic absorption methods and is even documented in text books from 50 years ago (Herrmann and Alkemade, 1963). In such analytical methods it was noted, for example, that all the Group IIA elements (Be, Mg, Ca, Sr, Ba) can interfere with aluminum's spectral analysis (Bitto et al., 1990; Luecke, 1992, 1992a, 1994). The nature of the aluminate formation mechanism is uncertain but one way to circumvent the difficulty and manage the analysis for Al with adequate sensitivity was to use high temperature C_2H_2/N_2O flames (2700°C) (De Galan and Samaey, 1970).

This would minimize any molecular or particle formation. A recent study examining the influence of various metal salts on soot formation noted that additions of $AlCl_3$ solution to an ethylene/air flame showed a negligible effect (Simonsson et al., 2018). This confirms early studies that rated its soot suppression as zero (Cotton et al., 1971; Bulewicz et al., 1975; Haynes et al., 1981). It also has a very limited flame ion chemistry (Goodings, 1992). It may have some catalytic ability in flames (Bulewicz and Padley, 1973; Vanpee and Shirodkar,1979), but is not listed as a flame inhibitor.

Combustion power plant data

With such a rich chemical understanding of this rather nonvolatile nature of aluminum it is interesting to examine the further observations that can be drawn from full-scale practical combustors. In coal combustion, mass balance studies for aluminum do indicate that all the element can be accounted for in the various ash samples (Nielsen et al., 2002; Goodarzi, 2006a). Utilizing solely ESPs for particulate control, stack emissions indicated Al loss of only 0.02%–0.06% to the environment. Moreover, this was aided by the fact that any submicron particles $<0.4\,\mu m$ emissions contain only low levels of Al. Due to its high levels in coal, the ash from such combustion has been extensively studied. It is found in bottom ash, fly ash and dust. China produced 620 million tons of coal ash in 2016 and now is tailoring the Al_2O_3 content to be in a crystal phase more compatible with recycling (Zhu et al., 2019). In recent years, due to an emphasis on small particulates, numerous interesting studies of fly ash have emerged where particle fractions were sized and analyzed. To keep things in perspective, it is to be remembered that ash of diameters that are $<2.5\,\mu m$ can constitute only about 3 wt.% of the total ash and similarly that of 2.5–5.0 μm about 7.5 wt.% (Martinez-Tarrazona and Spears, 1996). However, it is particularly this small ultrafine range that is medically of concern. Examining various size fractions from 0.3 to 100 μm (Querol et al., 1995) and from 2.5 to 65 μm (Martinez-Tarrazona and Spears, 1996), it was confirmed that Al showed no enrichment or particle size dependence as had been realized earlier (Coles et al., 1979; Smith et al., 1979). More recently this has been confirmed for coal fly ash (Lanzerstorfer, 2018; Zhao et al., 2018). This in fact is the reason why aluminum is used as a yardstick to gauge enrichment or depletion of other minerals in ash. This also emphasizes that aluminum normally is not influenced by its vaporization. Qualitative examination of fly ash by electron microscopy methods indicated that the aluminum is present with many elements such as Si, Fe, Ca, Mg, Na, K, Ti with which it might associate (Chen et al., 2005). However, Al and Si will generally always be in large excess over most of the other elements (Table 1.1).

Early ash studies confirmed overall that the concentrations of aluminum in the bottom ash and fly ash were similar. However, when collecting, sizing and analyzing the fine ash exiting an ESP the micron-sized particles show a bimodal distribution with peaks at about 0.05 μm (mass 6%) and 2 μm (mass 94%). The respective quantities of Al content in these were about 3% and 97% (Kauppinen and Pakkanen, 1990). It was also noted that 80 wt.% of sulfur in these two distributions was found in the smaller sized range. Ten years later, these submicron-sized ash particles were examined further and although the Al content was not measured it was shown that the ash actually appeared as a trimodal distribution of sizes, one of ultrafine particles at 0.1 μm, a central region of 0.3–3 μm and then larger sizes (Linak et al., 2000, 2002; Seames, 2003). Moreover, the two larger modes did not support being generated by a single broad fragmentation distribution and had a very different morphology indicating that there appeared to be three differing mechanisms involved. Now, although still not fully understood, two major efforts in Japan and China have confirmed this trimodal aspect of ash size formation essentially being of three types, ≤0.2, 0.2–3, and >3 μm. Interesting aspects are that there is very little Al in the smallest fraction confirming its nonvolatile nature. Moreover from the high levels of Na, K, Ca, Mg, Fe, P, and S, this fraction is clearly indicative of formation by a predominant chemical deposition mechanism (Liu et al., 2007a; Teramae and Takarada, 2009). Similarly, the fraction >3 μm has an Al concentration that is largely invariant with size obviously reflecting the basic coal Al/Si major core value. Consequently, the size range connecting these two regimes is intriguing by being quite different in not only displaying particles that have a very different morphology (Yu et al., 2008) but having a mechanism that displays Al concentrations that are smaller in the smaller particle sizes while at the same time sulfur and phosphorus content reflect just the opposite behavior (Yu et al., 2009; Xu et al., 2011). Clearly this reflects the nonvolatility of Al that is being affected by the volatility and deposition of some of the other elements. Whether there is slight loss of Al from ash remains uncertain. Ablation of Al_2O_3(s) by gaseous HCl present in flue gases is plausible at temperatures close to 1000°C (Bloem and Bosman, 1975). Interactions of aluminum in the ash are of course to be expected. That between Na_2SO_4(s) and various oxides such as Al_2O_3, SiO_2, Fe_2O_3 have been studied and do reflect weight losses at about 900°C and above indicative of mixed oxide formation with a loss of sulfur (Mobin et al., 1996). Nevertheless, what is noteworthy is that the small micron size particles that are more difficult to control and are of medical concern are seen to contain lower concentrations of Al and Si than the larger size distribution. Moreover, the presence of an FGD (flue gas desulfurization) system does help in retention of final traces of escaping fine particles. A detailed study of trace element retention in a coal/pet coke co-combustion plant suggested that traces of aluminum were further cut in half (Cordoba et al., 2012).

In summary, as a result of its nonvolatile nature, there is little chemical interest concerning aluminum in combustion. It behaves similarly in different modes of combustion or gasification whether with coal, biomass or blends with fuels such as sewage sludge or even in oxygen enriched combustion (Zhang et al., 2007; Jordan and Akay, 2012). However, the exact chemical nature of the aluminum in the various ashes remains largely unknown and will depend on conditions and temperatures. It is generally assumed to be present as the oxide, sulfate, phosphate, an aluminosilicate or aluminate. These of course are quite relevant to studies of leachability (Harris et al., 1988; Catalano et al., 2012), and are components that relate to the viscosity and nature of coal slags (Van Dyk et al., 2009; Yuan et al., 2012). Roasting municipal waste ash at 1000°C with either $CaCl_2$ or Cl_2 for 3 hours did not induce any Al volatilization at all (Chan and Kirk, 1999). The fact that Al concentrations are non-negligible in ash is finally suggesting that the recycling and leaching methods of ash may be a useful commercially viable source for this element (Yao et al., 2014; Singh et al., 2015; Unver and Terzi, 2018). In Northern China, a coal fly ash rich in aluminum is being mixed with ammonium sulfate in a fluidized bed reactor to recover the alumina content (Sun et al., 2017).

7.3 Antimony

Antimony gains its chemical symbol from the Latin name stibium and the name stibnite for its Sb_2S_3 ore. The element is located in group VA of the periodic table, namely N, P, As, Sb, and Bi. It is referred to as a metalloid, showing an increased metallic nature with chemistry very like that of As and Bi. It illustrates tri- and pentavalent molecules with the trivalency being more commonly encountered. Both are toxic and processed differently in the body. The valencies stem from its electronic configuration of ($5s^25p^3$) with a term symbol $^4S_{3/2}$. It is not an abundant element being ranked at about 66th in the Earth's crust with a concentration of about 0.2 ppmw, and is slightly more evident in the oceans at 46th. Nevertheless, it is found in many minerals, the most common and main ore being stibnite, Sb_2S_3, which can be roasted in air to the oxide. Currently, it is mined globally at a rate of 150,000 tons $year^{-1}$ and is becoming an element of critical raw materials as known reserves are only about 10 times larger than this. However, its uses are gradually being replaced and significant recycling is occurring. Current market costs are about \$3 lb^{-1}. More than a third is used globally as a fire retardant, followed by use as an alloy with Pb in acid batteries and munition bullets. It has many other lesser contributions in the semiconductor solid-state industry, uses in plastics [polyethylene terephthalate (PET)], ceramics, and glass but has very few bio- or medical applications. Also the use of Sb_2S_3 in brake lining pads and tires has added a concern for an environmental role (Varrica et al., 2013). A review of the global metallurgy summarizes some of these aspects (Anderson, 2012).

Mankind has always lived with natural atmospheric background levels of antimony that originate from volcanic emissions, air blown dusts, and sea aerosols. Nevertheless, it is now a polluting toxin as anthropogenic additions are approximately one and a half times larger than natural levels with combustion sources at a level at least that of nature. As of 1990 total atmospheric emissions approximated to about 6000 tons $year^{-1}$ (Nriagu, 1989, 1989a; Pacnya and Pacnya, 2001). About 40% of anthropogenic contributions result from metal smelting. In recent years, numerous studies have reexamined the anthropogenic contributions. In 2015 it was noted that China alone emitted 1250 tons $year^{-1}$, and from 1978 to 2012 their contribution had increased eightfold, 60% of which now is from combustion with a noticeable contribution from brake materials (Tian et al., 2011a, 2012a, 2014a, 2015). The same group of researchers has now predicted that this will double from 2010 to 2050 mainly as a result of incineration and metal refining (Zhou et al., 2015). One recent study of the Russian Cu-Ni industry concludes that previous Sb emissions in one region were actually very small and had been grossly overestimated (Boyd et al., 2009). The supporting evidence that might be acquired from ice-core, Antarctic snow, and lake sedimentary deposits can be difficult to accurately interpret. Ice cores from Russia and from Mt. Everest indicate historically low levels that grew with industrialization. These peaked around 1975 and have since fallen back to about 1950s levels due to numerous reasons (Hong et al., 2009; Eichler et al., 2014). The Antarctic snow deposits and lake sediments show similar trends although the latter work in Japan sees a continued rise after 1980 possibly due to increased coal and incineration in Asia (Hong et al., 2012; Kuwae et al., 2013).

Health concerns with antimony

Antimony generally under most conditions is noncarcinogenic but is a regulated toxin. It does occur in a wide range of foods and even tobacco but at very low levels such that an estimated daily diet consumes about 5 μg. The WHO (World Health Organization) has recommended a safe intake of about 0.6 μg kg^{-1} body wt. day^{-1} that is much below the toxic level. Two reviews of the toxicity and human exposure discuss in detail its hazards (ToxProfile Sb, 2017; Belzile et al., 2011). Whether antimony is alien to the body or has minimal tasks appears not to have been addressed but any consequences of deficiencies do not appear reported. Not all compounds are bioactive to the body and many are poorly absorbed by the intestines. Toxicity is not the usual concern because Sb compounds are emetics inducing violent vomiting and data for humans is lacking. A major medical use was as potassium antimony tartrate hydrate (tarter emetic) as an emetic and at certain doses to clear the body of certain parasites. This has now been replaced by alternate milder drugs. It appears that Sb(III) and Sb(V) can coexist in the body and do not interconvert. However, it is indicated that injected Sb(V) is mainly excreted in urine within 6 hours, whereas Sb(III) is excreted in feces after a day or so. Antimony has no affinity for carbon and organic forms are generally as organic acid salts such as the tartrate. Sb(III) has been found to biomethylate in nature and is known in plants and found in sediments and sludge (Filella, 2010). This appears to have a lower toxicity and little is known at present of other organics than the methyl and its substitutions. Attempts to methylate Sb(V) remain unsuccessful. There is no bioaccumulation noted in fish or foodstuffs. As a result, there is minimal reason for health concerns by the general public at present. Nevertheless, for those involved in the commercial production and applications of antimony products, databases have been established for monitoring employees. A level of 0.1 μg L^{-1} currently is recommended as a safe urine background value (Filella et al., 2013). The US testing that has included antimony has recorded low levels in blood, serum, and urine samples (Centers for Disease Control, US, 2009, 2018). One survey from a polluted area of China suggested an average for blood samples of about 2 μg L^{-1} and very low in hair (Li et al., 2014a). For some reason, antimony, which is rarely added to a survey test panel was added to a recent Norwegian study monitoring several other metals (Hg, Pb, Cd) for never-pregnant women in the ages of 18–40 years and indicated serum values in a range of about 3.2–4.1 μg L^{-1}, also hinting of a possible slight accumulation with age (Flotre et al., 2017). One Uganda school study showed high blood levels in all the children, those at the 95% point of the distribution registering a value of 8.8 μg L^{-1} that was difficult to explain (Cusick et al., 2018). However, it remains far too speculative to suggest possible deleterious roles for antimony such as in a recent study involving ADHD (attention deficit hyperactivity disorder) for Taiwanese children when its levels are very minor compared to other known neurotoxins also present such as lead and mercury (Lee et al., 2018a). For now it remains too early to establish sound medical risk levels but data is gradually accumulating in surveys to indicate normal human ranges (Goulle et al., 2005; Nisse et al., 2017). The most recent compared blood sample values in Morocco and the Canary Islands that found mean values of 2.56 and 2.53, but ranges of 1.4–8.9 and 0.7–4.4 μg L^{-1}, respectively (Henriquez-Hernandez et al., 2018). These appear to be in a range of minimum risk and are of no concern. Blood, serum, or urine biomarkers seem to be the better indicators. However, since we remain ignorant about long-term effects of even small toxin doses, the sources of such contamination would be interesting to identify at some point.

Antimony's basic chemistry

Antimony is a brittle metal (m.pt.631, b.pt.1587°C) and its atomic weight of 121.8 results from its two stable isotopes, 57% ^{121}Sb and 43% ^{123}Sb. A suitable radioactive ^{125}Sb can be used medically if needed and has the longest decay lifetime of its isotopes, 2.7 years. Antimony finds its uses alloying and strengthening other metals. Referring to the summaries of As and Bi chemistries given herein, it is seen from Table 7.6 listing condensed phase species for antimony, and Table 7.7 its gas phase trends, that antimony's does in fact lie between the two behaviors but is quite similar to arsenic. Along with phosphorus there is a tendency with its element and some compounds to favor cage-like molecules when they vaporize. Elemental Sb(s) vaporizes to predominantly $Sb_4(g)$ in an inert environment. The vapor distribution remains as this, still being 70% at 1000°C and then dissociates becoming 80% $Sb_2(g)$ by 1500°C with elemental Sb(g) contributions small until higher temperatures (Knacke et al., 1991; Barin and Platzki, 1995). Its oxide chemistry is dominant in air, trivalency being the more thermodynamically stable but Sb_2O_5 exists as do stable antimonates. One chemical thermodynamic calculation of coal combustion that included Sb noted that it had little affinity for other elements but favored $Sb_4O_6(g)$ formation (Yan et al., 2001).

TABLE 7.6 Properties of antimony's condensed phase molecules relevant to combustion, °C.

m.pt. b.pt. vapor pressure
Sb 631 1587, 100 ppmv/600°C, ins
Sb_4O_6 (Sb_2O_3) 656 1425, sublimes >400, sl s
Sb_2O_4 (III/V) 780, vaporizes, decomp 1230 to SbO(g)
Sb_2O_5 unstable, dissoc 525 to Sb_2O_4, sl s
Sb_2S_3 550, dissoc 1000 to SbS(g), reacts with air, water, ins
Sb_2S_5 decomp 75, ins
$SbCl_5$ 4 140, dissoc >100, reacts with water
$SbCl_3$ 73 223, dissoc 170, very sol, hydrolyzed
SbOCl decomp 170, reacts with water
$Sb_2(SO_4)_3$ decomp 100 to Sb_4O_6, hygroscopic, sl s
SbH_3 −88 − 17, unstable gas, sl s
$NaSbO_3$ ≈1000, dissoc to $NaSbO_2$, sl s
$KSbO_3$ ≈1250, dissoc to $KSbO_2$ sl s
$Ca(SbO_3)_2$ ≈1000 ins
$Mg(SbO_3)_2$ stable, ins
$Fe^{III}(SbO_3)_3$ ≈1000 + , stable, ins
$Fe^{II}(SbO_3)_2$ ≈1000 + , stable, ins

Decomp, decomposes; *dissoc*, dissociates; *sl s*, water slightly soluble; *ins*, water insoluble.

TABLE 7.7 Stability trends for relevant gaseous molecules of the N/P/As/Sb/Bi group, kJ mol^{-1}

Bond	N[a]	P[a,b]	As[b]	Sb[b]	Bi[b]
D_0(M-O)	626.8 ± 0.1	590 ± 3	476 ± 8	424 ± 10	333 ± 5
D_0(OM-O)	300.6 ± 0.6	497 ± 10	–	–	–
D_0(M-Cl)	260 ± 15[c,d,e,f]	300 ± 10	295 ± 6	288 ± 4	295 ± 10
D_0(ClM-Cl)	250 ± 50[d,e]	307 ± 10	300 ± 10	≈320	<240
D_0(Cl_2M-Cl)	160 ± 50[e]	354 ± 6	310 ± 15	≈330–356	≈300
D_0(M-S)	466 ± 10[g,h,i]	438 ± 10	380 ± 8	366 ± 5	311 ± 4
D_0(M-H)	328 ± 10	295 ± 8	270 ± 4	236 ± 4	210 ± 15
D_0(HM-H)	382 ± 10	310 ± 8	280 ± 4	244 ± 2.5	≈215
D_0(H_2M-H)	448 ± 10	351 ± 10[j]	311 ± 4	282 ± 2	≈240
D_0(M-M)	941.6 ± 0.6	485.6 ± 0.4	382 ± 2	296 ± 5[k,l]	197 ± 3
Bond ratios					
D_0 OM-O/M-O	0.48	0.84	–	–	–
D_0 M-S/M-O	0.74	0.74	0.80	0.86	0.93
D_0 M-Cl/ClM-Cl	1.04	0.98	0.95	0.90	1.2

[a]*Gurvich et al. (1989).*
[b]*see texts for phosphorus, arsenic, antimony and bismuth.*
[c]*Bettendorff and Peyerimhoff (1986).*
[d]*Papakondylis et al. (1995).*
[e]*Milburn et al. (1997).*
[f]*Xantheas et al. (1997).*
[g]*Peebles and Marshall (2002).*
[h]*Lodders (2004).*
[i]*Takeshita and Dunning (2015).*
[j]*Chase (1998).*
[k]*DeMaria et al. (1959).*
[l]*Kordis and Gingerich (1973).*

The pentavalent oxide dissociates with increasing temperature to the trivalent oxide. On melting at about 500°C the pentoxide loses oxygen and becomes a mix of Sb_2O_4 and Sb_6O_{13} then becoming all Sb_2O_4 which is regarded as a Sb^{III}/Sb^{V} mixed oxide, $Sb_2O_3.Sb_2O_5$ (Paoletti et al., 2001). This and also Sb_2O_3(s) sublime congruently above about 450°C to the Sb_4O_6(g) dimer. The oxides of this periodic table group are complex and well known for their differing crystal forms. Although Sb_4O_6(g) has been characterized spectroscopically no Sb_4O_{10}(g) has been reported. These complex forms remain difficult to model structurally by theory (Chrissanthopoulos and Pouchan, 2008). This polymorphism is displayed particularly by Sb_2O_3(s) which is stable at ambient temperatures in the crystal form as Sb_4O_6(s).

No SbO_2(g) is known probably as a result of its condensed phase mixed oxide nature. The Sb_4O_6(g) is unstable in the gas phase at higher temperatures and dissociates around 1000°C to SbO(g). A thermogravimetric vaporization of Sb_2O_3(s) in nitrogen reported full vaporization by 870°C (Paoletti et al., 2001), and above this temperature these gaseous oxides all fully dissociate to a dominance of SbO(g) (Asryan et al., 2003). It has also been reported when a crucible containing metallic antimony is heated in an air flow at 550°C that Sb_2O_3(s) nanoparticles can be deposited at 250°C downstream with SbO_2(s) (Sb_2O_4) remaining in the crucible (Xu et al., 2007). Similarly, if the sulfide ore Sb_2S_3(s) is heated in low O_2 flows (1%–5%) it volatilizes as Sb_4O_6(g). However, at higher oxygen partial pressures the vaporization can become limited by forming nonvolatile cervantite (SbO_2(s)) that finally requires 1150°C to dissociate to SbO(g) (Padilla et al., 2010). Nevertheless, although there are these complex crystal changes possible with the pentoxide and trioxide, they all dissociate to SbO(g) that becomes the main transport of antimony in combustion flue gases. No gaseous SbO_2 exists, and although its gaseous ion is known and various theoretical papers have appeared calculating the expected spectroscopy it remains undetected, possibly because of its III/V nature as mentioned above (Lee et al., 2006, 2007). Although these oxides can produce antimonates and antimonites, there is no ready acid formation. This is true in their volatility and it was seen that water had no effect on either Sb_2O_3(s) or Sb_2O_4(s) below 930°C with no evidence of hydroxide species (Furey and Bloore, 1970).

The sulfate $Sb_2(SO_4)_3$ is known but is hygroscopic and decomposes on melting at about 100°C to Sb_4O_6(s). Although present in one equilibrium calculation to higher temperatures, it can only play a minimal role if any in combustion (Danihelka et al., 2003). With antimony, no carbonate has been isolated, nor hydroxides in any phase. Additionally, the trihydride is known but is gaseous at normal conditions.

Chlorides for antimony exist but as is common in this group they react with water. The pentachloride dissociates at 120°C–260°C being complete at 300°C, and the trichloride hydrolyzes with water to SbOCl that also dissociates at 170°C. On heating, SbOCl(s) in N_2 disproportionates to Sb_2O_3 and $SbCl_3$, but in air produces a mix of Sb_2O_3/Sb_2O_4 (Costa et al., 1990). Although $SbCl_3$(s) was found to be a potential flame inhibitor (Babushok and Tsang, 2000), the mechanistic explanations for this now appear unreasonable and unresolved (Hastie, 1973; Hastie and McBee, 1975). However, this recently has been enlarged now to $SbBr_3$, calculating equilibrium distributions and introducing a very ambitious kinetic model that includes numerous gaseous Sb/O species (Babushok et al., 2017a). It tends to suggest that the major inhibition is via Sb/O species more so than the possible halide cycles, each being somewhat independent. To invoke mechanistic radical loops involving such species as SbOCl(g), BrSbO(g), SbO_2(g), SbOH(g) has to remain speculative at present solely from the uncertainties in the required thermochemical data. Also the corresponding unimportance of these species with As or Bi raises questions.

It is interesting that in nature the major ore for antimony is Sb_2S_3 indicating an affinity for sulfur but mainly under reducing conditions. This melts at about 550°C and in an inert environment will sublime unchanged but gradually dissociate to SbS(g) and other fragments (Faure et al., 1972; Piacente et al., 1992). In the presence of air it produces Sb_2O_3(s) and Sb_4O_6(g). At <630°C, Sb_2S_3(s) also reacts with water giving Sb_2O_3 and Sb, and above this temperature only Sb.

Like arsenic, antimony can produce a series of metallic antimonate and antimonite salts that remain poorly characterized or studied. They appear to be thermally quite stable, are regarded as mixed oxides to some degree and can have high melting points being partially refractory in nature. The sodium and potassium antimonates, $MSbO_3$ ($M_2O + Sb_2O_5$), are stable to about 1000°C and 1250°C, respectively, before breaking down at higher temperatures to the antimonites, $MSbO_2$ ($M_2O + Sb_2O_3$). The Sb^{III} antimonites appear more stable and those of Na, K, Rb, and Cs can even vaporize unchanged to the monomer and dimer such that their heats of formation in the gas phase have been measured by mass spectrometry (Shugurov and Lopatin, 2011). Calcium is also known to form the antimonate $CaSb_2O_6$ ($CaO + Sb_2O_5$) and the antimonites $CaSb_2O_4$ ($CaO + Sb_2O_3$), or $Ca_2Sb_2O_5$ ($2CaO + Sb_2O_3$). Antimonates of Fe, V, and even Si are also mentioned in the literature but not characterized.

Gas phase stabilities and transport

For antimony's gaseous molecules, SbO(g) is a dominant species. Spectroscopically it has been studied extensively and has numerous known electronic transitions (Huber and Herzberg1979; Rai and Rai, 1984). These have always tentatively indicated a value of D_0(Sb-O) $\leq$ 424 kJ mol^{-1} based on a predissociation limit assuming the state does dissociate to the same limits as the ground $X^2\Pi_r$ state. However, a theoretical assessment of its potential energy curves shows their density and complexity (Alekseyev et al., 1995). Only two mass spectrometric studies are available but do support this value. One studied the $Sb + H_2O/SbO + H_2$ equilibrium in hydrogen flame burned gases at 2000°C and on updating the data now suggests D_0(Sb-O) $\leq 430.5 \pm 20$ kJ mol^{-1} (Farber and Srivastava, 1975). The other measured the gaseous equilibrium using an effusion cell for $Sb + MoO_3/SbO + MoO_2$ at about 1560°C (Semenov, 1985). Updating this with improved molybdenum data indicates 426.6 ± 25 kJ mol^{-1} (Cordfunke and Konings, 1990). Although still with some error margin, a value of D_0(Sb-O) $= 424 \pm 10$ kJ mol^{-1} might be recommended. Only two databases include SbO(g). In one it may be noted that the listed heats of formation of both SbO and SbS have negative rather than the correct positive values and when corrected suggest a D_0 value of 406 kJ mol^{-1} (Barin and Platzki, 1995). The other, provides new tables but only considers one of these sources for D(SbO) (Pedley and Marshall, 1983). One further approximate estimate came from measuring the endothermic activation energy of the $Sb + O_2$ reaction. Assuming this activation energy reflects solely the reaction enthalpy, it indicates a D_0 value of 444 ± 28 kJ mol^{-1} at least confirming the correct magnitude (Cosic and Fontijn, 2000a).

Similarly, although the spectral bands of SbCl(g) are well established and characterized, its bond strength remains uncertain even though its spectrum is easily excited by photolyzing or having $SbCl_3$ in an activated nitrogen discharge. The bond strengths of the three bonds in the trichloride are more easily measured reliably from its condensed phase heat of formation coupled to its heat of vaporization, both of which are accurately measured. Quoted values are all in close agreement indicating $\Delta H_{f0} = -310.9 \pm 1$ kJ mol^{-1} (Knacke et al., 1991; Barin and Platzki, 1995). This implies $D_{0\text{Atomiz}} = 934 \pm 1$ kJ mol^{-1} for the sum of the three bonds. This is a useful guide to lend support to theoretical calculations that reproduce this value quite well (Skulan et al., 2006; Vasiliu et al., 2012). Of these theoretical predictions, that of Skulan et al. (2006) also calculated the heat of formation for the ground state $SbCl(X^3\Sigma^-)$ and its low lying singlet state. These indicated values lower than previously thought for the ground state, D_0(Sb-Cl) $= 288 \pm 4$ kJ mol^{-1}, and 205 ± 11 kJ mol^{-1} for a singlet SbCl state. An even lower theoretical value of 238 for the ground state had been suggested previously (Balasubramanian and Latifzadeh-Masoudipour, 1999). However, other quoted reference values had been suggested of $\approx 357 \pm 50$, 409 and 394 kJ mol^{-1}, respectively (Gaydon, 1968; Wagman et al., 1982; Barin and Platzki, 1995). Meanwhile, following these predictions, emission from the "spin-forbidden" SbCl transition indicated that the energy of this metastable low lying singlet state was $T_{0,0} = 78.3$ kJ mol^{-1} (Beutel et al., 1999). This is consistent with the theoretical prediction for a singlet state, with both the X- and a-states dissociating to ground state Sb and Cl. This lends support to now recommending the value D_0(Sb-Cl) $= 288 \pm 4$ kJ mol^{-1} that then implies the second and third bonds would have a summed strength of about 646 kJ mol^{-1}. Scaling the theoretical study for SbCl and $SbCl_2$ of Balasubramanian and Latifzadeh-Masoudipour (1999) to the above D_0(SbCl) value suggests the only known estimate of D_0(ClSb-Cl) $= 320$ kJ mol^{-1}. Coupling this to a measured D(Cl_2Sb-Cl) $= 356 \pm 15$ (Grob, 1963) would now indicate a sum for the three bonds of 964 slightly overestimating the required 934 value. This may possibly be suggesting a slightly too high an experimental value for D(Cl_2Sb-Cl). Consequently, it does appear for Sb that the first three chlorine bonds gradually gain strength (288, ≈ 320, $\approx 330-356$) somewhat like the trend for As (Table 7.7).

When these chloride values are compared to the high bond strength of SbO(g), D(Sb-O) $= 424$ kJ mol^{-1}, which has a similar stability to OH or H_2, and to that of HCl, it seems unlikely that the chlorides will play any role in fuel-lean combustion environments. Simply considering the enthalpy of the interchange,

$$SbCl + OH = SbO + HCl \quad \Delta H_0 \approx -140 \text{ kJ mol}^{-1}$$

this alone at 1000°C imposes a factor of 2×10^7 against its occurrence that could only be compensated for by a very high ratio of HCl/OH. Consequently, although it was suggested in equilibrium calculations that SbCl(g) might be present in incinerators above 750°C (Verhulst et al., 1996), if the theoretical predictions of a moderately weak bond for SbCl are correct, then chlorine can play no role with antimony.

Thermochemical values remain quite ill-defined for antimony, but now are reasonably well established for its gaseous hydrides. By ablating Sb in hydrogen in a matrix isolation study the infrared spectra of its three hydrides were observed providing their structures and frequencies (Wang et al., 2003a). Spectral systems have been identified and gaseous SbH has a ground state $X^3\Sigma^-$ configuration and a low-lying metastable $a^1\Delta$ state with 74 kJ mol^{-1} excitation energy (Beutel et al., 1996a). A useful theoretical study has calculated its numerous potential energy curves displaying many curve crossings (Alekseyev et al., 1998a). SbH_3 is more stable than AsH_3. From early measurements of the heat of dissociation of stibine, SbH_3, to its elements, its heat of formation was accurately determined implying $D_{0\text{Atomiz}} = 759 \pm 1$ kJ mol^{-1} for the sum of the three bond strengths (Barin and Platzki, 1995). A mass spectrometric photoionization study also provided an accurately measured distribution of these bond strengths with values of D_0(Sb-H) $= 236 \pm 4$, D_0(HSb-H) $= 244 \pm 2.5$, and D_0(H_2Sb-H) $= 282 \pm 2$ kJ mol^{-1} (Ruscic and Berkowitz, 1993). These values now are supported by several theoretical analyses (Dai and Balasubramanian, 1990; Skulan et al., 2006; Vasiliu et al., 2012).

The remaining gaseous species is the sulfide that will be favored only in reducing gasification and pyrolysis conditions. Reflecting the oxide, it will have similar spectroscopy and electronic nature. The ore stibnite, Sb_2S_3, melts at about 550°C but is volatile at and below this temperature. Mass spectrometric studies indicate a congruent vaporization possibly to Sb_4S_6(g) that fragments to many Sb_xS_y compounds with x and y in the range 2–4 (Faure et al., 1972). A later, possibly more detailed study, repeated this work at higher temperatures with more appropriate mass spectrometric ionizing potentials for the SbS, determining D_0(Sb-S) = 366 ± 5 kJ mol^{-1}, 10 kJ lower than previously indicated (Hino et al., 1986). Comparing this to SbO in Table 7.7, the ratio D(SbS)/D(SbO) = 0.86 is consistent with trends for this ratio with its neighboring periodic table column of C, Si, Ge, Sn, Pb. The four most significant comparisons for this ratio from the other tables in this review are Si, P (0.78, 0.74); Ge, As (0.84, 0.80), Sn, Sb (0.89, 0.86), and Pb, Bi (0.92, 0.93) confirming a trend of reasonable consistency.

Laboratory scale combustion studies

From all the listed chemical values for gaseous antimony compounds, SbO clearly is very dominant. Antimony can produce antimonates and antimonites as mentioned above, the latter being more stable and can evaporate unchanged above 1000°C. Nevertheless, in flame gases these will undoubtedly be dissociated to their oxides and not be major transporters of Sb at higher temperatures. Farber and Srivastava (1975) examined the behavior of antimony in atmospheric pressure H_2/O_2 flame gases at 2000°C, as also arsenic, by probe sampling into a mass spectrometer. Although the monitoring of the gaseous Sb and SbO derived a reasonable value for D(Sb-O), their additional claim for the existence of $HSbO_3$(g) and H_3SbO_4(g) now may be questioned and regarded more as a clustering with water during the sampling phase. Additions of potassium also appeared to result in KH_2SbO_4(g) formation. More recent studies of this possible probe perturbation of such sampling with the potential for inducing heterogeneous formation of such complex molecules now appears very plausible (Jiminez and Ballester, 2005). Their correct determination of D(SbO) probably resulted from the rapid kinetic coupling between these gaseous Sb and SbO major species that was approximately maintained through any heterogeneous loss of both antimony molecules in the sampling. Such bond strength values are extracted from an exponent term and an error of twofold in the concentration ratio of the two at these conditions would introduce only a 10 kJ mol^{-1} change. Moreover, to explain a homogeneous kinetic formation of KH_2SbO_4(g) in this flame system from gaseous KOH and SbO at these concentrations and timescales is beyond kinetic comprehension. There is no mention in the literature relating to antimony's flame ion chemistry, nor its interaction with soot chemistry or flame catalytic inhibition.

Combustion power plant data

As seen in Table 1.1; Sb is a trace component in all fuels and for most coals has concentrations in a 0.05–10 ppmw range (Xu et al., 2003). Commonly it is below 2 ppmw but was surprisingly quoted as 20–50 ppmw in a Spanish coal (Font et al., 2010). It is negligible in oils (Kitto, 1993), and biomass generally has levels below 1.5 ppmw (Chandrasekaran et al., 2012). Levels also are very low in pet coke and limestone

(Lind et al., 1994; Cordoba et al., 2012). Municipal wastes and sewage sludge have the potential for higher concentrations and respective levels of 10–60 ppmw (Paoletti et al., 2001; Poole et al., 2008) and 7 ppmw (Reed et al., 1997) have been noted. Generally, for most encountered fuel ash the levels of Sb are not of concern with regards to agricultural disposal (Zhang et al., 2002). However, in some cases of waste incineration where the input is from materials incorporating the use of Sb it can become elevated and require attention. Also in the antimony processing industry in particular, a waste slag is produced that is labeled as a hazardous waste. In this case the ash has to be treated by solidification/immobilization methods with gypsum or cement to enable even its land fill disposal (Salihoglu, 2014). Now in the age of energy transition, new interest has turned to producing biogases that remain high in methane content but as found can have a variable higher range of trace metals, significantly so for antimony, than those found in natural gas and may need monitoring and processing (Cachia et al., 2018). As discussed below, it seems very likely that any antimony oxide formed and held in the bottom ash as in grate fired or kiln burners will have sufficient time to interact there. If vaporized from pulverized fuels the antimony will deposit on cooling ash as the oxide with some possibility of reacting with ash components or blending into any silicate glass components of fly ash. This possibility has been raised in waste incinerators where more antimony is found in the bottom rather than the fly ash. It has also been shown that calcium can interact with antimony to form antimonates or antimonites in such cases and thermally stabilize the antimony at higher temperatures (Paoletti et al., 2001). There are no available chemical thermodynamic values for any antimonates or antimonites so they have never been considered in any thermodynamic calculation modeling. As a result distributions presented by such calculations are limited to the oxides and provide an example of the importance of requiring a complete database for any exactness (Frandsen et al., 1994; Verhulst et al., 1996). For a spectral analysis of antimony, it is common nowadays to utilize one of the numerous modes of separation or reduction to the gaseous hydride coupled to a spectroscopic or ICP in mass analysis mode (Gonzalvez et al., 2009; Simeonsson et al., 1997; Lachas et al., 1999). It poses few problems for its monitoring.

Although antimony has not been included in many full-scale combustion studies of trace metals, the ones reported do show a pattern of behavior that is now reasonably consistent and well understood. Care has to be taken in that differences arise as mentioned above relating to the mode of combustion. In spite of low concentrations in coals there has been renewed interest in better modeling the release mechanisms of the more volatile elements one of which is antimony. As a result, one model has considered the pyrite melt and the diffusion-vaporization features (Zeng et al., 2001). This has continued more recently to considerations of the chemical nature of the antimony, including particle fragmentation (James et al., 2014). For pulverized coal, one fly ash study reports antimony as having one of the largest enrichment factors on fly ash (Marrero et al., 2007). From an early date, it was noted that enrichment on fly ash was indicated by its increase to smaller particle size (Davidson et al., 1974; Klein et al., 1975; Kaakinen et al., 1975; Rowe and Steinnes, 1977; Coles et al., 1979; Ondov et al., 1989; Lind et al., 1994; Lanzerstorfer, 2018), even down to a 0.03–0.06 μm scale (Wang et al., 2008). Also with pulverized coal, there are small depleted amounts of antimony in bottom ash noted. Several studies also observed that antimony could be associated with large ash particles suggesting a calcium oxide/silicate interaction (Bool and Helble, 1995; Galbreath et al., 2000). This was confirmed by subsequent work that sampled ash particles at 1150°C concluding that antimony could have a range of volatility and in their case had to be retained in the fly ash by interactions with possibly Ca or Si (Senior et al., 2000). Incinerators have noted 50% of the antimony retained in the grate ash when previous thermodynamic calculations suggested it should be totally volatile (Paoletti et al., 2001). This latter study also clearly showed that a CaO/Sb_2O_3 mixture was much more thermally stable, to at least 1400°C, indicating the ability like arsenic to form antimonates/antimonites. This was also seen in cofiring of coal/sewage sludge treated with iron and lime where the ash retained antimony at 1100°C (Folgueras et al., 2007). Another rather complex study examined the co-combustion of coal with solid recovered fuels coupled to any effects observed by adding NaCl, PVC, $(NH_4)_2SO_4$, and kaolinite. Although very difficult to analyze it was concluded that PVC (polyvinyl chloride) with its high chlorine content did appear to vaporize Sb more, and sulfur additions decrease it (Wu et al., 2013). After initial release, depending on temperature all the antimony will be gaseous SbO with some Sb contribution. As ash particles cool, temperatures will be reached where antimony compounds cannot only form on the ash surface but also remain there due to a low vapor pressure or be held there by reaction with an ash component. This may be in the form of an antimonate/antimonite or a combination as a silicate glass form. On a complex ash particle this might involve a kinetic time constraint. With cooling, the levels of oxide deposit on the surfaces will increase and possibly modify from for example Sb_2O_3 to the more stable mixed valence Sb_2O_4 with enrichment on the smaller particle sizes. In other words deposition will cover a wide temperature range like arsenic, providing adequate time for essentially all the volatilized antimony to associate with particulate or surface matter. Recent Chinese coal combustor studies indicate that all the antimony was removed by the ESP units leaving little for the scrubbing flue gas desulfurization (FGD) system resulting in a zero stack emission (Tang et al., 2012; Cordoba et al., 2012; Zhao et al., 2016). As a result there should now be little concern regarding antimony release from combustion sources if units are equipped with efficient particulate control.

Few combustion studies have reported data for antimony concerning gasification, pyrolysis or oxygen-lean modes. Gasifiers associated with IGCC (integrated gasification combined cycle) systems operate in a 1200°C–1600°C range. They report for coal a high volatilization with antimony and little present in the slag. One study indicated 40% in the gas phase and 60% on the ash (Font et al., 2010). Another, that antimony was still in the vapor as atomic Sb at 500°C–600°C, which is the operating range of hot gas cleanup systems. Also that smaller ash particles were enriched (Helble et al., 1996). From the present analysis, with little oxygen, antimony's chemistry becomes limited. $Sb_2S_3(s)$ is stable but only below about 500°C, above this it congruently sublimes to Sb_2S_3 and also dissociates to SbS and Sb_4S_3 (Piacente et al., 1992). However, if traces of oxygen are present a distribution will arise between SbO(g) and SbS(g) and the assertive oxygen driven chemistry will dominate. The condensed phase sulfide also reacts with water and below 630°C produces Sb_2O_3 and Sb, while above this solely Sb(g). SbCl(g) may also be produced but whether it can compete with the sulfide will depend on some balance such as the following, which is heavily biased to the sulfide.

$$SbCl + SH = SbS + HCl \quad \Delta H_0 = -152\ kJ\ mol^{-1}$$

Other gases such as the hydrides are too weakly bound to contribute. It remains unknown if there is any degree of interaction with ash components under such low oxygen conditions.

In summary, although toxic, antimony is noncarcinogenic and an emetic. Although a heavy metal there is little concern naming it as a pollutant. As with chromium it can display two valences III and V both toxic. Known as a volatile element, SbO(g) will be a dominant carrier through a normal combustion train. However, combustion systems remain poorly characterized as to any resulting ash compositions. Stabilities are such that sufficient times will normally exist for the majority of the antimony to be present on ash deposits. Enrichment on smaller particles is clearly evident. Depending on conditions it is predominantly distributed between bottom and fly ash forms predictably as Sb_2O_3 or an antimonate/antimonite compound formed with other ash components such as calcium, iron or silicon. Chlorine or sulfur appear to play minor or no lasting roles in oxygen-rich modes but may become alternate gaseous carriers under gasification conditions.

7.4 Arsenic

Arsenic lies in the N, P, As, Sb, Bi group of the periodic table and is labeled as a metalloid. It shows an increased metallic nature being more like Sb and Bi rather than N and P. It has a similar $3d^{10}4s^24p^3$ electronic structure that controls its dominant tri- and pentavalent nature. It is not abundant having an average concentration of about 1.8 ppmw in the Earth's crust and ranked about 54th in the elements. It is more abundant in the oceans ranking there about 25th. Sulfide minerals of the type (Fe, Ni, Co, As)/S are the most common and it is extracted either as a by-product from Cu, Au, and Pb metal smelters or by roasting the ore arsenopyrite, FeAsS, from which the As sublimes. It is marketed as As_2O_3, global production currently at about 46,000 tons $year^{-1}$, with abundant reserves. It is regarded as a minor metal selling at 99% purity at about \$0.90 lb^{-1} for the metal and about \$0.30 lb^{-1} for the trioxide. Since the 1950s it has found major use as a wood preserver as chromated copper arsenate. This has now been largely phased out for residential usage. However, use has continued in nonresidential applications such as with railroad ties, utility poles, and marine timber. Otherwise its other uses are for alloys with Cu and Pb, in semiconductors and as GaAs in the optoelectronic industry. It is useful in pesticides, herbicides, and insecticides and is added in traces to some animal feeds. Its uses as pigments, in cosmetics, and medically have been largely phased out due to its toxicity. In the EU the safe level for arsenic in cosmetics is zero (Mohammed et al., 2017). It has been a military weapon in the form of its high trivalent toxicity and as defoliants.

Assessments of global natural and anthropogenic contributions to the atmosphere are difficult. This is especially so as controlling regulations now are beginning to drastically reduce previous emission levels. With As, fossil fuel combustion and metal smelting have always played the major anthropogenic role and estimates in the early 1990s tended to imply that As possibly was a pollutant being about twice the natural background (Nriagu, 1989; Buat-Menard, 1994; Han et al., 2003). The natural background is from dust, volcanoes, forest fires, and sea spray. Coal combustion and metal smelting are the major anthropogenic sources (Pacyna et al., 2007; Thurston et al., 2011). However, a further global summary now indicates much lower anthropogenic estimates with metal smelting remaining the main source. Totals appear to have been reduced with the indication that atmospheric anthropogenic levels now fall to only half those of the natural background, which would remove its pollutant tag in the atmosphere (Pacyna and Pacyna, 2001). Peat core analyses dating back 2000 years suggest an enrichment factor has occurred in the atmosphere over this time period (Shotyk et al., 1996). Also although this and related ice core analyses relating to historical atmospheric levels are fraught with difficulties, recent data also have been obtained from the Himalayas and the Antarctic. This is compatible with expectations by reflecting the dominance of the natural contributions to As during 700 years prior to the 20th century. Then, with industrialization, levels increased significantly and now are reflecting declines as regulation of the copper industry is becoming effective (Hong et al., 2009, 2012). Additionally, studies for China assessing heavy metal emissions from 1949 to 2012 illustrate a fivefold increase with As over the past 35 years. However, this increase has now been held constant for the past 10 years (Tian et al., 2015). Analysis of the Chinese atmospheric particle emission sizes and content confirmed that those particles of 1 μm diameter did contain peak As concentrations (Duan and Tan, 2013). However, analyses of the particle content indicated 80%−99% As(V), the less toxic arsenate form, with no indication of organoarsenates (Yang et al., 2012). Concerns for the anthropogenic cycle of arsenic continue in China, which remains now the largest supplier to the world. Emissions from its mining and the combustion of coal have increased significantly over the past two decades such that regulations now are being introduced (Shi et al., 2017). China shows concern with its heavy metal input to agricultural soils. For arsenic, atmospheric deposition from industrial emissions appears dominant over that from manures (Peng et al., 2019).

Health concerns with arsenic

Arsenic is a neurotoxin that is alien to the body. It is a poison, and now of concern possible more so by being carcinogenic. It appears to be needed in some living species and its elimination from the diets of rats, chickens, minipigs and goats did indicate health deterioration (World Health Organization Trace Elements, 1996). Recent experiments with rats indicate that even postnatal low concentrations of arsenic oxide in water can cause significant oxidative stress and neurological changes (Zhou et al., 2018b; Chandravanshi et al., 2018). It is automatically in the human diet but remains unknown to what extent, if any, is needed. It is unavoidable being ingested in water, dusts and food, the human natural health risks having been recently assessed (Rasheed et al., 2016). In history it has been used medically for diseases such as syphilis and was incorporated as a skin lightener in cosmetic wafers and soaps. As a cosmetic it still is used, although As is now banned from cosmetics in Europe and many countries (Mohammed et al., 2017; Michalek et al., 2019). Because of its toxicity several detailed environmental and toxicology reports have been published on the web. These list extensively experiments on animals, effects on the aquatic and biological environments and discuss the medical metabolism and suggested dietary levels (World Health Organization As, 2001; Tox Profile As, 2007). These indicate a minimum risk level (MRL) intake of 5 $\mu g\ kg^{-1}$ body wt. day^{-1} or an approximate adult dose in a broad range of

3–200 μg day^{-1}. An estimated US average daily adult dietary intake is about 10–50 μg day^{-1} and appears at such levels not to have any known lasting effects. Arsenic ingestion is complicated in that biological As can be methylated (also toxic), or in other organic forms is only mildly toxic. For example in fish it is present as arsenobetaine that is not metabolized or otherwise digested in the body and is excreted unchanged (Dopp et al., 2010). However, because total arsenic generally is what is generally monitored this can become deceptive and moves to measure also the inorganic As speciation are needed and are beginning to appear (Jeong et al., 2017; Phattamajintatamrong et al., 2017; Gao et al., 2018). Monitoring technology is rapidly developing and a recent review now assesses the advances in monitoring inorganic As in foods (Llorente-Mirandes et al., 2017). An additional recent speculation is that there may be a beneficial selenium interaction in the body with methylated arsenic forms (Skroder et al., 2018). Previous large scale testing surveys of populations are available (Schofield, 2016) to which eight additional studies now can be added. Blood surveys are available from Morocco, the Canary Islands and China each showing mean values of 2–3 μg L^{-1} but with extreme range values to even 29 μg L^{-1} (Henriquez-Hernandez et al., 2018; Zeng et al., 2019). One noteworthy observation is that arsenic hair values do not appear to change with age (Zhu et al., 2018). Also levels in hair appear uniformly low. Whether the body establishes stable baseline values for each toxin remains uncertain. However, as the environment changes, so do the average values recorded in surveys. Such testing has been intended to establish the distributions of values displayed by groups of people as potential biomarkers using blood, plasma, urine and hair samples and note averages and standard deviations. Several other recent studies provide data for blood, serum, and urine samples (Goulle et al., 2005; Nisse et al., 2017; Zeng et al., 2019). Recommended target values now for arsenic are: blood $<$12 μg L^{-1}, hair $\approx$ 0.9 ppmw and urine $<$20 μg/24 hours (Mayo Clinic, 2018). US National Health Survey testing urine levels have been tabulated to display their distributions for the populace (CDC 4th National Report on Human Exposure, US, 2009, 2018). Because the half-life of arsenic in blood is short, urine is regarded as a more reliable test. What was immediately noted, as with all the trace metals, is that people are all different and generally broad distributions are exhibited with a significant range of values. One recent analysis of a very large US database (7400 people) for their blood/urine levels of As, Pb, Cd and Hg showed that more than a half of those tested had all three or four of these present in measureable levels. Additionally, that when summed many had baseline values for these four neurotoxins that should be of concern for health (Shim et al., 2017). Although doctors use average values and distributions as their guide this emphasizes the majority and not those beyond this that may register outside the 95 percentile. Concern is now considering such people that appear to retain toxins, display extreme measures and are most probably at high risk for neurological consequences, but still may not be on the toxicity curve or showing any medical symptoms. As^{III} is indicated as being four- to fivefold more toxic than As^{V}. Arsenic is readily methylated in biology forming As-C bonding and is closer in nature to phosphorus rather than the metals. It is one metal that can methylate inorganic As in the body. Both As^{V} and As^{III} are equally absorbed by the body, are reduced to As^{III}, then methylated, and finally excreted. Some plants can accumulate arsenic but biomagnification is not evident. Inorganic As compounds are classified as carcinogenic but toxicity is also of great importance. Drinking water itself can be a very major concern (Sarkar and Paul, 2016). Levels have been recommended to be at $<10\ \mu g\ L^{-1}$ (WHO Guidelines, 2006a), but in some countries, such as Bangladesh, values can be $>50-3000\ \mu g\ L^{-1}$. Consequently, whereas acute arsenic poisoning is infrequent in the West, long-term exposure can become very severe in such places. It was estimated that there were 9000 deaths and 125,000 disabling years lost in 2001 in Bangladesh. Such long-term exposure also becomes more difficult to treat by chelation (Nielsen, 2017). Cases such as the Bangladesh situation can be of significant value not only in establishing toxicity levels but also indicating the symptoms of illness and the outcomes. In this manner it would appear to confirm that arsenic behaves as a bodily toxin and less than a neurological disabler. In one region of China, subject to high levels of As in the water, testing of hair and urine samples tentatively suggested that there was increased retention of Ba and Mn in hair but enhanced excretion of Cd, Cu and Mo in urine, showing the complexity of the body's balances (Wei et al., 2018). However, in one recent Chinese analysis of newborn cord blood, those with high As values did appear to correlate with later neurobehavioral measures (Wang et al., 2018). In cases of elevated arsenic in water, the additional transfers into maternal milk might be thought a risk for nursing children but babies appear protected by the body's efficient mammary gland filtering (Kazi et al., 2016; Schofield, 2016; Bastos et al., 2018; Altun et al., 2018). Recent testing of urine samples of children in Spain indicated 18% were above the accepted normal range, $<20\ \mu g\ L^{-1}$ (Roca et al., 2016) and the difficulty is assessing whether this has consequences. Additionally, attention has now been drawn heavily to a potential role for rice in the diet and control over its arsenic content (Saifullah et al., 2018; Wan et al., 2018; Xiao et al., 2018; Althobiti et al., 2018). To varying extents it does accumulate in the body and in the United States the FDA (Food and Drug Administration) now has set new limits of <100 ppbw for infant rice to match the European standard (Erickson, 2016). A recent global model also has indicated the atmospheric transport of arsenic containing particle emissions between the continents (Wai et al., 2016). In spite of extensive medical data collections there appear to be no specific correlations to areas of high arsenic concerning any of the current neurological diseases such as autism or Alzheimer's (Li et al., 2018g; Wei et al., 2018). Whether arsenic can contribute to these remains unknown.

Arsenic's basic chemistry

The chemistry of this Group VA of the periodic table is noteworthy in having a rather limited range of inorganic chemicals and that of As closely reflects Sb. The condensed phase molecules are summarized for As in Table 7.8. Neither has a stable carbonate, sulfate, nitrate or hydroxide. The major trend is for oxides, sulfides, halides, hydrides, and arsenates. Metallic arsenic occurs in nature as a single isotope ^{75}As, and like phosphorus has three allotropic structures formed on modifying its condensation. Gray is the most stable and common to which the other yellow and black forms will convert. It is insoluble in water but tarnishes in air and at 614°C sublimes to the $As_4(g)$ molecules below its official melting point. Above this temperature a distribution between As_4 and As_2 occurs in the vapor and at 630°C the two can be of similar magnitudes (Murray et al., 1973). It burns in air producing fumes of As_2O_3. Arsenic has few cationic condensed forms other than its two dominant deliquescent oxides.

TABLE 7.8 Properties of arsenic's condensed phase molecules relevant to combustion, °C

m.pt. b.pt	m.pt.
As 817 subl 614, ins	$AlAsO_4$ 600, dissoc 1700, ins
As_2O_3 274 460, cubic <180 s	$Ca_3(AsO_4)_2$ 1455, stable to 1500, s
313 460, monoclinic >180	$Fe_3^{II}(AsO_4)_2$ >930, dissoc 1000, ins
As_2O_5 315, stable to 530 s	$Fe^{III}AsO_4$ 830, dissoc 900, ins
H_3AsO_3 ($As(OH)_3$) stable only in soln	K_3AsO_4 440, dissoc 900, s
H_3AsO_4 ($AsO(OH)_3$) stable only in soln	$KAsO_2$ 300 decomp, slightly s
$AsCl_3$ −16 130, reacts with H_2O	Na_3AsO_4 300–500, dissoc 950, s
$AsCl_5$ −50 decomp in water	$NaAsO_2$ 550 decomp, s
AsH_3 −116 − 63 dissoc 230, slightly s	$Mg_3(AsO_4)_2$ 1315, decomp 1400, ins
As_2S_3 310 707, ins	$Ti_3(AsO_4)_4$ 900, dissoc 940
As_4S_4 320 565 stable to 300, ins	
As_2S_5 decomp	

s, water soluble; *ins*, insoluble; *decomp*, decomposes; *dissoc*, dissociates; *subl*, sublimes; *soln*, solution.

The condensed stable As_2O_3 oxide (m.pt.274, b.pt.460°C) is the dominant oxide occurring below 180°C with the cubic structure (arsenolite), and above this changing to monoclinic (claudelite). However, if in an open vessel it begins to sublime at about 200°C to As_4O_6(g). It is soluble in water being the anhydride for arsenous acid H_3AsO_3, stable only in solution. This is the precursor for arsenites and unlike phosphorus is actually tribasic with an As $(OH)_3$ structure. If As_2O_3 is adsorbed onto activated charcoal it gains stability and the major volatilization then occurs at 300°C–500°C (Cypers et al., 2009). The pentavalent oxide, As_2O_5 (m.pt.315°C), can be produced by oxidizing the trioxide, is less stable, but also soluble in water producing arsenic acid also only existing in solution. This is the foundation of arsenates. Thermodynamically the pentoxide is the more stable oxide but only at ambient temperatures and cannot be obtained directly from reaction of As and O_2. The acid formula is often written as $AsO(OH)_3$ but its condensed phase is regarded structurally as a mix of AsO_4/AsO_6 and it decompose to the trioxide above its melting point and vaporizes above 700°C to As_4O_6 and higher oxides (Brisdon et al., 1986; Konings et al., 1992). Mass spectrometric analyses of the vapor above these oxides at 100°C–1450°C is complex confirming this molecular dissociation as $As_4O_{6,5,3}$, $As_2O_{3,2}$ and AsO(g) at elevated temperatures (Brittain et al., 1983). The structural and vibrational nature of the M_4O_6 and M_4O_{10} molecules (M = P, As, Sb, Bi) have been addressed theoretically (Chrissanthopoulos and Pouchan, 2008). Of the corresponding chlorides the $AsCl_5$ is a liquid (m.pt. − 50°C), readily decomposes and is hydrolyzed by water. The trichloride, also liquid, is more stable, $AsCl_3$ (m.pt. − 16, b.pt.130°C) but is readily hydrolyzed by water to As_2O_3, and it has no oxyhalides. Reflecting the oxides there are known sulfides. As_2S_3 (m.pt.310, b.pt.707°C) is so insoluble in water that it is nontoxic. The pentavalent structure As_2S_5 is unstable. Indications are that it can be thermally stable to 280°C–500°C and sublimes as As_2S_3 or will be oxidized in air (Johri et al., 1968). Another mineral, Realgar, As_4S_4 is well known and characterized (m.pt.320, b.pt.565°C, stable to 300°C), it vaporizes as the molecule with numerous other As/S species. These sulfides react with air and convert to the oxide. Their vaporization has been examined in detail with various outcomes and like the oxide is molecularly complex showing numerous As/S structures that gradually dissociate as temperatures increase producing AsS(g) (Munir et al., 1971; Faure et al., 1973; Lau at al., 1982; Brittain et al., 1983a). The gaseous hydride, AsH_3 (m.pt. − 116, b.pt. − 63°C), is thermodynamically metastable to dissociation but kinetically constrained. It is pyrophoric, oxidized to As_2O_3, slightly soluble in water and decomposes to As/H_2 at 230°C. It arises in nature and microorganisms are noted as producing it in sewage sludge.

This is the limited range of arsenic's simple inorganic condensed phase compounds. To these can be added the numerous known mixed metal oxide arsenates. Although several arsenites are known (of anion AsO_3^{3-}), those that might play a role in combustion are the two trivalent meta-arsenites, $NaAsO_2$ (m.pt.550°C, decomposes) and $KAsO_2$ (m.pt.300°C, decomposes), being very or slightly water soluble, respectively. The slightly less toxic pentavalent arsenates are thermodynamically more stable than the arsenites and are of more importance to combustion. Those of interest are $AlAsO_4$ (m.pt.600°C, dissociates ≈1700°C, insoluble in water), $Ca_3(AsO_4)_2$ (m.pt.1455°C, stable below 1500°C, slightly soluble), $Fe^{III}AsO_4$ (m.pt.830°C, dissociates ≈900°C, insoluble), $Fe_3^{II}(AsO_4)_2$ (m.pt. >930°C, dissociates about 1000°C, insoluble), K_3AsO_4 (m.pt.440°C, dissociates 900°C, soluble), Na_3AsO_4 (m.pt.507°C, dissociates 950°C, soluble), $Mg_3(AsO_4)_2$ (m.pt.1315°C, dissociates 1400°C, insoluble), $Ti_3(AsO_4)_4$ (m.pt.900°C, dissociates 940°C). Arsenates of Na, K, Ca, Mg, and Al are indicated as dissociating below 1000°C. Consequently, it is noteworthy that although the Russian research group have mass spectrometrically studied the vaporization of about 200 or so mixed metal oxides over the past 50 years they report only having monitored the gaseous $NaAsO_2$ species for arsenic obtaining its heat of formation (Lopatin and Shugurov, 2013). Generally, arsenates are not sufficiently stable and vaporize dissociatively. However, one thermochemical equilibrium analysis has considered these potential interactions with ash components and indicates the pronounced stability of several arsenates over the arsenites and conveys the important role that these can play in cooling coal ash (Contreras et al., 2009). A more recent thermogravimetric study of several arsenates confirms the stability of those of Al, Ca, Fe, and Mg (Liu et al., 2016). As indicated in Table 7.8 several of the arsenates will have a longer time period during cooling for their ash formation and may play a role in As capture. With such a limitation on thermal stability of many of its condensed phase molecules during volatilization release, the variety of gaseous arsenic compounds is very restricted in combustion.

Gas phase stabilities and transport

Arsenic reflects P, Sb, and Bi in favoring tetrameric structures on vaporization. As a result the metal produces As_4(g) (vapor pressure 100 ppmv at 323°C) that then dissociated to a stable As_2(g) at higher temperatures. Due to this complexity care has to be taken in any reassessment of thermochemical modifications. This is particularly sensitive in this case in that even the basic $\Delta H_{f298}(As)_g$ remains uncertain. This is normally derived from the condensed phase heat of formation of the element coupled to its heat of sublimation. In the present case this involves three separate parameters, establishing the gaseous heats of formation of As_4, As_2 and the bond strength of As_2. Initially compiled by Hultgren et al. (1973), the value $\Delta H_{f298}(As)_g = 301.7$ kJ mol^{-1} was obtained and remains listed as such in currently used databases (Lias et al., 1988; Knacke et al., 1991; Barin and Platzki, 1995). However, in the photoionization mass spectrometric study of gaseous AsH_n ($n = 1-3$), from ionization potentials and ion appearance thresholds the bond strengths of these species were obtained and indirectly implied that $\Delta H_{f298}(As)_g = 288.7$ kJ mol^{-1} (Berkowitz, 1988). Over the years, many have preferably used this value. A recent theoretical study realized this dilemma and made a brief assessment of quoted values in the literature, then averaging these and favored using a midvalue (Mok et al., 2011). However, it is now possibly better to reassess the three measured parameters in the interpretation of Hultgen et al. (1973). The two aspects that have been better refined are $\Delta H_{f298}(As_4)$ (Herrick and Feber, 1968; Celmer et al., 1984), and $\Delta H_{f298}(As_2)$ (Murray et al., 1973; Rau, 1975; Drowart et al., 1978) supported also by other studies. When combined, these now reduce the initial value of Hultgren et al. (1973) to $\Delta H_{f298}(As)_g = 287.8$ kJ mol^{-1} agreeing with the independent value of Berkowitz (1988). Also with an earlier adjustment by Pabst et al. (1976) to 286.3 kJ mol^{-1}, a measurement of 286.6 kJ mol^{-1} (Drowart et al., 1978), and a theoretical prediction of 288.1 ± 3.3 (Feller et al., 2011). Consequently, this sends a ripple through all the thermochemistry of arsenic and all quoted values herein are adjusted to a value of $\Delta H_{f298}(As)_g = 288.5 \pm 3$ kJ mol^{-1}, 13 kJ mol^{-1} smaller than previously in use. There has always been less uncertainty with the bond strength of As_2 due to its rich spectral transitions that have been characterized (Huber and Herzberg, 1979). From a predissociation in its $A^1\Sigma_u^+$ state a value of D_0(As-As) = 382 kJ mol^{-1} was established. Mass spectrometric studies of the vapor obtained values indicating 379.1 and 384 (Kordis and Gingerich, 1973; Murray et al., 1973) and a recent theoretical calculation confirmed this magnitude as 383.5 (Feller et al., 2011). Further theoretical calculations now better understand the nature of the bonding and show the typical range of accuracy that can be realized with the present various basis sets but cannot yet specify any greater accuracy in this case (Xu and Truhlar, 2011; Xu and Dunning, 2015). Meanwhile, it is suggested that a value D_0(As−As) = 382 ± 2 kJ mol^{-1} be recommended. Care has to be taken in some databases that have modified the heat of formation of As_2 but not that for As (Knacke et al., 1991). The two values are coupled and require both to be adjusted for consistency.

As already indicated, the saturated vapor above As_2O_3(s) even at 100°C indicates As_4O_6(g) as dominant and stable (Brittain et al., 1982). If the temperature is raised to 780°C or 890°C fragmentation occurs through various oxides with stable AsO(g) starting to be dominant above about 650°C. On heating, As_2O_5(s) loses oxygen and similarly produces As_4O_6(g), which appears to be the most stable gaseous oxide together with other As_4O_x molecules in the region below about 700°C (Plies and Jansen, 1983; Barten and Cordfunke, 1984; Brisdon et al., 1986). For combustion, AsO(g) becomes the major arsenic transporter. As a result, this molecule has been studied spectroscopically quite extensively and many of its transitions identified (Rai and Rai, 1984), but only one experimental measurement has carefully assessed its thermochemistry. A mass spectrometric study measured the equilibrium between AsO and SnO obtaining a value D_0(AsO) = 476 ± 8 kJ mol^{-1} (Lau et al., 1981). A recalculation indicates this remains valid and is recommended. From a spectral analysis of predissociation in the $B^2\Sigma^+$ state a value of <480 had been suggested (Huber and Herzberg, 1979). A later theoretical calculation of the AsO low-lying states and their potential energy curves indicates an $X^2\Pi_{1/2}$ ground state with no others low-lying but suggests a too small ground state bond strength that remains questionable (Alekseyev et al., 1995a). An error in the sign of its listed heat of formation if corrected together with the newly recommended $\Delta H_{f298}(As)_g = 288.5 \pm 3$ kJ mol^{-1} indicates $D_0 = 476$ originating from the same experimental data (Barin and Platzki, 1995). This recommended value can be compared to the known strong bond of HCl (428 kJ mol^{-1}) implying adverse energies for any reactive formation steps for halides in the gas phase and a greater dominance by this AsO oxide. It is slightly weaker than $D_0(O_2) = 494$ kJ mol^{-1} implying a slight endothermicity for the As/O_2 reaction (18 kJ mol^{-1}). From an assessment of the thermochemistry of As_4O_6(g) it was illustrated that the average bond strength in this was much less than that of AsO (Lau et al., 1981). The indication was that on breaking the first As-O bond in the tetramer, the structure would not be stable and rapidly decay to AsO(g). Little is known of a dioxide. Those for Sb and Bi similarly are absent. Mass spectroscopically its ion was detected in an oxide vaporization study and thought to be from the parent AsO_2 (Brittain et al., 1982). Its ESR spectrum has been characterized in a matrix isolation study coupled to pulsed laser vaporization of As with O_2 (Knight et al., 1995). A chemiluminescence from As with O_2 was speculated to arise from AsO_2 but has never been examined further (Fraser, 1984). Additionally a theoretical description of its predicted structure and bonding concluded that its bent ground state is $^2B_2$ with a dioxide nature (Sun et al., 2004). Bonding would be predicted to be substantially weaker than that of As-O from the trends summarized already in Table 7.7. Recently, an (A-X) electronic spectrum has been assigned for the first time to HAsO and DAsO in a jet cooled beam (Grimminger and Clouthier, 2011). Now, high level theoretical calculations have confirmed the spectral measurements predicting the geometry, energy and frequencies of the (A, X) states (Mok et al., 2016; Xu et al., 2017a). Like the corresponding more stable HPO species this may play some slight transient role in the oxidation kinetics. Although $AlCl_3$ is listed as a combustion inhibitor, this is possibly due more to its chlorine content alone, than some As intermediate cycle (Babushok and Tsang, 2000). As indicated, arsenic has the two major sulfides, As_2S_3 and As_4S_4. These readily vaporize at low temperatures favoring the initial tetramer configuration but rapidly establishing a complex mix of As/S molecules. Utilizing mass spectrometric techniques a value of D_0(As-S) = 376 ± 6 kJ mol^{-1} was determined (Lau et al., 1982). The potential energy curves for 27 electronic states of AsS have been calculated confirming an $X^2\Pi_{1/2}$ ground state with no mixing of low lying states and reflecting the oxide (Shi et al., 2016). They also estimated D_0(As-S) = 392 kJ mol^{-1}. Previous approximate extrapolations of the vibrational energy levels had suggested values of 383 (Arafunnisa et al., 1975) and 400 kJ mol^{-1} (Reddy et al., 1985). The trend of the oxide/sulfide ratio of the group in Table 7.7 would suggest about 381 so there is an internal consistency. A recommended value of D_0(As-S) = 380 ± 8 kJ mol^{-1} seems currently appropriate. Of the two databases listing AsS(g), one lists heats of formation that are equally too high for both AsS and As so getting the recommended value (Knacke et al., 1991), while the other has reduced the heat of formation of AsS but left the As unchanged so this indicates D_0(As-S) = 393 but if corrected by the 13 kJ mol^{-1} smaller value for As is now correct (Barin and Platzki, 1995).

The trichloride is liquid at room temperatures but is thermally stable, very volatile (b.pt.130°C) and readily hydrolyzed by water. Its individual bond strengths remain uncertain but a reliable heat of formation of $AsCl_3$(g) is known from the heat of formation of the condensed phase coupled to a heat of vaporization. These imply $D_{0\text{Atomiz}} = 906$ kJ mol^{-1} (Knacke et al., 1991; Barin and Platzki, 1995), supported below by the latest theoretical estimate of 898 for the sum of the three bonds, an average of about 300 each. In the gas phase, AlCl(g) remains poorly characterized both spectrally and structurally. This has recently been improved by calculations of its low-lying electronic states and their potential energy curves (Pontes et al., 2015). Its system of triplet and singlet states has an $X^3\Sigma^-$ ground state with a low lying $a^1\Delta$ state at an energy 70 kJ mol^{-1} higher ($T_{0,0} = 6608$ cm^{-1}). They estimated D_0(As-Cl) = 296.6 kJ mol^{-1}. It had been noted in a mixture of $AlCl_3$(g) with O_2 in the presence of K vapor that this preferably produced AsO(g) chemiluminescent spectral bands (Kushawha, 1973, 1975). Attempts to formulate from these studies a bond strength for AsCl produced values that were too large. Previous theoretical calculations had suggested values that were too small, D_0(As-Cl) = 253 and D_0(ClAs-Cl) = 240 kJ mol^{-1} (Latifzadeh-Masoudipour and Balasubramanian, 1997). An older Russian study was useful in calculating the thermodynamic functions for these three arsenic chlorides but at that time could only estimate expected bond strength values (Shaulov and Mosin, 1973). The most recent theoretical effort to improve this situation now has reviewed prior theoretical work and provided more reliable estimates (Mok et al., 2011). Their heats of formation imply the values D_0(As-Cl) = 299, D_0(ClAs-Cl) = 307 and D_0(Cl_2As-Cl) = 293 kJ mol^{-1} totaling 898 kJ mol^{-1} and close to the measured total atomization value. However, examining Table 7.7 with corresponding values for the three chlorides of P, Sb and Bi their trend would suggest the strongest bond should be that of Cl_2As-Cl. Consequently, this raises some uncertainty and awaits clarification. For now, suggested recommended values might be D_0(As-Cl) = 295 ± 6, D_0(ClAs-Cl) = 300 ± 10 and D_0(Cl_2As-Cl) = 315 ± 15 kJ mol^{-1}. Nevertheless, when these expected magnitudes are compared with the strong bonds of D_0(H-Cl = 428) and D_0(As-O = 476 kJ mol^{-1}), it alone implies adverse energies for any reactive roles for chloride formation in flame gases.

Arsine, AsH_3, is the simplest compound of arsenic making it popular for theoretical structural modeling validations. It is pyrophoric and highly toxic but important to the semiconductor and metal refining industries. As a result its health risks are heavily documented (Fishbein, 2002). Mainly due to a disagreement between its independently measured bond strengths and its measured atomization energy it became clear that the old $\Delta H_{f298}(As)_g$ value was in error as discussed already and has indicated this need for a reevaluation of previous quoted bond strengths of gaseous arsenic molecules. In the case of AsH_3 by introducing photoionization mass spectrometry, from the ionization potential of AsH_2 and the ions appearance potential it was possible to derive an accurate value of D_0(H_2As-H) = 313.4 ± 2 kJ mol^{-1} free from other parameters. Similarly from the ionization potential of AsH and its appearance potential the value of D_0(HAs-H) = 278.2 ± 2 kJ mol^{-1} was obtained (Berkowitz, 1988). Only theoretical values are available but for the H_2As-H bond these are respectively 306.5, 308.9, 308.6, 307.3, and 306.0, spanning 20 years and possibly suggesting the experimental value may be at its lower error range (Dai and Balasubramanian, 1990; Binning and Curtiss, 1990; Mayer et al., 1997; Mok et al., 2011; Feller et al., 2011). Similarly, support for the HAs-H bond value come from numerous theoretical values that slightly overestimate the experimental value indicating 285.9, 288.1, and 285.7 (Binning and Curtiss, 1990; Mok et al., 2011; Feller et al., 2011). By determining the total dissociation energy of AsH_3 experimentally, its heat of formation was reliably measured with a 1.7 kJ mol^{-1} error bar (Gunn, 1972). This implied an atomization energy or sum of its three bond strengths of $D_0 = 860.8 \pm 4$ kJ mol^{-1}. This is supported by several theoretical calculations of 856, 860 and 857 kJ mol^{-1}. As a result, taking the now recommended values of $D_0(H_2As\text{-}H) = 311 \pm 4$ and $D_0(HAs\text{-}H) = 280 \pm 4$ kJ mol^{-1} this implies $D_0(As\text{-}H) = 270 \pm 4$ kJ mol^{-1} as suggested originally by Berkowitz (1988) and supported by theory (Ramakrishna and Duke, 2003; Shi et al., 2009; Mok et al., 2011; Feller, 2011). Interest lies mainly with AsH and its spectral systems and monitoring potential. The potential energy curves for AsH have been studied indicating the $X^3\Sigma^-$ ground state and other singlet and triplet states with none low-lying (Alekseyev et al., 1998, Shi et al., 2009). Compared to H_2 ($D_0 = 432$ kJ mol^{-1}), these magnitudes rule out any formation role for these hydrides in the majority of combustion systems. These molecular stabilities are summarized in Table 7.7 for comparison with those of P, Sb and Bi.

Laboratory scale combustion studies

Early laboratory studies using atmospheric pressure H_2/O_2 flames at about 2000°C, sampled these mass spectrometrically on adding traces of arsenic initially and then also together with potassium (Farber and Srivastava, 1975). They confirmed the dominance of AsO(g). However, they additionally reported observing gaseous $HAsO_3$, H_3AsO_4, and KH_2AsO_4 but whether these were true flame species now is very open to question and more likely reflects their formation in the probe sampling system. Meanwhile a significant amount of data now have been collected for arsenic from full-scale combustors and other bench type smaller scale experiments. However, measurement of gas phase arsenic components in a practical combustor remains a problem. LIF of atomic As can be studied in the laboratory utilizing an ICP but requires excitation at the more demanding 193.7, 197.2 nm vacuum ultraviolet wavelengths (Hueber et al., 1994; Simeonsson et al., 1997). In fact the 193.7 nm emission line was monitored by chance by the Hubble Space Telescope indicating atomic arsenic on a distant star (Bengtsson et al., 1992). AsO(g) is spectrally well characterized showing an (A-X) emission system at 317 nm and (B-X) emission and absorption at 250 nm but mainly has been studied in discharges (Huber and Herzberg, 1979). The collection of particles and their analysis is reliable but involves collecting and handling samples, often at a later date. Methods of sample preparation (Marrero et al., 2007) and speciation analysis concerning arsenite/arsenate have been reviewed (Butcher, 2007; Gonzalvez et al., 2009). However, although the As^{III}/As^{V} ratio can be measured, and is very important, it remains far easier to simply measure total As, which is done with an ICP, for example.

Laboratory measures that involve heating samples in a combustion or pyrolysis mode can indicate varying rates of release. This may relate to having longer interaction times for the releasing arsenic to interact with other fuel species. Recent work on volatilization shows a sensitivity also to blending different coals and also whether the coal has a low or high As content (Liu et al., 2016, 2016b). In one experiment heating a coal sludge mix with air at 10°C min^{-1}, arsenic was seen not to be released before even 1100°C. The sludge had been treated with lime and $FeCl_3$ so formation of Ca and iron arsenates certainly might have been possible also with the coal's high As content (Folgueras et al., 2007). Another study examined coal up to 1000°C in reducing atmospheres of either N_2 or H_2 noting the increased release to higher temperature but still fractional. It was suggested that retention was probably due to sulfide formation for such gasification conditions (Guo et al., 2004). As a result, the very recent report suggesting that chlorine enhanced the release of arsenic from municipal solid waste (MSW) incineration in an FBC system is unusual and unexpected (Shen et al., 2018). It is not currently supported by other studies. In a high arsenic Chinese coal at 1200–1400°C studying combustion, CO_2 gasification or N_2 pyrolysis, respective releases were 85%, 65%, and 45%. Also it was enriched on the fine particles 0.1–0.2 μm as As_2O_5 and calcium arsenate (Tian et al., 2016). A woodbark combustion measurement claimed about 75% release at 800°C but with a very low initial As content (Miller et al., 2003). The conditions of such experiments will control the result and the interpretation may not strictly relate to combustion of a fuel in real time at all. For example, this can be the case where the method involves heating the sample with air or some other gas in a furnace at 800°C–1100°C at a 10°C min^{-1} heating rate and then maintaining it at the temperature for a period of time. This is a different time scenario from what a fuel particle experiences when injected into a full-scale combustor, is rapidly heated and explosively fragmented by vaporization forces having little time to form any mixed metal arsenates or anything else until the cooling phase initiates later. In other words studies that talk of retention must be careful in their interpretation when relating to full-scale combustion. In an examination of this time history of a fuel particle, detailed models have formulated the stages through which the particle melts and species diffuse and vaporize (Zeng et al., 2001; James et al., 2014). There is no mention in the literature relative to arsenic's flame ion chemistry, nor any interaction with soot chemistry or flame catalytic inhibition. Its soot suppression capability was rated as zero (Cotton et al., 1971).

Combustion power plant data

Based on the indicated chemistry it is useful to review the full-scale combustion data that are available. This overviews what has been established with a certain reliability free of speculation. As this book goes to print an additional useful review of arsenic's combustion behavior has appeared in a journal publication and provides possibly a more in depth examination of current oxy-fuel effects and removal technologies. It also provides several thermochemical equilibrium expectations for various situations that are of value in indicating possible outcomes. However, as stressed throughout this book, these generally are limited through our minimal understanding of any potential kinetic constraints and also the still incomplete databases so should not be invoked without additional confirming experimental input (Wang et al., 2018f). As noted in Table 1.1, traces of As in coal can be quite variable, 0.5–30 ppmw and may average about 15 ppmw (Rowe and Steinnes, 1977; Xu et al., 2003). Pyrite (As^{III}) is often the most abundant source with organics, also trivalent, and some arsenate (As^{V}). A large fraction of arsenate is generated from pyrite oxidation (Kolker et al., 2000). Biomass, and oils contain minor levels (Kitto, 1993; Thurston et al., 2011). Refuse waste has levels that vary and can approach or exceed those of coals

(Werther and Ogada, 1999; Becidan et al., 2010). Sewage sludge can also be several fold larger (Werther and Ogada, 1999). Pet coke and limestone report very low levels (Lind et al., 1994; Cordoba et al., 2012). Synthetic biogases can have very significant levels, as much as 10-fold higher than in natural gas and can need processing before use (Cachia et al., 2018). Examining the potential chemistry of arsenic it is to be expected that its initial release will be significant if temperatures are above or about 1000°C as both normal inorganic molecules and mixed metal arsenates will all dissociate and vaporize. This is confirmed in the most recent study of As vaporization rates with six Chinese coals. Using an isothermal thermogravimetric tube reactor at 600°C–1300°C it was seen that rates of release increased from 700 to 1000 and were complete above 1200°C (Zou et al., 2017). AsO(g) will become the major transporter at elevated temperatures and will readily deposit on cooling and be open to reacting with ash species as temperature regimes become functional.

Due to its potentially complex gaseous oxide nature, care has to be taken in assessing the value of the numerous chemical thermodynamic equilibrium calculations and speculations that have appeared. One example even showed the limitations seen in the basic arsenic databases and the variations solely between different sources (Lundholm et al., 2007). If the effort is made and a variety of potential products are added to the database the calculations can be useful in indicating potential temperature ranges and thermal stabilities for one molecule compared to another. One effort showed potential roles for the Al and Mg arsenates (Yan et al., 2000, 2001). However, the outcome can also be dependent on concentrations and being in the database. In this case, they ruled out the importance of the calcium arsenate that was suggested in an earlier calculation, and as seen below can be predicted as important in combustion ash with various fuels (Frandsen et al., 1994). A recent ambitious thermochemical equilibrated-theoretic study assessed a possible role for innumerable As/H/O gaseous species to exist in the flue gas. Such molecules neither have the kinetic mechanisms nor time to exist and remain a pure speculation with little value (Du et al., 2016). Another study simulated co-combustion with higher levels of As and has included many of the arsenates appropriate to combustion (Contreras et al., 2009). Moreover, it is important to remember that such calculations relate only to equilibrium conditions and are mainly of value for condensed phase formation. In combustion with cooling gases, a dominant species such as AsO(g) at the high initial temperatures will have very significant kinetic constraints of time and low concentrations in the gas phase to produce any of its potential complex arrays of oxides from which it originated. However, via surface chemistry these may be able to readily form heterogeneously as temperatures cool. Nevertheless, AsO(g) will be the dominant gaseous species that is transporting the element through the combustion train. Examples of calculations by Frandsen et al. (1994) portray the regions of thermodynamic stability of simple As/O systems. Consequently, the known chemistry conveys that depending on the fuel and the maximum combustion temperature, this will determine if the arsenic is fully released. However, certainly for temperatures above about 1200°C, AsO(g) will play a dominant transport role for the element. It is interesting in combustion that an element can readily modify its valency. For example, in the present case As_2O_5(s) with its As(V) valency vaporizes to As_4O_6 with As (III) and then fully dissociates to AsO(g), which can be regarded as a reactive free radical rather than representing an As(II) molecule. Then this on depositing may form an arsenate, the AsO_4^{3-} structure reverts to As(V) which is the slightly less toxic valence.

In coal-fired power plants the arsenic is totally volatilized from the coal such that its molecular mode present in the coal is irrelevant (Huggins et al., 2007). Because of its volatility there is little in bottom ash and the distribution is between the gas and fly ash phases (Martinez-Tarazona and Spears, 1996; Llorens et al., 2001; Otero-Rey et al., 2003; Meng et al., 2005; Goodarzi, 2006). Monitoring flue gases in real time confirmed rapid release even at 850°C either in a combustion or pyrolysis mode (Shen et al., 2015). The release rate was more rapid for pyrolysis than combustion and remains not fully understood.

Mass balances have been obtained for As in coal and from the start indicated that generally only about 5% of the arsenic would be present in the bottom ash or slag and the majority would be on the fly ash with less in the gas phase as temperatures cool (Klein et al., 1975; Otero-Rey et al., 2003). The measurements are difficult and complicated by the fact that different utilities have either bag houses, ESPs and with no or various scrubbing methods for SO_2 control. As a result, mass balances ranged from 53% to 134% in one summary of nine plants. More recent mass balance studies of the full distribution for power plants with SCR, ESP (hot or cold), and better scrubbers now are available. These confirm the 5% in bottom ash, 83%–88% in the ESP ash, and 5%–7% removed by the FGD scrubbers with now less than 1% being emitted (Aunela-Tapola et al., 1998; Meng et al., 2005; Cheng et al., 2009; Nyberg et al., 2009; Tang et al., 2012). A more recent study of coal combustion noticed that the fly ash arsenic was present at 11 times or more than in bottom ash and was >90% As^{V}. Only about one-quarter of the As in the bottom ash was the more toxic As^{III}. They also noted enrichment in fly ash to smaller sizes (Zhao et al., 2018). Following this, a study added biomass to sludge combustion to convert the As^{III} to As^{V} in the bottom ash by forming more calcium salts (Zhao et al., 2018a). Additions of chlorine salt also appears to reduce As in bottom ash of coal/sludge combustion (Zhang et al., 2019). A Spanish study of FGD efficiencies indicated a 50% reduction solely across this device (Cordoba et al., 2012). It was shown that fabric filter bags were more efficient than ESPs and that the As in fly ash is present as As^{V} (Nelson et al., 2010). If fabric filter bags are used 99.2% of the arsenic can be captured (Goodarzi et al., 2008). Senior et al. (2000b) illustrated that ESPs can have up to a 5% penetration level by particles of the 0.1–1.0 μm diameter range. Moreover, it is these particles that are of a health concern and so do need to be further controlled. Consequently, supplemental particulate removal by an ESP plus scrubber or using a fabric filter bag is needed to achieve adequate control. Care also does have to be taken with lime FGD in that limestone itself can contribute to the arsenic content together with other trace metals. As a result any gypsum evaporating droplet release also needs to be minimized. A most recent pilot-plant test of a circulating FBC burning a mix of coals confirmed this general nature of As and illustrated the difficulty still of obtaining accurate mass balances. It was felt that small amounts of fine particulate arsenic still managed to escape the system (Duan et al., 2016). A more recent study by them burning pet-coke and coal now shows total control (Duan et al., 2017). A study of a modern large Chinese coal-fired plant equipped with an ESP, WFGD, and WESP indicated that most As was removed by the ESP with a final release of 0.12 mg ton^{-1} of coal, from initially 3.79 mg kg^{-1} or essentially a 100% retention (Zhao et al., 2016). This has now been followed by a flurry of reports from Chinese studies on coal-fired plants fitted with all available particulate devices confirming total retention as ash forms (Wang et al., 2017a; Zhang et al., 2017a; Zheng et al., 2017; Zhao et al., 2017a; Chang et al., 2019). The most recent of these quoted that As was captured to a 96% level. A final study obtaining a mass balance noted that without a hot ESP, arsenic can interact with the SCR catalysts introducing a difficulty (Wang et al., 2017c). Nevertheless, it is clear that if fully equipped with particulate control devices that emission release of arsenic now can be negligible. In an emerging different combustion mode such as oxy-pulverized coal fuel combustion, the recycling of the CO_2 laden flue gases provides lengthened time for the interactions with the fly ash. As a result in a recent study all of the arsenic can be captured on the fly ash (99%) without an FGD in the system (Cordoba and Diego, 2018). Numerous tests now have emerged concerning the stability against leaching of elements absorbed in the FGD system. For arsenic there appears to be minimum risk (Zhao et al., 2018b). Studies of leaching from other ash forms is more complex varying with the type of ash (Li et al., 2018).

Analysis of particle sizes from different regions of a combustion train indicate that arsenic concentrations increase with time (Sakulpitakphon et al., 2003), and that there is a significant enrichment that occurs as particle size decreases (Martinez-Tarazona and Spears, 1996; Marrero et al., 2007; Wang et al., 2008; Cheng et al., 2009; Wu et al., 2013; Lanzerstorfer, 2018). This long known enrichment relative to the coal is only second to that of selenium particularly for particles $<2.5\ \mu m$ (Coles et al., 1979), and confirms the vaporization/deposition mechanism. It was also noted that As concentrations in fly ash increased with row number in a cold-side ESP. Unexpectedly, a hot-side ESP indicated a 90% capture efficiency showing possible protection for SCR catalysts (Meng et al., 2005). The capture is a chemically induced deposition process that is controlled by surface area and in this case is even functional at higher temperatures. Also the presence of unburned carbon in the ash has been shown to have no effect for arsenic but there is a pronounced correlation between the retention and the ash Ca or Fe loading (Seames and Wendt, 2000; Senior et al., 2000a; Lopez-Anton et al., 2006). Seames and Wendt (2007) confirmed this by spiking various coals with calcium and iron observing enhanced retention. Another study of sewage sludge pyrolysis confirmed the retention by CaO and Fe_2O_3 up to 850°C (Han et al., 2016). A mix of sewage sludge and biomass in a FBC incinerator has confirmed that the Ca and Fe in the biomass retained the As and reduced ash leaching (Zhao et al., 2017b). Coal capture by ferrospheres separated from fly ash also has been seen effective at 600°C (Zhang et al., 2016).

Chemical analysis of coal fly ash samples from the start was difficult due to the mixed content of ingredients and the low concentration levels. Although an insensitive and poorly selective method, the X-ray absorption fine structure (XAFS) method was favored as an analysis method. This was capable of confirming a dominance of arsenate formation, but could not fully characterize it. Slight contributions from arsenites could not be fully ruled out but seemed minor (Huffman et al., 1994; Shoji et al., 2002; Galbreath and Zygarlicke, 2004; Huggins et al., 2004; Goodarzi and Huggins, 2005; Shah et al., 2007; Huggins et al., 2007). The consensus was that arsenic in ash is predominantly As^V. Requiring further confirmation, Luo et al. (2011) using an extended XAFS method suggested that the spectra were consistent with calcium pyroarsenate, $Ca_2As_2O_7$ in a crystalline rather than the less ordered glass phase. In other work, one study with oxy-fuel and the other with MSW incineration continues to illustrate the difficulties of clearly analyzing the ash (Low and Zhang, 2013; Hu et al., 2015). Such work continues now with X-ray photoelectron (XPS) and Auger electron spectroscopy (XAEF) methods on high As fly ash from a Stoker boiler. Analysis remains indefinite but they noted some ash surface enrichment with As, but less than for selenium. Also the presence of arsenates that appeared controlled by Fe content (Fu et al., 2018). As seen with alkali deposition listed herein, very complex molecules can be readily produced by chemical deposition on ash particles and surfaces so the true nature of the arsenic deposits still is not known. Arsenate formation with Fe or Ca is favored. It has been shown in a bench scale experiment that As_4O_6(g) reacts with CaO, $Ca_3(SiO_3)_2$ and $CaSiO_3$ at 600°C−1000°C to produce $Ca_3(AsO_4)_2$ and possibly $Ca_2As_2O_7$ (Jadhav and Fan, 2001; Sterling and Helble, 2003; Li et al., 2007a). In cases of high As-coals, Ca is often low and consumed as the sulfate. Al and Fe play a more significant role forming their arsenates (Yang et al., 2019). It was speculated in that study that the SCR may oxidize any vaporized arsenic to As_2O_5(s). If so this will be liquid (m.pt.315°C) and dissociate and vaporize as the gaseous As_2O_3 to be converted to the arsenate on the ash. On release from a fuel the arsenic above about 650°C will be AsO(g). This is the species that will deposit on ash surfaces, but as seen in other cases, its exact description seems irrelevant to the dominant thermodynamically driven chemical processes that only need any source of arsenic. As we see above, there are many metallic arsenates it can form. For MSW the thermodynamics clearly favor possible formation of $Ca_3(AsO_4)_2$ (Sorum et al., 2004). Although there is the possibility that high sulfur will preferentially favor $CaSO_4$ formation this does not appear to be the case. Because real systems contain numerous species, any model has to consider all possible interactions and preferences. One recent initial attempt to consider this modeled the possible partitioning when the three species As, Sb and Se might be present. It also permitted examining effects if Ca, Fe or Al was added (Raeva et al., 2011). It is unlikely in cases of high sulfur and low calcium that arsenic trioxide could deposit as previously was suggested (Seames and Wendt, 2007). The latter's vapor pressure is 5 ppmv at 125°C and 37 ppmv at 150°C (Behrens and Rosenblatt, 1972). At such low temperatures, if no alternatives are possible it is more probable that As_2O_5 would begin to form, with a much lower vapor pressure and is thermally stable (1 ppmv at 275°C) (Polukarov et al., 1971). Limestone as an absorbent has been considered in coal gasification. It was shown that a calcium arsenate can form and be retained (Diaz-Somoano et al., 2010). Again, the interaction of arsenic with some calcium lime form is difficult to resolve. By the time the gaseous arsenic reaches a lime desulfurization system in a full-scale combustor it is already on a particle and the interaction is difficult to assess (Taerakul et al., 2006). Also reacting As_4O_6(g) with mineral sorbents does certainly confirm $Ca(OH)_2$ as effective with the formation of the arsenate and its rate increases significantly up to 1000°C (Mahuli et al., 1997). Even though these laboratory scale experiments do not exactly relate to reality, it does appear with coal and biofuels that all the gaseous arsenic manages to deposit and probably react with the ash particle components and be retained somewhere in the flow train. It does not appear to suffer any time constraints. Arsenic's emission control becomes one of particle control.

For normally operating oxygen-rich combustion, there has been increased interest in arsenic due to its potential poisoning of SCR catalysts. As a result, preliminary modeling of its behavior has already been attempted (Senior et al., 2006; Li et al., 2015a). SCRs operate at about 370°C and so its catalytic surfaces are vulnerable to arsenic chemical deposition. The model is adequate in suggesting additions of calcium assuming a surface deposition of gaseous As_4O_6 (or more probably AsO) to reduce the arsenic levels ahead of the SCR. The model also allows for the competitive channels of arsenic and sulfur for the calcium. This is a good example of the difficulties and assumptions that have to be considered in formulating a model even when a reasonable understanding of its chemical behavior has been established. It also illustrates that an adequate engineering model can be created even though not chemically rigorous. After reviewing other work, simulations of oxy-burn conditions suggest little change in the behavior of arsenic and in fact release may be more rapid and the recycling will lengthen the fly ash interaction time and promote arsenate formation (Liu et al., 2016a).

Another group of data concerning arsenic centers on hazardous waste disposal. One special case is the combustion of chromated copper arsenate wood. Although now banned from residential use, large quantities of wood for nonresidential uses such as railroad ties, poles and marine timber still are produced. Landfill introduces leaching problems and combustion requires acceptable incineration. Domestic fire and stove burning has been seen to be very hazardous due to its high mineral content. A combustion study of the wood indicated 35%–60% of the As was released in the 500°C–1000°C range (Pedersen et al., 2010). Two laboratory studies confirmed that the wood could be incinerated at low temperatures (<1000°C). X-ray ash analysis indicated that Cr as Cr^{III} and Cu being nonvolatile are retained in the ash and only 11%–14% of the As remained volatized (Wasson et al., 2005; Lundholm et al., 2008). A more recent gasification study at 850°C indicated that most of the As was retained in the bottom ash and the small level of vaporization could be captured on a filter when the gas is cooled (Kramg et al., 2016). Another aspect of current concern relates to waste disposal in cement kilns. Incorporating the waste into the manufacture of cement can raise the level of arsenic in the clinker product fivefold and raise enhanced leaching rates (Huang et al., 2012). In the normal production of cement, the natural As content of the limestone is retained effectively in the clinker product (Yan et al., 2015).

When arsenic's chemistry is examined for fuel-rich conditions (gasification, pyrolysis, coking) it necessarily has to display a different behavior. Also it will become more sensitive to the functioning temperatures that control whether the arsenic is fully released or not from the fuel or reacts internally. The group V elements of the periodic table (P, As, Sb, Bi) all can form binary compounds with virtually every metallic element in the periodic table. As a result, in the absence of oxygen many arsenides may be formed and also the sulfide (As_4S_4, b.pt. ≈ 600°C) (Helble et al., 1996). These can be very stable, for example, AlAs (m.pt.1740°C), FeAs (m.pt.1030°C), $NiAs_2$, $ZnAs_2$, as well as those of Ca, Mg and Na. However, these latter three can react with water and produce gaseous arsine, AsH_3, which might be the mechanistic source for its presence that is reported under such conditions. For control and removal of arsenic from gasified fuels, suggestions have been absorption by copper, aluminum (Reiss and Muller, 2011), palladium (Poulston et al., 2011), zinc ferrite or titanate (Diaz-Somoano et al., 2004), effective at temperatures ranging from 200°C to 800°C. Only $PdAs_2$ has been identified as a product by its X-ray diffraction (XRD) spectrum. Use of activated charcoal has been examined but in one study seemed only partially effective (Lopez-Anton et al., 2007a). An examination of adsorption of $As_2O_3(g)$ on activated charcoal is interesting in that 29 wt.% appeared to be retained even at 600°C, when normally it should all have vaporized below 300°C (Cypers et al., 2009):

possibly an example of activated charcoals strong absorbent ability. Experimental observations imply that arsenic is volatile even under fuel-rich conditions but there appear to be few gaseous possibilities for it unless there are traces of oxygen to additionally form AsO (g). Heterogeneous formation of gaseous AsH_3 or even $AsCl_3$ (Wu and Biswas, 1993; Font et al., 2010) may be possible and has been speculated but requires examination. In underground coal gasification, arsenic is evident but possibly 60% can be retained. That emerging in the gaseous product can contain the condensable phases of As_2S_3 and elemental As (Liu et al., 2006). All suggestions tend to be based on equilibrium calculations that have limited and questionable data bases. Nevertheless, a sizable fraction close to 50% of arsenic can be retained in the ash together with these as yet uncharacterized gaseous components (Wang and Tomita, 2003; Bunt and Waanders, 2008; Font et al., 2010; Ma et al., 2010). It is to be remembered that the gas phase distribution can easily moderate as temperatures increase with dissociation, but on cooling the chemistry is not simply reversible. Kinetic and time constraints generally will freeze distributions in the gas phase and they no longer can track back equilibrium expectations. Of course the exact nature of the molecules on the ash or on any other absorber, however complex it becomes, is important. Capture elicits the requirement for disposal and involves detailed studies of whether this can be done adequately without introducing leaching difficulties that have to be addressed further before full and safe sequestration becomes possible. Due to the generally low concentrations of trace metals in fly ash, policy still permits fly ash not to be regarded as toxic. Many studies are available concerning leaching methods (Kim and Hesbach, 2009) and various options for various arsenic wastes have been reviewed (Leist et al., 2000; Skodras et al., 2009; Thorneloe et al., 2010). Recent studies have examined and discussed the mobility of the toxic elements in ash and shown the arsenic to be predominantly As^{V}, however implying it still could contaminate top soils and will probably be the critical element to be leached from coal ash (Wang et al., 2008b; Su and Wang, 2011; Flues et al., 2013; Deonarine et al., 2016). Most recently, concerns with high levels of As and Zn in sewage sludge have led to cofiring studies of this with biomass. The latter is richer in Ca, Al and Fe promoting the capture as stable arsenates that are less leachable. In this manner the wastes become more compatible with agricultural use (Zhao et al., 2017). Similar type experiments are reported concerning this possible conversion of the arsenic retained on ash but simply by heating it in a fluidized reactor to provide more time for interactions with other ash components such as Ca, Fe and Mg to promote this formation of less leachable final components (Zhou et al., 2017a).

In summary, arsenic is regarded as a volatile trace element with few stable molecules other than mixed metal arsenates. The major gaseous carrier is AsO. The majority of the arsenic is found on fly ash with very little in bottom ash. It is a neurotoxin so that deposition on ash and other retention methods is required. Modern combustors that are fully equipped are now seen to operate very efficiently with regards to arsenic. Questions then revolve about disposal and leaching. Arsenates are stable below 1000°C, so there can be sufficient interaction time in most combustors for arsenate or oxide formation on the ash. Calcium, aluminum and iron arsenates are favored. Chlorine appears to play a minimum role in its chemistry and sulfur only as an alternate sulfate product that can be formed on ash more favorably utilizing the calcium and iron components.

7.5 Barium

Barium is present at slightly higher levels than strontium in the Earth's crust and constitutes the 14th most abundant element. Its name comes from the Greek barys meaning heavy and although only 56th in the periodic table, heavy densities are becoming evident. Its two ores are barite, the sulfate $BaSO_4$, and witherite, the carbonate $BaCO_3$. Both are insoluble in water, and as a result, levels of barium salts in the oceans are reduced, it being the 22nd element most present. The sulfate is more important commercially, world mining production being about 8 million tons per year but with abundant reserves. Its main use (90%) is as the sulfate, ground to form a heavy mud, used extensively in oil well drilling. An alternate name is "heavy spar" with a density of 4.5 g cm^{-3}, but still less toxic than the heavier molecules of $HgSO_4$ or $PbSO_4$ of densities of about 6.4 g cm^{-3}. Other smaller uses involve alloying, the glass and ceramic industry, as a white pigment, in rubber and tires and the chemical industry. It is the green coloration for fireworks and has been used as a rat poison. Barium sulfate is commonly used to coat interior linings such as the intestines and stomach to aid X-ray diagnostics due to its insolubility in water or dilute acids. Otherwise all soluble barium salts are toxic.

Health concerns with barium

Medically, barium is similar to calcium and its potential health and environmental issues have been extensively reviewed in open publications by the World Health Organization and by the Agency for Toxic Substances and Disease Registry in the United States (World Health Organization Ba, 1990, 2001a; ToxProfile Ba, 2007). Barium, atomic number 56, is a heavy metal and toxic, soluble salts such as the acetate, nitrate and chloride are more toxic than the insoluble carbonate, sulfate or phosphate. Nonessential to the human body, a dietary intake of 0.3–1.7 mg day^{-1} is normally consumed from food and water. Minimal risk levels range from 0.02 to 0.2 mg Ba kg^{-1} body wt. day^{-1} in the above reports. It does not accumulate in the human body to any extent and is soon eliminated with only possibly small amounts retained by bone. Noncarcinogenic, with lethal doses being 10–60 mg kg^{-1} body wt. depending on its chemical form. As a result, in normal situations its toxicity is of little concern. Nevertheless, if so, it can be monitored by blood, serum, hair or urine sampling that displays the normally observed spread of values in a population over a several fold range (Goulle et al., 2005; Wei et al., 2018). Some blood testing facilities suggest that total Ba levels should be <10 μg L^{-1}, <7 μg L^{-1} for urine and for hair samples <2.5 ppmw. However, it is rarely included in health or population testing surveys. The distribution of urine values for the US population has been listed from recent National Health Surveys and is of no concern (Centers for Disease Control and Prevention, 2009, 2018). Barium does not appear to accumulate in plants but can affect growth as it does to some microorganisms. Release of barium to the environment from industry is in particulate form and has not risen to any significant level of discussion as a heath concern for the public. However, it is regulated in incinerators and included in the ten elements of concern to the Environmental Protection Agency (Carnes et al., 1992; Linak and Wendt, 1993). Barium's common use as an additive to diesel oil as a soot suppressant was banned in the United States. It is not included in natural and anthropogenic global emission studies. Due to its similar occurrence in nature to strontium it is not surprising in coal and biomass fuels that their levels are similar but much below those of calcium but still quite significant as a trace metal (Table 1.1). As seen it is also expected to be present in other waste fuels and sewage sludge at similar type levels. Only low ppmw traces are present in pet coke but it can range up to $\leq$ 40 ppmw in limestone (Lind et al., 1994; Cordoba et al., 2012). Recent considerations of producing alternate energy sources, such as synthetic biogases, report that these can display barium levels very much larger than the low levels seen with natural gas (Cachia et al., 2018).

Barium's basic chemistry

A brief summary of the physical properties of its condensed phase molecules relevant to combustion is listed in Table 7.9. Barium tends to be the first in the periodic table labeled as a heavy metal (m.pt.727, b.pt.1894°C, 10 ppmv vapor pressure at 683°C) divalent with a repertoire of chemicals and a chemistry that reflects those of calcium and strontium. On heating the carbonate begins to dissociate in the solid phase to the oxide (10 ppmv vapor pressure of CO_2 at 750°C) (Barin and Platzki, 1995; Stern, 2001). The carbonate melting ($\approx$810°C) and boiling points ($\approx$1360°C) are not well defined but controlled by the CO_2 ambient pressure (Gurvich et al., 1994). The sulfate $BaSO_4$ is thermally stable, melting without obvious decomposition at 1580°C. Its vaporization to SO_2 and O_2 has been studied mass spectrometrically and does indicate a vapor pressure of 10 ppmv at 1155°C (Mohazzabi and Searcy, 1976; L'vov, 2007). The sulfite, $BaSO_3$, is less stable dissociating to the sulfate and oxide at about 380°C (Stern, 2001). The thiosulfate BaS_2O_3 is even less stable dissociating at 220°C. BaO(s) (m.pt.2013, b.pt.3092°C) is a refractory oxide the most thermally stable of the alkaline earths. It differs in that it vaporizes molecularly so that a heat of sublimation is meaningful. Its vapor pressure for BaO(g) is about 10 ppmv at 1480°C (Gurvich et al., 1994). Dissociation of this to Ba(g) becomes more important only to higher temperature especially under reducing conditions (Lamoreaux et al., 1987). It is a more reactive oxide being hygroscopic, reacting with water to become the hydroxide and also oxidizing in air to the peroxide. Also in combustion environments, the volatility of BaO(s) is additionally enhanced by ablative evaporation with water vapor directly to $Ba(OH)_2$(g) (Sasamoto et al., 1979). With barium, the peroxide BaO_2 gains stability over the other alkaline earths with a melting point of 450°C, decomposing to the monoxide above about 600°C (Tribelhorn and Brown, 1995). Its infrared spectrum has been studied by matrix isolation methods (Ault and Andrews, 1975; Andrews et al., 1994, 1996). The anhydrous $Ba(OH)_2$ melts at 408°C and loses water at 780°C becoming the oxide. It is less reactive to CO_2 absorption than other hydroxides. The nitrate is not particularly thermally stable, dissociating at its 590°C melting point via the nitrite to the oxide (Stern, 2001). Its dissociating vapor pressure is 10 ppmv at 390°C. The nitrite, $Ba(NO_2)_2$ is less stable dissociating to the oxide above about 217°C (Stern, 2001). As with all the alkaline earth elements the chloride $BaCl_2$ is an important chemical being stable and water soluble. It vaporizes by sublimation (m.pt.962, b.pt.1560°C). It has an equilibrium vapor pressure of 10 ppmv at 1005°C (Gurvich et al., 1994).

TABLE 7.9 Properties of barium's condensed phase molecules relevant to combustion, °C.[a]

	m.pt. b.pt. vapor pressure
Ba	727 1894, reacts with water and air, 10 ppmv/683°C, s
BaO	2013 3092, sublimes, reacts with water, 10 ppmv/1474°C, s
BaO_2	450 800, dissoc >500, ss
$Ba(OH)_2$	408, dehydrates >780, 10 ppmv/780°C, s
$BaCO_3$	810 1360, dissoc in the melt, 10 ppmv CO_2/752°C, ins
$Ba(NO_3)_2$	590, decomp via nitrite to BaO, 10 ppmv/389°C, s
$Ba(NO_2)_2$	267, decomp >217 to BaO, s
$BaSO_4$	1580 1600, decomp, 10 ppmv/1155°C, ins
BaS_2O_3	220, decomp, ss
$BaSO_3$	>380 dissoc to sulfate/oxide, ins
$BaCl_2$	962 1560, sublimes, 10 ppmv/1005°C, s
BaS	2227, dissoc >1500, reacts with water >650, and O_2, s
BaH_2	1200 >675 dissoc, reacts with water and air, s
$BaAl_2O_4$	1815
BaB_2O_4	1095, ins
$BaFe_2^{III}O_4$	1420, ins
$BaMoO_4$	1450
$Ba(PO_3)_2$	1560, ins
$Ba_2P_2O_7$	1430, ins
$BaSiO_3$ (meta)	1605, ins
$BaSi_2O_5$ (di-)	1420
Ba_2SiO_4 (ortho)	1760
$BaTiO_3$	1664, ins
Ba_2TiO_4	1860
$Ba(VO_3)_2$	707, 0.2 ppmv/1730°C
$BaZrO_3$	2700, ins

[a]*Gurvich et al. (1994), see text.*
dissoc, dissociates; *decomp*, decomposes; *s*, water soluble; *ss*, slightly soluble; *ins*, water insoluble.

Its $BaCl_2(g)$ vapor begins to dissociate to the monochloride only at higher temperatures and a theoretical study of its geometrical structure is available (Levy and Hargittai, 2000). Of more interest to pyrolysis or oxygen-lean systems, the sulfide, BaS, although having a melting point of about 2200°C readily reacts with O_2 and water above 650°C producing $H_2S(g)$ and H_2 (Ota and Conger, 1977). A mass analysis of its vaporization indicated the formation of BaS(g) that dissociates to Ba(g) and $S/S_2(g)$ above about 1500°C (Colin et al., 1964). The hydride BaH_2

(m.pt.1200°C), reacts with water, is spontaneously flammable in air and dissociates in an inert environment at 675°C. Of the remaining compounds of potential interest to combustion, barium can form numerous ceramic type materials that could be produced at higher temperatures in ash but remain poorly characterized. Barium tends to take a cationic role and all are expressed as barium molecules. However, for barium they are generally not as thermally stable as those of the alkali metals and only the perrhenates, borates, molybdates, and tungstates can evaporate congruently without dissociation or partial dissociation. In combustion, the ones of interest include the aluminate $BaAl_2O_4$ (m.pt.1827°C), metaborate $Ba(BO_2)_2$ (m.pt.1095°C), ferrite $BaFe_2O_4$ (m.pt.1420°C), metaphosphate $Ba(PO_3)_2$ and pyrophosphate $Ba_2(P_2O_7)$ (m.pts.1560 and 1430°C), orthosilicate Ba_2SiO_4, metasilicate $BaSiO_3$, and disilicate $BaSi_2O_5$ (m.pts.1760, 1605 and 1420°C), and titanate $BaTiO_3$ (m.pt.1625°C). Generally such salts are insoluble in water. The gaseous $BaSiO_3$, $BaTiO_3$, $BaFe_2O_4$, $Ba(BO_2)_2$ and $BaPO_3$ molecules have been monitored as stable entities in high temperature mass spectrometric vaporization studies (Lopatin and Shugurov, 2013). Their potential lifetimes in flame gases remain unknown.

Consequently, from this array of condensed molecular forms, only BaO and $BaCl_2$ can vaporize directly, together with $Ba(OH)_2$ from water aided ablation of BaO(s), and possibly BaS in dry, oxygen-free environments. Whether other gaseous forms can be generated and act as barium element transporters in combustion such as BaO_2, BaOH, BaCl, Ba(OH)Cl, or BaH has to be considered.

The dissociation energy of BaO and its meaning

A major difference with barium from the other Group IIA periodic table elements is that BaO(s) vaporization is a simple sublimation to BaO(g). In such cases, because the heats of formation of the solid and its sublimation can be measured with good accuracy its gaseous heat of formation can be readily obtained. This was evaluated as $\Delta H_{f298} = -117.9 \pm 6$ kJ mol^{-1}(Gurvich et al., 1994). This transcribes to a reliable value of D_0(Ba-O) $= 542 \pm 8$ kJ mol^{-1} the most thermally stable oxide of the Group IIA elements indicating a bond stronger than that of O_2 or OC-O. Its infrared spectrum was additionally confirmed by matrix isolation using the reaction of the metal vapor with ozone in a nitrogen matrix (Ault and Andrews, 1975a). This thermochemical relationship is simple and the various parts are now accurately known:

$$D(\text{Ba-}O)(g) = \Delta \text{H}_f(\text{Ba})(\text{g}) + \Delta \text{H}_f(\text{O})(\text{g}) - \Delta \text{H}_f(\text{BaO})(\text{s}) - \Delta H_{\text{sublim}}(\text{BaO})(\text{s})$$

In fact a reasonable result, 538 ± 15 kJ mol^{-1}, was obtained 60 years ago by this method (Drummond and Barrow, 1951). Since then, numerous equilibrium type measurements have obtained values in support as also theoretical calculations (Langhoff et al., 1986; Stiakaki et al., 1994). Probably the most accurate recent theoretical calculation now indicates 548.5 ± 8 kJ mol^{-1} (Vasiliu et al., 2018). Nevertheless, due to the interest in BaO(g) there have been about 20 experimental studies reporting measured values of its bond strength by various alternate chemical systems. These were extensively reviewed by Gurvich et al. (1994), and Huntelaar and Cordfunke (1997). From this diverse range of methods, results were found to lie in a range of about 534–569 kJ mol^{-1}. Some were easily discarded due to interfering hydroxide formation but some lie beyond the limits of experimental error and obviously are displaying a system dependence that has to reflect to some extent as kinetic constraints on the actual dissociation dynamics. Such ranges of values are similarly observed for all the Group IIA diatomic oxides and sulfides and is a most interesting and very intriguing dilemma. As a result, the recommended value for D_0(Ba-O) over the years has varied away from the directly assessed thermochemical equilibrium value depending on the specific preferences selected. For example, in four major reviews published, quoted values are 558 kJ mol^{-1} (Pedley and Marshall, 1983), 552 (Cordfunke and Konings, 1990), 542 (Gurvich et al., 1994), and 536 (Huntelaar and Cordfunke, 1997). One is left in a quandary as to what this bond strength means. What is clear is that equilibrated studies such as Knudsen cell-mass spectrometric measures that have no time constraints, and also theoretical values, better reproduce the suggested thermochemical equilibrium cycle value (Newbury et al., 1968; Farber and Srivastava, 1975a). A possible resolution comes from the detailed study of the appropriate potential energy curves for BaO, Fig. 7.2. Due to spin-conservation, the ground state $X^1\Sigma^+$ of all the group IIA oxides and sulfides does not correlate to the separated ground state elements but connect to the next lowest energy pair. With BaO this is to the excited Ba(^3D_g) (111 kJ mol^{-1} additional energy content) and O(^3P_g). Based on potential energy concepts a direct connection to ground state products does not exist and is forbidden.

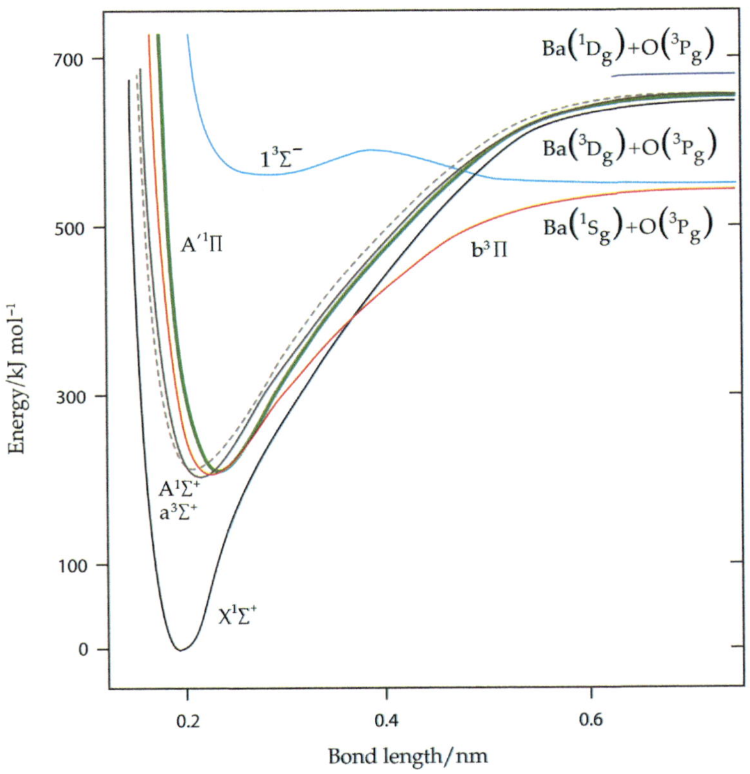

FIGURE 7.2 Low-lying potential energy curves for gaseous BaO. Source: *Redrawn from the data of Field, R.W., 1974. Assignment of the lowest $^3\Pi$ and $^1\Pi$ states of CaO, SrO and BaO. J. Chem. Phys. 60, 2400–2413; Gottscho, R.A., Koffend J.B., Field R.W. 1980. Optical-optical double resonance spectroscopy of BaO: the low-lying states. J. Mol. Spectrosc. 82, 310–338; Bender, D., Schaefer, S.H., Tiemann, E., 1986. Laser spectroscopy of perturbed states of BaO in the region above 32,000 cm^{-1}. J. Mol. Spectrosc. 116, 286–314.*

$$BaO(X^1\Sigma^+) = Ba(^1S_g) + O(^3P_g) \quad D_0 = 542 \pm 8 \text{ kJ mol}^{-1} \text{(spin forbidden)}$$

$$BaO(X^1\Sigma^+) = Ba(^3D_g) + O(^3P_g) \quad D_0 = 542 + 111 = 653 \text{ kJ mol}^{-1} \text{(allowed)}$$

It is the lower lying bound metastable ($b^3\Pi$) excited state of BaO, containing a 208 kJ mol^{-1} energy content that in this case rigorously correlates to ground state elements.

$$BaO(b^3\Pi) = Ba(^1S_g) + O(^3P_g) \quad D_0 = 542 - 208 = 334 \text{ kJ mol}^{-1} \text{(allowed)}$$

The complex spectroscopy of BaO now is well characterized and several of the lowest excited states in this case lie in this 197–210 region above the ground state quite different from other Group IIA oxides (Field, 1974; Gottscho et al., 1980; Bender et al., 1986). There is a significant overlapping of many vibrational levels of the first five bound excited electronic states. In this case these are currently labeled the $a^3\Sigma^+(T_{00}$ 197.3 kJ mol^{-1}), $A^1\Sigma^+$(200.0), $b^3\Pi$(208.0) and $A'^1\Pi$(209.5) and all have a negligible equilibrium total population of only about 0.006% that of the BaO(X) at 2000°C and so will be minimally involved in any normal chemical coupling to that of the ground state. Also from these only the $b^3\Pi$ bound state couples to the ground state dissociation products, the others all correlate to the second substantially higher limit. The question raised with all the Group IIA oxides and sulfides is the degree by which the dissociation mechanism proceeds unperturbed on its own curve to the second limit followed by quenching of the excited Ba(^3D_g) to Ba($^1S_0$) in this case. In fact, although metastable with a lifetime of 1.5 ms this is quite reactive. Although not readily quenched by inert gases (Adams et al., 2001), it may be quenched by other gases and does have reactive exothermic correlating surfaces for reactions with O_2, OH and H_2O to BaO(X), BaO(a,A,b,A′) and BaOH(X), respectively, and so probably has a shortened lifetime. In an equilibrium situation the reverse of a reaction between ground state Ba and O would produce BaO($^3\Pi$) which then also would be quenched to BaO ($X^1\Sigma^+$). Overall, such a full mechanism could effectively reproduce the forbidden transaction under equilibrium experiments with no kinetic time constraints. However, experiments where reaction times are short or in spectroscopic studies of reactions between specific states, these will be more indicative of the true mechanism proceeding and can modify individual concentrations slightly. Moreover, as earlier noted (Gaydon, 1968; Shukla et al., 1975), and now confirmed utilizing accurate spectral parameters for the $X^1\Sigma^+$ state (Li et al., 2000), a simple Birge-Sponer extrapolation of its vibrational energy levels does predict D_0(Ba-O) = 660 kJ mol^{-1} surprisingly close to the 650 kJ mol^{-1} now expected above for the allowed transition However, the largest experimental measure has indicated only 569 kJ mol^{-1}. Several attempts have been made to better resolve what we mean by this measured dissociation energy. These monitored certain reaction products. Under single-collision conditions, the reaction of barium with CO_2 was examined in detail (Dagdigian et al., 1974).

$$Ba(^1S_0) + CO_2(X^1\Sigma_g^+) = BaO(X^1\Sigma^+) + CO(X^1\Sigma^+) \quad \Delta H_0 = +D_0(OC\text{-}O) - D_0(Ba\text{-}O)$$

This reaction is fully spin allowed and with D_0(OC-O) = 526 kJ mol^{-1} and a value of D_0(Ba-O) = 542 or 650 will imply that the reaction is approximately exothermic by either 16 or 124 kJ mol^{-1}. By monitoring with LIF the maximum rotational level populated of the

$BaO(X^1\Sigma^+)$ product by this reaction, it was obvious that the lower value was appropriate. This was followed by a similar study on the reaction of Ba with ClO_2 (Engelke et al., 1976). Both established a bond strength value of 559 ± 5 kJ mol^{-1}.

Only two electronic states correlate with ground state products namely the bound $b^3\Pi$, and also a nonbonding repulsive $^3\Sigma^-$ illustrated in Fig. 7.2 the exact shape of which remains unknown and is speculated herein. These necessarily both have to cross the $X^1\Sigma^+$ curve. Spin–orbit coupling between the $^3\Pi/X^1\Sigma$ states was shown to be very small for BeO, MgO, and CaO and ruled out for these (Stagg and Richards, 1974). Although forbidden by Kronig's selection rules (Herzberg, 1963), induced predissociation utilizing the repulsive $^3\Sigma^-$ state has been shown possible for the analogous case of BeS dissociation (Larbi et al., 2010). As a result BaO dissociation may be occurring to different extents, partly on its allowed transition to the second limit, and also by curve crossings onto either the low lying $b^3\Pi$ or the repulsive $^3\Sigma^-$ state both of which can then produce ground state products. These mechanisms all could inflate the resulting value in those experiments with time constraints. This is an interesting aspect of reaction dynamics that needs further study to resolve this very old controversy that is more pronounced for some of the other alkaline earths. Consequently, it is seen that the recommended bond strength of BaO (542 kJ mol^{-1}) is valid for fully equilibrated studies. Under certain other conditions its apparent value might be elevated slightly. Nevertheless, in combustion if released into the gas phase this will only make slight adjustments. What is a more noteworthy question centers on whether it becomes hydrolyzed or chlorinated?

Gas phase stabilities and transport

With increasing atomic number the stability of the Group IIA condensed phase peroxides gradually increases. BaO_2(s) can in fact be obtained by heating BaO(s) in O_2. At room temperature it is stable and can be purified by further treatment with O_2. However, on heating it decomposes back to BaO(s) above about 500°C and can be used as a source of O_2 (Haller et al., 1996). It has been formed in the gas phase in crossed molecular-beam studies interacting gaseous Ba with O_3 (Davis et al., 1990). Although the reaction produced mainly gaseous BaO, 10% channeled to produce gaseous BaO_2 seen in this experimental work for the first time. From the interaction energies it was implied that it had quite strong bonding of $D_0(\text{Ba-O}_2) = 500 \pm 80$ or $D_0(\text{OBa-O}) = 452 \pm 80$ kJ mol^{-1}, but still a much weaker bond than that of BaO(g) and so unlikely to be stable in a H_2/O_2 system or combustion environment. Of the other gaseous molecules of interest BaH can be formed and its spectroscopy is well characterized (Pearse and Gaydon, 1976; Huber and Herzberg, 1979). Its bond strength of $D_0(\text{Ba-H}) = 190 \pm 3$ kJ mol^{-1} was obtained from spectral predissociation limits (Gurvich et al., 1994). This was supported by an early theoretical calculation that underestimated this value (Fuentealba et al., 1987). Its electronic structure and potential energy curves for the low-lying states were studied theoretically (Allouche et al., 1992). It can be expected to be readily oxidized and react with water. It is of little concern with regards to combustion speciation of products. The dihydride is only known from its characterization in a matrix isolation measurement (Wang and Andrews, 2004b). BaS(s) is stable and sublimes at the lower temperatures but then produces a vapor by dissociation of Ba, S/S_2, and BaS above 1500°C and does not permit obtaining a reliable value for a heat of sublimation (Colin et al., 1964; Gurvich et al., 1994). BaS is the analogue of BaO in having a similar spectroscopy, a ground state $X^1\Sigma^+$ and a $^3\Pi$ state that lies about 140 kJ mol^{-1} to higher energy (Cummins et al., 1981). As discussed with all the alkaline earth oxides and sulfides these are all intriguing and the reassessed dissociation energy, D_0(Ba-S)(g) = 420 ± 15 kJ mol^{-1} (Gurvich et al., 1994), is obtained from a Knudsen cell-mass spectrometric equilibrated experiment and relates to the optically spin-forbidden dissociation, namely

$$\mathrm{BaS}(\mathrm{X}^1\Sigma^+) = \mathrm{Ba}(^1\mathrm{S}_0) + S(^3\mathrm{P}_g) \quad D_0 = 420 \pm 15\ \mathrm{kJ\ mol^{-1}}(\text{spin forbidden})$$

It is the low-lying $^3\Pi$ excited state of BaS that correlates directly to ground state elements.

$$\mathrm{BaS}(^3\Pi) = \mathrm{Ba}(^1\mathrm{S}_0) + S(^3\mathrm{P}_g) \quad D_0 = 420 - 140 \approx 280\ \mathrm{kJ\ mol^{-1}}$$

If there is no curve hopping interaction between the $X^1\Sigma^+/^3\Pi$ states, a kinetic or spectroscopic measurement might derive a larger value by 111 kJ mol^{-1} as the $X^1\Sigma^+$ correlates in this case with either of the following overlapping energy state asymptotes which are spin allowed and hard to distinguish.

$$\mathrm{BaS}(\mathrm{X}^1\Sigma^+) = \mathrm{Ba}(^1\mathrm{S}_0) + S(^1\mathrm{D}_g) \quad D_0 = 420 + 111 = 531\ \mathrm{kJ\ mol^{-1}}$$

$$\mathrm{BaS}(\mathrm{X}^1\Sigma^+) = \mathrm{Ba}(^3\mathrm{D}_g) + S(^3\mathrm{P}_g) \quad D_0 = 420 + 111 = 531\ \mathrm{kJ\ mol^{-1}}$$

Transfer by curve hopping to the repulsive $^3\Sigma^-$ state, that is the only other curve that correlates to ground state products for these oxides and sulfides, Fig. 7.2, has also been suggested in the case of BeS via a spin–orbit induced predissociation mechanism (Larbi et al., 2010). As described above for the similar situation for BaO whether the above recommended bond strength is a true general representation will have to be determined from studies based on different methods of determination. What remains true is that the actual reaction dynamics of the dissociation remain uncertain. Trends in the ratio of MS/MO in the periodic table are not too helpful. Comparing this ratio for the Ca/Sr/Ba (Table 7.10), and Sc/Y/La (see Tables 7.15 and 7.19 for Sc,Y) adjacent groups suggest that D_0 420 kJ mol^{-1} for BaS is reasonable and should not be smaller by more than 20 kJ mol^{-1} if in error. Nevertheless, with regards to combustion, BaS(g) although stable can only be expected to survive in strongly reducing systems free of oxygen or water. The description of the role of the hydroxides was mainly studied in laboratory flames and is described below.

Laboratory combustion studies

The early flame studies immediately indicated that the hydroxides BaOH(g) and Ba $(OH)_2$(g) could both be important in flames (Ryabova and Gurvich, 1965; Sugden and Schofield, 1966; Cotton and Jenkins, 1968). In a later more detailed study of a $\Phi = 1.7$ fuel-rich $H_2/O_2/N_2$ flame at 2200°C the distribution of barium in the burned gases was shown to be 0.1% Ba, 17% BaO, 37% BaOH, 37% $Ba(OH)_2$, 9.5% $BaOH^+$, and 0.04% Ba^+

TABLE 7.10 Recommended bond strength values and trends for relevant Group IIA gaseous molecules, kJ mol^{-1}.

Bond	Be	Mg	Ca	Sr	Ba
D_0(M-O)	440 ± 10[a]	360 ± 10[b]	385 ± 10[a]	410 ± 10[a,c,d]	542 ± 8[a,c,e]
D_0(M-O_2)	368 ± 16[f]	105 ± 16[f]	226 ± 15[f]	226 ± 25[f]	500 ± 80[g]
D_0(OM-O)	420 ± 20[h]	–	>300	–	452 ± 80[g]
D_0(M-OH)	442 ± 15[i]	315 ± 12	386 ± 15	397 ± 10[j,d]	438 ± 8[d]
D_0(HOM-OH)	574 ± 40	456 ± 25	462 ± 20	435 ± 10[j,d]	440 ± 10[d]
D_0(M-Cl)	385 ± 5	320 ± 6	400 ± 6	409 ± 8	441 ± 5[k]
D_0(ClM-Cl)	536 ± 8	464 ± 8	502 ± 9	473 ± 8	480 ± 8[j]
D_0(M-H)	196 ± 4	124.0 ± 0.0[l]	163 ± 2	156 ± 2	190 ± 3
D_0(M-S)	350 ± 20[a,m,n,o]	270 ± 25[p]	330 ± 15[a,j]	330 ± 15[a]	420 ± 15[a]
Bond ratios					
D_0 OM-O/M-O	0.95	–	–	–	0.83
D_0 M-S/M-O	0.80	0.75	0.86	0.80	0.77
M-OH/HOM-OH	0.77	0.69	0.84	0.91	0.99
M-Cl/ClM-Cl	0.72	0.69	0.80	0.86	0.92
M-Cl/M-OH	0.87	1.02	1.04	1.03	1.01
HOM-OH/ClM-Cl	1.07	0.98	0.92	0.92	0.92

[a] *Apparently predissociating to ground state atoms.*
[b] *D_0 for an X/a mixture, $D_0(X^1\Sigma^+) = 447$, $D_0(a^3\Pi) = 226$, see text.*
[c] *Huntelaar and Cordfunke (1997).*
[d] *Vasiliu et al. (2018).*
[e] *Shown to dissociate to ground state atoms.*
[f] *Bauschlicher et al. (1992).*
[g] *Davis et al. (1990).*
[h] *Assumes $D(OM\text{-}O) = D(M\text{-}O_2) - D(M\text{-}O) + D(O_2)$; $D(O_2) = 493.6$ kJ mol^{-1} However, the nature of the bonding remains uncertain for MO_2.*
[i] *Vasiliu et al. (2010).*
[j] *Average value: Gurvich et al. (1994), Barin and Platzki (1995), Chase (1998).*
[k] *Average value (Huber and Herzberg, 1979; Gurvich et al., 1994; Barin and Platzki, 1995).*
[l] *Shayesteh et al. (2007).*
[m] *Partridge et al. (1988).*
[n] *Larbi et al. (2010).*
[o] *De Oliveirs-Filho et al. (2011).*
[p] *Uncertain, approximate value for a $D_0(X^1\Sigma^+) = 335$, $D_0(a^3\Pi) = 200$ mixture, see text.*
From Gurvich et al. (1994) unless otherwise indicated.

(Jensen and Jones, 1978). This shows several characteristic changes with barium. The greater stability of BaO enhances its presence, the two hydroxides can be of similar importance, and the chemiionization of barium is significant. Knudsen cell-mass spectrometric measurements from four experiments suggested a value of D_0(Ba-OH)(g) = 438 ± 15 kJ mol^{-1} (Gurvich et al., 1994) and from three 440 ± 30 kJ mol^{-1} (Chase, 1998). An early theoretical paper had suggested 444 (Bauschlicher et al., 1986). A very recent accurate theoretical estimate now suggests D_0(Ba-OH)(g) = 437 ± 8 kJ mol^{-1} and D_0(Ba-O)(g) = 549 ± 8 kJ mol^{-1} close to the values now recommended (Vasiliu et al., 2018). There appears to be reasonable agreement and the previous error bars can now be narrowed to a recommended D_0(Ba-OH)(g) = 438 ± 8 kJ mol^{-1}. It had been realized for a longtime that BaO(s) readily reacted with water vapor and appeared to vaporize to $Ba(OH)_2$(g). Measured by transpiration this was seen to be the most efficient for barium of the Group IIA elements (Sasamoto et al., 1979).

$$BaO(s) + H_2O(g) = Ba(OH)_2(g) \quad \Delta H_{298K} = 163 \text{ kJ mol}^{-1}, K_{1500K} = 1 \times 10^{-3} \text{ (Chase, 1998)}$$

In the presence of about 10% water vapor this would now suggest a vapor pressure of about 100 ppm, higher than had been reported by this transpiration experiment. However, the system appears to be more complex than assumed therein with gaseous BaO, BaOH, and $Ba(OH)_2$ contributing (Stafford and Berkowitz, 1964). Values for the strength of the second hydroxide bond remain very uncertain depending on the selected choice of experimental data. D_0(HOBa-OH)(g) = 419 ± 25 kJ mol^{-1} (8 studies, Gurvich et al., 1994), 439 ± 35 (3 studies, Chase, 1998), 419 ± 25 (4 studies, Cordfunke and Konings, 1990). However, the recent reportedly accurate theory that does correctly establish the magnitudes for BaO, BaOH, BaCl, $BaCl_2$ now estimates D_0(HOBa-OH)(g) = 442 ± 10 kJ mol^{-1}, a stronger value than its suggestion of 437 for the first bond (Vasiliu et al., 2018). Table 7.10 shows the overall trends in bond strengths for these Group IIA molecules. This can help as a guide in many cases. From this it is clear across these molecules that the second bond strength in the dichlorides compared to the dihydroxides is invariably the larger as now is being suggested by this data. The M-Cl/M-OH ratio is close to unity for Ca through Ba and the HOM-OH/ClM-Cl ratio is consistent with a recommended value of D_0(HOBa-OH) (g) = 440 ± 10 kJ mol^{-1}, the two bonds being of essentially equal strength.

Consequently, in combustion, a balance between BaO(g) and the gaseous hydroxides will occur due to their respective bond strengths, the presence of water and the flame conditions. The electronic structure of $Ba(OH)_2$ was noted in matrix isolation experiments ablating Ba with O_2 and H_2 or H_2O_2 mixtures, indicating a bent shape and its vibrational frequencies (Wang and Andrews, 2005). Calculations also indicated that the monohydroxide had no low-lying states (Allouche and Aubert-Frecon, 1994).

Additions of chlorine to flames clearly indicated it could modify a barium distribution. This was evident in a 50-year-old Russian paper where CCl_4 vapor was added to a C_2H_2/air flame containing traces of barium (Zhitkevich et al., 1963). Flame spectral emission from atomic Ba, Ba^+, BaO, BaCl, and BaOH is well characterized, is generally thermal in nature and has long been useful in monitoring such species (Pearse and Gaydon, 1976; Haraguchi et al., 1979). The Russian study noted on gradually adding chlorine that the Ba*, BaOH* emissions linearly decreased while that of BaCl* initially increased and then also began to decrease. Further work confirmed their interpretation with linear effects being observed with low chlorine levels consistent with monohalide formation becoming quadratic with increased chlorine and dichloride dominance. However, realized independently, the analysis of such data implied that more than one monohalide was evident and the Ba(OH)Cl species was suggested, which might be expected with such similar thermochemistry (Schofield and Sugden, 1971; Gurvich et al., 1971, 1974; Ryabova et al., 1971). The bond strength of the gaseous radical BaCl(g) has been extensively studied mass spectrometrically in Knudsen cells, spectroscopically in flames, by transpiration, and by electron impact. A value of D_0(Ba-Cl) = 441 ± 5 kJ mol^{-1} is recommended (Huber and Herzberg, 1979; Gurvich et al., 1994; Chase, 1998). It is based on mass spectral analyses of several gas phase equilibrium reactions and appears reliably measured. As a result the recent supposedly very accurate theoretical estimate of 453 ± 8 kJ mol^{-1} is unsettling (Vasiliu et al., 2018). This chloride is the most stable of the five Group IIA monochlorides and reflects the bond strength of the monohydroxide. It is seen to have no low-lying electronic states (Allouche et al., 1993). $BaCl_2$(g) also is very stable, and its heat of formation can be reliably determined from the independent measures of its heat of sublimation and the $BaCl_2$(s) heat of formation both of which are accurately known. Coupled with the value above for the monohalide this implies a recommended value of D_0(ClBa-Cl) = 480 ± 8 kJ mol^{-1} (Gurvich et al., 1994; Chase, 1998), the recent theory confirming this 485 ± 8 kJ mol^{-1}. This is one case where the dichloride second bond is stronger but the dihydroxide becomes slightly weaker. The subsequent Russian work indicated that the mixed hydroxychlorine molecule had approximate bond strengths of D_0(HOBa-Cl) = 435 kJ mol^{-1} and D_0(ClBa-OH) = 448 kJ mol^{-1}, the two being similar (Gurvich et al., 1974). So, as seen in Table 7.10, going through the Ca, Sr, Ba sequences, both the monohydroxides and monochlorides gain strength at the expense of the dihydroxides and dichlorides on moving to barium. Distributions of the gaseous $BaCl_2$, BaCl, BaO and the two hydroxides will be the most stable species in combustion and will depend on the combustion equivalence ratio and HCl/H_2O ratios. These will transport any barium that manages to be released and then deposit onto cooling particles or surfaces. Moving from Be through to barium the interplay between the hydroxides and chlorides changes. For beryllium in flames the hydroxide can be significantly in excess of the dichloride. However, considering the balanced reaction below for gaseous barium molecules, with 10% water vapor and 1% chlorine it suggests a ratio of about 250-fold of dichloride over the dihydroxide instead of the 70 fold excess of the dihydroxide that is seen with beryllium.

$$BaCl_2 + 2\,H_2O = Ba(OH)_2 + 2\,HCl \quad K_{p\ 2000K} \approx 4.0 \times 10^{-5} \text{ (Chase, 1998).}$$

In flame atomic absorption analytical spectroscopy it was noted that Al in a sample could interfere with analyses of Ba or Sr content (Luecke, 1994). It appeared that flame barium, for example, was reacting with the Al in the high temperature flame gases and producing a barium aluminate that was stable certainly in the condensed phase.

Barium is noteworthy for its chemiionization mechanism that invokes the $BaOH^+$ ion. Chemiionization was introduced to explain nonthermal ion formation of hydrocarbon radicals in flames (Gaydon and Wolfhard, 1979). It was then realized that such mechanisms could also explain the high levels of ionization observed even for metals with large ionization potentials (Alkemade et al., 1982; Goodings, 1992). This was the case particularly for the alkaline earths and in particular for Ba, ionization potential 5.21 eV, slightly larger than Na 5.14 eV, where for barium the chemical mechanism could extensively enhance the number of atomic and molecular ions in a flame (Schofield and Sugden, 1965). With the alkaline earths atoms, their ions are isoelectronic with the alkali metals. As a result, it is very novel that Ba and Ba^+, and the other alkaline earth atoms and atomic ions can be monitored simply by optical spectroscopy and their ratios established (Schofield and Sugden, 1965). It was through this, coupled with mass spectrometry that the ease of chemiionization was confirmed and later validated (Jensen, 1968; Cohen et al., 1971; Goodings, 1992). When it was found that barium salts were very effective in suppressing soot, both its coagulation and volume, a role for the ions was foremost. Of the Group II elements their effectiveness was scaled as Be 0; Mg 0, Ca 2; Sr 26, and Ba 100, that correlated with ease of ionization (Cotton et al., 1971; Bulewicz et al., 1975). All three alkaline earths were noted for reducing the amount of soot, but only Ba increased the number density of the particles. Reviewed along with other metal elements, it was noted that when used in diesel engines Ba was found to be emitted as the carbonate. Also it is evident that a full understanding of the mechanisms still is needed as to whether catalytic oxidation of the smaller soot particles also plays a significant role (Howard and Kausch, 1980).

Although not strictly connected, it was noted in this same time frame that numerous trace metal elements could become important in flame suppression (Linteris et al., 2008). The three alkaline earths were listed as showing a strong effect but less than Cr, Pb and Fe. Again they were in the order Ba > Sr > Ca (Cotton and Jenkins, 1971; Bulewicz and Padley). The still prevailing models utilize catalytic kinetic reaction cycles that invoke rapid conversions between the metal oxide and hydroxides that effectively recombine H and OH radicals so attenuating their propagation mechanisms. Such chemistries in the laboratory are easily induced. Whether they relate to full-scale combustion is another question.

Combustion power plant data

The chemistry and basic studies therefore indicate that barium inorganic salts can register vapor pressures of 10 ppmv at approximately 750°C ($BaCO_3$), 780°C ($Ba(OH)_2$), 1005°C ($BaCl_2$), 1155°C ($BaSO_4$), and 1480°C (BaO). However, except for the dichloride most of these are dissociating on vaporization to BaO so a release of gaseous barium will only occur from some fuels except at high temperatures. If this does occur, as temperatures cool, chemical deposition on ash or surfaces will initially favor these molecules in reverse order depending on the sulfur and chlorine levels in flue gases. Moreover, as temperatures cool, time permitting a deposit formed at higher temperatures can convert to one more thermodynamically stable. In this way, $BaCO_3$ could constitute the final deposit composition at low temperatures if it became the most thermodynamically stable. Additionally, it is to be remembered that a deposit or ash component of BaO(s) can interact with others in the ash and similarly be converted to the sulfate or chloride, time constraints permitting, as thermodynamic preferences change on cooling.

Information concerning barium's chemistry in combustion environments is not extensive but is found in various different disciplines. These involve its initial use as a soot suppressor in diesel fuel combustion, in flames used for analysis, with particle formation methods and for CVD, in addition to full-scale power plant data.

As already described, barium was found to be a very good soot suppressor. Now banned as an additive for environmental reasons in the United States, this centered mainly on the smaller size of the soot particles emitted rather than the insoluble $BaSO_4$ produced on the particles. The functioning mechanisms were never fully resolved but clearly involved increased ionization-induced nucleation rates of smaller soot particles that then could be oxidized by flame radicals more rapidly (Ndubizu and Zinn, 1982; Hirschler, 1986; Bonczyk, 1988). As mentioned, when used as an additive to diesel fuels, it was found to be emitted as $BaCO_3$ and also $BaSO_4$ particles (Turley et al., 1973; Truex et al., 1980).

In early flame atomic analytical methods the detection sensitivity depended on the level of free atoms in the flame. When assessed, it was realized that in the case of barium with a C_2H_2/air flame at about 2150°C, only 0.1% of the flame barium was in the atomic state (De Galan and Wineforder, 1967; Zeegers et al., 1969). It was also further confirmed on adding up to 2000 ppmv of chlorine that $BaCl_2$ formation became the dominant gaseous species (Kantor and Ernyei, 1994). There was a realization that high temperature combustion was needed for barium release.

Although BaO(s) has a very high melting point of about 2013°C and a low vapor pressure it is not useful as a ceramic material. It is too reactive, being hygroscopic and forming $Ba(OH)_2$(s) with water and absorbing oxygen to become BaO_2(s). As a result few efforts have been made to synthesize its nanoparticles in flames. One study using high temperature acetylene flames and more interested in aluminum and magnesium oxide nanoparticles did include Ba. However, the conditions were not tailored for its optimum formation (Fennell et al., 2007a). More interest has centered on fabricating thin films or particles of $BaTiO_3$ or $Ba_xSr_{1-x}TiO_3$ using CVD methods. These add aerosols of the two or three salt solutions simultaneously into a flame and either collect the high melting point nanoparticles (Brewster and Kodas, 1997; Maric et al., 2002), or deposit a film onto a 500°C–1000°C substrate (Ichinose et al., 1994; Olijaca et al., 2003). Again this illustrates that irrespective of the speciation of the barium, strontium and titanium in the flame a pure titanate deposit can occur compatible with thermodynamic preferences. The structural nature of such $BaTiO_3$ recently has been analyzed theoretically (Chibisov, 2015).

Although the level of barium is not one of the major elements it is present at nonnegligible levels in all fuels. Depending on the mode of combustion it can be labeled as either nonvolatile or volatile. This depends on temperature and conditions and is mainly volatile in coal combustion with its higher temperatures. In such cases, it is evident that barium can vaporize to some degree, become a component of the forming fume of gaseous elements and then deposit to contribute to the very small enriched submicron size particles. It appears that deposition is efficient and any gaseous barium molecule is short-lived. As a result, barium in a fuel is to be found in either the slag, bottom ash or any fly ash. Overall, no fractionation is noted between the bottom ash and the fly ash (Querol et al., 1995; Martinez-Tarrazona and Spears, 1996; Duan et al., 2016).

Evidence of enrichment of barium as a function of particle size particularly below 2.5 μm was noted in early studies and since reported in numerous other coal ash studies (Coles et al., 1979; Ondov et al., 1979; Meij, 1994; Querol et al., 1995; Martinez-Tarrazona and Spears, 1996; Nielsen et al., 2002; Goodarzi, 2006a). However, most recently, an ambitious study monitored 43 elements in size segregated coal fly ash samples but noted no pronounced enhancement for Ba in the particles for 2–20 μm diameters (Lanzerstorfer, 2018). This was true also for the Mg, Ca and Sr elements of Goup II included. Nevertheless, a bimodal distribution of the micron-sized particles has been observed with coal, reflecting the trend of increasing barium particle concentrations with smaller diameters. As indicated, this was noted in numerous early studies (Ragaini and Ondov, 1977; Quann et al., 1990), and particularly in greater detail in one (Markowski and Filby, 1985).

By examining particles in a size range of 0.04–10 μm upstream of the ESP this displayed particle and barium peaks at 0.05 and 2 μm. This behavior of barium reflects that of As, Sb Se and Ca clearly evidencing a contributing vaporization mechanism. Barium constituted 3.3% of the fume mass, and one-quarter of the total barium in particles <10.5 μm was in the submicron fume. As a result, any leakage and emission of small submicron particles will have the additional aspect of being severalfold enriched in barium over normally obtained ash. A British coal-fired plant reported that any slight accumulation on small size ash was a minor surface association, and that similar Ba concentrations were noted in all the ash forms (Spears and Martinez-Tarrazona, 2004).

A Danish mass balance of barium in two coal-fired power plants indicated the high efficiency of removal as fly ash with stack escaping emissions at a 0.18% level (Nielsen et al., 2002). Other studies support low releases using solely ESP or coupled to scrubber technology (Klein et al., 1975; Meij, 1994; Goodarzi, 2006, 2006a). More recent studies considering the overall emissions of trace elements from modern large coal-fired power plants in China show the effectiveness of particle control. Barium is effectively removed mainly by the ESPs and other particulate controls and the stack quantity emitted has become negligible being reduced five orders of magnitude below the quantity in the coal (Zhao et al., 2016, 2017a). Additional retention by modern scrubbers indicates these have close to a 50% efficiency for barium, removing final traces (Cordoba et al., 2012). The gypsum products manufactured from the sulfur scrubber extract has been extensively tested in recent years concerning leaching of its trace elements content and thermal stability. Levels of barium are reported to be low and minimally leached (Zhao et al., 2018b).

The normally found levels in ash are not sufficiently high for concerns with agricultural usage. Leaching studies of ash appear to differ. One study with ash from various coal blends burned at 950°C in a muffle furnace indicated that barium was readily leachable (George et al., 2008). However, two studies, one Chinese (Zhao et al., 2008) and one Japanese (Sukandar et al., 2006) involving incineration of hospital waste at below about 1100°C temperatures found the barium in the ash either minimally or significantly leachable possibly indicating a fuel compositional dependence. Roasting municipal waste ash at 1000°C with Cl_2 for 3 hours induced negligible volatilization of barium (Chan and Kirk, 1999). Few calculated equilibrium studies have included barium and have to be incomplete due to their current limited thermochemical databases. One for waste incineration implied a role for $BaSO_4$ below 955°C (Poole et al., 2008) and another for a coal gasifier indicated possible roles for $BaCO_3$, $BaCl_2$ and above 925°C the orthosilicate (Bunt and Waanders, 2010).

In summary, the nature of barium in combustion depends on temperature and is generally one of being retained in the ash, deposited on the ash, or contributing to the formation of submicron particles. Ash composition remains uncertain and could involve mixed metal oxide formation and depend on the presence of chlorine and sulfur. It reflects the behavior of calcium and strontium, volatility requiring higher temperatures. Escaping as particles from any combustion system will depend on the quality of the particulate control equipment. Leakage will generally result as a result of submicron particles surviving through these devices. At present for barium this does not appear to relate to any risk factor (Nelson et al., 2010) and on a modern power plant releases are negligible.

7.6 Beryllium

Beryllium is not entirely rare being about the 48th most abundant element in the Earth's crust. However, the element was not isolated in commercial quantities until about 1957. It is unusual in that although categorized with the alkaline-earth elements in the periodic table, chemically it has little in common with them and if anything is more similar to aluminum. Its differences are associated with it being atomic number 4, which implies a very small atomic and ionic radii and a charge to radius ratio that is similar to that of Al. It has a single stable isotope ^{9}Be and other radioactive masses raise no environmental concerns. However, ^{7}Be is a cosmogenic radionuclide formed in the atmosphere with a half-life of 53 days. This has found use in atmospheric studies considering solar activity, regional patterns and seasonal variations (Bas et al., 2016). With beryllium in either atomic or molecular forms all are extremely toxic especially if inhaled. Its major ore is beryl an aluminosilicate, $Be_3Al_2(SiO_3)_6$, that can exist in emerald, aquamarine and morganite gem mineral phases. Since its introduction into commercial application, beryllium's usage initially increased but now averages only about 260 tons year^{-1}, extracted from about 6500 tons year^{-1} of globally mined beryl. Main use is in aerospace and defense applications with much used in its elemental form in alloys with Cu, Al, or Ni. Some early potential uses were shown to be environmental hazards and so were replaced with other materials. Beryllium's main assets are its light weight, strength and corrosion resistance. If pure its cost is about \$7.50 g^{-1}.

Health concerns with beryllium

Medically, of course, beryllium is another of the metals that are alien to the body. However, due to low levels (<μg L^{-1} range) in drinking water and in edible plants (<1 μg dry wt.), a normal diet may still ingest small quantities thought to be <15 μg day^{-1}. Although absorbed, it is normally eliminated rapidly but some retention can occur in the skeleton with an average half-life of about 450 days with manifestations of possible consequences only after long periods. It is not beneficial to plant growth but deposition can contaminate leaves and affect growth. Due to its carcinogenic nature (lung cancer), strict controls were imposed on its manufacturing processes and on emission from incinerators (Carnes et al., 1992; Donnelly, 1993). Very extensive environmental reports, available on the Web have been published by the World Health Organization and an extensive toxicology report also is available (World Health Organization Be, 1990; Bruce and Odin, 2001; Tox Report Be, 2002). An MRL of about 2 μg day^{-1} kg^{-1} body wt. has been suggested. These reports and an earlier one from EPA (Anderson, 1973) indicate that natural levels of beryllium in the atmosphere from volcanic particles or dust constitute only about a 2.7% contribution to the environment. The main component now appears to be from coal combustion (93.5%) that remains unregulated. Fuel oil is a lesser component (3.7%) together with now regulated metal processors (0.2%) (Bruce and Odin, 2001). Nevertheless, there still appear to be significant health concerns for workplace exposure and whether there actually is a "safe level" (Brisson, 2009; Shay et al., 2013). In the United States, legislation by the Occupational Safety and Health Administration is in process to reduce the allowed beryllium dust levels in industrial locations by a further factor of ten to minimize any risk of lung cancer (Widener, 2015). Major concerns center on the dust levels connected with using coal slag as an abrasive blasting material. Although emissions from the cement industry and incinerators appear included in the US 1990 Clean Air Act and are now regulated in the United States, the emissions from coal combustion were excluded as not being sufficiently concentrated. However, individuals in one region burning coal with a high beryllium content did appear to show signs of possible exposure (Bencko et al., 1980). As a result, although present in coal on a par with mercury it is interesting to review such conclusions. It is known that biomagnification in the food chain is not evident for beryllium and that it has never entered the medical or pharmaceutical repertoire. However, anthropogenic levels remain higher than previously experienced by mankind. Not often included in test surveys, individuals do appear to reflect very low levels when tested in their blood or urine samples ($<0.1\ \mu g\ L^{-1}$) (Goulle et al., 2005; Cesbron et al., 2013; Nisse et al., 2017). The Mayo Clinic testing in the United States suggests blood values of $<1.0\ \mu g\ L^{-1}$ are acceptable (Mayo Clinic, 2018). However, one current survey comparing blood samples from the two populations of the Canary Islands and Morocco recorded similar high Be ranges of $0.8-2.8\ \mu g\ L^{-1}$, the possible source not being discussed (Henriquez-Hernandez et al., 2018).

Beryllium's basic chemistry

Due to early expectations that beryllium might be useful as a potential rocket propellant fuel, 50 years ago its then largely unknown basic chemistry began to be studied quite extensively. It was realized that it was a rather unique element that although predominantly divalent generally was never fully ionic nor covalent. For example, its rather simple BeO(g) molecule is neither ionic nor conventionally double-bonded but has a structure rather like BeOH(g) with a similar strength (Boldyrev and Simons, 1995). Other than the elemental metal with its melting point of 1287°C (b.pt.2470, vapor pressure 19 ppmv at 1230°C) the major molecule is again the inert ceramic oxide, BeO(s) (m.pt.2578°C). Its vapor under equilibrium conditions has been studied at high temperatures by the Knudsen cell-mass spectral analysis method displaying dissociative vaporization with the formation of gaseous Be, BeO, and $(BeO)_{n=2-4}$ polymers and a little Be_2O (Chupka et al., 1959; Theard and Hildenbrand, 1964; Lamoreaux et al., 1987). $Be_2O(g)$ was monitored by a mass spectrometer and has been explained by the unusual bonding evident in the BeO diatomic (Kalemos, 2017). However, it is not normally evident in beryllium's chemistry and shows little importance. Also this oxide form in the generally divalent group II elements is not evident for the other members. Temperatures can be above about 2200°C for the vapor pressure of BeO(s) to move onto a ppmv scale. However, it had earlier been noticed that vapor pressures were much lower in a vacuum (about 1 ppmv at 2000°C) than in the presence of water where volatility was noticed even at 1250°C (Erway and Seifert, 1951). Further studies indicated that beryllia, BeO(s), could be ablated by water vapor directly to $Be(OH)_2(g)$ with little evidence of BeOH(g) being formed (Altman, 1963; Blauer et al., 1966; Douglas, 1972).

$$BeO(s) + H_2O(g) = Be(OH)_2(g) \quad \Delta H_{298K} = +212/+172\ kJ\ mol^{-1},\ K_{1600K} = (0.8-1.4)\times 10^{-4}$$

The differing thermochemical values reflect the level of uncertainty of the two databases that contain $Be(OH)_2(g)$ (Gurvich et al., 1994; Chase, 1998). Such values imply under equilibrium conditions in the presence of 10% water vapor that the partial pressure of the dihydroxide will be about 10 ppmv at 1340°C ± 45°C. Looking at the repertoire of beryllium's condensed phase chemicals appropriate to combustion these are listed in Table 7.11. It is noted that other than BeO(s) few of its compounds are thermally stable. The peroxide, BeO_2(s), is unstable and has not been reported. $Be(OH)_2$(s), slightly soluble in water, dissociates to the oxide above 200°C. $BeSO_4$, insoluble in cold water is stable to about 600°C where its vapor pressure is about 10 ppmv and dissociates via a basic sulfate to the oxide (Stern and Weise, 1966; Vasil'ev and Ershova, 1972; Tagawa, 1984). It has no anhydrous carbonate but only a hydrated form that becomes a basic carbonate on heating, $Be_3(OH)_2(CO_3)_2$, is insoluble and dissociates to the oxide (Stern and Weise, 1969). The nitrate, $Be(NO_3)_2$, soluble in water, loses NO_2 at 125°C going through a basic nitrate to the oxide (Stern, 1972). The nitrite is even less stable. The dichloride, $BeCl_2$, is stable in dry air but is deliquescent and very soluble in water. It is very volatile (m.pt.415°C), subliming with a vapor pressure 7 ppmv at 230°C, boiling at 531°C and hydrolyzed quite readily to the dihydroxide in ambient air and so to the oxide as temperatures are raised. Particles can become coated with BeO(s) by covering them with a paste of $BeCl_2$ followed by heating at 1000°C (Alfredson and Cairns, 1964; Nagai, 1969). A solid dihydride exists

TABLE 7.11 Properties of beryllium's condensed phase molecules relevant to combustion, °C.[a]

	m.pt. b.pt. vapor pressure
Be	1287 2470, 19 ppmv/1230°C, ins
BeO	2507 3788, low volatility, but ablated by H_2O above 1300, ss
BeO_2	not known
$Be(OH)_2$	>200 dissoc to BeO, ss
$BeCO_3$	54 100, only basic hydroxyl-forms, ins
$Be(NO_3)_2$	>125 dissoc, s
$Be(NO_2)_2$	unstable
$BeSO_4$	1127, dissoc >600, 10 ppmv/700°C, s
$BeCl_2$	415 487, hydrolyzed by water, sublimes, 7 ppmv/230°C, s
BeS	>750 dissoc, slowly reacts w water, oxidized in air
BeH_2	>250 dissoc in air, reacts with water
$BeAl_2O_4$	1873
BeB_2O_4	–
$Be_3(BO_3)_2$	1495
$BeFe_2^{III}O_4$	>360 dissoc
$Be_3(PO_4)_2$	–, s
Be_2SiO_4	1560
$BeTiO_3$	1300–1650

[a]*Gurvich et al. (1994), see text.*
dissoc, dissociates; *s*, water soluble; *ss*, slightly soluble; *ins*, water insoluble.

but decomposes in air at 250°C to the oxide. The infrared spectra of BeCl, $BeCl_2$, BeH, BeH_2 molecules have been studied in low-temperature matrix isolation studies obtaining their vibrational frequencies (Yu et al., 2017a; Wang and Andrews, 2005h). Although thermally stable like the oxide, the corresponding sulfide, BeS, is stable only to about 750°C, is reactive, oxidizing in air to the sulfate, and decomposed by HCl to the dichloride and H_2S. It is slowly decomposed by water and is unstable in CO_2. Other mixed metal oxide compounds are known, for example, the aluminate, $BeAl_2O_4$ (m.pt.1873°C), silicate, Be_2SiO_4 (m.pt.1560°C), phosphates and borates, and a titanate, $BeTiO_3$ (m.pt.1300°C–1650°C) and a less stable ferrite (<360°C). Some of these may be volatile such as the metaborate, $Be(BO_2)_2$, shown to have a vapor pressure of about 0.2 ppmv at 1230°C (Blackburn and Buchler, 1965). The titanate and metaborate are sufficiently stable in the gas phase to have been monitored by mass spectrometry at high temperatures indicating some non-dissociative vaporization (Lopatin and Shugurov, 2013).

The dissociation energy of BeO and its meaning

Gaseous BeO has been monitored but vaporizes and dissociates only at high temperatures (Chupka et al., 1959). As is discussed in more detail for each of the Group IIA oxides, the gaseous bond strengths remain intriguing. In having an $X^1\Sigma^+$ ground state, for spin conservation, dissociation should proceed to the second energy limit, in this case the ground state $Be(^1S_0)$ and the excited $O(^1D_2)$ (energy content 189 kJ mol^{-1}), as indicated in Fig. 7.3. The two major mass spectrometric measurements, this mentioned and one by Theard and Hildenbrand (1964) agreed and indicated a recommended value of $D_0(\text{Be-O})_g = 440 \pm 10$ kJ mol^{-1}. This value was supported by numerous theoretical calculations that refer to ground state products suggesting values in a 410–450 kJ mol^{-1} range (Langhoff et al., 1986; Stiakaki et al., 1994; Sorensen and England, 2000; Sullivan et al., 2003; Vasiliu et al., 2010; Prascher et al., 2011; Kalemos, 2017). However, the two measures are equilibrated mass spectrometric measures and provide no information on the reaction dynamics. Moreover, although the $O(^1D_2)$ state is metastable with a long radiative lifetime it is rapidly quenched collisionally by many gases and should not affect the outcome of these equilibrium measurements (Schofield, 1978). However, the obtained experimental values and theory imply that the spin-forbidden dissociation relates to the overall outcome.

$$\text{BeO}(X^1\Sigma^+) = \text{Be}(^1S_0) + \text{O}(^3P_g) \quad D_0 = 440 \text{ kJ mol}^{-1} \text{(spin forbidden)}$$

$$\text{BeO}(X^1\Sigma^+) = \text{Be}(^1S_0) + \text{O}(^1D_2) \quad D_0 = 440 + 189 = 629 \text{ kJ mol}^{-1}$$

$$\text{BeO}(a^3\Pi) = \text{Be}(^1S_0) + \text{O}(^3P_g) \quad D_0 = 440 - 96 = 344 \text{ kJ mol}^{-1}$$

$$\text{BeO}(A^1\Pi) = \text{Be}(^1S_0) + \text{O}(^1D_2) \quad D_0 = 629 - 110 = 519 \text{ kJ mol}^{-1}$$

This poses a dilemma for the chemical kinetic dynamics of the dissociation, which are expected to proceed to the $O(^1D_2)$ product. As seen, the lowest lying excited state of BeO, the $a^3\Pi$ is metastable and lies at $T_0 \approx 96$ kJ mol^{-1} (Gurvich et al., 1994; Chambaud et al.,

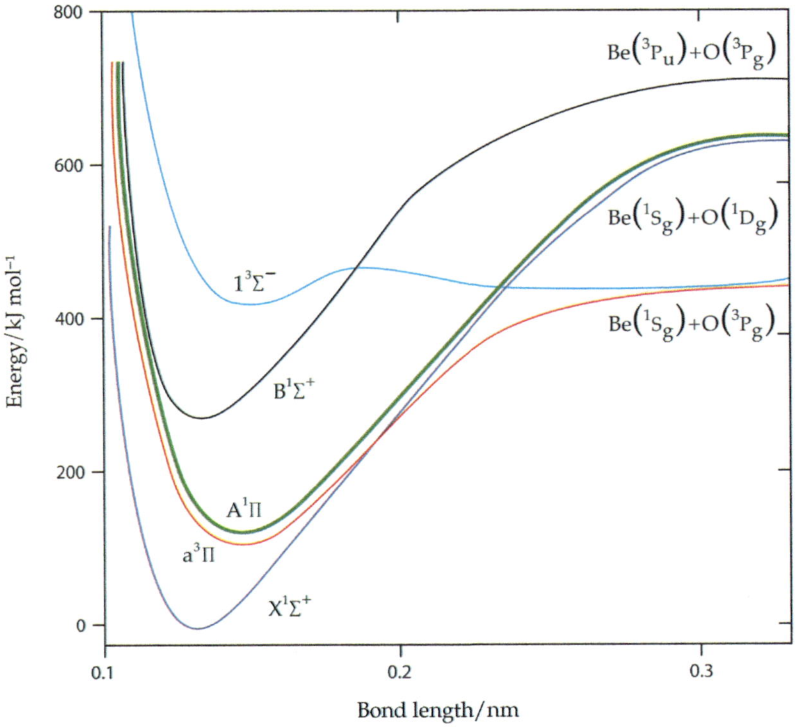

FIGURE 7.3 Approximate low-lying potential energy curves for gaseous BeO. Source: *Adapted and rescaled from Sorensen, TE, England, W.B., 2000. Valence states of BeO Feynman's way. Int. J. Quantum Chem. 76, 259–279.*

2008). Normally this will make only a 3.6% contribution even at 2000°C to an equilibrated BeO population. Also any potential energy curve hopping between the $a^3\Pi/X^1\Sigma^+$ states is expected to be negligible in this case (Stagg and Richards, 1974). In what appears to be the only option left for a direct channeling to $O(^3P_g)$ appears to be the possibility of spin–orbit induced predissociation. As shown, a repulsive $^3\Sigma^-$ state is predicted to exist and is the only other state that correlates to the ground state products. Its exact location remains uncertain but it has to cross the $X^1\Sigma^+$ curve at energies close to the first dissociation limit as illustrated recently for the analogous BeS (Larbi et al., 2010), MgS (Ribas et al., 2010), and CaO systems (Khalil et al., 2011). The general nature of these potential energy curves is illustrated in Fig. 7.3 adapted from a theoretical study. Even though a $^1\Sigma^+/^3\Sigma^-$ predissociation violates Kronig's selection rules based on both spin and $+/-$ symmetry (Herzberg, 1963) it has been observed to occur and is suggested as the efficient process for the analogous BeS system (Larbi et al., 2010). It is mentioned as a possibility by Lefebvre-Brion and Field (2004). Generally, such spin–orbit interactions occur mainly with increased atomic number elements. However, they indicate that it is known to occur for NH and PH molecules that are isoelectronic with BeO (Wayne and Colbourn, 1977, 1984). Also such a curve crossing predissociation has been measured in detail for a low-lying level of the ion BeO^+ (Ghalila et al., 2008). In that case, the bound $1^2\Sigma^+$ is crossed by a repulsive $1^4\Pi$, predissociating the $v = 18$ level and above, to the lower dissociation limit, and reducing its lifetime to a ps (10^{-12} seconds) timescale. This aspect of the nature of the potential energy curves now appears to be the most likely explanation for the reported divergent bond strength values particularly for the alkaline earth oxides and sulfides (Huber and Herzberg, 1979; Schofield, 1967). It has to be an efficient transfer. No alternate different type measures yet have been made for *D*(BeO), but methods based on nonequilibrated systems would be very interesting to establish its dominant reaction dynamics.

Gas phase stabilities and transport

The gas phase is interesting in that beryllium has little volatility. In most combustion, temperatures are such that any BeO(s) vaporization will have to be that facilitated by its ablation with water to $Be(OH)_2$(g), but this is not a rapid kinetic release. Such erosion rates are of importance for corrosion but generally not fast enough to contribute directly to combustion (Myers et al., 2019). A peroxide, BeO_2(g), remains undetected but theoretical calculations suggest a significant thermal stability (Bauschlicher et al., 1992). Values of $D_0(Be\text{-}O_2) = 368 \pm 16$ and $D_0(OBe\text{-}O) = 420 \pm 20$ kJ mol^{-1} are implied depending on the structural nature of the bonding. However, this was not observed in Knudsen cell-mass spectrometric vapor analyses of BeO(s) (Lamoreaux et al., 1987). A linear structure OBeO was suggested from the infrared spectra of laser-ablated Be atoms depositing with O_2 in a low-temperature Ar matrix isolation study (Thompson and Andrews, 1994). However, this could not be repeated in a N_2 matrix where more complex oxides resulted (Andrews et al., 1996). In combustion it would be vulnerable to conversion to BeO with possible hydrolysis to the dihydroxide.

In hydrogen systems, beryllium may be present as either BeOH or $Be(OH)_2$. The monohydroxide is in essence a radical and has been studied theoretically very extensively suggesting $D_0(Be\text{-}OH)_g$ values in the range $400 - 450$ kJ mol^{-1} (Bauschlicher et al., 1986; Stiakaki et al., 1994; Theodorakopoulos et al., 1999; Vasiliu et al., 2010; Heaven et al., 2011). The BeOH structure is about 143 kJ mol^{-1} more stable than HBeO and there is an additional activation energy of 131 kJ mol^{-1} isomerization barrier between them (Jain et al., 2000; Zaidi et al., 2006). Although theory still suggests a bent structure (Stiakaki et al., 1994; Jain et al., 2000; Zaidi et al., 2006; Vasiliu et al., 2010; Kalemos, 2017), experiment tends to indicate it is linear. At 1500°C, a mass analysis indicated that gaseous BeOH was negligible compared to $Be(OH)_2$ (Altman, 1963). It was finally successfully characterized by a mass spectrometric study of the equilibrium between Be_2O(g) and H_2O(g) that derived its heat of formation suggesting $D_0(Be\text{-}OH)_g = 458 \pm 40$ kJ mol^{-1} (Gurvich et al., 1994) now accepted as the best experimental value displacing earlier larger measures. However, further confirmation and accuracy is certainly still required. In this particular example, a case may be presented to prefer the theoretical estimate by Vasiliu et al. (2010) of 442 kJ mol^{-1}. This could be argued from their other accurate estimates obtained for BeO, BeCl, $BeCl_2$ and $Be(OH)_2$ all within 10 kJ mol^{-1} of presently recommended values. For now a value $D_0(Be\text{-}OH)_g = 442 \pm 15$ kJ mol^{-1} might be appropriate. The monohydroxide ESR and infrared spectra have been noted only in matrix isolation experiments (Brom and Weltner, 1976; Thompson and Andrews, 1996) with optical spectral emission bands assigned (Antic-Jovanovic et al., 1988). The dihydroxide has been studied in many effusion or transpiration experiments that indicate a reasonably accurate heat of formation. This indicates a sum value of 1032 ± 16 kJ mol^{-1} for the bond strength of the two Be-OH bonds in linear $Be(OH)_2(g)$ (Gurvich et al., 1994). The NIST-JANAF tables had suggested a value of 1067 ± 38 kJ mol^{-1} (Chase, 1998) assessing the same data. These at least indicate that the bonding will be strong for both bonds and accepting the value for BeOH this would imply $D_0(\text{HOBe-OH})_g = 574 \pm 40$ kJ mol^{-1} (Gurvich et al., 1994) or 21 kJ mol^{-1} stronger (Chase, 1998). However, both reviews of the available data agree that the second bond is 1.25 stronger than the monohydroxide. Several theoretical papers have included the dihydroxide. These supported the lower value indicated here (Schulz et al., 1999; Sullivan et al., 2003). One suggested both 442 kJ mol^{-1} for the first bond and 583 for the second, lending significant support for these magnitudes (Vasiliu et al., 2010). As a result, in combustion this very strong bonding in the dihydroxide ensures its dominant presence to high temperatures if gaseous beryllium is present with it finally dissociating to BeO.

The corresponding very volatile chloride has been examined by Knudsen cell mass spectrometry obtaining the heat of sublimation of $BeCl_2(s)$, which implies with its known condensed phase heat of formation an atomization energy for both bonds of $D_{0\text{Atomiz}} = 921.5 \pm 6$ kJ mol^{-1} (Gurvich et al., 1994). Theory indicated 922.6 providing support for its other beryllium estimates (Vasiliu et al., 2010). Gaseous $BeCl_2$ is linear with an $X^1\Sigma_g^+$ ground state. A detailed Knudsen cell study examined the equilibration of gaseous $BeCl_2$/Be/BeCl and also Be/AlCl/BeCl/Al. These unconnected studies were in very close agreement leading to the currently recommended value of $D_0(\text{Be-Cl})_g = 385 \pm 5$ kJ mol^{-1} for its $X^2\Sigma^+$ state (Hildenbrand and Theard, 1969; Gurvich et al., 1994). This then provides the quite accurate value of $D_0(\text{ClBe-Cl})_g = 536 \pm 8$ kJ mol^{-1}. These values show a slightly more pronounced difference than the ratio of the two hydroxides. Even so, when the trends of the two hydroxide and two chloride bond ratios are examined in Table 7.10, the two show consistent tracking over the five Group II elements within about 10%. Nevertheless, even this bonding level for beryllium suggests that the chloride concentrations in the gas phase will be held to the minimal in the presence of water when compared to the dihydroxide. For example, at 2000K with the gaseous equilibration of this reaction with 10% water and 1% chlorine the dihydroxide will be 100 times larger than the dichloride (Gurvich et al., 1994).

$$BeCl_2 + 2\,H_2O = Be(OH)_2 + 2\,HCl \quad K_p \approx 1$$

This is just the opposite of barium where the situation is totally reversed and the dichloride is about 400 times the larger. There appear to be no studies of beryllium and chlorine in flames. However, the formation of BeO thin films or coating of particles using a $BeCl_2/H_2O$ feed at 690°C–1200°C mentioned above is a strong confirmation of this hydrolysis of the dichloride through to the oxide (Alfredson and Cairns, 1964; Nagai, 1969).

Observed spectroscopically, a value of $D_0(\text{BeH(g)} = 196 \pm 4$ kJ mol^{-1}) is known from predissociation limits so the molecule will be too vulnerable to oxidation and dissociation to be of significance in combustion mixtures (Huber and Herzberg, 1979). Theoretically this five-electron molecule has been an important benchmark molecule in testing formalisms and also in modeling the nature of the Born–Oppenheimer breakdown in spectroscopy. Theory can adequately reproduce the bond strength (Fuentealba et al., 1987). A recent study has accurately calculated the ground $X^2\Sigma^+$ state potential energy curve with refined methods for the three isotopic varieties BeH, BeD, and BeT (Dattani, 2015). This now has been further confirmed by an accurate curve fitting that suggests $D_0(\text{BeH(g)}) \approx 200\ \text{kJ mol}^{-1}$ (Spirko, 2016).

BeS(g), like the oxide is expected to be quite stable and has been studied spectroscopically (Huber and Herzberg, 1979; Pouilly et al., 1982) and theoretically (Partridge et al., 1988; Chambaud et al., 2008; Larbi et al., 2010; De Oliveira-Filho et al., 2011). Its thermochemistry has not yet been measured experimentally but has theoretical recommendations of about $D_0(\text{Be-S}) = 314\ \text{kJ mol}^{-1}$. This differs from the estimate of 350 ± 20 suggested in the evaluation of Gurvich et al. (1994). The basis for this was the trend in the M-S/M-O bonding of these group II elements as apparent here in Table 7.10. Until resolved this value $D_0(\text{Be-S}) = 350 \pm 20\ \text{kJ mol}^{-1}$ remains currently suggested but uncertain. Like the oxide it is intriguing in that the unimolecular dissociation is required by spin conservation to couple to the second dissociation channel producing $S(^1D_2)$, adding an energy content of $111\ \text{kJ mol}^{-1}$. These potential energy curves image those of Fig. 7.3 for the oxide and have been well illustrated by Larbi et al. (2010). Although this $S(^1D_2)$ is a metastable state with a long radiative lifetime it is rapidly quenched collisionally by many gases (Schofield, 1979).

$$\text{BeS}(X^1\Sigma^+) = \text{Be}(^1S_0) + S(^3P_g) \quad D_0 = 350\ \text{kJ mol}^{-1} \ \text{(spin forbidden)}$$
$$\text{BeS}(X^1\Sigma^+) = \text{Be}(^1S_0) + S(^1D_2) \quad D_0 = 350 + 111 = 461\ \text{kJ mol}^{-1}$$

However, as outlined above with BeO, equilibrated reaction chemistry is expected to relate to the ground state products. It is the lowest exited state of BeS, $a^3\Pi$ ($T_0 \approx 87\ \text{kJ mol}^{-1}$) that correlates to ground state atoms (Pouilly et al., 1982; Chambaud et al., 2008; Larbi et al., 2010; De Oliveira-Filho et al., 2011).

$$\text{BeS}(a^3\Pi) = \text{Be}\left(^1S_0\right) + S\left(^3P_g\right) \quad D_0 = 314 - 87 = 227\ \text{kJ mol}^{-1}$$

However, its population will only be about 6% that of $\text{BeS}(X^1\Sigma^+)$ at 2000°C and so will usually play a small role in any chemistry. Consequently as with BeO, the higher lying repulsive $^3\Sigma^-$ state that correlates to the ground state products appears most probably to play the pivotal role in the actual reaction dynamics as indicated in the corresponding potential energy curves shown for the BeO case (Fig. 7.3). Theoretically, these curves have been illustrated (Larbi et al., 2010) indicating the possible curve hopping from the BeS ($X^1\Sigma^+$) but whether the suggested spin–orbit induced predissociation does occur needs further study especially as it violates Kronig's selection rules (Herzberg, 1963). Nevertheless, in connection with combustion, these sulfides readily oxidize and react with water and so generally will play no role in controlling observed products even under pyrolysis or gasification conditions (Frandsen et al., 1994). In a study of soot suppression by metal additives, on a scale of 100 listed for Ba, Be showed no effect given a value of zero (Cotton et al., 1971). It also has a very large ionization potential and appears neither to chemiionize nor participate in any catalytic cycles of recombination of flame radicals.

Combustion power plant data

With such a chemistry it is interesting to examine what has been observed in various combustion systems and see what positive conclusions can be made at present. It remains important to have an understanding of the mass balance for this element in the combustion cycle and its ultimate disposal aspects. With 7 billion tons $year^{-1}$ globally of coal combustion, even a low level of 0.1 ppmw content, implies a 700 tons $year^{-1}$ total quantity to capture (Table 1.1). An interesting UK study recently has measured the Be concentrations in the atmosphere at numerous locations in the country. Although levels remain low and below those of concern they are seen to be regional consistent with industrial sources (Goddard et al., 2016).

Like most metal elements, coals can have a diverse range of beryllium concentrations but are often below or on a low ppmw range (Wang et al., 2008c). Sewage sludge can be comparable but levels are generally smaller in biomass or wastes (Table 1.1). It appears to be very low in pet coke and limestone (Cordoba et al., 2012). Although there have been few quantitative studies of mass balances a reasonable display of expected behavior now is evident. There is little difference in concentrations between those found in bottom ash and fly ash. The latter has slightly higher values reflective of more complete burnout (Meij, 1994; Querol et al., 1995; Aunela-Tapola et al., 1998; Llorens et al., 2001; Font et al., 2010). Similarly, there is only slight enrichment to micron-size particles (Coles et al., 1979; Querol et al., 1995; Nielsen et al., 2002). The other elements of Group II are labeled as not enriching fly ash particles to smaller sizes (Lanzerstorfer, 2018). In a laboratory tube reactor studying both combustion and pyrolysis, no volatilization of beryllium was evident at the temperatures of $<1150°C$ (Wang and Tomita, 2003). Three detailed mass balance studies have been published. The average from two Finnish coal plants indicated a beryllium distribution between bottom ash (4.8%), fly ash (94%), FGD (1.0%), and release (0.2%) (Aunela-Tapola et al., 1998). This was confirmed by a similar Danish and a less detailed Dutch report illustrating that the ESP is the dominant control device and that although the FGD compliments this, a small fraction of the beryllium can transit the whole flow train (Meij, 1994; Nielsen et al., 2002). However, a coal-fired Canadian study with a plant equipped solely with an ESP indicated a 0.12% final release (Goodarzi, 2006). In more recent work, the supplemental role of the FGD system has been seen to retain many of the trace elements that evade capture upstream. As a result, at low concentrations in a coal-pet coke fired co-combustion power plant, even the final traces of Be were further reduced by this (Cordoba et al., 2012). Less work is reported on cofiring with biomass or sludge. The results from one suspension-fired reactor burning such mixtures confirmed a high retention in the ash with coal, sludge or straw (George et al., 2010). However, two other works are in conflict, one indicating that some beryllium appeared to be released from coal/straw mixtures but the beryllium levels in the biomass were very small (Miller et al., 2002). The other using a laboratory furnace at $\leq 1100°C$ implied sewage sludge might have a 30% volatilization, which is unlikely (Folgueras et al., 2007).

It is quite clear that hydrolytic vaporization of beryllium as $Be(OH)_2(g)$ may not be of concern in combustion but that most of the beryllium leaves a power plant in ash form. The use of ESP and FGD devices are capable of reducing atmospheric releases to below the accepted limits. As indicated, environmental assessments now relate not to stack release but to ash disposal (Aunela-Tapola et al., 1998). Leaching studies for ash that are available do show variations in co-combustion cases (George et al., 2008), but with equal water leaching from bottom and fly ash coal-fired samples (Llorens et al., 2001). A remaining question concerns the detailed nature of the beryllium in ash that remains unknown. Release appears unlikely but formation of the oxide could moderate during cooling to the water soluble sulfate or further interact with ash components to form for example either the silicate or aluminate. In this regard, when atomic flame analysis was common for the metal elements, it was noted that Al did interfere with the quantitative analysis of Be containing samples (Luecke, 1992). A method of avoidance and also for increasing the analytical sensitivity was to raise the temperature of the flame by switching to C_2H_2/N_2O (2700°C). Using such a high temperature flame ensured gas phase components with about 10% of atomic beryllium present (De Galan and Samaey, 1970).

In summary, the stable BeO(s) renders it as a nonvolatile element requiring high temperatures. It is found in ash forms reflecting little evaporative release. Water ablation to produce $Be(OH)_2$ appears a slow process, more appropriate to corrosion studies, and not apparent in presently reported power plant studies. Aspects of reactions within ash are unknown concerning its final chemical form and leaching remain uncertain. Chlorine and sulfur roles are not evident.

7.7 Bismuth

Atomic number 83, bismuth (m.pt.271, b.pt.1564°C) is chemically related to arsenic and antimony but differs in being one of the least toxic of the heavy metals. Due to this it is regarded as having a "green" environmental nature and for example is replacing the use of lead in its alloys and in bullets and as fishing sinkers. It is a rare element having an average concentration in the Earth's crust of about 9 ppbw, ranked about 69th in the elements and on par with Au and other low trace elements. It is slightly more evident in the oceans being 60th at a 20 pptw concentration. In nature it exists mainly as the sulfide, bismuthinite Bi_2S_3, with also the oxide bismite, Bi_2O_3. However, it is produced mainly as a byproduct of Cu/Pb ore smelting. Globally it is being produced at about 8500 tons year^{-1} with very adequate reserves and supply. Its price can fluctuate but at present is about \$6.00 lb^{-1}. It is the only nontoxic heavy metal and does not bioaccumulate nor is carcinogenic. As a result no toxicological reports have been published. Bi is the most diamagnetic metal, has a low thermal conductivity and a high electrical resistance. It has value in expanding up to 3.3% on cooling. It exists in nature primarily as one isotope ^{209}Bi. Although not yet accurately assessed it is considered that bismuth's atmospheric particulate loading is possibly 20% natural from volcanoes and rock/soil erosion and 80% anthropogenic (Ferrari et al., 2000). Earlier estimates have suggested 4000 or 1200–1700 tons year^{-1} input from volcanoes (Patterson and Settle, 1987; Candelone et al., 1995) and only 15 tons year^{-1} from coal and oil combustion. Current suggestions are that volcanoes may not be quite as dominant a natural source as previously thought but that metal smelting of lead that produces bismuth as a by-product is certainly a major anthropogenic component (Kaspari et al., 2009).

The pure metal has a very low melting point (m.pt.271, b.pt.1564°C), but is not volatile until higher temperatures. It possibly vaporizes as the tetramer Bi_4 but this is not stable and the vapor is found to be predominantly Bi_2 followed by Bi(g) with Bi_3 and Bi_4 negligible (Rovner et al., 1967). At about 750°C the Bi_2 and Bi have pressures at equilibrium of about 26 and 15 ppm, respectively (Barin and Platzki, 1995). It has many minor uses. It is a soft, brittle metal and with its low melting point can form useful very low melting point alloys. These find use in solders and in fire sensing detectors where alloys with melting points as low as 47°C exist. It has become an important metal to replace toxic lead, and is used in pigments, paint, metal type or molding. It can produce a powerful MnBi permanent magnet.

Health concerns with bismuth

Bismuth has throughout the ages found some use in medicine as Bi subsalicylate in Pepto-Bismol for gastrointestinal treatment. Colloidal bismuth subcitrate has also found use in gastric and stomach ulcer diseases. It is thought the heavy metal ion is toxic to bacteria. Historically it was one treatment for syphilis. Toxic poisoning is rare and the body excretes it rapidly. A full body half-life might be about 5 days. It generally is of less concern in the environment due to the insolubility of most of its compounds. Alkyl derivatives of bismuth have been detected in landfill areas and sewage fermentation but are very limited. They tend to be unstable with weak Bi-C bonding. Available information in the biological environment has been summarized and appears to remain nontoxic to humans but can be toxic to microbes (Filella, 2010a). As a biomarker for testing the body, whole blood is preferred. Values should be $<1\ \mu g\ L^{-1}$ for the unexposed (Mayo Clinic, 2018). It is commonly at levels of $<0.02\ \mu g\ L^{-1}$, but if subject to medical treatments with bismuth medications it may be $4-30\ \mu g\ L^{-1}$ but still of little concern. Few published surveys include it, one from France surveyed 100 healthy adults finding very low levels in all media (Goulle et al., 2005; Cesbron et al., 2013). However, a more recent one reported individuals in Morocco and the Canary Islands and displayed mean blood values of 1.4 and $4.8\ \mu g\ L^{-1}$, with 25%–75% percentile ranges of 1.0–2.5 and $1.3-8.4\ \mu g\ L^{-1}$, respectively (Henriquez-Hernandez et al., 2018). The difference was explained in terms of a higher level of economic development, which is to be remembered as a plausible cause of higher exposure. For bismuth it does appear that hair and urine levels are normally very low.

Bismuth's basic chemistry

Its chemistry reflects that of As and Sb but is simplified. The physical properties of its condensed phase Bi molecules are summarized in Table 7.12. Electronically its structure is $5d^{10}6s^{2}6p^{3}$ but no Bi(V) valent molecules are found in the bio-environment. Although similarly tri- and pentavalent, any five-valent molecules are unstable and not often encountered. The metal is stable in air and moisture, and needs to be hot to react with water to produce the Bi_2O_3 insoluble oxide. This has numerous solid phases depending on heating and cooling cycles. Again the metal has a very limited range of thermally stable molecules, none of which are soluble in water.

TABLE 7.12 Properties of bismuth's condensed phase molecules relevant to combustion, °C.

	m.pt. b.pt. vapor pressure
Bi	271 1564, 40 ppmv (Bi/Bi_2)/750°C
Bi_2O_3	825 1890
Bi_2O_4	305, reacts with water, unstable
Bi_2O_5	unstable
$(BiO)_2CO_3$	low stability dissoc
$Bi(OH)_3$	stable in solution
$BiCl_3$	234 439, reacts with water, vap. press. 550 ppmv/200°C
BiOCl	218 461, dissoc 700, reacts with water
$Bi(NO_3)_3$	260 440, dissoc dry air, reacts with water
$BiONO_3$	dissoc
$Bi_2(SO_4)_3$	405 decomp, reacts with water
BiH_3	−67 17, unstable gas
Bi_2S_3	775, sublimes 450, dissoc 1368, reacts with air
$BiAsO_4$	750 dissoc
$BiFeO_3$	940
$NaBiO_3$	vaporizes to $NaBiO_2$, dissoc to B_2O_3

decomp, decomposes; *dissoc*, dissociates; They are either insoluble or hydrolyzed by water.

In the condensed phase, the main oxide is Bi_2O_3 (m.pt.825, b.pt.1890°C) (Sidorov, 1980). A pentoxide can be made but is one of the least stable known, losing O_2 and becoming the trioxide. A mixed oxide B_2O_4(III/V) is known but not normally encountered (m.pt.305°C) and reacts with water. A trihydroxide exists but tends to be regarded as the hydrated oxide, $Bi_2O_3 \cdot 3H_2O$. Also a basic carbonate $(BiO)_2CO_3$, the subcarbonate is stable at ambient temperatures but dissociates to the trioxide on heating. The nitrate also is generally basic, $BiONO_3$, the subnitrate. The $Bi(NO_3)_3$ can be formed but is hygroscopic, dissociates at about 260°C and reacts with water. It is stable to about 440°C in dry air. The sulfate, $Bi_2(SO_4)_3$, begins to decompose above 405°C through oxysulfates to the Bi_2O_3 (m.pt.825°C) (Stern, 2001). The trichloride (m.pt.234, b.pt.447°C) is very volatile, 550 ppmv at 200°C, increasing rapidly with temperature. It is readily hydrolyzed by water to the oxyhalide BiOCl. This oxychloride (m.pt.218, b.pt.461°C) disproportionates at 700°C to $Bi_2O_3 + BiCl_3$ and is also hydrolyzed by water (Qi et al., 2016). Bi_2O_3 has been seen to react with Cl_2 starting at 38°C being first converted to BiOCl (100°C−400°C) and then to $BiCl_3$ (600°C−800°C). However, with HCl, Bi produces $BiCl_3$ in a single stage starting at 20°C (Ivashentsev et al., 1968). A more recent study heated MSW fly ash at 1000°C in flowing Cl_2 for 3 hours finding all the Bi was released, somewhat confirming it had been present as Bi_2O_3 (Chan and Kirk, 1999).

The trihydroxide, although trivalent can be oxidized to pentavalent bismuthates such as $NaBiO_3$. This is a strong oxidizer that is not stable and is hydrolyzed by warm water to NaOH and Bi_2O_3. It is unusual being one of the few cold water insoluble sodium salts. As a result, unlike arsenic there is no significant range of bismuthates or bismuthites that it can invoke. The alkali bismuthates are known and vaporize on heating to the gaseous bismuthites that show trivalency in the form of $NaBiO_2$(g) (Lopatin and Shugurov, 2013). Bismuth does have the mineral "rooseveltite" that is $BiAsO_4$. This dissociates at less than 800°C. A $BiPO_4$ exists and is usually incorporated with other molecules in catalysts. Recent interest has turned to Bi_2O_3/SiO_2 glasses and their potential future uses. Knudsen cell-mass spectrometric studies have modeled their melts and indicate at 700°C that the main vapor species are Bi and O_2 (Stolyarova et al., 2014). $BiFeO_3$ (m.pt.940°C) has been synthesized and found to be a very effective flame retardant (Qi et al., 2016).

Under reducing conditions potential molecules are BiH_3(m.pt. − 67, b.pt.17°C), which is gaseous and unstable (Dai and Balasubramanian, 1990). Bi_2S_3(s) is the main condensed phase stable molecule that sublimes in a vacuum or inert atmosphere at 450°C but dissociates to liquid Bi and S_2(g) at about 550°C−650°C, the Bi vaporizing at about 850°C−1100°C (Padilla et al., 2011). Heated in air it reacts with oxygen at about 440°C gradually becoming more oxidized and is Bi_2O_3 at 800°C (Johnson et al., 2004).

Gas phase stabilities and transport

The vaporization of these condensed phase molecules like those of As and Sb has been studied and shows a complex breakdown in air through various oxides to produce a dominant BiO(g) above about 800°C (Sidorov, 1980). This latter mass spectrometric study of Bi_2O_3(s) vaporization at 730°C−920°C again indicated the polymeric nature of the vaporization indicating the formation of Bi_4O_6(g) and numerous of its oxide dissociating fragments. A theoretical study has discussed this propensity with this group of elements (P, As, Sb and Bi) to form these cage-like structures in the gas phase of the M_4O_6, M_4O_{10} type (Chrissanthopoulos and Pouchan, 2008). Only the Bi_4O_6 structure is known for Bi and along with the Bi_4 component in the elements vaporization they are seen to be less stable than for As and Sb (Marschman and Lynch, 1984).

The bond strength of gaseous BiO now has been measured quite accurately. The review by Pedley and Marshall (1983) remains valid for the several mass spectrometric measures. Their recommended value was reproduced by a later Knudsen cell study of the Bi_2O_3 vaporization (Oniyama and Wahlbeck, 1998). As a result, D_0(Bi-O) = 333 ± 5 kJ mol^{-1} is suggested, and as seen from Table 7.7 that illustrates stability trends, is the weakest of this periodic table column. The value is adequately supported by theoretical calculations (Stoll et al., 2002). Also its spectroscopy is well developed with a $X^2\Pi_{1/2,3/2}$ doublet ground state and no others low-lying (Huber and Herzberg, 1979). Like antimony, the gaseous BiO_2 was not seen in any mass analyses of Bi/O vapors. It was reported though in an earlier matrix isolation study reacting Bi with O_2 displaying a bent structure and more recently as having an endo BiOO form (Konnov et al., 1981; Wang et al., 2013). If these can exist, it may be that any BiO_2^+ formed in the mass analysis studies is unstable to fragmentation. Nevertheless, it would be expected to be much weaker than the monoxide. A review of the thermodynamic data for Bi and Bi-O vapor systems summarized values for these molecules (Marschman and Lynch, 1984). With the metal vapor, as indicated Bi_2(g) is dominant, its bond strength accepted from the early mass spectrometric studies as D_0(Bi-Bi) = 197 ± 3 kJ mol^{-1} (Rovner et al., 1967; Knacke et al., 1991). For a brief period it acquired significant attention as a potential chemical laser candidate. Theoretical studies of the potential energy curves for its electronic states and their parameters were published (Balasubramanian and Liao, 1991; Das et al., 1995). Only in recent years has gaseous BiOH been characterized spectroscopically. In a low-temperature LIF spectroscopic study, Bi vapor was mixed with discharged O_2 and strong BiO(g) spectra recorded. On adding H_2O vapor a new spectrum identified as BiOH was apparent with a significant decrease in the BiO intensities (Fink et al., 1997; Shestakov et al., 1998). As with its family of elements, namely N, P, As, and Sb, it can occur in two isomeric forms BiOH and HBiO. Theory suggests that BiOH is the more stable by about 137 kJ mol^{-1} and estimates D(Bi-OH) ≈ 240 kJ mol^{-1} (Khandogin et al., 1997). The question is whether this is sufficiently strong to compete at flame temperatures with the other possible gas phase species that appear to be stronger. Consequently, like the corresponding HNO, POH, AsOH, and SbOH of its group it seems unlikely that BiOH will play any role. Additionally, boron halides are not seen to be strong flame inhibitors in spite of their halide content (Babushok and Tsang, 2000).

The bond strengths of the chlorides remains somewhat uncertain. From a heat of formation and sublimation of $BiCl_3$(s) these indicate an accurate heat of formation and $D_{0Atomiz} = 835 \pm 4$ kJ mol^{-1} for the gas phase (Cubicciotti, 1967; Barin and Platzki, 1995). The spectroscopy of the main electronic states of the monohalide is well established. The ground state is labeled X_10^+ (a mixed $^{3,1}\Sigma$) with the low lying $a^1\Delta(a_2)$ now assigned with an energy of about 141 kJ mol^{-1} larger (Beutel et al., 1996). Three very different methods agree on the bond strength of BiCl, 300 ± 4 (transpiration measurement) (Cubicciotti, 1967), 294 (ground state vibrational level extrapolation) (Venkateswarlu and Khanna, 1960), and 291 kJ mol^{-1} (theory) (Stoll et al., 2002). A database quotes 303 ± 8 kJ mol^{-1} (Barin and Platzki, 1995). Until further resolved a value of D_0(Bi-Cl) = 295 ± 10 kJ mol^{-1} might be suggested. This implies that the second and third bonds account for 540 kJ mol^{-1}. Two things may guide in their unraveling. Cubicciotti (1967) indicated that in his transpiration experiment only BiCl and $BiCl_3$ were major contributors and by setting an upper limit estimated D_0(ClBi-Cl) < 243 kJ mol^{-1}. This suggested D_0(Cl_2Bi-Cl) ≈ 292 kJ mol^{-1} and similar to the monochloride. Another guide, supporting these magnitudes are the periodic table trends of this group of elements (Table 7.7). For P, As and Sb the trihalide bond is consistently stronger with the second bond showing weakness. Based on this it is suggested that values of D_0(Cl_2Bi-Cl) ≈ 300 and D_0(ClBi-Cl) ≈ 240 kJ mol^{-1} be accepted. Whether they can play a role in combustion gases containing water remains uncertain. It is possible that they are readily hydrolyzed and effects of chlorine in combustion are not indicated for bismuth as transporters.

The sulfide reflects the oxide forming the trivalent Bi_2S_3 with a melting point of 775°C. At 450°C it sublimes but by 550°C this changes to a dissociative vaporization to Bi(l) and S_2(g) with low levels of BiS(g) present (Cubicciotti, 1963). This was studied further in a Knudsen cell-mass spectrometric measurement of the equilibrium of BiS with PbS(g) (Uy and Drowart, 1969). Adjusting their analysis to the current bond strength of PbS indicates a recommended value of D_0(Bi-S) = 311 ± 4 kJ mol^{-1}, the same as the single database listing (Knacke et al., 1991). Its electronic spectroscopy remains quite limited experimentally but its potential energy curves and parameters for the nine low-lying states have been treated theoretically (Lingott et al., 1999). Examining the trends in the ratio of this bond strength relative to the oxide (Table 7.7) illustrates the weakening of both bonds on moving from N to Bi with it being more pronounced with the oxide. This is reflected by the $D(MS)/D(MO)$ bond ratio gradually increasing toward unity.

The trihydrides of the group VA elements (N-Bi) are all gases (BiH_3 b.pt.17°C), weakening with atomic weight. Arsine, AsH_3 decomposes very slowly at room temperature and rapidly at 200°C. Bismuthine, BiH_3, is thermodynamically metastable at 0°C and readily decomposes. The three hydrides of bismuth were identified in a matrix isolation study condensing the products from laser-ablated Bi with H_2. Frequencies and structures were characterized (Wang et al., 2003a). Only in one database is a value quoted for the BiH_3 heat of formation (Knacke et al., 1991). This implies $D_{0Atomiz} = 669$ kJ mol^{-1} for the strength of the three bonds. An early attempt to calculate these, their geometries and electronic states suggested $D(H_2Bi\text{-}H) = 271$, $D(HBi\text{-}H) = 190$, $D(Bi\text{-}H) = 172$ kJ mol^{-1}, not particularly consistent with the recommended values below but following an expected trend (Dai and Balasubramanian, 1990). Most interest arose with BiH due to its rich spectroscopy and potential value for monitoring. The only experimental measure of its bond strength utilized a predissociation in the $E0^+$ state. The uncertainty of the dissociation limits of this state have now been resolved and indicate $D_0(Bi\text{-}H) = 210$ kJ mol^{-1} that remains the most reliable estimate (Alekseyev et al., 1994). Theoretical calculations report that BiH is the most complex of the heavy hydrides with its many low-lying states and spin–orbit couplings. They imply a nature for the ground $X0^+$ state that is 76% $^3\Sigma^-$ and 24% $^1\Sigma^+$ (Balasubramanian, 1989). A more recent calculation does suggest 220 kJ mol^{-1} (Stoll et al., 2002), and that the low-lying metastable $^1\Delta$ state has an energy of about 124 kJ mol^{-1} above the ground state (Setzer et al., 2015). As a result the value of $D_0(Bi\text{-}H) = 210 \pm 15$ kJ mol^{-1} is recommended and is consistent with the trends in Table 7.7. However, within each element of this group the trend is that the diatomic bond is the weakest with strengthening to the triatomic. An assessment of this trend accepting 210 for the monohydride would imply $D_0(HBi\text{-}H) \approx 215$ and $D_0(H_2Bi\text{-}H) \approx 240$ kJ mol^{-1} consistent with the total atomization value. Although bismuth was seen to possibly have some soot suppression capability, rated as 23 on a scale with barium at 100, it has never been investigated further (Cotton et al., 1971). Nor has its flame ion chemistry, and flame inhibition has never been mentioned.

Combustion power plant data

As noted in Table 1.1, bismuth is a trace metal generally being less than 3 ppmw in coals (Kronberg et al., 1981; Querol et al., 1995; Hu et al., 2006; Font et al., 2010). Levels are negligibly small in biomass, oils, and are very low in pet coke and limestone (Cordoba et al., 2012). Its presence can be more variable in municipal waste (Hasselriis and Licata, 1996), and in hospital wastes (Zhao et al., 2008). Although still not extensively documented, there are indications that it can be significant in sewage sludge possibly due to consumer product usage. As a result, because bismuth can be well controlled in incinerations (Hasselriis and Licata, 1996), there are no foreseeable considerations relating to regulation of bismuth with respect to combustion.

Among the potential gaseous Bi carriers in combustion, the strongest gas phase bonding is seen with the diatomic oxide (333) and the monochloride (295 kJ mol^{-1}) that can be compared in magnitude to that of NaOH(g) (D_0 334 kJ mol^{-1}). However, it is most probable that the gaseous chlorides are hydrolyzed. Both of these bismuth molecules are intriguing radical species with reactive available bonding. One spectral emission study did note in a flow system at low temperatures that when chlorine was added to a Bi/O_2 mixture showing strong BiO emission bands that these were largely reduced and replaced by BiCl bands (Fink et al., 1991). Obviously, gaseous BiO and BiCl in coal combustion where HCl is ever present could be in a dynamic balance.

$$BiO + HCl = BiCl + OH \quad \Delta H_0 = +42 \text{ kJ mol}^{-1}$$

The adverse enthalpy barrier could be readily compensated for by the HCl/OH concentration ratio. However, it does seem that the chloride is hydrolyzed by water. For oxygen-rich conditions the nature of the deposition on cooling would still favor the formation of the very stable Bi_2O_3(s). Even so, this deposit may be prone to ablative loss by HCl to $BiCl_3$(g) that can then be hydrolyzed back to the diatomic gas phase oxide. This cycling will depend on relative rates and may limit to some degree the time zone within which Bi_2O_3(s) can ultimately be removed by the ash. For oxygen-lean dry conditions, differences will be apparent with dominant BiS(g) and BiCl(g) probably coexisting under a similar scenario due to their closely related bond strengths and propensity for producing Bi_2S_3 deposits.

Monitoring capabilities have improved for bismuth as with other metals. Major techniques now include ICP-AES and MS, and atomic and absorption methods that invoke either hydride generation or graphite furnace atomization. Partly due to its nonregulation or environmental concern, the monitoring of bismuth has rarely been included in combustion studies and analyses. Although its concentrations are generally small in coal and biomass its presence can be quite significant in municipal wastes, especially those of hospitals and in sewage sludge incineration. Currently, four studies of interest are in the literature, one examining subbituminous coal combustion (Querol et al., 1995), one the 1200°C–1600°C gasification of bituminous coal/pet-coke/limestone mixes in an IGCC plant (Font et al., 2010), one with municipal waste incineration in a rotary kiln (Waterland et al., 1994), and one for Chinese incinerators disposing of hospital wastes (Zhao et al., 2008). Their data all appear to conform to the chemical expectations. Provided temperatures are not too low, that is about >800°C, bismuth is very volatile and has a high condensation potential on ash. There can be very little in slag or bottom ash and most of the bismuth is on the fly ash leading to very significant fly ash/slag ratios. Enrichment factors in the fly ash can be 10-fold especially in the finest size fractions (<10 μm). The behavior seen for incinerators is very sensitive to temperature. Bismuth's vaporization and transfer to fly ash increases with temperature. This is clearly seen in comparisons between low and high temperature incineration of hospital wastes where Bi levels can be quite significant and comparable to As or Sb in the fly ash (Zhao et al., 2008). Additionally, spiking incinerator experiments with added chlorine indicated no effect (Waterland et al., 1994). As a result, BiCl(g) may be hydrolyzed and the halides of less importance. A recent study has shown an effect on the fly ash composition of reducing the flue gas temperature from 250°C to 150°C through a heat exchanger. The fuel was biomass from metal contaminated soils burned in a small pilot-plant boiler. Boiler temperatures are not quoted but a very significant increase in ash concentration appeared to occur for bismuth and many other elements as the temperature was so reduced through the heat exchanger. As no further metal deposition was possible to increase the concentrations especially for elements such as Ti or Zr that are fully non-volatile it appears that particle segregation coupled to fragmentation may be occurring with the ash size distribution changing to favor the small enriched sizes at the lower temperature (Kovacs et al., 2018). It illustrates the difficulty in interpreting such observations without a more involved level of study.

In summary, bismuth's combustion behavior is as expected from its chemical nature. Somewhat like mercury it appears to be oblivious to the other mineral components either in the fuel or fly ash components. Its combustion centers on dissociations forming Bi_2O_3(s) that can vaporize to BiO(g) the main transporter. As temperatures begin to cool this gaseous bismuth readily deposits on the fly ash. Deposition products have not been analyzed but most likely will reflect dominant Bi_2O_3 if oxygen-rich and Bi_2S_3 if oxygen-lean with no formation of mixed metal oxides. As a result, any degree of atmospheric emission will depend simply on submicron particle leakage through the particulate control systems. A recent study of capture and retention of trace metals by an FGD system in a co-combustion coal/pet coke plant included bismuth. Although levels were low, there was the indication that the FGD would further lower any escaping flue gas Bi (Cordoba et al., 2012). Chlorides are prone to hydrolysis and most probably play no role. Bismuth remains unregulated.

7.8 Boron

Although boron is grouped with Al/Ga/In/Tl in the periodic table it is chemically quite different, a nonmetal and tends rather to be more reflective of Si especially in its complexity of acids. It is present in the Earth's crust at a concentration of about 10 ppmw making it the 36th most present. Its concentration in the oceans at about 4.4 ppmw places this much higher at 12th. The element itself is unusual and is described as a metalloid. This is defined as having properties between metallic and nonmetallic. They generally are electrical semiconductors. Boron is also interesting as it is said to have been formed by cosmic ray spallation and in supernovae and not by stellar nucleosynthesis. It is the fifth element in the periodic table placed between Be and C and has a $1s^22s^22p^1$ electronic configuration. It is thermally nonvolatile (m.pt.2075, b.pt.3866°C), with a vapor pressure of about 14 ppmv at its melting point. It has always been of interest as a rocket fuel or supplement due to its low weight and large enthalpy of oxidation. This might be renewed now that nanoparticle technology is being developed (Young et al., 2009). Another recent use of submicron particles of boron is to replace Al in thermite processes and fuse B/Bi_2O_3 or B/CuO into new ceramic materials (Comet et al., 2014). The atom cannot easily lose electrons due to high ionization potentials and so it has no cationic nature, is covalent in bonding and predominantly trivalent with planar structures. In nature the common ore is tourmaline, an aluminum borosilicate but this contains only about 3% boron. As a result, the commercial ores mined are kernite and tincal (sodium borates generally referred to as borax), colemanite (calcium borate), and ulexite (sodium/calcium borate). Global production is at about 4 million tons $year^{-1}$. It is abundant and reserves are quoted at 210 million tons. It is an inexpensive material priced at about \$0.30−0.45 lb^{-1}. Although precise numbers remain undocumented concerning natural and anthropogenic levels in the environment it seems clear that the human contributions to the global cycle have probably doubled the mobilization of boron in the environment and enhanced its transport in rivers (Park and Schlesinger, 2002). Borax and boron compounds are water soluble and estimates of the atmospheric cycle suggest rainfall magnitudes of 3 million tons $year^{-1}$ and dry deposition of 2 million tons $year^{-1}$. This is balanced by a major sea aerosol 1.5 million tons component, 0.2 million from rock weathering, 0.35 million from biomass, 0.2 million from coal combustion, and 0.3 million from its mining. As a result, the atmosphere shows an unexplained output that is greater than these suggested inputs. Boron has two major isotopes ^{10}B and ^{11}B that are easily fractionated by various natural processes to differing extents. In this manner hydrologists can qualitatively estimate the sources of boron. Using this method, a study of rainfall in China indicated that 20% of the boron in the rain originated from combustion (Zhao and Liu, 2010). Boron compounds have been known for thousands of years and were used in Chinese glazes. In more modern times a main use has been in borosilicate glass and fiberglass, ceramics, abrasives, detergent cleaners, fertilizer, insecticides and the semiconductor industry. It also is involved in the production of ferroboron steels, and in the synthesis of the second and third hardest materials after diamond, namely the ceramics BN (m.pt.3000°C) and B_4C (m.pt.2700°C). As a result it does appear that its role in combustion is minor and any environmental concern with boron call for an understanding of its combustion mass balance rather than issues of emissions control.

Health concerns with boron

Human's dietary requirements are not entirely known for boron. It is a necessary element and appears to be involved in steroid hormone balance and other functions not yet well defined. Deficiencies that are unusual to note, can create medical difficulties (World Health Organization Trace elements, 1996). Medically, it is a new therapeutic element finding use for arthritis and osteoporosis. It also appears to be useful in treating certain infections. A mean dietary intake has been assessed as 1.2 mg day^{-1} readily available from a broad variety of foods. Partly through a general lack of nutritional deficiencies, emphasis has concentrated on its noncarcinogenic and toxicity aspects. Two very detailed reports are available on the Web outlining the extensive toxicology with animals and humans (World Health Organization Boron, 1998; Toxicol Report Boron, 2010). From these an MRL of 0.2 mg kg^{-1} body wt. day^{-1} has been suggested for humans. It is minimally toxic and is not carcinogenic. Involved in the plant growth cycle it can be safely ingested in a diet at levels of 15 mg day^{-1}. Doses of a magnitude on a gram scale can indicate clinical symptoms and greater than 30 g becomes toxic. It is absorbed by the body easily but does not metabolize. This results from the chemical bond strength of borates in any biological system. Although there can be some accumulation in plants this does not magnify in the food chain or accumulate in humans. As a result there has been very little interest in monitoring biomarkers for boron in humans. Testing centers do not generally list and analyze for it. One noteworthy study was published in France when the four major possible biomarkers, blood, plasma, hair and urine levels were compared in 100 healthy individuals (Goulle et al., 2005). This indicated that a blood measure was reliable and should normally be $<100\ \mu g\ L^{-1}$.

Plasma was similar, with hair showing only traces of <2 ppmw. Urine is the major channel for excretion due to boron's general solubility, most being eliminated within a day, with only traces retained and stored in bone (Mayo Clinic, 2018). Cases of toxicity can be readily treated and result mainly from accidental exposure or relate to industrial processes.

Boron's basic chemistry

Several of boron's compounds are very volatile and hazardous requiring special handling. Their spontaneous combustion is appropriate as potential rocket propulsion fuels and relate to their significant energy content. Consequently, such molecules as B_2H_6, BH_3CO, B_2S_3 and BCl_3 are encountered in the field of propulsion and not in normal combustion systems. The hydrogen boranes, for example, exist up to a decaborane that finally has the mass to be in a condensed state with a melting point of 100°C. Generally, all these are readily hydrolyzed to boric acid and are readily oxidized.

From the point of view of combustion, boron's chemistry is defined by its B/H/O compounds and its quite stable metal borates. Its condensed phase compounds are discussed below and summarized in Table 7.13. As a result, sulfur and chlorine are not expected to play any direct roles in its combustion chemistry but can possibly be involved in their indirect effect on the availability of other potential metal interactions.

TABLE 7.13 Properties of boron's condensed phase molecules relevant to combustion, °C.

m.pt. b.pt. vapor pressure
B 2075 3866, 10 ppmv/2075°C, ins
B_2O_3 450 2066, hydrolyzed by water, 10 ppmv/1200°C, s
H_3BO_3 ($B(OH)_3$) 171 300, decomp to HBO_2 at 57, 50 ppmv/75°C, s
HBO_2 236, 5 ppmv/627°C, s
B_2H_6 −166 − 92, reacts with water
BH_3CO −137 − 64, reacts with water
BCl_3 −107 12.5, hydrolyzed by water
B_2S_3 563 1146, hydrolyzed by water, 5 ppmv/500°C
BP 1125 decomp, reacts with water
BPO_4 1400, slightly s
$NaBO_2$ 967, 5 ppm/818°C, s
KBO_2 947, 5 ppm/760°C, s
$Na_2B_4O_7$ 743, s
$K_2B_4O_7$ 515, s
$Ca(BO_2)_2$ 1154, slightly s
CaB_4O_7 986, reacts with water, slightly s

decomp, decomposes; *s*, water soluble; *ins*, water insoluble.

The major stable oxide is B_2O_3(s) (m.pt.450, b.pt.2066°C). If dry, this sublimes to a stable B_2O_3(g) with a vapor pressure of about 10 ppmv at 1200°C (Gurvich et al., 1994). Its gaseous structure is well characterized by electron diffraction and infrared spectra (Spiridonov et al., 1983; Serebrennikov et al., 1983). However, the oxide readily reacts with water vapor to produce orthoboric acid generally written as H_3BO_3(s) but has a $B(OH)_3$ structure (m.pt.171, b.pt.300°C). This is very volatile (vapor pressure 50 ppmv at 75°C) and can vaporize congruently at low temperatures to H_3BO_3(g), stable to about 600°C (Balasubramanian et al., 2008). However, its condensed phase is metastable and on heating begins to decompose about 57°C, loses water and becomes the more stable metaboric acid HBO_2(s) (m.pt.236°C). This also can vaporize, is less volatile (5 ppmv at 627°C), but produces HBO_2(g) trimer that has a very stable O = B-OH bent structure. However, at 427°C it converts to B_2O_3(l) which with water vaporizes as HBO_2(g). Consequently, any form of boron oxide on heating in the presence of water vapor will evaporate and convert to H_3BO_3(g) or HBO_2(g), which can become the major boron carriers below and above 650°C in some cases. Infrared studies of the vapor above B_2O_3(l)/H_2O(g) at 980°C−1180°C confirmed this dominance of HBO_2(g) (White et al., 1960). Boron differs from many elements in not having the normal range of inorganic categories such as a carbonate, nitrate or sulfate. Other than the two acidic hydroxides it does exhibit a trichloride that is normally encountered in its gaseous phase (BCl_3 m.pt. − 107, b.pt.12.5°C). This is hydrolyzed by water and readily oxidized and so is hazardous and no role player in combustion other than being a potential gaseous additive. This is also the case for the volatile hydrides, the well-known one being the dimer, diborane B_2H_6, (b.pt. − 92°C). A gaseous carbonyl also exists, BH_3CO (m.pt. − 137, b.pt. − 64°C). Reflecting the oxide, a sulfide is known, B_2S_3(s) (m.pt.567°C), that also sublimes to B_2S_3(g) (Sommer et al., 1960; Chen and Gilles, 1972). It is readily hydrolyzed by water and subject to oxidation. Consequently, it is seen in nature that boron is not normally found in its condensed oxide or hydroxyl forms but rather as metal borates. The alkali borates are the most studied due to their global source as sodium borate. However, the nature of their thermal dissociation still appears uncertain concerning the degree that the orthoborates, M_3BO_3, can either vaporize directly and then shed M_2O or form gaseous metaborates MBO_2 directly. Even so, it is found that the vapor produced is predominantly the thermally stable metaborate MBO_2(g) form, the molecular structures of which have been confirmed by electron diffraction (Ezhov and Komarov, 1978). No evidence for gaseous orthoborates has been observed. The metaborates are not refractory and for Li, Na, K, and Cs have melting points of 844°C, 967°C, 947°C, and 717°C, respectively. Similarly, alkaline-earth orthoborates are known and can have the structures $M_3(BO_3)_2$ or MB_4O_7 and their metaborates have the formulation $M(BO_2)_2$ in the condensed phase to satisfy valency requirements. Their melting points appear to be similar to the alkalis and around 1000°C. That for $Be_3(BO_3)_2$ was quoted as 1495°C (Blackburn and Buchler, 1965) and for $Ba(BO_2)_2$ 1095°C. Consequently, it is interesting to note that mass analyses of their vapors in the case of Sr and Ba only show an MBO_2(g) dominance in spite of this, implying a monovalent radical nature by the alkaline-earth (Kou and Asano, 1987; Asano and Kou, 1989). Boron is known for producing borides with many metals of the periodic table. These generally are refractory ceramics that are very stable. They are not easily produced and will play no role in normal combustion conditions.

Gas phase stabilities and transport

Examining the gas phase, the very basis of the thermochemical values centers on gaseous stabilities of B, BO, BO_2 and their hydrogen bearing derivatives. Their stabilities have already been listed in Table 7.5 and trends can be compared with its other periodic table elements Al/Ga/In and Tl. For a long time the sublimation energy of B(s) was not accurately known. This was finally better assessed and is now recommended as $\Delta H_{f298K}(B)_g = 564 \pm 2$, $\Delta H_{f0} = 559 \pm 2\ kJ\ mol^{-1}$ (Cordfunke and Konings, 1990; Nordine et al., 1991). This increased the previously used value by $5.2\ kJ\ mol^{-1}$ and so affected all prior analyses before that date. For gaseous BO, and the dioxide which is the only Group IIIA one observed, the evaluation by Gurvich et al. (1994) remains the most current database. As a result, changes have occurred over time and in this case the heat of formation for BO_2 has changed by $25\ kJ\ mol^{-1}$ from that listed earlier in NIST-JANAF (Chase, 1998) to the more current Gurvich et al. (1994) database. As a result, extra care has to be taken in comparing old and new measurements. From Knudsen cell-mass spectrometric equilibria of isomolecular exchange reactions with LaO, YO, and ZrO these led to a reliable value of $D_0(B\text{-}O) = 800 \pm 10\ kJ\ mol^{-1}$. A recent theoretical study constructed 38 potential energy curves for BO and calculates their spectral parameters. It also derived a theoretical bond strength of 798 in very close agreement (Magoulas and Kalemos, 2014). Similarly from several other studies of the dissociative vaporization of B_2O_3 and studies of the BO/BO_2 equilibration a value of $D_{0Atomiz}\ (BO_2) = 1364 \pm 20\ kJ\ mol^{-1}$ was derived. This implies $D_0(OB\text{-}O) = 564 \pm 22\ kJ\ mol^{-1}$. The gaseous oxides BO and BO_2 are important in the case of $B_2O_3(s)$ vaporization and unlike aluminum the dioxide replaces the role of a possible B_2O formation. The infrared spectra of BO and BO_2 were recorded in a low-temperature matrix isolation study obtaining the vibrational frequencies and suggesting a linear structure for the dioxide (Burkholder and Andrews, 1991). In fact the green chemiluminescence of boron in flames has always been attributed to electronically excited BO_2 (Kaskan et al., 1961). Recently, this emission has been suggested as a possible indirect method for monitoring thermal NO formation (Maligne et al., 2009). It is noteworthy that the BO bond strength is extremely strong and only exceeded by CO and the actinide ThO, and equal to that of LaO. However, in the presence of water, these oxides have to compete with the formation of other B/O/H molecules such as BOH, HBO, HBO_2, HBOH, etc. Bond strengths for these have been estimated, $D_0(H\text{-}BO) = 433 \pm 25\ kJ\ mol^{-1}$ and $D(B\text{-}OH) = 606 \pm 30\ kJ\ mol^{-1}$ but these are minor components in the B/H/O system even to high temperatures (Gurvich et al., 1994). A theoretical calculation confirms that the structure HBO(g) is about $200\ kJ\ mol^{-1}$ more stable than its isomeric form BOH (De Yonker et al., 2005). Moreover their isomerization also entails an additional energy barrier of about $110\ kJ\ mol^{-1}$ (Richards et al., 1995). $HBO_2(g)$ is the more important molecule being the vaporized metaboric acid from the B_2O_3/H_2O interaction. Its structure is HOBO conforming to trivalent B. It has been extensively studied by this liquid/gas equilibria and so is reliably determined as $D_{0Atomiz}\ (HBO_2) = 1827 \pm 10\ kJ\ mol^{-1}$ or $\Delta H_{f0}(HBO_2) = -557 \pm 10\ kJ\ mol^{-1}$. These imply $D_0(H\text{-}BO_2) = 463 \pm 20\ kJ\ mol^{-1}$. It can also be produced via kinetics such as

$$BO_2(g) + H_2O(g) = HBO_2(g) + OH(g) \quad \Delta H_{298K} = +30\ kJ\ mol^{-1}.$$

In fact, because the thermochemistry was more accurately known for $HBO_2(g)$, Uy et al. (1971) used this balance to better establish the bond strength of $BO_2(g)$.

The homogeneous gas phase kinetics have been extensively modeled for both oxygen-rich and oxygen-lean conditions and illustrate this trend of $HBO \rightarrow BO \rightarrow BO_2 \rightarrow HBO_2$ (Yetter et al., 1991; Pasternack, 1992; Brown et al., 1994).

$$HBO(g) + OH(g) = HBO_2(g) + H(g) \quad \Delta H_{298K} = -170 \text{ kJ mol}^{-1}$$

Under normal combustion conditions the gaseous HBO_2, B_2O_3, BO_2, HBO, the B_2O_2 dimer and BO tend to be the more important species. With such an H/B/O system it would seem that the boron might vaporize on heating and then on cooling the $HBO_2(g)$ would condense below about 600°C to $HBO_2(s)$. With water vapor this would probably convert to $H_3BO_3(s)$ and revaporize until temperatures were sufficiently cool to retain the two acid forms (Yetter et al., 1991). As a result, their sublimation becomes important with respect to being potential gaseous transporters of boron and their formation in ash. Vapor pressures will control stable deposition and the preference may certainly be to form a metal borate.

With boron's hydrides, formation of gaseous BH can occur as an intermediate in various processes making it of value as a nonperturbing spectral monitor. A recent heated W wire experiment with $B_2H_6/H_2/He$ mixtures is one example where B and BH were produced and monitored by LIF (Umemoto et al., 2014). BH has a rich molecular spectroscopy that has been documented (Huber and Herzberg, 1979). Recently, all the available spectroscopic data for the X, A and B states have been combined into a unified set (Shayesteh and Ghazizadeh, 2015). It has no low-lying states but from the establishment of predissociation limits of the $A^1\Pi$ state the ground state bond strength is accurately known as D_0(B-H) = 338.1 ± 0.4 kJ mol^{-1} (Persico, 1994). The dihydride has been less studied but a recent LIF study has now characterized in detail the bent X^2A_1 and low-lying linear A^2B_1(T_0 4200 cm^{-1}) states (Sunahori et al., 2015). The bond strength is only known from theory. Values approximately agree and a more recent estimate indicates ΔH_{f298K} = 328 ± 3 kJ mol^{-1} (Feller et al., 1998). Their corresponding value for BH agrees with that above, and only slightly varies with that of previous theory for BH_2 (Gurvich et al., 1994). It implies acceptable values of $D_{0Atomiz}(BH_2)$ = 664 or D_0(HB-H) = 331 ± 5 kJ mol^{-1}. As a result, the two bonds are of equal strength. The hydrides are only compatible with dry, oxygen free reducing environments and are of minor interest. The same applies to the chlorides. The M-Cl bonding can be strong and comparable to that of M-OH as clearly seen in the periodic table of trends for the boron periodic table column of elements, Table 7.5. However, boron itself differs from this trend. Diatomic gaseous BCl has been extensively characterized spectroscopically and seen to have no low-lying electronic states (Huber and Herzberg, 1979). The latest effusion cell-mass spectrometric study of equilibria in the B(s)/$MgCl_2$/$SrCl_2$ systems has refined the previous evaluation of D_0(B-Cl) = 510 ± 20 kJ mol^{-1} (Gurvich et al., 1994). A more precise recommended value of D_0(B-Cl) = 508 ± 5 kJ mol^{-1} now is indicated (Hildenbrand, 1996). This value now has been satisfactorily reproduced by theory together with its low-lying potential energy curves (Shi et al., 2011; Liu et al., 2012). Evaluations for the dihalide BCl_2 prefer a photoionization value of $D_{0Atomiz}(BCl_2)$ = 861 ± 12 implying D_0(ClB-Cl) = 353 ± 12 kJ mol^{-1} (Gurvich et al., 1994). The heats of formation and vaporization for BCl_3 are well characterized implying

$D_{0Atomiz}(BCl_3)_g = 1322 \pm 5$ and therefore $D_0(Cl_2B\text{-}Cl) = 461 \pm 14$ kJ mol^{-1} for the gaseous trichloride (Gurvich et al., 1994). The first bond is the strongest, the second the weakest and the third with a midstrength. Although strongly bound, BCl is vulnerable to oxygen and water destruction that will minimize its role in any H/O containing system. ClBO(g) has been studied in B_2O_3(l)/BCl_3 equilibrium effusion experiments. It is known to be linear with a $X^1\Sigma^+$ ground state. Structurally it is synonymous with HBO_2 (HOBO) shown to be the important gaseous acid of boron. Using the recommended heats of formation of Gurvich et al. (1994) it is seen like the boron chlorides this also will be readily hydrolyzed by water.

$$\mathrm{ClBO(g) + H_2O(g) = HOBO(g) + HCl(g)} \quad \Delta H_{0K} = -91 \text{ kJ mol}^{-1}$$

The gaseous monosulfide, BS, has been spectrally characterized with numerous electronic states observed, a $X^2\Sigma^+$ ground state with none low-lying. It has also been studied mass spectrometrically in equilibria of B/YS/BS/Y and BO/YS/BS/YO systems. A value of $D_0(B\text{-}S) = 575 \pm 30$ kJ mol^{-1} is recommended (Gurvich et al., 1994). This is 70% the strength of BO, and supported by a similar decrease for Al and other sulfides. BS_2(s) and B_2S_3(s) can vaporize congruently and molecularly at lower temperatures and a heat of sublimation of B_2S_3(s) was obtained (Chen and Gilles, 1972; Gurvich et al., 1994). However, at elevated temperatures a variety of sulfides can occur complicating analyses. From a mass spectrometric study of CeS/BS_2 and B/BS/BS_2 equilibria a value of $D_{0Atomiz}(BS_2) = 1060 \pm 20$ or $D_0(SB\text{-}S) = 485 \pm 25$ kJ mol^{-1} was obtained indicating a weaker second bond but by less than seen with the oxides assuming values are correct (Gurvich et al., 1994). Nevertheless, the sulfides are only compatible with dry, oxygen-free systems.

As mentioned already, metal borates are an important part of boron's chemistry. These also have significant stabilities in that they can vaporize congruently and molecularly. They have been extensively studied for the alkali and the alkaline earth elements. A vapor pressure of about 5 ppmv requires temperatures of 910°C, 818°C, 760°C, 692°C, 630°C for Li/Na/K/Rb/Cs, respectively (Gorokhov et al., 1971; Cordfunke et al., 1989; Chase, 1998). Relevant to combustion those for sodium and potassium are stable as $NaBO_2$(g) and KBO_2(g). Most previous data used the heat of formation of BO_2 from the JANAF-NIST tables (Chase, 1998). However, this has now been reassessed (Gurvich et al., 1994). It is a significant modification and has the effect of lowering previous bond strength values for M-BO_2 by 25 kJ mol^{-1}. Taking this into account, current magnitudes for $D_0(M\text{-}BO_2)$ are about 500 ± 20, 445 ± 15, 450 ± 10, 460 ± 15, 460 ± 10 kJ mol^{-1} for Li/Na/K/Rb/Cs updating the measurements from several sources (Jensen, 1969; Farber and Srivastava, 1985; Kou and Asano, 1987; Cordfunke et al., 1989; Asano and Kou, 1990a; Lopatin and Shugurov, 2013). These values are comparable to that of $D(H\text{-}BO_2) = 463$ kJ mol^{-1}. Consequently, it is seen with boron that gas phase chemistry with other metals can play a role in its speciation with reactions such as the close to energy neutral interchange

$$\mathrm{HBO_2(g) + Na(g) = NaBO_2(g) + H(g)}$$

Whether gaseous NaOH or NaCl can similarly invoke four-center reaction exchanges to contribute to the balance between HBO_2(g) and $NaBO_2$(g) or KBO_2(g) is plausible but unnecessary as these species are in fast balanced reactions with the Na(g). The chemical thermodynamic tendency will be to form the alkali metaborates. This was confirmed by adding traces of an alkali salt aerosol (Li, Na, or K) and trimethyl borate, $B(OCH_3)_3$ (b.pt.68°C) into a hydrogen flame and monitoring such balanced reactions to obtain reasonable measures of the bond strengths of the gaseous alkali metaborates (Jensen, 1969). Consequently, the gaseous metal borates will be evident in combustion systems with alkalis especially as HBO_2 along with B_2O_3 generally will be the major species (Yetter et al., 1991; Pasternack, 1992). It was noted by Blackburn and Buchler (1965) that the gaseous M-BO_2 molecules have a pseudochlorine nature in bond strengths with corresponding M-Cl. This in fact is quite pronounced and the respective borates are 30 ± 8 kJ mol^{-1} stronger for the five alkali elements, illustrated graphically in a previous review of alkali, alkaline earth flame chemistry (Schofield, 1992). It is similar with the alkaline earth elements but in this case the condensed phase $M(BO_2)_2$ molecule may vaporize unchanged but appears to then dissociate to the more stable MBO_2. In the earliest mass spectrometric studies with Be, Mg and Ba they report monitoring $M(BO_2)_2^+$ (Blackburn and Buchler, 1965; Gusarov, 1970; Il'in et al., 1974). However, this remains doubtful, and has not been confirmed in later work with Mg, Ca, Sr, or Ba. Moreover, this early work also did not appear to provide acceptable values for MBO_2. In two of the cases they indicated that the two bonds in $M(BO_2)_2$ were either weaker or stronger than each other, which is not an expected trend. With the exception of Be that was estimated from trends, bond strengths of D(M-BO_2) for the alkaline earths have been measured. Although still rather uncertain in most cases, an updated reassessment of the values indicates: D_0(Be-BO_2) = 490 ± 20 kJ mol^{-1} (Chase, 1998), D_0(Mg-BO_2) = 422 ± 20, D_0(Ca-BO_2) = 390 ± 20 (Lopatin and Shugurov, 2013). This measured value for calcium drops below D_0(Ca-Cl) = 400 ± 6 (Gurvich et al., 1994) and significantly breaks the borate-chloride trend. As a result an estimated value of D_0(Ca-BO_2) = 420 ± 30 would appear more appropriate. For strontium, the two published studies show a 65 kJ mol^{-1} discrepancy in their heats of formation and hence dissociation energy values. These imply either D_0(Sr-BO_2) = 485 (Asano and Kou, 1990) or 420 kJ mol^{-1} (Lopatin and Shugurov, 2013). With D_0(Sr-Cl) = 409 kJ mol^{-1} this suggests an average value 450 kJ mol^{-1} might be a better value at present. Barium is similar with the same two groups suggesting D_0(Ba-BO_2) = 498 (Asano and Kou, 1990a) and 459 kJ mol^{-1} (Lopatin and Shugurov, 2013). D_0(Ba-Cl) = 440 kJ mol^{-1}, so an average value of 478 ± 25 kJ mol^{-1}might be appropriate. Again, the borates are stronger than their corresponding chlorides and track each other in a definite trend of decrease in strength from Be to Mg and then gradually strengthening again with increasing atomic number. For normal combustion fuels, interactions with Mg and Ca components would seem to be of greatest interest but these are weaker than either the 463 kJ mol^{-1} value for HBO_2(g) or the alkali borates. Consequently in the gas phase, if available, Na and K will have a much greater affinity for boron than Mg or Ca due to the energetics of such exchange reactions such as:

$$MgBO_2(g) + Na(g) = NaBO_2(g) + Mg(g) \quad \Delta H \approx -23 \text{ kJ mol}^{-1}$$
$$CaBO_2(g) + K(g) = KBO_2(g) + Ca(g) \quad \Delta H \approx -30 \text{ kJ mol}^{-1}$$

Other metal borates that have been examined and might be of interest to combustion are those of Al, Fe and Pb. That with Al is in fact the strongest of all those mentioned here, D_0(Al-BO_2) = 559 ± 20 kJ mol^{-1} being similarly stronger than its chloride (498),

Table 7.5. The others are much weaker and of less interest, D_0(Fe-BO_2) = 267 and D_0(Pb-BO_2) = 246 kJ mol^{-1} (Lopatin and Shugurov, 2013). There is no mention of boron as a soot suppressor, but its trihalides, BF_3, BCl_3, and BBr_3 were mentioned as possible flame inhibitors (Rosser et al., 1966; Babushok and Tsang, 2000).

As a result, if vaporization occurs, there are numerous possibilities for gaseous molecular formation and transportation. Boron exhibits a flame ion chemistry based on $BO^+.H_2O$ interactions with other flame ions, both positive and negative (Goodings, 1992). Whether they participate in soot interactions remains unreported but there are suggestions that boron is a slight flame inhibitor (Miller, 1969; Babushok and Tsang, 2000).

Combustion power plant data

As seen in Table 1.1 of typical fuel content, boron is not negligible in most fuels. It is ever present in coals, less in biomass but has a presence in wastes and sewage sludge. It is less evident in pet coke and limestone being generally <20 ppmw (Cordoba et al., 2012). Due to its solubility, boron's history is one of moving from the land, dissolving into solution and finally returning to sedimentation processes in the rivers and oceans. If anything, the major municipal concern relates to ensuring that boron in drinking water is maintained at satisfactory levels. This is particularly sensitive in Japan where boron can be high in surface waters due to industrial emissions and efforts are made to import only low boron coals (Noda et al., 2013).

In full-scale combustion systems most of the data concerning boron have emerged in recent years partly due to an increased interest but also because of improved methods of analysis for boron. Various ICP modes now make it possible to include boron in trace element analyses. Prior to this, spectral atomic emission or absorption failed even in the hottest flames (De Galan and Samaey, 1970). Even so, full-scale combustion data do remain very scant especially for biomass, waste and sewage sludge fuels, which can reflect lower levels of boron. Few definitive studies exist and with regards to the boron speciation most conclusions still have to be regarded as speculative. One recent reliable full-scale study has reported an extensive mass balance for boron in a large coal power plant operating with SCR, ESP and wet FGD systems (Cheng et al., 2009). They noted removal efficiencies of about 8% by the ESP, 77% by the FGD with little boron in the bottom ash and <0.5% released as a gas from the stack. They implied that about 99% of the coal's boron was discharged in the waste water. It has been known for a long time that boron's emissions are predominantly gaseous (Fogg and Rahn, 1984). Meij (1994) also indicated that the boron appears to be so volatile that it passes through an ESP and is largely controlled by the FGD scrubbers that then may need the use of absorbers for the waste waters to reduce the boron content below effluent standards (Ohki et al., 2011; Cordoba et al., 2012). One drop tube furnace at 1400°C with a 2 or 4 seconds residence time examined seven coals with 16–134 ppmw B and ash quite rich in Al, Fe, and Ca. They measured a mass balance for the B utilizing ICP analysis and endeavored to analyze the distribution with equilibrium calculations. They assumed that B_2O_3 was mainly in the ash and that pyrolysis and combustion behaviors differed (Noda et al., 2013). A very similar Japanese pyrolysis study noted lower B release at 900°C if Si and Al content was higher. Also that some of the produced boron forms were soluble and some not (Mochizuki et al., 2014). In spite of such efforts, little understanding yet is realized. Studies that did examine collected fly ash indicate higher concentrations on smaller particle size fractions indicative of vapor deposition (Querol et al., 1995; Narukawa et al., 2003). One study of a coal power plant fit solely with a cold-side ESP indicated a boron final release of about 2% (Goodarzi, 2006). There is little retained in any bottom ash but the fly ash that is collected can be rich in boron with levels as high as 1500 ppmw (James et al., 1982). In Japan, its levels can exceed the regulated limits. As a result, they have done most of the research in endeavoring to analyze the nature of the boron in ash. Using mainly NMR analysis and leaching experiments they conclude that most of the boron in raw coal is mainly organic and that 50% can be released by 200°C and 80% by 400°C with possibly some inorganic contribution (Kashiwakura et al., 2011). For ash, their work still remains speculative but provides some indication for the presence of calcium orthoborates and pyroborate, and possibly contributions from magnesium, iron, sodium and potassium. There was no suggestion of silicate interaction (Kashiwakura et al., 2009; Hayashi et al., 2010). However, several leaching experiments with coal ash have indicated that rates for boron release can cover a wide range of 0%–85% in one Japanese study suggesting a dependence on the specific mineral content and type of coal and formation of insoluble borates (James et al., 1982). Consequently, it does appear that depositing boric acid does associate with various metals in the ash to produce insoluble borates such as possibly those of Al, Ca, Fe and Zn. Although involving higher concentrations, it is known that additions of boron to black liquor combustion catalyze the regeneration of NaOH and reduces the release of atomic sodium by forming $Na_3BO_3(s)$ (Saw et al., 2010).

It is apparent that boron exhibits a very volatile nature and does not have sufficient time to be scavenged and retained by the ash particles to a significant degree. As to why this occurs is partly apparent on assessing the plausible deposition chemistry. Initial release from a fuel does not seem limiting in the presence of oxygen and water vapor, moreover any inorganic borate or metaborate also will probably vaporize by 800°C–900°C. Although gaseous metal metaborates are stable and also can form in the flue gases it is quite possible that this is really irrelevant with respect to subsequent deposition. As flue gases cool, all the gaseous forms of boron will begin to interact with the surface of ash particles. The surface then assumes chemical control and will function according to its thermodynamic preferential options. For example, at 900°C temperatures will be too high for any B/H/O compounds that might form to be retained on the surface. Similarly, any depositing, gaseous $NaBO_2$ will be seen by the surface simply as a source both of Na and B. If sulfur is available it will instantly preferably form $Na_2SO_4(s)$ and the boron will be released back into the flue gases as $HBO_2(g)$. At even lower temperatures, if any $HBO_2(g)$ deposits on the surface and forms $NaBO_2(s)$ this also will still be at risk of conversion if sulfur is available. Consequently, only below about 600°C will $HBO_2(g)$ be able to remain on the surface as $HBO_2(s)$ and not be vaporized. Even then it may react further with ash components to form stable borates. Consequently, because boron has few molecules that are not volatile, or may lose its metallic component to sulfur or some other thermodynamically more stable compound, it appears that boron has to remain in the gas phase until such lower temperatures are reached. Moreover, although boron is a component of borosilicate glass there appears to be little indication at present for interaction between the gaseous boron and the ash silica. Rather like mercury in combustion, the boron appears to be constrained by temperatures and time. This is why little boron can be found on the ash captured by the ESPs and why it has to be controlled by FGD scrubbing. Moreover, this is another good example of ash deposits not necessarily relating to the gas phase speciation and in fact leaving the nature of the final deposited boron compounds uncertain. With present oxy-pulverized coal fuel combustion, the recycling of the CO_2 laden flue gases does provide lengthened time for such interaction with the fly ash. As a result in a recent study more boron is captured on the fly ash (63%) with an overall capture of about 82% without the previous FGD in the system (Cordoba and Diego, 2018).

Few reports have appeared concerning boron under reducing or gasification conditions. In oxygen deficient systems little release might be expected with the boron remaining in the char. Additions of SiO_2 and Al_2O_3 have been shown to suppress the release of boron during pyrolysis and CaO during gasification, which has to imply possible glass or calcium or aluminum borate formation (Mochizuki et al., 2009). Similarly in a gasifier at 960°C the use of a dolomite (Ca/Mg carbonate) hot gas filter at 580°C did effectively absorb all the gaseous boron (Reed et al., 2001). Any role of B_2S_3 under reducing conditions (m.pt.567°C, 5 ppmv vapor pressure at 500°C) is unlikely due to its hazardous reaction with water. Also boron phosphate BPO_4 (m.pt.1400°C) is known but not generally encountered. Films of it can be produced by CVD methods on substrates at 300°C–550°C simply from volatile organoboron and phosphorus molecular mixtures (Sato et al., 1991; Michael and Boland, 1997). However, it is an unusual covalent molecule that on vaporization at about 1100°C dissociates to B_2O_3 and P_2O_5 (Lopatin and Semenov, 1999). Moreover there is some hint that water at 280°C may hydrolyze BPO_4 to its oxides (Montoneri et al., 1986). Additionally, in a combustor the affinity of other elements for P may of course be much greater than that of B.

In summary, boron is an interesting element as it displays great volatility. It is readily released with a regimen of H/B/O volatile gaseous molecules. It does not appear in bottom or only slightly on ESP ash having insufficient interaction time. Its volatility is controlled by the flue gas temperatures preventing its permanent deposition onto fly ash. It is mainly removed by the FGD scrubbers. The transporters are gaseous HBO_2, H_3BO_3, and even $NaBO_2$ or other borates that may be formed. Most of boron's molecules are water soluble.

7.9 Cadmium

In between zinc and mercury in the periodic table, and at the end of the second row transition elements, cadmium's chemical behavior is more similar to that of zinc. However, like mercury it is one of the highly toxic heavier metals. It is a rare element being about 65th in the listing content of the Earth's crust (0.1–0.5 ppmw), and is ranked at about 50th in the oceans. Commercially it has been produced only for just over 100 years and became known as the metal of the 20th century. Global production of metallic Cd is about 22,000 tons year^{-1} and with no specific ore. It is found in nature associated with zinc and so is a byproduct of the latter's smelting, with about 20% now recycled. About 80% is produced from Zn processing (3 kg Cd produced per ton of Zn) and 20%

from Pb and Cu smelting. Present cost is about \$0.88 lb^{-1} not too dissimilar from that of Al, Pb and Zn and below the more expensive Cu, Sn, and Ni metals. Cadmium's melting point 321°C and boiling point 767°C are slightly lower than those of zinc. It has an equilibrium vapor pressure at atmospheric pressure equivalent to 100 ppmv at 310°C. It is used mainly in Ni/Cd batteries (86%), electroplating, fertilizers, pesticides, paints, and alloys. Although now 25 years old, the estimated values then of about 1300 tons $year^{-1}$ were assessed as the natural atmospheric emissions' background and are still generally accepted. Volcanoes are known to be a major contributor with forest fires, particles and volatiles the remainder (Nriagu, 1989; Malakhov and Makhon'ko, 1990). An additional Canadian report was published in 2001 questioning this value and estimating the natural background of numerous trace metals (Cd, Cu, Pb, Hg, Ni, Zn). All were assessed as being significantly larger by one to two orders of magnitude than previously considered (Richardson, 2001). That for Cd now became assessed at 40,000 tons $year^{-1}$. This is a very major change that remains unacceptable. It would dwarf the anthropogenic value and historically indicate no change over time. The previous such contribution now is thought to have been halved by regulations from the 25-year-old estimate of about 7600 tons $year^{-1}$ (Nriagu, 1989a; Malakhov and Makhon'ko, 1990). A value closer to 3000 tons $year^{-1}$ now is suggested and whether this is larger, smaller or comparable to the natural level remains unclear (United Nations Cd, 2010). Of this quantity, about a half may still result from combustion with the rest associated with metal smelting and industries. Russian ice core data show the significant increase of anthropogenic Cd from 1950 to 1990 but this appears to have fallen back now to its 1950 levels (Eichler et al., 2014). Cadmium has been assessed as one of the important environmental risks in Japan (Gamo et al., 2003). However, reports confirm decreasing atmospheric levels in Canada (Li et al., 2003), Greenland (Boutron et al., 1991), and Europe (Azimi et al., 2005; Pacyna et al., 2007). A concern remains in China where emissions from coal combustion have increased eightfold from 1980 to 2008 (Tian et al., 2012, 2015; Cheng et al., 2014, 2015a), and suggest combustion contributes about 45% to their anthropogenic emissions. Also problems were related to cadmium in China in those regions where high fluorine/cadmium-content coals were in use domestically for cooking and heating (Tang et al., 2009). An assessment of long-term accumulation has also been made for anthropogenic emissions of Cd when the ultimate potential environmental problem can result in soil pollution. Unless controlled, this can have a long half-life (Malakhov and Makhon'ko 1990). Atmospheric deposition to soils in China is of concern and they are analyzing control measures (Peng et al., 2019). The potential of using certain plants to remove Cd from polluted soils has been shown feasible (Sterckeman and Puschenreiter, 2018).

Health concerns with cadmium

Its compounds are very toxic and are rated as carcinogenic if inhaled. For this reason there is growing environmental concern with Cd for long-term effects. As a result, several very extensive environmental assessments have been published on the Web by the United Nations (United Nations Cd, 2010), and by the World Health Organization (WHO Cd 1992, 1992a) together with a detailed toxicological report (ToxProfile Cd, 2012). Like mercury, it appears to be a nonessential mineral to humans and is toxic. Nevertheless, it is unavoidable being in the diet and an average intake has been estimated at 10–40 μg day^{-1}. An MRL of 0.5 μg Cd kg^{-1} body wt. day^{-1} has been suggested, beginning to overlap with this average intake. This enhances the medical concerns for cadmium. The human diet is not as distant a source from the toxicity levels as is often hoped for. Recent testing of vegetable/crops near a zinc smelting region of China measured up to 2.2 μg g^{-1} in leafy vegetables, which for a 75 kg person would imply no more than 0.3 lb $week^{-1}$ of these to match the limit (Yang et al., 2011). Similarly, edible seaweed was seen to contain 0.59 mg kg^{-1} dw. This implies that 5 g of dried seaweed represents about 23% of the tolerable weekly intake (Paz et al., 2019). The MRL value has been converted into a blood level biomarker of about either 5 μg L^{-1} in one case (Mayo Clinic, 2018) or 1.0 μg L^{-1} (ToxProfile Cd, 2012), or even lower wishful recent suggestions. However, such levels are mainly useful approximate indicators for a physician. Lower levels only embrace a sense of judicious safety but can only be reached by modifying potential sources. One realistic study compared the four generally used human biomarkers directly, levels in blood, plasma, hair and urine samples from 100 healthy humans (Goulle et al., 2005). This indicated blood levels <2 μg L^{-1}, and urine <1 μg L^{-1} as being normal plausible levels to be favored over the traces seen in hair or blood plasma. Nevertheless, because long-term effects still remain uncertain, and are a concern for any toxin, it is useful to have this benchmark guide however crude. In the case of cadmium, many blood survey analyses are continuing to be published, at least five this year and averages do lie in or close to this desirable value (Schofield, 2016; Nisse et al., 2017; Dix-Cooper and Kosatsky, 2018; Lin et al., 2018; Oggiano et al., 2018; Cusick et al., 2018; Henriquez-Hernandez et al., 2018; Eom et al., 2018; Zeng et al., 2019). There are also notable outlier values in some of these distributions, these should be of concern but hopefully not sufficiently large to trigger toxicity. It now has been banned from cosmetics in most countries (Michalek et al., 2019). Cadmium remains inorganic and is not readily absorbed in the body (1%–10%). That absorbed is not easily excreted and can remain particularly in the kidneys where it can cause dysfunction. This retention is the major concern as the body half-life has been estimated as 10–40 years. As a result although generally not listed as a neurotoxic, the longer in the body the greater the probability of some blood/brain crossing or other damage. In this regard its neurotoxicity remains not understood and it is not listed as such (Wang and Du, 2013). One dilemma is the accumulation particularly in tobacco leaves. As a result tobacco smokers show elevated blood Cd levels ($\approx$0.5 μg L^{-1}) higher than the normal nonsmoking adults. However, the ongoing studies of blood and urine samples for Cd in the US population has displayed a continuing decline that is a reflection of reduced smoking (Riederer et al., 2013; CDC 4th National Report, 2009, 2018). Over a decade, at the 95% percentile mark, blood levels have declined from about 1.6 to 1.2, and urine 1.2–0.8 μg L^{-1} for nonsmokers compared to smokers. However, an assessment of the elements Cd, Pb, Hg, and As in a recent US National survey of blood and urine level testing now indicates that half of those tested had either three or all four of these elements present at measurable levels (Shim et al., 2017). Additionally, cadmium has been suspected as possibly being responsible in some cases of infertility, but this remains speculative compared to possible similar roles for Pb, As, or Hg (Tanrikut et al., 2014). Nevertheless, its risk for inducing cancer at levels below those currently regarded as safe now is being addressed

(Ali et al., 2015). Risks during pregnancy remain unknown although it is suspected at having several possible adverse outcomes. MRLs for pregnancy are suggested to be as small as possible, but few studies are available (Taylor et al., 2014). Fortunately, to some extent the body does have protective metallothioneins that are a family of cysteine-rich low molecular weight proteins. Cysteine contains the mercaptan, −CHSH, group, that has the ability to absorb numerous metals, particularly mercury and apparently also Cd. These act as transporters and detoxifiers. Selenium also has similar proteins that are even more effective as metal detoxifiers. A recent study with mice illustrated selenium's ability to prevent any lung damage by Cd (Hu et al., 2018). However, the field of synergistic effects of combinations of toxins in the body still remains largely untested due to the body's complex chemical balances. However, one such study with rats dosed with both Cd and As suggested this induced a faster elimination of the As (Diacomanolis et al., 2014). In nature, Cd does not have an organometallic nature as does Hg and does not normally methylate (World Health Organization Trace Elements, 1996). Organic associations tend solely to be ionic salts such as cadmium stearate. Any covalent organocadmium structure synthesized in the laboratory tends to be sensitive to air, light or moisture. Industrially it is accepted that airborne Cd can cause lung and possibly prostate cancer. It has not clearly been established whether Cd should be generally labeled carcinogenic. A recent edited volume has appeared covering all Cd's bioenvironmental aspects relating to plants and marine species (Sigel et al., 2013).

Cadmium's basic chemistry

The chemistry of cadmium reflects that of zinc showing a dominant divalent behavior resulting from its filled inner electron shells, $4p^6 4d^{10} 5s^2$. As summarized in Table 7.14 many of its condensed phase compounds that relate to combustion are not thermally stable as temperatures are raised and will generally decompose to CdO(s). For example, $Cd(OH)_2$ (decomposes 130°C), $CdCO_3$(decomposes 500°C), $Cd(NO_3)_2$ (decomposes above 300°C). CdH_2 is much less stable (decomposes −20°C) (Wang and Andrews, 2004a). CdO_2 is explosively unstable (decomposing exothermically at 195°C), whereas ZnO_2 similarly dissociates but endothermically (Hoffman et al., 1959; Ropp and Aia, 1962). The reason for this latter curious difference remains uncertain but for cadmium a role for traces of water has been suggested with possible initial formation of $Cd(OH)_2$ and atomic oxygen. Like that of zinc, $CdSO_3$ is known but dissociates above 400°C. A monohalide such as Cd_2Cl_2 has been shown theoretically to be unstable, decomposing disproportionally to the dihalide and the element (Liao et al., 1995). Those compounds that are more thermally stable are $CdSO_4$, dissociating like mercury via $2CdO \cdot CdSO_4$ above about 700°C (Ostroff and Sanderson, 1959; Pannetier et al., 1962; Tagawa, 1984; Tagawa and Kawabe, 1987). CdO extensively dissociates on vaporization above about 900°C and differs from zinc in producing Cd(g) and O_2(g) directly (L'vov, 2007). Knudsen cell vaporization measurements only detected very small quantities of CdO(g), and as discussed below is possibly even less significant than therein indicated (Lamoreaux et al., 1987). $CdCl_2$ is very stable, has a melting point of 564°C but sublimes below this, 100 ppmv (470°C), 725 ppmv (527°C) (Knacke et al., 1991). Although with a high melting point, CdS begins to dissociate above 700°C to Cd(g) and S_2(g), with Cd(g) vapor pressures of about 90 ppmv (750°C), 265 ppmv (800°C) (Nakamura and Fuwa, 1987).

TABLE 7.14 Properties of cadmium's condensed phase molecules relevant to combustion, °C[a]

	m.pt. b.pt. vapor pressure
Cd	321 767, 100 ppmv/310°C, ins
CdO	>750 dissoc, 20 ppmv/800°C, ins
CdO_2	>195 decomp
$Cd(OH)_2$	130 decomp, ins
$CdCO_3$	>287 decomp, ins
$Cd(NO_3)_2$	350, decomp 345, s
$CdSO_4$	1053, dissoc >700, 100 ppmv/700°C, s
$CdSO_3$	400 decomp, s
$CdCl_2$	568 963, 100 ppmv/470°C, s
Cd_2Cl_2	unstable, disproportionates
CdS	1475, dissoc >1360, 90 ppmv/750°C, ins
CdH_2	−20 decomp
Cd_3P_2	>350 dissoc, ins
CdP_2	>350 dissoc, ins
$Cd_3(PO_4)_2$	>1500 dissoc, ins
$CdAl_2O_4$	>1100 dissoc, ins
$CdFe_2O_4$	>800 dissoc, ins
$CdSiO_3$(meta)	1512, stable, ins
Cd_2SiO_4(ortho)	1250, stable, ins
$CdTiO_3$	stable >1150, ins

[a] *Review of sparingly water soluble cadmium salts (Clever et al., 1992);*
decomp, decomposes; *dissoc*, dissociates; *s*, water soluble; *ins*, water insoluble.

It is more volatile than ZnS and at 630°C the ratio of their vapor pressures is 180-fold (Steinbrunner and Edwards, 1991). Two phosphides are known CdP_2 and Cd_3P_2 but both of these dissociate above about 350°C (Lazarev et al., 1976). $CdSO_4$, $Cd(NO_3)_2$ and $CdCl_2$ are water soluble whereas $CdCO_3$, CdS and CdO are insoluble. Other stable molecules that are possibly formed in combustion deposits might be $Cd_3(PO_4)_2$ stable to 1500°C, $CdAl_2O_4$ formed at 1000°C–1300°C (Kurihara and Suib, 1993), $CdFe_2O_4$ formed at 500°C–800°C, $CdTiO_3$(stable at 1150°C) (Neil et al., 1971) and the metasilicate $CdSiO_3$(m.pt.1240°C) and orthosilicate Cd_2SiO_4(m.pt.1250°C), both of which can be formed at 600°C–900°C and all of which are insoluble in water. These latter molecules are less well characterized and their chemical thermodynamic properties not established. It is noteworthy that the Russian studies of many hundreds of mixed metal oxides have not encountered a Cd/metal oxide combination that manages to survive to higher temperatures to vaporize without dissociation (Lopatin and Shugurov, 2013).

Gas phase stabilities and transport

Gaseous species exist, but surprisingly are still poorly characterized, mainly due to their low stabilities and less important roles (Table 7.15). As with zinc and mercury, the experimental bond strength of cadmium's oxide, sulfide and hydroxide have not been reliably measured. This results from the complexities of the low-lying potential energy curves for CdO and CdS that are spin–orbit coupled and subject to curve crossing transitions, and the difficulty of measuring their gaseous low levels. CdO(s) dissociates immediately on vaporizing. At 750°C the CdO^+ was difficult to observe compared to Cd^+ in early mass spectrometric vaporization studies of CdO(s) (Berkowitz and Marquart, 1963; Marquart and Berkowitz, 1963). Although D_0(Cd-O) had been recommended as 228 kJ mol^{-1} (Lamoreaux et al., 1987) and 276 (Barin and Platzki, 1995), more recent reliable theoretical calculations now question these and suggest a much lower value of about 95 kJ mol^{-1} (Shepler and Peterson, 2006; Peterson et al., 2007; Song et al., 2007). This is supported by the Zn, Cd, Hg bond weakening trends (Table 7.16). It is to be remembered that their oxides have $X^1\Sigma^+$ ground states that strictly correlate to Cd($^1S_0$) and O,S($^1D_2$) dissociation products. They reflect the potential energy curves of the alkaline earth oxides and sulfides that have similar filled shells, shown in Figs. 7.2 and 7.3, the two valence electrons and similar dissociation correlations. However, in this case theoretical calculations now indicate that the low-lying $a^3\Pi$ state that does correlate to ground state products is extremely low and levels are mixed. The ground state essentially has a $X^1\Sigma^+/a^3\Pi$ nature for both CdO and CdS. Due to the latter's sixfold higher multiplicity this means that most of the population can readily dissociate directly to Cd($^1S_0$) and O,S($^3P_2$). The locations of the low lying CdO, CdS($a^3\Pi$) states has been confirmed theoretically (Peterson et al., 2007; Chambaud et al., 2008). Similarly, although the corresponding sulfide was previously quoted as D_0(Cd-S) $\leq$ 197 kJ mol^{-1} (Huber and Herzberg, 1979) and 169 (Barin and Platzki, 1995), theory now predicts a value of 108 kJ mol^{-1} (Peterson et al., 2007). Gaseous sulfide bonding is generally slightly weaker or at least equal to the corresponding oxide so some further adjustment may be necessary in the CdS value. However, this difference may be unique and a consequence of the potential energy locations. No gaseous dioxide or disulfide is known but CdO_2 and CdO were characterized in a Cd laser ablation/O_2 matrix isolation study (Chertihin and Andrews, 1997a), and CdS and CdSS similarly (Martin and Schaber, 1982).

Bond strengths for the chlorides are sparse, one spectral absorption method monitored the variation of Cd with temperature in a gaseous Cd/CdCl/$CdCl_2$ equilibrated system obtaining D_0(Cd-Cl) = 204 kJ mol^{-1} (Bruner and Corbett, 1964). This is consistent with approximate earlier spectral estimates (Huber and Herzberg, 1979). Three theoretical calculations have quoted values of 202, 181 and 155 kJ mol^{-1}, respectively (Liao et al., 1995; Shepler and Peterson, 2006; Cheng et al., 2007). Data for the dichloride is limited but is more reliable being derived from a heat of formation and sublimation.

TABLE 7.15 Stability trends for relevant gaseous molecules of the second/third row transition elements and uranium, kJ mol^{-1} [a]

Bond	Y	Zr	Mo	W	Ag	Cd	Hg	U
D_0(M-O)	712 ± 15	766 ± 10	527 ± 2	668 ± 20	200 ± 10	95 ± 10	18 ± 4	759 ± 10
D_0(OM-O)	374 ± 15	658 ± 20	636 ± 20	595 ± 30	–	–	–	730 ± 10
D_0(M-OH)	–	≈525	–	–	≈210	170 ± 20	40 ± 20	–
D_0(HOM-OH)	–	≈525	–	–	–	–	≈300	–
D_0(M-Cl)	≈500 ± 20	522 ± 25	370 ± 40	368.4	311 ± 2	180 ± 10	100 ± 3	439 ± 10
D_0(ClM-Cl)	≈480 ± 20	512 ± 25	444 ± 40	450 ± 40	–	370 ± 10	346 ± 7	481 ± 10
D_0(M-H)	275 ± 10	280 ± 30	≈207 ± 15	240 ± 20	211 ± 8	65 ± 10	36 ± 0.1	≈270 ± 25
D_0(M-S)	526 ± 10	570 ± 10	≈360 ± 30	476	210 ± 10	108 ± 15	24 ± 4	494 ± 15
Bond ratios								
OM-O/M-O	0.53	0.86	1.21	0.89	–	–	–	0.96
M-S/M-O	0.74	0.74	≈0.68	0.71	1.05	1.14	≈1.3	0.65
M-OH/HOM-OH	–	≈1	–	–	–	–	≈0.1	–
M-Cl/ClM-Cl	≈1.1	1.02	0.83	0.82	–	0.49	0.29	0.91
MCl/M-OH	–	≈1	–	–	≈1.48	1.06	≈2.5	–

[a] *See text for details.*

TABLE 7.16 Stability trends for gaseous Zn/Cd/Hg molecules, D_0 kJ mol^{-1} (see text)

	M-O	M-S	M-Cl	ClM-Cl	M-H	M-OH
Zn	155 ± 10	140 ± 10	229 ± 8	405 ± 10	82 ± 4	< 230
Cd	95 ± 10	108 ± 15	180 ± 10	370 ± 10	65 ± 10	170 ± 20[a]
Hg[b]	18 ± 4	24 ± 4	100 ± 3	346 ± 7	36.1 ± 0.1	40 ± 20

[a]*Estimated, see text.*
[b]D_0(*Hg-Br*) = 65 ± 2; D_0(*BrHg-Br*) = 303 ± 8; D_0(*HOHg-OH*) ≈ 300; D_0(*HHg-H*) ≈ 309.

These indicate a value of $D_{0\mathrm{Atomiz}}$ $(CdCl_2)_g$ = 555 or 545 ± 6 kJ mol^{-1} (Knacke et al., 1991; Kubaschewski et al., 1993). A reliable theoretical paper that reproduces this same heat of formation also indicates D_0(ClCd-Cl) = 381 kJ mol^{-1} (Shepler and Peterson, 2006).

As a result, reflective of Zn and Hg (Table 7.16) the second chloride bonding can be very significantly stronger, by about twofold with cadmium. Although uncertainties remain, values of D_0(Cd-Cl) = 180 and D_0(ClCd-Cl) = 370 kJ mol^{-1} presently appear appropriate possibly accurate to ± 10 kJ mol^{-1}.

CdH has been spectrally well characterized and its transitions can be a useful monitor. From a short extrapolation of the ground state vibrational levels a bond strength is suggested of D_0(Cd-H) = 65 kJ mol^{-1} (Huber and Herzberg, 1979). An earlier theoretical study confirmed the spectroscopic electronic state assignments and showed the shallowness of this ground $X^2\Sigma^+$ state, by far the weakest bound in the second row transition element hydrides (Balasubramanian, 1990). A more recent theoretical study reproduced the suggested bond strength (Shepler and Peterson, 2006). The weakening also is consistent with the Zn/Cd/Hg sequence. The infrared spectrum of gaseous CdH_2 has been characterized in a discharge and by matrix isolation and is considered to be more stable than the monohydride reflecting the mono- and dihalides but its strength remains unknown (Greene et al., 1995; Andrews, 2004; Shayesteh et al., 2005a). Little is known of the as yet uncharacterized *D*(CdOH). From the trends and expected similarity to the monohalide it might be approximately estimated at 170 ± 20 kJ mol^{-1}. Gaseous $Cd(OH)_2$, also expected from the trends to be more stable than the monohydroxide has only been studied by matrix isolation spectroscopy (Wang and Andrews, 2005g). It is predicted to be covalent unlike ionic $Sr(OH)_2$. A theoretical estimate in that study suggested an approximate value of about 417 kJ mol^{-1} for the combined strengths of the two Cd-OH bonds. Nevertheless, even in early mass spectrometric studies of CdO(s), O_2 and H_2O(g) at 970°C–1100°C the only gaseous cadmium species was Cd(g) with no hint of hydroxides (Glemser and Stocker, 1963).

Chemical kinetic constraints

The weakness of the CdO(g) bond again implies that although CdO(s) may be important, its vapor will be Cd(g) with a vapor pressure of the order of 5 ppmv (750°C), 20 ppmv(800°C), and 200 ppmv (900°C) (Coyle and Lewis, 1974; Miller, 1975). Consequently, reflecting zinc, the only potential gaseous cadmium molecule of relevance in cadmium's higher temperature chemistry will be the dichloride. However, like zinc and mercury, kinetic constraints can be expected concerning whether this can be formed via the weaker CdCl(g) or necessarily be confined to heterogeneous condensed phase formation with subsequent vaporization. Unfortunately, this commonality of Zn, Cd, and Hg has not been addressed from a kinetics viewpoint other than for Hg. There the difficulty of forming $HgCl_2$ in the gas phase was realized and heterogeneous chemistry was indicated.

$$\text{Zn, Cd, Hg} + \text{HCl} = \text{ZnCl, CdCl, HgCl} + \text{H} \quad \Delta\text{H}_0 = +199,\ 248,\ 328\ \text{kJ mol}^{-1}$$

$$\text{Zn, Cd, Hg} + \text{OH} = \text{ZnO, CdO, HgO} + \text{H} \quad \Delta\text{H}_0 = +269,\ 329,\ 406$$

Such reactions are obviously negligible in the time frame of combustion's imposing constraints and forcing the kinetics into equally slow three-body processes. This was initially invoked for Hg in full-scale coal combustion models and additionally could only be possible at low temperatures.

$$\text{Hg} + \text{Cl} + \text{N}_2 = \text{HgCl} + \text{N}_2 \quad \Delta H_0 = -100\ \text{kJ mol}^{-1}$$

Unfortunately, this kinetic constraint generally has been ignored and interpretations still accept the thermodynamic predictions that like Hg stress the gas phase importance of a dichloride dominance.

Laboratory combustion studies

Simple laboratory tests adding aerosol cadmium solutions to natural gas flames at about 1000°C confirmed several chemistries. Addition of cadmium nitrate produced predominantly $CdCO_3$ (96%) and CdO (4%) deposits on a glass fiber filter after sampling through a stainless steel probe. Additions of cadmium chloride produced deposits of the oxide and dichloride, all identified by their XRD spectra (Winter et al., 1994). A similar study examined the influence of steam, HCl and H_2S on evaporating Cd(s) in a heated tube with flowing H_2/Ar (Abascal et al., 2016). $CdCl_2$, CdO, and CdS were deposited on glass probes at different temperatures. The calculated equilibrium vapor pressures for $CdCl_2$ are about 42 ppmv (450°C) and 287 ppmv (500°C) indicating a very volatile nature but this does not resolve its formation mechanism whether heterogeneous, directly from the Cd(s) surface or created in the sampling lines or on the collection surface. The deposits were regarded as an elementary condensation process assuming CdO, $CdCl_2$, and CdS to be kinetically formed in the gas phase. This is now regarded as not possible and these species can be simply formed on reaching the surface and result heterogeneously from thermodynamic preferences. This applies also to any mass spectrometric monitoring. If something is detected it cannot reflect the gas phase unless there are kinetic processes in place to produce it there. Depending on the exact conditions, one thermodynamic calculation implied that $CdCl_2$(g) might be important only in the 580°C–880°C temperature range even if formed (Frandsen et al., 1994). Whether gas phase hydrolysis kinetics can convert any MCl_2 molecules stepwise into their correspondingly stable $M(OH)_2$(g) counterparts in the gas phase also is an intriguing speculation for future researchers. Again, it would now be expected when deposited heterogeneously onto a surface, these would modify to the oxide. A Japanese study of Cd thin films did note their easy corrosion. In the presence of HCl, this probably produces $CdCl_2$ on the surface directly. Its observation in leachate from fly ash in biomass combustion can either be significant (soluble $CdCl_2$) or minimal (insoluble CdO, $CdSiO_3$) (Hansen et al., 2001). Once Cd(g) deposits on ash the surface will control its preference for product formation. Also as it cools this might even convert to something that is more stable.

As a result, an overview of the available chemistry predicts that above 900°C the only gaseous form in combustion will be Cd(g) irrespective of conditions. This is interesting in that early flame atomic emission or absorption analysis methods were concerned about having a high degree of atomization for Cd. However, quantitative estimates of the degree of dissociation of Cd in analytical flames in this case always suggested a lower than unity value (De Galan and Samaey, 1970). This undoubtedly arose from the difficulty of the measurement especially as the resonant line of Cd is in the vacuum ultraviolet at 228.8 nm. It is now known with such flames that complete dissociation will be essentially instant. Rather surprisingly, such a spectral analysis still is being undertaken but now at low temperatures with a flame heated oven concept to actually try and monitor the rate and mechanism of dissociation of cadmium compounds when coupled to other instruments (Nagy et al., 2014). It has a high ionization potential and no reported ion chemistry in flames (Goodings, 1992). Similarly, undoubtedly due to its chemistry and being in the Zn, Cd, Hg periodic group, it has no effect on soot (Howard and Kausch, 1980), rated zero on a scale with Ba at 100 (Cotton et al., 1971). Nor does it appear to have any catalytic inhibition properties regarding combustion (Babushok et al., 2000; Linteris et al., 2008).

Combustion power plant data

As seen from the average fuel contents in Table 1.1, cadmium concentrations in coals are quite low and were not even measured in the initial NBS analyses of bituminous and subbituminous samples as standard reference materials (Rowe and Steinnes, 1977; Germani et al., 1980). A slightly increased presence is the case with biomass, but municipal waste and sewage sludge fuels have the potential for higher values which has raised environmental concerns for their combustion and is where more emphasis now is placed. Levels similarly are very low in pet coke and limestone (Lind et al., 1994; Cordoba et al., 2012). A recent study on the production of synthetic biogases illustrates that some of the trace metals can be present in much enhanced levels over those in natural gas. Cadmium can be one indicating several thousand-fold larger quantities (Cachia et al., 2018). In fuel combustion the main questions with trace components concern their volatility and distribution between the ash and gas phase at the higher temperatures and then their behavior as temperatures cool. The important question is what has transpired, what are they now, and how is the process concluded in the least disruptive and beneficial manner. Although the combustion literature on trace metals is extremely extensive, cadmium has not been of pivotal concern until recent years. As indicated already, it is only due to its higher potential levels in wastes and sewage sludge, and its apparent ease of leaching from ash that has now included it with other trace metals of concern. Nevertheless, although the form of cadmium's speciation present in fly ash or scrubber wastes remains unresolved, its overall distribution within combustion systems is now reasonably well defined. Coal combustion exhibits the widest range of temperatures and the fullest range of chemical opportunities for an element. With cadmium it was evident from initial work that it reflected a high volatility in 1000°C incinerators and showed a high 90% level of scavenging by the fly ash as it cooled with <0.1% stack emissions (Morf et al., 2000). Little is evident in the bottom ash or slag with the order of a 10-fold concentration increase on the fly ash (Klein et al., 1975; Huang et al., 2004). The enhanced release of Cd from the fuel at lower temperatures with higher chlorine content certainly does result in smaller amounts in the bottom ash (Zhang et al., 2008). This is compatible with $CdCl_2(s)$ formation in the fuel possibly via the oxide with subsequent release.

ESPs are very effective in fly ash collection (≈98/99%), and such samples indicate a predominant collection of cadmium (Coles et al., 1979; Galbreath et al., 2000) with increasing enrichment through the various collection fields (Huang et al., 2004). A recent analysis of five and six coal-fired plants in China with a variety of particle air pollution devices indicated overall collection efficiencies for Cd at 98.9%–99.9% (Deng et al., 2014a; Zheng et al., 2017). Whether such accuracy is possible may be questioned. Another such Chinese study was equipped with all the major air pollution control devices including an SCR, with the task of producing ultralow trace metal emissions of As, Se and Cd. Their overall efficiencies were listed as As (95.9%), Se (98.2%), and Cd (92.7%) numbers that are probably closer to reality (Chang et al., 2019). One modern Chinese large coal-fired plant recently indicated an even distribution of cadmium between the bottom ash, fly ash, and gypsum (Zhang et al., 2017a). With ESPs, there is always some slippage of the fine particles peaking in the 0.1–1 μm size range (Senior et al., 2000b). This resulted in an emission loss of 1.4% Cd in one Canadian coal-fired power plant (Goodarzi, 2006). If no further particle control devices are used, it is these fine particles that are released and of environmental concern. However, now with FGD scrubbing, one study with coal/pet coke co-combustion indicated minimal cadmium is emitted from the stack (Cordoba et al., 2012). One study of the proximity to a coal-fired plant equipped solely with an ESP indicated that local emission could be fivefold above the low natural background level for Cd (Jayasekher, 2009). This results from the fact that any small slippage fraction can be extensively enriched, concentrations increasing with smaller diameters. A study of the nano (0.03–0.06), ultrafine (0.06–0.1), and submicron (0.7–1.1 μm) fractions illustrated enrichment factors of 100-fold, approaching those of similarly volatile selenium (Wang et al., 2008). The micron size particles form a bimodal distribution with peaks at 0.05 and 2 μm. Although the smaller of these two peaks represents only about 6% of the slippage mass it was shown to contain one-third of the resulting cadmium (Kauppinen and Pakkanen, 1990). Waste incineration also reports that much of the Cd is found in the $PM_{1.0}$ size fraction (Lind et al., 2007). Addition of desulfurization scrubbing or fabric filter bags was reported to reduce this leakage to essentially zero (Meij, 1994). Other earlier reports had suggested final low levels of stack emission of 0.06% or 0.48% in two separate power plants (Aunela-Tapola et al., 1998). Recent medical testing of the population near a modern incinerator in Spain confirmed a lack of concern with regards to cadmium (Zubero et al., 2010). Nevertheless, the disposal of the various ash collections cannot be ignored as these display moderate leaching (George et al., 2008; Thorneloe et al., 2010). The resulting gypsum from the FGD control appears to have very low levels and be of no concern with minimal leaching risk (Zhao et al., 2018b). A recent study examined gypsum samples from 12 Chinese coal-fired plants confirming this (Hao et al., 2017). Nevertheless, it was found to be very mobile and as much ash should be captured as possible upstream on the other devices.

An examination of the potential chemistry suggests for the higher temperatures of coal combustion in furnaces or boilers that a total release of cadmium to Cd(g) occurs followed by chemical deposition and absorption/conversion on the ash as it cools with attempts to reflect its thermodynamic preferences. Deposition on ash from fluidized bed incineration of wastes with a freeboard temperature of 700°C–900°C was established by XRD, carbonate and oxide under normal operation and the dichloride with PVC (heavier Cl content) (Wey et al., 1999). Deposition of CdO(s) was noted at 1000°C onto silica/clay structures and remained as such (Lee et al., 2005). However, cadmium is absorbed by kaolin type materials at 1300°C indicating it can interact with Al/Si structures at the higher temperatures (Linak et al., 1995; Gale and Wendt, 2005; Yao and Naruse, 2005). Its theoretical interaction has been analyzed very recently (Wang et al., 2016f). Also absorption of $CdCl_2$(g) on various minerals indicated chemisorption on aluminum oxide and calcium oxide in the 750°C–950°C range (Masseron et al., 1999). However, this may not relate to full-scale combustion gases. In furnace-type ashing of coal samples at 850°C/950°C some unexpected retention of cadmium was noted and from its leaching resistance was possibly silicate formation due to the extended interaction time (George et al., 2008). Although other attempts to analyze ash particles by X-ray methods have hinted at the possible existence of the silicate, sulfate, oxide and chloride these could only examine large 40–200 μm particles from lower temperature waste incinerators and remain speculative (Camerani et al., 2004; Pinzani et al., 2002, 2004). Nevertheless, one might expect some Cd/Si/Al/Fe/P interactions depending on their concentrations and interaction time and efficiency. However, the ever presence of oxygen, chlorine and sulfur appears to induce deposition that ultimately converts to the stable sulfate with some possibly remaining as the dichloride. These are both water soluble and would easily leach.

Studies of cadmium with biomass combustion are scant and of less interest due to its low concentrations. One at 800°C did show that chlorine and sulfur enhanced Cd release and that some cadmium appeared retained in the ash even though the corresponding levels of Si/P/Al/Fe/Cl/S also were very low (Miller et al., 2003). Another noted 50% of the cadmium in the ash could be leached with water possibly also suggesting some insoluble oxide formation (Hansen et al., 2001). A recent study has shown an effect on the fly ash composition of reducing the flue gas temperature from 250°C to 150°C by passage through a heat exchanger. The fuel was biomass from metal contaminated soils burned in a small pilot-plant boiler. Boiler temperatures are not quoted but a 20% increase in cadmium's ash concentration appeared to occur and for many other elements as the temperature was so reduced through the heat exchanger. As to whether further metal deposition was possible for this increase in concentration remains questionable especially as similar larger increases were seen for such elements as Ti or Zr that are fully nonvolatile. It is possible that a particle segregation is occurring and the ash size distribution changing to favor the small enriched sizes at the lower temperature (Kovacs et al., 2018). It illustrates the care needed to resolve and interpret such observations.

Examples from incineration of wastes and sewage sludge with their higher Cd levels, and fluidized bed combustion (FBC) modes become a little more complex due to their functioning temperatures that hover in the 1000°C region and in the case of FBC incorporate limestone for SO_2 absorption that can have a greater Cd content than coal (Lind et al., 1994). With municipal waste incinerators, cadmium appears easily released from fuels above 300°C and can be complete at 400°C (Verhulst et al., 1996). The Cd release below 800°C with fluidized bed waste incineration now has been adequately modeled (Soria et al., 2013). On flue gas cooling, sulfur and chlorine have been shown to be important with a preferential formation of the water soluble sulfate (Jiao et al., 2011). If not formed directly, one can assume the possibility of oxide to chloride to sulfate ash conversion. One study obtained 92% of cadmium, leaching with hydrochloric acid (Tang and Steenari, 2016). Another study of ash from incinerators suggested no silicate formation and ash heat treatment indicates cadmium removal by 850°C (Jakob et al., 1995). One experiment heated waste incinerator ash at 1000°C for 3 hours with either a mix with $CaCl_2$ or with flowing Cl_2. In either case all the Cd in the ash volatized, probably confirming no silicate or complex binary metal oxide formation (Chan and Kirk, 1999). Incineration is obviously heavily controlled by chlorine and sulfur content (Jiao et al., 2013a). One recent incineration study examined the influence of added sulfur. This appeared to enhance the release of Cd at 750°C but the mechanism remains speculative and it may only be due to the formation of CdS(s) (Liu et al., 2017). Cadmium is readily leached from fly ash (Li et al., 2018), and in China the ash has to be processed before disposal (Wang et al., 2010a). There is little in bottom ash. However, when such ash was further mixed with additional Al/Si/Ca and heated to 760°C/940°C it appeared that the cadmium was retained (Jakob et al., 1995). Implications may be that silicate formation although plausible for cadmium is a slower process than other deposition formation chemistries.

Sewage sludge can typically have 10-fold more cadmium than coal, far more than allowed for agricultural usage. For this reason it is generally cofired with coal. Similarly, most of the cadmium is found in the ash and is effectively removed by high ESP efficiencies (Werther and Ogada, 1999; Amand and Leckner, 2004). There are clear indications that increased levels of chlorine induce more rapid release of the already volatile $CdCl_2$, but phosphorus can be an inhibitor (Luan et al., 2013). This was also seen with coal/waste cofiring, enhancements by chlorine, and reduced volatility by sulfur (Wu et al., 2013). Even in a FBC mode with bed temperatures of 800°C–850°C there is little cadmium in the bed, cyclones or scrubber and is concentrated on the small ash particles with about 0.4% emitted from the stack (Lopes et al., 2003; Van de Velden et al., 2008). Fired on a benchscale test at 800°C/900°C there appeared to be some retention of 25%–50% at the higher temperature. As to its nature remains speculative but sewage sludge is rich in Si/P/Al/Fe (Miller et al., 2004). In this connection, coal/sludge treated with Fe and Ca also was burned in an electric furnace at 800°C–1000°C. Higher levels of Cd retention were clearly apparent but suggested roles with the Si and possibly the Al remain speculative (Folgueras et al., 2007).

Consequently, until molecular speciation of ash particles are more fully analyzed for cadmium, the nature of combustion has to be assessed from the known chemistry and leaching experiments. It is clearly established that most of the cadmium in the fuel becomes present on the various fly ash components with the ultrafine $<2\ \mu m$ being non-negligible requiring removal by scrubbing or additional filter bags. In fact early incinerator work suggested that cadmium emissions could be reduced to near zero with such efforts

(Cesmebasi et al., 1988). Nevertheless, ash particularly from sewage sludge combustion should be regarded as toxic.

Although gasification/pyrolysis chemistry under reducing conditions has not been stressed above, it is expected to be very simple. Thermodynamic calculations have suggested that CdS(s) is stable but only to 380°C above which it is Cd(g) (Frandsen et al., 1994). More recent measurements of its dissociating vaporization indicate the vapor pressure of dissociated Cd(g) is only 2.8 ppmv at 700°C suggesting a slightly higher temperature for dissociation and possible chemideposition below such temperatures (Nakamura and Fuwa, 1987). As with Hg and Zn, a role for $CdCl_2$ formation would appear to be unlikely from Cd(g) and gaseous HCl under reducing conditions. Publications support such conclusions indicating cadmium is all Cd(g) above 425°C in a coal gasifier (Bunt and Waanders, 2008), above 600°C with a high sulfur coal (Scaccia and Mecozzi, 2012), and above 850°C in the steam fluidized bed gasification of biomass. Cadmium is suggested as being one of the more volatile of the metals and not redepositing above 380°C (Syc et al., 2011). In an IGCC gasifier fed coal/pet coke/limestone at 1200°C–1600°C cadmium was dominant on the fly ash with high enrichment factors of seven- to ninefold with evidence of CdS(s) on cooling (Font et al., 2010). Although kaolin type clays are known to absorb Cd under oxidizing conditions, it was noted that they do not function under reducing conditions (Tran et al., 2004).

As with Zn and Hg, published thermodynamic calculations have to be assessed very carefully for cadmium. It suffers a lack and completeness of thermochemical data and also ignores restrictive chemical kinetics that will most likely prevent the gas phase formation of species such as $CdCl_2$(g) and $Cd(OH)_2$(g) invoked in such calculations. One recent comparison of databases for cadmium helped to show the effects of inaccuracies and indicates the totally missing species (Lundholm et al., 2007). An interesting effort has been made to incorporate some data for the silicate and aluminate. Whether this is reliable or not remains to be seen but it produces an interesting pictorial display of the possible temperature ranges for formation of the plausible species for what appears to be a high chlorine coal (Liu et al., 2010).

In summary, reading the combustion data for Cd is difficult. There is a tendency to depend heavily on thermodynamic calculations to a point of accepting them as proven. Statements appear such as "$CdSiO_3$ is formed easily," remain no more than speculation (Liu et al., 2016c). Also because the gas phase components have not been measured, there remain large uncertainties concerning the details of Cd's release. Increased Cl fuel content tends to be reflected as a greater release but what this means is not clear, possibly enhanced rates, in that most of the Cd is released and ends on the fly ash anyway. An involvement of $CdCl_2$(g) has not been proven (Zhang et al., 2008) and its formation is constrained like Hg by the chemical kinetics, necessarily involving the weak CdCl(g) intermediary. On cooling, the situation is better understood with some information from ash particle analysis and leaching data. It has been suggested that the chloride is responsible for the observed leaching from waste incineration ash and is formed on cooling by surface heterogeneous deposition of any cadmium molecule (Jiao et al., 2016). Thermodynamic preferences will be dominant at this stage with potential conversions to the most stable deposits depending on available time and cooling temperature.

7.10 Calcium

Calcium is the first of the three main elements encountered in the periodic table labeled as the alkaline-earths that have very similar chemical natures. The other two are strontium and barium and all differ significantly from Be and Mg that are also in Group IIA with a similar electronic structure but not embraced by this term. Calcium is the most common being the fifth abundant element in the Earth's crust and similar in a level of occurrence to Fe, and sixth in the oceans being comparable to K in spite of its low solubility. In nature it occurs abundantly as the carbonate (limestone, calcite, dolomite) or sulfate (gypsum). Mined and processed in the many millions of tons year^{-1} its supply is relatively limitless. Uses are very many, including construction, agriculture, medical, glass production, in alloys and for sulfur absorption. There are few environmental issues concerning calcium, one has been the past use of $CaCl_2$ as a road deicer but this now seems to be found acceptable.

Health concerns with calcium

Calcium is important to living organisms, essential to leaves, shells, bones, teeth and many bodily mechanisms. It is the major metal element in the body (1.4% body mass) and the largest component of human incineration ash. An adult diet requires about 1–3 g day^{-1} of Ca and is nontoxic. Only with excessive or minimal intake does it produce any medical concern. The chloride is one compound that can be an irritant and has been used to sterilize male animals. Deficiencies can occur and there remain concerns with diets in many poor countries. One assessment in south east Africa concluded that the dietary Ca per capita was unlikely to be adequate (Broadley et al., 2012). Additionally, high Ca levels can interfere with body uptake of Zn (World Health Organization Trace elements, 1996). Because concerns for body Ca levels are somewhat rare, especially in the more developed countries, few health test centers include this in their regimen of elements. However, as a measure of health the US Mayo Clinic does suggest the calcium values: total serum ≈ 100 mg L^{-1}, serum ionized calcium ≈ 50 mg L^{-1}, and urine <200–250 mg/24 hour (Mayo Clinic, 2018). Calcium levels have appeared in several recent published surveys. Blood testing now appears favored in two recent findings. One reported a 5%–95% spread in values of 47–67 mg L^{-1} for 455 participants, ages 5–80 in central China (Zeng et al., 2019). The other monitored 30 healthy people in Italy finding average blood values of 54 mg L^{-1}, and also 430 ppmw in hair and 158 mg L^{-1} in urine (Oggiano et al., 2018). Other reported hair levels have indicated wide ranges, 130–4700 in Russia, and 2000–6000 ppmw in China questioning its value as a useful monitor (Drobyshev et al., 2017; Zhu et al., 2018). Because metal biology in the body remains a complex mix of elements, even calcium remains in the realm of speculation concerning various illnesses. It remains linked to possibly playing some role in Huntington's disease (Bichell et al., 2017).

Calcium's basic chemistry

Calcium's chemistry is predominantly divalent and ionic. It has a full repertoire of inorganic compounds many of which have an affinity for water and hydrated salt formation. However, the carbonate, oxide, hydroxide, sulfate, phosphate and sulfide are all only sparingly soluble in water. Its chloride and nitrate are very soluble. The physical properties of those most relevant to combustion are summarized in Table 7.17.

The metal (m.pt.842, b.pt.1484°C) is quite volatile (26 ppmv at 627°C) and does react with water. Like many elements, numerous of its basic inorganic molecules are not thermally stable and produce the stable CaO(s) known as lime. This is refractory (m.pt.2613, b.pt.3850°C) is nonvolatile (1 ppmv at 2200°C) but also reacts with water. Alkaline earth peroxides are known but like most are thermally unstable with regards to losing oxygen, stability increasing with atomic number. CaO_2(s) has various uses, including being a baking and bread making ingredient, but dissociates at 200°C–300°C, which is accelerated by the presence of water and is also converted to the dihydroxide by H_2. The sulfate, $CaSO_4$ decomposes below its melting point of 1150°C (Ostroff and Sanderson, 1959) having a vapor pressure of 10 ppmv at 925°C in a Knudsen cell-mass spectrometric study (Lau et al., 1977; Hildenbrand et al., 1990). Because of this, it is a controlling temperature with fluidized bed combustion (FBC) that is removing SO_2 by CaO(s). The sulfite, $CaSO_3$, is less stable and has also been studied mass spectrometrically dissociating above 450°C to SO_2(g) and the oxide (Cubicciotti et al., 1977). The carbonate, $CaCO_3$, an ore of the oxide, also dissociates most extensively above about 600°C having a CO_2 vapor pressure of 10 ppmv at 430°C and 500 ppmv at 540°C (Stern and Weise, 1969; Stern, 2001), by what is a reversible reaction. The nitrate, $Ca(NO_3)_2$ dissociates above 500°C (Mahadevan et al., 2002), and the nitrite, $Ca(NO_2)_2$ at about 280°C (Stern, 1972). The hydroxide that is only slightly soluble in water and is known as slaked lime losses water on heating at 580°C producing the oxide, also in a reversible process. Although the hydride CaH_2 is stable, it reacts with water changing to the dihydroxide and dissociates to its elements on vaporization (Ehlert et al., 1968). Of its thermally stable molecules appropriate to combustion CaO, $CaCl_2$, CaS, Ca_2SiO_4, and ortho $Ca_3(PO_4)_2$ or pyrophosphates $Ca_2P_2O_7$ are noteworthy. CaO(s) (m.pt.2613°C) heated alone vaporizes at high temperatures (1800°C for ppmv concentrations) predominantly by dissociation to the elements but with oxygen present produces a mixture of Ca(g) and CaO(g) in similar concentrations of about 1 ppmv at about 2200°C (Lamoreaux et al., 1987). The chloride (m.pt.772°C) sublimes having pressures of 10 ppmv at 930°C and is stable to higher temperatures (Gurvich et al., 1994). It is hygroscopic and the presence of water does reduce its stability. Like the oxide, the sulfide is very stable (m.pt.2525°C) and similarly dissociates on vaporization above 1400°C to a mixture of gaseous Ca, CaS, S, S_2 (Berkowitz and Marquart, 1963; Marquart and Berkowitz, 1963; Colin et al., 1964). However, on heating above 500°C it is oxidized in air to $CaSO_4$ that then starts to dissociate to CaO (Madarasz et al., 1996; Qiu et al., 1997). As with other metals the possibility of mixed metal oxide reactions occurring is always possible. Normally they will have a refractory or ceramic nature characterized by a high melting point. Calcium metasilicate, $CaSiO_3$ (m.pt.1540°C) is used as a high temperature insulator and a substitute for asbestos. $CaAl_2O_4$ has a similar melting point and can form glasses. The phosphates also can be encountered with melting points of the order of 1670°C.

TABLE 7.17 Properties of calcium's condensed phase molecules relevant to combustion, °C.[a]

m.pt. b.pt. vapor pressure
Ca 842 1501, reacts with water, 26 ppmv/627°C, s
CaO 2927 3850, reacts with water, nonvolatile 1 ppmv/2200°C, s
CaO_2 >156 dissoc, reacts with water, ss
$Ca(OH)_2$ 835, dissoc >450 to CaO, s
$CaCO_3$ (calcite, hexagonal) 1339, dissoc >897, 10 ppmv/430°C, ss
(aragonite, orthorhombic) 825, changes to calcite >450°C, ss
$Ca(NO_3)_2$ 561, dissoc >500, s
$Ca(NO_2)_2$ dissoc >280, s
$CaSO_4$ 1450, dissoc >950 to CaO, 10 ppmv/925°C, s
$CaSO_3$ dissoc >386, s
$CaCl_2$ 772 2006, hygroscopic, sublimes, 10 ppmv/930°C, s
CaS 2525, dissoc >1400 oxidized in air, ss
CaH_2 816, dissoc on vaporization, reacts with water
$CaAl_2O_4$ 1600, reacts with water
CaB_2O_4 (meta) 1160, s
CaB_4O_7 990, ss
$CaCr_2O_4$ 2097
$CaFe_2^{III}O_4$ 1216, ins
$CaMoO_4$ 1445
$Ca_3(PO_4)_2$ (ortho) 1810, ins
$Ca_2P_2O_7$ (pyro) 1358, ins
$CaSiO_3$ (meta) 1544, ins
Ca_2SiO_4 (ortho) 2130, ss
$CaTiO_3$ 1970, ins
$Ca(VO_3)_2$ (meta) 778, ss
$Ca_2V_2O_7$ (pyro) 1015, ins
$Ca_3(VO_4)_2$ (ortho) 1390, ss
$CaZrO_3$ 2325, ins
$CaWO_4$ 1580

[a]*Gurvich et al. (1994), see text.*
dissoc, dissociates; *s*, water soluble; *ss*, slightly soluble; *ins*, water insoluble.

The titanate, $CaTiO_3$, is even more stable (m.pt.1850°C), and a ferrite $CaFe_2O_4$ also has been mentioned. An interesting Russian study has recently characterized the ferrospheres that can be recovered from high calcium fly ash. This is ash rich in Ca, Fe and Si and contains calcium ferrate/ferrite phases (Sharonova et al., 2015). $Ca_2Fe_2O_5$ also is being considered to replace Fe_3O_4 in the "steam-iron" chemical looping mechanism for producing H_2 (Ismail et al., 2016). Some of these mixed oxides are strongly bound and thermally stable. In fact in high temperature mass spectrometric studies of their vapors, gaseous $CaSiO_3$, $CaTiO_3$, and $CaPO_2$, for example, have been monitored (Lopatin and Shugurov, 2008, 2013). In flame atomic or absorption spectral analyses, Al was found to be an interferent for the alkaline earth metals undoubtedly through a Ca/Al interaction in the flame gases (Luecke, 1992a). CaO(s) is quite reactive in that at lower temperatures it can absorb CO_2, HCl, NO_2, H_2S, phosphorus oxides and SiO_2 to reconstitute its appropriate molecules.

The dissociation energy of CaO and its meaning

In combustion, it is often not realized in cases of dissociation, that the ionic condensed phase structure generally always produces neutral atoms. The dissociation to ions takes significantly more energy due to the ionic attraction and so utilizes a changeover from an ionic potential energy curve to a lower energy one correlating to the neutrals. An avoided curve crossing occurs to facilitate this and is well established as an example by the alkali halides (Hosseini et al., 2009). Consequently, the ions monitored in flames do not result from ionic salt dissociation but are from mechanisms of either a thermal ionization or chemiionization nature. However, it is this ionic nature that has posed severe demands on theory and it is only in recent years that a better understanding of the gas phase behavior of ionic molecules has emerged. This has been particularly relevant to the dissociation energy of CaO(g), possibly one of the most studied diatomic molecules. It is more ionic than BeO or MgO but again being a singlet ground state $CaO(X^1\Sigma^+)$ on dissociating needs to correlate to allowed product states. Although this is the second limit $M(^1S_0) + O(^1D_2)$ for M = Be and Mg it now changes slightly for the alkaline earths. As seen with the potential energy curves for CaO illustrated in Fig. 7.4, the lowest spin allowed energy products for $CaO(X^1\Sigma^+)$ becomes $Ca(^3P_u) + O(^3P_2)$, in this case preferred, being 8.5 kJ mol^{-1} lower in energy than the $O(^1D_2)$ channel indicated for Be and Mg. It is the low-lying $CaO(a^3\Pi)$ that correlates to the ground state dissociation limits.

Moreover, the $Ca(^3P_1)$ is a metastable excited state ($\tau = 0.34$ ms) but can collisionally relax (Drozdowski et al., 1997). This was initially illustrated 44 years ago (Field, 1974) together with the fact that the lowest lying excited electronic state of CaO, the $a^3\Pi_u$, now lies at an energy ($T_0 = 98$ kJ mol^{-1}), much higher than that of MgO as also is confirmed by more recent theory (Khalil et al., 2011, 2012). In other words, at 2000°C, the metastable $CaO(a^3\Pi_u)$ (radiative lifetime seemingly still unknown), now can only constitute a 1.6% contribution to the CaO(g) population with this proportion decreasing rapidly to lower temperatures. Consequently, unlike magnesium, the $a^3\Pi_u$ state can play no kinetic population role in the dissociation chemistry. As a result, the uncertainties that center from the scattered measurement results for the CaO bond strength need an alternate explanation. The most recent evaluation of the data recommends D_0(Ca-O) =

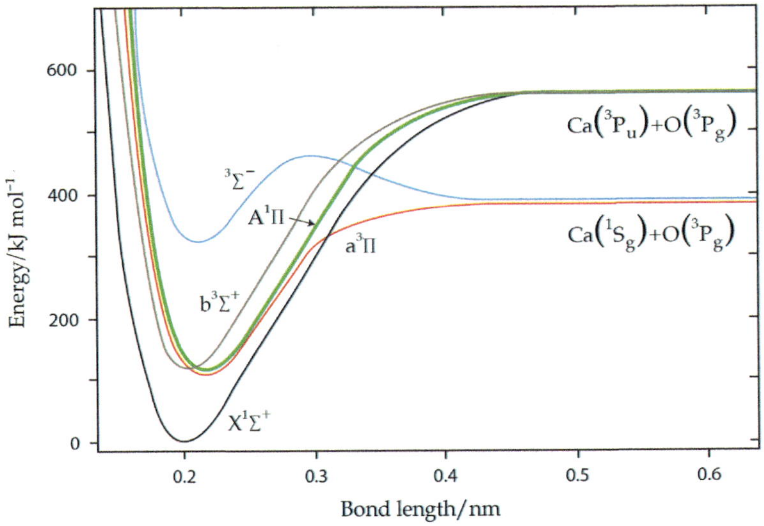

FIGURE 7.4 Approximate low-lying potential energy curves for gaseous CaO. Source: *Redrawn and scaled from data of Khalil, H., Le Quere, F., Brites, V., Leonard, C., 2012. Theoretical study of the rovibronic states of CaO. J. Mol. Spectrosc. 271, 1–9.*

385 ± 10 kJ mol^{-1}, which emphasizes the results from Knudsen cell, mass spectrometric analyses (Gurvich et al., 1994). However, it is to be remembered that these relate to a thermochemical equilibrium balance which is not bound by the more energetic spectroscopically forbidden transition that is written as

$$CaO(X^1\Sigma^+) = Ca(^1S_0) + O(^3P_g) \quad D_0 = 385 \text{ kJ mol}^{-1} (\text{spin forbidden})$$

However, this disguises the actual reaction dynamics that facilitate the process. This magnitude now has been confirmed by numerous theoretical studies that span values from 385 to 406 kJ mol^{-1} (Field, 1974; Langhoff et al., 1986; Harrison et al., 2002; Sullivan et al., 2003; Haworth et al., 2005; Vasiliu et al., 2010; Khalil et al., 2011). These generally either calculate the dissociation of the $X^1\Sigma^+$ state to the second limit then subtract the energy of excited Ca(^3P_u), or by calculating the dissociation value for the CaO($a^3\Pi$) and then add the known $X^1\Sigma^+ - a^3\Pi$ separation energy. They basically indicate that dissociation to other products will have the following related values:

$$CaO(X^1\Sigma^+) = Ca(^3P_u) + O(^3P_g) \quad D_0 = 385 + 181 = 566 \text{ kJ mol}^{-1} (\text{allowed})$$
$$CaO(a^3\Pi) = Ca(^1S_0) + O(^3P_g) \quad D_0 = 385 - 98 = 287 \text{ kJ mol}^{-1} (\text{allowed})$$

Efforts have been made to utilize thermochemical cycle values. These are based on knowing the heat of formation of CaO(s) and the heat of sublimation of Ca(s), which are well known, but it critically needs the heat of sublimation of CaO(s) which is not known due to its dissociative vaporization. Also the result of such a study is simply another equilibration exercise and failed for CaO and SrO (Drummond and Barrow, 1951). Early flame spectroscopic studies for CaO (Schofield, 1967) were large, for example, by 55–79 kJ mol^{-1} (Lagerqvist and Huldt, 1954) and 95 kJ mol^{-1} over the recommended equilibrium value raising concerns (Veits and Gurvich, 1956). These are still well below a 566 kJ mol^{-1} range for the allowed direct transition but were later seen to be unacceptable due to the presence of the gaseous dihydroxide in such flames as well as some CaOH(g) that was ignored (Sugden and Schofield, 1966). However, a larger than expected result is noted in numerous other cases and does appear to relate to the actual dynamics and is extremely important. Consequently, for thermochemical equilibrium calculations the recommended bond strength value appears reliable and relates to $CaO(X^1\Sigma^+)$. For real time measures the results can obviously indicate chosen controlling mechanistic channels. It had been calculated theoretically that the spin–orbit coupling element between the a/X states for CaO was very small so curve crossing to the lowest ground state products remains forbidden for $X^1\Sigma^+$ (Stagg and Richards, 1974). The most recent spectroscopic analysis examined the coupling and perturbations between these low-lying states noted no predissociative curve crossing (Khalil et al., 2012). They did note the intriguing way that the bond dipole moment changes with ion separation and bond breaking (Khalil et al., 2011). The $X^1\Sigma^+$ and $a^3\Pi$ do cross at the low vibrational levels but do not appear to perturb one another significantly. If so, some flame derived values would have produced a 385 kJ mol^{-1} value. However, there also remains the possibility of spin–orbit induced predissociation by the repulsive $^3\Sigma^-$ state that crosses the $X^1\Sigma^+$ curve and is the only other state that correlates with ground state products and was suggested for the similar case of BeS (Larbi et al., 2010). Depending on the curves exact and uncertain shape, as seen in Fig. 7.4, this could lead to values slightly higher than the first limit. If efficient this provides a mechanism but such a $^1\Sigma^+/^3\Sigma^-$ interaction appears to violate Kronig's selection rules concerning both spin and symmetry (Herzberg, 1963). Additional considerations considering complexities of forming the metastable $Ca(^3P_u)$ in the dissociation also appear unreasonable. Extensive relaxation studies for this state are available that indicate physical and reactive quenching on timescales of μs in atmospheric pressure systems (Pasternack and Dagdigian, 1978; Nicks et al., 1982; Husain and Schifino, 1983; Husain and Roberts, 1985, 1986). As a result it appears that the measurement of *D*(Ca-O) may be system dependent as seen with the other group II elements and further studies are needed to examine the extent of the possible $^3\Sigma^-$ curve hopping role.

Gas phase stabilities and transport

As seen from Table 7.17, CaO is a major oxide formed from many of its compounds on dissociation. However, it is not a volatile oxide needing temperatures of about 2000°C for vaporization. Calcium's gas phase is similar in nature to that of the alkali metals being controlled to a certain extent by its strongly ionic nature. Its $CaO(X^1\Sigma^+)$ ground state had been noted in matrix isolation measurements of its infrared spectrum (Ault and Andrews, 1975a; Andrews et al., 1994). As seen in the previous Table 7.10, it is the weakest of the three alkaline earth oxides with a bond strength of about $D_0(\text{Ca-O}) = 385 \pm 10$ kJ mol^{-1}.

The gaseous superoxide, CaO_2, has been examined by matrix isolation infrared spectroscopy indicating bent and cyclic structures (Andrews et al., 1994, 1996). Theoretical estimates based on a C_{2v} $^1A_1$ cyclic structure gives a bond strength of $D_0(\text{Ca-O}_2) = 226 \pm 15$ kJ mol^{-1}(Bauschlicher et al., 1992; Rosi et al., 1998). More recent calculations predict $D_0(\text{Ca-O}_2) = 216\ \text{kJ mol}^{-1}$ for the $^1A_1$ configuration and $171/182\ \text{kJ mol}^{-1}$ for $^3A_2$ (Nien et al., 1993; Broadley and Plane, 2010). Their values are supported by corresponding acceptable calculated values for the heats of formation of CaO, CaOH, and $Ca(OH)_2$. Being essentially a peroxide of a divalent metal atom it is unusually stable although the full nature of its bonding remains unclear yet probably is substantially ionic. Kinetic studies of the termolecular recombination of Ca with O_2 in He or N_2 indicate that reaction does occur and appears to be the only reasonable channel because reaction to CaO is $109\ \text{kJ mol}^{-1}$ endothermic. As a result rate constants have been measured at the lower temperatures where this termolecular reaction is dominant (−57°C to 830°C) (Nien et al., 1993), (30°C−460°C) (Vinckier and Remeysen, 1994), and (23°C−350°C) (Campbell and Plane, 2001), its occurrence in these ranges supporting its stability. A measurement suggesting the reaction below indicated a high efficiency.

$$CaO + NO_2 = CaO_2 + NO$$

For exothermicity, this would need $D_0(\text{OCa-O}) > 300\ \text{kJ mol}^{-1}$ but could be a different molecular structure for the CaO_2 from the termolecular insertion type interchange (Broadley and Plane, 2010). As a result, the nature of the second bond in CaO_2 remains uncertain. Nevertheless, although a potential player in upper atmospheric chemistry (Broadley and Plane, 2010), a role in combustion is unlikely due to its reactivity with water or hydrogen and easy conversion to the more stable CaO.

Fortunately, controversies with other gaseous species are more minimal. In isolation CaO(s) has a low volatility that is mainly dissociative as found to differing degrees with all the Group IIA oxides. However, in the presence of water vapor, vaporization is enhanced by ablation forming $Ca(OH)_2(g)$ directly as with other Group IIA oxides (Gurvich et al., 1994).

$$CaO(s) + H_2O(g) = Ca(OH)_2(g) \quad \Delta H_{298K} = 278 \pm 15\ \text{kJ mol}^{-1},\ K_{1980K} = 1.0 \times 10^{-4}$$

As a result, an equilibrium vapor pressure of $Ca(OH)_2(g)$ will be about 10 ppmv at 1700°C in the presence of water at 10% atmospheric pressure and has been sufficient for transpiration measurements to be made (Matsumoto and Sata, 1981; Sata et al., 1982). Hydrogen flame studies had previously established that the dihydroxide appeared to be the dominant gas phase component (Sugden and Schofield, 1966). A later more detailed study confirmed this in a 2200°C flame indicating a close to equilibrium gaseous distribution of 2% Ca, 0.3% CaO, 27% CaOH, 68% $Ca(OH)_2$ (Jensen and Jones, 1978). Values for CaOH's bond strength were obtained by monitoring thermal Ca* and CaOH* spectral emissions from flames and also by a Knudsen cell-mass spectroscopic method utilizing $CaO(s) + H_2$ equilibrations and from theoretical calculations. Based on these values $D_0(\text{Ca-OH})(g) = 386 \pm 15\ \text{kJ mol}^{-1}$ has been recommended (Gurvich et al., 1994) placing more weight on the two mass analysis studies that themselves differed by about $10\ \text{kJ mol}^{-1}$ (Murad, 1981; Farber et al., 1987). For some reason, flame values are slightly larger (Starovoitov et al., 1977) as are theoretical values (Bauschlicher et al., 1986; Haworth et al., 2005; Vasiliu et al., 2010) but fall within the error bounds. Being linear and rather like a diatomic, $CaOH(X^2\Sigma^+)$ is spectrally well characterized in emission and fluorescence (Human and Zeegers, 1975; Blackburn et al., 1978) and has no lying electronic states that might perturb analyses (Allouche and Aubert-Frecon, 1994; Taylor et al., 2005). It is regarded as an ionic radical. Also the possible alternate structure of HCaO is not favored being about 240 kJ mol^{-1} less stable (Kong and Boyd, 1996).

The dihydroxide is seen to be an important component of any volatile calcium. Its electronic structure and infrared spectrum were analyzed by matrix isolation (Wang and Andrews, 2005). Evaluations of its thermochemical stability indicate results from a transpiration measure, flame and mass spectrometric methods. A value of $\Delta H_{f298K} = -598$ kJ mol^{-1} has been recommended (Gurvich et al., 1994). Coupling this to the bond strength for CaOH(g) implies D_0(HOCa-OH) = 462 ± 20 kJ mol^{-1} a substantially strong molecule thermodynamically. Theoretical estimates of ΔH_{f298K} fall within 15 kJ mol^{-1} of that recommended but if coupled to their own calculated values for D_0(Ca-OH) come closer to confirming this D_0(HOCa-OH) value (Schulz et al., 1999; Sullivan et al., 2003; Haworth et al., 2005; Vasiliu et al., 2010).

Like the alkali chlorides, calcium chloride, $CaCl_2$, is volatile, stable and only hydrolyzed by water at elevated temperatures (≈1000°C). It has been shown to have a linear structure (Levy and Hargittai, 2000). At high temperatures the monohalide has also been characterized. It has a rich optical spectroscopy monitored in emission, absorption and fluorescence (Huber and Herzberg, 1979). It is seen to have no low lying electronic states (Allouche et al., 1993). Slightly, more stable than the corresponding monohydroxide, a value of D_0(Ca-Cl) = 400 ± 6 kJ mol^{-1} is recommended (Gurvich et al., 1994). Additional more recent studies support this utilizing a chemiluminescent threshold method, 405 ± 9 kJ mol^{-1} (Menendez et al., 1993), and theoretically a value of 406 kJ mol^{-1} (Haworth et al., 2005). From reliable measurements of the heats of formation and sublimation of $CaCl_2$(s), these imply an accurate value of $\Delta H_{f298K} = -485 \pm 7$ kJ mol^{-1} for $CaCl_2$(g), or a total atomization energy D_0 of 902 kJ mol^{-1} for the sum of the two bond strengths (Gurvich et al., 1994). As a result, the recommended bond strength for the second bond is D_0(ClCa-Cl) = 502 ± 9 kJ mol^{-1} accepting that above for the monochloride, and in combustion $CaCl_2$(g) will be very stable being comparable to the strong bonding of H–OH and O–O. The value is reasonably supported by theory (Wang and Schwarz, 1998; Haworth et al., 2005). Examining the trends in these values for the hydroxides and halides the ratios of the values for Mg through to Ba are remarkably similar further supporting the quoted values (Table 7.10).

Although CaH(g) has a rich spectroscopy (Huber and Herzberg, 1979), it is weakly bound D_0(Ca-H) = 163 ± 2 kJ mol^{-1} (Gurvich et al., 1994) and will play no role in combustion. The value is derived from a spectral predissociation limit. This was supported theoretically by an older calculation and quite accurately by more recent ones that also consider the various potential energy curves for numerous lower states (Fuentealba et al., 1987; Leininger and Jeung, 1995; Haworth et al., 2005; Shayesteh et al., 2017). The dihydride is only known from its characterization in a matrix isolation measurement (Wang and Andrews, 2004b). Theoretical calculations predict a stronger bonding than the monohydride, D(HCa-H) ≈250 (Haworth et al., 2005), and have analyzed its structure and potential energy surfaces (Koput, 2005). The other species, CaS(g), also more appropriate to reducing conditions, has been mentioned already as vaporizing from the solid phase like its oxide to Ca(g) and CaS(g). Its CaS structure was analyzed in a matrix isolation experiment (Martin and Schaber, 1982). As with all the other Group IIA sulfides and oxides, the important aspects tend to be where the low-lying $a^3\Pi$ state is located, and to what do the measured dissociation energy results relate. In this case, although not yet well fixed it appears that this first low-lying $a^3\Pi$ state that correlates to ground state atoms is at about 6600 cm^{-1} (79 kJ mol^{-1}) similar to BeS and SrS. This implies that at equilibrium at 2000°C the $a^3\Pi$ population will be about 9% that of the total CaS. A possible border line case value where it might make slight contributions in some cases to the dissociation dynamics. Nevertheless, at present, databases mainly have reworked the results of the two available Knudsen cell-mass spectrometric studies. These recommend a value of D_0(Ca-S) = 329 ± 10 kJ mol^{-1} (Gurvich et al., 1994; Barin and Platzki, 1995; Chase, 1998). This refers to dissociation to ground state atoms although this is spectroscopically spin-forbidden. The value is supported by theoretical calculations for such ground state products indicating $D_0 \approx 322$ kJ mol^{-1} (Partridge et al., 1988; Haworth et al., 2005).

$$CaS(X^1\Sigma^+) = Ca(^1S_0) + S(^3P_g) \qquad D_0 = 329\ kJ\ mol^{-1}\ (spin\ forbidden)$$

$$CaS(X^1\Sigma^+) = Ca(^1S_0) + S(^1D_g) \qquad D_0 = 329 + 111 = 440\ kJ\ mol^{-1} allowed)$$

$$CaS(a^3\Pi) = Ca(^1S_0) + S(^3P_g) \qquad D_0 = 329 - 79 = 250\ kJ\ mol^{-1} (allowed)$$

As seen from these relationships, the ground state correlates to the second dissociation limit which in this case is to $S(^1D_g)$ (energy content 111 kJ mol^{-1}). Although this atomic state is metastable with a long radiative lifetime it is rapidly quenched collisionally by many gases (Schofield, 1979). Consequently, fully equilibrated systems will relate to the first spin-forbidden reaction. Whether the actual dissociation dynamics proceed through any of these reactions remains unknown. Also as discussed with all the other group II diatomic oxides or sulfides, which share this dissociative dilemma, whether a role of potential energy curve hopping to the $a^3\Pi$ dissociative curve or spin–orbit induced predissociation utilizing the repulsive $^3\Sigma^-$ state occurs will only become apparent with further measurements. This has been discussed above and is illustrated in Fig. 7.4 for CaO. If the dynamics are invoking only the allowed dissociation transition a value of 440 kJ mol^{-1} remains possible under certain experimental conditions. As noted in Table 7.10, the recommended ratio of *D*(CaS)/*D*(CaO) follows the normal tracking of the alkaline earths in the periodic table (Hauge and Margrave, 1972). In this Group IIA, the CaS shows the largest strength relative to its oxide. Although of interest to chemistry, the sulfide will be of little importance to combustion but could arise in dry/oxygen free gasification and pyrolysis. Otherwise the CaS(g) will be vulnerable to conversion to the dihydroxide with water, and be readily oxidized and changed to the dichloride with HCl.

Laboratory combustion systems

As a result of these chemical stabilities, flame gases to which calcium salts were added were noted to contain calcium in a complex mixture of Ca, CaO, CaOH, $Ca(OH)_2$, CaCl, $CaCl_2$, with low levels of the ions Ca^+ and $CaOH^+$. The distribution would depend on temperature, water and chlorine content. Nevertheless, by utilizing a series of hydrogen flames in which the radical concentrations of H, OH and O were known and chlorine absent it was possible in numerous studies to obtain reasonable thermochemical values for the CaOH(g) and $Ca(OH)_2$(g) (Gurvich et al., 1994). Two detailed flame studies, but from different groups, examined the effects of added chlorine obtaining independent similar conclusions. At low levels of chlorine, linear effects were observed consistent with monohalide formation and as levels were increased the dichloride became more dominant (Schofield and Sugden, 1971; Gurvich et al., 1974). Also both analyses implied that more than one monohalide compound was present. When included, values were derived for D_0(Ca-Cl) and D_0(ClCa-Cl) close to those herein recommended. As a result, although not rigorously proven by mass spectrometry, it seems probable that the dominant dihydroxide can become partially chlorinated producing the speculated Ca(OH)Cl molecule. Consequently, the thermochemical values now known do adequately describe calcium's behavior in the gas phase in combustion. This was important because quantitative spectral analysis for calcium prior to the ICP had utilized flame photometry in atomic emission, absorption and fluorescence modes. With calcium it was realized that the amount of free atomic calcium even in C_2H_2/air flames at 2200°C was only about 3%–14% (De Galan and Samaey, 1970; Morozov et al., 1973). Moreover, many interferences could be encountered with such analyses that are now understood in terms of induced disturbances of the calcium distribution in a sample from that of the calibration solution. That calcium's distribution could be modified by chlorine was clearly evident in a now 55-year-old Russian publication that noticed how the Ca*, CaOH* and CaCl* spectral emissions changed on adding CCl_4 to a C_2H_2/air flame containing calcium (Zhitkevich et al., 1963). Also large additions of chlorine were seen to be linearly correlated with a decrease of Ca if interpreted as a result of dominant $CaCl_2$ formation (Kantor and Ernyei, 1994). As a result, in fuel combustion, any calcium that enters the gas phase will be predominantly transported as hydroxide or chloride forms with minor contributions of Ca and CaO. Ca, Sr and Ba salts all affect soot formation in flames. However, Ca is the weakest, rated 2.0 against Ba 100 (Cotton et al., 1971; Bulewicz et al., 1975; Haynes et al., 1979; Howard and Kausch, 1980). Tentatively explained by its lower level of ionization, the mechanism is thought to center on the chemiionization ions. A most recent study in this connection utilized laser-induced incandescence and elastic light scattering to monitor such soot formation suppression on adding seven metal salt solutions to premixed ethylene/air sooting flames. This confirmed the lower efficiency with calcium indicating only a smaller amount of light scattering soot formed (Simonsson et al., 2018). Additionally, the three alkaline earth elements are known to catalytically inhibit flames by free radical recombination mechanisms (Bulewicz and Padley, 1971; Cotton and Jenkins, 1971; Linteris et al., 2008). The effect by Ca is again the weakest of the three and still remains not fully understood.

The alkaline earth atomic ions are novel in that on losing an electron they become isoelectronic with the corresponding alkali atoms. It was partly through noting that Ca^+, Sr^+, and Ba^+, optical spectral emissions could be monitored in flames, in spite of their large thermal ionization potentials that their chemiionization mechanisms producing also their MOH^+ ions were discovered (Schofield and Sugden, 1965). Although much lower in concentration than the neutral species, this ionization that could also be a lower temperature source of electrons did find many uses and was extensively studied (Goodings, 1992). The kinetic rates were established (Jensen, 1968; Hayhurst and Kittelson, 1974; Cohen et al.,

1975). It was found that the efficiency was in the order Ba >Sr >Ca explaining barium's optimal role in soot suppression.

Some current concerns with calcium in fuels center on its release with sodium and iron in oxy-fuel combustion of high content alkali coals (Wang et al., 2018a). In the laboratory tube furnace experiment at 600°C–1300°C it was noted that calcium was in the ash as the carbonate, sulfate aluminate and silicate.

A remaining question concerning deposition of gaseous calcium as flue gases cool, concerns their condensed phase thermodynamic preferences and the composition of the resulting ash or substrate.

Combustion power plant data

As indicated a chemistry is seen where calcium may partly vaporize either as $Ca(OH)_2(g)$ by its ablation from CaO(s) by water and also as $CaCl_2(g)$ by its volatility. Within the gas phase the $Ca(OH)_2(g)$ is stable but even this may convert to the more stable dichloride, $CaCl_2(g)$, depending on the extent of chlorine in the system and its balance with water. Whether full-scale combustion follows such expectations will now be appraised. Due to its natural abundance it is not surprising that calcium is a major component of the noncombustibles in most fuels including municipal wastes and sewage sludge, but is slightly less in the biofuels (Table 1.1). It is generally at very low negligible levels in pet-coke (Cordoba et al., 2012). Questions to answer are its nature in ash; are there significant gaseous transport components in combustion, and are combustion emission controls adequate? It was realized from an early time that calcium could play an important role in defining the nature and physical properties of the ash product. Initially, particularly with coal combustion, problems centered on understanding the practical problems of fouling, sintering, slagging, and agglomeration in coal boilers. Calcium generally was labeled as nonvolatile and fully retained in some ash form (Raask, 1985). It is still a concern in some coal-fired boilers regarding the effects of Ca and Fe on modifying melt temperatures (Dai et al., 2013). Additionally there are Chinese coals rich in sodium and calcium that cannot be burned alone without severe fouling (Wang et al., 2017b). This has now been resolved in one case by adding kaolin to absorb some of the sodium and calcium. General studies have reflected this problem confirming the high concentrations of calcium in some ash distributions either with coal or biomass (Klein et al., 1975; Meij, 1994; Querol et al., 1995; Kauppinen and Pakkanen, 1990; Jensen et al., 1997). Moreover, there was no sensitivity to size for the then measured $>0.3\ \mu m$ ash particles (Coles et al., 1979; Querol et al., 1995). Equilibrium calculations reflected a condensed phase chemistry as discussed above involving the possible formation of calcium silicate or mixed metal silicates with either Fe, Al, Mg, or phosphates at the higher temperatures and the sulfate or oxide on cooling (Poole et al., 2008; Bourgel et al., 2011; Furimsky, 2000).

Mass balance studies in two Danish coal plants (Nielsen et al., 2002) and in the Netherlands' plants (Meij, 1994) clearly have indicated that most of the calcium is removed by the ESPs (99.9% efficiency) and depleted to very low levels by the FGD scrubbers (90%). Stack emissions were considered to be on a 0.1% scale and possibly a result of scrubber gypsum or lime particle loss rather than from the coal. In a Canadian study with a subbituminous coal and only a cold-side ESP, a 0.04% particle loss was reported (Goodarzi, 2006a). Meanwhile, health concerns particularly with emissions of submicron particle emissions have led to many more detailed new studies of the calcium in ash. Using a drop tube furnace, Sarofim's group examined a wide range of 20 or so coals. It was clear that calcium release is very low below 1600°C but increases significantly above this. This appears to suggest that volatilization is restricted to CaO either by normal or ablative mechanisms. There is no indication of a noteworthy role of release as volatile $CaCl_2$. Moreover, there is no clear correlation between coal calcium content and vaporization levels. This relates more to coal rank being greater in lower ranked coals. With a bimodal distribution of particle sizes, calcium was seen to be enriched in the submicron distribution peaking at 0.03 μm and obviously a result of vapor deposition (Neville et al., 1981; Quann and Sarofim, 1982; Quann et al., 1982, 1990). Calcium enrichment in the smaller sizes of the bimodal size distribution was also confirmed in further studies (Kauppinen and Pakkanen, 1990; Nielsen et al., 2002). In recent years more emphasis has been placed on examining this 0.03–1.0 μm size range. In a lignite-fired plant the particles exiting the ESP were size segregated in a 13-stage sampler. Calcium was clearly evident in the nanorange of 0.03–0.06 μm sizes (Wang et al., 2008). Burning a bituminous coal the bimodal distribution with calcium was present with a peak at 0.1 μm (Liu et al., 2007a). When cofired with 7% of a solid recovered fuel that enriched the calcium content, the concentration of calcium in the ash smaller than 1 μm increased fourfold at the 0.08 μm size (Wu et al., 2011a). The higher temperatures of coal combustion and its local reducing zones appear to be pivotal in the still quite small and short lived releases of calcium to the gas phase. A recent ambitious study that monitored very many elements in size segregated coal fly ash concluded no enrichment for calcium or the other Group II elements. However, the smallest size range was only 2.2 μm (Lanzerstorfer, 2018). Fuels that are incinerated or burned at lower temperatures are not expected to release calcium from the ash particles at all. This becomes possible if raised temperature co-combustion with coal is utilized and one such study with Ca-rich sewage sludge although at 1200°C did imply that calcium emitted in particles smaller than 1 μm was slightly reduced for the combination of fuels from their separate contributions (Wang et al., 2009a). The form of calcium in these following vapor depositions is considered to be mainly $CaSO_4$. Roasting municipal waste ash at 1000°C with Cl_2 for 3 hours, induced negligible volatilization of calcium (Chan and Kirk, 1999). In the bulk ash, if temperatures are not much above 1000°C not only will there be minimal releases but it is unlikely that silicates will be able to form. It is well documented that CaO converts to $CaCl_2$ with HCl at 600°C–800°C (Chyang et al., 2009), and to $CaSO_4$ with SO_2/O_2 at 600°C–900°C (Allen and Hayhurst, 1996). Moreover, $CaCl_2$ also converts to $CaSO_4$ with SO_2/O_2 at 350°C–650°C (Ozawa et al., 2003). Because of this, concerns have been raised as to whether gaseous $CaCl_2$ can play a role as a corrosive deposit especially in biomass combustion. However, there seems little evidence that $CaCl_2$ manages to form and so this remains an untested conjecture (van Lith et al., 2009). Questions relating to the extent that calcium may retain various toxic volatile metals in the ash such as arsenic (Sterling and Helble, 2003) or selenium (Seames and Wendt, 2007) have been suggested. It does appear that by cofiring biomass with sewage sludge, the calcium does help to capture more lower toxic As^V in the bottom ash as the arsenate, $Ca_3(AsO_4)_2$ (Zhao et al., 2018a). In the author's own work it was established that flames containing Ca, Na, W, S, and Cl at low ppmv levels would preferentially produce deposits at 500°C of $CaWO_4$ rather than $CaSO_4$, $CaCO_3$, CaO, or $CaCl_2$ (Schofield, 2003). Similarly, $CaMoO_4$ deposits were obtained if a salt precursor in the flame for Mo was substituted for that of W (Schofield, 2005). This confirmed that surface deposits or ash components will tend to reflect their most preferred thermodynamically stable molecules if at all possible.

The role of calcium in gasification/liquefaction under reducing conditions with low oxygen content has been studied in recent years to some degree. An interesting result with biomass appears to show a dependence on heating rate. If heating rates are very high 15%–20% releases of calcium can be noted. However, with slow rates loss is minimal. It appears that if sufficient time exists the calcium may react within and be retained by the ash particle (Okuno et al., 2005). However, fast heating rates may disintegrate particles into a spectrum of smaller sizes. A recent study with gasification of cane bagasse pyrolysis is somewhat confusing implying that elements such as Si/Ca/Al/Fe are released to the syngas but must in fact be present therein as small ash particles (Jordan and Akay, 2012). One might expect the stable CaS and $CaCl_2$ to be principal candidates in pyrolysis. However, the sulfide is readily oxidized (Torres-Ordonez et al., 1989, 1993) and reactive with water and also with gaseous HCl. Molecules such as hydroxysulfides, $Ca(SH)_2$, that correspond to their oxygen analogue are also unstable and hydrolyzed by water.

In summary, it is clear that calcium has a nonvolatile nature and, if temperatures never exceed much above 1000°C, calcium may not leave the fuel, nor be able to interact with fellow ingredients. The higher temperatures of coal are needed to finally witness some low release of CaO(g) with this becoming evident on the ultrafine ash particles. The complete nature of ash composition remains unknown. Calcium has the ability to form mixed oxide molecules but it will be very dependent on conditions. On cooling any deposition that may occur for calcium will probably finally be present as the sulfate. As a result, with efficient particle control devices all the calcium will be retained. There seems no documented role for chlorine in spite of its favorable volatility and stability. Disposal of various ash forms as agricultural fertilizers now has gained importance. It is now even possible to recover phosphorus compounds from sewage sludge ash by selective leaching to produce calcium phosphate fertilizer (Gorazda et al., 2012).

7.11 Chromium

Chromium, the fourth transition element in the first row of the periodic table is a light weight element that has critical importance due to its corrosion resistance properties and strategic role in stainless steel manufacturing. It occurs in nature primarily as the ore chromite, $FeCr_2O_4$ that exhibits divalent iron and trivalent chromium. In fact, the trivalency of chromium is thermodynamically the most stable form and is the main form found in nature and is nontoxic. Electronically it is one of the transition elements for which the $3d^5 4s^1$ is the ground state below the expected $3d^4 4s^2$ configuration. The mineral lead chromate is known but rare,

$PbCrO_4$, and contains the carcinogenic, toxic hexavalent element. Consequently, as will be seen, chromium is an environmentally interesting element that differs from other metals. It was largely nontoxic for early mankind but now can be toxic and carcinogenic as a result of industrialization with the introduction of chromate salts. It is of particular chemical interest in its ability to modify its valency between these two dominant states. The Cr^{VI} tends to be produced at moderate temperatures. Chromite (Cr^{III}) can be oxidized to form sodium chromate which is the important precursor for numerous commercially useful chromates (Cr^{VI}). These are used as wood preservers, in leather tanning, metal finishing, coloring pigments, plastics, glass and ceramic refractories and can be more water soluble than some Cr^{III} compounds. A study is available of the anthropogenic chromium global cycle (Johnson et al., 2006). Recently, a European Commission has added four Cr^{VI} compounds to its list of substances of very high concern, namely dichromium tris-chromate, strontium chromate, potassium hydroxyoctaoxodizincate-dichromate, and pentazinc chromate octa-hydroxide (Government Policy, 2014). It is the industrial release of such toxic and carcinogenic forms that have given chromium its reputation as an undesirable and highly toxic pollutant. The oxide has cosmetic use and the chloride and sulfate even for medical usage as dietary supplements.

Chromium is present in the Earth's crust with an average concentration of about 100 ppmw, being 19th in relative elemental concentration and somewhat similar in concentration to V and Zr. Its concentration in the oceans is closer to 0.3 ppbw lying 30th in magnitude and similar to that of Cs. Chromite, the major ore, currently is mined globally at levels of about 24 million tons $year^{-1}$ and has very long-term reserves. In the atmosphere, chromium is always in particulate form. Dust and volcanic sources are the main natural emissions and have been approximately assessed at about 44,000–60,000 tons $year^{-1}$ (Nriagu, 1989; Malakhov and Makhon'ko, 1990). Anthropogenic releases about 20 years ago were assessed as 30,000 tons $year^{-1}$ but are now possibly reduced due to regulation in recent decades in some countries (Nriagu and Pacyna, 1988). This emission's release results predominantly from coal combustion and the iron/steel industries. A more recent appraisal for Europe is that atmospheric emissions may now have been reduced by about 40% (Pacyna et al., 2007), to a global value of 15,000 tons $year^{-1}$ (Pacyna and Pacyna, 2001). These estimates remain uncertain and in a study of rainwater content it was stated that although natural atmospheric releases were about 54,000 tons $year^{-1}$, the anthropogenic amount was closer to 75,000 tons $year^{-1}$ with one-third of this being the toxic Cr^{VI} (Kieber et al., 2002). For the overall environment what is of greater concern than air releases are the industrial discharges to water and soil, which have been quoted with magnitudes of 140,000 and 900,000 tons $year^{-1}$, respectively (Nriagu and Pacyna, 1988). How chromium differs from some of the other polluting metals is that the speciation of Cr^{III} and Cr^{VI} still generally remains unknown and quoted analyzed levels refer to total Cr. As a result, China in particular is now becoming very concerned about its chromium emission, primarily from coal combustion, that increased about eightfold from 1980 to 2008 and remains large (Tian et al., 2012; Cheng et al., 2015a). However, although they now have extensive data the important speciation factor that concerns health is required (Tian et al., 2011, 2013; Duan and Tan, 2013). Analytical quantification of the valences remains difficult requiring an initial separation stage. This is the current problem in that two emission standards are essentially necessary, one for Cr^{III} and one for toxic Cr^{VI}. A recent review summarized very well the current analytical approaches (Namiesnik and Rabajczyk, 2012). One difficulty, as will be seen, is also ensuring that the analysis is true to the sample because evidence of oxidation of Cr^{III} can occur as well as reduction of Cr^{VI} depending on the conditions of separation and analysis (Tirez et al., 2011; Malherbe et al., 2011). One newly designed atmospheric sampler that reduces the humidity on a sampling filter measured average winter/summer values in one location in New Jersey. This reported a value of Cr^{VI} at 57/72% compared to the National Monitoring Station's value of 46% for both (Torkmahalleh et al., 2013a).

Health concerns with chromium

Although available, the current excess of publications from around the world on the level of chromium in vegetable foods or present in human blood, hair or urine remains of limited value without the known speciation, other than helping establish a preferred meaningful biomonitor for total chromium (Liang et al., 2018; Drobyshev et al., 2017; Arcega-Cabrera et al., 2018). One rainfall study, collecting and analyzing the chromium speciation was interesting in that care was taken to conserve the sample's nature. It concluded that approximately all the chromium released to the atmosphere is removed by rainfall in about equal quantities of Cr^{VI} and Cr^{III} (Kieber et al., 2002). Some level of Cr^{III} is needed by the human body. This is the natural molecular form and at low levels is essential for all mammals. It is more readily absorbed by the body via respiration rather than diet and is neither toxic nor an irritant. The normal content in humans is less than about 1 $\mu g\ L^{-1}$ of blood, urine or plasma (Cesbron et al., 2013; Nisse et al., 2017, CDC 4th 2018; Zeng et al., 2019) and this is the suggested safe level (Mayo Clinic, 2018). Hair levels also appear a good biomarker generally being about <1 ppmw (Goulle et al., 2005; Drobyshev et al., 2017). It plays various roles in the body, one involving glucose balance and another concerning cholesterol. A recent study of 60 autopsies in Poland indicated Cd blood levels in a range of 0.1–16.4 $\mu g\ L^{-1}$ indicating that high levels might accumulate in some people (Dudek-Adamska et al., 2018). However, such a high value also was recorded in a recent Moroccan survey (Henriquez-Hernandez et al., 2018). Nevertheless, concerns about contamination in food or agricultural soil levels have not been raised (Xiao et al., 2018; Liang et al., 2018; Huang et al., 2019). Nevertheless, the biogeochemistry of Cr in the soil/plant system is continually under study as chromium atmospheric deposition to soil is a major contributor (Shahid et al., 2017; Peng et al., 2019).

Radiolabeled studies of the body indicate that most Cr^{III} or Cr^{VI} is excreted within days or a week. Chromium indicates no evidence of birth defects but it may reflect into birth weights. Daily intake has been estimated at 20–45 $\mu g\ day^{-1}$ lying much below toxicity limits and recommended dose levels remain very uncertain (Goldhaber, 2003). It is poorly absorbed by the body and has low toxicity.

The problem arises somewhat due to the metastable nature of the Cr^{VI} state. Although generally stable, it is strongly oxidizing so readily reverts to the more stable Cr^{III} form if possible. It is difficult to measure once in the body as it does rapidly reduce to this trivalency. The toxicity for life forms appears to arise from the differing molecular coordination shapes of the two valences. Of these, Cr^{VI} compounds can more readily penetrate cell membranes. Once within the cell they are reduced to the lower valency forms but by then have caused oxidation damage. Toxicity generally appears to stem from this lifetime of the hexavalent form when released to nature. The biosphere is generally reducing for Cr^{VI} and as a result no bioamplification occurs but it is also toxic for plants, bacteria, fish, and animals. However, given sufficient time it is reduced in deeper groundwater or by plants and animals and can become harmless in sediments or soils (Kimbrough et al., 1999). Nevertheless, if deposited in quantities into fresh water or food supplies, it can have significant effects on nature. In the United States a current maximum value for total Cr in drinking water is recommended by EPA and the FDA as 100 $\mu g\ L^{-1}$. Various US States set more rigorous standards and California has a more restrictive value of 20 $\mu g\ L^{-1}$. My own district is currently trying to get this level reduced to 10 $\mu g\ L^{-1}$, but not really considering that the speciation is the more important question and that low levels of Cr^{III} are actually beneficial to health. However, recent testing for Cr(VI) in 918 California ground water wells did report a range from <1 to 32 $\mu g\ L^{-1}$, and was over the present limit in 4% of them. It was both of a natural and anthropogenic origin (Izbicki et al., 2015). A more recent report indicates that numerous other routes for oxidation of natural Cr^{III} now are arising to contaminate groundwater from agricultural and other uses of chemicals (Hausladen et al., 2018). In addition there is a potential problem of stability in sampling. Cr^{VI} has been seen to be unstable in anthropogenically contaminated water regardless of any water treatment methods. It is especially sensitive to the presence of Fe^{II}, Mn^{IV}, and SO_2, being reduced to Cr^{III} (Cadkova and Chrastny, 2015). Because of the well-documented pollution cases for Cr^{VI} that have occurred with levels in local drinking water as high as 600 $\mu g\ L^{-1}$ or even more in other legal cases, coupled to the need to protect workers in the chromium industries, very significant environmental studies have emerged and have been summarized for these two valence forms and are freely available on the Web (Kimbrough, 1999; World Health Organization Cr(III), 2009; World Health Organization Cr(VI), 2013). To these can be added a very extensive US toxicology report (Tox Report Cr, 2012). An interesting recent case in North Carolina, United States, questioned the source of Cr^{VI} being found in drinking water wells. The source could either be natural aquifers or leakage from nearby coal ash ponds. Unexpectedly, it was found that the natural sources were responsible and a health risk to that region (Vengosh et al., 2016). It is normally seen that high temperature systems that contain traces of chromium are more likely to produce Cr^{VI} but whether this survives or is reduced back to Cr^{III} is the major question. Naturally, this will include volcanoes and forest fires and normal combustion processes. Combustion can be expected to be one important anthropogenic atmospheric emitter and whether any Cr^{VI} retains its toxic nature or reduces to the nontoxic Cr^{III} in any particulate formation has to be established. Another aspect is whether chromium has a mainly condensed molecular nature or can have gaseous components as transporters.

Chromium's basic chemistry

Chromium has a very extensive chemistry. Its basic oxides and chromates open to a chemistry that is rich in a series of hydroxyoxides that with chlorine become chlorooxides. At the higher temperatures of combustion this can lead to a complexity of molecules that are either tri- or hexavalent. Thermal and thermodynamic stabilities play a major role in both the situations of increasing and decreasing temperatures. There is a lot of information available, some possibly unreliable, and still not entirely clear, but it does help to remove much speculation and helps to interpret observations and indicate more-so remaining questions. The condensed phase molecules of chromium are generally stoichiometric with well-defined physical properties. For convenience their physical properties are summarized in Table 7.18.

Trivalent chromium is more cationic, forming the oxide Cr_2O_3 and sulfate as the most important compounds. A basic sulfate, hydroxide, chloride, nitrate, and phosphate also exist, some water soluble some not. With trivalency, the trioxide is commercially important being thermally very stable and insoluble in water (m.pt.2330, b.pt. > 3000°C). The soluble sulfate is mainly basic, $Cr(OH)SO_4$, and may be stable to about 900°C dissociating to the oxide. The normal sulfate, $Cr_2(SO_4)_3$ is not stable above 300°C and is insoluble (Stern, 2001). No carbonate is indicated, and both the soluble hydroxide, $Cr(OH)_3$ and the nitrate, $Cr(NO_3)_3$, dissociate similarly to the oxide at low temperatures. The slightly soluble trihalide, $CrCl_3$, (m.pt.890°C) sublimes >1007°C and then dissociates to hygroscopic $CrCl_2$, is oxidized in air and hydrolyzes at higher temperatures. The insoluble phosphate, $CrPO_4$ (m.pt.1907), dissociates to oxides >1450°C and appears to be the only molecule other than the ore itself and the chromites and chromates that can relate to combustion. The main ore, $FeCr_2O_4$, ferrous chromite, is water insoluble and is also used as a refractory material melting at about 1850°C. A stable sulfide is known, Cr_2S_3 (m.pt.1350°C), but is insoluble and oxidizes on heating. Divalency is seen as the unstable CrO (dissociates at the 300°C m.pt.), and tetravalency with CrO_2 (dissociates at the 400°C m.pt.), that have minimal roles with chromium. However, hexavalency centers on the even lower melting point trioxide, CrO_3 (197°C). This is thermodynamically metastable, subliming congruently below its melting point and decomposing at about 250°C to the more stable trivalent Cr_2O_3. Nevertheless, CrO_3 is soluble and is the anhydride of chromic acid that exists only in solution but is the basis of the stable chromate ($^{VI}CrO_4^{2-}$) and dichromate ($^{VI}Cr_2O_7^{2-}$) chemistries. It is a strongly oxidizing acid as are its molecules and the combination of chromic and sulfuric acids was an important cleaning agent and metal surface patina and colorant agent for many years. Numerous metal chromates are known and rather interesting is the fact that sodium chromate the major one is produced directly from the trivalent ore chromite, $FeCr_2O_4$. Heated with Na_2CO_3 in air the chromium is oxidized to Cr^{VI} as Na_2CrO_4. Chromium's periodic table family includes Mo and W. They all form these mixed metal oxides displaying their hexavalency. However, as seen in the sections on molybdenum and tungsten, chromates do not reflect the same high thermodynamic stabilities seen with molybdates and tungstates. Nevertheless, numerous chromates, particularly with the alkali and alkali earths can be made by heating their compounds with the trivalent $Cr_2O_3(s)$ oxide. Some are water soluble, some not, but chromates are known that are quite stable but not to elevated temperatures. They tend to dissociate to their more stable trivalent chromites near or below their melting points and these become more ceramic. For example, Li_2CrO_4 (m.pt.485°C), Na_2CrO_4 (m.pt.794°C, dissociates to $Na_2Cr_2O_4$ at ≈1000°C), $Na_2Cr_2O_7$ (m.pt.357°C, decomposing at 400°C), K_2CrO_4 (m.pt.973°C, dissociates to $K_2Cr_2O_4$ at ≈ 1000°C), $K_2Cr_2O_7$ (m.pt.398°C, decomposing at 500°C), $MgCrO_4$ (m.pt.1065°C, dissociates to $MgCr_2O_4$ at ≈ 600°C–700°C that is insoluble with a m.pt. of 2350°C), $CaCrO_4$ (m.pt.1030°C, dissociates to $CaCr_2O_4$ above 1000°C), and $PbCrO_4$ (m.pt.844°C). Because of this conversion behavior it has been difficult to acquire chemical thermodynamic properties for gaseous chromates.

TABLE 7.18 Properties of chromium's condensed phase molecules relevant to combustion, °C.[a]

m.pt. b.pt. vapor pressure
Cr 1857 2679, 10 ppmv/1383°C, ins
CrO dissoc 300, readily oxidized
Cr_2O_3 2330 >3000, vap to CrO(g), CrO_2(g), CrO_3, 1 ppmv/1590°C, ins
CrO_2 dissoc 400 to Cr_2O_3, ins
CrO_3 197, metastable, dissoc >250 to Cr_2O_3, s
H_2CrO_4 exists in solution only
$Cr(OH)_3$ >400 dissoc to Cr_2O_3, s
$Cr(NO_3)_3$ >90 dissoc to Cr_2O_3, s
$Cr_2(SO_4)_3$ >737 dissoc to Cr_2O_3, ins
$Cr(OH)SO_4$ stable <900, dissoc to Cr_2O_3, s
$CrCl_2$ 815 1302 hygroscopic, hydrolyzes on heating, 5 ppmv/630°C, s
$CrCl_3$ 890, sublimes >1007 and dissoc to $CrCl_2$, oxidized in air, hydrolyzes, ss
CrO_2Cl_2 −97 117, fumes in air, reacts with water
$Cr^{III}PO_4$ 1907 2671, dissoc >1450 to oxides, ins
CrS 1565
Cr_2S_3 1350, oxidizes in air, ins
$Cr(CO)_6$ 152, decomp >130, ins
$CaCr_2{}^{III}O_4$ 2170, ins
$CaCr^{VI}O_4$ 1030, converts to $CaCr_2{}^{III}O_4$ >1000, s
$FeCr_2{}^{III}O_4$ 1850, ins
$K_2Cr_2{}^{III}O_4$ thermally stable
$K_2Cr^{VI}O_4$ 973, converts at 1000 to $K_2Cr_2{}^{III}O_4$, mol vapor observed, s
$K_2Cr_2{}^{VI}O_7$ 398, decomp 500, ins
$MgCr_2{}^{III}O_4$ 2350, ins
$MgCr^{VI}O_4$ 1065, converts at 600–700 to $MgCr_2{}^{III}O_4$
$Na_2Cr_2{}^{III}O_4$ high m.pt.
$Na_2Cr^{VI}O_4$ 794, converts at 1000 to $Na_2Cr_2{}^{III}O_4$, mol vapor observed, s
$Na_2Cr_2{}^{VI}O_7$ 357, decomp 400 to $Na_2Cr^{VI}O_4$ and Cr_2O_3, s

[a]*see text.*
dissoc, dissociates; *decomp*, decomposes; *s*, water soluble; *ss*, slightly soluble; *ins*, water insoluble; *vap*, vaporizes; *mol vap*, molecular vaporization.

Thermal treating of chromates also is a way to convert them to the nontoxic chromites. An old but very interesting mass spectrometric Knudsen cell study compared the thermodynamic stabilities of the chromates, sulfates, molybdates and tungstates (Choudary et al., 1975). In the condensed phase, stabilities decreased in the order tungstates > molybdates > sulfates ≈ chromates. By taking the limited data on the heats of formation of the gaseous alkali chromates these can be converted into atomization dissociation energies. Used in the case of lithium for which $Li_2O(g)$ data exists, the additional bonding energy gained for a specific metal oxide in forming the chromate with $CrO_3(g)$. With Li this bonding energy is $D_{298}(Li_2O\text{-}CrO_3 \approx 404\ kJ\ mol^{-1})$. Comparisons seen in later tables of gaseous molybdates and tungstates and with sulfate data indicate approximate values of about 585, 530, 436, and 404 kJ mol^{-1} for this same lithium type bonding with the tungstate, molybdate, sulfate, and chromate, respectively. This clearly indicates the relative weakness of the gaseous chromates compared to these other elements. As will be seen this is reflected in their vaporization and gas phase existence. The only other available data are for the vaporization of K_2CrO_4 that implies its atomization energy is the same as that for Li suggesting a similar overall chromate strength (Brittain et al., 1987). At about 950°C the total vapor pressure for the potassium chromate vapor was 1.5 ppmv and very similar to that of the sulfate. However, 83% maintained its molecular structure in the vapor for the chromate compared to 63% for the sulfate. In this case the chromate vapor is slightly more favored than the sulfate. Nevertheless, the increased stabilities gained by chromium converting to the chromites particularly with the alkaline earths makes for a significant difference from Mo and W and introduces more of a condensed refractory phase nature. The chromates generally are highly colored and extensively used in metal finishing, tanning and as pigments. Due to their toxicity, industries utilizing or producing them are now strictly regulated particularly as one estimate indicated that half a million workers are involved in such employ in the United States alone. Another toxic hexavalent compound is the oxychloride, CrO_2Cl_2 (m.pt. − 97, b.pt.117°C), a red liquid at room temperature, hydrolyzed by water to chromic acid ($CrO_2(OH)_2$) solution and HCl, and fumes in moist air. Chromium is encountered mainly in these valences of three and six in the condensed phase. Exceptions are, CrO_2 that dissociates at 400°C, and hygroscopic $CrCl_2$ (m.pt.824, b.pt.1302°C, vapor pressure 5 ppmv at 630°C) (Hildenbrand, 1995). Hydrides are regarded more as solid alloys and are poorly defined stoichiometrically as are numerous sulfides.

Gas phase stabilities and transport

An early Knudsen cell-mass spectrometric study of the vaporization of $Cr_2O_3(s)$ from 50 years ago clearly indicated chromium molecules or atoms in the gas phase with a total vapor pressure of about 1 ppmv at 1590°C in an equilibrated cell that then modified under slightly more oxidizing conditions (O_2 several ppmv) to a temperature of 1640 for the same 1 ppmv pressure (Grimley et al., 1961). The gaseous speciation was comprised mainly of Cr, CrO and CrO_2 with some evidence of CrO_3 under the more oxidizing conditions. The vapor pressures were sufficiently quantitative to approximately describe the individual bond strengths of the three oxides. Similar chromium vapor pressures of a few ppmv at 1750°C were also reported more recently from a Knudsen cell study of a Cr_2O_3-CaO-MgO-Al_2O_3-FeO-SiO_2 slag (Wang et al., 2009). In subsequent studies with sufficient oxygen it was proven, for example, that Cr_2O_3(s) at 900°C or 1200°C lost weight according to a dominant CrO_3(g) formation in dry air. Its formation rate also was correlated with a correct O_2 concentration dependence (Kim and Belton, 1974; Opila et al., 2007).

$$0.5\,Cr_2O_3(s) + 0.75\,O_2 = CrO_3(g) \quad \Delta H^{o}_{298} = +244 \pm 9\ \text{kJ mol}^{-1}$$

Although quite endothermic such a reaction is driven at measureable rates by higher temperatures and oxygen concentrations and exhibits this intriguing switch from tri- to hexavalency. On combustion scales, unless such very low levels have significant effects, these corrosive-ablative measurements do not compete with more major direct channels (Myers et al., 2019). Levels are generally at the level of mass spectrometric measurements. Under limited oxygen conditions the Knudsen cell indicates mono- and dioxide formation. A more recent Knudsen cell-mass spectrometric study on chromates also comments that the vapor over Cr_2O_3(s) was rich in CrO and CrO_2 in such a self-produced oxygen environment (Shugurov and Lopatin, 2006). This was also the case in fuel-rich H_2 flames to which was added either an aerosol of chromic acid solution or chromium carbonyl vapor (Bulewicz and Padley, 1971a). The oxides were also characterized by their matrix isolation FTIR spectra when deposited from laser ablation of Cr in 1% O_2 (Chertihin et al., 1997b). Although there has been a controversy over the exact stability of these oxides for many years, nine acceptable values for CrO(g) utilizing seven different methods have gradually improved accuracies. The past values and the current photofragmentation result now recommend D_0(Cr-O)g = 445 kJ mol^{-1}, $\Delta H^{o}_{f298} = +197.5$ kJ mol^{-1} for its bond strength (Cooper et al., 2017). Their quoted experimental error limit may be too optimistic due to the complexity of the analysis, but the value appears accurate to 445 ± 4 kJ mol^{-1}. This is seen to be more strongly bound than OH but displays a significant decrease from that of the first three transition element monoxides of Sc, Ti, and V (Table 7.19).

This bonding level and electronic structure have been adequately modeled theoretically together with its low-lying potential energy curves (Espelid and Borve, 1997; Espelid et al., 1998; Miliordos and Mavridis, 2010). The latter even indicated that one experimental value of 457 kJ mol^{-1} was probably too large. The ground state is $X^5\Pi$ with the lowest lying state being about 5000 cm^{-1} higher in energy. Whenever the accuracy of a fundamental value such as this is improved upon, it will necessarily send a rippling effect through the whole previous database range of values for related molecules. In this case all determinations invoking CrO now need to be correctly scaled to this new CrO value. The dioxide second bond strength still remains uncertain with only minimal data available. The early mass spectrometric study of vapor pressure data has been the major source (Grimley et al., 1961). Ebbinghaus (1993) in his assessment of thermochemical values questioned the previous analysis of this that had led to the database value of D_0(OCr-O)g = 509 ± 40 (Barin and Platzki, 1995; Chase, 1998). His reanalysis suggested a possible value of D_0(OCr-O)g = 525 kJ mol^{-1} but was based on D_0(Cr-O)g = 460 kJ mol^{-1}. Updating the data now to the latest value for CrO this becomes D_0(OCr-O)g = 540 ± 10 kJ mol^{-1}, $\Delta H^{o}_{f298}(CrO_2) = -98$ kJ mol^{-1} that is currently recommended. Farber and Srivastava (1973) examined the chemistry of chromium in hydrogen-rich flames at about 2250K, probe sampling into a mass spectrometer.

TABLE 7.19 Stability trends for relevant gaseous molecules of the first row transition elements, kJ mol^{-1}

Bond	Sc	Ti	V	Cr	Mn	Fe	Co	Ni	Cu	Zn
D_0(M-O)	668 ± 1	663 ± 4	621 ± 12	445 ± 4	370 ± 10	403 ± 2	373 ± 5	373 ± 5	290 ± 5	155 ± 10
D_0(OM-O)	380 ± 32	609 ± 8	564 ± 20	540 ± 10	484 ± 14	443 ± 20	≈410 ± 20	≈375	≈300	–
D_0(M-OH)	469 ± 20	≈400	–	355 ± 15	317 ± 15	342 ± 10	344 ± 10	375 ± 15	≈370 ± 10	<230
D_0(HOM-OH)	502 ± 20	–	–	–	436 ± 15	471 ± 10	424 ± 10	377 ± 15	–	–
D_0(M-Cl)	438 ± 15	420 ± 8	426 ± 8	377 ± 5	334 ± 5	332 ± 6	340 ± 6	369 ± 7	375 ± 4	229 ± 8
D_0(ClM-Cl)	560 ± 25	495 ± 8	473 ± 8	371 ± 7	447 ± 14	459 ± 7	420 ± 7	371 ± 10	≈370 ± 20	405 ± 10
D_0(M-H)	201 ± 8	200 ± 9	206 ± 7	186 ± 7	126 ± 18	147 ± 10	188 ± 10	249 ± 8	254 ± 8	82 ± 4
D_0(M-S)	470 ± 10	458 ± 10	438 ± 8	325 ± 15	282 ± 10	312.6	336 ± 15	352 ± 0.3	287 ± 10	140 ± 10
D_0(M-P)[a]	130 ± 30[b]	220 ± 20[b]	312 ± 20	211 ± 20	148 ± 20	197 ± 20[b]	218 ± 20[b]	265 ± 20	203 ± 20	≈60 ± 20b
Bond ratios										
OM-O/M-O	0.57	0.92	0.91	1.21	1.31	1.10	1.10	1.01	1.03	–
M-S/M-O	0.70	0.69	0.71	0.73	0.76	0.78	0.90	0.94	0.99	0.90
M-OH/HOM-OH	0.93	–	–	–	0.73	0.73	0.81	0.99	–	–
M-Cl/ClM-Cl	0.75	0.85	0.90	1.02	0.75	0.72	0.81	0.99	1.01	0.57
MCl/M-OH	0.91	1.05	–	1.06	1.05	0.97	0.99	0.98	1.01	≈1.0

[a] *Tong et al. (2003).*
[b] *See text.*

They observed strong signals relating to CrO, CrO_2, CrO_3, and H_2CrO_4 concentrations. Using this data they assumed these to be in balanced reactions, one with another on their millisecond timescale, and also that the probe sampling data did reflect the true concentrations in the flame. Taking a much higher value then for CrO, $\Delta H^o_{f298} = +222$ kJ mol^{-1}, they derived values for the other three components. However, recalculating their data to current values now produces D_0(OCr-O)g = 480 kJ mol^{-1} that remains unacceptably low. Theoretical attempts to validate such values remain uncertain for chromium compounds but do act as a guide on correct magnitudes and quote values of 532 (Espelid et al., 1998), and 579 (Gutsev et al., 2000a). The former did calculate an acceptable value for D_0(CrO). The value of about D_0(OCr-O)g = 540 ± 10 kJ mol^{-1}, $\Delta H^o_{f298} = -98$ kJ mol^{-1}, appears to be the best estimate currently. The second bond is seen to be significantly stronger, consistent with the trend seen in Table 7.19 for Cr and its following transition elements. Practical interest has centered on the gaseous trioxide due to its apparent role in corrosion and combustion in oxygen-rich environments. This centers on it being the acidic anhydride for chromic acid, H_2CrO_4, that is known only in solution, but as discussed below also can be present in the gas phase structurally as $CrO_2(OH)_2$. Although CrO_3(g) was observed mass spectrometrically in the vapor above Cr_2O_3(s) with added oxygen, the early data now appears less accurate and is no longer acceptable (Grimley et al., 1961). Ebbinghaus (1993) preferred and reanalyzed a later transpiration volatility measurements of Cr_2O_3(s) in flowing oxygen/argon where it seemed that gaseous CrO_3 was the major gaseous product (Kim and Belton, 1974). This was shown to correlate to the correct oxygen concentration dependence. More recently a further confirmation at 900°C (Opila et al., 2007) verified Ebbinghaus's recalculation of this data indicating $\Delta H_{f298}(CrO_3)_g = -323.5 \pm 4.2$ kJ mol^{-1}, now recommended, or D_0(atomization) = 1455 kJ mol^{-1}. This updates the previous database value of -293 ± 42 kJ mol^{-1} (Barin and Platzki, 1995; Chase, 1998). Unfortunately, we are still in a time when uncertainty often remains concerning even the most appropriate theoretical method best suited to address such a molecule (DeYonker et al., 2007; Handzlik and Kurleto, 2013). As a result, theoretical estimates are mainly of value in supporting the magnitude, and in this case have suggested $\Delta H_{f298}(CrO_3)_g = -334$ (Espelid et al., 1998) and -283 (Bross et al., 2013). The flame study indicated above derived a recalculated value now implying $\Delta H_{f298}(CrO_3)_g = -295$ kJ mol^{-1} also unacceptable (Farber and Srivastava, 1973). If the presently suggested values are stated as the strengths of the three bonds in CrO_3, the trend is D_0 (Cr-O, OCr-O, and O_2Cr-O) of values 445 ± 4, 540 ± 10, 470 ± 10 kJ mol^{-1}. The dioxide clearly has a slightly greater stability, a very pronounced trend reflected by Mo, its periodic table column relative.

Chromium's gaseous chemistry was found to change on the addition of hydrogen, water or chlorine becoming potentially more complex. One Knudsen cell-mass spectrometric study measured the vapor in equilibrium with Cr_2O_3(s) with added H_2 at high temperatures monitoring Cr, CrO, CrO_2, and CrOH (Gorokhov et al., 1990). They derived a heat of reaction for the following balance in the 1986K−2218K range.

$$CrOH(g) + O(g) = CrO_2(g) + H(g) \quad \Delta H^o_{298} = -206 \text{ kJ mol}^{-1} \text{(Ebbinghaus, 1993)}$$

Further recalculation now with the above recommended value for CrO_2 modifies its implication for CrOH to D_0(Cr-OH)g = 355 ± 15 kJ mol^{-1}, now recommended. Uncertainties remain due to the necessary calibration of their instrument and the values used for O and H concentrations. However, it remains the only experimental value. Ebbinghaus (1993) had suggested 358 from this data. The value is supported by theory that calculated a bond strength value of 361 or 365 (Espelid and Borve, 1997; Espelid et al., 1998). These calculations also had suggested $D_{0\mathrm{Atomization}}(CrO_2) = 985$ kJ mol^{-1} agreeing with the values recommended above for the sum of the two bonds, and an acceptable value for the three individual oxide bonds. As a result the value is accepted and is consistent with trends for the M-OH and M-Cl first row transition elements (Table 7.19) being slightly weaker than D_0(Cr-Cl). As indicated, additions of water and chlorine to chromium appear to have the ability to open up a complexity of hydroxyoxides and chlorooxy volatile compounds that will be discussed below in detail with the chlorides.

Turning otherwise to other potential gaseous molecules of chromium it is apparent these have little to contribute. Oxides have corresponding sulfides and for chromium only the gaseous diatomic has been studied by the Knudsen cell-mass spectrometric method. This indicated D_0(Cr-S)g $= 325 \pm 15$ kJ mol^{-1} (Drowart et al., 1967), as usual weaker and about 73% the strength of the monoxide, similar to its neighbors vanadium and manganese (Table 7.19). Theory still is developing to reproduce this magnitude, some methods predicting values slightly too low and others too high (Bauschlicher and Maitre, 1995; Wu et al., 2007; Yang et al., 2009). The pure rotational spectrum for its $X^5\Pi_r$ has been analyzed also suggesting a low-lying $^5\Sigma^+$ state (Pulliam and Ziurys, 2010). The LIF ($B^5\Pi - X^5\Pi$) transition was reported earlier (Shi et al., 2001). Other gaseous sulfides namely CrS_2 and CrS_3 have been isolated in matrix isolation studies by laser ablation of Cr in discharged sulfur vapor reporting their vibrational frequencies and geometries from FTIR infrared spectra (Liang and Andrews, 2002b). However, they can exist only in oxygen-free systems, and otherwise as temperatures rise are oxidized to form Cr_2O_3(s). A similar situation arises with the phosphides and has been studied in detail. On heating in the absence of air, CrP (s) vaporizes by dissociation producing gaseous P_2 (Myers et al., 1985). With mixtures of $CrCl_3/PCl_3/H_2/Ar$ heated at 700°C−1200°C, chemical deposition can occur onto substrates as CrP (<800°C) and as Cr_3P at >1100°C (Motojima and Higashi, 1985). In air these oxidize to $CrPO_4$ and Cr_2O_3. Theoretical assessments indicate a CrP($X^4\Sigma^-$) state, confirmed by LIF measurements, and with a dissociation energy to the first limits of about D_0(Cr-P)g $= 211 \pm 20$ kJ mol^{-1} (Table 7.19) (Tong et al., 2003; Li and Balfour, 2005).

Gaseous chromium hydride is of more interest to astronomers than to combustion. The most accurate experimental measure determined D_0(Cr-H)g $= 186 \pm 7$ kJ mol^{-1} from reaction threshold limits (Chen et al., 1993). An earlier mass spectrometric study had indicated an upper limit value of <188 kJ mol^{-1} (Kant and Moon, 1981). Goel and Masunov (2008) summarized the previous theoretical work on CrH, attempts mainly overestimating the bond strength value. However, extensive work on several transition metal hydrides successfully reproduced the bond strength for CrH and calculated its potential energy curves. No low-lying excited states were predicted (Koseki et al., 2006). A matrix isolation study of laser-ablated Cr with H_2 did show CrH and CrH_2 but with only other complex hydride structures (Wang and Andrews, 2003). These will all be readily oxidized and be of little importance even in reducing gasification processes.

Laboratory combustion studies and gas phase stabilities

In the gas phase, chromium is expected to be sensitive to experimental conditions, specifically temperature, H_2O, O_2, and chlorine content. Due to this possible complexity, a greater effort has been made to try and understand its behavior with various analyses and from the controllable laboratory studies that can better isolate chromium. In this regard, several experimental flame studies are of interest. One already mentioned above reported observing gaseous chromic acid, $CrO_2(OH)_2$, in mass spectroscopic sampling of atmospheric pressure fuel-rich hydrogen flames at 2250K containing chromium (Farber and Srivastava, 1973). However, very significant additions of $CrCl_2$ (1%–2%) vapor were made to the flame gases, much larger than usual in such flame studies making notable difficulties for obtaining unperturbed probe mass spectral sampling without inducing heterogeneous chemistry. The reported ions of CrO, CrO_2, CrO_3, and H_2CrO_4 appeared dominant with no signs of chlorine, but the study now has been critically questioned as being quantitative and reflecting their gas phase concentrations. The occurrence of CrO_3 also was not expected by Ebbinghaus (1993) for such reducing conditions and he discarded their data as questionable. Nevertheless, as seen below, the value obtained for the heat of formation of gaseous $CrO_2(OH)_2$ in the flame-probe study does appear acceptable and at least confirms its existence. Another early mass spectrometric study sampled the vapors also at atmospheric pressure that resulted from flowing O_2 saturated with H_2O at 21°C over Cr_2O_3(s) at 1100°C (Fryburg et al., 1977). Ions only of $CrO_2(OH)_2$ and $CrO_2(OH)$ were observed and the latter may have been real or a fragment ion. No ions of CrO, CrO_2, or CrO_3 were noted suggesting again that their $CrO_2(OH)_2$ was produced heterogeneously, directly on the surface and then vaporized. Although such corrosive ablation studies are important for observed corrosion, their rates generally are very low and not really appropriate to rapid combustion chemistry (Myers et al., 2019). Another quantitative flame study was that of Bulewicz and Padley (1971a) utilizing the analytical techniques developed by Sugden and his group. This involved using series of fuel-rich hydrogen flames, about 14 in this case, for which the content of flame species such as H, OH, O, H_2, H_2O were well characterized in each case, were quite varied and nonequilibrated at known temperatures (1390°C–2200°C) displaying noteworthy differences from flame to flame. Any chemical correlations observed with added species could then enable the dominant chemistry to be resolved and provided a sensitive method. In this case, chromium was added either as a nebulized spray of a weak chromic acid solution or as the vapor of chromium carbonyl, both hexavalent chromium. Flame concentrations of chromium were less than 1 ppmv. Generally, this technique is one of homogeneous chemistry with such trace additions. Consequently, one surprising result was that streaking Cr_2O_3(s) particles were present in the burned flame gases even with traces of additive as low as 5 ppbv. In fact, by adding a trace of sodium and comparing the spectral absorption intensity of gaseous atomic Cr to that of Na it was shown that the bulk of the chromium did remain in the gas phase. Results indicated under these reducing conditions that other than gaseous Cr, CrO, and CrO_2, a hydroxide was formed in balance with these oxides (Bulewicz and Padley, 1971a). Correlations implied $CrO_2(OH)$ an unusual pentavalent formation for Cr but the data apparently ruled out any other dominant forms. Ebbinghaus (1993) in his thermochemical analysis suggested that it may possibly have been

CrO(OH) that is also more compatible with trivalency under reducing conditions rather than this rare pentavalency. His recalculations provided a value of $\Delta H_{f298}(CrO(OH))_g = -247 \pm 5$ kJ mol^{-1}. Another aspect came from this early work on chromium (Bulewicz and Padley, 1970, 1971a). They noticed that very low levels as small as 1–10 ppmv additions of a chromium salt to hydrogen flames strongly catalyzed the recombination of the H and OH radicals. It had been noted earlier that CrO_2Cl_2 was effective at extinguishing flames (Miller, 1969). This was confirmed by later research (Vanpee and Shirodkar, 1979; Babushok and Tsang, 2000) and reviewed later (Linteris et al., 2008). There was an indication that this inhibition was sensitive to particular chemical form. Also what it appeared to imply from the beginning was that at such low levels the mechanism could not be heterogeneous on flame particles. It required gaseous chromium molecules that could exhibit efficient recombination kinetic cycles, such as the oxides, and hydroxides. Efficient also at higher concentration levels, it has been suggested there, that a heterogeneous mechanism on Cr_2O_3(s) particles may also occur (Vanpee and Shirodkar, 1979). Although connections between flame radical concentration recombinations and soot suppression remain tenuous and difficult to explain together, it is noteworthy that one study noted that Cr was possibly a slight soot promotor depending on conditions (Bulewicz et al., 1975). An earlier study rated it as a low soot suppressor 1.7, on a scale with Ba at 100 (Cotton et al., 1971). Also at a level of 0.1% in the flame, no downstream effect on soot was reported (Haynes et al., 1981). As a result Cr is largely ignored in soot control discussions (Howard and Kausch, 1980).

Due to chromium's importance in alloys it was realized that both oxygen alone or with steam could evaporate chromium at temperatures above 1000°C (Caplan and Cohen, 1961; Zaplatynsky, 1977). The presence of water was seen to enhance the rates over those with dry air (Graham and Davis, 1971). Consequently, for use in steam turbines and other purposes it was necessary to produce superalloys and try and resolve this functioning chemistry. The earliest experimental measurements invoked the transpiration method. An initial study used CrO_3(s) at the low temperatures of 135°C–185°C noting an evaporative loss with a linear dependence on water vapor concentration yet none on the oxygen level.

$$CrO_3(s) + H_2O(g) = CrO_2(OH)_2(g) \quad \Delta H^o_{298} = +71 \pm 10 \text{ kJ mol}^{-1}$$

They correctly suggested the formation of gaseous $CrO_2(OH)_2$ (chromic acid) even at such low temperatures (Glemser and Mueller, 1964). Later, a similar study then examined Cr_2O_3(s) in oxygen/water vapor flows at higher temperatures 1300°C–1585°C (Kim and Belton, 1974). This indicated a square root dependence on both water and oxygen concentrations but concluded now a dominant formation of the monohydroxide, $CrO_2(OH)(g)$ with a pentavalent Cr.

$$0.5\, Cr_2O_3(s) + 0.5\, H_2O + 0.5\, O_2 = CrO_2(OH)(g) \quad \Delta H^o_{298} = +250 \pm 15 \text{ kJ mol}^{-1}$$

From this Ebbinghaus (1993) quoted $\Delta H_{f298}(CrO_2(OH))_g = -438 \pm 6$ kJ mol^{-1}. A density functional study published the same year confirmed this within its error bars (Ziegler and Li 1995). At 1223K it was seen that $CrO_3(g)$ was present at very low levels of water but then $CrO_2(OH)_2(g)$ became dominant with higher humidity (Gindorf et al., 2005). By oxidation on the surface, Cr_2O_3(s) could convert to hexavalent CrO_3(g) that with water would then form the acid, $CrO_2(OH)_2$(g), a change in valence from III to VI (Fryburg et al., 1977).

Other work used transpiration but at 300°C–900°C, and extensively reexamined the data of Kim and Belton (1974) and questioned parts of Ebbinghaus's earlier recommendations (Opila et al., 2007). In this intermediate temperature range they noted a unit pressure dependence on water vapor and a three-quarters dependency on O_2 as required for a hexavalent Cr dihydroxy dominance with Cr_2O_3(s). It was also shown that contributions from CrO_3(g) or $CrO_2(OH)$ were negligible.

$$0.5\,Cr_2O_3(s) + H_2O + 0.75\,O_2 = CrO_2(OH)_2(g) \quad \Delta H^{o}_{298} = +49 \pm 10\ \text{kJ mol}^{-1}$$

They recalculated the Gibbs free energy functions for this dihydroxide with significant changes of 40 J $mol^{-1}\,K^{-1}$ and now recommend a second law value of $\Delta H_{f298}(CrO_2(OH)_2)_g = -760 \pm 7$ kJ mol^{-1}, over the previous value of -748.3 ± 4.3 (Ebbinghaus, 1993). Although the mass spectrometric flame-probe study with chromium mentioned already did not provide reliable data for the individual oxides a recalculation of the simpler balance

$$CrO(g) + 3H_2O(g) = CrO_2(OH)_2(g) + 2H_2(g) \quad \Delta H^{o}_{298} = -232 \pm 10\ \text{kJ mol}^{-1}$$

for which H_2O and H_2 are well defined, coupled to the new thermochemistry of CrO(g), now predicts a value of $\Delta H_{f298}(CrO_2(OH)_2)_g = -762$ kJ mol^{-1} in agreement with the above (Farber and Srivastava, 1973). All that is needed in such a determination is to obtain a reasonable estimate for the approximate ratio of the two gaseous chromium molecules in the flame from their ion signatures. This may be fortuitous here or maybe the balance between the two chromium compounds is roughly maintained by the fast chemistry even though being perturbed by the probe. Theoretical values indicate possible values of -787 (Espelid et al., 1998) and -792 ± 20 kJ mol^{-1} (Opila et al., 2007) indicating possible further needs for refinement. A theoretical mechanistic modeling also has been presented to describe the surface corrosive oxidation and vaporization for the Cr_2O_3(s) to $CrO_2(OH)_2$(g) involving H_2O and O_2 (Panas et al., 2004). The free energy functions now are much improved and a recalculation of the earliest transpiration study with CrO_3(s) provides a supporting value of -767 (Glemser and Mueller, 1964). This refinement now also removes previous questions concerning the analysis of Kim and Belton (1974), which are now accepted, and it appears that $CrO_2(OH)$(g) can become dominant at the higher temperatures although pentavalent. It is still not possible to accurately calculate any $CrO_2(OH)_2$(g) role at the higher temperatures but the thermochemical trend predicts a minimal contribution. A recalculated value of $\Delta H_{f298}(CrO_2(OH))_g = -450 \pm 7$ kJ mol^{-1} now is recommended. In the oxygen-rich high temperature transpiration experiments of Kim and Belton (1974) acceptance of a dominance of $CrO_2(OH)$ makes it rather unlikely that this same molecule would be dominant under reducing conditions where CrO_3 and $CrO_2(OH)_2$ are negligible. Another estimated value for a similar type molecule that may play a role under differing conditions is $\Delta H_{f298}(CrO(OH)_2)_g = -542 \pm 20$ kJ mol^{-1} (Ebbinghaus, 1995). As a result, Ebbinghaus's earlier thermochemical value estimates (1993) for all the simple hydroxyoxides of chromium remains a useful starting database of values, as now adjusted, but still remaining approximate. However, it has been of great value in illustrating the potential complexity in chromium's chemistry and the possible roles in combustion for gaseous $CrO(OH)$, $CrO(OH)_2$, $CrO_2(OH)$, and $CrO_2(OH)_2$. A preliminary representative thermodynamic equilibrium calculation by Ebbinghaus (1993) using the full database illustrated that many of the other possible oxyhydroxy structures were unlikely to play a role and that gaseous CrO_3 and $CrO_2(OH)_2$ would be dominant at equilibrium, which of course is always the pivotal question. Moreover, even if equilibrium is attained, a comparison of various databases has indicated that the magnitudes of the remaining uncertainties might still lead to differences (Gindorf et al., 2005; Lundholm et al., 2007). Some of these molecules have been identified in matrix isolation measurements where laser-ablated Cr reacted with H_2O_2 or H_2/O_2 mixtures and products condensed, vibrational frequencies and structures were obtained for CrOH and $Cr(OH)_2$ (Wang and Andrews, 2006). The earliest research flame studies showed several important aspects. In high temperature fuel-rich hydrogen flames a difference was noted between chromium salt additions and that of $Cr(CO)_6$. The former could be as low as 6 ppbv in the 2000°C flames and still produce instant streaks of Cr_2O_3(s) particle formation. However, with carbonyl addition, concentrations had to be more than 26 ppmv, 4000 times larger (Bulewicz and Padley, 1971a). Kennedy and his team followed this work with a detailed series of studies mainly with laminar diffusion or premixed hydrogen flames. By then, it was becoming realized that chromium did require kinetic consideration and could not be accepted as being thermodynamically equilibrated. Moreover, this sensitivity to the Cr source was becoming apparent. Their initial study, probe-sampled 2 cm downstream of the H_2 diffusion flame tip, diluting and cooling the gases, which then passed through a 20 nm pore filter (at about 100°C) and then through solution impingers. With chromium nitrate solution additions, the filter showed cenospheres of Cr_2O_3(s) with Cr(VI) evident in the vapor caught by the impingers. Use of $Cr(CO)_6$ appeared to lead to a thin deposit of CrO_3(s) (only stable below 250°C) and similar evidence of Cr(VI) in the vapor phase. In both cases the extent of Cr(VI) formation decreased with flame temperature quite strongly in this 1600°C–2180°C range. The carbonyl possibly produced Cr(VI) three times more effectively than the nitrate under similar conditions (Yoon et al., 1996). The behavior of $Cr(CO)_6$ was extended in a further study but with cooler laminar premixed H_2 flames at 1000°C, 1050°C, and 1100°C. Cr(VI) levels initially increased downstream to about 12% of the total chromium but then dropped to about 5% (Yu et al., 1998). The paper presented an initial chemical kinetic scheme for chromium that was developed further in a following paper and was suggested as being adequate. However, it contains many three-body and other unlikely four-center reactions and kinetically cannot be fully accepted (Yu et al., 1998; Yu and Kennedy, 2000). What it reinforced though was the realization that Cr, like Hg, cannot solely be modeled in the gas phase but needs to invoke heterogeneous components. An examination followed of the morphology of the oxide particles from laminar H_2 diffusion flames at 1600°C–2180°C (Kennedy et al., 1999). A bimodal size distribution was evident with many small particles (0.03 μm) of Cr_2O_3(s) (m.pt.2330°C) that appeared to originate from the nebulized nitrate droplet with subsequent breakup. Filtering at 105°C when the carbonyl source was used led to a very different deposit. It was consistent with CrO_3(s), which had to be a deposition process forming from CrO_3(g), $CrO_2(OH)_2$(g) or any other vaporized Cr^{VI} molecule present. In other words, when Cr is injected essentially in atomic form it will rapidly oxidize through to CrO_3(g). Whether it can form hydroxyoxides remains unproven but it appears not to nucleate or deposit as the stable Cr_2O_3(s) particles as found in reducing flames by Bulewicz and Padley (1971a). Kennedy's work had difficulty in accounting for all the chromium and some may have been in the CrO_3(s) deposit. A further study examined solely carbonyl additions sampling 52 cm downstream where the temperature had fallen to 190°C and chemistry had ceased (Guo and Kennedy, 2004). Unfortunately, too many parameter differences confuse the data. Residence time in the flame and temperature were important but the small particles were predominantly Cr_2O_3(s). Evidence of Cr^{VI} was less and it seems that when CrO_3(g) condenses on a particle or nucleates it has little choice at most temperatures but to reduce valency to Cr_2O_3(s). The extent of CrO_3(s) condensing on particles will always be very limited due to its metastability. An actual measurement of CrO(g) in a lean H_2 flame at about 1150°C fed by chromium carbonyl indicated a very low concentration of 1.6 ppbv indicating its limited lifetime (Wilcox et al., 2003). Having added about 130 ppmv of chromium this implies that only 0.001% was present as CrO(g) confirming its rapid oxidation toward CrO_3(g).

Ebbinghaus (1995) also extensively documented the possible chemical thermodynamic role in combustion of chlorine with chromium. This was equally important to assess as from an early period it was known that chlorine could leach CrO_2Cl_2(g) from chromite ore and be an alternate extraction method (Gertler, 1941). Additions to this 20-year-old compilation of data result from the more accurate measures now available for the bond strengths of CrCl(g), $CrCl_2$(g), CrO(g), and CrO_2(g) that throw a ripple through this database (Hildenbrand, 1995, 2008). These values now are D_0(Cr-Cl) = 377 ± 5 (ΔH_{f298} = 139.7 ± 7 kJ mol^{-1}) and D_0(ClCr-Cl) = 371 ± 7 (ΔH_{f298} = −110.0 ± 4 kJ mol^{-1}), the two bond strengths being very similar and need consideration together with the changes discussed above for the two oxides. Monomeric molecules are dominant in the $CrCl_2$ vapor phase (Schoonmaker et al., 1959). Theoretical calculations for CrCl reproduce this bond strength for the $X^6\Sigma^+$ state, indicate the nature of the low-lying potential energy curves and predict no low-lying states (Kardahakis and Mavridis, 2009). Although a high level ab initio theory reproduced the bond strength of the monochloride it overestimated that for the dichloride (Nielsen and Allendorf, 2005), that earlier was adequately described by a density functional study (Wang and Schwarz, 1998). The first row transition element dichlorides were shown to be linear in structure (Drake and Rosenblatt, 1979). Corresponding to the hydroxyoxides, the four most interesting and potentially important analogous gaseous oxychlorides appear to be CrOCl, $CrOCl_2$, CrO_2Cl, and CrO_2Cl_2. As an illustration of how tenuous thermodynamic information still remains, the present updating for these can be compared to their previously listed values (Ebbinghaus, 1995). Now that data for CrCl, CrO, and CrO_2 are quite accurate the present values change: ΔH_{f298}(CrOCl) = −93 ± 10 replaces −118 kJ mol^{-1}. With this value, ΔH_{f298}($CrOCl_2$) = − 294 ± 10 replaces −337 kJ mol^{-1}, and ΔH_{f298}(CrO_2Cl) = −290 ± 20 replaces −310 kJ mol^{-1}, while that for ΔH_{f298}(CrO_2Cl_2) = −519 ± 5 remains unaffected. Experimentally these four species were observed in different ways. CrOCl was monitored by a Knudsen cell-mass spectrometric method using a $Cr_2O_3/BaCl_2$ mixture at 1480°C (Milushin and Gorokhov,

1988). $CrOCl_2$ was proven similarly by mass analysis of $CrOCl(s)/Cl_2$ and $Cr_2O_3(s)/Cl_2$ mixtures at 630°C−1080°C (Plies, 1991), and by utilizing the mass spectrometric ionization efficiencies of the fragments from CrO_2Cl_2 vapor (Flesch et al., 1969). This latter study also monitored CrO_2Cl. The most detailed and accurate analysis was for $CrO_2Cl_2(g)$ utilizing the transpiration method (Sano and Belton, 1974). This examined dry flows of O_2 with Cl_2 over $Cr_2O_3(s)$ at 627°C−977°C. Vaporization was proven as being dominantly by the process:

$$Cr_2O_3(s) + 2\ Cl_2 + 0.5\ O_2 = 2\ CrO_2Cl_2(g)$$

Rates of formation confirmed this being directly proportional to the Cl_2 pressure and one-quarter that of O_2. Oxygen was varied from 1.2% to 90.8% of the flow and Cl_2 from 1.8% to 14%. In this way, roles in the study for $CrOCl_3$ and CrO_2Cl were ruled out and that for CrO_2Cl had to be <5%. Coupled to the known leaching of $CrO_2Cl_2(g)$ from chromium minerals, as with water and the dioxydihydroxy formation, it appears that Cl_2 and also possibly HCl can produce this molecule by heterogeneous surface chemistry with subsequent vaporization. As a result, it is noteworthy to compare these heterogeneous transpiration studies of $Cr_2O_3(s)$ with H_2O/O_2 (Opila et al., 2007) and dry Cl_2/O_2 (Sano and Belton, 1974) to illustrate the potential effect of chlorine. At 630°C with 1% Cl_2 and 5% O_2 the vapor pressure of $CrO_2Cl_2(g)$ is about 3×10^{-5} atm (30 ppmv). Correspondingly, with 10% H_2O and 5% O_2 the vapor pressure of $CrO_2(OH)_2(g)$ is about 4×10^{-8}, three orders of magnitude less. With Cl_2 the surface formation efficiency from chromium sources may control any observed chromium release. Although $CrO_2Cl_2(l)$ (b.pt.117°C) has a very high vapor pressure it is hydrolyzed to chromic acid fume in moist air. As a result whether it can be a surviving source would seem questionable and probably requires reformation in the gas phase at higher temperatures where it can regain stability against hydrolysis. Many nonchromate salts are known to dissociate on heating to the very stable $Cr_2O_3(s)$. This oxide though is very slowly vaporized in 5% O_2 at 1000°C to $CrO_3(g)$ with an equilibrium vapor pressure of about 0.02 ppmv; by 10% $H_2O/5\%O_2$ to $CrO_2(OH)_2$ with a pressure of about 0.3 ppmv but with 2% chlorine/5%O_2 to CrO_2Cl_2 with a pressure of about 350 ppmv. All rates of course would increase with temperature. Consequently, for the potential release of chromium in the vapor state from a fuel it is apparent that it requires the presence of chlorine or a temperature in excess of 1750°C (vapor pressure of 250 ppmv by 10% $H_2O/5\%O_2$ to produce $CrO_2(OH)_2(g)$). Consequently, examining these eroding type mechanisms utilizing particularly chlorooxy molecules it appears that they can provide a more efficient channel for chromium to be released and transported in the gas phase at lower temperatures (Nocker and Gruehn, 1993a). As with his study of the chromium oxyhydroxy stabilities, Ebbinghaus (1995) similarly displayed examples of the expected thermodynamically equilibrated chlorooxy distributions for several combustion cases at differing temperatures. At lower temperatures $CrO_2Cl_2(g)$ could be dominant with hexavalent $CrO_2(OH)_2$ gradually changing to CrO_2Cl, CrO_2OH along with $CrOCl_2$ and CrO_3 as temperatures increased. An important question not clearly addressed in all these studies was whether HCl(g) and Cl_2 behaved similarly (Nocker and Gruehn, 1993) especially as HCl is the dominant form of chlorine in combustion at all equivalence ratios. Although interesting, the research work of the Kennedy's group discussed above was insufficient, especially for incineration chemistry, as potential effects due to chlorine were ignored. As a result, a few years later the same group attempted to resolve this deficiency (Guo and Kennedy, 2001). As an extension of their previous work, a laminar CH_4/air flame at 1130°C was examined with added chromium hexacarbonyl and CH_3Cl. The probe was positioned at twice the flame length on its centerline and the sample diluted and cooled by flowing N_2. A filter similarly collected particles larger than 20 nm and vapors passed into solution impingers. Analysis was difficult and only emphasized Cr^{VI}, leaving interpretation difficult with no mass balance of chromium. The solid phase Cr^{VI} changed only slightly and remained about 5% of the total input. However, there was a large increase in Cr^{VI} in the gas phase caught by the impingers. The fraction of Cr^{VI} both on the filter and in the solutions relative to the total flame input rose from 5% with no chlorine to about 60% with about 2% addition of chlorine. That on the 105°C filter has to result from condensation of CrO_3(g) and cannot be CrO_2Cl_2(s) which has a boiling point of 117°C. What seems to be happening is that chlorine is the catalyst in kinetically producing CrO_2Cl_2(g) more easily from CrO_3(g), which then opens up the whole hydroxyoxy, chloroxy balances of Cr^{VI}. The chlorine itself is unlikely to appear in any ultimate product under such oxygen rich conditions. Additionally any formation of Cr_2O_3(s) particles in the flame might become vulnerable, converting to CrO_2Cl_2(g) certainly above 1000°C where its vapor pressure in equilibrium with the surface is about 350 ppmv. It does seem that the chlorine can maintain more of the chromium in the gas phase as Cr^{VI} for a longer period. Consequently these studies have left many unanswered questions but have indicated that Cr^{VI} can be created and survive in some particles but not necessarily in significant proportions. Also chromium hexacarbonyl or such an atomic chromium precursor will not be a commonly encountered input. From the chemistry above it has to be remembered, under oxidizing conditions, not only that all Cr^{III} salts dissociate to Cr_2O_3(s) but also does CrO_3(s) (Khilla et al., 1982). Also as indicated earlier, most chromates on heating convert from a Cr^{VI}, $CaCrO_4$, structure to the chromite, Cr^{III}, $CaCr_2O_4$. Above 1000°C, K_2CrO_4 was noted to vaporize initially molecularly and also fragment increasingly with temperature to Cr_2O_3(s) (Brittain et al., 1987). This is due to the fact that they initially can dissociate to CrO_3(s) but this is metastable and above 250°C changes to the trivalent Cr_2O_3(s). Infrared matrix isolation spectra and structures were in fact obtained by vaporizing Rb_2CrO_4 and Cs_2CrO_4 at 970°C above their melting points and shown to be consistent with the chromate structure (Bencivenni and Gingerich, 1982). Consequently even the combustion of wastes and sludge will tend only to produce Cr_2O_3(s) at temperatures up to about 1500°C unless chlorine can vaporize this and initiate chloro/hydroxy and CrO_3(g) chemistry. Additionally, the chromium in coal appears invariably to be present as Cr^{III} (Huggins et al., 2000). Opila et al. (2007) in a chlorine-free system observed liquid deposits of chromic acid below 600°C. Described later in full scale and flame data, chlorine as HCl is noted as enhancing hexavalent chromium vapor transport, which might confirm the supposition (Guo and Kennedy, 2001). In the mass spectrometric flame sampling study, adding up to 2% of $CrCl_2$ to a flame, the $CrO_2(OH)_2$(g) ion was reported but no ions containing Cl in spite of the 4% of HCl in the flame (Farber and Srivastava, 1973). Possibly, more important was a simple flame study by Bulewicz et al. (1961). This added a trace solution of a chromium salt into fuel-rich hydrogen premixed flames on less than a part per million scale. Chlorine addition to the flame was $<0.5\%$ of the burned gases. A clear balance was apparent between atomic Cr and CrCl in these flames leading to a very adequate estimate of its bond strength and confirming this chemistry. In spite of the presence of trace particles in the flame of presumably Cr_2O_3(s) the CrCl appeared to be dominant and unperturbed with negligible chromium oxychlorides or dihydroxides. If in the vapor state as CrO_3(g) or one of these other Cr^{VI} molecules, on cooling they will condense to reform Cr_2O_3(s) or possibly $CrO_2(OH)_2$(l) at low temperatures (Opila et al., 2007). If on an ash particle, time permitting, another option may be to form a chromate or add to a glass component. Its low level natural flame ion chemistry has been studied. Due to the high ionization potential of the transition metal elements, chemiionization depends on charge transfer from a natural flame ion such as H_3O^+ (Goodings, 1992). In the case of chromium the main ion observed is Cr^+ (>90%) together mainly with its monohydrated form. These are very minor components compared to neutral levels and have no properties that relate to combustion.

Chemical kinetic constraints

It is interesting to briefly assess the expected reaction chemistry of these oxychlorides and oxyhydroxides since it has never really been fully reviewed other than for assumed conditions of equilibrium. The bond strengths and heats of formation of expected participants now have been indicated above and appear to be reasonably established. Using any set of thermochemical values, it is clearly evident that chromium will have to encounter severe kinetic constraints in certain situations. This was first realized by Kashireninov and Fontijn (1998) and has only been cursorily addressed since in kinetic descriptions (Yu et al., 1998; Yu and Kennedy, 2000; Shah et al., 2012). It remains unsatisfactory. Equilibrium predictions can be useful, but as found for mercury, even though $HgCl_2$(g) is predicted by chemical thermodynamics to be totally dominant in any combustion system containing traces of Hg and Cl, reality is that there are no gas phase kinetics to facilitate this. Moreover, forming complex molecules in the gas phase, even $CrO_2(OH)_2$ the hydrate of CrO_3 and its chloro forms from atoms or smaller molecules often requires an efficient sequence of reactions and adequate time. As a result, like mercury it does appear that chromium's chemistry can be subject to kinetic constraints. In the case of chromium, the oxides should be able to form rapidly in oxygen-rich combustion, for example, with 5% O_2, 10% H_2O, at 1000°C and atmospheric pressure:

$$\mathrm{Cr + O_2 = CrO + O} \quad \Delta H_{298} = +49\ \mathrm{kJ\ mol^{-1}} \quad \tau_{1/2} \approx 1.2\ \mu s$$

$$\mathrm{CrO + O_2 = CrO_2 + O} \quad \Delta H_{298} = -44\ \mathrm{kJ\ mol^{-1}} \quad \tau_{1/2} \approx 11\ ns$$

$$\mathrm{CrO_2 + O_2 = CrO_3 + O} \quad \Delta H_{298} = +22\ \mathrm{kJ\ mol^{-1}} \quad \tau_{1/2} \approx 9\ \mu s$$

In other words atomic chromium can instantly become predominantly CrO_3(g), considering these representative reactions with the approximate half-lives indicated, assessed assuming reaction energy control. However, the major reactants in combustion that can hydrolyze these oxides sufficiently rapidly to the hydroxyoxides are H_2O, OH, and H_2, but these are strongly bound molecules, D_0 493, 424, and 432 kJ mol^{-1}, respectively.

Ebbinghaus (1993) indicated the strength of a single Cr-OH bond in any of the hydroxy-oxides to be in a range of values less than these, implying that any such proposed gas phase reactions will necessarily be quite endothermic. The larger concentration of H_2O will necessarily be needed to help overcome the adverse energy barriers. This severely reduces possibilities and isolates limited reactions such as the following to illustrate the point at 1000°C:

$$CrO_3 + H_2O = CrO_2(OH) + OH \quad \Delta H_{298=} + 155 \text{ kJ mol}^{-1} \quad \tau_{1/2} \approx 13 \text{ ms}$$

$$CrO_2(OH) + H_2O = CrO_2(OH)_2 + H \quad \Delta H_{298=} + 150 \text{ kJ mol}^{-1} \quad \tau_{1/2} \approx 8 \text{ ms}$$

In fact any reaction scheme has to further invoke chlorine to modify CrO_3 due to the encountered adverse kinetics. Consequently, in chlorine-free systems the O_2 and H_2O will undoubtedly maintain any free chromium as CrO_3, hexavalent, in the gas phase. Consequently, attainment of the type of equilibrated distribution illustrated by Ebbinghaus and others is certainly questionable at lower temperatures based solely on thermodynamic considerations unless molecules such as the hydroxyoxides are produced directly by vaporization. Of the experimental data, the transpiration studies bring some insight to this aspect. The fact that $CrO_2(OH)_2(g)$ was formed at temperatures of 135°C–185°C using $CrO_3(s)$ (Glemser and Mueller, 1964) has to imply that it had to be formed heterogeneously on the solid surface and vaporize. There are no homogeneous kinetic reactions that can hydrolyze $CrO_3(g)$ at these temperatures. The other studies used $Cr_2O_3(s)$ and different ranges of temperatures of 300°C–900°C (Opila et al., 2007) and 1300°C–1585°C (Kim and Belton, 1974). The mass spectroscopic H_2 flame study under reducing conditions did report the presence of $CrO_2(OH)_2(g)$ by adding $CrCl_2$ vapor to the flame, but a formation mechanism is not readily apparent (Farber and Srivastava, 1973). Its presence did confirm though that it was a major presence and not subject to kinetic constraints at 2250K. It is notable that Opila et al. (2007) proved a dominance of $CrO_2(OH)_2(g)$ at 900°C and reported that cool vapor collection surfaces indicated $CrO_2(OH)_2(l)$ below 600°C ($CrO_3(s)$ is reported to dissociate above 250°C) yet insoluble $Cr_2O_3(s)$ deposition at 600°C–900°C. This remains very strange in that if $Cr_2O_3(s)$ is depositing at the higher temperature there seems little reason to expect a valence change to Cr^{VI} at the lower temperature. Opila et al. (2007) did confirm the interpretation of Kim and Belton (1974) that at temperatures above about 1000°C the oxymonohydroxide could gain importance. This would probably be by the onset of the reverse reduction by H-atom, as above, which would still be faster than any thermal dissociation mechanism. It already seems that chromium might get an immediate jump start from some surface sources to instantly become $CrO_2Cl_2(g)$ or $CrO_2(OH)_2(g)$, which will obviously be very different from a gaseous Cr(g) origin. The latter has been briefly addressed and is more likely to encounter kinetic constraints. The former is interesting in that if chromium starts fully oxidized as Cr^{VI} in an oxidizing environment it will likewise be kinetically constrained to reduce its valency. In flames, with the presence of HCl (possibly at 1%) and H_2O (10%) there is little doubt that balances should couple the various oxychlorides and hydroxides not only between themselves but also with each other if chlorine is present. For example, low energy-barrier reactions clearly illustrate this.

$$CrO_2(OH)_2 + 2\,HCl = CrO_2Cl_2 + 2\,H_2O \quad \Delta H_{298} = -58\ \text{kJ mol}^{-1}$$

$$CrO_2(OH) + HCl = CrO_2Cl + H_2O \quad \Delta H_{298} = +16$$

$$CrO(OH) + HCl = CrOCl + H_2O \quad \Delta H_{298} = -8$$

$$CrO(OH)_2 + 2\,HCl = CrOCl_2 + 2\,H_2O \quad \Delta H_{298} = -57$$

$$CrO(OH)_2 + O_2 = CrO_2(OH)_2 + O \quad \Delta H_{298} = +31$$

$$CrO(OH) + O_2 = CrO_2(OH) + O \quad \Delta H_{298} = +46$$

$$CrOCl_2 + O_2 = CrO_2Cl_2 + O \quad \Delta H_{298} = +30$$

$$CrOCl + O_2 = CrO_2\,Cl + O \quad \Delta H_{298} = +69$$

The initial attempts to formulate a chemical kinetic model for the combustion of a Cr/H/Cl system were informative. In spite of the large number of possible reactions, constraints restricted products to be namely CrO_3, CrO_2, CrO_2Cl, and $CrO_2(OH)$ in initial efforts (Kashireninov and Fontijn, 1998). This disagreed with Wu and Bismas (1993) and Ebbinghaus's predictions (1993, 1995) of a dominance of $CrO_2(OH)_2$ and CrO_2Cl_2 for incinerators at equilibrium. A model for chromium in H_2/air flames also realized that kinetic results were differing from equilibrium expectations and that reaction kinetics had to be considered. Additionally by sampling experimental flames it was clearly seen that hexavalency existed at the higher temperatures but modified to the trivalency on cooling (Yu et al., 1998; Yu and Kennedy, 2000). No additional attempts to fully model the Cr/H/Cl system any further have yet been attempted although the necessary data are now much better characterized. In combustion, the corrosion of Cr based steels by alkalis has been a major problem for decades even at lower temperatures (Lehmusto et al., 2012). In a laboratory study, emphasizing more the nature of alkali deposition from flames onto surfaces, the aspect of corrosion of an Inconel-600 stainless steel probe (72% Ni, 14%–17% Cr, 6%–10% Fe) gave an interesting insight into this remarkably fast surface interaction between NaCl and the steel surface (Schofield, 2012a). Propane or hydrogen/air flat flames seeded with a nebulized sodium nitrate solution at a calibrated rate were used at 1700K–2700K with fuel rich or lean conditions to vary the alkali flame speciation of NaCl, NaOH, Na concentrations. What was surprising was that deposits were seen to be occurring above 900K when the vaporization pressure was too high for NaCl to remain on the surface. Analysis by ICP-AES and an assessment of vapor pressures indicated the small unexpected deposits were mainly soluble Na_2CrO_4. As a result, even when the probe was at a temperature too high for any NaCl to be retained, short-lived NaCl formation continued, induced rapid corrosion that could retain it while in the instantaneous process of vaporizing. It showed that metal elements depositing onto surfaces in combustion processes from flame gases is generic, and always occurring, simply controlled by influx/egress rates and thermodynamic/kinetic preferences that establish any permanent outcome. Otherwise it continues unseen and unnoticed.

Combustion power plant data

As listed in Table 1.1, coals, wastes, sewage sludge, and fuel oil can each have reasonable trace levels of chromium, generally more so in biomass, and can be significant in wastes and sludge. It is present but at much lower concentrations in pet coke and limestone (Lind et al., 1994; Cordoba et al., 2012). Biogases may reflect higher values than natural gas (Cachia et al., 2018).

Examining fuel combustion results for clues of behavior, the measures of chromium will be more reliable in waste and sludge type combustion where its concentration is larger. It still remains difficult to reliably measure the speciation of chromium. Care has to be taken due to the ease that can occur for Cr^{VI}/Cr^{III} interconversion during the handling or processing of samples. The ICP-AES or ICP/MS instruments have no difficulty monitoring total chromium (Boumans and Vrakking, 1987) but require a coupled chromatographic separation method for speciation (Bednar et al., 2009; Wolf et al., 2011). Nonchromatographic methods are available and their advantages and drawbacks have been reviewed (Gonzalvez et al., 2009). Sampling of atmospheric particles also has encountered problems on filters questioning accuracy and data bias (Huang et al., 2013a; Torkmahalleh et al., 2013). Another well used method invokes X-ray absorption, which now appears to have been validated and which clearly resolves the spectrum of Cr^{III} from that of Cr^{VI} but of course needs a synchrotron radiation laboratory and is testing old nonvirgin samples that may have changed (Huggins et al., 2004; Ohata and Matsubayashi, 2014). Unanswered questions in combustion also revolve around sampling. A filter can be a passive or an active device. For example, it is now found that fabric filter bags can enhance Hg oxidation. In the present case filtering at a lower temperature might produce a larger deposition of Cr_2O_3(s) or this might be eroded by chlorine and then elevate any Cr(VI) later impinger solution collection. However, evidence would seem to indicate that Cr(VI) has a difficult time condensing from the gas phase and maintaining its valency. The analytical difficulties have led to the current regulations that control solely total chromium, which now is seen not to be the real concern. Because of the difficulties, all earlier work solely measured the total chromium mass balance in combustors. It was found that essentially all of the chromium was in the various ash or slag forms and often roughly divided between bottom ash and fly ash but somewhat dependent on the nature of the fuel (Meij, 1994; Reddy et al., 2005; Goodarzi, 2006; Marrero et al., 2007; Shao et al., 2008; Swanson et al., 2013; Zhang et al., 2013; Wu et al., 2013; Duan et al., 2016). It did not appear to have any significant gas phase component. In pyrolysis of coal, no release of Cr was observed at 1150°C or 1000°C (Wang and Tomita, 2003; Guo et al., 2004), and in coal/biomass mixtures no Cr was observed below 1400°C (Stam et al., 2011). A small loss of Cr was noted when dried sewage sludge was treated with 0.5% Cl and heated at 850°C (Luan et al., 2013). One defect that arises in many of the earlier papers is that the combustion temperature and fuel chlorine levels often were not indicated impairing a complete assessment. A very recent study examined the chromium in 21 Chinese coals. Its concentrations varied from 0.1 to 179 ppmw with one Stone coal registering 1738 ppmw. Indications from leaching methods of the raw coals suggested about 39% was associated on average with silicates, 36% organic, 3% carbonate, and 9% sulfides (Liu et al., 2016e). In a vaporization tube with flowing O_2 it was noted that on average at 950°C 13% of the chromium was released, about 27% from a lignite and 2% from an anthracite. No chlorine levels in the coals were quoted. But the complication of coal rank was evidenced. Generally it has been accepted that chromium is only volatile above about 1500°C and essentially controlled by the onset of Cr_2O_3(s) dissociation. As a result, it is interesting that several studies have reported loss of chromium at temperatures as low as 800°C. However, a closer examination of such reports indicates that these may be questionable being mainly based on difficult mass balance differences (Miller et al., 2003; Spears and Martinez-Tarrazona, 2004; Huang et al., 2004; Miller et al., 2004; Zhang et al., 2013). Concentrations in fly ash have been seen to increase with temperature (Wu et al., 2014a) and be higher at the ESP inlet than in the bottom ash (Bhanarkar et al., 2008). A better understanding of chromium's potential volatility comes from examining fly ash particle concentration as a function of size. Numerous studies have confirmed an enrichment of chromium in coal ash particles as they approach micron or submicron sizes (Quann et al., 1990; Martinez-Tarrazona and Spears, 1996; Nielsen et al., 2002; Spears and Martinez-Tarrazona, 2004; Goodarzi et al., 2008; Shah et al., 2012). Also a change was noted between the collection fields of an ESP (Huang et al., 2004) and in the final escaping fine particles in the flue gas (Meij, 1994). Direct (in-flame) fly ash sampling studied four differing coals. Burning these in an entrained flow combustor at about 1500°C, after a residence time of about 1.6 seconds, ash was sampled at 1150°C, rapidly cooled, diluted and size segregated (Senior et al., 2000). Even with a coal of 20 ppm Cr and 0.17% chlorine, little chromium was found in the $<0.4\ \mu m$ particles indicating possibly low volatility under these conditions or too rapid a cooling for any deposition to occur. A more detailed study was reported of ash analyzed from the output side of an ESP in a lignite fired combustor (Wang et al., 2008). This ash was segregated by 13 stages and categorized as nano (0.03–0.06 μm), ultrafine (0.06–0.1), submicron (0.7–1.1) and fine sizes (1.8–2.7). Unlike the nonvolatile Ti and Zr that can also appear in such particles by forceful evacuation from larger particles, it does seem that the presence here of chromium probably did result by gaseous transport. In particular, the enrichment factors are quite significant in the nano- and ultrafine sizes. However, the mechanism of transport has not been resolved but seems to be gaseous.

In recent years, attention now has been directed more to factors controlling the speciation of the chromium. The environmental concern centers on chromium's potential ability to produce Cr^{VI} molecules particularly the chromates that are readily leachable and toxic. For coals and other fuels the natural input of chromium is Cr^{III} (Huffman et al., 1994; Huggins et al., 2000; Goodarzi and Huggins, 2005a). It is hard to have an input source of Cr^{VI} except with a molecule such as $Cr(CO)_6$ or possibly from the addition of heavy industrial waste taken by the cement industry (chromates), which are not otherwise encountered. As a result, some early work that assumed additions of CrO_3 solution could be compared to those of Cr^{III} did not then realize that these were essentially equivalent sources (Linak et al., 1996; Linak and Wendt, 1998). Also it was very quickly realized in spite of analytical difficulties that most combustion ash had a very low content of Cr^{VI} in FBC, incinerators or any combustor where temperatures remained below about 1600°C (Huffman et al., 1994; Huggins et al., 1999; Goodarzi and Huggins, 2005a; Shah et al., 2008, 2012; Goodarzi et al., 2008; Nelson et al., 2010; Jiao et al., 2011a; Hu et al., 2013). Bottom ash contained essentially none and fly ash <2% (Shah et al., 2012). One recent study indicated Cr was mainly in bottom ash with incineration and there simply by entrainment and a minimal problem (Li et al., 2018). In fact it was suggested that coal combustion should no longer be categorized as an air pollutant of Cr^{VI} (Huggins et al., 1999). However, there had been a few exceptions when more significant measures were recorded. One noticed a geographic effect in that using a down-fired combustor three Eastern America coals showed negligible Cr^{VI} while four Western coals were in a range of 9%–26%, this being tentatively ascribed to calcium chromate ash formation (Shoji et al., 2002). A similar ash study showed a dependency on coal type one reflecting 43% Cr^{VI} and an Illinois coal 6% even though furnace temperatures were <1300°C (Galbreath et al., 2000; Galbreath and Zygarlicke, 2004). A final example has come from cofiring coal with wood when levels of 12%–16% were noted (Stam et al., 2011). Although thermochemical equilibrium modeling for incineration conditions has been published it remains unacceptable. Kinetic constraints are present and the mechanisms need further detailed assessments (Chen et al., 1998).

From the known chemistry, the formation of Cr^{VI} might be expected either from the oxidation of Cr_2O_3(s) but only at quite high temperatures, from HCl oxidizing the ablative evaporation of this as CrO_2Cl_2(g) or from the formation of metal chromates in the ash. A recent study is the best illustration of such potential chromate formation (Verbinnen et al., 2013). They noted that heating Cr_2O_3(s) in air with alkali hydroxides (NaOH(s), KOH(s)) or alkaline earth oxides (MgO(s)-BaO(s)) promoted this formation of the Cr^{VI} molecule. Similar interactions with Fe_2O_3 and other oxides have been examined (Chen et al., 2013). Such interactions had been suggested earlier in two other studies that heated coal or ash with lime at 500°C–900°C for periods noting the presence of Cr^{VI} (Galbreath et al., 2000; Hu et al., 2013). Due to this apparent ease of alkali and alkaline earth chromate formation, their formation on ash might be expected. Studies now are refining this aspect by shaping the speciation with controlled quantities of alkali or Ca available to the Cr (Swietlik et al., 2014; Hu et al., 2015a). A most recent study has suggested adding Al_2O_3, Fe_2O_3, or SiO_2 to react with any alkali or alkaline earth element to minimize any chromate formation during the heating of solid waste materials containing chromium (Mao et al., 2016). However, as seen in the ash of residual oil combustion, although Na and K are present, the chromium tends to be present in its trivalent form $Cr_2(SO_4)_3$ (Galbreath et al., 1998). Chromate chemistry is assessed in ash thermal processing for agricultural usage. CaO levels should be low so that chromium transfers into any aluminosilicate glass rather than produce $CaCrO_4$. Nevertheless, above 1000°C this can be reduced back to the trivalent chromite $CaCr_2O_4$ (Hu et al., 2013, 2015a). However, because most of the chromium is retained as Cr^{III} in the ash, calcium levels should be kept low to avoid promoting calcium chromate formation and its potential leaching (Hu et al., 2013; Zhao et al., 2013a). Sewage sludge ash is often treated with Cl-donors ($CaCl_2$, $MgCl_2$, NaCl) and heated at 1000°C. Several heavy metals are released in this manner but not chromium. However, if Na_2CO_3 is added this is seen to promote a 10%–15% formation of Cr^{VI}, presumably Na_2CrO_4, but can be prevented by using $MgCO_3$ instead that promotes trivalent $MgCr_2O_4$ at 600°C (Vogel et al., 2014). An additional study of Cr containing sludge in the presence of CaO confirms chromate formation at 500°C–900°C converting to the trivalent chromite $CaCr_2O_4$ at 1000°C–1200°C but similarly preventable by adding Mg when the chromite is produced at lower temperatures, 900°C in their case (Mao et al., 2015). In a modern coal combustor retention of ash is now at very high levels. Nevertheless, 100% is rarely achievable however closely approached. One recent measure of the effectiveness of a FGD system in a co-combustion plant burning coal and pet-coke indicated that any stray Cr was further reduced significantly across this final control device (Cordoba et al., 2012). Recent Chinese studies of large coal-fired combustors have monitored the trace element behavior across the now numerous control devices. All are seen to be effectively captured within the power plant with efficiencies lowering their concentrations by more than three orders of magnitude below their ppmw levels in the coal. Ba can be removed most effectively (five orders of magnitude reduction) but surprisingly Cr the least by just more than the three orders (Zhao et al., 2016). The more recent Chinese power plants now are equipped with an SCR, ESP, fabric filters and a wet FGD removing chromium as ash or present in the scrubber gypsum (Wang et al., 2017a; Zhao et al., 2017a, 2018; Zhang et al., 2017a; Zheng et al., 2017). Leaching can occur from gypsum and fly ash implying care should be taken in its disposal and indicating a preference for more efficient precipitator ash removal (Hao et al., 2017; Zhang et al., 2017a; Zhao et al., 2018d). In one study, Cr was seen to have some slight water leaching from FGD gypsum, but was regarded as a low risk (Zhao et al., 2018b).

The history of chromium in combustion is mainly one of low volatility and retention as some ash form. Although the laboratory transpiration experiments indicated that chlorine induced evaporative vaporization of CrO_2Cl_2(g) from stable Cr_2O_3(s), this generally appears to play a minor role in practical combustors (Pedersen et al., 2010). Whether this is due to the chlorine being as HCl and not Cl_2 is not clearly indicated, the latter generally is more reactive. Hydrolysis to the dihydroxyoxides normally would be expected unless chlorine levels were very high. This could then dehydrate and deposit as CrO_3(s) at cooler temperatures. However, no such Cr^{VI} deposits or signs of volatility with chlorine generally are reported (Linak et al., 1996; Linak and Wendt, 1998; Jiao et al., 2011a; Wu et al., 2013). There have been suggestions that high chlorine content can locate more chromium into the $<10\,\mu m$ particles (Waterland et al., 1994), or remove it from particles $<5\,\mu m$ (Jiao et al., 2011a) but such results need confirmation as does mechanistic speculation as with recent modeling (Chen et al., 2012). Consequently, chromium's combustion reflects one needing high temperature and being constrained by kinetics. All the many published thermochemical equilibrium analyses now are generally of little value other than at hinting what might occur, they should not be invoked by further speculation. Such calculations still emerge but now should exclude chromium (Sorum et al., 2004; Lundholm et al., 2007; Poole et al., 2008; Roy et al., 2013). A recent study that is intriguing has shown an effect on the fly ash composition of reducing the flue gas temperature from 250°C to 150°C through a heat exchanger. The fuel was biomass from metal contaminated soils burned in a small pilot-plant boiler, and included monitoring 30 elements including chromium in the ash samples. Boiler temperatures are not quoted but a twofold increase in ash concentration appeared to occur for chromium over this temperature range and also numerous other elements as the temperature was so reduced through the heat exchanger. No further metal deposition was possible to explain this increase in concentration that was similarly reported for such elements as Ti or Zr that are known to be fully nonvolatile. Instead it seems apparent that particle segregation was occurring for chromium and the ash size distribution was changing favoring more of the small enriched sizes at the lower temperature (Kovacs et al., 2018). It illustrates the care still needed in interpreting data that is rather limited.

Another group of data concerning chromium centers on hazardous waste disposal. One special case is the combustion of chromated copper arsenate wood. Although now banned from residential use, large quantities of wood for nonresidential uses such as railroad ties, poles and marine timber still are produced. Landfill introduces leaching problems and combustion requires acceptable incineration. Domestic fire and stove burning has been seen to be very hazardous due to its high mineral content. Two laboratory studies confirmed that the wood could be incinerated at low temperatures (<1000°C). X-ray ash analysis indicated that

Cr as Cr^{III} and Cu being nonvolatile are retained in the ash and only 11%–14% of the more volatile As volatizes (Wasson et al., 2005; Lundholm et al., 2008). Another area of research relates to the cement industry. The rotary kilns used for forming the initial clinker cement product operate at about 1450°C but have a flame heater that can be a few hundred degrees hotter. Residence times are about 8 seconds and although fueled generally by coal they are increasingly being used globally as a method for disposing heavy industrial wastes (Huang et al., 2012). These can include materials such as electroplating sludge, red mud, automobile shredder waste, iron and steel slags, tanning wastes, coal fly ash and contaminated soil, some of which can be very rich in chromium probably as chromates. Although very well controlled, with 99.7% of the Cr becoming fixed in concrete products, concerns center mainly on their potentiality for excessive leaching. With additions of up to 5% red mud it is seen that about 35% of the chromium becomes water soluble as the hexavalent chromate (Vangelatos et al., 2009). In fact it has been suggested that chromium wastes should no longer be processed in such a manner (Sinyoung et al., 2011). From what is now known of coal combustion and incinerators with a less transit time it would seem that these might be a preferable mode of disposal cofiring with coal and the chromium being predominantly retained in the ash as Cr^{III}. One suggestion for the cement industry has been to sinter the clinker under reducing conditions which then extensively lowers the Cr^{VI} content (Chen et al., 2012a). Nevertheless, reports, for example, still emerge with cement dusts that are found to contain 20% of their chromium content as Cr^{VI} in the United States and 11% in a Nigerian plant (Ogunbileje et al., 2013). As mentioned above, better tailoring of conditions might modify the Cr into a chromite state. China is the largest producer of cement, 22 billion tons in 2012. However, a recent inventory probably not including any chromium waste disposal indicates low Cr emissions that have changed only twofold in the last 30 years, 74 tons $year^{-1}$ in 1980 and 166 in 2012 but with no indication of speciation (Hua et al., 2016).

In coal oxy-burn tests at 1273K and added HCl/SO_2, the coal Ca and Fe were noted as capturing the Cr. However, the higher levels of H_2O through HCl reactions appeared to oxidize some Cr^{III} to Cr^{V} in the ash, which needs further study. The SO_2 favored more retention as the trivalent sulfate (Li et al., 2018a).

With coal, ash mass balances for chromium now appear adequate, little gaseous release is expected and measured stack emission losses are very low with adequate particulate control (Meij, 1994; Nielsen et al., 2002; Goodarzi, 2006a). Nevertheless, urban atmospheric testing in some regions indicates Cr^{VI} in the aerosol particles: 30%–40% near transport roads and above 10% in most regions, rather surprising when biospecies in nature generally effectively reduce the Cr (Swietlik et al., 2011; Sillerova et al., 2014).

Little has appeared on the use of chromium for nanoparticle or thin film technologies. It would seem that if needed nontoxic $Cr_2O_3(s)$ nanoparticles could be readily formed by flame methods (Bulewicz and Padley, 1971a; Guo and Kennedy, 2004). A very high melting point film of lanthanum chromite, $LaCrO_3$ was reported (≈2450°C) using lanthanum and chromium nitrates by propane flame deposition and a substrate surface at 750°C (Ichinose et al., 1994a). Also plasma enhanced vapor deposition of $Cr(CO)_6$ produced $Cr_2O_3(s)$ thin films at 250°C–400°C. At that temperature any $CrO_3(g)$ would have no other formation choice (Wang et al., 2008a). Likewise the thermal decomposition of either $CrO_3(s)$ or $CrO_2Cl_2(s)$ in O_2 at 330°C–370°C can produce either metastable $CrO_2(s)$ or $Cr_2O_3(s)$ depending on the substrates (Cho and DeSisto, 2003). An intriguing study of mixed metal (Mn, Cr, Fe) oxide nanoparticles uncovered another aspect of potential interest. The particles were collected as their separate oxides or their oxide mixtures using a hydrogen diffusion flame to which the respective metal carbonyls were added. It was noted both in their formation and also later in their aging that the metal oxide mixture could influence each other's oxidation states and so possibly modify their toxicology or leachability (Nico et al., 2009).

In summary, there is now a reasonable understanding of chromium's potentially complex combustion chemistry. The nature of most combustion methods with their temperature and time constraints generally will leave the chromium in ash form in its nontoxic Cr^{III} form. Coupled with modern-day particulate control, combustion no longer should need to be considered an air pollutant for chromium. It is present in raw coal as Cr^{III} and its release from the fuel needs a high temperature of above 1000°C and seems not to be facilitated by chlorine. Volatile chromium most probably results from Cr_2O_3(s) dissociation and may lead to Cr^{VI} at high temperatures but then has a valence change on cooling. In coal ash, many of the other metal components appear to favor elements such as sulfur, chlorine or silicon rather than reacting with chromium. Other than Cr_2O_3(s), sulfate and metal chromates, chromites, few other stable condensed phase molecules in practical systems should be produced. Also because most chromates dissociate at high temperatures to trivalent chromites, time constraints probably inhibit their final detection. CrO_3(s) is known but is metastable, dissociating at low temperatures to Cr_2O_3(s). Although with water it forms chromic acid, $CrO_2(OH)_2$, possibly catalyzed by chlorine it will condense on cooling to Cr_2O_3(s). Future regulations concerning chromium should concern Cr^{VI} only. In combustion, it is one of few metallic elements that exhibits pronounced chemical kinetic constraints and requires an adequate homogeneous–heterogeneous model for meaningful engineering application. Time and kinetics limit its ability to exploit its array of Cr/O/H and Cr/O/Cl molecules that it can produce. Care has to be taken with the use of rotary cement kilns in disposal of chromium wastes, which differ having longer retention times and undoubtedly can produce leachable chromates within the ash, that then have to be controlled by further thermal heat treatments.

7.12 Cobalt

The nickname, the Goblin metal, arose from miners who were not at ease working with its minerals. This arose because its natural ores can be associated with arsenic in the forms $CoAs_2$ (smaltite) and CoAsS cobaltite, so resulting in a somewhat toxic environment. Various other sulfur based ores exist but now the majority of cobalt is produced as a by-product from Cu and Ni processing. This is because in nature, cobalt is widespread in the Earth's crust being mainly associated with these metals and also with As, Pb, and Fe. It is not common, being about the 33rd most abundant element in the Earth's crust and is at very low levels in the oceans. Although historically in use for more than 5 millennia to color ceramics and glass, only in the past century was it introduced around the 1940s as a pure metal, finding initial use mainly as an aircraft alloy and in magnet technology.

Global refined production now appears to be about 125,000 tons $year^{-1}$ with available resources approximated to be in a range of 10 (World Health Organization Co, 2006) to 30 million tons (Mudd et al., 2013) and probably more if possible recycling components were included. Basic information on the cobalt market and activities is listed by the Cobalt Development Institute located in the United Kingdom (CDI, 2015). Cobalt is not regarded as a critical mineral in current economic analyses (National Research Council, 2007). Its supply is tied to the production outputs of Cu and Ni. More importantly it is also related to the efficiency of the separation process for these metals which is also somewhat market driven. Extractions can be as low as 10% or as high as 80 + % introducing a significant variable. Prices for metallic cobalt have been much higher in previous years but now are quite low at \$15 lb^{-1}. An interesting recent life cycle analysis for cobalt also indicates an increase in recycling of the element, which may be needed more in the future (Harper et al., 2012; Kim et al., 2006). It is noteworthy that 11 tons $year^{-1}$ of all minerals are needed for every US citizen to maintain the current lifestyle so recycling may become necessary at some point for most elements (National Research Council, 2007). Although accurate assessments of atmospheric releases of cobalt have not appeared, there are indications of approximate levels. Cobalt's nature is nonvolatile so its presence in the atmosphere will be as particulates. Natural emissions are predominantly from blown dust, sea spray, volcanoes and forest fires, with global estimates totaling about 6000–7000 tons $year^{-1}$ (Nriagu, 1989; Malakhov and Makhon'ko, 1990; Kim et al., 2006). Anthropogenic sources are mainly from combustion of coal, from sewage sludge, phosphate fertilizers, mining and smelting. These have been listed as possibly being slightly below natural levels (Malakhov and Makhon'ko, 1990) or comparable (Kim et al., 2006). Recent sampling of snow-pits in the mountains of Tibet, although showing signs of anthropogenic contributions for other metals indicated low enrichment factors for Co (Huang et al., 2013). Although meteorites are mainly iron and nickel they can contain low levels of cobalt. Since the introduction of $LiCoO_2$ battery technology and now based on Li/Ni/Co/Al oxides, this has become a major end use of cobalt. However, current battery research aims to reduce or remove the role of cobalt due to its cost. The Cobalt Development Institute now indicates consumptions by the battery industry (38%), superalloys (17%), and hard metals (10%) (CDI, 2015). Other components would be as catalysts, semiconductor components (McCoy, 2014), and chemicals for ceramics, glass, paints, and ink.

Health concerns with cobalt

Cobalt is another of the elements critical to life. Needed only in trace amounts, 5–40 $\mu g\ day^{-1}$ it is the essential component of vitamin B_{12}, a coenzyme required by humans and animals. It has numerous roles in the brain and nervous system, in blood formation, cell metabolism and in DNA synthesis and regulation. It is a reminder of how the human body is so finely tuned. Humans acquire vitamin B_{12} through diet, and animals have to graze and have the necessary bacteria to produce it. Due to the importance of such trace amounts in the body, the vitamin is stored in substantial amounts in the liver, sufficient for several years. It is one example of the body's intelligence in housekeeping for future unforeseen events; somewhat contrary to natural evolutionary concepts. In parts of the world where animals cannot acquire sufficient components from the earth it is necessary to supplement their feed. B_{12} is the largest and structurally complex vitamin. The Co atom, mass 59 sits in the center of the molecule with molecular weight 1355. Only bacteria and archaea have the enzymes to synthesize it. It is still artificially produced today by bacterial fermentation synthesis, about 46 tons or so a year. It was discovered from its relationship with pernicious anemia. This illness is when certain gastric stomach cells involved in the vitamin's absorption are destroyed. Many people have died in the past due to this deficiency. Due to this requirement for B_{12}, agricultural use on the land of manures and sewage sludge are beneficial from the point of view of their cobalt content. Moreover, constant applications over long periods pose no risk with regards to this metal's accumulation (Eriksson, 2001).

Medically, two radioactive isotopes ^{60}Co (half-life 5 years) and ^{57}Co (272 days) have found major use in radiation therapy for cancer treatment. Cobalt is present in a broad spectrum of foods. Total body burden is about 1.1–1.5 mg scattered throughout the body. Its toxicology has been very extensively studied and it is not regarded as a toxin nor carcinogen (Lauwerys and Lison, 1994;ToxProfile Co, 2004; Kim et al., 2006; World Health Organization Trace elements 2006). Concerns that do arise have centered on industrial workers who encounter Co in particulate form. Inhalation over time can result in lung illnesses that have been quantified by many animal studies. Cobalt is rarely detected in drinking water so that the human food ingestion is the dominant source. In nature, plants are very tolerant of cobalt, absorbing it from soils. Some can even bioaccumulate the mineral which will be mainly as the divalent form, Co(II). It becomes toxic only at high levels. Its use in some joint and dental prosthetic devices can introduce Cr-Co alloy wear that is only a concern if medical test levels become elevated. Its favored biomonitoring methods appear to be by either blood or urine samples. The Mayo Clinic (2018) suggests $<1.0\ \mu g\ L^{-1}$ and $<2.0\ \mu g\ L^{-1}$ for these, respectively, with other test centers satisfied by slightly higher levels. Surveys, rarely include cobalt and in those published generally for western adult populations these levels are rarely exceeded (Goulle et al., 2005; Cesbron et al., 2013; Nisse et al., 2017; CDC US 4th National Report, 2009, 2018). In some countries, depending on geographical locations, they can display slightly more elevated levels in children but pose no health risks (Cusick et al., 2018). Its levels in hair are well below 1 ppmw for even outlier cases (Goulle et al., 2005; Drobyshev et al., 2017). Grazing animals have the necessary bacteria in their digestion stomachs to synthesize B_{12}. An interesting review of the nature of cobalt in terrestrial waters has recently appeared summarizing its soil/water chemistry. Divalent cobalt is dominant due to the general low solubility and oxidizing nature of Co(III) (Collins and Kinsela, 2010).

Cobalt's basic chemistry

Cobalt, ^{59}Co, atomic number 27, being a well-entrenched transition element is rich in electrons $3d^7 4s^2$. Nevertheless, the higher oxidation states are difficult to achieve and its basic chemistry is rather simple displaying predominantly divalency with few trivalent compounds. These latter are less stable due to their oxidizing tendencies. The pure metal (m.pt.1495, b.pt.2927°C, and vapor pressure 10 ppmv at 1509°C) is stable in air or water at ambient temperatures, is magnetic but in the metallic state finds little use outside of alloying with other metals. The oxides again are important, Co_3O_4 of mixed valence (II/III) is stable to about 840°C when it starts losing oxygen changing to CoO (m.pt.1933°C). The process is reversible in that CoO at low temperatures becomes metastable and absorbs oxygen. $CoCO_3$ dissociates to CoO above 130°C (vapor pressure 100 ppmv at 230°C) (Munir and Rice, 1981; Stern, 2001). A sulfate and its hydrates are known but dissociate to CoO and Co_3O_4 at about 700°C (Ostroff and Sanderson, 1959; Tagawa, 1984; Tagawa and Saijo, 1985). The nitrate decomposes to Co_3O_4 at about 100°C apparently in a single stage (Stern and Weise, 1966). A hydroxide can be precipitated from solution, $Co(OH)_2$, but is oxidized in air to $Co(OH)_3$ and dissociates at about 160°C to CoO. The latter monoxide also reacts with water to produce $Co(OH)_3$ which is similarly prone to dehydration on heating regenerating CoO. In recent years, nanoparticles of CoO have grown in importance. They generally are synthesized by thermal dissociation or solution combustion methods (Deng et al., 2014). They could be readily made in flame burned gases. A problem they may have is that their surface will gradually oxidize in air to the more stable Co_3O_4 oxide but this seems to produce a protective coating. The Co_3O_4 oxide now is used as a semiconductor and can form nanoparticles (Salavati-Niasari and Khansari, 2014). The dichloride (m.pt.740, b.pt.1049°C) is soluble in water and is hygroscopic. It readily forms hydrates and can be a desiccant. It is used in humidity sensors, where humidity modifies its color from sky blue to purple and then pink. It can also be a sensitive monitor for trace water leakage. It is volatile (1 ppmv 460°C, 10 ppmv 515°C) forming predominantly the gaseous monomer. However, when heated in air it decomposes to Co_3O_4 at about 400°C–500°C. Various sulfides are known, CoS, that can have fractional sulfur to Co_9S_8 and these are not well defined stoichiometrically. CoS_2 and Co_3S_4 can all be oxidized to $CoSO_4$ or the oxide and dissociate on heating. They are also vulnerable to hydrolysis at elevated temperatures but have melting points about 1200°C if in an inert environment. Hydrides are not easily formed and are poorly characterized. They have recently been synthesized at high pressures where they gain stability (Wang et al., 2018b). With the advent of semiconductor and slag compositional studies, interest has turned to the mixed metal oxides that might be important in cobalt's chemistry. Of these $CoTiO_3$ has been studied in detail with Knudsen cell-mass spectrometry (Balducci et al., 1984). Co_2TiO_4 is stable to 750°C then forming $CoTiO_3$ up to about 1100°C followed by more stable $CoTi_2O_5$ to about 1430°C. It becomes a question of thermodynamic stabilities of various structures in the condensed phase with a myriad of complexities that also include time constraints for formation. Other possibilities in combustion are the known $CoAl_2O_4$ (m.pt.1057°C), $CoSiO_3$ (m.pt.1420°C), Co_2SiO_4 (m.pt.1417°C), and $CoFe_2O_4$ (m.pt.1600°C) (Dyachenko et al., 2016). Cobalt blue glass is of course well known and can be K/Si/Co or Co/Al/Si based. The use of $LiCoO_2$ in solid-state batteries has helped to introduce the word amphoteric more so into cobalt's chemistry. Although Co_2O_3 does not readily exist in the condensed phase, the literature can be confusing in discussing it as another amphoteric oxide of cobalt that decomposes at 895°C. The term amphoteric is old and relates to mainly oxides and hydroxides in solution if they can display both an acidic and a basic character. In fact more than 60 years ago this was noted for $Co(OH)_2$ in that it can dissolve both in HCl and NaOH solutions (Gayer and Garrett, 1950). Alkali and alkaline earth cobalt mixed oxides are not yet well characterized but possibly exist. However, at present, cobalt is better known for its cationic oxide nature. Of other compounds carbonyls can form with $Co_2(CO)_8$ being the best known but is stable only to 50°C. Now other gaseous organocobalt compounds are available such as those invoking c-pentadienyl structures and with higher boiling points. Consequently, from the point of view of combustion, cobalt has a limited condensed phase chemistry. Its range of molecules relating to combustion and discussed above is summarized in Table 7.20.

Potential combustion outcomes were portrayed in an early equilibrium analysis (Frandsen et al., 1994). As temperatures are raised most cobalt salts dissociate to the oxides, any chloride vaporizes and becomes oxidized. As a result unless the cobalt is present

TABLE 7.20 Properties of cobalt's condensed phase molecules relevant to combustion, °C.[a]

m.pt. b.pt. vapor pressure
Co 1495 2927, 10 ppmv/1509°C, ins
CoO 1830, >1400 dissoc to Co(g), oxid to Co_3O_4, reacts with water to $Co(OH)_3$, ins
Co_2O_3 895, dissoc to CoO, ins
Co_3O_4 >840 losses O_2 becomes CoO(s), ins
$Co(OH)_2$ >143 dissoc to CoO, oxid in air to $Co(OH)_3$, ss
$Co(OH)_3$ decomp to CoO, ins
$CoCO_3$ >182 decomp to CoO, 100 ppmv CO_2/230°C, ins
$Co(NO_3)_2$ >100 decomp to Co_3O_4, s
$Co(NO_3)_3$ unstable, hygro, s
$CoSO_4$ >965 dissoc to CoO, Co_3O_4 s
$CoCl_2$ 721 1081, hygro, decomp in air 400–500 to Co_3O_4,10 ppmv monomer/515°C, s
Co_2P 1386, hygro, ins
$Co_3(PO_4)_2$ ins
CoS 1182, nonstoichiometric, >680 dissoc to Co_4S_3, oxid to $CoSO_4/Co_3O_4$, hydrol, ins
CoS_2 <190 stable, >270 decomp, ins
Co_2S_3 unstable
$Co_2(CO)_8$ 51, decomp, ins
$CoAl_2O_4$ 1967, stable, ins
$CoFe_2O_4$ 1600, ins
$CoSiO_3$(meta-) 1420, ins
Co_2SiO_4(ortho-) 1417, ins
$CoTiO_3$ 1463, stable 750–1100, converts gradually to finally $CoTi_2O_5$ at 1430, ins
Co_2TiO_4 1562, dissoc >750 to $CoTiO_3$, ins

[a]*See text.*
dissoc, dissociates; *decomp*, decomposes; *s*, water soluble; *ss*, slightly soluble; *ins*, water insoluble; *oxid*, oxidizes; *hygro*, hygroscopic; *hydrol*, hydrolyzes.

in the ash as a silicate or mixed metal component it becomes an oxide. Up to 657°C this will be the Co_3O_4; from 837°C to 1377°C CoO(s) will become dominant at which time this will begin to dissociate. Under reducing gasification conditions, options are even fewer with sulfides breaking down to CoS, which itself is only stable to about 680°C producing free Co(s) particles.

Gas phase stabilities and transport

In the gas phase, the stability of CoO(g) is reflected in the gaseous CoO/Co distribution above heated cobalt oxides. Vaporization is mainly dissociative with Co(g) being dominant over CoO(g) (Grimley et al., 1966). Above 1627°C most of the cobalt in the fuel will have been released as Co(g), if time permits for thermal equilibrium and no reactive retention occurs. The two early major Knudsen cell-mass analysis studies established its bond strength from such measurements (Grimley et al., 1966; Smoes et al., 1972). They showed that the condensed phase vaporization was predominantly a dissociative process with only about 1% remaining as gaseous CoO(g). However, the vapor pressure of Co(g) approximated to only about 1 ppmv at 1400°C under equilibrium conditions. These results were later reanalyzed, when knowledge of the ground electronic state of CoO still remained uncertain as also its low-lying states and a $^4\Sigma^-$ ground state was assumed (Pedley and Marshall, 1983). Since then, the very complex spectroscopy has been partially resolved (Merer, 1989). Now the $X^4\Delta_i$ has been proven to be the ground state with a $^4\Pi_i$ state 5514 cm^{-1} above this in energy (Ram et al., 1993). To these an expected $^4\Sigma^-$ state was assigned to a 3038 cm^{-1} energy and two other states to 5989, and 9105 cm^{-1} energies (Barnes et al., 1997). Since then absorption spectra of CoO isolated in solid neon also have hinted at the possible locations of four other low-lying states; $^4\Sigma^-$ 3390, $^4\Pi_i$ 5837, $a^6\Delta$ 7028, and $b^2\Delta$ 10,131 cm^{-1} (Danset and Manceron, 2003). Theoretical attempts at modeling cobalt molecule bonding and electronic nature, although extensive, have encountered significant difficulties and were reviewed by Harrison (2000). More recently it has been concluded that only very high level theory is adequate for CoO. In a very detailed study, this has been attempted examining the lowest 50 electronic states and predicting their potential energy curves (Sakellaris and Mavridis, 2012). They now predict an order and location of the four lowest states that might be important in contributing to its chemical thermodynamic behavior; $X^4\Delta_i$, $a^6\Delta$ 4400, $A^4\Phi$ 4900, $B^4\Sigma^-$ 5100, and $C^4\Pi$ 7100 cm^{-1}. Even though these may not be exact, their degeneracies and energies tend to show the effect they will have on thermochemical data obtained below 2000K. The electronic description of the ground state is most important since this doubles the degeneracy from the fourfold taken by Pedley and Marshall (1983), in their analysis, now to eight, and with the low-lying levels also making minor contributions. In a Third Law reduction of thermochemical data, the value obtained for D_0 will decrease as the degeneracy increases by an amount RTln $(New/Old)_{degeneracy}$. Consequently, at 1500K a twofold change can result in a 9 kJ mol^{-1} modification, indicating its potential magnitude. Actually the early data of Smoes et al. (1972) was quite accurate because they had simply assumed a degeneracy of 10 for low state contributions. Reevaluating these values with current thermochemical data now gives a recommended value of D_0(Co-O) $= 373 \pm 5$ kJ mol^{-1}. One additional mass spectrometric measurement not previously included in earlier evaluations suggests a value of about 378 (Balducci et al., 1984). These values appear acceptable so there is concern for the higher results of (394–398) ± 18 kJ mol^{-1} obtained in ion-beam reaction threshold studies (Fisher et al., 1990a; Liu and Armentrout, 2005). This is a particularly difficult molecule for theory which is looking for experimental values with which to compare. In the present case, the most sophisticated theoretical exercise (Sakellaris and Mavridis, 2012) recently chose to compare its calculations of the bond strength against the then reported larger magnitude. Their best theoretical estimate was 372 kJ mol^{-1}, so instead of their possible previous disappointment, I feel they should now be congratulated on an extremely good effort. Such is the progress of science, we build on an error laden foundation! The vibrational frequencies of CoO and CoO_2 were obtained from FTIR matrix isolation studies (Chertihin et al., 1997d). It was further confirmed by other IR matrix isolation work and ESR monitoring that the dioxide preferred only a linear symmetric structure (Van Zee et al., 1992; Danset et al., 2005). At ambient temperatures Co(g) does not react with O_2 being blocked by a significant activation energy barrier (Brown et al., 1991; Matsui et al., 1997). The electronic structure of CoO_2 has been examined theoretically (Uzunova et al., 2002, 2008). Such calculations remain difficult and introduce conflicts suggesting now reasonable values for D_0(Co-O) of 364–370 kJ mol^{-1} and 371 (Gutsev et al., 2003), yet conflicting values for D_0(OCo-O) of 453 (Gutsev et al., 2000a) and 307 kJ mol^{-1} (Uzunova et al., 2002). That a strong bond is suggested has to be considered as possible. As seen below, and summarized already in Table 7.19 for the first row transition elements, very strong second bonds in the gaseous hydroxides and dichloride appear to be the case for Mn, Fe, and Co as well as with the oxides. The next element, Ni, shows them all becoming similar. Additionally, the work by Uzunova et al. (2008) suggests that frequencies observed for CoO_2 may be modified by matrix interactions in such experiments. Until this dilemma is resolved a 10% stronger bond might be suggested as D_0(OCo-O) ≈ 410 ± 20 kJ mol^{-1}. Delaying a discussion of the chemistry of the hydroxides and chlorides to the section below on laboratory combustion studies where it is more appropriate, the remaining molecules of interest are the sulfides, phosphides and hydrides. As mentioned above, cobalt has a series of non-stoichiometric condensed phase sulfides. These are not well characterized but are known to oxidize at elevated temperatures and thermally dissociate on heating. Nevertheless, in the gas phase CoS is of interest due to its transition-state metal nature and its relationship to the oxide. Its spectroscopy remains still bare bones but two recent experimental efforts have confirmed the $X^4\Delta_i$ state similar to the oxide (Flory et al., 2005), and have monitored the $A^4\Phi_i$ (≈3700 cm^{-1}) and $B^4\Pi_i$ (≈5000 cm^{-1}) electronic states (Yu et al., 2006). Other similar states to the oxide can be expected. Only one mass spectrometric Knudsen cell study has measured its bond strength. This compared its equilibrium balance with that of MnS quoting a value of D_0 327 ± 15 kJ mol^{-1} (Drowart et al., 1967). Adjusting this almost 50-year-old result to present-day values for D_0(MnS) (+7 kJ mol^{-1}) and the heats of formation of Mn and Co(g) (+2 kJ mol^{-1}) would now give a recommended value of D_0(Co-S) = 336 ± 15 kJ mol^{-1}. Although the free energy functions used in the original analysis remain uncertain, 5 years later the same author did use 10-fold as the degeneracy for the ground state of CoO (Smoes et al., 1972) minimizing further adjustments. Although the value seems a little larger than might have been expected, the trend in the periodic table suggests some support. On advancing across the first row of transition elements, the ratio of $D(MS)/D(MO)$ gradually increases from about 70% for V, Cr, to 80% for Mn, Fe, and is now about 90% for Co, 94% for Ni, and 100% for Cu (Table 7.19). Until further studies emerge, it might remain accepted. Unfortunately, theory still has difficulty with these transition elements, one study obtaining very high bond strength values (526, 436 kJ mol^{-1}) for CoO and CoS, respectively (Bridgeman and Rothery, 2000). Nevertheless, this has a tempting ratio of $D(MS)/D(MO) = 83\%$ which could fit better in the trend and confuse the discussion. More recent work has positively confirmed the ground state electronic assignment and suggested a bond strength of 316 kJ mol^{-1} with earlier publications listed therein being in the range 284–436 (Wu et al., 2007). Although the CoP(s) and other transition phosphides are not relevant to combustion due to their sensitivity to air, in recent years they have illustrated a catalytic nature where they can effectively strip hydrogen out of water (Hu et al., 2016). Interest was raised concerning the gaseous transition phosphides and theory established their electronic nature. CoP(g) was indicated to have an $X(^5\Delta)$ state, several low lying states and an appraised value of D_0(Co-P) $\approx 218 \pm 20$ kJ mol^{-1} dissociating to the lowest energy limits (Tong et al., 2003). Gaseous CoH also is known, but is of more interest to astronomy and theory than to combustion. Although quite extensively studied, its spectroscopy is complex, experiments are difficult and the molecule has proved to be a theoretical challenge. Uncertainties remain. However, the ground state is proven as $X^3\Phi_4$, both experimentally (Beaton et al., 1988) and theoretically (Sakellaris and Mavridis, 2012a). Below 7000 cm^{-1} only the $a^5\Phi$ state (6625 cm^{-1}) has been observed (Miller et al., 1987), but numerous other low-lying states have been predicted by theory; $A^3\Delta$, $T_e = 1800–3700$ cm^{-1}; $B^3\Pi$ 1450–1850; $C^3\Sigma^-$ 1130–1840; $^5\Sigma^-$ 5300 (Freindorf et al., 1993; Varberg et al., 1989; Sakellaris and Mavridis, 2012a). Consequently, depending on temperature these will make slight contributions to any chemical thermodynamic data. It is not a thermally stable molecule and experimentally different approaches suggest a value of D_0(Co-H) of about 189 ± 12 kJ $mole^{-1}$ (Kant and Moon, 1981) and 185 ± 10 (Georgiadis et al., 1989; Fisher and Armentrout, 1990; Kickel and Armentrout, 1995a). Numerous theoretical descriptions have indicated the calculation difficulties, significantly over or underestimating this magnitude. The most recent and advanced theory calculating 27 of the lower potential energy curves now has derived a value of $D_0 = 194 \pm 4$ (Sakellaris and Mavridis, 2012a). As a result, until further studied a value of D_0(Co-H) $= 188 \pm 10$ kJ mol^{-1} is suggested. Nothing is known of a possible gaseous dihydride but one theoretical study has indicated a linear $X^4\Phi_g$ structure (Siegbahn et al., 1984) and another that produced a reasonable estimate for D(Co-H) suggested that the second bond D(HCo-H) might be about 100 kJ mol^{-1} stronger (Uribe et al., 2010). A matrix infrared study did obtain the vibrational frequencies and shape of the dihydride (Andrews, 2004).

Laboratory combustion studies

Many flame laboratory studies have continued to establish the nature of cobalt's basic chemistry. Using well controlled $H_2/O_2/N_2$ fuel-rich flames at 1800K–2615K, its behavior was studied from additions of traces of c-pentadienyl cobalt carbonyl vapor (1 ppmv) (Jensen and Jones, 1976). This involved monitoring the spectral emission of an atomic line (352.8 nm) and a molecular band transition (563.5 nm) that resulted from the Co addition.

Assuming the band emission was a reflection of the CoO flame concentration it was possible from the relative magnitude of these intensities to derive a second-order value of about D_0(Co-O) = 430 ± 20 kJ mol^{-1}. This is obviously too large by about 57 kJ mol^{-1} questioning the assumptions made. The complexity and unknown nature of the CoO spectral band monitored would now seem to be the most likely explanation. This visible spectral region is known to be very complex and its variation with temperature might also result from overlapping transitions within the bandwidth (Barnes et al., 1997; Guo et al., 2006). The flames were said to be homogeneous and it is apparent that Co(g) is dominant together mainly with CoO. The flames were reducing so CoO_2(g) would not be expected but hydroxides may have been present. This study was at a time when catalytic recombination of radicals in flames by low levels of trace metals was becoming reported (Bulewicz and Padley, 1971; Jensen and Jones, 1976; Vanpee and Shirodkar, 1979). In these studies, it was noted on increasing cobalt additions above about the 5 ppmv level in the burned flame gases that gas phase catalytic recombination of flame radicals was apparent and it could be a flame inhibitor. It was less efficient than Cr, the effect falling away rapidly across the transition element neighbors, Mn and Fe on one side to the no-effect seen with nickel (Linteris et al., 2008). Similar mechanisms generally were invoked as those being proposed for the effective elements. This tentatively implied the possible presence of CoO, CoOH, and $Co(OH)_2$ in flames because the presence of only gaseous Co and CoO would have no plausible kinetic way to promote radical recombination. Of course in such flame studies the cobalt was added as an easily dissociated salt solution and so became gaseous atomic cobalt essentially instantly in the reactive burned gases at about 2000K. Similar type studies examining its effects on soot formation and control showed only slight effects (Howard and Kausch, 1980). It had been rated earlier with a zero effect (Cotton et al., 1971) and possibly promoting it slightly (Bulewicz et al., 1975).

An early transpiration study had indicated that at 1275°C–1466°C, cobalt wire could be vaporized by gaseous water vapor to form $Co(OH)_2$(g) and H_2(g) (Belton and Jordan, 1967). The interaction was shown to be controlled by the reaction

$$\mathrm{Co(s)} + 2\,\mathrm{H_2O(g)} = \mathrm{Co(OH)_2(g)} + \mathrm{H_2(g)}$$

The degree of formation of the dihydroxide was shown to correlate with both the water and hydrogen concentrations. From this an estimate of the average Co–OH bond strength was obtained indicating 769 kJ mol^{-1} for the sum of both hydroxyl bonds. Similar ablations were noted for cobalt's neighboring transition elements Fe and Ni. Moreover, a strong implication was noted that the bonding of these hydroxides closely appeared to reflect those of the chlorides. Now, the bond energies for the mono- and dichlorides of these three elements are quite well established as are the two bonds in $Fe(OH)_2$. In all three chlorides (Fe, Co, Ni) and in the iron hydroxides the second bond in the molecule is consistently stronger. For Fe this is by 38%, for Co 24%, and for Ni the difference has dropped to almost equal (0.05%). Consequently, using this ratio for Co and the above sum of the two hydroxyl bonds a very good approximate estimate might be D_0(Co-OH) = 344 ± 10 kJ mol^{-1}, and D_0(HOCo–OH) = 424 ± 10 kJ mol^{-1}. The first bond has 81% the strength of the second. Laser-ablated Co reacting with H_2O_2, or H_2/O_2 mixtures, has identified the dihydroxide in matrix isolation (Wang and Andrews, 2006b). A bent structure and frequencies were established. Although such transpiration or ablative corrosion studies overtime have been useful to establish fundamental chemical values, they generally are concerned with small value loss rates (Myers et al., 2019). Consequently, when applied to combustion, the question to be noted is whether such trace amounts can compete or even contribute with the magnitudes and competition arising between species in full scale more practical combustion systems. The dichloride is a major salt of cobalt and is very volatile forming predominantly the monomeric vapor (Schoonmaker et al., 1959). However, it is readily oxidized and so can only be relevant to combustion either in a vaporizing release mode from a fuel or in gasification or other inert atmospheres. The thermochemistry of both gaseous CoCl and $CoCl_2$ are now reliably established but they await a more detailed analysis of its complex spectra and a still better theoretical description of its electronic nature. The ground state for CoCl has been proven to be $X^3\Phi_4$ (Wong et al., 2003; Ram et al., 2007), and it has been found that its low lying states correlate with those known for CoF and CoH. Using the mass spectroscopy-effusion beam method, the equilibration was measured in the Co/CoCl/Ag/AgCl system. Updating the thermochemical database, this provided a recommended value of $D_0(\text{Co-Cl}) = 340 \pm 6\ \text{kJ mol}^{-1}$ (Hildenbrand, 1995, 2008). Because $CoCl_2(s)$ vaporizes to the monomer it has a well-defined heat of sublimation. Hildenbrand in his study derived this more accurately (1995). Coupling this to the accurate heat of formation of $CoCl_2(s)$ (Chase, 1998) provides the best value of $\Delta H_{f298}(CoCl_2)_g = -94.6 \pm 2.6\ \text{kJ mol}^{-1}$. With the still slightly uncertain heat of evaporation of Co(s), $\Delta H(\text{subl})_{298}\text{Co(s)} = 424.7 \pm 2.0\ \text{kJ mol}^{-1}$ (Kubaschewski et al., 1993) this provides the most accurate value for the atomization energy of both bonds $D_{0\text{atm}}(CoCl_2)(g) = 760 \pm 3\ \text{kJ mol}^{-1}$. In the mass spectrometric Knudsen cell study of the Co/CoCl/$CoCl_2$ equilibration an experimental value of $776 \pm 10\ \text{kJ mol}^{-1}$ was obtained (Hildenbrand, 1995). Consequently, this provides support for taking the simpler experimental measure for CoCl and the value based on sublimation for $CoCl_2$: suggesting recommended values of $D_0(\text{Co-Cl}) = 340 \pm 6\ \text{kJ mol}^{-1}$ and $D_0(\text{ClCo-Cl}) = 420 \pm 7\ \text{kJ mol}^{-1}$. Theory does predict the triplet nature of gaseous CoCl (Jensen et al., 2007), and although it has low spin lying states, the various methods still appear to have difficulty adequately predicting its dissociation energy. The dichloride has been shown to be linear and its structure measured (Jacox and Milligan, 1969; Hargittai, 2000). Theoretically it has been studied quite extensively partly due to its later position in the first row of transition elements. Not known experimentally, due to its several low-lying electronic states, there was controversy concerning its ground state configuration. This does appear at present to favor the $X^4\Delta_g$ state with three other low-lying states (Sliznev et al., 2004; Bross et al., 2013). The close lying $^4\Sigma_g^-$ had been otherwise so favored (Bridgeman, 1997; Wang and Schwarz, 1998). Because the atomization energy of gaseous $CoCl_2$ has been quite well known for a longer time, it has been used as a benchmark molecule against which to test and develop theoretical models. The most recent efforts now show that certain models can reproduce the bond strengths adequately, almost within the error bars, with some understanding of the nature of the electronic structure (Jiang et al., 2012; Bross et al., 2013).

A major effort arose during the development of flame atomic absorption and emission methods as standardized methods of analysis. Either in absorption or emission these monitored the free atomic concentration in flame gases as compared to a known concentration of a standardized sample. It was a widespread method and levels of sensitivity centered on the degree of free atomic concentrations in the flame. Because of this, although not a necessity regarding its ability as an analyzer, there was interest in establishing this extent of free atoms for many metals in differing flames. Due to their reducing nature, high temperature fuel-rich flames were preferred using either H_2/O_2 or air, or C_2H_2/O_2 or N_2O depending on the necessary temperatures for required sensitivities. Several studies included cobalt. This was difficult to assess accurately as coarse spray nebulizers then were used and the degree of the cobalt nitrate solution vaporization had to be estimated. Also uncertain spectral parameters and partition functions were needed to convert a spectral emission or absorption into a quantitative atomic number density. Results were considered correct at best within a probable factor of 3. Nevertheless, several studies quoted results implying free cobalt atoms to be as low as 2% or as high as 41%, but results remained uncertain (De Galan and Samaey, 1970). A more quantitative and reliable laboratory study examined well defined $H_2/O_2/N_2$ flames at 1530°C–2340°C on a flat flame premixed burner to which c-pentadienyl cobalt dicarbonyl vapor was added at low levels (Co <3 ppmv) (Jensen and Jones, 1976). The cobalt additive, being an organometallic and a carbonyl would break down to atomic cobalt instantly at these temperatures. Atomic Co emission was monitored via its strong 352.8 nm line and cobalt molecular emission via its broad continuum peaking at 563.5 nm (Herrmann and Alkemade, 1963). Using 20 flames of differing composition the Co thermal emission closely tracked the flame temperatures confirming that gaseous atomic Co was dominant and molecular formation minor. A further study of 14 flames with the same temperature but differing composition suggested that the molecular emission was more probably from CoO(g) rather than the stable Co $(OH)_2$(g). Their conclusions appear to remain valid but the contribution of CoO(g) is now considered to be much lower than even they considered making any role for $Co(OH)_2$(g) unlikely at low cobalt concentrations. They did report a strong catalytic radical recombination effects with cobalt at higher flame concentrations (100 ppmv) (Jensen and Jones, 1976). The same effect, less pronounced, but apparent with lower concentrations was also studied (Bulewicz and Padley, 1971; Vanpee and Shirodkar, 1979; Linteris et al., 2008). Whether this arises from CoO(s) particle surfaces in the flame possibly producing the dihydroxide or heterogeneous mechanisms remains speculative. However, all these studies used reducing flames which would minimize oxide formation and promote atomization. As a result, a role at all for the hydroxides remains quite uncertain. Consequently it is of interest to examine more practical combustion that is oxygen rich. Additionally, cobalt appears to play a minimal role concerning soot formation and control. Compared to barium placed at 100 on a scale, cobalt was listed at zero (Cotton et al., 1971; Bulewicz et al., 1975; Howard and Kausch, 1980; Haynes et al., 1981). Its ionization in flames has been reported with the Co^+ being dominant but at low levels (Goodings, 1992). Due to its high ionization potential, this is regarded as resulting from charge transfer from the natural flame ion H_3O^+ rather than thermally.

As noted in the bond strength summary of Table 7.19, that for CoO(g) of 373 kJ mol^{-1} reflects a general weakening on crossing through the first row transition elements from ScO (668) to ZnO(155) but it may be stable enough to form directly in the gas phase at low concentrations under certain conditions. Cobalt's inorganic chemistry immediately reflects the limitation imposed on numerous other transition elements with regards to combustion. As temperatures are rapidly elevated, thermal dissociations or oxidations will occur producing a dominance of condensed phase oxide formation with some slight possibility of formation of silicate, ferrite, aluminate or titanate in the ash or slag. Rapid heating could cause disruptive disintegration of fuel particles producing a spectrum of sizes. Moreover, only with combustion at more elevated temperatures ($>1400°C$) will the presence of mainly gaseous Co becomes a possibility with traces of CoO(g). Even then, the latter is not stable being fully dissociated by about 1650°C. In fact published thermochemical tables concerning cobalt exist only for the gaseous element, its ion and the volatile chlorides. The very extensive mass spectrometric studies by the Russian group of the vaporization of mixed metal oxides has only reported gaseous $CaCoO_2$, $SrCoO_2$, $BaCoO_2$, and $CoPO_3$, $CoPO_2$ molecules for cobalt (Shugurov and Lopatin, 2006; Lopatin and Shugurov, 2013). Even these appear marginally stable against fragmentation. Consequently, free transport of gaseous cobalt as a fume will be possible where high temperatures are involved and deposition mechanisms then come into play on its cooling.

Combustion power plant data

Cobalt is a trace in most fuels with generally low levels, being present mainly in coal, municipal waste and sewage sludge (Table 1.1). It is found in biomass but at even lower levels. Levels in limestone and pet-coke are very low (Lind et al., 1994; Cordoba et al., 2012). As a result, and due to cobalt's benign nature, minimal information has emerged concerning its nature in combustion. Additionally, low concentrations additionally make quantitative assessments more difficult to meaningfully interpret. In coal and sewage sludge combustion, cobalt is predominantly found in the fly and bottom ash or slag. No enrichment is noted between these (Querol et al., 1995; Aunela-Tapola et al., 1998; Llorens et al., 2001; Shao et al., 2008). In one study of fly ash, particle analysis upstream of the ESP in a CFBC burning Venezuelan coal indicated that only 1% of the total mass of cobalt was in the $<1\mu m$ particles. This suggested minimal vaporization and/or fragmentation formation of small particles in this mode (Lind et al., 1994). In coal pyrolysis with lower temperature combustion no gaseous release of cobalt was noted below 1370°C (Galbreath et al., 2000; Wang and Tomita, 2003). Also in the reducing conditions of gasification, cobalt was labeled nonvolatile at 1450°C (Helble et al., 1996). However, there have been indications of cobalt release and enrichment on submicron particles at higher temperatures. Initial detailed work examined eleven different coals burned at a peak temperature of about 1730°C with a 30 ms residence time (Quann et al., 1982, 1990). These showed a bimodal size distribution with significant Co enrichment factors of from 3- to 32-fold in the submicron particles. No correlation with coal type was evident with differences seen both within one type or the lignite, bituminous and subbituminous coals tested. Another size segregated measurement used a 13-stage cascade impactor sampling at the ESP outlet on a 100 MW lignite fired plant (Wang et al., 2008). This indicated for Co enhancement factors particularly in the 0.03–0.06 and 0.06–0.10 μm ranges confirming a vaporization/deposition process. Other full-scale power plants did note cobalt enrichments in the small quantity of submicron particles that evaded capture and were emitted (Meij, 1994; Nielsen et al., 2002). In a recent study of the overall stack releases of 13 trace metals from a large modern coal-fired Chinese power plant, Co was included

(Zhao et al., 2016). With coals averaging 6 ppmw Co essentially 100% was retained within the various collection devices. A recent ambitious analysis reported concentrations of about 40 trace elements in coal fly ash segregated into five coarse sizes, the smallest being 2.2 μm diameter (Lanzerstorfer, 2018). This categorized Co into those elements displaying enrichment to lower particle size. With regards to possible chlorine effects, at 1100°C it was noted that the presence of high chlorine content sludge had no effect on Co release from blending with coal (Folgueras et al., 2007). This was seen also in a large entrained flow IGCC gasifier at 1600°C with chlorine contents of 1000–2000 ppmv (Font et al., 2010). In low-temperature (900°C) FBC combustion of sewage sludge rich in sulfur and chlorine, the cobalt was distributed between the bottom and fly ash (Shao et al., 2008). One study collected ESP fly ash from an MSW combustor. This was roasted with either $CaCl_2$ or with gaseous Cl_2 at 1000°C for 3 hours. The Cl_2 vaporized all the cobalt from the ash whereas the calcium chloride removed only one-third, indicating the presence of resistant molecular cobalt, either there originally or formed in the roasting process (Chan and Kirk, 1999). No explanation was sought but Ca/Co molecules appear to remain poorly characterized and as mentioned above $CaCoO_2$(g) has been monitored in a mass spectrometer. Moreover, the ease of vaporization by Cl_2 might even rule out the presence of mixed metal oxide formation. It was widely noted in coal or waste combustion that current particulate control devices reduce losses of cobalt in fine particles to a minimal level (Lind et al., 2003; Goodarzi, 2006, 2006a). A Spanish detailed study of the effectiveness of FGD retention of trace elements noted this system would reduce any stray levels of Co by a further factor of three- to fourfold in a co-combustion plant burning coal and pet-coke (Cordoba et al., 2012). Few equilibrium calculations of value are available due to the scarcity of cobalt's thermochemical tabulations. A recent study has shown an effect on the fly ash composition of reducing the flue gas temperature from 250°C to 150°C through a heat exchanger. The fuel was biomass from metal contaminated soils burned in a small pilot-plant boiler. Boiler temperatures are not quoted but a noteworthy 2.5-fold increase in the ash concentration appeared to occur for cobalt and numerous other elements as the temperature was so reduced through the heat exchanger. As no further metal deposition was possible to increase the concentrations, especially for such elements as Ti or Zr that are fully nonvolatile, it was apparent that particle segregation was mainly occurring and the ash size distribution was changing favoring collection of the smaller enriched sizes at the lower temperature (Kovacs et al., 2018). It illustrates the care needed in such interpretations.

In summary, for normal coal combustion, it is predicted that as temperatures increase, solid oxides will dominate in the fuel particle, dissociating to mainly Co(g) with a small fraction of CoO(g) above 1600°C. There are no other thermally stable molecules and vaporization depends on high temperatures. There is minimal transport, dependent on temperature coupled to possible fragmentation. Generally, volatility will be low. Chlorine appears to play no role, and cobalt plays a slight if any role with soot suppression, and catalyzed radical recombination aspects noted with cobalt remain unexplained. On cooling, the cobalt in the ash or depositing will begin to exhibit thermodynamic preferences with possible conversions to the sulfate as temperatures cool with only a minor if any role for the chloride (Frandsen et al., 1994). Under reducing conditions the sulfides will dominate up to about 600°C when these dissociate to elemental Co. Some calculations have attempted to include roles of mixed metal oxide formation and do hint of possible $CoFe_2O_4(s)$ or $Co_2SiO_4(s)$ existence around $\geq 800°C$ with wastes, but such formation remains speculative (Abanades et al., 2002; Poole et al., 2008).

7.13 Copper

Although not the last of the first row transition elements, copper favors a ground atomic state of $3d^{10}4s^1$. Following the trend of elements before it, only low valencies are favored being predominantly diatomic with few monovalent compounds. Trivalency is rare and would be strongly oxidizing. Its ores are numerous as sulfides or the mixed iron sulfide, as hydroxycarbonates or the oxide. It is about the 26th most abundant element in the Earth's crust at 60 ppmw a very similar concentration to its neighboring elements Ni and Zn. It is at a lower ranked concentration (33rd) in the oceans. Global mining is at about 20 million tons $year^{-1}$ with about 700 million tons in reserve. Nevertheless considerable efforts exist to recycle copper and possibly as much as 50% now is being achieved (Graedel et al., 2004). It is smelted or leached from ores and refined by electrolysis. It is the lowest cost coinage metal currently at about \$2.20 lb^{-1}. Copper's properties of high thermal and electrical conductivity and use in alloys have made it a critical mineral. This has led to an emphasis for promoting recycling and finding gradual replacement materials (National Research Council, 2007). It is difficult to assess the magnitude of the role it plays as a heavy metal in the environment. Early estimates have been made concerning the magnitudes of copper's natural background and current anthropogenic contributions and whether it is a pollutant. Natural atmospheric emissions were listed 25 years ago at about 20,000–28,000 tons $year^{-1}$, being mainly from dust and volcanoes, with anthropogenic estimates 35,000–140,000 tons $year^{-1}$, from metal processing and combustion (Nriagu, 1989a; Malakhov and Makhon'ko, 1990). A more recent report suggested that the natural emissions level had been underestimated by about 70- to 80-fold (Richardson, 2001). Such a large increase is unreasonable and unacceptable. It would be so large that it would dwarf observing any anthropogenic contributions and simply reflect the same approximately unchanged atmospheric level throughout all time. Moreover the corresponding estimates for natural emissions of Cd (38-fold increased), Pb (90), Hg (5), Ni (60), Zn (85-fold) are similarly very high, not indicative of any simple scalable error and so remains a dilemma. It has been indicated that anthropogenic contributions for copper in the 1900–80 period increased sixfold over the 1850–1900. Moreover, releases into rivers and water sources and to soil are now much larger (Nriagu, 1990). Some witness can come from ice core drilling analysis. From cores taken in Greenland and the Antarctic reflecting the 1770–1990 time period, it was concluded that the year 1770 was a good natural baseline with the 1990 levels about threefold higher but in decline (Boutron, 1995). Recent cores drilled in Siberia cover a period from 1680 to 1991. These indicate very low concentrations until about 1950 when there was significant growth until the 1980s. Copper smelting was regarded as the main contributor and volcanic eruptions in this period were not evident in the data. From the 1980s, when emission control measures started to be implemented in many countries, the levels became severely depleted such that they now appear to be back at the 1950 low levels (Eichler et al., 2014). There is no doubt that emission control regulations on industry are now modifying quite drastically these previous estimates of the anthropogenic contributions. This was apparent in measurements of the heavy metals in the atmosphere of Paris, France for the 1994–2002 period. In one district copper levels had decreased by 2.5-fold (Azimi et al., 2005). Controls are now being introduced in China where atmospheric particle emissions are still quite evident from metal smelters and combustors with particle sizes mainly around 1 μm (Duan and Tan, 2013). Nevertheless, there are still many regions close to metal smelters where downwind lakes show signs of heavy metal pollution besides copper (Gashkina et al., 2015). One recent reassessment of emissions from a smelting region in Northern Russia in 1994 indicated that earlier estimated values had been underestimated by a factor of sevenfold (Boyd et al., 2009). Emissions can arise also from the cement industry. Although on a lesser scale, one study of China's inventory showed a fivefold increase in Cu emissions in the past 30-year period with little diminution yet evident (Hua et al., 2016). Its main commercial use is in alloys, pipe, electrical wiring, fertilizers, fungicides, and for its chemical compounds. As will be noted, the use of sewage sludge ash as an agricultural fertilizer is of value partly for its copper content, which has regulated maximum levels for dispersal. China that has concerns regarding its soils notes that copper levels are increasing, resulting mainly from cattle manures (75%) together with 22% from atmospheric deposition, and suggesting a need for animal feed changes (Peng et al., 2019). The first row transition elements have atomic numbers 21–30. Even so, from Cr (24) through to Ga (31) the metals show high metal densities (all $>7\ g\ cm^{-3}$), peaking at Cu (29) with $9.0\ g\ cm^{-3}$ and are oft referred to as heavy metals.

Health concerns with copper

Nevertheless, there is less concern with copper concerning human health. It is actually the third most important metal in the body following Fe and Zn, also members of this first row of transition elements. It is an essential mineral involved in at least 12 major proteins involved in catalytic or structural purposes. It is readily transported across the blood/brain barrier supplying the brain (Jiang, 2017). The main human exposure is orally and a daily diet intake of $0.9–2.2\ mg\ day^{-1}$ is suggested for adults, but remains approximate, varying in different countries and for different groups (World Health Organization Trace elements, 1996; World Health Organization Guidelines, 2004, 2006a). The high tolerable level is considered as $<10\ mg\ day^{-1}$ that has been suggested as the no observable adverse effect level (NOAEL) limit (Goldhaber, 2003). Being tested with labeled ^{64}Cu, it has been shown to be absorbed (20%–60%) gastrointestinally and this study also indicated its body distribution. Most living organisms appear to have developed automatic level control mechanisms for the element. Excesses are expelled in feces or urine and overdoses have to be intentional. Normal human urine excretion can be $\leq 60\ \mu g\ day^{-1}$. Its toxicology has been reported in two detailed reports available on the Web and in a book (EHC Cu, 1998; ToxProfile Cu, 2004; Landner and Reuther, 2004). Its body half-life is several days, and it is regarded as a minor toxin and noncarcinogen. Various large surveys from around the world that measured blood levels of copper confirmed that these did appear somewhat regulated being generally in similar ranges of about 0.5–1.9 mg L^{-1} (5%–95% percentiles) (Cesbron et al., 2013; Schofield, 2016; Oggiano et al., 2018; Zeng et al., 2019). Recommended serum levels appear to track these same levels (Goulle et al., 2005; Mayo Clinic, 2018; CDC 4th National Report, 2018). As in human toxicology generally, deciding the best biomarker in the body to test as a measure of health has remained difficult (Goldhaber, 2003; Bornhorst et al., 2018). Other surveys that are useful compare blood, serum, urine and hair samples from the same groups of healthy people (Goulle et al., 2005; Oggiano et al., 2018). From these, urine also appears a good measure at <12 μg L^{-1}. Hair also reflects Cu levels but can display broader distributions that is less useful as a reliable monitor, but suggested by one test center to be best in a 13–35 ppmw range. Copper appears not to be an accumulative hazard in the body and there is no evidence of biomagnification in the food chain or on agricultural land (Huang et al., 2019). However, its deficiency and excesses become clearly evident in genetically driven diseases named as Menke's syndrome for extreme deficiencies and Wilson's disease, a recessive disorder leading to excessive accumulation and toxic death. Other lesser known genetically related illnesses have also invoked some role for copper. Still very much uncertain, numerous groups are examining the possible role for copper in the string of neurological illnesses such as Huntington's disease that involves uncontrolled movements and becomes evident usually at ages 35–40 years (Bichell et al., 2017). Also whether it has a role in Parkinson's and Alzheimer's diseases more commonly found above 65 years of age still remains highly speculative (Mot and Crouch, 2017; Squitti et al., 2017; Brewer, 2018). If these diseases are growing in prevalence, as does seem to be the case, it would need to correlate with some increase in copper dietary intake from water or food. In nature certain species are more sensitive and with fish they depend very much on the level of Ca in the water, soft water being more toxic (Landner and Reuther, 2004). Excessive Fe/Zn supplements can also inhibit human Cu absorption and be very detrimental. One recent environmental concern is with the collection of rainwater runoff for human consumption from copper surfaces such as roofs in some countries (Hedberg et al., 2014). Whether this introduces a dietary risk would have to be assessed. Risks from Cu generally are associated with too low an intake which tends to be infrequently encountered in developed countries due to the basic low requirements. Copper of course is used in water pipes that in some circumstances can lead to higher levels if not flushed and can be of some concern (World Health Organization Guidelines, 2004, 2006a). Use in cooking utensils generally is regarded as acceptable.

Copper's basic chemistry

Copper has an extensive inorganic condensed phase chemistry being enriched by having both monovalent and divalent molecules with the same element. These are briefly summarized in Table 7.21. Commercially, the oxide and sulfate are its most important salts. The metal has a slightly lower melting point (1085°C) and boiling point value (2562°C) than Co and Ni. Its vapor pressure is about 10 ppm at 1240°C and 100 ppm at 1390°C (Barin and Platzki, 1995).

It is only superficially oxidized in ambient air forming a protective green colored patina that is a hydroxycarbonate or sulfate. On heating with oxygen it forms black CuO (m.pt.1446°C), and is insoluble in water. Copper is interesting in that it has two oxides.

TABLE 7.21 Properties of copper's mono- and divalent condensed phase molecules relevant to combustion, °C.[a]

m.pt. b.pt. vapor pressure **Monovalent**	m.pt. b.pt. vapor pressure **Divalent**
Cu 1085 2569, 100 ppmv/1390°C, ins	–
Cu_2O 1244, <375 metastable, decomp 1800, ins	CuO 1446, converts 1125 to Cu_2O, ins
–	$Cu(OH)_2$ >140 dissoc to CuO, ins
–	$CuCO_3.Cu(OH)_2$ dissoc 315–350 to CuO, ins
–	$Cu(NO_3)_2$ subl 150–225, dissoc >225 to CuO, s
Cu_2SO_4 oxid to $CuO.CuSO_4$, >830 dissoc to CuO	$CuSO_4$ >800 dissoc to CuO via oxysulfate, s
–	$CuO.CuSO_4$ stable <860
CuCl 436 1209, volatile as $n = 1-4$ polymers, ss	$CuCl_2$ 589, dissoc to CuCl >200°C, hydrol, s
Cu_2S 1130, stable <1500, oxid to CuO/Cu_2O, ins	CuS 501, >500 converts to Cu_2S, oxid in air, ins
CuH >60 dissoc, unstable	CuP_2 900, ins
–	$Cu_3(PO_4)_2$ poorly characterized, ins
$CuAlO_2$ stable >1100, ins	$CuAl_2O_4$ 1280, converts >1100 to $CuAlO_2$, ins
$CuFeO_2$ 1197, ins	$CuFe_2O_4$, 1085, ins

[a]*dissoc*, dissociates; *decomp*, decomposes; *s*, water soluble; *ins*, insoluble; *ss*, slightly soluble; *oxid*, oxidizes; *hydrol*, hydrolyzes with water; *subl*, sublimes.

CuO is the more stable at normal temperatures but in a Knudsen cell-mass spectrometric study at 760°C this indicated only gaseous O_2 and a solid phase that consisted of the two solids CuO and Cu_2O (Ait-Hou et al., 1995). Water insoluble red-brown Cu_2O is metastable below 375°C dissociating to CuO and Cu. It becomes the more stable at higher temperatures and although the two oxides have different lattice types they can coexist in the condensed phase. As temperatures are increased, CuO converts fully by 1125°C in the solid phase to Cu_2O (m.pt.1244). Finally a melt with Cu_2O is formed that then will dissociate on vaporization at higher temperature to gaseous Cu and CuO. The Cu(g) vaporizes from the melt with the vapor pressures noted above. On cooling from high temperatures a mix of Cu(g) or CuO(g) in air will solidify at about 1065°C. This ability of CuO to lose O_2 on heating and then reform on cooling has spawned the "chemical looping" technique for CO_2 combustion recovery. This is now actively being developed as a new combustion method for all carbon based fuels (coal, CO), and also with H_2. The concept is to use CuO as the source of oxygen for the fuel and collect the combustion products (CO_2, H_2O). The formed Cu_2O is then rejuvenated with air back to CuO and recycled as the oxygen carrier (Chuang et al., 2009, 2011; Adanez-Rubio et al., 2014; Penthor et al., 2015; Coppola et al., 2015; Deutsch et al., 2017). A pure carbonate does not occur with copper but exists as a compound carbonate/hydroxide, malachite $CuCO_3.Cu(OH)_2$, and azurite $2CuCO_3.Cu(OH)_2$, that are water insoluble and decompose to CuO at about 315°C or 350°C, respectively, being complete at 420°C and 475°C (Beck, 1950; Shaheen and Selim, 1998). $Cu(OH)_2$

can be precipitated from solution but on heating dissociates to CuO above 140°C. The cuprous sulfate Cu_2SO_4 oxidizes in air to the basic $CuO.CuSO_4$ that dissociates to CuO at about 830°C (Stern, 2001). The more commonly encountered water soluble $CuSO_4$ is stable, decomposing at 598°C (Ostroff and Sanderson, 1959). The decomposition mechanisms of both $CuSO_4$ and $CuO.CuSO_4$ have been studied by torsion-effusion and mass spectrometric methods (Tagawa, 1984; Brittain et al., 1989; Hildenbrand et al., 1990). SO_3(g) is the primary decomposition product in both cases with a vapor pressure of about 2 ppm at 450°C and a CuO(s) product. $Cu(NO_3)_2$ is water soluble and hygroscopic. It is volatile and has been shown by mass spectrometry to sublime at 150°C–225°C to the monomer (Porter et al., 1959). Its vapor pressure is about 420 ppm at 160°C and at slightly higher temperatures dissociates to CuO, NO_2, and O_2 (Stern, 2001).

Copper has two stable chlorides. Monovalent CuCl (m.pt.430, b.pt. ≈ 1212°C) is water insoluble. On vaporization this favors formation of the trimer in closed cell configurations by rapid clustering but can exist also as the dimer and tetramer (Shelton, 1961). At low concentrations in open systems such as combustion the stable gaseous monomer will exist. The divalent $CuCl_2$ is hygroscopic (m.pt.630°C), is very water soluble and can be hydrolyzed. On heating, effusion studies of $CuCl_2$(s) indicate formation of CuCl(s) and Cl_2 in a situation rather similar to the oxides, univalency is favored at higher temperatures (Hammer and Gregory, 1964). This undoubtedly reflects the relative stabilities of the Cu ($3d^{10}$, or $3d^9$) electronic configurations. The solid dichloride starts to lose chlorine at low temperatures (> 200°C) forming CuCl(s). The two solids are not soluble in one another but are in an equilibrium balance that depends on the concentration of the chlorine. As temperatures reach the melting point of CuCl (423°C) a solution forms with $CuCl_2$(s) further dissociating from this mix as temperatures rise to its melting point, while the CuCl gradually vaporizes up to its boiling point. The presence of water in this system can also hydrolyze the $CuCl_2$ to CuCl with the formation of HCl(g) (Wang et al., 2015a).

The chemical thermodynamic values for CuS(s,g) and Cu_2S(s) have been tabulated; they are water insoluble (Barin and Platzki, 1995). In an oxygen free environment these reflect the behavior of the oxides. Above 500°C CuS loses sulfur and converts before melting to the more stable Cu_2S (m.pt.1130°C), which is thermally stable to about 1300°C–1500°C (Ferrante et al., 1981a). In air, the sulfides are unstable being oxidized to CuO or Cu_2O. A CuH is known but is unstable dissociating at 60°C. A phosphide has been reported CuP_2 (m.pt.900°C) and a poorly characterized phosphate $Cu_3(PO_4)_2$. As with other transition elements, numerous mixed oxide salts can exist such as $CuAlO_2$ (> 1100°C), $CuAl_2O_4$ (<1100°C), $CuFeO_2$ (m.pt.1197°C) $CuFe_2O_4$ (m.pt.1085°C) and $CuMoO_4$ (m.pt.500°C) all of which are water insoluble but largely uncharacterized. The monovalent structures appear more thermally stable than if divalent. $CuFe_2O_4$ recently has gained interest in possibly becoming a superior oxygen carrier in the chemical looping combustion of coal (Wang et al., 2016).

Gas phase stabilities and transport

Possibly due to its position at the end of the first row transition elements, coupled to its related chemical nature, the characterization of the gaseous molecules for copper has been a struggle both experimentally and particularly theoretically. Two gaseous oxides have been detected CuO and CuO_2. The dioxide is rarely encountered and has relevance more to atmospheric chemistry or low temperatures. CuO(g) has been quite extensively studied experimentally and theoretically. Spectroscopically it is a complex molecule (Merer, 1989). Its ground state is $X^2\Pi_{3/2,1/2}$ with no low-lying states but with the $Y^2\Sigma^+$ state higher in energy by about 92 kJ mol^{-1} (T_0 7715 cm^{-1}) (O'Brien et al., 1996). The electronic states and potential energy curves are now well established (Hippe and Peyerimhoff, 1992; Daoudi et al., 1999; Xian et al., 2000). Two equilibrium measures have studied CuO(g). The earlier compared its stability to that of NiO(g) (Smoes et al., 1972). Updating the data using their equilibrium constants indicates NiO, its neighbor in the periodic table, is 111 kJ mol^{-1} stronger than CuO. This now would suggest a value of D_0(Cu-O) = 262 ± 10 kJ mol^{-1} reflecting a pronounced bond weakening to the end of the first transition element row (Schofield, 2006). The later mass spectrometric attempt to repeat this same Knudsen cell equilibrium had trouble confirming equilibration but derived D_0(Cu-O) = 259 ± 30 kJ mol^{-1} from what seemed to be reliable data (Watson et al., 1993). Unfortunately, these values are at odds with those obtained utilizing threshold energy values by guided ion-beam mass analysis that quote, D_0(Cu-O) = 275 ± 14 (Clemmer et al., 1991) and 284 ± 12 kJ mol^{-1} (Rodgers et al., 1999; Metz et al., 2005). A more recent two-photon dissociation method measured the energetics of the CuO($F^2\Pi$) producing Cu ($^2D_{3/2}$) and O($^1D_2$) now indicates D_0(CuO) = 293 ± 3 and appears reliable (Parry et al., 2014). Using the same method they appear to have obtained a very reasonable value for D_0(AgO) (Cooper et al., 2015). Unfortunately theory still has considerable difficulty with CuO finding no universal functional that is consistent. Some methods produce low values, 239 (Uzunova et al., 2008), 257 (Bauschlicher and Maitre, 1995), or higher, 270 (Daoudi et al., 1999), 300 (Bridgeman and Rothery, 2000), 300 (Gutsev et al., 2003), and recent efforts that can be spread over 70 kJ mol^{-1} or more depending on the method used (Legge et al., 2001; Furche and Perdew, 2006; Jensen et al., 2007). However, as seen below, some support may come from CuS(g). It has generally been noted that metal sulfides are invariably weaker than their monoxides. However, CuS was shown to be closely equal in strength to MnS implying D_0(Cu-S) ≈ 287 ± 10 kJ mol^{-1} (Smoes et al., 1972; Kubashewski et al., 1993). Since trends invariably indicate that a gaseous oxide is generally stronger than its corresponding sulfide, it would appear that the mass spectrometric values are too low for CuO and a value of D_0(Cu-O) = 290 ± 5 kJ mol^{-1} seems more appropriate. Although not normally observed, there has been considerable interest in the possible Cu/O_2 molecule mainly through its biological connections and possible reactions in the atmosphere. Numerous studies have characterized it using matrix isolation methods, the latest reviewing the previous works (Chertihin et al., 1997). The CuOO and $Cu(O_2)$ structures have been noted that can modify with photolysis or added energy to the stable OCuO form. The latter has been observed as the product in anion photodetachment spectra of CuO_2^- (Wu et al., 1997). As seen also with Cr and Ni, but not with Fe, the Cu can react with O_2 in the gas phase at room temperature via a three body reaction.

$$Cu + O_2 + Ar = CuO_2 + Ar$$

This kinetic behavior is rather unexpected and still remains not fully understood. In these cases, reactivity seems to come from copper's $3d^{10}4s^1$ ($^2S_{1/2}$) ground electronic state. Ni appears to similarly use its $3d^94s^1$ state that overlaps the $3d^84s^2$ ground state, and with

Cr this has a $3d^5 4s^1$ ground state structure. Fe has a $3d^6 4s^2$ ground state and a much higher energy $4s^1$ state and displays no corresponding reaction. Reaction rates decrease in the order Cr, Ni, and Cu undoubtedly reflecting the falling away of bond strength. A weak $Cu(O_2)$ product is formed ($D_0(Cu\text{-}O_2) \approx 63 \pm 20$ kJ mol^{-1}) as there is insufficient energy to produce the normally expected OCuO structure (Vinckier et al., 1988, 1992; Brown et al., 1991; Mitchell, 1992). The $Cu(O_2)$ complex can have two structures, a side-on triangular $Cu\text{-}O_2$ or a bent end-on CuO-O. They appear to be close in energy (Hasegawa et al., 2001). The normally observed OM-O linear bonding is seen to gradually weaken across the transition elements (Table 7.19). All generally favor this OMO structure as the most stable, but Cu differs, favoring instead the $Cu(O_2)$ complex which lies very close in energy to the $Cu + O_2$ dissociation products, $\approx$ 40–80 kJ mol^{-1} (Gutsev et al., 2000a; Uzunova et al., 2008). The relative energies of the side-on and end-on forms lie within theoretical accuracy and energies are indistinguishable (Mochizuki et al., 1989; Bauschlicher et al., 1993; Hrusak et al., 1994; Pouillon and Massobrio, 2000; Hasegawa et al., 2001; Guell et al., 2009). CuO_2 has been studied theoretically very extensively being the most controversial in the first row transition elements, results showing a dependence on calculation method. Although the ground state of linear OCuO was labeled $^2\Pi_g$ in earlier papers (Ha and Nguyen, 1985; Mochizuki et al., 1991a; Deng et al., 1999), the latest work now suggests $^4\Sigma_g^+$ (Uzunova et al., 2008). The linear OCu-O bond strength has not been measured but was calculated as 338 kJ mol^{-1} (Gutsev et al., 2000a). Trends in Table 7.19 might suggest a value close to 300 kJ mol^{-1}. The additive complex is too weak as is the linear insertion to play any major role in combustion. The understanding of CuOH(g) is still very uncertain. It was noted in an early matrix isolation experiment after the initial adduct between the metal atoms with water was photolyzed (Kauffman et al., 1985a). They estimated a very approximate bond strength of $D(Cu\text{-}OH) = 326 \pm 80$ kJ mol^{-1}. Spectroscopically the molecule's structure and frequencies were obtained from its microwave spectrum (Whitham et al., 2000). LIF and chemiluminescent emission spectra have characterized the (B, A-X) spectra with an X^1A' bent shape of angle 111 degrees (Trkula and Harris, 1983; Jarman et al., 1991). Experimentally, its reaction chemistry has not been examined in any confirming detail. The low value had a significant impact on theory that was finding appropriate descriptive functionals to reflect this erroneous low bond strength estimate. Moreover, the Cu atom as a late row transition element introduces complexities with which theory is still trying to resolve. The most recent calculations for the bond strength now are in fact the largest at $D_0(Cu\text{-}OH) = 316$ kJ mol^{-1} (Jiang et al., 2012), with one earlier 309 value (Papai, 1995). All other earlier work suggest values no larger than 250 kJ mol^{-1} (Illas et al., 1985; Bauschlicher, 1986; Mochizuki et al., 1991; Ikeda et al., 2003). Further experimental and theoretical studies would appear needed to resolve this uncertainty. As seen below, theoretical estimates for D(CuCl) against the now accurately measured experimental value also underestimate the bond strength by 20–110 kJ mol^{-1}, supporting an argument for the herein suggested larger value for $D_0(Cu\text{-}OH)$. A gaseous CuSH has been observed by LIF with a X^1A' ground state and its structure compared to that of CuOH (Sunahori et al., 2006). The gaseous dihydroxide remains uncharacterized other than its structure and frequencies being noted in an infrared matrix isolation study of laser-ablated Cu with either H_2O_2, or H_2 and O_2 mixtures (Wang and Andrews, 2005b). The UV–visible range of spectroscopy for CuCl(g) is quite well documented and theory and experiment have now helped to correctly assign the transitions from the lower states to the ground state $X^1\Sigma^+$ (O'Brien et al., 2000; Parekunnel et al., 2001). The electronic states and potential energy curves have been adequately calculated (Guichemerre et al., 2002; Ramirez-Solis, 2006). Effusion beam mass analysis has studied the AgCl/CuCl/Ag/Cu system at equilibrium and further refined prior work to improve the accuracy now quoting a recommended value of D_0(Cu-Cl) = 375 ± 4 kJ mol^{-1} (Hildenbrand and Lau, 2006a). This is 10 kJ mol^{-1} larger than in the thermochemical data tables (Chase, 1998; Barin and Platzki, 1995). Theory again struggles to predict this magnitude. Only three studies appear to be within 20 kJ mol^{-1} and are on the low side (Mahe and Barthelat, 1997; van Wullen, 1998; Guichemerre et al., 2002), while others can be 30–110 kJ mol^{-1} too low (Nguyen et al., 1986; Huzinaga et al., 1986; Suzumura et al., 1999; Mayhall et al., 2009). $CuCl_2$(g) can be produced and has been extensively studied with respect to its ground $X^2\Pi_g$ and low-lying states using LIF (Ross et al., 1996; Bosch et al., 2000; Hodges et al., 2007). Vibrational frequencies and several electronic transitions also have been obtained by matrix isolation methods (Lorenz et al., 1997). Its electronic structure and these transitions now have been addressed theoretically in detail (Ramirez-Solis and Daudey, 2004, 2005; Zou and Boggs, 2009). The magnitude of the bonding still remains difficult to quantify with estimates of 267 (Wang and Schwarz, 1998) and 238 kJ mol^{-1} (Jiang et al., 2012) for D_0(ClCu-Cl). This would indicate a significant weakening from the accurate 375 value for the primary Cu-Cl bond. Trends in the periodic table for the whole first row of Sc through Zn transition elements display slight variations for the M-Cl bond but a consistently stronger second bond decreasing to unity for $NiCl_2$. Consequently, the theoretical estimates are in error and D_0(ClCu-Cl) ≈ 370 ± 20 kJ mol^{-1} seems more appropriate.

The spectroscopy of CuS(g) has been recorded and analyzed for the ($Y^2\Sigma^+-X^2\Pi$) transition (O'Brien et al., 1999). It was included in a Knudsen cell-mass spectrometric study initially comparing its distribution against that of MnS(g) in equilibrium with Cu and Mn. The gaseous concentrations implied that D_0(CuS) was 5.0 ± 10 kJ mol^{-1} larger than D_0(MnS = 282 ± 10 kJ mol^{-1}) (Drowart et al., 1967). On repeating this work several years later, looking at the CuS dissociation equilibrium, or that with AlS, these studies indicated D_0(CuS) = 271 ± 15 kJ mol^{-1} (Smoes et al., 1972; Uy and Drowart, 1971). As discussed above, these values of 287 and 271, need to be compared with the now somewhat better resolved value for D_0(CuO). Bond weakening is seen across the first row of transition metal monoxides. This reflects also into their sulfides but it has long been known that the ratio of D_0(M-S)/D_0(MO) <1.0 is consistent for these, including the end elements, namely from K through Se (16 elements) and also for many others (Hauge and Margrave, 1972). However, as seen in one plot of such values, the ratio is approaching 1.0 for this rather unique case of copper (Mn 0.76; Fe 0.81; Co 0.90; Ni 0.93) (Kretzschmar et al., 2001). As a result, in this case CuO and CuS should be essentially equally, strongly bonded and the value of D_0(CuS) = 287 ± 10 kJ mol^{-1} may be appropriate until improved. The Thermochemical Tables suggest a value of 291 kJ mol^{-1} but from an old reference source (Barin and Platzki, 1995). Numerous theoretical calculations have been published. Since, 2000, values for D_0(CuS) have been in the range 252–294 kJ mol^{-1} (Bridgeman and Rothery, 2000; Midda et al., 2006; Wu et al., 2007) and others display the variation depending on the calculation software used (Legge et al., 2001; Jensen et al., 2007). Throughout, five theoretical studies that have included both CuO and CuS obtain very similar values for each of them. One study predicted the largest values, 300 for CuO and 294 for CuS giving a CuS/CuO ratio of 0.98 (Bridgeman and Rothery, 2000). Little interest yet has centered on the CuS_2(g) molecule. One FTIR matrix isolation study observed the geometries and frequencies of linear symmetric CuS_2 and also the bent CuSS structures (Wang et al., 2009b). A simple gaseous phosphide CuP is known, and is possibly more stable in air and water compared to most other phosphides. Theoretical calculations indicate this has an $X^3\Sigma^-$ state that is weakly bound, D_0(CuP) ≈ 203 ± 20 kJ mol^{-1} (Tong et al., 2003). It will be of little significance in combustion.

The major spectroscopic transitions of CuH(g) have been observed and confirmed theoretically (Huber and Herzberg, 1979; Marian, 1991). The ground state is $X^1\Sigma^+$ with no low-lying states. A $^3\Sigma^+$ state is predicted but not yet identified in the vicinity of the known $A^1\Sigma^+$. Several differing experimental measures of the bond strength have been published but none recent. In this case, extrapolation of the potential energy vibrational levels appears reasonable indicating D_0(CuH) = 263 kJ mol^{-1} (Jenc and Brandt, 1985). A mass spectrometric Knudsen cell approach studying the Cu/H_2 system at 1120°C–1290°C obtained D_0(CuH) = 244 ± 8 (Third Law) and 250 ± 4 (Second Law) analyses (Kant and Moon, 1979). Following these, a series of studies using guided ion-beam mass detection of reaction energy thresholds came to a consensus of D_0(CuH) = 254 ± 6 kJ mol^{-1} (Georgiadis et al., 1989; Fisher and Armentrout, 1990; Kickel and Armentrout, 1995). Placing more weight on this later work, a value of D_0(CuH) = 254 ± 8 kJ mol^{-1} is suggested for the present and closely reflects its neighbor Ni, but indicating the strongest bound of this first row transition elements, which might seem unexpected (Table 7.19). Like other transition element hydrides, that for CuH has been regarded as a challenge for theory. It also has interest in astrophysics. As a result almost 20 efforts have been made to find an adequate structural description of this diatomic that appears to be simple but in fact is not. Results obtained before 2000 have been reviewed and span values for its bond strength of 232–292 (Harrison, 2000). Since then, six newer studies have continued this effort but still with varying values in the 250–300 kJ mol^{-1} range (Wang et al., 2003; Furche and Perdew, 2006; Daza et al., 2006; Goel and Masunov, 2008). A more recent work showed again a dependence on method and D_0 values from 274 to 298 (Mayhall et al., 2009). The latest publication using higher order correlation claims to have high accuracy giving D_0(CuH) = 259 ± 6 kJ mol^{-1} but confirming this strength (Bross et al., 2013) lying within the error bars of the recommended value above. An FTIR matrix isolation method using laser-ablated Cu with H_2 indicates formation of CuH and linear and bent CuH_2 structures with geometries and frequencies (Wang et al., 2003; Andrews, 2004). Earlier, a theoretical study had predicted that CuH_2 was metastable, lying about 30 kJ mol^{-1} in energy above Cu + H_2 products. Stabilization for this linear $^2\Sigma_g{}^+$ ground state was provided by a large energy barrier against dissociation (Siegbahn et al., 1984). Further calculations confirmed this, estimating the activation energy at about 138 kJ mol^{-1} (Guitou-Guichemerre and Chambaud, 2005). They also predicted a weaker second bond in CuH_2 calculating D_0(HCu-H) = 127 kJ mol^{-1}. These gaseous bond strength values for copper molecules are summarized in Table 7.19 and compared to the other first row transition elements. The extensive Russian mass spectrometric work examining any vaporization of mixed metal oxides has reported no such vapors containing copper (Lopatin and Shugurov, 2013). Although as seen above, copper can have a condensed phase aluminate or ferrite, as temperatures increase they generally convert to univalent copper structures but are not sufficiently stable to vaporize molecularly.

Laboratory combustion studies

Looking at this list of potential copper chemicals from the point of view of combustion it is apparent that there are simplifications. In the condensed phase only the oxide and chloride have any pronounced thermal stability. As temperatures are raised rapidly, Cu_2O and CuCl can dominate and finally vaporize to Cu, CuO, and CuCl. In the gas phase CuO (g) is not thermally stable and could produce CuOH. This will dominate unless chlorine is present and then a distribution between CuOH and CuCl will exist controlled by the chlorine level. The hydride is too weak to play any role but the sulfide CuS can exist under reducing (low oxygen) conditions. This is supported by the flame laboratory studies undertaken. For some reason although data was needed few studies emerged. In fuel-rich H_2 flames, Cu and weak CuOH, CuO, and CuH spectral emissions were noted that were replaced by Cu and CuCl emissions if chlorine was added. However, if free of chlorine, by simply measuring the spectral atomic Cu thermal emission in a series of flames of different equivalence ratios and temperatures of 1800K–2200K, scaling the data to equal concentrations of copper, it was apparent that the atomic copper was the dominant species. The Cu emission correlated exactly to its excitation energy. It clearly indicated that any CuH or CuOH was $\leq 5\%$ of the copper (Bulewicz and Sugden, 1956a). The flame photometric work comes from an early period when reaction chemistries were being resolved by new powerful multiflame comparison techniques. However, in this particular case questionable assumptions were made. The CuOH emission intensity was assumed as thermally excited and a reflection of the CuOH concentration, and formed by a balanced equilibrium, namely

$$Cu + OH + Z = CuOH + Z$$

This was considered because the data were not consistent with the normally expected balanced reaction with water when using the CuOH* emission as a measure of CuOH.

$$Cu + H_2O = CuOH + H$$

Present understanding now would interpret the CuOH* emission as having a chemiluminescent basis, formed by the three-body reaction, kinetically dependent as seen on the Cu and OH concentrations, but not a reflection or measure of the ground state CuOH concentration. The latter could be present from the water balance and thermally excited but the measured stronger chemiluminescent emission intensities did not correctly reflect its concentration. As a result their analysis of the importance of CuOH was invalid. In a brief and early flame mass spectrometric sampling study that introduced copper salicylate solution into a fuel-rich H_2 flame at 2030°C, the ions were recorded of the gaseous species Cu, CuO and $Cu(OH)_2$ in nearly equal concentrations (Farber and Srivastava, 1978). Such sampling now is questioned and whether $Cu(OH)_2$ plays a role in the flame or is an artifact of probe heterogeneous formation remains uncertain. In a corresponding flame photometric study as that above examining CuOH, H_2 fuel-rich flames monitored CuH emission finding it proportional to the very low Cu and H atom concentrations. The assumptions of equilibration between Cu, H, and CuH, and that the CuH* reflected that of CuH concentration similarly is invalid due to the chemiluminescent CuH emission. As a result, their estimate of $D_0(CuH) = 276 \pm 8$ is not acceptable (Bulewicz and Sugden, 1956a).

Extensive work during the development of flame atomic emission or absorption analysis methods also was interested in the release of the free Cu atoms in various fuel types. The flames used generally had high temperatures (≥2200°C) and copper was seen to be present predominantly as atoms with negligible ionization (De Galan and Samaey, 1970; Smyly et al., 1971). The very low levels of ionized copper were Cu^+ with negligible molecular forms and came simply from ion exchange with flame H_3O^+ (Goodings, 1992). In one study of atomization efficiencies, copper was used as the benchmark example of being fully dissociated (Willis, 1970). Another interesting related study actually added CuCl to the 1700K–2200K flame as a $CuCl_2$ solution (Kitagawa et al., 1974). They measured the flame distribution as between 70% and 90% dissociated confirming that with chlorine a slight distribution between Cu and a little CuCl could be present at the high temperature. Waste disposal is becoming a critical problem, with wastes and sludges being at higher trace metal concentrations. Decision to utilize incineration or cement kiln operations is still being quite extensively studied in the laboratory in lower temperature tube furnaces or in full scale. A recent study concerning incineration of coal/sludge mixtures reviews the previous work, extending it to examine the effects of added chlorine or FeS_2 at 1300°C (Zhang et al., 2019). It is generally noted that chlorine can release some copper from the bottom ash onto other ash forms. Fe and sulfur will retain more in the bottom ash as mixed metal compounds such as $CuFe_2O_4$. Another study added the chlorides of Al, Fe, Ca, and Na to a cement kiln co-combustion of sewage sludge. 1.6wt% of Cl volatilized 60% of Cu at 1350 °C. Pb, Zn and Ni also were quite significantly vaporized and molecular retention was a secondary feature (Yu et al., 2017).

The catalytic effect of metals on free radical recombination rates (inhibition) was first reported by Bulewicz and Padley (1971). For copper the effect was quite small, and in a fuel-rich H_2 flame at 1600°C, copper, as CuCl, was listed on a scale as 182, with 1000 for Fe and 900 for Pb (Babushok and Tsang, 2000; Linteris et al., 2008). Compared to the efficient catalytic inhibition by Cr and Ba this might support that Cu has very little molecular formation in such flames and so a minimal capability to have the necessary molecular mechanisms for reducing radical concentrations. When analyzed as to its effect on soot formation it can be more confusing. A slight increase was noted in acetylene/oxygen flames but shown to be dependent on fuel and equivalence ratio (Bulewicz et al., 1975). Others reported either very minimal effects (Haynes et al., 1981) or zero (Cotton et al., 1971). In a review on soot control, it was concluded that Cu was most probably ineffective (Howard and Kausch, 1980). It is now established that CVD methods using seeded flames and a cooled substrate can produce simple thin films of CuO (Yates et al., 2008), or even submicron YBaCuO superconductor deposits (Zachariah and Huzarewicz, 1991, 1991a). Consequently, in combustion deposition it would seem that the oxides, and mixed oxides such as the ferrite, $CuFe_2O_4$ or aluminates $CuAl_2O_4$, $CuAlO_2$ might form. It is a question of thermodynamic stabilities and the opportunity to form on or within the ash as temperatures cool. Copper appears to be one element that has little affinity for Si and silicate formation.

Combustion power plant data

Generally, combustion is not regarded as a significant pollutant source for Cu. Nevertheless, it is present in most fuels at nonnegligible levels and can be emitted in the submicron particle size of health concerns if not controlled. As seen in Table 1.1, coals generally have Cu concentrations in the region of 10–100 ppmw with lignites sometimes higher. Pet coke and limestone have low contents ($\leq$ 10 ppmw) (Klika et al., 2001; Font et al., 2010; Cordoba et al., 2012). Currently produced synthetic biogases can have higher copper level content than natural or biomethane gas (Cachia et al., 2018). Except for chromated copper arsenate treated wood, biomass content can be low (Demirbas, 2005). Fuel oils have low Cu fuel content but their small ash content becomes enriched (Huffman et al., 2000). In recent decades the fuels of environmental concern have become municipal wastes (average $\approx$1000 ppmw) (Lind et al., 2007) and sewage sludge (300–3000 ppmw) (Han et al., 2006; Van de Velden et al., 2008; Kruger et al., 2014; Jones et al., 2014a). These distributions among the fuels are important when it comes to utilizing the various ash components as agricultural supplements and their composition is regulated not to exceed certain levels (Eriksson, 2001; Poykio et al., 2010; Nurmesniemi et al., 2012; Chandrasekaran et al., 2012). Copper has been included in most quantitative combustion studies of trace metal distributions. As a result it became evident from initial studies that copper was essentially present in the various ash forms in either coal-fired power plants (Querol et al., 1995; Llorens et al., 2001; Klika et al., 2001; Nielsen et al., 2002; Spears et al., 2004), or with burning sewage sludge (Lopes et al., 2003; Van de Velden et al., 2008; Shao et al., 2008). Initially no pronounced enrichment was reported and concentrations were found to be similar (Miller et al., 2004). Copper displayed a nonvolatile nature in many of the lower temperature combustion conditions. Early mass balances had clearly indicated that most of the copper existed in the various ash streams (Kaakinen et al., 1975). Nevertheless, later more detailed studies in coal or cofueled plants did indicate that ash Cu content increased significantly downstream in the smaller ash particles and even across the stages of the ESP (Meij, 1994; Nielsen et al., 2002; Lopes et al., 2003; Swietlik et al., 2012). A recent study of trace element partitioning in a circulating FBC with a bed temperature of 880°C did report 80% of the Cu on the fly ash for anthracite or bituminous coals (Duan et al., 2016). Nevertheless, with particulate control and scrubbers, reported stack losses of Cu are very low from coal or sludge combustion and of the order of 0.1% or now even less (Nielsen et al., 2002; Goodarzi, 2006a; Van de Velden et al., 2008). In fact a recent Chinese study has measured the stack emissions of 13 trace metals from a modern and large coal-fired plant (Zhao et al., 2016). This indicated 99.98% of the Cu was retained within the plant's control features. A Spanish detailed study of the effectiveness of FGD retention of trace elements noted that such a system could reduce any stray levels of Cu by a further significant factor in a co-combustion plant burning coal and pet-coke (Cordoba et al., 2012). Although it is now common to cofire hazardous wastes with an appropriate fuel there remain sole fuel hazardous waste incinerators. One recent such study noted that most Cu was retained in the bottom slag that showed low leachability possibly through $CuFe_2O_4$ or aluminate formation. However, high chlorine released more to the fly ash (Li et al., 2018b). Pyrolysis experiments confirm that Cu release from coal is very temperature dependent taking several minutes at 900°C (Chen et al., 2008). In a suspension-fired reactor with Polish coal and sewage sludge, total retention was noted at 800°C but above this a slight loss due to the higher temperature and chlorine content (Miller et al., 2004). Even with sludge containing significant Cl levels, temperatures need to be about 1000°C before Cu is released (Amand and Leckner, 2004; Han et al., 2006; Latosinska and Gawdzik, 2012). Enhanced Cu release was noted on adding HCl(g) to biomass wood bark combustion in a suspension-fired reactor at 800°C (Miller et al., 2003). With sewage sludge and treatment at 850°C, it was noted that linear volatilization occurred with chlorine from zero to 0.5%, sulfur had only a minimal effect, but the presence of P could retain Cu (Luan et al., 2013). With sewage sludge/coal cofiring, the $FeCl_3$ used to treat the sludge enhanced Cu release above 830°C (Folgueras et al., 2007). Similarly with a mix of coal/straw/sludge, the combination of high Cu in the sludge coupled to the high Cl levels from the coal and biomass produced significant Cu release at 900°C (George et al., 2010). From studies that treat ash, it is found that Cl can remove the Cu. When municipal waste ash is roasted with either Cl_2 or $CaCl_2$, either will remove most of the Cu at 1000°C (Chan and Kirk, 1999). Sludge ash can either be treated with HCl(g) at 1000°C (Vogel and Adam, 2011), or with $CaCl_2$ or $MgCl_2$ at 1050°C to similarly remove 97% of the Cu (Nowak et al., 2012). A heat treatment of municipal waste fly ash from 670°C to 1300°C was found to remove almost all the copper as a result of the 5.6% Cl ash content (Jakob et al., 1995). Recently, heavily contaminated sewage sludge cofired with raw meal in a cement kiln did note added volatilization of the copper content with added chloride salts (Yu et al., 2017). Nevertheless, it seemed surprising that they still only recorded 60% vaporization of Cu at 1450°C, which would not be expected. Normally, with sufficient temperature and/or chlorine presence, Cu can be readily vaporized.

With the advent of particle sizing, it became evident that particles less than 2.5 μm were enriched in Cu reflecting a vaporization deposition mechanism (Querol et al., 1995; Martinez-Tarazona and Spears, 1996). In a combustion study of waste, rich in Cu and Cl, it was apparent that large fractions of 1 μm particles were present in the fly ash whether a grate or a FBC combustion mode was used (Lind et al., 2007). A recent ambitious study size segregated coal fly ash monitoring the concentrations of about 40 trace elements in five size ranges down to 2.2 μm (Lanzerstorfer, 2018). They noted strong enrichment to the smaller sizes for Cu. Utilizing an 11-stage impactor and collecting particles after the ESP it was noted that a bimodal distribution of sizes were present with peak quantities at 0.05 and 2.0 μm; 22% of the Cu was found in the finer size and 78% in the larger (Kauppinen and Pakkanen, 1990). More recently this work has been confirmed with a 13 stage impactor and a lignite coal combustor showing bimodal peaks at 0.06 and 1.8 μm (Wang et al., 2008). Particles in the 0.03–0.06 μm size range indicated enrichment factors of >10-fold in Cu; in the 0.06–0.1 μm range 10-fold; 10-fold in the 0.7–1.1 μm range and unity for 1.1–1.8 sizes. This was confirmed recently in a down-fired pilot-plant burning a Chinese anthracite coal at 1580°C (Zhang et al., 2017b). Copper was seen to be depleted in the bottom ash but was highly enriched on the fly ash. The ESP removed >98% of the copper-bearing particles. The presence of copper increased significantly for particles <5 μm and was very high in the 1 μm size. Copper was seen to behave rather like its neighbors Ni and Zn in the periodic table being about the eighth most volatile of the metals in combustion. In two cases examining ash particles from biomass combustion it was seen that the particles are small (Demirbas, 2005). For olive residue biomass combustion at 1450°C, an 11-stage sizing impactor illustrated that 3% of the total fly ash mass was in a fine (<1 μm) distribution with a 0.05 μm peak representing about 15% of the Cu (Jiminez et al., 2008). A recent study has shown an effect on the fly ash composition of reducing the flue gas temperature from 250 to 150°C through a heat exchanger. The fuel was biomass from metal contaminated soils burned in a small pilot-plant boiler. Boiler temperatures were not quoted but a general increase in ash concentration appeared to occur for copper, just over twofold, as also for many other elements as the temperature was so reduced through the heat exchanger. As no further metal deposition was possible to increase the concentrations especially for such elements as Ti or Zr that are fully nonvolatile it was apparent that another mechanism had to be apparent. In the case of copper any in the gas phase is rapidly deposited on the ash and should be fully complete by 250°C. As a result, it is possible that particle segregation was occurring through the heat exchanger and the ash size distribution was changing favoring at 150°C more of the small enriched sizes collected at the lower temperature (Kovacs et al., 2018). It illustrates the care needed in such interpretations.

The vapor transporter for copper is undoubtedly CuCl(g) in most combustion situations but the possibility of a contributing role for Cu(OH)(g) remains unclear. Deposition then will be controlled by the temperature of a cooling ash or nucleating center. Copper has few thermally stable condensed phase molecules at higher temperatures and major deposition can only occur when temperatures approach and fall below 1000°C. Several thermodynamic equilibrium calculations have been attempted that include Cu and only few other species and remains unsatisfactory (Lundholm et al., 2007; Konttinen et al., 2013). Moreover the data bases for copper compounds are limited so few species can be included. In the gas phase it would appear that the bond strengths for CuOH, CuCl, and $CuCl_2$ are essentially equal so their distribution will be critically controlled by the levels of water and HCl. Consequently, irrespective of fuels these will control volatility and then have a somewhat limited choice of what to form when temperatures cool and deposition can occur (Chen et al., 1998a; Yan et al., 2001a; Abanades et al., 2002; Danihelka et al., 2003; Sorum et al., 2004; Poole et al., 2008). Very few attempts have been made to identify the nature of the copper in the various ash forms. Residual fuel oil ash was analyzed by X-ray absorption concluding that the copper was possibly present as $CuSO_4$ (Huffman et al., 2000). Similarly, using X-ray absorption spectroscopy (XANES), filter ash from a waste FBC appeared compatible with CuO or $CuSO_4$ but not CuCl although the specificity of this method is poor (Lassesson and Steenari, 2013). An additional study used a small lab-scale waste rotary-kiln reactor examining the effects of Cl, S and water on the cooling flue gases below 600°C (Jiao et al., 2013a). The data were analyzed with the aid of a thermodynamic equilibrium program. It confirmed that deposition does occur at these low temperatures and will consist of $CuSO_4$ if sulfur is present otherwise a mix of CuO and $CuCl_2$ will be dominant. An interesting question rarely addressed is whether there will be sufficient time on cooling for the condensed phase to maintain its thermodynamic preferences. Also time for any significant interaction with other ash components such as Fe or Al. Although the ferrite is known, $CuFe_2O_4$, and aluminates, $CuAl_2O_4$ and $CuAlO_2$, it would seem these are likely to be produced more in the bottom ash or slag and time constraints may prevent their occurrence in fly ash. Leaching tests with water on bottom and fly ash from a coal-fired plant did indicate solubility of copper, comparable to arsenic and barium, so the soluble sulfate or chloride are possible end products (Llorens et al., 2001). Also ash thermal or other treating with chlorine or other reactants can provide answers. The question of whether ash can become a source of trace metals now is being considered more seriously. Solid waste incineration ash has in fact been shown to be easily leached of its Cu content (Tang and Steenari, 2016; Jiao et al., 2016). A recent monitoring study of emissions from six incinerator sites in the United Kingdom examined whether metal ratios might be used to fingerprint their source. Although, this was possible and might be useful in certain cases, emissions were seen not to be significant and rapidly blended into the environmental background confirming a negligible ambient contribution for copper (Font et al., 2015).

The only concerns with copper and biomass has centered on chromated copper arsenate treated wood that is rich in these elements and poses a modern-day disposal problem. Although now banned from residential use, large quantities of this wood for nonresidential uses such as railroad ties, poles and marine timber still are produced. Landfill introduces leaching problems and combustion requires acceptable incineration. Domestic fire and stove burning is very hazardous. Two laboratory studies confirmed that the wood could be incinerated at low temperatures (<1000°C) with the Cu and Cr being nonvolatile and retained in the ash while 11%–14% of the more volatile As concentrated in the fly ash (Wasson et al., 2005; Lundholm et al., 2008). X-ray analysis and modeling concluded that the copper in this case would likely form $CuCrO_2$, $CuCr_2O_4$ and $Cu(AsO_4)_2$ with their high ash metal content.

Detailed studies of copper's behavior under gasification or reducing conditions are limited. For steam gasification of biomass at 850°C, as expected copper is not volatile and is retained in the various ash forms (Syc et al., 2011). This is similarly the case for sewage sludge gasification at 770°C–960°C (Reed et al., 2005). In a coal gasifier Cu was described as semivolatile, 10% at 1025°C, 75% at 1125°C, and 100% at 1316°C (Bunt and Waanders, 2009). In an IGCC plant gasifying at 1200°C–1600°C, using coal, pet coke and limestone, it was noted that copper was equally dispersed between the slag and fly ash (Font et al., 2010). No studies appear to have addressed any possible roles being played by copper's condensed or gaseous phases. A thermodynamic equilibrium gasification calculation indicates that CuH(g), CuS(g), and Cu_2S(g) species with slag formation of Cu_2S(s) and $CuFeS_2$(s) might occur (Yan et al., 2001a).

In summary, copper is not rich in thermally stable molecules in the condensed or gaseous phase. As temperatures are increased, monovalency becomes dominant favoring CuCl(s) and Cu_2O(s). These then will vaporize to CuOH and CuCl with their distribution controlled by water and chlorine levels. They can be the dominant transport molecules readily depositing on the ash as it cools. Either a sufficient temperature or the presence of chlorine will promote copper release. Depending on the combustion mode, it can be found in bottom ash or will be present in some ash form. Coal fly ash becomes highly enriched especially on particles <5 μm being pronounced on particles of about 1 μm. Most ash is collected by an ESP with high efficiency (≥98%) and there is little if any stack emissions now from modern power plants with FGD scrubbers. The chemical nature in ash remains unknown but could reflect a thermodynamic preference of divalency at lower temperatures. If the ash can track thermodynamic preferences, there are hints of dominant formation of $CuSO_4$, and CuO possibly with $CuCl_2$. Any possible interaction with Al or Fe within the ash remains unreported and will depend on temperature, time and interaction constraints. Leaching from some ash is seen implying water solubility.

7.14 Iron

Iron is the 26th element in the periodic table: central in the first row of transition elements lying between Mn and Co. It is a mid-weight "heavy" metal with a solid density of 7.9 g cm^{-3}, heavier than the 2.7 of Al but lighter than that of tungsten 19.3. It is the most abundant metal in the solar system constituting most of the 2150 mile core radius of the earth. The composition of meteorites also confirms this universal occurrence. It is the fourth most abundant element by weight in the Earth's crust (5.6%) after O, Si, and Al, but like Al, has a concentration seven orders of magnitude lower in the oceans. It is mined mainly as its oxides and carbonate ores at an increasing global rate that is currently about 2.8 billion tons $year^{-1}$. Although abundant it has been questioned that sources may not be infinite and may require managing at some point. About 95% of the world's metal production is iron, primary due to its availability and low cost.

Iron has innumerable uses. The pure metal is too reactive, oxidizing and corroding in air. As a result, it has to be alloyed with carbon and other metals to form the large range of steels that have been developed. It is a metal that does not amalgamate with mercury and so has always been the metallic storage vessel for quicksilver. Its salts can have a range of colors for pigmentation for paints and even tattoo inks with diverse uses that include fertilizers, in animal food, and as a purifier for water. Magnetite, Fe_3O_4, is the most magnetic natural mineral known. Many of iron's compounds can show magnetic properties but hemoglobin in the blood is nonmagnetic. Natural sources are dust, volcanic emissions, and from forest fires. Anthropogenic sources from industrial and combustion particle losses are not regarded as pollutants but rather a possible fractional addition to the normal atmospheric burden. Few reviews include iron's natural atmospheric emissions but one suggests they are in the millions of tons per year (Malakhov and Makhon'ko, 1990). With a concentration of 5.6% in the Earth's mantle the estimates of dust and rock erosion are clearly obvious (Nriagu, 1989). Residence times in the atmosphere are estimated to be about 10–20 days (United States EPA Fe, 1984). It has been suggested that iron in ash from volcanic eruptions becomes soluble in the atmosphere before enriching the ocean waters (Hoshyaripour et al., 2015). Additionally there is now one extensive evaluation of the sources, transport and deposition of iron in the global atmosphere (Wang et al., 2015b). They estimate the emission of iron from coal combustion in the 1960–2007 period. This was 5.3 million tons $year^{-1}$ distributed over particle sizes of <1 μm (1%), 1–10 μm (27%) and >10 μm (72%). Iron as mineral dust 41 million tons $year^{-1}$. Atmospheric deposition controls Ocean concentrations and is about 8.4 million tons $year^{-1}$ of which 7% is the combustion contribution. Now that particle control efficiency from major combustors has significantly increased even this figure may be too large. As a result, iron is one commercial metal that cannot be listed as a pollutant, and combustion contributions are negligible.

Health concerns with iron

Not only is iron important to developing civilizations but it is ubiquitous to life forms with unique versatility being involved in a large range of bodily functions. Of these, 65% involve its role in blood hemoglobin but the biochemistry of iron in living organisms is a very extensive and an interesting field on which numerous books are available (Bezkorovainy, 1980; Crichton, 2001). In health, the body contains about 2.8–3.5 g Fe/adult. Free metal ions in the body are at low levels and the element is mainly bound in proteins or biomolecules. Unlike many metals where concerns center on toxicity, the major global concern for iron is a dietary deficiency. According to the World Health Organization, 30% of the world's population is anemic (World Health Organization Trace elements, 1996). Iron is present in most foods and a suggested dietary intake is about 15–30 mg day^{-1} for adults. The body does show different absorption rates from meats or grains so the nature of the diet does control the actual intake. Small amounts exist in drinking water. Generally there are no guideline values for this, levels often are ≤ 1 mg L^{-1}, higher values modify its taste and will stain. The body utilizes and can store iron (30%) in various organs for release on demand as needed, and if in excess is excreted or shed by hair growth. Excessive retention of iron is known as hemochromatosis, deficiencies produce anemia, both requiring medical attention. For adults, an onset of toxicity has been suggested for intakes of 70 mg day^{-1} with 50 mg day^{-1} regarded as a safe upper limit (Goldhaber, 2003). Clinical testing for human iron levels is not common. Few testing facilities quote methods or reference levels. The Mayo Clinic in the United States recommends a serum test that should normally register in the range 500–1500 μg L^{-1}. Only a few surveys have included it, giving ranges of values observed in whole blood, hair and urine samples. These are useful for establishing a preferred method that becomes a reliable indicator of normal good health. Hair has been listed as having a suitable range of 6–15 ppmw. However, because of its nature of being in an air environment it appears unreliable, one recent Russian study indicating a range of 7–1500 ppmw values with school children (Drobyshev et al., 2017). Two others indicate averages of about 50–75 and 5 ppmw introducing inconsistent uncertainties (Zhu et al., 2018; Oggiano et al., 2018). Whole blood (≈ 500 mg L^{-1}) appears to maintain more reliable narrow ranges (Oggiano et al., 2018; Zeng et al., 2019). The body releases only low levels of blood to urine (5–10 μg L^{-1}) so if monitored it is more a diagnostic for internal bleeding. There are risks in taking supplemental oral doses of iron (the sulfate or organic salts have been shown to be equivalent), in that the body mechanisms with other metals such as Al, Cu, Se, Zn are in complex balances that can then become modified. Usually such supplements become a high risk to children and are cases of household accidental overdosing. Often there can be concerns that mother's milk is not sufficient for a baby's iron needs and supplementary food has to be provided during the first 9 months of growth.

Being magnetic, iron nanoparticles such as Fe_3O_4 that are nontoxic are of particular interest in medicine where they can be monitored and directed to certain locations and then be inductively heated or used in other modes (Ficai et al., 2014). Iron is readily transported through the body and the brain (Jiang, 2017). It readily transfers into the brain and is essential. The brain maintains a regulated balance and its levels are essentially buffered. It does appear to increasingly accumulate in the brain with age (White, 2017). Whether this contributes to various neurodegenerative diseases remains highly debated. Along with other metals found present in brain autopsies it has the ability as do Cu and Cr to change valence and induce oxidative stress (Mot and Crouch, 2017; Kanninen and White, 2017; Bondy, 2017). Iron's main asset is that it has low toxicity to humans, is nontoxic to birds and only slightly toxic to rats and fish. It is also classified as noncancerous.

Atmospheric health concerns center more on the particle effects of inhalation rather than any iron content.

Iron's basic chemistry

As a transition element, iron's outer electronic atomic shells are $3d^6 4s^2$ implying eight possible valencies. However, the highest valence seen (6) is with K_2FeO_4, potassium ferrate, a stronger oxidant than the isostructures with Cr and Mn. It is purple in color with a melting point of 198°C at which it decomposes. Its use is limited in that it decomposes to Fe_2O_3 with water. The major valences exhibited for iron are simply 3 (ferric, d^5), and 2 (ferrous, d^6), with a large range of their compounds, which are commonplace and extensively used and studied. These are summarized in Table 7.22. All ferrous salts tend to be reducing agents in their attempt to become trivalent. Its chemistry closely mirrors that of Mn.

Pure iron (m.pt.1538, b.pt.2861°C) has a low vapor pressure of about 10 ppmv at 1455°C. Its major thermodynamically stable molecules again center about its oxides, all water insoluble. Three are readily available reflecting the valencies shown, FeO (II), Fe_2O_3 (III), and Fe_3O_4 (II/III). FeO is the least thermally stable and can be nonstoichiometric. On heating it can be oxidized by disproportionation to Fe(s) and Fe_3O_4(s). It has a melting point of 1377°C and vaporizes above 600°C to a partially dissociated FeO(g). This produces Fe(g) and O_2 (Balducci et al., 1971). A Knudsen cell-mass spectrometric study is reported at 1580°C–2480°C of the vaporization of a slag mixture of CaO/MgO/Al_2O_3/Cr_2O_3/FeO/SiO_2 (Wang et al., 2009). This noted that FeO(g) was the first to vaporize at >1300°C producing Fe(g). Fe_2O_3 is stable to its melting point of 1565°C. But from about 1450°C it converts to Fe_3O_4 that has a 1597°C melting point and is the most magnetic of all natural minerals. The two oxides Fe_2O_3 and Fe_3O_4 can readily be converted one into the other by oxidation or reduction. A Knudsen cell-mass analysis of vapors above the iron oxides at 1500°C–1594°C indicated only Fe and FeO gaseous species. At 1594°C FeO(g) was about one third that of Fe(g) (Washburn, 1963). Corresponding hydroxides are known, $Fe(OH)_2$, $Fe(OH)_3$, and FeO(OH), all water insoluble and can be made by precipitation from solution. The di- and trihydroxides are not thermally stable and above about 250°C conversions by dehydration occur to FeO(OH) and then Fe_2O_3.

The reversible nature of the thermodynamic stabilities describing exact distributions between these oxides and hydroxides is controlled mainly by the specific temperature, oxygen and water concentrations (Klissurski et al., 1984; Florindi et al., 1989). FeO(OH), also magnetic, is a useful yellow pigment and used in tattoo inks. The carbonate, $FeCO_3$ siderite, occurs naturally, is insoluble in water and above 500°C produces FeO that then oxidizes to Fe_2O_3 (Stern, 2001).

Iron's possible nitrites and nitrates are unstable but the deliquescent ferric salt does form a hydrated nitrate. Ferrous sulfate, $FeSO_4$, dissociates at 537°C to Fe_2O_3/Fe_3O_4 without melting (m.pt.680°C) (Ostroff and Sanderson, 1959) and oxidizes in air at ambient temperatures to ferric sulfate and Fe_2O_3. The ferric salt behaves similarly, dissociating at about 485°C to Fe_2O_3 (Tagawa, 1984; Tagawa and Saijo, 1985; Stern, 2001). The ferrous sulfate is very water soluble while the ferric salt only slightly so. Apparently, in 2009 the Spirit Rover on Mars stuck permanently in a patch of soft nontractable ferric sulfate. The

TABLE 7.22 Properties of iron's condensed phase molecules relevant to combustion, °C.[a]

m.pt. b.pt. vapor pressure
Fe 1538 2861, oxidizes in air, 10 ppmv/1455°C, ins
FeO 1377, unstable, vaporizes, >600 dissoc to FeO(g) and Fe(s), Fe_3O_4, ins
Fe_2O_3 1565, >1450 converts to Fe_3O_4, ins
Fe_3O_4 1597, stable, most magnetic, ins
$Fe(OH)_2$ 1085, unstable, >250 dehydrates, oxidizes to FeO(OH) then Fe_2O_3, ins
FeO(OH) >200 decomp to Fe_2O_3, ins
$Fe(OH)_3$ unstable, >250 dehydrates, oxidizes to FeO(OH) then Fe_2O_3, ins
$FeCO_3$ >500 dissoc to FeO then Fe_2O_3, ins
$Fe(NO_3)_3$ unstable, forms hydrate, s
$FeSO_4$ 680, dissoc >670 to Fe_2O_3/Fe_3O_4, oxidizes in air to ferric sulfate, s
$Fe_2(SO_4)_3$ dissoc on melting >490 to Fe_2O_3, ss
$FeCl_2$ 677 1023, vapor monomer but 25% dimer at m.pt., hydrol and oxidizes to Fe_2O_3, Fe_3O_4, 50 ppmv/510°C, s
$FeCl_3$ 304 330 sublimes to dimer, dissoc to $FeCl_2$, hydrol, 1000 ppmv/190°C, s
$Fe(C_5H_5)_2$ 173 249, sublimes >100, ins
$Fe(CO)_5$ −20 103, ins
$Fe_2(CO)_9$ 100, decomp, ins
FeS 1188 slowly oxidizes in air, dissoc >988 to Fe(s), ins
FeS_2 (pyrite) >743 dissoc to other sulfides, oxidizes in air, hydrol to Fe_3O_4, ins
FeP 1147, reacts with water, dissoc to Fe_2P(s), 5 ppmv P_2/870°C, ss
Fe_2P 1370, reacts with water, dissoc to Fe_3P(s), 5 ppmv P_2/650°C, ss
Fe_3P 1166, reacts with water, ss
FeH_n not stable
$FeMoO_4$ 1068
$Fe^{III}PO_4$ 1240, ins
$FeSiO_3$ (meta-) 1140, ins
Fe_2SiO_4 (ortho-) 1217, ins
$FeTiO_3$ 1385, ins
Fe_2TiO_4 1390, ins
$FeWO_4$ 989
$Al_2Fe^{II}O_4$ 1797, ins
$CaFe_2O_4$ 1215, ins
$KFeO_2$ unstable in air, s
$K_2Fe^{VI}O_4$ >198 unstable, decomp to $KFeO_2$, reacts with water forming Fe_2O_3, ss
$MgFe_2O_4$ 1750, stable, ins
$NaFe^{III}O_2$ 1347, >800 dissoc, s
$Na_2Fe_2O_4$ 1345, hygroscopic, s

[a]*dissoc,* dissociates; *decomp,* decomposes; *s,* water soluble; *ss,* slightly soluble; *ins,* water insoluble; *hydrol, hydrolyzes.*

chlorides are very water soluble. $FeCl_2$ (m.pt.677, b.pt.1023°C) is hydrolyzed by water at low temperatures producing Fe_3O_4 and Fe_2O_3 (Taylor, 1984). In a Knudsen cell the dichloride displayed its volatility and vaporized molecularly at low temperatures, predominantly to the monomer (4 ppmv at 450°C), the dimer gradually increasing with temperature being 25% present at the melting point (Schoonmaker and Porter, 1958). This may be a possible transport molecule for iron in volcanoes at the lower temperatures with low water content (Wahrenberger et al., 2002). It also is readily oxidized on heating to the oxides. The trichloride is more volatile and its vaporization more complex. Below 230°C it sublimes to the dimer with a vapor pressure of about 1000 ppmv at 190°C. Its melting point is 304°C and with increasing temperature the $FeCl_3$(s) can dissociate to $FeCl_2$(s) and the vaporizing dimer either dissociates to the monomer or forms an $FeCl_2$(s) aerosol (Blairs, 2006). It can similarly be hydrolyzed by water to FeO(OH) and is subject to oxidation. Both chlorides are corrosive and hazardous, $FeCl_3$ being used also as a water purifier.

The divalent sulfide (FeS, troilite) occurs in nature, is magnetic with a melting point of 1188°C at which it dissociates to iron and sulfur (Ferro et al., 1989). It tends to be non-stoichiometric and is insoluble in water. In powder form it is pyrophoric in air and slowly oxidizes in the bulk (Ajersch and Benlyamani, 1989). A disulfide occurs in nature in two structural forms marcasite and the more familiar pyrite (fool's gold). Marcasite exists below 430°C but is a metastable structure changing rapidly to pyrite above this temperature. Pyrite becomes unstable and above 640°C converts to pyrrhotite Fe_7S_8 and FeS(s) together with S_2, which with water or oxygen converts to Fe_3O_4 (Ferro et al., 1989). The trivalent sulfide Fe_2S_3 is unstable and little studied. Stable phosphides are known, FeP (m.pt.1147°C), Fe_2P (m.pt.1370°C), Fe_3P (m.pt.1100°C), but require moisture and oxygen-free environments for stability. An early Knudsen cell-effusion study showed that these gradually lose phosphorus on heating, changing to the next iron-rich formulation. Reflecting their relative thermodynamic stabilities, the phosphorus vapor pressure is about 5 ppmv above FeP_2(s) at 650°C, above FeP(s) at 870°C and only 0.5 ppmv at 1080°C above Fe_2P (Lewis and Myers, 1963). A more recent thermodynamic study adding mass analysis confirmed the values (Zaitsev et al., 1995). Although they can be detected in their vapor, these gaseous phosphides are not stable. Due to a complexity of low-lying states the FeP (g) ground state has not been confirmed. One theoretical appraisal suggests $X^6\Sigma^+$ and a bond strength D_0(Fe-P) $\approx 197 \pm 20$ kJ mol^{-1} (Tong et al., 2003). Like many phosphides these are hazardous due to the toxic PH_3(g) formation with water. Fe_2O_3 can react with phosphorus compounds under oxygen-rich conditions and form a complex range of phosphates but these are not expected to be stable to high temperatures, $FePO_4$ (m.pt.1240°C) and are insoluble in water. Iron has little affinity for hydrogen. Although molecular forms have been characterized there is no bulk solid phase hydride. Other molecules of commercial use are the carbonyls and ferrocene, ranked as the most effective metal based flame inhibitors (Babushok et al., 2000). The carbonyls are a useful surrogate for gaseous iron in combustion research, CVD or nanotechnology fields. The pentacarbonyl, $Fe(CO)_5$, is a gas (m.pt. − 20, b.pt.103°C), and is hazardous and toxic. The nonacarbonyl, $Fe_2(CO)_9$ decomposes at its 100°C melting point, sublimes and is less dangerous to handle. Use of ferrocene, $Fe(C_5H_5)_2$, often is a preferred choice (m.pt.173, b.pt.249°C) subliming above 100°C and is less toxic. All are water insoluble. With ash formation from solid fuels, iron's extensive mixed metal chemistry is possibly most relevant. From the major components that can be present in various ash forms it may for example produce compounds with Ca, Al, Si, and Ti. One class of compounds are the ferrates/ferrites terms loosely interchanged that can contain Fe(III) and are of the form $M^{II}Fe_2O_4$ such as $CaFe_2O_4$ (m.pt.1215°C). Others can be based on divalent iron such as the aluminate (written as $Fe(AlO_2)_2$ or Al_2FeO_4, m.pt.1800°C), orthosilicate (Fe_2SiO_4, m.pt.1217°C), titanate ($FeTiO_3$, m.pt.1385°C), orthovanadate ($FeVO_4$, m.pt.880°C) (Slobodin et al., 1976), and metavanadate ($Fe(VO_3)_3$), generally all water insoluble. Measurements have been made of the reaction of Fe_2O_3 and Fe_3O_4 with sodium salts and find that these do produce $NaFeO_2$ at temperatures close to 1000°C (Fotiev et al., 1996). This was evident also through weight loss of Fe_2O_3 with either NaCl (Mobin et al., 1990) or Na_2SO_4 (Mobin et al., 1996) at similar temperatures. Fe_2O_3 also reacts with V_2O_5 above 600°C to produce the orthovanadate. Consequently, with iron's nonvolatile nature, its oxides have many options to fuse with other ash components if temperatures are above 1000°C and there are no time constraints. Also condensing metals or metal oxides may readily react with iron oxides in ash particles as they cool.

Gas phase stabilities and transport

In the gas phase, iron can have a normal type chemical reaction nature. As indicated, FeO(g) is the predominant volatile oxide at high temperatures and is in a balance with Fe (g). Many measures of its gaseous bond strength have been published over the years and even the NIST-JANAF table, dated 1966 (Chase, 1998) managed to assess a closely accurate value of 409 ± 20 kJ mol^{-1}. Since then, methods involving equilibria, photofragmentation, and photoionization have better resolved its value to D_0(Fe-O) $= 403 \pm 2$ kJ mol^{-1} now suggested (Metz et al., 2005; Chestakov et al., 2005). The nonvolatility of the iron oxides, coupled to this dissociation energy and the complexity of the FeO spectroscopy delayed establishing $X^5\Delta_4$ as the ground state with a low lying $A^5\Sigma^+$ state at a still uncertain energy of 3905 or 2080 cm^{-1} ($T_0 = 47$, 25 kJ mol^{-1}) (Huber and Herzberg, 1979; Kim et al., 2015). The complexity of the spectral analysis for such diatomic transition metal oxides has been reviewed (Merer, 1989). Early theoretical assessments predicted a very low-lying $a^7\Sigma^+$ state (Dolg et al., 1987) now detected experimentally (Drechsler et al., 1997). This is located at 4965 cm^{-1} energy (59 kJ mol^{-1}) above the ground state and potentially affecting the thermochemistry at elevated temperatures (Kim et al., 2015). The most recent and thorough theoretical effort in this regard has estimated all the lower potential energy curves and summarized the data for FeO(g) (Sakellaris et al., 2011). This and the other studies still slightly underestimate the bond strength (Uzunova et al., 2008). However, before these, theory was heavily overestimating its value (Bridgeman and Rothery, 2000; Gutsev et al., 2003; Furche and Perdew, 2006; Jensen et al., 2007). In matrix isolation measurements both FeO and FeO_2 have been noted and analyzed for their basic parameters. However, in the condensed phase, interpretations of the formation of the various structures of FeO_2 remain unclear, but a symmetrical bent (150°) shape is the most stable (Chertihin et al., 1996; Andrews et al., 1996a; Gong et al., 2009). Experimentally, gaseous FeO_2 has only been reported in one Knudsen cell-mass analysis study with Fe_2O_3(s) at 1445°C–1490°C (Hildenbrand, 1975). Initially, by estimating the required thermochemical data for FeO_2, a value of D_0(OFe-O) $= 427 \pm 20$ kJ mol^{-1} was quoted, slightly stronger than the first bond.

In a recalculation this was raised to 443 kJ mol^{-1} but is based on a questionable thermochemical database for FeO_2(g) (Smoes and Drowart, 1984), which still has not been accurately reassessed. More recently, its chemistry has been invoked in the upper atmosphere. There is now interest in the demise of the gaseous iron ablated daily from the 120 tons of meteors and dust that enters the atmosphere from space. As a result, kinetic modeling has displayed the interplay between the oxides and hydroxides in the 70–110 km altitude regions of the atmosphere tentatively suggesting a dominant reservoir of FeOH and FeO (OH) formed via FeO and FeO_2 (Self and Plane, 2003). However, their work has raised questions not yet resolved concerning the bond energy of FeO_2. Their theoretical estimates for various structures suggest a value no greater than 375 ± 40 kJ mol^{-1} for D_0(OFe-O) (Plane and Rollason, 1999). A shock tube experiment did examine the reactions between Fe(g) and O_2 at 680°C–2230°C (Giesen et al., 2002). Fe and O atom concentrations were monitored simultaneously. It was found that FeO_2 formation was dominant below 1380°C but above this FeO(g) and O gained importance. Also a role for FeO_2 has always been assumed in iron's catalytic inhibition of combustion to be discussed later. This was analyzed again quite recently but did not address the question of this bond strength (Shvartsberg et al., 2010). Theoretical studies of the FeO_2(g) electronic structures remain very challenging partly because of the many low-lying and close levels that can indicate either triplet or quintet ground states depending on the method used (Garcia-Sosa and Castro, 2000; Gutsev et al., 2000a; Uzunova et al., 2008; Grein, 2009). Table 7.19 summarizes iron's and the first row transition elements' gaseous bond strengths. As seen, the periodic table trends fail to resolve this dioxide dilemma as iron's two prior neighbors Cr and Mn display a much stronger second bond in their dioxide while Co and Ni the following neighbors show a decreasing value. However, as with other transition elements, trends in the mono- and dichlorides and hydroxides can be very insightful. As will be seen, it appears quite apparent that in both these cases for iron their second bond is consistently stronger. Consequently, it appears quite reasonable to conclude that the D_0(OFe-O) bond may be larger than is currently being considered. As seen in Table 7.19 the trend in the D(OFe-O)/D(Fe-O) ratio does appear too low (1.06) with values of 427 and 403, raising this ratio would be more consistent with that of Co (1.10). This would support the higher suggested value of about 443 ± 20 kJ mol^{-1} for D_0(OFe-O) and is now recommended.

Leaving a discussion of the halides and hydroxides to the section below, there are other remaining gaseous molecules that may be encountered. FeS(g) has been of particular interest concerning gaseous sulfides particularly in comparison to the diatomic oxide. Experimentally, its stability has only been examined once in a Knudsen cell-mass analysis experiment (Drowart et al., 1967) and this remains the basis for the recommendation in the 1977 thermochemical table (Chase, 1998). The experiment established the stability of FeS relative to that of MnS. Because of the slight uncertainty of the latter, a value of D_0(Fe-S) = 326 ± 15 kJ mol^{-1} now appears more appropriate from that study. However, recent work from the accurate measurement of its predissociation threshold now has obtained a recommended value of D_0(Fe-S) = 312.6 kJ mol^{-1} with minimal error (Matthew et al., 2017). This bond strength of FeS is about 78% the value of the corresponding oxide and follows the trend seen with Mn (Table 7.19). Containing a transition element, FeS has been favored for a long time in helping to develop theory. In the last decades, methods have been refined to describe the electronic behavior and reproduce the magnitude of the bonding more closely but the overestimating of this value by about 10 kJ mol^{-1} has remained (Hubner and Sauer, 2002; Wu et al., 2007). Even though a diatomic, it is complex but now has been shown to have an $X^5\Delta_4$ ground state similar to the monoxide. It has very low-lying $^5\Sigma^+$, $^5\Delta$ and $^7\Sigma^+$ states that are not yet fully characterized, but theory consistently still indicates an $X^5\Sigma^+$ state. Coupled to this new experimental value, a further theoretical study has examined the latest preferred functional forms. The effort displays the continuing uncertainties in how to model this molecule with best efforts remaining no better than 17 kJ mol^{-1} from the accepted bond energy value and still predicting an $X^5\Sigma^+$ state (Sharkas et al., 2017). The various sulfide molecules created by laser ablating iron with sulfur in a matrix isolation method have been reported providing vibrational frequencies (Liang et al., 2009). Such sulfides will only play minor roles in combustion as they are vulnerable to oxidation and hydrolysis at elevated temperatures. They can play roles only in oxygen and water free pyrolysis or gasification. The gaseous hydrides are also of little interest to combustion being weakly bound and readily oxidized. They are of more interest to astrophysics and theory. In this respect, significant efforts were made to establish the bond strength of FeH and develop adequate theoretical descriptions of its electronic nature. Most of this extensive work is now more than 20 years old when the ground state was confirmed to be $X^4\Delta$ (Beaton et al., 1988). Some advances have been made over the years to more reliably anchor values and observe its spectroscopic transitions. Nevertheless, theory still appears to overestimate its bond strength but does identify the low-lying states. Experimentally it is difficult to measure. An early equilibrium-mass analysis study failed to monitor FeH and from this suggested an upper limit of about D $\leq$180 kJ mol^{-1} (Kant and Moon, 1981). Armentrout et al. in the 1988–95 utilized the method of establishing reaction thresholds of endothermic reactions between Fe^+ and alkanes and silane. A summary of their data recommended D_0(Fe-H) = 147 ± 5 kJ mol^{-1} (Kickel and Armentrout, 1995a). However, other ion-reaction hydride abstraction methods led to corresponding values of about 120 (Sallans et al., 1985) and 176 (Tolbert and Beauchamp, 1986). Further work utilizing a negative Fe^- ion-electron detachment reaction approach obtained 143 ± 16 kJ mol^{-1} supportive of a midrange value (Miller et al., 1993). For now, a value of D_0(Fe-H) = 147 ± 10 kJ mol^{-1} might be appropriate. Along with its neighbor Mn, these are the smallest values in the other first row transition element hydrides, exempting Zn the last member. Being a hydride, this has attracted many theoretical studies mainly for refining alternate model approaches. The work before 2000 was reviewed and indicated the difficulty encountered not only in overestimating the bond strength but in the difficulty of establishing the order of the lowest lying electronic states (Harrison, 2000). A more recent effort outlines the remaining theoretical difficulties (DeYonker and Allen, 2012). Their ordering of states $X^4\Delta$, $A^4\Pi$(1065 cm^{-1}), $a^6\Delta$(1530), and $b^6\Pi$(3580) is consistent with spectroscopic analyses that also remain uncertain in their relative positions, but the obtained bond strength of 180 kJ mol^{-1} appears to remains too large as do other recent efforts (Daza et al., 2006; Goel and Masunov, 2008). However, as indicated above from trends, this has yet to be resolved. Not observed in the gas phase until 1999 (Korsgen et al., 1999), the FeH_2 molecule has only been studied theoretically (Siegbahn et al., 1984) and with matrix isolation methods (Rubinovitz and Nixon, 1986; Chertihin and Andrews, 1995; Andrews, 2004; Wang and Andrews, 2009). These indicate a close to linear molecule with an $X^5\Delta$ state and are of more interest to catalysis than combustion.

Laboratory combustion studies

Early development of atomic emission/absorption analysis methods for metals utilizing high temperature flames did include iron (De Galan and Samaey, 1979). These methods depended on the atomization of the iron in the flame and measurements did note approximate estimated efficiencies of producing about 40%–90% atomic Fe. Consequently, if temperatures are sufficiently high it is apparent that atomic Fe will be evident if easily released into flame gases. An early laboratory study introducing $Fe(CO)_5$ or an iron salt solution at 0.9–3 ppmv concentration levels to hydrogen flames indicated that molecular formation other than gaseous Fe and FeO was dominant. At flame temperatures in the presence of water, FeO is hydrolyzed to its hydroxides (Jensen and Jones, 1973). The monohydroxide, FeOH, has been quite extensively studied due to its potential roles in combustion and astronomy. Initial Knudsen cell-mass analysis studies examined equilibria established in cells containing Fe_2O_3 with H_2 or H_2O. At the time, accurate thermochemical analysis was limited as the electronic nature of FeOH(g) was ill-defined. Later recalculation of these studies (Smoes and Drowart, 1984; Gorokhov et al., 1990) suggested a too high value of D_0(Fe-OH) = 380 ± 20 kJ mol^{-1} from the flame study (Jensen and Jones, 1973), and 344 ± 16, and 346 ± 16 from Knudsen cell-mass spectrometric measures (Murad, 1980a; Gorokhov et al., 1990). To these, now using an ion-molecule mass analysis and reaction bracketing method, an accurate value for the ionization potential of FeOH has been obtained (Schroder and Schwarz, 2003; Schroder, 2008). From this, a further value of 337 ± 10 kJ mol^{-1} can be derived. Although still a demanding task, several theoretical estimates have been published, 304 ± 20 (Glukhovtsev et al., 1997), 360 ± 20 (Kellogg and Irikura, 1999), 339 ± 25 (Rollason and Plane, 2000) and most recently 348 kJ mol^{-1} (Cao, 2005). Averaging the three lower experimental values suggests a recommended D_0(Fe-OH) = 342 ± 10 kJ mol^{-1} now consistent with much of the data. FeOH(A,X) states have been more fully characterized electronically and more accurate thermochemical analyses should soon become available (Hirano et al., 2009, 2010). A subsequent flame study involving $Fe(CO)_5$ addition and mass spectral analysis, although somewhat questionable due to the heavy flame seeding and probe sampling did report correct magnitudes for D(FeO) and $\Delta H_f(Fe(OH)_2)$ (Farber et al., 1974). Moreover, it recorded only these two iron species and showed that the dihydroxide was dominant over the monohydroxide. By taking the thermodynamic data for FeO and $Fe(OH)_2$ (Chase, 1998) it is now clear in the 2000K flames of Jensen and Jones (1973), that the gaseous ratio of $Fe(OH)_2$/FeO was >20-fold. This neglect of the dihydroxide explains their too high estimate for D(Fe-OH). The dihydroxide is in fact an important gaseous species of iron. It was first established in a chemical ablation study examining weight loss from Fe(s) or Fe_2O_3(s) in flowing H_2O/O_2 at 1300°C–1460°C (Belton and Richardson, 1962). A careful study eliminated alternate oxide or other hydroxide forms, the overall mechanism being defined by

$$Fe(s) + 2\,H_2O(g) = Fe(OH)_2(g) + H_2(g)$$

Measured dihydroxide concentrations can be on a ppmv at these temperatures and this helped explain the erosion that was being observed in steam boilers. Its thermochemical values remain based on this work and were supported by the above flame study (Farber et al., 1974) and theoretical estimates (Schroder and Schwatz, 2003). Matrix isolation studies reacting thermal metal atoms with water or laser-ablated atoms and condensing the products at low temperatures have noted dihydroxide formation. These have provided structural and frequency data (Kauffman et al., 1985a; Wang and Andrews, 2006b). From the heat of formation that is based on the above ablation study (Chase, 1998), it implies that the sum of the two Fe-OH bonds have a value of 813 $kJ\ mol^{-1}$. With the recommended value above for the first bond, this indicates D_0(HOFe-OH) = 471 ± 10 $kJ\ mol^{-1}$. The second bond is significantly more stable than the first. However, as noted, the quoted values of the two bonds are coupled. A recent theoretical paper hinted that the heat of formation of the dihydroxide could be larger (Jiang et al., 2012). If this is the case it would reduce the sum of the two bonds and weaken the second bond value slightly. However, as seen below, this ratio of the two recommended hydroxide bond values matches almost exactly the two measured bonds in the dichloride providing strong support at present. It is also reasonably consistent with its periodic table neighbors Mn and Co (Table 7.19).

Iron has volatile chlorides but these are vulnerable to oxidation or hydrolysis to the oxides Fe_2O_3(s)/Fe_3O_4(s). $FeCl_2$(s) (m.pt.677, b.pt.1023°C) sublimes predominantly to the monomer gas initially but forms 25% dimer at the melting point and has a vapor pressure of about 50 ppmv at 510°C (Kana'an et al., 1969; Barin and Platzki, 1995). $FeCl_3$(s) is even more volatile (m.pt.304, b.pt.330°C) subliming as the dimer, which becomes the monomer that then dissociates to the solid dichloride on heating (Blairs, 2006). From these, the gaseous FeCl($X^6\Delta$) and $FeCl_2$($X^5\Delta$) are its most stable chlorides at elevated temperatures, now confirmed theoretically (Bridgeman, 1997). Hildenbrand was the first to quite accurately establish their stabilities utilizing effusion-mass analysis methods and also obtain an accurate heat of sublimation (Hildenbrand, 1995). His results, later reanalyzed with updated thermochemical data, better including the contribution of low-lying states, still remain a little uncertain due to this continuing lack of spectroscopic data (Hildenbrand, 2008). However, bringing these values up to date they now indicate D_0(Fe-Cl) = 332 ± 6 $kJ\ mol^{-1}$ and D_0(ClFe-Cl) = 470 ± 10 $kJ\ mol^{-1}$ from the mass spectrometric equilibrium data. For $FeCl_2$(s) though an accurate heat of formation (Barin and Platzki, 1995) coupled to the more accurate heat of sublimation now permits an improved value for ΔH_f $(FeCl_2)_g$ and hence a value for both bonds of $D_0(FeCl_2)_{atomiz}$ = 791 ± 3 $kJ\ mol^{-1}$. This now implies with the current recommended value of D_0(Fe-Cl) = 332 ± 6 $kJ\ mol^{-1}$ the result D_0(ClFe-Cl) = 459 ± 7 $kJ\ mol^{-1}$. These reflect the ratio value for the corresponding hydroxides above. In the last decade, theoretical methods have been refined and now can reproduce these values very closely (Nielsen and Allendorf, 2005; De Yonker et al., 2007; Mayhall et al., 2009; Bross et al., 2013). Matrix isolation and electron diffraction studies established the frequencies and the structure of the dichloride was found to be linear (Jacox and Milligan, 1969; Hastie et al., 1971; Hargittai et al., 1991). The spectroscopy of FeCl(g) remains sparse but from comparisons it is expected to have numerous low-lying states and the ground state now has been proven to be a sextet (Allen et al., 1997). Corrosion of steel is accepted as involving $FeCl_2$(g). At elevated temperatures, deposits of chlorides such as NaCl on a surface can produce formation of $FeCl_2$(s) either directly or by the release of chlorine on its conversion to the alkali sulfate. The $FeCl_2$ vaporizes migrating through the scale and becomes oxidized, releasing the chlorine back into a somewhat catalytic cycle (Grabke et al., 1995). An alternate mechanism initially forming FeS(s) (presumably under reducing conditions), which is then converted to gaseous $FeCl_2$ has more recently been suggested but remains uncertain (Kung, 2014).

Major research topics for over 30 years and still of interest include two, seemingly at odds. These relate to its behavior of being a partial soot suppressor yet also a strong flame inhibitor. Generally by adding iron in the form of either the carbonyl, $Fe(CO)_5$, or ferrocene, $Fe(C_5H_5)_2$, many studies have shown that the iron does not appear to modify the soot precursor formation mechanism but may be encapsulated possibly as Fe_2O_3(s) in the soot particles and catalyzes its oxidation (Howard and Kausch, 1980; Bonczyk, 1991; Hahn and Charalampopoulos, 1992; Zhang and Megaridis, 1996; Kim et al., 2008; Kim and Kim, 2018). Iron also may produce many more small particles on which the soot can be formed leading to smaller average soot particle sizes that are more easily oxidized. The model remains unsatisfactory but stresses having the soot particles small to be more rapidly oxidized catalytically. However, the data on soot suppression is confusing and appears to be very dependent on combustion conditions and chemical form. The early studies indicated no effect or at the most an efficiency of 5% that of Ba (Cotton et al., 1971; Bulewicz et al., 1975; Haynes et al., 1981). Unlike Ba that similarly oxidizes and controls soot, no similar chemiionization role can be invoked for iron. The ions Fe^+ and its monohydrate are known to be the natural dominant iron ions, with about 2% of $FeOH^+$, but are insufficient (Goodings, 1992). As a result, no consistent explanation for the metal inhibition exists. It appears that under certain conditions Fe appears to control soot emission and is as effective as Ba or Mn without introducing the latter's toxicity issue (Howard and Kausch, 1980). Two recent studies have extended this work but still leave it in an unsatisfactory state concerning the question of soot. One monitored the soot formation on adding $FeCl_3$ solution spray into a C_2H_4/air flame (Simonsson et al., 2018). They used laser-induced incandescence and elastic light scattering imaging to monitor the effects. However, for iron they noted few effects. A second study introduced the pentacarbonyl into a methane/air diffusion flame. Utilizing a kinetic model of the methane flame and invoking the Fe, FeO, FeO_2, FeOH, $Fe(OH)_2$, and FeOOH species it attempted to relate the decline in NO and hydrocarbon emissions and suggests a perturbation of the soot precursors, but this has to remain very speculative (Raj et al., 2018).

The second topic involving iron in combustion is its long known flame inhibition quality (Miller, 1969; Bulewicz and Padley, 1971; Vanpee and Shirodkar, 1979). An inhibitor ranking was published in 2000 showing iron could be even superior to chromium under certain conditions (Babushok and Tsang, 2000). A significant kinetic modeling effort was undertaken to try and distinguish between the gas phase Fe/FeO/FeO_2/FeOH/$Fe(OH)_2$ plausible radical recombination cycles, or even a possible heterogeneous catalytic particle role to recombine the flame propagating radicals (Jensen and Jones, 1974; Kellogg and Irikura, 1999; Rumminger et al., 1999; Rumminger and Linteris, 2000, 2000a). They indicated that a gas phase kinetic model could be adequate in reproducing most of the data for added iron carbonyl in premixed or diffusion flames. However, the effect was seen to be concentration dependent and lose its affect above a certain additive level possibly through particulate formation (Linteris et al., 2008). Also when compared with other catalytic inhibitors such as Ba or Mn, iron's performance became combustion mode dependent and possibly affected by particles (Linteris et al., 2004). A general heterogeneous model though is suggested as too slow. At low concentrations iron can be the best of metal inhibitors but does not model correctly at higher levels. One study has noted and examined the variation with equivalence ratio (Shvartsberg et al., 2010). Since then a soft ionization molecular-beam mass spectrometer has finally monitored the iron species in an atmospheric pressure slightly fuel-rich hydrogen flame seeded with iron pentacarbonyl (Gerasimov et al., 2011). This displays a growing downstream dominance of $Fe(OH)_2$ in the flame, followed by FeOH and Fe, with FeO/FeO_2 small and FeO very small. FeO appears very important though and is kept low by fast kinetics. Downstream with a constant temperature of 1650K, $Fe/FeOH/Fe(OH)_2$ were close to their equilibrium values, with the dihydroxide representing about 80% of the iron in the flame. Carbonyl additions were kept low at about 0.01% to prevent flame particles clogging the probe sampler. Particles appear to be present, possibly Fe_2O_3 or Fe_2O_3 that will be marginally existing at this temperature. The kinetic model is approximate and needs further development but appears a good basis to explain the catalytic radical recombination cycling. As suggested here, addition of iron as the carbonyl or an aqueous solution spray will undoubtedly produce a background of nanoparticles of iron oxides condensing either homogeneously or probably on minute condensation nuclei in the flame gases, so both mechanisms are likely present to some degree. Also the fact that iron can possibly oxidize soot while also having its flame radicals reduced by recombination cycles almost appears contradictory. It might be that soot suppression is a downstream catalytic oxidation with O_2, while radical recombination is an upstream reaction zone mechanism. The mass spectrometer measurements did note that the iron in the reaction zone region started as fully elemental, so the chemistry is rapid. A recent theoretical model studied the fragmentation interaction of the Fe atom with forming flame polycyclic aromatic hydrocarbons. It appears that the atom can attach and fragment PAH, forming radicals that promote growth and soot nucleation (Mao and Luo, 2019). Yet other exercises now have been added to this discussion. One using a thermogravimetric analyzer (TGA) has examined several different ranks of coal combustion with or without added traces of Fe_2O_3 (Gong et al., 2010). The coals were ground, similarly mixed with graphite, and then some seeded with Fe_2O_3. Rather surprisingly the iron oxide appeared to catalyze the combustion. Ignition temperatures were lowered, the combustion rates were increased and burn out times reduced. No mechanism was suggested. Another experiment displayed this remarkable catalytic behavior of iron if present in flame gases. Introducing either ferrocene or an iron nitrate salt solution into a pyrolysis nonsooting CO/H_2 flame configuration at about 650K, it was discovered that the CO produced single-wall carbon nanotubes (Vander Wal and Hall, 2002; Vander Wal, 2002). It appeared that Fe nanoparticles, initially formed, can catalyze the CO to produce the carbon nanotubes.

Combustion power plant data

As a result, we see a chemistry that can be largely nonvolatile except at elevated temperatures. Dominated by stable iron oxides, the main gaseous species formed will be the monoxide with a very stable dihydroxide. Chlorides and sulfides might play short-lived roles but are too easily oxidized or hydrolyzed in most practical mixtures. Although as listed in Table 7.22 Fe can produce numerous mixed metal oxides, few vaporize molecularly or are stable. The extensive Russian mass spectrometric studies of such vaporizations have only recorded gaseous K_2FeO_2, $FePO_2$, $FePO_3$ together with $SrFeO_2$ and $BaFeO_2$ that contain iron (Lopatin and Shugarov, 2013). Laboratory studies differ from Full-scale combustion and can be very deceptive as they invariably invoke using iron carbonyl or ferrocene as the additive to the flame. This jump-starts any chemistry by immediately releasing gaseous Fe, raising the question of whether results obtained from such a format are at all meaningful to more practical situations. Consequently, it is interesting to examine what has been reported in truly full-scale systems. Depending on the coal, iron can be named as a major or minor component but as seen in the list of approximate fuel contents (Table 1.1) it is more of a trace element in biomass becoming important in some wastes and sludge. It has a low concentration in pet coke but can be a significant trace element in limestone if used (Lind et al., 1994; Cordoba et al., 2012). With coals it is important in that pyrite or marcasite (FeS_2) can account for about half the sulfur in coal. As a result, from 30 years ago it was realized that iron could play a major role in slag forming problems and numerous empirical slag indices and viscosity scales were introduced to minimize such slagging and fouling (Raask, 1985). There is a very rich literature available on the condensed phase behavior of iron. Bryers (1996) published an extensive review of this with various fuels indicating the sensitivity to composition and temperature. Since then, new technologies have emerged to better analyze the ash, examine its size segregation and begin to have a better understanding of these still complex interactions and their dependences. Coals rich in iron also can produce ash that consists of ferrospheres or cenospheres that are mainly its oxides, or aluminates or other ferrates such as $CaFe_2O_4$ or $MgFe_2O_4$ (Zhao et al., 2006a; Sharonova et al., 2015; Ranjbar and Kuenzel, 2017). These can be separated from the ash by magnetic and other methods and find use as catalysts and for other processes. In fact studies are underway to extract Fe and Al oxides from such ash by magnetic and leaching methods (Singh et al., 2015). High levels of recovery appear plausible. One study produced pure Fe_2O_3 cenospheres burning pulverized pyrite at 1000°C in a droptube furnace. It was noted that cooling rate was critical and if too rapid the thermal shock would fragment the particles (Wu and Li, 2019).

The refining of heavy crude oil can encounter severe deposition problems on its heat exchanger surfaces. Deposits of FeS and FeO(OH) are a common fouling problem due to the content of iron (Behranvand et al., 2018).

Iron can have a significant presence in coal, fuel oils and sludge on a percentage scale (Table 1.1) With these it is generally found that the iron content exists in the ash components with no noticeable enrichment (Meij, 1994; Martinez-Tarrazona and Spears, 1996; Llorens et al., 2001; Bhanarkar et al., 2008). It has been noted that any limestone in use must be considered in any Fe mass balance because its levels may be larger than those in coal (Meij, 1994; Lind et al., 1994). Levels in biomass are quite low in comparison being only slightly higher in solid wastes. With iron-rich coals the ash particle core is mainly iron (Neville and Sarofim, 1982). At temperatures below 800°C, iron will not interact with other components, but above 1000°C FeO can form and react with SiO_2 or Al_2O_3 forming glass structures to higher temperatures. The ratios of Fe/Si/Al/Ca become important in modifying the condensed phase controlling its melting and viscosity. This mineral behavior introduces the complexity depending on composition and temperature. It can be manipulated by blending rich and low Fe-coals (Qiu and Zheng, 1999). Unless temperatures are well above 1000°C it will be difficult to realize any gaseous release of iron. The concentration of iron in the ash from three stages of an ESP at 120°C showed only a slight decrease across them from a combustor with peak temperatures of 1800°C–1935°C (Swietlik et al., 2012). In fact, in one characterization of 28 elements in ESP sampled fly ash, the enrichment factors were normalized against that of Fe (Marrero et al., 2007). Nevertheless, in a recent study it was found that the speciation of iron in the ash can become dependent on oxygen concentration, temperature and particle size. At high temperatures where interaction with silica is possible, this affects the Fe^{III}/Fe^{II} ratios found to control melting point and viscosity (Zeng et al., 2009). Numerous ash particle size analyses now are available. In the majority of studies that do not have temperatures above about 1400°C, Fe is present in all particle sizes but without enrichment. With the smaller size ranges found after the ESP, mass concentrations favor coarser sizes of 2 or 5 μm but the iron is present due to fragmentation and coalescence and not by vaporization and deposition (Ondov et al., 1979; Kauppinen and Pakkanen, 1990; Wang et al., 2008; Teramae and Takarada, 2009; Wu et al., 2011a). An ambitious study analyzed 40 elements in coal fly ash, size segregating the ash into five categories. The smallest was 2.2 μm but displayed no variation in concentration with Fe over the differing sizes (Lanzerstorfer, 2018). A recent study reported an effect on the Fe fly ash composition by reducing the flue gas temperature simply from 250°C to 150°C through a heat exchanger. The fuel was biomass from metal contaminated soils burned in a small pilot-plant boiler. Boiler temperatures were not quoted but a general increase in ash concentration of twofold appeared to occur for iron and also many other elements, to varying degrees, as the temperature was so reduced through the heat exchanger. As no metal deposition is expected for iron during this time in the flue gases, to increase the concentrations for iron as well as such elements as Ti or Zr that are fully nonvolatile it seems apparent that particle segregation was occurring. The ash size distribution was changing, passing more of the small enriched sizes at the lower temperature, enriched simply arising by disintegration and combustion mass reduction (Kovacs et al., 2018). It illustrates the care needed in such interpretations.

At high temperatures though, there does appear to be adequate indications that in some cases iron can vaporize as Fe and/or FeO and be seen in a small-size fume component. The FeO will probably convert to the more stable $Fe(OH)_2$ if possible, but this really becomes irrelevant, both remaining transporters that will deposit on some particle. In studies above 1800°C with 11 different coals, Fe was present in the submicron part of the bimodal distribution. Not only was it a significant contributor but apparent with lignite, sub- and bituminous coals (Quann et al., 1982, 1990). This was explained by the suggestion that iron pyrite combustion coupled to the iron oxidation rapidly releases excessive energy that can both fragment particles and vaporize components (Baxter, 1990; Baxter and Mitchell, 1992). The mechanism of the release remains uncertain and further suggestions involving reduction of the oxides remain speculative (Lee et al., 2000). An earlier in situ Raman spectroscopy study clearly indicated that with sufficient oxygen FeS_2 rapidly oxidizes through FeS to Fe_2O_3 (Hamilton et al., 1986). Even at lower temperatures of 1400°C, with oxygen enrichment, it has been noted that Fe can be concentrated in the smaller particles (Zhang et al., 2007; Liu et al., 2007a). Moreover, such small particles containing iron now are becoming of environmental concern (Jayasekher, 2009). However, recent work indicates that coal blending to enhance the interaction with aluminosilicates can suppress Fe content in the submicron ash (Zhou et al., 2010a). In incineration or waste combustion there is no indication that chlorine or sulfur plays any role in releasing Fe (Chan and Kirk, 1999; Pedersen et al., 2010). However, as noted above, at higher temperatures the exothermic oxidation of iron pyrites in coal can elevate temperatures to vaporize iron as Fe(g). Ash melting can occur via aluminosilicate formation and is facilitated by the pyrites (Huffman et al., 1990; Bool et al., 1995; ten Brink et al., 1996; Zeng et al., 2009). It is the breakdown of the pyrite to form FeS, and possible pyrrhotite unusual sulfide minerals of structures such as Fe_7S_8, that is the basis of many of the slag problems (Raask, 1985; Bryers, 1996). The situation involving reducing conditions that exist in IGCC or other gasification methods can be a little different. The iron tends to be maintained in its Fe^{II} state which is beneficial for decreasing any melting points and viscosity considerations (McLennan et al., 2000; Font et al., 2005). Ash melting can be as low as 1000°C (Wang et al., 2012, 2012a). Due to the lack of oxygen the iron tends to be FeO and any pyrite will be similarly oxidized to this favoring acceptable slag conditions (Shannon et al., 2009). In some syngas producing facilities, particularly utilizing asphaltene, FeS(s) deposits appear on the cooling tubes probably from surface interaction between the oxide and sulfur (Wang et al., 2016a).

Under normal combustion conditions, the possibility of iron forming the thermally stable arsenate or selenite also remains possible in the ash (Seames and Wendt, 2007). Consequently, the iron will exist in the condensed phase and any release will depend on the quality of the particulate control systems, which now can perform at better than 99.9% efficiency (Nielsen et al., 2002; Goodarzi, 2006a; Bhanarkar et al., 2008; Jayaseker, 2009). A study on the ability of modern FGD systems to capture and retain any remaining traces of iron containing particulates indicated this device can reduce these levels by a further fourfold (Cordoba et al., 2012). In the case of residual oil combustion, efforts were made to characterize the speciation of the iron in the small $PM_{2.5}$ ash utilizing X-ray methods (Pattanaik et al., 2012). Although such methods have minimal selectivity it did appear in this case that the ultimate state of the iron was as $Fe_2(SO_4)_3$, most likely from SO_2 deposition interactions with the oxide on cooling.

Another aspect of iron with its three oxides has blossomed in recent years into a now substantial development. This involves the concept of chemical looping whereby the fuel is oxidized by an oxygen carrier so permitting a method of CO_2 capture. It is now being applied to all fuels such as coal, biomass and gas in fluidized combustion modes. It is based on the oxidizing power for example of Fe_2O_3(III) being reduced to Fe_3O_4(II/III) and then FeO(II) as it oxidizes a fossil fuel and then is rejuvenated by oxidation with air back to Fe_2O_3(III) (Bao et al., 2013; Chadeesingh and Hayhurst, 2014). It is facilitated by the ease with which iron can readily modify its oxide valence. Mixes coupling the iron with other metal oxides such as in $Ca_2Fe_2^{III}O_5$ now are being examined in this different mode of combustion (Ismail et al., 2016).

Pure nanoparticle formation from flames now has been of interest for many years. Using low pressure oxygen-rich hydrogen (Janzen and Roth, 2001) or with ethylene or acetylene flames (Emelianov et al., 2010), small diameter nanoparticles of Fe_2O_3 readily form of size 3–6 nm diameter. Recently, an ambitious kinetic model has been published simulating particle growth rates of the 5–10 nm sizes (Wlokas et al., 2013). A pilot plant scale flame reactor utilizing $Fe(CO)_5$ addition now has been tested for Fe_2O_3 production (Feroughi et al., 2015). The possible health effects of such iron oxide particles also has been examined in some detail (Guo and Kennedy, 2007; Kennedy, 2007). A recent laboratory study adding $Fe(CO)_5$ to a CO diffusion flame concluded little oxide formation and that elemental Fe played the crucial roles (Kim and Kim, 2018). Also the potential aging aspects have been considered for particles that may contain oxides of differing metals such as Fe, Cr, and Mn. Their oxidative natures can possibly modify, with time, valencies from the initial speciation (Nico et al., 2009). Also interesting for the case of iron is the ability to produce binary nanoparticles, such as Fe_2O_3/SiO_2, now known for 18 years (Ehrman et al., 1999). However, because iron has no ready gaseous transport mechanism it will produce particles which coalesce with others and in this case with SiO_2. The result will obviously depend on time and temperature as to how these two mix and determine the nature of the product nanoparticles. Also the possibility of chemical flame film deposition is not as readily available for iron but would involve the film formed from condensing nanoparticles.

In summary, for full-scale combustion, iron shows a nonvolatile nature. Above 1000°C it can begin to react with other ash components such as SiO_2 or Al_2O_3 forming glass like structures. Temperatures have to be high, ≈1800°C, to witness FeO(g)/Fe(g) contributions and evidence of enrichment in the smaller ash particles. Sulfur and chlorine play no roles. The full nature of the ash composition is controlled by the conditions and properties of the fuel. As indicated, mixed metal ferrates and other fused components can form in the ash particles. It will depend on time and temperature constraints. In some cases, if the oxide remains, this will tend to be finally converted to the sulfate. Laboratory studies with experimental flames totally differ from practical combustors. For these, the addition of iron carbonyls or ferrocene instantly introduces Fe(g) into the flame gases. As a result, the chemistry displayed is more unique, displaying a pronounced catalytic behavior for Fe and has been developed for numerous possible applications. These do relate to combustion but in the areas of soot suppression and flame inhibition. Although ferrocene was used in gasoline engines as an antiknock additive, its fouling of spark plugs restricted its use more to soot suppression in diesel combustion. In practical combustion of fuels, iron is rarely released from the ash particle other than by disintegration forming smaller particle sizes. It does indicate though that the iron oxides are very catalytic being able to suppress soot emissions and inhibit or accelerate combustion by mechanisms still not fully understood.

7.15 Lead

Lead is another heavy metal known since antiquity and present in combustible fuels. When pure it has a low melting point of 327°C (b.pt.1749°C) with ores of galena, PbS, cerussite or white lead, $PbCO_3$, used by Queen Elizabeth I as a white face cosmetic, and anglesite, $PbSO_4$, acquiring its name from its discovery in Anglesey, the island off the coast of north west Wales. It is not abundant in the Earth's crust or oceans being present at about 36th and 55th in these, respectively. A known neurotoxin, its organocompounds, mainly of anthropogenic origin are more toxic than its inorganic forms but are not carcinogenic. About 5.5 million tons year^{-1} of lead are mined globally, but now, due to a

40%–50% level of recycling, about 9.6 million tons year^{-1} are refined. There are significant global reserves. The removal of the use of a tetramethyl/ethyl group of lead additives from automobile gasoline, ended its then major use. Uses now are mainly lead-acid electrical batteries (80%), lead chemicals (8%), lead sheeting (5%), ammunition (2%), alloys (2%), and as a petroleum additive (1%). It is still permitted for use in leaded fuels for some older general aviation planes. In the United States this retention in fuel has been quoted as the largest current contributor to the atmosphere and its economic damage has been assessed (Wolfe et al., 2016). As a metal it is known for its corrosion resistance. It is present in alloys with Sn, Cu, and Sb. Previously, it was an important component of pewter and Ag/Pb pipe solder. Lead is a combustion inhibitor and for this reason from the 1920s became the predominant octane improver for gasoline fueled engines. Mainly due to the environmental aspect that 90% of the lead found in atmospheric aerosols originated from this source, and could concentrate above cities lying in basin-type geography, a phaseout was initiated in 1976. However, it was not until 1992/1996 that its addition to gasoline totally ended in the United States precipitated to a larger degree by the fact that lead was incompatible with automobile catalytic converters. Since then the rest of the world has followed such that now usage of this additive is low. Through this and also its previous use in paint (still in use in many developing countries at levels of 1%) (O'Connor et al., 2018), and leaded glass, it became one of the environmentally sensitive issues that helped spawn today's awareness of such heavy metal risks. The natural background level of lead emission to the atmosphere remains uncertain but has been estimated at about 20,000–27,000 tons year^{-1} resulting from wind erosion, volcanoes, forest fires and vegetation (Nriagu, 1979; Nriagu and Pacyna, 1988; Malakhov and Makhon'ko, 1990). A recent Canadian study has caused an uncertainty by suggesting much higher natural levels of emission for Cd, Cu, Hg, Ni, Zn, as well as Pb. In the case of lead this is now estimated as a 90-fold increase to 1.8 million tons which needs further examination and is unacceptably large (Richardson, 2001). The anthropogenic levels of lead use over time, recorded in ice core (Boutron et al., 1991; Boutron, 1995), or peat bog analyses (Shotyk et al., 1996) tend to illustrate three peak periods. If the natural levels were as large as estimated in the Canadian study this would dwarf this anthropogenic data which reflects changes with time and would not be apparent in the archival deposits, further proof for its unacceptance. Initially before its environmental problems were known, the Romans not only used lead pipe extensively but also inadvertently discovered that their food could be sweetened by basically adding old wine (acetic acid) in its preparation. Consequently, in any lead based cookware this effectively produced small toxic levels of the sweetener lead acetate. This is now considered as contributing to their downfall, for which lead pipe usage, previously blamed, appears now to have been discarded. The industrial revolution saw another increase in usage including production of leaded glass that contains 18%–40% PbO but can be a source of slow leaching as also pewter. It has since been considered a possible cause of gout in that period. The 20th century illustrated the largest peak usage with leaded gasoline. As a result, estimates of the ratio of anthropogenic to natural levels around 1980 could have been at least 15- to 30-fold (Nriagu, 1989; Malakhov and Makhon'ko, 1990). However, in recent years the removal of leaded gasoline coupled to better emission control devices on combustors and incinerators has seen a continued significant drop and the latest research possibly suggests current levels are approaching the natural ones of medieval times

(Boutron, 1995; Azimi et al., 2005; Pacyna et al., 2007; Kemsley, 2016; Cimova et al., 2016). However, recent assessments of lead emissions from coal combustion in China alone estimate levels of 14,000 tons in 2012, about a 50-fold increase from that of 1949 (Tian et al., 2012, 2015). Similarly, atmospheric emissions of Pb from cement plants in China have increased over the years and are approaching 1000 tons $year^{-1}$ (Hua et al., 2016). Nevertheless, even such numbers do suggest that global anthropogenic emissions of lead are in decline. Total anthropogenic emissions of Pb in China are about half from coal but also represent the highest emitter in China from among the other trace heavy metals (As, Cd, Cr, Hg) (Cheng et al., 2015a). Natural biomethylation or biological chemistry of Pb does not occur. Any in the environment is from anthropogenic sources. Lead is also emitted as dust in the sintering plant of steel industries. This is a process where iron ore, coke and limestone are ground, heated to 1300°C en route to the blast furnace. Off gases and dust are processed by ESPs and filtered, but fine particles that contain lead carbonate can be emitted to the environment (Sammut et al., 2010). Even so, it is reported that heavy metal input to soils in China arises predominantly from atmospheric deposition (Peng et al., 2019). With industrial controls it appears to be decreasing and there is no indication or concern with lead accumulation in Chinese soils (Huang et al., 2019).

Health concerns with lead

Lead is another element that is not essential to the human diet. Due to the seriousness of lead toxicity, numerous very extensive toxicity and environmental reports have been issued by the United Nations, the World Health Organization and the US Agency for Toxic Substances and Disease Registry. These are all readily available on the Web (World Health Organization Pb, 1989a, 1995; ToxProfile Pb, 2007; UN Review Pb, 2010). There is an enormous amount of information available on the health effects of lead.

Medical concerns tend to center on the nature of lead in that it can accumulate in the bone through a lifetime. Any nutritional essentiality has been difficult to assess for any of the elements beyond iodine (atomic number 53) in the periodic table, none of which play a role in life forms. With lead there is no evidence of a biological role in animals and no minimum requirement is suggested (World Health Organization Trace elements, 1996; World Health Organization Food additives, 2000). Lead also can transfer to a fetus through the placenta and the blood/brain barrier so maternal blood levels are important. As a result, no minimal dosage has been suggested by any agency. It is accepted now that even low levels in children affects their schooling and possibly even growth (Aizer et al., 2018; de Jesus et al., 2018; Deierlein et al., 2019). Both the brain and kidneys are susceptible to lead toxicity and risks have been reviewed. Chelation therapies now are standard for the body but are more involved for the brain (Aaseth et al., 2018). Such deleterious disorders can be apparent even with blood levels of 20 $\mu g\ L^{-1}$ and no safe minimum threshold is apparent. In other words the general consensus now is that the only safe level for lead is zero (Shefa and Heroux, 2017). Simply from analyzing human blood samples it is realized that there is still an average US dietary intake of the order of about 1–3 $\mu g\ kg^{-1}$ body weight day^{-1}. Crops grown near zinc smelting regions in China with a lead content of 35–91 $\mu g\ g^{-1}$ (dry weight) can certainly exceed such limits

(Yang et al., 2011; Li et al., 2018c). However, in one study, the levels of Se in the body appear to introduce a protection against the lead but at the price of possibly depleting the Se (Zhou et al., 2019). A Norwegian study of women with low blood levels of lead were considered to ingest this from tea and red wine (Flotre et al., 2017). In Western society, elimination of soldered food cans, use in house pipe plumbing, and removal from paints has helped to continually reduce environmental lead. In the body, Pb is not molecularly changed or metabolized and normally 99% will be excreted over a few weeks. Any retained is by the teeth and bone. Children retain it longer and are more sensitive to its toxicology. Lead tends to affect every organ in the body including the brain although it is not carcinogenic. It has significant effects on pregnancies and on fertility. Fortunately it does appear that the mammary glands do minimize the presence of lead in mother's breast milk to some extent (Bastos et al., 2018; Altun et al., 2018). However, in some cases it appears still too high (Schofield, 2016; Vahidinia et al., 2019). In 1970, 220,000 tons of lead-tetraethyl were burned and emitted by cars in the United States alone. In 2004 the World Health Organization estimates globally that 143,000 children died of lead poisoning. Mental retardation in children due to lead is still reported at a >600,000 level globally (O'Connor et.2018). A recent long-term study with rats considered a lifespan exposure to lead. They noted some damage in infancy, correcting in adulthood, but then intensified again in old age (Fenga et al., 2019). The critical period for intervention appeared to be in early life. As a consequence of such studies, there have been very many surveys of blood, urine, hair and even nail samples analyzed in the past decades. Blood and urine tend to be favored as biomarkers but hair and nails, that have very similar measures, also are adequate. Recommended levels for blood vary from one analysis clinic to another but prefer levels of 50 $\mu g\ L^{-1}$ or less (Mayo Clinic, 2018). Chelation for children is recommended if levels rise to 250, or to 450 $\mu g\ L^{-1}$ for adults, but these are just flexible guide numbers. Suggested levels in other body media tend to be <3 $\mu g\ L^{-1}$ for urine or about ≤3.9 ppmw for hair or nails. As seen in such surveys these are often wishful levels, but in numerous geographical locations such as one reported in French Guiana of 500 pregnant women tested, 25% had blood levels >50 $\mu g\ L^{-1}$ and 5% over 100 (Rimbaud et al., 2017). In China one area indicated 88% of the children with >50 $\mu g\ L^{-1}$ and affected their IQ (Xu et al., 2018a; Pan et al., 2018a). In a Ugandan blood survey of 100 healthy children, 10% had levels >100 $\mu g\ L^{-1}$ (Cusick et al., 2018). A recent Nigerian survey of 440 pregnant women indicated a median level of 640 $\mu g\ L^{-1}$ for the mother's blood and 393 $\mu g\ L^{-1}$ for the cord. Seventy percent of them were >50 $\mu g\ L^{-1}$ with potentially significant implications (Ladele et al., 2019). One analysis, concerned about the lack of recommended safe levels particularly for pregnancy, listed much of the earlier survey data for lead showing the extensive range of values listed in some studies (Taylor et al., 2014). Many populations now are addressing this risk. This is reflected in the data from the US National Surveys over recent years. For blood samples, the United States lead level at the 95% percentile point in the distribution measured 50 $\mu g\ L^{-1}$ in 2000, 32 in 2012, and 28 in 2016 confirming continued improvements. Similar gradual reductions are being noted in the German population (Goen et al., 2018). The fact that lead levels are in such decline tends to eliminate it as a possible single cause for autism as recently suggested (Alawad et al., 2019). Nevertheless, for pockets of people, even in New York City, one recent survey indicated 20% of the adults and 15% of the children had >50 $\mu g\ L^{-1}$ lead in their blood (Hore et al., 2017). Levels closer to zero are desirable from the many animal studies reported. One noteworthy recent observation analyzed the United States 2007–12 surveys. From the tested levels for Pb, Cd, Hg, and As they noted that more than 50% of the samples contained at least three of these elements at significant levels (Shim et al., 2017). It is recommended in houses built before 1950 in the United States or anywhere with older water line plumbing that either a filter for Pb and other impurities is used, or a flushing of standing drinking water lines should certainly be run before consumption. Even private water wells in the United States are reported with high lead from various sources (Pieper et al., 2018). The official WHO guideline for water quality is $<10\,\mu g\,L^{-1}$ (WHO Guidelines, 2006a). Another final US concern centers on use of leaded bullets or areas of military firing ranges that can exceed the presently suggested standards (Editorial: lead, 2012). Although lead among the heavy metals has been suggested as possibly contributing to Alzheimer's disease, the fact that lead in the environment has and continues to decrease seems not consistent with any epidemic-type growth rate (Costa, 2017). Approximate estimates are available of the cost of health damage from environmental lead, 80% as a result of increased mortality and 20% from IQ loss (Nedellec and Rabi, 2016).

Lead's basic chemistry

In comparing elements of the C/Si/Ge/Sn/Pb periodic table column group, there is a significant change from the tetravalent nonmetallic behavior for carbon, to that of a predominantly divalent and metallic nature for lead. Lead is a very heavy metal with atomic number 82. Its chemistry reflects a dominance of a true inorganic nature with a strong cationic presence displaying the normal range of inorganic molecules. The physical properties of its condensed phase molecules relevant to combustion are listed in Table 7.23.

Although the dioxide exists, PbO_2(s), this decomposes at 290°C to the more stable PbO (s). Similarly the Pb_2O_3(s) (II/IV) and the Pb_3O_4(s) (II/II/IV) oxides both decompose to PbO at about 500°C. The red (litharge) form of PbO(s) is stable at low temperatures but changes at 489°C to its yellow (massicot) more stable form that has a melting point of 897°C with a 20 ppmv vapor pressure of PbO(g) at 864°C (b.pt.1477). Below 489°C the yellow phase is metastable (Gurvich et al., 1991). All these oxides are insoluble in water. $PbSO_4$(s) is insoluble in water but is thermally stable, starting to dissociate at 803°C via its basic sulfate being complete and PbO(s) at 960°C (Ostroff and Sanderson, 1959). Its vapor pressure is about 10 ppmv at 690°C (Stern and Weise, 1966). $PbSO_3$(s) is known but dissociates to PbS and $PbSO_4$ before producing PbO(s) in a 400°C–700°C range. $PbCO_3$(s) is insoluble in water and dissociates to PbO(s) between 315°C and 510°C (Beck, 1959). The water soluble nitrate $Pb(NO_3)_2$(s) also dissociates to the oxide in the 300°C–535°C range, and its nitrite, $Pb(NO_2)_2$(s) is less stable starting to dissociate at 110°C and complete at 320°C–450°C (Stern, 1972). At low temperatures, chlorine can convert PbO(s) to the dichloride, $PbCl_2$(s). This dichloride is sparingly soluble in water (m.pt.510, b.pt.954°C) with a vapor pressure of 20 ppmv at 430°C; quite volatile, evaporating to stable $PbCl_2$(g) (Gurvich et al., 1991). The dihydroxide, $Pb(OH)_2$(s) decomposes at 145°C. Other relevant divalent compounds that are known are $Pb_3(PO_4)_2$(s) (m.pt.1014), $Pb(PO_3)_2$ (m.pt.670), $Pb_2P_2O_7$ (m.pt.824), $PbSiO_3$(s) (m.pt.764°C), basic $PbSiO_3.Pb(OH)_2$, and Pb_2SiO_4(s) (m.pt.743°C). Other ceramic compounds are $PbTiO_3$ (m.pt.1286°C), the piezoelectric $PbZrO_3$ and lead zirconium titanate that exhibits even greater thermal stability.

TABLE 7.23 Properties of lead's condensed phase molecules relevant to combustion, °C.[a]

m.pt. b.pt. vapor pressure
Pb 327 1749, 10 ppmv/705°C, ins
PbO(litharge) >489 converts to massicot, ins
PbO(massicot) 897 1477, 20 ppmv/863°C, ins
PbO_2 300, dissoc to Pb_3O_4, ins
Pb_2O_3(II/IV) decomp 530 to PbO, ins
Pb_3O_4(II/II/IV) 830, decomp >600 to PbO, ins
$Pb(OH)_2$ decomp 145, ins
$Pb(OH)_4$ hydrated PbO_2
$PbCO_3$ >315 dissoc to PbO, 440 ppmv/130°C, ins
$Pb(NO_3)_2$ 470, >266–540 dissoc through basic salt to PbO, 135 ppmv/220°C, s
$Pb(NO_2)_2$ >110 dissoc to PbO, complete 320–450, s
$PbSO_4$ 1087, dissoc 800–960 via oxysulfate to PbO, 10 ppmv/690°C, ins
$PbSO_3$ >400 dissoc to PbS/$PbSO_4$ then PbO, ins
$PbCl_2$ 510 954, 20 ppmv/430°C, slightly s
$PbCl_4$ −15, >50 dissoc, hydrolyzed by water
PbS 1113, decomp in air 1320 to $PbSO_4$ then PbO, 20 ppmv/663°C, ins
PbH_4 unstable
$Pb_3(PO_4)_2$ 1014, dissoc to Pb, PO_x, ins
$PbAl_2O_4$ >970 dissoc, ins
$PbFe_2O_4$ >725 dissoc, ins
$PbSiO_3$(meta) 764, dissoc to PbO, ins
Pb_2SiO_4(ortho) 743, dissoc to PbO, ins
$PbTiO_3$ 1200, dissoc, 4 ppmv PbO(g)/1000°C, ins
$PbZrO_3$ 1570, ins

[a]*Sparingly water soluble lead salts review (Clever and Johnston, 1980).*
decomp, decomposes; *dissoc*, dissociates; *s*, water soluble; *ins*, water insoluble.

Lead ferrite, $PbFe_2O_4$ of magnetic use dissociates above about 725°C and $PbAl_2O_4$ above 970°C. Under reducing conditions PbS(s) (m.pt.1113°C) exists, insoluble in water and vaporizes congruently to PbS(g) with a vapor pressure of 20 ppmv at 663°C (Gurvich et al., 1991). If heated in air PbS(s) oxidizes at 1320°C to $PbSO_4$ and PbO. It is also converted to $PbCl_2$ if heated above 600°C with other chlorides such as $CoCl_2$. Few tetravalent compounds are known and are not thermally stable. $PbCl_4$(s) dissociates at 100°C and is hydrolyzed by water. $Pb(OH)_4$ is regarded as a hydrated oxide and PbH_4 although now known, is expected to be a gas but unstable. The dihydride normally is not encountered. As a result, in combustion as temperatures increase, the only thermally stable molecules likely to be encountered in the condensed phase are going to be PbO, $PbSO_4$, and $PbCl_2$ with the possibility of some silicate or phosphate formation, and with PbS under reducing conditions.

Gas phase stabilities and transport

Early flame photometric atomic analysis methods were quite indicative in showing that high levels of atomic Pb could be generated on introducing any of its molecules even into 1750°C H_2/air flames (De Galan and Samaey, 1970). This in itself indicated that none of the gaseous forms of molecular lead are strongly bound. As with Ge, Sn and Pb, vaporization of the oxides form predominantly the MO(g) component, which as seen in Table 7.24 increasingly weakens with atomic number but still remains reasonable for PbO(g) yet less strong than many flame radicals.

TABLE 7.24 Bond strength trends for relevant gaseous molecules of the C/Si/Ge/Sn/Pb group, kJ mol^{-1}

Bond	C	Si	Ge	Sn	Pb
D_0(M-O)	1072 ± 0.4[a]	792 ± 5[a]	650 ± 5[a]	524 ± 4[a]	372 ± 4[a]
D_0(OM-O)	526 ± 0.4[a]	427 ± 18[b]	313 ± 20[a]	260 ± 30[a,c]	180 ± 40[a,d]
D_0(M-Cl)	404 ± 2[e,f]	414 ± 8[g]	387 ± 10[h]	385 ± 12[a]	305 ± 12[a]
D_0(ClM-Cl)	321 ± 3[e,f,i]	430 ± 10[g]	388 ± 11[h]	356 ± 14[a]	304 ± 13[a]
D_0(M-S)	710 ± 0.6[a,j]	614 ± 2.5[a]	548 ± 0.5[a]	464 ± 5[a]	341 ± 1.5[a]
D_0(M-H)	333 ± 1.2[a,j]	295 ± 0.5[a]	268 ± 8[k,l]	245 ± 10[l,m]	151 ± 10[j,n]
Bond ratios					
D_0 OM-O/M-O	0.49	0.54	0.48	0.50	0.48
D_0 M-S/M-O	0.66	0.78	0.84	0.89	0.92
D_0 M-Cl/ClM-Cl	1.26	0.96	1.00	1.08	1.00

[a] *Gurvich et al. (1991).*
[b] *Hildenbrand and Lau (1994b).*
[c] *Nielsen et al. (2003).*
[d] *Benjelloun et al. (2005a).*
[e] *Dixon and Peterson (2001).*
[f] *Csontos et al. (2010).*
[g] *Hildenbrand et al. (2003).*
[h] *Hildenbrand and Lau (1996).*
[i] *Tarczay et al. (2005).*
[j] *Huber and Herzberg (1979).*
[k] *Ricca and Bauschlicher (1999).*
[l] *Song et al. (2008).*
[m] *Li et al. (2013a).*
[n] *Zhao et al. (2017).*

From the quite accurate heat of formation of PbO(s) and its heat of sublimation, a reliable bond strength of D_0(Pb-O) = 372 ± 4 kJ mol^{-1} has been recommended (Gurvich et al., 1991). The ground state is $X^1\Sigma^+$ with no low-lying electronic states. In closed equilibrated systems the gaseous polymers can also be present, $(PbO)_{2-4}$, but decrease in relevance with size. $(PbO)_2$ is still present at about 800°C as a reasonable fraction of PbO(g) that itself starts dissociating above about 850°C (Drowart et al., 1965; Knacke and von Richthofen, 1994; Popovic et al., 1997). For MO_2(g), as noted in Table 7.24, in all cases in this periodic group the bond strength of the OM-O bond is very close to about half that of the M-O first oxygen bond. This explains the dominance of MO(g) and in fact for Ge, Sn, and Pb, the dioxides have never been observed in mass spectrometric studies of their vapors. The one exception is an early flame study (Farber and Srivastava, 1978). They added lead salicylate to a 2300K fuel-rich hydrogen flame, probe sampling into a mass spectrometer and noted similar magnitudes for the monitored ions of Pb, PbO, and PbO_2. Whether this reflected these species in the flame gases or whether they were heterogeneously formed during sampling is still questioned but they did appear to note a gaseous dioxide, which would in itself be rather suspicious. An additional study monitored the atomic Pb in flames spectroscopically by LIF (Buckley et al., 2002). The chemical thermodynamic value remains an estimate, D_0(OPb-O) = 180 ± 40 kJ mol^{-1}. One other study that reported measuring PbO_2^+ produced thermochemical values that implied that the two bonds of PbO_2 had equal strengths which is inconsistent with all other studies and is regarded as erroneous (Zmbov and Miletic, 1978). As shown by theory, PbO_2(g) will favor a cyclic structure with the linear OPbO shape lying about 50 kJ mol^{-1} higher in energy (Benjelloun et al., 2005a). The infrared structures of such shapes have been confirmed in low-temperature matrix isolation experiments (Chertihin and Andrews, 1996). The stabilities of $PbCl_2$(g) and PbCl(g) species are well characterized. Graphical extrapolation of the vibrational levels either of the ground PbCl or its $A^2\Sigma^+$ electronic states indicate a dissociation limit of D_0(Pb-Cl) = 305 ± 12 kJ mol^{-1}(Gurvich et al., 1991). This was confirmed by a recent theoretical estimate of D_0 299 kJ mol^{-1} and is regarded as acceptable (Li et al., 2013a). A value for the dichloride again is quite reliably derived from the solid's heat of formation and sublimation. This indicates a 0°C atomization value of 608.6 ± 5.1 kJ mol^{-1} or D_0(ClPb-Cl) = 304 ± 13 kJ mol^{-1}, essentially identical to the first bonding (Gurvich et al., 1991).

Theoretical calculations support this magnitude (Benavides-Garcia and Balasubramanian, 1994). Chlorine reactions such as the following will be endothermic and pose quite large kinetically limiting barriers, needing high levels of HCl and temperatures for reaction.

$$Pb + HCl = PbCl + H \qquad \Delta H_0 = +123 \text{ kJ mol}^{-1}$$

$$PbO + HCl = PbCl + OH \qquad \Delta H_0 = +70 \text{ kJ mol}^{-1}$$

$$PbCl + HCl = PbCl_2 + H \qquad \Delta H_0 = +124 \text{ kJ mol}^{-1}$$

The rate constant for the reaction of Pb + HCl has been measured and seen to be solely controlled by this endothermicity (Cosic and Fontijn, 2000). This confirms the bond strength for PbCl(g) which was assessed only from spectroscopy and not combustion flame studies. However, from the point of view solely of thermodynamic stabilities, calculations for combustion indicate $PbCl_2(g)$ might be fractionally important in the range of 680°C–1180°C depending on conditions and the quality of the database (Frandsen et al., 1994; Sorum et al., 2004; Lundholm et al., 2007). Again, although this displays an intriguing picture, whether the chlorides can be rapidly formed in the flame gases has in fact not been discussed. Thermodynamically it is sensitive to the chlorine concentration used and can be as small as 2% of the Pb content at 1100°C in one chemical thermodynamic calculation with 190 ppmv of Cl (Bankiewicz et al., 2012). Nevertheless, the formation of the chlorides in the gas phase, with bond strengths of about 305 kJ mol^{-1} will be subject to kinetic constraints as seen above due to the bond strength of HCl (D_0 428 kJ mol^{-1}) overriding the thermodynamics. A similar expectation of PbOH(g) in flame gases, reflecting the chloride also was not observed (Friswell and Jenkins, 1972). Instead, PbO was clearly dominant in flames and they derived a valid bond strength value of 382 kJ mol^{-1} for this, from this 46-year-old study with no refinement to their data. A bond strength for PbOH(g) has not been measured but would be expected to reflect that of PbCl. However, the enthalpies of the reactions to form this are clearly constraints.

$$PbO + H_2O = PbOH + OH \quad \Delta H_0 = +136 \text{ kJ mol}^{-1}$$

The high inhibition nature of lead compounds in combustion generally is explained by a model invoking an intermediate role played by PbOH(g) formation and destruction to catalyze radical recombination (Benson, 1988; Babushok and Tsang, 2000). From such a need, gaseous bond strengths of D(Pb-OH) = 285 and D(HOPb-OH) = 280 kJ mol^{-1} were estimated (Benson, 1988), with theoretical studies implying bent structures and D_0(Pb-OH) = 302 and D_0(HOPb-OH) = 340 kJ mol^{-1} (Benjelloun et al., 2004, 2005a). However, this depended on invoking HO_2 chemistry that remains very speculative. They did note that the PbOH structure was significantly more stable than the isomeric HPbO. Their infrared frequencies and structures were reported in matrix isolation studies (Wang and Andrews, 2005c; Teng et al., 2007). As a result, it cannot be fully accepted yet that gaseous PbCl or $PbCl_2$ and their corresponding hydroxides play a major role in the inhibition as modeled. As discussed below, there may be kinetic alternate routes to circumvent any apparent constraints. However, it remains strange that neither PbCl nor PbOH have yet been monitored in flames. Under reducing conditions PbS(g) and Pb(g) will be dominant species as evidenced in Knudsen cell-mass spectrometric measurements (Botor et al., 1989). A value of D_0(Pb-S) = 341 ± 1.5 kJ mol^{-1} has been assessed bringing together in a most consistent manner the value from the heats of formation and sublimation of the solid, the spectral limit observed in the (D-X) absorption spectrum and the mass spectral study of the PbS/Pb/S_2 Knudsen cell studies (Gurvich et al., 1991). PbS(g) starts dissociating to Pb(g) and $S_2(g)$ above about 750°C. It is the weakest sulfide in this C/Si/Ge/Sn/Pb group of elements, and only 8% weaker than the oxide, the smallest relative decrease in this group. $PbS_2(g)$ is thought to exist like the dioxide but similarly has not been observed. Also reflecting the ratio of D(OM-O)/D(M-O) an estimate of D_0(SPb-S) = 160 ± 10 kJ mol^{-1} has been suggested (Gurvich et al., 1991). The gaseous monohydrides are well established spectroscopically and theoretically, and as noted in Table 7.24, PbH($X^2\Pi_{1/2}$) is the weakest bound of the series with D_0(Pb-H) ≈ 151 ± 10 kJ mol^{-1}. This is based on an early spectral predissociation limit estimate, and the average of five theoretical estimates derived and listed together with potential energy curves in the most current study (Huber and Herzberg, 1979; Zhao et al., 2017d). Mass spectrometric thermodynamic studies have been reported recently for gaseous lead phosphates species emerging from a Knudsen cell effusion. This examined the vaporization of the above mentioned $Pb_3(PO_4)_2$, $Pb(PO_3)_2$ and $Pb_2P_2O_7$ salts (Emelyanova et al., 2016). With $Pb_3(PO_4)_2$ above 1100°C only gaseous PbO/Pb species were noted. However, in the other two cases above 750°C beside these, smaller amounts of gaseous $PbPO_2$, $PbPO_3$ and $Pb(PO_3)_2$ were evident. Although these may exist in such a controlled equilibrated environment they are displaying fragmenting vaporization and may not be present in the more hostile more open combustion conditions where the chemistry would drive them more rapidly to stable PbO(g).

Laboratory combustion studies

Other than the flame experiments mentioned above, that concerned more the determination of fundamental chemical stabilities, two other practical areas have been examined. These center on the subject of soot, formation and suppression, and also the observation of flame inhibition or catalytically enhanced flame radical recombination effects. The question of soot was not extensively examined as the early reports were not sufficiently positive. An initial study rated the effects of lead on soot suppression as zero on a scale with barium at 100 (Cotton et al., 1971). A follow-up study with numerous elements reported that the effects were slight, either enhancing or depleting soot but dependent on fuel and flame conditions (Bulewicz et al., 1975). This was reiterated in a review showing little interest concerning lead (Howard and Kausch, 1980). A far more important aspect was the realization that lead tetraethyl was a remarkable antiknock additive to automotive fuels. Although lead was listed as having only a slight effect as an inhibitor, Cr being far superior, it became apparent that this depended on the chemical nature of the lead and the flame conditions (Bulewicz and Padley, 1971; Linteris et al., 2008). In an organic form, the lead was seen to be only second to iron in its ability to inhibit combustion (Vanpee and Shirodkar, 1979; Babushok and Tsang, 2000). As a result, organic lead, formed the basis of an octane enhancer, or antiknock, additive for automobile gasoline for several decades. Fortunately, it was not compatible with the introduction of the catalytic converter on cars and so became banned to the benefit of the environment and human health. It was replaced by Mn, now in the process of being phased out due to its toxicity (Howard and Kausch, 1980). Although of little interest to combustion, the chemiionization of Pb particularly in hydrocarbon flames has been studied slightly (Goodings, 1992). It remains not fully understood and as indicated awaits further examination.

Combustion power plant data

As noted in Table 1.1, concentrations of lead can be significant in coals, low in biomass, but quite significant in municipal wastes and sewage sludge. Levels are negligible in pet coke and are low in limestone (Lind et al., 1994; Cordoba et al., 2012). Methods for analyzing all the inorganic contaminants in automobile fuels and biofuels now have been reported (Santos et al., 2012). The use of a blend of lead tetramethyl/ethyl additive compounds as an ignition control method for many years in gasoline engines promoted rapid combustion release raising the question of its emission. As expected, transport and deposition onto any particles as the gaseous Pb and PbO was noted with heterogeneous conversion on cooling with the available gaseous sulfur to $PbSO_4$ in accord with thermodynamic stabilities (Lundholm et al., 2007).

Due to the extensive incorporation of chemical thermodynamic modeling to help certain observations, the question as to whether such equilibrations can be achieved has never been seriously addressed. If the data readily conforms to the predictions, one can conclude that potential kinetic constraints that were envisioned to exist may still exist but the system has alternate mechanisms not considered. This was found to be the case with mercury. This now appears to be the case for Pb concerning its halides. As noted above gas phase kinetics are constrained in converting gaseous PbO into the gaseous halides or even hydroxides. However, it is now known that gaseous HCl can react with solid PbO, which has not been seriously considered.

$$PbO(s) + 2\,HCl(g) = PbCl_2(s) + H_2O(g) \quad \Delta H_{298K} = -199\ kJ\ mol^{-1}\ \text{(Chase, 1998)}$$

$$PbO(s) + H_2O(g) = Pb(OH)_2(s) \quad \Delta H_{298K} = -56 kJ\ mol^{-1}\ \text{(Bensen, 1988)}$$

Such heterogeneous reactions as these become a different channel for understanding such otherwise seemingly difficult gaseous transformations. The gaseous HCl can basically release lead from its condensed phase and vaporize it as the dichloride. Consequently the formation constraints in combustion may not exist and chemical thermodynamic expectations may be acceptable to some extent. From the known chemistry summarized above, combustion of traces of lead in fuels can be expected to be controlled by the levels of sulfur, chlorine and the combustion temperatures. As seen in Table 7.23 of lead's inorganic molecules, nearly all its condensed phase molecules dissociate on heating to the oxide. $PbCl_2(s)$ is the most volatile lead compound, and if formed, will vaporize. It may become oxidized to a certain degree, so with the oxide and the element these will be the major known transporters at higher temperatures. Although not characterized, PbOH may also exist at low levels. Nevertheless, the low bond strength of gaseous PbO and its chlorides will be affected by the more stable flame radicals, establishing some gaseous Pb/PbO balance. A large body of experimental studies is available but these remain limited in being qualitative, or based more on thermodynamic expectations for their modeling of lead's behavior, that depend on the nature of the database considered. Quite often, a role of sulfur or chlorine was not realized or assessed in some studies. The important aspects of lead concern its release behavior from fuels, the deposition on the ash, the relative levels of sulfur and chlorine and the temperature regimes. On heating coal samples in air or argon it was noted that lead vaporization initiated at about 750°C and was very significant above 950°C. Sulfur was present at 0.1%–0.8% levels but no chlorine content was stated (Wang and Tomita, 2003). From the initial studies, lead was seen to be quite volatile in combustion and immediately indicated that deposition onto fly ash was a dominant process that was surface controlled (Kaakinen et al., 1975; Galbreath et al., 2000; Van de Velden et al., 2008). Also due to the lower levels of lead found in some coal and biomass, emphasis has been placed more on characterizing waste and sewage sludge incineration, and also certain co-combustion modes with their potentially higher levels and more accurate analyses. Combustion temperature remains an important aspect. Whether the combustion process is one that produces larger quantities of bottom ash or not, the fly ash indicates enrichment factors for lead that can span from a two- to fivefold range or more, independent of fuel type (Klein et al., 1975; Korzun and Heck, 1990; Meij, 1994; Querol et al., 1995; Martinez-Tarrazona and Spears, 1996; Lopes et al., 2003; Huang et al., 2004; Goodarzi, 2006; Zhao et al., 2008). A recent ambitious study of coal fly ash, size segregated the ash into five coarse ranges, the smallest being 2.2 μm (Lanzerstorfer, 2018). Concentrations of about 40 elements have been analyzed and elements can be readily divided into those showing enrichment to smaller ash particle sizes and those not. That for lead is quite pronounced. It had long been noted that provided temperatures are above about 1000°C to vaporize the lead, there will be a range of fly ash diameters with concentrations significantly increasing to smaller particle sizes (Coles et al., 1979; Smith et al., 1979). It was later seen that the micron sizes indicated a bimodal distribution with the major peak centered at about 2 μm that is highly enriched in lead (Kaakinen et al., 1975; Querol et al., 1995; Martinez-Tarrazona and Spears, 1996; Swietlik et al., 2012). A finer mode of 0.05 μm can have about 6% and the coarser 2 μm 94% of the Pb mass in these small particles and together, even though these constitute only about 3%–6% of the total fly ash, they can contain 9% of the total available lead (Kauppinen and Pakkanen, 1990). Wang et al. (2008) illustrated this further by separating different ash fractions into nano (0.03–0.06 μm), ultrafine (0.06–0.1 μm), submicron (0.7–1.1 μm) and fine (1.1–1.8 μm) showing the high degrees of enrichment basically on very small ash nuclei. Consequently, this happens on the particle sizes that are more difficult to control by the ESP (Senior et al., 2000b). The enrichment that occurs on fly ash particles is clearly evident also in the later stages of an ESP or any escape release of the smaller particles (Klein et al., 1975; Huang et al., 2004; Bhanarkar et al., 2008). This represents the environmental concern, but with an ESP and filter bags and/or scrubbers it was earlier indicated that total lead releases as fine particles could be controlled and reduced to less than the 0.05%–0.5% Pb leakage range (Meij, 1994; Sandelin and Backman, 1999; Nielsen et al., 2002; Goodarzi, 2006; Van de Velden et al., 2008; Deng et al., 2014a). In large modern coal combustors in China equipped with SCR, ESP + FF, WFGD devices, studies of the stack emissions now indicate that for the 13 trace elements measured these are even better controlled (Zhao et al., 2016, 2017a; Wang et al., 2017a). In the case of Pb only a 5×10^{-5} fraction of that in the coal was emitted finally, or a reduction of 2×10^4 within the plant. It has also been indicated that the retention by a FGD system notably aides in fractionally reducing any final remaining carryover of lead (Cordoba et al., 2012).

Utilizing X-ray absorption, an initial study of lead on coal fly ash from five power plants did indicate an approximate 30%–70% $PbCl_2$ content and 30%–60% $PbSO_4$ (Shah et al., 2009). PbO was not detected in two of the ash samples. In the raw coals, lead at a 2–14 ppmw level was present 65%–70% as PbS, with sulfur levels of 0.1%–0.6% but unfortunately the chlorine content was not measured. It would certainly have been less than 1000 ppmw (0.01%) (Tillman et al., 2009) implying the following probable deposition ranking from any gaseous PbO, PbCl or Pb influx to the ash particle surface.

$$PbSO_4 > PbCl_2 > PbO$$

However, a more recent application of X-ray near-edge absorption monitoring of $PbCl_2$ indicates it can be a perturbing analysis, releasing chlorine and questions its quantitative nature (Schwartz and Prendergast, 2015). As a result, earlier analyses of dichloride content may be suspect and need confirmation but this is the expected thermodynamic preferential ranking.

One biomass combustion study that did carefully measure the Pb release from 550°C to 1150°C is quantitative (vanLith et al., 2008). It indicated early Cl release at 500°C but only slight Pb release below 700°C increasing above this and complete at 1150°C. The implication was that it was not related to $PbCl_2$ vaporization. A later study confirmed Pb retention at 500°C even with PVC present in waste fuels (Pedersen et al., 2010). From this chemical nature it suggests there is no formation of $PbCl_2$ in the fuel or with the release of PbO(g) at the higher temperature. If the fuel naturally contains the dichloride it could certainly vaporize. However, no studies have really answered the question whether any PbCl, $PbCl_2$ gases can form in the gas phase and the possibility remains uncertain due to the demanding kinetics. Nevertheless, with Pb and/orPbO in the gas phase with HCl, it can be expected that some slight balance will be established with PbCl that at high chlorine might extend to the dichloride. A recent study has shown an unexpected effect on the fly ash composition by reducing the flue gas temperature from 250°C to 150°C through a heat exchanger. The fuel was biomass from metal contaminated soils burned in a small pilot-plant boiler. Boiler temperatures are not quoted but a general increase in ash element concentrations appeared to occur for lead (almost twofold) and many other elements as the temperature was so reduced in this manner. No further metal deposition of lead is expected at these temperatures to increase its ash concentrations, and a similar effect is seen for Ti or Zr that are well known to be fully nonvolatile elements. As a result it appears that particle segregation was occurring and the monitored ash size distribution was changing, favoring the smaller enriched sizes at the lower temperature (Kovacs et al., 2018). It illustrates the care needed in such interpretations. Biomass fuels contain sulfur, but can have higher levels of chlorine and be low in silicon/aluminum. Although biomass and some wastes can have ratios of Cl >S, it is normal in coals and sewage sludges for S >Cl unless the sludge has been treated with $FeCl_3$ in its processing (Amand and Leckner, 2004). Consequently, when sludge is cofired with coal or biomass, generally both the levels of Pb and the S >Cl ratio can increase. As a result, in low-temperature co-combustion (800°C–900°C), the sludge appears to increase retention in the bottom ash partly through the low chlorine levels (Miller et al., 2004). However, with sludge treated with $FeCl_3$ such that the chlorine content can be twice that of sulfur, more vaporization of lead occurs above 400°C (Folgueras et al., 2007). A recent study with co-combustion with sludge in a cement kiln noted that more lead was released above about 1000°C on 1% by weight addition of either Al, Fe, Ca, or Na chloride salts. For example, 45% of the lead was normally released at 950°C but on adding $CaCl_2$ this increased to 75% at 1050°C (Yu et al., 2017). Coal/sludge co-combustion indicates that added chlorine as NaCl tends to release Pb from any bottom ash, added sulfur as FeS_2 retains any lead as $PbSO_4$ (Zhang et al., 2019). This could be supportive of a heterogeneous role for HCl ablating condensed phase PbO.

The major changes with sludge in lower temperature FBC modes (750°C–850°C) tends to be an increase in the bottom ash, less fly ash and a decrease in the very fine sizes, all indicators of more retention and less lead release (Lopes et al., 2003). A further recent study examining the effects of added sulfur to waste incineration below 900°C confirmed the enhanced retention presumably due to formation of more stable PbS with its low vapor pressure (Liu et al., 2017).

With municipal waste incineration, it was clearly indicated in the absence of sulfur, that $PbCl_2$(s) would begin to deposit below about 400°C consistent with its vapor pressure. However, on adding 0.1%, 0.2% SO_2 or 0.2% SO_2 + 8% H_2O the deposit would increasingly extend to higher temperatures as $PbSO_4$(s) (Jiao et al., 2011). Although it has not been measured in detail, it has been noted in cases when Cl >S by severalfold, that the flue gases can become more corrosive which is suggestive as being associated with $PbCl_2$(s) deposition (Vainikka et al., 2011; Enestam et al., 2011a). This can be modified to the less corrosive sulfate deposit simply by insuring that chlorine is less than sulfur when sulfate deposition becomes preferred above 550°C (Jiao et al., 2013). However, in the incineration of some wastes this S/Cl ratio can swing either way over orders of magnitude (Becidan et al., 2010; Enestam et al., 2011). Thermochemical calculations for such waste incineration conditions with a Cl/S ratio of 4.5 predicts that any gaseous lead chloride will be decreased by the presence of water, suggestive of hydrolysis but that this has little effect on the fly ash or slag composition (Poole et al., 2008). As largely confirmed by thermodynamic calculations it is clear that lead is sensitive to this S/Cl ratio (Frandsen et al., 1994; Furimsky, 2000; Sorum et al., 2004; Lundholm et al., 2007; Bankiewicz et al., 2012). One earlier XRD study doped metals into an incinerator operating at less than 900°C and noted $PbCl_2$ deposition under high levels of chlorine, and PbO/Pb_2O_3 deposition with normal lower levels. Sulfur's role appeared minimal in this case (Wey et al., 1999). A more recent X-ray study of waste incineration ash implies more a balance between PbO and $PbCl_2$ (Fujimori et al., 2013). It is extensively documented that most of the lead is concentrated on the fly ash even for waste incineration (Zhang et al., 2017a; Li et al., 2018b). In lower temperature FBC, the Pb is retained in the ash at 500°C − 1000°C and has the potential for interacting with the Al, Ca, Mg, Si components to be stably retained (Zhou et al., 2018a).

Biomass fuels contain sulfur but can have higher levels of chlorine and be low in silicon/aluminum. For these, lead release can start at 500°C and either be complete at 780°C (high chlorine) (Strand et al., 2004), or 1000°C in a case of S >Cl (van Lith et al., 2008). However, some levels of retention were suggested at 800°C with no effects shown by added SO_2 or HCl (Miller et al., 2003). In FBC incineration of solid wastes and sewage sludge combustion, it was confirmed that lead did not interact with the SiO_2 bed at <850°C (Lin et al., 2005a), but was predominantly on the fly ash (Shao et al., 2008). Also in vitrification of waste incineration ash at 1400°C − 1500°C the lead content vaporized with no Si/Al interaction (Wang et al., 2010a). Studies of waste incineration at 950°C − 1000°C confirmed that release was less sensitive to available time than to temperature (Youcai et al., 2004), and that rates decreased with lower levels of chlorine, but sulfur raised the temperature of volatility (Verhulst et al., 1996). The effect of water on waste incineration has been examined in a simulation system. The interpretation of the data is difficult but changes were noted that appeared to depress the lead volatilization (Youcai et al., 2004). Another study of pyrolysis of waste at 650°C noted only a slight effect (Raclavska et al., 2015). However, with sewage sludge at 850°C, the effects of 0.1%−0.5% Cl and 1%−5% S linearly increased the rate of lead volatilization, but phosphorus inhibited it (Luan et al., 2013). The presence of sulfur and water were shown to convert a dichloride deposit to lead sulfate (Jiao et al., 2013a). However, this is a typical study, symbolic of much of this research, and arises from the fact that it remains difficult to analyze the speciation of the elements in the ash or gas phase. When examined in detail, realistically such attempts fail to relate to waste incineration at all. This study takes readily dissociated acetate salts of Pb, Cd, Zn, and Cu, heats them in a rotary kiln oven with in turn gaseous HCl, SO_2, and H_2O passing over them. The emitted vapors are condensed as a function of temperature. Analysis is mainly aided by thermodynamic expectations and is extensively detailed. This raises doubts as to the reliability of the results. As a consequence, in reading any of these studies that purportedly relate to full-scale combustion for lead they need close inspection to see if they really can convey any real information or are simply publishing chemical thermodynamic expectations from an unspecified database.

Solely heating solid waste ash with a Cl/S content ratio of about twofold in air or argon, a 98% − 100% release was noted by 750°C also implying no presence of silicate formation (Jakob et al., 1995). However, with co-combustion of coal and waste, additions of Cl enhanced lead release but those of sulfur or kaolin reduced them by sulfate deposition or Al silicate absorption (Wu et al., 2013). Coal/sludge co-combustion indicates that added chlorine as NaCl tends to release Pb from any bottom ash, added sulfur as FeS_2 retains any lead as $PbSO_4$ (Zhang et al., 2019). Over the years, several experiments have examined the specific interactions between lead components and kaolin, an Al/Si clay, especially remembering that PbO is a major ingredient incorporated in leaded glass. By injecting kaolin powder and a lead acetate aerosol into a natural gas flame at about 1160°C in a downflow furnace the degrees of retention of sodium, cadmium, nickel or lead was examined (Linak and Wendt, 1993; Linak et al., 1995; Gale and Wendt, 2003, 2005; Yao and Naruse, 2005). These confirmed that the kaolin can absorb the gaseous PbO, PbCl, $PbCl_2$ and possibly Pb into its structure and that melting of the kaolin inhibits the process. Maximum interaction is at about 1000°C − 1100°C and could occur in some combustion systems. Of course ash is not kaolin and only creates a glass like core during combustion. Consequently, although remaining a possible interaction it may be time constrained. Another potential problem is that the sodium interaction with Si/Al/P is very dominant and might displace that of lead. Little seems to be known concerning lead phosphate, $Pb_3(PO_4)_2$. Its melting point is 1014°C and it decomposes to Pb(g) and P_xO_y but otherwise remains poorly characterized. Lead is also known for creating lead titanate (m.pt.1286°C) and with zirconium dioxide ceramic piezoelectric materials. Also $PbTiO_3$ films can be readily formed on substrates at 500°C–700°C by flame CVD methods (Ichinose et al., 1994; Choy et al., 1995). Due to such industrial usage its stability has been measured and it is known to dissociate to PbO(g) and TiO_2(s). At atmospheric pressure, it has a PbO(g) vapor pressure of about 5 − 10 ppmv at about 1000°C (Hardtl and Rau, 1969; Schmahl et al., 1979). However, with the low concentrations of Pb and Ti in most fuels such contributions of mixed metal interactions may have difficulty competing with those of oxygen or sulfur. $PbFe_2O_4$ has been characterized but is only thermally stable below about 725°C, but $PbAl_2O_4$ can form below about 1000°C. These would be in competition with $PbSO_4$ formation. Nevertheless, formation of the aluminate is being considered as one method for reducing the lead levels in lead-rich sludges (Lu and Shih, 2015). Injection of Al_2O_3 particles into cooling flue gases also appeared to promote $PbAl_2O_4$ formation (Jiao et al., 2013).

In reality, whether a chlorine level can control the distribution between PbO(g), PbCl(g) and $PbCl_2$(g) in the gas phase, it is somewhat irrelevant as the deposition can no longer be regarded as a simple condensation process but one of chemical deposition and thermodynamic preference. How else could one ever get a deposit of $PbSO_4$(s) from such a gaseous mix? Consequently, the deposits that are obtained will totally reflect the thermodynamic stabilities of PbO(s), $PbCl_2$(s), and $PbSO_4$(s) as a function of temperature for the specific operating conditions and whether there are any kinetic constraints that seem to be quite rare for condensed phase surface conditions. If chlorine levels are high, any PbO(s) that might start depositing below 800°C will be ablated off again as $PbCl_2$(g). It has been indicated that HCl, the dominant chlorine gaseous species in combustion, does not react with $PbSO_4$(s) but does so with the oxide, which is an important observation (Eddings and Lighty, 1992). As a result, below 650°C when $PbSO_4$(s) can form, although this may be thermodynamically equally vulnerable to conversion, HCl(g) cannot kinetically convert it. Then only below about 400°C when its condensed phase vapor pressure is sufficiently low will $PbCl_2$(s) deposition manage to compete with that of sulfate. Recently, it was noted that $PbSO_4$(s) deposition was favored in oxy-fuel combustion over the dichloride (Song et al., 2013). It should be noted that although PbO and $PbSO_4$ are insoluble in water, $PbCl_2$ is soluble. In fact as seen in Table 7.23, few of lead's compounds are water soluble. As a result, leaching studies can be useful and one recently indicated that HCl acid could leach 90% of the lead from waste incineration ash making it recoverable (Tang and Steenari, 2016). Although this picture is qualitatively supported by many thermodynamic calculations it is limited by our remaining lack of information on further possible interactions between gaseous lead forms and the Si/Al/P/Ti core content of ash. No analysis has yet attempted to include such interactions due to the lack of available thermodynamic data.

There appears to be some renewed interest in any potential combustion releases of radionuclides present in coal, in particular ^{210}Pb. This is a naturally occurring minor isotope of lead, formed by decaying radon and uranium isotopes. It has a half-life of about 23 years and has a very low concentration in the environment. The geographic distribution of such coals in the United States was documented (Lauer et al., 2015). In combustion it does not differ in behavior from the major stable isotopes and so will be found as a stable ash component. As concluded it poses no risks in coal-fired plants (Ozden et al., 2018; Li et al., 2019a). Additionally, any traces of lead, in any fine ash that is finally absorbed by the FGD scrubbers, will end being in the gypsum product. Numerous tests of recent have been concerned with potential leaching of trace elements from the gypsum. A recent study indicated that lead was stable in the gypsum and rated it as a medium risk leachant (Zhao et al., 2018b). A recent analysis of FGD gypsum samples from 12 coal-fired power plants in China noted the very low levels of lead confirming its high retention and control by the ash forms (Hao et al., 2017).

Due to the fact that ash particularly from municipal wastes and sewage sludge can be toxic, several programs have studied methods to reduce heavy metal components to levels where such waste can then be considered for agricultural purposes. With lead, solely heating to 1100°C is sufficient in vaporizing the lead components (Jakob et al., 1995). However, more general methods for also removing other metals of concern invoke heating the ash to 1000°C for 3 hrs with added chloride components such as $CaCl_2$ or treating it with Cl_2 or HCl and converting lead or its oxide to gaseous $PbCl_2$ that can be condensed and recycled (Chan and Kirk, 1999; Vogel and Adam, 2011; Nowak et al., 2012).

Less has been published on the behavior of lead under pyrolysis or gasification reducing conditions. For coal, thermodynamic calculations with S (0.3%), Cl (100 ppmv) (Frandsen et al., 1994; Helble et al., 1996), and S (1.6%), Cl (810 ppmv) (Bunt and Waanders, 2008), all represent cases of S >> Cl. Even without such excesses, PbS is very stable in both phases and has a vapor pressure of 1 ppmv at 570°C increasing to 20 ppmv at 660°C. However, comparisons of measurements and calculations suggested that the latter were overestimating the volatility by 15% (Bunt and Waanders, 2008). Other measurements of volatility to 1150°C indicated only about a 5% retention on slow pyrolysis in nitrogen yet higher values for rapid pyrolysis (Wang and Fomita, 2003). With biomass gasification and Cl, 2.5 times that of sulfur, volatilization was essentially complete at 855°C (Syc et al., 2011). For sewage sludge and S >Cl by factors of about 7- to 15-fold, at 770°C–960°C significant enrichments were noted in the micron size ash collected at 500°C indicative of significant vaporization and deposition of most probably PbS(s) and showing further enhanced removal if the hot gas filter was lowered to 450°C (Reed et al., 2001, 2005). This enrichment on submicron particles reflects that seen with coal (Helble et al., 1996; Font et al., 2010). Consequently, experimental information is scant but reaffirms that lead volatilization is significant with sulfur playing a far more significant role than chlorine. PbS(g) and Pb(g) are expected to be the dominant transporters of the lead and will deposit as PbS(s) below about 550°C where its vapor pressure drops below the surface influx of gaseous lead components. Under reducing conditions, any dichloride present or formed will be vaporized at the lower temperatures and any sulfur will produce PbS(s) that also vaporizes above about 600°C. Depending on the absence of oxygen and water, the gaseous sulfide and the chlorides are of similar thermal stability but will increasingly break down to Pb(g) above 1000°C. On cooling PbS(s) will be formed predominantly, any production of $PbCl_2$(s) being restricted to low flue gas temperatures and controlled by the S/Cl ratio. A recent mass spectral study of lead's speciation under gasification conditions and up to 1000°C indicates a distribution of Pb, $PbCl_2$ and PbS (Blasing et al., 2018).

In summary, although very volatile, gaseous atomic or molecular lead deposits readily on ash particles. Details of its release from fuels and whether the formation of gaseous chloride plays any role remain uncertain in oxygen-rich combustion. A new heterogeneous formation mechanism for gaseous HCl ablating condensed phase PbO is one possible route to resolve kinetic constraints. PbO(g) is the dominant transporter and all lead forms are vaporized as this by 1000°C. Many lead release studies have been published but product speciation cannot be measured. Chlorine appears to facilitate lead's release but it was noted in experiments that chlorine is released before lead and they do not correlate to a dichloride vaporization. Whether PbCl(g) or PbOH(g) play transport roles is not clear and dichloride or dihydroxide formation may be constrained. However, lead appears not to be constrained kinetically and approximate equilibrium expectations may be useful with a reasonable database, but will remain limited by the remaining lack of information on possible interactions between lead forms and the Si/Al/P/Ti core content of ash. The mechanism of lead's ability to catalytically recombine flame radicals tends to invoke chloride or hydroxide gaseous intermediates but they are significantly weaker than the monoxide. The gas phase chemistry appears to remain simple and as temperatures cool, efficient deposition onto ash proceeds, according to thermodynamic preferences, of initially PbO(s), then $PbCl_2$ and finally $PbSO_4$. It was noted that the sulfate once formed cannot react with

HCl. There is a pronounced enrichment on the smaller ash particle sizes. Although gaseous lead can be absorbed by kaolin, there appears to be little mixed metal interactions within ash, with lead preferring sulfur. Modern particulate capture should be able to retain all possible lead emissions. If necessary to satisfy agricultural ash-use programs, there are now simple methods to reduce the lead content of ash if required.

7.16 Magnesium

Magnesium is the seventh most abundant element in the Earth's crust being comparable in concentration to sodium. Rather surprisingly it is the fifth most abundant element in the oceans being only second to sodium in the metallic elements. As a result, as seen in Table 1.1, it is present as a major component (percentage scale) in all fuels and their ash. It is quite low in pet-coke but similarly can be a fractional percent in limestone (Cordoba et al., 2012). Its common ore is dolomite, $CaCO_3 \cdot MgCO_3$, but it exists as other minerals and is both mined and extensively extracted from sea water or brine lakes. The supply is inexhaustible and many millions of tons year^{-1} of MgO and other magnesium chemicals are produced directly. Globally, about 750,000 tons year^{-1} are reduced to the metal. The metal is extensively used in alloys due to its low weight and low melting 650°C and boiling point 1090°C. Its oxide is a valuable refractory and is also used as a small quantity additive to Portland cement. There are extensive medical applications and it is a required fertilizer.

Health concerns with magnesium

Magnesium's compounds are regarded as not toxic and are involved in many metabolic mechanisms in the body. A dietary requirement of about 300–400 mg day^{-1} is recommended and any excess is readily excreted from the body. Few studies are available showing any health or environmental concerns for magnesium. It is an important element in the body and brain. It transports readily into the brain and is found decreased in the brains of Alzheimer's disease patients. Whether it is involved in this remains speculative (Jiang, 2017). If blood levels fall below 10 mg L^{-1} it signals a deficiency and >90 can become toxic (Mayo Clinic, 2018). The inclusion in surveys and normal testing is somewhat rare for magnesium. Blood levels generally fall in the 30–55 mg L^{-1} range and seem a reliable measure as indicated in two surveys (Oggiano et al., 2018; Zeng et al., 2019). Hair the other biomarkers, display greater variation, values being anywhere from 150 to 1000 ppmw for hair and about 200 mg L^{-1} for urine making any diagnostic assessment difficult (Drobyshev et al., 2017; Zhu et al., 2018; Oggiano et al., 2018). Mother's breast milk appears comparable to her blood level of about 33 mg L^{-1} (Altun et al., 2018). In the majority of countries this is of little dietary concern but in parts of Africa it still is only being marginally met (Broadley et al., 2012).

Magnesium's basic chemistry

Magnesium has a full list of inorganic compounds. Those relevant to combustion are listed in Table 7.25, displaying a dominant divalent nature and its compounds generally are water soluble. However, in combustion most are not thermally stable and on heating will dissociate to the ceramic oxide MgO that has a 2850°C melting point (b.pt.3600°C) with a very low vapor pressure below 1800°C. For example, the carbonate with a melting point of about 990°C dissociates above 500°C (Beck, 1950), the sulfate above about 850°C, having a vapor pressure about 5 ppmv at 750°C (Lau et al., 1977; Tagawa, 1984; Hildenbrand et al., 1990; Ostroff and Sanderson, 1959), and the sulfite above 400°C.

TABLE 7.25 Properties of magnesium's condensed phase relevant to combustion, °C.[a]

m.pt. b.pt. vapor pressure
Mg 650 1090, 10 ppmv/430°C, ins
MgO 2827 3600, dissoc >1700 to Mg/O, 7 ppmv/1750°C, ss
MgO_2 223, decomp >100, reacts with water, ins
$Mg(OH)_2$ >350 dissoc, losses H_2O, ins
$MgCO_3$ 990, dissoc >500, ss
$Mg(NO_3)_2$ >330 dissoc, s
$Mg(NO_2)_2$ >100 dissoc, s
$MgSO_4$ 1140, dissoc >850, 5 ppmv/750°C, s
$MgSO_3$ >400 dissoc, s
$MgCl_2$ 707 1361, sublimes, 10 ppmv/630°C, s
MgS 2226, reacts with air, water
MgH_2 327, decomp >287, reacts with air, water
$MgAl_2O_4$ 2135, ins
$MgFe_2^{III}O_4$ 1750, ins
$Mg_3(PO_4)_2$ (ortho) 1348, ins
$MgSiO_3$ (meta) >1577 decomp, ins
Mg_2SiO_4 (ortho) 1898, ins
$MgTiO_3$ 1565, ss
$MgTi_2O_5$ 1695, ss
Mg_2TiO_4 1740, ss
$Mg(VO_3)_2$ (meta) 760
$Mg_2V_2O_7$ (pyro) 1010
$Mg_3(VO_4)_2$ (ortho) 1100

[a] *Gurvich et al. (1994), see text.*
decomp, decomposes; *dissoc*, dissociates; *s*, water soluble; *ss*, slightly soluble; *ins*, water insoluble.

Similarly the nitrate dissociates above 330°C, and the nitrite above 100°C (Stern, 1972). The well-known dihydroxide losses water above about 330°C (L'vov et al., 1998). The dichloride is stable and sublimes. Its melting point is 714°C and it has a vapor pressure of about 10 ppmv at 630°C. It can absorb water though (hygroscopic) with resulting destabilization. A peroxide exists with a melting point of 223°C dissociating at about 100°C. Moreover it reacts with water to form the hydroxide. The sulfide is known, is thermally stable with a high melting point (2226°C). However, it reacts with water producing H_2S and can be readily oxidized. A dihydride is known that is pyrophoric, dissociates at 287°C and reacts with water producing H_2. Again a chemistry is seen where in the condensed phase the oxide MgO can be expected to be stable and predominant.

MgO(s) is noted possibly for forming a wide range of mixed metal oxides listed in Table 7.25, including the refractory molecules with Al, Fe, P, Si, Ti, and V discussed further below with respect to their possible roles as a component of ash.

The dissociation energy of MgO and its meaning

Due to its importance as a refractory oxide the potential vaporization of MgO(s) has been extensively studied over the years especially due to intriguing disagreements concerning the dissociation energy of MgO(g). Vaporization experiments, especially those utilizing mass analysis, soon established the stability of the oxide condensed phase. All agreed that MgO(s) dissociated on vaporization mainly to its elements with a small amount of MgO(g). At temperatures of about 1750°C the total vapor pressure of the four gaseous species O, O_2, Mg and MgO is of the order of 7 ppmv with the Mg and oxygen species concentrations being similar, but that of MgO(g) an order of magnitude smaller (Farber and Srivastava, 1976; Kazenas et al., 1983; Lamoreaux et al., 1987). The earliest spectrophotometric study monitored the Mg emission from an acetylene/air flame at 2140°C and derived a value of 478 kJ mol^{-1} for the MgO bond strength (Huldt and Lagerqvist, 1950). However, from the mass spectrometrically measured equilibrium distributions, values for D_0(Mg-O) were in a range of 340–390 kJ mol^{-1} that compared reasonably well with transpiration flow-loss values of 370–390 kJ mol^{-1} (Gurvich et al., 1994). Values also were obtained from fuel-rich flames in which only gaseous Mg and MgO were expected to be dominant in the flame gases, controlled by a partial equilibration with the H and OH flame radicals, namely via reactions of the type

$$Mg + OH = MgO + O$$

Moreover, due to the partial equilibrium between H, OH, O, and O_2 in such fuel-rich flames the Mg/MgO distribution would become defined by the simple overall balance of

$$Mg + H_2O = MgO + H_2$$

Consequently, from a series of such well-defined flames, that did show this H_2 dependence, a value for D_0(MgO) should be reliable. However, efforts by several groups, with flame temperatures of 1300°C–2250°C, derived distributions implying 400–455 kJ mol^{-1} for this bond strength (Veits and Gurvich, 1956; Cotton and Jenkins, 1969; Gurvich et al.,

1994). This clearly introduced an uncertain dilemma of values and a quandary of what this implied. In addition to these experimental studies, reliable theoretical calculations emerged that reproduced the accepted bond strength values for BeO and CaO, but added further uncertainty by indicating an even lower value for D_0(MgO) of 250–265 kJ mol^{-1} (Langhoff et al., 1986; Stiakaki et al., 1994; Hwang and Mebel, 2000; Sullivan et al., 2003; Vasiliu et al., 2010; Prascher et al., 2011). Although the theory was adequate, it appears that the MgO($X^1\Sigma^+$) ground state is one of a very difficult electronic structure system and demands the highest quality of theory to acquire high accuracy for calculations of bond lengths, frequencies and energies (Bauschlicher, 2018). Fortunately, one additional independent experiment now has helped to begin to resolve this dilemma that still appears not to have been discussed or extensively reviewed in the literature. An accurate photodissociation threshold energy for MgO^+ was measured for the following allowed transition (Operti et al., 1989),

$$MgO^+(X^2\Pi_{3/2}) + h\upsilon = Mg^+\left({}^2S_{1/2}\right) + O\left({}^3P_g\right)$$

From the thermochemical energy cycle relationship,

$$D_0(MgO) = D_0(MgO^+) + IP_0(MgO) - IP_0(Mg)$$

using an accurate ionization potential for MgO (Bellert et al., 2000) and Mg, this indicated a value of D_0(MgO) = 257 ± 12 kJ mol^{-1} consistent with theoretical predictions. As a result, it now appeared that the range of different values obtained were not due to errors of measurement but rather to the experimental conditions, and the interpretation and meaning of the observed dissociation energy. This in fact is an excellent example in chemistry of illustrating the need for confirmation of values by utilizing differing methods of determination. The disagreements remain disturbing but now appear resolved to some degree by the nature of the potential energy curves for MgO and its electronic distribution. The lowest electronic energy states have been illustrated by Maatouk et al. (2010) and Bauschlicher and Schwenke (2017) for MgO, and have been redrawn here more simply in Fig. 7.5. They confirm the nature of the potential energy states predicted more than 40 years earlier (Schamps and Lefebvre-Brion, 1972).

Such curves are common to all the Group IIA oxides and sulfides, and also to ZnO, CdO, and HgO that have singlet ground states. In the present case of MgO, as common to all, due to spin conservation, the MgO($X^1\Sigma^+$) ground state correlates on dissociating to the second energy limit products of Mg($^1S_0$) + O($^1D_2$). The metastable O($^1D_2$) has a long radiative lifetime but with rapid quenching by many gases should not affect the outcome for measured equilibrated systems (Schofield, 1978). However, on examining the potential energy curves it is noted that only two excited states correlate to the ground state products, namely ($a^3\Pi$) and a repulsive unbound $e^3\Sigma^-$ state. MgO differs from BeO and the other Group IIA elements in having two much lower lying excited states, $A^1\Pi$ (T_0 3498 cm^{-1}, 42 kJ mol^{-1}) and $a^3\Pi$ (T_0 2552 cm^{-1}, 30.5 kJ mol^{-1}) (Murtz et al., 1995; Kim et al., 2001; Kagi and Kawaguchi, 2006). As seen in Fig. 7.5, these correlate to the second and first dissociation limits, respectively. As a result, these states provide alternate spin allowed dissociation mechanisms with the energies indicated in these reactions.

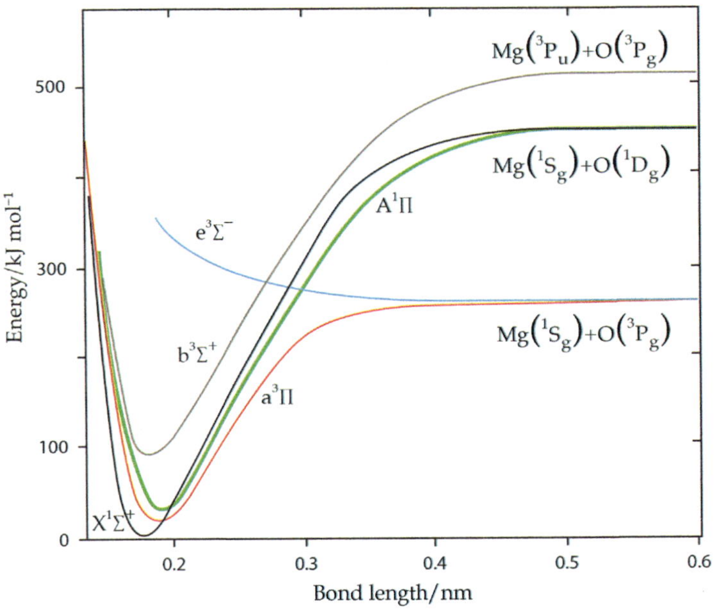

FIGURE 7.5 Approximate low-lying potential energy curves for gaseous MgO. Source: *Redraw from Maatouk, A., Ben Houria, A., Yazidi, O., Jaidane, N., Hochlaf, M., 2010. Electronic states of MgO: spectroscopy, predissociation and cold atomic Mg and O production. J. Chem. Phys. 133 (144302), 12 pp.*

$$MgO(X^1\Sigma^+) = Mg\left(^1S_0\right) + O\left(^3P_g\right) \quad D_0 = 257\ \text{kJ mol}^{-1}\ \text{(spin forbidden)}$$

$$MgO(X^1\Sigma^+) = Mg\left(^1S_0\right) + O(^1D_2) \quad D_0 = 257 + 190 = 447\ \text{kJ mol}^{-1}$$

$$MgO(a^3\Pi) = Mg\left(^1S_0\right) + O\left(^3P_g\right) \quad D_0 = 257 - 31 = 226\ \text{kJ mol}^{-1}$$

$$MgO(A^1\Pi) = Mg\left(^1S_0\right) + O(^1D_2) \quad D_0 = 447 - 42 = 405\ \text{kJ mol}^{-1}$$

Consequently, a measurement of D_0(MgO) appears to depend on the experimental method, its temperature and its experimental timescale. The ion-photodissociation measurement mentioned above indirectly established the energy of the first of these reactions, which may not occur naturally as suggested. Theoretically calculated values generally have been based on calculating the dissociation energy of MgO($a^3\Pi$) and then adding $T_0(a^3\Pi)$ and so they also indirectly obtained a value of about 257 kJ mol^{-1} for MgO(X). Often in spectroscopy, the smaller atomic number elements more rigorously comply with correlation rules. This appears to be the case here and very early calculations had concluded that the possible spin–orbit interaction between the $X^1\Sigma^+$ and $a^3\Pi$ electronic states in BeO, MgO, and CaO were all very weak. This prevents an avoided crossing whereby the X-state can easily hop onto the $a^3\Pi$-state potential energy curve and produce ground state products. Instead each potential energy curve seems to be rigorously independent (Stagg and Richards, 1974; Maatouk et al., 2010). However, recent theoretical work on CaO, BeS, and MgS that are analogues of MgO and have the same electronic states and symmetries also clearly show the repulsive $^3\Sigma^-$ state that crosses the $X^1\Sigma^+$ potential energy curve at energies slightly above this ground state limit (Larbi et al., 2010; Ribas et al., 2010; Khalil et al., 2011). This is clearly seen in Fig. 7.5 for MgO. It has now been suggested as a possible spin–orbit induced predissociation mechanism for BeS that could promote the mechanism to produce ground state products (Larbi et al., 2010). Such a transition would violate Kronig's selection rules for predissociation. Moreover, no experimental values in the case of MgO have obtained dissociation energies as small as this limit to support this suggestion. Also if predissociation were coupling the states the large range of values would not be expected. Although the MgO ($A^1\Pi$–$X^1\Sigma^+$) optical transition now has been reported, it is a weak transition, lengthening its radiative lifetime to 0.23 ms and further probably eliminating any possible role (Diffenderfer et al., 1983). The MgO($a^3\Pi$) state though is truly metastable with an estimated radiative lifetime of about 0.11 seconds (Thummel et al., 1989). Consequently, allowing for these state degeneracies, an approximate estimate of the MgO equilibrated populations of the $X^1\Sigma^+/a^3\Pi/A^1\Pi$ states at elevated temperatures indicates percentage distributions of 73/24/3 at 1000°C, 53/40/6 at 1500°C and 42/49/9 at 2000°C. The $A^1\Pi$ state will be a minor contributor at such temperatures but the $a^3\Pi$ can play a very major kinetic role. In other words, any thermally equilibrated system measurement will necessarily involve a mix of both $X^1\Sigma^+$ and $a^3\Pi$ states and this has to be at the basis for explaining the differing controversial values. The resulting dissociation energy can be expected to vary with temperature. Statistical thermodynamics allows for such population distributions in chemical analyses except in the present case it cannot allow for the consequences of differing dissociation limits and products. In other words the mass spectrometric and transpiration dissociation studies relate to measures of a balanced MgO(a/X) distribution. The flame studies that examine flames on millisecond timescales raise serious questions as to the state of equilibration of the MgO(a/X) populations. Whenever a long-lived metastable state is introduced into such a relatively fast kinetic system it has the potential to be a perturbation on the equilibrium distributions if relaxation processes are not sufficiently rapid. In the flame studies, it would seem that the metastable $a^3\Pi$ might be rapidly depleted and the system becomes biased upward toward a true value relating solely to D_0MgO(X) and singlet products. Collisional energy transfer rates between closely lying states in the $X^1\Sigma^+$ and $a^3\Pi$ manifolds have been studied theoretically but only minimally and for collisions with only N_2O (Pouilly et al., 1984). However, the fact that flame measurements tend to imply a large D_0(MgO) value close to that for the ground state dissociating to O($^1D_2$) tends to suggest that the $a^3\Pi$ state does struggle to maintain an equilibrium population and is depleted in the flame cases but can remain in equilibrium for Knudsen cell measurements. Chemically, it is a most interesting kinetic problem. Consequently, the above values based on the photodissociation of MgO^+ (Operti et al., 1989) and further confirmed by theoretical calculations on the ion (Partridge et al., 1986) are one step in resolving the previous extended controversy and pose the intriguing question of what we mean by a dissociation energy in the case of MgO. Currently recommended values are biased toward thermochemically equilibrated systems and suggest D_0(Mg-O) = 341 kJ mol^{-1} (Huber and Herzberg, 1979), 335 ± 25 kJ mol^{-1} (Chase, 1998), 360 ± 10 kJ mol^{-1} (Gurvich et al., 1994), with no value listed by Barin and Platzki (1995). This difficulty of measuring the group II diatomic oxides or similarly the diatomic sulfides applies to each member. Invoking the low-lying $a^3\Pi$ state appears valid in the case of Mg but cannot apply to the others. In their cases curve-crossing mechanisms appear to be involved, which retains the question whether these also contribute also with MgO to some degree.

Gas phase stabilities and transport

Potential gas phase molecules to consider in combustion other than the diatomic oxide are MgO_2(g), MgOH(g), $Mg(OH)_2$(g), MgCl(g), and $MgCl_2$(g). The gaseous superoxide MgO_2(g) has been studied theoretically suggesting $D_0(\text{Mg-O}_2) = 105 \pm 16$ kJ mol^{-1} (Bauschlicher et al., 1992), and further work now has confirmed the values (C_{2v} symmetry 88, linear 100 kJ mol^{-1}) also deriving acceptable values for MgO(g), MgOH(g), and Mg $(OH)_2$(g) (Plane and Whalley, 2012). The dioxide is much weaker than the trend seen for the other four elements of Group IIA and suggests a different bonding (Table 7.10). That MgO_2(g) can exist appears evident in independent kinetic studies of the direct reaction of Mg(g) with O_2 at 75°C–350°C (Plane, 1992; Nien et al., 1993) and 32°C–750°C (Vinckier et al., 1992). At these low temperatures, the three-body reaction appears to be the only kinetic mechanism possible due to the high endothermicity of reaction to MgO(g). Although such studies cannot determine specific bond strengths, a value of $D_0(\text{Mg-O}_2) =$ 82–106 kJ mol^{-1} has been suggested as necessary to be consistent with the kinetic data (Vinckier and Helaers, 1998). The nature of the bonding remains uncertain and may be considered as Mg^{2+} inserted into O_2^{2-}, linear or triangular. It cannot be compared with that of the diatomic oxide, Mg-O. It has been observed in matrix isolation studies by its infrared spectra and predicted to be linear (Andrews and Yustein, 1993; Andrews et al., 1996), and now is invoked in low-temperature atmospheric chemistry (Plane and Whalley, 2012) but is not expected to play a role in combustion.

As noted, pure MgO(s) does vaporize, but at elevated temperatures, mainly to gaseous Mg and MgO so interest has been additionally raised as to the possibility of the latter being hydrolyzed in the gas phase to MgOH(g) and $Mg(OH)_2$(g). With beryllium, the presence of water can aid ablation of BeO(s) directly to $Be(OH)_2$(g) and there does appear to be an indication that it can also occur for MgO(s) but to the least extent of all the Group IIA elements. Early transpiration experiments have been assessed and these do support this ablative interaction with MgO(s) (Gurvich et al., 1994).

$$\text{MgO(s)} + \text{H}_2\text{O(g)} = \text{Mg(OH)}_2\text{(g)} \quad \Delta H_{298\text{K}} = +291 \pm 20 \text{ kJ mol}^{-1},\ K_{2000\text{K}} = 4.7 \times 10^{-5}$$

This implies in the presence of 10% water vapor that at equilibrium a partial pressure of 10 ppmv of $Mg(OH)_2$(g) will require a temperature of about 1820°C (Alexander et al., 1963). However, the presence of $Mg(OH)_2$(g) has not been reported in flame gases although it has been observed in infrared matrix isolation studies of the products from laser-ablated Mg with O_2 and H_2 (Wang and Andrews, 2005). Two Knudsen cell-mass-spectrometric studies have characterized MgOH(g). To ensure no complications from any $Mg(OH)_2$(g) contributions, H_2 flowed over MgO(s) either at 1690°C–1730°C (Murad, 1980) or at 1750°C–1890°C (Hastie et al., 1980). No MgO(g) was detected and MgOH(g) was at concentrations 500 or more than a 1000-fold smaller then Mg(g). As recalculated, these indicate values in reasonable agreement suggesting $D_0(\text{Mg-OH}) = 315 \pm 12$ kJ mol^{-1} (Gurvich et al., 1994). The only other more recent independent measure comes from an $MgOH^+$ photodissociation threshold mass spectrometric determination (Operti et al., 1989). However, for the analysis this also requires the ionization potential of MgOH, which remains uncertain (Chen et al., 1999). The value of $D_0(\text{Mg-OH}) = 280 \pm 25$ kJ mol^{-1} so obtained is less accurate but consistent with these earlier results. Theoretical studies now also are fully consistent supporting 315 ± 12 kJ mol^{-1} (Bauschlicher et al., 1986; Stiakaki et al., 1994; Vasiliu et al., 2010). Such theoretical calculations again also have indicated that the MgOH structure is favored being about 250 kJ mol^{-1} more stable than that of HMgO, and their isomerization also includes an additional activation energy (Kong and Boyd, 1996; Jain et al., 2000; Li et al., 2012b). The ground and excited states of MgOH have also been appraised (Theodorakopoulos et al., 1999). Additionally, early flame studies of hydrogen-rich flames at 1300°C–2100°C showed that MgOH(g) and $Mg(OH)_2$(g) concentrations were negligibly small in the flame gases that contained mainly Mg(g), with 5%–40% MgO(g), and high 6%–44% H_2 and 18%–34% H_2O concentrations (Cotton and Jenkins, 1969). Consequently, although a large and acceptable value of D_0(HOMg-OH)$_g = 456 \pm 25$ kJ mol^{-1} has been recommended (Gurvich et al., 1994) and is supported by theoretical calculations (Schulz et al., 1999; Sullivan et al., 2003; Vasiliu et al., 2010), it appears that $Mg(OH)_2$(g) cannot be easily produced directly from MgO(g) due to kinetic constraints. Kinetically it may require the Mg(OH)(g) intermediate that is too weak to compete with MgO(g). Also because of its high stability, care may be needed in accepting thermochemical equilibrium calculations which incorrectly imply a dominance of Mg $(OH)_2$(g) that does not occur (Yan et al., 1999).

$MgCl_2$(s) is a common salt of magnesium and is water soluble. It is quite volatile, 10 ppmv vapor pressure at 630°C subliming to gaseous $MgCl_2$. Knudsen cell-mass spectrometric experiments measuring the equilibrium between gaseous MgCl and $MgCl_2$ established an accurate value of D_0(Mg-Cl) $= 320 \pm 6$ kJ mol^{-1} (Gurvich et al., 1994). Coupling the heat of sublimation and heat of formation of $MgCl_2$(s) this gives an accurate estimate of D_0(ClMg-Cl) $= 464 \pm 8$ kJ mol^{-1}, which values mirror almost exactly those above for the hydroxides and lends support to those values. Although thermally stable, due to the nature of the hydroxides, they would appear to exhibit a similar fate. Nanoparticles of MgO(s) can in fact be readily produced in flames by additions of $MgCl_2$ water solution aerosols (Fennell et al., 2006, 2007). Both optical and mass spectrometric monitoring methods are very sensitive and can observe trace levels on a ppbv or less scale. One such 50-year-old Russian study is quite interesting in that it examined the influence of carbon tetrachloride additions to an C_2H_2/air flame already containing a spray of $MgCl_2$ (Zhitkevich et al., 1963). Not normally observed, spectral radiation of MgCl(A-X) was seen in this case on adding excess chlorine. As chlorine levels increased, the weak emission radiations from Mg(g) and MgOH (g) linearly decreased while that of MgCl(g) continued to increase linearly. A similar type behavior was seen with Ca, Sr and Ba, except in those cases, their monochloride radiations were more intense, were more strongly affected, and also showed an onset of dichloride formation at higher chlorine levels. In an identical Hungarian study also at about 2130°C but using atomic absorption of Mg(g), a small decrease was monitored on adding up to 2000 ppmv equivalent of Cl (Kantor, 1994; Kantor and Ernyei, 1994). This was speculatively interpreted as formation of $MgCl_2$(g) but was a 40-fold smaller effect than that seen with calcium. Nevertheless, these two studies seem to suggest some slight formation of chloride, probably MgCl(g) at higher temperatures and chlorine content formed from the free Mg(g) or low traces of gaseous hydroxide. However, like the hydroxides, this still suggests that the magnesium chlorides play only a minimal role with magnesium in combustion.

Of other magnesium compounds, MgH_2(s) reacts vigorously with water, spontaneously ignites in air, decomposes at 287°C and can be used as a source of H_2. It is known to vaporize only to MgH(g) (Ehlert et al., 1968). Nevertheless, the FTIR spectrum of MgH_2(g) has been recorded in an electric discharge (Shayesteh et al., 2003), along with that of MgH by the matrix isolation method (Wang and Andrews, 2004). Gaseous MgH finds great importance in studying stellar atmospheres and has been analyzed spectroscopically. It has no low-lying electronic states. From Fourier transform emission spectra, the ground state $X^2\Sigma^+$ potential energy curve has recently been accurately determined implying a bond strength D_0(Mg-H)(g) = 124.0 ± 0.01 kJ mol^{-1} implying low thermal stability and no pertinent role in combustion (Shayesteh et al., 2007; Henderson et al., 2013). From this more accurate data, new partition functions and chemical thermodynamic tabulations have resulted mainly for astrophysical use (Szidarovszky and Csaszar, 2015). Most recently, the potential energy curves have been constructed theoretically, radiative lifetimes estimated and the accurate bond strength value confirmed (Chattopadhyaya, 2016).

MgS(s) appears to readily evaporate by dissociation. Its gaseous infrared absorption appears to have been monitored in a matrix isolation experiment (Martin and Schaber, 1982). It is thought to have a high melting point (2226°C) but reacts with water producing H_2S and is unstable in air. Due to the similar electronic configuration to MgO, the gaseous MgS($X^1\Sigma^+$) displays the same characteristic that its potential energy curve dissociates to Mg($^1S_0$) and S($^1D_2$) (additional excitation energy 111 kJ mol^{-1}) for spin conservation and it again also has an important very low-lying metastable $a^3\Pi$ state. These singlet and triplet potential energy curves were most recently illustrated in detail by Ribas et al. (2010). Theoretical values place the $a^3\Pi$ state at T_0 4170, ≈2000, 3120 and 3045 cm^{-1} above the X(v = 0) state (Partridge et al., 1988; Chase, 1998; Chambaud et al., 2008; Ribas et al., 2010). As also with the other four members of group IIA elements, the $a^3\Pi$ state of their gaseous diatomic sulfide is consistently lower than that of the oxide. As seen above, (T_0 2552 cm^{-1}, 30.5 kJ mol^{-1}) is well fixed for MgO(a), implying that the theoretical estimate of about 2000 cm^{-1} (24 kJ mol^{-1}) may be the most acceptable for MgS(a). Although the S($^1D_2$) state is metastable with a long radiative lifetime it is rapidly quenched collisionally by many gases (Schofield, 1979). The only mass spectrometric search for MgS(g) was negative, leading to an upper limit bound when recalculated with current thermochemical values of D_0(MgS) <255 kJ mol^{-1} (Colin et al., 1964). This is supported by two theoretical calculations indicating D_0(Mg-S)(g) = 224 kJ mol^{-1} (Partridge et al., 1988) and 219 (Ribas et al., 2010). The theoretical calculations relate to ground state products or the spin-forbidden reaction as indicated. Moreover, the value can be expected to be smaller than that of MgO.

$$\mathrm{MgS}(\mathrm{X}^1\Sigma^+) = \mathrm{Mg}\left(^1\mathrm{S}_0\right) + \mathrm{S}\left(^3\mathrm{P}_g\right) \quad D_0 = 224\ \mathrm{kJ\ mol^{-1}}\ \text{(spin forbidden)}$$

This would indicate that the following transitions have the energies indicated.

$$\mathrm{MgS}(\mathrm{X}^1\Sigma^+) = \mathrm{Mg}\left(^1\mathrm{S}_0\right) + \mathrm{S}(^1\mathrm{D}_2) \quad D_0 = 224 + 111 = 335\ \mathrm{kJ\ mol^{-1}}$$

$$\mathrm{MgS}(\mathrm{a}^3\Pi) = \mathrm{Mg}\left(^1\mathrm{S}_0\right) + \mathrm{S}\left(^3\mathrm{P}_g\right) \quad D_0 = 224 - 24 = 200\ \mathrm{kJ\ mol^{-1}}$$

Additionally, the theoretical papers above indicate the next state, the $A^1\Pi$ is about 1100 cm^{-1} higher than the $a^3\Pi$. Consequently, these low-lying excited states will become significantly populated at equilibrium especially at elevated temperatures. At 1500°C the MgS distribution will be 7% (A), 50% (a), and 43% (X), and at 2000°C, 9/57/34% (A/a/X). In other words a Knudsen cell-mass spectrometric value would be of interest to examine this dominant role of the MgS($a^3\Pi$) state and see if it reflected a bond strength value closer to 200 kJ mol^{-1}. This is rigorously metastable so the maintaining of the a/X distribution in some experimental methods could further illustrate details of the molecular transfer. As with all these Group IIA oxides and sulfides there also remains the possibility of curve hopping or even spin–orbit induced predissociation by the repulsive $^3\Sigma^-$ state as suggested and illustrated for BeS (Larbi et al., 2010) and mentioned elsewhere herein for all the Group II oxides and sulfides and further portrayed in Fig. 7.5 for the similar designated potential energy curves for MgO. A low-temperature photodissociation study could help resolve this and see if it can be restrained to go solely to the excited product. As a result, to recommend a value for the bond strength of MgS(g) poses a similar dilemma for combustion in our current understanding. That it was not seen in the only mass spectrometric study that did obtain values for CaS, SrS and BaS is interesting. Moreover the low-lying $a^3\Pi$ state will play a significant role making the gaseous MgS a mixed a/X state possibly reflecting two differing dissociation channels. Additionally, it may be expected to have a bond strength lower than the already low value for MgO(g). Quoted recommendations in databases at present are an estimate of 300 kJ mol^{-1} that may now seem to be too large (Gurvich et al., 1994), 275 kJ mol^{-1}(Barin and Platzki, 1995) and ≤ 276 kJ mol^{-1} (Chase, 1988). An estimate of D_0(Mg-S) = 270 ± 25 kJ mol^{-1} may be all that can be suggested, but the trends in Table 7.10 for the Ca, Sr, and Ba D_0MS/MO ratio are complicated by both molecules normally exhibiting this mix of $X^1\Sigma^+$ and $a^3\Pi$ states. With regards to combustion, whether MgS(g) is negligible like MgO(g) remains uncertain, but its vaporization does appear to be dissociative. It is unlikely to play a role in either combustion or pyrolysis/gasification if water or oxygen is present.

Laboratory combustion studies

Other than the flame experiments mentioned above, that concern more the determination of fundamental chemical stabilities, two other practical areas have been examined for magnesium and many of the other trace metal elements. These center on the subject of soot, formation and suppression, and also the observation of flame inhibition or catalytically enhanced flame radical recombination effects. With regards to soot suppression, magnesium is clearly not an alkaline earth (Howard and Kausch, 1980). In one of the earliest studies of forty metals, magnesium was rated with a zero effect (Cotton et al., 1971). Later there was a suggestion that it might even slightly promote soot formation (Bulewicz et al., 1975). Other studies have reported no effect (Haynes et al., 1981), as also a very recent one aspirating magnesium chloride solution into a C_2H_4/air flame (Simonsson et al., 2018). This used laser-induced incandescence and elastic light scattering mapping analysis but recorded no noticeable effects.

Surprisingly, although not catalytic with regards to soot, magnesium is rated quite high as an efficient flame radical recombiner and flame inhibitor. This was noted from early studies when magnesium levels of only 1 ppmv of the addition were very effective (Bulewicz and Padley, 1971; Vanpee and Shirodkar, 1979). It is as equally effective as Ca, better than Fe but not the equal of Cr, U, or Ba (Linteris et al., 2008). The intriguing question still not satisfactorily answered concerns the mechanism. Most favored is to invoke the metal forming the oxide or hydroxide that then is reduced back by flame radicals, forming a kinetic cycle and recombining the radicals to water or hydrogen. This is hard to invoke for magnesium that has only traces of hydroxide and would need a high kinetic efficiency. More complex kinetic models have been presented for inhibition efficiencies of Sn, Mn and Fe but contain kinetic difficulties that leave such models unsatisfactory or somewhat speculative (Linteris et al., 2002). Although the alkaline earths, Ca, Sr, and Ba, have well characterized chemiionization mechanisms, where the ions $MOH^+ > M^+$ are observed, magnesium behaves more like a transition element with $Mg^+ > Mg.H_2O^+ > MgOH^+$ (Goodings, 1992). Its ion chemistry has fostered little interest.

Combustion power plant data

Regarded as an important component in most fuels on a percentage scale, Table 1.1, magnesium is readily analyzed and so has been quite extensively studied in full-scale combustors and incinerators. In the rapid pyrolysis of Victorian brown coal at 900°C it was noted that 25% of the Mg could be released (Sathe et al., 2003). This seems rather extensive and might be physical ejection of an explosive nature but of course also possibly more a realistic reflection of ash combustion. Full-scale results generally confirm magnesium's very low volatility with concentrations being comparable in bottom ash and fly ash allowing for slight vaporization variations of other elements (Meij, 1994; Martinez-Tarrazona and Spears, 1996). Also early studies indicated no variation of concentration with ash particle size (Coles et al., 1979; Martinez-Tarrazona and Spears, 1996; Querol et al., 1995). Mass balance measurements at coal-fired plants in Europe confirmed that the magnesium is present totally in the ash forms and removed mainly by ESPs supplemented by an FGD (Meij, 1994; Nielsen et al., 2002). Early stack losses were estimated at about 0.1% and that possibly from the limestone in the FGD system or even 0.02% from a Canadian coal-fired plant (Goodarzi, 2006a). A more detailed study of absorption and retention of trace elements in a modern FGD system for a cofired coal/pet coke plant indicated more significant retention for Mg and in essence total ash control (Cordoba et al., 2012). Nevertheless, as a result of potential health concerns with submicron size particles there has been further interest in magnesium's behavior. Its variation of concentration in coal can be at least 20-fold, normally being higher in subbituminous and lignite coals and much lower in bituminous. As a result, particularly at high burning temperatures with coals such as lignites rich in Ca and Mg it was noted that these could play a part in submicron particle formation. Extensive studies by Sarofim's group indicated that above furnace temperatures of 1550°C slight vaporization of magnesium could in fact occur particularly with the subbituminous and lignite coals and rapidly increased to higher temperatures (Neville et al., 1981; Quann and Sarofim, 1982; Neville and Sarafim, 1982; Quann et al.,

1982, 1990; Helble et al., 1986). In these studies they examined the possible formation mechanisms for the ultrafine particles (0.005–0.1 μm) that constitute less than one percent of the total ash. These fractions were seen to be enriched by up to a factor of 10-fold more in magnesium relative to the ash as a whole. Obviously, partly because the burning particle can have a temperature hotter than its environment for a short time, and may even be reducing at subsurface layers, the magnesium can be released as free Mg(g) that would then oxidize and nucleate further particles or condense on such. Other studies have confirmed this, noting a twofold increase in Mg content as particles become smaller than 1.0 μm (Teramae and Takarada, 2009). Further analyses of <0.1 μm fractions of ash from coal (Wang et al., 2008) and sewage sludge combustion (Zhang et al., 2007) confirm low levels of magnesium regarded as originating from a vaporization/deposition mechanism. However, one analysis of fine particles (0.05 and 2.0 μm) leaking through an ESP that had removed 99.8% of the ash were found to contain magnesium at similar concentrations to the general fly ash (Kauppinen and Pakkanen, 1990), most probably a reflection of an unspecified low furnace temperature and a low level of coal magnesium. A recent ambitious study of coal fly ash, size segregated the ash into five coarse ranges, the smallest being 2.2 μm (Lanzerstorfer, 2018). Concentrations of about 40 elements have been analyzed and these can be readily divided into those showing enrichment to smaller ash particle sizes and those not. Possibly due to the coal combustion temperature and the coarse sizes of particles no enhancement to smaller particle was noted for magnesium, in fact for the 2.2 μm sizes, there was a depletion if anything. One study has in fact suggested adding magnesium at higher combustion temperatures as a method for reducing micron-sized particles (Ninomiya et al., 2009). That a small amount of magnesium manages to vaporize and leave a particle is not unreasonable if the temperatures are sufficiently high. One recent study has shown a strange effect on the fly ash composition of reducing the flue gas temperature from 250°C to 150°C through a heat exchanger. The fuel was biomass from metal contaminated soils burned in a small pilot-plant boiler. Boiler temperatures are not quoted but a more than twofold increase in ash concentration appeared to occur for magnesium and many other elements as the temperature was so reduced through the heat exchanger. As no further metal deposition was possible to increase the concentrations especially for such elements as Ti or Zr that are fully nonvolatile it was apparent that particle segregation was occurring and the sampled ash size distribution was changing, favoring the small enriched sizes at the lower temperature (Kovacs et al., 2018). It illustrated that the small size ash can become enriched by mechanisms other than deposition and that interpretations of even simple experiments can be difficult.

The fact that combustion does free some atomic magnesium was the basis of flame atomic absorption or emission analytical methods before the higher temperature ICP analyzer was developed. In order to obtain sufficient sensitivity for elements such as magnesium, H_2/O_2 flames were found to be marginal (Smyly et al., 1971) and the hotter C_2H_2/air or N_2O flames were needed. With these, studies showed that at 2100°C–2700°C the fraction of magnesium present in a sample injected into the flame could be about 60% atomic Mg providing adequate analytical measuring sensitivity (De Galan and Winefordner, 1967; Zeegers et al., 1969; De Galan and Samaey, 1970). CVD using solutions of organic compounds of Mg and Al dissolved in toluene and burned as a diffusion liquid-fueled flame torch can readily form $MgAl_2O_4$ ceramic deposits (m.pt.2135°C) on a substrate at 850°C–1150°C and such molecules are not difficult to form on surfaces (Stollberg et al., 1997). It was noted that Al interfered with flame analysis of Mg and the question remains whether gaseous or particulate aluminate formed in the gas phase depleting the gaseous magnesium by such particle and deposition formation (Luecke, 1992a).

In summary, any magnesium in the fuel of lower temperature combustors such as incineration or fluidized bed combustion will be fully retained in the original ash and can be controlled by particle removal methods. The form in which it finally remains is generally uncertain. High temperature combustion increasingly above 1550°C will begin to promote magnesium vaporization and facilitate enhanced submicron particle formation that although not large in mass becomes more difficult to control. There appears little evidence of hydroxide or chloride formation acting as transporters. Sulfur also appears to play only a minimal possibly deposition role. Roasting municipal waste ash at 1000°C with either $CaCl_2$ or Cl_2 for 3 hrs only induced about a 10% volatilization with Cl_2 volatilization suggesting some oxide presence (Chan and Kirk, 1999). Thermochemical equilibrium calculations suggest stabilities for various silicates or the phosphate, $Mg_3(PO_4)_2$ (m.pt.1348°C), at elevated temperatures (Bourgel et al., 2011) and $MgSO_4$, $MgCl_2$, and MgO at lower temperatures on the final ash (Poole et al., 2008). The data appears to imply that thermochemical predictions in this case may be a reasonable approximation of its behavior. Ash from biomass burning plants has been reported to be rich in calcium and magnesium phosphates (Lanzerstorfer, 2015). Other mixed metal oxide refractory molecules that might be produced in the cooling ash depending on chemical thermodynamic preferences are the silicates, $MgSiO_3$ (m.pt.1577°C), Mg_2SiO_4 (m.pt.1898°C), titanates, $MgTiO_3$ (m.pt.1630°C), $MgTi_2O_5$ (m.pt.1695°C), Mg_2TiO_4 (m.pt.1732°C), the ferrite $MgFe_2O_4$ (m.pt.1750°C), and vanadates, $Mg(VO_3)_2$ (m.pt.760°C), $Mg_2V_2O_7$ (m.pt.1010°C), $Mg_3V_2O_8$ (m.pt.1100°C). The last two vanadates being stable at 1300°C can mitigate V_2O_5 corrosion with heavy oils by adding magnesium sulfate or oxide (Di Salvia et al., 2015). Nevertheless, it should be noted that such predictions are solely preferences that may or may not be possible. Their formation on or in the ash during combustion or deposition will depend on these species managing to interact if temperatures and time constraints permit. Nevertheless, even though magnesium can produce many such mixed metal compounds in theory, it will also depend on the full ash composition and whether other metals take a precedence over the magnesium for a certain species. Modern-day combustors will retain all the magnesium in some ash form.

7.17 Manganese

In the center of the first row transition elements, manganese displays a rich multivalent chemistry due to its $3d^5 4s^2$ electron configuration. It lies between Cr and Fe and is more similar to the latter in some regards except it is less refractory (m.pt.1247, b.pt.2061°C) and oxidizes slowly in water. It is quite abundant in the Earth's crust (0.09%), being about 12th in quantity, and of the heavy metals only Fe is more abundant. In the oceans its concentration is more central in all the elements being comparable to that of Cu, Sb and Se. It has numerous ores, the main commercial ones being pyrolusite (MnO_2), rhodocrosite (carbonate), and rhodonite (silicate). Globally it is mined at about 18 million tons of ore year^{-1} and remains abundant. The United States imports its needs, which center on use in steel. With this, there is no adequate substitute and Mn is regarded as a strategic element (National Research Council, 2007). For 50% Mn ore content, present cost is about \$1.70 kg^{-1}, but can be 40 times higher for the pure element. It is present in the atmosphere in particulate form. Some compounds of manganese are soluble in water (nitrate, sulfate, chloride, alkali manganates, and permanganates) while others generally not. Hence in surface waters it can be in solution or solid phase but will modify and ultimately become a part of a sediment. Likewise in soil it can migrate and be leached. Initially, manganese was mainly of concern by Governmental Agencies with industrial occupational environments where sustained exposure could lead to neurological and respiratory consequences. However, more recent reports now name it as a developmental neurotoxin of major importance with children and pregnancy calling for tighter controls (Ritter, 2014). This arises mainly from its extensive use in sprayed fungicides (Mora et al., 2014; Gunier et al., 2014), and now being a major antiknock in gasoline. Its presence in raw food supplies (Nuapia et al., 2018), and having ferromanganese alloy plants in the vicinity of schools has given rise to increased monitoring (Rodriques et al., 2018). Otherwise, current interest is centered on its presence in atmospheric particulate emissions. Although there appear to be no recent studies, previous approximate assessments of natural and anthropogenic contributions to the atmosphere indicate that manganese is one of few trace metal elements that may not really be a pollutant. Anthropogenic releases, which hopefully are in fact decreasing as tighter regulations are constantly being applied to industry, had been indicated earlier at 215,000 or 40,000 tons year^{-1} (Malakhov and Makhon'ko, 1990; Nriagu and Pacyna, 1988). Now, this has been reduced to 11,000 tons year^{-1} (Pacyna and Pacyna, 2001). These atmospheric releases result mainly from the steel/metal industry, agriculture and from coal and sewage sludge combustion. Contributions from oil combustion and cement production appear minor but the current addition of 1/32 g gallon^{-1} of methyl cyclopentadienyl manganese tricarbonyl (MMT) to gasoline as the replacement antiknock ingredient needs to be included. In the United States with its annual consumption of about 140 billion gallons year^{-1} in automobiles will add 5000 tons year^{-1} of molecular Mn to anthropogenic particles, and will of course be present more in urban regions. Research on its use in combustion continues and a recent study indicated improved performance by the additive with biodiesel fuels (Celik et al., 2016). A recent report of US cement dust having Mn levels of 2526 ppmw compared to Nigerian levels of 370 ppmw might suggest further examination also of the cement industry contributions (Ogunbileje et al., 2013). Because of manganese's occurrence in the Earth's crust, the natural dust levels become a dominant source of manganese in the air. They dwarf contributions from volcanoes or forest fires and totals have been estimated at 320,000 or 600,000 tons year^{-1}, respectively (Nriagu, 1989; Malakhov and Makhon'ko, 1990). The corresponding anthropogenic disposals to water and soils are large but are monitored by water districts, and drinking water in California, United States, is regulated to being <0.5 mg L^{-1}. A recent study of the releases to the coastal waters from the Pearl River Delta in southern China noted this as 15,000 tons $year^{-1}$ from industrial waste (Geng et al., 2015).

Health concerns with manganese

Manganese is a necessary element for humans and animals playing innumerable roles in the human body chemistry (Chen et al., 2018). It has now come more to the forefront due to its possible neurotoxicity, even though needed by the brain. Like most elements, only small quantities are required of about 2–5 mg day^{-1} for adults. No MRL has been published but a value of <11 mg day^{-1} has been reported to be slightly below the level of observing adverse effects (Goldhaber, 2003). Dietary food is the main input source. The whole body has a content of about 12–20 mg Mn with a daily turnover of about 5–8 mg. Only about 5% in the diet is absorbed by the intestines and by using ^{54}Mn as a marker the half-life in the brain was measured as 53 days (dos Santos et al., 2017). The body, if not fully retaining it all, has the ability to regulate quantities by excretion mechanisms mainly by feces but also by bile, urine or sweat. There is a need for reliable biomarkers for health although risks from manganese are normally minimal and its body levels are rarely tested (Bornhorst et al., 2018). This could be measured by sampling blood, hair, urine or even toenails that have all been examined. Of them, blood stands out as being more reliable and consistent, being in the 5–20 μg L^{-1} range (Bocca et al., 2011; Cesbron et al., 2013). These values now are followed by most testing centers (Mayo Clinic, 2018). Quite a large number of monitoring survey results from around the world are available (Goulle et al., 2005; Schofield, 2006; Nisse et al., 2017; Oggiano et al., 2018; Zeng et al., 2019). These generally fall within this range. The US National testing data indicates for 2011–16 that 95% of the blood values are below 17 μg L^{-1}, with urine levels being very low (CDC 4th Report 2018). Nevertheless, numerous surveys do indicate cases of significant outlier values several fold higher, so whether some people are at risk remains unknown. Hair content has also been measured but is more difficult for a correlating accuracy. This arises from normal dust in the hair possibly biasing the data (Eastman et al., 2013). Other biomarkers of teeth and nails currently are being developed to examine their reliability particularly for pregnant women and young children (Mora et al., 2014; Gunier et al., 2014). The fact that urine is a minor excretion channel for the body makes it of less value. Any large excess intake by inhalation can cause adverse respiratory and nervous system problems rather than any main body-organ effects. Toxicity generally is not fatal and is treatable. It shows obvious symptoms similar to Parkinson's disease. The relative toxicity of manganese's compounds generally remains neither known nor the effect of its oxidation state in the body. Entry through the blood/brain barrier has been well studied and numerous influx transporters for Mn identified. The method of efflux is less well known and the question of possible accumulation has been raised (Jiang, 2017). In autopsies of Alzheimer patient's brains there is an accumulation, however, it is a critical element needed by the brain. Numerous studies speculate that Mn may play a role in other neurological illnesses of the elderly such as Parkinson's, Alzheimer's, or Huntington's disease (Bichell et al., 2017; dos Santos et al., 2017). The neurotoxic mechanisms of potential neural damage are being researched but remain unclear (Li et al., 2019b). Higher valency molecules such as heptavalent potassium permanganate, $KMnO_4$, are strong oxidizers and so react somewhat as bleach if consumed but are rarely encountered. The body does seem to modify valence though and in the brain can oxidize Mn^{2+} to trivalency. Both valences seem to be required and both are toxic. Their individual roles remain unclear. As suggested there is generally a minor concern with regards to manganese as a trace metal in the environment. However, whether some levels in drinking water may be affecting more susceptible people has been questioned (Simms, 2011). The WHO guidelines for drinking water are 0.4 mg L^{-1} for Mn (WHO Guidelines, 2006a). Nevertheless, there remain concerns in certain regions (Valcke et al., 2018). The British though, known for their frequent cup of tea, which can itself contain about 1 mg cup^{-1}, generally consume many cups per day, with rare instances of manganese overdoses so it is probably unlikely that drinking water places the public at too great a risk. In addition it is interesting to note that plants, algae and fish can bioaccumulate manganese, and fish can normally contain up to 20 mg kg^{-1} and up to 100 mg kg^{-1} in polluted waters. It appears that manganese has found no specific application in the medical profession and has not become a component of the health supplement industry. However, in cases of required parenteral nutrition that contains Mn, this is received intravenously, increasing its bioavailability to 100%. Care then has to be taken so that over supplementation is not provided (Livingstone, 2018; Jin et al., 2018). Its only medical aspect is either cases of manganese's rare deficiency or excess. Recently a World Health Organization report (World Health Organization Mn, 2004a) and one from the US Agency for Toxic Substances (ToxReport Mn, 2012), that are freely available on the web, document our present knowledge of the health and environmental aspects of manganese and its compounds. Currently, in the United States there are only advisory established levels suggested for its control by the Environmental Protection Agency. Nevertheless, there are health concerns, and as indicated, efforts continue to promote biomarkers that would indicate early signs of Mn toxicity that can be treated. The concern relates more to pregnancy and potential effects on the fetus that can be vulnerable as manganese does pass readily through the placenta. Speculative suggestions have been raised that it may relate to birth weight and growth effects (Ashley-Martin et al., 2018; Lee et al., 2018). Occupational long-term exposures remain of concern particularly from smelters (Cowan et al., 2009), and in Scandinavia through heavy biomass combustion emissions (Jumpponen et al., 2014). Also China has reported that their Mn levels in atmospheric particles do exceed World Health Organization suggested levels in many of its cities (Duan, 2013). A particulate source assessment of $PM_{2.5}$ in the United States using 2000–05 data indicated that Mn release was predominantly from the steel industry and minimally from coal, biomass or oil burning, and appears to discount Mn use in gasoline as being negligible (Thurston et al., 2011). The situation with using MMT in gasoline remains unclear. It is an additive in the United States (but not in California), banned by some oil refiners but still apparently in wide global use with an addition to gasoline of about 8.3 mg L^{-1}. Attempts to phase out its use may be in the near future.

Manganese's basic chemistry

Manganese has a rich chemistry that finds many uses. Its major application remains with the steel and metal industry. It is used as a hardener for steel and aluminum. Because the United States imports all its Mn needs it is regarded as critical for the economy (NRC Report, 2007). It has the ability also to add or remove colorations with glass and ceramics, forms ceramic materials, is used in advanced battery technology and can find wide use from being a fertilizer to a strong disinfectant (fungicide), oxidant or preserving agent. Although many valences can be invoked in its molecules, the divalent state is the most stable where it losses the two outermost 4s electrons and leaves a high-spin ion with the five free 3d-electrons each occupying one of the five d-orbitals. As summarized in Table 7.26 along with its other major condensed phase molecules it has five characterized oxides. These have stable ranges of temperature and on heating gradually lose oxygen and change structure (Shaheen and Selim, 1998).

As a result, MnO_2, found in nature and the most stable one that is richest in oxygen content, it starts losing oxygen above 220°C. Although Mn_2O_7(s) is known, and is the only water soluble oxide, it is an explosive and unstable molecule. Consequently, in the solid phase MnO_2 converts (535°C) to Mn_2O_3 which above 1080°C then has converted to Mn_3O_4 that then is quite stable to about 1300°C with a melting point of 1567°C (Shaheen and Selim, 1998). This likewise begins to dissociate into MnO(s), which vaporizes and begins to dissociate into a MnO(g) and Mn(g) mixture with oxygen and has a vapor pressure of about 1 ppmv at 1500°C (De Maria, 1984). The melting point of MnO(s) is about 1945°C. As a result manganese's oxides do not show any direct vaporization ability except for MnO at high temperatures (Brewer and Mastick, 1951). There is a dihydroxide, $Mn(OH)_2$, formed by precipitation from a soluble manganese salt such as the nitrate by alkali addition (McKenzie, 1971). It is water insoluble but dissociates on heating at 160°C–200°C through MnO(OH), which then forms MnO_2. The carbonate $MnCO_3$, also insoluble, dissociates to the dioxide above 300°C (Stern and Weise, 1969; Shaheen and Selim, 1998; Stern, 2001). The nitrate breaks down to the oxide above about 160°C (Stern, 1972, 2001). The sulfate melts and dissociates to Mn_3O_4 near 700°C (Ostroff and Sanderson, 1959; Stern and Weise, 1966; Tagawa, 1984; Stern, 2001). Except for the dichloride the chlorides are not thermally stable to higher temperatures. $MnCl_2$(s) is stable (m.pt.650, b.pt.1190°C), and readily formed on chlorination of the oxides (Fouga et al., 2009). At 500°C the vapor pressure is about 1 ppmv and 10 ppmv at 570°C increasing rapidly above its melting point (Barin and Platzki, 1995).

An early mass spectrometric Knudsen cell study indicated that monomeric $MnCl_2$(g) was the dominant vapor species (Schoonmaker et al., 1959). However, the chloride is covalent and susceptible to oxidation and hydrolysis to the oxide and HCl by steam above 600°C. Any oxychlorides are less stable and similarly hydrolyzed or oxidized. Of the other known single metal manganese compounds the sulfides are stable in reducing atmospheres. For example, MnS (m.pt.1530°C), insoluble in water, has been shown to vaporize by dissociation to Mn(g) and S_2(g) in Knudsen cell effusion and mass analysis studies (Wiedemeier and Gilles, 1965; Viswanadham and Edwards, 1975). They are vulnerable to oxidation through the sulfate to the oxide. Similarly, the corresponding phosphides are known, MnP (m.pt.1147°C), MnP_2 (m.pt.1327°C), and also vaporize congruently but to Mn (g) and P_2(g) (Myers et al., 1982). However, they are more stable against oxidation or hydrolysis. Phosphates are known such as $Mn_3(PO_4)_2$ (m.pt.1140°C), which is very water soluble, and the insoluble pyrophosphate $Mn_2P_2O_7$ (m.pt.1196°C). The former is extensively used as a metal protective coater and can be readily substituted by Li or Fe in its structure. The pyrophosphate also is thermally stable, a refractory ceramic, dissociating to $Mn_3(PO_4)_2$ above 1400°C. Both these phosphates and the corresponding silicates, meta-

TABLE 7.26 Properties of manganese's condensed phase molecules relevant to combustion, °C.[a]

m.pt. b.pt. vapor pressure

Mn 1247 2061, oxidizes slowly in water, 40 ppmv/1025°C, ins

MnO 1945, 1 ppmv/1500°C, ins

MnO_2 >200 losses O_2 converts >535 to Mn_2O_3, ins

Mn_2O_3 940, >850 converts to Mn_3O_4, ins

Mn_3O_4 1567 2847, stable to 1300, dissoc to MnO, ins

Mn_2O_7 6 95, unstable, >55 explodes, forms $HMnO_4$ with water, s

MnO(OH) >220 decomp to MoO_2, ins

$Mn(OH)_2$ dissoc 160–200 to MnO(OH) and MnO_2, ins

$MnCO_3$ >200 dissoc to MnO_2, >450 to Mn_2O_3, ins

$Mn(NO_3)_2$ >160 dissoc to MnO_2, hygroscopic, s

$MnSO_4$ 700 melts and dissoc to Mn_3O_4, s

$MnCl_2$ 650 1190, >600 hydrolyzes, oxidizes, 10 ppmv/570°C, s

$MnOCl_3$ <0 decomp to $MnCl_3$, hydrolyzes and oxidizes

$Mn_2(CO)_{10}$ 154, sublimes 60, dissoc 240–310 to Mn, ins

MnS 1530, oxidizes to sulfate and oxide, ins

MnP 1147, stable with water or air, dissoc on vaporization, ins

Mn_2P 1327, dissoc on vaporization, ins

$Mn_3(PO_4)_2$ 1140, s

$Mn_2P_2O_7$ (pyro-) 1196, dissoc >1400 to $Mn_3(PO_4)_2$, ins

$MnSiO_3$ (meta-) 1291, dissoc on vaporization, ins

Mn_2SiO_4 (ortho-) 1345, dissoc 1286 on vaporization, ins

$Mn^{II}Al_2O_4$ 1850, stable <1200

$MnFe_2O_4$ stable 1400

$Mn^{III}Ti^{III}O_3$ 1404, decomp 680, reforms >1000, ins

$Mn^{V}TiO_4$ 1420, ins

$Mn_2^{II/III}TiO_4$ 1460, ins

$CaMn^{VI}O_4$ produced at 1300

$Ca_2Mn^{IV}O_4$ produced at 1300

$KMn^{VII}O_4$ (permanganate) 240, decomp, photolyzed to manganate, s

$K_2Mn^{VI}O_4$ (manganate) 190, decomp by water to $KMnO_4$, s

$MgMnO_4$ formed at 1000

$NaMnO_4$ 36, stable 120–180, s

Na_2MnO_4 dissoc <600, s

[a]*dissoc,* dissociates; *decomp,* decomposes; *s,* water soluble; *ins,* water insoluble.

$MnSiO_3$ (m.pt.1291°C, insoluble), and ortho-Mn_2SiO_4 (m.pt.1345°C, insoluble, dissociates on vaporization) all reflect a divalent Mn atom. Like numerous other transition elements, manganese forms a solid carbonyl, $Mn_2(CO)_{10}$ (m.pt.154°C), that thermally dissociates to atomic Mn at slightly higher temperatures (240°C–310°C).

Turning to manganese's mixed metal oxide salts most appropriate to combustion formation, numerous have been characterized to some degree. $MnAl_2O_4$ is a spinel structure, a ceramic that is thermally stable to at least 1200°C but has a listed melting point of 1850°C. $MnFe_2O_4$ similarly is thermally stable to about 1300°C. $MnTiO_3$ (m.pt.1404°C) and Mn_2TiO_4 (m.pt.1460°C) have been reported. $CaMnO_4$ can be made at 1300°C but remains poorly characterized. Of this category, the best known molecule is potassium permanganate $KMnO_4$ (m.pt.240°C) that dissociates on heating to the manganate K_2MnO_4 (m.pt.190°C). This is a valency change from seven to six. On heating further above 540°C this reduces its valency to form K_3MnO_4 that remains stable in dry air dissociating at 930°C–1000°C. These are the better known molecules, and sodium and potassium can each form 12 such oxocompounds reflecting Mn valences that run from 2 to 7 (Gontarz et al., 1995). It is unusual, but the potassium salts are mostly encountered simply because their synthesis is simpler. In this case sodium and potassium do not exactly reflect their usual chemical similarities. The high valency molecules are very strong oxidizers and so are readily reduced producing an oxide.

This listing of condensed phase molecules indicates the low thermal stability and volatility for manganese. In fact on vaporization it appears that the mixed metal oxides dissociate. The only one reported in a very extensive study of such molecules is gaseous $MnTiO_3$ for which its heat of formation was obtained (Lopatin and Shugurov, 2013).

Gas phase stabilities and transport

Release and transport from a fossil fuel is obviously different from laboratory studies where weakly bound molecules can be added directly to flames. In the case of manganese elevated temperatures appear necessary. It has no easily volatile molecules other than the dichloride that is prone to being hydrolyzed by water vapor particularly in the condensed phase above about 600°C. Some ablative chlorination of condensed phase oxides has been reported which could lead to some gaseous species (Fouga et al., 2009). The gaseous monoxide MnO can only be produced from vaporization at high temperatures. It is the weakest bound of the first eight transition elements that progressively decrease in strength from Sc to Ni. Utilizing the Knudsen cell method, the equilibrium of the gaseous Fe/FeO/Mn/MnO balance was analyzed. From a reliable value for the FeO bond strength, a value of D_0(Mn-O) = 370 ± 8 kJ mol^{-1} now can be obtained (Smoes and Drowart, 1984). Using a guided ion beam system, the energy threshold for a somewhat complex reaction of Mn^+/c-C_2H_4O leading to MnO indicated a bond strength of 354 ± 25 kJ mol^{-1} (Sunderlin and Armentrout, 1990). No other acceptable data are available but the Knudsen cell study's agreement between second and third law analyses supports its acceptance. As a result a value of D_0(Mn-O) = 370 ± 10 kJ mol^{-1} is currently recommended. The periodic trend would suggest no lower value. Few of its spectral states have been characterized but the MnO($A^6\Sigma^+$–$X^6\Sigma^+$) transition in the 536–558 nm region has been confirmed and observed in emission and fluorescence in flames (Padley and Sugden, 1959; Blackburn et al., 1978; Westblom et al., 1994), but now relabeled as the $C^6\Sigma^+$ state (Miliordos and Mavridis, 2010). The spectroscopic difficulties arise from its many unpaired electrons that produce states of such high spin multiplicity (Merer, 1989). As a result, adequate theoretical treatments of its electronic structure have been difficult until recently (Bakalbassis et al., 1996; Gutsev et al., 2000; Gong et al., 2009; Miliordos and Mavridis, 2010). One interesting aspect is that the ground state $X^6\Sigma^+$ correlates to Mn(^{6}S) and O(^{1}D) on a rigorous spin conservation symmetry consideration and not the first dissociation limit of O(^{3}P). The suggested bond strength can be reproduced theoretically if the molecule is taken to be highly ionic with a $Mn^+–O^-$ structure (Miliordos and Mavridis, 2010). The sparse experimental values mentioned above refer to the first dissociation energy limit. Whether other techniques that are more sensitive to time constraints produce differing results, as observed with the alkaline earth oxides, has yet to be seen. In the present case three low-lying states correlate to the first limit and more options exist for curve crossings. These are of high multiplicity lying below 9000 cm^{-1} and now labeled $b^4\Pi$, $a^8\Pi$, and $A^6\Pi$. They illustrate the potential energy curves that show spectral complexity. There is also a repulsive $1^4\Sigma^-$ that also crosses the $X^6\Sigma^+$ state and connects to the first dissociation limit. Consequently the reaction dynamics that control the dissociation of MnO remain unknown and whether they invoke the low lying states or curve crossing onto the repulsive state remain unresolved. Both the monoxide and dioxide were characterized by matrix isolation methods and fixed the dioxide's OMnO bent structure (Chertihin and Andrews, 1997). This now is confirmed theoretically as also the low-lying states of MnO_2(g) and its possible structures (Gutsev et al., 2000, 2000a; Uzunova et al., 2008). In a Knudsen cell-mass spectrometric study of MnO(s) with a mixture of O_2 and H_2O at about 1600°C–1700°C, vapors of MnO, MnO_2, MnOH, Mn$(OH)_2$, and MnO(OH) were characterized (Hildenbrand and Lau, 1994a). Their concentrations were small and of a several tenths or less ppmv magnitude. Nevertheless thermochemical data were obtained. This experimental dioxide bond strength now recommended is D_0(OMn-O) = 484 ± 14 kJ mol^{-1} based on the current value for MnO. It is clearly much stronger than the monoxide following a Cr, Mn, Fe trend (Table 7.19). Values for the hydroxides and oxyhydroxide derived in the same study are D_0(Mn-OH) = 317 ± 15, D_0(HOMn-OH) = 436 ± 15, D_0(OMn-OH) = 365 ± 15 kJ mol^{-1}, D_0(HOMn-O) = 401 ± 15 kJ mol^{-1}. Much earlier, by nebulizing an aqueous spray of $MnCl_2$ into fuel-rich hydrogen flames at a concentration of about 10 ppmv and temperatures spanning 1500°C–2250°C, the thermal emission bands of MnO and MnOH were monitored. Values of their estimated bond strengths were close approximations, 400 and 305 kJ mol^{-1} to those derived above (Padley and Sugden, 1959). This confirmed their presence in the burned gases but not in dominant amounts otherwise there would have been interferences and modified values. Also it similarly indicated a minimal role for the dihydroxide in spite of its stability. The infrared spectra also have been monitored for MnOH and $Mn(OH)_2$ in matrix isolation studies resolving their frequencies and structures (Zhou et al., 2001a; Wang and Andrews, 2006b). The dihydroxide has the strongest bonding, and was suggested as being the possible species involved in the reported longtime gradual erosive loss in moist air at 1200°C of Mn from stainless steel alloys, such as Nichrome and Inconel (Zaplatynsky, 1977; Hildenbrand and Lau, 1994a). Whether it can form in flame gases without being subject to kinetic constraints has to be seen. Turning to the gaseous chlorides, manganese has two that are well characterized. $MnCl_2$ the predominant species on vaporizing $MnCl_2$(s), and MnCl($X^7\Sigma^+$) formed at higher temperatures with its X, A, B states spectrally characterized (Huber and Herzberg, 1979). In a detailed torsion-effusion mass spectrometric method, accurate measures indicated D_0(Mn-Cl) = 334 ± 5 kJ mol^{-1} and D_0(ClMn-Cl) = 447 ± 14 kJ mol^{-1} (Hildenbrand, 1995). MnCl(g) has a multiplet ground state but no others below about 8000 cm^{-1}(100 kJ mol^{-1}) in energy. As a result, their absence will not contribute to the chemical thermodynamics at temperatures below 2000°C. The quoted values were not modified in a later updating of the thermochemical analysis (Hildenbrand, 2008). Electron diffraction and theory indicate that $MnCl_2$(g) is linear and symmetric (Hargittai et al., 1991). Theoretical descriptions of this bonding now can reproduce these values quite closely (Wang and Schwarz, 1998; Nielsen and Allendorf, 2005; Nhat et al., 2012; Bross et al., 2013). The second bond formed on changing from the diatomic to triatomic molecule is consistently 31%–37% stronger on moving from the oxides to chlorides to hydroxides and the chlorides mirror the hydroxides closely. In two flame studies, the effect of chlorine addition was noted on either the emission or absorption of the free elemental Mn. In fuel-rich C_2H_2/air flames, substantial additions of CCl_4 vapor were interpreted as indicating predominant $MnCl_2$ formation (Kantor, 1994; Kantor and Ernyei, 1994). In fuel-rich $H_2/O_2/N_2$ flames at 1200°C–2300°C with ≤10 ppmv Mn salt and ≤0.5% HCl, it was assumed that MnCl was dominant but the too high value of D_0 = 357 kJ mol^{-1} then obtained now suggests an ignored contribution from $MnCl_2$ (Bulewicz et al., 1961). Consequently, even with this scant data it would seem that there is probably a gaseous balance in flames between the oxide, chloride and hydroxide, dependent on water and chlorine levels. For example,

$$\mathrm{MnCl + OH = MnO + HCl} \qquad \Delta H_{298} = -40\ \mathrm{kJ\ mol^{-1}}$$

$$\mathrm{MnOH + OH = MnO + H_2O} \qquad \Delta H_{298} = -122\ \mathrm{kJ\ mol^{-1}}$$

$$\mathrm{MnOH + HCl = MnCl + H_2O} \qquad \Delta H_{298} = -88\ \mathrm{kJ\ mol^{-1}}$$

Gaseous sulfides of manganese have been characterized but are limited to being in oxygen-free environments. Two experimental mass spectrometric studies examined the vaporization process of MnS(s) at 1200°C–1500°C, and although this was congruent but predominantly dissociative, sufficient MnS(g) was detectable for a thermochemical analysis (Colin et al., 1962; Wiedemeier and Gilles, 1965). The more reliable values indicated D_0(Mn-S) = 297 ± 16 kJ mol^{-1}. This has now been updated slightly to a current recommended value of 282 ± 10 kJ mol^{-1} (Knacke et al., 1991; Kubaschewski et al., 1993). It is about 76% the value of MnO and consistent with first row transition element MS/MO ratio trends (Table 7.19) (Hauge and Margrave, 1972). Numerous theoretical descriptions of the nature of this bonding are available with bond strengths supporting a 256–300 kJ mol^{-1} range (Anderson et al., 1987; Bauschlicher and Maitre, 1995; Wu et al., 2007; Mayhall et al., 2009; Gusarov and Yorish, 2011). The gaseous MnP is less studied but theoretical estimates indicate an $X^5\Pi$ state with D_0(Mn-P) ≈ 148 ± 20 kJ mol^{-1}, too weak for any significance in combustion (Tong et al., 2003). Theoretically, the hydrides have been of significant interest due to their high spin multiplicity, MnH($X^7\Sigma^+$), MnH_2($X^6\Sigma_g^+$), and the pronounced difficulty in developing appropriate methods to meaningfully model their electronic structures. The nature of the electronic ground and low-lying states now is verified but the bond strengths remain to be adequately calculated. Even experimentally these have been difficult both with their spectroscopy and thermochemistry. They are not listed in any standard volume of thermochemical values.

However, using guided ion-beam mass spectrometry a reliable experimental measure of D_0(Mn-H) = 126 ± 18 kJ mol^{-1} was established (Sunderlin and Armentrout, 1990) consistent with earlier predictions of such a magnitude (Kant and Moon, 1981). Other than ZnH, it has the weakest bonding in the first row of transition elements. Theoretical values before 2000 have been reviewed (Harrison, 2000), and although one estimate predicted low values (95,102 kJ mol^{-1}) (Barone and Adamo, 1997), the tendency was then, as even today, to overestimate the bond strength, values ranging from 131 to 210 with little improvement over time (Daza et al., 2006; Koseki et al., 2006; Furche and Perdew, 2006; Goel and Masunov, 2008; Gusarov and Yorish, 2011; Nhat et al., 2012). The hydride structural bonding differs in certain aspects and attempts to explain the calculation difficulties have been addressed (Goel and Masunov, 2008). The low-lying states and potential energy curves have been calculated (Langhoff et al., 1989a; Koseki et al., 2006; Tomonari et al., 2009) but similarly place the lowest state, the $a^5\Sigma^+$, much higher than the observed 1725 cm^{-1} (21 kJ mol^{-1}) range (Stevens et al., 1983; Balfour, 1988). The dihydride has been studied very minimally due to the theoretical difficulties concerning accuracies. It is now accepted that the molecule is linear with no low-lying electronic states (Demuynck and Schaefer, 1980; Fujii and Iwata, 1996; Balabanov and Boggs, 2002; Wang and Andrews, 2003a). These various bond strengths for the potential gaseous molecules of Mn are summarized in Table 7.19 and display their trends. Consequently, manganese in the center of the first row transition elements has a chemistry that might be expected to display a nonvolatile nature except at higher temperatures, and play mainly an ash ingredient role. Also manganese levels although listed as a trace element in most fuels are certainly not negligible and can be among the most important trace elements at times in ash particles.

Laboratory combustion studies

Early analysis by flame atomic emission or absorption had indicated adequate sensitivity limits for monitoring manganese in samples using a variety of high temperature acetylene or hydrogen flames (De Galan and Samaey, 1970; Parsons et al., 1975). Of course, a gaseous chemistry for manganese can be readily induced in flame gases by injecting low levels of a weakly bound molecule such as a gaseous carbonyl or the aqueous spray of a salt such as the nitrate (Westblom et al., 1994). Such studies indicated that the atom could at least persist in the flame gases with a reasonable lifetime and any oxide formation would not minimize it. As noted with several transition elements such as Mn, Co, Fe, and Cr, even at low concentrations, in combustion they can have curious catalytic effects that still remain not fully understood. Although not evident in full-scale combustion plants, these observed effects have commercial applications that have been developed. They are described as addition effects and find use in such diverse areas as soot suppression and flame inhibition. Along with barium and iron (ferrocene), manganese generally in the form of MMT can be effective for soot suppression. Although these may function by different mechanisms, manganese is felt to attach to the soot particles in oxygen-rich flames and catalytically oxidize the soot (Howard and Kausch, 1980). The metal does not appear to affect the soot formation mechanism but effectively oxidizes or reduces soot particle sizes downstream. The other effect, flame inhibition would almost seem a contrary behavior but is explained as a reaction zone cyclical mechanism where the element is involved in chemical reactions that involve flame propagating radicals, catalytically recombining them. This has been well documented and confirmed in several early studies that suggested possible molecular compound dependence (Bulewicz and Padley, 1971; Vanpee and Shirodkar, 1979; Linteris et al., 2008). Since then, and with its practical use, extensive kinetic modeling has been invoked, including all the gaseous Mn, MnO, MnO_2, MnOH, MnOOH, $Mn(OH)_2$, and MnH species (Linteris et al., 2002, 2004). Such a model is a first generation effort and must remain speculative at present due to possible kinetic constraints. MMT was found to be an adequate substitute for lead in gasoline as an octane improver and still is in use. There is concern regarding its toxicity and it is being phased out in some fuels and areas. Particles emitted from such use have been analyzed showing that manganese phosphate is a major constituent, followed by the sulfate with some oxide (Zayed et al., 1999). Such an additive to a fuel modifies the combustion in that the metal is automatically freed and open to its full range of chemistry. Release of manganese from fossil or other fuels is a different challenge. Manganese is a transition element, automatically having a high ionization energy. Nevertheless its chemiionization mechanisms in flames have been examined (Goodings, 1992). Analysis of one flame indicated a distribution of the ions in the flame gases as Mn^+ (93%), $Mn^+ \cdot H_2O$ (4.3%), $MnOH^+$ (1.3%), and $MnOH^+ \cdot H_2O$ (1.2%). Ion levels remain low and are of little interest.

Combustion power plant data

As noted in Table 1.1, manganese can be reasonably large on the ppmw concentration with coals, even higher in biomass and also significant in wastes and sludge. It is quite negligible in pet-coke. Additionally, particularly in FBC and scrubber technologies it has to be remembered that there can be more Mn in the limestone and sand than in the fuel, possibly complicating some interpretations (Lind et al., 1994; Cordoba et al., 2012). From early measurements, it was confirmed that manganese had negligible vapor phase transport mechanisms from solid fuels and has predominantly a condensed phase history in combustion. Nevertheless, numerous full-scale and laboratory studies are available to give a reasonable picture of its behavior with coal and the cofiring with various fuels. An earlier thermochemical equilibrium analysis for manganese in coal combustion (Yan et al., 1999) predicted oxide formation to 1500°C for both oxidizing and reducing conditions. Gaseous molecules appeared only above this temperature with the possibility of $MnCl_2(s)$ formation at low temperatures under reducing conditions. A more recent publication for gasification of biomass and recycled fuel (Konttinen et al., 2013) suggests a possible importance of MnS(s), MnO(s) and a low-temperature manganese aluminosilicate formation with a small gaseous $MnCl_2(g)$ component. However, other than being qualitative indicators of possibilities, without confirmation from realistic systems, little confidence can be placed on such equilibrium thermochemical predictions. In previous publications it is reported with coal that the concentration of Mn in bottom ash and fly ash is approximately the same with no indication of any particular enrichment (Klein et al., 1975; Kauppinen and Pakkanen, 1990; Meij, 1994; Llorens et al., 2001; Huang et al., 2004; Duan et al., 2016). Moreover, numerous studies that size segregated the fly ash found no noticeable enrichment on submicron size particles (Coles et al., 1979; Smith et al., 1979; Kauppinen and Pakkanen, 1990; Querol et al., 1995; Martinez-Tarrazona and Spears, 1996; Wang et al., 2008; Lanzerstorfer, 2018). This is also the case in coal gasification where very little Mn was found in the small particles (Helble et al., 1996). However, a detailed recent study examined Mn behavior in six coal-fired plants. Their coals differed with a range of Mn content from about 20–160 ppmw and all with similar Cl levels (213–345 ppmw) and S in the range 0.5%–4.0% (Deng et al., 2014a). In this case, some partitioning was noted between the fly ash and bottom ash; the more Mn in the coal, the more remained in the fly ash (partitioning ratio 2.6). The lowest Mn coals indicated the opposite trend with a slightly higher concentration in the bottom ash (ratio 0.9). This obviously is a reflection of a change from coal to coal and above 100 ppmw Mn content due to conservation. It appears that there is transport of material between the two ash forms. This is a rather noteworthy observation. A recent evaluation of fly ash in the three stages of an ESP has taken the understanding a step further. A coal/biomass fuel was burned at 1800°C–1935°C and it was noted that the Mn concentration did in fact decrease slightly in the ash from the first to third stages. This differed totally from Cd, Cu, Pb, and Zn which are known to follow a vaporization/deposition mechanism. The main element differences in the particles were measured using energy dispersive X-ray analysis. This clearly showed a morphological change in ash shape with particle size between the stages and suggested that element transport for Mn can also be a result of a mechanical mechanism in the condensed phase that does not invoke vaporization (Swietlik et al., 2012). Especially at higher temperatures if molecules melt, the possibility of phase partitioning within a burning particle can occur. As larger particles fragment the resulting distribution of a specific element in smaller particles will depend on whether the element has moved toward the surface or the core. This would appear to be the most acceptable explanation for Mn in coal combustion. In some cases, small or negligible amounts of Mn were detected in micron-sized or smaller ash particles from biomass or waste combustion (Lind et al., 2007; Jiminez et al., 2008). In coal combustion or pyrolysis below 1150°C no volatilization of Mn occurred (Wang and Tomita, 2003; Guo et al., 2004). A steam gasification of biomass at 800°C and 0.7% Cl reported no manganese release (Cui et al., 2013). As a result, the suggestion of some volatilization of Mn in a coal Sasol–Lurgi gasifier operating at or below 1000°C needs confirmation (Bunt and Waanders, 2010). One important question is whether chlorine content can play a role. A study burning waste on a fire grate with 1% Cl in the fuel at 1030°C noted no volatilization of the 130–350 ppmw Mn (Lind et al., 2007). Also burning coal with sewage sludge treated with $FeCl_3$ and lime at 1100°C and containing 1.1% Cl showed negligible vapor loss of manganese (Folgueras et al., 2007). A further series of studies utilized a suspension-firing reactor with wood bark co-combustion with various wastes. Initially at 800°C, additions of HCl and SO_2 were studied equivalent to a bark fuel content of 4% and 2.5%, respectively. This did appear to show some release of Mn in the chlorine case (Miller et al., 2003). When repeated later at 800°C and 900°C, cofiring with agricultural, sewage, and plastic wastes all four fuels having different levels of Mn and Cl ($\leq$0.26%), the data showed no similar consistent pattern of behavior and may be questionable (Miller et al., 2006). However, it is seen that if incinerator waste fly ash is roasted at

1000°C with a chlorinating agent such as $CaCl_2$ or Cl_2 then the manganese in the ash will be fully vaporized as $MnCl_2(g)$ (Chan and Kirk, 1999). Apparently in such a case the manganese was probably present as oxides that do react with chlorine if given sufficient time. As a result, it would not be surprising in high chlorine cases to possibly vaporize a little manganese as $MnCl_2$ but on the whole this does not appear to be observed. Mn generally is predominantly retained in the condensed phase. A blend of bituminous coal and sludge did report that Mn appeared to increase in the fly ash slightly on the addition of NaCl (Zhang et al., 2019). A recent study has reported an effect on the fly ash composition of reducing the flue gas temperature from 250°C to 150°C through a heat exchanger. The fuel was biomass from metal contaminated soils burned in a small pilot-plant boiler. Boiler temperatures are not quoted but a twofold increase in ash concentration appeared to occur for manganese and many other elements as the temperature was so reduced through the heat exchanger. As no deposition was possible to increase the concentrations especially for this, and such elements as Ti or Zr that are also nonvolatile, it was apparent that particle segregation was occurring in the heat exchanger and the ash size distribution was changing, favoring the smaller sizes at the lower temperature enriched possibly by a mechanical fragmentation mechanism (Kovacs et al., 2018). Any atmospheric release of manganese in particulate form will simply depend on the quality of the particulate controls invoked by the power plant. One Canadian coal-fired plant with solely an ESP reported an emission rate of 0.04% of the fuel manganese (Goodarzi, 2006a) and a Danish plant with ESP and a limestone FGD scrubber reported a value of 0.1% (Nielsen et al., 2002). It is to be noted that a limestone scrubber is not passive and can contribute slightly to the final emission of stack particles. A Spanish study did measure more recently the retention effectiveness of a FGD system for the trace elements with a cofired coal/pet coke power plant. In spite of the high content of Mn in the limestone, the stray emissions of Mn were further reduced to low levels (Cordoba et al., 2012). Similarly a Chinese study of a large modern coal-fired combustor monitored the actual stack emission levels of 13 trace metals. All were effectively captured within the power plant with efficiencies lowering their concentrations by more than three orders of magnitude below their ppmw levels in the coal. Ba was removed most effectively (five orders of magnitude reduction). Mn also was very effectively captured, the amount in the coal being reduced by a factor of 1.7×10^4. This implies a 99.99% capture (Zhao et al., 2016). However, due to water solubility, the Mn captured in the FGD device indicates high levels of Mn in the desulfurization wastewater (Zhao et al., 2017a) and also in the resulting gypsum samples that were seen to pose a high risk for Mn (Hao et al., 2017; Zhao et al., 2018b). Emission control for manganese is one of particulate capture, removing doubt over any Mn release to the atmosphere. However, as with other trace elements, final ash disposal becomes the major concern and involves the nature of the chemicals in the ash. For Mn, several leaching exercises have been reported with various ashes. In one coal study it was noted that although water leaching was small (<0.1%) it was twice the level with bottom ash than with fly ash (Llorens et al., 2001). Also with coal ash, or coal/sludge/straw co-combustion, acidic leaching was pronounced probably due to the presence of dissolvable oxides (Dutta et al., 2009; George et al., 2010; Flues et al., 2013). At higher combustion temperatures (1800°C–1935°C), larger quantities

($\approx$50%) of the Mn were locked in the ash most probably as an aluminosilicate indicating the complexity of leaching and its dependence on the details of combustion (Swietlik et al., 2012). A current leaching exercise examined Chinese ash from four coal-fired plants (Zhao et al., 2018d). They noted significant manganese leaching to the point of indicating it was a high risk to the environment. An intriguing study of mixed metal (Mn, Cr, Fe) oxide nanoparticle formation uncovered another aspect of potential interest. The particles were collected as their separate oxides and also as their three oxide mixture using a hydrogen diffusion flame to which the respective metal carbonyls were added. It was noted both in their formation and also later in their aging that the metal oxides could influence each other's oxidation states and so possibly modify their toxicology or leachability (Nico et al., 2009). Also Mn in particular has an ability to absorb oxygen and then the oxide can act as an oxidant of other metallic states. These pure manganese oxide nanoparticles so formed displayed an average 2.4 oxidation state and were a mixed II/III phase and regarded as hazardous.

In summary, although manganese has a rich potential chemistry, it is one of nonvolatility in full-scale fossil fuel combustion. Pyrolysis or gasification operates at too low a temperature to witness any release from fuels. The slight enrichment that is seen in some ash sizes is probably the result of mechanical fragmentation transport and internal migration rather than vaporization/deposition. Oxide chemistry will be dominant within the ash. Chlorine appears to play no role. Mixed metal oxide formation might occur and significant leaching is reported from some coal ash, gypsum and waste water to the point of being hazardous for the environment. Invoking more of manganese's chemistry, Mn additives such as MMT have been developed for use in areas of soot suppression or flame inhibition, and as an engine fuel octane improver. Its chemical mechanisms remain uncertain.

7.18 Mercury

Mercury is an unusual heavy metal in the periodic table. It is located along with Zn and Cd in the last column of transition elements all having a $d^{10}s^2$ electronic configuration. These are known as the volatile metals with mercury also being unusual as a liquid at room temperature. The reason for this remains uncertain theoretically, especially as to lower and higher atomic numbers it has condensed phase Au and Pb as neighbors. Mercury is a potent neurotoxin that is alien to living species but until recent decades did find very extensive commercial uses. Currently, it is regarded as the neurotoxin of most concern for health, and remains suspect as being a major contributor to the current wide spectrum epidemic of neurological illnesses. As to what finally triggered increased public concern is difficult to establish but it obviously was spawned initially by the mercury pollution from an acetaldehyde manufacturer in the Japanese fishing village of Minamata in Japan. Neurological and birth defects started to appear there in the 1950s but were of minor global concern until books and reports emerged in the 1970s and 1980s publishing the extent of this extensive disaster. Even so, regulation proceeded slowly, but now mercury has been removed almost totally from most of the world market products and is no longer mined. Its economic value has dropped to being essentially zero. It only remains unregulated in the combustion and cement industries, in medicine and dentistry, in artisanal gold and silver mining, and still as a skin bleacher cosmetic in some countries. It now seems that the ban on cosmetic use is essentially complete (Michalek et al., 2019). However, even these are all under extensive review as to the best methods of their control. Nevertheless, indications remain that Hg elemental emissions are still significant globally and, for example, China's cement industry has increased 15-fold in the past 30 years (now 75 tons Hg total $year^{-1}$, with 76% emitted as Hg^{2+}) and continues to increase (Hua et al., 2016). The published literature concerning mercury is overwhelming, with close to 5000 papers published a year. It is presently the most researched element in the periodic table, driven by medical and environmental interests. Numerous current reviews are available and concern its global anthropogenic releases, distribution, and atmospheric and oceanic modeling. Continuing improved controls are appearing, however approximations remain in any assessment, and it is still not clear whether global mercury contamination will be able to return to preindustrial levels given time (Semeniuk and Dastoor, 2017). Recent global and historical overviews are available (Streets et al., 2017, 2018, 2019; Obrist et al., 2018). Overviews also summarize the Chinese emissions over several time periods apportioning the contributions from combustion, metal smelting and cement industries (Wu et al., 2016c; Liu et al., 2016, 2018b; Huang et al., 2017). The sintering process in the iron and steel industry of China has been reviewed with regards to Hg emissions. In China it appears to be a significant contributor to anthropogenic releases, the source mainly being from the iron ore (Xu et al., 2017; Wu et al., 2017). However, a study in Japan indicated that, with adequate controls, emissions can become minimal (Fukuda et al., 2011). Emission budgets for the United States are available (Outridge et al., 2018), with better assessments now of the air/surface exchange flux (Zhu et al., 2016a; Wang et al., 2016j; Zhang et al., 2016e). Monitoring over time has shown positive signs of decreasing emissions in the Northeastern area of the United States but whether this results from coal plant closures or better FGD scrubbers and controls remains impossible to correlate (Zhou et al., 2017). Additionally, decreases in the United States and Europe appear to be balanced to some degree by the economic growth of China and Asia (Pacyna et al., 2006). Nevertheless, one positive sign of environmental improvement is the declining mercury concentrations in Bluefin Tuna in the North Atlantic that have been reported for the 2004–12 period (Lee et al., 2016). Tuna is the largest predatory fish that supplies the American diet with about 40% of its mercury intake.

Mercury is a rare element ranking about 70th in quantity in the Earth's crust, about similar in occurrence to Ag and Se, and found in comparable level concentrations as the sulfide ore HgS, cinnabar. This has extremely low water solubility and hence low toxicity, which is fortunate for peoples living next to lakes or the ocean adjacent to large geological outcrops. It is slightly higher in concentration in the oceans, about 57th and similar to those of Pb, Zr, or Tl. Historically, mercury gained its fame from its ability to amalgamate (alloy) with other metals, particularly Au and Ag providing a method of processing these metals from early days. The district of Almaden in Spain is the largest geochemical anomaly of mercury on Earth (Higueras et al., 2006, 2013). Its mines have been active from Roman times, uninterrupted until final recent closure. It facilitated what is now referred to as the colonial period of about 250 years (1570–1820) of Spanish-American silver mining (Camargo, 2002). Added to this was the American gold rush period (1850–1900), which appears to have been followed in the last century by an even larger industrial usage of mercury (Hylander and Meili, 2005). Concerns with the polluting of chemicals center on the levels of their natural and anthropogenic contributions to the environment. In order to understand mercury's distribution, the magnitude of all its source contributions are necessary if any model can be constructed to facilitate predictions of the extent and concerns with its current pollution levels. The natural, preindustrial level is difficult to assess. A very large number of studies are available, obtaining estimates either by summing the major assumed contributors. This also includes utilizing methods such as drilling and analyzing core samples taken from ice, peat bogs, and lake sediments all over the world that can illustrate trends with time. Although now much improved, it has been difficult to accurately monitor mercury and its speciation at the low levels encountered in the atmosphere or in combustion flue gases (Gonzalvez et al., 2009; Suvarapu et al., 2013; Jaffe et al., 2014; Huang and Gustin, 2015). As a result older data (more than 40 years) have to be examined carefully and more recent studies generally are preferentially favored. Even core drilling data can be difficult to accurately analyze and still can be somewhat of a quandary. It is agreed that natural emissions are dominated by volcanoes and geothermal venting, but now with significant contributions from soil and also from ocean and plant evasion (a term now in use for vaporizations that are somewhat unexpected and evading consideration). Suggestions examining previous work have indicated values for volcanic emissions up to about 1000 tons $year^{-1}$ (Pyle and Mather, 2003). However, a Canadian study implied 26,000 tons $year^{-1}$ creating a dilemma (Richardson, 2001). Such high values also were quoted by this for Cd, Cu, Pb, Ni, and Zn and are inconsistent with historical estimates of preindustrial levels that do register the evidence of earlier volcanic eruptions and compare these to current accepted anthropogenic data. With such large natural estimates, anthropogenic contributions would not even be apparent. As a result, estimates of the total mercury natural emission have been suggesting values of about 5000–5200 tons $year^{-1}$ and are currently accepted with oceans contributing about 36% (Pirrone et al., 2010; Pacyna et al., 2016), up from an early estimate of 2500 (Nriagu, 1989). The Canadian report indicated about 60,000 tons $year^{-1}$, a very significant disagreement that needs more detailed clarification. To scale all these estimates to global numbers is not easy and there remains doubt as to their accuracy even to an order of magnitude (Gustin, 2003). Anthropogenic mercury emissions now are in a total state of change with its removal from all commercial uses such as thermometers, barometers, switches, the chlor-alkali industry, pharmaceuticals, pesticides, fungicides, batteries, paints and inks and cosmetics. Additionally, its remaining presence in fluorescent lights (mg/tube) will undoubtedly soon be removed from the market place when LED usage becomes more globally general. Moreover regulations now are being imposed on the metal smelting and cement industries but globally they remain in various stages of implementation and improvement control (Wang et al., 2016d). Mercury is no longer a marketed commodity and has been reduced to having a minimal value. It is questionable if it can still be purchased without permits. As a result, anthropogenic emissions continue to be drastically reduced. One estimate of all-time anthropogenic releases from human activities implied that 39% were emitted before 1850 and 61% thereafter (Streets et al., 2011, 2017; Horowitz et al., 2014), now with a 3.2-fold increased atmospheric burden relative to 1850. At present, the remaining anthropogenic emissions arise from combustion, predominantly from coal, the continued use of thimerosal in vaccines in many countries of the world, and similarly a declining use in dentistry. Care has to be taken whether the evasive recycling of historical pollution from past years on the land and in the oceans that is very slowly being recycled before sedimentation is included in natural or anthropogenic listings. Studies have suggested this is a major 60%–75% contribution if added to current anthropogenic sources (Amos et al., 2013; Song et al., 2015). An interesting aspect that might explain this reemission of Hg to the atmosphere appears in recent publications that needs more careful confirmation. It has been reported that when mercury compounds are mixed into gypsum, alumina, TiO_2 or activated charcoal they appear to evaporate dissociatively as the free atom (Wu et al., 2011b; Zhu et al., 2016). This also appears to be the case for release from $HgCl_2$ in sand and soils and the reduction mechanism remains unknown (Xin et al., 2007; Carpi et al., 2009; Mann et al., 2016; Zhao et al., 2017c). Although this might be expected for many of the weakly bound mercury compounds, this is not the normal case with dihalides that are known to sublime unchanged and have measured vapor pressures that differ from that of dissociation.

In Minamata Bay, even 25 years after polluted sediments were dredged, concentrations of surface waters are little different, so a final demise for released mercury may have a very long life (Akito et al., 2014). Nevertheless, global atmospheric levels of mercury are showing significant declines that are expected to continue (Pacyna et al., 2009; Slemr et al., 2011; Brown et al., 2015; Weigelt et al., 2015; Castro and Sherwell, 2015; Zhang et al., 2016d). If current levels reflect only nonregulated sources, this anthropogenic channel may have fallen to about 2200 tons $year^{-1}$, now solely from combustion (two-thirds) and cement processes (Pacyna et al., 2006, 2010, 2016a; Pirrone et al., 2010). This would now be below natural levels and be improved even more by final regulation of these remaining emission sources. As indicated, it is difficult to accurately establish historical levels of environmental mercury and the reliability of core sampling has been questioned (Yang, 2015). Nevertheless, there is a large quantity of data that show consistent trends. It appears that current mercury levels in the environment may be about three- to fivefold higher than in precolonial times (Roos-Barraclough et al., 2002; Fitzgerald et al., 2005; Biester et al., 2007; Zhang et al., 2014a; Lamborg et al., 2002, 2014; De Simone et al., 2016, 2017). Moreover, levels before 1540 were very small, and in colonial days (1600–1850) were also much below present values (Uglietti et al., 2015). Ice core data suggested that overall, 8% of the anthropogenic Hg on record was released during 1600–1850, 14% in the American gold rush (1850–1900), and 78% in the 20th century (Beal et al., 2015; Zhang and Kang, 2015). These studies also noted an upturn after 1990 that reflected the increased rate of growth in Asia (Tian et al., 2015). A recent Northwestern tree ring analysis of atmospheric mercury over the past 400 years reproduces the trends. Levels were stable to 1750, increased steadily to 1950, paused from 1951 to 1975 but then increased to now record high levels (Clackett et al., 2018). Using archival data a model has reproduced these temporal trends showing the sensitivity to various scenarios and permitting expected consequences for future reductions (Amos et al., 2015). Hopefully with reduced sources below natural levels, this will begin to return mercury toward its preindustrial low level balance. Additionally, recent studies examining the isotopic composition of mercury has the potential for separating natural and anthropogenic sources (Yu et al., 2016). One recent study analyzed the soil levels in China of eight heavy metals, including mercury, concluding that Cd and Hg levels were increased above normal background (Huang et al., 2019). Atmospheric deposition to soils is the main mechanism in China but this is not much larger than from fertilizers, manure, or sludge and is of little concern (Peng et al., 2019). It is clear that there is a growing global consensus that a policy of control is necessary and the consequences of further delaying mitigation efforts have been studied (Angot et al., 2018; Chen et al., 2018c).

Health concerns with mercury

The health concern over mercury arose when it was realized that mercury displayed an even more toxic organic side, was bioaccumulated in carnivorous fish by many orders of magnitude and was becoming present to the human diet in larger quantities. Realized to be both a poisonous toxic to the body, and also a neurotoxin for the brain, it became and still is regarded as probably the major suspect in contributing to the global epidemic of neurological illnesses now clearly evident in society (Schofield, 2016, 2017a; Cariccio et al., 2019). Mercury is alien to the body, serves no use, and so adds to the toxin body-burden. It has always been in the environment through natural processes, but the anthropogenic increases from a fish diet and vaccines together, and its very slow attrition has spawned present concerns. As a result, a large number of reviews have analyzed the environmental and toxicology of mercury in its different chemical states and have been published on the Web by the World Health Organization (World Health Organization, 1989, 1991, 1996 Trace Elements), and the Agency for Toxic Substances and Disease Registry (Tox Profile Hg, 1999, 2013). Abbreviated reviews also has been published in the literature (Syversen and Kaur, 2012; Beckers and Rinklebe, 2017). The conclusions are that inorganic mercury can be regarded more as a poison, has a low absorption by the intestines, has difficulty crossing the blood–brain barrier, and if in not too large an excess can be processed from the body by the kidneys and liver or otherwise be chelated and medically treated. However, whether long-term damage from even small doses or quantities that are retained for lengths of time such that traces may cross into the brain are now being questioned (Al-Saleh et al., 2016; Dorea, 2018). One recent study with rats clearly showed that low doses of inorganic mercury, if a chronic exposure, did cause neural problems (Teixeira et al., 2018). The uptake of methylmercury from a pregnant rat into the fetal brain has been extensively studied (Sakamoto et al., 2018a). Organomercurials are totally different chemicals. Methyl mercury is biochemically formed and is the most important mercury form in nature with an ability to bioaccumulate in carnivorous fish; the larger and older the fish the higher the accumulated mercury concentration (Fitzgerald et al., 2007; Hintelmann, 2010). Currently it is also being found in rice and other water raised plants but remains at lower levels and is of less concern (Huang et al., 2018; Liao et al., 2019). In fact, recent research has suggested that some rice may have a self-detoxification mechanism, demethylating any methylmercury with no bioaccumulation (Xu et al., 2016). Both methylmercury and the medically used ethyl analogue (as thimerosal in vaccines) are potent neurotoxins equally dangerous to the body that can readily cross the blood–brain barrier (Aschner et al., 2010; Hirner and Rettenmeier,

2010). Extensive animal experiments have helped to resolve their interesting and closely similar chemistries, measured their brain half-lives, and illustrate the complexity and resilience of the protective mechanisms in the body. It appears that once in the body and also the brain, there are innumerable Se proteins such as selenocysteine and selenomethionine that together with the glutathione enzyme can attach the Hg and either sequester or transport it from the brain or remove it from the body (Yoneda and Suzuki, 1997; Sakamoto et al., 2013; Zhang et al., 2014; Bjorklund et al., 2017). Nevertheless, one complication and somewhat unique behavior is that the organomercury can dealkylate while in the brain, become the inorganic ion that is then more difficult to remove, and has a much longer brain half-life (Burbacher et al., 2005; Dorea et al., 2013). This lifetime in the brain will control the level of permanent damage that may not be repairable. A recent study examined the metabolism and elimination of methylmercury on 37 adults after consuming three fish meals (Caito et al., 2018). They found an average half-life of 44 days (28–62 range) displaying a significant human variation probably controlled by genetic susceptibility. It is now considered that a disruption of the selenium protein supply in the brain is a possible basis for mercury's neurotoxicity (Bjorklund et al., 2017; Schofield, 2017a; Ralston and Raymond, 2018). A recent study of whales provides evidence of Hg-Se cluster formation and nanoparticles in the whale's liver and brain. It has also been suggested that Se depletion in the effort to maintain sufficient Se-cysteine detoxicant products may become a hazard (Gajdosechova et al., 2016). This finding now has raised a nutritional complication. It appears that on an atomic basis, Se can neutralize the toxicity of Hg if in excess. Moreover, it is being noted that fish can have a Se content that is variable. A study on two shark species in the Baja Mexican Ocean found that one had Se>Hg while the other Hg>Se (Rafael et al., 2019). One study found evidence that Se induced demethylation of mercury in marine fish themselves (Wang and Wang, 2017). In other words, when making a dietary risk assessment concerning fish, it is now necessary to know both the fish Hg and Se content. The possible importance of selenium in connection with mercury and the other neurotoxins is covered more extensively in the section concerning selenium. This has become of significant importance for pregnancy where fish is regarded as very beneficial to the diet but also possibly toxic to a fetus. Guidelines have been suggested but do not assess the Se role and need reconsideration (Taylor et al., 2018). This results from organic mercury readily passing through the placenta and entering the fetal blood. The fact that many healthy babies are born illustrates the efficient protection and possible repair mechanisms within its newly forming brain. Two possibly important studies have measured the ratio of Se to Hg in the fetal birth cord blood. One indicated an average ratio of 21 from 54 Japanese births and the other a 5-fold to 626-fold ratio from 91 Polish births (Sakamoto et al., 2018; Kosik-Bogacka et al., 2018). This could be the answer as to how the fetus survives the constant onslaught of neurotoxins through pregnancy. Nevertheless, the overloading of the body-burden with organic mercury is becoming a recognized risk factor that should be avoided. Also it is noted globally that all women appear to have elevated mercury levels normally (Friedrich, 2017). The recent testing of newcomer women to Toronto showed slightly elevated blood levels of $HgCH_3$, with some above recommended Canadian levels for women of reproductive age (Wiseman et al., 2019). Additionally, it has to be remembered that mercury is only one of five hazardous neurotoxins found in the body (Schofield, 2017a). Moreover in a recent analysis of the 2007–12 US National Survey Data for Hg, Pb, Cd, and As blood content, half the participants had at least three of these four present at reasonable levels (Shim et al., 2017).

Also reports now are indicating possible synergistic effects between the various toxins. One Chinese study has suggested that autism may have a multielement cause (Wu et al., 2018a). Cases of mercury poisoning in adults are now more common but long-term consequences remain unknown (Hightower and Moore, 2003; Karimi et al., 2014). Dietary risk indicators are available on the Web that indicate the level of mercury intake from consuming certain fish type quantities. However, Se fish content is not yet available. Risks increase with the larger predatory fish. Concerns center predominantly on pregnancies when benefit/risk assessments are critical for diet, and care has to be taken with vaccinations that still can contain neurotoxic adjuvants (Schofield, 2016, 2017a). The medical profession has to realize that although vaccines may be individually safe, they are no longer the only player on the field. Neurotoxins can be significant and ingested from many other sources. Although MRLs for mercury have been established they are only about 10-fold below the onset level of toxicity. After decades long studies of the people that live on high seafood diets in either the Seychelles or the Faroe Islands, recommended biomarkers have been established for doctors. Blood is a favored measure and although set at $\leq 5.8\ \mu g\ L^{-1}$ in the United States, various test centers have lowered this to 2 or even 0.8 in some countries. Hair is an alternative with a recommendation of ≈ 1 ppmw. These are useful benchmarks and now possibly 50 or more large surveys have been published of people's blood levels around the world (CDC 4th Rpt, 2009, 2018; Cesbron et al., 2013; Taylor et al., 2014; Schofield, 2016; Nisse et al., 2017; Oggiano et al., 2018; Henriquez-Hernandez et al., 2018; Zeng et al., 2019). Many of these indicate distributions above the recommended values that stretch to almost an order of magnitude larger, always with a small percentage of outliers that certainly are beginning to be at risk. Germany reports falling mercury urine levels over the last 20 years (Goen et al., 2018). Taylor et al. (2014) was among the first to stress the need for concern with regards to children and pregnancy for which no recommended levels are established at all. Measures of mother's milk do indicate filtering of mercury by the mammary glands to lower values (Al-Saleh et al., 2015; Schofield, 2016; Vahidinia et al., 2019). As a result efforts are definitely necessary to continue decreasing anthropogenic releases or uses of mercury. Risk assessments and health benefit cost analyses now are appearing, that from China indicating a cost running into billions of dollars (Nedellec and Rabi, 2016; Zhang et al., 2017c; Bellinger et al., 2019). Emissions from fossil fuel combustion continue as one major source. Public concerns appear to be ever increasing. There is a move toward greater blood testing as a preventive approach (Budnik and Casteleyn, 2019).

Mercury's basic chemistry

An interesting aspect of mercury is its very limited repertoire of chemicals that mainly favor divalency (Table 7.27).

Also in the condensed phase these are not particularly thermally stable. Its ore, cinnabar HgS, is one of the more stable, sublimes and dissociates in air at about 450°C to Hg and SO_2. In the absence of air, like Zn and Cd, it sublimes dissociatively to Hg and sulfur before melting and has a measurable vapor pressure above 450°C (Mitchell and Munir, 1970). Its oxide has fame for the discovery of oxygen by Priestley, who observed this on focusing the sun on a sample of HgO(s). This dissociates to Hg and O_2 above 250°C and its TGA curve has been measured (Wendlandt, 1974).

TABLE 7.27 Properties of mercury's condensed phase molecules relevant to combustion, °C

m.pt. b.pt. vapor pressure Hg −39 357 370 ppmv/100°C, ins Divalent	m.pt. Monovalent
HgO 370 decomp, ins	Hg_2O unstable, 100 decomp, ins
HgS 825, dissoc >589, ins	–
$Hg(OH)_2$ none	–
$HgCO_3$ none	Hg_2CO_3 130, decomp, ins
$HgSO_4$ 425 decomp, s	Hg_2SO_4 unstable, 300 decomp, ss
$HgSO_4 \cdot 2HgO$ 410 decomp, ins	–
$Hg(NO_3)_2$ 79, decomp, s	$Hg_2(NO_3)_2$ unstable, ss
$HgCl_2$ 276 304, 1% atm/180°C, s	Hg_2Cl_2 525, sublimes 383 to $HgCl_2$, ins
$HgBr_2$ 241 318, s	Hg_2Br_2 407, ins
HgI_2 259 354, ins	Hg_2I_2 290
HgH_2 none	–
$Hg_3(PO_4)_2$ 680 decomp, ins	–
$HgSeO_3$ 460	–

mercury salts solubility (Clever et al., 1985); Hg_2Cl_2 solubility (Marcus, 1980); *s*. water soluble; *ins*, insoluble; *ss*, slightly soluble; *decomp*, decomposes.

A monovalent oxide, Hg_2O, is occasionally discussed in the literature but is stable only at low temperatures and is of little interest and not generally encountered. The sulfate has importance being thermally stable to about 350°C when it begins to lose sulfur forming the basic sulfate, $HgSO_4 \cdot 2\ HgO$ and continues to dissociate surviving to about 410°C. Hg_2SO_4 exists but at above about 300°C this disproportionates to the divalent sulfate (Tariq and Hill, 1981).

It has no known dihydroxide, dihydride or carbonate. The nitrate is known but is very hygroscopic, water soluble and dissociates below about 80°C. The dihalides are major chemicals. The difluoride is of little interest readily hydrolyzing to Hg and HF but the other three dihalides, $HgCl_2$, $HgBr_2$, and HgI_2 are stable and show similarities. They are almost equally volatile with melting points of 276°C, 241°C, 259°C and boiling points only slightly higher 304°C, 318°C, 354°C, respectively. They sublime molecularly with no dissociation evident (Nagayama and Takada, 1989). Vapor pressures of $HgCl_2$ have been determined (Bernard et al., 1997). The vapor pressures of Hg and $HgCl_2$ are about the same at 180°C and their two heats of sublimation differ by only 22 kJ mol^{-1} at ambient temperatures. The dihalides water solubility at 25°C does differ and by orders of magnitude, 7.31, 0.61, and 0.006 g/100 g H_2O, the dichloride being by far the most soluble. Their corresponding monohalides exist with melting points (F, Cl, Br, I) of 570°C, 525°C, 407°C, 290°C but are insoluble in water and dissociate on vaporization to produce the divalent halide and free metal element (Awitor et al., 2000). The vaporization behavior has been displayed as TGA mass loss curves for some of these di- and monohalides (Wendlandt, 1974; Nagayama and Takada, 1989). Mixed metal oxide compounds are plausible for mercury but have not been explored or characterized to any extent other than being theoretically screened as potential sorbents for mercury (Couling et al., 2012). Molecules such as $Hg_3(PO_4)_2$ (stable to 680°C), Hg_2CrO_4 (stable <260°C), $HgSeO_3$ (m.pt.460°C), or reduced state molecules such as HgSe (dissociative vaporization, vapor pressure 145 ppmv at 327°C) (Imperatori and Piacente, 1984), and others are mentioned in the literature but none are particularly thermally stable.

Gas phase stabilities and transport

The gas phase for mercury is remarkable in it being devoid of molecules. Most possibilities reflect the almost nonexistent bonding in this heaviest element of the Zn, Cd, Hg column of the periodic table. Values discussed below are summarized for mercury in Table 7.15, and Table 7.16 and emphasizes the observed trends for these three elements. Although early mass spectrometric studies implied a relatively strong bonding for gaseous HgO and HgS, that remains utilized in various thermochemical databases (Knacke et al., 1991; Barin and Platzki, 1995), these are incorrect and should no longer be considered. It can be credited to theoretical chemistry that several detailed high quality theoretical studies noted that these molecules have overlapping nearly isoenergetic ground state $^3\Pi$ and $^1\Sigma^+$ electronic states (Shepler and Peterson, 2003; Chambaud et al., 2008). The former of these is a shallow triplet ground state for HgO.

$$HgO(X^1\Sigma^+) = Hg\left(^1S_g\right) + O\left(^1D_g\right) \quad \text{Spin-allowed}$$

$$HgO(a^3\Pi) = Hg\left(^1S_g\right) + O\left(^3P_g\right) \quad \text{Spin-allowed}$$

Conserving spin, this triplet state correlates directly to Hg(^{1}S) and O(^{3}P) dissociation products with a minimal binding energy. The detailed calculations that have followed now all agree to a very low value of D_0(Hg-O) = 18 ± 4 kJ mol^{-1} (Shepler et al., 2005; Peterson et al., 2007; Cremer et al., 2008). The situation for isoelectronic HgS changes slightly with $X^1\Sigma^+$ becoming the ground state, which dissociates to Hg(^{1}S) and O(^{1}D). However, curve crossing with the $a^3\Pi_2$ state predissociates all levels above $X^1\Sigma^+(v = 11)$ or above 3797 cm^{-1} (45 kJ mol^{-1}) to the ground state products (Cressiot et al., 2007). Taking this into account a value of D_0(Hg-S) = 24 ± 4 kJ mol^{-1} now is accepted (Filatov and Cremer, 2004; Cressiot et al., 2007; Cremer et al., 2008). It is an usual and rare case that the sulfide is slightly stronger than the oxide but this is not the normal case of comparing spectroscopically similar ground states but instead mixed (a/X) states. This mirrors also into the oxides and sulfides of Zn and Cd. The disagreement with earlier experimentation has been explained for mercury possibly by the tendency of gaseous HgO and HgS to form ring type oligomers or chain polymers (Tossell, 2006). This disagreement for mercury between experiment and theory is not apparent in any of the other theoretical calculations involving stabilities of mercury diatomic molecules. In fact there generally is close agreement. The monohalides also are weakly bound. A photoionization study of $HgCl_2$ managed to obtain an approximate estimate of D_0(Hg-Cl) = 88.3 ± 9.2 kJ mol^{-1} (Linn et al., 1983). Other than this, the most reliable estimates are from vibrational extrapolations of the $X^2\Sigma^+$ state that suggest a value close to 100 kJ mol^{-1} (Wilcomb and Bernstein, 1976; Tellinghuisen et al., 1982). A value of D_0(Hg-Cl) = 100 ± 3 kJ mol^{-1} might be recommended, compatible with that suggested by Huber and Herzberg (1979) and Barin and Platzki (1995). Theoretical modeling is consistent with this magnitude (Khalizov et al., 2003; Shepler et al., 2005; Cremer et al., 2008; Kim et al., 2010). The situation is similar for HgBr where an accurate extrapolation of vibrational levels indicated D_0(Hg-Br) = 65 ± 2 kJ mol^{-1} that is recommended (Tellinghuisen and Ashmore, 1983), supported by theory (Shepler et al., 2005). The thermochemical values for the dihalides are more firmly established as these can be derived from the more reliably measured heats of formation of the condensed phase coupled to their heats of vaporization which are molecular in nature. As a result, the values of $D_{0\text{Atomiz}}$($HgCl_2$ = 446 ± 6) and ($HgBr_2$ = 368 ± 8 kJ mol^{-1}) are accurately known (Barin and Platzki, 1995). Consequently, coupled to the monohalide values these imply values of D_0(ClHg-Cl) = 346 ± 7 and D_0(BrHg-Br) = 303 ± 8 kJ mol^{-1}. They display two very disparate bond strengths within these molecules which have significant implications for mercury's combustion chemistry. The gaseous dichloride in particular is seen to be thermally stable in the gas phase certainly to about 450°C. As a result, the suggestion that $HgCl_2$(g) is thermally unstable decomposing back to Hg(g) starting at 80°C peaking at 200°C is a misinterpretation of its actual vaporization and relates to its behavior in soils and other materials (Ernest et al., 2014; Schofield, 2016a). Thermogravimetric measurements for $HgCl_2$(s) indicate its rapid sublimation (Wendlandt, 1974), and an attached atomic absorption analyzer confirmed the molecular nature of the vaporization (Nagayama and Takada, 1989). Heterogeneous conversions of Hg(g) to $HgCl_2$(g) at temperatures of 250°C–370°C on passage through base-steel inserts also validates this stability (Schofield, 2012, 2017), as are some conversions across the final stages of the SCR at similar temperatures (Cao et al., 2007). However, to confuse things it is reported that passage of $HgCl_2$(g) through various solid absorbent powders such as SiO_2, glass wools and activated charcoal above 200°C seem to have the ability to induce its dissociation (Wu et al., 2011b). Also recent analyses of the fate of mercury in FGD gypsum samples does appear to indicate that mercury has difficulty vaporizing in molecular form from solid matrices and appears as the elemental vapor (Zhu et al., 2016, 2017; Bian et al., 2018). The molecule HgOH, which might be expected to reflect the strength of HgCl has been minimally examined. It has been invoked in atmospheric and liquid phase chemistries. One theoretical study obtained values ranging from 29 to 66 kJ mol^{-1} for the gaseous D(Hg-OH), and a value of 316–374 for both dihydroxide bonds D(HO-Hg-OH) or D(HOHg-OH) ≈ 300 kJ mol^{-1} mirroring the dihalides with dissimilar first and second Hg–OH bond strengths (Tossell, 2003). A matrix isolation study did characterize the dihydroxide, indicating a linear O–Hg–O basic structure, obtaining its vibrational frequencies and indicating a very stable molecule (Wang and Andrews, 2005d). There appears to be no stable condensed phase dihydride at ambient temperatures (Wang and Andrews, 2005f). However, there has always been interest in the gaseous diatomic HgH due to its transition metal nature and richness of electronic states (Huber and Herzberg, 1979). The ground state $X^2\Sigma^+$ is shallow and has no low-lying electronic states. Spectral transitions have been noted and the potential energy curves and spectral constants calculated theoretically (Alekseyev et al., 1996a; Mosyagin et al., 2002). From accurate locations of the ground state vibrational levels, their extrapolation established an accurate value of D_0(Hg-H) = 36.1 ± 0.08 kJ mol^{-1} that is still recommended (Stwalley, 1975). The infrared spectrum was obtained for the gaseous linear HgH_2 molecule (Shayesteh et al., 2005b). Theoretical studies of its $X^1\Sigma_g^+$ potential energy surface estimate a thermodynamically more strongly bound structure with $D_0(\text{HHg-H}) \approx 309\ \text{kJ mol}^{-1}$, another molecule of mercury with dissimilar strengths of its two bonds (Li et al., 2005; Botschwina et al., 2007). In summary it is noted that the diatomic gaseous mercury compounds are weak and only $HgCl_2$ and $HgBr_2$ have the stability to be of interest in combustion systems. Although examined, it appears that Zn, Cd, and Hg show no indication of any gaseous ion chemistry either (Goodings, 1992), or have any chemistry relating to soot or flame inhibition.

Chemical kinetic constraints

Early chemical thermodynamic calculations for mercury predicted that in Hg/air systems, HgO(s) would be dominant below about 250°C. With the addition of sulfur, $HgSO_4$(s) then becomes dominant below 375°C. If chlorine is present, $HgCl_2$ will be dominant at all temperatures below 450°C, being in condensed form below about 100°C, but due to its high volatility then vaporizing. Above about 450°C, water vapor can hydrolyze the gaseous dichloride back to elemental mercury (Frandsen et al., 1994; Yan et al., 2001). This behavior at equilibrium relates to oxygen-rich equivalence ratios and becomes much more limited for reducing conditions when only elemental Hg will exist possibly with some HgS(s). Nevertheless, the equilibrium expectations are invalid. First, no stable gaseous HgO can exist even at reasonably low temperatures due to its very weak bonding. Second, the formation of $HgCl_2$ is kinetically constrained from forming under combustion conditions in the gas phase. In reality, this was soon realized for fuels such as coal containing both sulfur and chlorine. Most of the mercury was never equilibrated and exhibited a very unpredictable distribution between gaseous Hg and $HgCl_2$ that was not understood. Initially, full-scale data was based mainly on one measurement, the quantity and speciation of the flue gas mercury when it reached far downstream. Also it was noted that this distribution was coal-plant dependent, varying one from another, and that constraints prevented the predicted conversion to gaseous $HgCl_2$. From the beginning, accurate measurements were very difficult in power plant flue gases at the parts per billion by volume encountered, but nevertheless it was clear that all the mercury in the fuel was readily released in elemental form. Hardly any or none was found in bottom ash, but some could be collected on the fly ash. Generally, only low levels were "oxidized" in the gas phase to a gaseous molecular form. From the start, this was assumed correctly to be the only known gaseous molecule that was appropriate, namely $HgCl_2$. This is water soluble and so could be absorbed in a FGD system but its range of variation implied that most of the mercury remained in elemental form and was being emitted to the environment (Laudel et al., 2000). The plant-to-plant variations were unexpected and unexplained, but a rough correlation was illustrated showing the degree of conversion appeared to depend on coal-chlorine content (Senior and Johnson, 2005). If coal Cl was below 30–100 ppmw content, a negligible conversion was expected, and to obtain high conversions more than a two orders of magnitude increase in Cl was necessary. Primarily, because the mercury began and ended in the gas phase, a homogeneous gas phase kinetic model understandably was implied and developed. This model immediately illustrated temperature and time constraints for producing gaseous $HgCl_2$ from

Hg. It had to depend primarily on a slow three-body molecular reaction of Hg with atomic Cl at low flue gas temperatures to permit time for the required weakly bound HgCl(g) intermediate to form. Such models necessarily failed at low chlorine levels, where occasionally measureable conversions were reported in full-scale plants. The model was solely a fit to a simple decay curve and needed unrealistic rate constants and atomic Cl levels. Also the need for reactive Cl atoms so far downstream and for homogeneous chemistry to occur in an otherwise high heterogeneity of coal particle combustion soon provoked the models to add a role for unknown heterogeneous interactions on ash particles but also continued to invoke the Hg with atomic Cl reaction (Senior et al., 2000c; Sliger et al., 2000; Fujiwara et al., 2002; Norton et al., 2003; Wilcox et al., 2012). Meanwhile, laboratory measurements of the model's critical rate constant value indicated that the homogeneous chemical rates were too small for any homogeneous contribution at all (Donohoue et al., 2005; Van Otten et al., 2011). Additionally, decay rates in the flue gases displayed no indication of kinetic behavior such as varying with residency times or showing any concentration dependences necessary for a system dependent on chemical kinetics. Nevertheless, this uncertain empirical combined homo-, heteromodels became accepted as no other explanation appeared feasible. However, once a model is accepted and published in modern science it is hard to dispel even when the true chemistry is realized. As a result this outdated model persists to this day, still with no valid proof other than being empirically mathematically tailored to a power plant and is based on no chemical understanding (Krishnakumar and Helble, 2007; Sable et al., 2008; Gharebaghi et al., 2011; Smith et al., 2011; Wu et al., 2013a; Zhang et al., 2013a). The latest such model consists of a 279 reaction scheme that continues to invoke the reaction of Hg with Cl in the gas phase, and now also interactions of Hg with Fe_2O_3 and unburned carbon on the ash (Yang et al., 2016b). This remains pure speculation and is erroneous. Additional models also have appeared that invoke computational neural network algorithms based on no chemical understanding (Ticknor et al., 2014; Cheng et al., 2015; Ling et al., 2015; Wang et al., 2016b), and a model exists including Cl_2 (negligible in flames) that predicts $HgCl_2$(g) formation above 700°C, which is inconsistent with any coal combustion observations (Yang et al., 2013).

Laboratory combustion studies

The continuing use of such kinetic models now is very disturbing in that further research identified the unknown heterogeneous chemistry in 2004 that has since been further validated in pilot-plant coal combustors but is still generally ignored (Schofield, 2004, 2005a, 2008, 2012, 2017). From these later extensive studies it was instantly clear that mercury has a significant propensity for heterogeneous interactions, visually quite apparent as seen from the very first experiments portrayed in Fig. 2.1. This had been totally overlooked and not previously thought possible. It is quite understandable due to the previous acceptance that mercury was too volatile and could condense only at very low temperatures. This well-disguised mechanism invoked by atomic mercury in coal combustors arises as a result of the inertness of atomic mercury with HCl. It requires an alternate mechanism to convert to its thermodynamically most stable $HgCl_2$(g) molecule. In coal combustion any free chlorine is predominantly present ($>90\%$) as gaseous HCl under any

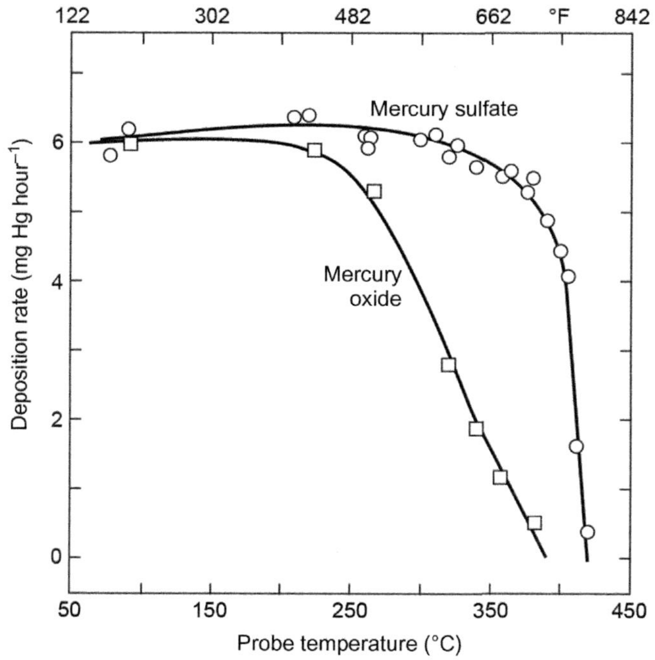

FIGURE 7.6 Accurate rates of deposition of $HgSO_4$ or HgO as a function of the steel probe surface temperature. From a fuel-lean propane/air flame containing low ppmv traces of an aspirated mercury salt solution with either no additives (HgO) or with low levels of SO_2 added ($HgSO_4$). Source: *From Schofield, K., 2017. Mercury emission control: phase II. Let's now go passive. Energy 122, 311–318.*

equivalence ratio and Cl_2 is negligible. This is simply a result of its high bond strength, D_0(H-Cl) = 428 kJ mol^{-1}, essentially equal to that of OH. Recent studies with bromine differ because D_0(H-Br) = 365 kJ mol^{-1}, a sufficient weakening, that permits a presence to occur of molecular Br_2 in flames that modifies mercury's options. What has now been fully proven is that within a temperature range, in combustion, atomic mercury can be chemically captured on noncatalytic metal surfaces to form thermodynamically stable HgO(s) or $HgSO_4$(s). The deposition rates are indicated in Fig. 7.6 displaying their operational temperature range and the dissociative falloff to higher temperatures. The formation mechanism involves the atomic Hg colliding with a surface but rather than being a condensation process it becomes a rapid surface chemical formation that retains the mercury in stable molecular form. The term chemideposition is more appropriate and now modifies the depositional concept of dew point. There is no balance between the species in the gas phase and the produced molecule on the surface. The mechanism becomes a two-dimensional collisional one that increases its efficiency by modifying the thermodynamics from gas phase to condensed phase. The deposition rate is the difference between the mercury influx to the surface and the egress from the surface that might be normal or dissociative vaporization. For combustion, the additional critical aspect is that either the formed oxide or sulfate is then vigorously and instantly attacked by gaseous HCl with conversion to $HgCl_2$(s). As known, this is extremely volatile with a boiling point of 304°C and sublimes off the surface. If chlorine had been present in the flame data of Figs. 2.1 or 7.6 no deposit would have been apparent.

What is clearly illustrated is that atomic mercury is colliding with the surface all the time. When the temperature conditions are appropriate, the mercury rapidly forms the sulfate or oxide. It does not react with HCl directly but chooses this alternate mechanism to

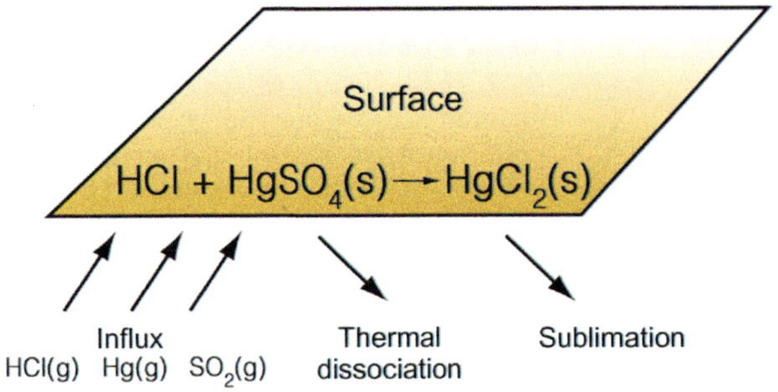

FIGURE 7.7 Overall portrayal of the heterogeneous conversion mechanism of elemental gaseous mercury to gaseous mercury dichloride. Source: *Reproduced with permission from the American Chemical Society, from Schofield, K., 2008. Fuel-mercury combustion emissions: an important heterogeneous mechanism and an overall review of its implications. Environ. Sci. Technol. 42, 9014–9030 (Suppl Inform S1–S16).*

circumvent the kinetic constraint. Analysis of Fig. 7.6 in greater detail shows two additional features. First, the boiling point of elemental Hg is 357°C, whereas recorded depositions of sulfate continue to about 420°C. This is a powerful indication of the surface affinity and extreme efficiency in being able to retain the mercury in chemical form before normal mercury vaporization. Also in the falloff region, Hg/S analyses ratios of the observed deposits show that starting at above about 350°C increasing sulfate dissociation is occurring through an intermediate formation of $HgSO_4 \cdot 2\,HgO(s)$ known as Schuetteite (Schofield, 2008).

As a result, a heterogeneous mechanism is clearly evident for converting elemental Hg to gaseous $HgCl_2$ in systems containing HCl in a specific temperature range. Rates were carefully measured on both Pt, quality stainless steels and base inexpensive carbon steel surfaces, all showing identical rates and proving the mechanism was noncatalytic. The fact that deposition rates for mercury mirrored those of the alkalis (Fig. 2.2) also implied this. In other words the mercury can start and end in the gas phase with the actual conversion mechanism fully disguised. A pictorial display of the mechanism is illustrated in Fig. 7.7 for the sulfate intermediate. In the functional temperature range, gaseous Hg and SO_2, instantly form the thermally stable sulfate on any available surface, this is then rapidly attacked by HCl(g), converted to $HgCl_2(s)$, which can then vaporize back into the gas phase. If temperatures are too high for the stable formation of the sulfate to survive, depending on the HCl conversion rate it will either thermally dissociate back to Hg(g) or be converted to the dichloride. Once converted, this gaseous dichloride is water soluble and also better absorbed by ash particles. As a result, its mitigation occurs by particle collection and/or by wet or dry FGD scrubbing. Because it uses a surface solely as an inert intermediate, this surrogate role is not readily apparent leaving the surface unaffected. Through the development of this new understanding, better monitoring instrumentation has been installed in many power plants. The instrumentation has become essentially real time and the sampling difficulties overcome. An example of such modern monitoring in a coal-fired pilot-plant experiment using such a modern Tekran detector is illustrated in Fig. 7.8.

Utilizing two instruments, in the case illustrated, measurements were made across a length of flue duct containing added metal surface area. These monitors closely indicated similar concentrations in and out for total mercury. However, the added surface did change the level of free atomic mercury, being oxidized to a high percentage across that section. What was most clearly indicated by this exercise was that mercury at these practical levels can now be reliably measured, with instrument stability over long periods. Nevertheless, it was also a reminder of words of wisdom from Professor Jost Wendt, then Chemical Engineering Professor at the University of Utah. His comment was that it is

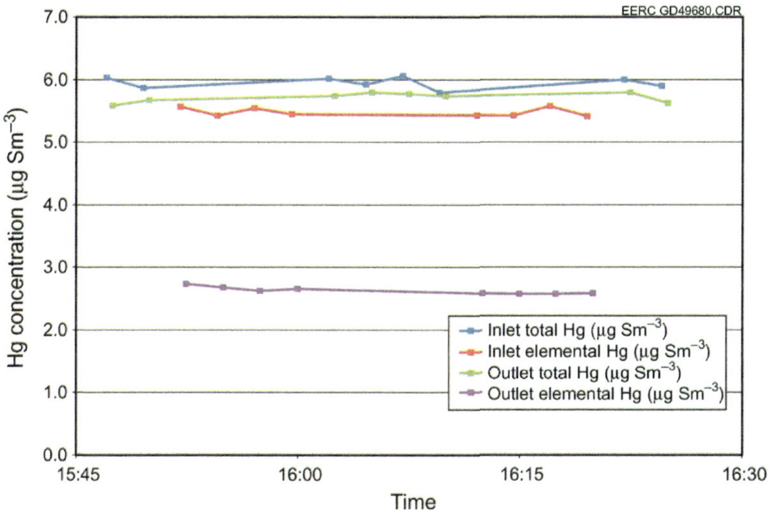

FIGURE 7.8 Observed changes of the mercury speciation passing through a slip steam test section held at 260°C containing base metal surface inserts in a coal-fired pilot plant. A 50% conversion was noted in this preliminary test of system reliability and stability. Source: *From Schofield, K., 2017. Mercury emission control: phase II. Let's now go passive. Energy 122, 311–318.*

hard to test in detail any strongly heterogeneous process in a pilot plant. This remains very true because although pilot plants are scaled down versions from full size in many ways, their ratio of volume to duct surface does not scale and pilot plants proportionally have much more available surface areas. This was illustrated by testing changes over other sections of the pilot plant and found that mercury has a natural propensity for this heterogeneous chemistry on any metal surface at the appropriate temperature. As a result, the advances now made have confirmed mercury's natural heterogeneous chemistry and given a better understanding of its combustion chemistry. The instrumentation has finally permitted establishing a much larger database, tracking mercury's history through the full combustor train and across each air pollution control devices present, for example, those installed for the mitigation of NO_x, SO_2, particulates, and for waste heat capture. As a result, in recent years it has now become evident that traces of mercury conversion are occurring everywhere, across the cooler end of the SCR, the cold-side ESPs, across fabric filter bags and the air preheater. It is also occurring on the flue gas ducting walls at the right temperature, time permitting. The problem for mercury is exactly this, it experiences time, temperature, and available surface areas constraints concerning its oxidation, so these contributions are generally each small. Nevertheless, a baseline value for the behavior of mercury is unique to each plant constituting the sum extent of these individual conversion contributions. Now, after more than a decade of intense research, the basic chemistry of mercury in combustion systems is better understood with the exception of the details concerning its capture and retention on particles and the chemistry occurring in the sulfur scrubbers. For oxygen-rich systems no lasting conversion of atomic mercury can occur until flue gas temperatures fall below about 400°C. The role of chlorine remains somewhat intriguing in that mercury is at ppbv in coal-fueled flue gases, so that chlorine even at 1 ppmv and predominantly as HCl in combustors, already has a concentration that is 1000 times larger than that of mercury. Consequently, the question is why does chlorine's concentration have to be these three orders of magnitude in excess for effectiveness as seen with bituminous coals? It cannot be a direct kinetic effect that would show a more sensitive and stronger dependence. It may simply be the required level of chlorine needed to find the sparsely populated oasis of condensed mercury on surfaces. The high level of chlorine may also initiate conversion to the dichloride at slightly higher temperatures than expected when the sulfate would normally just dissociate, broadening in this manner the functional temperature and lengthening the available time window. Conversion reaction rates are driven by concentration whereas thermal dissociation rates are controlled by an exponential energy term containing temperature. As a result there will be an automatic branching available to the mechanism at the high temperatures by this competition between either thermal dissociation or conversion.

Combustion power plant studies

As seen in Table 1.1, although the concentration of mercury in fuels is one of the smallest of trace metals, its volatility, toxicity and difficulty of control have made it a major environmental concern. Coal can have a range of concentrations of about 0.02–1.0 ppmw (Xu et al., 2003), but with the now approximately 7 billion tons $year^{-1}$ or more burned globally it is the major combustion source. Contributions from biomass combustion are less due to generally lower concentrations and volumes consumed (Friedli et al., 2001, 2003). A slight trace occurs in pet-coke and its levels in limestone are very low (Cordoba et al., 2012). Although gasoline and diesel fuels usage is high, analyses indicate their contribution also remains minor (Wilhelm, 2001; Conaway et al., 2005; Won et al., 2007). Natural gas does contain variable traces of mercury. Sometimes, but not always, these are removed by various methods in processing, mainly to minimize corrosion and catalyst poisoning. In some cases, mercury removal systems are needed to solely protect the metal components in the processing plant (Ezzeldin et al., 2016). Volumes of natural gas extracted are now about 0.4×10^{12} m^3 $year^{-1}$ in the United States alone, so a content of 1 μg m^{-3} would imply 0.4 ton $year^{-1}$. However, it can be 10-fold or larger in some locations and might require legislated control in such cases (Ryzhov et al., 2003). Concentrations of mercury are higher in municipal waste (van Velzen et al., 2002; Cheng and Hu, 2012), medical waste (Alvim-Ferraz and Afonso, 2003), and sewage sludge incineration (Werther and Ogada, 1999), but these operations function on smaller quantities of fuel and although their emitted flue gas concentrations are larger they are actually easier to control than the large volume/low ppbv emission levels realized in full-scale coal combustors (Takahashi et al., 2012). The cement industry also poses a significant problem as mercury is released not only from any possible coal usage but also in some cases from the limestone feed material and treating of other toxic wastes (Senior et al., 2010; Mlakar et al., 2010; Zemba et al., 2011; Wang et al., 2014). In coal combustion or incineration, due to the stability of mercury's compounds, atomic elemental mercury is readily released and none is found in bottom ash (Li et al., 2018b). It is gaseous and needs to be captured on some ash form or by the FGD scrubber otherwise it is emitted to the environment.

Initially, one of the air pollution control devices coming into major use was the SCR. This was to control NO_x emissions catalytically by conversion to nitrogen. As a result, because it has catalysts, it was hoped that it could also be modified to do double-duty by also oxidizing mercury to the water soluble dichloride. It was further considered that the heterogeneous Deakin Process might induce reaction between HCl and O_2 to produce Cl_2

(Hisham and Benson, 1995; Amrute et al., 2011). This is a known mechanism to create Cl_2 that could then heterogeneously convert Hg to $HgCl_2$ directly (Medhekar et al., 1979). However, such a process requires a rather specific catalyst such as copper and an optimum temperature of about 400°C, a temperature that is marginally too high for heterogeneous mercury chemistry. This temperature constraint tended to eliminate this possibility and additionally, experiments with the SCR indicated that the mercury oxidation observed depended solely on HCl ruling this idea out (Cao et al., 2007). However, this has not ended the ambition for an SCR contribution and much effort has continued. The problem would seem that the normal operating range for the SCR is above about 370°C. As seen in Fig. 7.6, this temperature lies on the thermal dissociation falloff parts of the curves for both HgO and $HgSO_4$ posing a major problem for their formation. Because of this, emphasis now appears to be concentrated on modifying the catalysts for NO_x/NH_3 conversion to operate at lower temperatures to not only control the NO_x, remove the excess NH_3 slip, and either convert or retain the mercury (Li et al., 2018d; Yang et al., 2018b; Zhou et al., 2018; Chen et al., 2018a,b; Cimino et al., 2018). Even so, this remains a demanding task to enable two or more processes to each have adequate operating efficiencies on the same system (Lv et al., 2018). One study did reduce the SCR operating temperature to 345°C reporting a 60% oxidation of the mercury, but did not mention its NO_x performance at that temperature (Su et al., 2017).

As emphasized, due to previous monitoring difficulties for mercury in combustion, care now should be taken in interpreting any earlier data as biased errors often could occur either in the sampling method or in the detector itself (Cauch et al., 2008). These problems now have become recognized and continuous emission monitors (CEMs) are available to measure the speciation (Hg and $HgCl_2$ proportions) of any mercury emission (Thiebaud et al., 2009). As indicated, reliability in monitoring in the last few years has led to a better analysis of the mass balance within combustion trains. The mentioned degrees of conversion now being noted across many of the air pollution control devices in power plants have illustrated two aspects. They have initially confirmed the heterogeneous mechanism for mercury and shown its propensity for reaction on surfaces in the appropriate temperature window. Second, it has illustrated that by modifying and optimizing these trace effects across each device they can aid and complement each other to optimize the natural baseline control of mercury emissions in a specific power plant (Pudasainee et al., 2009, 2012, 2017; Cheng et al., 2009; Wang et al., 2010b,c, Takahashi et al., 2010; Zheng et al., 2017). However, depending on coal type, ash nature and operating design, conditions will vary from plant to plant. For example, in some cases the ESP (cold-side) at temperatures of about 160°C may remove as low as 2%, or up to 43% Hg on the fly ash (Pudasainee et al., 2009; Tang et al., 2016), in this case controlled largely by the content of unburned carbon in the ash (Senior and Johnson, 2005; Hower et al., 2005, 2010; O'Dowd et al., 2006; Guo et al., 2007a). One recent Chinese study reported that the fly ash contained about equal amounts of $HgCl_2$ and HgS, the wet ESP particles were mainly $HgCl_2$ and that in the final gypsum were $HgCl_2$ and HgS (Zheng et al., 2018). A critical review of mercury's history through the coal combustor and in cement plants and metal smelters now summarizes the better data being collected (Zhang et al., 2016a). This shows the transforming contributions across all the devices but does leave most of the mercury still atomic and released. Some conversion can now be induced in the cooler final stages of an SCR (Cao et al., 2007; Eswaran and Stenger, 2005; Pudasainee et al., 2010). Conversion also can occur on fabric filter bags (Cao et al., 2008a; Scala and Clack, 2008; Choi et al., 2009), across air preheaters (Bilirgen and Romero, 2012) and on wall ducting itself. As a result, what used to be a very low conversion of mercury for low level chlorine coals can now be closer to 50% in some power plant designs before utilizing any additional remedial methods. One new power plant equipped with a cold-side ESP, a fabric filter baghouse, an air preheater and a lime spray absorber, fired with a low chlorine Powder River Basin coal indicated this a decade ago by showing a 59% baseline oxidation level on certification testing (Black Hills Corp. Report, 2009). Years earlier with no pollutant controls, it would have been expected to have an oxidized baseline of <10%. However, in all cases, there will remain a need for additional mitigation methods for total control as stack emissions often remain in the 50%–80% range, varying from coal to coal (Ancora et al., 2015; Chen et al., 2016; Tang et al., 2016; Liu et al., 2016d). Global social benefits greatly outnumber other considerations and it is nearing legislated controls that will gradually grow to worldwide acceptance (Giang and Selin, 2016; Sunderland et al., 2016). Delays have arisen through a lack of chemical understanding, difficulties in reliably monitoring its concentrations at very low levels, and in developing acceptable mitigation methods. However, as seen, most of these considerations now largely have been resolved. Nevertheless, because no Global Mandate is currently apparent to drastically control the anthropogenic emissions from coal combustion, the Chinese and other Asian power plant operators have stressed maximizing this collaborative cobenefit route coming from each of the air pollution control devices. They have optimized each device to elevate the overall baseline level of oxidation so reducing the resulting emission level at minimal cost (Zhang et al., 2017d; Pudasainee et al., 2017; Wu et al., 2018). Although not a final solution it can be an environmental benefit for now and reflects some interim progress, but even so, at least 20%–30% mercury will still be released. However, it is beneficial in that, the higher the baseline, the less mitigation processes will be finally necessary to meet any mandated target control. This nature of mercury is unique among the trace elements. As a result, for any power plant, its calibrated baseline behavior will be variable. This concept of optimizing the individual contributions from various control systems is now being considered in all anthropogenic mercury sources, such as power plants, incinerators, cement production and oil refiners, also monitoring the speciation of the mercury (Wu et al., 2018d; Noda and Ito, 2018; Takiguchi and Tamura, 2018; Chou et al., 2018; Mojammal et al., 2019).

As indicated, it was found that the particulate mercury on fly ash was a variable fraction of the total. Moreover, the nature of its interaction still is not clearly understood (Zhao et al., 2010; Wilcox et al., 2012; Fernandez-Miranda et al., 2015; Wang et al., 2016c). Atomic mercury has a minimal affinity for carbon or many other elements. Summaries have labeled its behavior as both inorganic and organic on ash. Even a decade ago, it was realized that no capture of elemental mercury occurred without its conversion (Dunham et al., 2003). Gaseous $HgCl_2$ is more readily absorbed by a physical interaction but only at temperatures below about 150°C (Chen et al., 2009). This is consistent with its volatility, which is similar to that of atomic mercury (Bernard et al., 1997). Recently, it was shown that most of the mercury that is collected on the ash decomposes below 300°C (He et al., 2016). This would tend to restrict possibilities to the dichloride, oxide or some other loose chemical binding through an oxygen or hydroxide on the surface. A desorption of Chinese coal fly ash suggested dominant $HgCl_2$ and HgO (Li et al., 2018e), and that the concentration in the gypsum was higher than on the fly ash (Zhang et al., 2017a; Zhao et al., 2017).

The $HgSO_4$ or sulfide would be quite stable at such a temperature (Tariq and Hill, 1981; Rubel et al., 2006; Lopez-Anton et al., 2011). Any release below 160°C is clearly indicative of $HgCl_2$. Consequently, time constraints limit the interaction and temperatures are low for the absorption of the oxidized Hg^{2+}. Also because Hg has no affinity for carbon it is probably bonded through oxygen or sulfur groups on the surface or within the ash. Some theories of carbon bonding have been suggested (Olson et al., 2009; Padak et al., 2006; Padak and Wilcox, 2009; Jew et al., 2015). Effects of HCl and SO_2 have been studied and note their important roles (Ochiai et al., 2009). Nevertheless, some mercury is retained and not readily leached (Luo et al., 2006) being shown to be dependent on particle size, degree of unburned carbon content and the inorganic constituents (Li et al., 2007; Lopez-Anton et al., 2009; Hower et al., 2010; He et al., 2015; Zhou et al., 2015a). Some appears to be reduced to the insoluble HgS. A Canadian study concluded that the quantity of carbon in the ash was not sufficient to explain the absorption (Goodarzi and Hower, 2008). An additional unknown parameter is that particles are not analyzed in situ but after sampling and a lapsed time period. Many studies of ash have been reported, but face the practical difficulty that the nature of fly ash differs not only from coal to coal but also by the combustion conditions through which it has traversed (Li et al., 2007). To enhance particulate capture, the addition of activated charcoal was examined. This improved capture performance but still needed C/Hg mass ratios of the order of 3000–20,000 showing a low efficiency (Choi et al., 2009; Wilcox et al., 2012). Absorption by injected activated charcoal also was shown to be a chemisorption process and X-ray or TGAs have suggested that the mercury may be present as $HgSO_4$, HgO, HgS, $HgCl_2$ and possibly the water insoluble Hg_2Cl_2 (Huggins et al., 2003; Takaoka et al., 2007). Nevertheless, early hopes had to be modified when efficiencies were seen not adequate for any governmental supported mitigation method (Sjostrom et al., 2010) unless the activated charcoal was doped with either a halogen (chloride or bromide) (Ghorishi et al., 2002; Hutson et al., 2007), or sulfur (Hsi et al., 2001). Also preferably collected on a fabric filter bag, which can be at an optimum temperature and also provide for a lengthened interaction time (Choi et al., 2009). Theoretical studies of the retention have suggested mechanisms but these have little chemical foundation and remain very speculative (Wilcox et al., 2011; Rungnim et al., 2016). There does appear to be a change in mechanism from physical to chemisorption as evidenced by the increasing efficiencies with temperature and also stability against leaching (Sasmaz et al., 2012; Bisson and Xu, 2015; Zhou et al., 2015b). Additionally with sulfur there is the suggestion that the carbon reduces it back to insoluble HgS (Graydon et al., 2009). Retention would be optimal if this or similarly insoluble Hg_2Cl_2 might be formed in the pores of the charcoal. The method, although developed to be adequate, unfortunately has become an expensive solution and was ruled as such by the US Supreme Court and consequently not possible to enforce at present by legislation on the coal and oil industries (Hess and McCoy, 2015). The cost of activated charcoal has been rated also too high for use by China (Zhang et al., 2017d). Other control methods such as amalgamation with gold are not commercially viable, but can be useful on a small scale as an absolute calibration standard (Zhao et al., 2006). Catalysts are generally assumed necessary to oxidize mercury (Presto and Granite, 2006), but closer inspection indicates this has never been rigorously illustrated and was solely a conceived suggestion in the literature that became a general belief (Brooks, 1997). A 1997 US Patent suggested using an absorption bed containing an unspecified commercial catalyst, most probably of Mn, V, Cr, Pb or Se oxides to successfully oxidize mercury. It was never reported tested. However, the current efficiencies of mercury chemideposition seen on base metals display rates essentially equal to those of alkali deposition, Fig. 2.2. As seen, this confirmed that mercury does not require a catalyst only an adequate surface (Schofield, 2005a). Catalysis implies the observation of an enhancement in rate. If rates are already high on base materials (carbon steel) and are similar to those on Pt surfaces, mercury's heterogeneous processes have to be noncatalytic and difficult to further enhance by a so-called catalyst (Medhekar et al., 1979; Schofield, 2005a, 2008, 2012, 2017). As a result, any oxidation effects seen in the SCR with Hg are unlikely catalytic but solely result from a region of sufficiently low temperature (Lee et al., 2008). Nevertheless, this has not minimized ambitious catalytic studies in attempts to facilitate the SCR as a two species modifier (Zhao et al., 2015; Zhang et al., 2015, 2015a; Fernandez-Miranda et al., 2016; Niksa et al., 2016; Huang et al., 2016). The mitigation method invoking bromine addition is very elegant in that it provides a more efficient method than chlorine for oxidizing mercury to the dihalide (Liu et al., 2007; Cao et al., 2007a, 2009). Unfortunately, studies with bromine have simply treated its chemistry as a more efficient extension of the previous chlorine interpretations. As a result homogeneous chemistry continues to be stressed but now clearly has no foundation (Van Otten et al., 2011; Gu et al., 2016). One such extensive chlorine model has been extended to include bromine (Niksa et al., 2010), and another model (Cl, Br, I) now involves 957 reactions and 202 species but neither of these have any meaningful chemical foundation and continue to be useful solely as empirical mathematical fits (Auzmendi-Murua and Bozzelli, 2016). The latest modeling effort does conclude that a significant role has to be heterogeneous but overlooks the previous work identifying this direct role with Br_2 (Yang et al., 2017a). Logically, it cannot be suggested that it involves HgBr in mechanisms similar to chlorine due to the even lower bond strength of both HgBr and $HgBr_2$ (Table 7.16). As a result, the greater efficiency of bromine simply has to imply a different mechanism from that of chlorine. This has to arise from the fact that the distribution of bromine in flames is very different from that of chlorine and under fuel-lean conditions about 50% can exist as Br_2. This drastically modifies the chemistry. The corresponding heterogeneous reaction between Hg and Cl_2 has long been known and occurs on any type surface, even Teflon or quartz, at about 250°C (Medhekar et al., 1979; Yan et al., 2005; Wang and Anthony, 2005). This offers an alternate and direct heterogeneous reaction in combustion between Hg and Br_2 to form similarly volatile but in fact less water soluble $HgBr_2$. This higher efficiency of bromine otherwise has been overlooked but suggests that in coals rich in bromine, such as in China, the role of the natural bromine concentration may be a significant contributor to that of chlorine. Very small ppmv levels of Br_2, HBr, or $CaBr_2$ appear to be capable of oxidizing a significant fraction of the mercury However, in a coal combustor, low levels of bromine can also be selectively removed by alkalis or other trace metals as their gaseous bromides and reduce the potential mercury control if this is not realized. Additionally, this heterogeneous chemistry between Hg and Br_2 now has been clearly evidenced in the sampling interferences noted with measurements of bromine's mercury speciation (Buitrago et al., 2013). As with chlorine, biased data can result unless the sampling method is fully validated. Acceptance of measurements with bromine before this date may require additional reanalysis. Few coal studies have ever quoted bromine concentration levels and some not even chlorine so assessments of certain studies are not possible. Iodine also has been shown to facilitate mercury conversion but is a less available material and not practical on any scale (Li et al., 2009; Chi et al., 2009). The difficulty with bromine and iodine can center on their low global abundance and current work with bromine reports its very high corrosive nature toward even stainless steel at low concentrations. As with chlorine, the interaction with bromine-activated charcoal has been extensively studied but the retention mechanisms remain unclear (Bisson et al., 2012; Sasmaz et al., 2012; Liu et al., 2013a).

The new technology of oxy-burn is promising for CO_2 capture. It consists of burning the fuel solely in oxygen with partial recirculation of the product CO_2, gradually withdrawing this with the water. This changes normal operating conditions and so raises questions also concerning the effects on the emissions of CO, NO_x, SO_2, and Hg. As a result there has already been a preliminary flurry of developmental papers addressing these questions and the parametric dependences. Rather surprisingly mercury has become of significant concern, but more with its corrosion effects (Stanger et al., 2014; Cordoba and Diego, 2018). In one large demonstration plant the flue gases were being partially recirculated over 2000 times (Cordoba et al., 2016). This had the effect of enhancing both the mercury and HCl concentrations, basically then oxidizing it all to $HgCl_2$. It was calculated that 44 mg Hg were being recirculated. Although this significantly enhanced the absorption on the fly ash (20%), the highly corrosive nature of the mercury became of concern for any Al alloys both in heat exchangers and for the CO_2 removal compressors (Belo et al., 2016; Li et al., 2018). This remains a pivotal problem with attempts to optimize the pollution control devices in an attempt to control mercury to the full in a specific plant. It was noted that increased formation of NO_2 also appeared to oxidize the gaseous elemental mercury. However, this may be questionable as gaseous HgO cannot exist in such conditions. It appears that the lengthened time does permit a high percentage of oxidized mercury (Wang et al., 2016e, 2016k). It still requires an additional mercury control method though and many models still in use are very speculative (Wang et al., 2017a). A further Chinese study examined the fly ash from such an oxy-fuel pilot plant (Yang et al., 2017). The greater accumulation of mercury on the ash was indicated as HgS and Hg binding to the surface utilizing oxygen binding, however the evidence remains speculative. Nevertheless, optimistic studies indicate pollutant levels can be highly abated utilizing gypsum absorption and filter ash bag technology (Font et al., 2011).

A major concern, not only in oxy-burn but in general, has been the ability to ensure no reemission of Hg(g) once captured, particularly from wet FGD systems (Ochoa-Gonzalez et al., 2011, 2013; Cordoba et al., 2012; Fernandez-Miranda et al., 2016a; Wu et al., 2016b). This is a very unusual aspect of mercury's chemistry in that although in molecular form in solutions, it can easily dissociate and vaporize as the element. This is a major concern in nature with any deposited mercury. It appears to recycle in the environment, lengthening the time until it is finally sequestered. Plans by the US EPA now are to more fully regulate any FGD wastewater in the direction of zero water discharges (Pakzadeh et al., 2014). This then would effectively control any reemission from the process, and from any resulting gypsum product that remains an additional concern (Zou et al., 2016; Zhao et al., 2017c; Liu et al., 2016; Zhu et al., 2017; Bian et al., 2018). However, even this regulated need is questioned by some (Reash, 2019). Due to its importance many reemission and leaching studies are available. Through additives and control of pH it does seem that these potential difficulties can be minimized in the FGD systems with the mercury being reduced to insoluble HgS (Wu et al., 2010; Tang et al., 2010; Omine et al., 2012; Ochoa-Gonzalez et al., 2013; Hao et al., 2016; Chang et al., 2017; Krzyzynska et al., 2018; Xu et al., 2018). Models now are being developed to mitigate this reemission (Lv et al., 2019). One recent Chinese study of six coal-fired plants equipped with ESP, fabric filters (FF) and an FGD reported that <12% of the mercury transferred into the gypsum (Gao et al., 2017). Other studies varying FGD parameters indicated it was possibly dissolved sulfite in the scrubber that reduced the mercury release (Heidel and Klein, 2017; Chang et al., 2017). A further attempt to identify the mercury species in FGD gypsum was not able to be exactly specific but did conclude that HgS was a major component. Its stability in the gypsum has been extensively tested with regards to thermal stability and leaching under various conditions (Zhu et al., 2016, 2017; Bian et al., 2018; Wang et al., 2018c; Zhao et al., 2018b). Only on high heating does it appear to vaporize any Hg, which needs confirmation due to its importance concerning the gypsum industry. The gypsum product is regarded as nontoxic. Similarly, there are concerns of leaching from coal ash in landfills where Hg and other trace metals require disposal considerations (Zhao et al., 2018d).

Numerous studies have addressed the behavior of mercury under fuel-rich conditions as can be encountered in pyrolysis, gasification and liquefaction processes for synthetic fuel production (Couling et al., 2012). A pyrolysis study of eight Chinese coals under N_2 up to 1200°C showed a normal release of mercury reflecting its chemical stabilities. The highest peak release at >750°C correlated with the mercury aluminosilicates in coals (Guo et al., 2018). The mercury balance in Polish coking ovens indicates the mercury transfer to final product forms but indicates a 14% emission release (Burmistrz et al., 2018). Iron oxides have been shown to absorb Hg but only at low temperatures, <140°C (Wang et al., 2013a). With IGCC (Pavlish et al., 2010) and Sasol–Lurgi gasifiers (Bunt and Waanders, 2008) it remains a significant problem as mercury is not compatible with high temperature gas cleaning (Reed et al., 2001a). Amalgamation has been examined with the noble metals, with Pd being possible up to 400°C (Granite et al., 2006). Efforts to reduce the mercury to HgS using sulfur treated activated charcoal appear to be useful at lower temperatures (Uddin et al., 2009), and especially for processing wellhead natural gas and for releases from underground coal gasification (Liu et al., 2006; Reddy et al., 2014).

In summary, although at low concentrations in fuels, mercury is very volatile, fully released in combustion and difficult to capture. Environmentally it has risen in importance with regards to health care, its organomercury compounds being highly neurotoxic. It has a simple chemistry with few thermally stable condensed phase molecules and only two of general importance in the gas phase, namely atomic Hg, $HgCl_2$, and $HgBr_2$. In normal fuel-lean combustion conditions, with flue gases containing trace metals, mercury released in its elemental state is the one element with the least available chemical reaction time available to it in a normal power plant. As a result, it is now apparent that any gas phase chemistry is kinetically too constrained. However, its potential heterogeneous chemistry is very efficient, but requires available surfaces in its brief reactive temperature window. Below about 370°C the elemental mercury can form either $HgSO_4$ or HgO on surfaces that then instantly reacts with the flue gas HCl. This converts it to the thermodynamically stable dichloride that vaporizes back into the flue gases as $HgCl_2(g)$. This is water soluble and can be controlled by the sulfur scrubber (FGD), ending in the gypsum product, becoming largely nonleachable and at low safe levels. It is an elegant mechanism with mercury starting and ending in the gas phase, circumventing the gas phase kinetic constraints. Alternatively, absorption by activated charcoal has been developed as a method for control, and bromine salts are an effective additive to induce an alternate heterogeneous mechanism forming $HgBr_2(g)$. Both these mitigation methods now are available but come with various operational difficulties, and are expensive solutions. In modern combustors with no mercury control a high percentage of the fuel atomic mercury continues to be released to the atmosphere. Efforts generally now are made to optimize its oxidation through the flue gas train to minimize naturally its emission level. It is still largely nonlegislatively controlled in most countries through the current high economic costs involved (Zheng et al., 2012).

7.19 Molybdenum

Molybdenum (second row transition element) and tungsten (third row) are chemically very similar but neither having a great deal of commonality to Cr the first member of their fourth column of the transition elements in the periodic table. Hexavalency now is more dominant and stable, and not as easily reduced. The three elements are of interest for theory in that of the six outer shell electrons their lowest energy configurations are Cr($3d^5 4s^1$), Mo($4d^5 5s^1$), yet W($5d^4 4s^2$). Molybdenum is ranked about the 58th in abundance in the Earth's crust and is essentially equal to that of tungsten. It is 22nd in seawater and the most abundant transition element. The main ore is the sulfide, molybdenite MoS_2, mined to an extent of 270,000 tons $year^{-1}$. Reserves are quoted as greater than 14 million tons. In air, the ore oxidizes to MoO_3 above 760°C and this can be further sublimed above 1100°C. It is refined from either pure MoO_3 or $(NH_3)_2MoO_4$ by H_2 reduction. Prices are currently about \$7.00 lb^{-1} for the trioxide. Although a refractory metal (m.pt.2623, b.pt.4680°C) with a high melting point, it does oxidize above 760°C. Its main use (85%) is in the production of steels and alloys. Also its chemicals find use as lubricants, catalysis and are a necessary component in agriculture. In recent years it has also gained interest as a wide band gap n-type semiconductor. Research in flame nanotechnology involving Mo has developed over many years. An early study showed that MoO_3 nanoparticles could be produced by adding the oxychloride, MoO_2Cl_2 to flames (Vergnon et al., 1980). Now, tubes and whiskers have been formed primarily of MoO_2 or MoO_3 using CVD methods at different temperatures (Gesheva et al., 1999; Merchan-Merchan et al., 2006). One recent study is reported of particle formation in ethylene or acetylene flames induced by the addition of $Mo(CO)_6$. This appears more complex in that the MoO_3 nanoparticles become embedded in soot particles and catalytically enhance its oxidation (Emelianov et al., 2010). Molybdenum also plays a role in current Li-ion battery technology (Xu et al., 2014).

Approximate estimates tend to indicate that natural additions to the atmosphere as particles are of a magnitude of about 3000 tons $year^{-1}$ (Nriagu, 1989, 1989a). Dust is a major component (1300), volcanoes and forest fires each contributing about 500 tons $year^{-1}$ and bio- and spray particles the rest. This is closely balanced by an anthropogenic source of about 3300 tons $year^{-1}$ making it not really a pollutant (Nriagu and Pacyna, 1988). This is mainly related to coal (2000) and oil (300) combustion and partly the steel industry. This figure was reduced slightly to 2640 in a later publication (Pacyna and Pacyna, 2001). Ice core samples from the Himalayas provide an interesting record of atmospheric and crustal concentrations of Mo over the past 800 years (Hong et al., 2009), as do similar analyses of snowpack in the Antarctic (Hong et al., 2012). These confirm effects of industrialization, apparent in the last century, but cannot fully rule out volcanic contributions (Wahrenberger et al., 2002). However, some slight enrichment of atmospheric particles and crustal concentrations is evident for Mo over the last century.

Health concerns with molybdenum

Molybdenum is biologically very active and is one of few heavy transition elements essential to life. It has been reported that a molybdenum deficiency in the oceans probably prevented any plant or animal life evolution there for possibly 2 billion years. It is essential in soils otherwise land is barren. One discussion queried how and why molybdenum had become an essential component of life (Williams and da Silva, 2002). The human body contains about 0.1 ppmw of molybdenum with a recommended daily dietary requirement of about 45–300 $\mu g\ day^{-1}$. This range is quite below a tolerable level of 2 $mg\ day^{-1}$ from animal studies. The Environmental Protection Agency suggested a Reference Dosage (RfD) of 5 $\mu g\ kg^{-1}$ body wt. day^{-1} for normal health (EPA Report, 1992). Serious illnesses occur with excesses of the order 10–15 $mg\ day^{-1}$, but its deficiency is quite rare in humans although it can be induced in animals (Goldhaber, 2003). It is readily absorbed by the intestine and biomarkers are now available for normal levels: blood 1–4 $\mu g\ L^{-1}$, urine $<$160 $\mu g\ L^{-1}$, hair 50–200 $ng\ g^{-1}$ (Mayo Clinic, 2018; World Health Organization Trace elements 1996). Few population surveys include this element but three that are published convey the validity of the blood and urine biomarkers (Goulle et al., 2005; Cesbron et al., 2013; CDC 4th Report, 2018). One testing hair levels appeared not to display a meaningful correlation (Wei et al., 2018). In life, Mo is involved in many enzyme chemical mechanisms, and with plants in the fixation of nitrogen. It has a very low toxicity but high levels of Mo or W can interfere with the body's utilization of copper. It is present in many grown foods and the content tends to be a reflection of soil quality. Tungsten can in fact be antagonistic to Mo's value as a nutrient (Fu and Tabatabai, 1988). Sewage sludge waste is utilized extensively in agriculture with molybdenum being one of the valuable components. However, in such use, due to the presence of other heavy metals (As, Cd, Cu, Pb, Hg, Ni, Se, Zn) all are regulated in numerous countries to be within certain safe limits (Poykio et al., 2010). Because the sludge wastes contain these trace metals, there is concern that those not depleted in agriculture will accumulate over time and be a potential future problem (Eriksson, 2001). As a result, agricultural use is limited, requires care in application (McBride and Cherney, 2004) and other methods of disposal for wastes have to be invoked (Fytili and Zabaniotou, 2008). Generally, the levels of Mo in sources of water are minimal and beneficial not requiring regulation. A guideline value is set at 0.07 $mg\ L^{-1}$ (WHO Guidelines, 2006a).

Molybdenum's basic chemistry

Appraising molybdenum's potential combustion chemistry is simplified restricting it to solely the condensed and gaseous phases. As with chromium, in the condensed phase this essentially involves a major oxide and molybdates. The physical properties of relevant molecules are discussed and summarized in Table 7.28.

TABLE 7.28 Properties of molybdenum's condensed phase molecules relevant to combustion, °C.[a]

	m.pt. b.pt. vapor pressure
Mo	2623 4680, 10 ppmv/2469°C, ins
MoO_2	dissoc 900 to Mo, MoO_3, 1 ppmv MoO_3/1380°C, ss
MoO_3	802 1105, vaporizes as $(MoO_3)_{3-5}$ polymers, ss
H_2MoO_4	exists as $MoO_2(OH)_2$ in gas phase only but thermally stable 1200°C.
$MoCl_2$	stable, >700 disprop to Mo, $MoCl_4$, r with wat
$MoCl_3$	1027, >500 disprop to $MoCl_2$, $MoCl_4$, r with wat
$MoCl_4$	317, sublimes, disprop >200 to $MoCl_3$, $MoCl_5$, r with wat
$MoCl_5$	194 268, dissoc to $MoCl_4$, Cl, r with wat
$MoCl_6$	unstable
MoO_2Cl_2	175, stable, sublimes 156, r with wat
$MoOCl_3$	297, sublimes, >215 disprop to $MoCl_3$, $MoOCl_4$, MoO_2Cl_2, r with wat
$MoOCl_4$	104 180, dissoc >200 to $MoOCl_3$, Cl, r with wat
Mo_2S_3	1807, dissoc 1867 in air to MoO_3, ins
MoS_2	>1100 sublimes, >760 dissoc in air to MoO_2, MoO_3, ins
$Mo(CO)_6$	148 155, decomp, ins
$Al_2(MoO_4)_3$	950, ss
$BaMoO_4$	1450
$CaMoO_4$	965 dissoc, mol vapor observed, ss
$CoMoO_4$	1040
$FeMoO_4$	1115, >800 dissoc to oxides, mol vapor observed
$Fe_2(MoO_4)_3$	940, >600 dissoc to oxides
K_2MoO_4	925, mol vapor observed, s
$K_2Mo_2O_7$	495, s
$MgMoO_4$	1240 stable, mol vapor observed, s
Na_2MoO_4	687, mol vapor observed, s
$Na_2Mo_2O_7$	615, s

[a]*dissoc*, dissociates; *decomp*, decomposes; *disprop*, disproportionates; *s*, water soluble; *ss*, slightly soluble; *ins*, water insoluble; *r with wat*, reacts with water; *mol vapor*, molecular vaporization.

The M = O bond formed with oxygen or water is strongly favored thermodynamically. With both Mo and W there is a minimal cationic nature. No carbonate, sulfate, nitrate or hydride. Hexavalency is dominant and, although as seen, $MoCl_n$ ($n = 2–5$), MoO_2, MoS_2 exist, they are not as thermally stable and can chemically convert to MoO_3. MoS_2 the major ore is readily oxidized by air and attacked by chlorine to produce MoO_3 and $MoCl_5$, respectively. The two oxides are MoO_2 and MoO_3. A trivalent $Mo_2O_3(s)$ is not known. The dioxide is stable to about 900°C when it disproportionates to Mo(s) and MoO_3(s) (Burns et al., 1960; Cordfunke and Konings, 1990). The trioxide is more volatile (m.pt.801, b.pt.1105°C) and vaporizes at lower temperatures than were seen with chromium and its dominant Cr_2O_3(s). The thermochemical data on the Mo-O and Mo-O-H systems have been reviewed due to their relevance to the nuclear industry (Cordfunke and Konings, 1990). Molybdenum can form a series of chlorides but these are all vulnerable to hydrolysis and sensitive to oxidation to the trioxide. $MoCl_2$, decomposes above 550°C, disproportionating at 700°C to $MoCl_4$(g) + Mo(s). $MoCl_3$, disproportionates above 500°C into $MoCl_4$(g) + $MoCl_2$(s). $MoCl_4$ sublimes and at >200°C disproportionating into $MoCl_5$(g) + $MoCl_3$(s). $MoCl_5$ (m.pt.194, b.pt.268°C) dissociates to $MoCl_4$(g) and chlorine. The hexachloride, unlike that of tungsten, has always been elusive but was recently synthesized and is unstable (Tamadon and Seppelt, 2013). Also of a complex nature, the oxychlorides are very volatile. MoO_2Cl_2, the most thermodynamically stable, sublimes having a low boiling point of 156°C. $MoOCl_4$, also hexavalent (m.pt.104, b.pt.180°C) dissociates above 200°C to $MoOCl_3$ and chlorine. This latter, less stable oxychloride $MoOCl_3$, disproportionates above 215°C to $MoCl_3$ + $MoOCl_4$ + MoO_2Cl_2. Consequently, all the chlorides and oxychlorides of molybdenum are very volatile and readily disproportionate into each other besides being vulnerable to oxidation and hydrolysis. As a result, their only possible role in combustion might be to aid volatile release. No simple phosphate is known and only complex formulations with other elements have been characterized. The disulfide, MoS_2, is stable but oxidizes on heating to >600°C, changing in one step to MoO_2/MoO_3 (Schaefer and Gokcen, 1980; Shigegaki et al., 1988). A nonstoichiometric Mo_2S_3 is seen in the same TGA study to also oxidize on heating, proceeding through various steps to the trioxide. Also as with chromium, a hexacarbonyl can be made (m.pt.148, decomposing at 155°C).

The trioxide gains additional importance in being the anhydride of molybdic acid, H_2MoO_4, structurally $MoO_2(OH)_2$, and is the basis for all metal molybdates, . However, MoO_3 is only slightly soluble in water and needs alkalinity for solution formation. The pure acid cannot be isolated in the condensed form. Molybdenum and tungsten similarly display not only numerous basic MoO_4^{2-} type anionic molecules with metals, but they have the ability more than chromium to absorb numerous additional trioxide units. Consequently, series of polymolybdates can exist, for example, increasing in thermodynamic stabilities with anions such as $Mo_3O_{10}^{2-}$, $Mo_6O_{19}^{2-}$, or $MoO_4(MO_3)_n^{2-}$. Their bonding is now modeled well by theory (Bridgeman and Cavigliasso, 2003). An assessment of those that might occur in combustion illustrates that these are not high temperature ceramic or refractory type molecules in the condensed phase, and vaporize or dissociate below about 1200°C, and that their water solubility can vary. $Al_2(MoO_4)_3$, slightly soluble, starts dissociating or vaporizing at about 700°C–800°C below its 950°C melting point. Ferric, $Fe_2(MoO_4)_3$, can be formed from Fe_2O_3/MoO_3 (m.pt.940°C) but starts dissociating in air above about 600°C (Zhukovskii et al., 1972). Other molybdates and these are discussed in detail later. Solubility varies, $MgMoO_4$ is water soluble, $CaMoO_4$ only slightly. The alkali and alkali-earth molybdates and polymolybdates are most studied and are mainly water soluble and thermally stable. With the higher polymolybdates, thermodynamic stability increases and melting points tend to lower, for example, $MoO_4(MO_3)_n^{2-}$, $n = 1-3$, 610°C, 540°C, 530°C for sodium and 495°C, 568°C, 546°C for potassium (Schofield, 2005). For the alkali metals, one Knudsen cell-mass spectrometric study of relative thermodynamic stabilities noted that the ranking was tungstates

>molybdates >sulfates >chromates (Choudary et al., 1975). With regards to coal combustion, an examination of their Gibbs free energy of formation in the condensed phase appears to indicate a thermodynamic preferential ranking for their formation of Ca > K > Na > Mg > Fe (Wagman et al., 1982; Iyer et al., 1990; Gavrichev et al., 2009). However, as temperatures are elevated their thermal stabilities against dissociate in the condensed phase to their respective oxides suggests an order K > Na > Ca > Mg > Fe with K_2MoO_4 being the most stable. The K_2O-MoO_3 bonding being $\Delta G_{298K} = 398$ kJ mol^{-1}compared to the FeO-MoO_3 bonding of 56 kJ mol^{-1} (Gavrichev et al., 2009; Chase, 1998). The trivalent ferric $Fe_2(MoO_4)_3$ molecule appears to be even more susceptible to reverting to its oxides while solid or liquid (Zhukovskii et al., 1972).

Gas phase stabilities and transport

Examining these chemicals in the gas phase with regards to molybdenum transport reduces the potential number of species. MoO_2(s) vaporizes mainly as MoO_3(g), the disproportionation in the solid phase controlling the balance (Burns et al., 1960). Studied by the Knudsen cell-mass spectrometric method, MoO_3(s) is seen to be very volatile with a low boiling point of 1105°C. However, it vaporizes initially as trimeric (major) and pentameric polymers (vapor pressure $(MoO_3)_3$(g) $\approx$ 4% atm at the m.pt.). There is no evidence of MoO(g), and MoO_2(g) is minor disproportionating to the trioxide (Berkowitz et al., 1957b; Blackburn et al., 1958; Ackermann et al., 1960). MoO_3(s) (m.pt.802, b.pt.1105°C) indicates a vapor pressure of 1 ppmv at 600°C and 170 ppmv at 700°C. At equilibrium in the confines of its own vapor, polymerization will be promoted. In an open combustion-gases environment, monomeric MoO_3(g) will most probably become dominant especially at low molybdenum concentrations. At sufficiently high temperatures this will ultimately equilibrate with MoO_2(g) and MoO(g). The three gaseous oxides have been studied and certainly have been shown to have strong bonding. However, the more recent measures of D_0(Mo-O) = 525, 520 kJ mol^{-1}, $\Delta H_{f298} = 380$, 384 kJ mol^{-1} did question some of the previous values from vaporization-mass spectrometric studies (Loock et al., 1998; Luo et al., 2016). These independent results were obtained by combining an accurately measured ionization potential to an apparent accurately known ion dissociation energy obtained from a guided ion-beam mass spectrometric method (Sievers et al., 1996). This is notable in that although the previous data are scattered, the three previous assessments had recommended values for D_0(Mo-O) of 556 (Pedley and Marshall, 1983), 547 (Cordfunke and Konings, 1990), and 594 (Chase, 1998). Other values that were obtained in this experimental study for ZrO and TiO using the same combination of ion data gave values in agreement with accepted magnitudes (Loock et al., 1998). More recently, a further examination has utilized a photofragmentation method and resolved this disagreement. This maps the velocities of the fragments and is able to identify the appropriate photodissociation channel. From this the value has been even more accurately established, confirms the other recent experimental value but tightens the error bar suggesting D_0(Mo-O) = 527 ± 2 kJ mol^{-1} (Cooper et al., 2017). This is currently recommended. Because experimental measurements have derived heats of formation for MoO_2(g) and MoO_3(g) these are essentially measures of the sum of their bond strengths. For MoO_2, a weakening of the first bond strength will imply a stronger second bond. Consequently, taking the recommended mass spectrometric vaporization data but now accepting this first bond strength of 527 kJ mol^{-1} changes the previously suggested values for the three MoO_3 bonds from 547/616/594 to now 527/636/594 (Cordfunke and Konings, 1990), or 594/563/583 to 527/630/583 (Chase, 1998). It moves the data into better agreement and suggests the bond strengths are in the order, second (strongest) and first (weakest). The latest constructed tables with D_0(OMo-O) = 636 ± 20 kJ mol^{-1} and $D_0(O_2$Mo-O) = 594 ± 20 kJ mol^{-1} are therefore recommended (Cordfunke and Konings, 1990). Older mass spectrometric studies can be used to test this new data from their measurements of the balance

$$2\,MoO_2(g) = MoO(g) + MoO_3(g)$$

For this, they had indicated a value of $\Delta H_{298} = -23$ kJ mol^{-1} (DeMaria et al., 1960), or −35 kJ mol^{-1} (Choudary et al., 1975). However, these current recommended values now indicate a change of +22 to now +42 kJ mol^{-1} (Cordfunke and Konings, 1990), or −20 to +47 (Chase, 1998), displaying how a new value for MoO can ripple through the chemical thermodynamics and illuminate existing remaining possible errors. These are very stable gaseous oxides, MoO_2 being comparable in bonding to that of NO (D_0 = 627 kJ mol^{-1}). As a result, if MoO_3 vaporizes it could form some balance with MoO_2. An FTIR matrix isolation study of laser-ablated molybdenum in oxygen did resolve the bond frequencies and geometries of MoO and MoO_2 (Bare et al., 1998). However, until recent years, optical spectroscopic measurements of any gaseous molybdenum molecule have been sparse due to spectral complexities and difficulties in obtaining theoretical guidance. At present MoO(g) is the most fully characterized, both experimentally and theoretically, and is shown to closely reflect the electronic states of CrO. In each of the three rows of transition elements, the monoxide bond strengths generally weaken with atomic number across each sequence but significantly increase with molecular weight in the first six columns of the transition elements. Consequently, gaseous MoO is seen to be weaker than ZrO and NbO but stronger than CrO, Tables 7.15 and 7.19. In three independent experimental studies, the ground state ($X^5\Pi$) has been confirmed and treated theoretically (Song et al., 2007). The molecule can display singlet, triplet and quintet spectral states. The quintet transitions have been most studied as they are connected to the ground state (Hamrick et al., 1991; Gunion et al., 1996). However, the triplet states that are the lowest lying ($a^3\Sigma^-$ and $^3\Delta$), both below 5000 cm^{-1}, are now well established (Gunion et al., 1996; Harms et al., 2014). The singlet states all lie to higher energies. This has been confirmed and aided by several theoretical studies that have resolved the nature of the bonding (Langhoff et al., 1989; Siegbahn, 1993a; Broclawik and Salahub, 1994; Broclawik, 1995; Broclawik and Borowski, 2001; Harms et al., 2014). Very little has appeared on the electronic structure of MoO_2(g), although it is intriguing that in this fourth column of the transition elements the second bond strength is larger than the first for both Cr and Mo but not W. Only one theoretical analysis appears to be available helping to explain its bent structure and enhanced bond strength (Siegbahn, 1993b). This also briefly addresses the trioxide geometry and structure.

Rather surprisingly the optical spectroscopy of molybdenum chlorides appears not studied experimentally. As seen above, molybdenum favors higher valencies and the chlorides readily vaporize and disproportionate to gaseous $MoCl_4$ and $MoCl_5$ which themselves are prone to oxidation and hydrolysis as are also the oxychlorides such as MoO_2Cl_2. This was shown by adding MoO_2Cl_2 to a flame and producing MoO_3 nanoparticles (Vergnon et al., 1980). Theoretically, the monochloride has been assessed with a bond strength of about 330 kJ mol^{-1} and a $X^6\Sigma^+$ ground state (Siegbahn, 1993; Cheng et al., 2007) but is listed also as 370 from vaporization studies (Knacke et al., 1991). Like the oxide, the second bond strength appears to increase in the $MoCl_2(X^5B_2)$ state, theoretically to about 345 kJ mol^{-1} and unlike its direct periodic table neighbors differs in having a bent structure (Siegbahn, 1994). The thermochemical values for the individual gaseous chlorides remain uncertain and are tabulated for the six in one database (Knacke et al., 1991) but only for $MoCl_4$, $MoCl_5$, and $MoCl_6$ in another (Chase, 1998). D_0(ClMo-Cl) and D_0(Cl_2Mo-Cl) are listed with values of 444 and 350 kJ mol^{-1} (Knacke et al., 1991). From the heat of formation and heat of vaporization of $MoCl_5$(s) this indicates an atomization or sum of its 5 bond strengths as $D_0 = 1700$ kJ mol^{-1} (Chase, 1998). This would suggest an average bond value of 340 kJ mol^{-1}, and with D_0(Cl_3Mo-Cl) 355 and D_0(Cl_4Mo-Cl) 262 kJ mol^{-1}. Knacke et al. (1991) estimate a stability of the five bonds as $D_0 = 1781$, in reasonable agreement. This pattern is repeated by tungsten's six chlorides but with a much stronger second bond. In both cases of Mo or W the MCl_5 and MCl_6 bonding weakens significantly.

As seen with chromium, the vaporization of MoO_3(s) is enhanced by the presence of water vapor. When condensed, this was found to be molybdic acid, $MoO_2(OH)_2$, hydrated MoO_3 (Wendlandt and Glemser, 1963). Quantified further by transpiration measurements at 600°C−690°C the direct proportionality to the water vapor concentration was indicated. These measurements indicated for MoO_3(s) in the presence of 10% H_2O in atmospheric pressure air that the vapor pressure of $MoO_2(OH)_2$ will be about 39 ppmv at 600°C and 320 ppmv at 700 (Glemser and Haeseler, 1962). These compare to the corresponding pressures of 1 and 170 ppmv MoO_3(g) in dry air quoted above. That the sole product was $MoO_2(OH)_2$ was further confirmed by a mass spectrometric study of MoO_3(s) vaporization by moist oxygen (Fryburg et al., 1977). A similar type volatilization by water was reported also for metallic Mo but at higher temperatures (1200°C−1500°C), also implying the thermal stability of $MoO_2(OH)_2$ to that temperature (Belton and Jordan, 1965). Although kinetically very complex this could proceed through an initial oxidation to the trioxide.

$$MoO_3(s) + H_2O(g) = MoO_2(OH)_2(g) \quad \Delta H_{1000K} = +27 \text{ kJ mol}^{-1}$$

$$Mo(s) + 4\,H_2O(g) = MoO_2(OH)_2(g) + 3\,H_2(g) \quad \Delta H_{1000K} = -284 \text{ kJ mol}^{-1}$$

This chemistry, common to Cr, Mo, and W relates directly to corrosion invoking formation of its hexavalent Mo oxydihydroxy molecule (Myers et al., 2019). Whether, corrosive rates are sufficient to compete with combustion rates is always debatable. However, one further indication of its more direct chemistry was indicated in a hydrogen flame study (Farber and Srivastava, 1973a). In that case, 1%−2% of MoF_6 vapor was added to oxygen-rich H_2/O_2 flames ($\Phi = 0.5-0.7$) at about 2300K. Probe-mass spectrometric sampling from the atmospheric pressure flame indicated the ions of MoO_2, MoO_3, and $MoO_2(OH)_2$. Although it can be questioned whether these accurately relate to the flame gases or are modified heterogeneously in the sampling probe, it does indicate, as with chromium, a relationship between gaseous MoO_3 and $MoO_2(OH)_2$. Also it suggests a mechanistic gaseous conversion of $MoF_6 \rightarrow Mo(g) \rightarrow MoO_3 \rightarrow MoO_2(OH)_2$. Although studied by FTIR matrix isolation methods, no simple hydroxide molecules have been observed with molybdenum (Wang and Andrews, 2006). Their structures relate to HMoO and H_2MoO_2 bonding.

Although the normal chlorides would not be expected to survive in combustion, MoO_2Cl_2(g) appears to be thermodynamically very stable to high temperatures. An analysis of stabilities indicates that the full atomization energy of MoO_2Cl_2(g) is 2032 kJ mol^{-1} compared to 1753 for MoO_3(g) (Pershina and Fricke, 1996; Chase, 1998). As a result, although MoO_3(g) hydrolysis will be occurring, depending on the concentration of HCl in a flame there might be a balance between the dihydroxy and dichloride molecules according to the following overall reaction, with MoO_3 itself becoming possibly minor if there are no reaction kinetic constraints:

$$MoO_2(OH)_2(g) + 2\,HCl(g) = MoO_2Cl_2(g) + 2\,H_2O(g) \quad \Delta H_{1000K} = -92\ \text{kJ mol}^{-1}, \quad K_{p,1000K} = 1.4 \times 10^3$$

In other words, if HCl is 1% and H_2O 10%, MoO_2Cl_2 could be 10-fold larger than the dihydroxide under such conditions. Thermodynamically, both molecules are stable to high temperatures, but on approaching 2000°C they will begin to dissociate and revert to MoO_3(g) with its own wide range of stability. Nevertheless, it does appear that molybdenum compounds may be more readily volatile than with chromium and be transported by any one of the three gaseous molecules, MoO_3, $MoO_2(OH)_2$, or MoO_2Cl_2. For the above calculations, recommended values have been assumed as $\Delta H_{f298}(MoO_2(OH)_2) = -851 \pm 4$ kJ mol^{-1} and $\Delta H_{f298}(MoO_2Cl_2) = -633 \pm 15$ kJ mol^{-1} (Chase, 1998).

Molecules with stable oxides invariably form corresponding sulfides. However, scarcely any information, experimental or theoretical, exists for molybdenum's gas phase sulfides. Being readily oxidized or hydrolyzed and of little astrophysical interest, this appears to have removed general interest. The three sulfides, MoS_n $n = 1-3$, were apparent in an FTIR matrix isolation study laser-ablating molybdenum in discharged sulfur vapor (Liang and Andrews, 2002b). This provided vibrational frequencies and structures. Also their DFT (Density Functional Theory) calculations suggested bond strengths of D(Mo-S $(X^5\Pi)) \approx 360$ kJ mol^{-1}, D(SMo-S) ≈ 450 kJ mol^{-1}, $D(S_2$Mo-S) ≈ 410 kJ mol^{-1}, weaker by 30% than the corresponding stronger oxide bonds. This was consistent with the earlier theoretical study that characterized the ground and low-lying states of MoO and suggested D(Mo-S) ≈ 344 kJ mol^{-1} (Langhoff et al., 1989). Molybdenum phosphate structures exist but as glass mixtures or catalysts of a complex nature. Such interactions remain possible in ash. Although not important in this study of combustion, the gaseous hydrides of molybdenum have been studied more extensively. This resulted on finding that transition metal ions could activate the C–H bond in hydrocarbons and introduced a need for M^+–H bond strengths. Also an FTIR matrix isolation experiment laser-ablating Mo with H_2 identified the MoH, MoH_2, MoH_4, MoH_6 species obtaining frequencies and structures (Wang and Andrews, 2005e). Two earlier ion-beam measurements had obtained values of 192 ± 12 (Sallans et al., 1985) and 222 ± 20 kJ mol^{-1} (Tolbert and Beauchamp, 1986) for D_0(Mo-H) suggesting an average value of 207 ± 15 kJ mol^{-1}. Theoretical calculations are in close agreement indicating values of 201 (Langhoff et al., 1987), 212 (Balasubramanian, 1990), 201 (Siegbahn, 1993), and 191 kJ mol^{-1} (Koseki et al., 2006) over the recent 30-year period. The latest work is also an in-depth study calculating potential energy curves and properties of low-lying states. Consequently, as expected, a suggested value of D_0(Mo-H) = 207 ± 15 kJ mol^{-1} reflects too weak a bonding to play any role in combustion. Only approximate theoretical calculations are available for the dihydride suggesting a slightly stronger second bond: D(HMo-H) ≈ 216 (Siegbahn, 1994) and 267 kJ mol^{-1} (Wang and

Andrews, 2005e). These gaseous recommended bond strengths for molybdenum species discussed here are summarized in Table 7.15.

Potential role for molybdates

As indicated above, molybdenum has a pronounced affinity with oxygen to produce mixed metal molybdates that are extremely stable in many cases, and also in the gas phase. Gaseous molybdates and tungstates standout as being among the most thermally stable of larger molecules and many display congruent vaporization that is molecular in nature. This was clearly demonstrated 50 years ago in Knudsen-type mass spectrometric studies that showed some gaseous molecular molybdates existing at 1200K–2400K (Verhaegen et al., 1965; Pupp et al., 1969; Johnson, 1975). Vaporization of polymolybdates can occur, but as seen such large complex structures would not be expected to have anything but a minimal lifetime in hot reactive gases before breaking down to simpler forms (Leonyuk et al., 1977). Since then the molybdate gas phase has been extensively studied. As listed in Tables 7.28 and 7.29 most molybdates have relatively low melting points and

TABLE 7.29 Thermochemistry of gaseous molybdates, kJ mol^{-1}

Molybdate m.pt.°C	Experimental Temp.°C	$-\Delta H_{f298}(g)^a$	$D_{298}{}^b$ Atomization	$D_{298}{}^c$ Metal oxide-MoO_3	References
Li_2MoO_4	1350–1430	1015	2990	502	Yamdagni et al. (1970)
705	–	989	2963	475	Choudary et al. (1975)
	1130–1230	1014	2989	501	Kazenas et al. (1998)
	–	1051	3025	537	Lopatin and Shugurov (2013)
Na_2MoO_4	1080–1380	1077	2947	?	Choudary et al. (1975)
687	–	1093	2964	?	Choudary et al. (1975)
	780–1030	1192	3062	?	Ermilova et al. (1976)
	–	1071	2941	?	Lopatin and Shugurov (2013)
K_2MoO_4	2000	1140	2974	?	Farber and Srivastava (1973a)
925	–	1129	2963	?	Choudary et al. (1975)
	900–1040	1158	2991	?	Kazenas et al. (2002)
	–	1134	2967	?	Lopatin and Shugurov (2013)
Rb_2MoO_4	–	1146	2963	?	Choudary et al. (1975)
938					
Cs_2MoO_4	–	1155	2964	?	Choudary et al. (1975)
957	800–900	1209	3018	?	Cordfunke and Konings (1990)
	860–950	1254	3063	?	Yamawaki et al. (1993)

(*Continued*)

TABLE 7.29 (Continued)

Molybdate m.pt.°C	Experimental Temp.°C	$-\Delta H_{f298}(g)^a$	$D_{298}{}^b$ Atomization	$D_{298}{}^c$ Metal oxide-MoO_3	References
$MgMoO_4$	1830	809	2605	492	Verhaegen et al. (1965)
1240	1380–1530	911	2714	601	Kazenas et al. (2004)
	–	803	2605	493	Lopatin and Shugurov (2013)
$CaMoO_4$	2130	866	2699	561	Verhaegen et al. (1965)
1445	1630–1900	859	2691	553	Lopatin et al. (2003)
$SrMoO_4$	1880	1013	2833	675	Verhaegen et al. (1965)
1040	1675–1810	912	2727	569	Lopatin et al. (2003)
	1400–1530	1068	2888	730	Kazenas et al. (2008)
$BaMoO_4$	1630–1820	970	2805	505	Cordfunke and Konings (1990)
1450	1370–1500	1065	2900	600	Kazenas et al. (2003)
	1770–1980	985	2818	518	Lopatin et al. (2003)
$TiMoO_4$	–	795	2924	508	Lopatin et al. (2014)
$VMoO_4$	1630–1830	676	2847	471	Gunina et al. (2012)
$CrMoO_4$	1640–1750	630	2683	473	Lopatin et al. (2003a)
$MnMoO_4$	1490–1630	638	2577	454	Stolyarova et al. (2004)
1130	–	655	2590	467	Lopatin and Shugurov (2013)
$FeMoO_4$	1600–1670	546	2618	462	Stolyarova et al. (2004a)
1068	–	560	2630	474	Lopatin and Shugurov (2013)
$CoMoO_4$	1440–1660	519	2599	452	Lopatin and Shugurov (2013)
1040					
$NiMoO_4$	–	432	2517	391	Lopatin and Shugurov (2013)
1275					
$BMoO_4$	–	744	2960	400	Lopatin et al. (2013)
Ga_2MoO_4	1310–1350	827	3021	376	Lopatin et al. (2009)
$SnMoO_4$	960–1100	699	2656	378	Shugurov et al. (2015)
$PbMoO_4$	890–900	676	2527	402	Kunkel et al. (2014)
1065	–	701	2551	426	Lopatin and Shugurov (2013)
$EuMoO_4$	–	962	2792	557	Lopatin and Shugurov (2013)

[a]*Known gaseous heat of formation.*
[b]*Heat of gaseous atomization.*
[c]*Dissociation energy into its two constituent oxides.*

their vapor pressures generally are quite low. For the alkali molybdates, temperatures need to be about 1200°C (alkaline earths about 1700°C) to move onto a ppm scale. As a result at high temperatures, although vaporization is congruent and initially molecular in nature, fragmentation to the base oxides can be the major component. As seen in Table 7.29 it is possible from the gaseous heats of formation to calculate the atomization energy. By subtracting the atomization energies of the two oxide components (MO and MoO_3) an approximate consideration of the additional stability gained by the MO-MoO_3 bonding can be assessed. As noted in the Table, this is particularly strong for the alkaline earth and the alkali molybdates and in the range of 500−700 kJ mol^{-1}. The one rare earth tested, $EuMoO_4$, also shows high gaseous stability, but others are weaker. It raises the question of whether these can have a pronounced molecular existence or formation in some combustion gases. It is hard to assess this for the alkali metals other than for lithium due to their lack of stable gaseous M_2O molecules, and thermochemical data.

Nevertheless, the trend is apparent and interesting. FTIR matrix isolation studies condensing the vapor from Knudsen cells with X-ray studies recorded all of the alkali molybdates and those of other elements, obtaining their frequencies and geometries (Bencivenni and Gingerich, 1982; Fomichev et al., 1992). Their thermodynamic properties have been established from very extensive studies over the years, primarily in Russia, either by observing vaporization or by the equilibration of the oxides at higher temperatures (Lopatin and Shugurov, 2013). However, it is to be remembered that these situations relate to Knudsen cell techniques where equilibrated situations can exist that are under no time constraints. This does not necessarily relate readily to combustion conditions where expected formation may not occur. Nevertheless it is obvious that Mo can form many molybdates that generally have melting points not much in excess of 1000°C. The early work indicated that the alkali molybdates occurred in the gas phase but probably fragmented above about 1300K (Johnson, 1975). The alkaline earth molybdates including Mg appear stable against fragmentation to possibly 2000K (Verhaegen et al., 1965; Pupp et al., 1969). Although questioned, but as listed in Table 7.29, the mass spectrometric probe sampling of a H_2/O_2 2000K flame containing both MoF_6 and a potassium salt solution did get an acceptable value for the heat of formation of $K_2MoO_4(g)$ (Farber and Srivastava, 1973a).

Laboratory combustion studies

The connection between laboratory combustion experiments and full-scale combustion varies. In the case of metals, and their general behavior in combustion, the two approaches can separate. In laboratory work there is greater flexibility and control. Being able to add thermally weak metal compounds directly into the flame gases tends to be directed more toward potential applications and chemical manipulation. In power plant combustion the behavior becomes more dependent on the chemical nature of the metal in the fuel, the chemicals and their possible release mechanisms. This is particularly important for the metals that are regarded as being nonvolatile. Molybdenum does have a volatile trioxide and important questions center on whether this can react in the fuel before or after release. More importantly, when vaporization does occur not only is this into a hostile environment full of radicals that will attempt to react with or break any molybdate into smaller components, but also that normal thermal fragmentation will occur at the higher temperatures.

The possibility of homogeneous molybdate formation within flame gases is an intriguing question. It has been invoked in two flame mass-spectrometric studies (Jensen and Miller, 1970; Farber and Srivastava, 1973a). In these cases, the molybdenum was supplied to the flame gases as either $Mo(CO)_6$ or MoF_6, implying these would rapidly dissociate and become gaseous MoO_3. In both cases their flames were also seeded with potassium to promote possible formation of gaseous $KHMoO_4$ or K_2MoO_4. As mentioned above, the derived heat of formation for the latter species from one of these flame studies is listed in Table 7.29 and is seen to agree with other accepted values. One thing mentioned already is that MoO_3 is the anhydride for H_2MoO_4, which may be a possible explanation of behavior in this case. Normally to form a larger gaseous molecule especially containing two K atoms such as $K_2SO_4(g)$ is a kinetic nightmare and in fact now has been shown unlikely. In the present case, the possibility of forming temporarily $H_2MoO_4(g)$ with the flame water is quite possible and provides a simpler mechanism of conversion to $KHMoO_4$ and K_2MoO_4, which are thermally very stable. These studies also are interesting in that the addition using MoF_6 appeared not to interfere, probably due to the dominant formation and nonparticipation of the very stable HF ($D_0 = 566$ kJ mol^{-1}) (Chase, 1998). Possible formation of KF(g), $D_0 = 489$ kJ mol^{-1} appears to be absent. Nevertheless, the possibility of forming such complex molecules in the gas phase will have kinetic constraints and a low probability. However, the two independent studies do lend some credence for this possibility, but the experiments are limited in scope and must remain speculative for now.

Various metallic molecules in gaseous forms have been added to fuel-rich $H_2/O_2/N_2$ flames and in one case measurements of the flame atomic hydrogen concentrations were monitored with and without their presence (Bulewicz and Padley, 1971). In many cases, this clearly displayed a catalytic recombination of the flame radicals with combustion inhibition. Chromium was seen to be the most effective in this screening of metals but molybdenum did display a more average level of effect. This was later studied in greater detail for molybdenum (Jensen and Jones, 1975). The additive was $Mo(CO)_6$ with a similar range of fuel-rich hydrogen flames with temperatures 1800K–2150K and additions of 1–140 ppmv metal. A noticeable reduction of H-atom concentrations was monitored. It was considered that a catalytic kinetic cycle invoking MoO_3, H_2MoO_3, H_2MoO_4 reactions in a cycle recombined the H-atoms. Whether this was in the gas phase or on particles remains speculative (Linteris et al., 2008). It is regarded though as a moderate flame inhibitor.

Although minimally mentioned as modifying soot formation in an overall authoritative review on combustion (Howard and Kausch, 1980), a paper published in that same time period indicated significant effects, depressing the soot volume fraction and coagulation, while increasing the particle number density (Haynes et al., 1981). Additions of ammonium molybdate, $(NH_4)_6Mo_7O_{24}$, that is quite water soluble, were found to be fractionally as effective as barium, the most efficient metal element known, and against which others are graded. This followed and confirmed previous work adding $(NH_4)_2MoO_4$ and Na_2MoO_4 that listed it half as effective as Ba (Cotton et al., 1971; Bulewicz et al., 1975). It was then believed that atomic or molecular metallic ionization might be the mechanism, but this is ruled out for molybdenum that has too high an ionization potential and does not readily chemiionize (Schofield, 2006). It was finally considered that the metal may become occluded in the soot particles with enhanced catalytic oxidation. However, the actual mechanism remains uncertain for most of the effective metals noted. One aspect of

Cr, Mo, and W that has been examined concerns their negative ion chemistry (Goodings, 1992). However, this relates more to electron attachment studies and controlling free electrons. In the power combustion industry it has little interest.

A further detailed, quantitative laboratory study did resolve the nature of molybdenum's deposition chemistry. The program's main aim was to find a method for preventing Na_2SO_4 deposition and corrosion, which can occur in any combustion system even with sub-ppmv concentrations of sodium and sulfur at temperatures below 1000°C. The condensed phase sulfate is very thermodynamically stable and it is difficult to disrupt its chemistry. After a long assessment of searching the periodic table, it was found that additions of molybdenum could be effective in modifying the observed deposition. Instead of depositing as the expected sulfate, any sodium or potassium in a flame was seen to preferentially favor formation of a polymolybdate deposit that was thermodynamically more stable (Schofield, 2005, 2009). These molecules are known as alkali bronzes and are benign at high temperatures with regards to corrosion. Both sodium and potassium behaved identically. To fuel-lean propane/O_2/N_2 flames at about 2000°C, quantitative additions of Na or K and Mo salt solutions could be nebulized into the flame gases together with SO_2 and Cl_2 all on ppmv scales. An air-cooled steel probe at about 500°C–600°C, located a few milliseconds downstream in the flame gases, collected deposited samples. It was noted on adding quantities of molybdenum in an exact atomic ratio to the alkali that changes were immediately apparent. The deposits were analyzed by Fourier transform Raman spectroscopy and X-ray powder diffraction. The deposits remained white as the sulfate but now were clearly different in nature. The Raman spectral analyses are indicated in Figs. 7.9 and 7.10, in this case for potassium with molybdenum, sulfur, and chlorine additions.

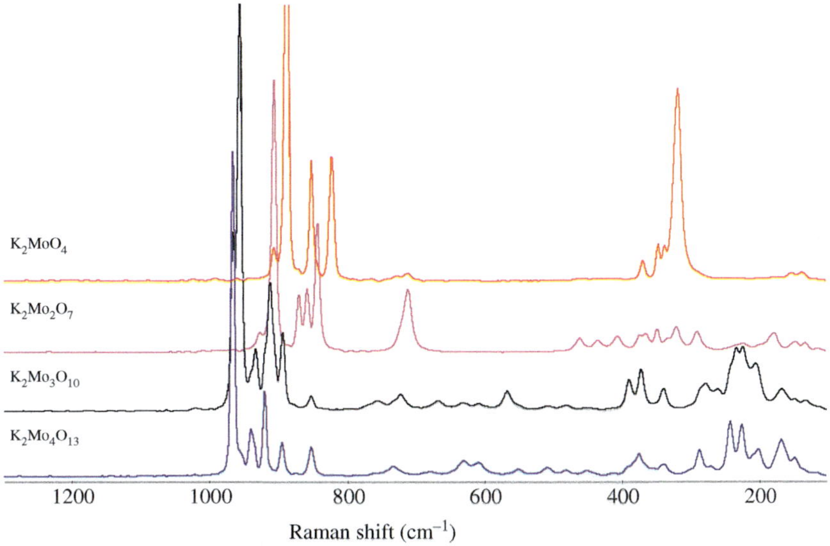

FIGURE 7.9 Fourier transform Raman spectra of pure samples of the first four potassium molybdates. Source: *Reproduced with permission from the American Chemical Society, from Schofield, K., 2005. A new method to minimize high-temperature corrosion resulting from alkali sulfate and chloride deposition in combustion systems. II. Molybdenum salts. Energy Fuels 19, 1898–1905.*

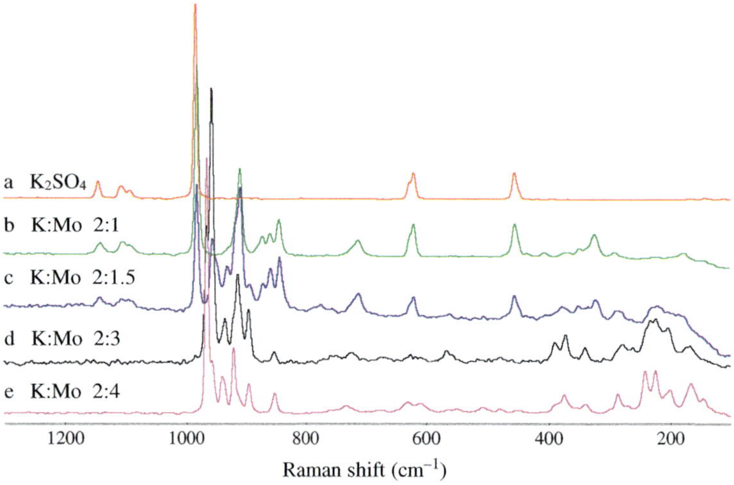

FIGURE 7.10 Spectra of deposits collected on a probe at about 500°C from a flame containing potassium, sulfur, and chlorine with Mo additions in the ratios indicated and discussed in the text. *Source: Reproduced with permission from the American Chemical Society, from Schofield, K., 2005. A new method to minimize high-temperature corrosion resulting from alkali sulfate and chloride deposition in combustion systems. II. Molybdenum salts. Energy Fuels 19, 1898–1905.*

Fig. 7.9 shows the clear analytical spectral differences that are expected for pure samples of the first four potassium molybdates. They each have very characteristic spectra. Fig. 7.10b–e are the spectra resulting from flame deposits for various experimental conditions, modifying the K:Mo atomic ratios. When the flame ratio of K:Mo was exactly 2:1 the spectrum (Fig. 7.10b) still is showing a pronounced K_2SO_4 deposit but changed slightly now indicating also some dimolybdate deposition. It required a ratio of 2:3 or 2:4 before the sulfate deposition fully ended and was replaced with one not of the simple molybdate but a pure polymolybdate, initially $K_2Mo_3O_{10}$ (Fig. 7.10d) and then pure $K_2Mo_4O_{13}$ (Fig. 7.10e) which are sufficiently thermodynamic stable to prevent any formation of the sulfate. These formations were additionally identified by X-ray analysis and indicate a high level of purity. Whether chlorine was present or not made no difference. Moreover, it was found if conditions are changed that any preformed or forming deposits will modify to their new thermodynamic preference in the new environment. In other words starting solely with potassium in a hydrocarbon flame a deposit of K_2CO_3 will be obtained. Then, if chlorine is introduced to the same flame the full deposit changes irreversibly to KCl and continues as such. Additions of sulfur can then convert all this to the sulfate, and as seen, further addition of molybdenum can convert this to a polymolybdate. Surfaces have preferential rankings that relate to thermodynamic stabilities.

Corresponding data are indicated in Fig. 7.11 for flame depositions with sodium and molybdenum. This more clearly shows the gradual reduction of sulfate formation as Mo is increased. In fact when the ratio of atomic Na to Mo is equal the dimolybdate is clearly evident.

This was an important discovery that resolved the nature of deposition from flame gases and was pertinent to ash chemistry in combustion. Moreover, this is heterogeneous surface chemistry, not gas phase related, and these very complex poly-molecules seen with molybdenum could not possibly form homogeneously (Schofield, 2005a). Relating to practical combustion with complex fuels such as coal, this illustrated that deposition can be modified continually as conditions change and with sufficient time. In the present program potassium or sodium clearly favored molybdenum over sulfur. Nevertheless, when a propane flame was further examined containing traces of K/Mo/Ca/S/Cl the potassium polymolybdate

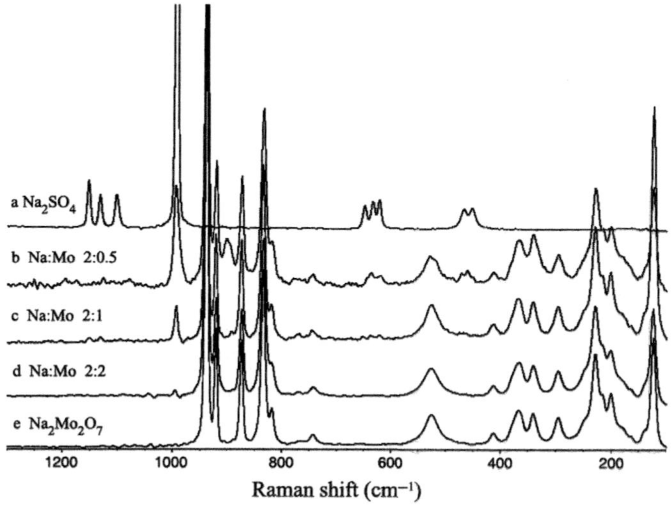

FIGURE 7.11 Corresponding spectra of deposits for sodium, sulfur, chlorine, molybdenum additions. Source: *Reproduced with permission from the American Chemical Society, from Schofield, K., 2005. A new method to minimize high-temperature corrosion resulting from alkali sulfate and chloride deposition in combustion systems. II. Molybdenum salts. Energy Fuels 19, 1898–1905.*

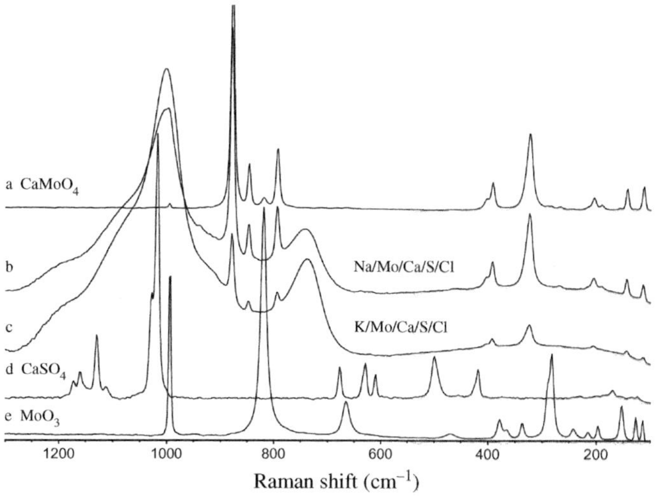

FIGURE 7.12 Raman spectra (b,c) of two flame generated deposits with equal flame ppm concentrations of alkali/Mo/Ca, compared to those of pure samples (a, d, e), confirming a preference for $CaMoO_4$ formation over alkali polymolybdates as discussed in the text. Source: *Reproduced with permission from the American Chemical Society, from Schofield, K., 2005. A new method to minimize high-temperature corrosion resulting from alkali sulfate and chloride deposition in combustion systems. II. Molybdenum salts. Energy Fuels 19, 1898–1905.*

previously deposited was no longer formed, but as apparent in Fig. 7.12, it now reflected $CaMoO_4$ formation. Calcium obviously can trump potassium by having a greater affinity for the molybdenum and the potassium then has to suffice with sulfur again producing K_2SO_4. Higher levels of Mo addition are necessary to overcome this Ca interference.

The thermodynamic values listed in Table 7.29 do not clearly predict this preference as strongly as seen later with similar behavior with tungsten and tungstates. Nevertheless it is hinted therein that alkaline earth molybdates do appear marginally more stable than those of alkali elements. This emphasizes the difficulty of assessing the products that will form in combustion ash because the full thermodynamic database remains incomplete and limits our predictive expectations of the order of the possible preferential rankings when numerous species are present. The pairings of all the species at their concentrations and temperature have to be assessed before a final correct thermodynamic ranking can be determined. What was clearly demonstrated in this study was that deposition is a surface formation mechanism that can modify if necessary to a more thermodynamically favored molecule. Also these deposits were noted as being individually chemically very pure if dominant.

Combustion power plant data

As indicated in the summary of Table 1.1, molybdenum is present in coals and wastes, is quite low in biomass and generally slightly more elevated in sludge. It is present at low levels in pet-coke and negligible in limestone (Cordoba et al., 2012). Nevertheless, it can show a slight presence even in natural gas and be higher in synthetic biogas (Cacchia et al., 2018). One study noted that attempts to preclean either Mo or U from some Chinese coals before combustion was very difficult (Duan et al., 2018). However, its nontoxic nature has tended to explain the lesser interest for examining its role in combustion. Its inclusion in studies often relates more to ash disposal and its implications for plants and aquatic organisms (Singh and Kolay, 2009). However, coal composition is becoming of greater interest as it becomes regarded as a possible source of trace elements. A recent discovery in China of an enriched Mo-coal has identified it as a natural source of value for Mo and Se (Long et al., 2019). In some cases it can be present at more significant concentrations in municipal wastes and sewage sludge and so is relevant to their ash disposal methods. More thermochemical data for molybdenum's chemicals now are available but have not generally been added to databases. As a result, equilibrium calculations involving Mo are almost nonexistent and if attempted indicate solely a dominance of $CaMoO_4$ if Ca is included (Poole et al., 2008). Consequently, although molybdenum can occasionally have a reasonable trace metal concentration in fuels, generally it has not been of interest (Xu et al., 2003). Occasionally, its presence is more noted in added pet-coke or limestone (Font et al., 2010). A few studies do exist of sewage sludge and wastes where it can be evident but some sludge is pretreated with $Ca(OH)_2$ and $FeCl_3$ which should certainly promote calcium molybdate formation (Folgueras et al., 2003). As a result, minimal full-scale combustion information has been reported. Even so, several studies have outlined its mass balance in combustion systems. Concentrations are found to be lowest in the bottom ash or slag and progressively increase in ash forms collected downstream (Kaakinen et al., 1975; Meij, 1994). In a suspension-fired reactor at 900°C with a coal/sludge/straw or coal/biomass mix the thermodynamically expected molybdenum hydroxyl release from the bottom ash was not apparent (George et al., 2010) and release was seen to vary from fuel to fuel and be modified by cofiring (Miller et al., 2002). In a coal gasifier it was noted that molybdenum release started at 725°C and was complete at 1025°C (Bunt and Waanders, 2009). In a steam gasifier of biomass at 800°C all the molybdenum was released (Cui et al., 2013), both probably indicative of MoO_3 or $MoO_2(OH)_2$. Most coal studies do report a high mobility (Flues et al., 2013) with gasification of the molybdenum and subsequent deposition on the fly ash. Behavior is reflective of As, Sb, or Zn with increasing enrichment noted on smaller diameter ash particles and clearly supporting release and deposition (Smith et al., 1979; Ondov et al., 1979; Coles et al., 1979; Meij, 1994; Querol et al., 1995; Martinez-Tarrazona and Spears, 1996; Spears and

Martinez-Tarrazona, 2004). No studies of submicron-sized ash have yet included Mo or analyzed its chemical composition. It does appear that the molybdenum is effectively collected by the ash and it is only due to the slippage of particulate control technologies that small quantities of a fraction of a percentage are emitted (Nielsen et al., 2002; Goodarzi, 2006) or 0.04% (Goodarzi, 2006a). In this regard, a recent measure of the effectiveness of a FGD system in a co-combustion plant burning coal and pet-coke indicated any stray Mo was further reduced significantly across this final control device (Cordoba et al., 2012). It is also at very low levels in limestone. In a more recent study of a state of the art large coal-fired Chinese plant, the stack emissions were monitored for 13 trace elements that succeeded in avoiding collection by the numerous control devices (Zhao et al., 2016). The levels of Mo exiting represented a 5000 reduction below the levels in the coal being burned or a capture rate of 99.98%. Molybdenum was found to be present more in the fly ash than the bottom ash and the molybdenum in the coal fly ash was one of the most water leachable (> 20%) in the 36 metals analyzed in one study (Llorens et al., 2001). Another study injected Trona (a sodium bicarbonate) into the flue gases, ahead of the air preheater and ESP, obtained ash with increased leaching for molybdenum (Dan et al., 2013). The normally high calcium coal ash apparently was overwhelmed by the sodium input creating soluble sodium molybdate over the insoluble calcium molybdate. Consequently, molybdenum's behavior is consistent with known chemistry but the finer details remain unstudied.

In summary, the combustion chemistry for molybdenum is somewhat restrictive. On heating, release from a fuel is most likely to be by volatile MoO_3 (b.pt.1105°C), the molecule most other molybdenum molecules will modify to. Whether its rate of release may be enhanced by water vapor or HCl under combustion conditions is uncertain. The gas phase may readily become a balance between MoO_3, $MoO_2(OH)_2$ and MoO_2Cl_2 at lower flame temperatures, but MoO_3 will be dominant at higher temperatures. Small ash particle enrichment is apparent. A role for chlorine or sulfur is not seen. If the fuel contains or forms molybdates these may vaporize molecularly and fragment depending on temperatures. However, on cooling, deposition on ash or other surfaces certainly will have the option of forming molybdates or polymolybdates as temperatures cool below 1000°C and provide time for deposition from the gas to surface phases. The molybdenum can be effectively retained predominantly on fly ash and by the FGD scrubber.

7.20 Nickel

Nickel is the eighth member of the first row of the transition elements having a d^8s^2 electronic structure. It follows the trend of the elements ahead of it showing a decreased stability of higher oxidation states favoring predominantly a divalent ionic character in its salts. In nature it occurs mainly as sulfides [millerite, NiS, pentlandite, $(Fe, Ni)_9S_8$] or silicate (Ni, Mg) ores. Mined globally by numerous countries at about 2.5 million tons $year^{-1}$ with abundant reserves. The United States finally commissioned a mine in 2014, having closed all its mines years ago even though the country is estimated to have 35% of the world reserves. Although the 24th most abundant element in the Earth's crust, its concentration increases toward the Earth's core which has been estimated to be about 6% Ni. It is rich in meteorites and exists with Mn as nodules on the Pacific Ocean floor; its presence in the water though is ranked about 31st of the elements. Now an important commodity, it markets for about \$4.70 lb^{-1}. Its processing over recent centuries, coupled to other sources, have added it as an anthropogenic component to the environment. Although difficult to appraise accurately, several older estimates of the natural atmospheric background nickel component indicated a magnitude of about 27,000–30,000 tons $year^{-1}$ globally, mainly from windblown dust (37%), volcanoes (47%), and forest fires (8%) (Nriagu, 1989a; Malakhov and Makhon'ko, 1990). Inflowing meteoric ablation has recently raised interest in Ni's chemistry in the upper atmosphere (Mangan et al., 2019). These values compare to approximate anthropogenic emissions now modified from about 56,000 to 80,000–95,000, mainly from oil combustion (48%), coal (25%), metal smelting (14%), incineration and wood combustion (3%) (Nriagu and Pacyna, 1988; Pacyna and Pacyna, 2001). Consequently, nickel is an atmospheric pollutant being probably two- to threefold greater in concentration than in preindustrial times. One review of natural emission levels included six trace metals (Cd, Cu, Hg, Pb, Zn) together with Ni (Richardson, 2001). Unfortunately this remains unacceptable indicating very large increases for all these from the previous estimates. That for Ni was increased by about 60-fold. The increases are so large that such natural levels would dwarf anthropogenic components and indicate no change from the preindustrialized period of time seen in core sampling or other time sequence analysis methods. Current values now should be decreasing with tighter regulations on metal smelting that previously may have been significantly underestimated (Boyd et al., 2009). One study for Europe indicated a drop in anthropogenic emissions of Ni by 42% from 2000 to 2010 due to improvements in controls (Pacyna et al., 2007). China, where about 63% of its trace metal emissions are from Ni, this originates from coal combustion. These should gradually decrease as more particulate control devices are installed (Tian et al., 2014). Its emissions from cement plants are less and in decline but still a factor of 2.3 above its 1980 output (Hua et al., 2016). In the United States nickel in particulate matter in air pollution is more closely related to oil combustion (Thurston et al., 2011). China has assessed its inventory of metal input to agricultural soils (Peng et al., 2019). For Ni, atmospheric deposition is the dominant enricher but remains of little concern.

Nickel's main use is in alloys and predominantly in stainless steel with Cr and Fe. It also alloys with Cu and Zn. Otherwise its other major uses are in nickel plating, to color (green) ceramics and glass, in batteries, Alnico permanent magnets, catalysts, and for its chemicals. It is not an expensive metal, about 1.0 US cent g^{-1} (\$4.70 lb^{-1}) such that the American 5¢ Nickel coin does contain about its value in the metal content. Several years ago, when nickel's market value rose quite high, there was a problem that coinage became worth more than its face value. It then was made illegal in America to melt coinage for its metal content.

Health concerns with nickel

In trace quantities, nickel is essential for humans, animals, bacteria and vegetation. Its main role is in various enzymes. Oral exposure to food is the main exposure, particularly chocolate, soybeans, nuts and oatmeal. A daily intake for adults is about 70–160 μg. Levels in water are low, a guideline being 0.07 mg L^{-1} (WHO Guidelines, 2006a). The leaching of Ni from stainless steel cookware can supply much of the body's needs. It is toxic at higher doses but these would be on a gram scale for an adult. It does not accumulate in animals, fish or humans but can in some plants. It is processed mainly by the kidneys and can be excreted in feces after only a short delay. Any absorbed into the blood is discharged in urine with a half-life of about a day. Its absorption depends on the food source. It is not a neurotoxin displaying more gastrointestinal discomforts or kidney failure. Toxicity or carcinogenic effects only relate to heavy industrial exposures. Otherwise nickel is a low risk metal with little medical concern. Industrially, impure nickel is further purified by conversion to $Ni(CO)_4$ (m.pt. − 19, b.pt.43°C) that can then be dissociated above 60°C. This is hazardous in that the tetracarbonyl is very toxic due to its CO content and it is this that is of most medical concern. In air the carbonyl is relatively unstable with a half-life of about 100 seconds and so can pose problems within the industry, but less so for any nearby community. The fact that it has been detected in sewage gas and could become a component in electronic waste recycling may modify these views (Thayer, 2010). A useful toxicology report for nickel is available on the Web from the Center for Disease Control (ToxProfile Ni, 2005). Generally nickel is not included in large human survey sampling, including the US Annual Human Exposure Surveys (CDC 4th Report, 2009, 2018). Biomarkers can be blood, plasma serum, urine or hair. Generally <4 μg L^{-1} in blood and <6 μg L^{-1} levels in urine samples are regarded as being healthy and of no concern. Levels in hair are low at about 1 ppmw normally (Drobyshev et al., 2017), but can be <6 ppmw if more industrialized (Li et al., 2014a; Zhu et al., 2018). Large surveys that involved from 100 to 2000 French healthy individuals indicated ranges of <4.2 for blood and <6.0 μg L^{-1} for urine (Goulle et al., 2005; Cesbron et al., 2013; Nisse et al., 2017). In heavily industrialized centers much larger ranges can be reported, <140 μg L^{-1} in blood of residents of the Pearl River Delta in China (Li et al., 2014a). A study of the population near a waste incinerator in Spain indicated that their body Ni levels decreased significantly from 1998 to 2007 with the introduction of better controls (Ferre-Huguet et al., 2009). A similar Spanish study compared autopsy values (1998, 2003) showing low levels of Ni in the body and mainly in the bone (Bocio et al., 2005). Concerns with metals often focus on potential pregnancy effects. In the case of nickel it can pass freely through the placenta to the fetus. Recent studies of pregnant women has indicated there may be a correlation with preterm low birth weight deliveries but this remains speculative (Chen et al., 2018d; Sun et al., 2018a).

Nickel's basic chemistry

Nickel is a biochemical with an enormous organometallic chemistry. Nevertheless, in combustion it is unlikely that their presence in fuels will be of significance other than being possible sources for the release of nickel. Its major roles generally are more commonly displayed in the nature of its inorganic molecules. These are predominantly divalent and cationic. The metallic nickel is stable and not prone to oxidation (m.pt.1455, b.pt.2913°C). The only major oxide is NiO (m.pt.1955°C), thermally stable and insoluble in water, but dissociates on vaporization. As will be seen, this stability implies a low vapor pressure on a ppm scale at about 1400°C. A trivalent Ni_2O_3 is reported but is not well characterized and may be a nonstoichiometric NiO, but is listed as decomposing at 600°C. A $Ni(OH)_2$ can be precipitated from solution but decomposes to the divalent oxide at about 230°C. The carbonate $NiCO_3$ also readily dissociates completely to the oxide by about 200°C (Stern, 2001). $NiSO_4$ is very water soluble but dissociates above 640°C–675°C directly to the oxide without melting (Ostroff and Sanderson, 1959; Tagawa, 1984; Stern, 2001). A $Ni(NO_2)_2$ is only stable in inert gases and below 260°C. The nitrate is stable and very water soluble dissociating at 260°C via the nitrite to the oxide (Stern, 2001). The $NiCl_2$ is thermally stable, water soluble and sublimes at its boiling point of 985°C. It begins to be hydrolyzed by water at elevated temperatures. NiO(s) can react with gaseous HCl and chlorine above about 350°C to form the dichloride which can sublime as temperatures are raised (Anufrieva et al., 2001). This might contribute to nickel's release and also be involved in changes that can occur in cooling ash. Elemental Ni can absorb and store quantities of hydrogen but no solid hydrides have been characterized and these are described more as an alloy between the two. $Ni_3(PO_4)_2$ is water insoluble and not well characterized. Nickel forms a series of sulfides, phosphides and silicides that have approximate stoichiometries, NiS (water insoluble, m.pt.976°C), NiS_2 (m.pt.1007°C) and Ni_3S_2 (m.pt.789°C). At higher temperatures in air these convert to NiS, then the oxide. Ni_2P (m.pt.1100°C), $NiSi_2$ (m.pt.993°C), Ni_2Si (m.pt.1255°C) are not well known and will be mainly favored under reducing conditions. It also forms a very useful but toxic carbonyl, $Ni(CO)_4$ (m.pt. − 19, b.pt.42°C). As with other transition elements, Ni will alloy with several metals and forms numerous mixed metal oxide compounds. These are of use as catalysts, nanoparticles or in battery technology (e.g., NiCd). A $NiAl_2O_4$ molecule of the two oxides has been characterized with a spinel type structure and can be produced as a nanopowder (Kurihara and Suib, 1993). Also in turbines it has been noted that NiO can sequester vanadium to suppress corrosiveness as $Ni_3V_2O_8$, and can also react with iron to form a stable ferrite, $NiFe_2O_4$ (Rocca et al., 2008; Di Salvia et al., 2015; Dyachenko et al., 2016). Synthesis of pure NiO nanoparticles by flame synthesis has become of importance especially as small particle diameters can be produced (Srikesh and Nesaraj, 2015). Because of its strong cationic nature, mixed metal oxides with the alkali and alkaline earth oxides are not expected to exist readily. As a result, in combustion considering this chemistry, the oxide NiO(s) is again dominant with only Ni(g), NiO(g) and $NiCl_2$(g) potentially playing gaseous roles. Otherwise nickel's other condensed phase molecules are not sufficiently thermally stable. This condensed phase nature of its compounds and their physical properties are summarized in Table 7.30 and provide several options for deposition or formation of mixed metal oxides in ash forms on cooling.

Gas phase stabilities and transport

Even from the early work it was concluded that NiO(s) vaporized predominantly by dissociation (Brewer and Mastick, 1951). This was confirmed by several further mass spectrometric Knudsen cell type experiments showing that NiO(g) was present on a low percentage ppmv scale (Grimley et al., 1961a). Also this was indicated as a Ni:NiO gaseous ratio of about 100:1 in equilibrium with the produced oxygen at 1440°C (Rutner and Haury, 1974), or 25 (1440°C), 76 (1730°C) (Farber and Srivastava, 1974), and 72 (1200°C), 36:1 (1350°C) (Watson et al., 1993).

TABLE 7.30 Properties of nickel's condensed phase molecules relevant to combustion, °C.[a]

m.pt. b.pt. vapor pressure
Ni 1455 2913, 100 ppmv/1677°C, ins
NiO 1955, extensive dissoc on vaporization, ≈10 ppmv Ni(g), O_2(g)/1440°C, ins
Ni_2O_3 dissoc 600 to NiO, ins
$Ni(OH)_2$ >230 dissoc to NiO, ins
$NiCO_3$ >150 decomp to NiO, ss
$Ni(NO_2)_2$ unstable, >260 decomp to NiO, s
$Ni(NO_3)_2$ dissoc >260 via nitrite to NiO, s
$NiSO_4$ dissoc >640 without melting to NiO, s
$NiCl_2$ sublimes, b.pt.980, deliq, resistant to oxid, 100 ppmv monomer/592°C, s
$Ni_3(PO_4)_2$ known but poorly characterized
Ni_2P 1110, decomp to phosphorus oxides at higher T, ins
NiS, NiS_2 ≈1000, oxid in air, ins
Ni_3S_2 635–806 melting range, oxid in air, ins
Ni_3S_4 >356 dissoc to lower oxides
NiH, NiH_2 unstable Ni/H_2 alloys
$Ni(CO)_4$ −19 42, volatile, dissoc >60, ins
$NiAl_2O_4$ 2110, stable, ins
$NiCr_2O_4$ known
$NiFe_2O_4$ known, ins
$NiSiO_3$ decomp 1545
Ni_2SiO_4(ortho-) 1545, ins
$NiTiO_3$ decomp 1394
$Ni_3V_2O_8$ known

[a]*dissoc*, dissociates; *decomp*, decomposes; *s*, water soluble; *ss*, slightly soluble; *ins*, *water insoluble*; *oxid*, oxidizes; *deliq*, deliquescent.

However, such closed cell experiments solely have the oxygen from the molecule in equilibrium. In open systems where oxygen is at larger concentrations this will even limit to some degree the dissociative vaporization and distribution. Recalculating the four mass spectrometric studies available concerning its bond strength, these indicate values of 376 ± 7 (Grimley et al., 1961a), 377 ± 12 (Smoes et al., 1972), 373 ± 3 (Farber and Srivastava, 1974), and 373 ± 3 kJ mol^{-1}(Watson et al., 1993). A recommended value of D_0(Ni-O) = 373 ± 5 kJ mol^{-1}, ΔH_{f298} = 302 ± 5 kJ mol^{-1} now is suggested for the $X^3\Sigma^-$ state. This reflects the strength of its periodic table neighbor, CoO exactly, but with differing electronic symmetries (Table 7.19).

This remains in agreement with Huber and Herzberg (1979), but upgrades other previous database recommendations 377 ± 17 (Pedley and Marshall, 1983), 366 kJ mol^{-1} (Knacke et al., 1991; Barin and Platzki, 1995). Since these evaluations, it has been noted that the lowest lying measured electronic state of NiO is the $A^3\Pi$ at T_0 4338 cm^{-1} (52 kJ mol^{-1}) above the ground state. Allowing for this, it modifies the chemical thermodynamic analyses slightly, but only reduces the bond strength value by about 0.4 kJ mol^{-1} (1500K) or 1.4 kJ mol^{-1} (2000K) even with its multiplicity (Ram and Bernath, 1992; Wu and Wang, 1997; Moravec and Jarrold, 1998). A recent theoretical paper reproduces this bond strength very well and also displays its lowest 54 potential energy curves (Sakellaris and Mavridis, 2013). It suggests that three other low lying states can exist, $a^5\Sigma^-$, $B^3\Phi$, $b^1\Delta$, one slightly below the $A^3\Pi$ and the others slightly above and so these may similarly have a small impact on the analyzed bond strength value. Final agreement between Second and Third Law data reductions as more data emerge will help resolve that issue. Consequently, NiO(s) is thermally stable and has low vapor pressures to well above 1000°C at which point depending on the ambient oxygen a distribution will occur between mainly Ni(g) with traces of NiO(g). The NiO and NiO_2 molecules have been isolated and their structures and frequencies characterized quite extensively by infrared matrix isolation methods (Citra et al., 1997; Danset et al., 2001; Allouti et al., 2006). There is some question that the observed structures may be dependent on the formation process energies. Also the dioxide implies a tetravalency for Ni, which is rare. However, for NiO_2, negative-ion photoelectron detachment measurements confirm both linear ($X^1\Sigma_g^+$) and Ni(O_2) triangular structures (Wu and Wang, 1997; Ramond et al., 2002). Reliable theoretical calculations for the dioxide have proven to be very difficult for a long time and still remain unsatisfactory. Numerous attempts (seven) have been published since 2000 and the consensus is that the symmetric linear structure is the most stable (Gutsev et al., 2000a; Deng et al., 2003; Bauschlicher, 2004; Allouti et al., 2006; Song et al., 2006; Uzunova et al., 2008; Hubner and Himmel, 2012). The most recent of these indicates that the linear structure is 148 kJ mol^{-1} lower in energy than Ni(O_2). From this study, calculated bond strengths for gaseous D(Ni-O_2) are 111 kJ mol^{-1} for the side-on triangular structure and 259 kJ mol^{-1} for the linear O-Ni-O. Based on the D(NiO) value above, this would suggest an even larger value for the second bond of D(ONi-O) = 380 kJ mol^{-1}. Although there has been no indication of gaseous NiO_2's importance, it is noteworthy that for Mn, Fe, Co, and Ni the dihydroxide and dichloride second bonds do appear to be consistently stronger than the corresponding mono-compound's bond (Table 7.19). This appears to be the case also for their gaseous oxides to lessening degrees. Using the ratio of the accurately known bond strengths below for the nickel chlorides, a value of D_0(ONi-O) ≈ 375 kJ mol^{-1} might be tentatively suggested. For nickel, second bond strengths appear to reflect that of the first. A further kinetic study of the reaction of ground state Ni with O_2 at 296K in an Ar buffer at total pressures of 5–700 Torr indicated a termolecular reaction with a measurable rate constant. Ground state Ni has the $a^3F_{4,3,2}$ configuration (energies 0, 1332, 2217 cm^{-1}) with the $3d^84s^2$ electronic structure. However, these levels are overlapped by the first excited state for the $3d^94s^1$ structure with an $a^3D_{3,2,1}$ configuration (energies 205, 880, 1713 cm^{-1}). As a result, even at room temperature, the apparently more reactive a^3D_3 level is significantly populated (22%) and remains in equilibrium as reaction occurs. Monitoring its decay by LIF, this was related to a termolecular rate constant as a function of O_2 and Ar pressures. From this an approximate estimate of D(Ni-O_2) of

≥ 238 kJ mol^{-1}was derived (Brown et al., 1991; Mitchell, 1992). This implies D(ONi-O) $\geq$ 359 kJ mol^{-1} for the linear structure, which may have the correct magnitude, but the experimental analysis and the nature of the interaction remains uncertain. What is remarkable is the nature of the required mechanism that at ambient temperatures might be able to break molecular oxygen, requiring 496 kJ mol^{-1} and recover this energy by forming the NiO_2 bonding. However, the nature of this bonding, linear or additive, remains uncertain. Nevertheless, in combustion, energies are available for such production and if kinetically allowed it might be able to compete thermodynamically with NiO(g).

As with iron and cobalt, a transpiration study indicated that at 1175°C–1422°C, nickel wire can be slowly vaporized by gaseous water vapor to form $Ni(OH)_2$(g) and H_2(g) (Belton and Jordan, 1967). The interaction was shown to be controlled by the reaction

$$Ni(s) + 2\,H_2O(g) = Ni(OH)_2(g) + H_2(g)$$

The formation of the dihydroxide varied proportionally with the water and hydrogen concentrations, but required quite high levels of H_2O(g) (0.4–0.5 atm) for measurements. An estimate of the average Ni-OH bond strength was made indicating 752 kJ mol^{-1} for the sum of both hydroxyl bonds. There was the implication that the bonding reflected trends of the gaseous hydroxides, chlorides and fluorides of Fe and Co. As seen below, reliable values for Ni-Cl and ClNi-Cl now indicate 369 and 371 kJ mol^{-1}, respectively, and a sum of 740 for the two. Consequently, this would seem to support the trend seen of a stronger second bond decreasing from Fe to Co and with Ni they are becoming of similar strength (Table 7.19). Scaling accordingly, this suggests the recommended values D_0(Ni-OH) $\approx$ 375 $\pm$ 15 kJ mol^{-1} and D_0(HONi-OH) $\approx$ 377 $\pm$ 15 kJ mol^{-1}. Laser-ablated Ni reacting with H_2O_2, or H_2/O_2 mixtures, has identified the monohydroxide and dihydroxide in matrix isolation (Wang and Andrews, 2006b). A linear O-Ni-O and bent Ni-O-H structure with frequencies were established.

The dichloride is a major salt of nickel and is very volatile, subliming and forming predominantly the monomeric vapor with negligible dimerization (Schoonmaker et al., 1959). At about 500°C its vapor pressure is 5 ppmv changing to 100 ppmv at about 600°C (Barin and Platzki, 1995). Electron diffraction measurements confirmed the linearity of the dichloride (Hargittai et al., 1991). Using the mass spectroscopic-effusion beam method, the equilibration was measured in the Ni/NiCl/Ag/AgCl and Ni/NiCl/$NiCl_2$ systems, together with a determination of an accurate value for the heat of sublimation of $NiCl_2$. Updating the thermochemical database for the five low-lying electronic states that can contribute to the thermochemical data (Muzangwa et al., 2011) this provided a recommended value of D_0(Ni-Cl) $= 369 \pm 7$ kJ mol^{-1} (Hildenbrand, 1995, 2008). Because $NiCl_2$(s) vaporizes to the monomer it has a well-defined heat of sublimation. Hildenbrand in his study derived this more accurately (Hildenbrand, 1995). Coupling it to the accurately known heat of formation of $NiCl_2$(s) (Chase, 1998) provides the value of $\Delta H_{f298}(NiCl_2)_g = -72.0 \pm 3.0$ kJ mol^{-1}. Together with the still slightly uncertain heat of evaporation of Ni(s), $\Delta H(\text{subl})_{298}$Ni(s) $=$ 430 $\pm$ 8 kJ mol^{-1} (Kubaschewski et al., 1993) this provides a value for the atomization energy of both bonds $D_{0atomiz}(NiCl_2)(g) = 740 \pm 9$ kJ mol^{-1}. In the mass spectrometric Knudsen cell study of the Ni/NiCl/$NiCl_2$ equilibration an adjusted confirming value of 740 ± 9 kJ mol^{-1} also was obtained (Hildenbrand, 1995). As a result the value D_0(ClNi-Cl) = 371 ± 10 kJ mol^{-1} can be recommended.

In a 55-year-old study, the behavior of atomic nickel thermal emission was monitored in a series of fuel-rich, reducing $H_2/O_2/N_2$ flames (Bulewicz et al., 1961). Low levels (0.1 ppm Ni) of an aqueous nickel salt spray were nebulized into the flames at temperatures of 1500–2600K with additions of chlorine up to 0.5% HCl. The optically thin atomic nickel emission could be directly related to the balance between Ni and NiCl in the reaction as a function of chlorine, namely

$$\mathrm{Ni + HCl = NiCl + H}$$

From a separate measure of the H-atom concentration, the equilibrium constant could be readily determined and converted to a value of D_0(Ni-Cl) = 368 ± 20 kJ mol^{-1}. This is in exact agreement with today's recommendation. This straightforward simple measurement and its agreement strongly confirms in such reducing flames that in such a case the Ni(g) and NiCl(g) are the only dominant Ni flame species. Chlorine levels were not sufficient to form $NiCl_2$ which would have been evident by curvature of the obtained straight line plots. Moreover, the chloride appears to be dominant over the equally strong monohydroxide due to the high chlorine addition and the more adverse enthalpy of the corresponding reaction with water. The energetics suggests a probable sensitivity to chlorine levels. Theory has had some difficulty with adequately modeling the electronic structural nature of the nickel chlorides. For many years now these have been benchmark test cases against which the adequacy of various theoretical functionals can be verified. With NiCl, a part of the problem arises from the five low-lying and overlapping electronic states below 2000 cm^{-1} with the $X^2\Pi_{3/2}$ ground state. The latest efforts by trial and error now can reproduce the bond strengths to be within about 5 kJ mol^{-1} for NiCl (Jiang et al., 2012) and 2 kJ mol^{-1} for $NiCl_2$ (Bross et al., 2013). Consequently, for nickel it is seen in Table 7.19 that the bond strengths have modified to the point that the oxides, hydroxides and chlorides are all almost of equal values. As a result, their distributions will be heavily controlled by water, chlorine and flame oxygen levels. It does appear that gaseous Ni and NiCl will be dominant even though the bond strengths are so similar.

As with other sulfides, those for Ni have been studied less than the oxide. It can form a range of stoichiometries with a melt around 1000°C that consists of NiS, NiS_2, Ni_3S_2, Ni(s) and S(g) species. Recent FTIR emission spectra observed the NiS ($A^3\Pi - X^3\Sigma^-$) bands in the 3300–4400 cm^{-1} region confirming that its spectral features reflect those of NiO (Ram et al., 2009a). Laser-ablated Ni mixed with the sulfur species from microwave discharged sulfur vapor noted NiS in a matrix isolation study, establishing its geometry and frequency (Liang et al., 2009a). Experimental studies concerning its bond strength consist of a reliable mass spectrometric Knudsen cell study comparing its bond energy with MnS (Drowart et al., 1967). Adjusting to the latter's current value this now indicates D_0(Ni-S) = 348 ± 15 kJ mol^{-1}. More recently, using a two-photon ionization spectral method, the predissociation threshold of gaseous NiS has been accurately measured (Mathew et al., 2017). This now indicates an apparently reliable value of D_0(Ni-S) = 352.3 ± 0.3 kJ mol^{-1}. This is consistent with the Fe/Co/Ni trends of MO and MS bond strengths that illustrate a gradual weakening of the MO bond and strengthening of MS with atomic number. The ratio of

D_0(M-S)/D_0(M-O) changes from 0.76: 0.78: 0.90: 0.94 for Mn, Fe, Co, Ni, respectively (Table 7.19). In the case of NiS it has a bond strength 94% that of the oxide, quite large for a metal sulfide. The next element Cu sees them essentially equal. Few attempts have been made to theoretically address the electronic structure and bond strength of NiS. Three earlier attempts span a 20-year period and leave many issues unresolved. The $X^3\Sigma^-$ ground state is confirmed but calculated bond strengths are D_0(Ni-S) = 325 (Bauschlicher and Maitre, 1995), 355 (Bridgeman and Rothery, 2000), and 301 kJ mol^{-1} (Wu et al., 2007) showing a disturbing range of variation. A current attempt still shows the sensitivity of the structure to the theoretical model used and although one method can get within 4 kJ mol^{-1}, the predicted range of values remains large (Sharkas et al., 2017).

Although there are no well-defined condensed phase hydrides, NiH(g) has been extensively studied spectroscopically and theoretically. The combination of a Ni $^3F(d^8s^2)$ ground state atom with an almost degenerate and more reactive $^3D(d^9s^1)$ state of the first row transition elements and a simple H atom introduces a significant challenge for both these fields. The spectroscopy is complex but indicates an $X^2\Delta_{5/2}$ ground state (Beaton et al., 1988) with low-lying $^2\Delta_{3/2}$, $^2\Pi_{3/2,1/2}$ and $^2\Sigma^+$ below 3000 cm^{-1} (Kadavathu et al., 1990; Marian, 1990; Gray et al., 1991). A matrix isolation study clearly noted absorptions from this ground state NiH (Li et al., 1997). Several experimental methods have measured its bond strength. From a mass spectrometric effusion cell equilibrium method a value of D_0(Ni-H) = 249 ± 8 kJ mol^{-1} was obtained from second and third law analysis methods (Kant and Moon, 1979). Using guided ion-beam mass spectrometry and reaction energy thresholds a value of 247 ± 10 kJ mol^{-1} was suggested (Fisher and Armentrout, 1990; Kickel and Armentrout, 1995a). A value of D_0(Ni-H) = 252 kJ mol^{-1} is currently listed in a database (Knacke et al., 1991) and is seen to be much stronger than the values for FeH (147) and CoH (188 kJ mol^{-1}) but similar to that of Cu. Theoretical estimates still have trouble establishing an adequate functional. Many attempts have been published. Those before 2000 have been summarized and of the seven considered they span a range of 208–269 and actually average to 248 kJ mol^{-1} (Harrison, 2000). Two studies have appeared since and give average values of 250 ± 7 (Furche and Perdew, 2006), and 269 ± 20 kJ mol^{-1} (Goel and Masunov, 2008) indicating that slight inconsistencies still exist. A value of D_0(Ni-H) = 249 ± 8 kJ mol^{-1} currently is recommended.

Of little interest to combustion, the gaseous NiH_2 also has been characterized. The matrix isolation study indicated an assignment of a singlet bent X^1A_1 state for NiH_2 with a low-lying linear $^3\Delta_g$ (Li et al., 1997). This disagreed with earlier theoretical attempts but has since been confirmed in three more recent theoretical calculations (Barron et al., 1998; Barysz and Papadopoulos, 1998; Zou and Boggs, 2008). No estimates are yet available concerning its binding energy. Although the phosphides are of little interest to combustion chemistry due to their reactive nature with oxygen, the electronic structure of those of the first-row transition elements have been analyzed theoretically (Tong et al., 2003). NiP($X^4\Delta$)(g), (D_0 265 ± 20), and VP($X^3\Delta$)(D_0 312 ± 20 kJ mol^{-1}) are exceptional as being the two most strongly bound (Table 7.19). Over a very extended period of time, the extensive Knudsen cell-mass spectrometric gas phase studies in Russia have only recorded nickel in the species $NiPO_2$, $NiPO_3$, $BaNiO_2$, $NiMoO_4$, and $NiWO_4$ (Shugurov and Lopatin, 2006; Lopatin and Shugurov, 2013). As mentioned above, due to its strong cationic nature, nickel is not known for readily forming mixed metal oxides and it appears that their bonding generally is not strong enough to allow molecular vaporization. Those known listed in Table 7.30 tend to decompose at temperatures of about 1500°C, with the exception of $NiAl_2O_4$ that is stable to 2100°C.

Laboratory combustion studies

Nickel's high temperature chemistry is limited and not easily monitored in flame gases. As a result it has not been studied extensively in flames. Early spectroscopic studies did not readily indicate analyzable spectral emission systems and nickel displays only weak emission from atomic nickel. The thermal dissociation of NiO was so pronounced that it initially prevented a determination of its bond strength (Huldt and Lagerqvist, 1954). Only one flame study using premixed hydrogen flames at about 1500–2600K containing nickel and chlorine obtained a bond strength for NiCl of 368 ± 20 kJ mol^{-1} in full agreement with recent recommendations (Bulewicz et al., 1961). This clearly confirmed that additions of $\leq 0.5\%$ Cl with nickel at about 10 ppmv produced solely Ni and NiCl forms in the flame gases. The oxide and hydroxide had to be negligible as also the dichloride. However, using combustion techniques it is still possible to synthesize NiO powders with varying shape, size and other characteristics (Hadke et al., 2015). Most metals have been examined in connection with their ability to inhibit combustion and also to suppress soot formation. The initial work adding a nickel salt solution to hydrogen flames indicated no catalytic radical recombination effects (Bulewicz and Padley, 1971). However, later work suggested that nickel chloride showed some effect and noted that the carbonyl was a reasonable anti-knock agent (Vanpee and Shirodkar, 1979). These comments appeared in a later review but also suggested that the carbonyl could even promote combustion under certain conditions (Linteris et al., 2008). As a result, nickel does not stand out as an inhibitor of merit other than suggesting that effects may be molecular compound, and combustion mode dependent. It probably is a reflection of Ni having no available kinetic looping mechanisms in the gas phase. The same kind of mixed results can be said for nickel's effects with soot. Studies showed either no or only slight downstream suppression (Cotton et al., 1971; Howard and Kausch, 1980; Haynes et al., 1981). A third study even suggested soot promotion on the addition of the nitrate to an acetylene/oxygen diffusion flame (Bulewicz et al., 1975). Its chemiionization in flames has been reported (Goodings, 1992). The Ni^+ (77%) and $Ni^+ \cdot H_2O$ (17%) ions formed from natural flame ion charge exchange reflects Co and is of little importance. This lack of chemical participation with Ni supports an absence of oxide and hydroxide in its combustion. It appears that the possible metal corrosion observed is too slight to affect flame compositions (Myers et al., 2019).

Combustion power plant data

Turning to the field of full-scale combustion and what has been noted, there is a realization that nickel has been of less interest until recent years. This partly arose from its lower chemical volatility and also low medical risk. Nevertheless, the measurements that indicated diesel and fuel oils can have reasonable levels of nickel in their ash and are present in roadside submicron particles has raised concerns (Wang et al., 2003b; Lin et al., 2005). Coal nickel content, like Co can be variable and nonnegligible (Table 1.1). Biomass levels can be even slightly higher and can vary as with municipal wastes. Sewage sludge has the potential for larger content. However, the two more significant fuel sources are residual oil and especially petroleum coke (Cordoba et al., 2012). Use of the latter has grown to very significant levels, now being a major fuel for cement kilns (35% of global demand) and in cogeneration power plants (Lee et al., 1999a). Its nickel concentrations can range from 200 to 3000 ppm. Limestone can contain measurable and variable levels (Lind et al., 1994; Cordoba et al., 2012). Present-day methods of producing biogas also find it can be quite enriched in nickel and other trace metals compared to natural gas (Cachia et al., 2018). With coal, it was initially noted that the major ash formations conveyed little enrichments (Martinez-Tarazona and Spears, 1996; Llorens et al., 2001; Reddy et al., 2005; Bhanarkar et al., 2008; Flues et al., 2013). A mass balance for power plants indicated nickel was retained in some form of ash (Aunela-Tapola et al., 1998; Nielsen et al., 2002). Measured stack emissions for escaping nickel as small particulates was reasonably less than 0.5% (Aunela-Tapola et al., 1998; Nielsen et al., 2002; Goodarzi, 2006). In fact now, a recent Chinese study has measured the stack emissions of 13 trace metals from a modern and large coal-fired plant (Zhao et al., 2016). This indicates 99.97% of the Ni can be retained within the plant's control features. A Spanish detailed study of the effectiveness of FGD retention of trace elements noted this device would reduce any stray levels of Ni by a further factor of 2- to 20-fold in a co-combustion plant burning coal and pet-coke (Cordoba et al., 2012). A study examining anthracite combustion in a down-fired furnace noted little in the bottom ash with enrichment in the fine fly ash. Ash removal efficiency by the ESP was >98% for Ni. Also that ash leaching was at low levels (Zhang et al., 2017b). Few pyrolysis studies are available but do indicate no release of nickel from coal, in one case noting no change at 1150°C (Wang and Tomita, 2003), but suggest an onset of volatilization above about 1225°C in full-scale gasifier measurements (Bunt and Waanders, 2009). In many coal studies that examined fly ash particle size distributions in greater detail, a noted enhancement was reported at sizes of 2.5 μm or less (Querol et al., 1995; Martinez-Tarazona and Spears, 1996; Nielsen et al., 2002; Goodarzi, 2006a). Two noteworthy studies used 11 or 13 stage impactors sampling the fly ash sizes after the ESP and analyzed the Ni content in the various submicron and micron ranges. The earlier work noted a bimodal particle distribution with a small peak centered at 0.05 μm and a larger one centered at 2 μm. Although the overall quantity of Ni was less in the fewer smaller particles, an enrichment of 1.2-fold was apparent over larger sizes (Kauppinen and Pakkanen, 1990). This was more clearly indicated in a more recent effort. Ni was one of the metals showing significant enrichments (50- to 100-fold) in the nano (0.03–0.06 μm), ultrafine (0.06–0.1 μm) and submicron (0.7–1.1 μm) fractions relative to the PM_{10} particles (Wang et al., 2008). Clearly, in a full-scale coal combustor, with temperatures sufficiently high, nickel can be released and redeposited. Because of the low nickel concentrations, and presence of other minerals, the chemical nature of the nickel in ash still is very uncertain. Experimental analysis methods are limited and only XANES has been pursued. This is a difficult technique with an ability mainly to characterize chemical type rather than specific identity (Galbreath and Zygarlicke, 2004). One study of ash from an Australian coal is the most recent for nickel (Shah et al., 2009). It suggested that 28%–80% of the nickel was $NiSO_4$, and 0%–15% as NiO, a difference that would show in leaching as the oxide is water insoluble and the sulfate soluble. However, other leaching studies of ash samples find that only about 1% of the Ni can be leached leaving the chemical nature uncertain (Llorens et al., 2001; Flues et al., 2013). A recent study has reported a threefold enhancement effect for Ni's fly ash concentration by simply reducing the flue gas temperature from 250°C to 150°C through a heat exchanger. The fuel was biomass from metal contaminated soils burned in a small pilot-plant boiler. Boiler temperatures are not quoted, but a general increase in ash concentration appeared to occur for nickel and many other elements as the temperature was so reduced through the heat exchanger. Further metal deposition is unlikely to increase these concentrations especially as increases were also noted for such elements as Ti or Zr that are fully nonvolatile. As a result, it is apparent that particle segregation was occurring in the heat exchanger and the ash size distribution was changing, favoring the smaller enriched sizes at the lower temperature (Kovacs et al., 2018). It illustrates the care needed in such interpretations.

Due to the higher nickel content, more studies have concentrated on residual fuel oil and pet coke. These produce less ash that makes it easier to analyze due to the higher metal concentrations. It was noticed that the characteristics of the particle formation depended on how efficiently the oil was burned. Fire tube boilers produced a broad coarse distribution of diameter sizes reflecting more char and unburned carbon. Most particles were >10 μm with 30%–40% <2.5 μm and only 1% in the submicron part of the bimodal distribution. However, if a refractory-lined combustor was used, to better simulate full-scale units, the distribution changed drastically to a unimodal peak centered on 0.1 μm. Clearly, nickel was only slightly vaporized in the first situation but now was displaying a higher temperature with 99% vaporization and depositing growth on the smallest of particles (Linak et al., 2000a). The work was repeated in a drop-tube furnace with high temperatures that similarly confirmed a dominance of small sizes (43% <0.1 μm), 90% <2.5 μm and 66% <0.5 μm (Jang et al., 2007). Higher temperatures favored a change to smaller particles. Attempts have been made to establish the chemical form of the nickel in these cases. The oils are chemically simpler than coal for their inorganic composition but can still vary considerably. Generally they will contain sulfur with low or negligible chlorine, with mainly Ni and V, and often lower levels of Na, Ca, Fe, and Si. X-ray absorption analysis methods remain the favored analysis. The most recent study used the $PM_{2.5}$ particles from both a fire-tube boiler and one refractory lined. The latter samples appeared to be mainly $NiSO_4$ whereas the fire tube furnace immobilized some of the nickel as the ferrate $NiFe_2O_4$, formed from reaction within the particle (Pattanaik et al., 2012). This conveys how little progress has been made over the past decade with the compositions still marginally known. Environmentally, the real concern was whether a carcinogenic sulfide could be produced but this was shown unlikely from the earliest work (Galbreath et al., 1998). However, simply due to the oxidizing combustion conditions this would not be expected. Due to the presence of Ni and V it is always possible that $Ni_3V_2O_8$ (m.pt.1310°C) also may be formed. This was recently confirmed in flameless oxidation of heavy fuels with added sprays of $VOSO_4$, $MgSO_4$ and $NiSO_4$ when inert $Ni_3V_2O_8$ was produced (Di Salvia et al., 2015). This addition of nickel had been indicated as a method for minimizing vanadium corrosion in gas turbines (Rocca et al., 2003). Using methods of examining the main particle size fraction, the water leached ash and the leached solution, it has become clear with high sulfur content that a mix of $NiSO_4$ and a nickel spinel such as $NiFe_2O_4$ can form. If sulfur is low, NiO may become a component (Huffman et al., 2000; Huggins et al., 2004; Galbreath et al., 2005). In corrosion studies it is becoming apparent that any codepositing NiO and Na_2SO_4 may react and possibly form Na_2NiO_2, an as yet uncharacterized nickelate. Under reducing gasification conditions, using a 50/50 mix of coal and pet coke (1700°C), 50% of the Ni was leachable, and there was the speculation that NiAs or arsenate may be present leaving the situation rather uncertain for such conditions (Font et al., 2005). Until a better analysis method is developed, the low specificity of the X-ray absorption spectra will leave any conclusions speculative. Gasification of oil-refining by-products such as asphaltene to produce syngas can run into corrosion and fouling due to the high content of Ni, V and S. For nickel this involves Ni_3S_2 deposit formation (Wang et al., 2016a). Moreover, mixing coal with pet coke makes for too complex a mix from which to obtain any fundamental data (Silva et al., 2013). More studies of the leaching of Ni may be an aid to resolve these compositional questions. They are also of interest in two other aspects relating to commerce. One is the possibility of recovering Ni from fuel oil combustion ash. Preliminary work suggested that 65% of the Ni can be extracted (Daous, 2005). Another is an environmental concern with cement kilns. These are now generally being used as a convenient method for coprocessing hazardous and other problem wastes along with the major fuel. In the case of nickel which is retained in the ash it becomes a component of a finished cement product. These now have been shown to have the potential for leaching the heavy metals and about 20% of the content for the case of Ni (Huang et al., 2012). A recent study examined the effects of chloride salt addition to the volatilization of nickel in co-combustion of sewage sludge in a cement kiln (Yu et al., 2017). They added up to 1.6 wt.% of either Al, Ca, Fe, or Na chlorides finding a dependence on additive and temperature. A 60% release of nickel was observed at 1450°C. One study examined the emissions from cement kilns noting that Ni emissions were controlled (Gupta et al., 2012). Due to its lower concentrations, there have been few studies considering biomass. With the generally higher concentrations of K, S, and Cl these will disguise any nickel effects to a large degree. However, one sampling of olive residues combustion ash with an 11-stage particle impactor and reaction temperatures up to 1450°C did note 11% of the nickel in submicron particles (Jimenez et al., 2008). For the case of combustion of wood bark and waste in a suspension firing reactor at 900°C, nickel was seen to be fully retained (Miller et al., 2006). Even 20 years ago it was also realized that nickel was not volatile in MSW incinerators and could be captured as ash with minimal emissions (Hasselriis and Licata, 1996). However, the incineration of sewage sludge alone or cofired can introduce higher levels of nickel and more importantly can have higher levels of chlorine content that may more easily volatize nickel as its dichloride. One dependence on chlorine concentration was noted that as Cl content was increased from 0.1 to 0.5 wt.% then 5% more Ni was released in sewage sludge incineration (Luan et al., 2013). This was also marginally observed for sludge combustion at 900°C with 0.06% Cl content but was reduced on cofiring with coal (Miller et al., 2004). However, combustion of a coal/sludge blend treated with $FeCl_3$ and CaO at 1100°C indicated no nickel release and hence no formation of either gaseous $NiCl_2$ or $Ni(OH)_2$ (Folgueras et al., 2007). Similarly, by adding NaCl, FeS_2, water and kaolin to co-combustion of a bituminous coal and sludge in a tube furnace at 1300°C, no effects of the Cl, S or water were noted on the nickel partitioning (Zhang et al., 2019). The nickel all remained in the ash. Passing gaseous HCl over sewage sludge ash heated to 1000°C also showed no effects (Vogel and Adam, 2011). However, when MSW ESP ash (207 ppm Ni content) was roasted at 1000°C for 3 hours with $CaCl_2$, a 30% loss occurred, and if similarly treated in flowing gaseous Cl_2 all the Ni vaporized, a behavior mirroring that of Co (Chan and Kirk, 1999). One experimental study that isolated Ni in a natural gas swirl flame incinerator was examined. This sprayed a $Ni(NO_3)_2$ aqueous solution into the flame at 1300°C with added chlorine such that chlorine was in excess of Ni by a 10-fold molar quantity (Linak et al., 1995; Linak and Wendt, 1998). The presence of chlorine noticeably modified the flame particle size distribution such that 80% of the nickel was in particles <1.1 μm with a distinct distribution of 0.05–0.1 μm clearly reflecting a vaporization/depositional growth mechanism. If kaolinite was present this retained the vaporizing nickel chloride, but not as the chloride, changing it to a nickel silicate or aluminate product. Recent concerns have centered on the FGD gypsum final product. One Chinese study measured the impurity content of gypsum collected from 12 coal-fired power plants (Hao et al., 2017). This showed a very low nickel content indicating no risk.

In summary, nickel requires either high temperatures or reasonable chlorine levels to aid vaporization. It has always been possible to monitor nickel by flame photometric atomic emission or absorption methods showing that the element is largely atomic in high temperature analysis flame gases (Fujiwara et al., 1975). Oil and coal combustion remain the major anthropogenic sources to the environment. Nickel is a low risk health metal raising few concerns. Its major oxide NiO is not volatile and also dissociates to the free atom. Other than this, the dichloride is thermally stable and volatile forming NiCl in the flame gases. Chlorine additions to fuels enhance nickel release and Ni and NiCl are most likely the gaseous transporters. Sulfur plays no role other than depositing on cooling ash as $NiSO_4$. The oxide may form mixed metal compounds with other metal oxides but options are limited. Nickel is at a low concentration in most fuels other than sludge, pet-coke or residual oil fuels. At higher temperatures, a vaporization/deposition mechanism is evident on fly ash with higher concentrations on the smaller particle size. Otherwise, if combustion temperatures are low, the nickel will be retained in the ash finally as the oxide, sulfate or some mixed metal spinel structure, for example with Al or Fe (Stollberg et al., 1997), or even react with alkalis to produce nickelate molecules (Mobin et al., 1990, 1996). Additionally, the fact that nickel displays no catalytic inhibition effects on combustion tends to imply its gas phase composition is simple. Inhibition of flame radicals requires cyclical mechanisms generally invoking hydroxide molecules, making any roles for gaseous nickel hydroxides appear unlikely in combustion. Capture on ash is very efficient with negligible atmospheric releases. Analyses of FGD gypsum find very low levels of nickel posing no disposal risk.

7.21 Phosphorus

Phosphorus is listed as the 13th element discovered, and it introduced the word phosphorescence into the English language from the glow emitted from the element. This remains, but with phosphorus the spectral emission now is more correctly known as chemiluminescence due to its short-lived molecular radiation from an oxidation chemical reaction. Phosphorus is common in the Earth's crust ranked at about 11th in concentration (0.1% by wt.) and is about 18th most common in the oceans (60 ppbw). It is always found in nature as an inorganic phosphate, the salts reflect pentavalency in the form of the $(PO_4)^{3-}$ anion. This reflects its electronic configuration of $P(3s^2 3p^3)$ with a $^4S_{3/2}$ ground state and a strong inorganic nature. The major ore is apatite, which is an impure form of tricalcium phosphate. About 230 million tons year^{-1} of the phosphate rock are processed globally driven mainly by the fertilizer market. Although reserves would appear to be large at about 67 billion tons, the growth of consumption suggests that the major commercial source areas may be depleted in about 100 years or less. This has raised major concerns for the future as the world's agriculture and food supply now have a total dependence on fertilizer application. If not adequately addressed, it could be the major problem in the next century, dwarfing any current global environmental aspects. Only about 1 million tons year^{-1} of ore is reduced to the pure element. Current costs are about \$122 ton^{-1} for the phosphate rock but much higher \$1.25–150 lb^{-1} for pure white/yellow phosphorus depending on its type and purity level, but generally sold as the more stable safer red allotrope. There are no gaseous forms of phosphorus in the atmosphere and any health risk concerns center on small particle inhalation. Anthropogenic contributions from combustion sources that include coal, oil, gasoline, diesel and incinerators have always been estimated as quite low and negligible, $\approx 5\%$ (Graham and Duce, 1979; Mahowald et al., 2008). However, recent concerns have questioned the atmospheric phosphorus budget model and now suggest a large difference between observed deposition rates and those assuming estimates for dust, sea aerosol, volcano, and biosources. This new analysis indicates that the difference has to result from underestimating the role of fuel combustion and also deforestation fires. The claim now is that the anthropogenic combustion contributions should be of a magnitude of 1.8 million tons year^{-1} and represent about 50% of the global atmospheric input (Wang et al., 2015). This work decreases significantly the several previous estimates for soil/particle erosion and for salt aerosol formation (Mahowald et al., 2008). There seems little justification for this, especially as biomass contains very low levels of phosphorus. As a result, until further data are collected the anthropogenic contribution to the atmospheric burden may remain in some doubt. Other work has suggested that river input to the oceans may be twice that of preindustrial levels due to fertilizer use, deforestation, soil loss and sewage sources (Savenko, 2001; Filippelli, 2008). River runoff is a major source of phosphorus to the oceans and is readily assimilated by the phytoplankton (Baturin, 2003). However, the overall concern from such data centers mainly on the loss of the phosphorus from the land to the oceans and whether it can be minimized to improve its agricultural usage. Also phosphorus in rivers has promoted unnecessary algal growth and oxygen deficiency in the form of a significant eutrophication problem in many fresh water sources. Several reviews emphasis this overall loss of phosphorus and outline the various approaches needed to capture phosphorus and enable global security (Cordell et al., 2011; Rittman et al., 2011). A more recent analysis emphasizes the range of benefits of recovering phosphorus and the need to seriously initiate this as a reality (Mayer et al., 2016). Moreover, recent measures of the significant lake and stream phosphorus concentrations in the United States would appear to make this imperative (Stoddard et al., 2016).

Phosphorus's main commercial usage is in fertilizers as the calcium dihydrogen phosphate, $Ca(H_2PO_4)_2$, that has low solubility but high bioactivity. The ore processing stage consumes a significant fraction of the sulfuric acid that is manufactured. Phosphorus has a wide spectrum of uses including its involvement in steel and Cu smelting, alloying as a phosphor bronze, in glass and chinaware, and as a fire retardant. Gaseous, highly toxic phosphine, PH_3, has importance for synthesizing the organophosphorus pesticides, compounds and weapon nerve agents. The isotopes ^{32}P and ^{33}P have a half-life suited to medical research. Initially present as the element in strike matches this was replaced by the sulfide for safety considerations.

One review gave an interesting history of the importance of phosphorus in agriculture and how we have traveled through different ages. Initially using animal manure, to now mainly fertilizers, to maintain a balance of nutrients in the soil. Soon a return to early methods using sewage and combustion ash may be needed to alleviate potential future shortages of natural phosphates (Ashley et al., 2011). As a result, research in this area is very active globally, particularly concerning recovery from sewage sludge (Sano et al., 2012; Havukainen et al., 2016; Mulligan and Sharifi-Nistanak, 2016; Sun et al., 2018; Fang et al., 2018). An extensive Swedish study analyzed 61 trace elements in sewage sludge and farmyard manure, including phosphorus, detailing the value and associated risks that can be encountered with its agricultural dispersion (Erikssen, 2001).

Health concerns with phosphorus

Phosphorus is essential to all life forms playing many major roles in the human body and DNA/RNA molecular mechanisms. An average body contains about 700 g mainly in the bones and teeth. A normal diet is suggested to contain about 3–4 g day^{-1}. It is a constant throughput element. It is absorbed into the blood from which the organs and soft tissues can be supplied and a little is stored in the bones. The kidneys then filter it from the blood but recycle it one time before excreting it in urine. By then it has been metabolized to the HPO_4^{2-} and $H_2PO_4^-$ salts. Historically it was from the analysis of urine that the element was originally discovered and one has to have great respect for their chemical analytical abilities. Urine contains the $(NH_4)NaHPO_4$ and the bones a hydroxyl-$Ca_3(PO_4)_2$. The latter is in teeth and water fluoridation modifies the phosphate-hydroxyl into the fluoroapatite form, which is a stronger material. Phosphorus generally is of low medical concern. Studies that are reported indicate adequate dietary intake of phosphorus in most countries (Olza et al., 2017; Trautvetter et al., 2018). Major sources are dairy, grain and meat products. Additionally maternal breast milk is found to be adequate (Kim et al., 2017; Altun et al., 2018). As a result phosphorus's levels in the body are not normally included in standardized tests. If there are concerns of dietary intake the recommended test is that of urine levels that should be <1.1 g/24 hours (Mayo Clinic, 2018). Hair samples can also be monitored but are very varied, in one study of children living in a remote area, a range of 6–240 ppmw was reported (Drobyshev et al., 2017). As will be seen, there now is renewed interest in recycling sewage wastes as a valuable resource of phosphorus. Dietary needs are essential but large excesses can begin to interfere with other mineral functions in the body especially with Ca, Mg, Fe, and Zn, and can enhance the loss of Cu. Also too little leads to disruptive and neurodysfunction. Most normally encountered phosphorus molecules are nontoxic and they control the rate of growth in plants and are involved in ecosystems. It is man-made organophosphorus structures that are highly toxic such as the fluorophosphate esters (ToxRpt Phosphate esters, 2012), and those used as pesticides and weapon nerve agents, and inorganic PH_3 with its related metal phosphides (World Health Organization Phosphine, 1988). An early review of phosphorus's air pollution aspects summarized phosphorus's chemicals and discussed their toxicities (Athanassiadis, 1969).

Phosphorus's basic chemistry

Phosphorus is present in most fuels and can be a nontrivial ash component, measured on a percentage scale. Because of this, and it being in the same periodic table group as As, Sb, and Bi, although a nonmetal, it is worthwhile to include it with the other trace metals in this review to examine its combustion behavior and how its chemistry may differ. Moreover, it is a good example of extrapolating its basic inorganic chemistry to explain and better understand the observations that are reported and are now seen as somewhat predictable. One aspect is its pronounced preference for pentavalency that weakens in this group of elements in the periodic table with increasing atomic number. Other valences do exist but so called reduced phosphorus compounds play few roles, but have been reviewed in depth (Morton and Edwards, 2005).

Elemental phosphorus is unusual in having different allotropic forms. The white/yellow structure (m.pt.44, b.pt.280°C) is the most reactive, highly toxic and least stable, reacting with air to produce a chemiluminescent glow and modifying slowly to a more usual yellow color. It is pyrophoric in air and stored under water due to its insolubility and for safety. It is a very effective white smoke munitions agent in warfare but is banned as an antipersonnel weapon. Its use as an incendiary weapon still is much debated in the laws of warfare. In sunlight and/or heating at 250°C in an inert atmosphere it changes to red phosphorus. This is denser, much more stable and less volatile. It sublimes at about 416°C. Also it is less toxic and not spontaneously flammable. It has been used in the later development of strike matches when it became the component of the strike strip. On heating to about 600°C in the absence of air and under pressure the phosphorus darkens to violet/black, the most stable and dense form but least reactive. This reverts to a stable orthorhombic structure at atmospheric pressure and is not easily ignited. The vapor pressures of the white and red allotropes are about 100 ppmv at 34°C and 216°C, respectively. They ignite in air at 34°C and 260°C. All forms melt to the same liquid state and if rapidly cooled produce the white form. Phosphorus vapor also condenses as the white form. Phosphorus reacts with oxygen and if this is restricted it produces the trivalent P_4O_6 (m.pt.24, b.pt.173°C). As the anhydride this can react readily with cold H_2O to form phosphorous acid (H_3PO_3) (structure $HPO(OH)_2/P(OH)_3$ but with an overall trivalent P bonding, m.pt.72, b.pt.200°C). Although P_4O_6(s) exists, its chemistry is not well characterized. It is less stable, decomposing above 210°C and is not a well-defined stoichiometry such that its heat of formation has not been measured. With excess oxygen, phosphorus oxidizes to the stable pentavalent P_4O_{10} (sublimes at 359°C, vap. press. 200 ppmv at 160°C, m.pt.426°C), soluble and similarly reacts with H_2O as the anhydride forming ortho-phosphoric acid (H_3PO_4) (structure $OP(OH)_3$, m.pt.42, b.pt.407°C). Its heat of sublimation and vapor pressure are well known (Gurvich et al., 1989). $P_4O_{10}(s)$ is one of the most effective drying agents known below 100°C due to its avidity for absorbing water. However, its ortho-acid, H_3PO_4, can be reversibly dehydrated on heating to produce pyro-, $H_4P_2O_7$, and metaphosphoric acids, HPO_3. In solution phosphorus has a very broad range of acids. Some such as hypophosphorous acid (H_3PO_2) have variable structures in equilibrium with one another, $HP(OH)_2$ and $(H)_2PO(OH)$ and illustrate the complexity. Those with trivalent phosphorus are thermally unstable and can oxidize to ortho-phosphoric with a release of toxic PH_3. Their acidic nature is controlled by their structures concerning whether they contain P-OH or HP = O bonding. Although quite complex in solution, these acids are the foundation for a wide range of molecules such as the very stable ortho- and pyrophosphates, K_3PO_4, $K_4P_2O_7$, metaphosphates, KPO_3, phosphites K_3PO_3, and hypophosphites KH_2PO_2. However, phosphites as one example, although the salts of phosphorous acid generally can exist in their condensed phase as mono- or disubstituted varieties such as NaH_2PO_3 due to these complexities of phosphorus solution chemistry.

Although phosphorus chemistry tends to emphasize its complex range of acids, other well defined phosphorus compounds such as its chlorides exist. PCl_5 (m.pt.167°C) is volatile and sublimes at 160°C with a measured heat of sublimation. It is toxic being hydrolyzed with water to ortho-phosphoric and HCl acids. Similarly, PCl_3 (m.pt. − 94, b.pt.76°C) is a liquid at room temperature, sublimes unchanged and is oxidized by Cl_2 to the pentachloride. It is very toxic and hydrolyzed by water to phosphorous and HCl acids. With O_2, the PCl_3 can be converted also to $POCl_3$ (m.pt. +1, b.pt.106°C), also toxic and hydrolyzed even in moist air to phosphoric and HCl acids.

Hundreds of simple and poly-phosphides are known with many elements of the periodic table and these are generally refractory with high melting points (Von Schnering and Honle, 1988). Some are indicated in Table 7.31. They are hydrolyzed by water producing PH_3 toxic fumes and so are dangerous.

Because they are finding application in numerous areas of science research, electronic structures of MgP, ZnP, CdP have been studied in matrix isolation (Fuller et al., 2010). Theoretical studies of the nature of their bonding and gas phase stabilities have been reported for the first row transition metal diatomic phosphides (Tong et al., 2003). These gas phase bond strengths are listed in Table 7.19 indicating low values and suggesting little importance in combustion even in oxygen-free systems. Similarly, there are many phosphorus sulfides of the type P_4S_n with $n \leq 10$. P_4S_{10} (m.pt.285, b.pt.515°C), and P_4S_6 (m.pt.290, b.pt.490°C), the analogues of the corresponding oxides and are of commercial significance as was P_4S_3 (m.pt.172, b.pt.408°C) in early strike match manufacturing. These are not as thermally stable and all are vulnerable to hydrolysis producing phosphoric acid but also with either toxic PH_3 or H_2S gases released. Because calcium phosphide can be an impurity in calcium carbide, hydrolysis of the latter in the manufacturing of acetylene can have this additional toxic hazard of PH_3. This potential trace impurity of PH_3 was noted as a background presence in a molecular fluorescence study of PO using an acetylene flame (Haraguchi et al., 1976). As seen, phosphorus chemistry is mainly dominated by its inorganic phosphates that are common for most metal elements and can be thermally stable with melting points and stabilities well above 1000°C. From the point of view of fossil fuel combustion this can involve potential formation within the ash of $AlPO_4$

TABLE 7.31 Properties of phosphorus's condensed phase molecules relevant to combustion, °C

m.pt. b.pt. vapor pressure	m.pt.
P(white) 44 280, 100 ppmv/34°C, ins	$AlPO_4$ 2000, ins
P(red) 44 431, subl 416, 100 ppmv/216°C, ins	$Al(PO_3)_3$ (meta) 1525, ins
P_4O_6 24 173, decomp 210, reacts w water	$Ca_3(PO_4)_2$ 1810, ins
P_4O_{10} subl 359 426, reacts w water, 200 ppmv/160°C	$Fe_3^{II}(PO_4)_2$ 1115, ins
H_3PO_4 (ortho) 42 407, $OP(OH)_3$, s	$Fe^{III}PO_4$ 1240, ins
H_3PO_3 (-ous) 72 200, $HPO(OH)_2/P(OH)_3$ mix, s	K_3PO_4 1640, s
H_3PO_2 (hypo) 27 130, s	$Mg_3(PO_4)_2$ 1348, ins
HPO_3 (meta) s	Na_3PO_4 1545, s
PCl_5 subl 160 167, decomp in water	
PCl_3 −94 76, decomp in water	AlP 2530, reacts w water
$POCl_3$ +1 106, decomp in water	Ca_3P_2 1600, reacts w water
PH_3 −134 −88, reacts with O_2, ins	Fe_3P 1100, reacts w water
PN decomp 830	Na_3P 650, reacts w water
P_4S_3 172 408, reacts w water	TiP 1950, reacts w water
P_4S_6 290 490, reacts w water	Ti_2P 1950, reacts w water
P_4S_{10} 285 515, reacts w water	

s, water soluble; *ins*, insoluble; *decomp*, decomposes; *reacts w water*, reacts with water; *subl*, sublimes.

(m.pt.2000°C), meta-$Al(PO_3)_3$ (1525°C), $FePO_4$ (1240°C), $Fe_3(PO_4)_2$ (1115°C), $Ca_3(PO_4)_2$ (1810°C), $Mg_3(PO_4)_2$ (1348°C), Na_3PO_4 (1545°C), and K_3PO_4 (1640°C) for many of which chemical thermodynamic data now are available (Knacke et al., 1991; Barin and Platzki, 1995). Other than the alkali phosphates these are generally insoluble in water. Although the melting point values are listed, many begin to vaporize dissociatively with emission of P_2(g) at much lower temperatures. In mixtures of elements, their presence will depend of course on the relative thermodynamic affinities of any metal present for phosphorus and whether chemical interreaction is possible. Table 7.31 summarizes the extent and physical properties of phosphorus's commonly encountered condensed phase molecules, and the limits encountered for combustion.

Gas phase stabilities

Phosphorus can show volatility in its chemistry. From early Knudsen cell-mass spectrometric measurements of vaporization the favoring of tetrameric structures was clearly evident for P, as also with As, Sb, and Bi elements in this periodic column. Elemental phosphorus with its low melting point vaporizes predominantly to a very stable P_4(g) with high vapor pressures of about 10^{-2} atm (120°C) and 0.3 atm (225°C). $P_4(g)$ is thermally very stable to at least 800°C due to the endothermicity of its breakdown first to $P_2(g)$ and finally P(g) at higher temperatures if in an inert environment.

$$P_4(g) = 2\,P_2(g) \quad \Delta H_0 = +229 \text{ kJ mol}^{-1}$$

$$P_2(g) = 2\,P(g) \quad \Delta H_0 = +485 \text{ (Barin and Platzki, 1995)}$$

Similarly with the major oxides $P_4O_{10}(s)$ and $P_4O_6(s)$, it was apparent in closed systems that these vaporized initially unchanged at low temperatures (Muenow et al., 1970). The vapor pressure of $P_4O_{10}(s)$ is about 200 ppmv at 160°C (Gurvich et al., 1989). Their structures are still demanding for theoretical description and they are considered as inorganic clusters (Chrissanthopoulos and Pouchan, 2008). However, if water vapor is present, the vapor becomes replaced by numerous H/P/O species. Although these large molecules are thermally stable to higher temperatures they are readily broken down to simpler gaseous molecules. Further mass spectrometric studies soon identified major gas phase roles for the PO, PO_2, HPO, HPO_2, and HPO_3 species (Hastie and Bonnell, 1980; Hildenbrand and Lau, 1994). It was noted that all the phosphorus compounds that were not thermally stable in moist air would dissociate to such gaseous species. These species began to be studied with respect to their thermodynamic stabilities. PO(g) had been extensively characterized spectroscopically and its lowest electronic states identified (Huber and Herzberg, 1979; Gurvich et al., 1989). Recent theoretical analyses have presented its potential energy curves for many of its states together with their parameters (Liu et al., 2013; Izzaouihda et al., 2014; Liu et al., 2017a). The ground state is $X^2\Pi_r$ and the $a^4\Pi_i$ is not low-lying. Moreover, a predissociation in the B′ $^2\Pi$ state provided a reliable value of D_0(P-O) = 590 ± 3 kJ mol^{-1}. Since then additional mass spectrometric studies of equilibria between PO and YO, GdO, and SnO oxides, measuring their relative strengths, has confirmed this recommendation (Pedley and Marshall, 1983; Gurvich et al., 1989). In addition theoretical analyses support such a magnitude (Bauschlicher, 1999; Haworth and Bacskay, 2002; Morgon, 2012). Less data are available for PO_2 and are all based on mass spectrometric equilibria either with Sn/SnO/P_2 or with PO/P_2 systems. Their data have been updated and now suggest a value of D_0(OP-O) = 497 ± 10 kJ mol^{-1} (Gurvich et al., 1989). This value also was supported by theoretical estimates (Haworth and Bacskay, 2002; Morgon, 2012). The frequencies and structures for both PO and PO_2 were noted in low-temperature matrix isolation experiments (Mielke et al., 1990; McCluskey and Andrews, 1991; Bauschlicher et al., 2000). The PO bonding is extremely strong, much stronger than the H-OH and O-O bonds, both of which are 493 kJ mol^{-1} and similar more to that of the phosphorus dioxide. The application of mass analysis to phosphorus in flames immediately conveyed the complex nature of the P/H/O interactions with not only a dominant gaseous PO but also PO_2, HPO, HPO_2, P_2, and possibly PN species (Hastie and Bonnell, 1980). As will be seen, although many of these P/H/O species have been invoked and included in kinetic models of phosphorus combustion only the HPO molecule has been studied in detail. This species was in fact attributed originally to the chemiluminescent emission observed from phosphorus in moist air and from flames. However, numerous later studies indicated a more complex situation where the emission can also originate from PO* and a $(PO)_2$* excimer (Van Zee and Khan, 1976, 1976a; Harris et al., 1985;

Hamilton and Murrells, 1986; El-Hag and Jano, 1989). Theoretical calculations confirm that HPO is the more stable geometric structure by about either 63 or 68–87 kJ mol^{-1}, respectively, rather than the alternate POH configuration (Luna et al., 1995; Haworth et al., 2002). Its bent structure has been resolved from its microwave spectrum and also the LIF of its (A-X) spectrum (Ozeki and Saito, 2003; Tackett and Clouthier, 2002). A recent theoretical study has confirmed its geometry and vibrational frequencies (Kitchens and Fortenberry, 2016). Theoretical studies also have appeared characterizing the two lowest electronic states (Nguyen et al., 1985; Lee et al., 2007a). There appears to be only one effusion cell-mass spectrometric experimental study that measures the stability of HPO. This implied $D_{0\text{Atomiz}} = 918$ kJ mol^{-1} (Hildenbrand and Lau, 1994). This value suggests that the bonding is essentially that of P-O and P-H as listed in Table 7.7 explaining its apparent importance. Since then, numerous theoretical calculations have been published indicating an atomization energy of 869 (Bauschlicher, 1999), 867 (Haworth and Bacskay, 2002), and 870 kJ mol^{-1} (Korobeinichev et al., 2005). This is an intriguing 50 kJ mol^{-1} discrepancy and appears to have been ignored. In fact this latter theoretical magnitude has been accepted and incorporated into the kinetic modeling of flame inhibition by phosphorus. Theoretical estimates also have been calculated for the heats of formation of other complex H/P/O gaseous molecules such as HPOH, H_2POH, H_3PO, HOPO, and $HOPO_2$ that may play kinetic roles (Haworth and Bacskay, 2002; Korobeinichev et al., 2005). Nevertheless, the gaseous PO and PO_2 species are extremely stable and will dominate the combustion gas phase chemistry. However, any stable reversion back to their condensed phase oxides will be possible but only at low temperatures due to their high volatility (P_4O_{10}(s) vapor pressure 200 ppmv at 160°C). In other words deposition only in the cool flue gas region.

Although phosphorus chlorides have commercial importance, their ease of hydrolysis and oxidation reduces their potential for roles in normal combustion systems. Gaseous PCl has been spectrally identified and PCl_3 and PCl_5 vapors result from their condensed forms. No experimental measures are available for the PCl bond strength but from comparisons a value of D_0(P-Cl) ≈ 300 kJ mol^{-1} was suggested (Gurvich et al., 1989). Theoretical calculations tend to support this magnitude with values of 296, 291, >298, 313, respectively (Gutsev, 1994; Papakondylis et al., 1995; Bravo and Machado, 1999; da Silva-Neto et al., 2004). A more detailed theoretical examination has calculated all the lower potential energy curves and now suggested D_0(P-Cl) = 299 kJ mol^{-1} in full agreement with the estimated value (Zhang et al., 2016b). For PCl the ground electronic state has a $X^3\Sigma^-$ configuration with a relatively low-lying metastable $a^1\Delta$ state with an energy of about 6500 cm^{-1} (78 kJ mol^{-1}). Zhao and Setser (1995) suggested a much larger value for D(P-Cl). Their statement that the PCl($A^3\Pi$) state dissociated to the same ground state limits as PCl(X) was in fact correct, but their thermochemical analysis not valid. The potential energy curve correlations now clearly show this is so, but that the $A^3\Pi$ state has a very shallow depth, <10 kJ mol^{-1} (da Silva-Neto et al., 2004; Zhang et al., 2016b). Gasous PCl_2, PCl_3, and PCl_5 do have accurately measured heats of formation (Gurvich et al., 1989) that can be converted into atomization energies summing their individual bond strengths. These indicate $D_{0\text{Atomiz}}$ values of 606.9 ± 6.1 (PCl_2), 960.8 ± 2.2 (PCl_3), 1284.7 ± 2.2 (PCl_5) kJ mol^{-1}. That for PCl_3 (957) was theoretically confirmed (Vasiliu et al., 2012). The theoretical calculations of Gutsev (1994) for PCl_n (n = 1–6) had indicated that the first four bonds were equally strong but the fifth was very weak and the sixth unstable (D_e 299, 283, 261, 263 kJ mol^{-1}). Consequently, it would appear reasonable to accept values close to 300 kJ mol^{-1} for each of the first three bonds in PCl_n with no stronger bonding possible in P/Cl molecules. For PCl_3 the values of Gurvich et al. (1989) are accepted and the individual bond strengths listed in Table 7.7.

Phosphine, the gaseous trihydride is thermally stable in the absence of water. It is flammable in air and becomes spontaneously so if traces of P_2H_4 are present. The molecule has been studied and the PH species is known from its spectral transitions. Like PCl its ground electronic configuration is $X^3\Sigma^-$, but it is unusual in that the "forbidden" PH ($a^1\Delta$-X) transition has been observed indicating a $T_{0,0}$ energy separation of 90 kJ mol^{-1} (7555 cm^{-1}) (Beutel et al., 1996b). An early photoionization mass spectrometric study reliably determined the PH_3 bond strengths as $D_0(H_2P\text{-}H) = 345 \pm 2$, $D_0(HP\text{-}H) = 310 \pm 8$, $D_0(P\text{-}H) = 295 \pm 8$ kJ mol^{-1} (Berkowitz et al., 1986). The values for these individual three bonds in PH_3 require such separate measurement for exactness due to their connectivity based solely on a heat of formation for PH_3. Otherwise, the value for one depends on values taken for the others. This factor does leave some uncertainty in this case. This arises from whether $D_0(H_2P\text{-}H)$ is 345 as reported above or an older value of 326 kJ mol^{-1} used by Gurvich et al. (1989). From the known heat of formation of PH_3, which is reliable, the atomization energy of all three bonds is 957 kJ mol^{-1}. The value for D_0(P-H) above is consistent with the assessment by Lodders (1999) and is only 5 kJ mol^{-1}smaller than that suggested by Gurvich et al. (1989). Theoretical calculations support a value in the range 294–305 (Haworth and Bacskay, 2002). Consequently accepting the 295 value the second and third bonds add up to 662 kJ mol^{-1}. Consequently, the 345 value above for the third bond strength indicates 317 for HP-H, and a 326 value would indicate 336 for HP-H. The theoretical calculations predict $D_{0\text{Atomiz}}(PH_2)$ in a range 607–618 and $D_0(HP\text{-}H) \approx 310$–320 (Gan et al., 1998; Ricca and Bauschlicher, 1998; Haworth and Bacskay, 2002). Consequently, this is an example of difficulties that arise in recommending data. In the present case, the analysis does tend to favor the work and values of Berkowitz et al. (1986) and differs slightly from current database values (Gurvich et al., 1989; Chase, 1998). In combustion such hydrides are unlikely to be encountered except in gasification or pyrolysis processes.

Corresponding to the oxide, phosphorus has a sulfide PS(g). Due to errors in its published thermodynamic table (Chase, 1998), this was revised and republished (Lodders, 2004). It now recommends $D_0(P\text{-}S) = 438 \pm 10$ and $\Delta H_{f298} = 152.4 \pm 10$ kJ mol^{-1} and the full thermodynamic table was recalculated. This is in agreement with the value of 440 ± 12 suggested by Gurvich et al. (1989) but both depend on a single experimental study, however, this is regarded as reliable (Drowart et al., 1973). The value compares to $D_0(P\text{-}O) = 590$ or 74% as strong, and is consistent with the sulfide/oxide ratio trend in its periodic column of N/P/As/Sb/Bi elements (Table 7.7). Corresponding values for this ratio with its direct periodic table neighbors indicate SiS/SiO (0.78) (Gurvich et al., 1991), S_2/SO (0.82) and SeS/SeO (0.86) (Table 7.34). PS(g) is very stable and could occur in oxygen/water free gasification or reducing conditions. Both PS and PS_2 have been analyzed in a matrix isolation experiment resulting from P_4 and S_8 in an Ar discharge. Structures and vibrational frequencies were obtained (Mielke et al., 1991). Theoretical descriptions still have difficulty reproducing this bond strength value for PS, but theory has calculated the potential energy curves for the 22 lowest lying states (Karna et al., 1988; Moussaoui et al., 1998).

As indicated above, Hastie and Bonnell (1980) sampled hydrogen flames seeded with the vapor of an organophosphorus compound monitoring the flame gases mass spectrometrically. The presence of small amounts of the species PN(g) was reported. In equilibrium calculations simulating oxygen-rich coal flue gases containing 50 ppmw of phosphorus it was also evident at a level of several percent of the phosphorus present (Frandsen et al., 1994). As temperatures elevate above 1100°C, PO(g) and PN(g) were indicated, but establishing equilibrium for such strongly bound molecules is very questionable at normal temperatures. Although uncertainty still surrounds the chemical thermodynamics of PN(g), there is little doubt this has similarly very strong bonding. For consistency, the previous thermochemical table (Chase, 1998) was adjusted (Lodders, 1999) and accepted the analysis of Gurvich et al. (1989). This now recommends D_0(P-N) = 614 ± 15 kJ mol^{-1}, a lower value from those taken previously: 681(Chase, 1998) and 676 (Barin and Platzki, 1995). It is based on a remaining rather uncertain mass spectrometric equilibrium study supported by a spectral extrapolation of the vibrational levels in the PN($A^1\Pi$) state. Nevertheless, PN is isoelectronic with SiO(g), D_0(Si-O) = 792, and CS(g), D_0(C-S) = 710 kJ mol^{-1} so a value closer to 700 certainly might be expected. It might be appropriate to reexamine the previous data. Experimentally, PN was produced from either the available fire retardant P_3N_5(s) or a mixture of UP(s) and BN(s) and then these were assumed to establish an equilibration between P_2(g)/N_2 and PN(g) (Gingerich, 1969; Ahlrichs et al., 1989). Having to form PN(g) from N_2, D_0(N−N) = 942, and P_2(g), D_0(P−P) = 486 kJ mol^{-1}, one has to question the nature of the necessary chemical kinetics especially in flames and whether its presence was actually an artifact of the mass analysis ion-formation sampling, or in equilibrium studies whether this can be plausibly attained. Most other studies forming PN(g) either utilize the dissociation of the solid nitride or have an electric discharge. It has a rich spectroscopy and its potential energy curves and spectral constants have been calculated (Grein and Kapur, 1983). This does not have any real significance for phosphorus combustion chemistry but it does stand out as anomalous in the flame/mass analysis data and requires further study before it can be fully accepted. A remaining gaseous species that exists is P_2 that has been very extensively studied spectroscopically with 16 stable electronic states identified (Huber and Herzberg, 1979). Due to the strength of PO it would be unlikely to play any role in an oxygen rich environment. Nevertheless, from a predissociation in the $C^1\Sigma_u^+$ state, supported by many mass spectrometric equilibrium studies with phosphorus vapor, an accurate value of D_0(P−P) = 485.6 ± 0.4 kJ mol^{-1} is recommended (Gurvich et al., 1989). If free phosphorus is vaporized, this explains its reluctance to fully atomize except at high temperatures.

Laboratory combustion studies

Phosphorus has been of significant interest to combustion particularly during the last three decades in various ways that have provided insights into its behavior. This has not related to pollution but to its fundamental chemistry. Numerous studies, about 30 years ago, reported interference by phosphorus with coal use in potassium-seeded magnetohydrodynamic (MHD) systems. From this it was concluded that the pivotal ionization mechanism was being perturbed by attachment of electrons to PO and PO_2 species

(Miller, 1978; Goodings and Hassanali, 1990), and that gaseous HPO, HPO_2 and HPO_3 were evident in the burned gases. Also that some of the potassium was possibly forming gaseous volatile phosphates with these such as KPO_2 or KPO_3 (Annen et al., 1981; Freedman et al., 1982; Unkel and Freedman, 1983). Interferences by phosphorus also began to be noted in flame photometric analyses of numerous metals involving either atomic absorption or emission modes (Long and Boss, 1985; Long et al., 1987; Maitra and Patsalides, 1987). This was assessed as arising from flame formation of metal phosphate aerosols. Since then, numerous Knudsen cell-mass spectrometric studies mainly by Russian groups have noted that many such MPO_3 and MPO_2 gaseous molecules can exist and have derived their heats of formation (Lopatin and Shugurov, 2013). From these values it is apparent that the alkali MPO_3 molecules are strongly bonded followed by their MPO_2 structures. Considering these as mixed oxides, from the MO and either the PO_2 or PO bond strengths coupled to the measured heats of formation, the strength of the additional bonding that results from their molecular formation can be calculated. For the five alkali metals (Li to Cs) values of $D_{298K}(MO\text{-}PO_2)$ are 687–720 kJ mol^{-1} and $D_{298K}(MO\text{-}PO)$ 647–652 kJ mol^{-1}. Corresponding values for the five Group 2 molecules (Be-Ba) are smaller, as are the known values for Fe, but these can exist in gaseous forms under certain conditions. As a result the strong interactions particularly between Na and K with P in flames are well evidenced and these species can exist in the harsh conditions of flame gases. Vapors above alkali phosphates have been mass analyzed at 1000°C–1300°C (Gingerich and Miller, 1975) and 860°C – 1180°C (Steblevskii et al., 1978). Such vapors also were condensed at 1200°C – 1330°C by matrix isolation and support a dominance of $NaPO_3(g)$ (Jenny and Ogden, 1979; Ogden and Williams, 1980). The gaseous alkaline earth phosphates also are seen but display dissociation. Mass spectrometrically analyzing the vapor above alkaline earth phosphates and phosphites clearly indicates these were producing in one case not only P_4O_{10} and PO_2 vapor species but also SrO(g) and $SrPO_2(g)$ (Lopatin and Semenov, 1986, 1987, 1990). Sodium metaphosphate produced 10 ppmv vapor pressures of $NaPO_3$ and $NaPO_2$ at 940°C and 950°C, respectively (Rudnyi et al., 1986). An earlier study with metaphosphates had also indicated vaporization by partial decomposition having MPO_3 vapor pressures of about 17 ppmv at 929 (Na) and 902°C (K) (Steblevskii et al., 1977). Such chemistry can exist in flames if the species are present and whether they can be adequately released in fuel combustion to participate in such a manner is another question.

Another area of phosphorus/combustion studies that has been of tremendous interest centers on phosphorus being a very effective flame inhibitor (Babushok and Tsang, 2000; van der Veen and de Boer, 2012). Due to this commercial use as a fire retardant, detailed studies of its chemical kinetic mechanism have been extensive. These invoke kinetic cycles of formation and loss of the gaseous HPO and HPO_2 species as the recombination steps for reducing the concentration levels of the propagating OH, H, and O radicals. Still ongoing, very extensive work by the Russian group of Korobeinichev (more than 20 publications) utilizing molecular beam mass spectrometric analyses of the species and other studies have now established and modeled this behavior of phosphorus compounds (Twarowski, 1995; Korobeinichev et al., 2000, 2005, 2007, 2009; Werner and Cool, 1999; Haworth et al., 2002; Mackie et al., 2002; Sullivan et al., 2004; Jayaweera et al., 2005; Shmakov et al., 2008; Shvartsberg et al., 2012). This rather ambitious undertaking extended the work further to organophosphorus compounds and their behavior in flames with even more elaborate kinetic reactions involving 41 phosphorus compounds and 202 reactions (Glaude et al., 2000; Korobeinichev et al., 2000a; MacDonald et al., 2001; Bolshova and Korobeinichev, 2006). This research is still very active with a rich data base (Mathieu et al., 2018; Li et al., 2019). The most recent study examined flame inhibition of phosphorus compounds in coflowing diffusion flames. Their kinetic model has been extended to 77 species and 886 reactions. They did note additions to the air stream to be much more effective (Takahashi et al., 2019). This obviously satisfies a modeling need for these effects. However, because most of the models are based on theoretical estimates of both heats of formation and kinetic rates they cannot be accepted as being chemically rigorous, but more of an engineering empirical modeling fit. Consequently, at present these cannot be seriously used as a basis for furthering chemical understanding or knowledge without additional insight into phosphorus's chemistry at a more accurate level of detail.

These flame laboratory type studies are very noteworthy in two respects. Although the majority of the flames were fuel-rich, one fuel-lean flame also confirmed the dominance of PO(g) and PO_2(g) with smaller contributions mainly from HPO_2 and less so from HPO. Also on using the so-called Na/Li method for measuring H-atom concentrations in fuel-rich flames various interferences were noted (Sugden, 1962). This method is based on the fact that in clean fuel-rich flames, sodium is present solely as atomic Na whereas lithium forms a distribution between Li and LiOH. Knowing the flame temperature and the equilibrium constant for the reaction

$$Li + H_2O = LiOH + H$$

a comparison of the relative concentrations of atomic Na and Li derived from emission or absorption measurements coupled to their transition probabilities is all that is needed to obtain the H-atom concentration absolutely. It is a simple and effective method for these ideal conditions. However, what was observed in the presence of phosphorus was its interference with the sodium and lithium. This was similar to that seen in MHD studies involving potassium where KPO_3 was positively identified (Freedman et al., 1982). Due to the high stability of HPO_2, Hastie and Bonnell (1980) considered $NaPO_2$(g) as the potential interfering species but could not rule out the other $NaPO_x$ possibilities.

If PO and PO_2 are dominant species in combustion gases rich in water it is kinetically reasonable to expect that reactions are plausible to produce their hydrogen bearing counterparts, if these are sufficiently stable, by mechanisms such as

$$\begin{aligned} PO + H_2O &= HPO + OH \\ &= HPO_2 + H \\ PO_2 + H_2O &= HPO_2 + OH \\ &= HPO_3 + H \end{aligned}$$

These are reactions involved in the observed phosphorus catalytic radical recombination and their rate constants have been calculated (Mackie et al., 2002). Similarly if other metal oxides or hydroxides are present it would seem equally reasonable to expect their interaction especially with the larger concentrations of PO or PO_2 to produce metal metaphosphate type molecules with an MPO_x formulation.

Molecular gas phase transport in combustion

As a result of this extensive chemistry, it is to be expected that phosphorus can be very volatile, depending on its chemical source. In the gas phase it will be transported predominantly as the gaseous forms PO, PO_2, possibly aided by water to HPO and managing to form fractional amounts of simple metal metaphosphates such as $MPO_3(g)$ if temperatures are not too high. As temperatures cool all these gaseous phosphorus species then will chemically deposit on ash or surfaces and form the thermodynamically preferred molecules, vapor pressures and time permitting. That such deposition is not simply a condensation process was actually realized half a century ago with phosphorus, but published in an Atomic Energy report with limited distribution (Grantham et al., 1961). Because of concerns relating to corrosion of metals by molten sodium polyphosphates, early Knudsen cell/mass spectrometric studies were undertaken. Above 700°C deposits of sodium trimetaphosphate were being formed on the cooler reaction walls. However, no evidence of such gaseous alkali phosphate species was detected above the molten sodium phosphates, only those of PO, PO_2, Na, NaO, and NaO_2. It was concluded that these species were "condensing and rearranging to form $Na_3(PO_3)_3$ on the cool surface." Even today this concept of chemical deposition still is struggling for acceptance in spite now of an extensive myriad of evidence. If metal phosphates exist in a fuel it can be expected that if temperatures are sufficient these will lose phosphorus and revert to the oxide and begin to evaporate dissociatively. A more recent Knudsen cell-mass spectrometric study of calcium phosphate at 1250°C − 1350°C noted such phosphorus loss from a mixed solid calcium oxide/phosphate mix (Nagai et al., 2009). Another study examined how to remove the solid calcium phosphate deposited on stainless steel surfaces in heat exchangers or flue gas cooling sections (Littlejohn et al., 2000).

Several thermodynamic equilibrium calculations have considered traces of phosphorus in coal combustion. These increase in complexity starting with a consideration of solely P/O species (Frandsen et al., 1994), and then in another adding Ca, Mg, and alkalis showing a potential role for their phosphates giving their possible operational temperature ranges under both oxidizing and reducing conditions (Yan et al., 1999). A more recent effort gives an interesting detailed portrayal of the condensed phases that may form in differing temperature ranges for two rather significant levels of phosphorus in the gasification of sewage sludge (Bourgel et al., 2011). This tends to confirm that calcium phosphate formation on ash can be favored with gaseous PO and PO_2 carriers. However, the calcium may have other competing preferences such as forming its sulfate or silicate. As noted throughout this book it has to be remembered that such thermodynamic equilibrium calculations have limited value in that they include no consideration of the availability of kinetic mechanisms being available to satisfy such expectations. They purely give an indication of relative stabilities and what may happen if there are no constraints. They are also only as good as the list of species included in their calculation data base.

As a result, from this known chemistry it is expected that phosphorus will be volatile. In oxygen or flame gases these will largely dissociate and be dominated by the very high stabilities of PO and PO_2.

Combustion power plant data

Phosphorus is present in most fuels and can be a nontrivial ash component. As seen from Table 1.1, its concentration is generally less than about 4% for coals and biomass. However, it can be variable in wastes and is of particular current interest with sewage sludge combustion where it can be a significant fraction. It occurs at low levels in pet-coke and appears negligible in limestone (Cordoba et al., 2012). In full-scale coal combustion most of the phosphorus can leave the fuel particle (Wu et al., 2011a). If it is associated with aluminosilicates this can vaporize even at lower temperatures. Otherwise, the dissociation of phosphate compounds will generally require elevated temperatures. Moreover there may be a small fraction that do not vaporize (Yu et al., 2009). As is usual, there will be a difference between phosphorus release from solid fuels, when it will depend on its chemical nature in the fuel, and the laboratory studies when it is added as a compound that readily disintegrates in the hot flame gases into simpler phosphorus species. With fuels, the phosphorus is found predominantly on the fly ash with small amounts, certainly less than 20%, depending on conditions, in the bottom ash (Meij, 1994; Martinez-Tarrazona and Spears, 1996). The dominant deposition onto fly ash facilitates its ready control. ESP measurements indicate 99.2%−99.5% collection efficiencies (Nielsen et al., 2002) that could be further raised to 100% utilizing a wet scrubber (Meij, 1994; Cordoba et al., 2012). Its chemical nature on ash has not been resolved in detail but Ca/Si phosphates are reported (Chen et al., 2005) and it has been noted in the aluminosilicate glass phase (Wee et al., 2005). In numerous ash particle size analyses it has been noted that the smaller sizes, $<1\ \mu m$, show a significant concentration enrichment indicative of vapor deposition (Martinez-Tarrazona and Spears, 1996; Nielsen et al., 2002; Yu et al., 2009; Teramae and Takarada, 2009; Wu et al., 2011a). The most recent ambitious study, sized coal fly ash into five categories, 2.2, 5.4, 9.7, 19, and 43 μm median diameters. Phosphorus was included in the 42 elements monitored and found to be increasingly enriched on the smaller particles (Lanzerstorfer, 2018).

Not really of concern for biomass combustion due to its low concentration levels, several studies of co-combustion with coal in FBC modes have noted a similar combustion behavior. Ca/Mg/K phosphate formation is apparent with total retention on the ash in the bed and negligible phosphorus emission (Valmari et al., 1999; Tissari et al., 2008; Piotrowska et al., 2010; Grimm et al., 2012).

Although shown to be of little concern for clinical waste incineration (Delay et al., 2001), this changes with combustion of sewage sludge, which relates at present to its main incinerator disposal method, used in particularly Europe and Asia. Not only does combustion significantly reduce its disposal volume but it can also be a source of energy. Moreover it is now being tailored, realizing that the ash is a useful source of minerals. It is generally rich in phosphorus with potential for agriculture after treatment to reduce any trace metals that may be excessive (Vogel and Adam, 2011). Due to the sludge levels it can be cofired with coal below about 1200°C to modify the ash size and trace element distributions (Wang et al., 2009a; Zhuo et al., 2012). Oxygen enrichment tends to promote phosphorus release as the oxides and enhances the small $<1\ \mu m$ particles due to the greater deposition on small nuclei (Zhang et al., 2007). Likewise above 1200°C and with 50% oxygen, phosphorus vaporizes readily and is increased in all size fractions <10 μm (Zhang and Ninomiya, 2007).

With the realization that we have only a finite supply of phosphorus in the world, there is a concern with recycling the phosphorus and its better management (Baker, 2011; Vaccari and Strigul, 2011; Dumas et al., 2011; Gupta et al., 2014). It is realized that sewage sludge, currently incinerated, can provide an ash rich in phosphorus and other trace minerals that is a promising large scale fertilizer source rather than a landfill waste (Mihelcic et al., 2011; Kruger et al., 2014; Peccia and Westerhoff, 2015; Westerhoff et al., 2015). However, it may be found in some cases that the phosphorus needs to be treated to become more bioactive as the phosphate (Herzel et al., 2016), and that some trace metals may need their concentrations modifying (Vogel et al., 2013; Ottosen et al., 2013). In comparison, biomass ash can be found to be too low for fertilizer use, not only in P but also in Ca and K (Poykio et al., 2010; Nurmesniemi et al., 2012). Consequently, cofiring of sludge with straw or biomass is useful and can better retain K in the ash K/Ca/P compounds produced (Li et al., 2013b; Ren and Li, 2015; Skoglund et al., 2016). Ca/Mg/Al silicates may be favored over $Ca_3(PO_4)_2$ even at high levels of phosphorus and if calcium is deficient, $P_4O_{10}(s)$ can be in the ash deposit. This may then dissolve in the silicate base and form a phosphorus glass as reported (Zhang et al., 2012).

In summary, although phosphorus is very volatile from fuels at higher temperatures it is volatile as predominantly gaseous PO, PO_2, HPO and possibly some alkali forms of type MPO_x. These have significant propensities for deposition on ash to form phosphates such as calcium phosphate. As a result, there will be lengthened transit time during combustion when the ash can produce predominantly phosphates as temperatures modify below their vaporization pressures. It will be controlled by the nature of the ash and the efficiency of the chemideposition of the various chemical options. They will have difficulty depositing as P_4O_{10} under oxygen conditions or P_4O_6 in low oxygen systems due to their high vapor pressures, and consequently these will be limited to low temperatures. Adequate particle control can ensure that minimal phosphorus is emitted to the environment in spite of its concentration in the fuel. Ash, rich in phosphorus, is increasing as an agricultural additive value.

7.22 Scandium

Scandium and yttrium are very similar in the first column of the transition elements but are chemically unlike the groups of elements (alkaline earths or transition elements) on either side of them in the periodic table. Yttrium is usually included with the lanthanides or rare earths and scandium is marginally on this border and sometimes included. Both are trivalent in nature. Scandium has the three electrons in its outer shells with a normal $3d^14s^2$ ground state electronic configuration, the first member of the transition elements. Scandium can occur in coals with concentrations up to about 10 ppmw and so is comparable to several other trace elements (Table 1.1). As a result it is interesting that the element has been minimally analyzed in coal or fuel research. Levels can be very small in biomass and wastes, but with sludge return to those seen with coals. Limestone and pet-coke levels are very low (Cordoba et al., 2012). It has been included in several coal analysis compilations but never discussed further in detail. Although the 31st most abundant element in the Earth's crust it has no concentrated deposits and is distributed in more than 100 mineral species. It is of interest to astronomy being more abundant in stars and the sun (23rd most abundant). Its commercial production is difficult and the global consumption tends to remain small (10 − 15 tons $year^{-1}$ globally of the oxide) being used in highly specialized applications. The growing demand is larger than the supply and it remains expensive, currently priced at \$4200 kg^{-1} for the pure oxide Sc_2O_3 and about \$15 g^{-1} for the pure metal. It is not mined directly but is a by-product extracted from the tailings of other mining processes. Its presence in coal ash like several other trace elements now has spawned the development of extraction mechanisms from this source. However, demand is spurring the supply, and Australia is estimating production of 65 tons $year^{-1}$ of the oxide (known as Scandia) at a lower future price of \$2000 ton^{-1}. This is from its first scandium mine in the world and is in the process of now beginning production. Being a lightweight metal, a main use is for alloying with Al, as an electronic ceramic, in high intensity light bulb technology and as an isotopic tracer in oil well drilling.

Health concerns with scandium

Its biology and biochemistry have been extensively summarized together with closely related yttrium in a two volume book series (Horovitz, 2000). These summarize the occurrence in nature, observing that annual tree ring analyses over 1200 years note only a few fold increase in environmental scandium. These also review dietary intake and nutritional studies examined by utilizing two of scandium's isotopes ^{46}Sc (84 day half-life) and ^{47}Sc (3.3 days half-life) with humans and animals. It illustrates the large volume of studies that have been made over the years, even with this element, that are necessary to obtain a reasonable understanding of any element for its behavior in living organisms. In fact it appears to be nonessential for life but being in the body is not totally inert but plays some minimal roles. There is no bioaccumulation in plants but its concentration is higher in roots rather than leaves and seeds. The metal is regarded as nontoxic and its compounds, although not studied to any degree, are listed as moderately toxic. Dietary intake is slight, estimated as 0.1–2.0 $\mu g\ day^{-1}$ (World Health Organization Trace elements, 1996) and has possibly not moderated over time by industrialization. No toxicology report has been released possibly indicating the low concern. It remains largely unknown by the medical profession, not commonly encountered, nor characterized yet with body biomarkers.

Scandium's basic chemistry

Scandium has a normal trivalent chemistry that is again dominated by a very stable refractory oxide, Sc_2O_3 (m.pt.2489°C). The pure metal has a m.pt. of 1540°C and b.pt.2836°C. Most of its compounds dissociate on heating to this oxide. A water insoluble trihydroxide can be formed, $Sc(OH)_3$ that dissociates at about 250°C as well as an oxy-hydroxide, ScO(OH), that also is not thermally stable. A hydrated carbonate exists, $Sc_2(CO_3)_3$, xH_2O, as well as a nitrate, $Sc(NO_3)_3$ and sulfate, $Sc_2(SO_4)_3$ which all dissociate at low temperatures. The trichloride is the only other stable compound of limited importance, $ScCl_3$ (m.pt.967°C) which sublimes to the polymers $n = 1-4$ and is deliquescent being soluble in water. This distribution of polymers simplifies as temperatures increase but the dimer can be important at low temperatures. By 580°C the monomer appears dominant with a vapor pressure of about 10 ppmv rapidly becoming more volatile (Patrikeev et al., 1973; Knacke et al., 1991). The oxide can react with chloride salts such as NaCl, KCl, $AlCl_3$ and rare earth chlorides in the condensed phase and such interaction has been used as a method of producing high purity $ScCl_3$. As a result, an option of a chloride/oxide balance might be plausible at high temperatures in the gas phase. However, when fly ash from a waste incinerator is heated at 1000°C for 3 hours with a chlorinating agent such as $CaCl_2$ or Cl_2 there was no sign of scandium's vaporization (Chan and Kirk, 1999). ScH_2 appears stable to about 1000°C in the absence of air but is not strictly stoichiometric and is described more as an alloy being poorly characterized. An ESR spectrum was obtained using matrix isolation of its vapor and suggested a nonlinear gas structure (Knight et al., 1981). A sulfide has been studied that reflects the oxide, Sc_2S_3, with a high melting point of 1775°C but dissociates to the metal and sulfur in the absence of air, otherwise it converts to the oxide. A phosphide ScP, also more appropriate to reducing conditions, similarly is very stable in the absence of air to temperatures up to 2000°C, but vaporizes congruently by dissociation to gaseous Sc and phosphorus (Franzen et al., 1980). Although minimally studied, Russian work by Rat'kovkii et al., published in Russian, did report examining the two trivalent phosphates $ScPO_4$ and $Sc(PO_3)_3$ by Knudsen cell effusion and mass analysis. They appear to be stable to >1400°C but then dissociate to Sc_2O_3(s), PO_2(g), and O_2. Sc compounds also are compatible with and can become incorporated into silicate glass structures. Otherwise although Sc is described as an amphoteric element, meaning it can appear as the cation or anion in molecules, in the present consideration it is strongly cationic. Few mixed metal oxide molecules are known and are not thermally stable to vaporization temperatures.

Consequently, other than the chloride, a nonvolatile nature is clearly evident. The physical properties of its rather limited range of condensed phase chemicals relevant to combustion are summarized in Table 7.32.

Gas phase stabilities and transport

Vaporization of the sesquioxide has been studied at high temperatures by Knudsen cell effusion and mass analysis methods (Ames et al., 1967; Rostai and Wahlbeck, 1999). It dissociates predominantly to ScO(g), which is the most stable diatomic oxide of the first row transition elements. Temperatures of 2280°C–2500°C were needed to create levels of vapor even for mass spectrometric measurements. Much lower levels of Sc(g) are present and the ratio of ScO(g)/Sc(g) was about 65. Like Y_2O_3 they appear to be the least volatile of the refractory oxides. Vapor pressures can be about 3 ppmv at 2250°C increasing to 60 ppmv at 2500°C. A significant effort to establish its bond strength has been reviewed

TABLE 7.32 Properties of scandium's condensed phase molecules relevant to combustion, °C.[a]

m.pt. b.pt. vapor pressure
Sc 1540 2836, 10 ppmv/1372°C, ins
Sc_2O_3 2489, refractory oxide, >2000 dissoc to ScO(g), ins
$Sc(OH)_3$ >250 dissoc to oxide, ins
ScO(OH) >400 dissoc to oxide, ins
$Sc_2(CO_3)_3$ >500 dissoc, ins
$Sc(NO_3)_3$ 125–150, dissoc to oxide, s
$Sc_2(SO_4)_3$ >600 dissoc to oxide, s
$ScCl_3$ 967 1086, sublimes to $n = 1–4$ polymers, 9 ppmv/580°C, s
Sc_2S_3 1775, dissoc to Sc(g), reacts with water
ScH_2 stable to 1000 in absence of air, nonstoichiometric metal/gas alloy
ScP >2000 dissoc to Sc/P, 50 ppmv/2000°C, oxidizes in air
$ScPO_4$ >1400 dissoc to oxide, stable in air
$Sc(PO_3)_3$ >1400 dissoc to oxide, stable in air

[a]*dissoc*, dissociation; *s*, water soluble: *ins*, water insoluble.

(Pedley and Marshall, 1983). Eight studies were evaluated and a value of 677 ± 11 kJ mol^{-1} was suggested. However, more recent work based on accurately measuring the ScO product energy distribution from the reaction of Sc with NO now provides the currently recommended value of D_0(Sc-O) = 667.7 ± 1.0 kJ mol^{-1} about 10 kJ mol^{-1} weaker (Luc and Vetter, 2001; Jeung et al., 2002). This measure differs in being independent of thermochemical analysis values. However, it is not easy to explain the difference, the correct degeneracy was used and there are no low-lying states to complicate the previous analysis. Until verified, some uncertainty has to remain. Recent theoretical estimates for this very stable $X^2\Sigma^+$ state support values in this 658–668 kJ mol^{-1} range (Mayhall et al., 2009). Potential energy curves have been designated indicating no low-lying states and lending close support to this bond strength (Miliordos and Mavridis, 2010). Previous theoretical descriptions of its electronic structure have been reviewed (Harrison, 2000), and the spectroscopy and the nature of its low-lying states predicted (Bauschlicher and Langhoff, 1986; Jeung and Koutecky, 1988; Wu and Wang, 1998). Although the oxides ScO and ScO_2 have been monitored with respect to geometries and frequencies in FTIR matrix isolation experiments, the dioxide is not readily produced in the gas phase (Chertihin et al., 1997a; Zhang et al., 2000). As a result experimental thermochemistry of the Sc, Y, La group dioxides is minimal. Values were finally obtained utilizing an indirect approach. This depended on obtaining the energy threshold of an endothermic ion-molecule reaction by ion-beam mass spectral analysis. The reaction studied was

$$ScO^+ + NO_2 = ScO_2 + NO^+$$

In this manner, the OSc-O bond strength could be obtained from known values of D (ON-O) and the ionization potentials of ScO and NO (Clemmer et al., 1992). The assessment appears reliable obtaining D_0(OSc-O) = 380 ± 32 kJ mol^{-1}, the error bar resulting mainly from that of the ionization potential value for ScO. The value initially was controversial as it was so much weaker (57%) than the first bond and unexpected at the time. The only value in a database had suggested 603 kJ mol^{-1}. This was an assessment based on the assumption that it might be similar to that of Ti or V with a value of about 90% of the first bond (Gurvich et al., 1982). Such trends are clearly apparent in Table 7.19 for the first row transition elements. However, theoretical calculations still struggle to adequately explain this structure which has two differing bond lengths (Siegbahn, 1993b; Gutsev et al., 2000a; Kim and Crawford, 2004; Uzunova et al., 2008). Even the most recent study screening a variety of density functional methods reports energy differences between possible C_{2v} and C_s structures and remains uncertain (Klawohn et al., 2018). The possibility of a hydroxide with scandium has been considered. In a matrix isolation FTIR experiment with laser-ablated Sc and water, both ScOH and HScO structures were characterized. The included theoretical analysis suggested that the ScOH structure was the more stable by about 28 kJ mol^{-1} (Zhang et al., 2000). A further theoretical paper confirmed that although close in energy, this ScOH structure was favored and that there could be a 170 kJ mol^{-1} energy barrier between the two structures: also that D_0(Sc-OH) ≈ 469 ± 8 kJ mol^{-1} with possibly too optimistic an error estimate (Miliordos and Hunt, 2011). This bond strength magnitude did reproduce that calculated earlier (Guo and Goodings, 2002). Laser ablation of Sc with H_2O_2 or H_2/O_2 produced $Sc(OH)_2$ and $Sc(OH)_3$ also in a matrix isolation FTIR study (Wang and Andrews, 2006a). Although it is rather unlikely that such structures could form and compete with ScO at high temperature, their thermochemical bond strengths and heats of formation have been estimated (Guo and Goodings, 2002).

The trichloride vaporization has been studied and it is quite volatile having a vapor pressure of 60 ppmv at 630°C, subliming extensively below its approximate 967°C melting point (Knacke et al., 1991). An electron diffraction measurement established the structure and indicated that 92% of the molecules were monomeric and 8% dimers at 630°C in the gas phase (Haaland et al., 1998). Gaseous ScCl is known to have a rich optical spectroscopy with an $X^1\Sigma^+$ ground state and three low-lying electronic states below 5000 cm^{-1} which will contribute to its higher temperature thermochemistry (Bencheikh, 1997). The lowest of these, $a^3\Delta$, lies at about 900 cm^{-1} (11 kJ mol^{-1}) and the potential energy curves for these states have been predicted theoretically (Taher-Mansour et al., 2008; Kardahakis and Mavridis, 2009). All these singlet and triplet states are seen to converge to the same ground state Sc(^{2}D) + Cl(^{2}P) limits so there is no confusion concerning its dissociation energy. Interestingly, the ScCl molecule is one to have missed the mass spectrometric technique and its dissociation energy remains unmeasured. Early spectroscopic extrapolation methods suggested a bond strength of 328 kJ mol^{-1} (Shenyavskaya et al., 1967), but this now is regarded as unreliable and too low. An earlier theoretical estimate obtained a value of D_0(Sc-Cl) = 439 kJ mol^{-1} and indicated a largely ionic structure (Langhoff et al., 1988). Using a Rittner ionic model a value of about 420 kJ mol^{-1} also was suggested (Struck and Baglio, 1991). Corresponding theoretical calculations in these two studies for the ScF bond strength indicated values of 579 and 578, respectively. However, in this case a reliable Knudsen cell-mass spectrometric measure is available indicating 595 ± 13 for ScF, so scaling the adequacy of the two theoretical analyses (Hildenbrand and Lau, 1995). More recently an additional theoretical value of D_0 (421 ± 20) for ScCl was indicated along with its potential energy curves (Kardahakis and Mavridis, 2009). As a result, uncertainty remains and from these three theoretical estimates it might be appropriate to suggest at present a value of D_0(Sc-Cl) = 438 ± 15 kJ mol^{-1}, which is somewhat below the ScOH(g) value. Over time these may move closer in value to one another. Little is known of the possible importance of the gaseous dichloride, the structure of which has been theoretically assessed together with its low-lying electronic states (Wang and Schwarz, 1998). An atomization energy of 999 kJ mol^{-1} was calculated for $ScCl_2$(g), clearly indicating strong bonding. Coupled to that of the first bond strength given above, this would indicate D_0(ClSc-Cl) ≈ 560 ± 25 kJ mol^{-1}, 28% stronger. A similar comparison of the two bonds with CaCl/$CaCl_2$ suggests a 25% difference while that for Ti is 12%. A stronger second chloride bond may therefore be expected for Sc but its exact magnitude needs confirmation especially as the second bond strength for YCl and YCl_2 suggests a weakening (Table 7.15). The heat of formation of $ScCl_3$(g) is known which indicates its atomization energy at 298 K is about 1412 kJ mol^{-1} (Knacke et al., 1991). This would suggest an average bond strength for each of the three bonds as 470 kJ mol^{-1}, which may remain consistent with these bond values above, but further confirmation is needed. As indicated already, although some erosion of Sc_2O_3(s) by HCl is possible, it appears to be a slow process and not likely to play a role in gaseous scandium release from a fuel. Its erosion in a very moist air environment (50% water) at 1000°C–1400°C examining the vaporization has also been reported (Courcot et al., 2010). Weight loss was very low but was linear with time and increased 10-fold in this temperature range. Loss was measured on a μg cm^{-2} h^{-1} scale and at 1400°C implied vaporization loss as ScOOH and $Sc(OH)_3$ at about 2 ppmv even at high water vapor concentrations. It will be negligible with respect to combustion.

Sc_2S_3(s) and ScS(s) are thermally very stable structures that vaporize at high temperatures (> 1600°C). Under reducing conditions, mass spectral analysis has indicated their dissociation forming a mixture of ScS(g), Sc(g), and S(g) and a value of D_0(Sc-S) = 460 ± 19 kJ mol^{-1} (Tuenge et al., 1976). This is slightly lower than the generally accepted value of 475 ± 10 kJ mol^{-1} obtained earlier in Knudsen cell-mass analyses of the equilibrium of the isomolecular exchange reaction between ScS(g) and Ge(g) (Coppens et al., 1967) and with Y(g) (Steiger and Cater, 1975a). The ScS bond strength is about 70% that of ScO which is normal for this section of the periodic table (Y, La, Ti, V, Cr) (Tables 7.15 and 7.19). Theoretical calculations established that the low-lying states of ScS reflect those of ScO and the $A^2\Pi - X^2\Sigma^+$ and $B^2\Sigma^+ - X^2\Sigma^+$ spectral transitions are well characterized (Huber and Herzberg, 1979; Bauschlicher and Langhoff, 1986). Later theoretical works have predicted values of 464 (Bauschlicher and Maitre, 1995) and 454 kJ mol^{-1} (Wu et al., 2007). Until further resolved the value of D_0(Sc-S) = 470 ± 10 kJ mol^{-1} might be suggested. Similarly in the absence of oxygen a thermally stable ScP(s) can exist. In a Knudsen cell-mass spectrometric study it was indicated that this vaporized congruently but dissociated to Sc(g) and P, P_2(g) (Franzen et al., 1980). The Sc(g) vapor pressure at 1800°C was about 3 ppmv increasing to 50 ppmv at 2000°C indicating the ceramic nature of the condensed phase. Nevertheless, by calculation it is seen that any gaseous ScP(g) is weakly bound with D_0(Sc-P) ≈ 146 kJ mol^{-1} and readily dissociates (Tientega and Harrison, 1994).

Its $X^1\Sigma^+$ state is interesting in that due to spin conservation, theoretically it will dissociate to the second limit of Sc(^{4}F) + P(^{4}S), 138 kJ mol^{-1} higher in energy than the first limit of Sc (^{2}D) + P(^{4}S). However, equilibrium studies and theory generally relate to the lower limit. Additional uncertainty appears to remain with later theoretical studies indicating a supporting value of 116 using multireference interaction methods (Daoudi et al., 2003), but a value of 264 kJ mol^{-1} for DFT methods, both for the ground state limit (Tong et al., 2003). As a result, a tentative value of D_0(Sc-P) = 130 ± 30 kJ mol^{-1} might be suggested until clarified further. Similarly a weakly bound ScH($X^1\Sigma^+$) (g) can exist. This has little importance in flames with D_0(Sc-H) = 201 ± 8 kJ mol^{-1} and was determined by Knudsen cell-mass spectral analysis adjusted for the low-lying $a^3\Delta$ state contribution to the thermochemistry (Kant and Moon, 1981; Armentrout and Sunderlin, 1992). ScH was characterized by the matrix isolation FTIR spectrum from laser-ablated Sc with H_2 (Wang et al., 2002). The nature of its electronic states and their potential energy curves now is well documented (Anglada et al., 1989; Jeung and Koutecky, 1988; Koseki et al., 2004; Goel and Masunov, 2008; Lodi et al., 2015), and indicates a low-lying $a^3\Delta$ state, as yet unobserved, situated about 2260 cm^{-1} (27 kJ mol^{-1}) above the ground state (Hubert et al., 2013). Numerous of the singlet state spectral transitions now have been characterized (Ram and Bernath, 1996, 1997). The nature of the molecular bonding in such transition metal hydrides has been reviewed (Harrison, 2000) with a further detailed study indicating a satisfactory theoretical description (Hubert et al., 2013). The dihydride appears not to have been observed experimentally. Nevertheless, theoretical descriptions of its potential nature are available (Balasubramanian, 1987; Fujii and Iwata, 1996). These values of stabilities of gaseous scandium compounds have been summarized with other first row transition elements in Table 7.19. Although flame ionization has little interest with scandium, its ions have been detected and mainly arise from charge exchange with the flames natural H_3O^+ ion (Goodings, 1992). In electric discharge systems, scandium atoms can react with oxygen atoms exothermically at low temperatures to produce electrons and ScO^+ (Schofield, 2006). This occurs due to large bond strength of ScO that is sufficient to chemiionize the oxide product.

Combustion power plant data

This survey of scandium's chemistry clearly indicates few options for scandium in any mode of combustion. It has a high temperature refractory nature based on its very stable oxide, which has a further option of interacting with other glass or ceramic forms. Consequently, the chemistry would indicate a nonvolatile nature. In fact, although scandium is the first of the transition elements, it is rarely mentioned in combustion. As noted in Table 1.1, this is a little surprising at first because it is present in most coals at concentration levels similar to numerous other trace metals that are more extensively studied. With the advent of ICP analysis it has been included in some studies when it seems a decision was made to monitor this or its related element yttrium. Moreover, due to the small quantities in industrial use, the levels that might arise in wastes and sewage sludge can be ≤ 10 ppmw making it of little concern if its ash is used in agriculture (Eriksson, 2001). Although limestone concentrations generally are low, care should be taken in interpreting

FBC results and stack loss which may result also from the lime in FGD cleaning (Lind et al., 1994; Font et al., 2010; Cordoba et al., 2012). From the earliest studies in combustion it was immediately noted that Sc had a nonvolatile nature. No enrichment occurs and concentrations were the same in the bottom ash, slag or fly ash as well as no variations were seen in the fly ash as a function of particle size (Coles et al., 1979; Meij, 1994). Even in a higher temperature drop tube furnace with lignite at 1780°C or 2180°C scandium showed no particle size effects and could be used like Al as a reference element for calculating enrichment factors of other trace metals (Quann et al., 1982). One study examined coal/sewage sludge pyrolysis at temperatures $\leq 1100°C$. The sludge had been treated with $Ca(OH)_2$ and $FeCl_3$ and contained 1.5 ppmw Sc and 1.2% Cl. The coal had a Sc content of 11.4 ppmw and 0.1% Cl (Folgueras et al., 2003). The nonvolatility was unchanged and obviously not affected by the chlorine. As a result, somewhat like its neighbor Ti in the periodic table, there is little concern regarding Sc in that it will be effectively removed by any particulate technology. The extent of global natural soil-dust levels has been estimated at about 3×10^{14} g (Nriagu, 1989). Based on this and an average soil content of 22 ppmw Sc, the natural content of Sc from such a natural source can be about 7000 tons year^{-1} to the atmosphere. Combustion of about 6 billion tons year^{-1} of coal globally with a content of 10 ppmw Sc could involve about 60,000 tons year^{-1} of Sc. Particulate control obviously collects most of this, and any system with 88% efficiency or greater overall ash removal will ensure no addition of Sc to the environment larger than the natural background. This is undoubtedly one reason along with its nontoxic nature why there are no current regulations involving scandium. Moreover, because yttrium and scandium are now regarded as strategic materials the possibility of leaching and recovering these from coal ash is seriously being considered (Taggart et al., 2016). In the case of Sc this remains marginal due to its low levels in coal and sludge ash (Folgueras et al., 2017).

In summary, the condensed phase Sc_2O_3 oxide is refractory in nature and the end dissociation product of most of its compounds in combustion. There is little evidence of its vapor in combustion or for any roles for chlorine and sulfur. As a result, although levels of scandium in fuels can be of a nonnegligible quantity, the scandium will never leave the fuel and remains in ash forms. It will be of no concern regards disposal of ash or use as an agricultural spread.

7.23 Selenium

Although a rare element, ranking about 69th in concentration in the Earth's crust (rather similar to Ag, Hg, and Pd) and 47th in the oceans, for 200 years physics has been intrigued by its electrical properties, and for 100 years medicine has increasingly realized its biological essentiality (Schrauzer and Surai, 2009). In the periodic table it is related to sulfur and as a result its minerals in nature are generally metal selenides. Such minerals are dispersed on the earth, and selenium is mainly produced as a by-product mainly of Cu smelting. Compared to some metals its global production remains quite low at about

2500 tons $year^{-1}$. Its cost also is quite variable being presently quite low at about \$26 lb^{-1} and has been declining in recent years. Reserves estimated at 120,000 tons remove concern over supplies. Most Se has been utilized in metallurgy, particularly in refining Mn with added SeO_2, but also in improving numerous alloys. It has a wide range of uses, now involved in thin film solar energy technology, with a call for greater recycling in the photovoltaic industry (Kavlak and Graedel, 2013). It is also encountered in Xerography, plastics, paints, and glass as a pigment, in cosmetics and medical use, in the semiconductor and optoelectronic industry, in fungicides and feed supplements and as a catalyst in rubber vulcanization. Natural sources contributing to the atmospheric content are difficult to assess but are estimated at about 9200 tons $year^{-1}$ from mainly volcanic and biosphere sources (Pacyna and Pacyna, 2001; Wen and Carignan, 2007). Anthropogenic sources of about 4600 tons $year^{-1}$ originate mainly from coal combustion and the remainder from waste incineration and metal smelting (Pacyna and Pacyna, 2001). However, a more recent assessment of trends in China indicate that selenium emissions increased from 635 tons $year^{-1}$ in 1980 to 2353 tons $year^{-1}$ in 2007 with coal being responsible for one-third of this with a 5% growth rate indicated (Tian et al., 2010). Even with advances in emissions control, difficulties occur with selenium due to its volatility. As a result, the man-made contributions are probably somewhat higher than previously estimated and similar to that of nature. Its distribution in the Earth's soils can be very varied and certain regions have concern over poor soil quality and food-crop selenium levels. One recent study in China assessed the value of adding straw to improve the soil selenium balance (Wang et al., 2019). An extensive model of the marine selenium cycle has been proposed from a Pacific Ocean study. It suggests that the reservoir of Se in the atmosphere both natural and anthropogenic has increased by 42% from preindustrial times. This is now balanced by the depletion rates over land and into the oceans. Overall, it is a very dynamic cycle that has increased Ocean surface concentrations possibly by 3% over the last 30 years due to anthropogenic emissions, and is being taken up by the phytoplankton. Final deep Ocean sedimentation rates are small and selenium has a very long marine residence lifetime (Mason et al., 2018).

The important health aspects of selenium

As indicated, selenium has a significant biological presence in the nature of life (Solovyev, 2015; Sobolev et al., 2018; Green, 2018). This is particularly of interest in that low levels in the human body are essential but too much is neurotoxic. A dietary intake in the range of 40–400 μg day^{-1} is required with an optimal range of 200–300 μg day^{-1} now being suggested for good health (Fordyce, 2007; Schrauzer and Surai, 2009; Sneddon, 2012). This is larger than earlier recommended levels. As a result of increased levels of analysis of Se in foods and in human bodies, in the last decade, renewed interest has arisen into just how important is it to our health? Although, as will be seen, selenium occurs mainly in Se^{IV} and Se^{VI} valences in a variety of compounds, their levels of toxicity for humans are not sensitive to this, but controlled by other aspects such as solubility and body absorption. As a result, regarding health, only the total Se content is of interest. Major dietary concerns arise due to the uneven occurrence of selenium in soils around the world leading to some deficiencies in dietary Se intake (Stoffaneller and Morse, 2015). Geology exerts its control over crops and animals and in China, for example, areas exist with either soil too rich or poor in selenium, both extremes raising medical concern (Li et al., 2012a). In regions of deficiencies there are supplementation programs (Bodnar et al., 2016), to try and introduce selenium into the food chain by fertilizing the soil (Alfthan et al., 2015), enriching animal feed for egg and meat production or for growing better foods crops. Recent studies have emphasized the possible enrichment of carrots, broccoli, strawberries and vegetables (Fisinin et al., 2009; Juniper et al., 2009; Gomez-Jacinto et al., 2012; Slupczynska et al., 2018; Yang et al., 2018a). The global biogeochemistry of selenium and its speciation in nature has been reviewed recently (Sharma et al., 2015). Also due to its role in living organisms, and the environmental health aspects for humans and animals, the toxicology profile for selenium and its compounds have been very extensively documented (World Health Organization Se 1987; Tox Profile Se, 2003). These large and detailed reports are available on the Web. They indicate that a normal diet in Western Society should usually provide about 70–150 $\mu g\ day^{-1}$ of selenium from mainly grains and cereals, followed by meat, fish and poultry. Although these reports previously recommended a daily intake of about 55 $\mu g\ day^{-1}$, suggestions as indicated above, now are that greater benefits result for health from levels about fourfold higher. In the body it appears to have numerous roles that could be summarized as forming antioxidant proteins such as selenocysteine and selenomethionine that strongly interact with numerous toxic metals such as Al, As, Cd, Pb, Mn, Hg, either sequestering or managing to purge these from the brain and the body (Ralston and Raymond, 2006; Rayman, 2012; Sadauskiene et al., 2016; Bjorklund et al., 2017; Schofield, 2017a; Hossain et al., 2018; Zhou et al., 2019). A suggested mechanism of methyl-mercury's potent neurotoxicity is that if its concentration is too high it can overpower the protective power of Se by inhibiting the selenoenzyme production rate (Ralston and Raymond, 2016, 2016a, 2018). There is an indication that a Se/Zn combination is even a stronger protection (Branca et al., 2018; Rahman et al., 2019). At least 25 selenium proteins are coded for production in the body, indicating their importance (Cardoso et al., 2015; Steinbrenner et al., 2016). To some degree this may arise from the fact that Se resists permanent oxidation in nature and can be readily oxidized and reduced (Reich and Hondal, 2016). It can readily substitute for its periodic family member sulfur and like the sulfur mercaptan, –SH group, can readily chelate toxic metals. As an antioxidant it is still being questioned in connection with inhibiting cancer formation, aiding heart problems, protecting the neural connections in the brain and possibly having a potential role in modifying Alzheimer's and other neurological diseases, but all these aspects remain controversial (Schrauzer, 2009; Solovyev, 2015; Meplan, 2015; Du et al., 2016a; Cardoso et al., 2017; Solovyev et al., 2018). Serious pilot trials are current concerning the benefits of selenium with regards to Alzheimer's disease (Cardoso et al., 2019). Similarly, other still incomplete studies indicate possible value for thyroid disorders in that the thyroid is the tissue involving the most Se (Santos et al., 2018; Negro et al., 2019). One recent study noted improved quality of life in the elderly by consuming a Brazil nut per day coupled to taking coenzyme Q10 but this is now being questioned (Johansson et al., 2015; Alehagen et al., 2015). Brazil nuts are rich in their Se content (15–200 $\mu g\ g^{-1}$). All plant growth absorbs selenium but some with accumulation (Pilon-Smits, 2017). These can cause poisoning in some animals but there is a buffer in the food chain to protect humans. Human death from poisoning is rare, needing 10- to 20-fold the usual daily dose and is generally accidental. Toxicity assessments now place the onset at about 1 mg day^{-1} consumption or 1 mg L^{-1} blood level (Goldhaber2003). However, as indicated, because food content of Se is variable around the world it is really difficult to assess intake without a biomarker. The common ones with scaled bioequivalents tend to be blood 100 $\mu g\ L^{-1}$, 80 $\mu g\ L^{-1}$ plasma or 10 $\mu g\ L^{-1}$ urine (Hays et al., 2014). Italy arrived at a reference value for blood samples of 140 (Bocca et al., 2011). Other than those values in surveys already listed (Schofield, 2016), other large blood measurements from France and Italy suggest values of 89–154, 85–142, and 153 indicating reasonably narrow distributions (Goulle et al., 2005; Cesbron et al., 2013; Oggiano et al., 2018). What is evident from the ranges of values in all these surveys is values are controlled to some degree at the upper range (253 and for pregnancy cases 374, 456) even with outliers, but some countries can have quite low values at the 5% percentile (14, 37, 41 $\mu g\ L^{-1}$) that could be dangerously low for cases of pregnancy. Now the medical profession is suggesting the selenoprotein level may be a preferred test but this makes for a more involved analysis (Bornhorst et al., 2018). Selenium also accumulates together with mercury in fish and this is now believed to alleviate to some degree the risk of mercury poisoning in some fish species. A recent intriguing study with fish now suggests that some of the CH_3Hg is demethylated by Se in the fish intestine (Wang and Wang, 2017). Recent co-administration of Se and Cd to mice indicated protection, and it evidently also provided protection from prenatal exposure to Ni avoiding the commonly resulting low birth weight in babies (Hossain et al., 2018; Hu et al., 2018; Sun et al., 2018a). Selenium compounds, inorganic or organic are readily absorbed by the body and are regarded as noncancerous. Tests with ^{75}Se labeling monitored selenium's metabolism in the body and indicated quite rapid excretion of any excess via urine and feces (Yim et al., 2018). Estimated from autopsies, the total body burden of Se is about 15 mg implying storage mainly in its protein and enzymic forms as no centralized storage mechanism in the body has been recognized. Selenium has a large organometallic chemistry forming both simple methylated compounds in nature with C-Se bonding, but with more complex structures in the body (Wallschlager and Feldmann, 2010). As indicated these behave as strong antioxidants and detoxicants. It is one of few elements that can bioaccumulate in the food chain. Currently, as outlined further under mercury, the presence of organic CH_3Hg in fish, and now a significant part of the human diet, has caused a new dilemma for the medical profession. The problem is that fish is nutritionally very beneficial but this now has to be assessed against the associated Hg toxicity. Many doctors still argue that the benefits outweigh the danger (Hibbein et al., 2018). However, this is not easily appraised, especially as Se now has entered into this discussion. It really requires a risk analysis assessment for every kind of fish type and still will assume that levels below the so-called safe level remain acceptable. A recent study has tried to categorize all fish in the Northeast Atlantic into their pelagic and demersal species. MeHg varied more than Se, but the Se:Hg molar ratio indicated all were >1, and showed a range from 1.9 to 43 in that study. Nevertheless certain fish should be eaten in limited quantities (Azad et al., 2019). This has been extended to two shark varieties, one having more Se and the other less Se than methyl-mercury (Rafael et al., 2019). It appears evident now that whales clearly have Hg $>$Se (Caceres-Saez et al., 2018). Lower levels of methyl-mercury are found in the smaller noncarnivorous varieties. Many studies of cultures living on high fish diets, such as in the Seychelles Islands, the Faroes Islands, northern Canada and the coast of Mexico, for example, have clearly shown adverse effects from mercury toxicity (Strain et al., 2008; Huang et al., 2018a; Ripley et al., 2018; Garcia-Hernandez et al., 2018). It is from such work, that the currently suggested MRL human blood value for mercury in the United States was set as $<5.8\,\mu g\,L^{-1}$. That for Se is suggested as $150-240\,\mu g\,L^{-1}$ (Mayo Clinic, 2018). Since it has now been found that selenium can neutralize mercury on an atom to atom basis, the units are rather confusing due to the two elements having different atomic weights of 200.6 (Hg) and 79.0 (Se). In other words, $5.8\,\mu g\,L^{-1}$ (Hg) and only $2.3\,\mu g\,L^{-1}$ (Se) contain the same number of metal atoms. Because normal blood levels of Se are generally in the range $50-300\,\mu g\,L^{-1}$ in 20 recorded blood monitoring surveys (Schofield, 2016), it is noted that the body selenium will normally be in significant excess. This is an important aspect. It clearly labels Se as the body's detoxer. All metals known to be toxic to the body have recommended safe blood levels in double digit or less in $\mu g\,L^{-1}$ units. As seen, and recommended, Se levels are normally found close to three digits. It would appear this is not by chance but programmed in the DNA Code. Countries appear now to be realizing this and are testing their populations and especially women in pregnancy to ensure adequate Se blood content levels (Schofield, 2016; Hu et al., 2018; CDC 4th Rept., 2018). This is very noteworthy as seen in two recent studies. The mercury and selenium blood levels were measured at birth in the placenta and umbilical cord. On an atomic basis, the Se to total Hg cord-ratio ranged from 5 to 626 in one study and was 60-fold in the other showing this pronounced dominance of selenium for the fetus (Kosik-Bogacka et al., 2018; Sakamoto et al., 2018). This is possibly the mechanism by which the tiny fetus, of minimal weight, survives pregnancy, it being saved by selenium. Otherwise with the other neurotoxins present, the fetus is always battling for survival, living with a mix of neurotoxins that are undoubtedly at toxic risk levels. As a result of selenium's antioxidant and chelating properties, there is now a general suggestion, by many, to add selenium to one's diet as a supplement. The natural form is best found in Brazil nuts, one nut/day. Selenium is an unusual element, essential to the body but becomes a neurotoxin at higher levels. However, a recent comment suggests that such supplementation should be done with careful consideration and only taken if necessary (Mazokopakis and Liontiris, 2018). Significant research followed the initial beneficial reports, and has now found that above a certain level in the blood, about $125\,\mu g\,L^{-1}$, the benefits plateau. Supplemental doses should therefore depend on geographic location and needs. In the United States where baseline values are already generally adequate (average about $190\,\mu g\,L^{-1}$) it may be unnecessary. Additionally long-term treatments with selenium even appear to possibly enhance mortality slightly, indicating no longevity benefit (Rayman et al., 2018; Giovannini et al., 2018). However, one aspect of the $HgCH_3$/Se interaction, now being suggested, is that excessive depletion of the Se by the mercury, and other toxins in the brain, in the form of the selenoenzymes may overwhelm their formation rate leaving the brain vulnerable (Ralston and Raymond, 2016, 2016a, 2018). In Canada recommended blood levels for selenium are $>100\,\mu g\,L^{-1}$, and for pregnancy, levels should be of a similar value or higher (Zidek et al., 2017; Varsi et al., 2017). Adequate levels of selenium appear to be very important in pregnancy, especially in those places with low selenium soils (Zachara, 2018).

Selenium's basic chemistry

Electronically, atomic selenium is described as $3d^{10}4s^{2}4p^{4}$ resulting in three major valences, divalent selenides, Se^{2-}, but four with SeO_2 and H_2SeO_3, and six, for example with H_2SeO_4 cationic type molecules. The pure metal can be formed in three phases, a red monoclinic structure, an amorphous solid and the most stable gray hexagonal crystal to which the others convert. It has six stable atomic isotopes that can complicate its mass analysis. Although reasonably volatile (m.pt.220, b.pt.685°C), it is insoluble in water and is not attacked by HCl. It has a vapor pressure of 10 ppmv at 230°C and 100 ppmv 280°C. Like sulfur it vaporizes to a complex mixture of Se_n polymers ($n=2$–8) that have been characterized mass spectrometrically (Berkowitz and Chupka, 1969). It reflects Hg in its condensed phase having few simple inorganic molecules (Table 7.33).

There is no SeO(s), and SeO_3(s) dissociates above 165°C. SeO_2(s) is the most frequently encountered molecule of selenium. It sublimes slightly below its melting point (m.pt.340, sublimes 315°C). It is very soluble in water forming selenous acid, H_2SeO_3 (m.pt.70°C, decomposes on heating) and is hygroscopic. This is the basis of many senenites.

TABLE 7.33 Properties of selenium's condensed phase molecules relevant to combustion, °C.

m.pt. b.pt. vapor pressure	m.pt.
Se 220 685, 10 ppmv/230°C, ins	$Al_2(SeO_3)_3$ 500, dissoc, ins
SeO_2 340, subl 315, 15 ppmv/125°C, s	$Al_2(SeO_4)_3$ 800, dissoc
SeO_3 120, >165 dissoc, s	$CaSeO_3$ 920, dissoc 1000, s
H_2SeO_3 70, decomp, s	$CaSeO_4$ dissoc 800, s
H_2SeO_4 58 260, decomp, s	$Fe_2(SeO_3)_3$ >500 dissoc, ins
Se_2Cl_2 −85 372, dissoc 130, s	$Fe_2(SeO_4)_3$ >500 dissoc
$SeCl_4$ 305, vap 196 to $SeCl_2$, s	K_2SeO_3 875, stable to 1000
$SeOCl_2$ 11 177, s	K_2SeO_4 stable to 1000
SeS_2 111 119, decomp, ins	$MgSeO_3$ 796, dissoc 730, ins
Se_2S_6 122, ins	$MgSeO_4$ dissoc 650, s
Se_4S_4 113, decomp	Na_2SeO_3 710, dissoc 900
Se_6S_2 122	Na_2SeO_4 777, dissoc 1100, s
Al_2Se_3 947, dissoc, s	
CaSe 1400, dissoc, s	
FeSe 458, dissoc, ins	
MgSe s	
P_2Se_3 245 380, s	
Na_2Se 875, s	
$SiSe_2$ b.pt.761, s	

s, water soluble; *ins*, insoluble; *decomp*, decomposes; *dissoc*, dissociates; *subl*, sublimes; *vap*, vaporizes.

Similarly SeO_3(s) (m.pt.120°C, sublimes, dissociating on heating), is hygroscopic dissolving readily in water to form selenic acid, H_2SeO_4 (m.pt.58, decomposes 260°C), and other selenates. It has no known sulfate, carbonate, nitrate, phosphate, or hydroxide. Its halides or oxyhalides either are volatile, unstable, or easily decomposed by water, Se_2Cl_2(l) (m. pt. − 85, b.pt.372°C, but dissociates at 130°C): $SeCl_4$(s) (m.pt.305, sublimes at 196°C to $SeCl_2 + Cl_2$), and $SeOCl_2$(l) (m.pt.11, b.pt.177°C) hydrolyzed by water. Condensed phase $SeCl_2$ has not been characterized. Selenium's chemistry with sulfur appears complex and numerous sulfides are known that are insoluble in water (Sedo et al., 2005). SeS_2 has a ring structure (m.pt.111, 119°C decomposes), Se_2S_6 (m.pt.122°C) is structurally like Se_8, as are Se_4S_4 (113°C decomposes) and Se_6S_2 (122°C). Selenium's major chemistry consists of its selenides, selenites, and selenates that it can form with many elements of the periodic table. In general the selenides have relatively low melting points below 1000°C (CaSe being exceptional, m.pt.1400°C) but all tend to react with water being hydrolyzed to the unstable hydroxide and releasing very toxic gaseous H_2Se (m.pt. − 66, b.pt. − 41°C). On heating the more complex selenides such as Al_2Se_3(s) may dissociate to Al_2Se(g), AlSe(g), and Al(g) (Bardi and Trionfetti, 1988). Selenates (VI) and selenites (IV) are not refractory salts, selenates generally dissociating through the selenite that can be easily reduced to Se. As indicated in Table 7.33, those appropriate to coal combustion dissociate below 1000°C with the Na and K salts being the most stable. These, Na_2SeO_4, Na_2SeO_3 and similarly for K are stable slightly above this temperature and even partly vaporize molecularly, their vapor having been monitored mass spectroscopically (Lopatin and Shugurov, 2013) and earlier by infrared matrix isolation (Brisdon et al., 1991). Generally, selenites are seen to be the more thermodynamically and thermally stable. With the exceptions of the alkalis, most metal selenites have a low solubility in water compared to the generally soluble selenates (Gospodinov et al., 1991).

Gas phase stabilities and transport

With combustion, the gas phase also has a very limited range of volatile Se species. Their bond strengths have been summarized in Table 7.34 and compared there with its periodic table column elements oxygen and sulfur. All its molecules show a weakening from those of sulfur. As indicated above, the metal vaporizes to numerous Se_n ($n = 2–8$) gaseous polymers.

It is considered that Se_8 puckered rings leave the surface and then dissociate as temperatures increase. Above 400°C, Se_2 and Se_5 are major, finally producing mainly a stable Se_2(g) dimer (Berkowitz and Chupka, 1966). The early determinations of the dimer bond strength were slightly low and seen to require a correction due to the electronic state splitting (Meschi and Searcy, 1969). The accepted value now of $D_0(Se–Se) = 327 \pm 8\ kJ\ mol^{-1}$ came from a Knudsen cell-mass spectrometric study (Drowart and Smoes, 1977), and a spectral predissociation limit (Huber and Herzberg, 1979). This is now supported by a theoretical confirmation (Hofener et al., 2012), and is the basis for the thermochemical tabulation (Barin and Platzki, 1995). Only two strongly bound gaseous oxides, SeO and SeO_2 exist at elevated temperatures. A trioxide exists and sublimes at lower temperatures (vapor pressure 229 ppmv at 92°C) (Ficalora et al., 1969), but dissociates to the

TABLE 7.34 Stability trends for gaseous molecules of O, S, and Se, D_0 kJ mol^{-1}

Oxygen[a]		Sulfur[a]		Selenium[b]	
Species	D_0	Species	D_0	Species	D_0
O-O	493.6 ± 0.2	S-S	421.566 ± 0.03	Se-Se	327 ± 8
–	–	S-O	517.0 ± 0.2	Se-O	426 ± 3
–	–	–	–	Se-S	367 ± 3[k]
O_2-O	102.3 ± 2	S_2-O	429.1 ± 1.4	Se_2-O	–
–	–	S_2-S	257.2 ± 4	–	–
–	–	S-SO	333.7 ± 1.4	–	–
–	–	OS-O	545.8 ± 0.4	OSe-O	411 ± 4
–	–	O_2-S	569.2 ± 0.4	O_2-Se	343 ± 5
–	–	O-SO_2	342.7 ± 0.8	O-SeO_2	201
O-H	423.7 ± 0.2	S-H	351.0 ± 3.5	Se-H	311 ± 1.0
H-OH	494.1 ± 0.2	H-SH	373.7 ± 3.5	H-SeH	330 ± 0.8
H-O_2	203.4 ± 3.0	H-SO	236 ± 10	H-SeO	Unstable
–	–	HS-O	402 ± 10[c–g]	–	–
HO-O	273.3 ± 3.0	HO-S	320 ± 10[c–f]	HO-Se	Unstable
–	–	H-OS	227 ± 10	–	–
O-Cl	265.38 ± 0.04	S-Cl	282 ± 5[h,i]	Se-Cl	260
ClO-Cl	139.8 ± 10	ClS-Cl	249 ± 5[j]	ClSe-Cl	≈ 246

[a] *Gurvich et al. (1989).*
[b] *See text.*
[c] *Wilson and Dunning (2004).*
[d] *Denis (2010).*
[e] *Nagy et al. (2011).*
[f] *Lindquist et al. (2013).*
[g] *Balucani et al. (2004).*
[h] *Leiding et al. (2011).*
[i] *Yang and Boggs (2005).*
[j] *Barin and Platzki (1995), Chase (1998).*
[k] *Lias et al. (1988).*

dioxide as temperatures are increased. These have been observed in infrared matrix isolation experiments and their frequencies observed and structures assessed theoretically (Brabson et al., 1996). An earlier study for SeO_2 obtained its infrared spectrum and calculated its thermodynamic functions (Cesaro et al., 1971). A recent theoretical study of numerous selenium molecules has calculated the three oxide atomization energies, frequencies, and heats of formation (Jackson et al., 2012). Their values for SeO_2 and SeO_3 provide the only estimate of the bond strength for the trioxide suggesting $D_0(\text{O-SeO}_2) = 201$ kJ mol^{-1} confirming its weakness and thermal instability. As seen from chemical thermodynamic calculations for fuel-lean combustion conditions (Frandsen et al., 1994), gaseous SeO_2 should be dominant below about 1500°C above which it begins to establish a partial equilibrium distribution with SeO that is controlled by the flame equivalence ratio and the prevailing radical concentrations. The oxides are both thermally very stable with strong molecular bonding. Although minimally studied, the Knudsen cell-mass spectrometric study of the balanced reaction $SeO + SeS = Se_2 + SO$ is considered reliable (Smoes and Drowart, 1984a; Drowart and Smoes, 1977). Although dependent on values for these other components, a reexamination indicates the value of $D_0(\text{Se-O}) = 426 \pm 3\ \text{kJ mol}^{-1}$ remains acceptable as published. It is consistent with a spectral predissociation value and replaces earlier indications of even larger values. Its magnitude is also supported by theoretical calculations (Bouchoux et al., 2012). $SeO_2(s)$ sublimes, so combining its heats of formation (Kubaschewski et al., 1993) and sublimation (Behrens et al., 1974) implies a gaseous accurate measure of $D_0(\text{OSe-O}) = 411 \pm 4\ \text{kJ mol}^{-1}$. The above mentioned theoretical study suggested 424 for SeO and 415 for O-SeO in close support (Jackson et al., 2012). To these, a value of 396 kJ mol^{-1} can be added from a study of the potential energy curves and structure of the low lying states of SeO_2 that confirmed its OSeO structure is far below the energy of a SeOO configuration (Grein, 2009a). Due to their favorable energetics, the preferred kinetic channels for its combustion chemistry will be via radical reactions rather than direct reaction with oxygen.

$$Se + OH = SeO + H \qquad \Delta H_0 = -2\ \text{kJ mol}^{-1}$$
$$SeO + OH = SeO_2 + H \qquad \Delta H_0 = +13\ \text{kJ mol}^{-1}$$
$$Se + O_2 = SeO + O \qquad \Delta H_0 = +68\ \text{kJ mol}^{-1}$$
$$SeO + O_2 = SeO_2 + O \qquad \Delta H_0 = +83\ \text{kJ mol}^{-1}$$

The rate constant for the direct reaction of Se with O_2 ($D_0(\text{O-O} = 494\ \text{kJ mol}^{-1})$) has been calculated and not only has this indicated the enthalpy energy barrier but also an additional activation energy (Urban and Wilcox, 2006, 2006a). As a result, if selenium is vaporized, in a short time period, equilibration of the gas phase should occur and a distribution of gaseous SeO_2, SeO, and Se will transport the selenium. In the laboratory, where nonequilibrated studies can examine the gas phase kinetics, it was noticed that as indicated in Fig. 7.13 for a fuel-lean hydrogen flame at 1390°C the expected dominance of SeO_2 was manipulated by the nonequilibrated radicals but decayed toward equilibrium on a millisecond timescale.

Indications are that at equilibrium $SeO_2(g)$ will be dominant to temperatures of about 1100°C when the equivalence ratio is oxygen-rich and will remain so in cooling gases. However, due to its very high volatility (approximately 110 ppmv at 150°C, 15 ppmv 125°C and 2 ppmv at 100°C), its physical absorption on surfaces generally will not be possible until very cool, unless chemideposition can effectively bind it molecularly on the surface at higher temperatures with some ash metal element (Ficalora et al., 1969; Behrens et al., 1974). The condensed phase oxides SeO_2 and SeO_3 are both very water soluble and are the basis of their corresponding acids, selenous (H_2SeO_3) and selenic (H_2SeO_4). These acids are not stable as gaseous species. Although also unstable, theory now indicates that the SeOH structure is preferred, being about 54 kJ mol^{-1} more stable than HSeO and the two are separated also by a 187 kJ mol^{-1} barrier (Viana, 2017). The corresponding gaseous

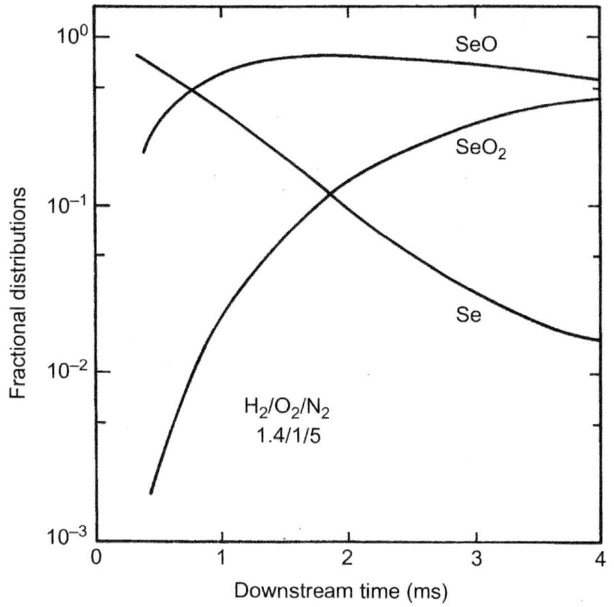

FIGURE 7.13 Predicted downstream nonequilibrated distribution of selenium in a 1670K oxygen-rich hydrogen flame, on a flat front burner, modifying with time (and downstream distance) toward equilibrium.

monosulfide also has been extensively studied by mass spectrometry. The three studies support a value of D_0(Se-S) = 367 ± 3 kJ mol^{-1} that reflects the normal weakening from the diatomic oxide by about 14% in this case (Drowart and Smoes, 1977; Huang et al., 1979; Grade et al., 1983). Extrapolation of the ground state vibrational levels had indicated a value of 357 kJ mol^{-1} (Huber and Herzberg, 1979). Like SeO and isoelectronic molecules O_2, S_2, and Se_2, these all have an $X^3\Sigma^-$ ground state and a potentially low-lying $a^1\Delta$ state that can contribute to its thermochemical nature. In the case of SeS this state lies 4058 cm^{-1} (48 kJ mol^{-1}) above the ground state (Setzer et al., 2008). Recent interest also has published two studies tabulating its spectral properties and low-lying potential energy curves (Zhu et al., 2013; Chattopadhyaya et al., 2016). Although the condensed phase SeS_2 is a useful molecule, on vaporization it appears to readily dissociate to the monosulfide.

The selenium chlorides are readily hydrolyzed by water and play no role in combustion as is clearly illustrated from incinerators where selenium shows no change with and without chlorine (Tillman, 1994). However, the monochloride is known, SeCl(g), that favors a Se_2Cl_2(g) structure. It has been spectrally characterized minimally but the electronic structure and potential energy curves of SeCl have been calculated theoretically (Woon and Dunning Jr 2009; Hermoso and Ornellas, 2012). These suggest a possible bond strength of about D_0(Se-Cl) = 260 kJ mol^{-1}.

$$\text{Se} + \text{HCl} = \text{SeCl} + \text{H} \quad \Delta H_0 = +161 \text{ kJ mol}^{-1}$$

Its direct formation from the atom will be difficult. Measured spectral absorption features ascribed to SeCl(g) were replaced by those of SeO(g) in the presence of O_2 confirming the latter's thermochemical dominance (Ciach et al., 1971). The dichloride, $SeCl_2$, has not been characterized but is suggested as a product in the dissociative sublimation of

$SeCl_4(s)$. The latter's heat of formation appears to have been estimated in thermochemical databases. This indicates an atomization value of 506 kJ mol^{-1} or D_0(ClSe-Cl) = 246 kJ mol^{-1}, implying two rather similar bond strengths in the $SeCl_2$ molecule (Knacke et al., 1991; Barin and Platzki, 1995).

As indicated above, of the potential metal selenites and selenates, only their alkali forms manage to molecularly vaporize to a limited extent but are not stable gaseous molecules to higher temperatures. Consequently the only other important gaseous species is the thermally stable hydride H_2Se that can be encountered mainly under reducing conditions. It is flammable, highly toxic but not affected by water, dissolving only slightly. A photoionization study reliably established its two similar bond strengths indicating D_0(Se-H) = 311 ± 1.0 and D_0(HSe-H) = 330 ± 0.8 kJ mol^{-1} (Gibson et al., 1986). A value of 309 kJ mol^{-1} for SeH had been obtained by extrapolation of its $A^2\Sigma^+$ electronic state levels (Huber and Herzberg, 1979). These values also have been adequately described by theoretical studies (Mayer et al., 1997; Duke and Radom, 1998; Ramakrishna and Duke, 2003; Bouchoux et al., 2012).

An important point of interest from reviewing its basic chemistry centers on the observed strong oxidizing power of gaseous SeO_2, a little like O_3. Although it is readily produced in the gas phase, in mixtures containing other species such as gaseous NH_3 or an aqueous solution of SO_2, for example, it will oxidize these, being reduced itself back to elemental Se. It is well known and used in organic chemistry as an oxidizing agent. As will be seen below, this is most probably the currently overlooked aspect of selenium combustion emissions and is noted here for the first time. It is the most likely explanation why selenium emissions remain difficult to totally control.

Combustion power plant data

On examining fuels (Table 1.1), selenium is present as a trace mineral in coal and occurs to varying extents in biomass and paper products. Coals can contain variable amounts that can average around 2 ppmw for US coals (Rowe and Steinnes, 1997; Germani et al., 1980; Hu et al., 2006) or possibly 4 ppmw in China (Wang et al., 2010) and even 8 ppmw or higher in some coals from South America (Lind et al., 1994). Fuel oils and gasoline have negligible traces on ppbw scales (Kitto, 1993; de Almeida et al., 2009). Municipal waste and sewage sludge can introduce higher input values for incinerators (Tillman, 1994). Its levels are very low in pet-coke but can be up to 2 ppmw in limestone (Cordoba et al., 2012). Synthetic biogases are seen to have much elevated levels over natural gas (Cachia et al., 2018).

Analysis of selenium and its speciation (distribution) in combustion remains difficult for real-time measurements. Generally, samples still require collection and generally total Se content analyzed mainly utilizing hydride generation-atomic absorption or ICP emission or mass analysis methods. A review details all the analysis methods in use (Gonzalvez et al., 2009). As mentioned before, the ICP-MS method has to take selenium's isotopes into account and its use also has been well documented (B'Hymer and Caruso, 2006; Bednar et al., 2009). The latest work has used it online in real-time on a coal gasification study measuring total Se (Edinger et al., 2016). The needed two-channel real-time atomic absorption or fluorescence analyzer for its speciation is still in development. Nevertheless, in spite of the demanding analytical difficulties, there are considerable data available detailing the behavior of selenium in combustion. At least ten or more mass-balance estimates concerning the selenium redistribution in coal combustion trains have been published that show a consistent trend (Andren et al., 1975; Kaakinen et al., 1975; Klein et al., 1975; Meij, 1984; Germani and Zoller, 1988; Ondov et al., 1989; Quick and Irons, 2002; Otero-Rey et al., 2003; Huang et al., 2004; Reddy et al., 2005; Cheng et al., 2009; Senior et al., 2010a; Tang et al., 2012). Such determinations are very difficult to obtain accurately, but the resulting approximations are very important in indicating the general distribution that occurs and the controlling aspects. Generally, little selenium is detected on slag or bottom ash (<2%) and its ready release from fuels is clearly evident. It is very volatile and its vaporization in coal combustion has been modeled (Zeng et al., 2001; James et al., 2014). The largest fraction is found on fly ash 50%–70%, which can be effectively removed by an electrostatic collector or fabric filter bag. Efficiencies for these are high except for the smallest sized fractions. Scrubbing by FGD devices can average about 60% efficiencies so removing another 15%–25% of the total selenium (Tian et al., 2010; Tang et al., 2012; Cordoba et al., 2012). However, these controls generally still leave an imbalance of volatiles that is emitted to the environment most probably as atomic Se on very fine particles. It varies with low sulfur coals and solely cold-side ESPs at about 5% (Otero-Rey et al., 2003; Goodarzi, 2006), or with coal-fired FBC, recently reported at about 2% (Zhou et al., 2018a). It has also been noted that this distribution is affected by the coal type and the sulfur level, which is discussed in detail later (Seames and Wendt, 2007; Senior et al., 2010a). One recent coal-fired fluidized bed measured the online selenium release. It was reported that the release was faster with gasification than for combustion possibly indicating different volatization mechanisms (Shen et al., 2016). A hazardous waste incinerator at <1100°C noticed that the selenium was distributed through the ash forms, on which they performed leaching tests. They concluded that the selenium became entrained within the ash particularly by Fe and Ca (Li et al., 2018b).

An important question is what is the nature of the selenium on ash? This has now been quite extensively examined but uncertainties remain. From early studies there has always been an interest in the relative enrichment of trace metals on fly ash both as a factor of time and also as a function of particle size. Even 30 years ago it was noted that in four different size fractions that not only did selenium's relative enrichment increase as particle size decreased but it reflected the greatest enrichment of all the trace elements (Coles et al., 1979). This is clearly seen again in recent ESP data where collections were made from various stages where temperature was decreasing as well as particle size (Suarez-Ruiz et al., 2007; Wang et al., 2008; Tang et al., 2012). Although a role of unburned carbon in ash has been discussed for mercury there seems little evidence for such a correlation in the case of selenium (Suarez-Ruiz et al., 2007; Lopez-Anton et al., 2007). It is interesting to note that there can be fly ash samples from certain coals, which are a richer source of selenium than any natural ore, and are considered of potential commercial value in this regard. It soon became evident for ash that the selenium was chemically absorbed and present as selenites showing a valency of Se(IV) (Ericzon et al., 1989). XANES analysis of ash confirmed the dominance of selenites but the presence of lesser contributions of the selenates has not been totally ruled out (Shah et al., 2007; Nelson et al., 2010). The speciation on the ash has only been examined using X-ray methods. The latest used XPS and XAEF. This examined ash from a coal-fired Stoker boiler. The Se was highly enriched on the ash surface, about 300-fold higher than in the core and mainly elemental (62%–83%), followed by selenite formation (17%–38%) (Fu et al., 2018). A slight difference was noted in the ash spectra between bituminous (iron rich) and subbituminous or lignite coals (calcium rich) for which no SeO_2 was present (Huggins et al., 2007). From the basic components of ash it has been surmised that Ca and Fe are probably the most likely selenites formed (Bool and Helble, 1995). It has been established that CaO is an efficient absorber of SeO_2 (Ghosh-Dastidar et al., 1996; Agnihotri et al., 1998; Li et al., 2006, 2006a). However, a dilemma is that this also absorbs SO_2 making for a competitive calcium sulfate formation channel. Whereas $CaSeO_3$ is only stable below 500°C in this mode (Ghosh-Dastidar et al., 1996) the corresponding sulfate melts at above 1000°C, is stable at FBC temperatures of 800°C–900°C but can dissociate above this, so its behavior is modified by flash ash temperature and other conditions (Jia et al., 2016). As a result, the sulfur has the advantage not only in concentration, but also in having a longer interaction time that has the ability to minimize calcium's effectiveness toward selenium. Generally sulfates will always be more stable than their corresponding selenites. As seen in Table 7.33, Al, Ca, Fe, K, Mg, and Na all have the potential for retaining the Se as a stable selenite below 500°C–900°C. This was tested directly by examining the retention of five of these, namely Fe, Ca, Mg, Na, and K by additions in FBC at 500°C–1000°C. It was noted that Fe, Ca, K, and Na were effectively retained at 600°C–700°C, with a smaller interaction with Mg (Zhou et al., 2018a). The ferric selenite, $Fe_2(SeO_3)_3$ is well characterized and also stable below about 500°C (Rai et al., 1988) but in this case the corresponding ferric or ferrous sulfates are only stable below 600°C. Consequently, a competition between sulfates and selenites is certainly expected to occur on the fly ash. It is unquestioned that the ash has the ability to remove the gaseous SeO_2 or even atomic Se by chemical deposition by this stabilizing selenite formation in a temperature range appropriate to the selenites. It is different from mercury only in retaining the selenium on the ash. Coals are generally always rich in Si and Al, so that ash in essence is glass-like, forming aluminosilicates, the melting point temperature of which will be very dependent on any incorporation of other constituents. This is another possibility for gaseous selenium in that selenite glasses are well known and its addition to produce red colored glass is a common practice. However, mass balance studies of bottom ash suggest only minimal retention there (Meij, 1994; Quick and Irons, 2002; Reddy et al., 2005; Cheng et al., 2009; Tang et al., 2012), and those of slag generally zero (Klein et al., 1975; Andren et al., 1975). Also data listed by Senior et al. (2010a) does indicate for high coal-sulfur situations that there is little retention of selenium on ash. Any interaction with the silicate component would provide sufficient time and be under no temperature constraint to be fully removed on the ash. Consequently, it might be concluded that selenium is chemically formed on the ash as selenites being limited by their stable temperature regimes and the competition with sulfur. This appears to be a constraint on selenium, minimizing like mercury its interaction time.

One detailed ash study analyzed the selenium/ash interaction for six differing coals. The correlations with coal calcium or iron are not entirely persuasive but do indicate the SO_2 interference. Also the suggestion therein that selenium either as Se or SeO_2 can condense on ash for low calcium/high sulfur coals is very unlikely due to its high vapor pressure without heterogeneous chemistry binding it (Seames and Wendt, 2007). Selenites are not highly thermally stable, as seen in Table 7.33, but are readily synthesized at temperatures below about 700°C. They are generally more stable than selenates in contrast to arsenic. Additionally, as seen in current alkali studies, this chemical formation/deposition can be reversible. In other words it is not inconceivable if K_2SeO_3 is produced on the ash surface that SO_2 then can convert this to the sulfate with the elimination of SeO_2(g). It is in fact surprising that the ash is as efficient as indicated from the data. The vapor pressure of SeO_2(s) is about 2 ppmv at 100°C and would need temperatures at this level or below for simple condensation, it obviously also has a propensity for chemideposition. An ambitious attempt by Senior et al. (2010a) modeled the data of six coals from Seames and Wendt (2007). This is consistent with some of the aforementioned chemistry and is a good illustration indicating the input information required and the assumptions that have to be made. In this present case it adequately reproduces the selenium ash content data and speculates that the role of iron relates to an interaction with the aluminosilicate and not from a direct iron selenite formation. Moreover, the suggestion of SeO_2 beginning to condense at below 200°C is far too high a limit without additional chemistry. A more recent model addresses the partitioning of Se, As and Sb and considers the numerous stages of dissociation, release, fragmentation and distribution of the elements in coal combustion (James et al., 2014).

The extent to which selenium is emitted from combustors to the atmosphere at present depends on three control devices if they are present, the SCR, the ESP or fabric filter, and the FGD scrubber. The role of the SCR enters the discussion due to its use of ammonia and the slight excess additions producing ammonia slip. Temperatures are at or falling below 400°C, too high for NH_4Cl formation, but not for the ammonia to be oxidized by the gaseous SeO_2 that is then reduced to gaseous or even particulate Se. Originally, additions of NH_3 were suggested as one control method for selenium in this manner and was patented. However, the formed atomic selenium condensation on cooler fabric filter bags as an orange deposit of the element was found to be held so tenaciously that it was not practical for its removal and recovery (Dismukes, 1994). The capture of ash particles by ESPs has always had high efficiencies invariably greater than 99.5% (Meij, 1994). As a result, the remaining volatile selenium leaving the ESP will be SeO_2 or Se and a low level of particulates that may be removed by a scrubber. However, recent mass balance studies still indicate stack emissions of 13% for an ESP/FGD combustor (Tang et al., 2012), 5% for one with an SCR/ESP/FGD (Cheng et al., 2009) and 2% in an ultralow coal-fired power plant (SCR, ESP, WFGD) (Chang et al., 2019). Indications are that following the ESP it is possible for some of the remaining selenium to survive the FGD. Most recently a detailed study has examined this question of the behavior of selenium across a wet FGD scrubber (Senior et al., 2015). They review previous studies that clearly indicate that although the scrubber can remove ≈98% of the SO_2, the cases they list show significant variations with only 22%–65% removal of Se and their own work displayed an average of 61%. They failed to adequately explain why this was so, but did note a Se enrichment in the submicron particles that pass through the scrubber and are difficult for it to retain. These fine particles are emitted by the stack. Ignored is the fact that SeO_2(g) can oxidize SO_2 in solution and itself be reduced back to insoluble elemental Se that will attach to any available particle. Consequently the chemistry in scrubbers can be complex with numerous competing channels. The SeO_2(g) goes into solution, some forming $CaSeO_3$ that is only slightly soluble. If strong oxidants in the FGD such as the peroxydisulfate ions are present,

($S_2O_8^{2-}$), these can oxidize some of this selenite to soluble $CaSeO_4$, which becomes more difficult to extract mainly through its similarity to $CaSO_4$ (Ohki et al., 2011). To minimize this, additions of a Mn^{2+} salt can be made (Akiho et al., 2010). Otherwise any SeO_2 can also oxidize anything it can find, including H_2SO_3 in solution and be reduced back to elemental Se. This potential for it being reduced back remains the one major problem for fully controlling selenium emissions. The temperatures in the flow train at this point are too low for it to reoxidize. The different rates of such processes will depend on scrubber conditions and helps to explain the variations seen between scrubbers in different power plants. In the FGD liquid residues, selenium is generally the selenite (Al-Abed et al., 2008; Zhong et al., 2011). However, a more detailed X-ray absorption method indicated selenite and some selenate that had become oxidized in solution (Akiho et al., 2012). The complex behavior of selenium in the wet limestone FGD systems have been extensively studied recently. The wastewater can become so enriched with selenium for certain coals that its recovery becomes commercially plausible (Cordoba, 2017; Cordoba and Staicu, 2018). However, this removal of selenium from wastewater is a major problem and various methods are becoming available (Stefaniak et al., 2018). In oxy-coal combustion it is now noted that the recycling of gases permits the fly ash to become further enriched (66% of the selenium) with 31% remaining in the flue gases (Cordoba and Diego, 2018). This study also indicated the difficulty of estimating a mass balance for Se in such a complex cycle. In conventional combustors, it may be that Se will have to be controlled upstream with, for example, alkaline earth or alkali based absorbers. At low temperatures the $SeO_2(g)$ is only weakly absorbed by activated carbon showing indications of being easily reduced to atomic Se (Jadhav et al., 2000). An alternate method might be to use a dry spray-dryer containing recovered fly ash as a more effective absorber. Numerous studies are reported of the effects on Se, As and Hg concentrations using a lime dry-spray FGD for SO_2 control and do show various effects on other elements (Taerakul et al., 2006; Gupta et al., 2007).

Of course disposal of ash, or with its agricultural use, introduces a need for additional information on leaching and possible transformations, which are being studied. Selenium content can be invaluable to agriculture. Because human toxicity is insensitive to valence, interest centers on total selenium content and how its chemical form will modify its migration or bioactivity (Su and Wang, 2011; Schwartz et al., 2016).

Like sulfur, with reducing or fuel-rich conditions, the chemistry of selenium changes, due to an absence of oxygen, and the dominant species becomes $H_2Se(g)$ that is slightly soluble in water but is one of the most toxic selenium compounds. Underground coal gasification indicates $H_2Se(g)$ as the main selenium gaseous species (Liu et al., 2006). This dominance is consistent with chemical thermodynamic expectations and is not affected by chlorine (Frandsen et al., 1994; Helble et al., 1996; Yan et al., 2000, 2001a). It results from its strong chemical bonding that is only about 10% weaker than for H_2S or SH species (Table 7.34). As a result, cold water scrubbing can be used to remove some $H_2Se(g)$. High temperature gas cleaning may be more difficult but can invoke selenide formation. Although very stable, CSe(g) ($D_0 = 587\ \text{kJ mol}^{-1}$) (Smoes and Drowart, 1977) and SeS(g) ($D_0 = 367\ \text{kJ mol}^{-1}$) (Drowart and Smoes, 1977) do exist. Also numerous stable condensed phase selenides already mentioned such as FeSe(s) (Yan et al., 2004) and even $SiSe_2(s)$ (O'Hare, 1995) or Al_2Se_3 (Uy and Drowart, 1971; Bardi and Trionfetti, 1988), have not yet been stressed in gasification conditions. Consequently, ash particles rich in minerals have the possibility of removing the $H_2Se(g)$ as chemically deposited selenides. Lime/limestone are again effective sorbents producing CaSe under these conditions (Diaz-Somoano et al., 2010). Activated charcoal as an absorber has been examined mainly in aqueous media where it is effective. It appears less effective for the capture of $H_2Se(g)$ when dry, as also is the unburned carbon in ash (Lopez-Anton et al., 2006). The use of zinc titanate or ferrite at 550°C has been seen to remove both sulfur and selenium as zinc sulfide or selenide (Dias-Somoano et al., 2004).

In summary, selenium displays a very volatile nature and is readily released from fuels. Transport through the flue gas train is dominated by the volatile Se and SeO_2. Removal is mainly on fly ash as a metal selenite, or as reduction to the elemental Se. Competition with sulfur concerning the preferred metal component is apparent, sulfates being favored over selenites. The gaseous SeO_2 is a strong oxidizer, comparable to ozone, and can oscillate between the oxide and free element depending on what it encounters in the flue gases. Ammonia slip will readily reduce it to its elemental form. Other than fly ash it is mainly controlled by the FGD scrubber in which it can dissolve. The chemistry therein is complex and the oxide can easily be reduced to the element that can escape as particulates. As with mercury, the other volatile element, it still remains difficult to control selenium totally and some of the element in small particle form will be emitted. The fly ash and FGD wastewater can become enriched in selenium for certain selenium-rich coals to the point of becoming a valuable resource for this relatively rare element. Selenium is an element with possible health benefits so its release to the environment may in fact be beneficial in some geographic locations, and be of less concern to control.

7.24 Silver

Although below copper in the periodic table family column of Cu, Ag, and Au, the three have few resemblances. With atomic number 47, silver lies at the central area of all the elements. Although lying ninth in the second row of transition elements its electronic structure favors a $4d^{10}5s^1$ structure (like Cu and Au) with no low-lying energy alternates. As a result it tends to favor monovalency. Its concentration in the Earth's crust is low at about 75 ppb being similar to that of Hg both in the crust and oceans in average concentration, with few metals at lower levels. Nevertheless, it is found as the ore argentite AgS, and horn silver AgCl, but nowadays about 75% is produced as a byproduct of Cu ore or from Pb, Zn smelters. It is mentioned in Genesis and has been purified from as early as 3000 BC. Its global market now is about 26,000 tons $year^{-1}$ with known reserves of 530,000 tons. Recycling now contributes to the numbers, being about 20% in the United States, and 57% of the silver is estimated recovered from discarded products globally (Eckelman and Graedel, 2007). Its role over the years has changed as also its value. In the United States its use in coinage was discontinued in 1970 and having a major use in photography has now become minimal with new digital technology. Currently silver prices are \$16 $oz.^{-1}$ (\$0.53 g^{-1}) but were as low as \$10 $oz.^{-1}$ before 2009, spiking to \$40 in 2011.

Few accurate assessments have been made as to silver's natural and anthropogenic levels in the atmosphere. Due to its low crustal concentrations the extent estimated from wind erosion, forest fires and volcanoes is low at about 110 tons $year^{-1}$. That in particulate form from predominantly metal smelting with smaller contributions from fuel combustion was estimated to be about 3100 tons $year^{-1}$ (Malakhov and Makhon'ko 1990). Air release in the United States in 1978 was estimated to be about 85 tons of which 40% was due to smelting, 28% the electroplating industry, and 12% from fuel combustion with some from cloud seeding usage (ToxProfile Ag, 1990). A model of silver's global anthropogenic release was made for the year 1997 (Eckelman and Graedel, 2007). This confirmed low atmospheric particulate release, with tailings and landfill being dominant disposal areas to land and water. As with other elements, records of historical deposition have been examined by analyzing ice core and other surface sampling methods around the globe. One such recent analysis concluded that 80% of the Ag in such deposits in the 1994–2004 time period were anthropogenic confirming the low natural background and also showing that deposition was in considerable decline from the 1960s (Krachler et al., 2008). Local water and soil components of the Earth were more heavily polluted during the height of the photographic manufacturing period with some river silver concentrations, close to plants, gradually declining from 260 $\mu g\ L^{-1}$ in the 1970s to $<10\ \mu g\ L^{-1}$ in the 1990s. Some concerns were raised about the possibly high levels of Ag in sewage sludge being dispersed for agricultural use due to its antimicrobial nature (CICADs Ag, 2002). However, levels now appear to be in a low ppm range and of little concern (Jones et al., 2014a). Silver has the highest known natural electrical and thermal conductivity and is stable, mainly tarnishing only in the presence of sulfur. Its uses are in alloys, photovoltaics, photography, solder, batteries, mirrors, dentistry, jewelry and sterling ware, cloud seeding, textiles and nanoparticles. It is valuable as a germicidal element.

Health concerns with silver

Elemental silver is surprisingly nontoxic and noncarcinogenic. It is hard to die from silver exposure. This also applies to many of its compounds and is due partly to their water insoluble nature. The nitrate differs being soluble and toxic. A toxic level of about 12 g day^{-1} dose for humans was estimated from animal studies. It is a nonessential element for the body with no known mechanisms. A normal dietary intake from water and food of 70 $\mu g\ day^{-1}$ was once estimated but this now has been reduced to about 4–7 $\mu g\ day^{-1}$ (WHO Trace Elements, 1996). An absorption rate in the body of about 10% is assessed with 2%–4% temporary retention in the tissues. Elimination is predominantly in feces with smaller amounts in urine and hair. One human study fed an individual a single dose of 6 mg of labeled ^{110}Ag acetate. Only 21% of this was seen to be retained after a week and then levels slowly declined with no effects. Silver is mildly toxic, but neither carcinogenic nor a neural toxin to humans. Deaths are not recorded and the lethally toxic levels has also been suggested as $50-500$ mg kg^{-1} body wt. (Nielsen, 2017). It is more toxic to biological systems being a strong antimicrobial element. Dining with silverware was by chance healthy for humans. Nevertheless, silver cutlery in acidic conditions does release small amounts (Soton et al., 2018). The low toxicity level can accumulate with constant use particularly with supplements, which should be taken after careful consideration. The illness symptom for excess is argyia or bluing of the skin. If in excess this reaches being severe when AgSe forms in the skin and then can no longer be chelated (Nielsen, 2017). Its different toxic nature with vegetation has been studied and mechanisms considered for cases of mixtures of metals that involve silver (Le et al., 2017). There is little concern with regards to health issues and its general use is not regulated. It is rarely tested in humans by any biotesting laboratory. The major biomarkers are blood and hair. Suggested safe levels are $<1\ \mu g\ L^{-1}$ for blood. A blood level of $<15\ \mu g\ L^{-1}$, becoming acute if $>1\ mg\ L^{-1}$ was regarded as normal at the height of production by the silver industry. Testing of workers with occupational exposures then could show blood levels of $0.1-20\ \mu g\ L^{-1}$ and higher with no medical effects. Few large scale testing surveys have included silver, and those published found blood levels (in 5%–95% percentiles) as $0.7-4.5\ \mu g\ L^{-1}$ (Goulle et al., 2005), and $0.15-0.49\ \mu g\ L^{-1}$ (Cesbron et al., 2013), the former also reported corresponding hair values of 0.02–1.31 ppmw. Now, with the advent of nanotechnology this has raised new questions concerning whether silver nanoparticles may be able to pass through the blood/brain barrier and pose a new increased risk (Karmakar et al., 2014). More concerns have centered on marine life due to the nature of silver. Testing shows it can bioaccumulate with significant variation over different marine species. However, the soluble Ag^{+} ion of concern, in natural environmental cases, generally complexes with sulfur or chlorine in water and becomes absorbed into sediments (Ratte, 1999). Nevertheless, the use of silver nanoparticles in clothing and textiles has spawned renewed concern to its environmental impact on marine organisms and its possible migration (Wang et al., 2014a; Wagener et al., 2016). Even so, in the marine environment, although complex and depending on the nature of the water, it is not regarded as a problem of current concern (CICADs Ag, 2002). Nevertheless, as a result of the past large silver industry, effects of silver and its compounds have been extensively reviewed for humans (ToxProfile Ag, 1990; Drake and Hazelwood, 2005).

Silver's basic chemistry

The extent of interest in the chemistry of an element and its compounds can often be judged by the degree of characterization of its chemical thermodynamics. The NIST-JANAF tabulations contain no listings for Ag or its compounds (Chase, 1998), that of Knacke et al. (1991) has 17 tables (4 for gaseous species), and Barin and Platzki (1995) has 25 (5 gaseous). These also clearly demonstrate the limited chemistry that silver displays having only the element and its halides and their ions known in the gas phase. An additional aspect of silver is that many of its compounds are photosensitive, producing a silver deposit, and consequently involve more difficult handling. This of course was the foundation of photography, with AgCl being the most sensitive photo-compound, then coupled to the use of the nitrate to circumvent its low water solubility. In the condensed phase, as indicated in Table 7.35, the metal (m.pt.962, b.pt.2162°C) has a vapor pressure of about 14 ppmv at 1000°C rapidly beginning to rise to 80 ppmv at 1100°C (Barin and Platzki, 1995).

TABLE 7.35 Properties of silver's condensed phase molecules relevant to combustion, °C.[a]

m.pt. b.pt. vapor pressure
Ag 962 2162, 80 ppmv/1100°C, ins
Ag_2O >200 dissoc, ss
AgO >100 dissoc, ins
AgOH possibly exists in solution only
Ag_2CO_3 218, >220 dissoc, ss
$AgNO_3$ 210, dissoc 200–300 to oxide then metal, light sens, s
$AgNO_2$ >140 dissoc via nitrate, ss
Ag_2SO_4 660, dissoc 1085, light sens, ss
Ag_2SO_3 >100 dissoc to sulfate, ss
AgCl 455 1547, vapor also contains trimer, 35 ppmv/700°C, light sens, ss
Ag_2S 837, >750 dissoc to Ag and S_2, ins
$AgPO_3$ 488, >600 unstable, ins
Ag_3PO_4 849, stable to 1000, light sens, ss

[a]*dissoc*, dissociates; *s*, water soluble; *ss*, slightly soluble; *ins*, water insoluble; *light sens*, light sensitive

It does form two oxides but these have low thermal stabilities and dissociate to the element: Ag_2O, dissociates at 200°C, and has very low water solubility: AgO dissociates above 100°C and is water insoluble. Most of its compounds are not thermally stable, generally dissociating through these oxides. Ag_2CO_3 dissociates above 220°C. For Ag_2SO_4 (m.pt.660°C), a TGA noted the melting and that dissociation to Ag, SO_2, and O_2 was 72% at 1100°C (Shkodin and Abisheva, 1986). Both the carbonate and sulfate have low water solubility. The sulfite, Ag_2SO_3, also of low solubility changes to the metal and sulfate above 100°C (Stern, 2001). The $AgNO_3$ (m.pt.210°C) dissociates in the 200°C–300°C range initially to Ag_2O and then the element, it is very water soluble, the one and only such silver compound considered herein (Brock and DeVore, 1991; Stern, 2001; Mahadevan et al., 2002). The nitrite, $AgNO_2$, is less stable (140°C) and only slightly soluble. The chloride indicates normal vaporization (m.pt.455, b.pt.1547°C) with a vapor pressure of about 2 ppmv (600°C) increasing to 35 ppmv (700°C), but at equilibrium, if contained, does indicate the trimer Ag_3Cl_3 structure (Barin and Platzki, 1995; Hildenbrand and Lau, 1996a). It also displays only low water solubility. A possible hydroxide is known only in solution but is poorly characterized. A sulfide, more thermally stable than the oxides is known (Ag_2S, m.pt.837°C), together with $AgPO_3$ (m.pt.488°C) and Ag_3PO_4 (m.pt.849°C), all water insoluble. No condensed phase AgH is available and would be unstable with water or air. The chloride, nitrate, sulfate and phosphate are all light sensitive being photolyzed to the element.

Of interest mainly to solid-state physics or catalysis, and other potential roles for new molecules with exotic properties, several mixed metal oxides have been synthesized with

Ag as well as Cu. An aluminate, ferrite and silicate are mentioned in the literature with silver and undoubtedly might form in cooling ash particles. However, due to its unstable oxides, there is no indication of a strong propensity for these. In other words interactions with other ash components appear doubtful. It will need low temperatures for molecular formation on ash and be subject to time constraints. Retention within the ash will need internal interactions. There is no evidence that mixed metal silver compounds have pronounced thermal stability or vaporize unchanged (Lopatin and Shugurov, 2013).

Gas phase stabilities and transport

In combustion, the gas phase chemistry of silver has to invoke in situ formation from Ag since other than AgCl there are no compounds that vaporize unchanged. Few studies exist, but the gaseous AgO, AgS, AgCl, and AgH diatomics have been characterized spectroscopically and mass spectrometrically. For AgO and AgS these also display the unusual divalency for Ag. From the early days of flame atomic emission or absorption analysis, it was realized that analytical accuracy and sensitivity could be obtained if the sample was either fully atomized or dissociated in the flame with a speciation that was insensitive to the sample matrix. Attempts to measure this degree of atomization for various elements in a variety of flames were quite common. On noting that silver was readily atomized, and also had a high ionization potential energy value, it became a standard for comparisons with other elements (Willis, 1971; Smyly et al., 1971). Recalculating the equilibrium data, measured in a Knudsen cell-mass spectrometric experiment for the gaseous balance between AgO/Ni/Ag/NiO, this now implies that D_0(NiO) is greater than D_0(AgO) by 160.6 ± 7 kJ mol^{-1} (Smoes et al., 1972; Pedley and Marshall, 1983). Coupling this to the presently recommended value for NiO (373) suggests $D_0(\text{AgO}) = 212 \pm 9$ kJ mol^{-1}. This compares to the approximate value of 209, estimated 48 years ago simply from the dissociation vapor pressure of AgO(s) (Brewer and Rosenblatt, 1969). The value indicates a weak bonding and so will be negligible in flame gases. Several optical transitions have been characterized in the visible and ultraviolet spectral ranges that facilitate its monitoring. The (D, C, A-$X^2\Pi$) transitions have been recorded (O'Brien et al., 1998) and have been predicted theoretically (Bauschlicher et al., 1990). However, electronic structural calculations for the binding energy of AgO still remain inadequate. Derived values are very dependent on the density functional method used and one study indicated values in the range 165–289 kJ mol^{-1} (Song et al., 2007). All other studies underpredict its value (Bauschlicher et al., 1990), or are far too large (Kharat et al., 2009). A recent new experimental measure has introduced a little uncertainty into this assessment by obtaining a value of 191 ± 1 kJ mol^{-1}. This is based on a two-photon photolysis of AgO($X^2\Pi$) dissociating to the two ground state atoms, namely Ag ($4d^{10}5s^1$, $^2S_{1/2}$) and O($2s^22p^4$, $^3P_2$), measuring their separation kinetic energy (Cooper et al., 2015). The published article suggesting involvement of the AgO($A^2\Sigma^+$) was a typographical error. This is a more demanding interpretation but by a similar study an apparently reliable measure of $D_0(\text{CuO}) = 293 \pm 3$ kJ mol^{-1} was obtained by the same group (Parry et al., 2014). As a result, until further confirmation, a value $D_0(\text{AgO}) = 200 \pm 10$ kJ mol^{-1} appears reasonable. Little is known of other gaseous oxides except a hint of an Ag(OO) bonded molecule was noted in a complex infrared spectral matrix isolation study (Wang and Andrews, 2001). Suggestions appear to indicate both a bent end-on and a side-on symmetric structure for this molecule (Kasai and Jones, 1986). Theoretically, both oxides are predicted to have very similar energies (Illas et al., 1986; Bravo et al., 1990). A gaseous AgOH has been observed in a discharge, sputtering silver vapor into moist air. With deuterium labeling its microwave rotational spectrum of AgOH was confirmed with a 108-degree bent structure (Whitham et al., 2000). This was predicted theoretically with an uncertain bond strength but possibly similar to that of AgO with D(Ag-OH) in the range 195–212 kJ mol^{-1} (Bauschlicher, 1986; Ikeda et al., 2003). The only mention of a dihydroxide was reported in a matrix isolation experiment with laser-ablated Ag into H_2O_2 or a H_2/O_2 mixture. A structure and frequencies were assigned to the dihydroxide (Wang and Andrews, 2005b).

Studies of AgCl(s) vaporization from a cell at equilibrium showed a dominance of AgCl (g), but also the gaseous trimer Ag_3Cl_3 and negligible amounts of other polymers. The mole fraction of the monomer varied from about 65% at 530°C to 80% at 680°C. This was studied accurately by a torsion-mass loss method to obtain the sublimation enthalpies and indicating $D_0(AgCl) = 311.3 \pm 2$ kJ mol^{-1}, close to previous values but now more exact (Hildenbrand and Lau, 1996a, 2006). Spectral transitions to the $X^1\Sigma^+$ ground state have been observed but not fully analyzed. Theoretical analyses of its potential energy curves for the 11 lowest energy lying states now have assigned these and predicted other singlet and triplet states (Guichemerre et al., 2002; Ramirez, 2006). None are sufficiently low-lying to contribute to the chemical thermodynamic analyses. Numerous theoretical attempts to calculate the AgCl bond strength have been reviewed (Guichemerre et al., 2002; Cheng et al., 2007). These generally always underestimate the value by about 16–68 kJ mol^{-1}, with one exception that ranges from being either lower by 27, or 4 kJ mol^{-1} larger depending on the level of sophistication invoked (Nasluzov and Rosch, 1996). It illustrates the continuing difficulty of modeling the transition element behavior in a reliable manner. The situation with silver dihalides is that divalent silver is rare and $AgCl_2$ is not known experimentally. One theoretical paper considered the bonding of the dihydrides, difluorides, and dichlorides of the second row transition elements but included elements only from Y to Pd (Siegbahn, 1994). Nevertheless three papers since then have analyzed the theoretical structure and electronic states in depth. They find the bonding differs considerably from that of Cu in spite of their similar $4d^{10}5s^1$ favored type configurations. If it exists, $AgCl_2$ will favor a linear ground state structure, but may have a low-lying excited electronic state. There appears to be debate as to whether the ground state is $X^2\Pi_g$ (Muller-Rosing et al., 2005; Ramirez-Solis et al., 2006) with a later study suggesting $X^2\Sigma_g{}^+$, which lies close in energy (Mishra, 2008). Such calculations are mainly of value in comparing the four halide differences.

Several studies have examined the nature of gaseous AgS. Like many, vaporization from Ag_2S is dissociative to elemental silver and sulfur. However, one Knudsen cell-mass spectrometric study examined the gaseous $AgS/Ag_2/S_2$ balance utilizing Ag(s) and MoS_2(s) in the 1140°C–1260°C range and derived the only measured bond strength of $D_0(AgS) = 213 \pm 15$ kJ mol^{-1} (Smoes et al., 1972). Their data indicated that the bonding of AgO and AgS becomes similar, rather like Cu and reflects the increase in the D_0(MS/MO) ratio to unity as seen in Table 7.19 across the first row of transition elements

(Kretzschmar et al., 2001). Two electronic structural studies show the theoretical limitations indicating binding about 15% (Bauschlicher et al., 1990) and 11% (Legge et al., 2001) below the experimental estimate. AgS is known to have an $X^2\Pi$ ground state, its rotational spectrum being analyzed (Okabayashi et al., 2016). Solely from their vibrational frequencies they estimate a very approximate value of $D_0(AgS) = 243$ kJ mol^{-1}. Until confirmed, considering also the MS/MO bond ratio and the experimental mass spectrometric result, a value of $D_0(AgS) = 210 \pm 10$ kJ mol^{-1} is currently recommended. This is an unusual area of the periodic table where the diatomic sulfide can become slightly stronger than its oxide. The spectral states of AgS are expected to reflect the states of AgO, and states lying up to 30,000 cm^{-1} above the ground state have been predicted with none of these doublet states energetically being low-lying (Bauschlicher et al., 1990). The $(A^2\Sigma^+-X^2\Pi_i)$ transition of AgS has been recorded and confirms its predicted location (Gupta et al., 2013). A bent AgSS molecule was reported in a matrix isolation experiment but remains otherwise uncharacterized (Wang et al., 2009b).

Studies of the transition metal hydrides are favored for developing theory, but quite often the necessary benchmark experimental value needed for comparison has uncertainty. In this case an equilibrated effusion cell-mass analysis method at 1100°C–1170°C did establish a value of $D_0(AgH) = 211.3 \pm 8$ kJ mol^{-1} (Kant and Moon, 1979). This was consistent with the early spectral extrapolation estimates of about 221 (Huber and Herzberg, 1979). A value of 199 ± 8 kJ mol^{-1} also was tentatively derived from a reaction energy threshold measure using a guided ion-beam mass spectral method examining the Ag^+ with c-C_3H_6 reaction, but this remains speculative (Chen and Armentrout, 1995). Spectroscopically, the molecule's electronic states have been quite extensively identified experimentally from earlier times (Huber and Herzberg, 1979). As a result, several detailed theoretical studies now have adequately described the potential energy curves for the lowest lying 12 singlet and triplet states and assigned these to the experimental spectral transitions (Witek et al., 2000, 2002; Li et al., 2004; Lo and Klobukowski, 2007). Reproduction of the bond strength generally has been more difficult and, in spite of the at least 10 or so attempts, variation still ranges from 182 to 236 kJ mol^{-1} with no obvious trend. The more recent theoretical values have implied $D_0(AgH) = 234$ (Witek et al., 2002), and 202 kJ mol^{-1} (Guitou-Guichemerre and Chambaud, 2005), so uncertainties remain. Although there has been interest in the coinage element gaseous dihydrides, AgH_2 has not been reported in any studies or seen in Ag/H_2 matrix isolation type experiments (Wang et al., 2003). Nevertheless, two theoretical analyses have described its probable structure and low-lying electronic states. Its ground state is predicted to be a bent (H-Ag-H) X^2B_2 structure that is stabilized by an energy barrier of about 75 kJ mol^{-1} against dissociating to Ag and H_2 (Balasubramanian and Liao, 1989; Guitou-Guichemerre and Chambaud, 2005). These gaseous silver molecule bond strengths are summarized in Table 7.15 together with several other second and third row transition elements, reflecting the overall weakness seen at the end of the transition element rows. Its gaseous molecules are weaker than those of Cu, with those of Cd being even weaker. As suggested, gaseous transport of silver at elevated temperatures will have to involve the element with possible traces of AgCl. Molecular formation will only be possible at lower temperatures by thermally stable chemidepostion on surfaces forming silver or its potential mixed metal compounds. Otherwise, silver may remain entrapped within the ash by chemical interaction.

Combustion power plant data

This summary of silver's chemistry displays its extreme limitations when considering any combustion role. Dissociation of most of the condensed phase silver compounds will favor formation of the elemental silver, which will only vaporize at higher temperatures (vapor pressure 100 ppmv at 1140°C) unless it can form the more volatile AgCl under the influence of chlorine (vapor pressure 100 ppmv at 775°C). No gaseous molecules are stable at elevated temperatures other than AgCl and even this will become minor in high temperature flames. The only other stable molecule is the sulfate below 1085°C and it is light sensitive. As a result, one might expect that full-scale combustion studies will reflect these predictions simply because few alternatives exist.

Silver is at very low levels in most fuels (Table 1.1). Coals and biomass contain concentrations up to 2.0 ppm (Xu et al., 2003; Hu et al., 2006), almost comparable to that of toxic mercury. It only becomes more significant in hospital and some waste incineration, and in sewage sludge combustion disposal where concentrations in resulting ash can occasionally rise higher (Zhao et al., 2008). A recent UK sampling of sludge from 28 wastewater treatment plants gave an average level of 2.7 ppmw (Jones et al., 2014a). It has been reported that incinerated sludge ash in computer technology areas such as Palo Alto, California, can contain recoverable amounts of Ag, Pd, and Au (Gabler and Neylan, 1977). Its presence in pet-coke and limestone are negligible. As a result, the scant information on silver in combustion comes mainly from incineration sources. Archival reviews of the trace elements in sewage sludge ash for the 1972–94 period do report a continuous decrease in concentrations probably via controls motivated by maximum permitted ash levels required for disposal as an agricultural supplement. Utilizing an electric furnace, sludge that had been treated with $Ca(OH)_2$ and $FeCl_3$ was heated at the rate of $10°C\,min^{-1}$ up to 800°C–1100°C and maintained there for an hour. The sludge contained about 10 ppm Ag, 1.2% Cl, and 0.5% S. Loss of silver was pronounced above about 400°C, 82% having vaporized by 800°C, and being total at 1000°C (Folgueras et al., 2003). Also if blended with coal (Ag 0.04 ppmw) to dilute it, volatility varied linearly with overall Cl content, evident at about 0.1% Cl, was 30% at 0.3% Cl, and essentially complete if $Cl \approx 1\%$. The vapor pressure of elemental Ag is about 14 ppm at 1000°C, whereas for AgCl it has risen to about 8500 ppmw at that temperature (Knacke et al., 1991). Clearly, chlorine becomes pivotal in liberating silver as gaseous AgCl, and the sulfur cannot retain it even though the sulfate is thermally stable to 1000°C. Nevertheless, they concluded that the transport of silver appeared to be more complex than simple vaporization release. For hospital waste incineration and municipal wastes, concentrations were quite low in bottom ash, but about 20-fold higher in the fly ash (Zhao et al., 2008). Sewage sludge incineration also was shown to have significant enrichment in the stack particles (Greenberg et al., 1981). Any vaporized silver form will deposit on the cooling fly ash with significant enrichment but will require low temperatures. However, one study took the waste incinerator ESP fly ash, mixed it with $CaCl_2$ or had $Cl_2(g)$ flowing over it in a tube furnace at 1000°C for 3 hours (Chan and Kirk,1999). Although other elemental forms such as K or Cu were fully volatilized from the ash by the chlorine, the 27 ppmv concentration of Ag in the ash was decreased only by

26% with the $CaCl_2$ and 37% with the molecular Cl_2. This tends to suggest that the speciation of Ag in the final ash was not AgCl, which would have vaporized, but during cooling the deposited silver became possibly Ag_2SO_4 or the phosphate that are not readily chlorinated. The nature of silver is one of particulates and any atmospheric stack emissions will be slight and in the form of small particles (0.17 g day^{-1} Ag from 16 tons incineration of sludge containing 29 ppm Ag), in other words 10 ppbw and quite negligible (Greenberg et al., 1981). In a paper assessing the retention effectiveness of trace metals by the FGD system in a coal-fired combustor, although 60 elements were included, Ag appeared negligible (Cordoba et al., 2012). One similar measure for coal stack emissions did include Ag. The coal contained only 0.06 ppmw and emissions were lowered even further by the FGD and fully negligible (Zhao et al., 2016). A study has reported a strange effect on the fly ash composition on reducing the flue gas temperature from 250°C to 150°C through a heat exchanger. The fuel was biomass from metal contaminated soils burned in a small pilot-plant boiler. Boiler temperatures are not quoted, but a two- to threefold increase in fly ash concentration appeared to occur for silver and many other elements as the temperature was so cooled through the heat exchanger. Deposition is unlikely to increase the concentrations for silver and it is noteworthy that similar increases were seen with such elements as Ti or Zr that are totally nonvolatile. It would appear that some particle segregation has to be occurring in the heat exchanger and the ash size distribution changing, favoring the smaller sizes at the lower temperature enriched possibly more by a mechanical fragmentation mechanism than by deposition (Kovacs et al., 2018). Renewed interest with silver has arisen partly from the use of nanoscale silver particles in the textile industry. This will ultimately affect silver levels in sewage sludge treatment plants and impact discharges to soil levels (Lombi et al., 2014; Johnson et al., 2014). A recent paper addressed such silver nanoparticles in sewage sludge. Such particles became Ag_2S nanoparticles during the treatment phase but then were converted back to Ag nanoparticles if incinerated at 800°C (Meier et al., 2016). Another area of interest is the low silver content in diesel fuels that results in the small nanoparticles in the roadside particles of heavily traveled roads, 37% of the Ag mass being found in particles $<0.1\ \mu m$ (Lin et al., 2005). Because AgCl is readily photolyzed to Ag, as with many of its other molecules, the final particulate speciation still remains unknown and may be the water insoluble sulfate or phosphate or even the free metal.

In summary, silver is not an element of concern for health or combustion. It is an antimicrobial that is toxic to agriculture and aquatics and does have to meet regulations concerning ash disposal. Coal has very low levels and waste and sludge fuels are the major concern. Chlorine content appears to catalyze gaseous release as atomic silver or AgCl. However, this may be more complex than simple vaporization. These are the only transport gaseous molecules in a very limited chemistry. Whether silver forms mixed metal compounds in or on any ash remains uncertain but chemideposition onto ash will have temperature and time constraints as only the sulfate or phosphate are solid and thermally stable at or below 1000°C. Modern-day combustors that include an ESP, bag filter and FGD scrubber will retain possible releases. Due to coal's low levels of silver, cofiring with sludge or wastes is one method of ash modification.

7.25 Strontium

Strontium is the central element of the three alkaline earths and this is reflected in the common divalent chemical properties that vary systematically from Ca to Ba. Being the 15th most abundant element in the Earth's crust its presence is less significant than that of calcium. In nature it occurs as the sulfate and carbonate both of which are insoluble in water. The sulfate is more abundant and favored for mining. Four stable nonradioactive isotopes exist naturally for strontium with ratios that vary slightly with location. This is useful at times particularly in geological and archeological studies. Although in previous centuries it was widely used by the sugar beet industry and was an early component in cathode ray tubes as an X-ray absorber strontium's usage today is in decline. It is no longer mined in the United States, but is readily available and can be imported for about \$1 kg^{-1}. World production is on a scale of about 350,000 tons $year^{-1}$. Current uses are mainly in ceramics, glass, ferrite magnets and basic chemicals. Its deep red spectral emission finds use for about 5% of its production for fireworks. There appears to be little interest in natural and anthropogenic levels of strontium in the atmosphere. However, one study does imply no present enrichment above the natural global abundance (Malakhov and Makhon'ko 1990).

Health concerns with strontium

The public became more aware of strontium when it was noted that Sr-90, one of its radioactive isotopes with a half-life of 29 years was a nuclear fallout product. Strontium's potential for being absorbed like calcium into bone stimulated interest and initiated human metabolic studies. It is not regarded as essential to the human body but is neither carcinogenic nor toxic except in large doses that can produce medical symptoms. Automatic daily intake of about 1–5 mg day^{-1} is consumed by the human diet in the food and water. About 11%–30% is absorbed if ingested and mainly excreted in feces. In nature, some mineral skeletons use it instead of calcium. It is present in human bone but generally Ca/Sr ratios are of the order 1000–2000. Nevertheless, in recent years it has been noted that low levels of strontium are beneficial and can add strength to bone. The medical profession now suggests strontium supplements for certain bone deficiencies. An NOAEL of about 10–40 mg kg^{-1} body wt. day^{-1} or a MRL (minimal risk level) of 2 mg kg^{-1} body wt. day^{-1} have been suggested. Two environmental-toxicity reports are available on the Web that tabulate in detail the medical and biological studies that involve strontium (ToxReport Sr, 2004; World Health Organization Sr, 2010). Consequently, like calcium there is no major health concern involving strontium and none involving combustion other than documenting the nature and possible magnitude of its emission in particulate form from differing fuels.

Few medical test centers measure for body strontium, but these indicate that adequate levels are 8–38 $\mu g\ L^{-1}$ for blood and 1–7.5 ppmw for hair. Similarly, few large surveys have included Sr but are compatible with these suggested ranges, 5–95 percentile results for blood being 9–41 (Goulle et al., 2005), 8–35 (Cesbron et al., 2013), and a full range of values 7–67 (Henriquez-Hernandez et al., 2018). Corresponding hair values are 0.2–4.6 ppmw (Goulle et al., 2005), urine 20–413 (Goulle et al., 2005), $\leq 299\ \mu g\ L^{-1}$ (CDC 4th Rept., 2018).

Strontium's basic chemistry

Like calcium, it has a broad spectrum of condensed phase chemicals that are slightly increased in their stability (Table 7.36). Many of these similarly favor dissociation on heating to the stable refractory low vapor pressure oxide, SrO (m.pt.2531, b.pt.3200°C).

TABLE 7.36 Properties of strontium's condensed phase molecules relevant to combustion, °C.[a]

m.pt. b.pt. vapor pressure
Sr 777 1412, reacts with water, 10 ppmv/520°C
SrO 2531 3200, reacts with water, nonvolatile 1 ppmv/2000°C, s
SrO_2 >215 dissoc, reacts with water
$Sr(OH)_2$ 535, decomp 710 to SrO, reacts with water, 10 ppmv/284°C, s
$SrCO_3$ >1494 dissoc, 10 ppmv/660°C, ins
$Sr(NO_3)_2$ 570, >645 dissoc, s
$Sr(NO_2)_2$ >260 dissoc, s
$SrSO_4$ 1606, >1374 dissoc, at 950 with Cl forms $SrCl_2$, 10 ppmv/930°C, ss
$SrSO_3$ >600 dissoc to sulfate/sulfite, ins
$SrCl_2$ 874 2056, sublimes, 10 ppmv/1000°C, s
SrS 2226, sublimes, >1500 dissoc, reacts with water and O_2, 10 ppmv/1640°C
SrH_2 1050, >675 dissoc, reacts with water and air
$SrAl_2O_4$ 1790
SrB_2O_4 known
$Sr_3(PO_4)_2$ ≈1300, ins
$Sr_2P_2O_7$ 1100, ins
$SrSiO_3$ (meta) 1580, ins
Sr_2SiO_4 (ortho) 2300, stable
$SrTiO_3$ 2040, ins
Sr_2TiO_4 1860, s
$SrZrO_3$ 2750, ss

[a]*Gurvich et al. (1994), see text. dissoc*, dissociates; *decomp*, decomposes; *s*, water soluble; *ss*, slightly soluble; *ins*, water insoluble.

The metal Sr (m.pt.777, b.pt.1412°C) has a vapor pressure of 10 ppmv at 520°C, slightly more volatile than either Ca or Ba. The carbonate dissociates with increasing temperature at its 1494°C m.pt. (CO_2 vapor pressure 10 ppmv at 610°C) (Gurvich et al., 1994; Barin and Platzki, 1995). The sulfate is stable to 1374°C dissociating to either SO_3 and or SO_2 which appears to remain uncertain (Ostroff and Sanderson, 1959). If heated with a chlorine salt at 950°C it converts to $SrCl_2$ (Castiel and Goldstein, 1947). As normal, the sulfite $SrSO_3$ is less stable decomposing to the sulfate and sulfide above 600°C. The nitrate is slightly more stable than calcium's, dissociating at 645°C (Mahadevan et al., 2002) and its nitrite above 260°C. Strontium oxide, SrO(s), is slightly soluble in water where it forms the hydroxide $Sr(OH)_2(s)$ that can be separated and dried as a powder (m.pt.535, b.pt.710°C). On heating this loses water and reverts to the oxide (water vapor pressure 10 ppmv at 284°C) (Barin and Platzki, 1995): SrO(s) (m.pt.2531, decomposes at b.pt.3200°C). If dry, it is a very stable refractory molecule requiring temperatures near 1800°C to begin to register vapors on a ppmv scale. Under neutral or oxidizing conditions it vaporizes both molecularly and by dissociation producing a mixture of gaseous Sr, SrO, and O_2 (Lamoreaux et al., 1987). However, as seen with the other Group IIA oxides, the direct ablative vaporization of SrO (s) by water vapor to stable gaseous $Sr(OH)_2$ is increasingly favored on changing from Mg through to Ba. The structure and vibrational frequencies have been recorded for the alkaline earth dihydroxides by matrix isolation experiments (Andrews and Wang, 2005; Wang and Andrews, 2005). The diatomic oxide also absorbs oxygen becoming the peroxide, SrO_2 (decomposes at 215°C m.pt.). This finds use as an oxidizing agent or in pyrotechnics but is not thermally stable dissociating above its melting point (Tribelhorn and Brown, 1995; Risold et al., 1996). It has been structurally characterized by infrared matrix isolation studies (Andrews et al., 1994, 1996). The dichloride is very stable and sublimes molecularly without dissociation (vapor pressure 10 ppmv at 993°C, m.pt.874, b.pt.2056°C) (Barin and Platzki, 1995). Similar to its oxide analogue, SrS(s) is very stable and appears to sublime congruently but dissociates extensively at high temperatures. At about 1830°C the total vapor pressure is about 10 ppmv but the Sr(g) product is about 25 times larger together with S and S_2 species than SrS(g) (Cater and Johnson, 1967; Gurvich et al., 1994; Barin and Platzki, 1995). Like other sulfides it readily reacts with water in this case producing the dihydroxide and $H_2S(g)$. Heated at 500°C with O_2 it converts directly to the sulfate (Madarasz et al., 1996). The dihydride (m.pt.1050°C) is known but similarly reacts with water or oxygen and is thermally stable only to 675°C. Its infrared spectrum has been characterized by matrix isolation methods (Wang and Andrews, 2004b). Also there are other mixed element oxides that relate to combustion and might form within or on ash particles. With strontium these become ceramic in nature having very high melting points, for example, the stable silicate, $SrSiO_3$ (m.pt.1580°C), and Sr_2SiO_4 (m.pt.2160°C). Strontium aluminate (m.pt. $\approx$1660°C) is very stable and unusual in being a long-lived phosphor. The titanate, $SrTiO_3$, has a high melting point (2040°C). The same applies to the known phosphor materials strontium phosphates, $Sr_3(PO_4)_2$ ($\approx$1300°C) and $Sr_2P_2O_7$ (1100°C). The ferrite $SrFe_2O_4$ is known but minimally studied. The gaseous molecules of some of these, namely $SrTiO_3$, $SrSiO_3$, $SrFeO_2$ have been characterized in high temperature mass spectrometric vaporization studies also confirming their gas phase molecular stabilities (Lopatin and Shugurov, 2013). If correct, for the latter this would imply that on vaporization the ferrite with Fe^{III} modifies to an Fe^{II} state.

Because the extent of free atomic Sr controls the sensitivity of flame atomic absorption/emission measurements, this was of particular interest in such analyses. One measure using absorption in a C_2H_2/air flame at 2140°C obtained a value of 7.5% free Sr, and another with $H_2/O_2/Ar$ flames at various stoichiometries and temperatures displayed no higher than a 9.4% proportion in the flame gases (Zeegers et al., 1969; Smyly et al., 1971). Clearly, the flame Sr was producing a molecular distribution. However, in such analysis work it is to be remembered that the strontium has been brought into solution and is injected directly into the analysis flame. This differs totally from fuel combustion where a release has to occur from the solid phase. From the array of condensed molecular forms discussed above it would seem that from a solid fuel, only SrO and $SrCl_2$ might vaporize directly, together possibly with $Sr(OH)_2$ from water aided ablation of SrO(s), and possibly SrS in dry, oxygen-free environments. Whether other gaseous forms can be generated in the gas phase and act as strontium transporters in combustion such as SrO_2, SrOH, SrCl, Sr(OH)Cl, or SrH has to be considered.

The dissociation energy of SrO and its meaning

On vaporizing, or being formed directly in the gas phase a simple question is what is the thermal stability of the gaseous molecules, or in particular in this case the nature of the SrO (g) bond? With all the Group IIA oxides and sulfides, this question still poses an intriguing enigma as is outlined individually in each of their separate discussions. It tends to arise partly from their ionic nature and also their spectroscopy. As clearly illustrated spectroscopically by Field (1974), and by matrix isolation spectra, the ground state is $SrO(X^1\Sigma^+)$ (Ault and Andrews, 1975a). As with all these oxides, although ionic, dissociation is to neutral atoms which is energetically strongly favored, taking less energy than to gaseous ions. Spectroscopically its ionic potential energy curve transfers as the atoms separate onto the covalent curve for $SrO(X^1\Sigma^+)$. However, as with all the other Group IIA oxides this ground state does not correlate to the ground state dissociation elements $Sr(^1S_0)$ and $O(^3P_g)$, being spin-forbidden. It is the first excited and metastable $SrO(a^3\Pi)$ state lying 108 kJ mol^{-1} above $X^1\Sigma^+$ that does (Field, 1974). Rigorously, the $X^1\Sigma^+$ always correlates to the second lowest energy pair of atoms on dissociation, which is spin allowed. Although all Group II elements experience this same dilemma, there is a slight difference between BeO and MgO, and the three alkaline earth oxides. In the case of Be and Mg the second dissociation limit invokes the excitation to $O(^1D_g)$ with about 190 kJ mol^{-1} and lies at the lowest energy, whereas for Ca, Sr, and Ba, excitation to the lowest excited states of these atoms now lies below that of $O(^1D_g)$, namely, $Ca(^3P_u)$ 181, $Sr(^3P_u)$ 171 and $Ba(^3D_g)$ 108 kJ mol^{-1}, and are all in fact metastable states. It satisfies spin conservation and correlation requirements.

For SrO, the potential energy curves mirror very closely those seen with CaO, already shown in Fig. 7.4 and so have not been redrawn here. The dilemma arises from the measured and theoretical values obtained for D_0(Sr-O)(g). These are from a variety of methods that involve mass transport, mass spectrometry, flame spectrometry, optical spectrometry, cross-beam measurements and theoretical calculations. The dilemma in fixing the bond strength arises from the fact that the values from these 21 or so experimental measures fall in the range of 342–482 kJ mol^{-1}. Theoretical predictions suggested either a range of about

408–420 or much higher or lower (Partridge et al., 1985; Langhoff et al., 1986). The latest most sophisticated calculations indicate 413 ± 8, that is supported, as will be seen, with very accurate values for the bond strengths of gaseous SrCl, $SrCl_2$, SrOH, and $Sr(OH)_2$ (Vasiliu et al., 2018). However, their results for SrO support the equilibrium studies and refer to the spin-forbidden transition. They either calculated the energy to the second dissociation limit and then subtracted the electronic excitation energy or calculated the dissociation energy of the low-lying $a^3\Pi$ state and added its own excitation energy. In other words, they calculated the energies of all three possible transitions and noted that the reported experimental values favored the forbidden exchange.

$$SrO(X^1\Sigma^+) = Sr(^1S_0) + O(^3P_g) \quad D_0 = 405\ \text{kJ mol}^{-1}\ (\text{spin forbidden})$$

$$SrO(X^1\Sigma^+) = Sr(^3P_0) + O(^3P_g) \quad D_0 = 405 + 171 = 576\ \text{kJ mol}^{-1}$$

$$SrO(a^3\Pi) = Sr(^1S_0) + O(^3P_g) \quad D_0 = 405 - 108 = 297\ \text{kJ mol}^{-1}$$

As with calcium, if dissociation does occur on the allowed curve, the metastable $Sr(^3P_0)$ is produced with a radiative lifetime of 21 μs (Drozdowski et al., 1997) and will be quite efficiently quenched by collisions such as with N_2 or CO_2 (Husain and Schifino, 1984). Nevertheless, such dynamics would be evident in some measures by showing a much larger activation energy. It would also need to maintain a balance between $Sr(^3P_0)$, $Sr(^1S_0)$, and SrO $(X^1\Sigma^+)$ in short measurement timescales. There is no doubt though that those results based on equilibrated systems, such as the mass spectrometric studies, better relate to the recommended forbidden transition value probably as they involve no time constraints, or reaction dynamic constraints. Obtaining the correct value is significant in that it is thought to be important for nuclear reactor modeling. Over the years, the available data have been extensively reviewed numerous times and a recommended value for D_0(SrO) picked from a considered favored list of values. An evaluation from 1974, based on four sources, was the basis for a listed value of 422 ± 20 kJ mol^{-1} (Chase, 1998), that by Pedley and Marshall (1983) obtained 422 ± 17 from eight measurements, Gurvich et al. (1994) listed 14 values, favored the 412–470 range suggesting 420 ± 10, Huntelaar and Cordfunke (1997) gave a very detailed analysis of 21 sources suggesting a later analysis of 405 ± 15 kJ mol^{-1}. This latest review implied that two main causes for the errant values was partly because the vapor concentrations are small and also can suffer interference from the gaseous dihydroxide if any water is present. This may be so in some cases, and the flame studies that suggested values of about 466 did drop to 416 if CO/O_2 flames were used. Nevertheless, one molecular-beam study measuring reaction threshold energy of the $Sr + O_2$ reaction suggested D_0 471 kJ mol^{-1} (Batalli-Cosmovici and Michel, 1972). All these reviews examine the same data base of values, largely from equilibrium systems, so the reaction dynamics are yet to be better tested by different type methods. It does seem that the nature of the potential energy curves certainly play a role in the exact reaction dynamics. None of the experiments have really invoked time constraints to examine this aspect for the divergent values. Like calcium, the normal equilibrium population of the $SrO(a^3\Pi)$ state will be very small (0.9% at 2000°C) probably too small to contribute to the dissociation or chemistry other than through this too small state mixing. From the corresponding curves for CaO (Fig. 7.4) it is seen necessary for a curve crossing to occur from the ground state $SrO(X^1\Sigma^+)$ to produce ground state atoms. Crossing has to be either to the $a^3\Pi$ or the repulsive $^3\Sigma^-$ as a possible solution. Recently, the electronic structure of SrO has been analyzed theoretically deriving the nature of 29 low-lying singlet and triplet states noting numerous curve crossings but none seems appropriate to these states: another theoretical value of D_0 of 444 kJ mol^{-1} was added to the data (Khatib and Korek, 2017). As a result, the unanswered spectroscopic questions really appear to involve this degree of curve hopping. That between the $X^1\Sigma^+$ and $a^3\Pi$ states would have the tendency to enhance the accepted value the lower its efficiency. It appears unobserved and unlikely. Whether the recent suggestion of spin–orbit induced predissociation utilizing the repulsive $^3\Sigma^-$ state, successfully applied to isoelectronic BeS, is the solution has to be seen (Larbi et al., 2010). If the curve crossings or relaxations are equally efficient, all measurements would have suggested the same value so the dilemma remains as to which dynamic is controlling and introducing possibly some system dependency. It should be remembered that even equilibrated systems must invoke some dynamic mechanism to achieve their goal.

Gas phase stabilities and transport

Although the solid peroxide is not thermally stable, the gaseous SrO_2(g) has been invoked as the product in the termolecular recombination of Sr with O_2 at low temperatures (30°C–695°C) (Vinckier and Helaers, 1998). As indicated above it also has been characterized by matrix isolation methods (Andrews et al., 1994, 1996). Theoretical calculations as with the other Group IIA elements suggest $D_0(\text{Sr-O}_2) = 226 \pm 25$ kJ mol^{-1}for a C_{2v} $^1A_1$ structure (Bauschlicher et al., 1992). Because the bonding relates more to an insertion of Sr^{2+} into O_2^{2-} the concept of the second OSr-O bond in relation to Sr-O has no rigorous meaning. With regards to combustion, even though thermally stable to a certain degree, any potential role is unlikely due to its reactivity with water or hydrogen and its tendency to convert to the more stable SrO(g).

Due to the low volatility of isolated SrO(s), its ablative evaporation directly to Sr $(OH)_2$(g) becomes potentially more important. However, this is driven by the thermal stability of the gaseous dihydroxide that is gradually weakening for strontium (Table 7.10).

$$\text{SrO(s)} + \text{H}_2\text{O(g)} = \text{Sr(OH)}_2\text{(g)} \quad \Delta H_{298\text{K}} = 237 \pm 30 \text{ kJ mol}^{-1},\ K_{1800\text{K}} = 1.4 \times 10^{-4}$$

Even so, and although uncertainties still exist between databases (Gurvich et al., 1994; Chase, 1998), in a gas containing 10% water vapor the equilibrium partial pressure of Sr $(OH)_2$(g) will be about either 36 or14 ppmv at 1800K, respectively. Volatility follows the trend of increasing from Ca to Sr, to Ba. The bond strengths now average to D_0(HOSr-OH)(g) = 435 ± 10 kJ mol^{-1} with that for D_0(Sr-OH) = 409 ± 8 kJ mol^{-1} from these sources, confirmed by accurate theoretical calculations (Vasiliu et al., 2018). As seen in Table 7.10 that summarizes these bond strengths, the monoxide gains strength in progressing from Mg to Ba. This is reflected in an analysis of a 2200°C hydrogen flame containing strontium that indicated a gaseous near equilibrium distribution of 3% Sr, 1.3% SrO, 38% SrOH, 49% $Sr(OH)_2$, and in this case a nonnegligible 8.3% $SrOH^+$ chemiionization contribution (Jensen and Jones, 1978). As a result, flames will most likely contain both of the hydroxides if there is a release of strontium. Although most of the thermochemical values for these hydroxides have generally been obtained from flame studies, the value for D(Sr-OH) also was confirmed by an additional Knudsen cell-mass spectrometric study that generated SrOH(g) by passing H_2 over SrO(s) (Murad, 1981). Additionally, as indicated recent theoretically calculated values are in agreement with previous values (Bauschlicher et al., 1986) and further calculations have established the nature of the lowest excited electronic states of SrOH that are not at low-lying energies (Allouche and Aubert-Frecon, 1994). The flame derived values are based on spectrometric measurements of either solely Sr or together with SrOH. The latter can be observed spectroscopically in emission, absorption and LIF, but the monohydroxide spectrum does consist of complex bands (Theo and Zeegers, 1975; Blackburn et al., 1978). Efforts resolved the excitation energies of these bands adequately (van der Hurk et al., 1974) and the spectral systems are now better characterized (Beardah and Ellis, 1999). Nevertheless, despite these difficulties, the flame derived values for D_0(Sr-OH) agree with the mass analysis study and are also the basis for the $Sr(OH)_2$ second bond strength value. This is also now confirmed by an accurate theoretical calculation. The electronic structure of the dihydroxide was determined in matrix isolation experiments using laser-ablated Sr either with H_2/O_2 or H_2O_2 mixtures (Wang and Andrews, 2005).

As with calcium or barium, the only molecules that can compete with the strontium hydroxides in the gas phase are the chlorides. This was noted 50 years ago on adding CCl_4 to a C_2H_2/air flame containing traces of strontium. Spectral emissions of Sr* and SrOH* decreased, with that for SrCl* initially rising and then also decreasing (Zhitkevich et al., 1963). Larger additions of chlorine were linearly correlated with a decrease of Sr and interpreted as dominant $SrCl_2$ formation (Kantor and Ernyei, 1994). The thermochemistry of the chlorides is well developed and SrCl(g) has been studied by mass spectrometry, flame photometry and molecular beam chemiluminescence methods. A reliable value of $D_0(\text{Sr-Cl}) = 409 \pm 8\ \text{kJ mol}^{-1}$ is recommended (Gurvich et al., 1994; Barin and Platzki, 1995). It is seen to have no low-lying electronic states (Allouche et al., 1993). Purely by monitoring the ratio of Sr and SrCl emissions from a series of flames, older values gave 400 ± 10 and $417 \pm 8\ \text{kJ mol}^{-1}$ for SrCl (Gurvich et al., 1974; Belyaev et al., 1982). Because $SrCl_2$(g) is stable, its heat of formation can be reliably determined from the measured heat of sublimation and the $SrCl_2$(s) heat of formation both of which are accurately known. As a result, this indicates the recommended value of $D_0(\text{ClSr-Cl}) = 473 \pm 8\ \text{kJ mol}^{-1}$ (Gurvich et al., 1994; Barin and Platzki, 1995). As noted in Table 7.10, these chlorides are more stable than the corresponding hydroxides and mirror calcium with the dichlorides being 9% stronger than the dihydroxides, and monochlorides about 3% stronger over monohydroxides. As a result their distribution in combustion will heavily depend on HCl/H_2O ratios. Flame studies readily see this with linear effects being observed with low chlorine levels consistent with monohalide formation becoming quadratic with increased chlorine and dichloride dominance. However, realized independently, an analysis of such data implied that more than one monohalide was evident and Sr(OH)Cl was suggested, which might be expected with such similar thermochemistry (Schofield and Sugden, 1971; Gurvich et al., 1971, 1974; Ryabova et al., 1971). The later Russian work indicated that the chlorine bond in this mixed molecule would be the stronger, $D_0(\text{HOSr-Cl}) = 448\ \text{kJ mol}^{-1}$ and $D_0(\text{ClSr-OH}) = 423\ \text{kJ mol}^{-1}$ (Gurvich et al., 1974). A theoretical study did investigate the geometrical structure of $SrCl_2$ (Levy and Hargittai, 2000).

SrH(g) can be formed in the gas phase and has been spectroscopically characterized. Predissociation limits establish a reliable value of D_0(Sr-H) = 156 ± 2.4 kJ mol^{-1} and confirm its thermal weakness (Gurvich et al., 1994). This was supported by an early but underestimated theoretical calculation (Fuentealba et al., 1987). It will also be readily oxidized and hydrolyzed with water. The remaining species SrS(g) is of interest in that it reflects in many ways SrO(g) and all the other Group IIA oxides and sulfides. It is stable, D_0(Sr-S) = 330 ± 15 kJ mol^{-1}, the same as CsS, but the value remains uncertain determined only by Knudsen cell-mass spectrometric equilibrium studies (Gurvich et al., 1994). The solid, that technically has a melting point of 2226°C, begins to sublime producing mainly Sr(g) and S, S_2(g) at temperatures as low as 1270°C with smaller amounts of SrS(g) (Colin et al., 1964; Cater and Johnson, 1967). Vaporization appears noncongruent and values are assessed from other Sr/S equilibria. Theoretical calculations confirm this bond strength magnitude (329 kJ mol^{-1}) that again relates to the thermodynamic balance for the spectrally spin-forbidden reaction (Partridge et al., 1988):

$$\mathrm{SrS}(\mathrm{X}^1\Sigma^+) = \mathrm{Sr}\left({}^1\mathrm{S}_0\right) + \mathrm{S}\left({}^3\mathrm{P}_g\right) \quad D_0 = 330 \pm 15 \text{ kJ mol}^{-1} (\text{spin forbidden})$$

However, without curve hopping and with spin conservation the $X^1\Sigma^+$ state correlates to an S($^1D_2$) product at the second dissociation limit. This differs slightly from SrO in this respect.

$$\mathrm{SrS}(\mathrm{X}^1\Sigma^+) = \mathrm{Sr}({}^1\mathrm{S}_0) + \mathrm{S}({}^1\mathrm{D}_2) \quad D_0 = 330 + 111 = 441 \text{ kJ mol}^{-1}$$

It is the lowest exited state of SrS (T_0 87 kJ mol^{-1}) that correlates directly to ground state atoms

$$\mathrm{SrS}(\mathrm{a}^3\Pi) = \mathrm{Sr}\left({}^1\mathrm{S}_0\right) + \mathrm{S}\left({}^3\mathrm{P}_g\right) \quad D_0 = 330 - 87 = 243 \text{kJ mol}^{-1}$$

The SrS(a$^3\Pi$) state is interesting in being metastable but lies too high in energy to have a significant population at equilibrium (6% of the total SrS at 2000°C). Although not accurately located it is thought to have an energy of about 7300 cm^{-1} (87 kJ mol^{-1}) similar to BeS and CaS (Partridge et al., 1988; Chase, 1998). Only in atomic recombination systems involving Sr + S might it be of importance. Current D_0 values are in agreement as they have all been derived from long time-equilibrated systems that reflect the spin-forbidden transition. As a result, and as extensively discussed herein for SrO, and in the similar cases of CaO and CaS, the mechanism of the dissociation remains uncertain whether it is actually one of curve hopping to the a$^3\Pi$ or proceeding via the S($^1D_2$) channel, which although metastable, is known to be collisionally relaxed quite rapidly (Schofield, 1979). Additionally another recent suggestion has been made for isoelectronic BeS of spin–orbit induced predissociation utilizing the repulsive $^3\Sigma^-$ state that is the only other state correlating with ground state products and has been illustrated for CaO in Fig. 7.4 (Larbi et al., 2010). This can only be resolved by further studies possibly invoking photodissociation of SrS(X$^1\Sigma^+$) or other time dependent methods to examine these possible differing dissociation channels that may imply different bond breaking energies. Consequently, depending on the method and the facilitating mechanism, values smaller or larger than that recommended still are conceivable. An early matrix isolation study characterized the SrS and

SrS_2 structures (Martin and Schaber, 1982). Even so, with respect to combustion it is unlikely that these sulfides can play any role except under pyrolysis or reducing gasification conditions. They are vulnerable to reactions with water and oxygen. Before the ICP, many metals in a sample were analyzed by flame atomic absorption or emission. A major concern was to resolve any interferences introduced by other trace metals that may exist in the sample. The most common was the presence of readily ionizable elements such as the alkali metals that would modify the ionic balances. Another interesting observation was the interference by Al salts on the analysis of Sr (Luecke, 1994). It clearly illustrated that the flame distribution of Sr was being perturbed by the presence of Al possibly by forming some gaseous strontium aluminate. Flame temperatures were high with C_2H_2/air flames ($\approx$2200°C) but a strong interaction was noted. Changing the flame to N_2O instead of air and raising its temperature by about 600°C did reduce the problem. Nevertheless, it remains unexplained as to whether the aluminate is being formed in the gas phase or being removed by chemideposition of known $SrAl_2O_4$ (Table 7.36) on any particle surfaces in the flame gases.

Laboratory combustion studies

Strontium, as with many trace metals can be easily examined in laboratory combustion flames through the ease of direct injection. Salt aqueous solutions such as with the soluble nitrate can be nebulized quantitatively. Mass spectrometric sampling can establish possible flame speciation of molecules. For strontium, such studies are aimed more toward examining applications of combustion other than on any environmental aspects based on energy generation. With the three alkaline earth elements, such interests generally can center on electron generation and the ease of chemiionization, and aspects of soot suppression or flame inhibition.

One early mass spectrometric sampling of the species in atmospheric pressure flames at 2000K–2100K, injected traces of a strontium salt, and monitored the gaseous Sr atoms and SrO concentrations (Farber and Srivastava, 1976a). Assuming an equilibrium balance with O_2 they derived a bond strength, now updated, to $D_0(\mathrm{SrO}) = 412 \pm 7\ \mathrm{kJ\ mol^{-1}}$, and from the vapor pressure data for SrO, $422 \pm 8\ \mathrm{kJ\ mol^{-1}}$. Although their flame-probe sampling into a low-temperature mass spectrometer has always been questioned concerning possible perturbations, their values remain quite accurate (Table 7.10).

During a period of high interest in electron generation and ionization it was realized that ionization of trace metal elements was occurring in flame gases at greater rates than possible thermally. Branching off from the field of chemiluminescence excitation, it was realized that some flame reactions had sufficient energy not only to electronically excite species but could actually ionize them, so called chemiionization (Schofield and Sugden, 1965). This developed into a large field as it had implications for producing electrons at lower temperatures, and also appeared to be involved in soot formation and control. A useful review on metal ions in flames includes the alkaline earth elements and illustrates how interest in the ion chemistry developed (Goodings, 1992). Subsequent work explained how Sr^+ and $SrOH^+$ were chemically formed, efficiencies increasing in the order Ca, Sr, Ba (Jensen, 1968; Hayhurst and Kittelson, 1974). In the area of soot suppression it was noticed that Sr lay between Ca and Ba for reducing the total amount of soot and reducing its coagulation. Unlike the other metals they appear to function at all equivalence ratios. Effectiveness appears to correlate with ion formation and when tested the three were on a scale of 2.0:26:100 for Ca:Sr:Ba (Cotton et al., 1971; Bulewicz et al., 1975). When reviewed, it was felt that the ionic explanation was not adequate for the lower levels of ions in flames for Ca and Sr (Howard and Kausch, 1980). Little has been done further to better explain this behavior. The two ions, Sr^+ and $SrOH^+$, appear to be in balance with flame H_2O and H, leading to an alternate method of obtaining absolute measures of H-atoms in flames from measuring this ion ratio (Jensen and Jones, 1975). The flame inhibition effects of trace metals is well known, but like soot suppression, has never been satisfactorily explained. It was shown that they were not on the same scale of effectiveness as Mn, Cr, Pb, and Fe, but the three alkaline earths could show quite strong effects (Bulewicz and Padley, 1971). Again Ba >Sr >Ca was their order and although catalytic kinetic cycles could invoke hydroxide participation in recombining flame radicals, it had limitations and the possibility of some contribution from heterogeneous surface reactions on particles appeared possible. As a result, a review left many loose ends but indicated that strontium was not a major player in any of these effects, showing no commanding presence for development (Linteris et al., 2008).

Combustion power plant data

As might be expected from its concentrations in fuels (Table 1.1), unlike calcium, an examination of strontium concentrations in ash is on a high parts per million scale even with waste and sludge samples but does remain a nonnegligible trace element. One benefit of now having the ICP analyzer is that strontium is readily monitored and included in some fuel analyses (Wang et al., 2008c), although chlorine content, which can be very important, is still often neglected. Strontium's levels in pet-coke are very low but it can have more of a presence in limestone, of the order of 200 ppmw in some cases (Cordoba et al., 2012).

As a result, with strontium levels already relatively low in fuels, coupled to its low volatility and nontoxicity, there have been few studies that include an examination of its behavior in full-scale commercial combustors. Nevertheless, these provide sufficient information to confirm various aspects. As with calcium, most of the strontium remains in ash form with approximately similar or slightly higher concentrations in fly ash over the slag or bottom ash whether for coal, waste or gasification systems (Klein et al., 1975; Meij, 1994; Querol et al., 1995; Martinez-Tarrazona and Spears, 1996; Spears and Martinez-Tarrazona, 2004; Zhao et al., 2008; Font et al., 2010). In FBC or scrubber systems using lime, it should be noted that this may probably contain more strontium in the lime than the fuel. In earlier studies with coal, no or very slight enrichment on smaller particles was noted (Coles et al., 1979; Querol et al., 1995; Martinez-Tarrazona and Spears, 1996). A later examination of coal ash concluded that only about 10% of any Sr was found on the surface, minimizing the role of vaporization (Spears and Martinez-Tarrazona, 2004). The lack of enrichment in the coal fly ash for any of the alkaline earth elements as well as Mg was reported again recently (Lanzerstorfer, 2018). They collected ash on the outlet of the ESP from a 420 MW coal-fired plant at 1480°C, size segregating it into five categories, the smallest being 2.2 μm. Mass balances in coal plants confirmed that the high efficiencies of ESPs removed most of the ash-transported strontium (Heinrichs, 1977; Ondov et al., 1979; Nielsen et al., 2002) and if coupled to scrubbers the stack leakage could be reduced to less than 0.1% (Nielsen et al., 2002).

Although the ultimate particle leakage to the stack is of a low level, their particle sizes are on the micron and submicron ranges. In a Danish study of two coal-fired power plants the bimodal distributions of diameters is clearly indicated with the larger peak at about 2 μm and the smaller submicron at 0.2 μm (Nielsen et al., 2002). This had been noticed much earlier for numerous elements (Markowski and Filby, 1985) with the most detailed study to date following soon thereafter (Kauppinen and Pakkanen, 1990). This examined particles collected downstream of the ESP, segregating the 0.01–11 μm sizes, and analyzing their element content. Of the two particle peaks at 0.5 and 2.0 μm in their study, strontium clearly reflected some vaporization and deposition enrichment on particles below 1 μm with concentrations linearly increasing to smaller diameter size. The quantities of these particles is of course very small but they noted that 11% of the strontium was in the submicron size and on average about twofold enriched over the micron size distribution. Although FBC methods do have lime additionally present, one such examination of fine particles upstream of the ESP (>0.01 μm) found no strontium in particles ($<0.1\%$) undoubtedly due to the lower combustion temperatures (Lind et al., 1994). The behavior of strontium seems to mirror that of calcium, both showing the evidence of some evaporative effects (Kauppinen and Pakkanen, 1990). No definitive studies have specifically considered chlorine effects. However, studies with sewage sludge blends rich in chlorine (1.2 wt.%) (Folgueras et al., 2003), incineration of hospital wastes (Zhao et al., 2008) and rotary kiln incineration (Waterland et al., 1994) show no indication of differences when significant levels of chlorine may be present. Roasting municipal waste ash at 1000°C with Cl_2 for 3 hours induced negligible volatilization of Sr (Chan and Kirk, 1999).

In summary, as with calcium there are still uncertainties as to how traces of strontium vaporize and enter the gas phase. The importance of any potential reducing mechanism in fuel particle surface regions starved of oxygen, or with unburned carbon present, remains unclear. Nevertheless, it is well established that any release to the environment is small and requires high temperatures to gain any noticeable level. Also any traces that are released will be deposited on the particles. Release from fuels will possibly be via $Sr(OH)_2(g)$ or $SrCl_2(g)$. Deposition of these could produce SrO(s) at high temperatures and $SrSO_4(s)$ as temperatures decrease. Below 1000°C either of these could then be converted to $SrCl_2(s)$, the final products depending on the thermodynamic rankings of these three for the conditions of sulfur and chlorine levels. Because of this, and the low concentrations there is little environmental concern with strontium. Particle control invoked for the more sensitive elements will also control particulate strontium. Whether SrO is sufficiently stable in the fuel particle, preventing release remains uncertain. It may form ceramic type mixed metal oxide structures that are stable. Its aluminate, phosphate, silicate, and titanate are possibilities. This will be relevant to ash leaching characteristics.

7.26 Thallium

As noted in Table 1.1, although thallium concentrations are small in coal, less in biofuels and wastes, it can be slightly richer in sludge waste but is very low in pet-coke and limestone (Cordoba et al., 2012). Nevertheless, its chemical nature is such that it should no longer be neglected. This stems from the fact that it is a heavy metal lying between Hg and Pb in the periodic table, the heaviest member of the B, Al, Ga, In, Tl column. With atomic number 81 it is rich in electrons but has a normal shell structure of $5d^{10}6s^{2}6p$ ($^{2}P_{1/2}$). This explains its dominant mono- and trivalencies. Most of its compounds are water soluble and more toxic than either of those of its neighbors. Moreover, as will be noted, its compounds are very volatile and even its carbonate is water soluble, the only such known heavy metal carbonate. There is a relatively small commercial demand for thallium of only about 15 tons $year^{-1}$. It appears not to be mined or refined in the United States. Its main uses are in the electronics industries (60%–70%) and for alloys and special glasses. With no particular viable ore it is extracted as needed as a by-product from metal smelters. As a result, this is one unusual case where there is no connection between its being manufactured, the estimated 3000–7000 tons $year^{-1}$ that are mobilized in the environment primarily by natural volcanic eruptions and forest fires, and also the anthropogenic emissions from combustion, smelting and particularly cement production (Nriagu and Pacyna, 1988). Combustion possibly contributes about 30%. Although largely ignored, recent outbreaks of serious thallium pollution have become apparent in China (Liu et al., 2018). The source has been ascribed to mining and smelting. It appears that enriched levels of thallium can result in the wastewaters from the steel industries. However, recent analysis of isotopic 205/203 ratios, from its two natural isotopes, indicate this can now help distinguish between natural and industrial sources. Although the emissions may be in decline due to suggested regulations, ice core data examining back many centuries estimate that general anthropogenic contributions to the environment are significant at about 90% (Krachler et al., 2008; McConnell and Edwards, 2008). Measurements of the Tl/S ratio of emissions from volcanoes suggests a global rate of about 600 tons $year^{-1}$ (Patterson and Settle, 1987). Ice core data clearly indicate this role of volcanic emission to the natural background levels (Hong et al., 2012) as also do forest fires. There has been little concern with thallium partly due to its low level emissions and the difficulty of monitoring such concentrations. Consequently these releases still remain largely approximate estimates (Pacyna and Pacyna, 2001). As a result, thallium emissions have remained largely uncontrolled except for possibly being included with other heavy metals and restricting particle emissions larger than $PM_{2.5}$. Now with new analytical instrumentation this monitoring difficulty has been removed. Nevertheless, one early study of the population living near a cement plant clearly indicated the need for environmental concerns (Brockhaus et al., 1981). A more recent study has reiterated this concern for areas close to smelters or coal burning facilities (Karbowska, 2016). Through the consumption of home grown vegetables and fruit their thallium intake can become significantly elevated. Also there are known regions of the world where thallium-rich soils are sufficient to affect drinking water and local vegetable growth that raise concerns (Zhang et al., 1998; Li et al., 2012a). Nevertheless, except for strict regulations on drinking water quality, the unregulated industrial releases are ignored. Like many developments, this new interest in thallium has arisen from new monitoring ability of isotopic tracing and fractionation, but is apparently needed in certain geographic and some industrial regions. Baseline measurements have recently been made comparing more virgin type soil areas with those affected in densely populated coal-fired plant regions (Vanek et al., 2016).

Health concerns with thallium

Several very detailed reports and papers have appeared that summarize in depth the environmental and health aspects of thallium some of which are now published on the web (ToxProfile Tl, 1992; Schaub, 1996; Kazantzis, 2000; Peter and Viraraghavan, 2005; Conerly and Rieth, 2009). In summary, thallium is a nonessential element for humans, but dietary intake from mainly vegetables and fruit can be about 5 $\mu g\ day^{-1}$. Its compounds are odorless and tasteless and have a half-life in humans of about 10 days which is also the length of time for a lethal dose to be effective. It has a narrow therapeutic dose range and is no longer in medical use. In earlier days it was used to treat syphilis and tuberculosis patients and kill rodents. Its salts are tasteless making it a choice as a poison until greater restrictions removed its availability. For a long time its radioactive tracer ($^{201}TlCl$) was valued for cardiac testing. As with most metals, any long time effects of low levels remains unknown. Concerns relate to water pollution and to major industries that may emit particles or vapors. Few medical studies exist but one speculated that prenatal thallium might affect female child growth in the first 2 years (Qi et al., 2019). It does pass through the placenta and is present in breast milk. Although not as widely publicized, some fish such as trout in the Great Lakes are now of concern regarding possible bioaccumulation (Nriagu, 2003). Little is mentioned of organothallium. However, its ions Tl^+ and $Tl(CH_3)_2{}^+$ can bioaccumulate in algae and plankton and could become of concern (Thayer, 2010). Blood, urine and hair samples can be used as biomarkers for Tl toxicity. The Mayo Clinic in the United States suggests blood levels of $<1\ \mu g\ L^{-1}$ with >10 implying a significant exposure. Urine should be $<1\ \mu g\ day^{-1}$ (Mayo Clinic, 2018). Few large surveys include thallium but those published all reflect low levels. Blood $<0.1\ \mu g\ L^{-1}$ (Goulle et al., 2005; Cesbron et al., 2013; Nisse et al., 2017; Zeng et al., 2018). Urine $<0.8\ \mu g\ L^{-1}$ (Goulle et al., 2005; Nisse et al., 2017), and hair very slight at <0.4 ppbw (Goulle et al., 2005).

Thallium's basic chemistry

Unlike the other members of its periodic table column (B, Al, Ga, In), thallium largely favors a univalent cationic chemistry (thallous) (m.pt.303, b.pt.1470°C). For a metal its vapor pressure is quite significant (10 ppmv 600°C, rising to about 250 ppmv at 750°C). It has trivalent compounds (thallic) but they are less stable and easily convert as temperatures are raised to monovalency. Its oxides are important but the stable yet volatile room temperature $Tl_2O_3(s)$ (m.pt.834°C) loses oxygen on heating, converting to $Tl_4O_3(s)$ (m.pt.717°C) and then becomes the water soluble $Tl_2O(s)$ (m.pt.580, b.pt. $\approx$1080°C). As a result, depending on the level of oxygen, a mixture of these three oxides can exist, as indicated by its phase diagram (Holstein, 1993a). These can interconvert and are quite volatile, all evaporating to the predominant $Tl_2O(g)$ (vapor pressure 1 ppmv at 400°C, 20 ppmv at 480°C) (Cubicciotti, 1970; Srivastava and Farber, 1978; Gurvich et al., 1994). This is stable to temperatures of at least 1300°C and is thought to fractionally fragment to Tl(g) and TlO(g). The gaseous diatomic oxide has only been studied theoretically and remains undetected. Vaporization distributions under different conditions assuming its presence have been published but show its minor role below 1330°C (Lamoreaux et al., 1987). Thallium has a very soluble hydroxide TlOH(s) that above 140°C dissociates to $Tl_2O(s)$ and also can absorb CO_2 to form the carbonate. The water soluble carbonate, Tl_2CO_3, can reversibly dissociate also to this oxide starting at about 270°C and becoming significant at >300°C (Stern and Weise, 1969). Vaporization of water soluble $TlNO_3$ was studied by the Knudsen cell-mass analysis method, and shown to lose NO above 300°C becoming $Tl_2O(s)$ with an increasing dissociative process (Cubicciotti, 1970a). The same researcher examined the water soluble sulfate Tl_2SO_4 noting this was more stable and vaporized congruently as $Tl_2SO_4(g)$ up to its melting point of about 632°C with a small fraction dissociating to the oxide (vapor pressure 10 ppmv at 620°C) (Cubicciotti, 1970b). At slightly higher temperatures dissociative vaporization becomes dominant. The water soluble trichloride $TlCl_3(s)$ (m.pt.155°C) is not stable and loses chlorine above 40°C forming the more stable TlCl(s) (m.pt.430, b.pt.720°C). This is also volatile subliming to thermally stable TlCl(g) (vapor pressure 1.0 ppmv at 260°C, 30 ppmv at 330°C) (Gurvich et al., 1994). Whether this or the sulfide is the form of thallium released from volcanoes appears uncertain. The condensed phase trivalent hydroxide, carbonate, nitrate and sulfate all dissociate readily at low temperatures to their monovalent molecule or the oxide. For example, the trivalent sulfate $Tl_2(SO_4)_3$ dissociates to Tl_2SO_4. Under reducing conditions the trivalent Tl_2S_3 similarly on heating readily loses S_2 producing thermally stable but water insoluble $Tl_2S(s)$ (m.pt.457, b.pt.1367°C). Although still somewhat uncertain, Knudsen cell-mass analysis studies show this to vaporize to $Tl_2S(g)$ to about 730°C then with dissociation to higher temperatures: a vapor pressure of about 20 ppmv at 575°C was indicated (Klanberg and Spandau, 1958). Few mixed metal thallates are known. $NaTl^{(III)}O_2$ and $Na_3Tl^{(I)}O_2$ are hygroscopic, soluble in water but little studied. Similarly, although phosphates such as $Tl^{(III)}PO_4$ and $Tl^{(I)}PO_3$(m.pt.438°C) are known they appear to have generated little interest. The monovalent form is more stable and congruently vaporizes without dissociation to higher temperatures (Lopatin et al., 1998). A Knudsen cell-mass spectrometric study of the vaporization of thallium metaborate is published indicating stable $TlBO_2(g)$ at 730°C with a possible strong bond strength $D(Tl\text{-}BO_2 = 417\ kJ\ mol^{-1})$ (Bagarat'yan et al., 1974). The physical properties of these condensed phase molecules are summarized in Table 7.37.

Gas phase stabilities and transport

On examining vapor phase species, some can be created by vaporization and some only by gas phase chemistry. Their stabilities and trends in comparison to B/Al/Ga and In are listed in Table 7.5 and further discussed here. Although values for D_0(Tl-O(g)) have been estimated from comparative trends as about 260 kJ mol^{-1} there is no evidence for its existence even spectroscopically and it is regarded as unstable (Gurvich et al., 1994).

TABLE 7.37 Properties of thallium's mono- and trivalent condensed phase molecules relevant to combustion, °C

m.pt. b.pt. vap. press. Monovalent	m.pt. b.pt. vap.press. Trivalent
Tl 303 1470, 18 ppmv/630°C, ins	–
Tl_2O 579 984, 20 ppmv/480°C, s	Tl_2O_3 834, dissoc 690 to Tl_4O_3/Tl_2O, ins
TlOH >140 dissoc to Tl_2O, s	$Tl(OH)_3$ unstable, dissoc to TlOH, Tl_2O, ins
Tl_2CO_3 522, >270 dissoc to Tl_2O, s	$Tl_2(CO_3)_3$ unstable, dissoc to Tl_2O
$TlNO_3$ 206 450, >300 dissoc, 63 ppmv/330°C, s	$Tl(NO_3)_3$ dissoc to Tl_2O, reacts w water
Tl_2SO_4 >632 dissoc, 10 ppmv/620°C, s	$Tl_2(SO_4)_3$ dissoc to Tl_2SO_4, reacts w water
TlCl 430 720, subl >260, 30 ppmv/330°C, ss	$TlCl_3$ 155 dissoc to TlCl >40°C, s
Tl_2S 457 1367, dissoc >730, 20 ppmv/575°C, ins	Tl_2S_3 losses S_2 on heating
$TlBO_2$ stable <730°C, ins	–
$TlPO_3$ 438, stable, vaporizes unchanged, ss	$TlPO_4$ hydrolyzed by water to $TlPO_3$, ss
Na_3TlO_2 hygroscopic	$NaTlO_2$ hygroscopic

s, water soluble; *ins*, insoluble; *ss*, slightly soluble; *decomp*, decomposes; *reacts w water*, reacts with water; *subl*, sublimes.

$Tl_2O(g)$ is the only gaseous oxide monitored and its structure was characterized in a matrix isolation study (Brom et al., 1971).

From the heats of formation and sublimation of $Tl_2O(s)$ a value for the atomization energy $D_{0Atomiz}(Tl_2O(g)) = 584 \pm 6$ kJ mol^{-1} has been determined (Gurvich et al., 1994). As a result, depending on the estimated value above for D(TlO), this would imply D_0(TlO-Tl) ≈ 324 kJ mol^{-1}, stronger than TlO but similar to that given below for TlOH. As a result, in combustion in the presence of water $Tl_2O(g)$ is hydrolyzed and forms a distribution with TlOH(g) by this thermoneutral reaction (Holstein, 1993).

$$Tl_2O(g) + H_2O(g) = 2\ TlOH(g) \quad \Delta H_{298K} = +0.1 \text{ kJ mol}^{-1},\ K_{p,1000K} = 1.6,\ K_{p2000K} = 3.2$$

These thermochemical values although still somewhat uncertain suggest in any combustion situation at 1000K, where water is present at about 10%, that the ratio of the hydroxide to the oxide will be about one-half and increasing to higher temperatures (Gurvich et al., 1994). However, in the development of flame atomic analysis of thallium it was noticed that sensitivities of detection were very good (50%–60% free atomic Tl) in a wide range of higher temperature flames, even H_2/air if temperatures were around 2000°C (De Galan and Samaey, 1970). This confirms the weaker bond strengths of the oxide and hydroxide. Early studies in fuel-rich H_2 flames at 1500°C–2200°C including the family members Ga and In had shown that these two indicated a balance in a distribution between the hydroxide and the atom. Namely by the reaction

$$M + H_2O = MOH + H$$

However, thallium was behaving as if solely free atoms and its thermal emission intensity (535 nm) varied correctly proportional to temperature indicating minimal molecular formation in the flame gases (Bulewicz and Sugden, 1958). Trends suggested a value of D_0(Tl-OH) ≈ 293 kJ mol^{-1} This was confirmed by subsequent similar later work that suggested a value of D_0(Tl-OH) <300 kJ mol^{-1} (Kelly and Padley, 1971). Now, from trends in other values for the fluorides, chlorides and hydroxides for Ga, In and Tl, Gurvich et al. (1994) indicate a recommendation of D_0(Tl-OH) ≤ 325 kJ mol^{-1}, which obviously has to be an upper limit or it would have been noticed in the simple early flame work. A recent theoretical paper concerning the electronic nature of TlOH suggests values close to 300 kJ mol^{-1} (Pershina and Ilias, 2018). Although the species $Tl(OH)_3$, $Tl(OH)_2$ and even TlO_2 have been created in matrix isolation experiments utilizing laser-ablated Tl with H_2O_2, H_2/O_2 or oxygen, there is no evidence of their existence as stable gas phase components (Kelsall and Carlson, 1980; Andrews et al., 1997; Wang and Andrews, 2007).

By utilizing a series of fuel-rich hydrogen flames (1200°C–2300°C) containing sub-ppm levels of metal additives and <0.5% HCl, a value of D_0(Tl-Cl = 364 ± 8) was obtained from the balance between Tl/HCl and TiCl/H (Bulewicz et al., 1961). This compares very well with the currently recommended accurate value of 367 ± 2 kJ mol^{-1} obtained from the heats of formation and sublimation of TlCl(s) (Gurvich et al., 1994). This fact confirmed the dominance of TlCl(g) over TlOH(g) in their flames and its obvious greater bond strength.

$$\mathrm{TlOH(g) + HCl(g) = TlCl(g) + H_2O(g)} \quad \Delta H_{298K} = -106\ \mathrm{kJ\ mol^{-1}},\ K_{p,1500K} = 5 \times 10^3$$

With this thermochemical value, TlCl(g) will become dominant for conditions of 10% H_2O if the HCl concentration is greater than 20 ppm and even lower if the bond strength of TlOH is weaker than currently considered. Although little is known of $TlCl_2$, on examining the bond strength values of this B to Tl group in the periodic table (Table 7.5), the consistent weakening of the first and second bond in the MCl_2 molecules across the first four elements of the Group strongly suggests a value of D_0(ClTl-Cl) ≈ 213 kJ mol^{-1} for this as yet unreported value.

The only remaining viable gaseous species are TlH and Tl_2S that both will relate to reducing conditions. Again, with fuel-rich hydrogen flames spanning a 1500°C–2100°C range, as noted Tl(g) was dominant in the flame gases with negligible ionization (ionization potential 589 kJ mol^{-1}) and displayed an equilibrated spectral thermal emission. However, weak spectral bands of TlH were evident indicating a slight presence. Assuming this was also in a thermal formation/dissociation balance with Tl(g) and H(g), a Second Law data reduction of the variation of the TlH emission intensities with conditions produced a reasonable value for its weak bond strength of 184 ± 28 kJ mol^{-1} (Bulewicz and Sugden, 1958). The accepted value from its spectral predissociation limit is now D_0(Tl-H) = 190.5 ± 0.2 kJ mol^{-1} confirming the reliability of the early flame analysis techniques (Huber and Herzberg, 1979; Gurvich et al., 1994). As indicated above, Tl_2S(s) does vaporize congruently and unchanged to its gas phase molecule but above about 730°C it dissociates to its elements. Invariably sulfides are weaker than their corresponding oxides by about 25% in this section of the periodic table so that the potential gaseous TlS and Tl_2S molecules will be of little importance. As seen from Table 7.5, these bond strengths for gaseous thallium compounds are consistent with the weakening trend seen when moving from Ga to In and to Tl, the three displaying a consistent pattern with their chemistries. Thallium displays no strongly bound gaseous forms other than the chloride. As a result, if any Tl is released it will quickly become the gaseous atom or chloride in hot gases. These will be the major transporters and will redeposit on cooling fly ash or surfaces as their most stable thermodynamic preferences depending on the conditions.

Combustion power plant data

As a result, thallium's combustion chemistry has the potential to be intriguing. Thallium has few thermally stable condensed phase molecules. Rather like mercury, its condensation chemistry cannot begin until temperatures fall, introducing possible time constraints. Consequently, at first sight it is surprising that it is not of some concern. Very few full-scale combustion studies have included thallium in their analysis. Two efforts to display expectations through thermodynamic equilibrium calculations are useful in reflecting predictions. One by Yan et al. (2001) taking a whole coal database indicates condensed phase thallium compounds only below temperatures of about 280°C for oxidizing conditions, and only Tl_2S below 80°C if reducing, and confirming the general dominance of TlCl (g) to about 1500°C. A second study raises the upper temperature limit for Tl_2SO_4(s) to exist to possibly 500°C (Danihelka1 et al., 2003). It indicated a dominance of TlCl(g) from about 430°C–1230°C above which atomic Tl exists.

As noted in Table 1.1, thallium's levels in coal are small and almost negligible in biomass (Chandrasekaran et al., 2012). Similarly, its levels normally also can be expected to be small in wastes but can be present in sewage sludge. However, occasionally, depending on the other materials used in the processing of sludge this can become even 10-fold more concentrated than mercury (Elled et al., 2007). Fortunately, although still difficult to accurately monitor due to its low levels, several studies have examined the chemical balance of thallium in combustors. These tend to agree that there is little thallium in the bottom ash but rather surprisingly indicate that most is removed on particles by the ESP. For coal, the fly ash to slag concentration enrichment can be about threefold (Heinrichs, 1977; Meij, 1994; Querol et al., 1995). There is the clear observation that concentrations increase as particle sizes decrease, indicative of vapor deposition (Davison et al., 1974; Querol et al., 1995). Ash from the second stage (smaller sizes) of an ESP is seen to be at least twice richer in thallium (Lind et al., 2003). On heating coal and coal/sludge blends in a laboratory furnace, all the samples lost their Tl content by about 830°C whether coal or sludge (Folgueras et al., 2007). A similar finding was reported for the pyrolysis of coal/biomass at 900°C (Miller et al., 2002). One mass balance study on two coal-fired power plants indicated average approximate capture rates for thallium of about 1.1% bottom ash, 94.2% fly ash, 4.6% FGD, with 0.1% release (Aunela-Tapola et al., 1998). Meij (1994) also confirmed that most thallium was captured on the fly ash with the remainder being removed by the FGD scrubber. Additional analyses of FGD system's retention ability indicate that any surviving thallium will be retained quite well (Cordoba et al., 2012). An older study with only an ESP control had indicated possibly a 7% stack loss of thallium from coal (Heinrichs, 1977). However, with the improvements in ESP collection and solely this control it now appears adequate for FBC of biomass and waste combustion (Lind et al., 2003).

A recent study in a coal-limestone fluidized bed plant tended to show a more uniform distribution of the thallium over the various ash forms. Enrichment on the smaller particles was noted with little dependence on the ash compositions and was related to surface area (Lopez-Anton et al., 2015).

As a result of this field data, there appears to be little concern at present with thallium emission control in spite of its potential volatility. However, it remains strange that something so similar to mercury is being successfully removed by the fly ash. Mercury does need temperatures <400°C and encounters its severe emission control problems as a result of this time/temperature constraint that leaves it insufficient time to interact with the fly ash (Schofield, 2008). This has to imply that thallium is depositing at such temperatures much more efficiently on the ash in spite of the limited time. With thallium in the gas phase it might deposit as the oxide Tl_2O or even Tl_2O_3(s) that will have a sufficiently low vapor pressure below 600°C (m.pt.834°C) to be retained (Holstein, 1993a). Also in that same work it was noted that Tl_2O reacted with an alumina boat above 700°C to form the aluminate, $AlTlO_2$. Another option might be to form the sulfate at temperatures below 600°C. With chlorine present, TlCl(s) could equally become the deposit. Even so, one of these or a similar interaction is operating to permit the thallium sufficient time to deposit on the ash and be controlled. Obviously, the nature of the thallium in the ash needs to be determined.

In summary, although thallium displays low thermal stability and high volatility in its condensed phase molecules, the limited quantity of full-scale combustion data confirms that most is retained by the ESP on the fly ash with evidence of enrichment to smaller diameter size. Because this can only happen in the cooling flue gases region it implies an efficient deposition interaction with the ash particles. Only those molecules stable at low temperatures will be chemideposited with the possibility of mixed metal molecules such as the aluminum thallates. Coupled to the FGD scrubber total capture is possible with low levels in a product bauxite. Although in a time constraining situation like mercury, it appears that the mercury/ash interaction is much less efficient.

7.27 Tin (Germanium)

Tin is atomic number 50 in the periodic table and is regarded as a heavy metal. It is also about 51st in abundance in the Earth's crust and on par with Be, Br, or U. It is also rare in the oceans, about 65th in order and somewhat similar to the presence of Au or the rare earths. Although neither toxic when inorganic nor of significant importance in combustion it can be present in nonnegligible concentrations in some coals and waste materials. Being chemically related to lead it is interesting to examine its combustion behavior to note differences with other metals and also whether it may interfere with other potential chemistries. Unlike lead, its organometallic chemistry has been much more extensively developed, its compounds are far more toxic and biomethylation and bioaccumulation in nature is more apparent. As indicated, although relatively rare in the Earth's crust, tin has been known from Bronze Age times. Its main ore is cassiterite (SnO_2) and current global production of tin is about 250,000 tons year^{-1} with about 10%–15% recycling. Adequate reserves of at least 4.7 million tons are known but sources are no longer regarded as unlimited and there now is more emphasis on recycling. It remains more expensive than the other major metals at \$6.60 lb^{-1}, 10 times that of Al. For a metal it does have a low melting point of 232°C but a high boiling point of 2600°C. Its equilibrium vapor pressure is about 10 ppmv at 1224°C. Uses are decreasing but it still is a component of solder, used in metal plating, in alloys such as pewter and bronze, in superconductivity technology, glass, ceramics, as a food preserver and in dentistry as SnF_2. Interestingly, the organo-forms favor tetravalency like carbon and the inorganic molecules divalency like lead. The organometallic uses of tin now tend to be greater than its inorganic role. These include uses as biocides, antifouling agents, sterilizers, catalysts and as a stabilizer in polymerized products such as PVC. There are now more than 800 organotin molecules that have been synthesized and characterized. Some have already caused problems in the environment and have been replaced (Gajda and Jancso, 2010).

Anthropogenic atmospheric levels are not well ascertained but estimates of about 6000 tons year^{-1} are quoted with about 5000 tons year^{-1} of this resulting from combustion (1000 coal, 3300 oil and 800 incineration tons year^{-1}) (Nriagu and Pacyna, 1988; Nriagu, 1989a; Tillman, 1994). The natural background level from dust, forest fires and volcanoes (emitting SnS) remains uncertain but is thought to be no more than 10%–20% of this anthropogenic level. Analysis of Himalayan ice-core samples dating the past 800 years indicates that since 1970, tin concentrations and crustal enrichment have significantly increased (Hong et al., 2009).

Health concerns with tin

Tin appears to have no known essential function in humans but is ingested in low levels from diet. In testing rats it did seem that low levels might be needed (World Health Organization Trace Elements, 1996). It is not easily absorbed in the body and easily excreted. Absorption appears to be dose related with larger ingestion values being less absorbed. Inorganic tin is nontoxic or carcinogenic and in the past resulted in diet from its use for the tin plated canning of foods. This had its problems of lid soldering and the use of lead containing solder. However, before this became widely realized, Al canning became dominant due to its fractional cost. In reality tin containers were very good due to tin's resistance to acids or alkalis. As a result the levels of tin resulting in diet now have become quite low. However, organic forms of tin are very toxic and are finding many industrial uses. The toxicity varies with structure but the triorganoquadravalent tin molecules are the most toxic and banned for agriculture. Due to this, numerous toxicological reports are available in the open literature of the Web covering animal and human findings concerning the risks and metabolism in the body. Safe dose levels of about 0.3 mg kg^{-1} body wt. day^{-1} have been suggested. Medical effects do become noticeable with intakes of Sn greater than about 15 mg day^{-1}. Inorganic Sn does not appear to cross the blood–brain barrier but industrial accidents with organotin compounds can have very significant medical effects that can be long lasting and affect the brain and nervous system (ToxProfile Sn, 2005; World Health Organization Sn, 2005). As a result, although inorganic tin is relatively harmless, the many organometallic forms are now permeating the environment, infiltrating food and could become a concern for humans. No long-term effects are known for such molecules. The actual toxicity range has not been assessed. Few surveys have been taken but do indicate blood levels of $<1.75\,\mu g\,L^{-1}$ (Goulle et al., 2005), and $<0.65\,\mu g\,L^{-1}$ (Cesbron et al., 2013). Some test centers that are infrequently used indicate that $<2.0\,\mu g\,L^{-1}$ is acceptable. One study, comparing blood samples for populations in Morocco and the Canary Islands reported rather high blood levels of about $16\,\mu g\,L^{-1}$, which appears unusually high (Henriquez-Hernandez et al., 2018). In the US National Survey, urine testing is favored and in 2014 indicated values of $<4.0\,\mu g\,L^{-1}$. Closer examination displayed wider ranges with age or race ($2-8\,\mu g\,L^{-1}$). Some testing has been reported next to waste incinerators in Spain indicating low hair levels, 0.37 ppmw in 1998 and 0.16 in 2007 (Ferre-Huguet et al., 2009). Russian workers in an aluminum plant showed slightly elevated but still totally safe hair levels (Skalny et al., 2018). Autopsies of people living near a hazardous waste incinerator indicated low levels and a reduction from 1998 to 2003, and were interesting to show the distribution of the tin throughout the body (Bocio et al., 2005). Its presence now is at low levels in the environment and only displayed more in heavily industrialized places (Li et al., 2014). The medical industry shows little concern regarding tin.

Tin's basic chemistry

Tin is another element with a very limited chemistry displaying a mixture of divalent and tetravalent tendencies. Many of its potential condensed phase molecules are thermally unstable and so will not be encountered in combustion. A critical analysis of the thermochemical values for the condensed phases of Sn, SnO, SnO_2, $SnCl_2$, $SnCl_4$, SnS, and SnS_2 is extremely useful (Gurvich et al., 1991). The physical properties of these relevant condensed phases are summarized in Table 7.38.

Tin has no known carbonate, the tetravalent nitrate, $Sn(NO_3)_4$, decomposes slightly above its 91°C melting point and the divalent nitrate is not stable (Stern, 1972). Its divalent sulfate, $SnSO_4$, can react with water, and above 380°C self-oxidizes to the tetravalent oxide, SnO_2 (Stern and Weise, 1966). The tetravalent sulfate, $Sn(SO_4)_2$, is hygroscopic and unstable above 100°C, similarly producing the dioxide. The hydroxide $Sn(OH)_2$ exists but is unstable and there is little evidence for $Sn(OH)_4$, both being regarded as acid anhydrides but readily dehydrated to their oxides. They have been formed and their infrared spectra characterized in matrix isolation studies (Wang and Andrews, 2005c). SnH_4 decomposes to Sn at 0°C. It has two chlorides, both easily hydrolyzed by water. $SnCl_4$ is liquid at room temperature (m.pt. − 33°C), is very volatile with a boiling point of 109°C (vapor pressure 0.04 atm at 25°C) (Pankratz, 1984). Its vapor is stable but dissociates to the oxide at higher temperatures and has found use in CVD forming SnO_2 films at 370°C–700°C when mixed with O_2 or H_2O vapor (Kleps et al., 1997; Koutsogianni and Tsamakis, 2001; Takeuchi et al., 2004). The dichloride, $SnCl_2$, also is volatile with a boiling point of 612°C, at which temperature it decomposes. Its stable vapor composition was confirmed by a Knudsen cell-mass spectrometric study (O'Brien and Mucklejohn, 1987). It has a vapor pressure of about 10 ppmv at 220°C (Pankratz, 1984), and has also been used in CVD as the precursor with O_2 or H_2O in

TABLE 7.38 Properties of tin's condensed phase molecules relevant to combustion, °C.

	m.pt. b.pt. vapor pressure
Sn	232 2600, 10 ppmv/1224°C, ins
SnO	977, decomp 1080, converts to SnO_2 >530, ins
SnO_2	1630, dissoc >1150 to SnO(g), 10 ppmv/1160°C, ins
$Sn(OH)_2$	unstable
$Sn(OH)_4$	dehydrates to SnO_2
$SnCO_3$	not known
$Sn(NO_3)_2$	>200 unstable, s
$Sn(NO_3)_4$	91, decomp 98°C to SnO_2, s
$SnSO_4$	378, decomp to SnO_2, s
$Sn(SO_4)_2$	>100 decomp to SnO_2, hygroscopic, s
$SnCl_2$	247 612, reacts w water, 10 ppmv/220°C, s
$SnCl_4$	−34 109, reacts w water, volatile, s
SnS	880 1210, 10 ppmv/595°C, ins
SnS_2	765, decomp 600, ins
SnH_4	−146 − 52, dissoc 0°C
$Sn_2{}^{II}P_2O_7$	>400 dissoc, ins
$K_2Sn^{IV}O_3$	>850 dissoc, hygroscopic, s
$K_2Sn_2{}^{II}O_3$	less stable, formed at 550°C, s
$K_4Sn^{IV}O_4$	>650 hygroscopic, dissoc to K_2SnO_3, s
$Na_2Sn^{IV}O_3$	stable >800, s
$Na_4Sn^{IV}O_4$	>800 dissoc to $Na_2Sn^{IV}O_3$, s
$MgSn^{IV}O_3$	>700 dissoc, ins
$Mg_2Sn^{IV}O_4$	stable at 1100, ins
$Mg_2Sn_2{}^{II}O_4$	ins
$SrSn^{IV}O_3$	stable at 1000, ceramic, ins
$Sr_2Sn^{IV}O_4$	>900 dissoc to $SrSn^{IV}O_3$, ceramic, ins

decomp, decomposes; *dissoc*, dissociates; *s*, water soluble; *ins*, water insoluble.

SnO_2 film formation (Saikia and Deori, 1999; Dogan et al., 2011). Tin appears to have little difficulty changing from divalent to tetravalent. The tetravalent oxide is one of the thermally stable molecules and is well studied, dissociating above about 1100°C to predominantly SnO(g) and lesser amounts of its polymers $(SnO)_{2-4}$(g) (Zimmermann et al., 1996, 1999). Its vaporization behavior under different conditions has been illustrated and under normal conditions will have a vapor pressure of about 10 ppmv at 1160°C.

The divalent oxide SnO also is stable, but above 530°C it converts to the dioxide vaporizing then as SnO(g). Since both oxides produce a very similar vapor composition it was argued that there is little difference between the two (Tumarev and Baimakov, 1972; Kostelov et al., 1979). In fact, the monoxide has been referred to as a metastable form and on heating SnO(s) disproportionates into a mixture of Sn(l) and SnO_2(s) (Tumarev and Baimakov, 1972; Mallika et al., 2001). Consequently, it is in essence SnO_2(s) that is evaporating in both cases. However, a major point of interest is that SnO_2(g) has never been detected and appears to be short-lived, instantly dissociating if formed to the gaseous monoxide, which survives to higher temperatures. Under reducing conditions the sulfides are stable, SnS and SnS_2. SnS to about 1250°C (m.pt.880, b.pt.1210°C). However, it sublimes with a vapor pressure of 10 ppmv at about 595°C (Nakamura and Fuwa, 1987; Wiedemeier and Csillag, 1979; Pankritz et al., 1987). SnS_2(s) is stable to about 400°C and below its melting point of 765°C it forms a mixed II/IV sulfide of Sn_2S_3(s) that begins to dissociates to SnS above 525°C (Wiedemeier and Csillag, 1980). Gaseous SnS_2 has not been reported and available data for this have been estimated from periodic table trends (Gurvich et al., 1991). In CVD experiments mixing gaseous $SnCl_4$ and H_2S produces deposits of SnS_2 on glass at <500°C, but between 525°C and 545°C Sn_2S_3, and above this temperature SnS(s) (Price et al., 1999). Similarly, using $SnCl_2$ and NH_3 in a CVD mode at 550°C–675°C, deposits on quartz of SnN_x are formed showing the thermodynamic power of such chemical deposition chemistry (Takahashi et al., 2000a). Tin can form mixed metal molecules, particularly the sodium or potassium stannates with tetravalent structures such as K_2SnO_3, K_4SnO_4. These appear to be somewhat unstable at higher temperatures, dissociating back to their respective oxides, are hygroscopic and react with CO_2 (Hagenmuller and Tournoux, 1963). The latter orthostannate dissociates above 650°C to the metastannate, K_2SnO_3, reported to be stable to at least 830°C (Devalette and Hagenmuller, 1967). The corresponding sodium stannates are similarly Na_4SnO_4 formed at 500°C disproportionating to Na_2SnO_3 above 800°C (Hagenmuller, 1967). The corresponding divalent stannites that would have formulas of $K_2Sn_2O_3$ and $K_4Sn_2O_4$ appear thermally less stable and remain uncharacterized, only the former of these has been reported as having been prepared at 550°C. Orthostannates with tetravalency can be very stable with some alkaline earth molecules including Mg, such as $SrSnO_3$ and Sr_2SnO_4 being ceramic materials stable to higher temperatures. Tin/phosphorus chemistry is not well defined, but like many metals, tin can react with phosphorus to form complex glass like structures. Tin pyrophosphate is one component that has been indicated as vaporizing with dissociation at 400°C but producing stable gaseous $SnPO_3$ and $SnPO_2$ (Lopatin and Semenov, 1996; Lopatin and Shugurov, 2013). Although these Russian studies over a 50 years period have examined numerous gaseous mixed metal oxides they have reported no others for Sn.

Gas phase stabilities and transport

Few tin species can exist in the gas phase especially at elevated temperatures. Additionally few are listed in thermochemical tables. The NIST-JANAF tables list none (Chase, 1998). SnO(g) and SnS(g) stand apart due to their very strong bonding. Gaseous SnO has been studied spectroscopically and its singlet and triplet electronic states assigned partly through its chemiluminescent Sn reactions with oxidizers. The ground state is $X^1\Sigma^+$ with no lying states. These have been illustrated theoretically with potential energy curves (Giri et al., 2002). An evaluation of the numerous experimental measures of its vaporization and mass analysis studies indicates a reliable value of $D_0(\text{Sn-O})_g = 524 \pm 4$ kJ mol^{-1} (Gurvich et al., 1991; Barin and Platzki, 1995). This bond strength is seen to exceed even that of water (D_0(H-OH) = 493 kJ mol$^{-1)}$ and is comparable to that of OC-O. Although SnO_2(g) has never been directly monitored, a shock tube experiment observed reaction between SnO(g) and O_2 at 1430°C−2030°C. SnO(s) nanoparticles were shock vaporized in <50 μs and the resulting rate of O-atom formation monitored (Herzler and Roth, 2003). Experimental care was taken to use pure reactants and remove traces of water or other background impurities from the shock tube. A strong activation energy was observed eliminating any role of the exothermic reaction of O_2 with any Sn that may have been formed. Assuming this activation energy corresponded to the adverse reaction enthalpy it provides an estimate of the SnO_2(g) heat of formation consistent with other estimates (Lamoreaux et al., 1987). These suggest $D(\text{OSn-O})_g \approx 310$ kJ mol^{-1} and imply that SnO_2(g) is relatively unstable to conversion to the far more stable SnO(g). Even so, when examining the trends in this periodic group concerning the ratio of D(Sn-O)/D(OSn-O) as in Table 7.24, a lower value of about 260 would appear more appropriate as was also suggested elsewhere (Gurvich et al., 1991). In fact theoretical calculations of the heats of formation for SnO and SnO_2 imply a supporting value of D_0(OSn-O) = 243 kJ mol^{-1} (Nielsen et al., 2003). As a result, this introduces more of a bias concerning any role for SnO_2(g) in the analyses of the gas phase compositions with temperature for SnO_2(s) vaporization, suggesting an even lesser role for SnO_2 than was considered (Lamoreaux et al., 1987). In reality, SnO(g) will be dominant, with lesser quantities of Sn(g) and oxygen to temperatures above 2000°C. Observations in a low-temperature matrix isolation infrared characterization also accept that the SnO_2(s) monitored therein was chemically formed during condensation and did not result from SnO_2(g) (Bos and Ogden, 1973). The trend of the M-O and OM-O bonding in this periodic table group is quite striking with the first bond being uniformly about twice as strong for C through Pb.

The questionable existence of the hydroxide SnOH(g) was analyzed in hydrogen flames containing tin but shown to be negligible with SnO(g) dominant and implying D(Sn-OH < 350 kJ mol^{-1}) (Bulewicz and Padley, 1971b). Nevertheless, it may be present in small quantities. No volatile hydroxide species were detected up to 930°C in a transpiration experiment with SnO_2 and H_2O vapor (Furey and Bloore, 1970). It is invoked to explain the flame inhibition properties of tin explained by the necessary catalytic reactions of SnO with H and OH radicals using temporary SnOH formation that then regenerates SnO (Bulewicz and Padley, 1971; Babushok and Tsang, 2000; Linteris et al., 2002). Although at higher temperatures the $SnOH^+$(g) ion can be present in flame gases this is not related to SnOH but produced either from the chemiionizing reaction of Sn with OH or from ion exchange between SnO and H_3O^+ (Goodings, 1992). A matrix isolation study examining the deposition of SnOH or HSnO tentatively assigned a frequency to HSnO but the simple hydroxides are not readily abundant or apparent (Teng et al., 2007). A theoretical paper also has estimated the vibrational frequencies for these structures (Ball, 2003).

Gaseous $SnCl_4$, $SnCl_2$, and SnCl are known and the latter spectrally well characterized. This provides a reliable value for the $X^2\Pi_{1/2}$ ground state of D_0(Sn-Cl) = 385 ± 12 kJ mol^{-1}

obtained from a short extrapolation of the energy levels of its high lying $a^4\Sigma^-$ state (Gurvich et al., 1991). A recent theoretical analysis indicated a supporting value of about 378 ± 8 kJ mol^{-1} (Li et al., 2013a). A value for $SnCl_2$ can be derived from its quite accurate heats of formation of the solid coupled to its heat of sublimation. This indicates a value of $D_{0Atomiz}(SnCl_2)_g = 741.3 \pm 7$ kJ mol^{-1} or D_0(ClSn-Cl) = 356 ± 14 kJ mol^{-1} displaying only a slight weakening of the second bond. Similarly, appraising the available data on the heats of formation of liquid $SnCl_4$ and its sublimation at low temperatures a value of $D_{0Atomiz}(SnCl_4)_g = 1256 \pm 2.7$ kJ mol^{-1}was derived (Gurvich et al., 1991). This would indicate that the combined strength of the third and fourth bonds is 515 ± 7 kJ mol^{-1} or weakening to an average of 258 kJ mol^{-1} each. A detailed experimental and theoretical kinetic study of the thermal dissociation of gaseous $SnCl_4$ implied an energy barrier of 326 kJ mol^{-1} to break the $SnCl_3$-Cl bond, which might suggest an additional activation energy in addition to the thermal barrier (Takahashi et al., 2000). Nevertheless, as seen in Table 7.24 the bonding in the mono- and dichlorides for the Si, Ge, Sn, and Pb are remarkably similar. As mentioned these gaseous chlorides are convenient sources of Sn and can be used in CVD formation methods for SnO_2 deposits without any chlorine effects. However, the chlorides are vulnerable to hydrolysis by water and oxidation to SnO(g).

Another possible gaseous species vulnerable to oxidation is SnH. No thermochemical table assessments are available for this. It has been studied spectroscopically and theoretically. The spectral data using predissociation limits in the $A^2\Delta$ state suggest a bond strength of ≤ 263 (Huber and Herzberg, 1979) and 224 kJ mol^{-1} (Alekseyev et al., 1996), whereas the theoretical studies now quote 241 and 250 kJ mol^{-1} (Song et al., 2008; Li et al., 2013a). As a result a value of D_0(Sn-H) = 245 ± 10 kJ mol^{-1} is tentatively suggested. Only recently has the gaseous SnH_2 been detected and analyzed spectroscopically in the gas phase for the first time (Smith and Clouthier, 2018).

Under reducing conditions SnS(g) can be stable to about 1000°C and has been detected as emitted from volcanoes. Its bond strength is accurately known from thermochemistry via accurate heats of formation and sublimation of the solid SnS, and also by a short extrapolation of vibrational levels in the $E^1\Sigma^+$state. These both suggest D_0(Sn-S) = 464 ± 5 kJ mol^{-1} (Gurvich et al., 1991). The electronic states and potential energy curves have been treated theoretically also confirming this bond strength magnitude deriving 454 kJ mol^{-1} (Giri and Das, 2005). As seen in Table 7.24, the ratio of the MS to MO bond strengths increase more drastically in this periodic table group than is normally observed, from 0.66 for carbon to 0.92 for lead and in this case 0.89 for tin (Hauge and Margrave, 1972). Reflecting the gaseous dioxide, there is no indication of a gaseous disulfide and vaporization of SnS_2(s) will produce SnS(g).

Consequently, tin's basic chemistry tends to imply that its combustion will be dominated by the oxides if oxygen-rich and by the sulfides if reducing. Moreover the degree of vaporization will be very sensitive to temperature whether above or below 1000°C. Effects of chlorine or sulfur may only facilitate initial release and probably not be involved in final ash products other than possible sulfate at low temperatures. Experience gained from flame atomic analysis methods indicated that to obtain a sufficient sensitivity, or in other words have an adequate concentration of free Sn atoms required a hot C_2H_2/N_2O flame. This would be close to 2700°C and able to dissociate the dominant SnO (De Galan and Samaey, 1970).

Combustion power plant data

There can be variable levels of Sn in coals but consistently less in biomass. Wastes and sludge can have more reasonable levels (Demirbas, 2005; Table 1.1). Both pet coke and limestone have negligibly low levels (Cordoba et al., 2012). However, mainly because tin emissions remain unregulated there are few detailed measurements concerning its speciation or behavior in combustors. Recent reviews on FBC incinerators and the co-combustion of coal/sewage sludge do not even include tin (Lopes et al., 2003; Van Caneghem et al., 2012). Also because tin is sensitive to combustion temperatures it is necessary to examine separately behavior below 1000°C in FBC and incinerators, and above this in coal boilers. With incineration of hospital wastes fired with diesel fuel below 1100°C, most ash is reported as bottom ash but the small amount of fly ash was found to contain a fourfold higher concentration of tin that is not leachable (most likely SnO_2) (Zhao et al., 2008). In a power plant fired with a subbituminous coal the ratio of tin in the fly ash to the slag was quoted as twofold (Querol et al., 1995) and in a FBC with a bed temperature of 580°C, burning a high ash lignite coal, the ash/slag ratio was fourfold (Selcuk et al., 2005). A study examining the efficiencies of retention of the FGD system in co-combustion of coal and pet-coke indicated that any stray Sn reaching the stacks was reduced a further 50% by this device (Cordoba et al., 2012). Under reducing conditions, the 1200°C–1600°C gasifier of an IGCC plant indicated a very high enrichment on the ash and small amounts of slag or gaseous component (Font et al., 2010).

The combination of the chemistry and two extensive equilibrium calculations (Frandsen et al., 1994; Poole et al., 2008) suggest that under oxidizing combustion conditions the distribution of the tin will be simple and will start vaporizing and/or be oxidized to the stable dioxide. Above about 400°C this will be complete and any chlorides will have vaporized or been hydrolyzed by water and retained. The SnO_2 formed will have a low vapor pressure until above 1000°C–1100°C and so will be retained. Above this temperature it will fully vaporize solely as SnO(g), stable to about 2000°C and be the main vaporized form of tin. No deposition on the ash can occur until temperatures fall again to below about 1100°C when the SnO(g) will deposit as the more thermodynamically stable SnO_2(s). This will continue until very cool temperatures of about 250°C when possible conversion or deposition as stannic sulfate occurs, which may become the more stable product in the presence of sulfur. The thermodynamic calculations suggest that the presence of alkalis or chlorine should not modify this behavior.

Under fuel-rich reducing conditions the presence of any oxygen in the system will continue a dominance of SnO_2/SnO. In its absence the chemistry changes to that of SnS_2(s)/SnS(s) that starts vaporizing above about 525°C to SnS(g), stable to about 1250°C or above. These are all so stable that other molecules are not expected to interfere. When leaching tests are made, both SnO_2(s) and SnS(s) are known to be insoluble, whereas the chlorides are soluble and other molecules such as the sulfate are hydrolyzed to a dioxide precipitate. Fly ash from a municipal waste incinerator was heated at 1000°C for 3 hours either with $CaCl_2$ or with Cl_2. The chlorine in the salt had no effect but the molecular Cl_2 vaporized all the tin in the ash sample (Chan and Kirk, 1999). Any molecular formation on the ash would be converted to SnO_2(s) at 1000°C and certainly chlorinated by Cl_2 so it conveys little information for Sn without additional leaching data. In one water leaching test it seemed that fly ash leached more efficiently than bottom ash (Llorens et al., 2001). Sulfate is soluble and the dioxide not, so it is difficult to interpret such data without special care being taken concerning the full history of the handling of such samples being analyzed.

The CVD experiments mentioned above, depositing films of SnO_2(s) by invoking the hydrolysis or deposition of either gaseous $SnCl_2$ or $SnCl_4$ are intriguing from the point of view of the mechanistic details that still remain largely unknown. Generally carried out with a substrate at 400°C–700°C one study created deposits even at 900°C–1300°C from $SnCl_4$ and O_2 (Nagano, 1984). Another nebulized an aerosol of $SnCl_4$ dissolved in methanol into an oxyhydrogen flame torch depositing nanocrystalline SnO_2(s) onto a heated substrate (Kim et al., 2009). In this case the tin clearly would be SnO(g) in the flame gases and depositing as the stable SnO_2(s). At the low temperatures with either $SnCl_4$ or $SnCl_2$, the SnO_2(s) cannot result from SnO_2(g) which would be difficult to form. One is forced to conclude that this has to be a fast chemical deposition mechanism that occurs directly at the surface with continued film growth simply driven by thermodynamic preferences. Moreover as outlined above, mixtures of a chloride and H_2S can produce SnS(s) deposits. With NH_3, deposits of SnN_x suggest these do not reflect differing gas phase kinetics but are general heterogeneous mechanisms allowed on the hot two dimensional surfaces.

Nanoparticles and particles or powders also can be flame synthesized without a deposition surface (Pratsinis, 1998; Roth, 2007). In this case a volatile tin precursor is added to either a hydrogen or fossil fuel in any type of flame configuration either as the chlorides or for example as tetramethyl tin. Concentrations can be large, typically ranging up to 0.5%. Under such conditions the flame gases become saturated with SnO(g). Whether this then homogeneously nucleates at appropriate temperatures to initiate the SnO_2(s) particle growth or requires nucleating centers remains unclear (Zhu and Pratsinis, 1997; Lindackers et al., 1998). Nothing is known of whether mixed metal compounds are formed in the ash but as indicated in Table 7.38 concerning possible Sn compounds, there are alkali and particularly alkaline earth stannates that have ceramic natures.

In summary, tin is of interest particularly as to how it fits into its periodic column of elements. It is again dominated by oxide (in oxygen environments) or sulfide chemistries (pyrolysis or oxygen-free) that are exceptional in having a very limited range of thermally stable chemicals. In combustion, temperatures control its gaseous release and transport. Its concentration in fuels is low and coupled to its nontoxicity, this removes most interest in its behavior. Its combustion inhibition qualities raise uncertainties concerning its exact mechanisms. The implication is that low levels of hydroxide must exist. With sulfur, deposition may change from the oxide to the soluble sulfate at low temperature. Whether mixed metal oxide compounds form in or on the ash remains uncertain.

Supplemental comments on Germanium (relative of Tin)

Although germanium has not been included in this book, its chemical behavior supports the conclusions seen above for tin, primarily due to its familial relationship in the periodic table. Its thermochemical values are listed in Table 7.24 along with those for C, Si,

Sn, and Pb. There is little concern in practical combustors or incinerators with regards to germanium and its fuel content is rarely analyzed even though it can have a range of about 0.5–50 ppmw in coals. Its chemistry mirrors that of tin. Although the NIST-JANAF database does not include its molecules (Chase, 1998), the other three useful sources do list most of its condensed and gaseous forms that relate to combustion (Gurvich et al., 1991; Knacke et al., 1991; Barin and Platzki, 1995). GeO(s) is seen to be even more metastable than SnO(s) and disproportionates to GeO_2(s) above 415°C and does not exist above 800°C (Zagorodnyuk et al., 1972; Ramakrishnan and Chandrasekharaiah, 1975). Germanium is similarly used in CVD and in flame synthesis methods to create GeO_2(s) films and particles. These vaporize predominantly to GeO(g) and as shown in Table 7.24 along with some of its other gaseous molecules, and is even more stable than SnO(g) with D_0(GeO)(g) = 650 ± 5 kJ mol^{-1} (Gurvich et al., 1991). It has vapor pressures of 6 ppmv (1100°C), 27(1150) and 110(1200) (Shimazaki et al., 1957). Similarly D_0(GeS)(g) = 548 ± 0.5 kJ mol^{-1} implies this is the most stable metal sulfide in the upper part of the periodic table. That flame generated particles of GeO_2(s) can be controlled by an electric field might suggest a role for electrical charge in any nucleation step (Katz and Hung, 1990). As a result, the oxides or sulfides will dominate any Ge transport, and if oxygen-rich, deposit on cooling ash initially as GeO_2 with possible conversion to the sulfate at lower temperatures. Now that interest is turning to ash as a possible source of materials, there may be renewed interest in extracting germanium (Unver and Terzi, 2018). In the absence of air, GeS(g) will be dominant forming GeS_2 deposits. The element is still being characterized and recently a spin-forbidden $a^4\Sigma^- - X^2\Pi_{1/2}$ transition of GeH(g) has located the energy of this low-lying electronic state for the first time (Harms et al., 2018).

7.28 Titanium

Being the second member of the first row of transition elements, titanium (atomic number 22) is still a lightweight metal. It has four valence electrons $3d^24s^2$ and displays a pronounced tendency to +4 valency. Although di- and trivalent molecules are known they are generally readily hydrolyzed or oxidized to the +4 state. In fact aqueous chemistry of +2 ions is not evident. It is relatively abundant in the Earth's crust (0.6%) ranking ninth behind elements such as Al, Ca, Fe, Si, Mg, Na, K, and O. Possibly due to its general insolubility, it is 5 million times less dense in the oceans with a concentration of 1 ppbw and so is about the 30th most abundant. Its main ore is ilmenite $FeTiO_3$, with some mining of rutile TiO_2. Commercially, the ilmenite is more extensively processed at about 6 million tons year^{-1} globally with 0.8 million tons year^{-1} of rutile. Global reserves for these ores are abundant being estimated at about 2 billion tons. About 95% of that mined is utilized as TiO_2 due to its white pigmentation. This is found in paint, paper, cosmetics, sunscreens, and as a nontoxic inert food colorant particularly in cheese. Due to titanium's lightweight, strength, and corrosion-resilient alloys, it finds extensive use as the metal in the aircraft, machinery and defense industries and for medical implants that are well tolerated in the human body. Its major commercial materials are the metal, TiO_2 and $TiCl_4$. Over the years its cost has fallen and currently is about \$8.00 lb^{-1} for Ti ingot. An interesting report on the critical minerals to the US economy states the rather shocking fact that each American consumes 11 tons $year^{-1}$ of mineral supplies. In the report, titanium is included. Although 95% is used as a pigment, substitutes for this could be found. However, the other 5% is critical to the aviation and high tech industries making it a material that should be regarded as critical to the economy (National Research Council, 2007).

Its natural and anthropogenic levels in the environment have not been assessed. Because its chemistry is not one of gaseous components, ambient environmental aspects center on its presence in particles. Natural rock-worn dust levels will establish a baseline which appears to be of little concern. There is a realization that local contamination can result from fossil fuel combustion and incineration particularly of painted items, paper, inks and plastics if the emissions are not effectively controlled.

As a result, although one would expect little interest in titanium or TiO_2, the *Journal of Physical Chemistry* did 1-year note that the number of papers it published on TiO_2 was about 500 $year^{-1}$ (Kamat, 2011). However, the papers published relating to combustion centered mainly on nanoparticle formation or CVD film deposition.

Health concerns with titanium

The World Health Organization has published on the Web an extensive report on titanium and its compounds with respect to its human health concerns (World Health Organization Ti, 1982). More recently this has been extended by a review of the bioinorganic chemistry of titanium as it relates to the environment (Buettner and Valentine, 2012). On the whole, concern remains minor in that TiO_2, titanium's major compound is not regarded as toxic or carcinogenic. Although not fully proven, titanium seems not to be an essential element for man or animals. No titanium enzyme has ever been identified. It is not readily absorbed by plant life but enters their cycle through enriched or contaminated soil. Some algae are indicated as bioaccumulating titanium and so can be introduced into the food chain. A typical diet can contain about 300–400 $\mu g\ day^{-1}$. Absorption by the body is slight (3%) but it can cross the blood/brain and placenta barriers and an average body composition will contain about 10–20 mg. Its half-life in the body if retained is thought to be about 1 year. Extensive studies of animals and workers in the titanium industry report no long-time repercussions of significant exposures either to the body or brain from inhalation, ingestion or inoculation. However, recent reports do indicate some effects of TiO_2 on the brain of Zebrafish (Palaniappan and Pramad, 2011). Moreover, the use of nanoparticles in a growing number of commercial products may now require a further assessment of its toxicity. TiO_2 nanoparticles also have been reported to cause damage to the brains of mice (Ze et al., 2013). This aspect may also affect plant forms and whether nanoparticles facilitate their uptake with possible greater introduction into the food chain (Larue et al., 2011). There is a growing usage of the nanoparticles in food and care products. One study measured the content in these using a rather intriguing unit of $\mu g\ mg^{-1}$. Some children's candies contained 1 $\mu g\ mg^{-1}$ which would appear small until changed to 0.1%. A realistic estimate of exposure to TiO_2 in the United Kingdom is 2–3 mg kg^{-1} body wt. day^{-1} for children under 10 years old and similarly 1 mg for the general public (Weir et al., 2012). Whether this should now be flagged as of possible concern requires further study. The man-made volatile chemical $TiCl_4$, liquid at normal ambient temperatures (m.pt. − 24, b.pt.136°C), is very hazardous due to its ready hydrolysis or oxidation to TiO_2 releasing toxic HCl. $TiCl_4$ finds importance as the precursor molecule in the synthesis of pure metallic Ti and TiO_2. A toxicology analysis has been published for this but will be encountered mainly in the industrial environment (Tox Profile TiCl4, 1997).

The lack of concern with titanium is apparent in that hardly any human testing of body biomarkers is published or its inclusion in any survey testing. The Mayo Clinic in the United States suggests a blood serum level of $<1\ \mu g\ L^{-1}$ but really this is as a marker for prosthetic implant wear. One testing clinic suggests hair levels should be about 0.1−0.7 ppmw. Only one survey has been found comparing the populations of Morocco and the Canary Islands (Henriquez-Hernandez et al., 2018). They indicted median values of $23/27\ \mu g\ L^{-1}$ for blood but also gave an unusual high range of values for tin, placing doubt. Even so, at present, there are no health concerns for titanium.

Titanium's basic chemistry

On examining titanium's condensed phase molecules a rather limited chemistry for titanium is apparent that emphasizes tetravalency. It has both a cationic nature (dioxide, tetrachloride, sulfate), and an extensive mixed metal oxide tendency to form many ceramic or refractory titanate molecules such as ortho- (Ba_2TiO_4) and meta-titanates ($BaTiO_3$) with numerous other metals such as Al, Ca, Co, Fe, Mg, Mn, Ni, Sr, and Zn. These are generally insoluble in water and some are true molecular titanates, while others, like that of Al, are in effect a solution of the two oxides one in the other namely Al_2O_3 and TiO_2 that are stable at high temperatures. Also the ortho-titanates tend to dissociate into their respective oxides or react with CO_2 at high temperatures to convert to the more stable metatitanates.

$$Ba_2TiO_4 + CO_2 = BaTiO_3 + BaCO_3$$

A compilation of titanium's condensed phase chemicals is listed in Table 7.39 indicating that few will be involved in combustion.

Metallic titanium is a refractory (m.pt.1668, b.pt.3287°C). It has no stable carbonate and the sulfate $Ti(SO_4)_2$ is hydrolyzed by water forming TiO_2. On heating above 150°C it dissociates initially forming the basic sulfate $TiOSO_4$ and then the dioxide (Stern and Weise, 1966; Stern, 2001). Similarly, the nitrate is not thermally stable and decomposes at its 58°C melting point to TiO_2 (Stern, 1972, 2001). A divalent oxide TiO(s) (m.pt.1750°C) can be formed by heating equal molecular quantities of Ti and TiO_2 at about 1500°C. However, it is a nonstoichiometric molecule and needs careful annealing to acquire a monoclinic structure. The same can be said for Ti_2O_3(III) (s) (m.pt.1840°C) formed by similarly mixing the appropriate proportions of Ti and TiO_2 and heating at about 1600°C. Heating Ti and TiO_2, or TiO_2 alone can produce a liquid above 1800°C that can contain a mixture of TiO, TiO_2 and Ti_2O_3 in equilibrium with one another (Banon et al., 1982). The thermodynamically stable oxide is TiO_2(s), which exists as rutile at all temperatures (m.pt.1857°C), but it also has anatase and Brookite phases.

TABLE 7.39 Properties of titanium's condensed phase molecules relevant to combustion, °C.[a]

m.pt. b.pt. vapor pressure
Ti 1668 3287, refractory, 10 ppmv/1709°C, ins
TiO 1750, nonstoichiometric, heating in air oxidizes to Ti_2O_3, Ti_3O_5, TiO_2
TiO_2 (rutile) 1857 2972, 1 ppmv/1815°C, ins
(anatase) 1676, metastable, >627 converts to rutile, ins
$Ti_2^{III}O_3$ 1842 3000, less encountered, ins
$Ti_3^{III/IV}O_5$ 1777
$TiO(OH)_2$, unstable, hydrated TiO_2
$Ti(CO_3)_2$, unstable, not known
$Ti(NO_3)_4$ 58, decomp to TiO_2
$Ti(SO_4)_2$ >150 dissoc via basic sulfate, hydrolyzed by water, ins
$TiCl_2$ 1035 dissoc, reacts with water and air
$TiCl_3$ >425 disproportionates, reacts with water and air
$TiCl_4$ −24 136, dissoc >700, reacts with water, 1.3×10^{-2} atm/20°C, s
TiS 1780, dissoc >1000, oxidized in air, ins
TiS_2 1930, dissoc >500 to TiS, oxidized by air, ins
TiH_2 decomp >450, reacts with water and air, ins
TiH_4, unstable, reacts with water and air
TiP 1990, dissoc >1400, oxidized in air, ins
Al_2TiO_5 1860, unstable 800−1280, ins
$CaTiO_3$ 1975 3000, molecular vaporization observed, ins
$Fe^{II}TiO_3$ (ilmenite) 1385, ins
K_2TiO_3 1615, ins
K_4TiO_4 >650 changes to K_2TiO_3, hydrolyzed by water, ins
$MgTiO_3$ 1630, ins
$MgTi_2O_5$ 1690, ins
Mg_2TiO_4 1740, ins
$NaTiO_3$ stable at 1000, ins
$TiSiO_4$ −, silicate glass, ins

[a]*See text.*
dissoc, dissociates; *decomp*, decomposes; *s*, water soluble; *ss*, slightly soluble; *ins*, water insoluble.

The latter two also exist in nature but are metastable and can occasionally be the product of oxidation or hydrolysis of titanium salts. They convert to the more stable rutile on heating to 550°C or 750°C, respectively. Although $TiO_2(s)$ is thermally very stable and insoluble in water, it is subject to a rather slow surface hydration by water vapor. If heated to 1200°C–1400°C, transpiration studies now have indicated it is slightly eroded by the formation of gaseous $TiO(OH)_2$, a basic hydroxide also labeled as the stable hydrate $TiO_2.H_2O$ (Nguyen, 2008; Myers et al., 2019). Titanium has several chlorides but the di- (m.pt.1035°C) and trivalent form (this disproportionates above 425°C to $TiCl_2$ and $TiCl_4$) are readily oxidized by air and hydrolyzed by water. The important $TiCl_4$ is a volatile liquid at room temperature (m.pt. − 24, b.pt.136°C) and is also converted to the dioxide by reaction with oxygen or water. The latter in fact converts it to aqueous $Ti(OH)_4$, its hydroxide, known as titanic acid, H_4TiO_4. However, this is not stable, losing water to become TiO_2. Sulfides are known, $TiS_2(s)$ and TiS(s), and although insoluble in water they are subject to surface hydrolysis to TiO_2. They are readily oxidized in air above about 400°C, and at 600°C TiS_2 can produce the metastable anatase structure. They do have high melting points (1930°C) (Franzen and Gilles, 1965) and vaporize to TiS(g) and $Ti(g)/S,S_2$ mixtures with $TiS_2(g)$ comprising <1% of the vapor (Edwards et al., 1971). With the hydrides, TiH_2 (decomposes above 450°C) (Rasooli et al., 2013) and its decomposition and ignition properties have been studied recently (Young et al., 2015a). TiH_4 is unstable and both of these hydrides are oxidized in air and hydrolyzed by water. The infrared spectra of condensed TiH, TiH_2, TiH_3, TiH_4 have been studied by matrix isolation of products from laser-ablated Ti in H_2 (Chertihin and Andrews, 1994). A titanium phosphate recently has been synthesized but is not normally encountered (Cumba et al., 2012). Similarly, although stable under reducing conditions, Ti(III)P is known, insoluble in water and stable to >1400°C but oxidized by air. In combustion and ash chemistry, the potential formation of titanates with other major ash constituents of Al, Ca, Fe, K, Mg, Na requires consideration. The alkali titanates are known for Na and K but are readily hydrolyzed by water (Devalette and Hagenmuller, 1967). When TiO_2 and either Na_2SO_4 or NaCl are mixed and heated at 950°C slight mass losses have been noted with evidence of some $Na_2TiO_3(s)$ and titanium chloride formation (Mobin et al., 1990; Mobin and Malik, 1996). The other titanates are all refractory materials and insoluble in water. Aluminum titanates have the problem of being unstable in the 800°C–1280°C range when they can dissociate back to their individual oxides. Otherwise they can be produced by combustion synthesis (Segadaes et al., 1998). Of the others, the meta-titanate forms TiO_3^{2-} are more stable. Numerous have been studied and have approximate melting points as indicated, $BaTiO_3$(1620°C), $CaTiO_3$(1975°C), $FeTiO_3$(ilmenite, the ore, 1385°C), $MgTiO_3$(1630°C) (Barin and Platzki, 1995). Many Russian studies have examined these meta-titanates at high temperatures and recently have reviewed their 50-year study of examining their gas phase with mass spectrometric methods (Lopatin and Shugurov, 2013). What is most interesting is that some of these ceramic type molecules particularly with the alkaline earth elements can vaporize molecularly and species such as $CaTiO_3(g)$ are so strongly bound $D(CaO\text{-}TiO_2) = 558\ kJ\ mol^{-1}$ that they can exist in high temperature gases. Beryllium titanate appeared to be the strongest bound but there has been no indication of any alkali titanate vaporizing unchanged.

Gas phase stabilities and transport

With such condensed phase chemistry, gaseous release of titanium from a fuel will only be possible at very high temperatures ($\approx 2000°C$) unless it is injected separately into a flame as a volatile additive. The fact that Ti does not atomize easily was abundantly clear for any mode of flame atomic analysis (Parsons et al., 1975). Approximate measures of Ti atom concentrations in high temperature C_2H_2/air flames indicated negligible dissociation of the dominant TiO(g) and TiO_2(g) species and reducing fuel-rich higher temperature C_2H_2/N_2O flames (2700°C) were required (Rasmuson et al., 1973; Fujiwara et al., 1975). The characterization of the gaseous titanium species that might be expected in combustion gases has been mainly studied by the Knudsen cell-mass spectrometric method due to its otherwise difficult monitoring sensitivity and species identification. For example, the oxides are refractory and so need high temperatures (vapor pressures ≈ 0.3 ppmv, 1750°C) (Semenov, 1969; Wu and Wahlbeck, 1972; Sheldon and Gilles, 1977). Due to their commercial importance, the vaporization of Ti(s), TiO(s), TiO_2(s), Ti_2O_3(s) mixtures has been extensively studied. It indicated that the high temperatures would define the composition balance between these. A 50/50 mix of Ti and TiO_2 could produce mainly TiO at high temperatures and enable it to be characterized. The gaseous vapor was shown to consist predominantly of TiO(g) and TiO_2(g). As many as 15 such studies were reviewed to assess values for the bond strengths of these, recommending 668 ± 9 kJ mol^{-1} for TiO (Pedley and Marshall, 1983), and 663 ± 8 (TiO) and 604 ± 14 (TiO_2) from an earlier review (Chase, 1998). Two further different type studies for TiO(g) have since appeared. One a crossed-beam study of the Ti + NO = TiO(a) + N reaction monitored the energy of the product by LIF and determined the reaction's enthalpy indicating D_0(Ti-O) $= 663 \pm 6$ kJ mol^{-1} (Naulin et al., 1997). Another accurately measured the ionization potential of TiO(g), which when coupled to the known dissociation energy of TiO^+ gave the same value of 663 ± 8 kJ mol^{-1} (Loock et al., 1998). As a result, a value of D_0(Ti-O) $= 663 \pm 4$ kJ mol^{-1} now seems appropriate and is supported by theoretical estimates (Bauschlicher and Maitre, 1995; Pan et al., 2017). Although it has two relatively low-lying states, $a^1\Delta$ (3444 cm^{-1}, 41 kJ mol^{-1}) and $d^1\Sigma^+$ (5663 cm^{-1}, 68 kJ mol^{-1}), due to the degeneracy of the ground $X^3\Delta_1$, even at 2000°C the population of the $a^1\Delta$ state will only constitute 3.6% of the total TiO concentration and be negligible (Kaledin et al., 1995; McKemmish et al., 2017). The electronic structure, bonding and potential energy curves of all the lower lying singlet and triplet states of TiO have been calculated (Miliordos and Mavridis, 2010). The dioxide, TiO_2, has a bent structure with two double bonds (Rosi et al., 1998). The ground X^1A_1 and nine low-lying singlet states have been studied theoretically (Chou, 2015). It has also been studied experimentally by mass spectrometric methods leading to values of $D_0(TiO_2)_{atomiz} = 1267 \pm 12$ kJ mol^{-1} (Chase, 1998), 1272 ± 12 (Hildenbrand, 1976), 1285 (Gurvich et al., 1982) and 1264 (Balducci et al., 1985). An average value tends to support $D_0(TiO_2)_{atomiz} = 1272$ kJ mol^{-1} or D_0(OTi-O) $=$ 609 ± 8 kJ mol^{-1}. The most recent theoretical predictions indicate about 604 kJ mol^{-1} consistent with these values (Pan et al., 2017). The second bond in the dioxide follows the pattern seen for the second and third columns of the transition elements of being slightly weaker than the monoxide bond (like Zr, V, Nb, Ta) (Tables 7.15, 7.19 and Schofield, 2006). The structures of these two oxides have also been characterized by matrix isolation methods at low temperatures (McIntyre et al., 1971; Chertihin and Andrews, 1995a).

The oxides' stability is extremely strong implying this refractory/ceramic nature. It appears to minimize any possible role in combustion for potential hydroxides of titanium. One study examined Cr, Mo and Ti metals at 1050°C in O_2 flows that contained NaCl(g) and H_2O(g). Monitoring the gaseous products by mass spectrometry, although Cr and Mo indicated corrosive removal, Ti was unaffected (Fryburg et al., 1977). As a result gaseous hydroxides have not been observed other than by laser ablation of Ti in H_2O_2 with low-temperature matrix isolation methods with infrared analysis. Molecules such as HTiOH, $H_2Ti(OH)_2$, HTiO, HTiO(OH), and $Ti(OH)_2$ have been tentatively identified (Kauffman et al., 1985; Zhou et al., 2000a; Wang and Andrews, 2005a). Theoretical calculations have predicted the structures of some of these (Ignatyev et al., 2007) together with the energies of the linear TiOH and bent HTiO (Miliordos et al., 2011). This latter study suggested an approximate value of about $D(\text{Ti-OH}) \approx 400$ kJ mol^{-1} with an H-TiO almost half the strength.

The thermochemistry of titanium's chlorides remained very uncertain until a recent reanalysis of previous experimental work with updated thermodynamic quantities that accounted for the effects of the low-lying electronic states. Hildenbrand's mass spectrometric measurements now indicate values of $D_0(\text{Ti-Cl}) = 420 \pm 8$ kJ mol^{-1}, $D_0(\text{ClTi-Cl}) = 495 \pm 8$, $D_0(\text{Cl}_2\text{Ti-Cl}) = 420 \pm 8$ and $D_0(\text{Cl}_3\text{Ti-Cl}) = 376 \pm 8$ (2009). These values are well supported by theoretical calculations of the electronic states and their potential energy curves for the mono- and dichloride (Wang and Schwarz, 1998; Kardahakis and Mavridis, 2009). In combustion, this bonding will be insufficient to prevent oxidation and hydrolysis (John et al., 2010). Also it is unlikely that chlorine can facilitate release of titanium utilizing any volatile nature. Because the combustion of $TiCl_4$ is one major method of producing TiO_2 particles, recent theoretical studies have considered all the potential intermediates in the process. In an initial attempt to create a kinetic model of the process, as one might imagine, this included the $TiCl_{4-x}(OH)_x$, $TiCl_2(OH)_2$ and hydroxy reduction and oxidation of $Ti(OH)_4$ breakdown species, ten in all (Totton et al., 2011). This now has been extensively increased to characterize 2328 possible intermediate species and is coupled to an equilibrium compositional calculation (Buerger et al., 2019). However, such large kinetic models lose chemical rigor and become engineering computer models. As a result, although such exercises are now in the realm of theory and computerization, they have to retain significant uncertainties and have to be accepted as speculations. In a mass spectrometric effusion cell study of a $TiCl_4/AlCl_3/H_2O$ mixture at about 700°C the $TiOCl_2$ molecule was evident. It was speculated that this might also be involved in the oxidation of $TiCl_4$ and in the chlorination of TiO_2 (Hildenbrand et al., 1991), and does appear to have significant stability (Buerger et al., 2019).

Reactions of laser-ablated Ti atoms with H_2 have identified the structures of the TiH_{1-4} species in low-temperature matrix isolation experiments (Chertihin and Andrews, 1994). Thermochemically, only TiH(g) has been studied in depth. Utilizing ion-reaction threshold energy measurements, from $D(\text{TiH}^+)$ and the ionization potential of TiH this produced a value of $D_0(\text{Ti-H}) = 200 \pm 9$ kJ mol^{-1} (Chen et al., 1991). Being a simple transition element hydride, numerous theoretical calculations have refined their methods on this molecule and have obtained values in a wide range from 180 to 280 kJ mol^{-1} (Bauschlicher, 1988; Anglada et al., 1990; Harrison, 2000; Goel and Masunov, 2008). Its numerous low-lying electronic states are identified (Anglada et al., 1990; Koseki et al., 2002). Little is yet known of the other three possible gaseous hydrides but the electronic structure of TiH_2 has been theoretically described with a second bond similar to the first (Kudo and Gordon, 1995; Fujii and Iwata, 1996) and no unusual bond strength changes are expected for the others. Moreover, with respect to combustion they are themselves combustibles and so will not exist in the burned gases being readily oxidized and hydrolyzed.

The sulfides reflect the oxides in being refractory with high melting points. Laser ablation of Ti reacting with discharged sulfur vapor was cocondensed in a matrix isolation experiment. It identified the TiS and TiS_2 species establishing their structural features (Liang and Andrews, 2002). In the gas phase, Knudsen cell type experiments confirmed the partial dissociation on vaporization to TiS(g), Ti(g), S/S_2(g) with no contribution from TiS_2(g) (Edwards et al., 1971). A later vaporization study obtained the heat of vaporization that indicated D_0(Ti-S) = 413 ± 3 kJ mol^{-1} (Pelino et al., 1979). This differs quite considerably from the earlier and generally reliable mass spectrometric studies that indicated 458 ± 12 kJ mol^{-1} (Smoes et al., 1969). Theoretical estimates have suggested 442 (Bauschlicher and Maitre, 1995) and 445 (Wu et al., 2007). For oxides and sulfides it has been noted that the ratio D(MS)/D(MO) can be very similar for families of elements in the periodic table (Hauge and Margrave, 1972). In the present situation this is quite pronounced, the sulfide/oxide bond strength ratio being 0.70 (Sc), 0.71 (V), and 0.74 (Y), 0.74 (Zr), Tables 7.15 and 7.19. Taking a ratio of 0.7, 0.705 and 0.71 for Ti would indicate D_0(Ti-S) = 464, 467, and 471 kJ mol^{-1}, certainly favoring the higher experimental value. Vaporization measurements of solids, in this case TiS(s) that derive a heat of vaporization are not always reliable in implying a valid heat of formation and hence bond strength. The assumption made is that the composition of the TiS(s) is constant. It seems very likely that in the above case this was not so and led to a biased misleading value. Consequently, until a confirming value is obtained, D_0(Ti-S) = 458 ± 10 kJ mol^{-1} is suggested rather than the lower value that has also been listed in databases (Kubaschewski et al., 1993; Barin and Platzki, 1995). The phosphides are of less interest in combustion due to their sensitivity to air and moisture. Nevertheless the gaseous diatomic TiP has been identified spectrally by its $B^2\Sigma^+-X^2\Sigma^+$ LIF spectrum (Li and Balfour, 2005). As a result, the earlier theoretical estimate of D_0(Ti-P) ≈ 220 ± 20 kJ mol^{-1} that correctly assigned the state is preferred over the later study suggesting 255 kJ mol^{-1} for an $X^5\Delta$ state (Glezakou et al., 1996; Tong et al., 2003).

Laboratory combustion studies

The main interest with titanium in combustion has resulted from injecting titanium salts into burned flame gases to invoke this dominant TiO/TiO_2 role that ends in TiO_2(s) formation. Several reviews of nanoparticle formation or CVD film deposition are available and not only involve TiO_2 but also formation of many mixed metal particles of this with Si, V, Al, Ba/Sr, Li, Sn, to list a few (Wooldridge, 1998; Pratsinis, 1998; Roth, 2007). Syntheses can involve creating powders (Brewster and Kodas, 1997; Jung et al., 1998; Ismail et al., 2015), specific tailored nanoparticles (Chen et al., 1998b; Ehrman et al., 1999; Maric et al., 2002; Zhao et al., 2005), V doped TiO_2 nanoparticles (Ren et al., 2015) and even ball-in-sphere nanospheres (Liu et al., 2009). Many papers center on the flame aspects that can control particle parameters such as size, number density, temperatures, agglomeration, and residence times (Katz and Hung, 1990; Singhal et al., 1999; Katzer et al., 2001; Rosner, 2005; Zhao et al., 2009). Monitoring methods also are being developed, for example, to monitor in situ the particles by Raman spectroscopy and characterize the crystal phase such as anatase versus rutile (Liu et al., 2010a). A new diagnostic measures the TiO_2 volume fraction in the flame using laser-induced breakdown spectroscopy (Zhang et al., 2015b). An attempt has been made to kinetically model the nanoparticle formation on adding Ti tetra-iso-propoxide to a hydrogen flame measuring some species flame profiles with a molecular beam mass spectrometer (Shmakov et al., 2013). Also by adding size resolved Ti nanoparticles (20–150 nm) into the burned gases of a laminar 1500°C–2200°C flame, their burning rates have been measured and shown to be surface reaction controlled (Zong et al., 2015). CVD film-forming methods can equally be used in a flame mode either for TiO_2 or titanate deposition (Ichinose et al., 1994; Wiedmann et al., 1994; Oljaca et al., 2003; Korobeinichev et al., 2012). The lack of other molecular formation prevents titanium from participation in any flame inhibition chemistries. It does exhibit mainly the $TiOH^+$ ion in flames (Goodings, 1992). The chemiionization reaction of Ti atoms with oxygen atoms to produce TiO^+ is essentially close to being thermoneutral and can produce electrons at low temperatures (Schofield, 2006).

Combustion power plant data

The chemistry of titanium therefore reflects few options in combustion processes that involve fuels. As noted in Table 1.1, with potential fuel content of the elements, titanium is generally listed with the major elements. This is on the percentage scale and can be as high as 3% in coal ash. It can be lower with biomass but does have a universal presence. With pet-coke and limestone it remains a trace element on a low ppmw scale (Cordoba et al., 2012). The titanium oxides have very low vapor pressures and their gaseous forms are so stable that they dominate any gas phase chemistry. Their bond strengths outlined above are summarized for titanium and other first row transition elements in Table 7.19. As a result any examination of full-scale combustion can expect to see retention of titanium in the ash with possibly some titanate ash formation if temperatures are sufficiently high with minimal transport of gaseous TiO/TiO_2 unless temperatures are well above 2000°C. With an increasing interest in ash as a source of materials, titanium may be a candidate for extraction (Unver and Terzi, 2018).

As a result of titanium's minimal chemical options, the general conception, which is correct, is that it will be retained in the ash in combustion, be readily controlled and be of little interest. However, as often occurs in research, surprises occur and nothing was lost by such studies. Equilibrium calculations for coal combustion imply that under all conditions from 25°C to 2000°C the rutile form of TiO_2 will be dominant and possibly form titanates in the ash (Frandsen et al., 1994; Yan et al., 1999). As expected, slag, bottom ash or fly ash indicate similar titanium concentrations with no enrichments apparent (Kauppinen and Pakkanen, 1990; Meij, 1994; Querol et al., 1995; Martinez-Tarrazona and Spears, 1996; Font et al., 2010). A recent study has shown an effect on the fly ash composition of reducing the flue gas temperature from 250°C to 150°C by passage through a heat exchanger. The fuel was biomass from metal contaminated soils burned in a small pilot-plant boiler.

Boiler temperatures are not quoted but a fivefold increase in titanium's ash concentration appeared to occur, and for many other elements as the temperature was so reduced through the heat exchanger. Further metal deposition is impossible for such an increase in concentration especially as similar increases were seen also for elements such as Zr, another nonvolatile element. If the data are correct, it must be that particle segregation is occurring in the heat exchanger and the ash size distribution is now favoring the small enriched sizes at the lower temperature (Kovacs et al., 2018). Mass balances indicate all the titanium is removed as ash with very minimal stack leakage of about 0.1% or less depending on the quality of particulate control and on a par with such elements as Ba, Mn, Ni, and Zn (Ondov et al., 1979; Nielsen et al., 2002; Goodarzi, 2006a). One study of retention by the FGD system with the co-combustion of coal and pet-coke indicated that this device was effective at removing at least 50% of any remaining stray Ti in small particles in the flue gas (Cordoba et al., 2012). MSW heated to 1000°C indicated no titanium loss (Pedersen et al., 2010) and even if mixed with $CaCl_2$ or Cl_2 and similarly heated for 3 hours no release was noted (Chan and Kirk, 1999). Several studies that examined potential particle enrichment as a function of size found that none was particularly apparent for titanium from the smallest nanosizes (0.03–0.06 μm) to larger (Coles et al., 1979; Querol et al., 1995; Martinez-Tarrazona and Spears, 1996; Wang et al., 2008; Wu et al., 2011a). It was noted though by using electron microscopy and other techniques that coal ash particles <0.1 μm indicated a presence of Fe, Al, and Ti with suggestions of titanate formation (Chen et al., 2005). However, the study by Baxter et al. (1997) of titanium behavior in coal particles was particularly noteworthy in relating to this and was a study that seems to have been overlooked for its far reaching implications. They simply examined the behavior of the titanium in coal particles during volatization. A small fractional weight loss of titanium was seen at temperatures of only 780°C–1000°C. This was found also to depend on the rate of particle heating. The only plausible explanation for such a release was that during coal volatilization numerous species are rapidly escaping from the particle. Depending on the conditions, this can actually physically induce other mineral matter loss. With titanium, losses could vary from zero to 16% of the original titanium mass. In other words, this confirmed the concept initially suggested by Helble et al. (1986) that 200–500 micron-sized particles could be generated from a single coal particle by such forceful shedding of mass. Now, taking a closer examination of the bimodal micron and submicron ash particle distributions this appears to explain titanium's presence in the very smallest ultrafine particles (Quann et al., 1982, 1990; Wang et al., 2008). Two studies examined particle sizes in the bimodal distribution that had peaks at about 0.05 and 2 μm and found similar behaviors with one sampling before and one after the 99.8% efficient removal ESP. The fine submicron particles (0.04–0.15 μm) were noted to contain about 4.5%–5.0% of the total titanium in all the particles <10 μm (Markowski and Filby, 1985; Kauppinen and Pakkanen, 1990). Also there was a slight depletion of titanium by a factor of 0.7 in the concentration of the fine relative to the coarse particles. Consequently, these studies of the nonvolatile titanium have helped to resolve a common source of nanosize particle formation from fuel particles in flames. These can act as the required nuclei for deposition of other volatile material and will contain a core of nonvolatile molecules from the original particle.

In summary, the known refractory thermal stability of TiO_2 dominates the behavior of titanium in combustion. It is retained in the ash and any extent of forming mixed metal titanates remains uncertain. Because there are no pollution or health concerns with titanium there has been little interest in its full-scale combustion behavior. A very major outcome has been the observation that disintegration of fuel particles can redistribute such nonvolatile elements. This explains how titanium can be detected in the submicron particles and become an effective nucleating center. Otherwise titanium is not affected by sulfur or chlorine interactions. It is not involved in flame or soot suppression chemistry, but surprisingly the atom can readily chemiionize with an O-atom to produce TiO^+ and an electron at ambient low temperatures. In flames it is seen to produce low levels of mainly the $TiOH^+$ molecular ion.

7.29 Tungsten

Tungsten (third row transition element) and molybdenum (second row) are chemically very similar but neither have a great deal of commonality to Cr the first member of their fourth column of the transition elements in the periodic table. Hexavalency is dominant and stable. The three elements are of interest for theory in that of the six outer shell electrons their lowest energy configurations are Cr($3d^5 4s^1$), Mo($4d^5 5s^1$), yet W($5d^4 4s^2$). Tungsten is ranked about the 58th in abundance in the Earth's crust (1.25 ppmw) and is essentially equal to that of molybdenum. Its levels in seawater are low, 100-fold below that of molybdenum. As indicated in Table 1.1, tungsten exists in normal fuels at levels much lower than molybdenum, and it is very sparse in biomass, and has never fostered much interest in combustion studies (Farber and Srivastava, 1973a; Schofield, 2003). It is at very low levels in petcoke and limestone (Cordoba et al., 2012). Its chemistry is a total reflection of molybdenum except displaying a greater thermodynamic stability overall. For comparison the physical properties of molybdenum's condensed and gaseous molecules have been indicated briefly in Tables 7.15, 7.28, and 7.29. For tungsten, some of its data were collected 50 years ago and although a little uncertain have not been much modified since. It is found in several ores, scheelite $CaWO_4$, and wolframite, $(Fe,Mn)WO_4$ being favored. About 87,000 tons year^{-1} are mined with 3.3 million in reserve. Cost is about \$25 kg^{-1} WO_3. China has about 75% of the world reserves. It has found many uses having the highest melting point of all the metals (3422°C) and boiling point (5930°C). Its thermal expansion matches that of borosilicate glass and so it is useful for metal-glass seals. Half the available tungsten is used to produce tungsten carbide an extremely hard material, stable to high temperatures, and also in alloys and steels. It is of value widely in heating elements and still exists as the filament in incandescent light bulbs. Its disulfide is a useful higher temperature lubricant.

Health concerns with tungsten

Tungsten is alien to the body but not known for toxicity nor a carcinogenic nature. It is the heaviest element known to be essential to some living organisms and is found in several bacteria and archaea. An extensive toxicological report has been published summarizing its occurrence in nature, the food supply and reported effects on animals and humans (Tox Report W, 2005). This raises little concern for humans other than for some industrial workers. Tungsten is indicated as interfering with Mo and Cu metabolism and it is somewhat toxic to animals. It can have elevated concentrations in drinking water in some geographical areas. There is little concern with health issues and it is rarely ever tested in surveys or at test centers. However, one survey of 100 healthy volunteers in France did compare body biomarkers finding blood $<0.08\ \mu g\ L^{-1}$, plasma $<0.75\ \mu g\ L^{-1}$, urine $<0.09\ \mu g\ L^{-1}$ and hair <7 ppbw (Goulle et al., 2005). It is included in the US Annual National Test Surveys as a urine test. Over the years, there have been only slight changes noted and for the 95% percentile of the population, the 2010 urine levels indicated $\leq 0.46\ \mu g\ L^{-1}$ (CDC 4th National Report, 2009, 2018). Levels in the United States appeared slightly higher for smokers but also a decline with age. It is obviously present in the human diet but apparently at negligible levels of no concern. A 1996 World Health Organization report on trace metals in nutrition and health chose not to include it. However, due to its presence in soils, plants and sewage sludge there remains an agricultural interest in its quantity (Fu and Tabatabai, 1988). No estimates appear of the natural or anthropogenic tungsten levels in the atmosphere.

Tungsten's basic chemistry

As with Mo its chemistry centers on the trioxide, and also its tungstates. Otherwise it is very limited, closely reflecting molybdenum. Hexavalency is favored. The WO_2(s) disproportionates completely to WO_3 and W below its melting point with vaporization then to $(WO_3)_n$ polymers (Blackburn et al., 1958). The WO_3(s) solid vaporizes congruently producing these polymers that was further confirmed (Berkowitz et al., 1957; Marushkin et al., 1985).

Very similar to Mo, the chlorides and oxychlorides, including WCl_6 that does exist in this case, are all volatile, but subject to dissociation and disproportionation one into another on heating, and are readily oxidized and hydrolyzed. The sequence for this modification generally is WCl_6, $WOCl_4$, WO_2Cl_2 to WO_3. A disulfide exists and is stable to 500°C before dissociating to W and sulfur. No hydride is stable at ambient pressures and temperatures. Forms of WH_n have been found stable but only at elevated pressures. A useful carbonyl $W(CO)_6$ can be made that is in condensed form to 170°C when it sublimes and dissociates.

As summarized in Table 7.40 and discussed below, these molecules and numerous tungstates similar to molybdates exist that exhibit significant stability. Studies of gasification of the ore Hubnerite, $MnWO_4$ with molecular Cl_2, and its reaction with Na_2WO_4 produced gaseous tungsten oxychlorides at >700°C (Fouga et al., 2012; Saeki et al., 1975). However, gaseous HCl, more relevant to combustion, generally has no effect on metal tungstates. As a result tungsten has a rather limited condensed phase chemistry relative to combustion interests and is centered mainly on the trioxide and potential tungstates.

TABLE 7.40 Properties of tungsten's condensed phase molecules relevant to combustion, °C.[a]

	m.pt. b.pt. vapor pressure
W	3422 5930, 10 ppmv/3204°C, ins
WO_2	1700, disprop 1530 below m.pt. to $W/(WO_3)_n$, ins
WO_3	1473 1840, 13 ppmv/1120°C, ins
H_2WO_4	120 decomp, ins
WCl_2	volatile, disprop 589, r with wat
WCl_3	sublimes, disprop 550, oxidizes to WO_3, r with wat
WCl_4	498, disprop to WCl_2, WCl_5, r with wat
WCl_5	230 288, disprop, r with wat
WCl_6	275 347, disprop >240 to WCl_5 and Cl_2, r with wat
WO_2Cl_2	volatile, >265 sublimes, disprop, r with wat
$WOCl_3$	240 disprop, r with wat
$WOCl_4$	210 228, disprop, r with wat
WH	unstable at atm. press.
WS_2	1250, dissoc >500 to W/sulfur, air oxidizes, ss
$W(CO)_6$	170, sublimes, decomp, ins
$Al_2(WO_4)_3$	stable to 850, ins
$CaWO_4$	1580, mol vapor observed, ss
Ca_3WO_6	2250, ins
$FeWO_4$	990, mol vapor observed, ins
$Fe_2(WO_4)_3$	dissoc >600 to Fe_2WO_6, s
Fe_2WO_6	1100
K_2WO_4	916, mol vapor observed, s
$MgWO_4$	1410, mol vapor observed, ss
Na_2WO_4	698, mol vapor observed, s

[a] *See text.*
dissoc, dissociates; *decomp*, decomposes; *disprop*, disproportionates; *vap*, vaporizes; *s*, water soluble; *ss*, slightly soluble; *ins*, water insoluble; *r with wat*, reacts with water; *mol vapor*, molecular vaporization.

Gas phase stabilities and transport

Due to the complex vaporization of WO_3 to polymers, the thermodynamic values for WO (g), WO_2(g) and WO_3(g) were obtained from reaction balances in Knudsen cell-mass spectrometric studies. The values clearly show the greater bonding strength of W over Mo and Cr. They were evaluated in thermodynamic tables (Chase, 1998) and were based on the three pivotal studies (Chupka et al., 1959; DeMaria et al., 1960; Ackermann et al., 1960). Matrix isolation and gas phase studies isolated WO, WO_2, and WO_3 obtaining infrared bond frequencies and geometric data (DeVore and Gallaher, 1986; Bare et al., 1998). Although early theory had indicated a possible $X^5\Pi$ state for WO(g) (Nelin and Bauschlicher, 1985), experimentally this has now been shown to be $X^3\Sigma^-$ with only its spin components as the low-lying states (Ram et al., 2001, 2001a). Several low-lying states of WO_2(g) have been assessed from the ion photo-electron spectrum of WO_2^- (Davico et al., 1999).

The single mass spectrometric value for WO(g) (DeMaria et al., 1960) remains the only early value but was accepted and reassessed by Pedley and Marshall (1983) as D_0(W-O) = 668 ± 40 kJ mol^{-1}. Utilizing a reaction bracketing mass spectrometric method, new measures were obtained for the ionization potential of WO and the WO^+ dissociation energy. Coupling these data indicated D_0(W-O) = 716 ± 70 kJ mol^{-1} (Blagojevic et al., 2004). Also using the same technique, from the onset of the reaction of W^+ with CO, coupled to the ionization potential of W, this derived the value D_0(W-O) = 676 ± 30 kJ mol^{-1} (Hinton et al., 2009). This confirms the approximate magnitude and probably reduces the error limit range. In any measure of D(W-O) care has to be taken if ΔH_f(W) is involved. This is true for all the basic metal elements in that uncertainties still remain even in the basic element thermochemistry. In the present case the heat of formation of W(g) is the enthalpy of vaporization and even now still carries an error bar. Pedley and Marshall (1983) used a value of $\Delta H_0 = 848 \pm 4$, Chase (1998) 851 ± 6, Plante and Sessoms (1973) 859 ± 5, so it is surprising to find Knacke et al. (1991) quoting 828 kJ mol^{-1}, which appears to be a printing error. The Chase value has been taken herein. The heats of formation of WO_2 and WO_3, respectively, had been obtained from the two and three independent mass spectrometric studies (Chase, 1998), indicating $D_{0\text{Atomiz}}(WO_2) = 1263 \pm 30$ and $D_{0\text{Atoms}}(WO_3) = 1877 \pm 30$ kJ mol^{-1}. One additional independent mass spectrometric study (Balducci et al., 1981) compared the O_2W-O and CrO bonding, which with the latest D_0(Cr-O) = 445 ± 4 now implies D_0(O_2W-O) = 612 ± 5, an independent measure that compares to its database value of 614 ± 40 kJ mol^{-1}. This now lends credence to this body of thermochemical data, suggesting that the thermo-database values for WO_2 and WO_3 may be acceptable. From their atomization data the recommended bond strengths now are suggested as D_0(W-O) = 668 ± 20 kJ mol^{-1}, D_0(OW-O) = 595 ± 30 kJ mol^{-1}. D_0(O_2W-O) = 612 ± 5 kJ mol^{-1}. Any reduction in the value for WO will be reflected into an increase in OW-O. That for the third bond is now reliably measured independently. All three are strong bonds and might be expected to form a balance with each other in combustion.

As with molybdenum, metallic tungsten and its oxide can experience ablative enhanced rates of evaporation at higher temperatures with water due to the direct surface formation of $WO_2(OH)_2$, the monohydrate of WO_3 and known as gaseous tungstic acid, H_2WO_4 (Myers et al., 2019). With WO_3(s), in air and 10% H_2O vapor, the vapor pressure of $WO_2(OH)_2$ (g) was measured as 1 ppmv at 700 and 45 ppmv at 900°C (Glemser and Ackermann, 1963). For W(s), higher temperatures were needed and studied at 1200°C–1500°C (Belton and McCarron, 1964). However, as seen, in the presence of low levels of H_2, the tungsten metal will favor reaction with water. Vaporization of metal tungsten with steam has been reported at ≥ 800°C.

$$WO_3(s) + H_2O(g) = WO_2(OH)_2(g) \quad \Delta H_{1000K} = +178\ \text{kJ mol}^{-1},\ K_p = 1.5 \times 10^{-5}$$

$$W(s) + 4\,H_2O(g) = WO_2(OH)_2(g) + 3\,H_2(g) \quad \Delta H_{1500K} = +109\ \text{kJ mol}^{-1},\ K_p = 2.1 \times 10^{-3}$$

A mass spectrometric study also reported the slow oxidation of tungsten wire with traces of oxygen producing gaseous WO_2, WO_3, and polymer products that gradually dissociated at higher temperatures to W and WO (Schissel and Trulson, 1965). The hydroxides have been observed in matrix isolation studies showing a variety of structures (Wang and Andrews, 2006). The monohydroxide indicates an HWO structure, the dihydroxide is H_2WO_2 with four independent bonds about W, and the dioxydihydroxy the very stable $WO_2(OH)_2$ structure.

As with Mo, the chlorides and oxychlorides similarly, including WCl_6 exist, are all volatile, but subject to disproportionation and dissociation on heating, and are readily oxidized and hydrolyzed. On heating in air the progression tends to follow $WCl_6 \rightarrow WOCl_4 \rightarrow WO_2Cl_2 \rightarrow WO_3$. Disproportionality is involved in the rearrangements and occurs at low temperatures, <600°C. Information for these individual gaseous chlorides is difficult to isolate and bond strengths remain uncertain. It was noted that gaseous HCl can react with WO_3(s) to produce WO_2Cl_2(g) (Funaki and Uchimura, 1962). Values for the chloride bond strengths (WCl_n, $n = 1$, 2, and 4–6) are mainly found in the data base evaluations (Knacke et al., 1991; Chase, 1998). Until recently there was only the one estimate of D_0(W-Cl) $= 416 \pm 40$ kJ mol^{-1}, a rough estimate from the other WCl_n values. Now, a two-photon ionization spectroscopy method was used to accurately measure the predissociation threshold for gaseous WCl. This obtained an accurate measure of D_0(W-Cl) $= 368.4$ kJ mol^{-1}, $\Delta H_{f298=}$ $+600.6$ kJ mol^{-1} that is now recommended (Sevy et al., 2017). The known heat of formation for WCl_6(g) is based on the heat of formation of WCl_6(s) coupled to its enthalpy of vaporization. The latter now has been confirmed and is accurate (Cordfunke and Konings, 1992). As a result, this heat of formation is a reasonable measure of the sum of all the six bond strengths. From it, the atomization energy is 2059 ± 25 kJ mol^{-1} implying an average strength of 343 kJ mol^{-1} for each of the six bonds. The database intermediate values at present depend on either estimated heats of vaporization or energies of disproportionation reactions that remain questionable. A theoretical paper on the gas phase structures of WCl_3 through WCl_6 has raised new interest in the bonding of these chlorides (Schimmelpfenig et al., 2001). Coupling their calculated heats of formation for WCl_3 and WCl_4 to those in the database (Chase, 1998) the following approximate D_{0K} bond strengths are implied: 416, 686, 165, 376, 216, and 201 for the strength of the first (W-Cl) through sixth bonds (WCl_5-Cl). Even though their sum is consistent with the hexachloride atomization energy their individual values may not be meaningful especially as the distribution is unexpected. The corresponding six bonds in molybdenum are 370, 444, 350, 355, 222, and 84 kJ mol^{-1}. A slight strengthening of the second bond might be acceptable, coupled to a weakening for the final two bonds. With tungsten the implied values for the second and third bond strengths currently appears unacceptable. Taking the theoretical heats of formation for WCl_3 and WCl_4 indicates atomization energies for these of 1273 and 1630, respectively, but these cannot help to resolve the individual values. It does suggest though that the second and third bond strengths add to about 900 kJ mol^{-1}. Comparing the changes to molybdenum, it would not be unreasonable for them to have similar ratios for the first and second bond strengths, making such current estimates for the six bonds in WCl_6 now suggests 368, 450, 455, 360, 220, and 205.

It was noted that gaseous HCl or chlorine gas can ablate WO_3(s) more efficiently than water with the formation of WO_2Cl_2(g) (Funaki and Uchimura, 1962).

$$WO_3(s) + 2\,HCl(g) = WO_2Cl_2(g) + H_2O(g) \quad \Delta H_{1000K} = +103\ \text{kJ mol}^{-1},\ K_p = 3.4 \times 10^{-2}$$

This indicates a pressure of 34 ppmv for the dichloride at this temperature with 1% HCl and 10% H_2O. Both WO_3(g) and WO_2Cl_2(g) are seen to be more stable than their structures with Mo or Cr (Pershina and Fricke, 1996; Chase, 1998). As with molybdenum, if $WO_2(OH)_2$ (g) forms in flame gases with chlorine present, then a distribution between this and the dichloride will undoubtedly be established:

$$WO_2(OH)_2(g) + 2\,HCl(g) = WO_2Cl_2(g) + 2\,H_2O(g) \quad \Delta H_{1000K} = -75\ \text{kJ mol}^{-1},\ K_p = 2.3 \times 10^3$$

In other words, if HCl is 1% and H_2O 10%, WO_2Cl_2 could be 20-fold larger than the dihydroxide in this case at such conditions. Thermodynamically, both molecules appear to be thermally stable to high temperatures, but on approaching 2000°C they will begin to dissociate and revert to WO_3(g) with its own range of stability. For the above calculations, recommended values have been assumed as $\Delta H_{f298}(WO_2(OH)_2) = -906 \pm 4$ kJ mol^{-1} and $\Delta H_{f298}(WO_2Cl_2) = -672 \pm 25$ kJ mol^{-1} (Chase, 1998).

The disulfide is stable only to about 500°C, dissociates to W(s) but also can be oxidized and hydrolyzed on heating. The three sulfides, WS_{1-3}, have been observed in matrix isolation and the bond strengths of the gaseous mono- and disulfide were approximately estimated theoretically and are proportionally weaker than their oxides as expected, D(W-S) $\approx$ 420, D(SW-S) $\approx$ 484 kJ mol^{-1} (Liang and Andrews, 2002b). Recently a two-photon ionization spectroscopy method was used to more accurately measure the predissociation threshold for gaseous WS. This has now obtained the reliable value of D_0 (W-S) = 476.2 kJ mol^{-1} (Sevy et al., 2017). As noted in Table 7.15, from the trend, this might imply a stronger MoS bond than there indicated. At high temperatures these oxidize to the oxides. There has been theoretical interest in the possible WH(g) molecule due to its heavy-transition metal nature. A matrix isolation study has noted the WH_n ($n = 1-4$, 6) species deriving their frequencies, structures and estimating bond strengths (Wang and Andrews, 2002). Theoretical studies have now established the electronic states and potential energy curves for WH suggesting $X^6\Sigma^+$ and few low-lying states (Ma and Balasubramanian, 1991; Koseki et al., 2006). A value of D(W-H) = 240 ± 20 kJ mol^{-1} appears appropriate. Also like Mo a hexavalent carbonyl is known, $W(CO)_6$, but decomposes on heating above its melting point. Bond strengths for these gaseous tungsten molecules are summarized in Table 7.15.

Potential role for tungstates

As shown in Table 7.41, a list of gaseous tungstate stabilities is illustrated for comparison with those of molybdates (Table 7.29). They display the same behavior with low vapor pressures until above their melting points: Cs_2WO_4 (5 ppm 1000°C), $CaWO_4$ (0.3 ppm 1730°C), $BaWO_4$ (2 ppm 1890°C), $CrWO_4$ (1 ppm 1730°C). Vaporization is generally congruent and molecular but depending on the conditions may fragment.

TABLE 7.41 Thermochemistry of gaseous tungstates, kJ mol^{-1}

Molybdate m.pt.°C	Experimental Temp.°C	$-\Delta H_{f298}(g)^a$	$D_{298}{}^b$ Atomization	$D_{298}{}^c$ Metal oxide-MoO_3	References
Li_2WO_4	1100–1260	1010	3176	550	Yamdagni et al. (1970)
735	–	1044	3211	584	Choudary et al. (1975)
	–	1048	3214	588	Lopatin and Shugurov (2013)
Na_2WO_4	–	1149	3211	?	Choudary et al. (1975)
698	960–1110	1157	3219	?	Kazenas et al. (2001)
	–	1142	3205	?	Lopatin and Shugurov (2013)
K_2WO_4	2000	1164	3190	?	Farber and Srivastava (1973a)
916	–	1185	3210	?	Choudary et al. (1975)
	–	1174	3200	?	Lopatin and Shugurov (2013)
Rb_2WO_4	–	1201	3211	?	Choudary et al. (1975)
955	–	1225	3235	?	Lopatin and Shugurov (2013)
Cs_2WO_4	–	1207	3207	?	Choudary et al. (1975)
962	740–940	1234	3235	?	Lopatin et al. (2000)
$BeWO_4$	–	795	2967	636	Lopatin and Shugurov (2013)
$MgWO_4$	1830	891	2885	634	Verhaegen et al. (1965)
1410	1590–1860	867	2861	609	Kazenas et al. (2004a)
	–	773	2768	517	Lopatin and Shugurov (2013)
$CaWO_4$	2030	971	2996	720	Verhaegen et al. (1965)
1580	1850–1950	898	2924	648	Lopatin et al. (2003a)
	1590–1720	989	3015	738	Kazenas et al. (2005)
$SrWO_4$	1930	1044	3056	759	Verhaegen et al. (1965)
1535	1955–2135	962	2971	675	Lopatin et al. (2003)
	1530–1620	1018	3030	734	Kazenas et al. (2007)
$BaWO_4$	1730–1840	1032	3059	621	Pupp et al. (1969)
1575	1840–1980	1025	3052	614	Lopatin et al. (2003)
$TiWO_4$	–	750	3071	517	Lopatin et al. (2014)
VWO_4	1630–1830	706	3069	555	Gunina et al. (2012)
$CrWO_4$	1650–1770	617	2862	514	Lopatin et al. (2003)
	–	645	2891	543	Lopatin and Shugurov (2013)
$MnWO_4$	1580–1710	613	2744	482	Stolyarova et al. (2004)
	–	635	2763	502	Lopatin and Shugurov (2013)

(Continued)

TABLE 7.41 (Continued)

Molybdate m.pt.°C	Experimental Temp.°C	$-\Delta H_{f298}(g)^a$	$D_{298}{}^b$ Atomization	$D_{298}{}^c$ Metal oxide-MoO_3	References
$FeWO_4$	1726–1800	521	2785	490	Stolyarova et al. (2004a)
990	–	543	2807	513	Lopatin and Shugurov (2013)
$CoWO_4$	1440–1660	477	2750	486	Lopatin and Shugurov (2013)
1280					
$NiWO_4$	–	479	2757	493	Lopatin and Shugurov (2013)
1270					
BWO_4	–	1051	3460	761	Lopatin and Shugurov (2013)
$GeWO_4$	–	760	2980	432	Lopatin and Shugurov (2013)
> 1000					
Ga_2WO_4	1285–1390	819	3211	428	Lopatin et al. (2009)
$SnWO_4$	980–1130	778	2927	512	Bulova et al. (2002)
$PbWO_4$	–	750	2793	530	Nalivaiko and Rat'kovskii (1981)
1123	–	736	2779	516	Lopatin and Shugurov (2013)
$EuWO_4$	–	986	3009	636	Lopatin and Shugurov (2013)

[a]*Known gaseous heats of formation.*
[b]*Heats of full gaseous atomization.*
[c]*Dissociation energy into the two constituent oxides.*

The gaseous tungstates are unusual in being extremely stable, even more so than the molybdates. As noted in the Table, from the atomization energy and the two oxide individual bond strengths, the added strength of the union bonding between the two oxides explains this stability.

Comparing molybdates (Table 7.29) to tungstates, and allowing for the difference in bond strengths of WO_3 and MoO_3, it is noteworthy that for alkali and alkaline earth molecules the tungstates gain on average an extra 90–100 kJ mol^{-1} of union bond strength in the metal oxide-WO_3 structure over what is seen for the molybdates. With tungstates this gaseous union bond strength can be 600–700 kJ mol^{-1}, remarkably large and implying that gaseous tungstates can certainly exist in flame gases if produced, and more so than molybdates. The most stable, yet found, appears to be BWO_4 followed by the alkaline earths. The one rare earth measured, $EuWO_4$, also is quite noteworthy. In combustion, the role of tungstates and molybdates may be more useful in providing ways to modify chemical flame processes. This will be seen below in laboratory flames by preventing alkali sulfate/chloride corrosive deposition. Additions of tungsten salts modified any such corrosive deposition by changing this to benign tungstates and polytungstates (Schofield, 2003). The composition of ash will also be heavily controlled by thermodynamic preferences and if possible tungstates will certainly form, even transforming other species.

Laboratory combustion studies

There has been little interest in tungsten in combustion. An initial study was mainly concerned with its molecular electron attachment ability (Jensen and Miller, 1970a; Goodings, 1992). $W(CO)_6$ or WF_6 were added to fuel rich $H_2/O_2/N_2$ flames at 1815K–2475K, seeded with a potassium salt solution or for their mass analysis study directly as KWO_4 solution to avoid probe sampling problems. The presence of tungsten reduced the levels of free potassium atoms. They suggested the presence of gaseous WO_3, H_2WO_4, and $KHWO_4$ in the flames. A similar study using probe mass spectrometric sampling of a flame at 2300K repeated this work (Farber and Srivastava, 1973a). The flame species WO, WO_2, WO_3, H_2WO_4, $KHWO_4$, and K_2WO_4 were recorded. The reliability of the probe sampling has always been questioned and whether heterogeneous deposition in the probe biased the data is unclear. Homogeneous formation of gaseous K_2WO_4 will be kinetically demanding. However, as seen in Tables 7.29 and 7.41, the values derived in their studies for the heats of formation of K_2WO_4 and also K_2MoO_4 in the same study were perfectly acceptable. The answer may be that these tungsten species are so dominant in the flame gases that their concentrations can only be slightly perturbed and still maintain a meaningful measure of their speciation. It is to be noted in such systems that the enthalpy is extracted from an exponent term so an error of twofold in an equilibrium constant changes the heat of formation by only 12 kJ mol^{-1}.

Various metallic molecules in gaseous forms have been added to fuel-rich $H_2/O_2/N_2$ flames and, in many cases, this has clearly displayed a catalytic recombination of the flame radicals with combustion inhibition. WF_6 was seen to show some effect in an early screening of metals (Miller, 1969). This was later studied in greater detail for tungsten (Jensen and Jones, 1975). The additive then was $W(CO)_6$ with a range of fuel-rich hydrogen flames spanning temperatures 1800K–2150K. A noticeable reduction of H-atom concentrations was monitored with observed effects even with only 10 ppmv additions. A still questionable catalytic kinetic cycle invoking $WO_3/HWO_3/H_2WO_4$ recombining H-atoms was proposed. Whether this is totally homogeneous gas phase or heterogeneous on particles remains speculative (Linteris et al., 2008). It is regarded though as a moderate flame inhibitor.

Although minimally mentioned as modifying soot formation in an overall review on combustion soot control (Howard and Kausch, 1980), and one paper indicating no effect with additions of 1 ppm (Haynes et al., 1981), earlier publications had indicated noticeable effects with Na_2WO_4 additions (Cotton et al., 1971). With Ba as the benchmark species rated at 100; Mo was 43 and W 18 in diffusion flames. Later similar work raised this to 37 (Bulewicz et al., 1975). It was then believed that atomic or molecular metallic ionization might be the mechanism, but this is ruled out for tungsten that has too high an ionization potential and does not readily chemiionize (Schofield, 2006). It was finally considered that the metal may become occluded in the soot particles with enhanced catalytic oxidation. However, the actual mechanism remains uncertain for most of the effective metals noted.

As with other metals the formation of nanoparticles is of important interest. Using WF_6 addition to low pressure hydrogen flames the synthesis of WO_3 is possible, one study even examining the slight stoichiometric changes resulting at different flame equivalence ratios (Gupta et al., 2011). Additionally, complex nanostructures can be produced by tungsten doping of MoO_3 flame formation (Cai et al., 2013).

A further detailed, quantitative laboratory study did resolve the nature of tungsten's deposition chemistry mirroring that done with molybdenum. The program's aim was to find a method for preventing Na_2SO_4 deposition and corrosion, which can occur in any combustion system even with sub-ppmv concentrations of sodium and sulfur at temperatures below 1000°C. The condensed phase sulfate is very thermodynamically stable and it is difficult to disrupt its chemistry. After a long assessment of searching the periodic table it was found that additions of tungsten could be effective in modifying the observed deposition. Instead of depositing as the sulfate, any sodium or potassium in a flame was seen to preferentially favor formation of a polytungstate deposit that was thermodynamically more stable (Schofield, 2003). These molecules are known as alkali bronzes and are benign at high temperatures with regards to corrosion. Both sodium and potassium behaved identically. To fuel-lean propane/O_2/N_2 flames at about 2000°C, quantitative additions of Na or K, and W salt solutions could be nebulized into the flame gases together with SO_2 and Cl_2 all on ppmv scales. An air-cooled steel probe at about 500°C−600°C, located a few milliseconds downstream in the flame gases, collected deposited samples. The deposits were analyzed by FT Raman spectroscopy and X-ray powder diffraction. The deposits remained white as the sulfate but now were clearly different in nature. FT Raman is a powerful method of analysis if there are no interferences in the spectrum. All the sodium and potassium polytungstates were synthesized and together with pure samples of acquired Na_2SO_4 and WO_3 all their individual spectra were recorded. These are illustrated in Fig. 7.14 for the potassium case and in Fig. 7.15 for three of the sodium tungstates that are easily resolved spectrally. Strangely the tritungstate, $Na_2W_3O_{10}$ is unstable and unknown yet the potassium salt can be made.

What is immediately noted is that these molecules have few frequencies that are Raman active giving each molecule a clear fingerprint, for comparisons. The spectrum for sulfates is often simple with one strong frequency at about 1000 cm^{-1}. Flames were burned that contained sulfur and chlorine and for which the atomic alkali/tungsten atomic ratio was

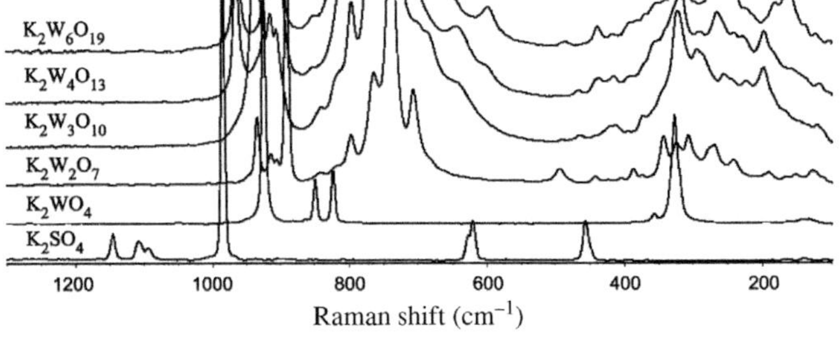

FIGURE 7.14 Fourier transform Raman spectra of purchased samples of pure K_2SO_4 and K_2WO_4 and synthesized samples of four polytungstates. Source: *Reproduced with permission from the American Chemical Society, from Schofield, K., 2003. A new method to minimize high-temperature corrosion resulting from alkali sulfate and chloride deposition in combustion systems. I. Tungsten salts. Energy Fuels 17, 191–203.*

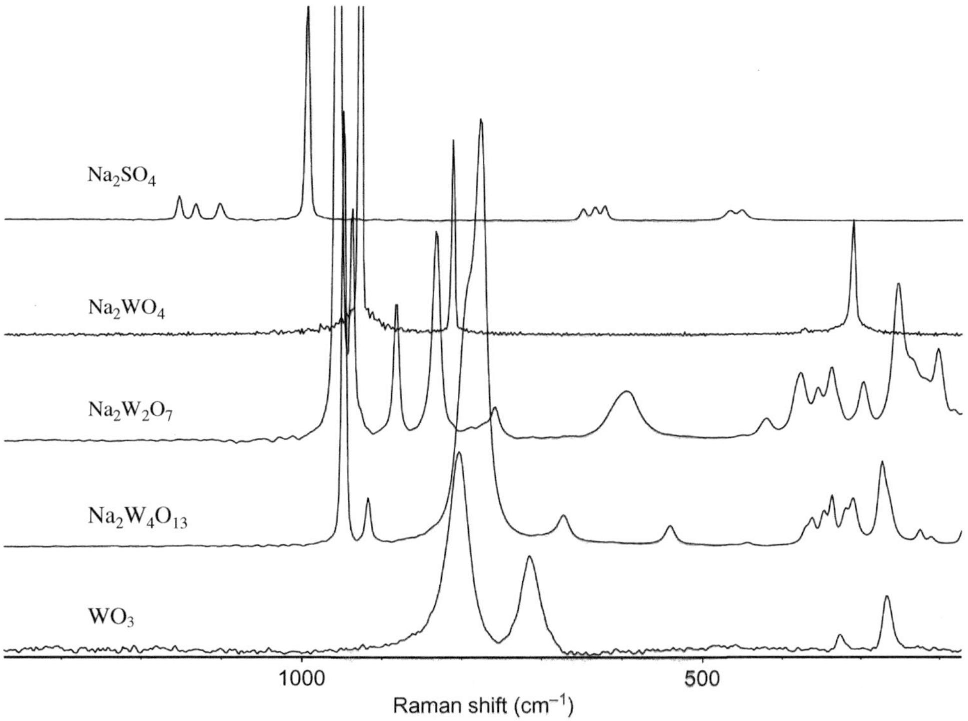

FIGURE 7.15 Fourier transform Raman spectra of purchased samples of pure Na_2SO_4, Na_2WO_4, and WO_3 and synthesized samples of the polytungstates, $Na_2W_2O_7$, $Na_2W_4O_{13}$; $Na_2W_3O_{10}$, is unstable. Source: *Reproduced with permission from the American Chemical Society, from Schofield, K., 2003. A new method to minimize high-temperature corrosion resulting from alkali sulfate and chloride deposition in combustion systems. I. Tungsten salts. Energy Fuels 17, 191–203.*

accurately controlled. What was noted was that as the quantity of tungsten salt was increased to equal that of the sodium, the sodium sulfate deposit was reduced to near zero (Fig. 7.16) and the morphology of the deposit changed in appearance. It became predominantly the ditungstate, $Na_2W_2O_7$.

When the tungsten was twice that of the sodium on an atomic basis, a pure deposit of $Na_2W_4O_{13}$ is resulted. The same behavior was observed with molybdenum. These have a range of melting points, tungstates higher than the corresponding molybdates. They gradually decrease with increased complexity for molybdenum dropping to about 550°C for the tetramolybdate, but are variable for tungsten ending at about 800°C–870°C, but these are all lower than the sodium, potassium sulfate values, 884°C, 1069°C, respectively (Schofield, 2009). These formations were additionally identified by X-ray analysis and indicate a high level of purity. Whether chlorine was present or not made no apparent difference. Moreover, it was found if conditions are changed that any preformed or forming deposits will modify to the new thermodynamic preference in the new environment. In other words starting solely with sodium in a hydrocarbon flame a deposit of Na_2CO_3 will be obtained. Then, if chlorine is introduced to the same flame the full deposit changes

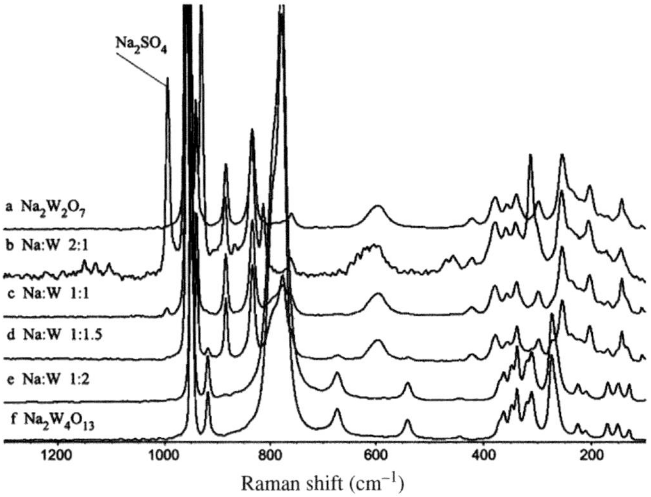

FIGURE 7.16 Fourier transform Raman spectra of synthesized samples of (a) pure $Na_2W_2O_7$ and (f) $Na_2W_4O_{13}$. Spectra (b)–(e) are deposits from flames containing Na (25 ppmv), SO_2 (75 ppmv), and W salts in the atomic proportions indicated, illustrating the inhibition of Na_2SO_4 formation and polytungstate formation. Source: *Reproduced with permission from the American Society of Mechanical Engineers, from Schofield, K., 2009. Alkali control for gas turbines by a combined hot gas clean-up approach. In: Proceedings of ASME Turbo Expo 2009: Power for Land, Sea and Air held in Orlando FL. American Society of Mechanical Engineers, New York. Paper No. GT2009-60254, 8 pp.*

irreversibly to NaCl and continues as such. Additions of sulfur can then convert all this to the sulfate, and as seen, further addition of tungsten or molybdenum will convert this to a polytungstate or polymolybdate. Surfaces have preferential rankings that relate to their thermodynamic stabilities. This was an important discovery that resolved the nature of deposition from flame gases and was pertinent to ash chemistry in combustion. Moreover, this is heterogeneous surface chemistry, not gas phase related, and these very complex poly-molecules seen with tungsten could not possibly form homogeneously (Schofield, 2005a). They also appear to form rapidly with no apparent kinetic constraints. Relating to practical combustion with complex fuels such as coal, this illustrated that deposition can be modified continually as conditions change time permitting. In the present program potassium or sodium clearly favored tungsten over sulfur. Nevertheless, when a propane flame was examined containing traces of Na/W/Ca/S/Cl the sodium polytungstate previously deposited was no longer formed, but as apparent in Fig. 7.17, now one reflecting $CaWO_4$. Calcium obviously can trump sodium or potassium thermodynamically by having a greater affinity for the tungsten. It has first claim to the tungsten, and the alkali then has to suffice with sulfur again producing K_2SO_4. Higher levels of W addition then are necessary to overcome such a Ca interference. This is clearly apparent from Table 7.41 that lists the relative thermochemical stabilities of the tungstates, where it was clearly displayed that the alkaline earth tungstates were more stable than those of the alkali elements.

This emphasizes the difficulty of assessing the final products that will form in combustion ash because the full thermodynamic database often still remains incomplete. This limits our predictive expectations of the order of the possible preferential rankings when all the species present require consideration together. The pairings of the species, their concentrations and the temperature have to be assessed before a final correct thermodynamic ranking can be determined. What was clearly demonstrated in this study was that deposition is a surface formation mechanism that can readily modify if necessary to a more thermodynamically favored molecule. Also these deposits if dominant were noted as being

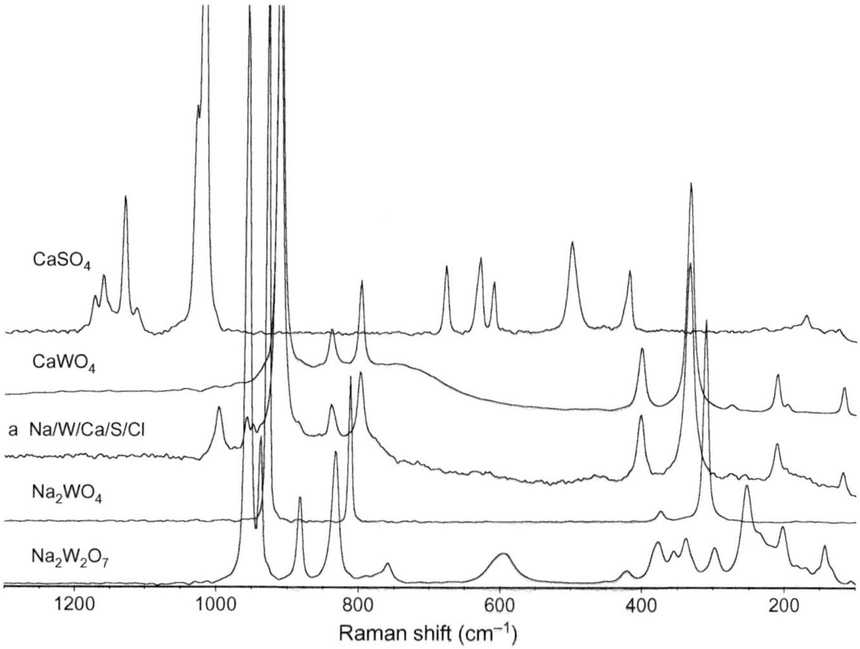

FIGURE 7.17 Fourier transform Raman spectra of purchased samples of anhydrous $CaSO_4$, $CaWO_4$, Na_2WO_4, and synthesized $Na_2W_2O_7$. (a) is a spectrum of a deposit from a flame containing Na:W:Ca in the atomic ratio 1:2:2 (Na = 10 ppmv) and 75 ppmv SO_2 and 50 ppmv Cl_2, showing dominant formation of $CaWO_4$. *Source: Reproduced with permission from the American Society of Mechanical Engineers, from Schofield, K., 2003b. A new method to control high temperature corrosion in gas turbines resulting from alkali sulfates. In: Proceedings of ASME Turbo Expo 2003: Power for Land, Sea and Air held in Atlanta Georgia, USA: American Society of Mechanical Engineers, Paper No. GT2003-38782, 7 pp.*

individually chemically very pure. However, in connection with flames, the main commercial interest remains in formation of, for example, WO_3 nanoparticles (Gupta et al., 2011; Kluge et al., 2017).

Combustion power plant data

As indicated in Table 1.1, tungsten exists in normal fuels at levels much lower than molybdenum, is very sparse in biomass, and not much higher in wastes or sludge. It is at very low levels in pet-coke and limestone (Cordoba et al., 2012). It has never fostered much interest in combustion studies other than the several laboratory studies outlined above (Farber and Srivastava, 1973a; Schofield, 2003). Due to tungsten's small presence in fuels there is little mention of it in full-scale combustion data. But surprising, from very early studies of coal combustion it was included in their analyses, and immediately noted to be volatile, being enriched in the 0.2 μm size particles (Coles et al., 1979; Ondov et al., 1979). In other detailed studies the tungsten was seen to be distributed evenly between the fly ash and slag with strong enrichment in the 0.2–2.5 μm fly ash size (Querol et al., 1995; Font et al., 2010). Also any particles after the ESP collection were further enriched downstream (Meij, 1994). Due to the use of coal fly ash in ultrafine cement this enrichment of various elements in coal combustion was recently repeated showing this enrichment for tungsten in the 2.2 μm size fraction (Lanzerstorfer, 2018). Any volatile tungsten appears to have sufficient time to be collected onto some ash in the flue trail. Recent studies of the FGD scrubbers also indicated this device is very useful as a final fine particle control device removing any remaining gaseous fine particles (Cordoba et al., 2012). With tungsten it was noted that more than half of such final remaining tungsten traces are removed indicating negligible stack release of any tungsten. Lower temperature fuel pyrolysis studies involving tungsten appear not available and the few combustion studies all involve the higher temperatures of coal combustion. This does appear to vaporize any tungsten transport into the combustion gases and may involve some formation and probable balance between gaseous WO_3, $WO_2(OH)_2$, and WO_2Cl_2 that remains unresolved. That tungsten is released from fuels is a little surprising due to its low vapor pressures but it may invoke some ablative corrosion by water or HCl gases, but certainly at higher temperatures only. These will deposit on cooling ash and either form tungstates that are refractory, and thermodynamically very stable, or remain as stable WO_3(s). Deposition can start at elevated temperatures so available time is lengthened. With such non-volatile elements, as with titanium, whether a transport role can be played by particle disintegration remains unclear. One leaching study did note that there was water leaching of tungsten from coal fly ash (Llorens et al., 2001). The oxides are insoluble so whether this was from an alkali tungstate formation or other water soluble metal tungstate remains unknown. The content of traces of tungsten in sewage sludge in connection with agricultural usage is of some importance due to the antagonistic effect of tungsten on molybdenum plant nutrition (Fu and Tabatabal, 1988; Eriksson, 2001). Otherwise there appears to be little interest in leaching tungsten from either coal or ash.

In summary, little is known concerning tungsten's full-scale combustion behavior. It does seem to closely mirror that of molybdenum showing significant volatility and transport onto fly ash deposition. This seems hard to explain, being expectingly non-volatile. However, this will only be apparent in higher temperature coal combustion, when even then, a fraction remains in any bottom ash and any final atmospheric release centers only on the efficiency of the ash control devices. The oxide or tungstates similarly will control the chemistry but due to its low presence in fuels there remains little interest in even any minor role it may play.

7.30 Uranium

Although outside the realm of most chemists, with no listings in the NIST-JANAF thermochemical tables (Chase, 1998), uranium is simply another element, but one whose molecules are radioactive. It is not rare, being about the 47th most present in the Earth's crust and about 25th and quite abundant in the oceans. A recent report indicates the world's oceans contain about 4 billion tons of dissolved uranium. The original source of uranium in the Earth and universe appears to remain unknown as is the volume of water on earth at the other end of the molecular weight scale. Uranium is widespread and mined globally in those areas where concentrations are conducive for marketable extraction. Due to its decay product radon, mining tends to be either open cast or by leaching methods. It occurs in many ores, pitchblende, the UO_2/U_3O_8 oxide form marketed as "yellowcake" is the most common but other ores that consist of Si/U, Ti/U, K/V/U, Ca/V/U, and Cu/P/U structures are known. Its detection initially was apparent using Geiger Counters but now can be mapped aerially by sensitive gamma-ray sensors. Coals can contain about 0.5–5 ppmw and certain U-rich lignites have even been used as a source of U_3O_8 in their ash. Moreover, active research fields exist, that now can extract by filtration/absorption the uranium content of sea water almost on a financially competitive basis. A new chelator/adsorber has recently emerged that is highly selective and easily regenerated (Jacoby, 2019). At the present, about 78,000 tons of the ore are marketed annually, predominantly for electricity production by the nuclear industry. Its price varies as a result of the public's acceptance or rejection of this energy source but growth worldwide does continue. Since 2000, prices have varied very significantly from about \$10–130 lb^{-1} and currently are at about \$20, a price required for continuing commercial production. Demand is predicted to continue its growth. It has many radioactive isotopes being in nature 99.3% as ^{238}U, the most stable of its atoms with a half-life of 4.5 billion years. The remainder is mainly the ^{235}U (0.7%, half-life 0.7 billion years) that is suited for energy capture. Gained knowledge of the dangers of radiation has largely eliminated other uses of uranium such as its coloration of glass, ceramics and in porcelain. So-called spent uranium with levels of radioactivity below the natural ore is generated as a by-product by the nuclear industry. This has become favored in replacing lead as the heavy metal needed in certain armament shells, denser by about 1.7-fold. Although such use still is questioned, it remains in the armaments of at least 20 countries and is regarded as legal in war, particularly by the United States, France and United Kingdom. Because uranium is present in many sources, its natural levels are ever present in the environment. Its composition approximates to the three main isotopes ^{234}U 0.0055%, ^{235}U 0.72%, and ^{238}U 99.27%. However, depleted uranium is correspondingly about ^{234}U 0.001%, ^{235}U 0.2%, and ^{238}U 99.8%. As a result, depending on the question being asked, it can be very difficult to distinguish between these two sources (Sahoo et al., 2014). Certain isotopes such as ^{236}U are only produced in nuclear reactors and so this can become very useful for confirming an anthropogenic source and identify that of earlier nuclear tests (Quinto et al., 2013). Atmospheric anthropogenic deposition pollution can arise from nuclear power plant accidental releases such as with Fukushima, Japan, from coal power plant emissions and from the use of depleted uranium as discussed already. However, accidental nuclear power plant releases generally will involve other volatile radionuclides and not uranium. These could be ^{137}Cs (half-life 30 years) and ^{134}Cs (2 years) and can be readily measured, shown to be pollutants and tested for, with regards to food safety (Merz et al., 2013). As noted, depleted uranium is rather hard to distinguish from a natural background unless a significantly raised level is apparent. Emissions from coal plants would be expected to pollute the neighborhood of a plant location by deposition. One recent study in Spain monitored many soil samples to a depth of 30 cm and in an area of 64 km^2 about the plant location finding no evidence of any kind of uranium enrichment (Charro et al., 2013). Another report, concerned about a possible pocket of child leukemia, studied coal-fired plant radioactive emissions in the United Kingdom. The discharges were found to be negligibly low (Wan and Wrixon, 1988). One recent analysis of ice core records from Mount Everest has shown a slight increase since the 1950s in the environmental atmospheric uranium content that possibly reflects increased mining output (Kaspari et al., 2009). In some countries, there is growing concern with the natural levels of minerals in drinking and wellwater. The excessive use of groundwater can modify its transport and aggravate the situation. One recent report of groundwater measurements that included uranium, raised significant concerns in the area of north-west India (Coyte et al., 2018). There also appear more concerns now concerning possible deposition of uranium from large scale coal combustion and more monitoring is beginning to emerge.

Health concerns with uranium

Although some bacteria have been noted that can accumulate uranium, most food plants have low content. It tends to concentrate in plant roots and so is quite evident in root crops such as potatoes, yams, radish, parsnips and turnips. Human diets have been estimated to contain about 0.9–1.5 $\mu g\ day^{-1}$. The World Health Organization has suggested a tentative drinking water guideline value of 15 $\mu g\ L^{-1}$, but this can vary geographically (WHO Guidelines, 2006a). A very extensive toxicological report on uranium and its compounds has been published (ToxProfile U, 2013). Water does vary and a recent study in Kabul, Afghanistan found it was at least an order of magnitude above recommended levels in some drinking water wells (Kato et al., 2016). An additional review illustrates that although uranium is not readily absorbed by the body this is in fact very sensitive to particulate size. Inhalation of UO_2 (0.2–0.6 μm) small particles can enter the blood stream readily, and in excess, lead to kidney failure (Lawrence et al., 2014). Uranium's target organs are the kidneys and the bone, and it is mainly excreted. Retention in bone has a half-life of about 70–220 days. If required, chelation by sodium citrate and other agents is effective. It is not rated as carcinogenic nor a neurotoxin but is toxic. Over the years, it became necessary to develop monitoring methods particularly for workers connected to mining operations, for communities living near or on geologically rich uranium sites and also for the army to monitor soldiers subject to possible depleted uranium contamination. Blood, urine and hair became the obvious simple biomarkers. Of these, urine appears to be favored and currently is recommended as $<0.1\ \mu g\ L^{-1}$ for good health (Mayo Clinic, 2018). The US National Monitoring Survey also includes a urine test. This has not varied significantly over the last decade and is close to $<0.04\ \mu g\ L^{-1}$ at the 95% percentile (CDC 4th Natl. Rpt. 2018). It is included in few surveys but early ones of the general public conveyed the generally very low levels, blood <0.006, urine $<0.01\ \mu g\ L^{-1}$, hair <0.03 ppmw (Goulle et al., 2005; Cesbron et al., 2013). If you are not exposed in the course of employment there is little concern with dietary intake. Hair was seen to indicate very significant levels but did not correlate linearly with blood or urine samples (Byrne and Benedik, 1991). This was confirmed in a more recent study in the Balkans where hair samples ranged from 0.9 to 449 $ng\ g^{-1}$ and they quoted a range of $5-1280\ ng\ g^{-1}$ in other studies (Sahoo et al., 2014). It seems that with uranium, loss through the hair varies at individual rates. However, exposed workers did indicate elevated levels with any monitoring method, becoming larger the more exposed they were in the commercial process (Byrne and Benedik, 1991). Hopefully, industrial regulations have been imposed since that study as some process workers had 50 $\mu g\ g^{-1}$ in their hair and 5 $\mu g\ L^{-1}$ in their urine, compared to the control group with 10 $ng\ g^{-1}$ (hair) and 10 $ng\ L^{-1}$ (urine). One recent survey comparing the inhabitants of Morocco and the Canary Islands did indicate slightly higher blood levels of ≤ 0.2 and 0.13 $\mu g\ L^{-1}$, respectively, which might correlate to their separate water supplies (Henriquez-Hernandez et al., 2018).

Many medical studies have been made concerning two aspects of uranium. One is whether any possible health effects may arise from living adjacent to a mining area or in fact on top of a region rich in natural uranium deposits (Au et al., 1998; Guimaraes et al., 2010). The other involves army veterans who suffered contamination from depleted uranium in warfare and may still have fragments in them (McDiarmid et al., 2011, 2011a). These are difficult and generally inconclusive studies because any effects that may be present are not sufficiently severe to be apparent. Nevertheless, such peoples are being subject to low levels of radiation and do appear to show some evidence of DNA damage and recovery. Whether these imply consequences concerning overall health problems or longevity remains uncertain and such studies continue.

Uranium's basic chemistry

As with any element, uranium has an inorganic chemistry that compares to other elements and depends on its molecular properties at various temperatures and conditions. Uranium is a large atom with 92 electrons and many shells. It is the fourth (Ac, Th, Pa, U) in the actinide series that are all radioactive and it has a ground state configuration of $5f^36d7s^2$. Complexities arise because these f, d, and s electrons have very similar energies and can contribute in differing ways to chemical binding. For U, valences favored are IV and VI. The metal (m.pt.1132, b.pt.4131°C) has a vapor pressure of 10 ppmv, 2050°C and 50 ppmv at 2215°C. Concerning its condensed phase chemistry, the oxide UO_2 is metastable and if stored or heated in air, gradually converts to the most stable oxide structure U_3O_8 at normal temperatures. This has disordered oxidation states in its crystal lattice. There is a UO_3 molecule but this is unstable dissociating at 200°C–650°C. On heating, U_3O_8 melts at 1150°C and dissociates to UO_2(s) at 1300°C. The dioxide although metastable melts at 2842°C with boiling at about 3530°C. Below this melting point, vaporization is dominantly to UO_2(g) (vapor pressure 1 ppmv at 1860°C; 50 at 2150 and 100 ppmv at 2210°C). At 1900°C the vapor is 94% UO_2 with about 0.6% UO(g) and 4.7% UO_3(g) (Babelot and Hoch, 1989). Only at much higher temperatures (>2500°C) do gaseous UO and UO_3 begin to contribute significantly (Fischer, 1989; Ackermann et al., 1980). A hydroxide is known but only in solution, $UO_2(OH)_2$, exhibiting the stable hexavalency of U. A valence difference is displayed in the two known sulfates, UO_2SO_4, (U(VI)) that is water soluble, and the less stable and less soluble $U(SO_4)_2$, (U(IV)). On heating, the latter dissociates at 460°C–665°C to the uranyl sulfate that in dry air and in SO_2 can be stable to 710°C before converting to U_3O_8 (Tridot, 1966).

A water soluble uranyl nitrate exists, $UO_2(NO_3)_2$, known originally through its early use as a photosensitive compound in early photography. It melts at 60°C and dissociates to UO_3 above 400°C.

The chlorides UCl_{3-6} are well characterized but all react with air and are hygroscopic. UCl_4 appears to be the most thermodynamically stable: UCl_3 (m.pt.841, b.pt.1578°C), UCl_4 (m.pt.590, b.pt.791°C), UCl_5 (m.pt.327), UCl_6 (m.pt.179, decomposes 120°C – 150°C). UCl_3 is oxidized to UCl_4 by water and both UCl_3 and UCl_4 are oxidized in air to UO_2Cl_2. The latter is very soluble in water but otherwise not thermally stable. UCl_5 decomposes to UCl_4 and Cl_2 and is hydrolyzed by water. UCl_6 is only thermally stable in the narrow 12°C–82°C window and reacts vigorously with water.

Ceramic type sulfides are known but also are prone to oxidation, US (m.pt.2462°C), US_2 (m.pt.1847°C, loses S at low temperatures), U_2S_3 (m.pt.2027°C). For example, US_2 reacts with O_2 at 270°C–350°C becoming initially UO_2SO_4. Similarly a TGA study of US to 900°C in oxygen or air indicated transition from US_2, through UOS, UO_2SO_4 to UO_2 (Nakai et al., 1969). The sulfides have found use in the nuclear industry together with their corresponding phosphides (UP, m.pt.2850°C) that appear to have been less developed. This does begin to dissociate to gaseous U, P and P_2 above about 1800°C. Heating uranium in hydrogen at 250°C–300°C produces the UH_3 hydride that can be decomposed at higher temperatures. It can be a useful compound from which to synthesize other uranium compounds.

Mixed metal compounds in the form of urinates can be readily formed and uranium's wide range of ores indicates it can assimilate into many crystal and glass forms. They tend to be less stable than molybdates and tungstates. In terms of possible formation in ash or slag numerous compounds are known; Na_2UO_4 (m.pt.1635°C), $CaUO_4$ (m.pt.1800°C), Ca_3UO_6 (m.pt.2175°C), $FeUO_4$, K_2UO_4 (m.pt.1620°C), and $MgUO_4$ (m.pt.1700°C). These all display the hexavalent preference and generally can modify below their melting points if heated in air to the $U_2O_7^{2-}$ structure. Due to its relevance to the nuclear industry the interactions between sodium and uranium oxides have been extensively studied (Smith et al., 2015, 2015a). These show the wide range of stoichiometric mixtures and urinates that can exist even in this one case. This brief behavior of uranium in the condensed phase is summarized in Table 7.42.

Gas phase stabilities and transport

As can be construed from the above, the gaseous nature of uranium has to be very limited in connection with combustion at normally encountered temperatures. Nevertheless, and as a result of the nuclear industry, extensive studies have been published on oxide and sulfide vaporization and the strengths of their respective bonding. Numerous matrix isolation studies identified the UO, UO_2, and UO_3 species in experiments vaporizing uranium in the presence of O_2. The more recent has confirmed their appropriate vibrational frequencies and the linearity of UO_2 (Zhou et al., 2000). This linearity of UO_2 also is predicted by DFT theory (Yang et al., 2016a). Numerous Knudsen cell-mass spectrometric studies examined the equilibrium distributions between U, UO, UO_2, UO_3, BO, SiO, LaO, and YO. These were summarized in the reassessment by Pedley and Marshall for the

TABLE 7.42 Properties of uranium's condensed phase molecules relevant to combustion, °C.[a]

m.pt. b.pt. vapor pressure
U 1132 4131, 50 ppmv/2215°C, ins
UO_2 2842 3530, metastable, slowly converts to U_3O_8(s), vapor to UO_2(g), 50 ppmv/2150°C, ins
UO_3 unstable, dissoc 200–650, ins
U_3O_8 1150, dissoc 1300 to UO_2(s), ins
$UO_2(OH)_2$ exists only in solution
UO_2CO_3 dissoc 136, ss
$UO_2(NO_3)_2$ 60, dissoc 400 to UO_3, s
$U(SO_4)_2$ dissoc 460–665 to UO_2SO_4 and then U_3O_8, ss
UO_2SO_4 dissoc 700–800 to U_3O_8, s
UCl_3 841 1578, r with wat to UCl_4, with air to UO_2Cl_2, 10 ppmv/730°C, s
UCl_4 590 791, r with wat, oxidized in air to UO_2Cl_2, s
UCl_5 327, dissoc to UCl_4, r with wat, s
UCl_6 179, only stable 12–82, dissoc 120–150, r with wat, s
UO_2Cl_2 577 1253, dissoc to UO_2(s), r with wat, s
US 2462, oxidizes in air at 900 to US_2, UOS, UO_2SO_4 then to UO_2, ins
US_2 1847, readily losses S, reacts with air 270–350 to UO_2SO_4, ins
U_2S_3 2027, thermally very stable, ins
UP 2850, dissoc 1800 to U, P, P_2, ins
UH_3 >500 dissoc, r with water giving H_2, s
$CaUO_4$ 1800, stable at 1350, ins
Ca_3UO_6 2175, stable, ins
$FeUO_4$ – , ins
K_2UO_4 1620, changes at 1200 to $K_2U_2O_7$, ins
$MgUO_4$ 1700, stable to 1250, decomposes in air, ins
Na_2UO_4 1635, changes at 1300 to $Na_2U_2O_7$, ins

[a]*dissoc*, dissociates; *s*, water soluble; *ss*, slightly soluble; *ins*, water insoluble; *vapor*, vaporizes; r with wat, reacts with water.

diatomic oxides (1983). At the time, the electronic spectroscopy of the low-lying states of UO was unknown and they assumed an electronic degeneracy in their thermochemical analysis of 10. Since then the major low-lying contributing states have been better resolved experimentally (Kaledin et al., 1994; Kaledin and Heaven, 1997). Additional theoretical electronic structural calculations have appeared assigning the electronic configurations and role of the 5f electrons in the low lying states of UO (Tyagi et al., 2014) and for both

UO and UO_2 (Heaven, 2006; Infante et al., 2007; Kovacs et al., 2015). For UO these now suggest a value of at least 12 may be more appropriate for the electronic degeneracy factor. In gaseous chemical thermodynamics the "fef" function is proportional to Rln*g* where "*g*" is the degeneracy and used in "Third Law" analyses of equilibrated thermodynamic reactions such as those involving UO. Derived values for D_0 will decrease as degeneracy increases by an amount RT ln $(\mathrm{New/Old})_{\mathrm{degeneracy.}}$ Consequently, for accuracy the correct degeneracy and basic fef values are very important. In the case of the uranium oxides, although much data have been collected, the thermochemical tables needed for such analysis have remained uncertain until recent years. For example, the main four data sources available for UO(g) indicated fef_{298K} values of 283.3 J $\mathrm{mol}^{-1}\,\mathrm{K}^{-1}$ (Green, 1980, 1980a, 1980b), 300.6 (Gurvich, 1982), 297.3 (Pedley and Marshall, 1983) and 293.7 (Fischer, 1989; Cordfunke and Konings, 1990). Such variations can introduce differences in enthalpies of the order 35 kJ mol^{-1}. Of these, the first used UO(g) vapor pressure data in the 1400°C–1900°C to adjust the required degeneracy factor (Green, 1980a). However, due to its small contribution and the dominance of UO_2(g) it appears that these values were not sufficiently sensitive for such an analysis. Also for the last source, a degeneracy of about threefold was used whereas a value of 10–12 now is more appropriate, implying a need to increase their values by about 9 J $\mathrm{mol}^{-1}\,\mathrm{K}^{-1}$. From the previous reassessment review of Pedley and Marshall (1983) and adding one additional study (Younes et al., 1981) it had been concluded that their value of $D_0(\mathrm{U\text{-}O}) = 755 \pm 10$ kJ mol^{-1} remains adequate (Schofield, 2006). Any "Second Law" values derived in these studies tended to be consistently larger than the "Third Law" analyses indicating that statistical degeneracies appear adequate; their increase would lower the recommended value further. However, this confusing situation now has been changed by a recent comprehensive review of the lanthanide and actinide oxides updating their thermochemical data and recalculating the heat of formation of UO(g). This placed most emphasis on the mass spectrometric studies and now recommends a recalculated value of $D_0(\mathrm{U\text{-}O}) = 758.9 \pm 10$ kJ mol^{-1} (Konings et al., 2014). The small change from the above documented arguments as to a preferred value is encouraging, and illustrates to some degree the ruggedness and approximate correctness of our chemical thermodynamic basis of understanding. The most recent theoretical attempts to effectively reproduce such a bonding strength still tend to overestimate with a prediction of about 811 kJ mol^{-1} (Kovacs et al., 2015).

The situation for UO_2(g) is simpler in that due to its congruent vaporization its heat of vaporization is reasonably well measured. However, although previously thought to be accurately known as $\Delta H_{\mathrm{sub,\,298K}}(UO_2)_s = 618.4 \pm 0.5$ kJ mol^{-1}(Cordfunke and Konings, 1990), this has now been replaced with 622.9 ± 12 kJ mol^{-1} in this recent comprehensive review of the lanthanide and actinide oxides (Konings et al., 2014). This also is a useful reminder to question error bars that in this case were rather misleading. Coupling this new value to the accurate heat of formation of $UO_2(s)_{298K}$ (-1085 ± 1) provides the value $\Delta H_{\mathrm{f,\,298K}}(UO_2)_g = -462.1 \pm 12$ kJ mol^{-1}. From this with the heat of vaporization of U(s) now indicates $D_0(UO_2)_{\mathrm{g,\,atomiz}} = 1489 \pm 15$ kJ mol^{-1}. Previously it was considered that the two bonds of UO_2(g) were essentially of similar strength, but now one value depends on the exactness of the other. At present this now indicates $D_0(\mathrm{OU\text{-}O}) = 730 \pm 10$ kJ mol^{-1}. Consequently what is seen with uranium is very strong bonding to oxygen in the gas phase that will not be easy to modify. Such strong metal bonding is only seen in the early lanthanides and actinides. There are no consistent trends in the four actinides Th, Pa, U, and Np, with weaker dioxide bonding in Th, U, and Np but equal strengths for Pa (Konings et al., 2014). Recent quantum chemical calculations do predict OU-O values in a correct magnitude of about 712–723 kJ mol^{-1} (Kovacs et al., 2015). Uranium has the corresponding sulfides US(g) and US_2(g) that have been extensively studied using vaporization and mass spectrometric methods. All this data have been critically reviewed and little has appeared for its modification (Gronvold et al., 1984). US(s) is thermally stable having a vapor pressure of about 3 ppmv at 2000°C. The vapor consists of U(g), S(g), US(g) with very low levels of US_2(g). The evaluation suggested a value of D_0(U-S) = 494 ± 15 kJ mol^{-1}. US_2(s) is more volatile, losing sulfur readily and modifying its solid phase stoichiometry. It displays a vapor pressure of sulfur of about 2 ppmv at 400°C. Nevertheless, the US_2(g) molecule shows thermal stability with a recommended value of D_{298}(SU-S) = 498 ± 15 kJ mol^{-1}, essentially of equal strength to the monosulfide. These values display the significant weakening (65%) of the sulfur bonding to uranium compared to that of oxygen that is normally expected. This is reflected by numerous elements in this region of the first few transition element columns of the periodic table (Sc, Ti, V, Y, Zr, Mo, W), Tables 7.15 and 7.19. Matrix isolation measurements have confirmed the bent structure of US_2 and examined its configuration theoretically with respect to that of linear UO_2 (Andrews et al., 2011). Otherwise, limited theoretical studies have addressed these sulfides. One has attempted to develop an acceptable approach encompassing all the actinide monosulfides, but calculated values for *D*(US) of 544 and 571 remain too high as they are also for *D*(UO) (Pereira et al., 2011).

The sensitivity to air and water removes most of any interest in the uranium chlorides with respect to combustion. They are certainly volatile and may be relevant to gasification and air free systems. An early matrix isolation study identified the infrared absorption of UCl_2 obtaining its frequencies and 100° bent structure (Hastie et al., 1971a). UCl_4(s) was known to be the most thermodynamically stable chloride and also vaporize unchanged. Its sublimation had been accurately measured and reviewed along with the scant data on the other chlorides (Fuger et al., 1983; Cordfunke and Konings, 1991). Soon thereafter a very detailed Knudsen cell-mass spectral analysis established the thermochemical properties of the UCl_{1-5} molecules (Lau and Hildenbrand, 1984). Due to the uncertainty of the electronic nature of these molecules, careful Second Law data were obtained. The enthalpy of sublimation of UCl_4(s) had been measured, which when coupled to its heat of formation provided a reliable measure for the heat of formation of UCl_4(g) and a value for its atomization energy and the sum of its four bond strengths. Preferentially utilizing this, the Knudsen cell data were slightly modified to build a consistent set of values. With this adjustment, it indicated values of D_{298}(U-Cl) = 439 ± 10 kJ mol^{-1}, D_{298}(ClU-Cl) = 481 ± 10 kJ mol^{-1}, D_{298}(Cl_2U-Cl) = 485 ± 10 kJ mol^{-1}, D_{298}(Cl_3U-Cl) = 420 ± 10 kJ mol^{-1}, D_{298}(Cl_4U-Cl) = 209 ± 10 kJ mol^{-1} (Hildenbrand et al., 1985; Hildenbrand and Lau, 1992). Thermochemical values are ever being improved and in this case uncertainties still exist even in the sublimation enthalpies of U(s), now $\Delta H_{sub,298K}$ = 533 ± 8 kJ mol^{-1}(Konings and Benes, 2010), and of UCl_4(s) (Fuger et al., 1983; Cordfunke and Konings, 1991), requiring some future readjustment of these recommended values. For example, in the present case, the values of $D_{298}(Cl_4U\text{-}Cl) = 238$ kJ mol^{-1}and $D_{298}(Cl_5U\text{-}Cl) = 207$ kJ mol^{-1} indicated in Barin's 1995 edition of Thermochemical Values illustrate a 20–30 kJ mol^{-1}uncertainty from the values being suggested here and were either estimates or from older tabulations no longer acceptable (Barin and Platzki, 1995). In the last decade or so, theoretical efforts have turned to these actinide halides. With UCl_3(g) two studies have effectively modeled its geometry, frequencies and low-lying electronic states, both suggesting a pyramidal shape with about 110-degree angle (Joubert and Maldivi, 2001; Kovacs et al., 2013). Recently, the UCl_4 molecule was modeled with DFT that closely reproduced its experimental enthalpy of formation. Studying also a corresponding $U(OH)_4$(g) structure this is predicted to have an atomization energy with U-OH bond strengths similar to any in UCl_4 (Johnson et al., 2013). Two density functional studies by the same group have approximately reproduced the five experimentally measured bond strengths in UCl_{1-5}. Moreover they predict a value of $D_{298}(Cl_5U\text{-}Cl) = 180–185$ kJ mol^{-1} for the unknown sixth bond apparently continuing the weakening trend observed in the fifth bond (Batista et al., 2004; Peralta et al., 2005). As a result these UCl_{1-4} chlorides display bond strengths stronger than HCl but are readily oxidized to the even stronger oxides.

The hydrides UH and UH_2 have been characterized in matrix isolation experiments with ablated uranium, one study being recent (Vent-Schmidt et al., 2015). These have established vibrational frequencies and noted that the dihydride is bent with an HUH structure. However, it seems rather strange in this day and age that the bond strength of UH has never been measured experimentally nor its spectral systems characterized. In recent years, its electronic structure has been examined theoretically and many low-lying states predicted (about 12 states below 5000 cm^{-1}) (Cao et al., 2008; Dolg and Cao, 2009), and also for both systems (Balasubramanian et al., 2003). The earliest study identified the fact that depending on the level of theory, the ground state of UH could be either a quartet or a sextet affecting the quoted bond strength. The more recent studies now favor the quartet for the ground state UH and predict $D(U\text{-}H) = 269/275$ kJ mol^{-1}, which is the only estimate of its stability at present other than a similar nontranslated Chinese study that confirms this and suggests $D(U\text{-}H) = 278$ and $D(HU\text{-}H) = 228$ kJ mol^{-1}. A summary of these gas phase bonding strengths and their comparison to some of the heavier transition-state elements is summarized in Table 7.15.

Although only slightly connected to combustion, the chemiionization of atomic uranium in oxygen discharges as a low-temperature source of electrons has one unique characteristic (Schofield, 2006). About 23 elements in the periodic table, including uranium, can react exothermically with atomic oxygen at low temperatures to produce the diatomic ion MO^+ and an electron. In the case of uranium this is stable and can react further with oxygen to the dioxide ion.

$$U + O = UO^+ + e^- \quad \Delta H_0 = -165 \text{ kJ mol}^{-1}$$
$$UO^+ + O_2 = UO_2^+ + O \quad \Delta H_0 = -245 \text{ kJ mol}^{-1}$$

An examination of such dioxide ions for the periodic table indicates these are generally vulnerable to rapid dissociative electron recombination back to the monoxide MO and so are short lived. Uranium is unique, its oxides are so stable that the normal decay,

$$UO_2^+ + e^- = UO + O \quad \Delta H_0 = +149 \text{ kJ mol}^{-1}$$

has this thermal barrier at lower temperatures. As a result, this appears to be the only dioxide ion that can produce electrons and be long-lived. It is forced into decaying like an atomic ion through a slower three body interaction.

$$UO_2^+ + e^- + Z = UO_2 + Z$$

Combustion power plant data

As indicated in Table 1.1, uranium can be consistently found in most fuels (≤3 ppmw), being lower in biomass. It is generally negligible in pet-coke and possibly ≤2 ppmw in limestone (Cordoba et al., 2012). It occasionally occurs that certain US and Chinese coals are enriched with uranium and other radioactive compounds. If burned, extra care has to be taken with their ash and any other disposable materials (Lauer et al., 2015, 2017). Attempts to preclean by leaching such high uranium coal have found it difficult to extensively reduce its levels and its use is not recommended (Duan et al., 2018). As a result, in combustion systems, if any uranium manages to get into the burned gas phase it is not hard to predict from its chemistry that it will be in the form of the condensed phase particles of the oxides U_3O_8 and UO_2 with only negligible levels of anything gaseous. As a result it is surprising that the behavior of uranium in early developments of atomic absorption and emission, with high temperature N_2O/C_2H_2 flames, was regarded as anomalous when no atoms could be detected (Stephens and Stevenson, 1975). However, it had been reported from early flame studies that additions of a nebulized aqueous solution of a salt such as uranyl nitrate even with flame concentrations of about 1 ppmv would produce streaking in the burned flame gases (Bulewicz and Padley, 1971). It was also noted that the presence of such UO_2 particles in the flame gases was very effective in enhancing the recombination rates of the H, OH radicals in fuel-rich hydrogen flames. In other words, like Cr, it was a very effective inhibitor of combustion (Bulewicz and Padley, 1971). Although many such effects have been explained by efficient gas phase chemical cycles the mechanism with uranium remains uncertain (Tischer and Scheller, 1970; Linteris et al., 2008). Now from the analysis of its chemistry it probably arises simply from a heterogeneous particle surface enhanced recombination of radicals. Uranium appears to play a minimal or no role in soot suppression (Bulewicz et al., 1975; Howard and Kausch, 1980). It is to be remembered that laboratory flame studies do not suffer from release considerations as do full-scale fuel combustion and so do not relate.

One major interest in combustion with uranium largely has centered on whether problems arise in potential radioactive emissions when burning coal or sewage sludge. As a result, several studies have been published indicating that uranium is retained in the ash forms, emission levels are very low and minimal risk exists with any radionuclide (Aly et al., 1994; ISCORS, 2005), although even ^{210}Po emission has been questioned (Yan et al., 2012). In the United States, 9000 facilities are licensed users of radioactive materials and can release them to sanitary sources. Consequently, disposal into sewage waste can become a fuel source, is normally monitored and under normal combustion conditions should be of little concern and can be suitable for agricultural dispersion (Jones et al., 2014a). From early days it was realized that the uranium stayed in ash forms and the concentrations in bottom or fly ash were closely the same (Coles et al., 1979; Ondov et al., 1979). The ash concentration enrichment over crustal occurrence was solely due to combustion material volume reduction (Marrero et al., 2007). Moreover, neither ash form was leachable with water (Llorens et al., 2001). Some enrichment was suggested on the smallest particles but was probably through material loss of other components and was not supportive of a uranium condensation process (Querol et al., 1995; Lanzerstorfer, 2018). Coal gasification under reducing conditions at 1450°C indicated only very low uranium levels in the small particles (Helble et al., 1996). Power plants burning subbituminous coals with solely ESP particle control indicated uranium emissions loss as small particulates of 0.18% (Goodarzi, 2006) and 0.02% (Goodarzi, 2006a) in two case studies. With added scrubbing facilities these levels are further reduced to be negligible but may need adjustment for any low level of uranium occasionally in the limestone (Meij, 1994; Cordoba et al., 2012). In one of the few laboratory combustion studies using sewage sludge it appeared that the uranium in sludge treated with $FeCl_3$ and $Ca(OH)_2$ remained involatile at 1100°C and no chlorine induced release was apparent (Folgueras et al., 2003). A renewed interest now appears to have occurred in China concerning the uranium in coal combustion. A detailed investigation examined coal from 18 mines finding a general level of 1–3 ppm presence but that coals rich in uranium can also exist (Zhang et al., 2016c). They confirmed that the uranium is to be found in the ash forms testing samples up to 1100°C in a tube furnace and to 1600°C with a drop tube. Very little (<0.1%) was detected in the flue gases most likely in particulate form. Another study mixed uranium and alkaline earth salts with coal in a quartz reactor up to 1200°C (Yang et al., 2016). It appears to have confirmed that urinates, particularly that with calcium can be formed. These do not appear to be stable above 1000°C leaving the question whether formation can occur in the cool down time period of the ash.

In summary, with adequate particulate control, uranium is removed from combustion systems in the form of particulates. Emission levels are negligible compared to those of nature. There is no information concerning their chemical nature. Whether time permits interactions between the uranium oxide and the particle ash minerals remains unknown, but formation of a calcium, magnesium or alkali urinate remains plausible.

7.31 Vanadium

Vanadium, the third of the first row transition elements lies between Ti and Cr. As with these, it is a valuable lightweight metal mainly used in the form of alloys with Fe, U, Ti, or Al. Its average concentration in the Earth's crust is about 120 ppmw, slightly less than that of Zr, but has slightly larger concentrations in the oceans due to its increased solubility. It is widespread, abundant but with few concentrated deposits. About 65 minerals are known that contain it, including about 40 vanadates. Of these, a complex K/U/VO mineral (carnotite) and a Pb/VO/Cl lead chlorovanadate (vanadinite) are the most important that are mined primarily for their other metal content. Vanadium also occurs more in crude oil (3–260 ppmw) and the refining tends to concentrate the vanadium into the heavier residual fractions (0.2–160 ppmw). Their combustion or further processing now is an additional source of the element. In nature it exists in two isotopes but ^{51}V is dominant (99.75%) with only a small contribution from the slightly radioactive ^{50}V, which is very stable with a half-life of $>10^{17}$ years.

Depending on the data accepted, vanadium may or may not be an atmospheric pollutant. Such releases to the atmosphere are very difficult to assess and earlier estimates suggested a threefold larger anthropogenic source over the natural levels (Nriagu and Pacyna, 1988; Nriagu, 1989). The largest natural level estimate is about 69,000 tons $year^{-1}$ originating from mainly soil dust, sea spray, volcanoes (possibly 23%) and forest fires (Malakhov and Makhon'ko, 1990; Hope, 1994). This was a factor of three above earlier estimates indicating the levels of uncertainty. Anthropogenic levels also show a large range of uncertainty and now is raised a factor of two- to threefold to 240,000 tons $year^{-1}$ mainly the result of fuel oil combustion, coal and sludge combustion and to a lesser extent the metal industry and the growth of Asian economies (Pacyna and Pacyna, 2001). Detailed global mass balances, considering all the various reservoirs and flux changes between them disagreed indicating that anthropogenic loading of the atmosphere was only about 30% of its content (Hope, 1994, 1997). These studies suggested larger roles for dust (48%) and volcanoes (23%) with no suggested concern for a global health threat. This global cycle model was repeated later but does not appear to take into account the efficient capture and handling of vanadium on ash. Its estimates cannot be currently accepted (Hope, 2008). A more recent assessment of the global biogeochemical cycle for vanadium suggests that anthropogenic releases to the atmosphere may be about 1.7-fold larger than natural sources (Schlesinger et al., 2017). However, in one study of the current metal pollution in China, this reports atmospheric levels of 18 ng m^{-3}, which is negligible compared to the World Health Limit of 1000 ng m^{-3}. This low level was assumed to be mainly from oil combustion (85%) and coal (11%) (Duan and Tan, 2013). Moreover, such amounts might be expected to decrease with an advent of increased natural gas usage and better oil refinement or combustion particulate control. Although rated with a low level toxicity, vanadium has fewer uses than might be expected. Its main application is to produce much improved alloys with iron and several other metals. It has useful catalytic qualities, is used in glazes, paints, inks and glass, and its hydride is a neutron moderator in nuclear reactors. Some salts are fungicides or insecticides but it has limited medical use. The world production of vanadium was about 80,000 tons $year^{-1}$ in 2017, sources being mainly China, South Africa and Russia, with very adequate reserves of about 63 million tons available. Recycling also is beginning to add to supply. Cost depends on product. Ferrovanadium steel which is about 85% of its use is currently \$20 kg^{-1}. It can contain from 0.15% to 5% V depending on the steel quality. Sold as the pentoxide, V_2O_5 it can be \$2.70 kg^{-1} whereas the less used pure metal can be expensive at \$135 kg^{-1}. The United States is totally depend on imports and it is not currently mined in the United States. Although a very important ingredient in steel and catalysts there is an abundance and it is not regarded as a critical at risk element in economies (National Research Council, 2007). Whether anthropogenic inputs into the atmosphere actually exceed the amount produced for commerce appears possible. China alone estimates its emission levels to the atmosphere from oil and coal combustion at 32,000 tons $year^{-1}$ (Duan and Tan, 2013).

Health concerns with vanadium

Whether vanadium is a required trace in the human body appears to remain uncertain but is quite possible. Lower forms of life do seem to incorporate it into various essential enzymes (World Health Organization Trace elements, 1996). Due to responses in animals, there is current research interest in vanadyl sulfate V^{IV}, $VOSO_4$, and sodium metavanadate V^V,$NaVO_3$, currently sold as supplements claiming possible benefits for diabetes and cancer. However, with suggested doses of 100 mg day^{-1} a doctor's advice is certainly suggested and testing is needed to see if the body can tolerate such a drastic change in levels from what is normally experienced. Medical research generally uses organovanadium salts. Vanadium's toxicology has been extensively reviewed in two interesting and detailed reports freely available on the Web (World Health Organization Vanadium, 2001b; Tox Profile V, 2012). A more recent appraisal raises questions again concerning its possible neurotoxicity (Fatola et al., 2019). Reviews outlining the history, uses and metallurgy, releases to the atmosphere, its aquatic chemistry, its toxicology and health aspects also have been published. These appeared in "Vanadium in the Environment," a two-volume special issue of *Advances in Environmental Science and Technology* (Nriagu, 1998).

Vanadium is naturally present in the atmosphere and soil and enters the food chain through plant growth. It appears to be more important in marine species and is minimal in animals and humans. It can be toxic to some fish. Levels in drinking water can vary and few water quality standards appear regulated. One in the Netherlands suggested 1.2 and 3.0 $\mu g\ L^{-1}$ for long- and short-term exposures. A British Research group has renewed concern over the environmental emerging risks associated with vanadium (Watt et al., 2018). However, at low levels there may be beneficial effects to health but this still is uncertain as also whether it is essential for life. It does appear so in chicken and rat studies, but its role remains obscure. Normal average human intake is about 11 $\mu g\ day^{-1}$, which is tolerated by the body. It does not noticeably accumulate in the body and is easily eliminated. It is known that it can transfer through the blood/brain and placenta barriers and is very low in human breast milk (<1 ppmw) (World Health Organization Vanadium, 2001b; Tox Profile V, 2012). It is ingested normally in diet as V^{IV} (vanadyl) and V^V(vanadate) compounds. These appear not to be readily absorbed and metabolized but are excreted with a half-life of 15–40 hours. General survey testing generally has not included vanadium. It is not in the US National testing reports (CDC 4th Natl. Report. 2018). If it has been included in population surveys it indicates blood levels of $<0.4\ \mu g\ L^{-1}$ (Cesbron et al., 2013; Nisse et al., 2017), urine 1.4–10 $\mu g\ L^{-1}$ (Goulle et al., 2005) and $<1.3\ \mu g\ L^{-1}$ (Nisse et al., 2017), and hair 1–51 ppbw (Goulle et al., 2005), <0.2 ppmw (Drobyshev et al., 2017). A recent comparison of the populations of Morocco and the Canary Islands indicated blood levels of <2 and $<3\ \mu g\ L^{-1}$, respectively (Henriquez-Hernandez et al., 2018). One interesting autopsy study of people living in the vicinity of a hazardous waste incinerator showed an even distribution over all the main body organs at a low density of <0.25 ppmw (Bocio e al.2005). Although regarded as a toxin, vanadium is only observed in industrial exposures. Extensive animal research has been reported and vanadium is not totally inert but does appear to interact with various enzymes and mechanisms. Certain bacteria and enzymes do need it, but functions remain unknown in humans. It mimics insulin in animals and this is the basis for possible human interest. Concern may also arise from the fact that V_2O_5 is a lung irritant and can induce lung cancer in mice (Rondini et al., 2010). Although there is this wealth of animal data, extrapolation to humans appears to still lack confidence. Moreover, water soluble compounds are expectedly more toxic than those nonsoluble. Long-term data from studies of industrial workers remains unclear but have produced minor concern about health. Whether long-term effects can ensue from gradual accumulation with age still remains unknown. One study on mice fed excessive vanadium doses did display induced oxidative stress in their brains. However, when stopped it appeared reversible (Folarin et al., 2018). There is significant interest for any future possible medical benefits of vanadium. A current medical research review emphasizes its observed anti-diabetic properties of lowering blood sugar (Korbecki et al., 2012), and possibly for aiding in controlling diseases found in certain tropical countries (Rehder, 2012). Also there are several suggestions now that in certain organic forms vanadium may in fact have inhibiting effects on numerous types of cancer (Bishayee et al., 2010; Chi et al., 2012). Vanadium is a good example of how difficult it is for the medical profession to establish with any degree of certainty whether a metallic element is beneficial to life or may have long-term consequences that go unnoticed. It is to be remembered that mankind has only known this particular element and had it readily available for 100 years or less.

Vanadium's basic chemistry

Vanadium, the third of the first row transition elements starts showing higher valences and a range of molecules that can be nonstoichiometric. Valences of 2–5 are seen with four-valent molecules generally containing the VO^{2+} component. In combustion, especially with oxygen-rich systems, the chemistry will mainly concern oxides and vanadate formation (VO_3^-, VO_4^{3-}). The range of compounds is limited with a basic sulfate, halides, sulfides and hydrides characterized, as will be briefly outlined here and are summarized in Table 7.43.

However, these remain minor to the dominance again of a more complex stable metal oxide chemistry. In the condensed phase, vanadium oxides show a pronounced ability to show stoichiometries that range from a 1:1 ratio (VO(s)) to 1:2.5 (V_2O_5(s)). VO, V_2O_3, VO_2(V_2O_4), and V_2O_5 are all dark in color and commercially available. From the earliest vaporization measurements it was noted that the highest valences can readily start losing oxygen when heated. For V_2O_5(s) with a melting point of 679°C this loss has been detected even at 400°C. The first Knudsen cell-mass spectral analysis of its vaporization indicated that on heating, the solid phase composition changed with time. The vapor over V_2O_5(liquid) was complex and gradually changed with its own liquid/solid mix of oxides as it lost oxygen. This was not amenable to extract thermochemical characteristics (Berkowitz et al., 1957a). However, their corresponding experiment starting with VO(s) (m.pt.1790°C) produced a vapor of only gaseous V, VO, and VO_2 enabling a very good value for *D*(V-O). Over the years, further work clarified the melting and vaporization of different mixtures (Frantseva and Semenov, 1969; Farber et al., 1972; Banchorndhevakul et al., 1985). These confirmed that condensed phase trivalent V_2O_3 or mixtures where the

TABLE 7.43 Properties of vanadium's condensed phase relevant to combustion, °C.[a]

m.pt. b.pt. vapor pressure

V 1902 3406, 10 ppmv/1555°C, ins

VO 1790 2627, nonstoichiometric, vap to gaseous V, VO, VO_2

VO_2 (V_2O_4) 1545, ins

V_2O_3 1967, vap to VO(g), VO_2(g), 1–5 ppmv/1600°C–1800°C, ins

V_2O_5 679, dissoc >1690 to V_2O_4, starts >400, s

$V_2(SO_4)_3$ >400 dissoc, s

$VOSO_4$ 105, decomp >400 to V_2O_4,V_2O_5, s

VCl_2 1350, sublimes at 1027, s

VCl_3 >500 disproportionates to VCl_2,VCl_4, s

VCl_4 −26 150, dissoc to VCl_3, reacts with water

VOCl 127

$VOCl_2$ 380, reacts with water

$VOCl_3$ −78 127, hydrolyzed and reacts with water, air

V_nS_m exists with $n/m = 0.6–1.25$, unstable with water or air, ins

V_2S_3 unstable, ins

VS >1700 dissoc to V(g), VS(g), and S(g), ins

$VH_{0.95-2}$ >650 dissoc, nonstoichiometric

$V(CO)_6$ >50 sublimes, decomp, ins

$AlVO_4$ 695, ins

$CaV_2^{IV}O_5$ 724, ss

$Ca(VO_3)_2$ (meta-) >780 dissoc to pyro-, ss

$Ca_2V_2O_7$ (pyro-) >1012 dissoc to ortho-, ss

$Ca_3(VO_4)_2$ (ortho-) >1390 dissoc to CaO, ss

$Fe^{III}VO_4$ 860, ins

$Fe^{II}(VO_3)_2$ >750 dissoc to $FeVO_4$, ins

$Fe_3^{III}VO_7$ 840, ins

KVO_3 (meta-) 495, s

K_3VO_4 (ortho-) >1000 dissoc, s

$Mg(VO_3)_2$ (meta-) 760, ss

$Mg_2V_2O_7$ (pyro-) 1010, ss

$Mg_3(VO_4)_2$ (ortho-) 1100, ss

$NaVO_3$ (meta-) 630, s

$Na_4V_2O_7$ (pyro-) 660, s

Na_3VO_4 (ortho-) 1200, s

[a] *dissoc*, dissociates; *decomp*, decomposes; *s*, water soluble; *ss*, slightly soluble; *ins*, water insoluble; *vap*, vaporizes.

V/O ratio was similarly 1.5 did vaporize congruently in a certain temperature and oxygen pressure range and that the composition was stable and initially VO(g) and VO_2(g). At temperatures of 1600°C–1800°C vapor pressures for these are about 1–5 ppmv and so permitted their thermochemical values to be established by mass spectrometry. The chemical thermodynamic nature describing the condensed phase was assessed for the various states associated with the solid and liquid phases as they changed with temperature (Brewer and Ebbinghaus, 1988). Consequently, at lower temperatures, complex gaseous molecular structures such as V_4O_{10} may vaporize but will rapidly fragment to VO and VO_2, the most stable gaseous species at elevated temperatures and low concentration conditions. The dissociation process appears reversible, and on cooling and condensing, vanadium oxides can reabsorb oxygen modifying their valency and structure with increasing complexity in the condensed phase. Moreover, although the melting point of V_2O_5(s) is only 679°C, as oxygen is lost the melting points of the lower oxides remain high, V_2O_4, V_2O_3, VO_2 all >1500°C, and nonstoichiometric VO 1790°C. Consequently, on heating the vanadium oxides go through a series of condensed phase mixtures as they lose oxygen but retain a high melting point and gradually reduce the valence of the vanadium from five down to two. As the temperature becomes elevated the continuing vaporization simplifies and only VO (g) and VO_2(g) are the stable gaseous molecules along with V(g). The highest valency seen at the elevated temperatures is V^{IV}.

Although apparently not pursed in detail, there are indications in the literature that the rate of vaporization of the vanadium oxide melts of V_2O_5/VO_2 at about 1000°C is significantly enhanced by water vapor (Hackert and Gruehn, 1998; Sanchez and Hager, 1994). It was suggested that the water vapor possibly produced gaseous VO_2(OH) and $VO(OH)_2$ Whether these or other hydroxides form reflecting 3 or 5 valency (VO(OH), $VO(OH)_3$) remains totally speculative. A valency of four is probably most favored. Two studies have cocondensed the products of the reaction between the vapor of either thermally or laser-ablated vanadium with water and found evidence for the HVOH molecule (Kauffman et al., 1985; Zhou et al., 2001). Also very recently a theoretical assessment concerning the HVO molecule indicated this to be bent and very slightly more stable (20 kJ mol^{-1}) than its bent VOH structure (Miliordos et al., 2013). Otherwise little is known of possible hydroxides but in light of the great oxide stability their lasting presence in combustion gases is probably unlikely but might still mirror the nature of its neighbor Cr to some degree.

Vanadium displays no carbonates or nitrates but several sulfates and their hydrates are known (Stern, 2001). $VOSO_4$ is stable and soluble in water but decomposes to the oxides V_2O_4, V_2O_5 above 450°C. These can revert to the sulfate in the presence of SO_2 or SO_3 (Tridot, 1966). The chlorides are hygroscopic, volatile and are readily oxidized by water or air: VCl_4(liquid) with a boiling point of 15°C dissociates on vaporization, $VOCl_3$(liquid) boiling point 127°C, $VOCl_2$(s) melts and decomposes at 380°C. No halides are compatible with combustion and the presence of chlorine will have no major effects on vanadium's combustion chemistry. The sulfides reflect a similar nonstoichiometric behavior as their related oxides. Structural XRD studies display phases with V/S ratios running from 0.6 to 1.25 representing stoichiometries close to V_3S_5, V_5S_8, V_2S_3, V_3S_4, VS, and V_5S_4 (De Vries and Jellinek, 1974). Thermogravimetric studies of such sulfides in the presence of air or water show their rapid degradation to the oxide chain leading to V_2O_5 (Basu and Taniguchi, 1984). One Knudsen cell-mass spectrometric study of VS(s) at 1700°C–2000°C confirmed its congruent dissociating vaporization to VS(g), V(g), and S(g) at such temperatures (Owzarski and Franzen, 1974). The hydride also is nonstoichiometric indicating $VH_{0.95-2}$ phases. Considered as a means of storing hydrogen, on heating it does release H_2, losing it totally at 650°C–800°C. It is not compatible with combustion.

A broad spectrum of vanadates can form with a wide range of stoichiometric molecules forming in molten mixes (Mittal and Elliott, 1984). Unlike titanium they tend not to be refractory in nature and like Mo and W generally have melting points below or about 1000°C, reflect valences of 4 or 5 and some vaporize incongruently and dissociate to their oxides on heating. A recent review of Russian work over the past 50 years indicates their studies of the vaporization of mixed metal/vanadium oxides and the behavior especially of the alkali and alkaline earth V^{IV} vanadates in the gas phase. These mass spectrometric studies generally can show congruent molecular evaporation with stable gaseous molecules then dissociating to their oxides at higher temperatures (Lopatin and Shugurov, 2013). Trends of their gas phase heats of formation in 15 cases permit estimates for a further ten and reflect their stabilities. Purchased stable vanadates normally have the fully oxidized V^V structure such as ammonium vanadate, NH_4VO_3, water soluble and decomposes at 200°C to V_2O_5. The alkalis form meta- and ortho-forms, tend to be hygroscopic, water soluble and have low melting points, $NaVO_3$ (630°C), Na_3VO_4 (1200°C), corresponding values for potassium being 495, >1000°C. The alkaline earth vanadates illustrate both tetravalency, $CaVO_3$, CaV_2O_5(m.pt.724°C), and V^V, CaV_2O_6 (780°C), $Ca_2V_2O_7$ (1012°C). However, on comparing the strengths of the gaseous binding between the two metal oxides the ranking is clearly tungstates >molybdates >vanadates each being close to 100 kJ mol^{-1} stronger than the next (Lopatin and Shugurov, 2013). What this implies is that although molecular vaporization of vanadates undoubtedly occurs they will not be stable in the gas phase as temperatures are elevated. Vanadate formation may be of particular importance though in deposition from burned gases when the stability of the solid becomes important and if there are no time constraints. For example, magnesium salts now are used commercially as an additive to alleviate sodium vanadate corrosion in boilers or gas turbines. This reflects the greater stability of its vanadates, 760°C (MgV_2O_6), 1010°C ($Mg_2V_2O_7$), 1100°C ($Mg_3V_2O_8$) over that of the alkalis (Barbooti et al., 1988; Tiwari and Prakash, 1998). Vanadates other than those of the alkalis and alkaline earths that could deposit from combustion gases depending on conditions are for example $AlVO_4$ (m.pt.695°C), $FeVO_4$ (860°C), and Fe_3VO_7 (840°C). As a result, any vanadates present in fuels will readily melt and vaporize molecularly but as the more stable V^{IV} format which then can dissociate to its gaseous oxide depending on stability and temperature. Other vanadium molecules will dissociate to the oxides and if temperatures are high with vanadium release then as VO(g) and VO_2(g) that will be dominant. On cooling these can then deposit as an oxide or preferentially form a vanadate with some ash component. In order to have sufficient sensitivity in the past, attempts to analyze for V content using atomic absorption or emission methods required the use of C_2H_2/N_2O flames at about 2700°C to ensure adequate dissociation (De Galan and Samaey, 1970). This fact by itself conveyed the low volatility to be expected in normal combustion systems and clearly illustrated the stabilities of these condensed phase oxides.

Gas phase stabilities of vanadium compounds are reflected by their bond strengths. Those of the oxides are now well characterized. Pedley and Marshall (1983) reviewed the available data and recommended D_0(V-O) = 621 ± 12 kJ mol^{-1} based on two Knudsen cell-mass spectrometric equilibria involving either AlO or GeO (Coppens et al., 1967; Farber et al., 1972). These remain valid and the suggested recommendation, still is surprisingly close to the initial value of 618 suggested from the early vaporization study of VO(s) (Berkowitz et al., 1957a). Two further studies have emerged since this previous evaluation. One examined the vaporization of V_2O_3(s) to VO(g) and VO_2(g) (Banchorndhevakul et al., 1985). Their second and third law analyses for D_0(VO) ranged from 598 to 620 kJ mol^{-1} arising from the vaporization complexities. The other examined the VO/EuO equilibrium, but on recalculating this to the present accepted value of D_0(EuO) = 482 kJ mol^{-1} their value for VO changes to 641 questionably too large (Balducci et al., 1983). The thermochemical databases have previously recommended 631 ± 20 kJ mol^{-1} (Barin and Platzki, 1995; Chase, 1998). The situation for the VO_2(g) bond strength was assessed (Schofield, 2006). Four studies were regarded as the most reliable and these now benefit from better thermochemical functions (fefs) being available for VO_2(g) (Balducci et al., 1983). Utilizing these and updating supplemental bond strengths the following values can be obtained for the following gaseous equilibria

	$\Delta H_{f298}(VO_2)$	D_0(Atomiz)	D_0(OV-O), kJ mol^{-1}	
$V + VO_2 = 2\ VO$	−180	1182	561*	Berkowitz et al. (1957a)
	−254	1256	635	Farber et al. (1972)
	−166	1167	546	Balducci et al. (1983)
	−187	1189	568*	Drowart (1985)
$V + 2\ AlO = 2\ Al + VO_2$	−248	1250	629	Farber et al. (1972)
$VO_2 + Eu = EuO + VO$	−184	1186	565*	Balducci et al. (1983)
$VO_2 + P = PO + VO$	−190	1192	571*	Drowart (1985)

A recent paper has measured the relative ionization cross sections for VO and VO_2 appropriate for high temperature mass spectrometric use, indicating the relative values used by Drowart were the most appropriate (Lopatin et al., 2013a). As a result biasing those starred values indicated, now suggests a recommended value of D_0(OV-O) = 564 ± 20 kJ mol^{-1}, 10% weaker than that of the first bond, 621 kJ mol^{-1}, and reflecting the behavior of its neighbor Ti (Table 7.19). This need to modify the thermochemical value especially for VO_2(g) was realized also by Gunina et al. (2012) through inconsistencies seen in their equilibria of vanadium molybdates and tungstates. They also reanalyzed all the available data and likewise now recommend values of D_0(V-O) = 624 ± 10 kJ mol^{-1} and D_0(OV-O) = 563 ± 15 kJ mol^{-1}, which is most gratifying. The two bonds together are now seen to be 48 kJ mol^{-1} weaker than previously suggested by the data bases (Barin and Platzki, 1995; Chase, 1998). Although the interesting vaporization studies of Frantseva and Semenov (1969) and Banchorndhevakul et al. (1985) also managed to derive similarly adequate values for VO_2 from their studies, they were not included in my own assessment due to complexities of their melt compositions and resulting uncertainties. They were included in the Russian assessment but appear not to have modified the outcome.

The potential energy curves for the ground $VO(X^4\Sigma^-)$ and 20/13 of its electronic states have been studied theoretically and compared to spectral observations (Miliordos and Mavridis, 2007; McKemmish et al., 2016). The lowest lying state is at about a 7000 cm^{-1} energy. Matrix isolation experiments of condensed products from laser-ablated V atoms reacting with O_2 have measured the geometries and vibrational frequencies of VO and VO_2 (Knight et al., 1996; Chertihin et al., 1997c). Most recently, a combined matrix isolation and theoretical study has analyzed and assigned the electronic absorption spectra of VO in Ne matrices (Hubner et al., 2015). Computational thermochemistry can now adequately reproduce the reported VO, VO_2 gaseous bond strengths (DeYonker et al., 2007). Several papers have examined the electronic structure of VO_2 and considered the relative stabilities of the OVO, VOO and $V(O_2)$ geometries (Gutsev et al., 2000a; Uzunova et al., 2008; Bande and Luchow, 2008). Although still questionable whether $V_2O_3(g)$ plays an initial role in the vaporization of condensed vanadium oxide melts, the thermochemical properties of this species recently have been calculated for inclusion in such modeling (Gusarov and Osina, 2012).

Few measures of stabilities are available for the corresponding sulfides partly aggravated by the deficiency of thermochemical data. An early value of D_0(V-S) = 445 ± 15 kJ mol^{-1} was obtained from comparisons with GeS and SnS (Drowart et al., 1967). A later mass spectrometric study of the congruent vaporization of VS(s) at 1430°C−1730°C indicated D_0(V-S) = 485 ± 17 kJ mol^{-1} and D_0(SV-S) = 477 ± 20 kJ mol^{-1} (Owzarski and Franzen, 1974). This disagreement now appears to be resolved by an accurate predissociation threshold value in its two-photon ionization spectrum. This indicates D_0(V-S) = 438 ± 8 kJ mol^{-1} supporting the lower value and is now recommended (Johnson et al., 2016). This is also consistent with the trend in the value of the ratio of D_0(V-S)/D_0(V-O) in this section of the periodic table. Values for this ratio for Sc, Ti, V, Cr are about 0.70, 0.69, 0.71, 0.73, respectively (Table 7.19), with those for second row transition elements Y, Zr, Mo and La, 0.74, 0.74, 0.68, 0.72 (Table 7.15). Similarly, assuming a trend in the D_0(SV-S)/D_0(V-S) ratio as that seen for the vanadium oxides, would predict a 0.908 ratio which then suggests D_0(SV-S) = 398 ± 30 kJ mol^{-1} and seems most appropriate at the present time. Theoretical studies adequately describe the electronic structures of the sulfides but are not yet sufficiently accurate to resolve these discrepancies (Bauschlicher and Langhoff, 1986; Anderson et al., 1987; Bauschlicher and Maitre, 1995; Wu et al., 2007). Matrix isolation studies of the VS and VS_2 molecules have recorded their geometries and vibrational frequencies such that better thermochemical properties may now be possible to tabulate (Liang and Andrews, 2002a). Phosphides are not encountered in combustion due their tendency to oxidize and be hydrolyzed with water producing toxic PH_3. However, gaseous VP has been studied theoretically and is expected to display an $X^3\Delta$ ground state and have a bond strength of about 312 ± 20 kJ mol^{-1} (Tong et al., 2003). Thermochemical tables are almost nonexistent for gaseous vanadium molecules with the possible exception of VCl_2, VCl_3, and $VOCl_3$ (Knacke et al., 1991; Barin and Platzki, 1995). However, the vanadium chlorides can only be of minor relevance to combustion due to their ready oxidation and hydrolysis to the oxides. Nevertheless, for comparisons, their electronic structures and bonding are of chemical interest. Recently their uncertain thermochemistry has been resolved. By Knudsen cell-mass spectrometry the values of D_0(V-Cl) = 426 ± 8 kJ mol^{-1} and D_0(ClV-Cl) = 473 ± 8 kJ mol^{-1} have been reliably established with much weakened bonding seen in the higher VCl_3, VCl_4 chlorides (Hildenbrand et al., 2008). This behavior of the second bond having more strength is not unusual for the metal chlorides and is reflected by its neighbor titanium. Theoretical studies of the electronic structure of VCl ($X^5\Delta$) and its potential energy curves indicate a very close low-lying $A^5\Pi$ state with several others below 5000 cm^{-1} in a complex array of states (Kardahakis and Mavridis, 2009). Similar theoretical studies of VCl_2 establish its geometries, frequencies and states (Wang and Schwarz, 1998; Vogel and Wenzel, 2005), as does an earlier infrared matrix isolation study (Hastie et al., 1971). Although not relevant to combustion, in systems containing molecular Cl_2, erosion of V_2O_5(s) has been noted at temperatures of about 300°C resulting in the formation of $VOCl_2$(g) (Hackert et al., 1996).

With vanadium, metal hydrides continue their low significance in trace metal combustion. A gaseous monohydride has been characterized with a bond strength D_0(V-H) = 206 ± 7 kJ mol^{-1} (Chen et al., 1993). This experimental value is supported by theory indicating complex electronic states with an $X^5\Delta$ ground state. The potential energy curves have been outlined with numerous low-lying states below 5000 cm^{-1} (Walch and Bauschlicher, 1983; Barone and Adamo, 1997; Koseki et al., 2004; Daza et al., 2006; Goel and Masunov, 2008). Theory has addressed the possible dihydride indicating its electronic states and suggesting the second bond could be slightly stronger than the first as with the chlorides (Fujii and Iwata, 1996). With this limited repertoire of condensed and gas phase chemistry it is interesting to examine what has been observed in laboratory and full-scale combustion systems.

Laboratory combustion studies

One such preliminary laboratory study added 0.5% VCl_4 to a H_2/O_2 laminar flame at 2000°C (Farber and Srivastava, 1973). Molecular beam-mass spectrometric sampling analyzed flame gaseous components. The tetrachloride instantly produced the oxides which were only partially vaporized at this temperature but did indicate the presence of gaseous V, VO, and H_2VO_3 ($VO(OH)_2$ or $VO_2.H_2O$). Subsequent simultaneous additions of a potassium salt appeared then to also produce $KHVO_3$(g) with traces of K_2VO_3(g), both tetravalent V. In light of the discussion above concerning the stabilities of gaseous vanadates, such a finding relating to the 2000°C flame gases would appear very debatable. It seems quite possible with this heavy seeding level that the probe sampler perturbed the system. The observed H_2VO_3 may be representing VO_2(g) that hydrated in the probe. Additionally, since no evaporative release from a fuel was involved in this case, the potassium vanadates would have to be formed in the flame gases. Gas phase formation is questionable, but deposition on a VO_2(s) particle in the flame or probe might heterogeneously produce and even vaporize the vanadate. Nevertheless, this remains a complex quandary. In the same study chromium was examined, behaved similarly, providing gaseous heats of formation for H_2CrO_4 and K_2CrO_4 that closely agreed with other previous values. Unfortunately, their values for H_2VO_3 and K_2VO_3 remain unconfirmed. It is interesting to compare the implied stabilities of such gaseous molecules, for example, K_2WO_4, K_2MoO_4, K_2CrO_4, K_2VO_3, and K_2SO_4. From their known heats of formation and hence atomization energies, subtracting the atomization energies of WO_3, MoO_3, CrO_3, VO_2, and SO_3, this provides a measure of the K_2O oxide-oxide binding in these molecules. Not knowing the atomization energy for K_2O (taken as x), since this is the same in all cases, indicates the energies between the two respective oxides as $1309 - x$, $1214 - x$, $1175 - x$, $1106 - x$, $1107 - x$ kJ mol^{-1}. This exercise in itself illustrates the gradual bond weakening across the series. There is no solid evidence that K_2SO_4 is present in 2000°C flames, so one has to seriously question the validity in this case of the vanadates existence and the conclusions by Farber and Srivastava (1973). Vanadium appears to have no effects on soot suppression (Howard and Kausch, 1980), and displayed no ability with respect to flame inhibition (Linteris et al., 2008). Its induced ion chemistry was examined but not noteworthy (Goodings, 1992).

Combustion pilot plant data

Vanadium is significantly more present in crude oil in which its concentration can be variable from around the world but tends to be especially large in Venezuelan oils. Refining now can reduce it extensively from the light fractions such as kerosene, gasoline, diesel and domestic heating oils (20–30 ppmw) but it can remain more concentrated in the heavier residual oil used for large heating and power systems and can approach 1 mg g^{-1} (1000 ppmw) in crude oil itself. In fact the ash from heavy oils is now an economically viable source of the metal. Much Russian work is available that discusses among other things briquetting the ash for metallurgical processing (Daous, 2005; Navarro et al., 2007; Mambote et al., 2008). In oil gasification and coking processes, one product now in large volumes is petroleum coke (pet coke). Its production tends to concentrate trace metals into the coke which is increasingly being used as a useful solid fuel but now has to readdress vanadium's historical corrosive nature in combustion. Moreover, it produces an ash that verges on being listed as either a hazardous waste or a useful by-product as it can contain as much as 20% weight composition. A useful review comparing coal and pet coke use was published a few years ago (Lee et al., 1999a). Other solid fuels such as coal, biomass, municipal wastes or sewage sludge generally have much lower vanadium content, and the ash from which is of less concern (Table 1.1) (Demirbas, 2005; Font et al., 2007; Chandrasekaran et al., 2012). Natural gas is essentially free of vanadium. Limestone can contain low levels ($\leq$3 ppmw) (Lind et al., 1994; Cordoba et al., 2012). A recent study has analyzed the nature of vanadium in 21 Chinese coals. It was present at concentration levels of 3–62 ppmw with one sample of a stone coal displaying 3000 ppmw (Liu et al., 2016e). Like all coals, the vanadium compositions varied widely with silicates being 40%–70%, organics 20%–60%, and sulfides 2%–25%. Volatility testing at 950°C showed an overall average release of 9%. However, this varied with coal type, lignites being as high as 27% and anthracites 2%. The coals also had varying levels of Al and Fe which could have played some role. However, it did show that vanadium could be transported to a certain degree at such a low temperature. Various thermochemical calculations have been published that include vanadium even with its rather sparse and unsatisfactory thermochemical database. Although limited they do suggest that above about 1200°C–1300°C gaseous VO and VO_2 may begin to exist otherwise it will remain in a condensed form. Three have examined primarily coal, emphasizing solely the oxides and oxysulfate

(Frandsen et al., 1994; Danihelka et al., 2003; Jano-Ito et al., 2014). One, the cofiring of coal/sludge/straw introducing the possibility of Fe and Ca vanadates (George et al., 2010). Two others have emphasized either gasification of solid wastes in a fluidized bed or petroleum coke gasification (Konttinen et al., 2013; Li et al., 2016). The most recent of these is the most detailed, concentrating solely on vanadium. It indicates little vaporization below 1300°C with the possibility of Ca, Fe, and Na vanadate formation within the ash particle. It does not favor Al, Si, or Ni and if sulfur concentrations are high this will preferably react with the Ca and Fe so removing these from possible vanadate formation. These of course are all predictions, assuming no constraints and obviously require validation. What is observed with coal, biomass or sludge combustion is that the vanadium distributes itself between the bottom ash and the fly ash with very slight enrichment apparent in the latter and with no gaseous component obviously apparent (Querol et al., 1995; Martinez-Tarrazona and Spears, 1996; Sandelin and Backman, 1999; Nielsen et al., 2002; Miller et al., 2004; Amand and Leckner, 2004; Goodarzi, 2006). However, due to the normal slippage of small particles through an ESP, and even with FGD control, there remain low levels of very fine particle release to the stack emission that was about 0.7% 40 years ago (Klein et al., 1975), and can now be about 0.15% or less today (Nielsen et al., 2002; Goodarzi, 2006, 2006a). A more recent measure of the effectiveness of a FGD system in a co-combustion plant burning coal and pet-coke indicated any stray V was further reduced threefold across this control device (Cordoba et al., 2012). Bench-type combustion and pyrolysis experiments confirm no release of vanadium from coal below temperatures of about 1150°C (Wang and Tomita, 2003; Chen et al., 2008). Fluidized bed combustion of coal also can show no sign of vanadium vaporization due to the lower temperatures (Lind et al., 1994). Similarly, no effects of chlorine were noted in any study with coal/sewage sludge samples at temperatures up to 1100°C (Folgueras et al., 2007). Several reports have been made of the fine particles that pass through ESPs. It was apparent from initial measurements that the smaller ESP outlet particles were enriched in vanadium (Klein et al., 1975; Coles et al., 1979) as were stack emitted particles (Meij, 1994). Also with biomass/peat/refuse fuels a similar behavior was noted as enrichment in the second field of ESPs (Lind et al., 2003). Other coal studies reported such enrichment in particles as the small micron sizes are approached (Querol et al., 1995; Galbreath et al., 2000), less than 5 μm (Martinez-Tarrazona and Spears, 1996), <2.5 μm (Nielsen et al., 2002) or in the submicron fume of 11 selected coals (Quann et al., 1990). Even more detailed studies of size segregated coal ash have examined the bimodal distributions of the micron sizes peaking at 2 and 0.05 μm. Three studies are noteworthy in confirming the volatile nature of vanadium at the higher temperatures with deposition and enrichment on these very fine coal ash particles (Quann et al., 1982; Kauppinen and Pakkanen, 1990; Wang et al., 2008). These are particularly intriguing in clearly showing the rich composition of elements even in particles <0.06 μm illustrating this wide range some of which have deposited from the vapor phase and some that have remained in their condensed forms but have been expelled from larger particles and become the basis of condensation nuclei. Deposition enrichment from the vapor is characterized by its increase to smaller sizes and is seen for vanadium.

The burning of heavy, residential or diesel oils, or pet coke differs from the burning of other solid fuels. Depending on the degree of refining, vanadium and nickel can become more important and rather than having ash particles heavily based on Si/Al/Fe/Ca, oils will have much lower levels of these. The lighter oils will produce cleaner burned gases possibly with only some carbon char. The combustion gases can be less heavily loaded with condensation nuclei surfaces that induce heterogeneous chemistry and begin to approach the laboratory clean homogeneous-type research flames to which controlled additions can be made.

Relating such a discourse to oil combustion, species such as VO(g) and VO_2(g) would be expected to form on oil combustion due to their easier release from a liquid fuel. Temperature then will determine vanadium's transport, and if carbon or other trace species form potential condensation nuclei they can deposit as an oxide or may even homogeneously nucleate if their concentrations are sufficient, which is possible in heavy, less refined oils. What has been recorded is that particles are formed that are small, $<2.5\ \mu m$ and have a unimodal distribution that is narrow and centered on about 0.1 μm (Linak and Wendt, 1993; Linak et al., 2000, 2000a). More recent drop-tube furnace work has confirmed this potentially significant concentration of vanadium in these very small particles (23% in those $<0.1\ \mu m$) (Jang et al., 2007). This tends to illustrate that in many countries any health concerns with vanadium should be directed more to oil combustion and now also to pet coke ash, that can be $\leq 20\%$ vanadium, rather than to coal. Even the more refined diesel oil for vehicular use still contains vanadium levels that seem to be of some concern. In gasification studies involving pet coke the lack of detecting vanadium was thought to be due to its presence as Ca and Mg vanadates (Blasing et al., 2014). However, in a high pressure (60 atm) gasification of asphaltene, deposits of V_2O_3 were noted below about 1500°C in the syngas cooling tubes (Wang et al., 2016a). A Taiwanese study indicated diesel vehicle exhaust was the largest anthropogenic source of vanadium in that country (Wang et al., 2003b). Another report from the same group, size-segregated the particles into nano (0.01−0.06 μm), ultrafine (0.1), fine (<2.5) and coarse (>2.5) and confirmed that about 38% of the emitted vanadium mass was in the $<0.1\ \mu m$ sizes (Lin et al., 2005).

One aspect that remains uncertain with vanadium concerns its final chemical nature in any type of ash particle. This remains difficult to resolve due to the complexity that arises from the presence of the many other components (Wang et al., 2008). In the case of vanadium, even with some heavy oil combustion ash or pet coke ash that is significantly enriched in vanadium it is still not in isolation but in the presence of other chemistries. Numerous attempts utilizing mainly X-ray absorption methods such as EXAFS and XANES spectroscopy, or electron beam methods to establish the valences of vanadium in ash simply have illustrated the lack of sensitivity and potential difficulty for such molecular characterization (Silk et al., 1989; Huffman et al., 2000; Silva et al., 2013). This was attempted also for IGCC fly ash produced at 1700°C under reducing conditions (Font et al., 2005). Attempts to analyze pet-coke ash in a FBC mode may have detected calcium vanadate formation using XRD (Jia et al., 2002). Such studies suggest the presence of V^{III} and V^{IV} valences but remain unpersuasive. Leaching methods might give clues and a chromatographic/ICP mass analysis method identified V^{IV} and V^{V} in coal ash but of course whether this is an oxide or a vanadate is not clear (Wang et al., 1995). Solubility can help a little in that the oxides are insoluble but some vanadates (Na, K, Ca, Mg) are soluble in water. However, even this is obviously complex and is mainly of value in establishing which elements leach under different pH conditions (George et al., 2008; Dan et al., 2013). Currently there is no easy solution for exact molecular identification of ash forms.

Consequently, there remains only speculation as to whether a stable oxide of the V_xO_y structure is produced especially in the case of oil combustion, or a metal vanadate. Moreover once in the condensed phase also whether this gradually modifies over time by oxidation or if sulfur is present by even conversion to the sulfate. With coal ash the ever presence particularly of higher concentrations of Al, Fe, Ca, Na, and K, for example, can undoubtedly facilitate vanadate formation on cooling from oxide deposition. Boilers and gas turbines using heavy oils have been plagued for more than half a century with alkali vanadate corrosion (Bryers, 1996). These molecules have low melting points ($NaVO_3$, 630°C) and are effective liquid flux materials that induce corrosion. Remediative methods developed over the years include washing the salt out of the oil, and/or adding magnesium or nickel in excess of the alkali impurity to preferentially deposit benign magnesium or nickel vanadate. These have higher melting points (1010°C–1100°C) and remain condensed and are less corrosive. Also on turbine blades these are friable and less accumulating (Rocca et al., 2003, 2008). A recent study noted the possible preferential formation of $Mg_2V_2O_7$ and $Ni_3V_2O_8$ in a laboratory flameless oxy-combustion mode with fuels containing vanadium at 1200°C–1600°C (Di Salvia et al., 2015).

Although vanadium is not yet a major player in nanotechnology, there is growing interest in the properties of oxide particles such as VO_2 (Kamat, 2011). Such oxides can be readily synthesized by introducing, for example, volatile vanadium halide molecules into flames (Wooldridge, 1998), or formation of mixed oxide composites is possible such as with V_2O_5/Al_2O_3 or V_2O_5/TiO_2 (Roth, 2007; Ren et al., 2015). Recent studies find that particle size can be controlled by modifying the combustion conditions (Wu et al., 2018b).

In summary, our knowledge of vanadium's chemistry facilitates a still rather qualitative understanding of observations in combustion systems. Nevertheless, there is sufficient information to illustrate the importance of temperature and combustion mode in the possible role of some probably short-lived transport by volatile vanadium oxides at higher temperatures. Chlorine or sulfur appear to be of little importance. The chemistry is predominantly one of oxides or vanadates. Environmental concerns tend to mainly involve its leaching from petroleum coke and ash deposits rich in vanadium.

7.32 Yttrium

Yttrium is chemically very similar to lanthanum and is mainly found in rare earth minerals as well as uranium ores. It is traded as one of the rare earths and is the second most abundant of these in the Earth's crust after cerium. It has an average concentration of about 33 ppmw in the Earth's crust and is the 29th most abundant element being about the same as Co and more than As or Pb. Nevertheless, it is mined with rare earth sources, which at present are mainly located in China. Its concentration in a suitable source can range from minor to 65% (Chen, 2011). Four rare earth elements are dominant in sources, namely Y, Ce, La, Nd, and these constitute about 85% of the rare earths mined. Demand for Y has increased in recent years and production is about 12,000 tons $year^{-1}$ of the oxide,

Y_2O_3 (Du and Graedel, 2011). Reserves are abundant and estimated at 10 million tons. Use of the oxide is dominant and only small tonnage of the pure metal are produced. Relative to its lighter weight periodic-table member scandium it is more abundant and less costly. Current prices are about \$80 kg^{-1} for the metal and \$20–40 kg^{-1} for the oxide. Similarly, yttrium is not regarded as a pollutant and is not currently included in any drinking water standards. If global wind borne soil/dust particles are of a magnitude 2×10^{14} g $year^{-1}$ as estimated (Nriagu, 1989), one might expect a natural global atmospheric background of yttrium in dust particles of about 7000 tons $year^{-1}$.

Health concerns with yttrium

The element appears to have no critical role in life forms. Its biology and biochemistry have been extensively summarized along with closely related scandium in a two-volume book series (Horovitz, 2000). These review diet intake and nutritional studies that utilized several of yttrium's isotopes ^{88}Y(107 days half-life), ^{91}Y (59 days half-life), and ^{90}Y (2.7 days half-life) with humans and animals. These efforts illustrate the large volume of studies that have been made over the years even with this element that are necessary to obtain a reasonable understanding of behavior in living organisms. It appears to be nonessential for life but being in the body is not totally inert but plays minimal roles. Total quantities in the human body have been estimates at <0.5 mg. It occurs in edible plants and seeds on a ppmw basis but does not appear to accumulate in the body. Dietary intake has been approximated and is very low in children 1.0 μg day^{-1} increasing to about 10 μg day^{-1} for adults. Water insoluble compounds are rated nontoxic and soluble ones mildly toxic. Its toxicity is more of a concern with inhalation and certain exposures can produce lung diseases. Industries are not regulated but recommendations for air quality have been established by US Occupational Safety and Health establishments. With the advent of nanoparticle production of Y_2O_3 such concerns now may have to be readdressed (Gojova et al., 2007). At present it is not too surprising that no testing for yttrium is undertaken by the available test centers or even included in any surveys. It has been found only in one recent blood survey of just over 100 inhabitants each in a comparison between Morocco and Canary Islands' populations (Henriquez-Hernandez et al., 2018). Yttrium was detected in just under 10% of each group with 25%/75% values in the range 0.3–0.7 μg L^{-1} for each location, indicating its trace quantities in human diets.

Yttrium's basic chemistry

The metal, yttrium is relatively stable in air (m.pt.1522, b.pt.3334°C). It can react with water at 750°C due to oxide pacifying formation on its surface. Small particles can ignite in air at 400°C and very fine nanoparticles can become hazardous in air. Its major molecule is the oxide, which is a refractory material with low thermal expansion and shock proof properties. This has wide uses as a phosphor, in alloys with Al or Mg, in garnets, ceramics and glass, as a superconducting component and for jewelry. Many have used

YAG (yttrium aluminum garnet) lasers. Medically, one of its isotopes finds use in cancer treatments. Yttrium chemistry is again trivalent reflecting very closely that of scandium so a nonvolatile nature seems predictable from the start.

The refractory oxide, yttria, Y_2O_3, is the most studied of yttrium's compounds with a high melting point of 2439°C and boiling point of about 4300°C and is insoluble in water. It is one of the most stable refractory oxides. As a result of its nature, its vaporization is difficult to examine and has invoked high temperature Knudsen cell-mass analysis and effusion rates in one case. Vaporization is dissociative to YO(g) and O(g) and is congruent. The fraction of Y(g) observed is always one to two orders of magnitude smaller and regarded as negligent. Vapor pressures of YO(g) are about 2–3 ppmv at 2250°C, only increasing to about 30 ppmv at 2430°C (Walsh et al., 1960; Ackermann et al., 1964; Ames et al., 1967; Liu and Wahlbeck, 1974; De Maria, 1984; Marushkin et al., 1990; Kablov and Folomeikin, 2015). A trihydroxide can be precipitated from solution, $Y(OH)_3$. This is insoluble and stable at normal ambient temperatures. On heating to about 300°C–400°C, TGA measurements indicate it loses water and forms YO(OH), which at higher temperatures becomes the oxide at about 700°C (Sato et al., 1988; Titov et al., 2008). A matrix isolation study with laser-ablated Y reacting with water has confirmed that hydroxide structures can be produced with assignable shapes and vibrational frequencies (Zhang et al., 2001). A $Y(OH)_2$ was noted and that HYO appeared favored over YOH. The carbonate, $Y_2(CO_3)_3$, similarly insoluble in water decomposes via certain oxycarbonates to the oxide on heating to 600°C–700°C (Titov et al., 2008). TGA measurements similarly follow the decomposition of the nitrate, $Y(NO_3)_3$, to the oxide being complete by about 450°C–500°C. Yttrium, like the rare earths, has a sulfate, $Y_2(SO_4)_3$ that at 920°C losses sulfur forming an oxysulfate, $Y_2O_2(SO_4)$ that by 1248°C has changed to the oxide (Nathans and Wendlandt, 1962). The chloride YCl_3 is one of the few yttrium compounds that displays volatility and is not prone to hydrolysis. Both condensed and gaseous phases are tabulated in one thermochemical database indicating a melting point of 721°C and boiling point of 1484°C (Knacke et al., 1991). Early mass analysis indicated a predominantly monomer vapor (McKinley, 1965). The vapor pressure is now accepted to be about 2 ppmv at 700°C rapidly rising to 60 ppmv at 800°C. The condensed phase can absorb and desorb water with no chemical change to the initial YCl_3. However, at elevated temperatures in the gas phase some hydrolysis to the YOCl molecule with loss of HCl has been suggested. Additionally, heating Y_2O_3 in the presence of CCl_4, or heating the oxide in chlorine, or when mixed with certain chloride salts such as KCl or NaCl, can induce chlorination of the oxide producing high purity YCl_3 for electronic applications (Miller et al., 1959). In other cases, when YCl_3 is gaseous in reducing H_2/CO_2 mixtures at about 1000°C–1300°C, CVD of pure Y_2O_3 is reported onto a substrate at 1090°C (Sipp et al., 1992). The condensed oxide can be chlorinated and the gaseous trichloride can be readily reduced back to the oxide. These three common compounds, the nitrate, sulfate and chloride are all water soluble. Of the sulfides both YS and Y_2S_3 are known. A Knudsen cell-mass spectrometric study of YS(s) at 1780°C–2080°C indicated congruent dissociative vaporization to YS(g), Y(g) and S(g) (Steiger and Cater, 1975, 1975a). Y_2S_3 also has a high melting point (1925°C) and although not studied is expected to vaporize similarly by dissociation. These sulfides are readily oxidized in air above 500°C. The phosphide, YP, was studied similarly at 1830°C–2030°C in a Knudsen cell and found to vaporize solely to Y(g) and P, $P_2(g)$ (Franzen and Hariharan, 1980).

TABLE 7.44 Properties of yttrium's condensed phase molecules relevant to combustion, °C.[a]

m.pt. b.pt. vapor pressure
Y 1522 3334, 100 ppmv/1800°C, ins
Y_2O_3 2439 4300, refractory oxide, 30 ppmv YO/2430°C, ins
$Y(OH)_3$ dissoc 400 to YO(OH), ins
YO(OH) dissoc 700 to oxide, ins
$Y_2(CO_3)_3$ dissoc 600–700 to oxide, ins
$Y(NO_3)_3$ dissoc 450–500 to oxide, s
$Y_2(SO_4)_3$ dissoc 920 to $Y_2O_2(SO_4)$ and >1250 to the oxide, s
YCl_3 721 1484, monomeric vapor, 60 ppmv/800°C, hydrol higher temperatures to YOCl, s
YS >2000 dissoc to Y(g), S(g), oxidized in air, ins
Y_2S_3 1925, dissoc on vaporization, oxidized in air, ins
YH_2 dissoc on vaporization, 30 ppmv/416°C
YH_3 >100 dissoc on vaporization, readily oxidized, 250 ppmv/180°C
YP >2000 dissoc to Y(g), P, $P_2(g)$
YPO_4 (mineral xenotime), >200 dissoc to oxides
$Y_3Al_5O_{12}$ (YAG) synthetic refractory garnet
$Y_3Fe_5O_{12}$ (YIG) synthetic refractory garnet

[a]*dissoc*, dissociates; *s*, water soluble; *ins*, water insoluble; *hydrol*, hydrolyzes.

The phosphate, YPO_4, essentially the mixed oxides of Y_2O_3 and P_2O_5 is not thermally stable and begins to dissociate by 200°C (Bamzai et al., 2013). As with other transition elements, metal hydrides are known for yttrium. YH_3 is unstable in air and dissociates to the dihydride on heating and then to the metal above about 200°C (Gabis et al., 2013). Like Sc, an ESR spectrum of the dihydride was obtained using matrix isolation of its vapor and suggested a nonlinear gas structure (Knight et al., 1981). It is volatile and similarly dissociates on vaporization. Consequently, yttrium's condensed phase chemistry reflects that of scandium with only the chloride showing any signs of volatility. Like scandium, yttrium has a strong cationic nature and few mixed metal oxides are known, two synthetic ones are refractory garnets, YAG with Al and YIG with Fe. The physical properties of these condensed phase molecules relevant to combustion are summarized in Table 7.44.

Gas phase stabilities and transport

As noted in the gas phase, vaporization of $Y_2O_3(s)$ is to YO(g), which is very stable. About 19 equilibrium type studies of this before 1980 were critically evaluated indicating a slight scatter of the data with a recommendation of $D_0(\text{Y-O}) = 715 \pm 11$ kJ mol^{-1}

($\Delta H_{f298K} = -46 \pm 10$ kJ mol^{-1}) indicative of very strong bonding (Pedley and Marshall, 1983). Since then, an unconnected approach obtained a very accurate measure of the ionization potential of YO that implies D_0(Y$^+$-O) − D_0(Y-O) = 10.04 ± 0.01 kJ mol^{-1} (Linton et al., 1999). Two values are available for D_0(Y$^+$-O), one from guided ion-beam mass spectrometric reaction cross section measurements (Sievers et al., 1996), and one utilizing collision-induced dissociation in an ion trap (Jackson et al., 2002). Their respective values were 699 ± 17 (suggested as a possible lower limit due to potential small contributions from low-lying states of Y$^+$), and 718 ± 25 kJ mol^{-1}. The latter value would imply D_0(Y-O) = 708 ± 25 kJ mol^{-1}, consistent with earlier trends. Consequently, a current recommendation of D_0(Y-O) = 712 ± 15 kJ mol^{-1} appears appropriate.

Until the guided ion-beam mass spectrometric study of the threshold energy of the reaction

$$YO^+ + NO_2 = YO_2 + NO^+$$

no information was available concerning the gaseous dioxide. This threshold energy involves only a need for D(NO-O), and the ionization energies of NO and YO to obtain a value for D(OY-O). Consequently, it is now quite accurately known updating the data with the more accurate YO ionization energy (Linton et al., 1999). This gives a value of D_0(OY-O) = 374 ± 15 kJ mol^{-1} now recommended (Clemmer et al., 1992), reflecting like scandium a very similar much weaker second bond. The YO and YO_2 molecules have been observed by their FTIR spectra in a matrix isolation study of laser-ablated Y in oxygen (Andrews et al., 1999).

Theoretically, the nature of their structural bonding now appears to be reasonably described and the bond lengths, shape and energies reproduced (Siegbahn, 1993b; Song et al., 2007). The YO(g) molecule has doublet and quartet spectral systems with $X^2\Sigma^+$, no low-lying states and as yet the quartet transitions remain unobserved (Huber and Herzberg, 1979; Langhoff and Bauschlicher, 1988). Similarly, the low-lying states of YO_2 have been observed in photoelectron spectroscopy (Wu and Wang, 1998) and characterized theoretically (Lee and al.2008a). Due to the highly exothermic oxide reactions that can be seen with the lanthanide type elements they are well known for being precursors for chemiluminescence and chemiionization. Reactions such as

$$Y + O_2 = YO + O \qquad \Delta H_0 = -216 \text{ kJ mol}^{-1}$$
$$Y + O = YO^+ + e^- \qquad \Delta H_0 = -109 \text{ kJ mol}^{-1}$$

have been extensively studied due to their ability to form products in electronically excited states and as a potentially low-temperature stable source of electrons without invoking thermal ionization (Schofield, 2006). In the present case the thermal ionization potential of Y is 600 kJ mol^{-1} clearly showing the value of chemiionization. This behavior is common to all six elements in the first two periodic table columns of the transition elements. With interest in chemical lasers and electron chemical sources these became significant fields of study in their own right for many years (Fontijn, 1992; Schofield, 2006). Ionization of gaseous yttrium has been studied in this connection in fuel-rich and lean $H_2/O_2/N_2$ flames at temperatures of 1900°C–2425°C (Patterson and Goodings, 1995). Nebulizing a fine aqueous spray of yttrium nitrate solution into the flames at ppmv levels produced red-dish/pink chemiluminescence from electronically excited YO and, although most of the yttrium produced oxide particles, ions were also detected utilizing a probe and mass spectrometer. Hydrated ions of yttrium oxide, $YO^+ \cdot nH_2O$ ($n = 1-4$) were monitored and whether YO^+H_2O or $Y(OH)_2^+$ are true flame ions requires confirmation as does their formation mechanisms. Nevertheless, such ions are at low concentrations and negligible in comparison to the main neutral reactions. Whether gaseous hydroxide chemistry with lanthanides plays any role in the overall combustion remains unknown. Nevertheless, pure rotational spectra of YOH and YOD($X^1\Sigma^+$) have recently been recorded indicating frequencies and bond lengths. A supersonic expansion method was used with laser-ablated yttrium with water (Halfen et al., 2015).

As indicated above, YCl_3(s) vaporizes intact with a melting point of 721°C, where it has a vapor pressure on a ppmv scale. At higher temperatures it dissociates and the spectroscopy of YCl(g) has been extensively studied. Its ground state is $X^1\Sigma^+$ and many of its low-lying states have been identified (Ye et al., 2007) and agree with earlier theoretical predictions (Langhoff et al., 1988). One interesting aspect is that the lowest $A^1\Delta$ at about 10,000 cm^{-1} (120 kJ mol^{-1}) is metastable, and not coupled radiatively to the ground state. Another aspect is that the additional triplet states that have been studied less, also have a low-lying metastable $a^3\Delta$. Its energy relative to $X^1\Sigma^+$ still remains somewhat uncertain. From a singlet/triplet intercombination band it has been suggested to lie about 6500 cm^{-1} (78 kJ mol^{-1}) above the ground state (Xin et al., 1996). No experimental measure of the bond strengths of YCl or YCl_2 appear to have been made. Several theoretical papers, spread over 25 years, seem to give an adequate description of the monochloride bonding suggesting values of 517 (Langhoff et al., 1988), 518 (Siegbahn, 1993), 472 (Cheng et al., 2007), and 504 kJ mol^{-1} (Ma et al., 2012). It is agreed that the molecule is largely ionic but when the bond strength is calculated with a Rittner polarizable-ion model it only suggests 452 kJ mol^{-1} (Struck and Baglio, 1991). Until measured, a value of D_0(Y-Cl) = 500 ± 20 kJ mol^{-1} might be tentatively accepted. The dichloride is less studied but it is predicted to be bent and have a second bond about 60 kJ mol^{-1} weaker than the first (Siegbahn, 1994). However, one mass spectrometric study of chlorine ablating heated yttrium surfaces observed only YCl_2(g) with no mono- or trihalide, but may have been modified by the heterogeneous nature of the experiment (McKinley, 1964). The heat of formation of YCl_3(g) is known and indicates an atomization energy at 298 K of 1490 kJ mol^{-1} (Kubaschewski et al., 1993). This indicates an average bond strength of about 500 kJ mol^{-1} for each of the three bonds. This level of magnitude is shown by its neighbor zirconium. Consequently the value for the dihalide second bond should not be too much lower or it pushes the values for the other two bonds probably too high. The periodic table values are of little help having shown that the second bond is the strongest for Sc, Ti, and V in the first row of transition element chlorides but displaying some semblance of a trend (Table 7.19). However, in the second row of corresponding elements, Y, Zr, and Nb, the average bond strength of the first three chloride bonds is even larger for each, being 500, 522, and 474 kJ mol^{-1}, respectively (Keneshea and Cubicciotti, 1969). Consequently, any weakening in the second bond for these would imply a very strong third bond. At present the first two bonds in Zr are closely equal (Table 7.15). Consequently a value of D_0(ClY-Cl) = 480 ± 20 kJ mol^{-1} might be more appropriate. As a result these chloride bond strengths remain somewhat uncertain, obviously strong, but in need of further study. Nevertheless, such bond strengths still remain far below that of YO(g) and as observed, additions of YCl_3 to flames immediately produces the oxide with no indication of halide spectral emission.

As indicated above, vaporization of YS(s) and Y_2S_3(s) is mainly dissociative. As a result, Knudsen cell-mass spectrometric studies aimed at measuring the YS(g) bond strength invoked the isomolecular method where YS(g) for example would be in equilibrium with GeS(g) and the two metal elements. In this way, D_0(Y-S) = 526 ± 10 kJ mol^{-1}was obtained and is still accepted (Coppens et al., 1967; Steiger and Cater, 1975, 1975a). Although readily oxidized in air above about 500°C, there has been significant interest in characterizing its spectroscopy. The ground state has $X^2\Sigma^+$ symmetry and other states observed are the $A^2\Pi_{1/2,3/2}$, $B^2\Sigma^+$ and $^4\Pi$. Most recently the potential energy curves of 55 low-lying states have been calculated and the spectroscopy reviewed (Farhat et al., 2013). There are no low-lying states below 10,000 cm^{-1} that might contribute to its thermodynamic behavior at combustion temperatures. An LIF study observed a transition from a high lying state to the ground $X^2\Sigma^+$ (Zang et al., 2015). Theory appears to describe the bonding well and reproduces the bond strength (Siegbahn, 1996). The ratio of D(YS)/D(YO) is 74% the same as its neighbor Zr and similar to Mo so displays a consistent trend (Table 7.15). The transition metal hydrides are of little interest in most fuel combustion but have been extensively studied generally as potential sources of hydrogen. Also because yttrium hydride only has four valence electrons its structure is of particular interest for theorists but is more complicated than expected and also is experimentally difficult to examine. In fact in the present case although YH(g) now has been characterized both theoretically and experimentally in detail no measured bond strength yet exists for this or a possible dihydride. An FTIR matrix isolation study of laser-ablated Y in H_2 established the structures and vibrational frequencies of $YH_{1,2,3}$ (Wang et al., 2002). YH ($X^1\Sigma^+$) is the ground state and the excited singlet and triplet states with their potential energy curves now have been well predicted (Langhoff et al., 1987; Balasubramanian et al., 1989; Balasubramanian, 1990; Koseki et al., 2004). Experimentally, many of these lower lying states also have been analyzed and the singlet and triplet manifolds coupled (Ram and Bernath, 1994; Jakubek et al., 2002; Ram et al., 2011). The lowest lying state is $a^3\Delta_1$ with an energy of 6205 cm^{-1}(74 kJ mol^{-1}) and so can play a role in high temperature thermodynamics. No other states lie below 10,000 cm^{-1}. Four theoretical calculations are in substantial agreement that the bond strength is about D_0(Y-H) = 275 ± 10 kJ mol^{-1} (Langhoff et al., 1987; Balasubramanian, 1990; Siegbahn, 1993; Koseki et al., 2004). One theoretical study tentatively indicated a 47 kJ mol^{-1} weaker second bond in YH_2(g) that may be reasonable (Siegbahn, 1994). It correctly predicted a bent structure but requires confirmation of the bonding. The gas phase stabilities of these yttrium molecules are summarized in Table 7.15 where trends with other second and third row transition elements can be compared. Consequently, this chemistry confirms minimal options for yttrium in combustion. Supplemental information from such exercises as CVD experiments confirm this. If salt solutions such as yttrium nitrate or organometallic solutions are sprayed into a flame then Y_2O_3 particles result. The method is used to produce nanoparticle powders (Karthikeyan et al., 1997), or together with zirconium can produce film deposits of ceramic thermal barrier coatings of Y_2O_3/ZrO_2 on gas turbine blades (Choy and Vyas, 1999). An examination of what is reported in practical combustors also seems to support this.

Combustion power plant data

Yttrium occurs in coal generally at levels of about ≤ 25 ppmw (Linak and Wendt, 1993; Llorens et al., 2001), mined in somewhat larger quantities than its related element scandium, it is evident at similar concentrations in sewage sludge (Eriksson, 2001). There is little mention of it in biomass or wastes and although negligible in pet-coke it can be present in limestone (≤ 5 ppmw) (Cordoba et al., 2012). Although generally regarded as a rare-earth element it is at concentrations in most fuels larger than many other trace metals (Table 1.1). Its behavior in combustion closely mirrors that of scandium showing a nonvolatile nature. Few coal analyses include both Sc and Y but tend to favor one or the other. Even from the earliest studies it was apparent that no enrichment occurred on combustion or pyrolysis and that bottom ash, slag and fly ash had similar compositions (Smith et al., 1979; Querol et al., 1995; Martinez-Tarrazona and Spears, 1996; Llorens et al., 2001; Spears et al., 2004). They noted also that there was no enrichment on ash particles as a function of size. It was reported that bottom ash leached with water five times more than fly ash but this still remained extremely low with $\leq 0.01\%$ yttrium loss (Llorens et al., 2001). One study examined coal/sewage sludge pyrolysis at temperatures $\leq 1100°C$. The sludge had been treated with $Ca(OH)_2$ and $FeCl_3$ and contained 4.2 ppmw Y and 1.2% Cl. The coal had a Y content of 18.5 ppmw and 0.1% Cl (Folgueras et al., 2003). The nonvolatility was unchanged and obviously not affected by the chlorine. This was reported also when it was noted that coal ash released no yttrium when heated at 1000°C with the addition of either Cl_2 or $CaCl_2$ (Chan and Kirk, 1999). Currently, there are no regulations concerning yttrium emissions. With suitable particulate control and FGD, the stack loss has been noted as being very low (Kaakinen et al., 1975; Cordoba et al., 2012). As with Sc, coal ash now is being more closely examined as a possible resource for the rare earth elements as these become of more strategic interest (Taggart et al., 2016). Numerous studies are available in attempts to speciate the nature of the yttrium in the ash to facilitate leaching (Taggart et al., 2018; Pan et al., 2018; Bartonova et al., 2018). Coals that have higher concentrations of Y (95–126 ppmw) have been suggested as viable for such resource extraction (Folgueras et al., 2017).

In summary, yttrium mirrors its neighboring elements in the first columns of transition elements in the periodic table. They display refractory or ceramic type oxides, volatile only at high temperatures with dissociation. The temperatures needed for release from fuels generally imply coal combustion with the yttrium remaining mainly in the ash with limited signs of release. Any such sign of enrichment to smaller diameters is more a reflection of particle fragmentation. Chlorine or sulfur appears to be of little importance. The natural reduction in fuel volume now is making fuel ash a new possible element source. This is the case with yttrium which is classed with the rare earths. There are no concerns about health or environmental issues and yttrium raises no combustion issues but remains a very important element in commercial applications and nanotechnology. It is mostly known in chemistry for its chemiluminescent and chemiionization properties. In combustion it plays no role in soot suppression or flame inhibition

7.33 Zinc

Zinc is quite a common element being 23rd in ranking of concentrations in the Earth's crust. Rather unusual, it has about the same ranking of presence in the oceans. With atomic number 30, zinc is a lightweight metal, the last in the first row of transition elements. However, because the elements Zn, Cd, and Hg have no molecules where the *d* electronic shell is not filled they are generally regarded chemically as nontransition elements simply having two electrons in their outer shell, ionized zinc resembling ionized magnesium in some ways. Industrially it is the fourth most common metal in use with current global production of about 13.5 million tons $year^{-1}$ with adequate reserves. Some recycling occurs (about 150,000 tons $year^{-1}$ in the United States) and its commercial value of about \$1.00 lb^{-1} is similar to that of Al and Pb and is less expensive than Cu, Ni or Sn. It is found in nature in numerous minerals but ZnS remains the common ore also with some carbonate and oxide. Zinc has a low melting point of 420°C and boils at 906°C. It has a wide spectrum of uses such as in galvanizing, producing alloys of brass or bronze, is used in currencies, in tires as a vulcanizing agent, ZnO as a pigment in paints and ceramics, in batteries and also in cosmetics and medically. The latter stems from its low toxicity and noncarcinogenic nature. It is in fact an important ingredient for life. Atmospheric pollution is in the form of particulates. It is difficult to assess the natural background level for most metals and Zn is no different. The main assessments were made almost 30 years ago and accepted. One group suggested about 45,000 tons $year^{-1}$ with 20% from volcanoes, 16% from forest fires, and the remainder various dust aerosols (Nriagu, 1989, 1989a). This was confirmed, but at about 90,000 tons $year^{-1}$ by a Russian study (Malakhov and Kakhon'ko, 1990). Arguing that there is now much more data available, a Canadian review has very significantly raised these estimates for Zn to 5.8 million, a 60-fold increase over the Russian estimate (Richardson, 2001). For the other metals considered in this Canadian study all the natural background levels were increased by factors for Cd (40-fold), Cu (80), Pb (90), Hg (5), and Ni (60) over earlier estimates. These values are very large and not acceptable. They would dwarf anthropogenic levels and suggest no noticeable change from preindustrial times. As seen below, the fact that China is now thought to emit about 22,000 tons $year^{-1}$ may possibly indicate from its magnitude that any global atmospheric emission may be no higher than 100,000 tons $year^{-1}$ for Zn, further questioning these large new natural estimates. Anthropogenic sources were thought 25 years ago to be about three times larger than the then considered background (Nriagu, 1989, 1989a). These now probably are much reduced as with cadmium by regulation (Boyd et al., 2009), with decreases of 2.5-fold seen in northern latitudes over the past 30-year period in ice and snow analyses (Boutron et al., 1991). Even so, only 17% was then considered to be due to combustion. The major contribution may remain with the extensive metal smelting and refining industries. The fraction from combustion alone may possibly remain at about 50% in magnitude of the natural background level (Nriagu, 1989, 1989a; Malakhov and Kakhon'ko, 1990). As noted in Table 1.1 listing approximate levels of fuel contents it can have a significant presence in some cases. More appropriately, an interesting recent review has assessed the Zn content in about 20 different fuel types (Jones et al., 2014). Tire-derived fuels are top of the list and can contain Zn up to levels of 17 g kg^{-1} (1.7%). Sludges can vary from an average of 2 g kg^{-1} to much higher. Coal content is lower averaging about 0.1 g kg^{-1} (100 ppmw) with biomass often less. Shale oil though can have up to 2.7 g kg^{-1}. As a result, the major fuels of coal and biomass have generated a less concern as they do not usually display the corrosive problems encountered at higher levels. Pet-coke can contain low levels (≤20 ppmw) while limestone is variable having a similar or negligible amount (Lind et al., 1994; Cordoba et al., 2012). Ice core drilling data in Russia did note the rise in Zn emissions mainly due to metal smelting from 1950, peaking at 1980 from which time they have slowly returned to 1955 values (Eichler et al., 2014). From 1949 to 2012 in China the anthropogenic emissions of Zn appear to have increased 100-fold to a value of 22,000 tons $year^{-1}$. This is the largest of 12 such different metal element emissions of concern for them (Tian et al., 2015). Slightly less than half of this value currently arises from coal combustion. However, a size distribution of the atmospheric particles shows a health concern distribution peaking at 1–2 μm in China (Duan and Tan, 2013). As a result, accurate global natural and anthropogenic atmospheric releases remain poorly defined for Zn at present. An attempt to summarize them appears in one extensive free environmental report published on the Web for zinc (World Health Organization Zn, 2001c). The interface between combustion and soil quality in agriculture is becoming an important study. Ash from fuels and also sewage sludge, raw or burned, are now regarded as potentially enriching methods for soils. A Swedish study did note that Cu, Zn, and Hg can gradually accumulate in soils from sewage sludge use but it did not transfer into the crops. The conclusion was that it did improve the soil fertility (Borjesson and Katterer, 2018). Similarly in China the study was concerned about natural anthropogenic deposition of heavy metals onto agricultural soils (Huang et al., 2019). This noted accumulation of Cd and Hg but found these to be decreasing over time. Zn showed no major accumulation. It is classed along with the heavy metals but is regarded as a valuable ingredient. It is to be expected that more dispersal policies will emerge due to this value both for agriculture and for a method of disposal. Combustion can become significant by specifically tailoring the fuel, cofiring of sewage sludge and biomass, for example, to better balance the beneficial against possible toxic elements in the resulting ash (Zhao et al., 2017b).

Health concerns with zinc

Due to its ever presence in the environment, zinc has always been a component of dietary intake. It is a very important trace metal in the body involved in innumerable mechanisms. In the body it does participates in the formation of over 300 Zn-ligand based enzymes (World Health Organization Zn, 2001c). It is the second most important trace metal in the body after Fe, being extensively involved in brain functions. In fact Zn deficiency is an important risk factor for human health, and deficiencies are found more than excesses (Landner and Reuther, 2004). The fact that zinc is regarded as nontoxic and noncarcinogenic under normal circumstances has reduced the level of concern for its presence in the environment. A normal dietary intake of 10–15 mg day^{-1} has been recommended.

This is generally easily satisfied by meat eaters. Fish can sequester zinc but there is no bioaccumulation. Oysters and mussels are enriched. Vegetarians can acquire their needs from grains, rice, legumes, and cheese. A 10-fold excess of the recommended diet would begin to produce clinical concerns. One study in a zinc smelting region of China did note elevated levels in leafy vegetables such as lettuce and cabbage grown in the vicinity. An estimate of dietary intake there increased to about 50 mg day^{-1} with no adverse effects but has to be close to borderline (Yang et al., 2011; Golghaber, 2003). The nutritional needs for health have been summarized in a detailed World Health Organization free report on the Web (World Health Organization Trace Elements, 1996). In addition to these extensive reports a toxicology report also is available detailing zinc and the body chemistry, including animal and human test results (ToxProfile Zinc, 2005). The body can store Zn and excrete any excess in urine and feces. The interactions between Zn and other metals in the body have been quite extensively examined in this particular case. It can interact and interfere with roles of Cu, Pb, Fe, Cd, and Ca in the body illustrating that care should be taken with trace metal supplements as certain balances can be disrupted. Nevertheless, there are indications that Zn is a beneficial element, especially if coupled to Se, and appears to provide some protection from Cd, Pb, As, and Hg toxicity (Branca et al., 2018; Rahman et al., 2019). Additionally there is evidence that it may inhibit tumor growth in animals. It is well known to be of value for wound healing. However, as it is important in the brain, it is under suspicion for being involved either alone or with another element as involved in the current epidemic of neural illnesses. Numerous reviews have appeared relating zinc to autism through the range of illnesses to Alzheimer's (Mot and Crouch, 2017; Grabrucker and Grabrucker, 2017; del Barrio et al., 2017; Jiang, 2017; Kanninen and White, 2017). These extensively list the known roles of zinc in the brain. These are intriguing hypotheses but they cannot explain why this has the appearance of being triggered especially with autism and some of the other diseases. No environmental changes, either natural or man-made have occurred concerning this element. These illnesses also are not local but global. Moreover, there are no reasons for genetic modifications to occur abruptly, As a result, unlike mercury, zinc should not be so categorized.

It is of value to test for zinc in the human body. The biomarkers of blood, serum, urine and hair now have been standardized as fully acceptable and available (Bornhorst et al., 2018). Test centers recommend acceptable ranges for blood as 4–8 mg L^{-1}, urine <1 mg L^{-1} and hair, which can be more variable as <250 ppmw. The fact that the unit is now mg and not μg reflects the importance of zinc in the body. For interest sake numerous surveys have been published that compare levels for large population samples in at least two biomarkers (Goulle et al., 2005; Nisse et al., 2017; Oggiano et al., 2018). One study on blood testing established the reference level for Italians (Sardinia only), 4.7–8.6 mg L^{-1} for the 5%–95% range (Bocca et al., 2011). Individual values can vary slightly, country to country, but median values remain close, as one comparison between Morocco and the Canary Islands indicated (Henriquez-Hernandez et al., 2018).

Zinc's basic chemistry

Of zinc's compounds normally encountered, only $ZnCl_2$ tends to be a hazardous compound, deliquescent and corrosive as a skin irritant, but is not normally lethal. In practical combustion it is this compound that is of most interest. As will be seen, it can be a component under certain conditions in boilers for producing sticky, corrosive deposits (Jones et al., 2013).

The chemistry of zinc is interesting in that it is predominantly inorganic and divalent, and as will be displayed tends to foretell its behavior in combustion. The physical properties of its condensed phase molecules are discussed herein and summarized in Table 7.45.

More so than Cd, zinc does not biomethylate in nature as does Hg. It has no specific organic nature and is in the environment essentially as Zn^{2+}. It is true that dimethyl (m.pt. − 42, b.pt.46°C) and diethyl zinc (m.pt. − 28, b.pt.117°C) liquids were possibly the first organometallic molecules ever to be synthesized. Nevertheless these are pyrophoric and handled by their solution in alkane solvents. Otherwise the Zn-C bonding is rare and not found. In nature zinc has a full range of condensed phase molecules but many are not thermally stable at raised temperatures dissociating to the more stable ZnO(s).

TABLE 7.45 Properties of zinc's condensed phase molecules relevant to combustion, °C.[a]

m.pt. b.pt. vapor pressure
Zn 420 906, 100 ppmv/397°C, ins
ZnO 1974, dissoc >1100, ins
ZnO_2 212, dissoc >150
$Zn(OH)_2$ 125 decomp, ins
$ZnCO_3$ >425 dissoc, ins
$Zn(NO_3)_2$ >180 decomp, s
$ZnSO_4$ >370 decomp via oxysulfate, s
$ZnSO_3$ >200 decomp, s
$ZnCl_2$ 290 732, 76 ppmv/350°C, reacts w water, s
Zn_2Cl_2 unstable, disproportionates
ZnS >400 decomp, 11 ppmv/850°C, ins
ZnH_2 >90 decomp
Zn_3P_2 stable 350–550, reacts w water, PH_3 release
$Zn_3(PO_4)_2$ 900, ins
$ZnAl_2O_4$ 1950, ins
$ZnFe_2O_4$ 1590, ins
$ZnSiO_3$ 1437, ins
Zn_2SiO_4 1509, ins
Zn_2TiO_4 1549

[a]*Sparingly water soluble zinc salts review (Clever et al., 1992): decomp*, decomposes; *dissoc*, dissociates; *s*, water soluble; *ins*, water insoluble.

For example, $Zn(OH)_2$ is insoluble in water (decomposes 125°C). Its structure and bond frequencies have been characterized by matrix isolation (Wang and Andrews, 2005g). Also $ZnCO_3$, as insoluble Smithsonite, decomposes beginning at 425 and ends at 550°C (Beck, 1959), and zinc nitrate (basic salt, stable to about 180°C) (Malecka et al., 2003). The water soluble $ZnSO_4$ is stable to about 370°C then modifies to the more stable basic sulfate $ZnO.2ZnSO_4$ up to 650°C–700°C above which it forms the oxide (Ostroff and Sanderson, 1959; Stern and Weise, 1966; Tagawa, 1984; Tagawa and Saijo, 1985; Hildenbrand et al., 1990; Jones et al., 2013). $ZnSO_3$ decomposes at 200°C and ZnO_2 at 225°C. Produced from solution, this ZnO_2 was found on dissociation to produce ZnO nanoparticles (Uekawa et al., 2003).

Similarly, ZnH_2 (unstable, decomposes 90°C) has been characterized by matrix isolation (Greene et al., 1995; Wang and Andrews, 2004a). Zn_2Cl_2(s) is unstable and disproportionates to $ZnCl_2$(s) (Liao et al., 1995). Those compounds that are more thermally stable are ZnO (m.pt.1974°C), but this extensively dissociates on vaporization to Zn(g) and O(g) products (L'vov, 2007). An earlier Zn(g) vapor pressure of about 1 ppm was recorded at 1000°C above ZnO increasing 1000-fold by 1300°C (De Maria, 1984). Like the oxide, ZnS vaporizes to Zn(g) and S_2(g), its zinc vapor pressure over a powdered form being about 11 ppmv (850°C), 97 ppmv (950°C), and 600 ppmv (1050°C) (Munir and Mitchell, 1969). The formation of ZnS and ZnS_2 were tentatively analyzed in an early matrix isolation study (Martin and Schaber, 1982). $ZnCl_2$ (m.pt.290, b.pt.732°C) is very volatile, 76 ppmv (350°C), 515 ppmv (400°C) (Barin and Platzki, 1995), with its vapor stable to about 1300°C but both phases are subject to hydrolysis by water and oxidation in air (Jones et al., 2013). Other molecules of potential interest are $Zn_3(PO_4)_2$ (m.pt.900°C) and Zn_3P_2(stable to 350°C–550°C) (Schoonmaker et al., 1967). ZnP was characterized by an ESR matrix isolation and theoretical study (Fuller et al., 2010). It was predicted to be very weakly bound in the gas phase (Table 7.19), and being unstable in air or water is of little interest for any combustion mode. Zinc is not known for refractory or ceramic mixed oxide molecules. However, it can produce Zn_2SiO_4 (m.pt.1509°C), $ZnAl_2O_4$(ceramic spinel m.pt. ≈1200°C) and recently $ZnFe_2O_4$ (formed at 900°C–1100°C), all insoluble in water (Suetens et al., 2015).

Gas phase stabilities

The gas phase chemistry is very limited by stabilities that were quite uncertain until recent years. Its gas phase molecular stabilities are summarized in Tables 7.16 and 7.19 where they are compared and trends observed with its periodic table members Zn/Cd/Hg and with the other members of the first row of transition elements. The bond strength values for ZnO(g) derived from Knudsen cell-mass spectrometric equilibria were in dispute but a reliable experimental value based on the guided ion-beam mass spectrometric method indicated a low value now recommended of D_0(Zn-O) = 155 ± 10 kJ mol^{-1}. The method used, established the kinetic threshold of the reaction of $Zn^+ + NO_2$ producing $ZnO + NO^+$. It assumed no activation energy for the reaction, that is normally expected for such reactions, but if present implies the value will be an upper limit (Clemmer et al., 1991). Previous theoretical analyses had predicted the need for a low value and this has now been confirmed further by numerous more sophisticated calculations (Bauschlicher and Partridge, 1998; Harrison et al., 2002; Gusarov and Iorish, 2006; Peterson et al., 2007; Jiang et al., 2012). Previous confusion had arisen in two ways. First its vaporization is extensively dissociative:

$$ZnO(s) = Zn(g) + O(g)$$

One reliable Knudsen cell/mass spectrometric study could not even detect ZnO(g) at 1470°C, implying in their system a vapor pressure of $<2\times10^{-10}$ atm (Watson et al., 1993). A more recent attempt illustrates the experimental difficulty of monitoring any ZnO(g). It could only estimate that its pressure had to be a factor of 10^{-5} smaller than those recorded for Zn(g) (Makarov et al., 2003). These recent vapor pressure measurements with ZnO(s) also indicated that the resulting pressure of Zn(g) is about 3 ppmv at 1000°C and 135 ppmv at 1160°C. Such values suggest for zinc that in combustion systems above temperatures of 1200°C the only gaseous species will be Zn(g). The experimental difficulties also have been aggravated by the fact that the $X^1\Sigma^+$ ground state potential energy curve is crossed by that of the low-lying $a^3\Pi$ excited state, reminiscent of all the alkaline earth oxides. The potential energy curve for the ground state $X^1\Sigma^+$ correlates to Zn(^{1}S) and O(^{1}D) dissociation products whereas $a^3\Pi$ relates directly to ground state Zn(^{1}S) and O(^{3}P) products. This $a^3\Pi$ state is very low-lying in the case of ZnO ($T_0 \approx 27 \pm 4$ kJ mol^{-1}) and modifies the chemical thermodynamic analyses extensively as states will be mixed with both contributing to any chemistry especially with the $a^3\Pi$'s higher multiplicity. This low-lying state has only been observed experimentally by indirect anion photoelectron spectra (Francher et al., 1998; Kim et al., 2001; Moravec et al., 2001), but also predicted theoretically with this energy magnitude (Bauschlicher and Partridge, 1998; Harrison et al., 2002; Peterson et al., 2007; Chambaud et al., 2008). One important review on the vaporization distributions of ZnO utilized a much higher D_0(Zn-O)$_g$ = 285 kJ mol^{-1} value now making their published results for Zn(g) and ZnO(g) meaningless (Lamoreaux et al., 1987). It in fact makes any equilibrium calculation questionable if data for ZnO(g) is implied to be of any importance. No electronic spectral systems of ZnO(g) have been characterized. The mixed ground state implies that any measured dissociation energy will relate to ground state products.

The electronic and geometric structures of the first row gaseous transition metal dioxides have been examined theoretically. All are thermodynamically stable with the one exception of ZnO_2, which is predicted to spontaneously dissociate to gaseous Zn and O_2 (Gutsev et al., 2000a). However, as mentioned above, ZnO_2(s) was found to produce ZnO nanoparticles on dissociation (Uekawa et al., 2003). Also a linear OZnO structure was observed and characterized in laser-ablated Zn/O_2 matrix isolation studies (Ropp and Aia, 1962; Chertihin and Andrews, 1997a).

Other important gaseous molecules are the chlorides. $ZnCl_2$ sublimes and from its heat of formation and sublimation the heat of formation of $ZnCl_2$(g) is accurately known as $\Delta H_{f298K} = -265.7 \pm 2.5$ kJ mol^{-1} (Kubaschewski et al., 1993). This implies a value of $D_{0Atomiz} = 634 \pm 3$ kJ mol^{-1} for the sum of its two bond strengths. By mass spectrometrically studying the vapor equilibrium between Zn/ZnCl/$ZnCl_2$ from a heated effusion cell, a value was obtained of D_0(Zn-Cl(g)) = 229 ± 8 kJ mol^{-1} close to spectral estimates (Hildenbrand et al., 1999). This magnitude has been confirmed theoretically and indicates the monochloride of Zn is the weakest bound across the first row of transition elements

(Wang and Schwarz, 1998). This implies D_0(ClZn-Cl) = 405 ± 10 kJ mol^{-1}, an approximately doubling in strength over the first bond, which is seen to be the common trend for the Zn/Cd/Hg elements (Table 7.16). The zinc monochloride although weakly bound does have various doublet spectral systems that have been characterized (Huber and Herzberg, 1979). A recent theoretical study of its four monohalides calculates their potential energy curves (Elmoussaoui et al., 2016). It confirms an $X^2\Sigma^+$ ground state with no low-lying states. It summarizes experimental and theoretical estimates that have been published concerning the bond strength suggesting a value of 220 kJ mol^{-1} in its own study. ZnOH(g) has been observed mass spectrometrically in a laser-ablated molecular beam (Cannavo et al., 2006), and by its millimeter spectrum in a Broida-type oven (Zack et al., 2012). A recent detailed study has examined theoretically the electronic structure and rovibrational properties of the X^2A' ground state (Hirano et al., 2014). However, its stability remains unknown. From M-Cl and M-OH trends it should be comparable to that of ZnCl. It can be no stronger than this as noted in the ZnCl experimental study above. There, the effusion cell equilibration between solely Zn and H_2O was examined but could observe neither ZnOH(g) nor $Zn(OH)_2$(g), which would have been expected under the conditions studied (Hildenbrand et al., 1999). Consequently, even if gaseous ZnOH is comparable in bond strength to ZnCl, formation of these from either reactions of Zn(g) and flame gases or from each other would seem unlikely due to the strong binding of H_2O and HCl.

Gaseous ZnS is of interest partly because of its possible formation under reducing conditions but mainly through its comparison with the oxide. Values for its structure and bonding are based on theory but can be judged by the comparison with ZnO. Its potential energy curves and electronic states will reflect those of ZnO and so interest also lies on the location of the contributing $a^3\Pi$ electronic state. Numerous theoretical estimates studies have been made for its bond strength and all are consistently slightly weaker than the oxide as expected. All such calculated dissociation values refer to ground state products. The four chosen values for ZnS indicate 140, 143.5, 140.7, and 137, respectively, implying D_0(Zn-S) = 140 ± 10 kJ mol^{-1} that can be compared in a meaningful manner with D_0(Zn-O) = 155 ± 10 kJ mol^{-1}. This reflects the narrowing of the D(MS)/D(MO) energy gap toward unity at the end of this row of transition elements (Table 7.19) (Bridgeman and Rothery, 2000; Peterson et al., 2007; von Szentpaly, 2008; Jiang et al., 2012). In fact the value for ZnS would remain consistent at about 147 kJ mol^{-1}. Only two studies appear to have estimated the $a^3\Pi/X^1\Sigma^+$ energy separation for ZnS, calculating values of 55 and 47 kJ mol^{-1} (Peterson et al., 2007; Chambaud et al., 2008). These average to about twice the value for the oxide, reducing the level of state mixing. As with the oxide the electronic state spectroscopy has not yet been resolved.

The diatomic hydride has been characterized spectroscopically and its four lowest doublet states identified. The ground state $X^2\Sigma^+$ has a shallow potential energy curve allowing a simple extrapolation of its energy levels and indicating $D_0(\text{Zn-H})_{(g)}$ = 82 kJ mol^{-1} (Huber and Herzberg, 1979). The value is supported by theory (Daza et al., 2006; Bross et al., 2013), and the potential energy curves for the low-lying states calculated (Jamorski et al., 1994). As indicated above, although solid ZnH_2 is unstable it has been recorded in matrix isolation studies (Greene et al., 1995; Wang and Andrews, 2004a). In addition its gas phase emission spectrum in the infrared has been analyzed and like $ZnCl_2$/ZnCl and corresponding molecules of Cd and Hg, it appears that the dihydride may be much more stable than the monohydride (Shayesteh et al., 2005, 2005a). A recent theoretical study has examined its linear $X^1\Sigma_g^+$ ground state to construct an accurate potential energy surface (Sebald et al., 2010). Thermochemical tables for gaseous ZnO, ZnCl, or ZnH are not yet available and that available for ZnS uses too large a bond strength (Knacke et al., 1991; Barin and Platzki, 1995).

Chemical kinetic constraints

The trends in the bond values for the Zn, Cd, and Hg periodic table family members are tabulated in Table 7.16. The general bond weakness reduces any importance for the oxides, sulfides and monohalides in combustion. This is important because any thermochemical equilibrium calculation will imply the dominance of the dihalides, dihydroxides or dihydrides if they are in the database based purely on their larger bond strengths. As is apparent, in these chloride cases the stability of HCl ($D_0 = 428$ kJ mol^{-1}) dwarfs that of ZnCl(g) ($D_0 = 229$ kJ mol^{-1}) and the other monochlorides more so. It also extends to the formation of the weak gaseous oxide and sulfide.

$$\text{Zn, Cd, Hg} + \text{HCl} = \text{ZnCl, CdCl, HgCl} + \text{H} \quad \Delta H_0 = +199,\ 248,\ 328\ \text{kJ mol}^{-1}$$

$$\text{Zn, Cd, Hg} + \text{OH} = \text{ZnO, CdO, HgO} + \text{H} \quad \Delta H_0 = +269,\ 329,\ 406$$

$$\text{Zn, Cd, Hg} + \text{SH} = \text{ZnS, CdS, HgS} + \text{H} \quad \Delta H_0 = +211, 243, 327$$

As a result the initial formation of ZnCl(g) poses the crucial kinetic constraint limiting the chemical formation of $ZnCl_2$(g) via the required subsequent reaction

$$\text{ZnCl(g)} + \text{HCl} = \text{ZnCl}_2\text{(g)} + \text{H} \quad \Delta H_0 \approx +23\ \text{kJ mol}^{-1}$$

This is common for Zn, Cd, and Hg and remains so for zinc, even though its two chloride bond strengths, 229 and 405 kJ mol^{-1}, pose less of a difficulty than in the well-known case of mercury, D(Hg-Cl) = 100 kJ mol^{-1} and D(ClHg-Cl) = 346 kJ mol^{-1}. Consequently, zinc still has no easy gaseous formation reactions for the chlorides from HCl and must invoke direct heterogeneous ablative reactions such as with ZnO(s).

$$\text{ZnO(s)} + 2\text{HCl} = \text{ZnCl}_2\text{(s)} + \text{H}_2\text{O} \quad \Delta H_{298} \approx -121\ \text{kJ mol}^{-1}$$

$$\text{ZnCl}_2\text{(s)} = \text{ZnCl}_2\text{(g)}\ \Delta H_{298} \approx +149\ \text{kJ mol}^{-1}\text{(Kubaschewski et al., 1993)}$$

Gaseous transport of zinc

Zinc has a very large ionization potential and the weakness of its gaseous bonding rules out any flame chemiionization for any of the Zn, Cd, Hg group. Also the laboratory studies of flame inhibition can often be a useful indicator of possible gas phase chemistry. In the present case, the question of hydroxide formation for zinc remains intriguing. One study did indicate measurable slight effects in an 1860K, H_2-rich flame, seeded with only 1 ppmw of a zinc salt (Bulewicz and Padey, 1971; Linteris et al., 2008). Such effects need either flame particles for surface interactions or catalytic recombination kinetics for an explanation. Consequently it does leave some doubt as to whether only gaseous Zn is present in such flames. The other intriguing laboratory studies, that are still not fully understood but useful, revolve around soot suppression or growth in flames. Little interest has been shown on the Zn, Cd, Hg element group in this connection (Howard and Kausch, 1980). Nevertheless, in one study Cd was rated as zero, and Hg 18 on a scale with Ba at 100 (Cotton et al., 1971). Soon thereafter it was noted that zinc appeared to in fact slightly promote soot formation. Finally, this has been resolved to some degree in a study based on measuring soot by laser incandescence and elastic light scattering methods in premixed ethylene/air flame (Simonsson et al., 2018). A slight decrease in soot volume and growth was noted but only at flame additions of about 700 ppmw. The current theory being that the element becomes occluded in the soot and catalytically oxidizes it. However, in these laboratory studies, due to its availability, the zinc salt atomized into the flame was the dichloride, and not the nitrate, which might be a compounding problem in the latest study.

Returning to questions of chlorine effects, as noted below, it is now apparent that $ZnCl_2(g)$ is oxidized in air at 800°C–1100°C back to ZnO(s) or Zn(g) releasing the chlorine as Cl or Cl_2 with its potential for inducing corrosion. As a result, although not proven, it would seem reasonable to assume that any reported presence of $ZnCl_2(g)$ in normal fuel combustion must result from heterogeneous formation and vaporization. Consequently, the recent study of the effects of HCl, H_2S and H_2O on zinc release requires reevaluation (Abascal et al., 2016a). Their assumption of ZnO, $ZnCl_2$, or ZnS being in the gas phase and simply condensing onto probes is unacceptable. It may satisfy thermochemical predictions but ignores kinetic constraints and also the probable heterogeneous formation on the probes. This applies also to their mass spectrometric sampling. Mercury encounters similar difficulties. In its case it was shown to invoke a heterogeneous deposition forming its sulfate or oxide that then were converted by HCl to the volatile dichloride (Schofield, 2008). Two observations related to this are noteworthy. First, it seems well documented by experiment that high levels of chlorine enhance the release of zinc from various fuels (Verhulst et al., 1996; Miller et al., 2003; Sippula et al., 2008; Pedersen et al., 2010; George et al., 2010; Luan et al., 2013). Also in one experiment, waste incinerator ash was heated at 1000°C for 3 hours in either a mix with $CaCl_2$ or with flowing Cl_2. In either case all the Zn in the ash volatized, confirming dichloride formation and also probably suggesting no silicate or complex binary metal oxide formation in that particular ash (Chan and Kirk, 1999). It was also evidenced by noting an enhanced release of Zn if solid recovered waste was cofired with coal. Zinc also was noted in the finer particles in that study (Wu et al., 2013). In such cases of gaseous $ZnCl_2$ release, heavy corrosion of any iron-based alloys in incinerators is evidenced by the formation of $ZnFe_2O_4$ (Ma et al., 2003; Bankiewicz et al., 2009, 2013). Also the role of the dichloride was confirmed by testing steel surfaces heated in the presence of $ZnCl_2/K_2SO_4$ mixtures (Bankiewicz et al., 2009). This corrosion also is reported in boilers that consume painted demolition wood that is contaminated by the zinc in the paint (Elled et al., 2008). $ZnCl_2$ is known to be a flux component for removing a surface oxide layer and now has been shown responsible for such corrosion (Vainikka et al., 2011). X-ray analyses studies have possibly identified $ZnSO_4$, $ZnAl_2O_4$, Zn_2SiO_4 and even K_2ZnCl_4 as products.

Equilibrium calculations also have been attempted to better understand observations (Enestam et al., 2011, 2011a, 2013). Also it is reported that addition of moisture can alleviate this corrosion possibly suggesting that the dichloride is hydrolyzed by water to ZnO at temperatures in the range 800°C–1100°C with the less corrosive HCl being released (Verhulst et al., 1996; Stucki and Jakob, 1997). With oxidation of $ZnCl_2$ it is more likely to release Cl or Cl_2, known to be highly corrosive (Jones et al., 2013). This probably occurs also in the condensed phase during the release mode with any dichloride formed therein being involved in both vaporization and conversion to the oxide channels, the relative rates of which will depend on the heating rate and conditions (Stucki and Jakob, 1997; Youcai et al., 2004). Such hydrolysis should be reflected in thermodynamic equilibrium calculations, which are a display of relative stabilities but appear absent in some cases possibly resulting from the relative concentrations of water and chlorine (Danihelka et al., 2003). Additionally, raised levels of sulfur can promote $ZnSO_4$ deposits over the dichloride but only at low temperatures when it becomes the more stable (<400°C). At high temperatures, deposition on ash will be limited to formation of a compound utilizing an ash component such as Al, Fe, or Si. Only when temperatures drop to about 1100°C when its vapor pressure is low enough can ZnO remain on the ash once it is formed. This will continue until it is cool enough either for conversions to occur or formations of $ZnCl_2$, the sulfate or its oxysulfate are sufficiently stable and not vaporize. This competition between the dichloride and sulfate has been quite extensively studied experimentally (Jiao et al., 2013, 2013a; Song et al., 2013). Both can be extensively produced below 650°C depending on Cl or S levels. On adding particulates of Fe_2O_3 or Al_2O_3 it was also possible to produce at a higher temperature (800°C) deposits of $ZnFe_2O_4$ and $ZnAl_2O_4$ with a little sulfate. A comparison of zinc deposition in air or oxy-fueled combustion noted some changes in whether sulfate (oxy-burn) or dichloride (air-fired) resulted in the cooled ash.

These initial assessments therefore suggest a dominance of thermal dissociation in the fuel of the less stable zinc compounds to ZnO(s) that then dissociates on vaporization to Zn(g) at temperatures above 1000°C or is vaporized by interaction with fuel chlorine. This single species then transports the zinc through a combustor with the ash particles and will be the depositing species on the ash as temperatures decrease. For coal combustion, maximum temperatures can be in the 1500°C–2000°C range with a very high release of zinc to the gas phase. This will facilitate a range of molecular formation on the ash surface as temperatures permit and will provide a reasonable gas/particle interaction time period. For lower temperature incinerator operations, although a more variable complex fuel, similar chemistry is expected but the extent of zinc release will become a greater factor. An interesting effort to examine this with a chemical thermodynamic model conveys this possible role of Al, Fe and Si but depends heavily on the accuracy of the thermodynamic database (Liu et al., 2010).

Combustion power plant data

In coals and biomass, zinc's presence can vary and be at levels up to 200 ppmw (Table 1.1). It can be much higher in wastes and sewage sludge. Many studies have been reported concerning the behavior of most of the trace metals in combustion. Although generally regarded of lesser importance, observations concerning zinc have been noted in numerous of these. A recent study of Zn release from single particles of biomass confirmed that at 1000°C all the Zn was released in gaseous atom form (Sommersacher et al., 2016). Due to zinc's more significant levels in municipal waste and sewage sludge incineration these fuels now are the ones receiving greater emphasis. Also this is particularly apparent in Europe and Asia where such ash is now regulated and increasingly being used in agriculture (Werther and Ogada, 1999; Han et al., 2006; Poykio et al., 2010). Japan now has set standards for zinc for waste incinerators (Yuan et al., 2005). In Europe, the level of incinerated waste is sufficiently high that the mass flow indicates a significant tonnage of zinc in the ash. This is encouraging the development of ash recovery methods (Purgar et al., 2016). Additionally, a complete survey of sewage sludge ash in Germany stresses the importance of establishing its composition from the viewpoint of resource recovery. The median ash content of Zn has been noted as 2.5 g kg^{-1} (Kruger et al., 2014). In spite of this, few studies yet are available that have analyzed the compositions of zinc in the bottom or fly ash. With incineration of municipal wastes at 1000°C only 54% of the zinc was thought to be in the fly ash with the remainder in the bottom ash with <0.1% released to the atmosphere (Morf et al., 2000). Now, studies of FGD systems indicate significant removal of any such final traces of Zn (Cordoba et al., 2012). Leaching from the gypsum FGD product from 12 coal-fired power plants in China was reported. This indicated very high levels of zinc leaching indicating it posed high risk (Hao et al., 2017). This possibly implied sulfate or chloride deposition on the fine ash particles. However, what it clearly showed was that power plants now collect various ash forms from different regions of the power plant. There is no reason for them to be the same and if the wish is to know a mass distribution correctly, each ash form needs independent analysis. Most combustion studies rely heavily on thermochemical calculations for their interpretation. Numerous examples are available of thermochemical calculations for coal (Frandsen et al., 1994; Danihelka et al., 2003) and municipal waste incineration (Chen et al., 1998a; Sorum et al., 2004; Poole et al., 2008). These may be useful in providing a qualitative overview of the relative thermodynamic stabilities at equilibrium but as already conveyed fail at considering the kinetic constraints for dihalide formation and the now corrected bond strength values. Obviously, with zinc, depending on their concentrations, if the silicate, aluminate, phosphate or ferrate are not included in the database an incomplete prediction will result. For incineration at 930°C of municipal waste, X-ray analysis and leaching of ash tentatively suggested the possible presence of ZnO(s), $ZnSO_4$(s), and $ZnAl_2O_4$(s) (Abanades et al., 2002). Examining ash particles from the ESP of a subbituminous Powder River Basin coal, the X-ray absorption spectra were very similar but not exactly the same with Zn_2SiO_4(s) and higher silicates (Luo et al., 2011). A similar study of fine fly ash noted possible $ZnFe_2O_4$ from an Eastern coal (Shoji et al., 2002). In co-combustion of coal and secondary fuels at 950°C, low leaching was observed and was of little concern (George et al., 2008). In another high temperature combustion of coal and coal/biomass an analysis of fly ash from the 120°C ESP stages, there are adequate Zn levels and it clearly increases across the stages to the smaller size particles confirming vaporization/deposition. Nevertheless, further analysis is complex and provides no reliable information (Swietlik et al., 2012). Due to differences in the various combustion parameters and fuel composition, besides the lack of adequate analysis methods, generalities concerning ash compositions remain speculative. However, because the sulfate and chloride are the only species soluble in water with the oxide and silicate requiring more acidic solutions a little information might be expected to emerge from such leaching data. For coal ash in one case, leaching was interpreted as indicating the presence of the aluminate (Catalano et al., 2012). It is plausible with coal and higher temperature conditions that this will necessarily provide not only this higher temperature component, but also will lengthen the interaction time between the flue gas and ash to promote such mixed metal oxide formation in the fly ash. Nevertheless, because all the zinc in incineration is present in the ash in some form, although compositions remain unknown in detail, it is now practice to use thermal treatments of ash. This is to either vitrify it to the silicate (Werther and Ogada, 1999; Wang et al., 2010a), or use evaporation (Jakob et al., 1995) as a method of mineral reduction control to permit its use commercially in other end products. A recent leaching test on waste incineration fly ash indicated that HCl acid treatment removed 80% of the zinc (Tang and Steenari, 2016). In waste incineration, the leaching behavior of fly ash suggested a mainly chloride component (Jiao et al., 2016).

For coal, the ratio of zinc in the fly ash can be several fold larger than in the bottom ash (Klein et al., 1975). Ash is enriched and even though the zinc is totally accumulated on the ash and none remains in the gaseous form it remains low on the list for enrichments with respect to the Earth's crustal average values (Marrero et al., 2007). As a result, zinc emissions are easily controlled by particulate removal and ESPs generally collect about 99%, scrubbers most of the remainder and little slip exists, less than about 0.3% of the zinc escaping as ultrafine particles (Nielsen et al., 2002; Yuan et al., 2005). It can have the highest removal efficiency of the trace metals. A recent study of trace element emissions from a modern large coal combustor in China confirmed the major removal by the ESP with very minimal stack emission on a several mg ton^{-1} of coal scale (Zhao et al., 2016). Zinc concentrations in coal ash generally follow the inverse particle diameter correlation, increasing with smaller particle size (Coles et al., 1979; Martinez-Tarrazona and Spears, 1996; Nielsen et al., 2002; Huang et al., 2004; Yuan et al., 2005; Wang et al., 2008; Swietlik et al., 2012; Lanzerstorfer, 2018) and with downstream time (Kaakinen et al., 1975). This is a reflection of high volatility followed by deposition but can depend at times on the mineralogy of the coal (Bool and Helble, 1995). One particle size analysis collected the fly ash downstream of the ESP for a coal combustor. The ash was collected in two size modes, fine (0.05 μm) and coarse ($\approx$2.0 μm), both indicating the presence of Zn but with no significant enrichment noted on the fine particles, which is a little unexpected (Kauppinen and Pakkanen, 1990). Parameters affecting the zinc are temperature, fuel composition, chlorine and sulfur levels. Normally, coal combustion temperatures appear to vaporize most of the zinc with little indication of interactions with the bulk elements on release (Senior et al., 2000). One study indicated high volatility with rapid heating but slight volatility in slow heating combustion, possibly indicative of reaction with bulk elements: also little vaporization occurred below 750°C increasing to significant at 950°C (Wang and Tomita, 2003).

Biomass is burned at lower temperatures and for wood, zinc release has been measured as 60% at 800°C (Miller et al., 2003), 85%–100% at 850°C in a different study but starting at 500°C (vanLith et al., 2008). Similarly in sewage sludge or waste incineration that is often cofired with coal more retention is evident below about 950°C (Miller et al., 2004; George et al., 2010). In these temperature ranges any $ZnCl_2$ formed (b.pt.732°C) will either be hydrolyzed or vaporize. However, as seen with other metals, on depositing onto ash the most thermodynamically stable molecule will form irrespective of the vapor composition.

The chemistry significantly simplifies for gasification or pyrolysis for reducing conditions and will be dependent on the extent of oxygen or chlorine that is in the fuel. Only a few studies have been published but indicate that emphasis centers more on ZnS(s) which should be stable under such conditions having a steady-state Zn(g) vapor pressure of 10–100 ppmv in the 850°C–950°C range. It has been reported stable in a gasifier system to 725°C dissociating then to Zn(g) (Bunt and Waanders, 2009). Although phosphides exist they can only form at lower temperatures and dissociate on vaporizing and react with water releasing toxic PH_3 (Schoonmaker et al., 1967). As mentioned above, the gaseous dihydride might be quite stable but will have formation difficulties due to a weak monohydride and unstable condensed phase dihydride. In a coal/waste gasifier fired at 960°C with a hot-gas filter operating at 450°C – 580°C there was no indication of downstream gaseous zinc suggesting that this was retained as the sulfide on the filter (Reed et al., 2001). The formation of the dichloride under reducing conditions remains uncertain as Zn(g) like Hg may not be able to chemideposit directly with HCl even heterogeneously. Whether the ash incorporates the zinc as the sulfide, silicate or refractory remains unknown.

With zinc, questions that remain center on the degree of zinc/bulk element interactions relative to surface chemical deposition together with establishing the exact nature of the final particle composition. Zn(g) is the dominant gaseous form and whether this interacts with a particle surface much above zinc's boiling point of 906°C is an interesting question. Formation of the silicate or aluminate is possible if the surface chemistry is sufficiently fast to secure the zinc. Experience with mercury and sodium has clearly indicated that surface chemistry can be faster than the collisional time of a species reaching and leaving the surface. In cases such as ZnO(s) deposition, its own vaporization is large until temperatures fall below 1100°C when deposition might then begin to establish. Depending on availability in the ash, some silicate, aluminate, ferrate or oxide may certainly be expected in the ash from such chemical deposition. Whether the sulfate and the dichloride form (via the oxide) at lower temperatures will depend on sulfur and chlorine levels. One interesting deposition study managed to grow ZnO nanowires of differing diameters using a flame method (Xu et al., 2007a). A zinc plated steel substrate was inserted in the methane/air flame such that the Zn coating was molten and slowly vaporizing. With a substrate temperature below about 1100°C the thermally stable ZnO nanowire could form with such stabilization. Additionally, it does appear that zinc dichloride is formed in the fuel release period especially in incineration with PVC wastes, is very volatile, and can induce corrosion if not controlled. Its formation mechanism remains uncertain Under what conditions $ZnCl_2$(g) survives, and is oxidized on a surface where it releases the chlorine, depends on numerous parameters such as oxygen and water content.

In summary, zinc is of the Zn, Cd, Hg group in the periodic table but behaves differently from the transition metals and has only a cationic chemistry based on its Zn^{2+} nature. Two aspects of its chemistry are noteworthy; one its stable ZnO(s) dissociates and above 1100°C is Zn(g). Its gaseous oxide is not thermally stable. Second, exactly like Hg, it has no gas phase mechanism for forming $ZnCl_2$(g), due to kinetic constraints, although predicted thermodynamically. Such formation needs heterogeneous ablation of possibly the oxide by HCl or water. Gaseous zinc can transport and deposit efficiently onto ash with the option of forming one of its thermally stable molecules at the lower temperatures, or reacting with an ash component to form a thermally stable mixed metal oxide. Each ash form in a combustor may differ due to its combustion history and may need individual analysis. Its overall behavior in combustion still requires information on its final speciation.

7.34 Zirconium

Zirconium lies below titanium in the periodic table, second column, second row. Second row transition elements can show slightly different properties from their first row lighter congeners. However, in this case zirconium and its very closely related third row element hafnium differ only slightly from titanium. However, they are heavier and reactively more stable. This is noted in their use in the nuclear industry and is also why in the combustion of metals as missile fuels they have the most energetically formed solid phase dioxides with ZrO_2 ($\Delta H_f = -1100$ kJ mol^{-1}, $D_{0atoms} \approx 1424$ kJ mol) and $HfO_2 = -1145$ compared to $TiO_2 = -945$ and $CO_2 = -400$ (Barin and Platzki, 1995). Zirconium is reasonably abundant, averaging an Earth's crustal concentration of about 165 ppmw. This is 30-fold less than Ti but more than Pb and Zn and is about the 18th element in a list of crustal abundances. Its concentration in the oceans is very low due to the insolubility of its major compounds. It is present widely over the earth and so not surprisingly also is present in coals generally being <100 ppmw but can reach to 200 (Kronberg et al., 1981; Xu et al., 2003). It can be of similar magnitudes in sewage sludge and municipal wastes but is not expected to be significant in biomass (Table 1.1). Levels in pet-coke and limestone are minimal (Cordoba et al., 2012). Its principal commercial ore is the orthosilicate $ZrSiO_4$ (zircon) with some occurrences of ZrO_2. These are widespread over the earth and rarely in concentrated deposits. The ore is not subject to weathering so there is little concern about its content in natural dusts. Global resources indicate abundant reserves and close to 1.5 million tons year^{-1} of ore are mined from which about 8000 tons year^{-1} of pure metal are extracted and large quantities of pure ZrO_2 are produced. As with titanium, the dioxide is the major commercial product. Estimates indicate adequate world reserves of at least 80 million tons measured as ZrO_2. Partly because the atomic and ionic radii of the Zr and Hf elements are the same, as well as their chemistries, deposits of zirconium invariably contain a low (1%–2.5%) of hafnium. Due to the difficulty of their separation most zirconium will be left with this slight very similar component. Only in the nuclear industry is there a real need for its separation and the pure metal and this application accounts for 90% of the production. The metal and its stable dioxide find major uses as a nuclear fuel rod cladder, as a refractory and strongly abrasive material. Also in jewelry as cubic zirconia, as a replacement in paints for lead, in alloys with Sn, Cr and Fe, in ceramics and also medically for noncorrosive implants that are compatible with the body. The price depends on purity and for

99.6% pure rod it may cost about \$80–200 kg^{-1}. Pure zirconium metal was first produced only about 100 years ago. It is evident in certain stars and in the sun and meteorites. Rocks on the moon appeared to be richer in Zr than on Earth.

Health concerns with zirconium

No known biological role is known for zirconium and it is regarded as nontoxic. It is abundant in nature and absorbed by plants. The body contains traces on a low mg scale acquired by accumulation from a daily intake, approximately assessed as 50 $\mu g\ day^{-1}$ for adults, and apparently much lower in one preschool children study. It can cross the blood/brain barrier, the placenta, and mammary glands. Nevertheless it is not associated with any metabolic function. An overall review of its biology is available (Ghosh et al., 1992). It does seem that the body finds it harmless. Utilizing injected radioactive Zr, it is seen that any retained is stored mainly in the bone skeleton with slow gradual leaching. Although of low toxicity, Zr salts can cause skin and lung irritation. A basic carbonate was once marketed as a treatment for poison ivy but was discontinued due to skin reactions. Also although controversial, zirconium and aluminum salts are used in antiperspirant products where the mechanism is to physically block the skin pores. Due to its apparent harmless nature there has been little interest in setting up biomarkers for monitoring and no toxicity reports have been issued. It cannot be found included in any of the normal population surveys but has been reported in blood and hair. One hair report was very low at 0.046 ppmw, with a test center indicating range of 0.02–0.5 ppmw was acceptable.

Zirconium's basic chemistry

The zirconium element resembles stainless steel, melting at 1855°C and only burns in air at very high temperatures forming the dioxide and even nitrides. It boils at 4430°C (10 ppmv vapor pressure at 2366°C). It is favored as a fuel rod cover as it only reacts with water at high temperatures but this is the mechanism of hydrogen release and explosions in nuclear accidents. It has a cationic nature and covalent bonding is preferred in its salts that are generally hydrolyzed by water to the insoluble dioxide. Tetravalency is preferred with some divalency possible. This results from its electronic configuration of $4p^6 4d^2 5s^2$. The dioxide plays a dominant role in its combustion chemistry being a refractory oxide (m.pt.2715, b.pt.4270°C, vapor pressure 1 ppmv/2318°C). A tetravalent sulfate and nitrate are known but are hygroscopic and dissociate directly to the oxide at low temperatures on heating (Stern and Weise, 1966; Stern, 1972; Tagawa, 1984; Tagawa and Kawabe, 1987). A tetrahydroxide is known but has the structure of the hydrated dioxide. Although chlorides exist, $ZrCl_4$ (sublimes 330°C) and $ZrCl_2$ decomposes (m.pt.730°C), these also are readily oxidized and hydrolyzed by water. The tetrachloride forms the $ZrOCl_2$ oxychloride that is not thermally stable. Only a basic carbonate is known that dissociates at about 135°C. ZrH_2 exists and can be used as a source of hydrogen to which it dissociates at about 800°C. Reflecting the oxide high stability, the stable disulfide can be formed that converts at 1100°C–1500°C to the sesquisulfide Zr_2S_3. This also is vulnerable to oxidation. Any phosphate formation is of a complex nature. These values for the physical properties of its condensed phase molecules are summarized in Table 7.46.

The silicate ore, $ZrSiO_4$, is well characterized and dissociates at about 1540°C into the oxides. Like titanium, its zirconates result from the mixed metal oxides. However, it seems that zirconates are weaker bound in the gas phase. In spite of many mass spectrometric studies of mixed oxide salts no volatile zirconate has been reported and dissociation appears to occur before vaporization (Lopatin and Shugurov, 2013). With the alkali zirconates such as K_4ZrO_4, these reflect their K_2O component hydrolyzing readily and volatize this oxide at low temperatures (> 650°C) (Devalette and Hagenmuller 1967). However, in recent years it has been noted that the alkaline earth, lanthanum zirconates and a few others do have structural modifications. They are of important use as high temperature refractories, $CaZrO_3$ (m.pt.2550°C) which indicates ionic Ca-O and covalent Zr-O bonding, $BaZrO_3$ (m.pt.2540°C), and with the rare earth elements primarily lanthanum zirconate

TABLE 7.46 Properties of zirconium's condensed phase relevant to combustion, °C.[a]

m.pt. b.pt. vapor pressure
Zr 1855 4430, refractory, 10 ppmv/2366°C, ins
ZrO_2 2715 4270, refractory, 1 ppmv/2318°C, ins
$Zr(OH)_4$, ($ZrO_2.2H_2O$), >550 dissoc to ZrO_2, ins
$Zr(CO_3)_2$ >135 dissoc, basic forms only, ins
$Zr(NO_3)_4$ >100 dissoc to ZrO_2, s
$Zr(SO_4)_2$ >400 dissoc to ZrO_2, s
$ZrCl_2$ 730 1290, decomp, reacts with water
$ZrOCl_2$ >300 dissoc, s
$ZrCl_4$ 437, sublimes 330, dissoc to $ZrCl_n$, reacts with air, and hydrol to $ZrOCl_2$, 680 ppmv/175°C, s
ZrS_2 1550, converts1150–1500 to Zr_2S_3, ins
ZrH_2 decomp >800 to H_2, 100 ppmv H_2/367°C, ins
$CaZrO_3$ 2550, ins
K_2ZrO_3 >600 decomp, hydrol by water
K_4ZrO_4 hydrol by water
$MgZrO_3$ 2150, ins
Na_2ZrO_3 1180, dissoc >1200, readily hydrol by water
$ZrSiO_4$ dissoc >1540, ins
$ZrTiO_4$ dissoc >1100°C
$Zr(WO_4)_2$ stable <777, shrinks on heating, ins

[a]*dissoc*, dissociates; *decomp*, decomposes; *s*, water soluble; *ss*, slightly soluble; *ins*, water insoluble; *hydrol*, hydrolyzes.

(m.pt.2300°C) being of use as thermal barrier coatings for gas turbine blades but having nonstoichiometric compositions that approximate to $La_2Zr_2O_7$, for example. Zirconium tungstate has been added due to its curious nature. Although not involved in combustion it is an unusual molecule in its condensed phase. It is seen to shrink in three dimensions on heating.

Gas phase stabilities and transport

Because of zirconium's refractory nature and industrial use, much interest was placed in the 1970s on quantifying particularly the dioxide thermal stability, establishing its vaporization and thermochemical properties. The early Knudsen cell-mass spectrometric measures (Ackermann and Rauh, 1974; Murad and Hildenbrand, 1975) indicated that ZrO_2(s) was very stable having very low vapor pressures that only attained values of a few ppmv even at 2400°C. At this temperature the gaseous dioxide starts dissociating to the more stable monoxide at that temperature and the two coexist. The vaporization was shown to be fully congruent (Rauh and Garg, 1981). Reassessing these data, using updated thermochemical data tables (Cordfunke and Konings, 1990), this has now provided a reliable value of D_0(Zr-O) = 766 ± 10 kJ mol^{-1}. This value for ZrO(g) was used in a recent overview of metal oxides (Schofield, 2006) and is close to that recommended previously, 772 ± 13 (Pedley and Marshall, 1983) and 763 ± 12 (Cordfunke and Konings, 1990). It was also confirmed independently and exactly (± 10 kJ mol^{-1}) by accurately measuring the ionization potential of ZrO and coupling this to a reliable value for D_0(Zr-O$^+$) (Loock et al., 1998). As confirmed by theory, it is the strongest bound diatomic oxide of the second row transition elements and has very few stronger in the periodic table (La, Hf, Ta, Ac, Ce, Th, Pa). However, theory initially underestimated the bond strength by about 10% (Siegbahn, 1993a). This continues, but several methods now have been developed that are in closer agreement (Song et al., 2007). Uncertainty centers more on the gaseous dioxide, which has most recently been recommended as D_0(OZr-O) = 658 ± 20 kJ mol^{-1} (Schofield, 2006). This is larger by about 68 kJ mol^{-1} than the values listed in current data bases (Barin and Platzki, 1995; Chase, 1998) and now is based on recalculating four major earlier works with updated thermochemical functions (Chupka et al., 1957; Ackermann et al., 1975; Murad and Hildenbrand, 1975; Marushkin et al., 1990). Moreover, although having produced the most current thermochemical tables (Cordfunke and Konings, 1990), their quoted value of 727 kJ mol^{-1} is clearly too large and too similar to that of ZrO to be correct. The ratio of D(OZr-O)/D(Zr-O) = 0.86 from the above recommended values is supported by the trend of the other early members of the transition elements, Ti 0.92, Hf 0.80, V 0.91, Nb 0.90, Ta 0.83. A value of 727 would imply a ratio of 0.95. A theoretical estimate for this ratio was 0.78 (Siegbahn, 1993b). The magnitude of the strength of these oxide bonds implies that in combustion the gaseous oxides will be dominant species in the gas phase, but even so only at very high temperatures, due to their small vapor pressures and condensed phase stability that will limit any release. The gaseous chlorides are far more volatile but are still strongly bound. $ZnCl_4$(s) is very volatile with a vapor pressure of at least 680 ppmv at 175°C, subliming, disproportionating and dissociating to $ZrCl_2$ above 330°C. This breaks down further to ZrCl and Zr at higher temperatures. Bond strengths are D_0(Zr-Cl) = 522 kJ mol^{-1} and D_0(ClZr-Cl) = 512 kJ mol^{-1} based on two early mass spectrometric equilibration studies but remains uncertain (± 25 kJ mol^{-1} or by ± 5%) (Chase, 1998). The monochloride is stronger than D_0(H-Cl) = 428 kJ mol^{-1}. However, with regards to combustion, zirconium's chlorides are incompatible with oxygen and readily oxidized. The weakening trends of the gaseous monochlorides and monohydrides across this second row of transition elements has been assessed theoretically (Siegbahn, 1993). However, this and a later theoretical study severely underestimate the ZrCl bond energy (Cheng et al., 2007). However, a close value for the dichloride was obtained (Siegbahn, 1994).

The bond strength for ZrH appears not to have been measured but was estimated as D_0 = 307 ± 20 kJ mol^{-1} (Chase, 1998). Several theoretical estimates have suggested values of about 220 (Langhoff et al., 1987), 255, 230 (Balasubramanian and Wang, 1989; Balasubramanian, 1990), 290 (Siegbahn, 1994), and 200 (Koseki et al., 2002) but with time have not even approached a consensus. Also because the trend is for theory to underestimate by 15% with these still difficult transition elements, a value of D_0 = 280 ± 30 kJ mol^{-1} might be the best estimate at present. Studying twelve potential energy surfaces of ZrH_2 in connection with the Zr/H_2 reaction, it was noted that the ground $ZrH_2(X^1A_1)$ state was 109 kJ mol^{-1} more stable than the two ground state Zr and H_2 reactants (Wang et al., 1990). This provides a heat of formation for the dihydride and a sum value for its two Zr-H bonds of 545 kJ mol^{-1}. This is consistent with the *D*(Zr-H) value above suggesting that the first and second bonds are similar and of order of about 275 kJ mol^{-1} each. Again these and any other potential polyhydride will not be stable under combustion conditions. Strong bonding seen in the gaseous sulfide is a reflection of its oxide, but is less characterized indicating D_0(Zr-S) = 570 ± 10 kJ mol^{-1} (Steiger and Cater, 1975a; Barin and Platzki, 1995). The ratio of MS/MO bond strengths is the same as its periodic table neighbor yttrium, providing some support for the values (Table 7.15). Nevertheless, ZrS is readily oxidized and hydrolyzed to the dioxide. The gaseous hydroxides, which often mimic the chlorides in bond strength have been tentatively suggested as being strongly bound, *D*(Zr-OH) = 531 kJ mol^{-1} and *D*(HOZr-OH) = 577. These values had been estimated and are listed by Cordfunke and Konings (1990). Another estimate of the interaction energy of gaseous Zr with H_2O_2 implies a value for the dihydroxide of *D*(HO-Zr-OH) = 1026 (Wang and Andrews, 2005a). Consequently, although uncertain it does seem that the single and second Zr–OH bonds could have strengths close to 525 kJ mol^{-1}. Whether these manage to play a part in steam vaporization/corrosion loss of certain alloys has been hinted in the literature but never confirmed. These gas phase molecular stabilities are summarized in Table 7.15 together with the trends seen with other second row transition elements.

Zirconium molecular spectroscopy

With zirconium molecules, the optical spectroscopy has mainly stressed gaseous ZrO due to its importance in stellar astronomy and for acquiring reliable chemical thermodynamic tabulations. The ground state of $ZrO(^1\Sigma^+)$ now has been established experimentally but it also exhibits a very low-lying $a^3\Delta$-state (1080 cm^{-1}, 12.9 kJ mol^{-1}) implying a mixture of these states at most temperatures. This produces a rich spectroscopy of singlet and triplet transitions that now are quite well characterized (Huber and Herzberg, 1979; Davis and Hammer, 1988; Afaf, 1995; Kaledin et al., 1995a). It has also been adequately described theoretically (Langhoff and Bauschlicher, 1990; Siegbahn, 1996). LIF studies are still ongoing to observe the remaining theoretically predicted systems (Balfour and Chowdhury, 2010). Being triatomic, spectroscopy of ZrO_2 is more complex but an optical transition has been reported recently (Le et al., 2011) and structures have been assessed theoretically (Siegbahn, 1993b). Also matrix isolation methods at low temperatures have condensed the ZrO and ZrO_2 oxides and established their structures and vibrational frequencies (Chertihin and Andrews, 1995a). ZrS spectroscopy would be expected to reflect that of the oxide, which it does. This is also important in astronomy and similarly has been characterized experimentally (Jonsson, 1995; Jonsson and Lindgren, 1995; Jonsson et al., 1998) and adequately modeled theoretically (Langhoff and Bauschlicher, 1988). It also exhibits a $^1\Sigma^+$ ground state and singlet and triplet spectral manifolds. The location of the low lying $a^3\Delta$-state still appears not to have been exactly anchored relative to the singlet system but the theory has indicated it to be even lower lying at $500 \pm 200\ cm^{-1}$ and so has a significant contribution to its thermochemical distribution. Gaseous ZrS_2 appears unreported but it is present in condensed form in matrix isolation experiments providing a structure and frequencies (Liang and Andrews, 2002). ZrCl, although a diatomic molecule, has a complex electronic structure. The combination of theory (Siegbahn, 1993; Sakai et al., 1999) and experiment (Ram et al., 2001) now have established its $X^2\Delta$ ground state but with very low lying $a^4\Phi$ ($616\ cm^{-1}$), $b^4\Sigma^-$ ($963\ cm^{-1}$), and six states in all lying below $5000\ cm^{-1}$ energy. More current two-photon spectral studies of ZrCl still are continuing (Martinez and Morse, 2011). Experimental spectroscopy of ZrH appears minimal and most emphasis has been theoretical due to the challenge of trying to model such second row transition element hydrides that display relativistic and spin–orbit effects in their low-lying spin mixed states (Langhoff et al., 1987; Balasubramanian, 1990; Koseki et al., 2002). One study calculated the potential energy curves for the 36 electronic states that lie below $14{,}000\ cm^{-1}$ (Balasubramanian and Wang, 1989), and a more recent analysis also has examined the locations of these curves (Koseki et al., 2002). ZrH appears to reflect ZrCl with a similar ground and low lying electronic state. Infrared studies of the matrix isolated hydrides have derived structures and vibrational frequencies for the hydrides, ZrH_n, $n = 1-4$ (Chertihin and Andrews, 1995b). The infrared spectra of the dihydroxide molecule has been obtained by laser ablation of Zr in H_2O, H_2O_2 or H_2/O_2 mixtures and characterized with regards to vibrations and geometry (Zhou et al., 2000a; Wang and Andrews, 2005a).

Laboratory combustion studies

In the author's laboratory, a series of experiments introduced zirconium at ppmv concentrations as a nitrate salt aerosol into a fuel-rich or fuel-lean propane flame. The aim of the program was to examine whether the presence of zirconium could modify the 2000K flame chemistry of an alkali (sodium or potassium) when sulfur and chlorine were also present. Without zirconium, deposits of Na_2SO_4 or NaCl were normally obtained on steel probes inserted in the flame and cooled to temperatures in the 400K–1200K range.

The question was whether Na_2ZrO_3 formation was possible and sufficiently more thermodynamically stable under these conditions to prevent sodium sulfate/chloride formation (Schofield, 2003). However, in all cases, it was found that the sodium and zirconium flame chemistry remained independent of one another and that the zirconium solely deposited monoclinic ZrO_2. With present knowledge, if an alkaline earth salt had been used instead of sodium it is possible, for example, that deposits would have modified to $BaZrO_3$. Such behavior is in fact now the basis for particle formation and CVD of ZrO_2 and zirconates utilizing flame and other CVD methods. Flame particle synthesis has been reviewed by Rosner (2005) and Roth (2007). Details of producing ZrO_2 nanoparticle powders and mixed ZrO_2/Y_2O_3 powders and film deposits are well documented (Karthikeyan et al., 1997; Choy and Vyas, 1999). CVD methods describe formation of ZrO_2, $BaZrO_3$, $PbZrO_3$, $PbZrTiO_3$, and others, generally comparing different zirconium precursors to modify volatility, handling of these chemicals, and substrate temperatures, and other variable parameter effects (Jones et al., 1998). As mentioned earlier, the zirconates differ from the vanadates and chromates in being more thermodynamically stable in the condensed phase. This results from the ceramic nature of zirconium oxide with its high melting point (2715°C) that is continued into numerous of its mixed metal zirconates, the alkaline earths for example, $MgZrO_3$(2150°C), $CaZrO_3$(2550°C), $SrZrO_3$(2750°C), and $BaZrO_3$(2705°C). In a recent study, nanoparticles of Zr metal were produced by laser ablation and then size selected to be in the 20–150 nm diameter range. These were then directly injected into a laminar flame at 1450°C–2250°C and the mechanism of particle oxidation to ZrO_2 monitored. This was seen to be controlled by surface reaction (Zong et al., 2015).

One noteworthy aspect of zirconium mirrors its second row transition elements, Y and Nb, and also Sc, Ti, La, Hf, and Ta elements in the first three columns. This relates to their very high thermal ionization potentials and their potential to ionize by energetic chemi-ionization reactions (Schofield, 2006). Reactions such as these can produce electrons at lower temperatures.

$$Zr + O_2 = ZrO + O \qquad \Delta H_0 = -272 \text{ kJ mol}^{-1}$$

$$Zr + O = ZrO^+ + e^- \qquad \Delta H_0 = -109 \text{ kJ mol}^{-1}$$

$$Zr + O_2 = ZrO_2^+ + e^- \qquad \Delta H_0 = -14 \text{ kJ mol}^{-1}$$

$$ZrO^+ + O_2 = ZrO_2^+ + O \qquad \Delta H_0 = +95 \text{ kJ mol}^{-1}$$

$$Zr^+ + O_2 = ZrO^+ + O \qquad \Delta H_0 = -255 \text{ kJ mol}^{-1}$$

In the case of zirconium, the ZrO^+ ion is energetically favored and the electrons quite long-lived with respect to recombination.

Combustion power plant data

The combustion chemistry of zirconium compounds would seem very restricted and one might conclude a limited release from ash and a rapid conversion to ZrO_2(s). It was soon realized in flame photometric analyses using hydrogen or acetylene fuels that these methods had no sensitivity to zirconium. Only with the introduction of the ICP with its high temperature plasma was a quantitative sensitive analysis possible by a spectral method monitoring its ionic or neutral atomic lines (Boumans and Vrakking, 1987). As a result, zirconium is expected to reflect titanium and be a nonvolatile component of fuels. This is confirmed in those studies mainly of coal and sewage sludge that have included the analysis of zirconium. The concentrations in ash are larger than in the raw coal simply due to combustion mass reduction and can be about 250 ppmw in the various ash forms and rather like its concentration in soil (Rowe and Steinnes, 1977). Consequently, zirconium is of little concern as a specific pollutant from combustion. Its content can vary in differing fuels (Table 1.1), generally being less than 100 ppmw in Western coals and possibly being lower in subbituminous than bituminous (Germani et al., 1980) and at low levels in biomass. It distributes itself totally in the bottom and fly ash or stack dusts (Xu et al., 2003) and no enrichment is apparent in these (Martinez-Tarrazona and Spears, 1996; Querol et al., 1995; Font et al., 2010; Lanzerstorfer, 2018). Similarly, in these latter studies no variation or fractionation was reported in particle size dependences for $<1.6\,\mu m$ and larger particles. A detailed mass balance of trace metal emissions was made on two Danish coal-fired plants. These confirmed that the bottom ash and fly ash contained most of the zirconium with no enrichments (Nielsen et al., 2002). However, as in other cases, particle collection is never 100% and there was evidence of 0.02 and 0.06% penetration of zirconium that was in the fine particle stack emissions. This also had been shown in an early study of two US plants burning Western coals with particulate controls (Ondov et al., 1979). However, now in a modern co-combustion with coal and pet-coke fuel it was shown that the FGD system was quite effective retaining any stray leakage of zirconium by halving it further (Cordoba et al., 2012). Specific studies aimed at examining the releasing capabilities of Cl_2 or $CaCl_2$ on fly ash (Chan and Kirk, 1999) and $FeCl_3$ treated coal/sewage sludge mixtures (Folgueras et al., 2003) also concluded that zirconium is negligibly volatized.

In summary, as with titanium, although there are limited data it clearly reflects the nonvolatility and fractionation of zirconium in fuels. However, there is zirconium present in the smallest submicron particles that elude collection by particulate control and are present in stack emissions. It appears that like titanium these may be produced not by a vapor nucleation mechanism but simply by physical fragment ejection from particles as they forcefully gasify.

Bibliography

Abanades, S., Flamant, G., Gagnepain, B., Gauthier, D., 2002. Fate of heavy metals during municipal solid waste incineration. Waste Manage. Res. 20, 55–68.

Abascal, M.B., Blasing, M., Ninomiya, Y., Muller, M., 2016. Influence of steam, hydrogen chloride and hydrogen sulfide on the release and condensation of cadmium in gasification. Energy Fuels 30, 943–953.

Abascal, M.B., Blasing, M., Ninomiya, Y., Muller, M., 2016a. Influence of steam, hydrogen chloride and hydrogen sulfide on the release and condensation of zinc in gasification. Ind. Eng. Chem. Res. 55, 6911–6921.

Abrahamson, D., 1992. Aluminum and global warming. Nature 356, 484.

Ackermann, R.J., Rauh, E.G., 1974. Thermodynamic properties of ZrO(g) and HfO(g); a critical examination of isomolecular oxygen-exchange reactions. J. Chem. Phys. 60, 2266–2271.

Ackermann, R.J., Thorn, R.J., Alexander, C., Tetenbaum, M., 1960. Free energies of formation of gaseous uranium, molybdenum and tungsten trioxides. J. Phys. Chem. 64, 350–355.

Ackermann, R.J., Rauh, E.G., Thorn, R.J., 1964. Thermodynamic properties of gaseous yttrium monoxide: correlation of bonding in group III transition metal monoxides. J. Chem. Phys. 40, 883–889.

Ackermann, R.J., Rauh, E.G., Alexander, C.A., 1975. The thermodynamic properties of $ZrO_2(g)$. High Temp. Sci. 7, 304–316.

Ackermann, R.J., Rauh, E.G., Rand, M.H., 1980. A re-determination and re-assessment of the thermodynamics of sublimation of uranium dioxide. In: Thermodynamics of Nuclear Materials 1979. Proceedings of International Symposium, Julich, Germany. International Atomic Energy Agency, Vienna, vol. 1, pp. 11–27.

Adams, J.W., Lei, J., Husain, D., 2001. The collisional behavior of $Ba[6s5d(^3D_J)]$ generated following pulsed dye-laser excitation at $\lambda = 553.5$ nm $\{Ba(6s6p\ ^1P_1) \leftarrow Ba(6s^2\ ^1S_0)\}$ in the presence of noble gases and ground state atomic barium by atomic spectroscopic marker emission measurements in the time-domain. Z Phys. Chem. 9, 1205–1230.

Adanez, J., Abad, A., 2019. Chemical-looping combustion: status and research needs. Proc. Combust. Inst. 37, 4303–4317.

Adanez-Rubio, I., Gayan, P., Abad, A., et al., 2014. Kinetic analysis of a Cu-based oxygen carrier: relevance of temperature and oxygen partial pressure on reduction and oxidation reaction rates in chemical looping with oxygen uncoupling. Chem. Eng. J. (Amsterdam) 256, 69–84.

Afaf, M., 1995. Assignment of omitted ultraviolet systems of zirconium oxide. Astrophys. J. 447, 980–983.

Agnihotri, R., Chauk, S., Mahuli, S., Fan, L.-S., 1998. Selenium removal using Ca-based sorbents: reaction kinetics. Environ. Sci. Technol. 32, 1841–1846.

Ahlrichs, R., Bar, M., Plitt, H.S., Schnockel, H., 1989. The stability of PN and $(PN)_3$: *ab initio* calculations and matrix infrared investigations. Chem. Phys. Lett. 161, 179–184.

Ahmaruzzaman, M., 2010. A review on the utilization of fly ash. Prog. Energy Combust. Sci. 36, 327–363.

Aho, M., Ferrer, E., 2005. Importance of coal ash composition in protecting the boiler against chlorine deposition during combustion of chlorine-rich biomass. Fuel 84, 201–212.

Aho, M., Silvennoinen, J., 2004. Preventing chlorine deposition on heat transfer surfaces with aluminium/silicon rich biomass residue and additive. Fuel 83, 1299–1305.

Ait-Hou, A., Rais, T., Chatillon, C., Allibert, M., 1995. Vaporization thermodynamics of the copper-oxygen system by mass spectrometry. Ann. de Chim. (Paris) 20 (6), 345–354.

Aizer, A., Currie, J., Simon, P., Vivier, P., 2018. Do low levels of blood lead reduce children's future test scores? Am. Econ. J. Appl. Econ. 10 (1), 307–341.

Ajersch, F., Benlyamani, M., 1989. Thermogravimetric identification and analysis of reaction products during oxidation of solid or liquid sulfides. Thermochim. Acta. 143, 221–237.

Akbar, S., Schnell, U., Scheffknecht, G., 2010. Modeling potassium release and the effect of potassium chloride on deposition mechanisms for coal and biomass-fired boilers. Combust. Theory Model. 14, 315–329.

Akiho, H., Ito, S., Matsuda, H., 2010. Effect of oxidizing agents on selenate formation in a wet FGD. Fuel 89 (9), 2490–2495.

Akiho, H., Yamamoto, T., Tochihara, Y., et al., 2012. Speciation and oxidation reaction analysis of selenium in aqueous solution using X-ray absorption spectroscopy for management of trace element in FGD liquor. Fuel 102, 156–161.

Akito, M., Shinichiro, Y., Akihiro, H., et al., 2014. Reevaluation of Minamata Bay, 25 years after the dredging of mercury-polluted sediments. Mar. Pollut. Bull. 89 (1–2), 112–120.

Al-Abed, S.R., Jegadeesan, G., Scheckel, K.G., Tolaymat, T., 2008. Speciation, characterization and mobility of As, Se and Hg in flue gas desulfurization residues. Environ. Sci. Technol. 42, 1693–1698.

Alawad, Z.M., Al-Jobouri, S.M., Majid, A.Y., 2019. Lead among children with autism in Iraq. Is it a potential factor? J. Clin. Anal. Med. 10 (2), 215–219.

Alehagen, U., Aaseth, J., Johansson, P., 2015. Reduced cardiovascular mortality 10 years after supplementation with selenium and co-enzyme Q10 for four years: follow-up results of a prospective randomized double-blind placebo-controlled trial in elderly citizens. PLoS One 10 (12), e0141641.

Alekseyev, A.B., Buenker, R.J., Liebermann, H.-P., Hirsch, G., 1994. Spin-orbit configuration interaction study of the potential energy curves and radiative lifetimes of the low-lying states of bismuth hydride. J. Chem. Phys. 100 (4), 2989–3001.

Alekseyev, A.B., Liebermann, H.-P., Buenker, R.J., Hirsch, G., 1995. Theoretical study of the electronic spectrum of antimony oxide employing relativistic effective core potentials. J. Chem. Phys. 102 (6), 2539–2550.

Alekseyev, A.B., Sannigrahi, A.B., Liebermann, H.-P., et al., 1995a. *Ab initio* potential curves, dipole moments and transition probabilities for the low-lying states of arsenic oxide. J. Chem. Phys. 103 (1), 234–244.

Alekseyev, A.B., Liebermann, H.-P., Buenker, R.J., Hirsch, G., 1996. *Ab initio* study of the low-lying states of SnH. Mol. Phys. 88, 591–603.

Alekseyev, A.B., Liebermann, H.-P., Buenker, R.J., Hirsch, G., 1996a. Spin-orbit configuration interaction study of potential energy curves and transition probabilities of the mercury hydride molecule and tests of relativistic effective core potentials for Hg, Hg^+ and Hg^{2+}. J. Chem. Phys. 104 (12), 4672–4684.

Alekseyev, A.B., Liebermann, H.-P., Hirsch, G., Buenker, R.J., 1998. The spectrum of arsenic hydride: an *ab initio* configuration interaction study employing a relativistic effective core potential. J. Chem. Phys. 108 (5), 2028–2040.

Alekseyev, A.B., Liebermann, H.-P., Lingott, R.M., et al., 1998a. The spectrum of antimony hydride: an *ab initio* configuration interaction study employing a relativistic effective core potential. J. Chem. Phys. 108 (18), 7695–7706.

Alemozafar, A.R., Madix, R.J., 2005. Surface reorganization accompanying the formation of sulfite and sulfate by reaction of sulfur dioxide with oxygen on Ag(111). J. Chem. Phys. 122, 214718.

Alexander, C.A., Ogden, J.S., Levy, A., 1963. Transpiration study of magnesium oxide. J. Chem. Phys. 39, 3057–3060.

Alfredson, P.G., Cairns, R.C., 1964. Development of coated fuel pellets for beryllia-based fuels. J. Nucl. Mater. 14, 459–464.

Alfthan, G., Eurola, M., Ekholm, P., Venäläinen, E.-R., Root, T., Korkalainen, K., et al., 2015. Effects of nationwide addition of selenium fertilizes on foods, and animal and human health in Finland: from deficiency to optimal selenium status of the population. J. Trace Elem. Med. Biol. 31, 142–147.

Ali, I., Damdimopoulou, P., Stenius, U., Halldin, K., 2015. Cadmium at nanomolar concentrations activates Raf-MEK-ERK 1/2 MAPKs signaling via EGFR in human cancer cell lines. Chem.-Biol. Interact. 231, 44–52.

Alkemade, C.Th.J., Hollander, Tj, Snelleman, W., Zeegers, P.J.Th, 1982. Metal Vapours in Flames. Pergamon Press, Oxford, UK.

Allen, D., Hayhurst, A.N., 1996. Reaction between gaseous sulfur dioxide and solid calcium oxide: mechanism and kinetics. J. Chem. Soc. Faraday Trans. 92, 1227–1238.

Allen, M.D., Li, B.Z., Ziurys, L.M., 1997. A millimeter/sub-millimeter wave spectroscopic study of the FeCl radical ($X^6\Delta_i$). Chem. Phys. Lett. 270, 517–526.

Allendorf, M.D., Melius, C.F., Cosic, B., Fontijn, A., 2002. BAC-G2 predictions of thermochemistry for gas phase aluminum compounds. J. Phys. Chem. A 106, 2629–2640.

Allouche, A.R., Aubert-Frecon, M., 1994. A ligand field approach for the low-lying states of Ca, Sr and Ba monohydroxides. J. Mol. Spectrosc. 163, 599–603.

Allouche, A.R., Nicolas, G., Barthelat, J.C., Spiegelmann, F., 1992. Theoretical study of the electronic structure of the BaH molecule. J. Chem. Phys. 96 (10), 7646–7655.

Allouche, A.R., Wannous, G., Aubert-Frecon, M., 1993. A ligand-field approach for the low-lying states of Ca, Sr and Ba monohalides. Chem. Phys. 170, 11–22.

Allouti, F., Manceron, L., Alikhani, M.E., 2006. The $Ni + O_2$ reaction: a combined infrared matrix isolation and theoretical study of the formation and structure of NiO_2. Phys. Chem. Chem. Phys. 8, 448–455.

Al-Naiema, I., Estillore, A.D., Mudunkotuwa, I.A., et al., 2015. Impacts of co-firing biomass on emissions of particulate matter to the atmosphere. Fuel 162, 111–120.

Al-Saleh, I., Abduljabbar, M., Al-Rouqi, R., et al., 2015. The extent of mercury (Hg) exposure among Saudi mothers and their respective infants. Environ. Monit. Assess. 187, 678. 29 pp.

Al-Saleh, I., Elkhatib, R., Al-Rouqi, R., et al., 2016. Alterations in biochemical markers due to mercury (Hg) exposure and its influence on infant's neurodevelopment. Int. J. Hyg. Environ. Health 219, 898–914.

Althobiti, R.A., Sadiq, N.W., Beauchemin, D., 2018. Realistic risk assessment of arsenic in rice. Food Chem. 257, 230–236.

Altman, R.L., 1963. Thermochemistry of the beryllia-water reaction. J. Chem. Eng. Data 8, 534–536.

Altun, S.K., Dinc, H., Temamogullari, F.K., et al., 2018. Analyses of essential elements and heavy metals by using ICP-MS in maternal breast milk from Sanhurfa, Turkey. Int. J. Anal. Chem. 2018, 1784073.

Alvarez-Barcia, S., Flores, J.R., Granucci, G., Persico, M., 2013. Theoretical study of the chemiluminescence of the $Al + H_2O$ reaction. J. Phys. Chem. A 117, 67–74.

Alvim-Ferraz, M.C.M., Afonso, S.A.V., 2003. Incineration of different types of medical wastes: emission factors for particulate matter and heavy metals. Environ. Sci. Technol. 37, 3152–3157.

Aly, A.I.M., Abdel-Aal, M.M., Ahmed, M.A., Sabek, M.G., 1994. Radiation doses resulting from a proposed coal fired power plant in the Suez Canal area. Energy 19, 55–61.

Amand, L.-E., Leckner, B., 2004. Metal emissions from co-combustion of sewage sludge and coal/wood in fluidized bed. Fuel 83, 1803–1821.

Amand, L.-E., Leckner, B., Eskilsson, D., Tillin, C., 2006. Deposits on heat transfer tubes during co-combustion of biofuels and sewage sludge. Fuel 85, 1313–1322.

Ames, L.L., Walsh, P.N., White, D., 1967. Rare earths. IV. Dissociation energies of the gaseous monoxides of the rare earths. J. Phys. Chem. 71, 2707–2718.

Amos, H.M., Jacob, D.J., Streets, D.G., Sunderland, E.M., 2013. Legacy impacts of all-time anthropogenic emissions on the global mercury cycle. Global Biogeochem. Cycles 27 (2), 410–421.

Amos, H.M., Sonke, J.E., Obrist, D., et al., 2015. Observational and modeling constraints on global anthropogenic enrichment of mercury. Environ. Sci. Technol. 49, 4036–4047.

Amrute, A.P., Mondelli, C., Hevia, M.A.G., Perez-Ramirez, J., 2011. Temporal analysis of products study of HCl oxidation on copper and ruthenium-based catalysts. J. Phys. Chem. C 115, 1056–1063.

Ancora, M.P., Zhang, L., Wang, S.-X., et al., 2015. Economic analysis of atmospheric mercury emission control for coal-fired power plants in China. J. Environ. Sci. China 33, 125–134.

Andersen, K.H., Frandsen, F.J., Hansen, P.F.B., Wieck-Hansen, K., Rasmussen, I., Overgaard, P., et al., 2000. Deposit formation in a 150 MW_e utility pulverized fuel-boiler during co-combustion of coal and straw. Energy Fuels 14, 765–780.

Anderson, D., 1973. Emission factors for trace substances. U.S. Environmental Protection Agency EPA-450/2-73-001, 91 pp.

Anderson, C.G., 2012. The metallurgy of antimony. Chem. Erde 72 (4), 3–8.

Anderson, A.B., Hong, S.Y., Smialek, J.L., 1987. Comparison of bonding in first transition-metal series: diatomic and bulk sulfides and oxides. J. Phys. Chem. 91, 4250–4254.

Andren, A.W., Klein, D.H., Talmi, Y., 1975. Selenium in coal-fired steam plant emissions. Environ. Sci. Technol. 9, 856–858.

Andrews, L., 2004. Matrix infrared spectra and density functional calculations of transition metal hydrides and dihydrogen complexes. Chem. Soc. Rev. 33, 123–132.

Andrews, L., Wang, X., 2005. Infrared spectra of the group 2 metal dihydroxide molecules. Inorg. Chem. 44, 11–13.

Andrews, L., Yustein, J.T., 1993. Reaction of pulsed laser evaporated magnesium atoms with oxygen. Infrared spectra of linear OMgO and MgOMgO in solid argon. J. Phys. Chem. 97, 12700–12704.

Andrews, L., Yustein, J.T., Thompson, C.A., Hunt, R.D., 1994. Reactions of pulsed-laser evaporated Ca, Sr and Ba atoms with O_2. Infrared spectra of the metal oxides, oxide dimers, dioxides and peroxides in solid argon. J. Phys. Chem. 98, 6514–6521.

Andrews, L., Chertihin, G.V., Thompson, C.A., Dillon, J., Byrne, S., Bauschlicher Jr, C.W., 1996. Infrared spectra and quantum chemical calculations of Group II MO_2, O_2MO_2 and related molecules. J. Phys. Chem. 100, 10088–10099.

Andrews, L., Chertihin, G.V., Ricca, A., Bauschlicher Jr, C.W., 1996a. Reactions of laser-ablated iron atoms with oxygen molecules: matrix infrared spectra and density functional calculations of OFeO, FeOO and $Fe(O_2)$. J. Am. Chem. Soc. 118, 467–470.

Andrews, L., Kushto, G.P., Yustein, J.T., Archibong, E., Sullivan, R., Leszczynski, J., 1997. Reactions of pulsed-laser-evaporated thallium atoms with oxygen. Matrix infrared spectra of new TlO_2 species. Trends in Group 13 dioxides and dioxide anions. J. Phys. Chem. A 101, 9077–9084.

Andrews, L., Zhou, M., Chertihin, G.V., Bauschlicher Jr, C.W., 1999. Reactions of laser-ablated Y and La atoms, cations and electrons with O_2. Infrared spectra and density functional calculations of the MO, MO^+, MO_2, MO_2^+ and MO_2^- species in solid argon. J. Phys. Chem. A 103, 6525–6532.

Andrews, L., Wang, X., Liang, B., et al., 2011. Matrix infrared spectroscopy and a theoretical investigation of SUO and US_2. Eur. J. Inorg. Chem. 4457–4463.

Anglada, J., Bruna, P.J., Peyerimhoff, S.D., 1989. The electronic spectrum of ScH: an MRD-CI study. Mol. Phys. 66, 541–563.

Anglada, J., Bruna, P.J., Peyerimoff, S.D., 1990. Theoretical investigation of the low-lying electronic states of TiH. Mol. Phys. 69, 281–303.

Angot, H., Hoffman, N., Giang, A., et al., 2018. Global and local impacts of delayed mercury mitigation efforts. Environ. Sci. Technol. 52, 12968–12977.

Annen, K.D., Kuzmenko, P.J., Keating, R., Self, S.A., 1981. Electron and positive ion measurements in MHD combustion plasmas with phosphorus addition. J. Energy 5, 31–38.

Antic-Jovanovic, A., Bojovic, V., Pesic, D., 1988. Electronic bands of the BeOH and BeOD. Spectrosc. Lett. 21, 757–765.

Anufrieva, T.A., Derlyukova, L.E., Vinokurova, M.V., 2001. Interaction of nickel oxide with chlorine. Zhur Neorg. Khim. 46 (1), 21–24.

Anyanwu, B.O., Ezejiofor, A.N., Igweze, Z.N., Orisakwe, O.E., 2018. Heavy metal mixture exposure and effects in developing nations: an update. Toxics 6 (4), 65.

Arafunnisa, Rao, D.V.K., Rao, P.T., 1975. Dissociation energies, r-centroids and Frank-Condon factors of AsS molecule. Indian J. Phys. 49, 580–585.

Arcega-Cabrera, F., Fargher, L., Quesadas-Rojas, M., et al., 2018. Environmental exposure of children to toxic trace elements (Hg, Cr, As) in an urban area of Yucatán, Mexico: water, blood, and urine levels. Bull. Environ. Contam. Toxicol. 100, 620–626.

Armentrout, P.B., Sunderlin, L.S., 1992. Gas phase organometallic chemistry of transition metal hydrides. In: Dedieu, A. (Ed.), Transition Metal Hydrides. VCH Publishers Inc., New York, pp. 1–64. Chapter 1.

Asano, M., Kou, T., 1989. Thermochemical properties of $BaBO_2(g)$ and $Ba_3B_2O_6(s)$. J. Chem. Thermodyn. 21, 837–845.

Asano, M., Kou, T., 1990. Thermochemical properties of SrSiO(g) and $SrBO_2(g)$. J. Chem. Phys. 22, 1223–1230.

Asano, M., Kou, T., 1990a. Vaporization and thermochemical properties in the BaO-B_2O_3 system. High Temp. Sci. 29, 171–187.

Aschner, M., Onishchenko, N., Ceccatelli, S., 2010. Toxicology of alkylmercury compounds. Chapter 12. In: Sigel, A., Sigel, H., Sigel, R.K.O. (Eds.), Metal Ions in Life Sciences, vol. 7. Organometallics in Environment and Toxicology. Royal Society of Chemistry, London, UK, pp. 403–434.

Ashley, K., Cordell, D., Mavinic, D., 2011. A brief history of phosphorus: from the philosopher's stone to nutrient recovery and reuse. Chemosphere 84, 737–746.

Ashley-Martin, J., Dodds, L., Arbuckle, T.E., et al., 2018. Maternal and cord blood manganese (Mn) levels and birth weight: the MIREC birth cohort study. Int. J. Hyg. Environ. Health 221 (6), 876–882.

Ashraf, M.A., Akib, S., Maah, M.J., Yusoff, I., Balkhair, K.S., 2014. Cesium-137: radiochemistry, fate and transport, remediation and future concerns. Crit. Rev. Environ. Sci. 44 (15), 1740–1793.

Asryan, N.A., Alikhanyan, A.S., Nipan, G.D., 2003. Specifics of sublimation of antimony oxides. Doklady Phys. Chem. 392, 221–226.

Athanassiadis, Y.C., 1969. 86 pp. Air Pollution Aspects of Phosphorus and its Compounds. National Air Pollution Control Administration.

Au, W.W., McConnell, M.A., Wilkinson, G.S., et al., 1998. Population monitoring: experience from residents exposed to uranium mining/milling waste. Mutat. Res. 405 (2), 237–245.

Ault, B.S., Andrews, L., 1975. Matrix reactions of alkaline earth metal atoms with oxygen molecules: infrared spectra of the metal superoxide and metal oxide dimer species. J. Chem. Phys. 62, 2312–2319.

Ault, B.S., Andrews, L., 1975a. Nitrogen matrix reactions of alkaline earth metal atoms with ozone: infrared spectra of the alkaline earth metal oxide molecules. J. Chem. Phys. 62 (6), 2320–2327.

Aunela-Tapola, L., Hatanpaa, E., Hoffren, H., Laitinen, T., Larjava, K., Rasila, P., et al., 1998. A study of trace element behavior in two modern coal-fired power plants. II. Trace element balances in two plants equipped with semi-dry flue gas desulfurization facilities. Fuel Process. Technol. 55, 13–34.

Auzmendi-Murua, I., Bozzelli, J.W., 2016. Gas phase mercury oxidation by halogens (Cl, Br, I) in combustion effluents: influence of operating conditions. Energy Fuels 30, 603–615.

Awitor, K.O., Bernard, L., Coupat, B., et al., 2000. Measurement of mercurous chloride vapor pressure. New J. Chem. 24, 399–401.

Azad, A.M., Frantzen, S., Bank, M.S., et al., 2019. Effects of geography and species variation on selenium and mercury molar ratios in North East Atlantic Marine fish communities. Sci. Total Environ. 652, 1482–1496.

Azimi, S., Rocher, V., Garnaud, S., Varrault, G., Thevenot, D.R., 2005. Decrease of atmospheric deposition of heavy metals in an urban area from 1994 to 2002 (Paris, France). Chemosphere 61, 645–651.

Babelot, J.-F., Hoch, M., 1989. The constant vaporizing composition of UO_2 at 2150 and 2800K. J. Nucl. Mater. 167, 25–29.

Babushok, V., Tsang, W., 2000. Inhibitor rankings for alkane combustion. Combust. Flame 123, 488–506.

Babushok, V.I., Linteris, G.T., Hoorelbeke, P., Roosendans, D., van Wingerden, K., 2017. Flame inhibition by potassium-containing compounds. Combust. Sci. Technol. 189 (12), 2039–2055.

Babushok, V.I., Deglmann, P., Kramer, R., Linteris, G.T., 2017a. Influence of antimony-halogen additives on flame propagation. Combust. Sci. Technol. 189 (2), 290–311.

Bagarat'yan, N.V., Il'in, M.K., Nikitin, O.T., 1974. Mass spectrometric study of thallium metaborate. High Temp. USSR 11, 888–893.

Baijot, V., Ducere, J.-M., Rouhani, M.D., Rossi, C., 2016. Effect of temperature and O_2 pressure on the gaseous species produced during combustion of aluminum. Chem. Phys. Lett. 649, 88–91.

Bajamundi, C.J.E., Vainikka, P., Hedman, M., et al., 2015. Searching for a robust strategy for minimizing alkali chlorides in fluidized bed boilers during burning of high solid recovery-energy-share fuel (SRF). Fuel 155, 25–36.

Bakalbassis, E.G., Stiakaki, M.-A.D., Tsipis, A.C., Tsipis, C.A., 1996. Ground and low-lying excited state properties of the first-row transition metal oxide diatomics calculated by an improved ASED-MO model. Chem. Phys. 205, 389–399.

Baker, L.A., 2011. Can urban P conservation help to prevent the brown devolution? Chemosphere 84, 779–784.

Bakar, C., Karaman, H.I.O., Baba, A., Sengunalp, F., 2010. Effect of high aluminum concentration in water resources on human health, case study: Biga Peninsula, Northwest Turkey. Arch. Environ. Contam. Toxicol. 58, 935–944.

Bakkali, K., Martos, N.R., Souhail, B., Ballesteros, E., 2012. Determination of heavy metal content in vegetables and oils from Spain and Morocco by inductively coupled plasma mass spectrometry. Anal. Lett. 45, 907–919.

Balabanov, N.B., Boggs, J.E., 2002. *Ab initio* study of structure and spectra of MnH_2, MnH_2^- and MnH_3. J. Phys. Chem. A 106, 6839–6843.

Balan, G., Losurdo, M., Spliethoff, H., 2013. Experimental study of high temperature chlorine-induced corrosion in dependence of gas velocity. Energy Fuels 27, 5628–5639.

Balasubramanian, K., 1987. Electronic states and potential energy surfaces of ScH_2. Chem. Phys. Lett. 135, 288–293.

Balasubramanian, K., 1989. Spectroscopic properties and potential energy curves for heavy p-block diatomic hydrides, halides and chalconides. Chem. Rev. 89, 1801–1840.

Balasubramanian, K., 1990. The low-lying states of the second-row transition metal hydrides (YH-CdH). J. Chem. Phys. 93, 8061–8072.

Balasubramanian, K., Latifzadeh-Masoudipour, L., 1999. Spectroscopic properties and potential energy surfaces of electronic states of $SbCl_2$, $SbBr_2$, $SbCl_2^+$, $SbBr_2^+$. J. Phys. Chem. A 103, 3044–3053.

Balasubramanian, K., Liao, D.-W., 1991. Spectroscopic constants and potential energy curves of Bi_2 and Bi_2^-. J. Chem. Phys. 95 (5), 3064–3073.

Balasubramanian, K., Liao, M.Z., 1989. Electronic states and potential energy surfaces of AgH_2: comparison with AuH_2. J. Phys. Chem. 93, 89–94.

Balasubramanian, K., Pitzer, K.S., 1984. Relativistic quantum calculations of low-lying states of lead hydride: comparison with experimental spectra. J. Phys. Chem. 88, 1146–1148.

Balasubramanian, K., Wang, J.Z., 1989. Spectroscopic properties and potential energy curves of thirty-six electronic states of ZrH. Chem. Phys. Lett. 154, 525–530.

Balasubramanian, K., Wang, J.Z., 1989a. Spectroscopic properties and potential energy curves of 29 electronic states of YH. J. Mol. Spectrosc. 133, 82–89.

Balasubramanian, K., Siekhaus, W.J., McLean II, W., 2003. Potential energy surfaces for the uranium hydriding reaction. J. Chem. Phys. 119 (12), 5889–5900.

Balasubramanian, R., Narasimhan, T.S.L., Viswanathan, R., Nalini, S., 2008. Investigation of the Vaporization of boric acid by transpiration thermogravimetry and Knudsen effusion mass spectrometry. J. Phys. Chem. B 112, 13873–13884.

Balducci, G., De Maria, G., Guido, M., Piacente, V., 1971. Dissociation energy of FeO. J. Chem. Phys. 55, 2596–2598.

Balducci, G., Gigli, G., Guido, M., 1981. Dissociation energies of the molecules $CrPO_2$(g) and CrO(g) by high temperature mass spectrometry. J. Chem. Soc. Faraday Trans. II 77, 1107–1114.

Balducci, G., Gigli, G., Guido, M., 1983. Thermochemical properties of the gaseous molecules VO, VO_2 and V_2O_4. J. Chem. Phys. 79, 5616–5622.

Balducci, G., Gigli, G., Goldenberg, M.M.F., Guido, M., 1984. Vaporization of $CoTiO_3$. J. Chem. Thermodyn. 16, 207–217.

Balducci, G., Gigli, G., Guido, M., 1985. Mass spectrometric study of the thermochemistry of gaseous $EuTiO_3$ and TiO_2. J. Chem. Phys. 83, 1909–1912.

Balducci, G., Gigli, G., Meloni, G., 1998. Dissociation energies of the Ga_2, In_2 and GaIn molecules. J. Chem. Phys. 109 (11), 4384–4388.

Balfour, W.J., 1988. The electronic spectrum of manganese hydride. J. Chem. Phys. 88, 5242–5243.

Balfour, W.J., Chowdhury, P.K., 2010. Laser induced fluorescence and dispersed fluorescence spectra of zirconium monoxide between 460 and 560 nm. Chem. Phys. Lett. 485, 8–10.

Ball, D.W., 2003. The structure and vibrational spectra of isotopomers of SnOH and OSnH. J. Mol. Struct. Theochem. 626, 217–221.

Balucani, N., Stranges, D., Casavecchia, P., Volpi, G.G., 2004. Crossed beam studies of the reactions of atomic oxygen in the ground ^{3}P and first electronically excited ^{1}D states with hydrogen sulfide. J. Chem. Phys. 120, 9571–9582.

Bamzai, K.K., Kachroo, N., Singh, V., Verma, S., 2013. Synthesis, characterization and thermal decomposition of pure and dysprosium doped yttrium phosphate system. J. Mater. 359514, 8 pp.

Banchorndhevakul, W., Matsui, T., Naito, K., 1985. Vaporization study on V_2O_3 and two-phase mixture of V_2O_3 and VO by mass spectrometric method. Thermochim. Acta 88, 301–306.

Bande, A., Luchow, A., 2008. Vanadium oxide compounds with quantum Monte Carlo. Phys. Chem. Chem. Phys. 10, 3371–3376.

Bankiewicz, D., Yrjas, P., Hupa, M., 2009. High temperature corrosion of superheater tube materials exposed to zinc salts. Energy Fuels 23, 3469–3474.

Bankiewicz, D., Vainikka, P., Lindberg, D., Frantsi, A., Silvennoinen, J., Yrjas, P., et al., 2012. High temperature corrosion of boiler waterwalls induced by chlorides and bromides. II. Laboratory scale corrosion tests and thermodynamic equilibrium modeling of ash and gaseous species. Fuel 94, 240–250.

Bankiewicz, D., Enestam, S., Yrjas, P., Hupa, M., 2013. Experimental studies of Zn and Pb induced high temperature corrosion of two commercial boiler steels. Fuel Process. Technol. 105, 89–97.

Banon, S., Chatillon, C., Allibert, M., 1982. High temperature mass spectrometric study of thermodynamic activities in solid titanium monoxide-titanium dioxide and liquid titanium trioxide-titanium dioxide systems. High Temp. Sci. 15, 105–128.

Bao, J., Li, Z., Sun, H., Cai, N., 2013. Experiment and rate equation modeling of Fe oxidation kinetics in chemical looping combustion. Combust. Flame 160, 808–817.

Barbooti, M.M., Al-Madfai, S.H., Nassouri, H.J., 1988. Thermochemical studies on hot ash corrosion of stainless steel 304 and inhibition by magnesium sulfate. Thermochim. Acta 126, 43–49.

Bardi, G., Trionfetti, G., 1988. The vapor pressure over Al_2Se_3. Thermochim. Acta 136, 313–314.

Bare, W.D., Souter, P.F., Andrews, L., 1998. Reactions of laser ablated molybdenum and tungsten atoms with dioxygen: resolved infrared spectra of natural molybdenum and tungsten isotopic oxides in argon matrices. J. Phys. Chem. A 102, 8279–8286.

Barin, I., Platzki, G., 1995. Thermochemical Data of Pure Substances, third ed., vol. 1, pp. 1–924; vol. 2, pp. 925–1885. VCH Verlagsgesellschaft mbH, Weinheim, Germany.

Barnes, M., Clouthier, D.J., Hajigeorgiou, P.G., Huang, G., Kingston, C.T., Merer, A.J., et al., 1997. The electronic spectrum of CoO in the visible region. J. Mol. Spectrosc. 186, 374–402.

Barone, V., Adamo, C., 1997. First row transition metal hydrides: a challenging playground for new theoretical approaches. Int. J. Quantum Chem. 61, 443–451.

Barron, J.R., Kelley, A.R., Liu, R., 1998. NiH_2 has a singlet ground state. J. Chem. Phys. 108 (1), 1–3.

Bartels, M., Lin, W., Nijenhuis, J., Kapteijn, F., van Ommen, J.R., 2008. Agglomeration in fluidized beds at high temperatures: mechanisms, detection and prevention. Prog. Energy Combust. Sci. 34, 633–666.

Barten, H., Cordfunke, E.H.P., 1984. The vapor pressure of diarsenic pentoxide, As_2O_5. Thermochim. Acta 80, 221–229.

Bartonova, L., Serencisova, J., Cech, B., 2018. Yttrium partitioning and associations in coal-combustion ashes prior to and after their leaching in HCl. Fuel Process. Technol. 173, 205–215.

Barysz, M., Papadopoulos, M.G., 1998. On the ground state of NiH_2. J. Chem. Phys. 109 (9), 3699–3700.

Bas, M.C., Ortiz, J., Ballesteros, L., Martorell, S., 2016. Analysis of the influence of solar activity and atmospheric factors on ^{7}Be air concentration by seasonal-trend decomposition. Atmos. Environ. 145, 145–157.

Bashir, M.S., Jensen, P.A., Frandsen, F., Wedel, S., Dam-Johansen, K., Wadenback, J., et al., 2012. Suspension-firing of biomass. I. Full-scale measurements of ash deposit build-up. Energy Fuels 26, 2317–2330.

Bastos, W.R., Vieira, S.M., Manzatto, A.G., et al., 2018. Heterogeneity of multimedia exposures to neurotoxic elements (Al, As, Cd, Pb, Mn, and Hg) in breast-fed infants from Porto Velho, Brazil. Biol. Trace Elem. Res. 184, 7–15.

Basu, N., Lanphear, B.P., 2019. The challenge of pollution and health in Canada. Can. J. Public Health-Rev. Can. Sante Publ. 110 (2), 159–164.

Basu, S.K., Taniguchi, M., 1984. Thermal analysis of the oxidation of the vanadium sulfides V_5S_8 and V_3S_4. J. Thermal. Anal. 29 (6), 1209–1223.

Batalli-Cosmovici, C., Michel, K.W., 1972. Dissociation energy of SrO from a crossed been experiment. Chem. Phys. Lett. 16 (1), 77–80.

Batista, E.R., Martin, R.L., Hay, P.J., 2004. Density functional investigations of the properties and thermochemistry of UF_n and UCl_n (n = 1-6). J. Chem. Phys. 121 (22), 11104–11111.

Baturin, G.N., 2003. Phosphorus cycle in the ocean. Lithol. Miner. Resourc. 38 (2), 101–119.

Bauschlicher Jr, C.W., 1986. OH-transition metal bonding. Int. J. Quantum Chem. Quantum Chem. Symp. 20, 563–572.

Bauschlicher Jr, C.W., 1988. Full configuration interaction benchmark calculations for TiH. J. Phys. Chem. 92, 3020–3023.

Bauschlicher Jr, C.W., 1999. Heats of formation for PO_n and PO_nH (n = 1-3). J. Phys. Chem. A 103, 11126–11129.

Bauschlicher Jr, C.W., 2004. What is the ground state of $Ni(O_2)$? J. Phys. Chem. A 108, 2871–2873.

Bauschlicher Jr, C.W., 2018. The convergence of the coupled cluster approach for MgO. Chem. Phys. Lett. 711, 27–31.

Bauschlicher Jr, C.W., Langhoff, S.R., 1986. Theoretical studies of the low lying states of ScO, ScS, VO and VS. J. Chem. Phys. 85, 5936–5942.

Bauschlicher Jr, C.W., Maitre, P., 1995. Theoretical study of the first transition row oxides and sulfides. Theor. Chim. Acta 90, 189–203.

Bauschlicher Jr, C.W., Partridge, H., 1998. A comparison of ZnO and ZnO^-. J. Chem. Phys. 109, 8430–8434.

Bauschlicher Jr, C.W., Schwenke, D.W., 2017. Low-lying electronic states of MgO. Chem. Phys. Lett. 683, 62–67.

Bauschlicher Jr, C.W., Langhoff, S.R., Partridge, H., 1986. *Ab initio* study of the alkali and alkaline-earth monohydroxides. J. Chem. Phys. 84, 901–909.

Bauschlicher Jr, C.W., Partridge, H., Langhoff, S.R., 1990. An *ab initio* study of the low-lying doublet states of AgO and AgS. Chem. Phys. 148, 57–68.

Bauschlicher Jr, C.W., Partridge, H., Sodupe, M., Langhoff, S.R., 1992. Theoretical study of the alkaline-earth metal superoxides BeO_2 through SrO_2. J. Phys. Chem. 96, 9259–9264.

Bauschlicher Jr, C.W., Langhoff, S.R., Partridge, H., Sodupe, M., 1993. Determination of the structure and bond energies of NiO_2 and CuO_2. J. Phys. Chem. 97 (4), 856–859.

Bauschlicher Jr, C.W., Zhou, M., Andrews, L., 2000. A study of the products of the reaction of phosphorus and dioxygen. J. Phys. Chem. A 104, 3566–3571.

Baxter, L.L., 1990. The evolution of mineral particle size distributions during early stages of coal combustion. Prog. Energy Combust. Sci. 16, 261–266.

Baxter, L.L., Mitchell, R.E., 1992. The release of iron during the combustion of Illinois no. 6 coal. Combust. Flame 88, 1–14.

Baxter, L.L., Mitchell, R.E., Fletcher, T.H., 1997. Release of inorganic material during coal devolatilization. Combust. Flame 108, 494–502.

Beal, S.A., Osterberg, E.C., Zdanowicz, C.M., Fisher, D.A., 2015. Ice core perspective on mercury pollution during the past 600 years. Environ. Sci. Technol. 49, 7641–7647.

Beardah, M.S., Ellis, A.M., 1999. Observation of several new electronic transitions of the SrOH free radical. J. Chem. Phys. 110, 11244–11254.

Beaton, S.P., Evenson, K.M., Nelis, T., Brown, J.M., 1988. Detection of the free radicals FeH, CoH, and NiH by far infrared laser magnetic resonance. J. Chem. Phys. 89, 4446–4448.

Becidan, M., Sorum, L., Frandsen, F., Pedersen, A.J., 2009. Corrosion in waste-fired boilers: a thermodynamic study. Fuel 88, 595–604.

Becidan, M., Sorum, L., Lindberg, D., 2010. Impact of municipal solid waste quality on the behavior of alkali metals and trace elements during combustion: a thermodynamic equilibrium analysis. Energy Fuels 24, 3446–3455.

Becidan, M., Houshfar, E., Khalil, R.A., Skreiberg, O., Lovas, T., Sorum, L., 2011. Optimal mixtures to reduce the formation of corrosive compounds during straw combustion: a thermodynamic analysis. Energy Fuels 25, 3223–3234.

Beck, C.W., 1950. Differential thermal analysis curves of carbonate minerals. Am. Mineral. 35, 985–1013.

Beckers, F., Rinklebe, J., 2017. Cycling of mercury in the environment: sources, fate, and human health implications: a review. Crit. Rev. Environ. Sci. Technol. 47 (9), 693–794.

Bednar, A.J., Kirgan, R.A., Jones, W.T., 2009. Comparison of standard and reaction cell inductively coupled plasma mass spectrometry in the determination of chromium and selenium species by HPLC-ICP-MS. Anal. Chim. Acta 632, 27–34.

Behranvand, E., Mozdianfard, M.R., Diaz-Bejarano, E., et al., 2018. The comprehensive investigation of refinery preheats foulant samples originated by heavy crude oil fractions as heating fluids. Fuel 224, 529–536.

Behrens, R.G., Rosenblatt, G.M., 1972. Vapor pressure and thermodynamics of octahedral arsenic trioxide, arsenolite. J. Chem. Thermodyn 4, 175–190.

Behrens, R.G., Lemons, R.S., Rosenblatt, G.M., 1974. Vapor pressure and thermodynamics of selenium dioxide: the enthalpy of atomization of $SeO_2(g)$. J. Chem. Thermodyn. 6, 457–466.

Bellert, D., Burns, K.L., Wampler, R., Breckenridge, W.H., 2000. An accurate determination of the ionization energy of the MgO molecule. Chem. Phys. Lett. 322, 41–44.

Bellinger, D.C., 2018. Applying methods of the global burden of diseases, injuries, and risk factors study to the developmental neurotoxicants: a commentary. Environ. Health 17, 53.

Bellinger, D.C., Devleesschauwer, B., O'Leary, K., Gibb, H.J., 2019. Global burden of intellectual disability resulting from prenatal exposure to methylmercury, 2015. Environ. Res. 170, 416–421.

Belo, L.P., Elliott, L.K., Stanger, R.J., et al., 2014. High temperature conversion of SO_2 to SO_3: homogeneous experiments and catalytic effect of fly ash from air and oxy-fuel firing. Energy Fuels 28, 7243–7251.
Belo, L.P., Elliott, L.K., Stanger, R.J., Wall, T.F., 2016. Impacts of sulfur oxides on mercury speciation and capture by fly ash during oxy-fuel pulverized coal combustion. Energy Fuels 30, 8658–8664.
Belton, G.R., Jordan, A.S., 1965. The volatilization of molybdenum in the presence of water vapor. J. Phys. Chem. 69, 2065–2071.
Belton, G.R., Jordan, A.S., 1967. The gaseous hydroxides of cobalt and nickel. J. Phys. Chem. 71, 4114–4120.
Belton, G.R., McCarron, R.L., 1964. The volatilization of tungsten in the presence of water vapor. J. Phys. Chem. 68, 1852–1856.
Belton, G.R., Richardson, F.D., 1962. A volatile iron hydroxide. Trans. Faraday Soc. 58, 1562–1572.
Belyaev, V.N., Lebedeva, N.L., Krasnov, K.S., Gurvich, L.V., 1982. Determination of the dissociation energies of molecules from absolute intensity measurements of their electronic transitions in SrCl flame spectra. High Temp. USSR 20, 844–848.
Belzile, N., Chen, Y.-W., Filella, M., 2011. Human exposure to antimony. I. Sources and intake. Crit. Rev. Environ. Sci. Technol. 41 (14), 1309–1373.
Benavides-Garcia, M., Balasubramanian, K., 1994. Bond energies, ionization potentials and the singlet-triplet energy separations of SnX_2 and PbX_2 (X = Cl, Br, I) and their positive ions. J. Chem. Phys. 100 (4), 2821–2830.
Bencheikh, M., 1997. Semiempirical calculations on scandium monohalides. J. Phys. B: At. Mol. Opt. Phys. 30, L137–L139.
Bencivenni, L., Gingerich, K.A., 1982. Infrared spectra of matrix isolated gaseous ternary oxides M_2XO_4. J. Chem. Phys. 76, 53–56.
Bencivenni, L., Pelino, M., Ramondo, F., 1992. *Ab initio* study on the Al_2O, Al_2O_2, Si_2O_2 and $SiAlO_2$ oxides and on the $LiAlO_2$ and $NaAlO_2$ molecules. J. Mol. Struct. Theochem. 253, 109–120.
Bencko, V., Vasilieva, E.V., Symon, K., 1980. Immunological aspects of exposure to emissions from burning coal of high beryllium content. Environ. Res. 22, 439–449.
Bencze, L., Rys-Matejczuk, M., Yazhenskikh, E., et al., 2016. Investigation of vaporization of alkali metals from solidified gasifier slags by Knudsen effusion mass spectrometry. Energy Fuels 30, 657–665.
Bender, D., Schaefer, S.H., Tiemann, E., 1986. Laser spectroscopy of perturbed states of BaO in the region above 32,000 cm^{-1}. J. Mol. Spectrosc. 116, 286–314.
Bengtsson, G.J., Berzinsh, U., Larsson, J., Svanberg, S., 1992. Determination of radiative lifetimes in neutral arsenic using time-resolved laser spectroscopy in the VUV region. Astron. Astrophys. 263, 440–442.
Benjelloun, A.T., Daoudi, A., Chermette, H., 2004. Interaction of lead atoms with atmospheric hydroxyl radical. An *ab initio* and density functional theory study of the resulting complexes PbOH and HPbO. J. Chem. Phys. 121 (15), 7207–7221.
Benjelloun, A.T., Daoudi, A., Chermette, H., 2005. *Ab initio* and density functional theory study of lead complexes of atmospheric interest PbH_2, PbOH, $Pb(H_2O)$, $Pb(HO_2)$ and $Pb(OH)_2$. Mol. Phys. 103, 317–335.
Benjelloun, A.T., Daoudi, A., Chermette, H., 2005a. Interaction of lead atom with atmospheric dioxygen and ozone: quantum study of the structure and the stability of the resulting $Pb(O_n)$ (n = 1,2,3) compounds. J. Chem. Phys. 122, 154304.
Benson, S.W., 1988. The mechanism of inhibition of knock by lead additives, a chain debranching reaction. J. Phys. Chem. 92, 1531–1533.
Berkowitz, J., 1988. Photoionization mass spectrometric studies of AsH_n(n = 1-3). J. Chem. Phys. 89, 7065–7076.
Berkowitz, J., Chupka, W.A., 1966. Equilibrium composition of selenium vapor: the thermodynamics of the vaporization of HgSe, CdSe and SrSe. J. Chem. Phys. 45 (11), 4289–4302.
Berkowitz, J., Chupka, W.A., 1969. Photoionization of high temperature vapors. VI. S_2, Se_2 and Te_2. J. Chem. Phys. 50 (10), 4245–4250.
Berkowitz, J., Marquart, J.R., 1963. Equilibrium composition of sulfur vapor. J. Chem. Phys. 39, 275–283.
Berkowitz, J., Chupka, W.A., Inghram, M.G., 1957. Polymeric gaseous species in the sublimation of tungsten trioxide. J. Chem. Phys. 27, 85–86.
Berkowitz, J., Chupka, W.A., Inghram, M.G., 1957a. Thermodynamics of the V-O system: dissociation energies of VO and VO_2. J. Chem. Phys. 27, 87–90.

Berkowitz, J., Inghram, M.G., Chupka, W.A., 1957b. Polymeric gaseous species in the sublimation of molybdenum trioxide. J. Chem. Phys. 26, 842–846.

Berkowitz, J., Curtiss, L.A., Gibson, S.T., Greene, J.P., Hillhouse, G.L., Pople, J.A., 1986. Photoionization mass spectrometric study and *ab initio* calculations of ionization and bonding in P-H compounds; heats of formation, bond energies and the $^3B_1$-$^1A_1$ separation in PH_2^+. J. Chem. Phys. 84, 375–384.

Bernard, L., Awitor, K.O., Badaud, J.P., Bonnin, O., Coupat, B., Fournier, J.P., et al., 1997. Determination of the vapor pressure of $HgCl_2$ by the Knudsen effusion method. J. Phys. III France 7, 311–319.

Beryllium, 1990. Environmental Health Criteria EHC106. World Health Organization; International Program on Chemical Safety, Geneva. Available from: www.inchem.org.

Bettendorff, M., Peyerimhoff, S.D., 1986. Electronic structure of the radicals NF and NCl. II. Potential energy curves for NCl. Chem. Phys. 104, 29–39.

Beutel, M., Setzer, K.D., Shestakov, O., Fink, E.H., 1996. The $a^1\Delta(a_2)$ states of BiCl, BiBr and BiI. J. Mol. Spectrosc. 175 (1), 48–53.

Beutel, M., Setzer, K.D., Shestakov, O., Fink, E.H., 1996a. The $a^1\Delta(a_2)$-$X^3\Sigma^-(X_11)$ transitions of SbH and SbD. J. Mol. Specrosc. 179, 79–84.

Beutel, M., Setzer, K.D., Shestakov, O., Fink, E.H., 1996b. The $a^1\Delta \rightarrow X^3\Sigma^-$ transitions of PH and PD. Chem. Phys. Lett. 249, 183–190.

Beutel, M., Setzer, K.D., Fink, E.H., 1999. The $b^1\Sigma^+(b0^+)$-$X^3\Sigma^-(X_10^+)$ and $a^1\Delta(a_2)$-$X^3\Sigma^-(X_10^+)$ transitions of SbF, SbCl, SbBr, SbI. J. Mol. Spectrosc. 195 (1), 147–153.

Bezkorovainy, A., 1980. Biochemistry of Nonheme Iron. Plenum Press, New York.

Bhanarkar, A.D., Gavane, A.G., Tajne, D.S., Tamhane, S.M., Nema, P., 2008. Composition and size distribution of particles emissions from a coal fired power plant in India. Fuel 87, 2095–2101.

B'Hymer, C., Caruso, J.A., 2006. Selenium speciation analysis using inductively coupled plasma-mass spectrometry. J. Chromatogr. A 1114 (1), 1–20.

Bian, S., Wu, J., Zhang, Y., Wang, T., et al., 2018. Increasing recovery ratios with an improved European Community Bureau of reference method for Mercury analysis in flue gas desulfurization gypsum. Energy Fuels 32, 8340–8347.

Bichell, T.J.V., Halbesma, T.C., Tipps, K.G., Bowman, A.B., 2017. Metal biology associated with Huntington's disease, Chapter 13, pp. 231–263. In: White, A.R., Aschner, M., Costa, L.G., Bush, A.I. (Eds.), Biometals in Neurodegenerative Diseases. Academic Press, Elsevier, London, 415 pp.

Biester, H., Bindler, R., Martinez-Cortizas, A., Engstrom, D.R., 2007. Modeling the past atmospheric deposition of mercury using natural archives. Environ. Sci. Technol. 41 (14), 4851–4860.

Bilirgen, H., Romero, C., 2012. Mercury capture by boiler modifications with sub-bituminous coals. Fuel 94, 361–367.

Binning Jr, R.C., Curtiss, L.A., 1990. Theoretical study of GeH_n, AsH_n and SeH_n: bond dissociation energies. J. Chem. Phys. 92, 1860–1864.

Bishayee, A., Waghray, A., Patel, M.A., Chatterjee, M., 2010. Vanadium in the detection, prevention and treatment of cancer: the in vivo evidence. Cancer Lett. 294, 1–12.

Bisson, T.M., Xu, Z.-H., 2015. Potential hazards of brominated carbon sorbents for mercury. Environ. Sci. Technol. 49 (4), 2496–2502.

Bisson, T.M., MacLean, L.C.W., Hu, Y.-F., Xu, Z.-H., 2012. Characterization of mercury binding onto a novel brominated biomass ash sorbent by X-ray absorption spectroscopy. Environ. Sci. Technol. 46 (21), 12186–12193.

Bitto, A., Posta, J., Tomcsanyi, L., 1990. About interelement effects of alkaline-earth metals and aluminum in chloride medium in the air/acetylene flame in atomic absorption spectroscopy. Acta Chim. Hung. 127, 407–415.

Bjorklund, G., Aaseth, J., Ajsuvakova, O.P., Nikonorov, A., et al., 2017. Molecular interaction between mercury and selenium in neurotoxicity. Coord. Chem. Rev. 332, 30–37.

Blackburn, P.E., Buchler, A., 1965. The thermodynamics of vaporization in the beryllium oxide-boron oxide system. J. Phys. Chem. 69, 4250–4255.

Blackburn, P.E., Hoch, M., Johnston, H.L., 1958. The vaporization of molybdenum and tungsten oxides. J. Phys. Chem. 62, 769–773.

Blackburn, M.B., Mermet, J.M., Winefordner, J.D., 1978. CW-laser excited molecular fluorescence of species in flames. Spectrochim. Acta A 34, 847–852.

Black Hills Corporation, 2009. Golden CO. Mercury optimization study: Wygen II Final Report. 68 pp.

Blagojevic, V., Koyanagi, G.K., Lavrov, V.V., et al., 2004. Ion-molecule reactions of W^+ and WO^+: new and improved values for IE(WO), $D_{298}(W^+\text{-}O)$, $D_{298}(W\text{-}O)$ and $\Delta_f H_{298}{}^{\circ}(WO)$. Chem. Phys. Lett. 389, 303–308.

Blairs, S., 2006. Sublimation study of anhydrous ferric chloride. J. Chem. Thermodyn. 38, 1484–1488.

Blasing, M., Muller, M., 2010. Mass spectrometric investigations on the release of inorganic species during gasification and combustion of German hard coals. Combust. Flame 157, 1374–1381.

Blasing, M., Muller, M., 2010a. Mass spectrometric investigations on the release of inorganic species during gasification and combustion of Rhenish lignite. Fuel 89, 2417–2424.

Blasing, M., Melchior, T., Muller, M., 2010. Influence of the temperature on the release of inorganic species during high temperature gasification of hard coal. Energy Fuels 24, 4153–4160.

Blasing, M., Nazeri, K., Muller, M., 2014. Fate of sulfur, chlorine, alkali metal and vanadium species during high temperature gasification of Canadian tar sand products. Energy Fuels 28, 6345–6350.

Blasing, M., Nazeri, K., Muller, M., 2014a. Release of alkali metal, sulfur and chlorine species during high temperature gasification and co-gasification of hard coal, refinery residue and petroleum coke. Fuel 126, 62–68.

Blasing, M., Abascal, M.B., Ninomiya, Y., Muller, M., 2018. Partitioning of lead and lead compounds under gasification-like conditions. Energy Fuels 32, 651–657.

Blauer, J., Greenbaum, M.A., Farber, M., 1966. The thermodynamic and physical properties of beryllium compounds. IX. The free energy of formation of beryllium hydroxide(g). J. Phys. Chem. 70, 973–977.

Blissett, R.S., Rowson, N.A., 2012. A review of the multi-component utilization of coal fly ash. Fuel 97, 1–23.

Bloem, J., Bosman, B., 1975. Etching of aluminum oxide in an hydrogen chloride-hydrogen ambient at 1200–1500 K. Philips Res. Rep. 30 (4), 206–217.

Bocca, B., Madeddu, R., Asara, Y., et al., 2011. Assessment of reference ranges for blood Cu, Mn, Se and Zn in a selected Italian population. J. Trace Elem. Med. Biol. 25, 19–26.

Bocio, A., Nadal, M., Garcia, F., Domingo, J.L., 2005. Monitoring the metals in the population living in the vicinity of a hazardous waste incinerator: concentrations in autopsy tissues. Biol. Trace Elem. Res. 106, 41–50.

Bodnar, M., Szczyglowska, M., Konieczka, P., Namiesnik, J., 2016. Methods of selenium supplementation: bioavailability and determination of selenium compounds. Crit. Rev. Food Sci. Nutr. 56 (1), 36–55.

Bojer, M., Jensen, P.A., Frandsen, F., Dam-Johansen, K., Madsen, O.H., Lundtorp, K., 2008. Alkali/chloride release during refuse incineration on a grate: full-scale experimental findings. Fuel Process. Technol. 89, 528–539.

Boldyrev, A.I., Simons, J., 1995. *Ab initio* study of the strong binding of BeO to Li, Be and B atoms in the hyperstoichiometric LiOBe, BeOBe, and BeOB molecules. J. Phys. Chem. 99, 15041–15045.

Bolshova, T.A., Korobeinichev, O.P., 2006. Promotion and inhibition of a hydrogen-oxygen flame by the addition of trimethyl phosphate. Combust. Explos. Shock Waves 42 (5), 493–502.

Bonczyk, P.A., 1988. Suppression of soot in flames by alkaline earth and other metal additives. Combust. Sci. Technol. 59, 143–163.

Bonczyk, P.A., 1991. Effect of Ferrocene on soot in a pre-vaporized iso-octane/air diffusion flame. Combust. Flame 87, 233–244.

Bondy, S.C., 2017. Metals and neuroinflammation, Chapter 6, pp. 83–93. In: White, A.R., Aschner, M., Costa, L.G., Bush, A.I. (Eds.), Biometals in Neurodegenerative Diseases. Academic Press, Elsevier, London, 415 pp.

Bool III, L.E., Helble, J.J., 1995. A laboratory study of the partitioning of trace elements during pulverized coal combustion. Energy Fuels 9, 880–887.

Bool III, L.E., Peterson, T.W., Wendt, J.O.L., 1995. The partitioning of iron during the combustion of pulverized coal. Combust. Flame 100, 262–270.

Boonsongsup, L., Iisa, K., Frederick Jr, W.J., 1997. Kinetics of the sulfation of NaCl at combustion conditions. Ind. Eng. Chem. Res. 36, 4212–4216.

Borjesson, G., Katterer, T., 2018. Soil fertility effects of repeated application of sewage sludge in two 30- year-old field experiments. Nutr. Cycl. Agroecosyst. 112 (3), 369–385.

Bornhorst, J., Kipp, A.P., Haase, H., Meyer, S., Schwerdtle, T., 2018. The crux of inept biomarkers for risks and benefits of trace elements. Trends Anal. Chem. 104, 183–190.

Bos, A., Ogden, J.S., 1973. Matrix isolation infrared study of the reaction between tin vapor and molecular oxygen: the characterization of molecular tin dioxide. J. Phys. Chem. 77, 1513–1519.

Bosch, E., Crozet, P., Ross, A.J., Brown, J.M., 2000. Fourier transform spectra of the $E^2\Pi_u\text{-}X^2\Pi_{g(3/2)}$ system of $CuCl_2$. J. Mol. Spectrosc. 202, 253–261.

Botor, J., Milkowska, G., Konieczny, J., 1989. Vapor pressure and thermodynamics of PbS(s). Thermochim. Acta 137, 269–279.

Botschwina, P., Sebald, P., Figgen, D., Stoll, H., 2007. Calculated spectroscopic properties of HgH_2. Mol. Phys. 105 (9), 1193–1205.

Bouchoux, G., Rashad, A.M., Helal, A.I., 2012. Theoretical investigation of selenium interferences in inductively coupled plasma mass spectrometry. J. Phys. Chem. A 116, 9058–9070.

Boumans, P.W.J.M., Vrakking, J.J.A.M., 1987. Detection limits of about 350 prominent lines of 65 elements observed in 50 and 27 MHz inductively coupled plasmas (ICP): effects of source characteristics, noise and spectral bandwidth. "Standard" values for the 50 MHz ICP. Spectrochim. Acta B 42, 553–579.

Bourgel, C., Veron, E., Poirier, J., Defoort, F., Seiler, J.-M., Peregrina, C., 2011. Behavior of phosphorus and other inorganics during the gasification of sewage sludge. Energy Fuels 25, 5707–5717.

Boutron, C.F., 1995. Historical reconstruction of the earth's past atmospheric environment from Greenland and Antarctic snow and ice cores. Environ. Rev. 3, 1–28.

Boutron, C.F., Gorlach, U., Candelone, J.-P., Bolshov, M.A., Delmas, R.J., 1991. Decrease in anthropogenic lead, cadmium and zinc in Greenland snows since the late 1960s. Nature 353, 153–156.

Boyd, R., Barnes, S.-J., De Caritat, P., Chekushin, V.A., Melezhik, V.A., Reimann, C., et al., 2009. Emissions from the copper-nickel industry on the Kola Peninsula and at Noril'sk, Russia. Atmos. Environ. 43, 1474–1480.

Brabson, G.D., Andrews, L., Marsden, C.J., 1996. Reactions of selenium and oxygen: matrix infrared spectra and density functional calculations of novel Se_xO_y molecules. J. Phys. Chem. 100, 16487–16494.

Branca, J.J.V., Morucci, G., Maresca, M., Tenci, B., et al., 2018. Selenium and zinc: two key players against cadmium-induced neuronal toxicity. Toxicol. In Vitro 48, 159–169.

Bravo, R., Machado, F.B.C., 1999. Dissociation energy of the ground state of PCl. Chem. Phys. Lett. 307, 511–517.

Bravo, G., Flores-Riveros, A., Novaro, O., 1990. A CI pseudopotential-based description of the low-lying states of AgO_2. Int. J. Quantum Chem. 37, 785–796.

Brett, B., Schrader, D., Rauchle, K., et al., 2015. Recovery of valuable metals from power plant ashes. I. Characterization of lignite ashes for recovery of strategic metals. Chem. Ingenieur Tech. 87 (10), 1383–1391.

Brett, B., Schrader, D., Rauchle, K., et al., 2015a. Recovery of valuable metals from power plant ashes. II. Thermal and chemical treatment of lignite ashes for recovery of strategic metals. Chem. Ingenieur Tech. 87 (11), 1514–1526.

Brewer, G., 2018. 140 pp. Environmental Causes and Prevention Measures for Alzheimer's Disease. Academic Press, Elsevier, London.

Brewer, L., Ebbinghaus, B.B., 1988. The thermodynamics of the solid oxides of vanadium. Thermochim. Acta 129, 49–55.

Brewer, L., Mastick, D.F., 1951. The stability of gaseous diatomic oxides. J. Chem. Phys. 19, 834–843.

Brewer, L., Rosenblatt, G.M., 1969. Dissociation energies and free energy functions of gaseous monoxides. In: Eyring, L. (Ed.), Advances in High Temperature Chemistry, vol. 2. Academic Press, New York, pp. 1–83.

Brewster, J.H., Kodas, T.T., 1997. Generation of unagglomerated, dense $BaTiO_3$ particles by flame-spray pyrolysis. AIChE J. 43 (11A), 2665–2669.

Bridgeman, A.J., 1997. Bonding in the monochlorides and dichlorides of iron and cobalt. J. Chem. Soc. Dalton Trans. 24, 4765–4772.

Bridgeman, A.J., Cavigliasso, G., 2003. Computational analysis of Mo and W oxoanions through bond order and bonding energy approaches. J. Phys. Chem. A 107, 4568–4577.

Bridgeman, A.J., Rothery, J., 2000. Periodic trends in the diatomic monoxides and monosulfides of the 3d transition metals. Dalton 2, 211–218.

Brisdon, A.K., Gomme, R.A., Ogden, J.S., 1986. Matrix isolation studies on the vaporization of arsenic(V) oxide, As_2O_5: the characterization of molecular species As_4O_n (n = 7-10) by infrared spectroscopy. J. Chem. Soc. Dalton Trans. 12, 2725–2730.

Brisdon, A.K., Gomme, R.A., Ogden, J.S., 1991. Matrix isolation and mass spectrometric studies of the vaporization of alkali metal oxoselenium salts: characterization of molecular M_2SeO_4, M_2SeO_3 and $MSeO_2$. J. Phys. Chem. 95, 2927–2931.

Brisson, M.J., 2009. Overview of beryllium sampling and analysis: occupational hygiene and environmental applications. In: Brisson, M.J., Ekechukwu, A.A. (Eds.), Beryllium: Environmental Analysis and Monitoring. Royal Society of Chemistry Publishing, Cambridge, UK, pp. 1–16.

Brittain, R.D., Lau, K.H., Hildenbrand, D.L., 1982. Molecular species in the As-O system. J. Phys. Chem. 86, 5072–5075.

Brittain, R.D., Lau, K.H., Hildenbrand, D.L., 1983. Molecular species in the arsenic-oxygen system. Proc. Electrochem. Soc. (Proc. 2nd Symp. High Temp. Mater. Chem.) 83–7, 1–3.

Brittain, R.D., Lau, K.H., Hildenbrand, D.L., 1983a. Vaporization of Realgar, arsenic sulfide, As_4S_4. J Electrochem. Soc. 130 (5), 1206–1210.

Brittain, R.D., Lau, K.H., Hildenbrand, D.L., 1987. Mechanism and thermodynamics of the vaporization of K_2CrO_4. J. Electrochem. Soc. 134, 2900–2904.

Brittain, R.D., Lau, K.H., Hildenbrand, D.L., 1989. Effusion studies of the decomposition of $CuSO_4$ and CuO. $CuSO_4$. J. Phys. Chem. 93, 5316–5319.

Broadley, S.L., Plane, J.M.C., 2010. A kinetic study of reactions of calcium-containing molecules with O and H atoms: implications for calcium chemistry in the upper atmosphere. Phys. Chem. Chem. Phys. 12, 9094–9106.

Broadley, M.R., Chilimba, A.D.C., Joy, E.J.M., Young, S.D., Black, C.R., Ander, E.L., et al., 2012. Dietary requirements for magnesium, but not calcium, are likely to be met in Malawi based on National food supply data. Int. J. Vitamin Nutr. Res. 82 (3), 192–199. SI.

Brock, L.R., DeVore, T.C., 1991. Vacuum thermal decomposition of $AgNO_3$. High Temp. Sci. 32, 101–112.

Brockhaus, A., Dolgner, R., Ewers, U., Kramer, U., Soddemann, H., Wiegand, H., 1981. Intake and health effects of thallium among a population living in the vicinity of a cement plant emitting thallium containing dust. Int. Arch. Occup. Environ. Health 48, 375–389.

Broclawik, E., 1995. Density functional theory and transition metal oxides. In: Seminario, J.M., Politzer, P. (Eds.), Modern Density Functional Theory-a Tool for Chemistry. Theoretical and Computational Chemistry, vol. 2. Elsevier Science Publishers, Amsterdam, pp. 349–370.

Broclawik, E., Borowski, T., 2001. Time-dependent DFT study on electronic states of vanadium and molybdenum oxide molecules. Chem. Phys. Lett. 339, 433–437.

Broclawik, E., Salahub, D.R., 1994. On the electronic structure of MoO: spin-polarized density functional calculations of spectroscopic properties of low-lying quintet, triplet and septet states. Int. J. Quantum Chem. 52, 1017–1026.

Brom Jr, J.M., Weltner Jr, W., 1976. ESR spectrum of the BeOH molecule. J. Chem. Phys. 64, 3894–3895.

Brom Jr, J.M., Devore, T., Franzen, H.F., 1971. Infrared spectrum and structure of matrix isolated thallous oxide. J. Chem. Phys. 54 (6), 2742–2749.

Brooks, R.J., 1997. Removal of mercury from a combustion gas stream and apparatus. US Patent 5607496.

Bross, D.H., Hill, J.G., Werner, H.-J., Peterson, K.A., 2013. Explicitly correlated composite thermochemistry of transition metal species. J. Chem. Phys. 139, 094302. 12 pp.

Brostrom, M., Kassman, H., Helgesson, A., Berg, M., Andersson, C., Backman, R., et al., 2007. Sulfation of corrosive alkali chlorides by ammonium sulfate in a biomass fired CFB boiler. Fuel Process. Technol. 88, 1171–1177.

Brown, C.E., Mitchell, S.A., Hackett, P.A., 1991. Dioxygen complexes of 3d transition-metal atoms: formation reactions in the gas phase. J. Phys. Chem. 95, 1062–1066.

Brown, R.C., Kolb, C.E., Cho, S.Y., Yetter, R.A., Dryer, F.L., Rabitz, H., 1994. Kinetic model for hydrocarbon-assisted particulate boron combustion. Int. J. Chem. Kinet. 26, 319–332.

Brown, R.J.C., Goddard, S.L., Butterfield, D.M., et al., 2015. Ten years of mercury measurement at urban and industrial air quality monitoring stations in the UK. Atmos. Environ. 109, 1–8.

Bruce, R.M., Odin, M., 2001. 76 pp. Beryllium and Beryllium Compounds. Concise International Chemical Assessment Document 32. World Health Organization, Geneva.

Bruner, B.L., Corbett, J.D., 1964. A spectrophotometric study of the gaseous equilibrium between cadmium and cadmium dichloride. J. Phys. Chem. 68, 1115–1121.

Bryers, R.W., 1996. Fireside slagging, fouling, and high-temperature corrosion of heat-transfer surface due to impurities in steam-raising fuels. Prog. Energy Combust. Sci. 22, 29–120.

Bu, Y., Song, X., 2000. Structural analysis of the cyclic AlO_2 and AlS_2 systems in doublet and quartet states at density functional theory and the electron correlation levels. J. Chem. Phys. 113 (10), 4216–4229.

Buat-Menard, P., 1994. Global change in atmospheric metal cycles. In: Hewitt, C.N., Sturges, W.T. (Eds.), Global Atmospheric Chemical Change. Chapman & Hall, London, pp. 271–311. Chapter 7.

Buckley, S.G., Sawyer, R.F., Koshland, C.P., Lucas, D., 2002. Measurements of lead vapor and particulate in flames and post-flame gases. Combust. Flame 128, 435–446.

Budnik, L.T., Casteleyn, L., 2019. Mercury pollution in modern times and it socio-medical consequences. Sci. Total Environ. 654, 720–734.

Buerger, P., Akroyd, J., Kraft, M., 2019. Extended first-principles thermochemistry for the oxidation of titanium tetrachloride. Combust. Flame 199, 441–450.

Buettner, K.M., Valentine, A.M., 2012. Bioinorganic chemistry of titanium. Chem. Rev. 112, 1863–1881.

Buha, J., Mueller, N., Nowack, B., et al., 2014. Physical and chemical characterization of fly ashes from Swiss waste incineration plants and determination of the ash fraction in the nanometer range. Environ. Sci. Technol. 48, 4765–4773.

Buitrago, P.A., Van Otten, B., Senior, C.L., Silcox, G.D., 2013. Impinger based mercury speciation methods and gas phase mercury oxidation by bromine in combustion systems. Energy Fuels 27, 6255–6261.

Bulewicz, E.M., Padley, P.J., 1971. Catalytic effect of metal additives on free radical recombination rates in $H_2/O_2/N_2$ flames. Symp. (Int.) Combust. Proc. 13, 73–80.

Bulewicz, E.M., Padley, P.J., 1971a. Photometric investigations of the behavior of chromium additives in premixed $H_2/O_2/N_2$ flames. Proc. R. Soc. Lond. A 323, 377–400.

Bulewicz, E.M., Padley, P.J., 1971b. Photometric observations on the behavior of tin in premixed $H_2/O_2/N_2$ flames. Trans. Faraday Soc. 67, 2337–2347.

Bulewicz, E.M., Padley, P.J., 1973. Catalytic effects as the origin of some interferences in flame spectrometry. Spectrochim. Acta 28B, 125–137.

Bulewicz, E.M., Sugden, T.M., 1956. Determination of the dissociation constants and heats of formation of molecules by flame photometry. III. Stability of gaseous cuprous hydroxide. Trans. Faraday Soc. 52, 1481–1488.

Bulewicz, E.M., Sugden, T.M., 1956a. Determination of the dissociation constants and heats of formation of molecules by flame photometry. II. Heat of formation of gaseous cuprous hydride. Trans. Faraday Soc. 52, 1475–1481.

Bulewicz, E.M., Sugden, T.M., 1958. Determination of the dissociation constants and heats of formation of molecules by flame photometry. IV. The stability of GaOH, InOH and TlH. Trans. Faraday Soc. 54, 830–837.

Bulewicz, E.M., Phillips, L.F., Sugden, T.M., 1961. Determination of dissociation constants and heats of formation of simple molecules by flame photometry. VIII. Stabilities of the gaseous diatomic halides of certain metals. Trans. Faraday Soc. 57, 921–931.

Bulewicz, E.M., Evans, D.G., Padley, P.J., 1975. Effect of metallic additives on soot formation processes in flames. Symp. (Int.) Combust. Proc. 15, 1461–1470.

Bulova, M.N., Alikhanyan, A.S., Gas'kov, A.M., 2002. High temperature thermodynamic study of micro- and nanocrystalline SnO_2-WO_3 systems. Inorg. Mater. 38 (7), 688–693.

Bunt, J.R., Waanders, F.B., 2008. Trace element behavior in the Sasol-Lurgi Mark IV Fixed-Bed Dry-Bottom gasifier. I. The volatile elements Hg, As, Se, Cd and Pb. Fuel 87, 2374–2387.

Bunt, J.R., Waanders, F.B., 2009. Trace element behavior in the Sasol-Lurgi MK IV FBDB gasifier. II. The semi-volatile elements Cu, Mo, Ni and Zn. Fuel 88, 961–969.

Bunt, J.R., Waanders, F.B., 2010. Trace element behavior in the Sasol-Lurgi fixed-bed dry-bottom gasifier. III. The non-volatile elements: Ba Co, Cr, Mn and V. Fuel 89, 537–548.

Buonanno, G., Scungio, M., Stabile, L., Tirler, W., 2012. Ultrafine particle emission from incinerators: the role of the fabric filter. J. Air Waste Manage. Assoc. 62 (1), 103–111.

Burbacher, T.M., Shen, D.D., Liberato, N., et al., 2005. Comparison of blood and brain mercury levels in infant monkeys exposed to methylmercury or vaccines containing thimerosal. Environ. Health Perspect. 113 (8), 1015–1021.

Burkholder, T.R., Andrews, L., 1991. Reactions of boron atoms with molecular oxygen. Infrared spectra of the BO, B_2O_2, B_2O_3, BO_2^- in solid argon. J. Chem. Phys. 95 (12), 8697–8709.

Burmistrz, P., Kogut, K., Marczak, M., Dziok, T., Gorecki, J., 2018. Mercury in Polish coking bituminous coals. Energy Fuels 32, 5677–5683.

Burns, R.P., DeMaria, G., Drowart, J., Grimley, R.T., 1960. Mass spectrometric investigation of the sublimation of molybdenum dioxide. J. Chem. Phys. 32, 1363–1366.

Butcher, D.J., 2007. Environmental applications of arsenic speciation using atomic spectrometry detection. Appl. Spectrosc. Rev. 42, 1–22.

Byrne, A.R., Benedik, L., 1991. Uranium content of blood, urine and hair of exposed and non-exposed persons determined by radiochemical neutron activation analysis, with emphasis on quality control. Sci. Total Environ. 107, 143–157.

Cabus, N., Oguz, E.O., Tufan, A.C., Adiguzel, E., 2015. A histological study of toxic effects of aluminum sulfate on rat hippocampus. Biotechnol. Histochem. 90 (2), 132–139.

Caceres-Saez, I., Haro, D., Blank, O., Lobo, A.A., et al., 2018. High status of mercury and selenium in false killer whales (*Pseudorca crassidens*, Owen 1846) stranded on Southern South America: a possible toxicological concern? Chemosphere 199, 637–646.

Cachia, M., Bouyssiere, B., Carrier, H., Garraud, H., et al., 2018. Characterization and comparison of trace metal compositions in natural gas, biogas, and biomethane. Energy Fuels 32, 6397–6400.

Cadkova, E., Chrastny, V., 2015. Isotope evidence of hexavalent chromium stability in ground water samples. Chemosphere 138, 74–80.

Cai, M., Carter, C.C., Miller, T.A., Bondybey, V.E., 1991. Fluorescence excitation and resolved emission spectra of supersonically cooled aluminum oxide, Al_2O. J. Chem. Phys. 95 (1), 73–79.

Cai, L., Rao, P.M., Feng, Y., Zheng, X., 2013. Flame synthesis of 1-D complex metal oxide nanomaterials. Proc. Combust. Inst. 34, 2229–2236.

Caito, S.W., Jackson, B.P., Punshon, T., Scrimale, T., et al., 2018. Variation in methylmercury metabolism and elimination status in human following fish consumption. Toxicol. Sci. 161 (2), 443–453.

Camargo, J.A., 2002. Contribution of Spanish-American silver mines (1570–1820) to the present high mercury concentrations in the global environment: a review. Chemosphere 48, 51–57.

Camerani, P., Maria, C., Somogyi, A., et al., 2002. Direct determination of cadmium speciation in municipal solid waste fly ashes by synchroton radiation induced μ-X ray fluorescence and μ-X ray absorption spectroscopy. Environ. Sci. Technol. 36 (14), 3165–3169.

Camerani, M.C., Golosio, B., Somogyi, A., Simionovici, A.S., Steenari, B.-M., Panas, I., 2004. X-ray fluorescence tomography of individual municipal solid waste and biomass fly ash particles. Anal. Chem. 76, 1586–1595.

Cameron, J.H., Goerg-Wood, K., 1999. Role of thermophoresis in the deposition of fume particles resulting from the combustion of high inorganic containing fuels with reference to Kraft black liquor. Fuel Process. Technol. 60, 49–68.

Campbell, M.L., Plane, J.M.C., 2001. Kinetic study of the gas phase reaction of $Ca(^1S_0)$ with O_2 from 296–623 K. J. Phys. Chem. A 105, 3515–3520.

Candelone, J.-P., Bolshov, M.A., Rudniev, S.N., Hong, S., Boutron, C.F., 1995. Bismuth in recent snow from Central Greenland: preliminary results. Atmos. Environ. 29, 1843–1848.

Cannavo, D., Knopp, G., Radi, P., Beaud, P., Tulej, M., Bodek, P., et al., 2006. Neutral molecular ZnX (X = O, OH, N) compounds in a molecular beam. J. Mol. Struct. 782, 67–72.

Cao, X., 2005. *Ab initio* studies on the electronic structure of FeOH. Chem. Phys. 311, 203–208.

Cao, Y., Chen, B., Wu, J., Cui, H., Smith, J., Chen, C.-K., et al., 2007. Study of mercury oxidation by a Selective Catalytic Reduction catalyst in a pilot-scale slipstream reactor at a utility boiler burning bituminous coal. Energy Fuels 21, 145–156.

Cao, Y., Wang, Q., Chen, C.-w, Chen, B., Cohron, M., Tseng, Y.-c, et al., 2007a. Investigation of mercury transformations by HBr addition in a slipstream facility with real flue gas atmospheres of bituminous coal and Powder River Basin coal. Energy Fuels 21, 2719–2730.

Cao, X., Moritz, A., Dolg, M., 2008. All-electron Douglas-Kroll-Hess and pseudopotential study on the low-lying states of uranium hydride UH. Chem. Phys. 343, 250–257.

Cao, Y., Cheng, C.-M., Chen, C.-W., Liu, M., Wang, C., Pan, W.-P., 2008a. Abatement of mercury emissions in the coal combustion process equipped with a Fabric Filter Baghouse. Fuel 87, 3322–3330.

Cao, Y., Wang, Q.-H., Li, J., Cheng, J.-C., Chan, C.-C., Cohron, M., et al., 2009. Enhancement of mercury capture by simultaneous addition of hydrogen bromide and fly ashes in a slipstream facility. Environ. Sci. Technol. 43, 2812–2817.

Cao, Y., Zhou, H., Jiang, W., Chen, C.-W., Pan, W.-P., 2010. Studies of the fate of sulfur trioxide in coal-fired utility boilers bases on modified selected condensation methods. Environ. Sci. Technol. 44, 3429–3434.

Capablo, J., 2016. Formation of alkali salt deposits in biomass combustion. Fuel Process. Technol. 153, 58–73.

Capablo, J., Ballester, J., 2015. Experimental study of the kinetics of sulfation of alkali chloride deposits. Fuel Process. Technol. 140, 215–221.

Caplan, D., Cohen, M., 1961. The volatilization of chromium oxide. J. Electrochem. Soc. 108, 438–442.

Carbone, F., Beretta, F., D'Anna, A., 2010. Multimodal ultrafine particles from pulverized coal combustion in a laboratory scale reactor. Combust. Flame 157, 1290–1297.

Cardoso, B.R., Roberts, B.R., Bush, A.I., Hare, D.J., 2015. Selenium, selenoproteins and neurodegenerative diseases. Metallomics 7 (8), 1213–1228.

Cardoso, B.R., Hare, D.J., Bush, A.I., 2017. The role of selenium in neurodegenerative diseases, Chapter 3, pp. 35–49. In: White, A.R., Aschner, M., Costa, L.G., Bush, A.I. (Eds.), Biometals in Neurodegenerative Diseases. Academic Press, Elsevier, London, 415 pp.

Cardoso, B.R., Roberts, B.R., Malpas, C.B., Vivash, L., et al., 2019. Supra-nutritional sodium selenate supplementation delivers selenium to the central nervous system: results from a randomized controlled pilot trial in Alzheimer's disease. Neurotherapeutics 16 (1), 192–202.

Cariccio, V.L., Sama, A., Bramanti, P., Mazzon, E., 2019. Mercury involvement in neuronal damage and in neurodegenerative diseases. Biol. Trace Elem. Res. 187 (2), 341–356.

Carnes, R.A., Santoleri, J.J., McHale, S.H., 1992. Metals and incinerators: the latest regulatory phase. J. Hazard. Mater. 30, 343–353.

Carpi, A., Petraco, N.D.K., Cocris, D., et al., 2009. Mechanisms of the reduction and emission of $HgCl_2$ from surfaces. In: 238th American Chemical Society National Meeting, Abstracts of Papers. August 2009:Geoc-041.

Castiel, N., Goldstein, D., 1947. Transformation of alkaline earth sulfates into alkaline earth halides by ammonium halides. Dept. Nacl. Producao Mineral Brasil Ministerio Agric. 27, 125–129.

Castillo, J.L., Rosner, D.E., 1989. Theory of surface deposition from a unary dilute vapor-containing stream allowing for condensation within the laminar boundary layer. Chem. Eng. Sci. 44 (4), 925–937.

Castro, M.S., Sherwell, J., 2015. Effectiveness of emission controls to reduce the atmospheric concentrations of mercury. Environ. Sci. Technol. 49 (24), 14000–14007.

Catalano, J.G., Huhmann, B.L., Luo, Y., Mitnick, E.H., Slavney, A., Giammar, D.E., 2012. Metal release and speciation changes during wet aging of coal fly ashes. Environ. Sci. Technol. 46, 11804–11812.

Cater, E.D., Johnson, E.W., 1967. Sublimation of strontium sulfide and the dissociation energy of SrS. J. Chem. Phys. 47, 5353–5357.

Catoire, L., Legendre, J.-F., Giraud, M., 2003. Kinetic model for aluminum-sensitized ram accelerator combustion. J. Propuls. Power 19, 196–202.

Cauch, B., Silcox, G.D., Lighty, J.S., Wendt, J.O.L., Fry, A., Senior, C.L., 2008. Confounding effects of aqueous phase impinger chemistry on apparent oxidation of mercury in flue gases. Environ. Sci. Technol. 42, 2594–2599.

CDC, 2009. 4th National report on human exposure to environmental chemicals. Department of Health and Human Services: Centers for Disease Control and Prevention, 530 pp. Updated Tables 2018, vol. 1, 861 pp. vol. 2, 779 pp. <www.cdc.gov/exposurereport>.

Celik, M., Yucesu, H.S., Guru, M., 2016. Investigation of the effects of organic based manganese addition to biodiesel on combustion and exhaust emissions. Fuel Process. Technol. 152, 83–92.

Celmer, R., Yamamoto, M., Toguri, J.M., 1984. Vapor pressure of arsenic. Can. Metal. Q. 23 (2), 169–172.

Cernuschi, S., Giugliano, M., Ozgen, S., Consonni, S., 2012. Number concentration and chemical composition of ultrafine and nanoparticles from waste-to-energy plants. Sci. Total Environ. 420, 319–326.

Cesaro, S.N., Spoliti, M., Hinchcliffe, A.J., Ogden, J.S., 1971. Infrared spectrum and thermodynamic functions of matrix isolated SeO_2. J. Chem. Phys. 55 (12), 5834–5835.

Cesbron, A., Saussereau, E., Mahieu, L., Couland, I., et al., 2013. Metallic profile of whole blood and plasma in a series of 106 healthy volunteers. J. Anal. Toxicol. 37, 401–405.

Cesmebasi, E., Magee, R.S., Shafai, N.A., 1988. Metal emissions from municipal solid waste incinerators. Particulate Sci. Technol. 6, 365–380.

Chadeesingh, D.R., Hayhurst, A.N., 2014. The combustion of a fuel-rich mixture of methane and air in a bubbling fluidized bed of silica sand at 700°C and also with particles of Fe_2O_3 or Fe present. Fuel 127, 169–177.

Chadwick, B.L., Domazetis, G., Morrison, R.J.S., 1995. Multiwavelength monitoring of photofragment fluorescence after 193 nm photolysis of NaCl and NaOH: application to measuring the sodium species released from coal at high temperatures. Anal. Chem. 67, 710–716.

Chambaud, G., Guitou, M., Hayashi, S., 2008. Specific electronic properties of metallic molecules MX, correlated to piezoelectric properties of solids MX. Chem. Phys. 352, 147–156.

Chan, C.C.Y., Kirk, D.W., 1999. Behavior of metals under the conditions of roasting MSW incinerator fly ash with chlorinating agents. J. Hazard Mater. B 64, 75–89.

Chandrasekaran, S.R., Hopke, P.K., Rector, L., Allen, G., Lin, L., 2012. Chemical composition of wood chips and wood pellets. Energy Fuels 26, 4932–4937.

Chandravanshi, L.P., Gupta, R., Shukla, R.K., 2018. Developmental neurotoxicity of arsenic: involvement of oxidative stress and mitochondrial functions. Biol. Trace Elem. Res. 186 (1), 185–198.

Chang, L., Zhao, Y., Li, H., Tian, C., et al., 2017. Effect of sulfite on divalent mercury reduction and re-emission in a simulated desulfurization aqueous solution. Fuel Process. Technol. 165, 138–144.

Chang, L., Yang, J.P., Zhao, Y.C., Liu, H., et al., 2019. Behavior and fate of As, Se and Cd in an ultra-low emission coal fired power plant. J. Clean. Prod. 209, 722–730.

Charro, E., Pardo, R., Pena, V., 2013. Chemometric interpretation of vertical profiles of radionuclides in soils near a Spanish coal fired power plant. Chemosphere 90, 488–496.

Chase Jr, M.W., 1998. J. Phys. Chem. Ref. Data Mono No 9, Parts I and II. 1963 pp NIST-JANAF Thermochemical Tables, fourth ed. American Institute of Physics, New York.

Chattopadhyaya, S., 2016. Electronic states and spectroscopic properties of MgH in absence and presence of spin-orbit coupling: a configuration interaction study. Mol. Phys. 114 (20), 3026–3039.

Chattopadhyaya, S., Nath, A., Das, K.K., 2016. Effects of spin-orbit coupling on the electronic states and spectroscopic properties of diatomic SeS. J. Phys. B: At. Mol. Opt. Phys. 49, 065101. 16 pp.

Chen, Z., 2011. Global rare earth resources and scenarios of future rare earth industry. J. Rare Earths 29, 1–6.

Chen, H.-Y., Gilles, P.W., 1972. High molecular weight boron sulfides. VIII. Vapor pressures of B_2S_3(g) and B_4S_6(g) over stoichiometric B_2S_3. J. Phys. Chem. 76, 2035–2038.

Chen, Y.-M., Armentrout, P.B., 1995. Guided ion beam studies of the reactions of Ag^+ with C_2H_6, C_3H_8, HC $(CH_3)_3$ and c-C_3H_6. J. Phys. Chem. 99, 11424–11431.

Chen, Y.-M., Clemmer, D.E., Armentrout, P.B., 1991. The gas-phase thermochemistry of TiH. J. Chem. Phys. 95, 1228–1233.

Chen, Y.-M., Clemmer, D.E., Armentrout, P.B., 1993. Gas phase thermochemistry of VH and CrH. J. Chem. Phys. 98, 4929–4936.

Chen, J.-C., Wey, M.-Y., Chiang, B.-C., Hsieh, S.-M., 1998. The simulation of hexavalent chromium formation under various incineration conditions. Chemosphere 36 (7), 1553–1564.

Chen, J.-C., Wey, M.-Y., Su, J.-L., Hsieh, S.-M., 1998a. Two-stage simulation of the major heavy metal species under various incineration conditions. Environ. Int. 24, 451–466.

Chen, Y.-J., Glumac, N.G., Skandan, G., Kear, B.H., 1998b. High-rate production of high-purity, nonagglomerated oxide nanopowders in flames. ACS Symp Ser: Synth. Character. Adv. Mater. 681, 158–169.

Chen, Q.-F., Milburn, R.K., Hopkinson, A.C., Bohme, D.K., Goodings, J.M., 1999. Magnesium chemistry in the gas phase: calculated thermodynamic properties and experimental ion chemistry in $H_2/O_2/N_2$ flames. Int. J. Mass Spectrom. 184, 153–173.

Chen, Y., Shah, N., Huggins, F.E., Huffman, G.P., 2005. Transmission electron microscopy investigation of ultrafine coal fly ash particles. Environ. Sci. Technol. 39, 1144–1151.

Chen, Y., Liu, G., Wang, L., Kang, Y., Yang, J., 2008. Occurrence and fate of some trace elements during pyrolysis of Yima coal, China. Energy Fuels 22, 3877–3882.

Chen, W.-C., Lin, H.-Y., Yuan, C.-S., Hung, C.-H., 2009. Kinetic modeling on the adsorption of vapor phase mercury chloride on activated carbon by thermogravimetric analysis. J. Air Waste Manage. Assoc. 59, 227–235.

Chen, J., Jiao, F., Zhang, L., Yao, H., Ninomiya, 2012. Use of synchrotron XANES and Cr-doped coal to further confirm the vaporization of organically bound Cr and the formation of chromium (VI) during coal oxy-fuel combustion. Environ. Sci. Technol. 46, 3567–3573.

Chen, Y.-L., Chang, J.-E., Lai, Y.-C., Ko, M.-S., 2012a. Effects of sintering atmosphere on cement clinkers produced from chromium-bearing sludge. J. Air Waste Manage. Assoc. 62, 587–593.

Chen, J., Jiao, F., Zhang, L., Yao, H., Ninomiya, Y., 2013. Elucidating the mechanism of Cr(VI) formation upon the interaction with metal oxides during coal oxy-fuel combustion. J. Hazard. Mater. 261, 260–268.

Chen, B.-Y., Liu, G.-J., Sun, R.-Y., 2016. Distribution and fate of mercury in pulverized bituminous coal-fired power plants in coal energy-dominant Huainan City, China. Arch. Environ. Contamin. Toxicol. 70 (4), 724–733.

Chen, P., Bornhorst, J., Aschner, M., 2018. Manganese metabolism in humans. Front. Biosci. Landmark 23, 1655–1679.

Chen, C., Jia, W., Liu, S., Cao, T., 2018a. Simultaneous NO removal and Hg^o oxidation over CuO doped V_2O_5-WO_3/TiO_2 catalysts in simulated coal-fired flue gas. Energy Fuels 32, 7025–7034.

Chen, C.M., Cao, Y., Liu, S.T., Chen, J.M., Jia, W.B., 2018b. SCR catalyst doped with copper for synergistic removal of slip ammonia and elemental mercury. Fuel Process. Technol. 181, 268–278.

Chen, C.Y., Driscoll, C.T., Eagles-Smith, C.A., Eckley, C.S., et al., 2018c. A critical time for mercury science to inform Global policy. Environ. Sci. Technol. 52, 9556–9561.

Chen, X., Li, Y., Zhang, B., Zhou, A., Zheng, T., et al., 2018d. Maternal exposure to nickel in relation to preterm delivery. Chemosphere 193, 1157–1163.

Cheng, H., Hu, Y., 2012. Mercury in municipal solid waste in China and its control: a review. Environ. Sci. Technol. 46, 593–605.

Cheng, L., Wang, M.Y., Wu, Z.J., Su, Z.M., 2007. Electronic structures and chemical bonding in 4d transition metal monohalides. J. Comput. Chem. 28, 2190–2202.

Cheng, C.-M., Hack, P., Chu, P., Chang, Y.-N., Lin, T.-Y., Ko, C.-S., et al., 2009. Partitioning of mercury, arsenic, selenium, boron and chloride in a full-scale coal combustion process equipped with selective catalytic reduction, electrostatic precipitation and flue gas desulfurization systems. Energy Fuels 23, 4805–4816.

Cheng, K., Tian, H., Zhao, D., et al., 2014. Atmospheric emission inventory of cadmium from anthropogenic sources. Int. J. Environ. Sci. Technol. 11 (3), 605–616.

Cheng, J.-R., Zhao, J.-Y., Li, S.-J., 2015. An improved Alopex-based evolutionary algorithm (AEA) using local research and its application to parameter estimation for homogeneous mercury oxidation. Chemometrics Intell. Lab. Syst. 141, 68–78.

Cheng, K., Wang, Y., Tian, H.Z., et al., 2015a. Atmospheric emission characteristics and control policies of five precedent-controlled toxic heavy metals from anthropogenic sources in China. Environ. Sci. Technol. 49 (2), 1206–1214.

Chertihin, G.V., Andrews, L., 1993. Reactions of pulsed-laser ablated Al atoms with H_2. Infrared spectra of AlH, AlH_2, AlH_3 and Al_2H_2. J. Phys. Chem. 97, 10295–10300.

Chertihin, G.V., Andrews, L., 1994. Reactions of laser ablated Ti atoms with hydrogen during condensation in excess argon: infrared spectra of the TiH, TiH_2, TiH_3 and TiH_4 molecules. J. Am. Chem. Soc. 116, 8322–8327.

Chertihin, G.V., Andrews, L., 1995. Infrared spectra of FeH, FeH_2 and FeH_3 in solid argon. J. Phys. Chem. 99, 12131–12134.

Chertihin, G.V., Andrews, L., 1995a. Reactions of laser ablated titanium, zirconium and hafnium atoms with oxygen molecules in condensing argon. J. Phys. Chem. 99, 6356–6366.

Chertihin, G.V., Andrews, L., 1995b. Reactions of laser-ablated Zr and Hf atoms with hydrogen: matrix infrared spectra of the MH, MH_2, MH_3 and MH_4 molecules. J. Phys. Chem. 99, 15004–15010.

Chertihin, G.V., Andrews, L., 1996. Infrared spectra of the reaction products of laser ablated lead atoms and oxygen molecules in condensing argon and nitrogen. J. Chem. Phys. 105, 2561–2574.

Chertihin, G.V., Andrews, L., 1997. Reactions of laser ablated manganese atoms with dioxygen: infrared spectra of MnO, OMnO, $Mn(O_2)$, $(MnO)_2$ and higher oxide complexes in solid argon. J. Phys. Chem. A 101, 8547–8553.

Chertihin, G.V., Andrews, L., 1997a. Reactions of laser ablated Zn and Cd atoms with O_2: infrared spectra of ZnO, OZnO, CdO and OCdO in solid argon. J. Chem. Phys. 106, 3457–3465.

Chertihin, G.V., Saffel, W., Yustein, J.T., Andrews, L., Neurock, M., Ricca, A., et al., 1996. Reactions of laser-ablated iron atoms with oxygen molecules in condensing argon. Infrared spectra and density functional calculations of iron oxide product molecules. J. Phys. Chem. 100, 5261–5273.

Chertihin, G.V., Andrews, L., Bauschlicher Jr, C.W., 1997. Reactions of laser-ablated copper atoms with dioxygen: infrared spectra of the copper oxides CuO, OCuO, CuOCuO and OCuOCuO and superoxide CuOO in solid argon. J. Phys. Chem. A 101 (22), 4026–4034.

Chertihin, G.V., Andrews, L., Rosi, M., Bauschlicher Jr, C.W., 1997a. Reactions of laser-ablated scandium atoms with dioxygen: infrared spectra of ScO, OScO,(O_2)ScO, $(ScO)_2$ and $Sc(O_2)_2$ in solid argon. J. Phys. Chem. A 101, 9085–9091.

Chertihin, G.V., Bare, W.D., Andrews, L., 1997b. Reactions of laser-ablated chromium atoms with dioxygen: infrared spectra of CrO, OCrO, CrOO, CrO_3, $Cr(OO)_2$, Cr_2O_2, Cr_2O_3 and Cr_2O_4 in solid argon. J. Chem. Phys. 107, 2798–2806.

Chertihin, G.V., Bare, W.D., Andrews, L., 1997c. Reactions of laser-ablated vanadium atoms with dioxygen: infrared spectra of VO, VO_2, $OOVO_2$ and V_2O_2 in solid argon. J. Phys. Chem. A 101, 5090–5096.

Chertihin, G.V., Citra, A., Andrews, L., Bauschlicher Jr, C.W., 1997d. Reactions of laser-ablated cobalt atoms with O_2: infrared spectra of cobalt oxides in solid argon. J. Phys. Chem. A 101, 8793–8802.

Chervonnyi, A.D., 2010. Composition of the gas phase over Al_2O_3 and the enthalpies of atomization of AlO, Al_2O and Al_2O_2. Russ. J. Inorg. Chem. 55, 556–559.

Chervonnyi, A.D., Piven, V.A., Kashireninov, O.E., Manelis, G.B., 1977. Mass spectrometric investigations of gas phase equilibria over Al_2O_3 at high temperatures. High Temp. Sci. 9, 99–108.

Chestakov, D.A., Parker, D.H., Baklanov, A.V., 2005. Iron monoxide photodissociation. J. Chem. Phys. 122, 084302. 3 pp.

Cheung, A.S.-C., Zyrnicki, W., Merer, A.J., 1984. Fourier transform spectroscopy of CrO: rotational analysis of the $A^5\Sigma$-$X^5\Pi$ (0,0) band near 8000 cm^{-1}. J. Mol. Spectrosc. 104, 315–336.

Chi, Y., Yan, N., Qu, Z., Qiao, S., Jia, J., 2009. The performance of iodine on the removal of elemental mercury from the simulated coal-fired flue gas. J. Hazard. Mater. 166, 776–781.

Chi, Z.X., Zhu, L.L., Lu, X.M., Yu, H.T., Liu, B., 2012. Comparative study of three mononuclear vanadium aromatic 1,2-diol complexes: structure, characterization and anti-proliferating effects against cancer cells. Zeit. Anorg. Allgem. Chem. 638, 1523–1530.

Chibisov, A.N., 2015. Effect of stoichiometry on the atomic and electronic structure of $BaTiO_3$ nanoparticles: a first-principles study. Mol. Phys. 113 (21), 3291–3295.

Chin, C.-H., Mebel, A.M., Hwang, D.-Y., 2004. Theoretical study of the reaction mechanism of BO, B_2O_2 and BS with H_2. J. Phys. Chem. A 108, 473–483.

Cho, Y.-N., DeSisto, W.J., 2003. Phase selective CVD of chromium oxides from chromyl chloride, CrO_2Cl_2. Chem. Vapor Deposition 9, 121–124.

Choi, H.K., Lee, S.H., Kim, S.S., 2009. The effect of activated carbon injection rate on the removal of elemental mercury in a particulate collector with fabric filters. Fuel Process. Technol. 90, 107–112.

Chou, Y.-C., 2015. The low-lying singlet electronic excited states of TiO_2: a symmetry adapted cluster-configuration interaction (SAC-CI) study. Comput. Theor. Chem. 1059, 51–59.

Chou, C.P., Chang, T.C., Chiu, C.H., Hsu, H.C., 2018. Mercury speciation and mass distribution of cement production process in Taiwan. Aerosol Air Qual. Res. 18 (11), 2801–2812.

Choudary, U.V., Gingerich, K.A., Kingcade, J.E., 1975. Mass spectrometric determination of the thermodynamic stabilities of gaseous molybdenum oxides and sodium molybdates and predicted stabilities of gaseous group 1A tungstates, molybdates, chromates and sulfates. J. Less-Common Metals 42, 111–126.

Choy, K.-L., Vyas, J.D., 1999. Processing and structural characterization of Y_2O_3-ZrO_2 films deposited using flame assisted vapor deposition technique. Key Eng. Mater. (Ceram. Soc. Japan) 161–163, 653–656.

Choy, K.-L., Wiedmann, I., Derby, B., 1995. Flame assisted deposition of functional ceramic thin films. In: Vincenzini, P. (Ed.), Advances in Inorganic Films and Coatings. Techna Srl, Faenza, Italy, pp. 299–305.

Chrissanthopoulos, A., Pouchan, C., 2008. A density functional investigation of the structural and vibrational properties of the highly symmetric molecules M_4O_6, M_4O_{10} (M = P, As, Sb, Bi). Vib. Spectrosc. 48, 135–141.

Christensen, K.A., Livbjerg, H., 1996. A field study of submicron particles from the combustion of straw. Aerosol Sci. Technol. 25, 185–199.

Christensen, K.A., Livbjerg, H., 2000. A plug flow model for chemical reactions and aerosol nucleation and growth in an alkali containing flue gas. Aerosol Sci. Technol. 33, 470–489.

Chu, C.C., Huang, H.Y., Whang, T.J., Tsai, C.C., 2018. Observation of double-well potential of NaH ($C^1\Sigma$) state: deriving the dissociation energy of its ground state. J. Chem. Phys. 148, 114301. 7 pp.

Chuang, S.Y., Dennis, J.S., Hayhurst, A.N., Scott, S.A., 2009. Kinetics of the chemical looping oxidation of CO by a co-precipitated mixture of CuO and Al_2O_3. Proc. Combust. Inst. 32, 2633–2640.

Chuang, S.Y., Dennis, J.S., Hayhurst, A.N., Scott, S.A., 2011. Kinetics of the chemical looping oxidation of H_2 by a co-precipitated mixture of CuO and Al_2O_3. Chem. Eng. Res. Design 89 (9), 1511–1523.

Chupka, W.A., Berkowitz, J., Inghram, M.G., 1957. Thermodynamics of the Zr-ZrO_2 system: the dissociation energies of ZrO and ZrO_2. J. Chem. Phys. 26, 1207–1210.

Chupka, W.A., Berkowitz, J., Giese, C.F., 1959. Vaporization of beryllium oxide and its reaction with tungsten. J. Chem. Phys. 30, 827–834.

Chyang, C.-S., Han, Y.-L., Zhong, Z.-C., 2009. Study of HCl absorption by CaO at high temperature. Energy Fuels 23, 3948–3953.

Ciach, S., Power, G., Thistlethwaite, P., 1971. The electronic absorption spectrum of the SeCl radical. Chem. Phys. Lett. 9, 349–350.

CICADs, 2002. 42 pp. Report 44: Silver and Silver Compounds: Environmental Aspects. World Health Organization, Geneva, Switzerland.

Cimino, S., Mangone, C., Scala, F., 2018. Combined mercury removal and low temperature NH_3-SCR of NO with MnO_x/TiO_2 sorbents/catalysts. Combust. Sci. Technol. 190 (8), 1488–1499.

Cimova, N., Novak, M., Chrastny, V., et al., 2016. Lead fluxes and $^{206}Pb/^{207}Pb$ isotope ratios in rime and snow collected at remote mountain-top locations (Czech Republic, Central Europe): patterns and sources. Atmos. Environ. 143, 51–59.

Citra, A., Chertihin, G.V., Andrews, L., Neurock, M., 1997. Reactions of laser-ablated nickel atoms with dioxygen. Infrared spectra and density functional calculations of nickel oxides NiO, ONiO, Ni_2O_2 and Ni_2O_3, superoxide NiOO, peroxide $Ni(O_2)$ and higher complexes in solid argon. J. Phys. Chem. A 101, 3109–3118.

Clackett, S.P., Porter, T.J., Lehnherr, I., 2018. 400-year record of atmospheric mercury from tree-rings in Northwestern Canada. Environ. Sci. Technol. 52, 9625–9633.

Clements, A.L., Buzcu-Guven, B., Fraser, M.P., et al., 2013. Role of particulate metals in heterogeneous secondary sulfate formation. Atmos. Environ. 75, 233–240.

Clemmer, D.E., Dalleska, N.F., Armentrout, P.B., 1991. Reaction of Zn^+ with NO_2. The gas phase thermochemistry of ZnO. J. Chem. Phys. 95 (10), 7263–7268.

Clemmer, D.E., Dalleska, N.F., Armentrout, P.B., 1992. Gas phase thermochemistry of the Group 3 dioxides: ScO_2, YO_2 and LaO_2. Chem. Phys. Lett. 190, 259–265.

Clery, D.S., Mason, P.E., Rayner, C.M., Jones, J.M., 2018. The effects of an additive on the release of potassium in biomass combustion. Fuel 214, 647–655.

Clever, H.L., Johnston, F.J., 1980. The solubility of some sparingly soluble lead salts: an assessment of the solubility in water and aqueous electrolyte solution. J. Phys. Chem. Ref. Data 9 (3), 751–784.

Clever, H.L., Johnson, S.A., Derrick, M.E., 1985. The solubility of mercury and some sparingly soluble mercury salts in water and aqueous electrolyte solutions. J. Phys. Chem. Ref. Data 14 (3), 631–680.

Clever, H.L., Derrick, M.E., Johnson, S.A., 1992. The solubility of some sparingly soluble salts of zinc and cadmium in water and in aqueous electrolyte solutions. J. Phys. Chem. Ref. Data 21 (5), 941–1004.

Cobalt Development Institute, 2015. Guildford, Surrey, UK. <www.cobaltinstitute.org/statistics>.

Cobos, C.J., 2002. Heats of formation for AlH, AlOH, OAlH and OAlOH and their monocations. J. Mol. Struct. (Theochem.) 581, 17–29.

Cohen, R.B., Majeres, P., Raloff, J.K., 1975. Chemi-ionization reactions of Ba, Sr, Ca with OH. Chem. Phys. Lett. 31 (1), 176–180.

Coles, D.G., Ragaini, R.C., Ondov, J.M., Fisher, G.L., Silberman, D., Prentice, B.A., 1979. Chemical studies of stack fly ash from a coal-fired power plant. Environ. Sci. Technol. 13, 455–459.

Colin, R., Goldfinger, P., Jeunehomme, M., 1962. Existence of gaseous sulfides of the transition elements: dissociation energy of gaseous MnS. Nature 194, 282–283.

Colin, R., Goldfinger, P., Jeunehomme, M., 1964. Mass spectrometric studies of the vaporization of the sulfides of calcium, strontium and barium: the dissociation energy of S_2 and SO. Trans. Faraday Soc. 60, 306–316.

Collins, R.N., Kinsela, A.S., 2010. The aqueous phase speciation and chemistry of cobalt in terrestrial environments. Chemosphere 79, 763–771.

Comet, M., Schnell, F., Pichot, V., et al., 2014. Boron as fuel for ceramic thermites. Energy Fuels 28, 4139–4148.

Conaway, C.H., Mason, R.P., Steding, D.J., Flegal, A.R., 2005. Estimate of mercury emission from gasoline and diesel fuel consumption, San Francisco Bay area, California. Atmos. Environ 39, 101–105.

Conerly, O., Rieth, S., 2009. Toxicological review of thallium and compounds. U.S. Environmental Protection Agency EPA/635/R-08/001F. <www.epa.gov/iris>.

Contreras, M.L., Arostegui, J.M., Armesto, L., 2009. Arsenic interactions during co-combustion processes based on thermodynamic equilibrium calculations. Fuel 88, 539–546.

Cooper, G.A., Kartouzian, A., Gentleman, A.S., et al., 2015. Dissociation energies of Ag-Rg (Rg = Ar, Kr, Xe) and AgO molecules from velocity map imaging studies. J. Chem. Phys. 143, 124302. 6 pp.

Cooper, G.A., Gentleman, A.S., Iskra, A., Mackenzie, S.R., 2017. Photo-fragmentation dynamics and dissociation energies of MoO and CrO. J. Chem. Phys. 147, 013921. 8 pp.

Coppens, P., Smoes, S., Drowart, J., 1967. Mass spectrometric determination of the dissociation energies of the molecules GeS, ScS, YS, LaS and CeS. Trans. Faraday Soc. 63, 2140–2148.

Coppola, A., Solimene, R., Bareschino, P., Salatino, P., 2015. Mathematical modeling of a two-stage fuel reactor for chemical looping combustion with oxygen uncoupling of solid fuels. Appl. Energy 157, 449–461.

Cordell, D., Rosemarin, A., Schroder, J.J., Smit, A.L., 2011. Towards global phosphorus security: a systems framework for phosphorus recovery and reuse options. Chemosphere 84, 747–758.

Cordfunke, E.H.P., Konings, R.J.M. (Eds.), 1990. Thermochemical Data for Reactor Materials and Fission Products. Elsevier Science Publishers, Amsterdam.

Cordfunke, E.H.P., Konings, R.J.M., 1991. The vapor pressure of UCl_4. J. Chem. Thermodyn. 23, 1121–1124.

Cordfunke, E.H.P., Konings, R.J.M., 1992. The vapor pressure of tungsten hexachloride. J. Chem. Thermodyn. 24 (3), 329–331.

Cordfunke, E.H.P., Konings, R.J.M., Westrum Jr, E.F., 1989. Recent thermochemical research on reactor materials and fission products. J. Nucl. Mater. 167, 205–212.

Cordoba, P., 2015. Status of flue gas desulfurization systems from coal-fired power plants: overview of the physic-chemical control processes of wet limestone FGDs. Fuel 144, 274–286.

Cordoba, P., 2017. Partitioning and speciation of selenium in wet limestone flue gas desulfurization system: a review. Fuel 202, 184–195.

Cordoba, P., Diego, R., 2018. Mass balance and partitioning of trace elements under oxy-coal combustion: first experiences. Fuel 215, 892–903.

Cordoba, P., Staicu, L.C., 2018. Flue gas desulfurization effluents: an unexploited selenium source. Fuel 223, 268–276.

Cordoba, P., Font, O., Izquierdo, M., et al., 2012. The retention capacity for trace elements by the flue gas desulfurization system under operational conditions of a co-combustion power plant. Fuel 102, 773–788.

Cordoba, P., Maroto-Valer, M., Delgado, M.A., et al., 2016. Speciation, behavior and fate of mercury under oxy-fuel combustion conditions. Environ. Res. 145, 154–161. Corrigendum 2016;150:665.

Cosic, B., Fontijn, A., 2000. Kinetics of Pb reactions with N_2O, Cl_2, HCl, and O_2 at high temperatures. J. Phys. Chem. A 104, 5517–5524.

Cosic, B., Fontijn, A., 2000a. Transitions in order and molecularity with temperature in gaseous metal oxidation reactions. The Sb-O_2 system. Zeit. Phys. Chem. 214 (11), 1521–1531.

Costa, L.G., 2017. Developmental exposure to metals and its contribution to age-related neuro degeneration, Chapter 12, 217–229. In: White, A.R., Aschner, M., Costa, L.G., Bush, A.I. (Eds.), Biometals in Neurodegenerative Diseases. Academic Press, Elsevier, London, 415 pp.

Costa, L., Paganetto, G., Bertelli, G., Camino, G., 1990. Thermal decomposition of antimony oxyhalides. I. Oxychlorides. J. Thermal Anal. 36 (3), 1141–1153.

Cotton, D.H., Jenkins, D.R., 1968. Dissociation energies of gaseous alkaline earth hydroxides. Trans. Faraday Soc. 64, 2988–2997.

Cotton, D.H., Jenkins, D.R., 1969. Bond dissociation energy of gaseous magnesium oxide. Trans. Faraday Soc. 65, 376–379.

Cotton, D.H., Jenkins, D.R., 1971. Catalysis of radical-recombination reactions in flames by alkaline earth metals. Trans. Faraday Soc. 67, 730–739.

Cotton, D.H., Friswell, N.J., Jenkins, D.R., 1971. The suppression of soot emission from flames by metal additives. Combust. Flame 17, 87–98.

Couling, D.J., Nguyen, H.V., Green, W.H., 2012. Screening of metal oxides and metal sulfides as sorbents for elemental mercury at elevated temperatures. Fuel 97, 783–795.

Courcot, E., Rebillat, F., Teyssandier, F., 2010. Thermochemical stability of rare earth sesquioxides under a moist environment at high temperature, Ceram. Trans. (Design Dev. Appl. Eng. Ceram. Compos.), 215. pp. 257–264.

Cowan, D.M., Zheng, W., Zou, Y., Shi, X., Chen, J., Rosenthal, F.S., et al., 2009. Manganese exposure among smelting workers: relationship between blood manganese-iron ratio and early onset neurobehavioral alterations. Neurotoxicology 30, 1214–1222.

Coyle Jr, R.T., Lewis, G., 1974. Knudsen effusion study of the vaporization of CdO(s). J. Am. Ceram. Soc. 57, 398–400.

Coyte, R.M., Jain, R.C., Srivastava, S.K., Sharma, K.C., et al., 2018. Large-scale uranium contamination of groundwater resources in India. Environ. Sci. Technol. Lett. 5, 341–347. Rathore, D.P.S., Comments 591–592, Response 593–594.

Cremer, D., Kraka, E., Filatov, M., 2008. Bonding in mercury molecules described by the normalized elimination of the small component and coupled cluster theory. ChemPhysChem 9, 2510–2521.

Crepeaux, G., Eidi, H., David, M.-O., Tzavara, E., et al., 2015. Highly delayed systemic translocation of aluminum-based adjuvant in CD-1 mice following intramuscular injections. J. Inorg. Biochem. 152, 199–205.

Cressiot, C., Guitou, M., Mitrushchenkov, A., Chambaud, G., 2007. Stability of the HgS molecule and spectroscopy of its low lying electronic states. Mol. Phys. 105 (9), 1207–1216.

Crichton, R., 2001. Inorganic Biochemistry of Iron Metabolism, third ed. Wiley & Sons, Ltd, New York.

Crichton, R.R., 2017. Metal toxicity: an introduction.RSC Metallobiology Series No. 8. Chapter 1 In: Crichton, R., Ward, R.J., Hider, R.C. (Eds.), Metal Chelation in Medicine. Royal Society of Chemistry, Cambridge England, pp. 1–23.

Crovisier, P.N., Horton, J.H., Hassanali, C.S., Goodings, J.M., 1992. Ion chemistry of aluminum and boron in methane-oxygen flames. Can. J. Chem. 70, 839–848.

Csontos, J., Rolik, Z., Das, S., Kallay, M., 2010. High accuracy thermochemistry of atmospherically important fluorinated and chlorinated methane derivatives. J. Phys. Chem. A 114, 13093–13103.

Cubicciotti, D., 1963. Thermodynamics of liquid solutions of bismuth and sulfur. J. Phys. Chem. 67 (1), 118–123.

Cubicciotti, D., 1967. The dissociation energy of bismuth chloride. J. Phys. Chem. 71 (9), 3066–3068.

Cubicciotti, D., 1970. The thermodynamics of vaporization of thallous oxide. High Temp. Sci. 2, 213–220.

Cubicciotti, D., 1970a. The vaporization of thallous nitrate. High Temp. Sci. 2, 131–139.

Cubicciotti, D., 1970b. Vaporization of thallous sulfate. High Temp. Sci. 2, 389–397.

Cubicciotti, D., Sanjurjo, A., Hildenbrand, D.L., 1977. The thermal dissociation of $CaSO_3$ and its enthalpy of formation. J. Electrochem. Soc. 124, 933–936.

Cui, H., Turn, S.Q., Keffer, V., Evans, D., Tran, T., Foley, M., 2013. Study on the fate of metal elements from biomass in a bench-scale fluidized bed gasifier. Fuel 108, 1–12.

Cumba, L.R., de Oliveira Bicalho, U., do Carmo, D.R., 2012. Preparation and voltammetric studies of titanium(IV) phosphate modified with silver hexacyanoferrate to a voltammetric determination of l-cysteine. Int. J. Electrochem. Sci. 7, 4465–4478.

Cummins, P.G., Field, R.W., Renhorn, I., 1981. Argon ion laser-induced BaS ($B^1\Sigma^+$ - $A^1\Sigma^+$, $A'\,^1\Pi$, $a^3\Pi$, $X^1\Sigma^+$) fluorescence spectra: analysis of $A \sim A'$, $A \sim a$, and $A' \sim a$ perturbations. J. Mol. Spectrosc. 90, 327–352.

Cusick, S.E., Jaramillo, E.G., Moody, E.C., Ssemata, A.S., et al., 2018. Assessment of blood levels of heavy metals including lead and manganese in healthy children living in the Katanga settlement of Kampala, Uganda. BMC Public Health 18, 717.

Cypers, F., De Dobbelaere, C., Hardy, A., Van Bael, M.K., Helsen, L., 2009. Thermal behavior of arsenic trioxide adsorbed on activated carbon. J. Hazard. Mater. 166 (2–3), 1238–1243.

Dagdigian, P.J., Cruse, H.W., Schultz, A., Zare, R.N., 1974. Product state analysis of BaO from the reactions $Ba + CO_2$ and $Ba + O_2$. J. Chem. Phys. 61, 4450–4465.

Dai, D., Balasubramanian, K., 1990. Geometries and energies of electronic states of AsH_3, SbH_3 and BiH_3 and their positive ions. J. Chem. Phys. 93, 1837–1846.

Dai, B.-Q., Low, F., De Girolamo, A., et al., 2013. Characteristics of ash deposits in a pulverized lignite coal-fired boiler and the mass flow of major ash-forming inorganic elements. Energy Fuels 27, 6198–6211.

Dan, Y., Zimmerman, C., Liu, K., Shi, H., Wang, J., 2013. Increased leaching of As, Se, Mo and V from high calcium coal ash containing trona reaction products. Energy Fuels 27, 1531–1537.

Danihelka, P., Volna, Z., Jones, J.M., Williams, A., 2003. Emission of trace toxic metals during pulverized fuel combustion of Czech coals. Int. J. Energy Res. 27, 1181–1203.

Danset, D., Manceron, L., 2003. Mid- and near-infrared electronic absorption spectrum of CoO isolated in solid neon: vibronic data for low-lying electronic states. J. Phys. Chem. A 107, 11324–11330.

Danset, D., Manceron, L., Andrews, L., 2001. Vibrational spectra of nickel and platinum dioxide molecules isolated in solid argon. J. Phys. Chem. A 105, 7205–7210.

Danset, D., Alikhani, M.E., Manceron, L., 2005. Reactivity of atomic cobalt with molecular oxygen: a combined infrared matrix isolation and theoretical study of the formation and structure of CoO_2. J. Phys. Chem. A 109, 97–104.

Daoudi, A., Benjelloun, A.T., Flament, J.P., Berthier, G., 1999. Potential energy curves and electronic structure of copper nitrides CuN and CuN^+ versus CuO and CuO^+. J. Mol. Spectrosc. 194, 8–16.

Daoudi, A., Baba, M.F., Elkhattabi, S., et al., 2003. Electronic structure and molecular spectroscopic constants of ScN and ScP investigated by several quantum chemistry methods. Mol. Phys. 101 (19), 2929–2951.

Daous, M.A., 2005. Recovery of vanadium and nickel from fly ash produced in power generation and desalination plants in Saudi Arabia. Proc. Int. Conf. Solid Waste Technol. Manage. 20, 971–980.

Das, K.K., Liebermann, H.-P., Buenker, R.J., Hirsch, G., 1995. Relativistic configuration interaction study of the low-lying electronic states of Bi_2. J. Chem. Phys. 102 (11), 4518–4530.

da Silva-Neto, A.G., Roberto-Neto, O., Ornellas, F.R., Machado, F.B.C., 2004. The low-lying electronic states of PCl. Chem. Phys. Lett. 395, 239–245.

Dattani, N.S., 2015. Beryllium monohydride, BeH: where we are now, after 86 years of spectroscopy. J. Mol. Spectrosc. 311, 76–83.

Davico, G.E., Schwartz, R.L., Ramond, T.M., Lineberger, W.C., 1999. An experimental study of the low-lying electronic states of WO_2. J. Phys. Chem. A 103, 6167–6172.

Davidsson, K.O., Amand, L.-E., Steenari, B.-M., Elled, A.-L., Eskilsson, D., Leckner, B., 2008. Countermeasures against alkali-related problems during combustion of biomass in a circulating fluidized bed boiler. Chem. Eng. Sci. 63, 5314–5329.

Davis, S.P., Hammer, P.D., 1988. Energy levels of zirconium oxide. Astrophys. J. 332, 1090–1091.

Davis, H.F., Suits, A.G., Hou, H., Lee, Y.T., 1990. Reactions of Ba atoms with NO_2, O_3 and Cl_2: dynamic consequences of the divalent nature of barium. Ber. Bunsenges. Phys. Chem. 94, 1193–1201.

Davison, R.L., Natusch, D.F.S., Wallace, J.R., Evans Jr, C.A., 1974. Trace elements in fly ash: dependence of concentration on particle size. Environ. Sci. Technol. 8, 1107–1113.

Dayton, D.C., Frederick Jr, W.J., 1996. Direct observation of alkali vapor release during biomass combustion and gasification. II. Black liquor combustion at 1100°C. Energy Fuels 10, 284–292.

Dayton, D.C., French, R.J., Milne, T.A., 1995. Direct observation of alkali vapor release during biomass combustion and gasification. I. Application of molecular beam-mass spectrometry to switchgrass combustion. Energy Fuels 9, 855–865.

Daza, M.C., Restrepo, G., Uribe, E.A., Villaveces, J.L., 2006. Quantum chemical and chemotopological study of fourth row monohydrides. Chem. Phys. Lett. 428, 55–61.

de Almeida, C.M.S., Ribeiro, A.S., Saint'Pierre, T.D., Miekeley, N., 2009. Studies on the origin and transformation of selenium and its chemical species along the process of petroleum refining. Spectrochim. Acta B 64, 491–499.

De Galan, L., Samaey, G.F., 1970. Measurement of degrees of atomization in premixed laminar flames. Spectrochim. Acta B 25, 245–259.

De Galan, L., Winefordner, J.D., 1967. Measurement of the free-atom fraction of 22 elements in an acetylene-air flame. J. Quant. Spectrosc. Radiat. Transfer 7, 251–260.

Deierlein, A.L., Teitelbaum, S.L., Windham, G.C., Pinney, S.M., et al., 2019. Lead exposure during childhood and subsequent anthropometry through adolescence in girls. Environ. Int. 122, 310–315.

De Jesus, L.D.F., Moreira, M.D.R., de Azevedo, S.V., Borges, R.M., et al., 2018. Lead and mercury levels in an environmentally exposed population in the central Brazil. Cadernos de Saude Publica 34 (2), e00034417.

Delay, I., Swithenbank, J., Argent, B.B., 2001. Prediction of the distribution of alkali and trace elements between the condensed and gaseous phases generated during clinical waste incineration. J. Alloys Compds. 320, 282–295.

del Barrio, M., Borghesani, V., Hureau, C., Faller, P., 2017. Metal binding to Amyloid-β peptide: coordination, aggregation, and reactive oxygen species production, Chapter 14, pp. 265–281. In: White, A.R., Aschner, M., Costa, L.G., Bush, A.I. (Eds.), Biometals in Neurodegenerative Diseases. Academic Press, Elsevier, London, 415 pp.

De Maria, G., 1984. Vaporization studies by Knudsen cell mass spectrometry and thermodynamic stability of high temperature gaseous species. In: Ribeiro da Silva, M.A.V. (Ed.), Thermochemistry and its Applications to Chemical and Biochemical Systems, C119. NATO Advanced Study Institute Series, pp. 157–181.

De Maria, G., Drowart, J., Inghram, M.G., 1959. Thermodynamic study of InSb with a mass spectrometer. J. Chem. Phys. 31 (4), 1076–1081.

De Maria, G., Burns, R.P., Drowart, J., Inghram, M.G., 1960. Mass spectrometric study of gaseous molybdenum, tungsten and uranium oxides. J. Chem. Phys. 32, 1373–1377.

Demirbas, A., 2005. Heavy metal content of fly ash from selected biomass samples. Energy Sources 27, 1269–1276.

Demuynck, J., Schaefer III, H.F., 1980. The gas phase structure of transition metal dihydrides. J. Chem. Phys. 72, 311–315.

Deng, K., Yang, J., Yuan, L., Zhu, Q., 1999. A theoretical study of the linear OCuO species. J. Chem. Phys. 111 (4), 1477–1482.

Deng, K., Yang, J., Zhu, Q., 2003. A theoretical study of the NiO_2 species. J. Chem. Phys. 118 (15), 6868–6873.

Deng, J., Kang, L., Bai, G., Li, Y., Li, P., Liu, X., et al., 2014. Solution combustion synthesis of cobalt oxides (Co_3O_4 and Co_3O_4/CoO) nanoparticles as supercapacitor electrode materials. Electrochim. Acta 132, 127–135.

Deng, S., Shi, Y., Liu, Y., et al., 2014a. Emission characteristics of Cd, Pb and Mn from coal combustion: field study at coal-fired power plant in China. Fuel Process. Technol. 126, 469–475.

Denis, P.A., 2010. Theoretical investigation of ionized HSO and SOH. Mol. Phys. 108 (13), 1739–1747.

de Oliveira, L.C.C., de Oliveira, J.F.G., de Melo, F.L.G., et al., 2015. Mineral sorbents for downstream sodium capture in biomass gasifiers. Fuel Process. Technol. 138, 629–636.

De Oliveira-Filho, A.G.S., Alves, T.V., Ribas, V.W., Ferrao, L.F.A., Roberto-Neto, O., Machado, F.B.C., et al., 2011. A CASSCF/MRCI study of the low-lying electronic states of the BeS molecule. Int. J. Quantum Chem. 111, 1694–1700.

Deonarine, A., Kolker, A., Foster, A.L., et al., 2016. Arsenic speciation in bituminous coal fly ash and transformations in response to redox conditions. Environ. Sci. Technol. 50, 6099–6106.

De Simone, F., Gencarelli, C.N., Hedgecock, I.M., Pirrone, N., 2016. A modeling comparison of mercury deposition from current anthropogenic mercury emission inventories. Environ. Sci. Technol. 50, 5154–5162.

De Simone, F., Hedgecock, I.M., Carbone, F., Cinnirella, S., et al., 2017. Estimating uncertainty in global mercury emission source and deposition receptor relationships. Atmosphere 8 (12), 236.

Deutsch, M., Horvath, F., Knoll, C., Lager, D., et al., 2017. High temperature energy storage: kinetic investigations of the CuO/Cu_2O reaction cycle. Energy Fuels 31 (3), 2324–2334.

Devalette, M., Hagenmuller, P., 1967. Potassium orthosalts of the Group IVA and IVB elements. Bull. Soc. Chim. Fr. 265, 3457–3458.

De Vries, A.B., Jellinek, F., 1974. Structural study of vanadium sulfides. Rev. Chim. Miner. 11 (5), 624–636.

DeVore, T.C., Gallaher, T.N., 1986. The vibrational infrared spectra of the tungsten oxide vapor molecule. High Temp. Sci. 21, 85–95.

De Yonker, N.J., Allen, W.D., 2012. Taming the low lying electronic states of FeH. J. Chem. Phys. 137, 234303. 16 pp.

De Yonker, N.J., Li, S., Yamaguchi, Y., Schaefer III, H.F., Crawford, T.D., King, R.A., et al., 2005. Application of equation-of-motion coupled-cluster methods to low-lying singlet and triplet electronic states of HBO and BOH. J. Chem. Phys. 122, 234316. 13 pp..

De Yonker, N.J., Peterson, K.A., Steyl, G., Wilson, A.K., Cundari, T.R., 2007. Quantitative computational thermochemistry of transition metal species. J. Phys. Chem. A 111, 11269–11277.

Diacomanolis, V., Noller, B.N., Ng, J.C., 2014. Bioavailability and pharmacokinetics of arsenic are influenced by the presence of cadmium. Chemosphere 112, 203–209.

Diaz-Somoano, M., Lopez-Anton, M.A., Martinez-Tarazona, M.R., 2004. Retention of arsenic and selenium during hot gas desulfurization using metal oxide sorbents. Energy Fuels 18, 1238–1242.

Diaz-Somoano, M., Lopez-Anton, M.A., Huggins, F.E., Martinez-Tarazona, M.R., 2010. The stability of arsenic and selenium compounds that were retained in limestone in a coal gasification atmosphere. J. Hazard. Mater. 173 (1–3), 450–454.

Diffenderfer, R.N., Yarkony, D.R., Dagdigian, P.J., 1983. Radiative lifetimes and transition moments in MgO. J. Quant. Spectrosc. Radiat. Transfer 29, 329–331.

Di Salvia, N., Malavasi, M.P., Di Salvia, G., Vaccari, A., 2015. Formation of Ni and Mg vanadates during the flameless oxy-combustion of heavy fuels. Fuel Process. Technol. 138, 534–539.

Dismukes, E.B., 1994. Trace element control in electrostatic precipitators and fabric filters. Fuel Process. Technol. 39, 403–416.

Dix-Cooper, L., Kosatsky, T., 2018. Blood mercury, lead and cadmium levels and determinants of exposure among newcomer South and East Asian women of reproductive age living in Vancouver, Canada. Sci. Total Environ. 619–620, 1409–1419.

Dixon, D.A., Peterson, K.A., 2001. Heat of formation of CCl and CCl_2 from *ab initio* quantum chemistry. J. Chem. Phys. 115, 6327–6329.

Dogan, F., Ozacar, M., Akbulut, H., Guler, M.O., 2011. Effect of temperature on SnO_2 thin films deposited by atmospheric pressure chemical vapor deposition. Adv. Sci. Eng. Med. 3, 52–57.

Dolg, M., Cao, X., 2009. Accurate relativistic small-core pseudopotentials for actinides: energy adjustment for uranium and first application to uranium hydride. J. Phys. Chem. A 113, 12573–12581.

Dolg, M., Wedig, U., Stoll, H., Preuss, H., 1987. *Ab initio* pseudopotential study of the first row transition metal monoxides and iron monohydride. J. Chem. Phys. 86, 2123–2131.

Donnelly, J.R., 1993. Metal emissions control technologies for waste incineration. Chapter 14, pp. 174–188. In: Khan, M.R. (Ed.), Clean Energy from Waste and Coal, American Chemical Society Symposium Series, 515 pp.

Donohue, D.L., Bauer, D., Hynes, A.J., 2005. Temperature and pressure dependent rate coefficients for the reaction of Hg and Cl and the reaction of Cl with Cl: a pulsed laser photolysis-pulsed laser induced fluorescence study. J. Phys. Chem. A 109, 7732–7741.

Dopp, E., Kligerman, A.D., Diaz-Bone, R.A., 2010. Organoarsenicals: uptake, metabolism and toxicity. Chapter 7 In: Sigel, A., Sigel, H., Sigel, R.K.O. (Eds.), Metal Ions in Life Sciences, vol. 7. Organometallics in Environment and Toxicology. Royal Society of Chemistry, London, UK, pp. 231–265.

Dorea, J.G., 2018. Low-dose Thimerosal (ethyl-mercury) is still used in infants' vaccines: should we be concerned with this form of exposure? J. Trace Elem. Med. Biol. 49, 134–139.

Dorea, J.G., Farina, M., Rocha, J.B.T., 2013. Toxicity of ethylmercury (and thimerosal): a comparison with methylmercury. J. Appl. Toxicol. 33, 700–711.

Doshi, V., Vuthaluru, H.B., Korbee, R., Kiel, J.H.A., 2009. Development of a modeling approach to predict ash formation during co-firing of coal and biomass. Fuel Process. Technol. 90, 1148–1156.

dos Santos, D.M., Aschner, M., dos Santos, A.P.M., 2017. Manganese and neuronal degeneration. Chapter 8, pp. 117–151. In: White, A.R., Aschner, M., Costa, L.G., Bush, A.I. (Eds.), Biometals in Neurodegenerative Diseases. Academic Press, Elsevier, London, 415 pp.

Dou, B., Pan, W., Ren, J., et al., 2007. Single and combined removal of HCl and alkali metal vapor from high temperature gas by solid sorbents. Energy Fuels 21, 1019–1023.

Douglas, T.B., 1972. An alternative to accurate pyrometry in distinguishing concurrent vaporization reactions. Postulated results using the vaporization of gold in analyzing the vaporization of beryllium oxide in water vapor. J. Res. Natl. Bur. Std. 76A, 511–515.

Douglas, M.A., Hauge, R.H., Margrave, J.L., 1982. Electronic absorption spectroscopy of aluminum metal atom hydration products isolated in rare gas matrices at 15 K. ACS Symp. Ser. 179, 347–354.

Douglas, M.A., Hauge, R.H., Margrave, J.L., 1983. Electronic absorption and emission studies of the Group IIIA metal suboxides isolated in cryogenic rare gas matrices. High Temp. Sci. 16 (1), 35–54.

Drake, P.L., Hazelwood, K.J., 2005. Exposure-related health effects of silver and silver compounds: a review. Ann. Occup. Hyg. 49 (7), 575–585.

Drake, M.C., Rosenblatt, G.M., 1979. Trends in structure and vibrational frequencies of MX_2 and MX_3. J. Electrochem. Soc. 126 (8), 1387–1396.

Drechsler, G., Boesl, U., Bassmann, C., Schlag, E.W., 1997. Mass selected anion-zero kinetic energy photoelectron spectroscopy (anion-ZEKE): ground and low excited states of FeO. J. Chem. Phys. 107, 2284–2291.

Drobyshev, E.J., Solovyev, N.D., Ivanenko, N.B., Kombarova, M.Y., Ganeev, A.A., 2017. Trace element biomonitoring in hair of school children from a polluted area by sector field inductively coupled plasma mass spectrometry. J. Trace Elem. Med. Biol. 39, 14–20.

Drowart, J., 1985. Mass spectrometric studies at high temperature. Adv. Mass Spectrom. 10A, 185–214.

Drowart, J., Smoes, S., 1977. Determination by the mass spectrometric Knudsen cell method and discussion of the dissociation energies of the molecules Se_2(g), SSe(g) and SeTe(g). J. Chem. Soc. Faraday Trans. II 73, 1755–1767.

Drowart, J., Colin, R., Exsteen, G., 1965. Mass spectrometric study of the vaporization of lead monoxide. Trans. Faraday Soc. 61, 1376–1383.

Drowart, J., Pattoret, A., Smoes, S., 1967. Mass spectrometric studies of the vaporization of refractory compounds. Proc. Br. Ceram. Soc. 8, 67–89.

Drowart, J., Myers, C.E., Szwarc, R., et al., 1973. The dissociation energies of the molecules PS, PSe and PTe. High Temp. Sci. 5, 482–488.

Drowart, J., Smoes, S., Vanderauwera-Mahieu, A., 1978. The atomization enthalpies of AsSb(g), As_4(g), Sb_4(g), As_3Sb(g), As_2Sb_2(g), $AsSb_3$(g), As_3SbO_6(g), $As_2Sb_2O_6$(g) and $AsSb_3O_6$(g): a thermodynamic study by the mass spectrometric Knudsen cell method. J. Chem. Thermodyn. 10, 453–464.

Drozdowski, R., Ignaciuk, M., Kwela, J., Heldt, J., 1997. Radiative lifetimes of the lowest $^3P_1$ metastable states of Ca and Sr. Z. Phys. D 41, 125–131.

Drummond, G., Barrow, F.R., 1951. Thermochemical dissociation energies of gaseous calcium, strontium and barium oxides. Trans. Faraday Soc. 47, 1275–1277.

Du, H., Yu, F., 2009. Kinetic modeling of nucleation experiments involving SO_2 and OH: new insights into the underlying nucleation mechanisms. Atmos. Chem. Phys. 9, 7913–7922.

Du, X., Graedel, T.E., 2011. Global in-use stocks of the rare earth elements: a first estimate. Environ. Sci. Technol. 45, 4096–4101.

Du, X., Tang, J., Gao, X., et al., 2016. Molecular transformations of arsenic species in the flue gas of typical power plants: a density functional theory study. Energy Fuels 30, 4209–4214.

Du, X.B., Wang, C., Liu, Q., 2016a. Potential roles of selenium and selenoproteins in the prevention of Alzheimer's disease. Curr. Topics Med. Chem. 16 (8), 835–848.

Duan, J., Tan, J., 2013. Atmospheric heavy metals and Arsenic in China: situation, sources and control policies. Atmos. Environ. 74, 93–101.

Duan, L., Sun, H., Jiang, Y., et al., 2016. Partitioning of trace elements, As, Ba, Cd, Cr, Cu, Mn, and Pb in a 2.5 MW_{th} pilot-plant circulating fluidized bed combustor burning an anthracite and a bituminous coal. Fuel Process. Technol. 146, 1–8.

Duan, L., Cui, J., Jiang, Y., Zhao, C., Anthony, E.J., 2017. Partitioning the behavior of arsenic in circulating fluidized bed boilers co-firing petroleum coke and coal. Fuel Process. Technol. 166, 107–114.

Duan, P.P., Wang, W.F., Sang, S.X., Tang, Y.G., et al., 2018. Geochemistry of toxic elements and their removal via the preparation of high-uranium coal in Southwestern China. Minerals 8 (3), 83.

Dudek-Adamska, D., Lech, T., Konopka, T., Koscieiniak, P., 2018. Chromium in postmortem material. Biol. Trace Elem. Res. 186 (2), 370–378.

Duke, B.J., Radom, L., 1998. Gaussian-2 (G2) theory for third-row elements: a systematic study of the effect of the *3d* orbitals. J. Chem. Phys. 109 (9), 3352–3359.

Dumas, M., Frossard, E., Scholz, R.W., 2011. Modeling biogeochemical processes of phosphorus for global food supply. Chemosphere 84, 798–805.

Dunham, G.E., DeWall, R.A., Senior, C.L., 2003. Fixed-bed studies of the interactions between mercury and coal combustion fly ash. Fuel Process. Technol. 82, 197–213.

Dutta, B.K., Khanra, S., Mallick, D., 2009. Leaching of elements from coal fly ash: assessment of its potential for use in filling abandoned coal mines. Fuel 88, 1314–1323.

Dyachenko, S.V., Martinson, K.D., Cherepkova, I.A., Zhernovoi, A.I., 2016. Particle size, morphology and properties of transition metal ferrospinels of the MFe_2O_4 (M = Co, Ni, Zn) type, produced by glycine-nitrate combustion. Russ. J. Appl. Chem. 89 (4), 535–539.

Eastman, R.R., Jursa, T.P., Benedetti, C., Lucchini, R.G., Smith, D.R., 2013. Hair as a biomarker of environmental manganese exposure. Environ. Sci. Technol. 47, 1629–1637.

Ebbinghaus, B.B., 1993. Thermodynamics of gas phase chromium species: the chromium oxides, the chromium oxyhydroxides and volatility calculations in waste incineration processes. Combust. Flame 93, 119–137.

Ebbinghaus, B.B., 1995. Thermodynamics of gas phase chromium species: the chromium chlorides, oxychlorides, fluorides, oxyfluorides, hydroxides, oxyhydroxides, mixed oxyfluorochlorohydroxides and volatility calculations in waste incineration processes. Combust. Flame 101, 311–338.

Eckelman, M.J., Graedel, T.E., 2007. Silver emissions and their environmental impacts: a multilevel assessment. Environ. Sci. Technol. 41, 6283–6289.

Eddings, E.G., Lighty, J.S., 1992. Fundamental studies of metal behavior during solids incineration. Combust. Sci. Technol. 85, 375–390.

Edinger, P., Tarik, M., Hess, A., et al., 2016. Online detection of selenium and its retention in reducing gasification atmosphere. Energy Fuels 30, 1237–1247.

Editorial, 2012. Lead exposure limits are inadequate. Chem. Eng. News 90 (50), 36.

Edwards, J.G., Franzen, H.F., Gilles, P.W., 1971. High temperature mass spectrometry, vaporization and thermodynamics of titanium monosulfide. J. Chem. Phys. 54, 545–554.

EHC-Copper: Environmental Health Criteria 200, 1998. Report. World Health Organization, Geneva, Switzerland, 182 pp. <www.who.int/ipcs/publications/ehc>.

Ehlert, T., 1977. Mass spectrometric investigations of the oxides of potassium. High Temp. Sci. 9, 237–241.

Ehlert, T.C., Hilmer, R.M., Beauchamp, E.A., 1968. Mass spectrometric study of the vaporization of magnesium, calcium, magnesium hydride and calcium hydride. J. Inorg. Nucl. Chem. 30, 3112–3115.

Ehrman, S.H., Friedlander, S.K., Zachariah, M.R., 1999. Phase segregation in binary SiO_2/TiO_2 and SiO_2/Fe_2O_3 nanoparticle aerosols formed in a premixed flame. J. Mater. Res. 14, 4551–4561.

Eichler, A., Tobler, L., Eyrikh, S., et al., 2014. Ice core based assessment of historical anthropogenic heavy metal (Cd, Cu, Sb, Zn) emissions in the Soviet Union. Environ. Sci. Technol. 48, 2635–2642.

Ekvall, T., Andersson, K., Leffler, T., Berg, M., 2017. K-Cl-S chemistry in air and oxy-combustion atmospheres. Proc. Combust. Inst. 36 (3), 4011–4018.

El-Hag, I.H., Jano, I., 1989. Study of the excited states of HPO and its emission spectrum in hydrogen based flames. Anal. Chim. Acta 226, 305–313.

Elled, A.-L., Amand, L.-E., Leckner, B., Andersson, B.-A., 2007. The fate of trace elements in fluidized bed combustion of sewage sludge and wood. Fuel 86, 843–852.

Elled, A.-L., Amand, L.-E., Eskilsson, D., 2008. Fate of zinc during combustion of demolition wood in a fluidized bed boiler. Energy Fuels 22, 1519–1526.

Ellison, W., Weiler, H., Sanyal, A., McIlvaine, R.W., 2002. Experience in high sulfur service to date in management of air preheater fouling and fly ash by-product contamination due to SCR operation. Annu. Int. Pittsburgh Coal Conf. Proc. 19, 308–318.

Elmoussaoui, S., El-Kork, N., Korek, M., 2016. Electronic structure of the ZnCl molecule with rovibrational and iconicity studies of the ZnX (X = F, Cl, Br, I) compounds. Comput. Theor. Chem. 1090, 94–104.

Emelianov, A., Eremin, A., Jander, H., Wagner, H.Gg, 2010. Formation of condensed particles in premixed flames catalyzed by metal carbonyls. Z. Phys. Chem. 224, 715–727.

Emelyanova, K.A., Shugurov, S.M., Panin, A.I., Lopatin, S.I., 2016. Thermodynamic properties of the gaseous lead phosphates. J. Chem. Thermodyn. 101, 337–342.

Enestam, S., Backman, R., Makela, K., Hupa, M., 2013. Evaluation of the condensation behavior of lead and zinc in bubbling fluidized bed combustion of recovered waste wood. Fuel Process. Technol. 105, 161–169.

Enestam, S., Boman, C., Niemi, J., Bostrom, D., Backman, R., Makela, K., et al., 2011. Occurrence of zinc and lead in aerosols and deposits in the fluidized bed combustion of recovered waste wood. II. Thermodynamic considerations. Energy Fuels 25, 1970–1977.

Enestam, S., Makela, K., Backman, R., Hupa, M., 2011a. Occurrence of zinc and lead in aerosols and deposits in the fluidized bed combustion of recovered waste wood. I. Samples from boilers. Energy Fuels 25, 1396–1404.

Engelke, F., Sander, R.K., Zare, R.N., 1976. Crossed-beam chemiluminescent studies of alkaline earth atoms with ClO_2. J. Chem. Phys. 65 (3), 1146–1155.

Eom, S.-Y., Lee, Y.-S., Lee, S.-G., Seo, M.-N., Choi, B.-S., et al., 2018. Lead, mercury and cadmium exposure in the Korean general population. J. Korean Med. Sci. 33 (2), e9. 13 pp.

EPA Report, 1992. Integrated Risk Information System: Molybdenum. CASRN 7439-98-7, Environmental Protection Agency.

Erickson, B., 2016. FDA targets arsenic in baby cereal. Chem. Eng. News 94 (15), 15.

Ericzon, C., Pettersson, J., Andersson, M., Olin, A., 1989. Determination and speciation of selenium in end products from a garbage incinerator. Environ. Sci. Technol. 23, 1524–1528.

Eriksson, J., 2001. Concentrations of 61 trace elements in sewage sludge, farmyard manure, mineral fertilizer, precipitation and in oil and crops. Swedish Environmental Protection Agency Report 5159, 69 pp.

Ermilova, I.O., Kazenas, E.K., Zviadadze, G.N., 1976. Study of the thermodynamics of sodium molybdate vaporization. Zhur Fiz Khim 50 (8), 2172–2173.

Ernest, C.T., Donohoue, D., Bauer, D., Ter Schure, A., Hynes, A.J., 2014. Progammable thermal dissociation of reactive gaseous mercury, a potential approach to chemical speciation: results from a field study. Atmosphere 5 (3), 575–596.

Erway, N.D., Seifert, R.L., 1951. Vapor pressure of beryllium oxide. J. Electrochem. Soc. 98, 83–88.

Escobar, I., Muller, M., 2007. Alkali removal at about 1400°C for the pressurized pulverized coal combustion combined cycle. II. Sorbents and sorption mechanisms. Energy Fuels 21, 735–743.

Espelid, O., Borve, K.J., 1997. Accurate enthalpies of formation for CrX(g), X = O, OH and F: a computational study. J. Phys. Chem. A 101, 9449–9456.

Espelid, O., Borve, K.J., Jensen, V.R., 1998. Structure and thermodynamics of gaseous oxides, hydroxides and mixed oxohydroxides of chromium: $CrO_m(OH)_n$ (m,n = 0-2) and CrO_3: a computational study. J. Phys. Chem. A 102, 10414–10423.

Eswaran, S., Stenger, H.G., 2005. Understanding mercury conversion in selective catalytic reduction catalysts. Energy Fuels 19, 2328–2334.

Ezhov, Yu.S., Komarov, S.A., 1978. Molecular structures of inorganic compounds of low volatility. II. Metaborates of alkali metals. J. Mol. Struct. 50, 305–312.

Ezzeldin, M.F., Gajdosechova, Z., Masod, M.B., et al., 2016. Mercury speciation and distribution in an Egyptian natural gas processing plant. Energy Fuels 30 (12), 10236–10243.

Fan, Q., Sun, W., Feng, H., 2009. Studies on the full vibrational energies and dissociation energies of some heteronuclear diatomic molecules. Spectrochim. Acta A 74, 911–916.

Fancher, C.A., de Clercq, H.L., Thomas, O.C., et al., 1998. Zinc oxide and its anion: a negative ion photoelectron spectroscopic study. J. Chem. Phys. 109 (19), 8426–8429.

Fang, L., Li, J.-S., Guo, M.Z., Cheeseman, C.R., Tsang, D.C.W., et al., 2018. Phosphorus recovery and leaching of trace elements from incinerated sewage sludge ash. Chemosphere 193, 278–287.

Farber, M., Srivastava, R.D., 1973. A mass spectrometric investigation of reactions involving vanadium and chromium with potassium seeded H_2/O_2 flames. Combust. Flame 20, 43–49.

Farber, M., Srivastava, R.D., 1973a. A mass spectrometric investigation of reactions involving tungsten and molybdenum with potassium seeded H_2/O_2 flames. Combust. Flame 20, 33–42.

Farber, M., Srivastava, R.D., 1974. The dissociation energy of NiO and vaporization and sublimation enthalpies of Ni. In: Porter, R.S., Johnson, J.F. (Eds.), Analytical Calorimetry, vol. 3. Plenum Press, New York, pp. 731–741.

Farber, M., Srivastava, R.D., 1975. An investigation of arsenic and antimony additives in unseeded and potassium-seeded H_2/O_2 flames. Combust. Flame 25, 101–106.

Farber, M., Srivastava, R.D., 1975a. The dissociation energy of barium oxide. High Temp. Sci. 7, 74–80.

Farber, M., Srivastava, R.D., 1976. The dissociation energy of magnesium oxide. High Temp. Sci. 8, 195–201.

Farber, M., Srivastava, R.D., 1976a. The dissociation energies of calcium oxide and strontium oxide. High Temp. Sci. 8 (2), 73–80.

Farber, M., Srivastava, R.D., 1978. A mass spectrometric investigation of the chemistry of plateau burning propellants. Combust. Flame 31, 309–323.

Farber, M., Srivastava, R.D., 1985. Electron impact and thermodynamic studies of potassium metaborate. J. Chem. Soc. Faraday Trans. I 81, 913–918.

Farber, M., Uy, O.M., Srivastava, R.D., 1972. Effusion mass spectrometric determination of the heats of formation of the gaseous molecules V_4O_{10}, V_4O_8, VO_2 and VO. J. Chem. Phys. 56 (11), 5312–5315.

Farber, M., Harris, S.P., Srivastava, R.D., 1974. An investigation of iron and rhenium additives in unseeded and potassium-seeded H_2/O_2 flames. Combust. Flame 22, 191–196.

Farber, M., Srivastava, R.D., Frisch, M.A., Harris, S.P., 1974a. Mass spectrometer studies of Al + H_2O reactions in effusion cells and in atmospheric H_2/O_2 flames. Faraday Symp. Chem. Soc. 8, 121–130.

Farber, M., Srivastava, R.D., Moyer, J.W., 1983. Effusion-mass spectrometric determination of thermodynamic properties of gaseous potassium sulfide species. High Temp. Sci. 16, 153–160.

Farber, M., Srivastava, R.D., Moyer, J.W., Leeper, J.D., 1987. Effusion mass spectrometric determination of thermodynamic properties of the gaseous mono- and di-hydroxides of calcium and KCaO(g). J. Chem. Soc. Faraday Trans. I 83, 3229–3236.

Farhat, A., Korek, M., Abdul-Al, S.N., Marques, M.A.L., 2013. Electronic structure with spin-orbit calculations of the low-lying electronic states of the molecule YS. Chem. Phys. 412, 109–116.

Fatchi, H., He, Y., Wang, Z., et al., 2015. Laser-induced breakdown spectroscopy (LIBS) measurements and numerical studies of potassium release during biomass gasification. Proc. Combust. Inst. 35, 2389–2396.

Fatehi, H., Li, Z.S., Bai, X.S., Alden, M., 2017. Modeling of alkali metal release during biomass pyrolysis. Proc. Combust. Inst. 36 (2), 2243–2251.

Fatola, O.I., Olaolorun, F.A., Olopade, F.E., Olopade, J.O., 2019. Trends in vanadium neurotoxicity. Brain Res. Bull. 145, 75–80.

Faure, F.M., Mitchell, M.J., Bartlett, R.W., 1972. Vapor pressure study of stibnite, Sb_2S_3. High Temp. Sci. 4, 181–191.

Faure, F.M., Mitchell, M.J., Bartlett, R.W., 1973. Vapor pressure study of arsenic trisulfide. High Temp. Sci. 5, 128–137.

Feller, D., Dixon, D.A., Peterson, K.A., 1998. Heats of formation of simple boron compounds. J. Phys. Chem. A 102 (35), 7053–7059.

Feller, D., Vasiliu, M., Grant, D.J., Dixon, D.A., 2011. Thermodynamic properties of arsenic compounds and the heat of formation of the As atom from high level electronic structure calculations. J. Phys. Chem. A 115, 14667–14676.

Fenga, C., Liu, S.S., Zhou, F.K., Gao, Y.Y., Li, Y.S., et al., 2019. Oxidative stress in the neuro-degenerative brain following lifetime exposure to lead in rats: changes in lifespan profiles. Toxicology 411, 101–109.

Fenimore, C.P., 1972. Oxidation of sulfur dioxide in stainless steel or Inconel sampling probes. Combust. Flame 18, 469–470.

Fennell, P.S., Dennis, J.S., Hayhurst, A.N., 2006. Production of nanoparticles of MgO, BaO and Al_2O_3 in a premixed flame and its relation to the flame structure. Combust. Explos. Shock Waves 42 (6), 642–648.

Fennell, P.S., Dennis, J.S., Hayhurst, A.N., 2007. The sampling of nanoparticles of MgO formed when doping an oxygen-rich flame with magnesium: the measurement of the concentrations and size-distributions of these nanoparticles. Combust. Flame 151, 560–572.

Fennell, P.S., Dennis, J.S., Hayhurst, A.N., 2007a. The size distributions of nanoparticles of the oxides of Mg, Ba and Al in flames: their measurement and dependence on the concentrations of free radicals in the flame. Proc. Combust. Inst. 31, 1939–1945.

Fernandez-Miranda, N., Rumayor, M., Lopez-Anton, M.A., et al., 2015. Mercury retention by fly ashes from oxy-fuel processes. Energy Fuels 29, 2227–2233.

Fernandez-Miranda, N., Lopez-Anton, M.A., Diaz-Somoano, M., Martinez-Tarazona, M.R., 2016. Mercury oxidation in catalysts used for selective reduction of NO_x (SCR) in oxy-fuel combustion. Chem. Eng. J. 285, 77–82.

Fernandez-Miranda, N., Lopez-Anton, M.A., Torre-Santos, T., et al., 2016a. Impact of oxy-fuel conditions on elemental mercury re-emission in wet flue gas desulfurization systems. Environ. Sci. Technol. 50 (13), 7247–7253.

Feroughi, O.M., Hardt, S., Wlokas, I., et al., 2015. Laser based in situ measurements and simulation of gas phase temperature and iron atom concentration in a pilot-plant nanoparticle synthesis reactor. Proc. Combust. Inst. 35, 2299–2306.

Ferrante, M.J., Stuve, J.M., Ko, H.C., Brown, R.R., 1981. Thermodynamic properties of aluminum sulfide, Al_2S_3. High Temp. Sci. 14, 91–101.

Ferrante, M.J., Stuve, J.M., Pankratz, L.B., 1981a. Thermodynamic properties of cuprous and cupric sulfides. High Temp. Sci. 14, 77–90.

Ferrari, C.P., Hong, S., Van de Velde, K., Boutron, C.F., Rudniev, S.N., Bolshov, M., et al., 2000. Natural and anthropogenic bismuth in Central Greenland. Atmos. Environ. 34, 941–948.

Ferre-Huguet, N., Nadal, M., Schuhmacher, M., Domingo, J.L., 2009. Monitoring metals in blood and hair of the population living near a hazardous waste incinerator: temporal trend. Biol. Trace Elem. Res. 128, 191–199.

Ferro, D., Piacente, V., Scardala, P., 1989. Decomposition enthalpies of iron sulfides. J. Chem. Thermodyn. 21, 483–494.

Ficalora, P.J., Hastie, J.W., Margrave, J.L., 1968. Mass spectrometric studies at high temperatures. XXVII. The reactions of aluminum vapor with $S_2(g)$, $Se_2(g)$, and $Te_2(g)$. J. Phys. Chem. 72 (5), 1660–1663.

Ficalora, P.J., Thompson, J.C., Margrave, J.L., 1969. Mass spectrometric studies at high temperatures. XXVI. The sublimation of SeO_2 and SeO_3. J. Inorg. Nucl. Chem. 31, 3771–3774.

Ficia, D., Oprea, O., Ficia, A., Holban, A.M., 2014. Metal oxide nanoparticles: potential uses in biomedical applications. Curr. Proteomics 11 (2), 139–149.

Field, R.W., 1974. Assignment of the lowest $^3\Pi$ and $^1\Pi$ states of CaO, SrO and BaO. J. Chem. Phys. 60, 2400–2413.

Fielder, W.L., Stearns, C.A., Kohl, F.J., 1984. Reactions of NaCl with gaseous SO_3, SO_2 and O_2. J. Electrochem. Soc. 131, 2414–2417.

Filatov, M., Cremer, D., 2004. Revision of the dissociation energies of mercury chalcogenides: unusual types of mercury bonding. ChemPhysChem 5, 1547–1557.

Filella, M., 2010. Alkyl derivatives of antimony in the environment. In: Sigel, A., Sigel, H., Sigel, R.K.O. (Eds.), Metal Ions in Life Sciences, vol. 7. Organometallics in Environment and Toxicology. Royal Society of Chemistry, London, UK, pp. 267–301. Chapter 8.

Filella, M., 2010a. Alkyl derivatives of bismuth in environmental and biological media. In: Sigel, A., Sigel, H., Sigel, R.K.O. (Eds.), Metal Ions in Life Sciences, vol. 7. Organometallics in Environment and Toxicology. Royal Society of Chemistry, London, UK, pp. 303–317. Chapter 9.

Filella, M., Belzile, N., Chen, Y.-W., 2013. Human exposure to antimony. III. Contents in some human excreted biofluids (Urine, milk, saliva). Crit. Rev. Environ. Sci. Technol. 43 (2), 162–214.

Filippelli, G.M., 2008. The global phosphorus cycle: past, present and future. Elements 4 (2), 89–95.

Fink, E.H., Setzer, K.D., Ramsay, D.A., Vervloet, M., 1991. The $X_21 \rightarrow X_10^+$ electronic band systems of bismuth monohalides in the near infrared. Chem. Phys. Lett. 179, 95–102.

Fink, E.H., Shestakov, O., Setzer, K.D., 1997. Fine structure transitions of BiOH in the near-infrared region. J. Mol. Spectrosc. 183, 163–167.

Fischer, E.A., 1989. A new evaluation of the urania equation of state based on recent vapor pressure measurements. Nucl. Sci. Eng. 101, 97–116.

Fishbein, L., 2002. Arsine: human health aspects. 36 pp Concise International Chemical Assessment Document 47. World Health Organization, Geneva.

Fisher, E.R., Armentrout, P.B., 1990. Reactions of Co^+, Ni^+ and Cu^+ with c-propane and ethylene oxide: metal-methylidene ion bond energies. J. Phys. Chem. 94, 1674–1683.

Fisher, E.R., Elkind, J.L., Clemmer, D.E., Georgiadis, R., Loh, S.K., Aristov, N., et al., 1990a. Reactions of fourth-period metal ions (Ca^+-Zn^+) with O_2: metal-oxide ion bond energies. J. Chem. Phys. 93, 2676–2691.

Fisinin, V.I., Papazyan, T.T., Surai, P.F., 2009. Producing selenium-enriched eggs and meat to improve the selenium status of the general population. Chem. Rev. Biotechnol. 29 (1), 18–28.

Fitzgerald, W.F., Engstrom, D.R., Lamborg, C.H., et al., 2005. Modern and historical atmospheric mercury fluxes in Northern Alaska: global sources and Arctic depletion. Environ. Sci. Technol. 39, 557–568.

Fitzgerald, W.F., Lamborg, C.H., Hammerschmidt, C.R., 2007. Marine biogeochemical cycling of mercury. Chem. Rev. 107, 641–662.

Flagan, R.C., Taylor, D.D., 1981. Laboratory studies of submicron particles from coal combustion. Proc. Combust. Inst. 18, 1227–1237.

Fleig, D., Vainio, E., Andersson, K., et al., 2012. Evaluation of SO_3 measurement techniques in air and oxy-fuel combustion. Energy Fuels 26 (9), 5537–5549.

Fleig, D., Alzueta, M.U., Normann, F., et al., 2013. Measurement and modeling of sulfur trioxide formation in a flow reactor under post-flame conditions. Combust. Flame 160, 1142–1151.

Flesch, G.D., White, R.M., Svec, H.J., 1969. The positive and negative ion mass spectra of chromyl chloride and chromyl fluoride. Int. J. Mass Spectrom. Ion Phys. 3, 339–363.

Florindi, A., Petroni, M., Pelino, M., 1989. Thermodynamic study of the dehydration process of α-goethite in vacuo. Thermochim. Acta 138, 47–53.

Flory, M.A., McLamarrah, S.K., Ziurys, L.M., 2005. High resolution spectroscopy of $CoS(X^4\Delta_i)$: examining 3d transition metal sulfide bonds. J. Chem. Phys. 123, 164312. 9 pp.

Flotre, C.H., Varsi, K., Helm, T., Bolann, B., Bjorke-Monsen, A.-L., 2017. Predictors of mercury, lead, cadmium and antimony status in Norwegian never-pregnant women of fertile age. PLoS One 12 (12), e0189169. 12 pp.

Flues, M., Sato, I.M., Scapin, M.A., Cotrim, M.E.B., Camargo, I.M.C., 2013. Toxic elements mobility in coal and ashes of Figueira coal power plant, Brazil. Fuel 103, 430–436.

Fogg, T.R., Rahn, K.A., 1984. Boron as a tracer of aerosol from combustion of coal. Geophys. Res. Lett. 11, 854–857.

Folarin, O.R., Adaramoye, O.A., Akanni, O.O., Olopade, J.O., 2018. Changes in the brain antioxidant profile after chronic vanadium administration in mice. Metab. Brain Dis. 33 (2), 377–385. Special Issue 1.

Folgueras, M.B., Diaz, R.M., Xiberta, J., Prieto, I., 2003. Volatilization of trace elements for coal-sewage sludge blends during their combustion. Fuel 82, 1939–1948.

Folgueras, M.B., Diaz, R.M., Xiberta, J., Alonso, M., 2007. Effect of inorganic matter on trace element behavior during combustion of coal-sewage sludge blends. Energy Fuels 21, 744–755.

Folgueras, M.B., Alonso, M., Fernandez, F.J., 2017. Coal and sewage sludge ashes as sources of rare earth elements. Fuel 192, 128–139.

Fomenko, E.V., Anshits, N.N., Vasilieva, N.G., et al., 2015. Characterization of fly ash cenospheres produced from the combustion of Ekibastuz coal. Energy Fuels 29, 5390–5403.

Fomenko, E.V., Anshits, N.N., Vasilieva, N.G., et al., 2016. Composition and structure of the shells of aluminosilicate microspheres in fly ash formed on the combustion of Ekibastuz coal. Solid Fuel Chem. 50 (4), 238–247.

Fomichev, V.V., Poloznikova, M.E., Kondratov, O.I., 1992. Structural features and spectroscopic and energy characteristics of alkali metal molybdates and tungstates. Russ. Chem. Rev. 61, 877–888.

Font, O., Querol, X., Huggins, F.E., Chimenos, J.M., Fernandez, A.I., Burgos, S., et al., 2005. Speciation of major and selected elements in IGCC flyash. Fuel 84, 1364–1371.

Font, O., Querol, X., Juan, R., Casado, R., Ruiz, C.R., Lopez-Soler, A., et al., 2007. Recovery of gallium and vanadium from gasification fly ash. J. Hazard. Mater. A 139, 413–423.

Font, O., Querol, X., Izquierdo, M., Alvarez, E., Moreno, N., Diez, S., et al., 2010. Partitioning of elements in an entrained flow IGCC plant: influence of selected operational conditions. Fuel 89, 3250–3261.

Font, O., Cordoba, P., Leiva, C., et al., 2011. Fate of mercury and other trace elements in an oxy coal combustion pilot plant. In: Proceedings of 4th World of Coal Ash, Denver CO, a39, pp. 1–10.

Font, A., de Hoogh, K., Leal-Sanchez, M., et al., 2015. Using metal ratios to detect emissions from municipal waste incinerators in ambient air pollution data. Atmos. Environ. 113, 177–186.

Fontijn, A. (Ed.), 1985. Gas-Phase Chemiluminescence and Chemi-Ionization. North-Holland, Amsterdam, 19 chapters, 372 pp.

Fontijn, A. (Ed.), 1992. Gas Phase Metal Reactions. Elsevier Science Publishers, Amsterdam, 28 chapters, 700 pp.

Fordyce, F., 2007. Selenium geochemistry and health. Ambio 36 (1), 94–97.

Fotiev, A.A., Surat, L.L., Danilov, N.F., 1996. Reactions of Fe_2O_3 and Fe_3O_4 with sodium salts under heating in air. Russ. J. Inorg. Chem. 41, 1710–1712.

Fouga, G.G., De Micco, G., Bohe, A.E., 2009. Chlorination of manganese oxides. Thermochim. Acta 494, 141–146.

Fouga, G.G., Taddeo, R.M., Bosco, M.V., Bohe, A.E., 2012. Kinetic study of hubnerite ($MnWO_4$) chlorination. Thermochim. Acta. 536, 30–40.

Frandsen, F.J., 2009. Ash research from Palm Coast, Florida to Banff, Canada: entry of biomass in modern power boilers. Energy Fuels 23, 3347–3378.

Frandsen, F., Dam-Johansen, K., Rasmussen, P., 1994. Trace elements from combustion and gasification of coal: an equilibrium approach. Prog. Energy Combust. Sci. 20, 115–138.

Frantseva, K.E., Semenov, G.A., 1969. Mass spectrometer study of the evaporation of certain vanadium oxides. High Temp. USSR 7, 52–57.

Franzen, H.F., Gilles, P.W., 1965. High temperature vaporization and thermodynamic properties of titanium monosulfide. J. Chem. Phys. 42, 1033–1038.

Franzen, H.F., Hariharan, A.V., 1980. The high temperature vaporization thermodynamics of lanthanum and yttrium monophosphides as determined by Knudsen effusion. J. Chem. Thermodyn. 12, 975–984.

Franzen, H.F., Hariharan, A.V., Merrick, J.A., 1980. The high temperature vaporization and thermodynamics of scandium monophosphide. High Temp. Sci. 12, 11–20.

Fraser, M.E., 1984. Kinetic and chemiluminescence studies of ground state arsenic atom reactions. I. J. Phys. Chem. 88 (16), 3383–3386.

Fredriksson, E., Carlsson, J.O., 1993. Chemical vapor deposition of aluminum oxides from various gas mixtures. J. Chem. Vapor Depos. 1 (4), 333–417.

Freedman, A., Wormhoudt, J.C., Kolb, C.E., 1982. Studies of alkali phosphate and phosphorus chemistry important to magnetohydrodynamics combustion. Am. Chem. Soc. Symp. Ser. 179, 601–613.

Freindorf, M., Marian, C.M., Hess, B.A., 1993. Theoretical study of the electronic spectrum of the CoH molecule. J. Chem. Phys. 99, 1215–1223.

Friedemann, J., Wagner, A., Heinze, A., et al., 2016. Direct optical observation of coal particle fragmentation behavior in a drop-tube reactor. Fuel 166, 382–391.

Friedli, H.R., Radke, L.F., Lu, J.Y., 2001. Mercury in smoke from biomass fires. Geophys. Res. Lett. 28, 3223–3226.

Friedli, H.R., Radke, L.F., Lu, J.Y., Banic, C.M., Leaitch, W.R., MacPherson, J.I., 2003. Mercury emissions from burning of biomass from temperate North American forests: laboratory and airborne measurements. Atmos. Environ. 37, 253–267.

Friedrich, M.J., 2017. High mercury levels found in women around the world. J. Am. Med. Assoc. 318 (19), 1857.

Friswell, N.J., Jenkins, D.R., 1972. Identification of lead compounds in flames and determination of the PbO bond energy. Combust. Flame 19, 197–201.

Froben, F.W., Schulze, W., Kloss, U., 1983. Raman spectra of matrix-isolated group IIIA dimers: Ga_2, In_2 and Tl_2. Chem. Phys. Lett. 99 (5–6), 500–502.

Fryburg, G.C., Miller, R.A., Kohl, F.J., Stearns, C.A., 1977. Volatile products in the corrosion of Cr, Mo, Ti and four superalloys exposed to O_2 containing H_2O and gaseous NaCl. J. Electrochem. Soc. 124, 1738–1743.

Fryburg, G.C., Miller, R.A., Stearns, C.A., Kohl, F.J., 1978. Formation of Na_2SO_4 and K_2SO_4 in flames doped with sulfur and alkali chlorides and carbonates. In: Hildenbrand, D.L., et al., (Eds.), Proceedings of Symposium on High Temperature Metal Halide Chemistry., vol. 78-1. The Electrochemical Society, Pennington, NJ, pp. 468–483.

Fryda, L., Sobrino, C., Cieplik, M., van de Kamp, W.L., 2010. Study on ash deposition under oxyfuel combustion of coal/biomass blends. Fuel 89, 1889–1902.

Fu, M.H., Tabatabai, M.A., 1988. Tungsten content of soils, plants and sewage sludges in Iowa. J. Environ. Qual. 17, 146–148.

Fu, B., Hower, J.C., Dai, S., Mardon, S.M., Liu, G., 2018. Determination of chemical speciation of arsenic and selenium in high arsenic coal combustion ash by X-ray photoelectron spectroscopy: examples from a Kentucky stoker ash. ACS Omega 3, 17637–17645.

Fuentealba, P., Reyes, O., Stoll, H., Preuss, H., 1987. Ground state properties of alkali and alkaline-earth hydrides. J. Chem. Phys. 87 (9), 5338–5345.

Fuger, J., Parker, V.B., Hubbard, W.N., Oetting, F.L., 1983. 267 pp. The Chemical Thermodynamics of Actinide Elements and Compounds. VIII. The Actinide Halides. International Atomic Energy Agency, Vienna.

Fujii, T.S., Iwata, S., 1996. Theoretical studies of the ground and low-lying excited electronic states of the early transition metal dihydrides with state averaged MC SCF method. Chem. Phys. Lett. 251, 150–156.

Fujimori, T., Tanino, Y., Takaoka, M., 2013. Thermochemical behavior of lead adjusting formation of chlorinated aromatics in municipal solid waste fly ash. Environ. Sci. Technol. 47, 2169–2176.

Fujiwara, K., Haraguchi, H., Fuwa, K., 1975. Profiles of distribution of atoms in the nitrous oxide-acetylene flame. Bull. Chem. Soc. Jpn. 48, 857–862.

Fujiwara, N., Fujita, Y., Tomura, K., Moritomi, H., Tuji, T., Takasu, S., et al., 2002. Mercury transformations in the exhausts from lab-scale coal flames. Fuel 81, 2045–2052.

Fukuda, N., Takaoka, M., Doumoto, S., Oshita, K., et al., 2011. Mercury emission and behavior in primary ferrous metal production. Atmos. Environ. 45, 3685–3691.

Fuller, R.O., Chandler, G.S., Davis, J.R., McKinley, A.J., 2010. Matrix isolation ESR and theoretical studies of metal phosphides. J. Chem. Phys. 133, 164311. 11 pp.

Funaki, K., Uchimura, K., 1962. Chlorine metallurgy of tungsten. IV. Equilibrium between tungsten oxide and hydrogen chloride gas. Denki Kagaku 30 (4), 249–255.

Furche, F., Perdew, J.P., 2006. The performance of semilocal and hybrid density functionals in 3d transition metal chemistry. J. Chem. Phys. 124, 044103. 27 pp.

Furey, D.A., Bloore, E.W., 1970. Effect of water on the volatility of fission product oxide. US Clearinghouse Feder. Sci. Tech. Inform. Report, AD702884, 13 pp.

Furimsky, E., 2000. Characterization of trace element emissions from coal combustion by equilibrium calculations. Fuel Process. Technol. 63, 29–44.

Fytili, D., Zabaniotou, A., 2008. Utilization of sewage sludge in EU application of old and new methods: a review. Renew. Sustain. Energy Rev. 12, 116–140.

Gabbard, A., 1995. Can coal combustion breed plutonium in the sky? Phys. Today 48 (5), 88–90.

Gabis, I.E., Chernov, I.A., Voyt, A.P., 2013. Decomposition kinetics of metal hydrides: experiments and modeling. J. Alloys Compds. 580, s243–s246.

Gabler, Jr R.C., Neylan, D.L., 1977. Incinerated municipal sewage sludge as a secondary resource for metals and phosphorus. In: Sludge Manage Disposal Utility Proceedings of National Conference, 3rd conference, pp. 197–200.

Gajda, T., Jancso, A., 2010. Organotins: formation, use, speciation and toxicology. Chapter 4 In: Sigel, A., Sigel, H., Sigel, R.K.O. (Eds.), Metal Ions in Life Sciences, vol. 7. Organometallics in Environment and Toxicology. Royal Society of Chemistry, London, UK, pp. 111–151.

Gajdosechova, Z., Lawan, M.M., Urgast, D.S., et al., 2016. In vivo formation of natural HgSe nanoparticles in the liver and brain of pilot whales. Sci. Rep. 6, 34361.

Galbreath, K.C., Zygarlicke, C.J., 2004. Formation and chemical speciation of arsenic, chromium and nickel-bearing coal combustion $PM_{2.5}$. Fuel Process. Technol. 85, 701–726.

Galbreath, K.C., Zygarlicke, C.J., Toman, D.L., Huggins, F.E., Huffman, G.P., 1998. Nickel and chromium speciation of residual oil combustion ash. Combust. Sci. Technol. 134 (1–6), 243–262.

Galbreath, K.C., Toman, D.L., Zygarlicke, C.J., Pavlish, J.H., 2000. Trace element partitioning and transformations during combustion of bituminous and subbituminous U.S. coals in a 7 kW combustion system. Energy Fuels 14, 1265–1279.

Galbreath, K.C., Schulz, R.L., Toman, D.L., et al., 2005. Nickel and sulfur speciation of residual oil fly ashes from two electric utility steam-generating units. J. Air Waste Manage. Assoc. 55, 309–318.

Gale, T.K., Wendt, J.O.L., 2003. Mechanisms and models describing sodium and lead scavenging by a kaolinite aerosol at high temperatures. Aerosol Sci. Technol. 37, 865–876.

Gale, T.K., Wendt, J.O.L., 2005. In-furnace capture of cadmium and other semi-volatile metals by sorbents. Proc. Combust. Inst. 30, 2999–3007.

Gallagher, N.B., Peterson, T.W., Wendt, J.O.L., 1996. Sodium partitioning in a pulverized coal combustion environment. Proc. Combust. Inst. 26, 3197–3204.

Gamo, M., Oka, T., Nakanishi, J., 2003. Ranking the risks of twelve major environmental pollutants that occur in Japan. Chemosphere 53, 277–284.

Gan, Z., Su, K., Wang, Y., Wen, Z., 1998. Several excited states and thermochemical properties of PH_n (n = 1-3): a high level *ab initio* study. Chem. Phys. 228, 31–38.

Gao, Q., Li, S., Yuan, Y., Zhang, Y., Yao, Q., 2015. Ultrafine particle matter formation in the early stage of pulverized coal combustion of high sodium lignite. Fuel 158, 224–231.

Gao, Q., Li, S., Yuan, Y., Zhao, Y., Yao, Q., 2016. Role of minerals in the evolution of fine particulate matter during pulverized coal combustion. Energy Fuels 30 (3), 1815–1821.

Gao, L., Wang, Y., Huang, Q., Guo, S., 2017. Emission of mercury from six low calorific value coal-fired power plants. Fuel 210, 611–616.

Gao, Y., Baisch, P., Mirlean, N., da Silver Junior, F.M.R., et al., 2018. Arsenic speciation in fish and shellfish from the North Sea (Southern bite) and Acu Port area (Brazil) and health risks related to seafood consumption. Chemosphere 191, 89–96.

Garba, M.U., Ingham, D.B., Ma, L., Porter, R.T.J., Pourkashnian, M., Tan, H.Z., et al., 2012. Prediction of potassium chloride sulfation and its effect on deposition in biomass-fired boilers. Energy Fuels 26 (11), 6501–6508.

Garcia-Hernandez, J., Ortega-Velez, M.I., Contreras-Paniagua, A.D., Aquilera-Marquez, D., et al., 2018. Mercury concentrations in seafood and the associated risk in women with high fish consumption from coastal villages of Sonora, Mexico. Food Chem. Toxicol. 120, 367–377.

Garcia-Sosa, A.T., Castro, M., 2000. Density functional study of FeO_2, FeO_2^+ and FeO_2^-. Int. J. Quantum Chem. 80, 307–319.

Gashkina, N.A., Tatsii, Yu.G., Udachin, V.N., et al., 2015. Indication of environmental contamination: a case study of a large copper smelter. Geochem. Int. 53 (3), 253–264.

Gavrichev, K.S., Ryumin, M.A., Tyurin, A.V., Gurevich, V.M., Komissarova, L.N., Khoroshilov, A.V., et al., 2009. A calorimetric study of the thermodynamic properties of potassium molybdate. Russ. J. Phys. Chem. 83, 327–333.

Gaydon, A.G., 1968. Dissociation Energies and Spectra of Diatomic Molecules, third ed. Chapman and Hall, London.

Gaydon, A.G., 1974. The Spectroscopy of Flames, second ed. Chapman & Hall, London.

Gaydon, A.G., Wolfhard, H.G., 1979. Flames: Their Structure, Radiation and Temperature, fourth ed. Chapman & Hall, London.

Gayer, K.H., Garrett, A.B., 1950. The solubility of cobalt hydroxide, $Co(OH)_2$, in solutions of hydrochloric acid and sodium hydroxide at 25°C. J. Am. Chem. Soc. 72, 3921–3923.

Geng, J.J., Wang, Y.-P., Luo, H.-J., 2015. Distribution, sources and fluxes of heavy metals in the Pearl River Delta, South China. Marine Pollut. Bull. 101 (2), 914–921.

George, A., Dugwell, D.R., Kandiyoti, R., 2008. Leaching hierarchies in co-combustion residues. Energy Fuels 22, 1641–1649.

George, A., Larrion, M., Dugwell, D., Fennell, P.S., Kandiyoti, R., 2010. Co-firing of single, binary and ternary fuel blends: comparing synergies within trace element partitioning arrived at by thermodynamic equilibrium modeling and experimental measurements. Energy Fuels 24, 2918–2923.

Georgiadis, R., Fisher, E.R., Armentrout, P.B., 1989. Neutral and ionic metal-hydrogen and metal-carbon bond energies: reactions of Co^+, Ni^+ and Cu^+ with ethane, propane, methylpropane and dimethylpropane. J. Am. Chem. Soc. 111, 4251–4262.

Gerasimov, I.E., Knyazkov, D.A., Shmakov, A.G., Paletsky, A.A., et al., 2011. Structure of an atmospheric pressure $H_2/O_2/N_2$ flame doped with iron pentacarbonyl. Combust. Explos. Shock Waves 47 (1), 1–11.

Germani, M.S., Zoller, W.H., 1988. Vapor phase concentrations of arsenic, selenium, bromine, iodine and mercury in the stack of a coal-fired power plant. Environ. Sci. Technol. 22, 1079–1085.

Germani, M.S., Gokmen, I., Sigleo, A.C., Kowalczyk, G.S., Olmez, I., Small, A.M., et al., 1980. Concentrations of elements in the National Bureau of Standards' bituminous and subbituminous coal standard reference materials. Anal. Chem. 52, 240–245.

Gertler, S.E., 1941. Chromium oxide. United States Patent, US 2263623.

Gesenhues, U., Wendt, H., True, E.I., 1986. Mass spectra of the vapor molecules of aluminum chloride and aluminum chloride dimer. Int. J. Mass Spectrom. Ion Process. 70, 225–229.

Gesheva, K.A., Ivanova, T., Iossifova, A., Gogova, D., Porat, R., 1999. Deposition and characterization of CVD MoO_3 thin films. J. Phys. IV France 9, 453–459.

Ghalila, H., Lahmar, S., Lakhdar, Z.B., Hochlaf, M., 2008. Spectroscopy and metastability of BeO^+. J. Phys. B At. Mol. Opt. Phys. 41, 205101. 7 pp.

Gharebaghi, M., Hughes, K.J., Porter, R.T.J., Pourkashanian, M., Williams, M., 2011. Mercury speciation in air-coal and oxy-coal combustion: a modeling approach. Proc. Combust. Inst. 33, 1779–1786.

Ghorishi, S.B., Keeney, R.M., Serre, S.D., Gullett, B.K., Jozewicz, W.S., 2002. Development of a Cl-impregnated activated carbon for entrained-flow capture of elemental mercury. Environ. Sci. Technol. 36, 4454–4459.

Ghosh, S., Sharma, A., Talukder, G., 1992. Zirconium: an abnormal trace element in biology. Biol. Trace Elem. Res. 35, 247–271.

Ghosh-Dastidar, A., Mahuli, S., Agnihotri, R., Fan, L.-S., 1996. Selenium capture using sorbent powders: mechanism of sorption by hydrated lime. Environ. Sci. Technol. 30, 447–452.

Giang, A., Selin, N.E., 2016. Benefits of mercury controls for the United States. Proc. Natl. Acad. Sci. USA 113 (2), 286–291.

Gibson, S.T., Greene, J.P., Berkowitz, J., 1986. A photoionization study of SeH and H_2Se. J. Chem. Phys. 85, 4815–4824.

Giesen, A., Woiki, D., Herzler, J., Roth, P., 2002. Oxidation of Fe atoms by O_2 based on Fe- and O-concentration measurements. Proc. Combust. Inst. 29, 1345–1352.

Gindorf, C., Singheiser, L., Hilpert, K., 2005. Vaporization of chromia in humid air. J. Phys. Chem. Solids 66, 384–387.

Gingerich, K.A., 1969. Gaseous phosphorus compounds. III. Mass spectrometric study of the reaction between diatomic nitrogen and phosphorus vapor and dissociation energy of phosphorus mononitride and diatomic phosphorus. J. Phys. Chem. 73, 2734–2741.

Gingerich, K.A., Miller, F., 1975. Thermodynamic study of gaseous sodium-phosphorus-oxygen ternary molecules by high temperature mass spectrometry. J. Chem. Phys. 63, 1211–1217.

Gingerich, D.B., Grol, E., Mauter, M.S., 2018. Fundamental challenges and engineering opportunities in flue gas desulfurization wastewater treatment at coal fired power plants. Environ. Sci.-Water Res. Technol. 4 (7), 909–925.

Giovannini, S., Onder, G., Lattanzio, F., Bustacchini, S., et al., 2018. Selenium concentrations and mortality among community-dwelling older adults: results from ilSIRENTE study. J. Nutr. Health Aging 22 (5), 608–612.

Giri, D., Das, K.K., 2005. Electronic states of SnS and SnS^+: a configuration interaction study. J. Phys. Chem. A 109, 7207–7215.

Giri, D., Buenker, R.J., Das, K.K., 2002. Electronic spectrum of tin oxide: MRDCI study. J. Phys. Chem. A 106, 8790–8797.

Glarborg, P., Marshall, P., 2005. Mechanism and modeling of the formation of gaseous alkali sulfates. Combust. Flame 141, 22–39.

Glaude, P.A., Curran, H.J., Pitz, W.J., Westbrook, C.K., 2000. Kinetic study of the combustion of organophosphorus compounds. Proc. Combust. Inst. 28, 1749–1756.

Glazer, M.P., Khan, N.A., de Jong, W., Spliethoff, H., Schurmann, H., Monkhouse, P., 2005. Alkali metals in circulating fluidized bed combustion of biomass and coal: measurements and chemical equilibrium analysis. Energy Fuels 19, 1889–1897.

Glemser, O., Ackermann, H., 1963. Gaseous hydroxides. VI. Supplementary research concerning gaseous $WO_2(OH)_2$. Zeit Anorg. Allgem Chem. 325, 281–286.

Glemser, O., Haeseler, R.J., 1962. Gaseous hydroxides. IV. Gaseous hydroxides of molybdenum and tungsten. Zeit Anorg. Allgem Chem. 316, 168–181.

Glemser, O., Mueller, A., 1964. Gaseous hydroxides. VIII. Gaseous chromium hydroxide. Zeit Anorg. Allgem Chem. 334, 150–154.

Glemser, O., Stocker, U., 1963. The dissociation reaction $CdO(s) = Cd(g) + 0.5\ O_2(g)$. Ber Bunseng Phys. Chem. 67, 505–509.

Glezakou, V.-A., Mavridis, A., Harrison, J.F., 1996. Electronic structure of the ground and low-lying excited states of TiP. J. Phys. Chem. 100 (33), 13971–13975.

Glukhovtsev, M.N., Bach, R.D., Nagel, C.J., 1997. Performance of the B3LYP/ECP DFT calculations of iron containing compounds. J. Phys. Chem. A 101, 316–323.

Goddard, S.L., Brown, R.J.C., Ghatora, B.K., 2016. Determination of beryllium concentrations in UK ambient air. Atmos. Environ. 147, 320–329.

Goel, S., Masunov, A.E., 2008. Potential energy curves and electronic structure of 3d transition metal hydrides and their cations. J. Chem. Phys. 129, 214303. 14 pp.

Goen, T., Lermen, D., Hildebrand, J., Bartel-Steinbach, M., et al., 2018. Discovering time-trends of the German populations exposure to contaminants by analysis of human samples of the German Environmental Specimen Bank. Toxicol. Lett. 298, 194–200. Spec Iss S1.

Gojova, A., Guo, B., Kota, R.S., Rutledge, J.C., Kennedy, I.M., Barakat, A.I., 2007. Induction of inflammation in vascular endothelial cells by metal oxide nanoparticles: effect of particle composition. Environ. Health Perspect. 115, 403–409.

Goldhaber, S.B., 2003. Trace element risk assessment: essentiality vs. toxicity. Regul. Toxicol. Pharmacol. 38, 232–242.

Gomez-Jacinto, V., Garcia-Barrera, T., Garbayo, I., et al., 2012. Metallomic study of selenium biomolecules metabolized by the microalgae *Chlorella sorkiniana* in the biotechnological production of functional foods enriched in selenium. Pure Appl. Chem. 84 (2), 269–280.

Gong, Y., Zhou, M., Andrews, L., 2009. Spectroscopic and theoretical studies of transition metal oxides and dioxygen complexes. Chem. Rev. 109, 6765–6808.

Gong, X., Guo, Z., Wang, Z., 2010. Reactivity of pulverized coals during combustion catalyzed by CeO_2 and Fe_2O_3. Combust. Flame 157, 351–356.

Gontarz, Z., Grzybowska, R., Sabalinski, R., 1995. Solid phase reactivity of sodium oxosalts of manganese. J. Thermal Anal. 45, 1125–1133.

Gonzalvez, A., Cervera, M.L., Armenta, S., de la Guardia, M., 2009. A review of non-chromatographic methods for speciation analysis. Anal. Chim Acta 636, 129–157.

Goodarzi, F., 2006. Assessment of elemental content of milled coal, combustion residues and stack emitted materials: possible environmental effects for a Canadian pulverized coal-fired power plant. Int. J. Coal Geol. 65, 17–25.

Goodarzi, F., 2006a. Morphology and chemistry of fine particles emitted from a Canadian coal-fired power plant. Fuel 85, 273–280.

Goodarzi, F., Hower, J.C., 2008. Classification of carbon in Canadian fly ashes and their implications in the capture of mercury. Fuel 87, 1949–1957.

Goodarzi, F., Huggins, F.E., 2005. Speciation of arsenic in feed coals and their ash byproducts from Canadian power plants burning sub-bituminous and bituminous coals. Energy Fuels 19, 905–915.

Goodarzi, F., Huggins, F.E., 2005a. Speciation of chromium in feed coals and ash byproducts from Canadian power plants burning subbituminous and bituminous coals. Energy Fuels 19, 2500–2508.

Goodarzi, F., Huggins, F.E., Sanei, H., 2008. Assessment of elements, speciation of As, Cr, Ni and emitted Hg for a Canadian power plant burning bituminous coal. Int. J. Coal Geol. 74, 1–12.

Goodings, J.M., 1992. Metal ion chemistry in flames. In: Fontijn, A. (Ed.), Gas Phase Metal Reactions. Elsevier Science Publishers, Amsterdam, pp. 493–527.

Goodings, J.M., Hassanali, C.S., 1990. Ion chemistry of phosphorus in hydrocarbon flames. I. Electron scavenging by negative ion formation. Int. J. Mass Spectrom. Ion Process. 101, 337–354.

Gorazda, K., Kowalski, Z., Wzorek, Z., 2012. From sewage sludge ash to calcium phosphate fertilizers. Pol. J. Chem. Technol. 14 (3), 54–58.

Gornostayev, S.S., Tanskanen, P.A., Heikkinen, E.-P., Kerkkonen, O., Harkki, J.J., 2007. An example of alkalization of SiO_2 in a blast furnace coke. Energy Fuels 21, 2637–2641.

Gorokhov, L.N., Gusarov, A.V., Makarov, A.V., Nikitin, O.T., 1971. Mass spectrometric study of the evaporation of alkali metaborates. High Temp. USSR 9, 1082–1085.

Gorokhov, L.N., Kilushin, M.I., Emelyanov, A.M., 1990. Knudsen effusion mass spectrometric determination of metal hydroxide stabilities. High Temp. Sci. 26, 395–403.

Gospodinov, G., Slavcheva, Y.U., Popova, E., 1991. Determination of the thermodynamic parameters of some slightly soluble selenites. Thermochim. Acta 181, 337–339.

Gottscho, R.A., Koffend, J.B., Field, R.W., 1980. Optical-optical double resonance spectroscopy of BaO: the low-lying states. J. Mol. Spectrosc. 82, 310–338.

Gottwald, U., Monkhouse, P., Bonn, B., 2001. Dependence of alkali emissions in PFB combustion on coal composition. Fuel 80, 1893–1899.

Gottwald, U., Monkhouse, P., Wulgaris, N., Bonn, B., 2002. In situ study of the effect of operating conditions and additives on alkali emissions in fluidized bed combustion. Fuel Process. Technol. 75, 215–226.

Goulle, J.-P., Mahieu, L., Castermant, J., Neveu, N., et al., 2005. Metal and metalloid multi-elementary ICP-MS validation in whole blood, plasma, urine and hair: reference values. Forensic Sci. Int. 153, 39–44.

Government Policy, 2014. Nine chemicals added to high-concern list. Chem. Eng. News 92 (35), 38.

Grabke, H.J., Reese, E., Spiegel, M., 1995. The effects of chlorides, hydrogen chloride and sulfur dioxide in the oxidation of steels below deposits. Corros. Sci. 37, 1023–1043.

Grabrucker, S., Grabrucker, A.M., 2017. Zinc in autism. pp, 153–173, Chapter 9 In: White, A.R., Aschner, M., Costa, L.G., Bush, A.I. (Eds.), Biometals in Neurodegenerative Diseases. Academic Press, Elsevier, London, 415 pp.

Grade, M., Wienecke, J., Rosinger, W., Hirschwald, W., 1983. Electron impact investigation of the molecules SeS (g) and TeSe(g) under high temperature equilibrium conditions. Ber. Bunsenges. Phys. Chem. 87, 355–361.

Graedel, T.E., Van Beers, D., Bertram, M., et al., 2004. Multilevel cycle of anthropogenic copper. Environ. Sci. Technol. 38, 1242–1252.

Graham, H.C., Davis, H.H., 1971. Oxidation/vaporization kinetics of Cr_2O_3. J. Am. Ceram. Soc. 54, 89–93.

Graham, W.F., Duce, R.A., 1979. Atmospheric pathways of the phosphorus cycle. Geochim Cosmochim Acta 43, 1195–1208.

Grandjean, P., Herz, K.T., 2015. Trace elements as paradigms of developmental neurotoxicants: lead, methylmercury and arsenic. J. Trace Elem. Med. Biol. 31, 130–134.

Granite, E.J., Myers, C.R., King, W.P., Stanko, D.C., Pennline, H.W., 2006. Sorbents for mercury capture from fuel gas with application to gasification systems. Ind. Eng. Chem. Res. 45, 4844–4848.

Grantham, L.F., Hiller, M.A., Young, T.L., Ginell, W.S., 1961. Molten phosphate reactor fuels. III. Mass spectrometric identification of gaseous corrosion products of metals in molten sodium polyphosphate. United States Atomic Energy Commission, Report Number NAA-SR-5927, 30 pp.

Gray, J.A., Li, M., Nelis, T., Field, R.W., 1991. The electronic structure of NiH: the ($Ni^{+}3d^{9}$ ^{2}D) supermultiplet. J. Chem. Phys. 95 (10), 7164–7178.

Graydon, J.W., Zhang, X., Kirk, D.W., Jia, C.Q., 2009. Sorption and stability of mercury on activated carbon for emission control. J. Hazard. Mater. 168, 978–982.

Green, D.R., 2018. An element of life. Cell 172, 389–390.

Green, D.W., 1980. Calculation of the Thermodynamic Properties of Fuel Vapor Species From Spectroscopic Data. Argonne National Laboratory, Argonne, IL, Report ANL-CEN-RSD-80-2, 85 pp.

Green, D.W., 1980a. Relationship between spectroscopic data and thermodynamic functions: application to uranium, plutonium and thorium oxide vapor species. J. Nucl. Mater. 88, 51–63.

Green, D.W., 1980b. Standard enthalpies of formation of gaseous thorium, uranium and plutonium oxides. Int. J. Thermophys. 1 (1), 61–71.

Greenberg, R.R., Zoller, W.H., Gordon, G.E., 1981. Atmospheric emissions of elements on particles from the Parkway Sewage-Sludge Incinerator. Environ. Sci. Technol. 15 (1), 64–70.

Greene, T.M., Brown, W., Andrews, L., Downs, A.J., Chertihin, G.V., Runeberg, N., et al., 1995. Matrix infrared spectroscopic and *ab initio* studies of ZnH_2, CdH_2 and related metal hydride species. J. Phys. Chem. 99, 7925–7934.

Greger, F., Hartinger, K.T., Monkhouse, P.B., Wolfrum, J., Baumann, H., Bonn, B., 1996. In situ alkali concentration measurements in a pressurized fluidized-bed coal combustor by excimer laser induced fragmentation fluorescence. Proc. Combust. Inst. 26, 3301–3307.

Grein, F., 2009. Ground and low-lying excited C_{2v} states of FeO_2: a challenge to computational methods. Int. J. Quantum Chem. 109, 549–558.

Grein, F., 2009a. Theoretical studies on the electronic spectrum of selenium dioxide: comparison with ozone and sulfur dioxide. Chem. Phys. 360, 1–6.

Grein, F., Kapur, A., 1983. The electronic spectrum of PN: a configuration interaction study. J. Mol. Spectrosc. 99, 25–34.

Grimley, R.T., Burns, R.P., Inghram, M.G., 1961. Thermodynamics of the vaporization of Cr_2O_3: dissociation energies of CrO, CrO_2 and CrO_3. J. Chem. Phys. 34, 664–667.

Grimley, R.T., Burns, R.P., Inghram, M.G., 1961a. Thermodynamics of the vaporization of nickel oxide. J. Chem. Phys. 35 (2), 551–554.

Grimley, R.T., Burns, R.P., Inghram, M.G., 1966. Mass spectrometric study of the vaporization of cobalt oxide. J. Chem. Phys. 45, 4158–4162.

Grimm, A., Skoglund, N., Bostrom, D., Boman, C., Ohman, M., 2012. Influence of phosphorus on alkali distribution during combustion of logging residues and wheat straw in a bench-scale fluidized bed. Energy Fuels 26, 3012–3023.

Grimminger, R., Clouthier, D.J., 2011. The electronic spectrum of the previously unknown HAsO transient molecule. J. Chem. Phys. 135, 184308. 8 pp.

Grob, V.E., 1963. Formation of negative ions in $SbCl_2$. J. Chem. Phys. 39 (4), 972–978.

Gronvold, F., Drowart, J., Westrum Jr, E.F., 1984. 269 pp. The Chemical Thermodynamics of Actinide Elements and Compounds. IV. The Actinide Chalcogenides (excluding Oxide). International Atomic Energy Agency, Vienna.

Gu, Y., Zhang, Y., Lin, J., et al., 2016. Homogeneous mercury oxidation with bromine species released from HBr-modified fly ash. Fuel 169, 58–67.

Guell, M., Luis, J.M., Rodriguez-Santiago, L., et al., 2009. Structure, bonding and relative stability of the ground and low-lying electronic states of CuO_2: the role of exact exchange. J. Phys. Chem. A 113, 1308–1317.

Guichemerre, M., Chambaud, G., Stoll, H., 2002. Electronic structure and spectroscopy of monohalides of metals of group I-B. Chem. Phys. 280, 71–102.

Guimaraes, A.C., Antunes, L.M., Ribeiro, H.F., et al., 2010. Cytogenetic biomonitoring of inhabitants of a large uranium mineralization area: the municipalities of Monte Alegre, Prainha and Alenquer in the State of Para, Brazil. Cell Biol. Toxicol. 26 (5), 403–419.

Guitou-Guichemerre, M., Chambaud, G., 2005. Electronic structure, reactivity and spectroscopy of dihydrides of group I-B metals. J. Chem. Phys. 122, 204325. 11 pp.

Gunier, R.B., Mora, A.M., Smith, D., et al., 2014. Biomarkers of manganese exposure in pregnant women and children living in an agricultural community in California. Environ. Sci. Technol. 48, 14695–14702.

Gunina, A.O., Lopatin, S.I., Shugurov, S.M., 2012. Gaseous vanadium molybdate and tungstates: thermodynamic properties and structures. Inorg. Chem. 51 (9), 4918–4924.

Gunion, R.F., Dixon-Warren, S.J., Lineberger, W.C., Morse, M.D., 1996. Ultraviolet photoelectron spectroscopy of molybdenum and molybdenum monoxide anions. J. Chem. Phys. 104, 1765–1773.

Gunn, S.R., 1972. The enthalpies of formation of arsine and biarsine. Inorg. Chem. 11 (4), 796–799.

Guo, B., Kennedy, I.M., 2001. The impact of chlorine on chromium speciation in a laminar diffusion flame. Combust. Flame 126, 1557–1568.

Guo, B., Kennedy, I.M., 2004. The speciation and morphology of chromium oxide nanoparticles in a diffusion flame. Aerosol Sci. Technol. 38, 424–436.

Guo, B., Kennedy, I.M., 2007. Gas phase flame synthesis and characterization of iron oxide nanoparticles for use in a health effects study. Aerosol Sci. Technol. 41, 944–951.

Guo, J., Goodings, J.M., 2002. Scandium ionization mechanisms in $H_2/O_2/N_2$ flames supported by calculated thermodynamic data. Int. J. Mass Spectrom. 214, 339–348.

Guo, R., Yang, J., Liu, Z., 2004. Behavior of trace elements during pyrolysis of coal in a simulated drop-tube reactor. Fuel 83, 639–643.

Guo, J., Wang, T., Zhang, Z., Chen, C., Chen, Y., 2006. Identification of new bands in the red system of CoO. J. Mol. Spectrosc. 240, 45–51.

Guo, J., Goodings, J.M., Hayhurst, A.N., 2007. What are the main gas phase species formed by aluminum when added to a premixed flame? Combust. Flame 150, 127–136.

Guo, X., Zheng, C.-G., Xu, M., 2007a. Characterization of mercury emissions from a coal-fired power plant. Energy Fuels 21, 898–902.

Guo, S., Jiang, Y.F., Li, J.Z., Yu, Z.L., et al., 2017. Correlations between coal compositions and sodium release during steam gasification of sodium-rich coals. Energy Fuels 31 (6), 6025–6033.

Guo, S., Zhang, L., Niu, X., Gao, L., et al., 2018. Mercury release characteristics during pyrolysis of eight bituminous coals. Fuel 222, 250–257.

Gupta, H., Thomas, T.J., Park, A.-H.A., et al., 2007. Pilot scale demonstration of the OSCAR(Ohio state carbonation ash reactivation) process for high temperature multipollutant control of coal combustion flue gas, using carbonated fly ash and mesoporous calcium carbonate. Ind. Eng. Chem. Res. 46 (14), 5051–5060.

Gupta, A., Ifeacho, P., Schulz, C., Wiggers, H., 2011. Synthesis of tailored WO_3 and WO_x ($2.9 < x < 3$) nanoparticles by adjusting the combustion conditions in a $H_2/O_2/Ar$ premixed flame reactor. Proc. Combust. Inst. 33 (2), 1883–1890.

Gupta, R.K., Majumdar, D., Trivedi, J.V., Bhanarkar, A.D., 2012. Particulate matter and elemental emissions from the cement kiln. Fuel Process. Technol. 104, 343–351.

Gupta, V., Mazzotti, F.J., Rice, C.A., et al., 2013. The $A^2\Sigma^+$-$X^2\Pi_i$ electronic transition of AgS. J. Mol. Spectrosc. 286/287, 52–55.

Gupta, D.K., Chatterjee, S., Datta, S., Veer, V., Walther, C., 2014. Role of phosphate fertilizers in heavy metal uptake and detoxification of toxic metals. Chemosphere 108, 134–144.

Gurvich, L.V., Ryabova, V.G., Khitrov, A.N., Starovoitov, E.M., 1971. Spectrophotometric determination of dissociation energies of molecules. I. SrCl and SrOH. High Temp. USSR 9, 261–264.

Gurvich, L.V., Ryabova, V.G., Khitrov, A.N., 1974. Determination of dissociation energies of alkaline earth chlorides, bromides and hydroxides by flame photometry. Faraday Symp. Chem. Soc. 8, 83–106.

Gurvich, L.V., Veits, I.V., Medvedev, V.A., Bergman, G.A., Yungman, V.S., Khachkuruzov, G.A., et al., 1982. Thermodynamic properties of individual substances, Cr, Mo, W, V, Nb, Ta, Ti, Zr, Hf, Sc, Y, La, Th, U, Pu, Li, Na, K, Rb, Cs elements and their compounds, Bk. 1.: Calculation of thermodynamic properties. Bk. 2.: Table of thermodynamic properties, vol. 4. Nauka, Moscow, 622, 560 pp. (in Russian).

Gurvich L.V., Veyts I.V., Alcock C.B. (Eds.), 1989, Thermochemical properties of individual substances. fourth ed. vol. 1. Elements O, H (D,T), F, Cl, Br, I, He, Ne, Ar, Fr, Xe, Rn, S, N, P and their compounds. Part 1. Methods and computation. Part 2. Tables. New York: Hemisphere Publishing Corporation; 1989, Part 1: 551 pp., Part 2: 448 pp.

Gurvich L.V., Veyts I.V., Alcock C.B., Iorish V.S. (Eds.) 1991. Thermochemical properties of individual substances. vol. 2. Elements C, Si, Ge, Sn, Pb and their compounds. Part 1. Methods and computation. Part 2. Tables, Hemisphere Publishing Corporation, New York, Part 1, 553 pp., Part 2, 372 pp.

Gurvich L.V., Veyts I.V., Alcock C.B., Iorish V.S. (Eds.) 1994. vol. 3. Thermochemical properties of individual substances. Elements B, Al, Ga, In, Tl, Be, Ca, Sr, Ba and their compounds. Part 1. Methods and computation. Part 2. Tables. CRC Press, Inc., Boca Raton, Florida, Part 1, 707 pp., Part 2, 448 pp.

Gurvich, L.V., Bergman, G.A., Gorokhov, L.N., Iorish, V.S., Leonidov, V.Ya, Yungman, V.S., 1996. Thermodynamic properties of alkali metal hydroxides. I. Lithium and sodium hydroxides. J. Phys. Chem. Ref. Data 25, 1211–1276.

Gurvich, L.V., Bergman, G.A., Gorokhov, L.N., Iorish, V.S., Leonidov, V.Ya, Yungman, V.S., 1997. Thermodynamic properties of alkali metal hydroxides. II. Potassium, rubidium and cesium hydroxides. J. Phys. Chem. Ref. Data 26, 1031–1110.

Gusarov, A.V., 1970. Determination of the stability of gaseous magnesium metaborates. High Temp. USSR 8, 1113–1118.

Gusarov, A.V., Iorish, V.S., 2006. Estimation of *a priori* errors in *ab initio* calculations of thermochemical values for the example of the dissociation energies of the ZnO and ZnS molecules. Russ. J. Phys. Chem. 80, 1864–1867.

Gusarov, A.V., Osina, E.L., 2012. Anharmonicity in the V_2O_3 molecule and thermodynamic properties of V_2O_3 in the gas phase. High Temp. 50, 56–60.

Gusarov, A.V., Yorish, V.S., 2011. Quantum chemical calculations of dissociation energies of a series of diatomic manganese compounds. High Temp. 49, 620–622.

Gustin, M.S., 2003. Are mercury emissions from geological sources significant? A status report. Sci. Total Environ. 304, 153–167.

Gutsev, G.L., 1994. A theoretical study on the structure and stability of the PCl_n and PCl_n^- series, n = 1-6. Chem. Phys. 179, 325–339.

Gutsev, G.L., Rao, B.K., Jena, P., Li, X., Wang, L.-S., 2000. Experimental and theoretical study of the photoelectron spectra of MnO_x^-(x = 1-3) clusters. J. Chem. Phys. 113, 1473–1483.

Gutsev, G.L., Rao, B.K., Jena, P., 2000a. Systematic study of oxo, peroxo and superoxo isomers of 3d-metal dioxides and their anions. J. Phys. Chem. A 104, 11961–11971.

Gutsev, G.L., Andrews, L., Bauschlicher Jr, C.W., 2003. Similarities and differences in the structure of 3*d*-metal monocarbides and monoxides. Theor. Chem. Acc. 109 (6), 298–308.

Ha, T.-K., Nguyen, M.T., 1985. An *ab initio* calculation of the electronic structure of copper dioxide. J. Phys. Chem. 89 (26), 5569–5570.

Haaland, A., Martinsen, K.-G., Shorokhov, D.J., Girichev, G.V., Sokolov, V.I., 1998. Molecular structure of monomeric scandium trichloride by gas electron diffraction and density functional theory calculations on $ScCl_3$ and Sc_2Cl_6. J. Chem. Soc. Dalton Trans. Inorg. Chem. 17, 2787–2792.

Hackert, A., Gruehn, R., 1998. $V_2O_3(OH)_4$(g): proof of existence, thermochemical characterization and chemical transport calculations for V_2O_5(s) in the presence of water. Zeit Anorg. Allgem Chem. 624 (11), 1756–1762.

Hackert, A., Plies, V., Gruehn, R., 1996. Proof of existence and thermochemical characterization of the gaseous molecule $VOCl_2$(g). Zeit Anorg. Allgem Chem. 622 (10), 1651–1657.

Hadke, S., Kalimila, M.T., Rathkanthiwar, S., Gour, S., et al., 2015. Role of fuel and fuel-to-oxidize ratio in combustion synthesis of nano-crystalline nickel oxide powders. Ceram. Int. 41 (10), 14949–14957.

Hagenmuller, P., 1967. Alkali metal stannates. Silicates Industriels 32, 333–336.

Hagenmuller, P., Tournoux, M., 1963. Two new compounds in the tin oxygen-potassium system. Comp. Rend. 257, 154–156.

Hahn, D.W., Charalampopoulos, T.T., 1992. The role of iron additives in sooting premixed flames. Proc. Symp. (Int.) Combust. 24, 1007–1014.

Halfen, D.T., Keogh, J.P., Ziurys, L.M., 2015. The Fourier transform microwave/millimeter-wave spectrum of YOH and YOD($X^1\Sigma^+$). J. Mol. Spectrosc. 314, 79–82.

Halle, J.C., Stern, K.H., 1980. Vaporization and decomposition of Na_2SO_4: thermodynamics and kinetics. J. Phys. Chem. 84, 1699–1704.

Haller, K., Lunsford, J.H., Laane, J., 1996. Temperature dependence of the Raman spectrum of barium peroxide. J. Phys. Chem. 100, 551–555.

Hamilton, P.A., Murrells, T.P., 1986. Mechanism for the chemiluminescence in oxygen-phosphorus systems. J. Phys. Chem. 90, 182–185.

Hamilton, J.C., McCarty, K.F., Nagelberg, A.S., 1986. In situ Raman spectroscopy of high temperature pyrite reactions related to deposit formation from coal. In: Shores, D.A., Yurek, G.J. (Eds.), Fundamental Aspects of High Temperature Corrosion II. Electrochemical Society, Inc, Pennington NJ, pp. 251–258.

Hammer, R.R., Gregory, N.W., 1964. Vaporization reactions in the copper chloride-chlorine system. J. Phys. Chem. 68 (11), 3229–3233.

Hamrick, Y.M., Taylor, S., Morse, M.D., 1991. Optical spectroscopy of jet-cooled MoO. J. Mol. Spectrosc. 146, 274–313.

Han, Y.-K., Hirao, K., 2000. On the ground-state spectroscopic constants of Tl_2. J. Chem. Phys. 112 (21), 9353–9355.

Han, F.X., Su, Y., Monts, D.L., Plodinec, M.J., Banin, A., Triplett, G.E., 2003. Assessment of global industrial-age anthropogenic arsenic contamination. Naturwissenschaften 90, 395–401.

Han, J., Xu, M., Yao, H., Furuuchi, M., Sakano, T., Kanchanapiya, P., et al., 2006. Partition of heavy and alkali metals during sewage sludge incineration. Energy Fuels 20, 583–590.

Han, H.D., Hu, S., Lu, C.F., et al., 2016. Inhibitory effects of CaO/Fe_2O_3 on arsenic emission during sewage sludge pyrolysis. Bioresour. Technol. 218, 134–139.

Handy, N.C., Carter, S., Yamaguchi, Y., Li, S., Turney, J.M., Schaeffer III, H.F., 2006. Rovibrational energy levels for the electronic ground state of AlOH. Chem. Phys. Lett. 427, 14–17.

Handzlik, J., Kurleto, K., 2013. Assessment of density functional methods for thermochemistry of chromium oxo compounds and their application in a study of chromia-silica system. Chem. Phys. Lett. 561/562, 87–91.

Hansen, L.A., Nielsen, H.P., Frandsen, F.J., Dam-Johansen, K., Horlyck, S., Karlsson, A., 2000. Influence of deposit formation on corrosion at a straw-fired boiler. Fuel Process. Technol. 64, 189–209.

Hansen, H.K., Pedersen, A.J., Ottosen, L.M., Villumsen, A., 2001. Speciation and mobility of cadmium in straw and wood combustion fly ash. Chemosphere 45, 123–128.

Hao, Y., Wu, S., Pan, Y., et al., 2016. Characterization and leaching toxicities of mercury in flue gas desulfurization gypsum from coal-fired power plants in China. Fuel 177, 157–163.

Hao, Y., Li, Q., Pan, Y., Liu, Z., et al., 2017. Heavy metals distribution characteristics of FGD gypsum samples from Shanxi province 12 coal-fired power plants and its potential environmental impacts. Fuel 209, 238–245.

Haraguchi, H., Fowler, W.K., Johnson, D.J., Winefordner, J.D., 1976. Molecular fluorescence spectroscopy of phosphorus monoxide in flames studied by a silicon intensified target-optical multichannel analyzer system. Spectrochim. Acta A 32, 1539–1544.

Haraguchi, H., Weeks, S.J., Winefordner, J.D., 1979. Selective excitation of molecular species in flames by laser-excited molecular fluorescence. Spectrochim. Acta A 35, 391–399.

Hardtl, K.H., Rau, H., 1969. PbO vapor pressure in the $Pb(Ti_{1-x}Zr_x)O_3$ system. Solid State Commun. 7, 41–45.

Hargittai, M., 2000. Molecular structure of metal halides. Chem. Rev. 100, 2233–2301.

Hargittai, M., Subbotina, N.Y., Kolonits, M., Gershikov, A.G., 1991. Molecular structure of first-row transition metal dihalides from combined electron diffraction and vibrational spectroscopic analysis. J. Chem. Phys. 94, 7278–7286.

Harms, J.C., Womack, K.A., O'Brien, L.C., Zou, W., 2014. Analysis of a new MoO transition in the near-infrared: a combined theoretical and experimental study. J. Chem. Phys. 141, 134310. 9 pp.

Harms, J.C., O'Brien, L.C., O'Brien, J.J., 2018. The spin-forbidden $a^4\Sigma$-$X^2\Pi_{1/2}$ transition of GeH detected in absorption using intra-cavity laser spectroscopy. J. Chem. Phys. 148, 204306. 7 pp.

Harper, E.M., Kavlak, G., Graedel, T.E., 2012. Tracking the metal of the Goblins: cobalt's cycle of use. Environ. Sci. Technol. 46, 1079–1086.

Harris, D.G., Chou, M.S., Cool, T.A., 1985. Experiments concerning phosphorus chemiluminescence. J. Chem. Phys. 82 (8), 3502–3515.

Harris, W.R., Silberman, D., Sears, D.R., 1988. Solubility of trace metals from fly ash from the combustion of lignite in an atmospheric fluidized bed combustor. Int. J. Environ. Anal. Chem. 34, 239–252.

Harrison, J.F., 2000. Electron structure of diatomic molecules composed of a first-row transition metal and main-group element (H-F). Chem. Rev. 100, 679–716.

Harrison, J.F., Field, R.W., Jarrold, C.C., 2002. Comparison of CaF, ZnF, CaO and ZnO: their anions and cations in their ground and low-lying excited states. Am. Chem. Soc. Symp. Ser. 828, 238–259.

Hasegawa, J.-Y., Pierloot, K., Roos, B.O., 2001. Ground state structure of CuO_2: a CASPT2 study. Chem. Phys. Lett. 335, 503–509.
Hasselriis, F., Licata, A., 1996. Analysis of heavy metal emission data from municipal waste combustion. J. Hazard. Mater. 47, 77–102.
Hastie, J.W., 1973. Mass spectrometric studies of flame inhibition: analysis of antimony trihalides in flames. Combust. Flame 21, 49–54.
Hastie, J.W., Bonnell, D.W., 1980. Molecular chemistry of inhibited combustion systems. National Bureau of Standards, US Department of Commerce, Report No. NBSIR 80-2169, 173 pp.
Hastie, J.W., McBee, C.L., 1975. Mechanistic studies of halogenated flame retardants: the antimony/halogen system. Am. Chem. Soc. Symp. Ser. 16, 118–148.
Hastie, J.W., Hauge, R.H., Margrave, J.L., 1971. Infrared spectra and geometries for the dichlorides of Ca, Sc, Ti, V, Cr, Mn, Fe and Ni. High Temp. Sci. 3, 257–274.
Hastie, J.W., Hauge, R.H., Margrave, J.L., 1971a. Infrared spectra and geometries of heavy metal halides: strontium chloride, barium chloride, europium chloride, europium difluoride, lead chloride and uranium dichloride. High Temp. Sci. 3 (1), 56–72.
Hastie, J.W., Bonnell, D.W., Plante, E.R., 1980. Slag and metal oxide vaporization in reactive atmospheres. High Temp. Sci. 13, 257–277.
Hastie, J.W., Zmbov, K.F., Bonnell, D.W., 1984. Transpiration mass spectrometric analysis of liquid KCl and KOH vaporization. High Temp. Sci. 17, 333–364.
Hauge, R.H., Margrave, J.L., 1972. Stabilities of gaseous diatomic sulfides. High Temp. Sci. 4, 170–177.
Hausladen, D.M., Alexander-Ozinskas, A., McClain, C., Fendorf, S., 2018. Hexavalent chromium sources and distribution in California groundwater. Environ. Sci. Technol. 52, 8242–8251.
Havukainen, J., Nguyen, M.T., Hermann, L., et al., 2016. Potential of phosphorus recovery from sewage sludge and manure ash by thermochemical treatment. Waste Manage. 49, 221–229.
Haworth, N.L., Bacskay, G.B., 2002. Heats of formation of phosphorus compounds determined by current methods of computational quantum chemistry. J. Chem. Phys. 117, 11175–11187.
Haworth, N.L., Bacskay, G.B., Mackie, J.C., 2002. The role of phosphorus dioxide in the H + OH recombination reaction: *ab initio* quantum chemical computation of thermochemical and rate parameters. J. Phys. Chem. 106, 1533–1541.
Haworth, N.L., Sullivan, M.B., Wilson, A.K., Martin, J.M.L., Radom, L., 2005. Structures and thermochemistry of calcium-containing molecules. J. Phys. Chem. A 109, 9156–9168.
Hayashi, S.-i, Takahashi, T., Kanehashi, K., Kubota, N., Mizuno, K., Kashiwakura, S., et al., 2010. Chemical state of boron in coal fly ash investigated by focused-ion-beam time-of-flight secondary ion mass spectrometry and satellite-transition magic angle spinning nuclear magnetic resonance. Chemosphere 80, 881–887.
Hayhurst, A.N., Kittelson, D.B., 1974. Ionization of alkaline earth additives in hydrogen flames. II. Kinetics of production and recombination of ions. Proc. R. Soc. Lond. A 338, 175–195.
Haynes, B.S., Jander, H., Wagner, H.G.G., 1979. The effect of metal additives on the formation of soot in premixed flames. Proc. Symp. (Int.) Combust. 17, 1365–1374.
Haynes, B.S., Jander, H., Matzing, H., Wagner, H.G.G., 1981. The influence of various metals on carbon formation in premixed flames. Combust. Flame 40, 101–103.
Hays, S.M., Macey, K., Nong, A., Aylward, L.L., 2014. Biomonitoring equivalents for selenium. Regul. Toxicol. Pharmacol. 70, 333–339.
He, P., Jiang, X.-M., Wu, J., Pan, W.-G., Ren, J.-X., 2015. Characterization of fly ash from coal fired power plant and their properties of mercury retention. Surf. Rev. Lett. 22 (2), 1550018. Available from: https://doi.org/10.1142/S0218625X15500183.
He, P., Zhang, X.-B., Peng, X., Wu, J., Jiang, X., 2016. Effect of fly ash composition on the retention of mercury in coal combustion flue gas. Fuel Process. Technol. 142, 6–12.
Heaven, M.C., 2006. Probing actinide electronic structure using fluorescence and multiphoton ionization spectroscopy. Phys. Chem. Chem. Phys. 8, 4497–4509.
Heaven, M.C., Bondybey, V.E., Merritt, J.M., Kaledin, A.L., 2011. The unique bonding characteristics of beryllium and the Group IIA metals. Chem. Phys. Lett. 506, 1–14.
Hedberg, Y.S., Hedberg, J.F., Herting, G., Goidanich, S., Wallinder, I.O., 2014. Critical review: copper runoff from outdoor surfaces at atmospheric conditions. Environ. Sci. Technol. 48, 1372–1381.

Hedgecock, I.M., Naulin, C., Costes, M., 1996. Crossed molecular beam study of the $Cr(a^7S_3) + O_2(X^3\Sigma_g^-) \rightarrow CrO(X^5\Pi_\Omega) + O(^3P_J)$ reaction. Chem. Phys. Lett. 207, 379–387.
Heidel, B., Klein, B., 2017. Reemission of elemental mercury and mercury halides in wet flue gas desulfurization. Int. J. Coal Geol. 170 (S1), 28–34.
Heinrichs, H., 1977. Emissions of 22 elements from Brown-coal combustion. Naturwissenschaften 64, 479–481.
Helble, J., Neville, M., Sarofim, A.F., 1986. Aggregate formation from vaporized ash during pulverized coal combustion. Symp. Int. Combust. 21, 411–417.
Helble, J.J., Srinivasachar, S., Boni, A.A., Charon, O., Modestino, A., 1992. Measurement and modeling of vapor-phase sodium chloride formed during pulverized coal combustion. Combust. Sci. Technol. 81, 193–205.
Helble, J.J., Mojtahedi, W., Lyyranen, J., Jokiniemi, J., Kauppinen, E., 1996. Trace element partitioning during coal gasification. Fuel 75, 931–939.
Henderson, R.D.E., Shayesteh, A., Tao, J., et al., 2013. Accurate analytic potential and Born-Oppenheimer breakdown functions for MgH and MgD from a direct-potential-fit data analysis. J. Phys. Chem. A 117, 13373–13387.
Henriksson, M., Warnqvist, B., 1979. Kinetics of formation of HCl(g) by the reaction between NaCl(s) and SO_2, O_2 and H_2O(g). Ind. Eng. Chem. Process Des. Dev. 18, 249–254.
Henriquez-Hernandez, L.A., Gonzalez-Antuna, A., Boada, L.D., Carranza, C., et al., 2018. Patterns of blood concentrations of 47 elements in two populations from the same geographical area but with different geological origin and lifestyles: Canary Islands (Spain) vs Morocco. Sci. Total Environ. 636, 709–716.
Hermoso, W., Ornellas, F.R., 2012. The radical SeCl: a theoretical contribution to the characterization of its low-lying electronic states. Chem. Phys. 397, 98–101.
Herrick, C.C., Feber, R.C., 1968. Vaporization studies on arsenic. J. Phys. Chem. 72 (4), 1102–1110.
Herrmann, R., Alkemade, C.T.J., 1963. Chemical Analysis by Flame Photometry, second rev. ed. Interscience Publishers, New York.
Herzberg, G., 1963. Molecular Spectra and Molecular Structure. I. Spectra of Diatomic Molecules, second ed. Van Nostrand, New Jersey USA.
Herzel, H., Kruger, O., Hermann, L., Adam, C., 2016. Sewage sludge ash: a promising secondary phosphorus source for fertilizer production. Sci. Total Environ. 542B, 1136–1143.
Herzler, J., Roth, P., 2003. High temperature gas phase reaction of SnO(g) with O_2. PhysChemChemPhys. 5, 1552–1556.
Hess, G., McCoy, M., 2015. Supreme Court blocks EPA mercury limits. Chem. Eng. News 92 (27), 5.
Hibbein, J., Gregory, S., Iles-Caven, Y., Taylor, C.M., et al., 2018. Total mercury exposure in early pregnancy has no adverse association with scholastic ability of the offspring particularly if the mother eats fish. Environ. Int. 116, 108–115.
Hightower, J.M., Moore, D., 2003. Mercury levels in high-end consumers of fish. Environ. Health Perspect. 111 (4), 604–608.
Higueras, P., Oyarzun, R., Lillo, J., et al., 2006. The Almaden district (Spain): anatomy of one of the world's largest Hg-contaminated sites. Sci. Total Environ. 356, 112–124.
Higueras, P., Oyarzun, R., Lillo, J., Morata, D., 2013. Intraplate mafic magmatism, degasification and deposition of mercury: the giant Almaden mercury deposit (Spain) revisited. Ore Geol. Rev. 51, 93–102.
Hildenbrand, D.L., 1975. Thermochemistry of molecular FeO, FeO^+ and FeO_2. Chem. Phys. Lett. 34, 352–354.
Hildenbrand, D.L., 1976. Mass spectrometric studies of the thermochemistry of gaseous TiO and TiO_2. Chem. Phys. Lett. 44, 281–284.
Hildenbrand, D.L., 1995. Dissociation energies of the monochlorides and dichlorides of Cr, Mn, Fe, Co and Ni. J. Chem. Phys. 103, 2634–2641.
Hildenbrand, D.L., 1996. Dissociation energies of the molecules BCl and BCl^+. J. Chem. Phys. 105 (23), 10507–10510.
Hildenbrand, D.L., 2008. Low-lying electronic states and dissociation energies of the monochlorides of Cr, Mn, Fe, Co and Ni. J. Phys. Chem. A 112, 3813–3815.
Hildenbrand, D.L., 2009. Low-lying electronic states and revised thermochemistry of TiCl, $TiCl_2$ and $TiCl_3$. J. Phys. Chem. A 113, 1472–1474.
Hildenbrand, D.L., Lau, K.H., 1991. Mass spectrometric searches for gaseous sodium carbonates. J. Phys. Chem. 95, 8972–8975.

Hildenbrand, D.L., Lau, K.H., 1992. Trends and anomalies in the thermodynamics of gaseous thorium and uranium halides. Pure Appl. Chem. 64 (1), 87–92.

Hildenbrand, D.L., Lau, K.H., 1993. Mass spectrum and sublimation pressure of sodium oxide vapor: stability of the superoxide molecule NaO_2. J. Chem. Phys. 98, 4076–4081.

Hildenbrand, D.L., Lau, K.H., 1994. Thermochemical properties of gaseous POBr and some H-P-O species. J. Chem. Phys. 100 (11), 8373–8376.

Hildenbrand, D.L., Lau, K.H., 1994a. Thermochemistry of gaseous manganese oxides and hydroxides. J. Chem. Phys. 100, 8377–8380.

Hildenbrand, D.L., Lau, K.H., 1994b. Thermochemistry of gaseous SiO(OH), $SiO(OH)_2$ and SiO_2. J. Chem. Phys. 101, 6076–6079.

Hildenbrand, D.L., Lau, K.H., 1995. Thermochemical properties of the gaseous scandium, yttrium and lanthanum fluorides. J. Chem. Phys. 102, 3769–3775.

Hildenbrand, D.L., Lau, K.H., 1996. Dissociation energy of the molecule GeCl. Chem. Phys. Lett. 263, 145–147.

Hildenbrand, D.L., Lau, K.H., 1996a. Thermochemistry of gaseous AgCl, Ag_3Cl_3 and CuCl. High Temp. Mater. Sci. 35, 11–20.

Hildenbrand, D.L., Lau, K.H., 2006. Dissociation energies of the Cu and Ag monohalides and of Ni monofluoride. J. Phys. Chem. A 110, 11886–11889.

Hildenbrand, D.L., Murad, E., 1970. Dissociation energy of NaO(g) and the heat of atomization of Na_2O(g). J. Chem. Phys. 53, 3403–3408.

Hildenbrand, D.L., Theard, L.P., 1969. Mass spectrometric measurement of the dissociation energies of BeCl and $BeCl_2$. J. Chem. Phys. 50, 5350–5355.

Hildenbrand, D.L., Gurvich, L.V., Yungman, V.S., 1985. 187 pp. The Chemical Thermodynamics of Actinide Elements and Compounds. XIII. The Gaseous Actinide Ions. International Atomic Energy Agency, Vienna.

Hildenbrand, D.L., Lau, K.H., Brittain, R.D., 1990. Mechanistic aspects of metal sulfate decomposition processes. High Temp. Sci. 26, 427–440.

Hildenbrand, D.L., Lau, K.H., Mastrangelo, S.V.R., 1991. Gaseous species in the Ti-Al-Cl system and reaction with H_2O. J. Phys. Chem. A 95, 3435–3437.

Hildenbrand, D.L., Lau, K.H., Roos, J.W., 1999. Thermochemistry of ZnCl(g). J. Chem. Phys. 111, 1337–1338.

Hildenbrand, D.L., Lau, K.H., Sanjurjo, A., 2003. Experimental thermochemistry of the SiCl and SiBr radicals: enthalpies of formation of species in the Si-Cl and Si-Br systems. J. Phys. Chem. A 107, 5448–5451.

Hildenbrand, D.L., Lau, K.H., Perez-Mariano, J., Sanjurjo, A., 2008. Thermochemistry of the gaseous vanadium chlorides VCl, VCl_2, VCl_3 and VCl_4. J. Phys. Chem. A 112, 9978–9982.

Hindiyarti, L., Frandsen, F., Livbjerg, H., Glarborg, P., Marshall, P., 2008. An exploratory study of alkali sulfate aerosol formation during biomass combustion. Fuel 87, 1591–1600.

Hino, M., Nagamori, M., Toguri, J.M., 1986. Thermodynamics of gaseous SbS. Metal. Trans. B 17, 913–914.

Hintelmann, H., 2010. Organomercurials: their formation and pathways in the environment. Chapter 11 In: Sigel, A., Sigel, H., Sigel, R.K.O. (Eds.), Metal Ions in Life Sciences: Volume 7. Organometallics in Environment and Toxicology. Royal Society of Chemistry, London, UK, pp. 365–401.

Hinton, C.S., Li, F., Armentrout, P.B., 2009. Reactions of Hf^+, Ta^+ and W^+ with O_2 and CO: metal carbide and metal oxide cation bond energies. Int. J. Mass Spectrom. 280, 226–234.

Hippe, D., Peyerimhoff, S.D., 1992. Theoretical spectroscopy of copper monoxide. Mol. Phys. 76 (2), 293–318.

Hirano, T., Bunker, P.R., Patchkovskii, S., Nagashima, U., Jensen, P., 2009. The predicted spectrum of FeOH in its Renner-degenerate X^6A' and A^6A'' electronic states. J. Mol. Spectrosc. 256, 45–52.

Hirano, T., Nagashima, U., Winnewisser, G., Jensen, P., 2010. Electronic structures and rovibronically averaged geometries of the X^6A' and A^6A'' states of FeOH. J. Chem. Phys. 132, 094303. 10 pp.

Hirano, T., Andaloussi, M.B.D., Nagashima, U., Jensen, P., 2014. Electronic structure and rovibrational properties of ZnOH in the X^2A' electronic state: a computational molecular spectroscopy study. J. Chem. Phys. 141, 094308. 14 pp.

Hirner, A.V., Rettenmeier, A.W., 2010. Methylated metal(loid) species in humans. Chapter 14 In: Sigel, A., Sigel, H., Sigel, R.K.O. (Eds.), Metal Ions in Life Sciences: Volume 7. Organometallics in Environment and Toxicology. Royal Society of Chemistry, London, UK, pp. 465–521.

Hirschler, M.M., 1986. Soot from fires. III. Soot suppression. J. Fire Sci. 4, 42–72.

Hisham, M.W.M., Benson, S.W., 1995. Thermochemistry of the Deacon Process. J. Phys. Chem. 99, 6194–6198.

Ho, P., Burns, R.P., 1980. A mass spectrometric study of the AlO_2 molecule. High Temp. Sci. 12, 31–39.

Ho, T.H., Pham-Tran, N.-N., Kawazoe, Y., Le, H.M., 2016. *Ab initio* investigation of O-H dissociation from the Al-OH_2 complex using molecular dynamics and neural network fitting. J. Phys. Chem. A 120, 346–355.

Hodges, P.J., Ross, A.J., Crozet, P., et al., 2007. On the spin-orbit splitting of $CuCl_2$ in its $^2\Pi_g$ ground state. J. Chem. Phys. 127, 024309. 9 pp.

Hofener, S., Ahlrichs, R., Knecht, S., Visscher, L., 2012. Relativistic and non-relativistic electronic molecular structure calculations for dimers of 4p-, 5p- and 6p-block elements. ChemPhysChem 13, 3952–3957.

Hoffman, C.W.W., Ropp, R.C., Mooney, R.W., 1959. Preparation, properties and structure of cadmium peroxide. J. Am. Chem. Soc. 81, 3830–3834.

Holstein, W.L., 1993. Thermodynamic properties of thallium(I) hydroxide in the vapor state. J. Chem. Thermodyn. 25, 1287–1294.

Holstein, W.L., 1993a. Thermodynamics of the volatilization of Tl_2O from Tl_2O, Tl_4O_3 and Tl_2O_3. J. Phys. Chem. 97, 4224–4230.

Hong, S., Lee, K., Hou, S., Hur, S.D., Ren, J., Burn, L.J., et al., 2009. An 800-year record of atmospheric As, Mo, Sn and Sb in Central Asia in high altitude ice cores from Mt. Qomolangma(Everest), Himalayas. Environ. Sci. Technol. 43, 8060–8065.

Hong, S., Soyol-Erdene, T.-O., Hwang, H.J., Hong, S.B., Hur, S.D., Motoyama, H., 2012. Evidence of Global-Scale As, Mo, Sb and Tl atmospheric pollution in Antarctic snow. Environ. Sci. Technol. 46, 11550–11557.

Hope, B.K., 1994. A global biogeochemical budget for vanadium. Sci. Total Environ. 141, 1–10.

Hope, B.K., 1997. An assessment of the global impact of anthropogenic vanadium. Biogeochemistry 37, 1–13.

Hope, B.K., 2008. A dynamic model for the global cycling of anthropogenic vanadium. Global Biogeochem. Cycles 22, GB4021, 16 pp.

Hore, P., Ahmed, M.S., Sedlar, S., Saper, R.B., et al., 2017. Blood lead level and potential risk factors for lead exposures among South Asians in New York City. J. Immigrant Minor. Health 19 (6), 1322–1329.

Horner, D.A., Allen, W.D., Csaszar, A.G., Schaefer III, H.F., 1991. The sodium superoxide radical: X $^2A_2$ and A $^2B_2$ potential energy surfaces. Chem. Phys. Lett. 186 (4/5), 346–355.

Horovitz, C.T., 2000. Biochemistry of Scandium and Yttrium. I. Physical and Chemical Fundamentals. II. Biochemistry and Applications. Kluwer Academic-Plenum Publishers, New York, vol. 1, 324 pp., vol. 2, 303 pp.

Horowitz, H.M., Jacob, D.J., Amos, H.M., Streets, D.G., Sunderland, E.M., 2014. Historical mercury releases from commercial products: global environmental implications. Environ. Sci. Technol. 48, 10242–10250.

Hoshyaripour, G.A., Hort, M., Langmann, B., 2015. Ash iron mobilization through physicochemical processing in volcanic eruption plumes: a numerical modeling approach. Atmos. Chem. Phys. 15, 9361–9379.

Hossain, K.F.B., Rahman, M.M., Sikder, M.T., Saito, T., et al., 2018. Inhibitory effects of selenium on cadmium-induced cytotoxicity in PC12 cells via regulating oxidative stress and apoptosis. Food Chem. Toxicol. 114, 180–189.

Hosseini, B.H., Weck, P.F., Sadeghpour, H.R., Kirby, K., Stancil, P.C., 2009. Photodissociation dynamics of lithium chloride: contribution of interferometric predissociation. J. Chem. Phys. 130, 054308. 5 pp.

Howard, J.B., Kausch Jr, W.J., 1980. Soot control by fuel additives. Prog. Energy Combust. Sci. 6, 263–276.

Hower, J.C., Robl, T.L., Anderson, C., Thomas, G.A., Sakulpitakphon, T., Mardon, S.M., et al., 2005. Characteristics of coal combustion products from Kentucky power plants, with emphasis on mercury content. Fuel 84, 1338–1350.

Hower, J.C., Senior, C.L., Suuberg, E.M., Hurt, R.H., Wilcox, J.L., Olson, E.S., 2010. Mercury capture by native fly ash carbons in coal-fired power plants. Prog. Energy Combust. Sci. 36, 510–529.

Hrusak, J., Koch, W., Schwarz, H., 1994. An *ab initio* molecular orbital study of the structures and energetics of the neutral and cationic CuO_2 and CuNO molecules in the gas phase. J. Chem. Phys. 101 (5), 3898–3905.

Hsi, H.-C., Rood, M.J., Rostam-Abadi, M., Chen, S., Chang, R., 2001. Effects of sulfur impregnation temperature on the properties and mercury adsorption capacities of activated carbon fibers. Environ. Sci. Technol. 35, 2785–2791.

Hsu, L.-J., Alwahabi, Z.T., Nathan, G.J., et al., 2011. Sodium and potassium released from burning particles of brown coal and pine wood in a laminar premixed methane flame using quantitative laser-induced breakdown spectroscopy. Appl. Spectrosc. 65 (6), 684–691.

Hu, X.F., Chan, H.M., 2018. Factors associated with the blood and urinary selenium concentrations in the Canadian population: results of the Canadian health measures survey (2007–2011). Int. J. Hyg. Environ. Health 221 (7), 1023–1031.

Hu, J., Zheng, B., Finkelman, R.B., Wang, B., Wang, M., Li, S., et al., 2006. Concentration and distribution of sixty-one elements in coals from DPR Korea. Fuel 85, 679–688.

Hu, H., Luo, G., Liu, H., Qiao, Y., Xu, M., Yao, H., 2013. Fate of chromium during thermal treatment of municipal solid waste incineration fly ash. Proc. Combust. Inst. 34, 2795–2801.

Hu, Z., Wang, X., Wang, Z., Wang, Y., Tan, H., 2014. Segmented kinetic investigation on condensed KCl sulfation in $SO_2/O_2/H_2O$ at 523-1023 K. Energy Fuels 28, 7560–7568.

Hu, H., Liu, H., Chen, J., et al., 2015. Speciation transformation of arsenic during municipal solid waste incineration. Proc. Combust. Inst. 35, 2883–2890.

Hu, H., Xu, Z., Liu, H., et al., 2015a. Mechanism of chromium oxidation by alkali and alkaline earth metals during municipal solid waste incineration. Proc. Combust. Inst. 35, 2397–2403.

Hu, G., Tang, Q., Jiang, D., 2016. CoP for hydrogen evolution: implications from hydrogen adsorption. PhysChemChemPhys 18, 23864–23871.

Hu, X., Chandler, J.D., Fernandes, J., Orr, M.L., et al., 2018. Selenium supplementation prevents metabolic and transcriptomic responses to cadmium in mouse lung. Biochim. Biophys. Acta-Gen. Top. 1862 (11), 2417–2426.

Hu, G., Liu, G., Wu, D., Fu, B., 2018a. Geochemical behavior of hazardous volatile elements in coals with different geological origin during combustion. Fuel 233, 363–376.

Hua, S., Tian, H., Wang, K., et al., 2016. Atmospheric emission inventory of hazardous air pollutants from China's cement plants. Temporal trends, spatial variation characteristics and scenario projections. Atmos. Environ. 128, 1–9.

Huang, J., Gustin, M.S., 2015. Uncertainties of gaseous oxidized mercury measurements using KCl-coated denuders, cation-exchange membranes and Nylon membranes: humidity influences. Environ. Sci. Technol. 49, 6102–6108.

Huang, M.-D., Meschi, D.J., Searcy, A.W., 1979. Thermodynamic properties of SeS gas. J. Electrochem. Soc. 126, 1825–1826.

Huang, Y., Jin, B., Zhong, Z., Xiao, R., Tang, Z., Ren, H., 2004. Trace elements (Mn, Cr, Pb, Se, Zn, Cd, Hg) in emissions from a pulverized coal boiler. Fuel Process. Technol. 86, 23–32.

Huang, Q., Yang, Y., Wang, Q., 2012. Potential for serious environmental threats from uncontrolled co-processing of wastes in cement kilns. Environ. Sci. Technol. 46, 13031–13032.

Huang, J., Kang, S., Zhang, Q., Guo, J., Chen, P., Zhang, G., et al., 2013. Atmospheric deposition of trace elements recorded in snow from the Mt. Nyainqentanglha region, Southern Tibet plateau. Chemosphere 92, 871–881.

Huang, L., Zhihua, T.F., Yu, C.H., Hopke, P.K., Lioy, P.J., Buckley, B.T., et al., 2013a. Interconversion of chromium species during air sampling: effects of O_3, NO_2, SO_2, particle matrices, temperature and humidity. Environ. Sci. Technol. 47, 4408–4415.

Huang, W.-J., Xu, H.-M., Qu, Z., et al., 2016. Significance of Fe_2O_3 modified SCR catalyst for gas phase elemental mercury oxidation in coal-fired flue gas. Fuel Process. Technol. 149, 23–28.

Huang, Y., Deng, M., Li, T., Japenga, J., et al., 2017. Anthropogenic mercury emissions from 1980 to 2012 in China. Environ. Pollut. 226, 230–239.

Huang, L., Li, B., Tam, N.F.Y., Wang, X., Ye, Z.H., 2018. Effects of environment and genotype on mercury and methylmercury accumulation in rice (*Oryza sativa* L.). Plant Soil 427 (1-2), 269–280.

Huang, L.S., Cory-Slechta, D.A., Cox, C., Thurston, S.W., et al., 2018a. Analysis of nonlinear associations between prenatal methylmercury exposure from fish consumption and neurodevelopmental outcomes in the Seychelles Maine Cohort at 17 years. Stochast. Environ. Res. Risk Assess. 32 (4), 893–904.

Huang, Y., Wang, L.Y., Wang, W.J., Li, T.Q., He, Z.L., Yang, X.E., 2019. Current status of agricultural soil pollution by heavy metals in China: a meta-analysis. Sci. Total Environ. 651, 3034–3042.

Huber, K.P., Herzberg, G., 1979. Molecular Spectra and Molecular Structure. IV. Constants of Diatomic Molecules. Van Nostrand Reinhold Company, New York.

Hubert, M., Olsen, J., Loras, J., Fleig, T., 2013. General active space commutator-based coupled cluster theory of general excitation rank for electronically excited states: implementation and application to ScH. J. Chem. Phys. 139, 194106. 11 pp.

Hubner, O., Himmel, H.-J., 2012. Cyclic and linear NiO_2: a multireference configuration interaction study. J. Phys. Chem. A 116, 9181–9188.

Hubner, O., Sauer, J., 2002. Structure and thermochemistry of $Fe_2S_2^{-/0/+}$ gas phase clusters and their fragments: B3LYP calculations. PhysChemChemPhys 4, 5234–5243.

Hubner, O., Hornung, J., Himmel, H.-J., 2015. The electronic structure of VO in its ground and electronically excited states: a combined matrix isolation and quantum chemical (MRCI) study. J. Chem. Phys. 143, 024309. 11 pp.

Hueber, D., Smith, B.W., Madden, S., Winefordner, J.D., 1994. Argon fluoride laser-excited atomic fluorescence of arsenic in a H_2/air flame and in an Ar ICP. Appl. Spectrosc. 48, 1213–1217.

Huffman, G.P., Huggins, F.E., Shah, N., Shah, A., 1990. Behavior of basic elements during coal combustion. Prog. Energy Combust. Sci. 16, 243–251.

Huffman, G.P., Huggins, F.E., Shah, N., Zhao, J., 1994. Speciation of arsenic and chromium in coal and combustion ash by XAFS spectroscopy. Fuel Process. Technol. 39, 47–62.

Huffman, G.P., Huggins, F.E., Shah, N., Huggins, R., Linak, W.P., Miller, C.A., et al., 2000. Characterization of fine particulate matter produced by combustion of residual fuel oil. J. Air Waste Manage. Assoc. 50, 1106–1114.

Huggins, F.E., Najih, M., Huffman, G.P., 1999. Direct speciation of chromium in coal combustion by-products by X-ray absorption fine-structure spectroscopy. Fuel 78, 233–242.

Huggins, F.E., Shah, N., Huffman, G.P., Kolker, A., Crowley, S., Palmer, C.A., et al., 2000. Mode of occurrence of chromium in four US coals. Fuel Process. Technol. 63, 79–92.

Huggins, F.E., Yap, N., Huffman, G.P., Senior, C.L., 2003. X-ray absorption Fine Structure characterization of mercury captured from combustion gases on sorbents at low temperatures. Fuel Process. Technol. 82, 167–196.

Huggins, F.E., HuffmanGP, Linak, W.P., Miller, C.A., 2004. Quantifying hazardous species in particulate matter derived from fossil fuel combustion. Environ. Sci. Technol. 38, 1836–1842.

Huggins, F.E., Senior, C.L., Chu, P., Ladwig, K., Huffman, G.P., 2007. Selenium and arsenic speciation in fly ash from full-scale coal-burning utility plants. Environ. Sci. Technol. 41, 3284–3289.

Huldt, L., Lagerqvist, A., 1950. The dissociation energies of the alkaline earth oxides. Ark. Fys. 2, 333–336.

Huldt, L., Lagerqvist, A., 1954. The dissociation energy of gaseous nickel oxide. NiO. Zeitschrift fur Naturforsch 9a, 358–359.

Hultgren, R., Desai, P.D., Hawkins, D.T., et al., 1973. Selected Values of Thermodynamic Properties of the Elements. American Society for Metals, Metals Park OH.

Human, H.G.C., Zeegers, P.J.T.H., 1975. Molecular fluorescence of CaOH, SrOH and BaCl in flames. Spectrochim. Acta B 30, 203–209.

Hung, H.-M., Hoffmann, M.R., 2015. Oxidation of gas phase SO_2 on the surfaces of acidic microdroplets: implications for sulfate and sulfate radical anion formation in the atmospheric liquid phase. Environ. Sci. Technol. 49, 13768–13776.

Huntelaar, M.E., Cordfunke, E.H.P., 1997. The standard molar enthalpies of formation of BaO(g) and SrO(g). J. Chem. Thermodyn. 29, 817–826.

Husain, D., Roberts, G., 1985. Collisional quenching of $Ca(4^3P_J)$ by H_2 and D_2 studied over the temperature range 850-1075 K by time-resolved atomic resonance emission at $\lambda = 657.3$ nm $[Ca(4\ ^3P_1 \rightarrow Ca(4\ ^1S_0) + h\upsilon]$ following pulsed dye-laser excitation. J. Chem. Soc. Faraday Trans. II 81, 527–538.

Husain, D., Roberts, G., 1986. Kinetic studies of the collisions with CO_2 of metastable states of Ca atoms generated by pulsed dye-laser excitation. Z. Phys. Chem. 150, 133–161.

Husain, D., Schifino, J., 1983. Kinetic study of $Ca(4^3P_1)$ by time-resolved emission, $4^3P_1 \rightarrow 4^1S_0 + h\upsilon$ ($\lambda = 657.3$ nm), following dye-laser excitation: spontaneous emission, diffusion and collisional quenching. J. Chem. Soc. Faraday Trans. II 79, 1265–1281.

Husain, D., Schifino, J., 1984. Collisional quenching of electronically excited strontium atoms $Sr(5\ ^3P_J)$, studied by time-resolved emission at $\lambda = 689.3$ nm ($5^3P_1 \rightarrow 5^1S_0 + h\upsilon$) following dye-laser excitation. J. Chem. Soc. Faraday Trans. II 80, 647–657.

Husain, D., Marshall, P., Plane, J.M.C., 1985. Temperature dependence of the absolute third-order rate constant for the reaction between $Na + O_2 + N_2$ over the range 571–1016 K studied by time-resolved atomic absorption spectroscopy. J. Chem. Soc. Faraday Trans. II 81, 301–318.

Hutson, N.D., Attwood, B.C., Scheckel, K.G., 2007. X-ray absorption spectroscopy and X-ray photoelectron spectroscopy characterization of mercury binding on brominated activated charcoal. Environ. Sci. Technol. 41, 1747–1752.
Huzinaga, S., Klobukowski, M., Barandiaran, Z., Seijo, L., 1986. A systematic study of basis set effects in *ab initio* SCF calculations: well-tempered GTF basis set calculations on transition metal halides. J. Chem. Phys. 84 (11), 6315–6327.
Hwang, D.-Y., Mebel, A.M., 2000. Theoretical study on the reaction mechanism of CO_2 with Mg. J. Phys. Chem. A 104, 7646–7650.
Hwang, S.M., Cooke, J.A., De Witt, K.J., Rabinowitz, M.J., 2010. Determination of the rate coefficients of the $SO_2 + O + M = SO_3 + M$ reaction. Int. J. Chem. Kinet. 42, 168–180.
Hylander, L.D., Meili, M., 2005. The rise and fall of mercury: converting a resource to refuse after 500 years of mining and pollution. Crit. Rev. Environ. Sci. Technol. 35, 1–36.
Hynes, A.J., Steinberg, M., Schofield, K., 1984. The chemical kinetics and thermodynamics of sodium species in oxygen-rich hydrogen flames. J. Chem. Phys. 80, 2585–2597.
Ichinose, H., Shiwa, Y., Nagano, M., 1994. Synthesis of $BaTiO_3/LaNiO_3$ and $PbTiO_3/LaNiO_3$ multilayer thin films by spray combustion flame technique. Jpn. J. Appl. Phys. 33, 5903–5906.
Ichinose, H., Shiwa, Y., Nagano, M., 1994a. Deposition of $LaMO_3$ (M = Ni, Co, Cr, Al) oriented films by spray combustion flame technique. Jpn. J. Appl. Phys. 33, 5907–5910.
Ignatyev, I.S., Montejo, M., Gonzalez, J.J.L., 2007. Structure and vibrational spectra of Ti(IV) hydroxides and their clusters with expanded titanium coordination: density functional theory study. J. Phys. Chem. A 111, 7973–7979.
Ikeda, S., Nakajima, T., Hirao, K., 2003. A theoretical study of transition metal hydroxides: CuOH, AgOH and AuOH. Mol. Phys. 101 (1/2), 105–110.
Il'in, M.K., Makarov, A.V., Nikitin, O.T., 1974. A study of the vaporization products of barium metaborate. Moscow Univ. Chem. Bull. 29 (4), 39–41.
Illas, F., Rubio, J., Barthelat, J.C., 1985. Cu as a one-electron atom: molecular structure and dissociation energy of CuOH. Chem. Phys. Lett. 119 (5), 397–402.
Illas, F., Rubio, J., Ricart, J.M., Cabot, P.L., 1986. Theoretical evidence for two geometrical isomers of AgO_2. J. Am. Chem. Soc. 108, 7893–7897.
Imperatori, P., Piacente, V., 1984. Vapor pressure of condensed HgSe from torsion-effusion measurements. J. Chem. Thermodyn. 16, 949–954.
Infante, I., Eliav, E., Vilkas, M.J., et al., 2007. A Fock space coupled cluster study on the electronic structure of the UO_2, UO_2^+, U^{4+} and U^{5+} species. J. Chem. Phys. 127, 124308,12 pp.
Ip, L.-T., Baxter, L.L., Mackrory, A.J., Tree, D.R., 2008. Surface temperature and time-dependent measurements of black liquor droplet combustion. AIChE J. 54 (7), 1926–1931.
ISCORS, 2005. Assessment of Radioactivity in Sewage Sludge: Recommendations on Management of Radioactive Materials in Sewage Sludge and Ash at Publicly Owned Treatment Works. Interagency Steering Committee on Radiation Standards Technical Report 2004-04, EPA 832-R-03-002B, 166 pp.
Ismail, M.A., Memon, N.K., Mansour, M.S., et al., 2015. Curved wall-jet burner for synthesizing titania and silica nanoparticles. Proc. Combust. Inst. 35, 2267–2274.
Ismail, M., Liu, W., Chan, M.S.C., et al., 2016. Synthesis, application and carbonation behavior of $Ca_2Fe_2O_5$ for chemical looping H_2 production. Energy Fuels 30 (8), 6220–6232.
Ivashentsev, Y.A.I., Kutakova, L.I., Ketov, A.N., 1968. Reaction of bismuth oxide with chlorine and hydrogen chloride. Vop. Khim. Khim. Tekhnol. 54–60. In Russian.
Iyer, V.S., Agarwal, R., Roy, K.N., Venkateswaran, S., Venugopal, V., Sood, D.D., 1990. Thermal properties of Na_2MoO_4(s) and $Na_2Mo_2O_7$(s) by high temperature Calvet calorimetry in the temperature range 335 to 760 K. J. Chem. Thermodyn. 22, 439–448.
Izbicki, J.A., Wright, M.T., Seymour, W.A., et al., 2015. Cr(VI) occurrence and geochemistry in water from public-supply wells in California. Appl. Geochem. 63, 203–217.
Izzaouihda, S., El Makarim, H.A., Komiha, N., et al., 2014. *Ab initio* potential energy curves of valence and Rydberg electronic states of the PO radical. Comput. Theor. Chem. 1049, 102–108.
Jackson, G.P., King, F.L., Goeringer, D.E., Duckworth, D.C., 2002. Collision-induced dissociation of lanthanide oxide ions in quadrupole ion traps: effects of bond strength and mass. Int. J. Mass Spectrom. 216, 85–93.

Jackson, V.E., Dixon, D.A., Christe, K.O., 2012. Thermochemical properties of selenium fluorides, oxides and oxofluorides. Inorg. Chem. 51, 2472–2485.

Jacoby, M., 2019. Chelator selectively grabs uranium. Chem. Eng. News 97 (8), 6.

Jacox, M.E., Milligan, D.E., 1969. Matrix isolation study of the infrared and ultraviolet spectra of several first series transition metal dichlorides. J. Chem. Phys. 51, 4143–4155.

Jadhav, R.A., Fan, L.-S., 2001. Capture of gas phase arsenic oxide by lime: kinetic and mechanistic studies. Environ. Sci. Technol. 35, 794–799.

Jadhav, R.A., Agnihotri, R., Gupta, H., Fan, L.-S., 2000. Mechanism of selenium sorption by activated carbon. Can. J. Chem. Eng. 78, 168–174.

Jaffe, D.A., Lyman, S., Amos, H.M., et al., 2014. Progress on understanding atmospheric mercury hampered by uncertain measurements. Environ. Sci. Technol. 48, 7204–7206.

Jager, T., 2013. All individuals are not created equal: accounting for individual variation in fitting life-history responses to toxicants. Environ. Sci. Technol. 47, 1664–1669.

Jain, S.K., Rout, C., Rastogi, R.C., 2000. Density functional study of the isomerization of MOH (M = Be and Mg). Chem. Phys. Lett. 331, 547–552.

Jaisankar, S., Patadiya, D.M., Sheshadri, T.S., 2017. Shockwave induced primary thermal fragmentation of coal particles. Combust. Explos. Shock Waves 53 (3), 329–339.

Jakob, A., Stucki, S., Kuhn, P., 1995. Evaporation of heavy metals during the heat treatment of municipal solid waste incineration fly ash. Environ. Sci. Technol. 29, 2429–2436.

Jakubek, Z.J., Nakhate, S.G., Simard, B., Balfour, W.J., 2002. Laser-induced fluorescence molecular beam investigation of new singlet and triplet electronic states of yttrium monohydride. J. Mol. Spectrosc. 211, 135–146.

James, W.D., Graham, C.C., Glascock, M.D., Hanna, A.-S.G., 1982. Water leachable boron coal ashes. Environ. Sci. Technol. 16, 195–197.

James, D.W., Krishnamoorthy, G., Benson, S.A., Seames, W.S., 2014. Modeling trace element partitioning during coal combustion. Fuel Process. Technol. 126, 284–297.

Jamorski, Ch, Dargelos, A., Teichteil, Ch, Daudey, J.P., 1994. Theoretical determination of spectral lines for the Zn atom and the ZnH molecule. J. Chem. Phys. 100 (2), 917–925.

Jang, H.-N., Seo, Y.-C., Lee, J.-H., Hwang, K.-W., Yoo, J.-I., Sok, C.-H., et al., 2007. Formation of fine particles enriched by V and Ni from heavy oil combustion: anthropogenic sources and drop-tube furnace experiments. Atmos. Environ. 41, 1053–1063.

Jano-Ito, M.A., Reed, G.P., Millan, M., 2014. Comparison of thermodynamic equilibrium predictions on trace element speciation in oxy-fuel and conventional coal combustion power plants. Energy Fuels 28, 4666–4683.

Janzen, C., Roth, P., 2001. Formation and characteristics of Fe_2O_3 nanoparticles in doped low pressure $H_2/O_2/Ar$ flames. Combust. Flame 125, 1150–1161.

Jarman, C.N., Fernando, W.T.M.L., Bernath, P.F., 1991. Laser spectroscopy of the A^1A'-X^1A' system of CuOH and CuOD. J. Mol. Spectrosc. 145, 151–163.

Jayasekher, T., 2009. Aerosols near by a coal-fired thermal power plant: chemical composition and toxic evaluation. Chemosphere 75, 1525–1530.

Jayaweera, T.M., Melius, C.F., Pitz, W.J., et al., 2005. Flame inhibition by phosphorus containing compounds over a range of equivalence ratios. Combust. Flame 140, 103–115.

Jenc, F., Brandt, B.A., 1985. The ground state reduced potential curves of the metal hydrides CuH and MgH. Spectrochim. Acta A 41 (10), 1211–1214.

Jenny, S.N., Ogden, J.S., 1979. Matrix isolation studies on alkali metal phosphates. I. The characterization of molecular sodium phosphate. J. Chem. Soc. Dalton Trans. 10, 1465–1467.

Jensen, D.E., 1968. Production of electrons from alkaline earths in flames: equilibrium and kinetic considerations. Combust. Flame 12, 261–268.

Jensen, D.E., 1969. Electron attachment and compound formation in flames. I. Electron affinity of BO_2 and heats of formation of alkali metal metaborates. Trans. Faraday Soc. 65, 2123–2132.

Jensen, D.E., Jones, G.A., 1972. Flame photometric determination of the standard enthalpies of formation of Al$(OH)_2$ and AlO. J. Chem. Soc. Faraday Trans. I 68, 259–268.

Jensen, D.E., Jones, G.A., 1973. Iron compounds in flames: relative stabilities of Fe, FeO, FeOH and $Fe(OH)_2$. J. Chem. Soc. Faraday Trans. I 69, 1448–1454.

Jensen, D.E., Jones, G.A., 1974. Catalysis of radical recombination in flames by iron. J. Chem. Phys. 60 (9), 3421–3425.
Jensen, D.E., Jones, G.A., 1975. Mass spectrometric tracer and photometric studies of catalyzed radical recombination in flames. J. Chem. Soc. Faraday Trans. I 71, 149–160.
Jensen, D.E., Jones, G.A., 1976. Aspects of the flame chemistry of cobalt. J. Chem. Soc. Faraday Trans. I 72, 2618–2630.
Jensen, D.E., Jones, G.A., 1978. Alkaline earth flame chemistry. Proc. R. Soc. Lond. A 364, 509–535.
Jensen, D.E., Miller, W.J., 1970a. Electron attachment and compound formation in flames. III. Negative ion and compound formation in flames containing tungsten and potassium. J. Chem. Phys. 53 (8), 3287–3292.
Jensen, D.E., Miller, W.J., 1970. Electron attachment and compound formation in flames. IV. Negative ion and compound formation in flames containing potassium and molybdenum. Proc. Combust. Inst. 13, 363–370.
Jensen, P.A., Stenholm, M., Hald, P., 1997. Deposition investigation in straw-fired boilers. Energy Fuels 11, 1048–1055.
Jensen, J.R., Nielsen, L.B., Schulz-Moller, C., Wedel, S., Livbjerg, H., 2000. The nucleation of aerosols in flue gases with a high content of alkali: a laboratory study. Aerosol Sci. Technol. 33, 490–509.
Jensen, P.A., Frandsen, F.J., Dam-Johansen, K., Sander, B., 2000a. Experimental investigation of the transformation and release to gas phase of potassium and chlorine during straw pyrolysis. Energy Fuels 14, 1280–1285.
Jensen, K.P., Roos, B.O., Ryde, U., 2007. Performance of density functionals for first row transition metal systems. J. Chem. Phys. 126, 014103. 14 pp.
Jeong, S., Lee, H., Kim, Y.-T., Yoon, H.-O., 2017. Development of a simultaneous analytical method to determine arsenic speciation using HPLC-ICP-MS: arsenate, arsenite, mono-methyl arsonic acid, dimethyl arsinic acid, dimethyldithio arsinic acid and dimethyl mono-thio arsinic acid. Microchem. J. 134, 295–300.
Jeung, G.H., Koutecky, J., 1988. Molecular bonding with scandium: diatomics ScH, ScO, ScC and ScN. J. Chem. Phys. 88, 3747–3760.
Jeung, G.-H., Luc, P., Vetter, R., Kim, K.H., Lee, Y.S., 2002. Experimental and theoretical study on the reaction Sc + NO = ScO + N. PhysChemChemPhys 4 (4), 596–600.
Jew, A.D., Rupp, E.C., Geatches, D.L., et al., 2015. Mercury interaction with the fine fraction of coal combustion fly ash in a simulated coal power plant flue gas stream. Energy Fuels 29, 6025–6038.
Jia, L., Anthony, E.J., Charland, J.P., 2002. Investigation of vanadium compounds in ashes from a CFBC firing 100% petroleum coke. Energy Fuels 16, 397–403.
Jia, X., Wang, Q., Cen, K., Chen, L., 2016. An experimental study of $CaSO_4$ decomposition during coal pyrolysis. Fuel 163, 157–165.
Jiang, H., 2017. Metal transporters in neurodegeneration. Chapter 16, pp. 313–347. In: White, A.R., Aschner, M., Costa, L.G., Bush, A.I. (Eds.), Biometals in Neurodegenerative Diseases. Academic Press, Elsevier, London, 415 pp..
Jiang, W., DeYonker, N.J., Determan, J.J., Wilson, A.K., 2012. Toward accurate theoretical thermochemistry of first row transition metal complexes. J. Phys. Chem. 116, 870–885.
Jiao, F., Cheng, Y., Zhang, L., Yamada, N., Sato, A., Ninomiya, Y., 2011. Effects of HCl, SO_2 and H_2O in flue gas on the condensation section of municipal solid waste incineration. Proc. Combust. Inst. 33, 2787–2793.
Jiao, F., Wijaya, N., Zhang, L., Ninomiya, Y., Hocking, R., 2011a. Synchroton based XANES speciation of chromium in the oxy-fuel fly ash collected from lab-scale drop-tube furnace. Environ. Sci. Technol. 45, 6640–6646.
Jiao, F., Zhang, L., Song, W., et al., 2013. Effect of inorganic particles on the condensation behavior of lead and zinc vapors upon flue gas cooling. Proc. Combust. Inst. 34, 2821–2829.
Jiao, F., Zhang, L., Yamada, N., et al., 2013a. Effect of HCl, SO_2 and H_2O on the condensation of heavy metal vapors in the flue gas cooling section. Fuel Process. Technol. 105, 181–187.
Jiao, F.C., Zhang, L., Dong, Z.B., et al., 2016. Study on the species of heavy metals in MSW incineration fly ash and their leaching behavior. Fuel Process. Technol. 152, 108–115.
Jimenez, S., Ballester, J., 2005. A comparative study of different methods for the sampling of high temperature combustion aerosols. Aerosol Sci. Technol. 39, 811–821.
Jimenez, S., Ballester, J., 2005a. Effect of co-firing on the properties of submicron aerosols from biomass combustion. Proc. Combust. Inst. 30, 2965–2972.
Jimenez, S., Perez, M., Ballester, J., 2008. Vaporization of trace elements and their emission with submicrometer aerosols in biomass combustion. Energy Fuels 22, 2270–2277.

Jin, J., Saqui, O., Allard, J.P., 2018. Effect of discontinuation of manganese supplementation from home parenteral nutrition solutions on whole-blood levels and magnetic resonance imaging of the brain: a five-year cohort study. J. Parenteral Enteral Nutr. 42 (1), 164–170.

Johansen, J.M., Jakobsen, J.G., Frandsen, F.J., Glarborg, P., 2011. Release of K, Cl, and S during pyrolysis and combustion of high-chlorine biomass. Energy Fuels 25, 4961–4971.

Johansson, L.S., Tullin, C., Leckner, B., Sjovall, P., 2003. Particle emissions from biomass combustion in small combustors. Biomass Bioenergy 25, 435–446.

Johansson, P., Dahlstrom, O., Dahlstrom, U., Alehagen, U., 2015. Improved health-related quality of life, and more days out of hospital with supplementation with selenium and coenzyme Q10 combined: results from a double blind, placebo-controlled prospective study. J. Nutr. Health Aging 19 (9), 870–877.

John, A.K., Savthri, S., Vijayalakshmi, K.P., Suresh, C.H., 2010. Density functional theory study of gas phase hydrolysis of titanium tetrachloride. Bull. Chem. Soc. Jpn. 83, 1030–1036.

Johnson, I., 1975. Mass spectrometric study of the vaporization of cesium and sodium molybdates. J. Phys. Chem. 79 (7), 722–726.

Johnson, V.S., White, R.J., Dann, S.E., Hellgardt, K., 2004. Structural investigation of the high temperature oxidation of bismuth sulfide using temperature programed oxidation TPO-MS and in-situ X-ray diffraction techniques. Ind. Eng. Chem. Res. 43 (12), 3127–3132.

Johnson, J., Schewel, L., Graedel, T.E., 2006. The contemporary anthropogenic chromium cycle. Environ. Sci. Technol. 40, 7060–7069.

Johnson, D.F., Bhaskaran-Nair, K., Bylaska, E.J., de Jong, W.A., 2013. Thermodynamics of tetravalent thorium and uranium complexes from first-principles calculations. J. Phys. Chem. A 117, 4988–4995.

Johnson, A.C., Jurgens, M.D., Lawler, A.J., et al., 2014. Particulate and colloidal silver in sewage effluent and sludge discharged from British wastewater treatment plants. Chemosphere 112, 49–55.

Johnson, E.L., Davis, Q.C., Morse, M.D., 2016. Predissociation measurements of bond dissociation energies: VC, VN and VS. J. Chem. Phys. 144, 234306. 9 pp.

Johri, K.N., Kaushik, N.K., Singh, K., 1968. Thermogravimetric analysis of arsenic (III,V) sulfides precipitated with potassium thiocarbonate. Curr. Sci. 37 (24), 699–700.

Joller, M., Brunner, T., Obernberger, I., 2007. Modeling of aerosol formation during biomass combustion for various furnace and boiler types. Fuel Process. Technol. 88, 1136–1147.

Jones, A.M., Harrison, R.M., 2016. Emission of ultrafine particles from the incineration of municipal solid waste: a review. Atmos. Environ. 140, 519–528.

Jones, A.C., Leedham, T.J., Wright, P.J., Crosbie, M.J., Lane, P.A., Williams, D.J., et al., 1998. Metal-organic chemical vapor deposition (MOCVD) of zirconia thin films by direct liquid injection using a new class of zirconium precursor. Chem. Vap. Depos. 4 (2), 46–49.

Jones, F., Tran, H., Lindberg, D., et al., 2013. Thermal stability of zinc compounds. Energy Fuels 27, 5663–5669.

Jones, F., Bankiewicz, D., Hupa, M., 2014. Occurrence and sources of zinc in fuels. Fuel 117, 763–775.

Jones, V., Gardner, M., Ellor, B., 2014a. Concentrations of trace substances in sewage sludge from 28 wastewater treatment works in the UK. Chemosphere 111, 478–484.

Jonsson, J., 1995. The $b^3\Phi$-$a^3\Delta$ and $d^3\Delta$- $a^3\Delta$ systems of ZrS. J. Mol. Spectrosc. 169, 18–29.

Jonsson, J., Lindgren, B., 1995. Infrared spectroscopy: the $b'^3\Pi$- $a^3\Delta$ system of ZrS. J. Mol. Spectrosc. 169, 30–37.

Jonsson, J., Wallin, S., Lindgren, B., 1998. On the singlet spectrum of ZrS. J. Mol. Spectrosc. 192, 198–214.

Jordan, C.A., Akay, G., 2012. Speciation and distribution of alkali, alkali earth metals and major ash forming elements during gasification of fuel cane bagasse. Fuel 91, 253–263.

Joubert, L., Maldivi, P., 2001. A structural and vibrational study of uranium(III) molecules by density functional methods. J. Phys. Chem. A 105, 9068–9076.

Jumpponen, M., Ronkkomaki, H., Pasanen, P., Laitinen, J., 2014. Occupational exposure to solid chemical agents in biomass-fired power plants and associated health effects. Chemosphere 104, 25–31.

Jung, C.H., Park, J.Y., Oh, S.J., Park, H.K., Kim, Y.S., Kim, D.K., et al., 1998. Synthesis of Li_2TiO_3 ceramic breeder powders by the combustion process. J. Nucl. Mater. 253, 203–212.

Juniper, D.T., Phipps, R.H., Ramos-Morales, E., Bertin, G., 2009. Effects of dietary supplementation with selenium enriched yeast or sodium selenite on selenium tissue distribution and meat quality in lambs. Anim. Feed. Sci. Technol. 149 (3–4), 228–239.

Kaakinen, J.W., Jorden, R.M., Lawasani, M.H., West, R.E., 1975. Trace element behavior in coal-fired power plant. Environ. Sci. Technol. 9, 862–869.

Kablov, E.N., Folomeikin, Yu.I., 2015. Mass spectrometric study of vaporization of high refractory ceramics. Dokl. Phys. Chem. 463 (1), 150–153.

Kadavathu, S.A., Scullman, R., Gray, J.A., Li, M., Field, R.W., 1990. New low-lying states in NiH: rotational analysis with the aid of laser-induced dispersed fluorescence spectroscopy. J. Mol. Spectrosc. 140, 126–140.

Kagi, E., Kawaguchi, K., 2006. Microwave spectroscopy of MgO in the $a^3\Pi$ and $X^1\Sigma^+$ states. J. Mol. Struct. 795, 179–184.

Kaledin, L.A., Heaven, M.C., 1997. Electronic spectroscopy of UO. J. Mol. Spectrosc. 185, 1–7.

Kaledin, L.A., McCord, J.E., Heaven, M.C., 1994. Laser spectroscopy of UO: characterization and assignment of states in the 0–3 eV range with a comparison to the electronic structure of ThO. J. Mol. Spectrosc. 164, 27–65.

Kaledin, L.A., McCord, J.E., Heaven, M.C., 1995. Laser spectroscopy of TiO: accurate term energies for the singlet states and ligand field assignment of states in the range 0 to 4 eV. J. Mol. Spectrosc. 173, 499–509.

Kaledin, L.A., McCord, J.E., Heaven, M.C., 1995a. Laser spectroscopy of ZrO: accurate term energies for the triplet states. J. Mol. Spectrosc. 174, 93–99.

Kalemos, A., 2017. The nature of the chemical bond in $BeO^{0,-}$, $BeOBe^{+,0,-}$, and in their hydrogenated products $HBeO^{0,-}$, BeOH, HBeOH, $BeOBeH^{+,0,-}$, and HBeOBeH. J. Chem. Phys. 146, 104307. 10 pp.

Kamat, P.V., 2011. Dominance of metal oxides in the era of nanotechnology. J. Phys. Chem. Lett. 2, 839–840.

Kana'an, A.S., McCreary, J.R., Peterson, D.E., Thorn, R.J., 1969. Enthalpy and entropy of sublimation of ferrous chloride: thermochemical properties of gaseous $FeCl_2$. High Temp. Sci. 1, 222–231.

Kang, H., Beauchamp, J.L., 1986. Gas phase studies of alkene oxidation by transition metal oxides: ion-beam studies of CrO^+. J. Am. Chem. Soc. 108, 5663–5668.

Kanninen, K.M., White, A.R., 2017. Abnormal function of metalloproteins underlies most neurodegenerative diseases. Chapter 21, 415-438 In: White, A.R., Aschner, M., Costa, L.G., Bush, A.I. (Eds.), Biometals in Neurodegenerative Diseases. Academic Press, Elsevier, London, 451 pp.

Kant, A., Moon, K.A., 1979. Dissociation energies of gaseous CuH, AgH, AuH and NiH. High Temp. Sci. 11, 55–62.

Kant, A., Moon, K.A., 1981. Mass spectrometric determination of the dissociation energies of gaseous AlH, GaH, InH, ScH and CoH and estimates of the maximum dissociation energies of TlH, CrH, MnH and FeH. High Temp. Sci. 14, 23–31.

Kantor, T., 1994. Atomization efficiencies in halocarbon-loaded acetylene-air flames. I. Monochlorides of readily volatilized elements. Spectrochim. Acta B 49, 1717–1732.

Kantor, T., Ernyei, L., 1994. Atomization efficiencies in halocarbon-loaded acetylene-air flames. II. Formation of dihalides and combined effects on vaporization and gas phase atomization processes. Spectrochim. Acta B 49, 1733–1744.

Karbowska, B., 2016. Presence of thallium in the environment: sources of contaminations, distribution and monitoring methods. Environ. Monitor Assess. 188 (11), 640.

Karimi, R., Silbernagel, S., Fisher, N.S., Meliker, J.R., 2014. Elevated blood Hg at recommended seafood consumption rates in adult seafood consumers. Int. J. Hyg. Environ. Health 217, 758–764.

Karlsson, S., Jonsson, T., Hall, J., et al., 2014. Mitigation of fireside corrosion of stainless steel in power plants: a laboratory study of the influences of SO_2 and KCl on initial stages of corrosion. Energy Fuels 28, 3102–3109.

Karmakar, A., Zhang, Q., Zhang, Y., 2014. Neurotoxicity of nanoscale materials. J. Food Drug Anal. 22 (1), 147–160.

Karna, S.P., Bruna, P.J., Grein, F., 1988. Configuration interaction studies on the electronic states and structure of PS. J. Phys. B: At., Mol. Opt. Phys. 21, 1303–1313.

Karthikeyan, J., Berndt, C.C., Tikkanen, J., Wang, J.Y., King, A.H., Herman, H., 1997. Nanomaterial powders and deposits prepared by flame spray processing of liquid precursors. Nanostruct. Mater. 8, 61–74.

Kasai, P.H., Jones, P.M., 1986. Cu, Ag and Au atom-molecular oxygen complexes: matrix isolation ESR study. J. Phys. Chem. 90, 4239–4245.

Kashireninov, O.E., Fontijn, A., 1998. Modeling of chromium combustion in incineration: thermochemistry of Cr-C-H-Cl combustion in air and selection of key reactions. Combust. Flame 113, 498–506.

Kashireninov, O.E., Chervonnyi, A.D., Piven, V.A., 1982. On the detection and thermodynamic characteristics of AlO_2. High Temp. Sci. 15, 79–91.

Kashiwakura, S., Kubo, H., Kumagai, Y., Kubo, H., Matsubae-Yokoyama, K., Nakajima, K., et al., 2009. Removal of boron from coal fly ash by washing with HCl solution. Fuel 88, 1245–1250.

Kashiwakura, S., Takahashi, T., Nagasaka, T., 2011. Vaporization behavior of boron from standard coals in the early stage of combustion. Fuel 90, 1408–1415.

Kaskan, W.E., Mackenzie, J.D., Millikan, R.C., 1961. Origin of the green bands in the boron-oxygen system. J. Chem. Phys. 34, 570–574.

Kaspari, S., Mayewski, P.A., Handley, M., et al., 2009. Recent increases in atmospheric concentrations of Bi, U, Cs, S and Ca from a 350 year Mount Everest ice core record. J Geophys Res D Atm 114 (4), 011088. 14 pp.

Kassman, H., Bafver, L., Amand, L.-E., 2010. The importance of SO_2 and SO_3 for sulfation of gaseous KCl: an experimental investigation in a biomass fired CFB boiler. Combust. Flame 157, 1649–1657.

Kassman, H., Brostrom, M., Berg, M., Amand, L.-E., 2011. Measures to reduce chlorine in deposits: application in a large-scale circulating fluidized bed boiler firing biomass. Fuel 90, 1325–1334.

Kassman, H., Normann, F., Amand, L.-E., 2013. The effect of oxygen and volatile combustibles on the sulfation of gaseous KCl. Combust. Flame 160, 2231–2241.

Kathirvel, P., Chandrasekaran, J., Manoharan, D., Kumar, S., 2014. Preparation and characterization of alpha alumina nanoparticles by in-flight oxidation of flame synthesis. J. Alloys Compds. 590, 341–345.

Kato, M., Azimi, M.D., Fayaz, S.H., et al., 2016. Uranium in well drinking water of Kabul, Afghanistan and its effective, low cost depuration using Mg-Fe based hydrotalcite-like compounds. Chemosphere 165, 27–32.

Katz, J.L., Hung, C.-H., 1990. Initial studies of electric field effects on ceramic powder formation in flames. Proc. Combust. Inst. 23, 1733–1738.

Katzer, M., Weber, A.P., Kasper, G., 2001. The effects of electric fields on growth of titania particles formed in a CH_4-O_2 diffusion flame. Aerosol Sci. 32, 1045–1067.

Kauffman, J.W., Hauge, R.H., Margrave, J.L., 1985. Reactions of atomic scandium, titanium and vanadium with molecular water at 15 K. J. Phys. Chem. 89, 3547–3552.

Kauffman, J.W., Hauge, R.H., Margrave, J.L., 1985a. Studies of reactions of atomic and diatomic Cr, Mn, Fe, Co, Ni, Cu and Zn with molecular water at 15 K. J. Phys. Chem. 89, 3541–3547.

Kaufmann, H., Nussbaumer, Th, Baxter, L., Yang, N., 2000. Deposit formation on a single cylinder during combustion of herbaceous biomass. Fuel 79, 141–151.

Kauppinen, E.I., Pakkanen, T.A., 1990. Coal combustion aerosols: a field study. Environ. Sci. Technol. 24, 1811–1818.

Kavlak, G., Graedel, T.E., 2013. Global anthropogenic selenium cycles for 1940–2010. Resourc. Conserv. Recycl. 73, 17–22.

Kazantis, G., 2000. Thallium in the environment and health effects. Environ. Geochem. Health 22, 275–280.

Kazenas, E.K., Samoilova, I.O., Zviadadze, G.N., 1983. Thermodynamics of the dissociation and sublimation of magnesium oxide. Russ. J. Phys. Chem. 57, 1571–1572.

Kazenas, E.K., Samoilova, I.O., Astakhova, G.K., Ovchinnikova, O.A., 1998. Mass spectrometric study of the thermodynamics of lithium molybdate evaporation. Metally (1), 34–36. In Russian.

Kazenas, E.K., Tsvetkov, Yu.V., Samoilova, I.O., et al., 2001. Mass spectrometric study of the thermodynamics of evaporation of sodium tungstate. Metally (4), 22–24. In Russian.

Kazenas, E.K., Tsvetkov, Yu.V., Samoilova, I.O., et al., 2002. Mass spectrometric study of the thermodynamics of evaporation of potassium molybdate. Metally (5), 33–35. In Russian.

Kazenas, E.K., Tsvetkov, Yu.V., Samoilova, I.O., et al., 2003. Mass spectrometric study of sublimation thermodynamics of barium molybdate. Metally (1), 14–17. In Russian.

Kazenas, E.K., Tsvetkov, Yu.V., Samoilova, I.O., et al., 2004. Thermodynamics of evaporation of magnesium molybdate. Metally (2), 83–85. In Russian.

Kazenas, E.K., Tsvetkov, Yu.V., Samoilova, I.O., et al., 2004a. Thermodynamics of evaporation of magnesium tungstate. Metally (4), 23–28. In Russian.

Kazenas, E.K., Tsvetkov, Yu.V., Samoilova, I.O., et al., 2005. Study of the thermodynamics of evaporation of calcium tungstate. Metally (3), 26–28. In Russian.

Kazenas, E.K., Tsvetkov, Yu.V., Astakhova, G.K., et al., 2007. Study of the thermodynamics of evaporation of strontium tungstate. Metally (2), 19–23. In Russian.

Kazenas, E.K., Tsvetkov, Yu.V., Samoilova, I.O., et al., 2008. Study of the thermodynamics of strontium molybdate sublimation. Metally (2), 16–19. In Russian.

Kazi, T.G., Brahman, K.D., Afridi, H.I., et al., 2016. The effects of arsenic contaminated drinking water of livestock on its total levels in milk samples of different cattle: risk assessment in children. Chemosphere 165, 427–433.

Kellogg, C.B., Irikura, K.K., 1999. Gas phase thermochemistry of iron oxides and hydroxides: portrait of a super-efficient flame suppressant. J. Phys. Chem. A 103, 1150–1159.

Kelly, R., Padley, P.J., 1971. Photometric studies in hydrogen + oxygen + carbon dioxide flames. I. Metal hydroxide dissociation energies. Trans. Faraday Soc. 67, 740–749.

Kelsall, B.J., Carlson, K.D., 1980. Matrix reactions of molecular oxygen with thallium atoms. J. Phys. Chem. 84, 951–959.

Kemsley, J., 2016. Lead pollution approaches natural background levels. Chem. Eng. News 94 (40), 11.

Keneshea, F.J., Cubicciotti, D., 1969. The thermodynamic properties of gaseous niobium chlorides. J. Phys. Chem. 73 (9), 3054–3062.

Kennedy, I.M., 2007. The health effects of combustion generated aerosols. Proc. Combust. Inst. 31, 2757–2770.

Kennedy, I.M., Zhang, Y., Jones, A.D., Chang, D.P.Y., Kelly, P.B., Yoon, Y., 1999. Morphology of chromium emissions from a laminar hydrogen diffusion flame. Combust. Flame 116, 233–242.

Khadri, F., Hochlaf, M., 2012. Theoretical calculations on the NaS and NaS^+ radicals: electronic structure, spectroscopy and spin-orbit couplings. Eur. Phys. J. D 66 (145), 14 pp.

Khalil, H., Brites, V., Le Quere, F., Leonard, C., 2011. *Ab initio* study of the low-lying electronic states of the CaO molecule. Chem. Phys. 386, 50–55.

Khalil, R.A., Houshfar, E., Musinguzi, W., Becidan, M., Skreiberg, O., Goile, F., et al., 2011a. Experimental investigation on corrosion abatement in straw combustion by fuel mixing. Energy Fuels 25, 2687–2695.

Khalil, H., Le Quere, F., Brites, V., Leonard, C., 2012. Theoretical study of the rovibronic states of CaO. J. Mol. Spectrosc. 271, 1–9.

Khalizov, A.F., Viswanathan, B., Larregaray, P., Ariya, P.A., 2003. A theoretical study on the reactions of Hg with halogens: atmospheric implications. J. Phys. Chem. A 107, 6360–6365.

Khandogin, Y., Alekseyev, A.B., Liebermann, H.-P., Hirsch, G., Buenker, R.J., 1997. *Ab initio* relativistic CI calculations of the spectroscopic constants and transition probabilities for the low-lying states of the BiOH/HBiO isomers. J. Mol. Spectrosc. 186, 22–33.

Kharat, B., Deshmukh, S.B., Chaudhari, A., 2009. 4d transition metal monoxides, monocarbides, monoborides, mononitrides and monofluorides: a quantum chemical study. Int. J. Quantum Chem. 109, 1103–1115.

Khardahakis, S., Mavridis, A., 2009. First-principles investigation of the early 3d transition metal diatomic chlorides and their ions, $ScCl^{0\pm}$, $TiCl^{0\pm}$, $VCl^{0\pm}$ and $CrCl^{0\pm}$. J. Phys. Chem. A 113, 6818–6840.

Khatib, M., Korek, M., 2017. Electronic structure of SrO molecules with dipole moments and ro-vibrational calculations. Can. J. Phys. 95, 610–620.

Khilla, M.A., Hanafi, Z.M., Mohamed, A.K., 1982. Physicochemical properties of chromium trioxide and its suboxides. Thermochim. Acta 59, 139–147.

Kickel, B.L., Armentrout, P.B., 1995. Guided ion-beam studies of the reactions of Mn^+, Cu^+ and Zn^+ with silane. M^+-SiH_x (x = 0-3) bond energies. J. Phys. Chem. 99, 2024–2032.

Kickel, B.L., Armentrout, P.B., 1995a. Reactions of Fe^+, Co^+ and Ni^+ with silane: electronic state effects, comparison to reactions with methane, and M^+-SiH_x (x = 0-3) bond energies. J. Am. Chem. Soc. 117, 764–773.

Kieber, R.J., Willey, J.D., Zvalaren, S.D., 2002. Chromium speciation in rainwater: temporal variability and atmospheric deposition. Environ. Sci. Technol. 36, 5321–5327.

Kilian, A., Morse, T.F., 2001. A novel aerosol combustion process for the high rate formation of nanoscale oxide particles. Aerosol Sci. Technol. 34, 227–235.

Kim, Y.-W., Belton, G.R., 1974. The thermodynamics of volatilization of chromic oxide. I. The species CrO_3 and $CrO_2(OH)$. Metall. Trans. 5, 1811–1816.

Kim, S.-J., Crawford, T.D., 2004. Hartree-Fock orbital instabilities and symmetry-breaking in ScO_2: is a C_s equilibrium structure viable? J. Phys. Chem. A 108, 3097–3102.

Kim, A.G., Hesbach, P., 2009. Comparison of fly ash leaching methods. Fuel 88, 926–937.

Kim, K., Kim, H.K., 2018. Characterization of products from $Fe(CO)_5$ seeded CO diffusion flame. Energy 148, 802–808.

Kim, J.H., Li, X., Wang, L.-S., et al., 2001. Vibrationally resolved photoelectron spectroscopy of MgO^- and ZnO^- and the low-lying electronic states of MgO, MgO^- and ZnO. J. Phys. Chem. A 105, 5709–5718.

Kim, J.H., Gibb, H.J., Howe, P.D., 2006. Cobalt and inorganic cobalt compounds. Concise International Chemical Assessment Document 69. World Health Organization.

Kim, K.B., Masiello, K.A., Hahn, D.W., 2008. Reduction of soot emissions by iron pentacarbonyl in iso-octane diffusion flames. Combust. Flame 154, 164–180.

Kim, Y., Yoon, Y., Shin, D., 2009. Fabrication of Sn/SnO_2 composite powder for anode of lithium ion battery by aerosol flame deposition. J. Anal. Appl. Pyrol. 85, 557–560.

Kim, J., Ihee, H., Lee, Y.S., 2010. Spin-orbit density functional and *ab initio* study of HgX_n (X = F, Cl, Br, I; n = 1,2,4). J. Chem. Phys. 133, 144309. 7 pp.

Kim, J.B., Weichman, M.L., Neumark, D.M., 2015. Low-lying states of FeO and FeO^- by slow photoelectron spectroscopy. Mol. Phys. 113 (15–16), 2105–2114.

Kim, H., Jung, B.M., Lee, B.N., Kim, Y.J., et al., 2017. Retinol, α-tocopherol, and selected minerals in breast milk of lactating women with full-term infants in South Korea. Nutr. Res. Pract. 11 (1), 64–69.

Kimbrough, D.E., Cohen, Y., Winer, A.M., Creelman, L., Mabuni, C., 1999. A critical assessment of chromium in the environment. Crit. Rev. Environ. Sci. Technol. 29, 1–46.

Kitagawa, K., Yanagisawa, M., Takeuchi, T., 1974. Spectroscopic determination of atomization efficiency ($CuCl \rightarrow Cu + Cl$) in an air-hydrogen flame. Anal. Chim. Acta 68, 212–216.

Kitchens, M.J.R., Fortenberry, R.C., 2016. The rovibrational nature of closed-shell third-row triatomics: HOX and HXO, X = Si^+, P, S^+ and Cl. Chem. Phys. 472, 119–127.

Kitto, M.E., 1993. Trace element patterns in fuel oils and gasolines for use in source apportionment. J. Air Waste Manage. Assoc. 43, 1381–1388.

Klanberg, F., Spandau, H., 1958. The thermal behavior of thallium sulfide. Naturwissenschaften 45, 209–210.

Klawohn, S., Kaupp, M., Karton, A., 2018. MVO-10: a gas phase oxide benchmark for localization/delocalization in mixed valence systems. J. Chem. Theory Comput. 14, 3512–3523.

Klein, D.H., Andren, A.W., Carter, J.A., Emery, J.F., Feldman, C., Fulkerson, W., et al., 1975. Pathways of thirty-seven trace elements through coal-fired power plant. Environ. Sci. Technol. 9, 973–979.

Kleps, A.I., Battiston, G.A., Gerbasi, R., Guerriero, P., Podaru, C., 1997. Electrical and structural characterization of SnO_2 atmospheric pressure chemical vapor deposition thin films. Proc. Electrochem. Soc. 97–25, 788–795.

Klika, Z., Bartonova, L., Spears, D.A., 2001. Effect of boiler output on trace element partitioning during coal combustion in two fluidized-bed power stations. Fuel 80, 907–917.

Klissurski, D.G., Subrt, J., Blaskov, V.N., Lipka, J., Hanousek, P., Bechine, K., 1984. Effect of the partial pressure of water vapor on the thermal decomposition of γ-FeOOH and on the dispersity of the solid decomposition products. J. Mater. Sci. 19, 183–188.

Kluge, S., Wiggers, H., Schulz, C., 2017. Mass spectrometric analysis of clusters and nanoparticles during the gas phase synthesis of tungsten oxide. Proc. Combust. Inst. 36 (1), 1037–1044.

Knacke, O., von Richthofen, A., 1994. Oligomerization of PbO vapor. Z. Phys. Chem. 187, 257–264.

Knacke, O., Kubaschewski, O., Hesselmann, K. (Eds.), 1991. second ed. Thermochemical Properties of Inorganic Substances, vols. 1 and 2. Springer-Verlag, Berlin, pp. 1–1113. , 1114–2412.

Knight Jr, L.B., Wise, M.B., Fisher, T.A., Steadman, J., 1981. The high temperature generation and ESR investigation of ScH_2, YH_2 and YD_2 radicals in rare gas matrices: a nonequilibrium vaporization technique for solid samples. J. Chem. Phys. 74, 6636–6643.

Knight Jr, L.B., Jones, G.C., King, G.M., et al., 1995. Electron spin resonance and theoretical studies of the PO_2 and AsO_2 radicals in neon matrices at 4 K: laser vaporization and X-irradiation radical generation techniques. J. Chem. Phys. 103 (2), 497–505.

Knight Jr, L.B., Babb, R., Ray, M., Banisaukas III, T.J., Russon, L., Dailey, R.S., et al., 1996. An electron spin resonance investigation of vanadium dioxide ($^{51}V^{16}O_2$ and $^{51}V^{17}O_2$) and $^{51}V^{17}O$ in neon matrices with preliminary assignments for VO_3 and V_2^+: comparison with *ab initio* theoretical calculations. J. Chem. Phys. 105, 10237–10250.

Knudsen, J.N., Jensen, P.A., Dam-Johansen, K., 2004. Transformation and release to the gas phase of Cl, K, and S during combustion of annual biomass. Energy Fuels 18, 1385–1399.

Kolker, A., Huggins, F.E., Palmer, C.A., et al., 2000. Mode of occurrence of arsenic in four US coals. Fuel Process. Technol. 63, 167–178.

Kong, J., Boyd, R.J., 1996. The $^2\Sigma^+$ states of HBeO, HMgO and HCaO. J. Chem. Phys. 104, 4055–4060.

Konings, R.J.M., Benes, O., 2010. The thermodynamic properties of the *f*-elements and their compounds. I. The lanthanide and actinide metals. J. Phys. Chem. Ref. Data 39 (4), 043102. 47 pp.

Konings, R.J.M., Cordfunke, E.H.P., Booij, A.S., 1992. The infrared spectra of gaseous phosphorus oxide (P_4O_{10}) and arsenic oxide (As_4O_6 and As_4O_{10}). J. Mol. Spectrosc. 152 (1), 29–37.

Konings, R.J.M., Benes, O., Kovacs, A., et al., 2014. The thermodynamic properties of the *f*-elements and their compounds. II. The lanthanide and actinide oxides. J. Phys. Chem. Ref. Data 43 (1), 013101. 95 pp.

Konnov, S.A., Serebrennikov, L.V., Mal'tsev, A.A., 1981. Infrared spectra of the products of the interaction of bismuth atoms with oxygen in an argon matrix. Russ. J. Phys. Chem. 55 (11), 1644–1646.

Konttinen, J., Backman, R., Hupa, M., Moilanen, A., Kurkela, E., 2013. Trace element behavior in the fluidized bed gasification of solid recovered fuels: a thermodynamic study. Fuel 106, 621–631.

Koput, J., 2005. *Ab initio* prediction of the potential energy surface and vibrational-rotational energy levels of calcium dihydride, CaH_2. J. Phys. Chem. A 109, 4410–4414.

Korbecki, J., Baranowska-Bosiacka, I., Gutowska, I., Chlubek, D., 2012. Biochemical and medical importance of vanadium compounds. Acta Biochim. Polon. 59 (2), 195–200.

Kordis, J., Gingerich, K.A., 1973. Atomization energy and standard heat of formation of gaseous diatomic arsenic. J. Chem. Eng. Data 18 (2), 135–136.

Kordis, J., Gingerich, K.A., 1973a. Mass spectrometric investigation of the equilibrium dissociation of gaseous Sb_2, Sb_3, Sb_4, SbP, SbP_3 and P_2. J. Chem. Phys. 58 (11), 5141–5149.

Korobeinichev, O.P., Chernov, A.A., Bolshova, T.A., 2000. The chemistry of the destruction of organophosphorus compounds in flames. IV. Destruction of diisopropyl-methylphosphonate in a flame of $H_2/O_2/Ar$. Combust. Flame 123, 412–420.

Korobeinichev, O.P., Ilyin, S.B., Bolshova, T.A., et al., 2000a. The chemistry of destruction of organophosphorus compounds in flames. III. The destruction of dimethyl methyl phosphonate (DMMP) and trimethyl phosphate (TMP) in a flame of hydrogen and oxygen. Combust. Flame 121, 593–609.

Korobeinichev, O.P., Shvartsberg, V.M., Shmakov, A.G., et al., 2005. Flame inhibition by phosphorus-containing compounds in lean and rich propane flames. Proc. Combust. Inst. 30, 2353–2360.

Korobeinichev, O.P., Shvartsberg, V.M., Shmakov, A.G., 2007. The chemistry of combustion of organophosphorus compounds. Russ. Chem. Rev. 76, 1094–1121.

Korobeinichev, O.P., Rybitskaya, I.V., Shmakov, A.G., Chernov, A.A., Bolshova, T.A., Shvartsberg, V.M., 2009. Inhibition of atmospheric pressure $H_2/O_2/N_2$ flames by trimethylphosphate over a range of equivalence ratio. Proc. Combust. Inst. 32, 2591–2597.

Korobeinichev, O.P., Shmakov, A.G., Maksyutov, R.A., Tereshchenko, A.G., Knyazkov, D.A., Bolshova, T.A., et al., 2012. Synthesis of mesoporous nanocrystalline TiO_2 films in a premixed $H_2/O_2/Ar$ flame. Combust. Explos. Shock Waves 48, 49–56.

Korsgen, H., Urban, W., Brown, J.M., 1999. The infrared spectrum of FeH_2, studied in the gas phase by laser magnetic resonance. J. Chem. Phys. 110, 3861–3869.

Korzun, E.A., Heck, H.H., 1990. Sources and fates of lead and cadmium in municipal solid waste. J. Air Waste Manage. Assoc. 40, 1220–1226.

Koseki, S., Ishihara, Y., Umeda, H., Fedorov, D.G., Gordon, M.S., 2002. Dissociation potential curves of low-lying states in transition metal hydrides. I. Hydrides of group IV. J. Phys. Chem. A 106, 785–794.

Koseki, S., Ishihara, Y., Fedorov, D.G., Umeda, H., Schmidt, M.W., Gordon, M.S., 2004. Dissociation potential curves of low-lying states in transition metal hydrides. II. Hydrides of groups 3 and 5. J. Phys. Chem. A 108, 4707–4719.

Koseki, S., Matsushita, T., Gordon, M.S., 2006. Dissociation potential curves of low-lying states in transition metal hydrides. III. Hydrides of groups 6 and 7. J. Phys. Chem. A 110, 2560–2570.

Kosik-Bogacka, D., Lanocha-Arendarczyk, N., Kot, K., Malinowski, W., et al., 2018. Concentrations of mercury and selenium in afterbirth and their relations with various factors. Environ. Geochem. Health 40 (4), 1683–1695.

Kosminski, A., Ross, D.P., Agnew, J.B., 2006. Influence of gas environment on reactions between sodium and silicon minerals during gasification of low-rank coal. Fuel Process. Technol. 87, 953–962.

Kosminski, A., Ross, D.P., Agnew, J.B., 2006a. Reactions between sodium and kaolin during gasification of a low-rank coal. Fuel Process. Technol. 87, 1051–1062.

Kosminski, A., Ross, D.P., Agnew, J.B., 2006b. Reactions between sodium and silica during gasification of a low-rank coal. Fuel Process. Technol. 87, 1037–1049.

Kosminski, A., Ross, D.P., Agnew, J.B., 2006c. Transformations of sodium during gasification of low-rank coal. Fuel Process. Technol. 87, 943–952.

Kostelov, V.V., Grebenshchikov, V.A., Bukhbinder, A.I., Baimakov, A.Yu, 1979. Vapor pressure and heat of vaporization of tin monoxide. J. Appl. Chem. USSR 52, 1882–1886.

Kosugi, T., 1972. Measurements of the vapor and dissociation pressures of potassium sulfate and carbonate at high temperatures. Bull. Chem. Soc. Jpn. 45, 15–19.

Kou, T., Asano, M., 1987. Mass spectrometric study of vaporization in the SrO-B_2O_3 system. High Temp. Sci. 24, 1–19.

Koutsogianni, A., Tsamakis, D., 2001. Tin oxide atmospheric pressure chemical vapor deposition thin films grown by $SnCl_4$ oxidation on glass and Si substrates in a cold wall reactor. J. Phys. IV 11 (3), 377–383.

Kovacs, A., Konings, R.J.M., Varga, Z., Szieberth, D., 2013. Structure and other molecular properties of actinide trichlorides $AnCl_3$ (An = Th-Cm). J. Phys. Chem. A 117, 11357–11363.

Kovacs, A., Konings, R.J.M., Gibson, J.K., et al., 2015. Quantum chemical calculations and experimental investigations of molecular actinide oxides. Chem. Rev. 115, 1725–1759.

Kovacs, H., Dobo, Z., Koos, T., Gyimesi, A., Nagy, G., 2018. Influence of the flue gas temperature on the behavior of metals during biomass combustion. Energy Fuels 32, 7851–7856.

Krachler, M., Zheng, J., Fisher, D., Shotyk, W., 2008. Atmospheric inputs of Ag and Tl to the Arctic: comparison of a high resolution snow pit (AD 1994–2004) with a firn (AD 1860–1996) and an ice core (previous 16,000 years). Sci. Total Environ. 399, 78–89.

Kramb, J., Konttinen, J., Backman, R., et al., 2016. Elimination of arsenic-containing emissions from gasification of chromated copper arsenate wood. Fuel 181, 319–324.

Kretzschmar, I., Schroder, D., Schwarz, H., Armentrout, P.B., 2001. The binding in neutral and cationic 3d and 4d transition-metal monoxides and -sulfides. Adv. Metal Semiconduct. Clusters 5, 347–395.

Krishnakumar, B., Helble, J.J., 2007. Understanding mercury transformations in coal-fired power plants: evaluation of homogeneous Hg oxidation mechanisms. Environ. Sci. Technol. 41, 7870–7875.

Kronberg, B.I., Brown, J.R., Fyfe, W.S., Peirce, M., Winder, C.G., 1981. Distributions of trace elements in Western Canadian coal ashes. Fuel 60, 59–63.

Kruger, O., Grabner, A., Adam, C., 2014. Complete survey of German sewage sludge ash. Environ. Sci. Technol. 48, 11811–11818.

Krzyzynska, R., Hutson, N.D., Zhao, Y., Szeliga, Z., Regucki, P., 2018. Mercury removal and its fate in oxidant enhanced wet flue gas desulfurization slurry. Fuel 211, 876–882.

Kubaschewski, O., Alcock, C.B., Spencer, P.J., 1993. Materials Thermochemistry, sixth ed. Pergamon Press, Oxford, 363 pp.

Kudo, T., Gordon, M.S., 1995. Molecular and electronic structure of TiH_2. J. Chem. Phys. 102, 6806–6811.

Kulmala, M., Petaja, T., Ehn, M., et al., 2014. Chemistry of atmospheric nucleation: on the recent advances on precursor characterization and atmospheric cluster composition in connection with atmospheric new particle formation. Ann. Rev. Phys. Chem. 65, 21–37.

Kung, S.C., 2014. Further understanding of furnace wall corrosion in coal-fired boilers. Corrosion 70 (7), 749–763.

Kunkel, K., Milke, E., Binnewies, M., 2014. Formation of ternary lead-molybdenum oxides $PbMoO_4$, $PbMo_2O_7$, Pb_2MoO_5, $PbMo_3O_{10}$ in the gas phase: a mass spectrometric quantum chemical investigation. Int. J. Mass Spectrom. 374, 12–19.

Kurihara, L.K., Suib, S.L., 1993. Sol-gel synthesis of ternary metal oxides. I. Synthesis and characterization of MAl_2O_4 (M = Mg, Ni, Co, Cu, Fe, Zn, Mn, Cd, Ca, Hg, Sr, Ba) and $Pb_2Al_2O_5$. Chem. Mater. 5, 609–613.

Kurosaki, Y., Yokoyama, K., 2012. *Ab initio* MRSDCI study on the low-lying electronic states of the lithium chloride molecule, LiCl. J. Chem. Phys. 137, 064305(1–9).

Kushawaha, V.S., 1973. Dissociation energy of AsCl. Spectrosc. Lett. 6 (10), 647–650.

Kushawaha, V.S., 1975. Mechanism of excitation of AsO in the reaction of $AsCl_3$ + O_2. Chem. Phys. Lett. 30, 130–136.

Kuwae, M., Tsugeki, N.K., Agusa, T., et al., 2013. Sedimentary records of metal deposition in Japanese alpine lakes for the last 250 years: recent enrichment of airborne Sb and In in East Asia. Sci. Total Environ. 442, 189–197.

Kyi, S., Chadwick, B.L., 1999. Screening of potential mineral additives for use as fouling preventatives in Victorian brown coal combustion. Fuel 78, 845–855.

Lachas, H., Richaud, R., Jarvis, K.E., Herod, A.A., Dugwell, D.R., Kandiyoti, R., 1999. Determination of 17 trace elements in coal and ash reference materials by ICP-MS applied to milligram sample sizes. Analyst 124, 177–184.

Ladele, J.I., Fajolu, I.B., Ezeaka, V.C., 2019. Determination of lead levels in maternal and umbilical cord blood at birth at the Lagos university teaching hospital, Lagos. PLoS One 14 (2), e211535.

Lagerqvist, A., Huldt, L., 1954. The dissociation energy of CaO, SrO and BaO. Z. Naturforsch A 9, 991–992.

Lam, J., Amans, D., Dujardin, C., et al., 2015. Atomistic mechanisms for the nucleation of aluminum oxide nanoparticles. J. Phys. Chem. A 119, 8944–8949.

Lamborg, C.H., Fitzgerald, W.F., Damman, A.W.H., Benoit, J.M., Balcom, P.H., Engstrom, D.R., 2002. Modern and historic atmospheric mercury fluxes in both hemispheres: global and regional mercury cycling implications. Glob. Biogeochem. Cycles 16, 51. 11 pp.

Lamborg, C.H., Hammerschmidt, C.R., Bowman, K.L., et al., 2014. A global ocean inventory of anthropogenic mercury based on water column measurements. Nature 512, 65–68.

Lamoreaux, R.H., Hildenbrand, D.L., 1984. High temperature vaporization behavior of oxides. I. Alkali metal binary oxides. J. Phys. Chem. Ref. Data 13, 151–173.

Lamoreaux, R.H., Hildenbrand, D.L., Brewer, L., 1987. High temperature vaporization behavior of oxides. II. Oxides of Be, Mg, Ca, Sr, Ba, B, Al, Ga, In, Tl, Si, Ge, Sn, Pb, Zn, Cd and Hg. J. Phys. Chem. Ref. Data 16, 419–443.

Landner, L., Reuther, R., 2004. Metals in society and in the environment: a critical review of current knowledge on fluxes, speciation, bioavailability and risk for adverse effects of copper, chromium, nickel and zinc 406 pp. Environ. Pollut. Ser., vol. 8. Kluwer Academic, Dordrecht.

Langhoff, S.R., Bauschlicher Jr, C.W., 1988. Theoretical studies of the monoxides and monosulfides of Y, Zr and Nb. J. Chem. Phys. 89, 2160–2169.

Langhoff, S.R., Bauschlicher Jr, C.W., 1990. Theoretical study of the spectroscopy of ZrO. Astrophys. J. 349, 369–375.

Langhoff, S.R., Bauschlicher Jr, C.W., Partridge, H., 1986. Theoretical study of the diatomic alkali and alkaline earth oxides. J. Chem. Phys. 84, 4474–4480.

Langhoff, S.R., Pettersson, L.G.M., Bauschlicher Jr, C.W., Partridge, H., 1987. Theoretical spectroscopic parameters for the low-lying states of the second-row transition metal hydrides. J. Chem. Phys. 86, 268–278.

Langhoff, S.R., Bauschlicher Jr, C.W., Partridge, H., 1988. Theoretical study of the scandium and yttrium halides. J. Chem. Phys. 89, 396–407. Erratum 7649.

Langhoff, S.R., Bauschlicher Jr, C.W., Pettersson, L.G.M., Siegbahn, P.E.M., 1989. Theoretical spectroscopic constants for the low-lying states of the oxides and sulfides of Mo and Tc. Chem. Phys. 132, 49–58.

Langhoff, S.R., Bauschlicher Jr, C.W., Rendell, A.P., 1989a. The spectroscopy of MnH. J. Mol. Spectrosc. 138, 108–122.

Lanzerstorfer, C., 2015. Chemical composition and physical properties of filter fly ashes from eight grate-fired biomass combustion plants. J. Environ. Sci.-China 30, 191–197.

Lanzerstorfer, C., 2018. Fly ash from coal combustion: dependence of the concentration of various elements on the particle size. Fuel 228, 263–271.

Larbi, T., Khadri, F., Ghalila, H., Lahmar, S., Hochlaf, M., 2010. Theoretical spectroscopy and metastability of BeS and its cation. Chem. Phys. 373, 193–202.

Larson, E.D., Consonni, S., Kreutz, T.G., 2000. Preliminary economics of black liquor gasifier/gas turbine cogeneration at pulp and paper mills. ASME J. Eng. Gas Turbines Power 122, 255–261.

Larue, C., Khodja, H., Herlin-Boime, N., Brisset, F., Flank, A.M., Fayard, B., et al., 2011. Investigation of titanium dioxide nanoparticles toxicity and uptake by plants. J. Phys. Conf. Ser. 304, 012057. 7 pp.

Lassesson, H., Steenari, B.-M., 2013. Speciation of copper in ash from a fluidized bed boiler fired with municipal solid waste. Energy Fuels 27, 3891–3897.

Latifzadeh-Masoudipour, L., Balasubramanian, K., 1997. Geometries and energy separations of electronic states of $AsCl_2$, $AsBr_2$, $AsCl_2^+$ and $AsBr_2^+$. J. Chem. Phys. 106, 2695–2701.

Latosinska, J., Gawdzik, J., 2012. Effect of incineration temperature on the mobility of heavy metals in sewage sludge ash. Environ. Protect. Eng. 38 (3), 31–44.

Lau, K.H., Hildenbrand, D.L., 1984. Thermochemical studies of the gaseous uranium chlorides. J. Chem. Phys. 80 (3), 1312–1317.

Lau, K.H., Cubicciotti, D., Hildenbrand, D.L., 1977. Effusion studies of the thermal decomposition of magnesium and calcium sulfates. J. Chem. Phys. 66, 4532–4539.

Lau, K.H., Cubicciotti, D., Hildenbrand, D.L., 1979. Effusion studies of the vaporization/decomposition of potassium sulfate. J. Electrochem. Soc. 126, 490–495.

Lau, K.H., Brittain, R.D., Hildenbrand, D.L., 1981. The dissociation energy of AsO from gaseous equilibrium measurements. Chem. Phys. Lett. 81, 227–229.

Lau, K.H., Brittain, R.D., Hildenbrand, D.L., 1982. Vaporization of arsenic trisulfide and the dissociation energy of arsenic monosulfide. J. Phys. Chem. 86 (22), 4429–4432.

Lau, K.H., Brittain, R.D., Lamoreaux, R.H., Hildenbrand, D.L., 1985. Studies of the vaporization/decomposition of alkali sulfates. J. Electrochem. Soc. 132, 3041–3048.

Laudal, D.L., Pavlish, J.H., Graves, J., Stockdill, D., 2000. Mercury mass balances: a case study of two North Dakota power plants. J. Air Waste Manage. Assoc. 50, 1798–1804.

Lauer, N.E., Hower, J.C., Hsu-Kim, H., Taggart, R.K., Vengosh, A., 2015. Naturally occurring radioactive materials in coals and coal combustion residues in the United States. Environ. Sci. Technol. 49, 11227–11233.

Lauer, N.E., Vengosh, A., Dai, S., 2017. Naturally occurring radioactive materials in uranium-rich coals and associated coal combustion residues from China. Environ. Sci. Technol. 51, 13487–13493.

Lauwerys, R., Lison, D., 1994. Health risks associated with cobalt exposure: an overview. Sci. Total Environ. 150, 1–6.

Lawrence, G.D., Patel, K.S., Nusbaum, A., 2014. Uranium toxicity and chelation therapy. Pure Appl. Chem. 86 (7), 1105–1110.

Lazarev, V.B., Greenberg, J.H., Shevchenko, V.J., Marenkin, S.F., Kozlov, S.E., 1976. Sublimation thermodynamics of Cd_3P_2. J. Chem. Thermodyn. 8, 61–65.

Le, A., Steimle, T.C., Gupta, V., Rice, C.A., Maier, J.P., Lin, S.H., et al., 2011. The visible spectrum of zirconium dioxide, ZrO_2. J. Chem. Phys. 135, 104303. 11 pp.

Le, T.T.Y., Peijnenburg, W.J.G.M., 2017. Modeling toxicity of metal mixtures: a generalization of new advanced methods, considering potential application to terrestrial ecosystems. Crit. Rev. Environ. Sci. Technol. 47 (7), 409–454.

Lee, E.P.F., Wright, T.G., 2002. The ionization energy of $KO_2(X^2A_2)$ and dissociation energies of KO_2 and KO_2^+. Chem. Phys. Lett. 363, 139–144.

Lee, E.P.F., Wright, T.G., 2004. Spectroscopy and thermodynamics of LiS/NaS($X^2\Pi$ and $A^2\Sigma^+$) and LiS^+/NaS^+($X^3\Sigma^-$ and $A^3\Pi$). Chem. Phys. Lett. 397, 194–199.

Lee, E.P.F., Wright, T.G., 2005. Heavier alkali metal monosulfides (KS, RbS, CsS and FrS) and their cations. J. Chem. Phys. 123, 144309-1-8.

Lee, S.H.D., Teats, F.G., Swift, W.M., Banerjee, D.D., 1992. Alkali vapor emission from PFBC of Illinois coals. Combust. Sci. Technol. 86, 327–336.

Lee, E.P.F., Soldan, P., Wright, T.G., 1999. The heat of formation of $NaO_2^+(X^3\Sigma^-)$ and $NaO_2(X^2A_2)$. Chem. Phys. Lett. 301, 317–324.

Lee, J.M., Baker, J.J., Rolle, J.G., Lierena, R., 1999a. Comparison of fuel properties of petroleum cokes and coals used in power generation. Am. Chem. Soc. Symp. Div. Fuel Chem. 44 (1), 80–89. www.anl.gov/PCS/acsfuel.

Lee, C.M., Davis, K.A., Heap, M.P., Eddings, E., Sarofim, A., 2000. Modeling the vaporization of ash constituents in a coal fired boiler. Proc. Combust. Inst. 28, 2375–2382.

Lee, E.P.F., Soldan, P., Wright, T.G., 2002. What is the ground electronic state of KO? J. Chem. Phys. 117 (18), 8241–8247.

Lee, M.-H., Cho, K., Shah, A.P., Biswas, P., 2005. Nanostructured sorbents for capture of cadmium species in combustion environments. Environ. Sci. Technol. 39, 8481–8489.

Lee, E.P.F., Dyke, J.M., Chau, F.-T., et al., 2006. An *ab initio* study on the ground state and low-lying doublet electronic states of SbO_2. J. Chem. Phys. 125, 064307, 12 pp.

Lee, E.P.F., Dyke, J.M., Mok, D.K.W., Chau, F.-T., Chow, W., 2007. *Ab initio* calculations on low-lying electronic states of SbO_2 and Franck-Condon simulation of its photodetachment spectrum. J. Chem. Phys. 127, 094306-1-10.

Lee, E.P.F., Mok, D.K.W., Chau, F.-T., Dyke, J.M., 2007a. *Ab initio* calculations on the X^1A' and A^1A'' states of HPO and Franck-Condon simulation of the single vibronic level emission spectra of HPO and DPO. J. Chem. Phys. 127, 214305. 12 pp.

Lee, C.W., Serre, S.D., Zhao, Y., Lee, S.J., Hastings, T.W., 2008. Mercury oxidation promoted by a selective catalytic reduction catalyst under simulated Powder River Basin coal combustion conditions. J. Air Waste Manage. Assoc. 58, 484–493.

Lee, E.P.F., Dyke, J.M., Mok, D.K.W., Chau, F.-T., 2008a. An *ab initio* study of the low-lying electronic states of YO_2 and Franck-Condon simulation of the first photodetachment band of YO_2^-. J. Phys. Chem. A 112, 4511–4520.

Lee, C.-S., Lutcavage, M.E., Chandler, E., et al., 2016. Declining mercury concentrations in Bluefin tuna reflect reduced emissions to the North Atlantic Ocean. Environ. Sci. Technol. 50, 12825–12830.

Lee, J.J., Valeri, L., Kapur, K., Hasan, M.O.S.I., et al., 2018. Growth parameters at birth mediate the relationship between prenatal manganese exposure and cognitive test scores among a cohort of 2-to-3 year old Bangladeshi children. Int. J. Epidemiol. 47 (4), 1169–1179.

Lee, M.-J., Chou, M.-C., Chou, W.-J., Huang, C.-W., et al., 2018a. Heavy metals'effect on the susceptibility to attention-deficit/hyperactivity disorder: implication of lead, cadmium, and antimony. Int. J. Environ. Res. Public Health 15 (6), 1221.

Leffler, T., Brackmann, C., Alden, M., Li, Z., 2017. Laser-induced photofragmentation fluorescence imaging of alkali compounds in flames. Appl. Spectrosc. 71 (6), 1289–1299.

Legge, F.S., Nyberg, G.L., Peel, J.B., 2001. DFT calculations for Cu-, Ag-, and Au-containing molecules. J. Phys. Chem. A 105 (33), 7905–7916.

Lehman, R.L., Gentry, J.S., Glumac, N.G., 1998. Thermal stability of potassium carbonate near its melting point. Thermochim. Acta 316, 1–9.

Lehmusto, J., Yrjas, P., Skrifvars, B.-J., Hupa, M., 2012. High temperature corrosion of superheater steels by KCl and K_2CO_3 under dry and wet conditions. Fuel Process. Technol. 104, 253–264.

Leiding, J., Woon, D.E., Dunning Jr, T.H., 2011. Bonding and isomerism in $SF_{n-1}Cl$ (n = 1-6): a quantum chemical study. J. Phys. Chem. A 115, 329–341.

Leininger, T., Jeung, G.-H., 1995. *Ab initio* calculation of ro-vibronic transition spectra of CaH. J. Chem. Phys. 103 (10), 3942–3949.

Leist, M., Casey, R.J., Caridi, D., 2000. The management of arsenic wastes: problems and prospects. J. Hazard. Mater. B 76, 125–138.

Lemly, A.D., 2014. An urgent need for an EPA standard for disposal of coal ash. Environ. Pollut. 191, 253–255.

Leonyuk, N.I., Pashkova, A.V., Gokhman, L.Z., 1977. Volatility and thermal decomposition of molten potassium trimolybdate. Russ. J. Inorg. Chem. 22, 1175–1177.

Leppa, J., Anttila, T., Kerminen, V.-M., Kulmala, M., Lehtinen, K.E.J., 2011. Atmospheric new particle formation: real and apparent growth of neutral and charged particles. Atmos. Chem. Phys. 11, 4939–4955.

Levy, J.B., Hargittai, M., 2000. Unusual dimer structures of the heavier alkaline earth dihalides: a density functional study. J. Phys. Chem. A 104, 1950–1958.

Lewis, G., Myers, C.E., 1963. Vaporization properties of iron phosphides. J. Phys. Chem. 67, 1289–1292.

Li, R., Balfour, W.J., 2005. Transition metal monophosphides: electronic spectra of TiP, VP and CrP. Chem. Phys. Lett. 403, 102–106.

Li, S., Wei, X., 2011. Behavior of alkali metal hydroxides/chlorides for NO reduction in a biomass reburning process. Energy Fuels 25, 3465–3475.

Li, S., Van Zee, R.J., Weltner Jr, W., et al., 1997. Magneto-infrared spectra of matrix-isolated NiH and NiH_2 molecules and theoretical calculations of the lowest electronic states of NiH_2. J. Chem. Phys. 106 (6), 2055–2059.

Li, H., Focsa, C., Pinchemel, B., Le Roy, R.J., Bernath, P.F., 2000. Fourier transform spectroscopy of BaO: new ground state constants for the $A^1\Sigma^+$-$X^1\Sigma^+$ chemiluminescence. J. Chem. Phys. 113, 3026–3033.

Li, C., Cornett, J., Ungar, K., 2003. Long-term decrease of cadmium concentrations in the Canadian Arctic air. Geophys. Res. Lett. 30 (5), 1256 (4 pp.).

Li, Y., Libermann, H.-P., Buenker, R.J., Pichl, L., 2004. A coupled treatment of $^1\Sigma^+$ and $^3\Pi$ states of the AgH molecule. Chem. Phys. Lett. 389, 101–107.

Li, H., Xie, D., Guo, H., 2005. A reliable *ab initio* potential energy surface and vibrational states for the ground electronic state of $HgH_2(X^1\Sigma_g^+)$. J. Chem. Phys. 122, 144314. 7 pp.

Li, Y., Li, J., Jin, Y., Cheng, S., Wu, Y., Gao, J., 2005a. Study on alkali metal vapor removal for high temperature cleaning of coal gas. Energy Fuels 19, 1606–1610. 2583.

Li, Y., Tong, H., Zhuo, Y., Chen, C., Xu, X., 2006. Simultaneous removal of SO_2 and trace SeO_2 from flue gas: effect of product layer on mass transfer. Environ. Sci. Technol. 40, 4306–4311.

Li, Y., Tong, H., Zhuo, Y., Wang, S., Xu, X., 2006a. Simultaneous removal of SO_2 and trace SeO_2 from flue gas: effect of SO_2 on selenium capture and kinetics study. Environ. Sci. Technol. 40 (24), 7919–7924.

Li, S., Cheng, C.-M., Chen, B., Cao, Y., Vervynckt, J., Adebambo, A., et al., 2007. Investigation of the relationship between particulate-bound mercury and properties of fly ash in a full-scale 100 MWe pulverized coal combustion boiler. Energy Fuels 21, 3292–3299.

Li, Y., Tong, H., Zhuo, Y., et al., 2007a. Simultaneous removal of SO_2 and trace As_2O_3 from flue gas: mechanism, kinetic study and effect of main gases on arsenic capture. Environ. Sci. Technol. 41, 2894–2900.

Li, Y., Daukoru, M., Suriyawong, A., Biswas, P., 2009. Mercury emissions control in coal combustion systems using potassium iodide: bench-scale and pilot-scale studies. Energy Fuels 23, 236–243.

Li, L., Yu, C., Huang, F., Bai, J., Fang, M., Luo, Z., 2012. Study on the deposits derived from a biomass circulating fluidized bed boiler. Energy Fuels 26, 6008–6014.

Li, S.H., Xiao, T.F., Zheng, B.S., 2012a. Medical geology of arsenic, selenium and thallium in China. Sci. Total Environ. 421, 31–40.

Li, W.-Z., Geng, F.-F., Pei, Y.-W., Cheng, J.-B., Li, Q.-Z., Gong, B.-A., 2012b. CASPT2 study on low lying states of HMgO and HOMg. Int. J. Quantum Chem. 112, 1209–1214.

Li, B., Sun, Z., Li, Z., Alden, M., Jakobsen, J.G., Hansen, S., et al., 2013. Post-flame gas phase sulfation of potassium chloride. Combust. Flame 160, 959–969.

Li, H., Feng, H., Sun, W., et al., 2013a. Diatomic silylynes, germylynes, stannylynes and plumbylynes: structure, dipole moments, dissociation energies and quartet-doublet gaps of EH and EX (E = Si, Ge, Sn, Pb; X = F, Cl, Br, I). Inorg. Chem. 52, 6849–6859.

Li, L., Ren, Q., Li, S., Lu, Q., 2013b. Effect of phosphorus on the behavior of potassium during the co-combustion of wheat straw with municipal sewage sludge. Energy Fuels 27, 5923–5930.

Li, Y., Gao, X., Wu, H., 2013c. Ash cenosphere from solid fuels combustion. II. Significant role of ash cenosphere fragmentation in ash and particle matter formation. Energy Fuels 27, 822–829.

Li, Y., Gao, X., Wu, H., 2013d. Further investigation into the formation mechanism of ash cenospheres from an Australian coal fired power station. Energy Fuels 27, 811–815.

Li, R.D., Kai, X.P., Yang, T.H., et al., 2014. Release and transformation of alkali metals during co-combustion of coal and sulfur-rich wheat straw. Energy Convers. Manage. 83, 197–202.

Li, J., Cen, D., Huang, D., Li, X., et al., 2014a. Detection and analysis of 12 heavy metals in blood and hair samples from the general population of Pearl River Delta area. Cell. Biochem. Biophys. 70, 1663–1669.

Li, W., Wang, L., Qiao, Y., et al., 2015. Effect of atmosphere on the release behavior of alkali and alkaline earth metals during oxy-fuel combustion. Fuel 139, 164–170.

Li, X., Li, J., Peng, Y., et al., 2015a. Regeneration of commercial SCR catalysts: probing the existing forms of arsenic oxide. Environ. Sci. Technol. 49, 9971–9978.

Li, J., Zhao, J., Guo, S., et al., 2016. Predicting the vanadium speciation during petroleum coke gasification by thermodynamic equilibrium calculation. Fuel 176, 48–53.

Li, X., Dong, H., Chen, J., Lu, C., et al., 2018a. Transformation of organically bound chromium during oxy-coal combustion: the influence of steam and mineral. Energy Fuels 32, 1992–1998.

Li, W., Ma, Z., Huang, Q.-X., Jiang, X., 2018b. Distribution and leaching characteristics of heavy metals in a hazardous waste incinerator. Fuel 233, 427–441.

Li, X.Y., Li, Z.G., Lin, C.J., Bi, X.Y., et al., 2018c. Health risks of heavy metal exposure through vegetable consumption near a large-scale Pb/Zn smelter in central China. Ecotoxicol. Environ. Safety 161, 99–110.

Li, H., Zhao, J., Zhang, W., Yang, J., et al., 2018d. NH3 inhibits mercury oxidation over low-temperature MnOx/TiO_2 SCR catalyst. Fuel Process. Technol. 176, 124–130.

Li, C.F., Duan, Y.F., Tang, H.J., Zhu, C., et al., 2018e. Study on the mercury emission and migration characteristics in a coal-fired power plant in China with an ammonia desulfurization process. Fuel 211, 621–628.

Li, X., Huang, Q., Luo, C., Zhou, D., et al., 2018f. Effect of acid gases on elemental mercury removal in an oxy-fuel CO_2 compression process. Energy Fuels 32, 4334–4340.

Li, H., Li, H., Li, Y., Liu, Y., Zhao, Z., 2018g. Blood mercury, arsenic, cadmium, and lead in children with autism spectrum disorder. Biol. Trace Elem. Res. 181, 31–37.

Li, W., Jiang, Y., Jin, Y., Wang, L., Xu, W., 2019. Experimental study of the influence of dimethyl- methylphosphonate on methane/air coflow diffusion flames using OH-PLIF. Fuel 235, 39–44.

Li, J., Wang, C., Pan, Z., Jiang, Z., et al., 2019a. Analysis of 210Pb and 210Po emissions from coal-fired power plants. Fuel 236, 278–283.

Li, J., Cen, Y.Y., Li, Y., 2019b. The research advances in the mechanism of manganese-218 eight the same as to 18 a system journal induced neurotoxicity. Toxin Rev. 38 (1), 54–60.

Liang, B., Andrews, L., 2002. Infrared spectra and density functional theory calculations of group IV transition metal sulfides. J. Phys. Chem. A 106, 6295–6301.

Liang, B., Andrews, L., 2002a. Infrared spectra and density functional theory calculations of Group V transition metal sulfides. J. Phys. Chem. A 106, 3738–3743.

Liang, B., Andrews, L., 2002b. Infrared spectra and density functional theory calculations of group VI transition metal sulfides in solid argon. J. Phys. Chem. A 106, 6945–6951.

Liang, B., Wang, X., Andrews, L., 2009. Infrared spectra and density functional calculations of group 8 transition metal sulfide molecules. J. Phys. Chem. A 113, 5375–5384.

Liang, B., Wang, X., Andrews, L., 2009a. Infrared spectra and density functional theory calculations of group 10 transition metal sulfide molecules and complexes. J. Phys. Chem. A 113, 3336–3343.

Liang, H., Wu, W.L., Zhang, Y.H., Zhou, S.J., et al., 2018. Levels, temporal trend and health risk assessment of five heavy metals and fresh vegetables marketed in Guangdong province of China during 2014–2017. Food Control 92, 107–120.

Liao, M.-S., Zhang, Q.-E., Schwarz, W.H.E., 1995. Properties and stabilities of MX, MX_2 and M_2X_2 compounds (M = Zn, Cd, Hg; X = F, Cl, Br, I). Inorg. Chem. 34, 5597–5605.

Liao, W., Wang, G., Li, K.M., Zhao, W.B., Wu, Y., 2019. Effective cooking on speciation and in vitro bioaccessibility of mercury and arsenic from rice, using ordinary and pressure cookers. Biol. Trace Elem. Res. 187 (1), 329–339.

Liaquat, L., Sadir, S., Batool, Z., Tabassum, S., et al., 2019. Acute aluminum chloride toxicity revisited: study on DNA damage and histopathological, biochemical and neurochemical alterations in rat brain. Life Sci. 217, 202–211.

Lias, S.G., Bartmess, J.E., Liebman, J.F., et al., 1988. Gas phase ion and neutral thermochemistry. J. Phys. Chem. Ref. Data 17 (Suppl. 1), 861 pp.

Lin, C.-C., Chen, S.-J., Huang, K.-L., Hwang, W.-I., Chang-Chien, G.-P., Lin, W.-Y., 2005. Characteristics of metals in nano/ultrafine/fine/coarse particles collected beside a heavily trafficked road. Environ. Sci. Technol. 39, 8113–8122.

Lin, C.-L., Wey, M.-Y., Yu, W.-J., 2005a. Emission characteristics of organic and heavy metal pollutants in fluidized bed incineration during the agglomeration/defluidization process. Combust. Flame 143, 139–149.

Lin, Z.Q., Chen, X., Xi, Z., Lin, S.B., et al., 2018. Individual heavy metal exposure and birth outcomes in Shenqiu County along the Huai River basin in China. Toxicol. Res. 7 (3), 444–453.

Linak, W.P., Wendt, J.O.L., 1993. Toxic metal emissions from incineration: mechanisms and control. Prog. Energy Combust. Sci. 19, 145–185.

Linak, W.P., Wendt, J.O.L., 1998. Partitioning of the refractory metals, nickel and chromium, in combustion systems. Combust. Sci. Technol. 134, 291–314.

Linak, W.P., Srivastava, R.K., Wendt, J.O.L., 1995. Sorbent capture of nickel, lead and cadmium in a laboratory swirl flame incinerator. Combust. Flame 100, 241–250.

Linak, W.P., Ryan, J.V., Wendt, J.O.L., 1996. Formation and destruction of hexavalent chromium in a laboratory swirl flame incinerator. Combust. Sci. Technol. 116/117, 479–498.

Linak, W.P., Miller, C.A., Wendt, J.O.L., 2000. Comparison of particle size distributions and elemental partitioning from the combustion of pulverized coal and residual fuel oil. J. Air Waste Manage. Assoc. 50, 1532–1544.

Linak, W.P., Miller, C.A., Wendt, J.O.L., 2000a. Fine particle emissions from residual fuel oil combustion: characterization and mechanisms of formation. Proc. Combust. Inst. 28, 2651–2658.

Linak, W.P., Miller, C.A., Seames, W.S., Wendt, J.O.L., Ishinomori, T., Endo, Y., et al., 2002. On trimodal particle size distributions in fly ash from pulverized coal combustion. Proc. Combust. Inst. 29, 441–447.

Linak, W.P., Yoo, J.-I., Wasson, S.J., Zhu, W., Wendt, J.O.L., Huggins, F.E., et al., 2007. Ultrafine ash aerosols from coal combustion: characterization and health effects. Proc. Combust. Inst. 31, 1929–1937.

Lind, T., Kauppinen, E.I., Jokiniemi, J.K., Maenhaut, W., 1994. A field study on the trace metal behavior in atmospheric circulating fluidized-bed coal combustion. Proc. Combust. Inst. 25, 201–209.

Lind, T., Hokkinen, J., Jokiniemi, J.K., Saarikoski, S., Hillamo, R., 2003. Electrostatic precipitator collection efficiency and trace element emissions from co-combustion of biomass and recovered fuel in fluidized-bed combustion. Environ. Sci. Technol. 37, 2842–2846.

Lind, T., Kauppinen, E.I., Hokkinen, J., Jokiniemi, J.K., Orjala, M., Aurela, M., et al., 2006. Effect of chlorine and sulfur on fine particle formation in pilot-scale CFBC of biomass. Energy Fuels 20, 61–68.

Lind, T., Hokkinen, J., Jokiniemi, J.K., 2007. Fine particle and trace element emissions from waste combustion: comparison of fluidized bed and grate firing. Fuel Process. Technol. 88, 737–746.

Lindackers, D., Janzen, C., Rellinghaus, B., Wassermann, E.F., Roth, P., 1998. Synthesis of Al_2O_3 and SnO_2 particles by oxidation of metalorganic precursors in premixed $H_2/O_2/Ar$ low pressure flames. Nanostruct. Mater. 10, 1247–1270.

Lindberg, D., Backman, R., Hupa, M., 2007. Thermodynamic modeling of the stability and melting properties of sodium borates relevant to black liquor combustion and gasification. Int. J. Mater. Res. 98 (10), 1012–1018.

Lindquist, B.A., Takeshita, T.Y., Woon, D.E., Dunning Jr, T.H., 2013. Bonding in sulfur-oxygen compounds HSO/SOH and SOO/OSO: an example of recoupled pair π bonding. J. Chem. Theor. Comput. 9, 4444–4452.

Ling, L.-X., Fan, M.-H., Wang, B.-J., Zhang, R.-G., 2015. Application of computational chemistry in understanding the mechanisms of mercury removal technologies: a review. Energy Environ. Sci. 8 (11), 3109–3133.

Lingott, R.M., Liebermann, H.-P., Alekseyev, A.B., Buenker, R.J., 1999. Electronic states and transitions of bismuth sulfide. J. Chem. Phys. 110 (23), 11294–11302.

Linn, S.H., Brom Jr, J.M., Tzeng, W.-B., Ng, C.Y., 1983. Molecular beam photoionization study of $HgCl_2$. J. Chem. Phys. 78 (1), 37–45.

Linteris, G.T., Knyazev, V.D., Babushok, V.I., 2002. Inhibition of premixed methane flames by manganese and tin compounds. Combust. Flame 129, 221–238.

Linteris, G.T., Katta, V.R., Takahashi, F., 2004. Experimental and numerical evaluation of metallic compounds for suppressing cup-burner flames. Combust. Flame 138, 78–96.

Linteris, G.T., Rumminger, M.D., Babushok, V.I., 2008. Catalytic inhibition of laminar flames by transition metal compounds. Prog. Energy Combust. Sci. 34, 288–329.

Linton, C., Simard, B., Loock, H.P., Wallin, S., Rothschopf, G.K., Gunion, R.F., et al., 1999. Rydberg and pulsed field ionization-zero electron kinetic energy spectra of YO. J. Chem. Phys. 111, 5017–5026.

Littlejohn, F., Grant, C.S., Saez, A.E., 2000. Mechanisms for the removal of calcium phosphate deposits in turbulent flow. Ind. Eng. Chem. Res. 39, 933–942.

Liu, M.B., Wahlbeck, P.G., 1974. Knudsen effusion and mass spectrometric studies of the vaporization of $Y_2O_3(s)$: dissociation energy of YO(g). High Temp. Sci. 6, 179–189.

Liu, F., Li, F.-X., Armentrout, P.B., 2005. Guided ion-beam studies of the reactions of Co^{+}_{n} (n = 2-20) with O_2: cobalt cluster-oxide and -dioxide bond energies. J. Chem. Phys. 123, 064304. 15 pp.

Liu, S., Wang, Y., Yu, L., Oakey, J., 2006. Volatilization of mercury, arsenic and selenium during underground coal gasification. Fuel 85, 1550–1558.

Liu, S.-H., Yan, N.-Q., Liu, Z.-R., Qu, Z., Wang, H.P., Chang, S.-G., et al., 2007. Using bromine gas to enhance mercury removal from flue gas of coal-fired power plants. Environ. Sci. Technol. 41, 1405–1412.

Liu, X., Xu, M., Yao, H., Yu, D., Gao, X., Cao, Q., et al., 2007a. Effect of combustion parameters on the emission and chemical composition of particulate matter during coal combustion. Energy Fuels 21, 157–162.

Liu, J., Hu, Y., Gu, F., Li, C., 2009. Flame synthesis of ball-in-shell structured TiO_2 nanospheres. Ind. Eng. Chem. Res. 48, 735–739.

Liu, J., Falcoz, Q., Gauthier, D., Flamant, G., Zheng, C.Z., 2010. Volatilization behavior of Cd and Zn based on continuous emission measurements of flue gas from laboratory-scale coal combustion. Chemosphere 80, 241–247.

Liu, X., Smith, M.E., Tse, S.D., 2010a. In situ Raman characterization of nanoparticle aerosols during flame synthesis. Appl. Phys. B 100, 643–653.

Liu, Y., Zhang, X., Yu, K., 2012. *Ab initio* calculation on the low-lying excited states of BCl radical. Comput. Theor. Chem. 991, 82–87.

Liu, H., Shi, D., Sun, J., Zhu, Z., 2013. Calculations on thirteen Λ-S states of PO radical: electronic structure, spectroscopy and spin-orbit coupling. J. Quant. Spectrosc. Radiat. Transfer 121, 9–22.

Liu, J., Qu, W., Zheng, C., 2013a. Theoretical studies of mercury-bromine species adsorption mechanism on carbonaceous surface. Proc. Combust. Inst. 34, 2811–2819.

Liu, H., Pan, W.-P., Wang, C., Zhang, Y., 2016. Volatilization of arsenic during coal combustion based on isothermal thermogravimetric analysis at 600–1500°C. Energy Fuels 30, 6790–6798.

Liu, H., Wang, C., Sun, X., et al., 2016a. Volatilization of arsenic in coal during isothermal oxy-fuel combustion. Energy Fuels 30, 3479–3487.

Liu, H.M., Wang, C.B., Zhang, Y., et al., 2016b. Experimental and modeling study on the volatilization of arsenic during co-combustion of high arsenic lignite blends. Appl. Thermal Eng. 108, 1336–1343.

Liu, J.-Y., Huang, L.-M., Sun, S.-Y., et al., 2016c. Thermodynamic equilibrium calculations on Cd transformation during sewage sludge incineration. Water Environ. Res. 88 (6), 548–556.

Liu, Q.-Z., Sun, Y., Bai, Y., et al., 2016d. Characteristics of mercury emissions from a coal-fired power plant. J. Resid. Sci. Technol. 13 (Suppl 1), S175–S180.

Liu, Y., Liu, G., Qi, C., Cheng, S., Sun, R., 2016e. Chemical speciation and combustion behavior of chromium and vanadium in coals. Fuel 184, 42–49.

Liu, M., Zhang, W., Wang, X., Chen, L., et al., 2016f. Mercury release to aquatic environments from anthropogenic sources in China from 2001 to 2012. Environ. Sci. Technol. 50, 8169–8177.

Liu, C.Q., Huang, Y., Wang, X., Zhang, S.Y., et al., 2017. Dynamic experimental investigation on the volatilization behavior of lead and cadmium in the simulated municipal solid waste influenced by sulfur compounds during incineration. Energy Fuels 31 (1), 847–853.

Liu, H., Shi, D., Sun, J., Zhu, Z., 2017a. Accurate potential energy curves and spectroscopic properties of the 27Λ-S states and 73Ω states of the PO radical. Mol. Phys. 115 (6), 714–730.

Liu, J., Wang, J., Tsang, D.C.W., Xiao, T., Chen, Y., Hou, L., 2018. Emerging thallium pollution in China and source tracing by thallium isotopes. Environ. Sci. Technol. 52, 11977–11979.

Liu, Y., Wang, Z., Lv, Y., Wan, K., He, Y., Xia, J., et al., 2018a. Inhibition of sodium release from Zhundong coal via the addition of mineral additives: the combination of online multipoint LIBS and off-line experimental measurements. Fuel 212, 498–505.

Liu, K., Wang, S., Wu, Q., Wang, L., Ma, Q., et al., 2018b. A highly resolved mercury emission inventory of Chinese coal-fired power plants. Environ. Sci. Technol. 52, 2400–2408.

Liu, Y., Wang, Z., Xia, J., Vervisch, L., et al., 2019. Measurement and kinetics of elemental and atomic potassium release from a burning biomass pellet. Proc. Combust. Inst. 37, 2681–2688.

Livingstone, C., 2018. Manganese provision in parenteral nutrition: an update. Nutr. Clin. Pract. 33 (3), 404–418.

Llorens, J.F., Fernandez-Turiel, J.L., Querol, X., 2001. The fate of trace elements in a large coal-fired power plant. Environ. Geol. 40, 409–416.

Llorente-Mirandes, T., Rubio, R., Lopez-Sanchez, J.F., 2017. Inorganic arsenic determination in food: a review of analytical proposals and quality assessment over the last six years. Appl. Spectrosc. 71 (1), 25–69.

Lo, J.M.H., Klobukowski, M., 2007. Relativistic calculations on the ground and excited states of AgH and AuH in cylindrical harmonic confinement. Theor. Chem. Acc. 118, 607–622.

Lodders, K., 1999. Revised thermochemical properties of phosphinidene (PH), phosphine (PH_3), phosphorus nitride (PN), and magnesium phosphate ($Mg_3(PO_4)_2$). J. Phys. Chem. Ref. Data 28, 1705–1712.

Lodders, K., 2004. Revised and updated thermochemical properties of the gases, mercapto (HS), disulfur monoxide (S_2O), thiazyl (NS) and thioxophosphino (PS). J. Phys. Chem. Ref. Data 33 (1), 357–367.

Lodi, L., Yurchenko, S.N., Tennyson, J., 2015. The calculated rovibronic spectrum of scandium hydride, ScH. Mol. Phys. 113 (13–14), 1998–2011.

Lombi, E., Donner, E., Scheckel, K.G., et al., 2014. Silver speciation and release in commercial antimicrobial textiles as influenced by washing. Chemosphere 111, 352–358.

Long, G.L., Boss, C.B., 1985. Effect of droplet size on the phosphine depression of calcium atomic emission signals in flame spectrometry. Anal. Chim. Acta 174, 191–201.

Long, G.L., Newman Jr, W.J., Klunder, G.L., Mahaney, J., 1987. The phosphine depression in flame atomic spectrometry. Appl. Spectrosc. 41, 255–260.

Long, J., Zhang, S.X., Luo, K.L., 2019. Discovery of anomalous molybdenum enrichment in lower carboniferous coal and its availability and origin. J. Geochem. Explor. 200, 104–111.

Loock, H.-P., Simard, B., Wallin, S., Linton, C., 1998. Ionization potentials and bond energies of TiO, ZrO, NbO and MoO. J. Chem. Phys. 109, 8980–8992.

Lopatin, S.I., Semenov, G.A., 1986. Gaseous calcium, strontium and barium phosphites. Dokl. Phys. Chem. 287, 249–250.

Lopatin, S.I., Semenov, G.A., 1987. A mass spectrometric study of thermal dissociation for strontium phosphates. Inorg. Mater. USSR 23, 1067–1070.

Lopatin, S.I., Semenov, G.A., 1990. Mass spectrometric study of the alkaline earth metal phosphates thermal dissociation. Phosphorus Sulfur Silicon 51, 423.

Lopatin, S.I., Semenov, G.A., 1996. Mass spectrometric study of vaporization of $Sn_2P_2O_7$. Russ. J. Gen. Chem. 66, 180–183.

Lopatin, S.I., Semenov, G.A., 1999. A mass spectrometric study of the vaporization of boron phosphate, BPO_4. Rapid Commun. Mass Spectrom. 13, 1398–1400.

Lopatin, S.I., Shugurov, S.M., 2013. Mass spectrometric study of stability, thermochemistry and structures of the gaseous oxyacid salts. Open Thermodyn. J. 7 ((Suppl.1), 35–56. Web available.

Lopatin, S.I., Semenov, G.A., Selevich, A.F., 1998. Mass spectrometric study of vaporization of thallium phosphates. Russ. J. Gen. Chem. 68, 844–846.

Lopatin, S.I., Semenov, G.A., Pilyugina, T.S., 2000. Thermodynamic study of gaseous salts of oxygen containing acids. V. Cesium tungstate. Russ. J. Gen. Chem. 70 (10), 1494–1496.

Lopatin, S.I., Semenov, G.A., Shugurov, S.M., 2003. Thermodynamic study of gaseous salts of oxygen containing acids. XIII. Molybdates and tungstates of alkaline earth metals. Russ. J. Gen. Chem. 73 (2), 169–175.

Lopatin, S.I., Shugurov, S.M., Semenov, G.A., Baranovskii, V.I., 2003a. The thermodynamic characteristics of gaseous chromium molybdates and tungstates. Russ. J. Phys. Chem. 77 (6), 879–884.

Lopatin, S.I., Shugurov, S.M., Stolyarova, V.L., 2008. Thermodynamics of gaseous calcium silicates. Dokl. Phys. Chem. 418 (1), 5–6.

Lopatin, S.I., Shugurov, S.M., Gunina, A.O., 2009. Thermodynamic properties of the gaseous gallium molybdates and tungstates. J. Phys. Chem. A 113, 13469–13474.

Lopatin, S.I., Panin, A.I., Shugurov, S.M., 2013. Thermodynamic properties and structure of gaseous BMoO4. Dalton Trans. 42 (4), 1210–1214.

Lopatin, S.I., Shugurov, S.M., Emelyanova, K.A., 2013a. High temperature mass spectrometric determinations of relative ionization cross sections of gaseous TiO, TiO_2, VO, VO_2, YO, HfO, and GeO molecules. Rapid Commun. Mass Spectrosc. 27, 2338–2342.

Lopatin, S.I., Panin, A.I., Shugurov, S.M., Emelyanova, K.A., 2014. Gaseous titanium molybdates and tungstates: thermodynamic properties and structures. Rapid Commun. Mass Spectrom. 28, 2636–2644.

Lopes, M.H., Abelha, P., Lapa, N., Oliveira, J.S., Cabrita, I., Gulyurtlu, I., 2003. The behavior of ash and heavy metals during the co-combustion of sewage sludges in a fluidized bed. Waste Manage 23, 859–870.

Lopez-Anton, M.A., Diaz-Somoano, M., Spears, D.A., Martinez-Tarazona, M.R., 2006. Arsenic and selenium capture by fly ashes at low temperature. Environ. Sci. Technol. 40, 3947–3951.

Lopez-Anton, M.A., Diaz-Somoano, M., Abad-Valle, P., Martinez-Tarazona, M.R., 2007. Mercury and selenium retention in fly ashes: influence of unburned particle content. Fuel 86, 2064–2070.

Lopez-Anton, M.A., Diaz-Somoano, M., Fierro, J.L.G., Martinez-Tarazona, M.R., 2007a. Retention of arsenic and selenium compounds present in coal combustion and gasification flue gases using activated carbons. Fuel Process. Technol. 88, 799–805.

Lopez-Anton, M.A., Abad-Valle, P., Diaz-Somoano, M., Suarez-Ruiz, I., Martinez-Tarazona, M.R., 2009. The influence of carbon particle type in fly ashes on mercury adsorption. Fuel 88, 1194–1200.

Lopez-Anton, M.A., Perry, R., Abad-Valle, P., Diaz-Somoano, M., Martinez-Tarazona, M.R., Maroto-Valer, M.M., 2011. Speciation of mercury in fly ashes by temperature programmed decomposition. Fuel Process. Technol. 92, 707–711.

Lopez-Anton, M.A., Spears, D.A., Diaz-Somoano, M., et al., 2015. Enrichment of thallium in fly ashes in a Spanish circulating fluidized-bed combustion plant. Fuel 146, 51–55.

Lorenz, M., Caspary, N., Foeller, W., et al., 1997. Electronic structure of triatomic copper (II) chloride. Mol. Phys. 91 (3), 483–493.

Low, F., Zhang, L., 2013. Arsenic emissions and speciation in the oxy-fuel fly ash collected from lab-scale drop-tube furnace. Proc. Combust. Inst. 34, 2877–2884.

Lu, X., Shih, K., 2015. Formation of lead aluminate ceramics: reaction mechanisms in immobilizing the simulated lead sludge. Chemosphere 138, 156–163.

Lu, H., Robert, W., Peirce, G., Ripa, B., Baxter, L.L., 2008. Comprehensive study of biomass particle combustion. Energy Fuels 22, 2826–2839.

Lu, H., Ip, E., Scott, J., et al., 2010. Effects of particle shape and size on devolatilization of biomass particle. Fuel 89, 1156–1168.

Lu, P., Hao, J.T., Yu, W., et al., 2016. Effects of water vapor and Na/K additives on NO reduction through advanced biomass reburning. Fuel 170, 60–66.

Luan, J., Li, R., Zhang, Z., Li, Y., Zhao, Y., 2013. Influence of chlorine, sulfur and phosphorus on the volatilization behavior of heavy metals during sewage sludge thermal treatment. Waste Manage. Res. 31 (10), 1012–1018.

Luc, P., Vetter, R., 2001. The Sc + NO = ScO + N reaction: rotational state distribution in ScO($X^2\Sigma^+$, $v'' = 0$). J. Chem. Phys. 115, 11106–11117.

Luecke, W., 1992. Systematic investigation of aluminum interferences on the alkaline earth elements in flame atomic absorption spectrometry. J. Anal. At. Spectrom. 7, 765–768.

Luecke, W., 1992a. Systematic investigation of aluminum interferences with the alkaline earths in flame atomic absorption spectrometry. II. The behavior of magnesium and calcium. Fresenius J. Anal. Chem. 344, 242–246.

Luecke, W., 1994. Systematic investigation of aluminum interferences on the alkaline earth elements in flame atomic absorption spectrometry. III. The behavior of strontium and barium. J. Anal. At. Spectrom. 9, 105–109.

Luna, A., Merchan, M., Roos, B.O., 1995. A theoretical analysis of the lowest excited states in HNO/NOH and HPO/POH. Chem. Phys. Lett. 196, 437–445.

Lundholm, K., Nordin, A., Backman, R., 2007. Trace element speciation in combustion processes: review and compilations of thermodynamic data. Fuel Process. Technol. 88, 1061–1070.

Lundholm, K., Rodgers, J.M., Haynes, B.S., et al., 2008. Fate of Cu, Cr and As during the combustion stages of CCA-treated wood fuel particles. Energy Fuels 22, 1589–1597.

Luo, Z., Hu, C., Zhou, J., Cen, K., 2006. Stability of mercury on three activated carbon sorbents. Fuel Process. Technol. 87, 679–685.

Luo, Y., Giammar, D.E., Huhmann, B.L., Catalano, J.G., 2011. Speciation of selenium, arsenic and zinc in Class C fly ash. Energy Fuels 25, 2980–2987.

Luo, Z., Chang, Y.-C., Pan, Y., et al., 2016. Rotationally resolved state-to-state photoelectron study of molybdenum monoxide cation, MoO^+. J. Phys. Chem. A 120, 4643–4654.

Lupion, M., Gutierrez-Ortiz, F.J., Navarrete, B., Cortes, V.J., 2010. Assessment performance of high temperature filtering elements. Fuel 89, 848–854.

Lv, Q., Cai, M., Wang, C.A., Zhang, Z.H., Che, D.F., 2018. Investigation on elemental mercury removal and anti-deactivation performance of modified SCR catalysts. Asia-Pacific J. Chem. Eng. 13 (4), e2208.

Lv, Y., Romero, C.E., Charles, J., Pauvlinch, K.A., Watkins, R.W., 2019. Developing steady and dynamic ORP models for mercury emissions control in power plants using WFGD operating data. Fuel 235, 54–62.

L'vov, B.V., 2007. Thermal Decomposition of Solids and Melts: New Thermochemical Approach to the Mechanism, Kinetics and Methodology. Springer, New York.

L'vov, B.V., Novichikhin, A.V., Dyakov, A.O., 1998. Mechanism of thermal decomposition of magnesium hydroxide. Thermochim. Acta 315, 135–143.

Lyons-Weiler, J., Ricketson, R., 2018. Reconsideration of the immunotherapeutic pediatric safe-dose levels of aluminum. J. Trace Elem. Med. Biol. 48, 67–73.

Ma, Z., Balasubramanian, K., 1991. Electronic states of WH. Chem. Phys. Lett. 181, 467–473.

Ma, H., Wang, L., Li, Y., 2003. Accelerated corrosion of pure Fe induced by gaseous $ZnCl_2$ at high temperatures. J. Mater. Sci. Technol. 19 (5), 456–458.

Ma, J.-J., Yao, H., Luo, G.-Q., Xu, M.-H., Han, J., He, X.-M., 2010. Distribution of Hg, As, Pb, and Cr in a coke oven plant. Energy Fuels 24, 5289–5290.

Ma, T., Tong, G.S.-M., Cheung, A.S.-C., 2012. Density functional theory study of yttrium monohalides. Mol. Phys. 110, 1407–1414.

Ma, S., Guo, M., Song, H., et al., 2014. Formation mechanism and influencing factors of ammonium bisulfate during the selective catalytic reduction process. Reliab. Fadian 43 (2), 75–78 (Chinese).

Maatouk, A., Ben Houria, A., Yazidi, O., Jaidane, N., Hochlaf, M., 2010. Electronic states of MgO: spectroscopy, predissociation and cold atomic Mg and O production. J. Chem. Phys. 133, 144302. 12 pp.

MacDonald, M.A., Gouldin, F.C., Fisher, E.M., 2001. Temperature dependence of phosphorus-based flame inhibition. Combust. Flame 124, 668–683.

Mackie, J.C., Bacskay, G.B., Haworth, N.L., 2002. Reactions of phosphorus-containing species of importance in the catalytic recombination of H + OH: quantum chemical and kinetic studies, J. Phys. Chem. A, 106. 10825–10830.

Madarasz, J., Leskelae, T., Rautanen, J., Niinistoe, L., 1996. Oxidation of alkaline earth metal sulfide powders and thin films. J. Mater. Chem. 6 (5), 781–787.

Magoulas, I., Kalemos, A., 2014. An *ab initio* study of the electronic structure of the boron oxide neutral (BO), cationic (BO +) and anionic (BO-) species. J. Chem. Phys. 141, 124308. 10 pp.

Mahadevan, R., Lee, D., Sakurai, H., Zachariah, M.R., 2002. Measurement of condensed phase reaction kinetics in the aerosol phase using single particle mass spectrometry. J. Phys. Chem. A 106, 11083–11092.

Mahe, L., Barthelat, J.-C., 1997. Pseudopotentials including part of the atomic correlation energy. J. Mol. Struct. Theochem. 401, 93–105.

Mahowald, N., Jickells, T.D., Baker, A.R., Artaxo, P., Benitez-Nelson, C.R., et al., 2008. Global distribution of atmospheric sources, concentrations and deposition rates, and anthropogenic impacts. Global Biogeochem. Cycles 22, GB4026. 19 pp.

Mahuli, S., Agnihotri, R., Chauk, S., et al., 1997. Mechanism of arsenic sorption by hydrated lime. Environ. Sci. Technol. 31, 3226–3231.

Maitra, A.M., Patsalides, E., 1987. Organophosphorus acid interferences in flame atomic absorption spectrometry. Anal. Chim. Acta 193, 179–191.

Makarov, A.V., Zbezhneva, S.G., Kovalenko, V.V., Rumyantseva, M.N., 2003. Mass spectrometric study of nanocrystalline ZnO vaporization. Inorg. Mater. 39, 594–598.

Malakhov, S.G., Makhon'ko, E.P., 1990. Emission of toxic metals into the atmosphere and their accumulation in the surface layer of the Earth. Russ. Chem. Rev. 59, 1037–1050.

Malecka, B., Gajerski, R., Malecki, A., Wierzbicka, M., Olszewski, P., 2003. Mass spectral studies on the mechanism of thermal decomposition of $Zn(NO_3)_2.nH_2O$. Thermochim. Acta 404, 125–132.

Malherbe, J., Isaure, M.-P., Seby, F., Watson, R.P., Rodriguez-Gonzalez, P., Stutzman, P.E., et al., 2011. Evaluation of hexavalent chromium extraction method EPA Method 3060A for soils using XANES spectroscopy. Environ. Sci. Technol. 45, 10492–10500.

Maligne, D., Cessou, A., Stepowski, D., 2009. Chemiluminescence of BO_2 to map the creation of thermal NO in flames. Combust. Flame 156, 348–361.

Mallika, C., Raj, A.M.E.S., Nagaraja, K.S., Sreedharan, O.M., 2001. Use of SnO for the determination of standard Gibbs energy of formation of SnO_2 by oxide electrolyte e.m.f. measurements. Thermochim. Acta 371, 95–101.

Maly, P.M., Zamansky, V.M., Ho, Loc, Payne, R., 1999. Alternative fuel reburning. Fuel 78 (3), 327–334.

Mambote, C.R., Xiao, Y., Schuiling, R.D., 2008. Vanadium recovery from fly ash: a review. Proc. Int. Mineral Process Congress 24th Beijing, China 3, 3697–3707.

Mangan, T.P., McAdam, N., Daly, S.M., Plane, J.M.C., 2019. Kinetic study of Ni and NiO reactions pertinent to the Earth's upper atmosphere. J. Phys. Chem. A 123, 601–610.

Mann, E., Khusial, R., Carpi, A. Mechanistic investigation of the reduction and volatilization of mercury in soil. 251st American Chemical Society National Meeting, Abstracts of Papers. March 2016: Geoc-92.

Mao, Q., Luo, K.H., 2019. Trace metal assisted polycyclic aromatic hydrocarbons fragmentation, growth and soot nucleation. Proc. Combust. Inst. 37, 1023–1030.

Mao, L., Gao, B., Deng, N., et al., 2015. The role of temperature on Cr(VI) formation and reduction during heating of chromium-containing sludge in the presence of CaO. Chemosphere 138, 197–204.

Mao, L.-Q., Deng, N., Liu, L., et al., 2016. Effects of Al_2O_3, Fe_2O_3 and SiO_2 on Cr(VI) formation during heating of solid waste containing Cr(III). Chem. Eng. J. 304, 216–222.

Marcus, Y., 1980. Compilation and evaluation of solubility data in the mercury (1) chloride-water system. J. Phys. Chem. Ref. Data 9 (4), 1307–1329.

Marian, C.M., 1990. Quasirelativistic calculation of the vibronic spectra of NiH and NiD. J. Chem. Phys. 93 (2), 1176–1186.

Marian, C.M., 1991. Theoretical study of the spectra of CuH and CuD. J. Chem. Phys. 94 (8), 5574–5585.

Maric, R., Sundell, S., Oljaca, M., Hunt, A., 2002. Nanoparticle synthesis via nanomiser device atmospheric pressure flame. Adv. Powder Metall. Part Mater. 1, 129–136.

Markowski, G.R., Filby, R., 1985. Trace element concentration as a function of particle size in fly ash from a pulverized coal utility boiler. Environ. Sci. Technol. 19, 796–804.

Marquart, J.R., Berkowitz, J., 1963. Dissociation energies of some metal sulfides. J. Chem. Phys. 39, 283–285.

Marrero, J., Polla, G., Rebagliati, R.J., Pla, R., Gomez, D., Smichowski, P., 2007. Characterization and determination of twenty-eight elements in fly ashes collected in a thermal power plant in Argentina using different instrumental techniques. Spectrochim. Acta B 62, 101–108.

Marschman, S.C., Lynch, D.C., 1984. Review of the Bi and Bi-O vapor systems. Can. J. Chem. Eng. 62, 875–879.

Marshall, P., Narayan, A.S., Fontijn, A., 1990. Kinetics and thermochemical studies of the recombination reaction $Na + O_2 + N_2$ from 590 to 1515 K by a modified high temperature photochemistry technique. J. Phys. Chem. 94 (7), 2998–3004.

Martin, T.P., Schaber, H., 1982. Matrix isolated II-VI molecules: sulfides of Mg, Ca, Sr, Zn and Cd. Spectrochim. Acta A 38 (6), 655–660.

Martinez, A., Morse, M.D., 2011. Spectroscopy of diatomic ZrF and ZrCl: 760-555 nm. J. Chem. Phys. 135, 024308. 9 pp.

Martinez-Tarazona, M.R., Spears, D.A., 1996. The fate of trace elements and bulk minerals in pulverized coal combustion in a power plant. Fuel Process. Technol. 47, 79–92.

Marushkin, K.N., Alikhanyan, A.S., Greenberg, J.H., Lazarev, V.B., Malyusov, V.A., Rozanova, O.N., et al., 1985. Sublimation thermodynamics of tungsten trioxide. J. Chem. Thermodyn. 17, 245–253.

Marushkin, K.N., Alikhanyan, A.S., Orlovskii, V.P., 1990. The thermodynamic properties of the oxides of zirconium, hafnium and yttrium. Russ. J. Inorg. Chem. 35, 1181–1185.

Maryutina, T.A., Soin, A.V., 2009. Novel approach to the elemental analysis of crude and diesel oil. Anal. Chem. 81, 5896–5901.

Mason, P.E., Darvell, L.I., Jones, J.M., Williams, A., 2016. Observations on the release of gas phase potassium during the combustion of single particles of biomass. Fuel 182, 110–117.

Mason, P.E., Jones, J.M., Darvell, L.I., Williams, A., 2017. Gas phase potassium release from a single particle of biomass during high temperature combustion. Proc. Combust. Inst. 36 (2), 2207–2215.

Mason, R.P., Soerensen, A.L., DiMento, B.P., Balcom, P.H., 2018. The global marine selenium cycle: insights from measurements and modeling. Global Biogeochem. Cycles 32, 1720–1737.

Masseron, R., Gadiou, R., Delfosse, L., 1999. Study of the adsorption of $CdCl_2$ vapor on various minerals using a drop tube furnace. Environ. Sci. Technol. 33, 3634–3640.

Masson, J.-D., Crepeaux, G., Authier, F.-J., Exley, C., 2018. Critical analysis of reference studies on the toxicokinetics of aluminum-based adjuvants. J. Inorg. Biochem. 181, 87–95.

Mathieu, O., Kulatilaka, W.D., Petersen, E.L., 2018. Experimental and modeling study on the effects of dimethyl methylphosphonate (DMMP) addition on H_2, CH_4 and C_2H_4 ignition. Combust. Flame 191, 320–334.

Matsui, R., Senba, K., Honma, K., 1997. Kinetics of Co(a^4F, b^4F, a^2F) and Ni(a^3F, a^3D, a^1D) depletion by O_2, NO and N_2O. J. Phys. Chem. A 101, 179–186.

Matsumoto, K., Sata, T., 1981. A study of the calcium oxide-water vapor system by means of the transpiration method. Bull. Chem. Soc. Jpn. 54, 674–677.

Matthew, D.J., Tieu, E., Morse, M.D., 2017. Determination of the bond dissociation energies of FeX and NiX (X = C, S, Se). J. Chem. Phys. 146, 144310. 10 pp.

Mavrodineanu, R., Boiteux, H., 1965. Flame Spectroscopy. Wiley & Sons, New York.

Mayer, P.M., Gal, J.-F., Radom, L., 1997. The heats of formation, gas-phase acidities and related thermochemical properties of the third-row hydrides GeH_4, AsH_3, SeH_2 and HBr from G2 *ab initio* calculations. Int. J. Mass Spectrom. Ion Process. 167/168, 689–696.

Mayer, B.K., Baker, L.A., Boyer, T.H., et al., 2016. Total value of phosphorus recovery. Environ. Sci. Technol. 50, 6606–6620.

Mayhall, N.J., Raghavachari, K., Redfern, P.C., Curtiss, L.A., 2009. Investigation of Gaussian theory for transition metal thermochemistry. J. Phys. Chem. A 113, 5170–5175.

Mayo Clinic, Rochester, 2018. Interpretive Handbook. Medical Laboratories, Rochester MN, 2388 pp.

Mazokopakis, E.E., Liontiris, M.I., 2018. Health concerns of Brazil nut consumption. J. Altern. Complement. Med. 24 (1), 3–6.

McBride, M.B., Cherney, J., 2004. Molybdenum, sulfur and other trace elements in farm soils and forages after sewage sludge application. Commun. Soil Sci. Plant Anal. 35, 517–535.

McCarty, K.F., Anderson, R.J., 1987. Real time measurements of deposit formation from sodium sulfate seeded flames. Combust. Sci. Technol. 54, 51–60.

McCluskey, M., Andrews, L., 1991. Matrix infrared spectra of the products of the P_2 and O_3 reaction. J. Phys. Chem. 95, 2988–2994.

McConnell, J.R., Edwards, R., 2008. Coal burning leaves toxic heavy metal legacy in the Arctic. Proc. Natl. Acad. Sci. 105, 12140–12144.

McCoy, M., 2014. A home for cobalt in computer chips. Chem. Eng. News 92 (34), 16–17.

McDiarmid, M.A., Albertini, R.J., Tucker, J.D., et al., 2011. Measures of genotoxicity in Gulf War I veterans exposed to depleted uranium. Environ. Mol. Mutagen. 52 (7), 569–581.

McDiarmid, M.A., Engelhardt, S.M., Dorsey, C.D., et al., 2011a. Longitudinal health surveillance in a cohort of Gulf War veterans 18 years after first exposure to depleted uranium. J. Toxicol. Environ. Health A 74 (10), 678–691.

McIntyre, N.S., Thompson, K.R., Weltner Jr, W., 1971. Spectroscopy of titanium oxide and titanium dioxide molecules. J. Phys. Chem. 75, 3243–3249.

McKemmish, L.K., Yurchenko, S.N., Tennyson, J., 2016. *Ab initio* calculations to support accurate modeling of the rovibronic spectroscopy calculations of vanadium monoxide, VO. Mol. Phys. 114 (21), 3232–3248.

McKemmish, L.K., Masseron, T., Sheppard, S., et al., 2017. MARVEL (Measured active rotational-vibrational energy levels) analysis of the measured high resolution rovibronic spectra of $^{48}Ti^{16}O$. Astrophys. J., Suppl. Ser. 228 (15), 1–22.

McKenzie, R.M., 1971. The synthesis of birnessite, cryptomelane and some other oxides and hydroxides of manganese. Mineralog. Mag. 38, 493–502.

McKinley, J.D., 1964. Mass spectrometric investigation of the yttrium chloride surface reaction. J. Chem. Phys. 41, 2814–2817.

McKinley, J.D., 1965. Mass spectrum of yttrium chloride vapor. J. Chem. Phys. 42, 2245–2246.

McLennan, A.R., Bryant, G.W., Stanmore, B.R., Wall, T.F., 2000. Ash formation mechanisms during pf combustion in reducing conditions. Energy Fuels 14, 150–159.

Medhekar, A.K., Rokni, M., Trainor, D.W., Jacob, J.H., 1979. Surface catalyzed reaction of $Hg + Cl_2$. Chem. Phys. Lett. 65, 600–604.

Meier, C., Voegelin, A., del Real, A.P., et al., 2016. Transformation of silver nanoparticles in sewage sludge during incineration. Environ. Sci. Technol. 50, 3503–3510.

Meij, R., 1994. Trace element behavior in coal-fired power plants. Fuel Process. Technol. 39, 199–217.

Menasha, J., Dunn-Rankin, D., Muzio, L., Stallings, J., 2011. Ammonium bisulfate formation temperature in a bench-scale single channel air preheater. Fuel 90 (7), 2445–2453.

Menendez, M., Garay, M., Verdasco, E., Urena, A.G., 1993. On the determination of D_0(CaCl) from translational energy threshold measurements. J. Chem. Phys. 99, 2760–2764.

Meng, L., Zhao, Z., Cao, Y., et al., 2005. Arsenic partitioning at three USA electric power plants. Proc. Int. Tech. Conf. Coal Utiliz. Fuel Syst. 30, 1227–1235. vol. 2.

Meplan, C., 2015. Selenium and chronic diseases: a nutritional genomics perspective. Nutrients 7 (5), 3621–3651.

Merchan-Merchan, W., Saveliev, A.V., Kennedy, L.A., 2006. Flame synthesis of molybdenum oxide whiskers. Chem. Phys. Lett. 422, 72–77.

Merer, A.J., 1989. Spectroscopy of the diatomic 3d transition metal oxides. Ann. Rev. Phys. Chem. 40, 407–438.

Merz, S., Steinhauser, G., Hamada, N., 2013. Anthropogenic radionuclides in Japanese food: environmental and legal implications. Environ. Sci. Technol. 47, 1248–1256.

Meschi, D.J., Searcy, A.W., 1969. Investigation of the magnetic moments of S_2, Se_2, Te_2, Se_6 and Se_5 by the Stern-Gerlach magnetic deflection method. J. Chem. Phys. 51 (11), 5134–5138.

Metz, R.B., Nicolas, C., Ahmed, M., Leone, S.R., 2005. Direct determination of the ionization energies of FeO and CuO with VUV radiation. J. Chem. Phys. 123, 114313. 6 pp.

Michael, L.H., Boland, J.T., 1997. Chemical vapor deposition of polycrystalline BPO_4 thin films. Surf. Coat. Technol. 94–95, 451–455.

Michalek, I.M., Benn, E.K.T., dos Santos, F.L.C., Gordon, S., Wen, C., Liu, B., 2019. A systematic review of global legal regulations on the permissible level of heavy metals in cosmetics with particular emphasis on skin lightening products. Environ. Res. 170, 187–193.

Midda, S., Bera, N.C., Bhattacharyya, I., Das, A.K., 2006. *Ab initio* and density functional study of spectroscopic properties of CuO and CuS. J. Mol. Struct. Theochem. 761, 17–20.

Mielke, Z., McCluskey, M., Andrews, L., 1990. Matrix reactions of P_2 and O_3 molecules. Chem. Phys. Lett. 165 (2–3), 146–154.

Mielke, Z., Brabson, G.D., Andrews, L., 1991. Matrix infrared spectra of the phosphorus sulfides, PS, P_2S and PS_2. J. Phys. Chem. 95, 75–79.

Mihelcic, J.R., Fry, L.M., Shaw, R., 2011. Global potential of phosphorus recovery from human urine and feces. Chemosphere 84, 832–839.

Milburn, R.K., Rodriquez, C.F., Hopkinson, A.C., 1997. Theoretical enthalpies of formation of NH_mCl_n: neutral molecules, cations and anions. J. Phys. Chem. B 101, 1837–1844.

Miliordos, E., Hunt, K.L.C., 2011. First principles calculations of the electronic and geometrical structures of neutral [Sc,O,H] molecules and the monocations, $ScOH^{0,+}$ and $HScO^{0,+}$. J. Phys. Chem. A 115, 4436–4447.

Miliordos, E., Mavridis, A., 2007. Electronic structure of vanadium oxide: neutral and charged species VO, VO^+ and VO^-. J. Phys. Chem. A 111, 1953–1965.

Miliordos, E., Mavridis, A., 2010. Electronic structure and bonding of the early 3d-transition metal diatomic oxides and their ions: $ScO^{0,\pm}$, $TiO^{0,\pm}$, $CrO^{0,\pm}$, and $MnO^{0,\pm}$. J. Phys. Chem. A 114, 8536–8572.

Miliordos, E., Harrison, J.F., Hunt, K.L.C., 2011. *Ab initio* investigation of titanium hydroxide isomers and their cations, TiOH, $TiOH^+$ and HTiO, $HTiO^+$. J. Chem. Phys. 135, 144111. 13 pp.

Miliordos, E., Harrison, J.F., Hunt, K.L.C., 2013. Ground and excited states of vanadium hydroxide isomers and their cations, VOH, VOH^+ and HVO, HVO^+. J. Chem. Phys. 138, 114305, 14 pp.

Miller, A.R., 1975. The vaporization of cadmium oxide by Knudsen effusion. High Temp. Sci. 7, 126–130.

Miller, W.J., 1969. Inhibition of low-pressure flames. Combust. Flame 13, 210–212.

Miller, W.J., 1978. Electron attachment and compound formation in flames. VII. Electron attachment in flames containing phosphorus and potassium. J. Chem. Phys. 69 (8), 3709–3713.

Miller, J.F., Miller, S.E., Himes, R.C., 1959. Preparation of anhydrous rare earth chlorides for physicochemical studies. J. Am. Chem. Soc. 81, 4449–4451.

Miller, A.E.S., Feigerie, C.S., Lineberger, W.C., 1987. Laser photoelectron spectroscopy of CrH^-, CoH^-, and NiH^-: periodic trends in the electronic structure of the transition metal hydrides. J. Chem. Phys. 87, 1549–1556.

Miller, A.E.S., Miller, T.M., Morris, R.A., Viggiano, A.A., Van Doren, J.M., Paulson, J.F., 1993. Reactions of Fe^- with acids: gas phase acidity and bond energy of FeH. Int. J. Mass Spectrom. Ion Process. 123, 205–216.

Miller, B.B., Dugwell, D.R., Kandiyoti, R., 2002. Partitioning of trace elements during the combustion of coal and biomass in a suspension-firing reactor. Fuel 81, 159–171.

Miller, B., Dugwell, D.R., Kandiyoti, R., 2003. The influence of injected HCl and SO_2 on the behavior of trace elements during wood-bark combustion. Energy Fuels 17, 1382–1391.

Miller, B.B., Kandiyoti, R., Dugwell, D.R., 2004. Trace element behavior during co-combustion of sewage sludge with Polish coal. Energy Fuels 18, 1093–1103.

Miller, B., Dugwell, D., Kandiyoti, R., 2006. The fate of trace elements during the co-combustion of wood-bark with waste. Energy Fuels 20, 520–531.

Milushin, M.I., Gorokhov, L.N., 1988. Mass spectrometric determination of the enthalpies of formation of chromium chloride and oxide chloride molecules. Russ. J. Phys. Chem. 62, 387–388.

Milushin, M.I., Emel'yanov, A.M., Gorokhov, L.N., 1986. Mass spectrometric determination of the enthalpy of formation of AlOH(g). High Temp. USSR 24, 347–352.

Miner, S., 1969. Air pollution aspects of barium and its compounds. National Air Pollution Control Administration Report, September 1969.

Miner, S., 1969a. Air pollution aspects of radioactive substances. National Air Pollution Control Administration Report, September 1969.

Minissale, M., Congiu, E., Baouche, S., et al., 2013. Formation of nitrogen oxides via NO + O_2 gas-solid reaction on cold surfaces. Chem. Phys. Lett. 565, 52–55.

Mintz, B., Chan, B., Sullivan, M.B., et al., 2009. Structures and thermochemistry of the alkali metal monoxide anions, monoxide radicals and hydroxides. J. Phys. Chem. A 113 (34), 9501–9510.

Mishra, S., 2008. Structural and spectroscopic study of the excited electronic states of silver dihalides by quantum chemical methods. PhysChemChemPhys 10, 3987–3991.

Mitchell, S.A., 1992. Association reactions of transition metal atoms with simple molecules near room temperature. In: Fontijn, A. (Ed.), Gas Phase Metal Reactions. Elsevier Science Publishers, Amsterdam, pp. 227–252.

Mitchell, M.J., Munir, Z.A., 1970. Studies on the sublimation of IIB-VIA compounds. III. Equilibrium and free surface pressures and enthalpies of sublimation of red mercury sulfide (cinnabar). High Temp. Sci. 2, 265–273.

Mittal, S.K., Elliott, J.F., 1984. Thermodynamic properties of Na_2O-V_2O_5 melts, 1030-1210 K. J. Electrochem. Soc. Solid State Sci. Technol. 131, 1194–1199.

Mlakar, T.L., Horvat, M., Vuk, T., Stergarsek, A., Kotnik, J., Tratnik, J., et al., 2010. Mercury species, mass flows and processes in a cement plant. Fuel 89, 1936–1945.

Moallemi, A., Kazemimanesh, M., Kostiuk, L.W., Olfert, J.S., 2018. The effect of sodium chloride on the nanoparticles observed in a laminar methane diffusion flame. Combust. Flame 188, 273–283.

Mobin, M., Malik, A.U., 1996. Studies on the interactions of transition metal oxides and sodium sulfate in the temperature range 900–1200 K in oxygen. J. Alloys Compds. 235, 97–103.

Mobin, M., Malik, A.U., Ahmad, S., 1990. High temperature interactions of metal oxides with NaCl. J. Less Common Metals 160, 1–14.

Mobin, M., Malik, A.U., Ahmad, S., Hasan, S.K., Ajmal, M., 1996. Studies on the interactions of metal oxides and Na2SO4 at 1100 and 1200 K in oxygen. Bull. Mater. Sci. 19, 807–821.

Mochizuki, Y., Nagashima, U., Yamamoto, S., Kashiwagi, H., 1989. A theoretical study of the bent form of CuO_2. Chem. Phys. Lett. 164 (2/3), 225–230.

Mochizuki, Y., Takada, T., Murakami, A., 1991. A theoretical study of the CuOH molecule. Chem. Phys. Lett. 185 (5/6), 535–543.

Mochizuki, Y., Tanaka, K., Kashiwagi, H., 1991a. Electronic structure of the linear form of OCuO. Chem. Phys. 151, 11–20.

Mochizuki, Y., Sugawara, K., Enda, Y., 2009. Effect of pretreatment and additives on boron release during pyrolysis and gasification of coal. Energy Fuels 23, 4502–4506.

Mochizuki, Y., Tsubouchi, N., Sugawara, K., 2014. Behavior of boron release and change in the occurrence of boron during fixed-bed pyrolysis of coals. Fuel 130, 54–59.

Moghadamnia, A.A., 2012. An update on toxicology of aluminum phosphide. Daru-J. Pharmaceut. Sci. 20, 25.

Mohammed, T., Mohammed, E., Bascombe, S., 2017. The evaluation of total mercury and arsenic in skin bleaching creams commonly used in Trinidad and Tobago and their potential risk to the people of the Caribbean. J. Public Health Res. 6 (3), 184–189.

Mohazzabi, P., Searcy, A.W., 1976. Kinetics and thermodynamics of decomposition of barium sulfate. J. Chem. Soc. Faraday Trans. I 72, 290–295.

Mohn, H., Wendt, H., 1995. Molecular thermodynamics of molten salt evaporation. IV. The evaporation of molten carbonates in atmospheres containing CO_2 and water vapor. Zeitschrift Phys. Chem. 192, 101–119.

Mojammal, A.H.M., Back, S.K., Seo, Y.C., Kim, J.H., 2019. Mass balance and behavior of mercury in oil refinery facilities. Atmos. Pollut. Res. 10 (1), 145–151.

Mok, D.K.W., Lee, E.P.F., Chau, F.-t, Dyke, J.M., 2011. The enthalpies of formation of AsX_n molecules, where X = H, F or Cl, and n = 1-3 by RCCSD(T) and UCCSD(T)-F12x calculations. Phys. Chem. Chem. Phys. 13, 9540–9553.

Mok, D.K.W., Lee, E.P.F., Dyke, J.M., 2016. Simulation of the single-vibronic-level emission spectra of HAsO and DAsO. J. Chem. Phys. 144, 184303. 10 pp.

Monkhouse, P., 2011. On-line spectroscopic and spectrometric methods for the determination of metal species in industrial processes. Prog. Energy Combust. Sci. 37, 125–171.

Montoneri, E., Salzano, F.J., Findl, E., Kulesa, F., 1986. High temperature proton conductors: boron and aluminum phosphates. Solid State Ionics 18–19, 994–1002.

Mora, A.M., de Joode, B.V.W., Mergler, D., et al., 2014. Blood and hair manganese concentrations in pregnant women from the infant's environmental health study in Costa Rica. Environ. Sci. Technol. 48, 3467–3476.

Moravec, V.D., Jarrold, C.C., 1998. Study of the low-lying states of NiO^- and NiO using anion photoelectron spectroscopy. J. Chem. Phys. 108 (5), 1804–1810.

Moravec, V.D., Klopcic, S.A., Chatterjee, B., Jarrold, C.C., 2001. The electronic structure of ZnO and ZnF determined by anion photoelectron spectroscopy. Chem. Phys. Lett. 341, 313–318.

Morf, L.S., Brunner, P.H., Spaun, S., 2000. Effect of operating conditions and input variations on the portioning of metals in a municipal solid waste incinerator. Waste Manage. Res. 18, 4–15.

Morgon, N.H., 2012. Enthalpies of formation of phosphorus and oxygen compounds determined by the correlation consistent composite approach. Int. J. Quantum Chem. 112, 3256–3260.

Morozov, V.N., Popyalkovskaya, L.K., Gracheva, G.P., Krest'yaninov, A.G., 1973. Degree of calcium atomization in acetylene/air and propane/butane/air flames. J. Appl. Spectrosc. USSR 18, 9–13.

Morton, S.C., Edwards, M., 2005. Reduced phosphorus compounds in the environment. Crit. Rev. Environ. Sci. Technol. 35, 333–364.

Mosyagin, N.S., Titov, A.V., Buenker, R.J., Liebermann, H.-P., Alekseyev, A.B., 2002. Generalized relativistic effective core potential (GRECP/MRD-CI) calculations on the Hg atom and HgH molecule. Int. J. Quantum Chem. 88, 681–686.

Mot, A.I., Crouch, P.J., 2017. Biometals and Alzheimer's disease. Chapter 1, 1–17. In: White, A.R., Aschner, M., Costa, L.G., Bush, A.I. (Eds.), Biometals in Neurodegenerative Diseases, 415 pp. Academic Press, Elsevier, London.

Motojima, S., Higashi, T., 1985. Deposition and hollow crystal growth of chromium phosphide (CrP) and trichromium monophosphide (Cr_3P) by CVD process. J. Crystal Growth 71 (3), 639–647.

Motzfeldt, K., 1955. The thermal decomposition of sodium carbonate by the effusion method. J. Phys. Chem. 59, 139–147.

Moussaoui, Y., Ouamerali, O., De Mare, G.R., 1998. Theoretical study of the properties of the radical PS, its anion and cation. J. Mol. Struct. (Theochem.) 425, 237–247.

Mudd, G.M., Weng, Z., Jowitt, S.M., Turnbull, I.D., Graedel, T.E., 2013. Quantifying the recoverable resources of by-product metals: the case of cobalt. Ore Geol. Rev. 55, 87–98.

Muenow, D.W., Uy, O.M., Margrave, J.L., 1970. Mass spectrometric studies of the vaporization of phosphorus oxides. J. Inorg. Nucl. Chem. 32, 3459–3467.

Mulholland, A.J., Gomatam, J., Jenkins, A.M., 1999. Modeling high temperature salt condensation: Fick and Soret diffusion effects. Int. J. Therm. Sci. 38, 436–441.

Muller III, C.H., Schofield, K., Steinberg, M., Broida, H.P., 1979. Sulfur chemistry in flames. Proc. Combust. Inst. 17, 867–879.

Muller-Rosing, H.-C., Schulz, A., Hargittai, M., 2005. Structure and bonding in silver halides. A quantum chemical study of the monomers: Ag_2X, AgX, AgX_2 and AgX_3 (X = F, Cl, Br, I). J. Am. Chem. Soc. 127, 8133–8145.

Mulligan, C.N., Sharifi-Nistanak, M., 2016. Conversion of sludge from a wastewater treatment plant to a fertilizer. Int J Geomate 11 (23), 2194–2199.

Munir, Z.A., Mitchell, M.J., 1969. Sublimation of IIB-VIA compounds. I. Sublimation coefficient and activation energy for sublimation of single crystalline zinc sulfide. High Temp. Sci. 1, 381–387.

Munir, Z.A., Rice, J.L., 1981. The thermal decomposition of cadmium and cobalt carbonates. High Temp. Sci. 14, 171–180.

Munir, Z.A., Street, G.B., Winters, H.F., 1971. Mass spectrometric and vapor pressure studies on the sublimation of Realgar (As_4S_4). J. Chem. Phys. 55 (9), 4520–4527.

Mupambwa, H.A., Dube, E., Mnkeni, P.N.S., 2015. Fly ash composting to improve fertilizer value: a review. South Africa J. Sci. 111 (7–8), 26–31.

Murad, E., 1980. Mass spectrometric study of the thermochemical properties of gaseous MgOH. Chem. Phys. Lett. 72, 295–296.

Murad, E., 1980a. Thermochemical properties of gaseous FeO and FeOH. J. Chem. Phys. 73, 1381–1385.

Murad, E., 1981. Thermochemical properties of the gaseous alkaline earth monohydroxides. J. Chem. Phys. 75, 4080–4085.

Murad, E., Hildenbrand, D.L., 1975. Thermodynamic properties of gaseous ZrO and ZrO_2. J. Chem. Phys. 63, 1133–1139.

Murray, J.J., Pupp, C., Pottie, R.F., 1973. Heat of dissociation of $As_4(g)$ and the heat of formation of $As_2(g)$. J. Chem. Phys. 58 (6), 2569–2578. Erratum ibid. 59(8):4572.

Murtz, P., Thummel, H., Pfelzer, C., Urban, W., 1995. New bands of the MgO($A^1\Pi$-$X^1\Sigma^+$) and ($a^3\Pi_{0,1}$-$X^1\Sigma^+$) systems by Faraday laser magnetic resonance spectroscopy. Mol. Phys. 86, 513–534.

Muzangwa, L.G., Ayles, V.L., Nyambo, S., Reid, S.A., 2011. Probing the electronic structure of the nickel monohalides: spectroscopy of the low-lying electronic states of NiBr and NiCl. J. Mol. Spectrosc. 269, 36–40.

Muzio, L., Bogseth, S., Himes, R., Chien, Y.-C., Dunn-Rankin, D., 2017. Ammonium bisulfate formation and reduced load SCR operation. Fuel 206, 180–189.

Myers, C.E., Kisacky, G.A., Patterson, E.L., Mevorach, R.A., 1982. The congruent vaporization of manganese phosphide, $Mn_2P(s)$. J. Electrochem. Soc. 129, 2343–2346.

Myers, C.E., Kisacky, G.A., Klingert, J.K., 1985. Vaporization behavior of chromium phosphides. The solid two-phase regions CrP-$Cr_{12}P_7$, $Cr_{12}P_7$-Cr_3P and Cr_3P-Cr. J. Electrochem. Soc. 132 (1), 236–238.

Myers, D.L., Jacobson, N.S., Bauschlicher Jr, C.W., Opila, E.J., 2019. Thermochemistry of volatile metal hydroxides and oxy-hydroxides at elevated temperatures. J. Mater. Res. 34 (3), 394–407.

Nagai, H., 1969. Preparation and properties of beryllium oxide thin films. Jpn. J. Appl. Phys. 8, 1221–1228.

Nagai, T., Miyake, M., Maeda, M., 2009. Thermodynamic measurement of calcium phosphates by double Knudsen cell mass spectrometry. Metall. Mater. Trans. B Process. Metall. Mater. Process. Sci. 40 (4), 544–549.

Nagano, M., 1984. Chemical vapor deposition of tin oxide thin films on rutile single crystals. J. Crystal Growth 67, 639–644.

Nagayama, K., Takada, T., 1989. Direct combination of thermogravimetric analyzer and atomic absorption spectrometer for detection of atomic vapor in thermal analysis. Thermochim. Acta. 156, 11–19.

Nagy, B., Szakacs, P., Csontos, J., et al., 2011. High accuracy theoretical thermochemistry of atmospherically important sulfur-containing molecules. J. Phys. Chem. A 115, 7823–7833.

Nagy, D., Nagy, T., Balogh, A., Falussy, C., Posta, J., 2014. Atomization of cadmium compounds under reactive and inert high temperature environment with stationary sample introduction. Appl. Spectrosc. 68 (3), 272–279.

Nakai, E., Kanno, M., Mukaibo, T., 1969. Oxidation behavior of uranium monosulfide. J. Nucl. Sci. Technol. 6 (3), 138–142.

Nakamura, S., Fuwa, A., 1987. Vapor pressure measurement of germanium sulfide, tin sulfide and cadmium sulfide by the transpiration-mass spectrometer system. Nippon Kinz Gakk 51, 124–130.

Nalivaiko, A.G., Rat'kovskii, I.A., 1981. Mass spectrometric study of the vaporization of lead tungstate. Izv. Akad. Nauk SSSR. Neorg. Mater. 17 (6), 1132–1133. In Russian.

Namiesnik, J., Rabajczyk, 2010. The speciation of aluminum in environmental samples. Crit. Rev. Anal. Chem. 40, 68–88.

Namiesnik, J., Rabajczyk, A., 2012. Speciation analysis of chromium in environmental samples. Crit. Rev. Environ. Sci. Technol. 42, 327–377.

Narayan, V., Jensen, P.A., Henriksen, U.B., et al., 2016. Behavior of alkali metals and ash in a low-temperature circulating fluidized bed gasifier. Energy Fuels 30, 1050–1061.

Narukawa, T., Riley, K.W., French, D.H., Takatsu, A., Chiba, K., 2003. Investigation into the relationship between major and minor element contents and particle size and leachability of boron in fly ash from coal fuel thermal power plants. J. Environ. Monit. 5, 831–836.

Nasluzov, V.A., Rosch, N., 1996. Density functional based structure optimization for molecules containing heavy elements: analytical energy gradients for the Douglas-Kroll-Hess scalar relativistic approach to the LCGTO-DF method. Chem. Phys. 210, 413–425.

Nathans, M.W., Wendlandt, W.W., 1962. The thermal decomposition of the rare earth sulfates: thermogravimetric and differential thermal analysis studies up to 1400°C. J. Inorg. Nucl. Chem. 24, 869–879.

National Research Council, 2007. Minerals, Critical Minerals and the U.S. Economy. Washington DC, 170 pp.

Naulin, C., Costes, M., 1999. Crossed beam study of the $Al(^2P_{1/2,3/2}) + O_2(X^3\Sigma_g^-) \rightarrow AlO(X^2\Sigma^+) + O(^3P_J)$ reaction at low and very low energies. Chem. Phys. Lett. 310, 231–239.

Naulin, C., Hedgecock, I.M., Costes, M., 1997. The dissociation energy of TiO determined from a crossed-beam study of the $Ti + NO = TiO + N$ reaction. Chem. Phys. Lett. 266, 335–341.

Navarro, R., Guzman, J., Saucedo, I., Revilla, J., Guibal, E., 2007. Vanadium recovery from oil fly ash by leaching, precipitation and solvent extraction processes. Waste Manage. 27, 425–438.

Ndubizu, C.C., Zinn, B.T., 1982. Effects of metallic additives upon soot formation in polymer diffusion flames. Combust. Flame 46, 301–314.

Nedellec, V., Rabi, A., 2016. Costs of health damage from atmospheric emissions of toxic metals: part II. Analysis for mercury and lead. Risk Anal. 36 (11), 2096–2104.

Negro, R., Hegedus, L., Attanasio, R., Papini, E., Winther, K.H., 2019. The 2018 European Thyroid Association Survey on the use of selenium supplementation in Graves' hyperthyroidism and Grave's orbitopathy. Eur. Thyroid J. 8 (1), 7–15.

Neil, J.M., Navrotsky, A., Kleppa, O.J., 1971. The enthalpy of Ilmenite-Perovskite transformation in cadmium titanate. Inorg. Chem. 10 (9), 2076–2077.

Nelin, C.J., Bauschlicher Jr, C.W., 1985. On the low-lying states of WO: a comparison with CrO and MoO. Chem. Phys. Lett. 118 (2), 221–225.

Nelson, P.F., Shah, P., Strezov, Z., Halliburton, B., Carras, J.N., 2010. Environmental impacts of coal combustion: a risk approach to assessment of emissions. Fuel 89, 810–816.

Neville, M., Sarofim, A.F., 1982. The stratified composition of inorganic submicron particles produced during coal combustion. Symp. (Int.) Combust. 19, 1441–1449.

Neville, M., Quann, R.J., Hayes, B.S., Sarofim, A.F., 1981. Vaporization and condensation of mineral matter during pulverized coal combustion. Symp. (Int.) Combust. 18, 1267–1274.

Newbury, R.S., Barton Jr, G.W., Searcy, A.W., 1968. Vapor species of the barium-oxygen system. J. Chem. Phys. 48, 793–800.

Nguyen, Q.N., 2008. High Temperature Volatility and Oxidation Measurements of Titanium and Silicon Containing Ceramic Materials. Thesis, Chemistry Department, Cleveland State University.

Nguyen, M.T., Hegarty, A.F., Ha, T.-K., Brint, P., 1985. An *ab initio* study of the ground and excited states of HPO. Chem. Phys. 98, 447–453.

Nguyen, M.T., McGinn, M.A., Fitzpatrick, N.J., 1986. A theoretical characterization of some diatomic copper species. J. Chem. Soc. Faraday Trans. II 82, 1427–1443.

Nhat, P.V., Cuong, N.T., Duy, P.K., Nguyen, M.T., 2012. Theoretical study of manganese hydrides and halides, MnX_n with X = H, F, Cl, Br and n = 1-4. Chem. Phys. 400, 185–197.

Nicks, C.M., Malins, R.J., Benard, D.J., 1982. Collisional relaxation of metastable calcium by molecules. Chem. Phys. Lett. 89, 523–526.

Nico, P.S., Kumfer, B.M., Kennedy, I.M., Anastasio, C., 2009. Redox dynamics of mixed metal (Mn, Cr and Fe) ultrafine particles. Aerosol Sci. Technol. 43, 60–70.

Nielsen, P., 2017. Chelation therapy for heavy metals. RSC Metallobiology Series No.8. Chapter 3 In: Crichton, R., Ward, R.J., Hider, R.C. (Eds.), Metal Chelation in Medicine. Royal Society of Chemistry, Cambridge England, pp. 56–105.

Nielsen, I.M.B., Allendorf, M.D., 2005. High level *ab initio* thermochemical data for halides of chromium, manganese and iron. J. Phys. Chem. A 109, 928–933.

Nielsen, H.P., Baxter, L.L., Sclippab, G., Morey, C., Frandsen, F.J., Dam-Johansen, K., 2000. Deposition of potassium salts on heat transfer surfaces in straw-fired boilers: a pilot-scale study. Fuel 79, 131–139.

Nielsen, H.P., Frandsen, F.J., Dam-Johansen, K., Baxter, L.L., 2000a. The implications of chlorine-associated corrosion on the operation of biomass-fired boilers. Prog. Energy Combust. Sci. 26, 283–298.

Nielsen, M.T., Livbjerg, H., Fogh, C.L., Jensen, J.N., Simonsen, P., Lund, C., et al., 2002. Formation and emission of fine particles from two coal-fired power plants. Combust. Sci. Technol. 174, 79–113.

Nielsen, I.M.B., Janssen, C.L., Allendorf, M.D., 2003. *Ab initio* predictions for thermochemical parameters for tin-oxygen compounds. J. Phys. Chem. A 107, 5122–5127.

Niemi, J., Kinnunen, H., Lindberg, D., Enestam, S., 2018. Interactions of $PbCl_2$ with alkali salts in ash deposits and effects on boiler corrosion. Energy Fuels 32, 8519–8529.

Nien, C.-F., Rajasekhar, B., Plane, J.M.C., 1993. Unusual kinetic behavior of the reactions $Mg + O_2 + M$ and $Ca + O_2 + M$ ($M = N_2$, He) over extended temperature ranges. J. Phys. Chem. 97, 6449–6456, 12422.

Niksa, S., Padak, B., Krishnakumar, B., Naik, C.V., 2010. Process chemistry of Br addition to utility flue gas for Hg emissions control. Energy Fuels 24, 1020–1029.

Niksa, S., Krishnakumar, B., Ghoreishi, F., 2016. Analytical management of SCR catalyst lifetimes and multipollutant performance. J. Air Waste Manage. Assoc. 66 (2), 215–223.

Ninomiya, Y., Wang, Q., Xu, S., Mizuno, K., Awaya, I., 2009. Effect of additives on the reduction of PM2.5 emissions during pulverized coal combustion. Energy Fuels 23, 3412–3417.

Nisse, C., Tagne-Fotso, R., Howsam, M., Richeval, C., et al., 2017. Blood and urinary levels of metals and metalloids in the general adult population of northern France: the IMEPOGE study, 2008–2010. Int. J. Hyg. Environ. Health 220, 341–363.

Nitodas, S.F., Sotirchos, S.V., 2002. Development and validation of a mathematical model for the chemical vapor deposition of alumina from mixtures of aluminum trichloride, carbon dioxide and hydrogen. J. Electrochem. Soc. 149, C130–C141.

Niu, Y., Tan, H., Hui, S., 2016. Ash related issues during biomass combustion: alkali induced slagging, silicate melt-induced slagging (ash fusion), agglomeration, corrosion, ash utilization and related countermeasures. Prog. Energy Combust. Sci. 52, 1–61.

Niu, Y., Wang, Z., Zhu, Y., et al., 2016a. Experimental evaluation of additives and K_2O-SiO_2-Al_2O_3 diagrams on high temperature silicate melt-induced slagging during biomass combustion. Fuel 179, 52–59.

Nocker, K., Gruehn, R., 1993. Chemical transport of chromium oxychloride (CrOCl) and chromium oxide (Cr_2O_3): experiments and model calculations for participation of $CrOCl_2$(g). Zeit Anorg. Allgem Chem. 619, 699–710.

Nocker, K., Gruehn, R., 1993a. Chemical transport of chromium (III) oxide with chlorine molecule and with mercuric chloride: experiments and model calculations. Zeit Anorg. Allgem Chem. 619, 1530–1542.

Noda, N., Ito, S., 2018. Mercury partitioning in coal-fired power plants in Japan. J. Japan Inst. Energy 97 (11), 342–347.

Noda, N., Ito, S., Nunome, Y., et al., 2013. Volatilization characteristics of boron compounds during coal combustion. Proc. Combust. Inst. 34, 2831–2838.

Noor, N., Zong, G., Seely, E.W., Weisskopf, M., James-Todd, T., 2018. Urinary cadmium concentrations and metabolic syndrome in US adults: the National Health and Nutrition Examination Survey 2001–2014. Environ. Int. 121, 349–356, part 1.

Nordine, P.C., Weber, J.K.R., Krishnan, S., Schiffman, R.A., 1991. The vapor pressure and sublimation enthalpy of pure boron. High Temp. Sci. 30 (2/3), 163–171.

Norton, G.A., Yang, H., Brown, R.C., Laudal, D.L., Dunham, G.E., Erjavec, J., 2003. Heterogeneous oxidation of mercury in simulated post combustion conditions. Fuel 82, 107–116.

Novakovic, A., van Lith, S.C., Frandsen, F.J., Jensen, P.A., Holgersen, L.B., 2009. Release of potassium from the systems K-Ca-Si and K-Ca-P. Energy Fuels 23, 3423–3428.

Nowak, B., Wegerer, H., Aschenbrenner, P., Rechberger, H., Winter, F., 2012. Sewage sludge ash to phosphate fertilizer by chlorination and thermal treatment: residence time requirements for heavy metal removal. Environ. Technol. 33, 2375–2381.

Nriagu, J.O., 1979. Global inventory of natural and anthropogenic emissions of trace metals to the atmosphere. Nature 279, 409–411.

Nriagu, J.O., 1988. A silent epidemic of environmental metal poisoning? Environ. Pollut. 50, 139–161.

Nriagu, J.O., 1989. A global assessment of natural sources of atmospheric trace metals. Nature 338, 47–49.

Nriagu, J.O., 1989a. Natural versus anthropogenic emissions of trace metals to the atmosphere. In: Pacyna, J.M., Ottar, B. (Eds.), Control and Fate of Atmospheric Trace Metals. Kluwer Academic Publishers, Dordrecht, The Netherlands, pp. 3–13.

Nriagu, J.O., 1990. Human influence on the global cycling of trace metals. Palaeogeogr. Palaeoclim. Palaeoecol. 82, 113–120.

Nriagu, J.O. (Ed.), 1998. Vanadium in the atmosphere. I. Chemistry and biochemistry. II. Health effects. Adv. Environ. Sci. Technol. 30, 1–410. 1998;31:1–403.

Nriagu, J.O., 2003. Thallium. Chem. Eng. News 81 (36), 153.

Nriagu, J.O., Pacyna, J.M., 1988. Quantitative assessment of worldwide contamination of air, water and soils by trace metals. Nature 333, 134–139.

Nuapia, Y., Chimuka, L., Cukrowska, E., 2018. Assessment of heavy metals in raw food samples from open markets in two African cities. Chemosphere 196, 339–346.

Numata, M., Sugiura, M., Fuwa, A., 1993. Standard Gibbs energies of formation for AlCl(g) and $AlCl_2$(g) species at temperatures from 600 to 1300 K. Mater. Trans. JIM 34 (6), 511–516.

Nurmesniemi, H., Makela, M., Poykio, R., Manskinen, K., Dahl, O., 2012. Comparison of the forest fertilizer properties of ash fractions from two power plants of pulp and paper mills incinerating biomass-based fuels. Fuel Process. Technol. 104, 1–6.

Nyberg, C.M., Thompson, J.S., Zhuang, Ye, Pavlish, J.H., Brickett, L., Pletcher, S., 2009. Fate of trace element hazardous air pollutants when applying mercury control technologies. Fuel Process. Technol. 90, 1348–1353.

O'Brien, N.W., Mucklejohn, S.A., 1987. A study of the vapor phase equilibria over tin (II) chloride. J. Chem. Thermodyn. 19, 1065–1077.

O'Brien, L.C., Kubicek, R.L., Wall, S.J., et al., 1996. Fourier transform spectroscopy of the $Y^2\Sigma^+$-$X^2\Pi_i$ transition of CuO. J. Mol. Spectrosc. 180, 365–368.

O'Brien, L.C., Wall, S.J., Henry, G.L., 1998. The $A^2\Sigma^+$-$X^2\Pi_i$ (0,1) band and reanalysis of the blue and ultraviolet transitions of AgO. J. Mol. Spectrosc. 191, 218–220.

O'Brien, L.C., Dulick, M., Davis, S.P., 1999. The near-infrared $Y^2\Sigma^+$-$X^2\Pi$ transition of CuS. J. Mol. Spectrosc. 195, 328–331.

O'Brien, L.C., Cao, H., O'Brien, J.J., 2000. The near-infrared transition of CuCl observed by intracavity laser spectroscopy. J. Mol. Spectrosc. 199, 100–108.

Obrist, D., Kirk, J.L., Zhang, L., Sunderland, E.M., Jiskra, M., Selin, N.E., 2018. A review of global environmental mercury processes in response to human and natural perturbations: changes of emissions, climate and land-use. Ambio 47, 116–140.

Ochiai, R., Azhar Uddin, M., Sasaoka, E., Wu, S., 2009. Effects of HCl and SO_2 concentration on mercury removal by activated carbon sorbents in coal-derived flue gas. Energy Fuels 23, 4734–4739.

Ochoa-Gonzalez, R., Cordoba, P., Diaz-Somoano, M., et al., 2011. Differential partitioning and speciation of Hg in wet FGD facilities of two Spanish PCC power plants. Chemosphere 85, 565–570.

Ochoa-Gonzalez, R., Cuesta, A.F., Cordoba, P., et al., 2011a. Study of boron behavior in two Spanish coal combustion power plants. J. Environ. Manage. 92, 2586–2589.

Ochoa-Gonzalez, R., Diaz-Somoano, M., Martinez-Tarazona, M.R., 2013. Control of Hg^0 re-emission from gypsum slurries by means of additives in typical wet scrubber conditions. Fuel 105, 112–118.

O'Connor, D., Hou, D.Y., Ye, J., Zhang, Y.H., et al., 2018. Lead-based paint remains a major public health concern: a critical review of global production, trade, use, exposure, health risk, and implications. Environ. Int. 121, 85–101. Part 1.

O'Dowd, W.J., Pennline, H.W., Freeman, M.C., Granite, E.J., Hargis, R.A., Lacher, C.J., et al., 2006. A technique to control mercury from flue gas: the Thief Process. Fuel Process. Technol. 87, 1071–1084.

Ogden, J.S., Williams, S.J., 1980. The identification of molecular $NaPO_3$ and $NaPO_2$ by matrix isolation infrared spectroscopy. J. Chem. Phys. 73, 2007–2008.

Oggiano, R., Solinas, G., Forte, G., Bocca, B., et al., 2018. Trace elements in ALS patients and their relationships with clinical severity. Chemosphere 197, 457–466.

Ogunbileje, J.O., Sadagoparamanujam, V.-M., Anetor, J.I., Farombi, E.O., Akinosun, O.M., 2013. Lead, mercury, cadmium, nickel, copper, zinc, calcium, iron, manganese and chromium (VI) levels in Nigeria and United States of America cement dust. Chemosphere 90, 2743–2749.

O'Hare, P.A.G., 1995. Thermochemical studies of inorganic chalcogenides by fluorine-combustion calorimetry: binary compounds of germanium and silicon with sulfur, selenium and tellurium. Thermochim. Acta 267, 1–13.

Ohata, M., Matsubayashi, N., 2014. Determination of hexavalent chromium in plastic certified reference materials by X-ray absorption fine structure analysis. Spectrochim. Acta B 93, 14–19.

Ohki, A., Yamada, K., Furuzono, T., Nakajima, T., Takanashi, H., 2011. Analysis of trace elements in flue gas desulfurization water in the coal combustion system and the removal of boron and mercury from the water. Energy Fuels 25, 3568–3573.

Ohrman, O., Haggstrom, C., Wiinikka, H., Hedlund, J., Gebart, R., 2012. Analysis of trace components in synthesis gas generated by black liquor gasification. Fuel 102, 173–179.

Okabayashi, T., Oya, A., Yamamoto, T., et al., 2016. Rotational spectrum of the AgS radical in the $X^2\Pi_i$ state. J. Mol. Spectrosc. 329, 13–19.

Okuno, T., Sonoyama, N., Hayashi, J.-i, Li, C.-Z., Sathe, C., Chiba, T., 2005. Primary release of alkali and alkaline earth metallic species during the pyrolysis of pulverized biomass. Energy Fuels 19, 2164–2171.

Oleschko, H., Muller, M., 2007. Influence of coal composition and operating conditions on the release of alkali species during combustion of hard coal. Energy Fuels 21, 3240–3248.

Oljaca, M., Luten, H.A., Tomov, T., Sundell, S., Hunt, A., 2003. Deposition of $Ba_XSr_{1\text{-}X}TiO_3$ in atmospheric pressure flame: combustion monitoring and optimization of thin film properties. Surf. Eng. 19, 51–57.

Olson, E.S., Azenkeng, A., Laumb, J.D., Jensen, R.R., Benson, S.A., Hoffmann, M.R., 2009. New developments in the theory and modeling of mercury oxidation and binding on activated carbons in flue gas. Fuel Process. Technol. 90, 1360–1363.

Olsson, J.G., Jaglid, U., Pettersson, J.B.C., Hald, P., 1997. Alkali metal emission during pyrolysis of biomass. Energy Fuels 11, 779–784.

Olza, J., Aranceta-Bartrina, J., Gonzalez-Gross, M., Ortega, R.M., et al., 2017. Reported dietary intake, disparity between the reported consumption and the level needed for adequacy and food sources of calcium, phosphorus, magnesium and vitamin D in the Spanish population: findings from the ANIBES study. Nutrients 9 (2), Art. No. e-168.

Omine, N., Romero, C.E., Kikkawa, H., Wu, S., Eswaran, S., 2012. Study of elemental mercury re-emission in a simulated wet scrubber. Fuel 91, 93–101.

Ondov, J.M., Ragaini, R.C., Biermann, A.H., 1979. Emissions and particle-size distributions of minor and trace elements at two Western coal-fired plants equipped with cold-side electrostatic precipitators. Environ. Sci. Technol. 13, 946–953.

Ondov, J.M., Choquette, C.E., Zoller, W.H., Gordon, G.E., Biermann, A.H., Heft, R.E., 1989. Atmospheric behavior of trace elements on particles emitted from a coal fired power plant. Atmos. Environ. 23, 2193–2204.

Oniyama, E., Wahlbeck, P.G., 1998. Phase equilibria in the bismuth-oxygen system. J. Phys. Chem. B 102, 4418–4425.

Operti, L., Tews, E.C., MacMahon, T.J., Freiser, B.S., 1989. Thermochemical properties of gas phase MgOH and MgO determined by Fourier Transform mass spectrometry. J. Am. Chem. Soc. 111, 9152–9156.

Opila, E.J., Myers, D.L., 2004. Alumina volatility in water vapor at elevated temperatures. J. Am. Ceram. Soc. 87 (9), 1701–1705.

Opila, E.J., Myers, D.L., Jacobson, N.S., Nielsen, I.M.B., Johnson, D.F., Olminsky, J.K., et al., 2007. Theoretical and experimental investigation of the thermochemistry of $CrO_2(OH)_2(g)$. J. Phys. Chem. A 111, 1971–1980.

Ostroff, A.G., Sanderson, R.T., 1959. Thermal stability of some metal sulfates. J. Inorg. Nucl. Chem. 9, 45–50.

Ota, K., Conger, W.L., 1977. Thermochemical hydrogen production via a cycle using barium and sulfur: reaction between barium sulfide and water. Int. J. Hydrog. Energy 2 (2), 101–106.

Otero-Rey, J.R., Lopez-Vilarino, J.M., Moreda-Pineiro, J., Alonso-Rodriguez, E., Muniategui-Lorenzo, S., Lopez-Mahia, P., et al., 2003. As, Hg and Se flue gas sampling in a coal fired power plant and their fate during coal combustion. Environ. Sci. Technol. 37, 5262–5267.

Ottosen, L.M., Kirkelund, G.M., Jensen, P.E., 2013. Extracting phosphorus from incinerated sewage sludge ash rich in iron or aluminum. Chemosphere 91, 963–969.

Outridge, P.M., Mason, R.P., Wang, F., Guerrero, S., Heimburger-Boavida, L.E., 2018. Updated Global and Oceanic mercury budgets for the United Nations Global Mercury Assessment 2018. Environ. Sci. Technol. 52, 11466–11477.

Owzarski, T.P., Franzen, H.F., 1974. High temperature mass spectrometry, vaporization and thermodynamics of vanadium monosulfide. J. Chem. Phys. 60, 1113–1117.

Ozawa, S., Naruse, K., Ito, K., Kojima, Y., Matsuda, H., Mishima, S.-I., et al., 2003. Evolution of HCl from KCl, $CaCl_2$ and NaCl in an SO_2- O_2-H_2O atmosphere under a simulated condition of municipal wastes incineration. J. Chem. Eng. Jpn. 36, 867–873.

Ozden, B., Guler, E., Vaasma, T., Horvath, M., Kiisk, M., Kovacs, T., 2018. Enrichment of naturally occurring radionuclides and trace elements in Yatagan and Yenikoy coal-fired thermal power plants, Turkey. J. Environ. Radioact. 188, 100–107.

Ozeki, H., Saito, S., 2003. Microwave spectra of HPO and DPO: molecular structure. J. Mol. Spectrosc. 219, 305–312.

Pabst, R.E., Bennett, S.L., Margrave, J.L., Franklin, J.L., 1976. Negative ion electron impact studies of arsenic trihalides: AsF_3, $AsCl_3$ and $AsBr_3$. J. Chem. Phys. 65, 1550–1560.

Pacyna, J.M., Pacyna, E.G., 2001. An assessment of global and regional emissions of trace metals to the atmosphere from anthropogenic sources worldwide. Environ. Rev. 9 (4), 269–298.

Pacyna, E.G., Pacyna, J.M., Steenhuisen, F., Wilson, S., 2006. Global anthropogenic mercury emission inventory for 2000. Atmos. Environ. 40, 4048–4063.

Pacyna, E.G., Pacyna, J.M., Fudala, J., Strzelecka-Jastrzab, E., Hlawiczka, S., Panasiuk, D., et al., 2007. Current and future emissions of selected heavy metals to the atmosphere from anthropogenic sources in Europe. Atmos. Environ. 41, 8557–8566.

Pacyna, J.M., Pacyna, E.G., Aas, W., 2009. Changes of emissions and atmospheric deposition of mercury, lead and cadmium. Atmos. Environ. 43, 117–127.

Pacyna, E.G., Pacyna, J.M., Sundseth, K., Munthe, J., Kindbom, K., Wilson, S., 2010. Global emission of mercury to the atmosphere from anthropogenic sources in 2005 and projections to 2020. Atmos. Environ. 44, 2487–2499.

Pacyna, J.M., Sundseth, K., Pacyna, E.G., 2016. Sources influxes of harmful metals. In: Pacyna, J.M., Pacyna, E.G. (Eds.), Environ Determin Human Health. Molecular and Integrative Toxicology Series, pp. 1–25. Available from: https://doi.org/10.1007/978-3-319-43142-0_1.

Pacyna, J.M., Travnikov, O., De Simone, F., et al., 2016a. Current and future levels of mercury atmospheric pollution on a global scale. Atmos. Chem. Phys. 16, 12495–12511.

Padak, B., Wilcox, J., 2009. Understanding mercury binding on activated carbon. Carbon 47, 2855–2864.

Padak, B., Brunetti, M., Lewis, A., Wilcox, J., 2006. Mercury binding on activated carbon. Environ. Prog. 54, 319–326.

Padilla, R., Ramirez, G., Ruiz, M.C., 2010. High temperature volatilization mechanism of stibnite in nitrogen-oxygen atmospheres. Metall. Mater. Trans. B Proc. Metall. Mater. Process. Sci. 41, 1284–1292.

Padilla, R., Villa, R., Ruiz, M.C., 2011. Decomposition and oxidation of bismuthinite in nitrogen-oxygen atmospheres. In: 2nd International Symposium on High Temperature Metallurgical Process, pp. 221–228.

Padley, P.J., Sugden, T.M., 1959. Determination of the dissociation constants and heats of formation of molecules by flame photometry. VI. Stabilities of MnO and MnOH and their mechanism of formation. Trans. Faraday Soc. 55, 2054–2061.

Pak, M.V., Gordon, M.S., 2001. The potential energy surfaces for AlO_2 using multireference wave functions. Chem. Phys. Lett. 344, 236–240.

Pakzadeh, B., Wos, J., Renew, J., 2014. Flue gas desulfurization wastewater treatment for coal-fired power industry. In: Proceedings of ASME Power Conference 2014; vol. 2, article V002T10A010.

Palaniappan, P.R., Pramad, K.S., 2011. The effect of titanium dioxide on the biochemical constituents of the brain of Zebrafish (*Danio rerio*): an FTIR study. Spectrochim. Acta A Mol. Biomol. Spectrosc. 79, 206–212.

Pan, Y., Luo, Z., Chang, Y.-C., et al., 2017. High level *ab initio* predictions for the ionization energies, bond dissociation energies and heats of formation of titanium oxides and their cations (TiO_n/TiO_n^+, n = 1, 2). J. Phys. Chem. A 121, 669–679.

Pan, J., Zhou, C., Liu, C., Tang, M., Cao, S., et al., 2018. Modes of occurrence of rare earth elements coal fly ash: a case study. Energy Fuels 32, 9738–9743.

Pan, S., Lin, L., Zeng, F., Zhang, J., Dong, G., et al., 2018a. Effects of lead, cadmium, arsenic, and mercury co-exposure on children's intelligence quotient in an industrialized area of southern China. Environ. Pollut. 235, 47–54.

Panas, I., Svensson, J.-E., Asteman, H., et al., 2004. Chromic acid evaporation upon exposure of Cr_2O_3(s) to H_2O (g) and O_2(g): mechanism from first principles. Chem. Phys. Lett. 383, 549–554.

Pankratz, L.B., 1984. Bulletin 674, 826 pp. Thermodynamic Properties of Halides. United States Department of the Interior. Bureau of Mines.

Pankratz, L.B., Mah, A.D., Watson, S.W., 1987. Bulletin 689, 427 pp. Thermodynamic Properties of Sulfides. United States Department of the Interior. Bureau of Mines.

Pannetier, G., Bregeault, J.M., Guenot, J., 1962. Study of the thermal dissociation of sulfates and intermediate basic sulfates. I. Pyrolysis of cadmium sulfate: the basic sulfate $CdSO_4.2CdO$. Bull. Soc. Chim. Fr. 2158–2162.

Paoletti, F., Sirini, P., Seifert, H., Vehlow, J., 2001. Fate of antimony in municipal solid waste incineration. Chemospere 42, 533–543.

Papai, I., 1995. Theoretical study of the $Cu(H_2O)$ and $Cu(NH_3)$ complexes and their photolysis products. J. Chem. Phys. 103 (5), 1860–1870.

Papakondylis, A., Mavridis, A., Metropoulos, A., 1995. Theoretical study of the electronic and geometrical structure of the ground and low-lying states of NCl_2, PCl_2, NCl and PCl radicals. J. Phys. Chem. 99, 10759–10765.

Parekunnel, T., O'Brien, L.C., Kellerman, T.L., et al., 2001. Fourier transform emission spectroscopy of CuCl. J. Mol. Spectrosc. 206, 27–32.

Park, H., Schlesinger, W.H., 2002. Global biogeochemical cycle of boron. Global Biogeochem. Cycles 16 (4), 20-1. 11 pp.

Parry, I.S., Hermes, A.C., Kartouzian, A., Mackenzie, S.R., 2014. Imaging the photodissociation dynamics of neutral metal clusters: copper dimer, Cu_2, and copper oxide, CuO. PhysChemChemPhys 16, 458–466.

Parsons, M.L., Smith, B.W., Bentley, G.E., 1975. Handbook of Flame Spectroscopy. Plenum Press, New York.

Partridge, H., Langhoff, S.R., Bauschlicher Jr, C.W., 1985. The dissociation energy of SrO. Chem. Phys. Lett. 121, 163–168.

Partridge, H., Langhoff, S.R., Bauschlicher Jr, C.W., 1986. *Ab initio* calculations on the positive ions of the alkaline-earth oxides, fluorides and hydroxides. J. Chem. Phys. 84, 4489–4496.

Partridge, H., Langhoff, S.R., Bauschlicher Jr, C.W., 1988. Theoretical study of the alkali and alkaline-earth monosulfides. J. Chem. Phys. 88, 6431–6437.

Partridge, H., Bauschlicher Jr, C.W., Sodupe, M., Langhoff, S.R., 1992. Comment on "An *ab initio* study of the ionization of sodium superoxide, NaO_2. J. Chem. Phys. 96 (10), 7871. Marshall P. "Reply". ibid. 7872.

Partridge, H., Bauschlicher Jr, C.W., Sodupe, M., Langhoff, S.R., 1992a. Theoretical determination of the alkali metal superoxide bond energies. Chem. Phys. Lett. 195, 200–206.

Pasternack, L., 1992. Gas phase modeling of homogeneous boron/oxygen/hydrogen/carbon combustion. Combust. Flame 90, 259–268.

Pasternack, L., Dagdigian, P.J., 1978. The reaction of metastable Ca atoms with O_2 and CO_2. Chem. Phys. 33, 1–11.

Patrascu, A.T., Hill, C., Tennyson, J., Yurchenko, S.N., 2014. Study of the electronic and rovibronic structure of the $X^2\Sigma^+$, $A^2\Pi$ and $B^2\Sigma^+$ states of AlO. J. Chem. Phys. 141, 144312. 10 pp.

Patrick, R., Golden, D.M., 1984. Termolecular reactions of alkali metal atoms with O_2 and OH. Int. J. Chem. Kinet. 16, 1567–1574.

Patrikeev, Yu.B., Morozova, V.A., Dudchik, G.P., et al., 1973. Saturated vapor pressure of scandium trichloride. Zhur Fiz Khim 47 (1), 266.

Pattanaik, S., Huggins, F.E., Huffman, G.P., 2012. Chemical speciation of Fe and Ni in residual oil fly ash fine particulate matter using X-ray absorption spectroscopy. Environ. Sci. Technol. 46, 12927–12935.

Patterson, C.C., Settle, D.M., 1987. Magnitude of lead flux to the atmosphere from volcanoes. Geochim. Cosmochim. Acta 51 (3), 675–681.

Patterson, P.M., Goodings, J.M., 1995. Ionization of the group III metals La, Y and Sc in $H_2/O_2/Ar$ flames. Int. J. Mass Spectrom. Ion Process. 148, 55–76.

Pavlish, J.H., Hamre, L.L., Zhuang, Y., 2010. Mercury control technologies for coal combustion and gasification systems. Fuel 89, 838–847.

Paz, S., Rubio, C., Frias, I., Gutierrez, A.J., et al., 2019. Toxic metals (Al, Cd, Pb, and Hg) in the most consumed edible seaweeds in Europe. Chemosphere 218, 879–884.

Pearse, R.W.B., Gaydon, A.G., 1976. The Identification of Molecular Spectra, fourth ed. Chapman & Hall, London.

Peccia, J., Westerhoff, P., 2015. We should expect more out of our sewage sludge. Environ. Sci. Technol. 49, 8271–8276.

Pedersen, A.J., van Lith, S.C., Frandsen, F.J., Steinsen, S.D., Holgersen, L.B., 2010. Release to the gas phase of metals, S and Cl during combustion of dedicated waste fractions. Fuel Process. Technol. 91, 1062–1072.

Pedley, J.B., Marshall, E.M., 1983. Thermochemical data for gaseous monoxides. J. Phys. Chem. Ref. Data 12, 967–1031.

Peebles, L.R., Marshall, P., 2002. A coupled-cluster study of the enthalpy of formation of nitrogen sulfide, NS. Chem. Phys. Lett. 366, 520–524.

Pelino, M., Viswanadham, P., Edwards, J.G., 1979. Vaporization chemistry and thermodynamics of titanium monosulfides by the computer-automated simultaneous Knudsen-torsion effusion method. J. Phys. Chem. 83, 2964–2969.

Pelucchi, M., Frassoldati, A., Faravelli, T., Ruscic, B., Glarborg, P., 2015. High temperature chemistry of HCl and Cl_2. Combust. Flame 162, 2693–2704.

Peng, B., Wu, D., 2015. Study on bromine release from bituminous coal during combustion. Fuel 157, 82–86.

Peng, B., Wu, D., Lai, J., Xiao, H., Li, P., 2012. Simultaneous determination of halogens (F, Cl, Br, and I) in coal using pyrohydrolysis combined with ion chromatography. Fuel 94, 629–631.

Peng, H., Chen, Y.L., Weng, L.P., Ma, J., Ma, Y.L., et al., 2019. Comparisons of heavy metal input inventory in agricultural soils in North and South China: a review. Sci. Total Environ. 660, 776–786.

Penthor, S., Zerobin, F., Mayer, K., et al., 2015. Investigation of the performance of a copper based oxygen carrier for chemical looping combustion in a 120 kW pilot plant for gaseous fuels. Appl. Energy 145, 52–59.

Peralta, J.E., Batista, E.R., Scuseria, G.E., Martin, R.L., 2005. All-electron hybrid density functional calculations on UF_n and UCl_n (n = 1-6). J. Chem. Theory Comput. 1, 612–616.

Percival, C.J., Welz, O., Eskola, A.J., et al., 2013. Regional and global impacts of Criegee intermediates on atmospheric sulfuric acid concentrations and first steps of aerosol formation. Faraday Discuss. 165, 45–73.

Pereira, C.C.L., Marsden, C.J., Marcalo, J., Gibson, J.K., 2011. Actinide sulfides in the gas phase: experimental and theoretical studies of the thermochemistry of AnS(An = Ac, Th, Pa, U, Np, Pu, Am, Cm). PhysChemChemPhys 13, 12940–12958.

Pershina, V., Fricke, B., 1996. Group 6 dioxydichlorides MO_2Cl_2 (M = Cr, Mo, W and element 106, Sg): the electronic structure and thermodynamic stability. J. Phys. Chem. 100, 8748–8751.

Pershina, V., Ilias, M., 2018. Electronic structure and properties of MAu and MOH, where M = Tl and Nh: new data. Chem. Phys. Lett. 694, 107–111.

Persico, M., 1994. Tunneling in the $A^1\Pi$ state and the dissociation energy of BH. Mol. Phys. 81 (6), 1463–1471.

Peter, A.L.J., Viraraghavan, T., 2005. Thallium: a review of public health and environmental concerns. Environ. Int. 31, 493–501.

Peterson, K.A., Shepler, B.C., Singeton, J.M., 2007. The group 12 metal chalcogenides: an accurate multireference configuration interaction and coupled cluster study. Mol. Phys. 105, 1139–1155.

Petrie, S., 1998. Thermochemistry of aluminum halides: a theoretical appraisal. J. Phys. Chem. A 102, 7828–7834.

Phattamajintatamrong, N., Puanngam, M., Imyim, A., 2017. Speciation analysis of arsenite, arsenate and dimethyl arsenic acid using silica-based columns and ICP-OES. Curr. Anal. Chem. 13 (2), 137–143.

Piacente, V., Scardala, P., Ferro, D., 1992. Study of the vaporization behavior of Sb_2S_3 and Sb_2Te_3 from their vapor pressure measurements. J. Alloys Compds. 178, 101–115.

Pieper, K.J., Nystrom, V.E., Parks, J., Jennings, K., Faircloth, H., et al., 2018. Elevated lead in water of private wells poses health risks: case study in Macon County, North Carolina. Environ. Sci. Technol. 52, 4350–4357.

Pilon-Smits, E.A.H., 2017. Mechanisms of plant selenium hyperaccumulation. In: Pilon-Smits, E.A.H., Winkel, L. H.E., Lin, Z.Q. (Eds.), Selenium in Plants: Molecular, Physiological, Ecological and Evolutionary Aspects. Plant Ecophysiology Book Series, Springer International Publishing, Springer Nature, vol. 11, pp. 53–66.

Pinzani, M.C.C., Ansell, S., Somogyi, A., Steenari, B.-M., Lindqvist, O., 2004. Microextended X-ray absorption fine structure studies of cadmium speciation within single municipal solid waste fly ash particles. Anal. Chem. 76, 1596–1602.

Piotrowska, P., Zevenhoven, M., Davidsson, K., Hupa, M., Amand, L.-E., Barisic, V., et al., 2010. Fate of alkali metals and phosphorus of rapeseed cake in circulating fluidized bed boiler. II. Co-combustion with coal. Energy Fuels 24, 4193–4205.

Pirrone, N., Cinnirella, S., Feng, X., Finkelman, R.B., Friedli, H.R., Leaner, J., et al., 2010. Global mercury emissions to the atmosphere from anthropogenic and natural sources. Atmos. Chem. Phys. 10, 5951–5964.

Plane, J.M.C., 1992. A comparison between the oxidation reactions of the alkali and alkaline earth atoms. In: Fontijn, A. (Ed.), Gas Phase Metal Reactions. Elsevier Science Publishers, Amsterdam, pp. 29–56.

Plane, J.M.C., Rajasekhar, B., 1989. Kinetic study of the reactions $Na + O_2 + N_2$ and $Na + N_2O$ over an extended temperature range. J. Phys. Chem. 93, 3135–3140.

Plane, J.M.C., Rollason, R.J., 1999. A study of the reactions of Fe and FeO with NO_2, and the structure and bond energy of FeO_2. PhysChemChemPhys 1, 1843–1849.

Plane, J.M.C., Whalley, C.L., 2012. A new model for magnesium chemistry in the upper atmosphere. J. Phys. Chem. A 116, 6240–6252.

Plane, J.M.C., Rajasekhar, B., Bartolotti, L., 1990. Kinetic study of the reaction $K + O_2 + M$ ($M = N_2$, He) from 250 to 1103 K. J. Phys. Chem. 94 (10), 4161–4167.

Plante, E.R., Sessoms, A.B., 1973. Vapor pressure and heat of sublimation of tungsten. J. Res. Natl. Bur. Stand. A Phys. Chem. 77 (2), 237–242.

Plies, V., 1991. Mass spectrometric investigations of the reactions of chlorine with chromium (3 +) and chromium (3 +) oxychloride at high temperatures: the heat of formation of gaseous chromium (4 +) dichloride oxide. Zeit. Anorg. Allgem. Chem. 602, 97–104.

Plies, V., Jansen, M., 1983. Mass spectrometric studies of the gaseous phase above arsenic pentoxide. Zeit. Anorg. Allgem. Chem. 497, 185–190.

Politzer, P., Lane, P., Grice, M.E., 2001. Energetics of aluminum combustion. J. Phys. Chem. A 105, 7473–7480.

Polukarov, A.N., Makhmetov, M.Z.H., Buketov, E.A., 1971. Dissociation of arsenic pentoxide. Zhur. Fiz. Khim. 45, 2688.

Pontes, M.A.P., de Oliveira, M.H., Ferrao, L.F.A., et al., 2015. AsCl radical: the low-lying electronic states and the (1) $^3\Pi$- $X^3\Sigma^-$ electronic transition. Chem. Phys. Lett. 634, 66–70.

Poole, D., Argent, B.B., Sharifi, V.N., Swithenbank, J., 2008. Prediction of the distribution of alkali and trace elements between the condensed and gaseous phases in a municipal solid waste incinerator. Fuel 87, 1318–1333.

Popkov, O.S., Semenov, G.A., 1971. Mass spectrometric study of the vaporization of lithium and sodium aluminates. Zhur. Fiz. Khim. 45 (2), 476–477.

Popovic, A., Lesar, A., Gucek, M., Bencze, L., 1997. Mass spectrometric investigation of the evaporation properties of lead oxide. Rapid Commun. Mass Spectrom. 11, 459–468.

Porter, R.F., Schoonmaker, R.C., Addison, C.C., 1959. Mass spectrum of gaseous cupric nitrate. Proc. Chem. Soc. 11–12.

Pouillon, Y., Massobrio, C., 2000. A density functional study of CuO_2 molecules: structural stability, bonding and temperature effects. Chem. Phys. Lett. 331, 290–298.

Pouilly, B., Robbe, J.-M., Schamps, J., Field, R.W., Young, L., 1982. Perturbations in the $A^1\Pi$-$X^1\Sigma^+$ system of the BeS molecule: the $a^3\Pi_i$ state. J. Mol. Spectrosc. 96, 1–55.

Pouilly, B., Robbe, J.-M., Alexander, M.H., 1984. Cross sections for collisional energy transfer between low-lying electronic states of MgO($X^1\Sigma^+$, $a^3\Pi$, $A^1\Pi$) in collisions with N_2O. J. Phys. Chem. 88, 140–148.

Poulston, S., Granite, E.J., Pennline, H.W., Hamilton, H., Smith, A.W.J., 2011. Palladium based sorbents for high temperature arsine removal from flue gas. Fuel 90, 3118–3121.

Poykio, R., Nurmesniemi, H., Keiski, R.L., 2010. Heavy metals in flue gas cleaning residue. Environ. Chem. Lett. 8, 295–300.

Prascher, B.P., Woon, D.E., Peterson, K.A., Dunning Jr, T.H., Wilson, A.K., 2011. Gaussian basis sets for use in correlated molecular calculations. VII. Valence, core-valence and scalar relativistic basis sets for Li, Be, Na and Mg. Theor. Chem. Acc. 128, 69–82.

Pratsinis, S.E., 1998. Flame aerosol synthesis of ceramic powders. Prog Energy Combust Sci 24, 197–219.

Presto, A.A., Granite, E.J., 2006. Survey of catalysts for oxidation of mercury in flue gas. Environ. Sci. Technol. 40, 5601–5609.

Price, L.S., Parkin, I.P., Hardy, A.M.E., Clark, R.J.H., Hibbert, T.G., Molloy, K.C., 1999. Atmospheric pressure chemical vapor deposition of tin sulfides (SnS, Sn_2S_3 and SnS_2) on glass. Chem. Mater. 11, 1792–1799.

Pudasainee, D., Kim, J.-H., Seo, Y.-C., 2009. Mercury emission trend influenced by stringent air pollutants regulation for coal fired power plants in Korea. Atmos. Environ. 43, 6254–6259.

Pudasainee, D., Lee, S.J., Lee, S.-H., Kim, J.-H., Jang, H.-N., Cho, S.-J., et al., 2010. Effect of Selective Catalytic Reactor on oxidation and enhanced removal of mercury in coal-fired power plants. Fuel 89, 804–809.

Pudasainee, D., Kim, J.-H., Yoon, Y.-S., Seo, Y.-C., 2012. Oxidation, re-emission and mass distribution of mercury in bituminous coal-fired power plants with SCR, CS-ESP and wet FGD. Fuel 93, 312–318.

Pudasainee, D., Seo, Y.-C., Sung, J.-H., Jang, H.-N., Gupta, R., 2017. Mercury co-beneficial capture in air pollution control devices of coal-fired power plants. Int. J. Coal Geol. 170, 48–53.

Pueschel, R.F., 1976. Aerosol formation during coal combustion: condensation of sulfates and chlorides on fly ash. Geophys. Res. Lett. 3, 651–653.

Pulliam, R.L., Ziurys, L.M., 2010. The pure rotational spectrum of the CrS radical in its $X^5\Pi_r$ state. J. Chem. Phys. 133 (17), 174313. 6 pp.

Pupp, C., Yamdagni, R., Porter, R.F., 1969. Mass spectrometric study of the evaporation of $BaMoO_4$ and $BaWO_4$. J. Inorg. Nucl. Chem. 31, 2021–2029.

Purgar, A., Winter, F., Blasenbauer, D., et al., 2016. Main drivers for integrating zinc recovery from fly ashes into the Viennese waste incinerator cluster. Fuel Process. Technol. 141, 243–248.

Pyle, D.M., Mather, T.A., 2003. The importance of volcanic emissions for the global atmospheric mercury cycle. Atmos. Environ. 37, 5115–5124.

Pyykonen, J., Jokiniemi, J., 2003. Modeling alkali chloride superheater deposition and its implications. Fuel Process. Technol. 80, 225–262.

Qi, Y., Wu, W., Han, L., et al., 2016. Using TG-FTIR and XPS to understand thermal degradation and flame-retardant mechanism of flexible poly(vinyl chloride) filled with metallic ferrites. J. Therm. Anal. Calorim. 123, 1263–1271.

Qi, J., Lai, Y.P., Liang, C.M., Yan, S.Q., Huang, K., et al., 2019. Prenatal thallium exposure and poor growth in early childhood: a prospective birth cohort study. Environ. Int. 123, 224–230.

Qiu, J.-R., Li, F., Zheng, C.-G., 1999. Mineral transformation during combustion of coal blends. Int. J. Energy Res. 23, 453–463.

Qiu, K., Mattisson, T., Steenari, B.-M., Lindqvist, O., 1997. Thermogravimetric combined with mass spectrometric studies on the oxidation of calcium sulfide. Thermochim. Acta 298, 87–93.

Quann, R.J., Sarofim, A.F., 1982. Vaporization of refractory oxides during pulverized coal combustion. Symp. (Int.) Combust. 19, 1429–1440.

Quann, R.J., Neville, M., Janghorbani, M., Mims, C.A., Sarofim, A.F., 1982. Mineral matter and trace element vaporization in a laboratory pulverized-coal combustion system. Environ. Sci. Technol. 16, 776–781.

Quann, R.J., Neville, M., Sarofim, A.F., 1990. A laboratory study of the effect of coal selection on the amount and composition of combustion generated submicron particles. Combust. Sci. Technol. 74, 245–265.

Querol, X., Fernandez-Turiel, J.L., Lopez-Soler, A., 1995. Trace elements in coal and their behavior during combustion in a large power station. Fuel 74, 331–343.

Quick, W.J., Irons, R.M.A., 2002. Trace element partitioning during the firing of washed and untreated power station coals. Fuel 81, 665–672.

Quinto, F., Hrnecek, E., Krachler, M., et al., 2013. Measurement of ^{236}U in ancient and modern peat samples and implications for post-depositional migration of fallout radionuclides. Environ. Sci. Technol. 47 (10), 5243–5250.

Raask, E., 1985. Mineral Impurities in Coal Combustion: Behavior, Problems and Remedial Measures. Hemisphere Publishing, Washington, DC.

Raclavska, H., Corsaro, A., Hlavsova, A., et al., 2015. The effect of moisture on the release and enrichment of heavy metals during pyrolysis of municipal solid waste. Waste Manage. Res. 33 (3), 267–274.

Raeva, A.A., Klykov, O.V., Kozliak, E.I., et al., 2011. In situ evaluation of inorganic matrix effects on the partitioning of three trace elements (As, Sb, Se) at the outset of coal combustion. Energy Fuels 25, 4290–4298.

Rafael, T.L., Laura, A.M., Felipe, G.M., Sujitha, S.B., Jonathan, M.P., 2019. Understanding the antagonism of mercury and selenium in two shark species Baja California South, Mexico. Sci. Total Environ. 650, 202–209. Part 1.

Ragaini, R.C., Ondov, J.M., 1977. Trace element emissions from Western United States coal-fired power plants. J. Radioanal. Chem. 37, 679–691.

Rahman, M.M., Hossain, K.F.B., Banik, S., Sikder, M.T., Akter, M., et al., 2019. Selenium and zinc protections against metal-(loids)-induced toxicity and disease manifestations: a review. Ecotoxicol. Environ. Safety 168, 146–163.

Rai, S.B., Rai, D.K., 1984. Electronic spectra of diatomic oxides of Group 5A elements. Chem. Rev. 84, 73–88.

Rai, D., Mattigod, S.V., Moore, D.A., 1988. Characterization of $Fe_2(SeO_3)_3$ $6H_2O$, a new synthetic ferric selenite. Mater. Res. Bull. 23, 1621–1628.

Raj, A., Croiset, E., Wen, J.Z., 2018. Numerical analysis of effects of iron pentacarbonyl as fuel additive for reducing NO and soot precursors from methane/air diffusion flame. Fuel 216, 768–780.

Ralston, N.V.C., Raymond, L.J., 2006. Selenium's protective effect against mercury toxicity. Neurotoxicology 27 (6), 1172.

Ralston, N.V.C., Raymond, L.J., 2016. Functional deletion of brain selenoenzymes by methylmercury. pp.71–72. In: Banuelos, G.S., et al., (Eds.), Global Advance in Selenium Research from Theory to Application. Proceedingsv of 4th International Conference on Se, Brazil, 2015. CRC Press, Taylor and Francis Group, Leiden, The Netherlands, 207 pp.

Ralston, N.V.C., Raymond, L.J., 2016a. The "SOS" mechanisms of methylmercury toxicity. pp. 73–74. In: Banuelos, G.S., et al., (Eds.), Global Advance in Selenium Research from Theory to Application. Proceedings of 4th International Conference on Se, Brazil, 2015. CRC Press, Taylor and Francis Group, Leiden, The Netherlands, 207 pp.

Ralston, N.V.C., Raymond, L.J., 2018. Mercury's neurotoxicity is characterized by its disruption of selenium biochemistry. Biochim. Biophys. Acta-General Subj. 1862 (11), 2405–2416.

Ram, R.S., Bernath, P.F., 1992. Fourier transform infrared emission spectroscopy of a new $A^3\Pi_i$-$X^3\Sigma^-$ system of NiO. J. Mol. Spectrosc. 155, 315–325.

Ram, R.S., Bernath, P.F., 1994. High resolution Fourier transform emission spectroscopy of YH. J. Chem. Phys. 101, 9283–9288.

Ram, R.S., Bernath, P.F., 1996. Fourier transform emission spectroscopy of the $B^1\Pi$-$X^1\Sigma^+$, $C^1\Sigma^+$- $X^1\Sigma^+$, and $G^1\Pi$-$X^1\Sigma^+$ systems of ScH and ScD. J. Chem. Phys. 105, 2668–2674.

Ram, R.S., Bernath, P.F., 1997. Fourier transform emission spectroscopy of ScH and ScD: the new singlet electronic states $A^1\Delta$, $D^1\Pi$, $E^1\Delta$ and $F^1\Sigma^-$. J. Mol. Spectrosc. 183, 263–272.

Ram, R.S., Jarman, C.N., Bernath, P.F., 1993. Fourier transform infrared emission spectroscopy of a new $A^4\Pi_i$-$X^4\Delta_i$ system of CoO. J. Mol. Spectrosc. 160, 574–584.

Ram, R.S., Adam, A.G., Sha, W., Tsouli, A., Lievin, J., Bernath, P.F., 2001. The electronic structure of ZrCl. J. Chem. Phys. 114, 3977–3987.

Ram, R.S., Lievin, J., Li, G., et al., 2001a. The $X^3\Sigma^-$ ground state of WO. Chem. Phys. Lett. 343, 437–445.

Ram, R.S., Gordon, I., Hirao, T., Yu, S., Bernath, P.F., Pinchemel, B., 2007. Fourier transform emission spectroscopy of the $C^3\Delta$- $X^3\Phi$, $D^3\Delta$- $X^3\Phi$, $G^3\Phi$-$X^3\Phi$ and $G^3\Phi$-$C^3\Delta$ systems of CoCl. J. Mol. Spectrosc. 243, 69–77.

Ram, R.S., Lievin, J., Bernath, P.F., 2009. Fourier transform emission spectroscopy and *ab initio* calculations on WO. J. Mol. Spectrosc. 256, 216–227.

Ram, R.S., Yu, S., Gordon, I., Bernath, P.F., 2009a. Fourier transform infrared emission spectroscopy of new systems of NiS. J. Mol. Spectrosc. 258 (1–2), 20–25.

Ram, R.S., Li, G., Bernath, P.F., 2011. Fourier transform emission spectroscopy of YH and YD: observation of new $A^1\Delta$ and $B^1\Pi$ electronic states. J. Chem. Phys. 135, 194308. 6 pp.

Ramakrishna, V., Duke, B.J., 2003. Extension of complete basis set model chemistries to molecules containing third row atoms Ga-Kr. J. Chem. Phys. 118 (14), 6137–6143.

Ramakrishnan, E.S., Chandrasekharaiah, M.S., 1975. Thermodynamic stability of amorphous germanium monoxide. Trans. Indian Inst. Metals 28, 58–63.

Ramirez-Solis, A., Daudey, J.P., 2004. *Ab initio* study on the spectroscopy of $CuCl_2$. I. Benchmark calculations on the $X^2\Pi_g$-$^2\Sigma_g^+$ transition. J. Chem. Phys. 120 (7), 3221–3228.

Ramirez-Solis, A., Daudey, J.P., 2005. *Ab initio* study on the spectroscopy of $CuCl_2$. II. Benchmark calculations on the $X^2\Pi_g$-$C^2\Delta_g$ and $X^2\Pi_g$-$D^2\Delta_g$ transitions. J. Chem. Phys. 122, 014315, 7 pp.

Ramirez-Solis, A., 2006. The spectroscopy of copper and silver monohalides: what modern quantum chemistry can and cannot do. Theor. Chem. Acc. 116, 641–654.

Ramirez-Solis, A., Poteau, R., Daudey, J.P., 2006. The electronic spectrum of $AgCl_2$: *ab initio* benchmark versus density-functional theory calculations on the lowest ligand-field states including spin-orbit effects. J. Chem. Phys. 124, 034307. 10 pp.

Ranjbar, N., Kuenzel, C., 2017. Cenospheres: a review. Fuel 207, 1–12.

Rasheed, H., Slack, R., Kay, P., 2016. Human health risk assessment for arsenic: a critical review. Crit. Rev. Environ. Sci. Technol. 46 (19–20), 1529–1583.

Rasmuson, J.O., Fassel, V.A., Kniseley, R.N., 1973. An experimental and theoretical evaluation of the nitrous oxide-acetylene flame as an atomization cell for flame spectroscopy. Spectrochim. Acta B 28, 365–406.

Rassooli, A., Boutorabi, M.A., Divandari, M., Azarniya, A., 2013. Effect of high heating rate on thermal decomposition behavior of titanium hydride, TiH_2, powder in air. Bull. Mater. Sci. 36, 301–309.

Ratte, H.T., 1999. Bioaccumulation and toxicity of silver compounds: a review. Environ. Toxicol. Chem. 18 (1), 89–108.

Rau, H., 1975. Vapor composition and van der Waals constants of arsenic. J. Chem. Thermodyn. 7, 27–32.

Rauh, E.G., Garg, S.P., 1981. The congruently vaporizing compositions and thermodynamic properties of the cubic phase of substoichiometric zirconium dioxide. High Temp. Sci. 14, 121–134.

Rayman, M.P., 2012. Selenium and human health. Lancet 379 (9822), 1256–1268.

Rayman, M.P., Winther, K.H., Pastor-Barriuso, R., Cold, F., et al., 2018. Effect of long-term selenium supplementation on mortality: results from a multiple-dose, randomized controlled trial. Free Radical Biol. Med. 127, 46–54.

Raymond, T.M., Davico, G.E., Hellberg, F., et al., 2002. Photoelectron spectroscopy of nickel, palladium, and platinum oxide anions. J. Mol. Spectrosc. 216, 1–14.

Reash, R.J., 2019. Bioavailability of mercury in powerplant wastewater and ambient river samples: evidence that the regulation of total mercury is not appropriate. Integr. Environ. Assess. Manage. 15 (1), 142–147.

Rechtsteiner, G.A., Ganske, J.A., 1997. The adsorption and oxidation of sulfur dioxide gas on alkali halide surfaces. Abstracts of Papers of the 213th National Meeting of the American Chemical Society, 213, p. 430.

Reddy, R.R., Reddy, A.S.R., Rao, T.V.R., 1985. Estimation of dissociation energy of the AsS molecule. Pramana (India) 25, 187–190.

Reddy, M.S., Basha, S., Joshi, H.V., Jha, B., 2005. Evaluation of the emission characteristics of trace metals from coal and fuel-oil fired power plants and their fate during combustion. J. Hazard. Mater. B 123, 242–249.

Reddy, K.S.K., Al Shoaibi, A., Srinivasakannan, C., 2014. Elemental mercury adsorption on sulfur-impregnated porous carbon: a review. Environ. Technol. 35 (1), 18–26.

Reed, G.P., Brain, S., Cahill, P., Fantom, I., 1997. Control of trace elements in gasification: a measurement methodology to validate thermodynamic predictions. In: Preto, F.D.S. (Ed.), 14th International Conference on Fluidized Bed Combustion. The American Society of Mechanical Engineers, New York, pp. 1285–1293.

Reed, G.P., Dugwell, D.R., Kandiyoti, R., 2001. Control of trace elements in gasification: distribution to the output streams of a pilot-plant gasifier. Energy Fuels 15, 794–800.

Reed, G.P., Ergudenler, A., Grace, J.R., Watkinson, A.P., Herod, A.A., Dugwell, D., et al., 2001a. Control of gasifier mercury emissions in a hot gas filter: the effect of temperature. Fuel 80, 623–634.

Reed, G.P., Paterson, N.P., Zhuo, Y., Dugwell, D.R., Kandiyoti, R., 2005. Trace element distribution in sewage sludge gasification: source and temperature effects. Energy Fuels 19, 298–304.

Rehder, D., 2012. The potentiality of vanadium in medicinal applications. Fut. Med. Chem. 4, 1823–1837.

Reich, H.J., Hondal, R.J., 2016. Why nature chose selenium. Am. Chem. Soc. Chem. Biol. 11 (4), 821–841.

Ren, Q., Li, L., 2015. Co-combustion of agricultural straw with municipal sewage sludge in a fluidized bed: role of phosphorus in potassium behavior. Energy Fuels 29, 4321–4327.

Ren, Y., Zhang, Y., Li, S., Law, C.K., 2015. Doping mechanism of vanadia/titania nanoparticles in flame synthesis by a novel optical spectroscopy technique. Proc. Combust. Inst. 35, 2283–2289.

Ribas, V.W., Ferrao, L.F.A., Robert-Neto, O., Machado, F.B.C., 2010. Transition probabilities and molecular constants of the low-lying electronic states of the MgS molecule. Chem. Phys. Lett. 492, 19–24.

Ricca, A., Bauschlicher Jr, C.W., 1998. Accurate heats of formation for PH_n, PH_n^+ and PH_n^-. Chem. Phys. Lett. 285, 455–458.

Ricca, A., Bauschlicher Jr, C.W., 1999. Heats of formation of GeH_n (n = 1-4) and Ge_2H_n (n = 1-6). J. Phys. Chem. A 103, 11121–11125.

Richards Jr, C.A., Vacek, G., DeLeeuw, B.J., Yamaguchi, Y., Schaefer III, H.F., 1995. An *ab initio* study on the ground state HBO-BOH system. J. Chem. Phys. 102, 1280–1287.

Richardson, G.M., 2001. Report, 59 pp. Critical Review on Natural Global and Regional Emissions of Six Trace Metals to the Atmosphere. Risk Logic Scientific Services, Inc., Ottawa, Canada.

Riederer, A.M., Belova, A., George, B.J., Anastas, P.T., 2013. Urinary cadmium in the 1999-2008 US National Health and Nutrition Examination Survey (NHANES). Environ. Sci. Technol. 47, 1137–1147.

Riess, M., Muller, M., 2011. High temperature sorption of arsenic in gasification atmosphere. Energy Fuels 25, 1438–1443.

Rimbaud, D., Restrepo, M., Louison, A., Boukhari, R., et al., 2017. Blood lead levels and risk factors for lead exposure among pregnant women in western French Guiana: the role of manioc consumption. J. Toxicol. Environ. Health A Curr. Issue 80 (6), 382–393.

Ripley, S., Robinson, E., Johnson-Down, L., Andermann, A., et al., 2018. Blood and hair mercury concentrations among Cree First Nations of Eeyou Istchee (Quebec, Canada): time trends, prenatal exposure and links to local fish consumption. Int. J. Circumpolar Health 77 (1), 1474706.

Risold, D., Hallstedt, B., Gauckler, L.J., 1996. The strontium-oxygen system. Calphad 20, 353–361.

Ritter, S.K., 2014. Study fuels toxicity debate: scientists call neurotoxic effects of industrial chemicals a global pandemic. Chem. Eng. News 92 (11), 40–41.

Ritter, S.K., 2016. A new life for coal ash. Chem. Eng. News 94 (7), 10–14.

Rittmann, B.E., Mayer, B., Westerhoff, P., Edwards, M., 2011. Capturing the lost phosphorus. Chemosphere 84, 846–853.

Robinson, A.L., Junker, H., Baxter, L.L., 2002. Pilot scale investigation of the influence of coal-biomass cofiring on ash deposition. Energy Fuels 16, 343–355.

Roca, M., Sanchez, A., Perez, R., et al., 2016. Biomonitoring of 20 elements in urine of children: levels and predictors of exposure. Chemosphere 144, 1698–1705.

Rocca, E., Steinmetz, P., Moliere, M., 2003. Revisiting the inhibition of vanadium-induced hot corrosion in gas turbines. J. Eng. Gas Turbine Power Trans. ASME 125, 664–669.

Rocca, E., Aranda, L., Moliere, M., 2008. Chemistry of ash-deposits on gas turbines hot parts: reactivity of nickel, zinc and iron oxides in (Na,V,S) molten salts. Mater. Sci. Forum 595/598: Part I High Temp. Corros. Prot. Mater. 7, pp. 169–176.

Rodgers, M.T., Walker, B., Armentrout, P.B., 1999. Reactions of Cu^+(^1S and ^3D) with O_2, CO, CO_2, N_2, NO, N_2O and NO_2. Int. J. Mass Spectrom. 182/183, 99–120.

Rodrigues, J.L.G., Araujo, C.F.S., dos Santos, N.R., Bandeira, M.J., et al., 2018. Airborne manganese exposure and neurobehavior in school-aged children living near a ferro-manganese alloy plant. Environ. Res. 167, 66–77.

Rollason, R.J., Plane, J.M.C., 2000. The reactions of FeO with O_3, H_2, H_2O, O_2 and CO_2. PhysChemChemPhys 2, 2335–2343.

Rondini, E.A., Walters, D.M., Bauer, A.K., 2010. Vanadium pentoxide induces pulmonary inflammation and tumor promotion in a strain-dependent manner. Particle Fibre Toxicol. 7, article 9.

Roos-Barraclough, F., Martinez-Cortizas, A., Garcia-Rodeja, E., Shotyk, W., 2002. A 14,500 year record of the accumulation of atmospheric mercury in peat: volcanic signals, anthropogenic influences and a correlation to bromine accumulation. Earth Planet Sci. Lett. 202, 435–451.

Ropp, R.C., Aia, M.A., 1962. Thermal analysis of phosphor raw materials. Anal. Chem. 34, 1288–1291.

Rosi, M., Bauschlicher, C.W., Chertihin, G.V., Andrews, L., 1998. The vibrational frequencies of CaO_2, ScO_2 and TiO_2: a comparison of theoretical methods. Theor. Chem. Acc. 99, 106–112.

Rosner, D.E., 2005. Flame synthesis of valuable nanoparticles: recent progress/current needs in areas of rate laws, population dynamics and characterization. Ind. Eng. Chem. Res. 44, 6045–6055.

Ross, A.J., Crozet, P., Bacis, R., et al., 1996. Laser induced fluorescence from $CuCl_2$ in a free-jet expansion: high resolution Fourier transform spectra. J. Mol. Spectrosc. 177, 134–142.

Rosser Jr, W.A., Inami, S.H., Wise, H., 1966. The quenching of pre-mixed flames by volatile inhibitors. Combust. Flame 10, 287–294.

Rostai, M., Wahlbeck, P.G., 1999. Dissociation energy of gaseous scandium monoxide. J. Chem. Thermodyn. 31, 255–261.

Roth, P., 2007. Particle synthesis in flames. Proc. Combust. Inst. 31, 1773–1788.

Rovner, L., Drowart, A., Drowart, J., 1967. Mass spectrometric determination of dissociation energies of molecules Bi_2, Bi_3, Bi_4 and BiPb. Trans. Faraday Soc. 63, 2906–2912.

Rowe, J.J., Steinnes, E., 1977. Determination of 30 elements in coal and fly ash by thermal and epithermal neutron activation analysis. Talanta 24, 433–439.

Roy, B., Choo, W.L., Bhattacharya, S., 2013. Prediction of distribution of trace elements under oxy-fuel combustion condition using Victorian brown coals. Fuel 114, 135–142.

Rubel, A.M., Hower, J.C., Mardon, S.M., Zimmerer, 2006. Thermal stability of mercury captured by ash. Fuel 85, 2509–2515.

Rubinovitz, R.L., Nixon, E.R., 1986. The photochemical Fe + H_2 reaction in Ar and Kr matrices by irradiation in the visible region. J. Phys. Chem. 90, 1940–1944.

Rudnyi, E.B., Vovk, O.M., Sidorov, L.N., et al., 1986. Enthalpy of formation of PO_2^-, PO_3^- and $NaPO_2$. High Temp. USSR 24, 56–63.

Rumminger, M.D., Linteris, G.T., 2000. Inhibition of pre-mixed carbon monoxide- hydrogen-oxygen-nitrogen flames by iron pentacarbonyl. Combust. Flame 120, 451–464.

Rumminger, M.D., Linteris, G.T., 2000a. The role of particles in the inhibition of premixed flames by iron pentacarbonyl. Combust. Flame 123, 82–94.

Rumminger, M.D., Reinelt, D., Babushok, V., Linteris, G.T., 1999. Numerical study of the inhibition of premixed and diffusion flames by iron carbonyl. Combust. Flame 116, 207–219.

Rungnim, C., Promarak, V., Hannongbua, S., et al., 2016. Complete reaction mechanisms of mercury oxidation on halogenated activated carbon. J. Hazard. Mater. 310, 253–260.

Ruscic, B., Berkowitz, J., 1993. Determination of consecutive bond energies by photoionization of SbH_n (n = 1-3). J. Chem. Phys. 99, 5840–5848.

Rutner, E., Haury, G.L., 1974. Nickel-NiO thermodynamic cycle: values of ΔG^o and ΔH^o. J. Chem. Eng. Data 19 (1), 19–27.

Ryabova, V.G., Gurvich, V.L., 1965. Investigation of the energy of the metal-hydroxyl bond in the CaOH, SrOH and BaOH molecules. High Temp. USSR 3, 284–287.

Ryabova, V.G., Gurvich, L.V., Khitrov, A.N., 1971. Spectrophotometric determination of molecular dissociation energies. II. $SrCl_2$ and SrOHCl. High Temp. USSR 9, 686–689.

Ryzhov, V., Mashyanov, N.R., Ozerova, N.A., Pogarev, S.E., 2003. Regular variations of the mercury concentration in natural gas. Sci. Total Environ. 304, 145–152.

Sable, E.R., 1990. Trace metals in combustible municipal solid waste. Proceedings of the 1990 National Waste Processing Conference: Integrated Waste Management, The Challenge Increases. 14th biennial meeting held in Long Beach CA. American Society of Mechanical Engineers, New York, pp. 229–233.

Sable, S.P., de Jong, W., Spliethoff, H., 2008. Combined homo- and heterogeneous model for mercury speciation in pulverized fuel combustion flue gases. Energy Fuels 22, 321–330.

Sadauskiene, I., Staneviciene, I., Liekis, A., Zekonis, G., 2016. Catalase activity in mouse brain: the effects of selenium and aluminum ions. Trace Elem. Electrolytes 33 (2), 64–69.

Saeki, Y., Kato, A., Matsuzaki, R., 1975. Reaction process between sodium tungstate and chlorine. Chem. Lett. Jpn. 935–938.

Sahoo, S.K., Mishra, S., Zunic, Z.S., et al., 2014. Distribution of uranium and selected trace metals in Balkan human scalp hair using inductively coupled plasma mass spectrometry. Int. J. Mass Spectrom. 373, 15–21.

Saifullah, Dahlawi, S., Naeem, A., Iqbal, M., Farooq, M.A., Bibi, S., et al., 2018. Opportunities and challenges in the use of mineral nutrition for minimizing arsenic toxicity and accumulation in rice: a critical review. Chemosphere 194, 171–188.

Saikia, P.K., Deori, K.L., 1999. Optical and electrical properties of tin oxide films prepared by atmospheric pressure chemical vapor deposition. Asian Chem. Lett. 3, 66–69.

Sakai, Y., Mogi, K., Miyoshi, E., 1999. Theoretical study of low-lying electronic states of TiCl and ZrCl. J. Chem. Phys. 111, 3989–3994.

Sakamoto, M., Yasutake, A., Kakita, A., et al., 2013. Selenomethionine protects against neuronal degeneration by methylmercury in the developing rat cerebrum. Environ. Sci. Technol. 47, 2862–2868.

Sakamoto, M., Chan, H.M., Domingo, J.L., Koriyama, C., Murata, K., 2018. Placental transfer and levels of mercury, selenium, vitamin E, and docosahexaenoic acid in maternal and umbilical cord blood. Environ. Int. 111, 309–315.

Sakamoto, M., Tatsuta, N., Chan, H.M., Domingo, J.L., et al., 2018a. Brain methylmercury uptake in fetal, neonate, weanling and adult rats. Environ. Res. 167, 15–20.

Sakellaris, C.N., Mavridis, A., 2012. Electronic structure and bonding of cobalt monoxide, CoO, and its ions CoO^+ and CoO^-: an *ab initio* study. J. Phys. Chem. A 116, 6935–6949.

Sakellaris, C.N., Mavridis, A., 2012a. First principles study of cobalt hydride, CoH, and its ions CoH^+ and CoH^-. J. Chem. Phys. 137, 034309. 13 pp.

Sakellaris, C.N., Mavridis, A., 2013. First principles exploration of NiO and its ions NiO^+ and NiO^-. J. Chem. Phys. 138, 054308. 15 pp.

Sakellaris, C.N., Miliordos, E., Mavridis, A., 2011. First principles study of the ground and excited states of FeO, FeO^+ and FeO^-. J. Chem. Phys. 134, 234308. 16 pp.

Sakulpitakphon, T., Hower, J.C., Trimble, A.S., Schram, W.H., Thomas, G.A., 2003. Arsenic and mercury partitioning in fly ash at a Kentucky power plant. Energy Fuels 17, 1028–1033.

Salavati-Niasari, M., Khansari, A., 2014. Synthesis and characterization of Co_3O_4 nanoparticles by a simple method. Comptes Rendus Chim. 17, 352–358.

Salihoglu, G., 2014. Immobilization of antimony waste slag by applying geopolymerization and stabilization/solidification technologies. J. Air Waste Manage. Assoc. 64 (11), 1288–1298.

Sallans, L., Lane, K.R., Squires, R.R., Freiser, B.S., 1985. Generation and reactions of atomic metal anions in the gas phase. Determination of the heterolytic and homolytic bond energies for VH, CrH, FeH, CoH and MoH. J. Am. Chem. Soc. 107, 4379–4385.

Sammut, M.L., Noack, Y., Rose, J., et al., 2010. Speciation of Cd and Pb in dust emitted from sinter plant. Chemosphere 78, 445–450.

Sanchez, J.L., Hager, J.P., 1994. Investigation of volatile hydroxide formation as the basis for the enhanced vapor transport of refractory metal oxides. In: Sohn, H.Y. (Ed.), Metallurgical Processes Early Twenty First Century. Proceedings of 2nd International Symposium Metallurgical Processes, vol. 1, pp. 523–542.

Sandelin, K., Backman, R., 1999. A simple two-reactor method for predicting distribution of trace elements in combustion systems. Environ. Sci. Technol. 33, 4508–4513.

Sankar, G., Kumar, D.S., Balasubramanian, K.R., 2019. Computational modeling of pulverized coal-fired boilers: a review on the current position. Fuel 236, 643–665.

Sano, N., Belton, G.R., 1974. The thermodynamics of volatilization of chromic oxide. II. The species CrO_2Cl_2. Metall. Trans. 5, 2151–2154.

Sano, A., Kanomata, M., Inoue, H., et al., 2012. Extraction of raw sludge containing iron phosphate for phosphorus recovery. Chemosphere 89, 1243–1247.

Santos, A.L., Takeuchi, R.M., Munoz, R.A.A., et al., 2012. Electrochemical determination of inorganic contaminants in automotive fuels. Electroanalysis 24 (8), 1681–1691.

Santos, L.R., Neves, C., Melo, M., Soares, P., 2018. Selenium and selenoproteins in immune mediated thyroid disorders. Diagnostics 8 (4), article 70.

Saqib, N., Backstrom, M., 2016. Chemical association and mobility of trace elements in 13 different fuel incineration bottom ashes. Fuel 172, 105–117.

Sarkar, A., Paul, B., 2016. The global menace of arsenic and its conventional remediation: a critical review. Chemosphere 158, 37–49.

Sasamoto, T., Mizushima, K., Sata, T., 1979. Transpiration study of the reaction of water vapor with barium oxide. Bull. Chem. Soc. Jpn. 52, 2127–2129.

Sasmaz, E., Kirchofer, A., Jew, A.D., et al., 2012. Mercury chemistry on brominated activated charcoal. Fuel 99, 188–196.

Sata, T., Sasamoto, T., Matsumoto, K., 1982. High temperature vaporization of calcium oxide. High Temp. High Pressures 14 (4), 399–408.

Sathe, C., Hayashi J-i, Li. C.-Z., Chiba, T., 2003. Release of alkali and alkaline earth metallic species during rapid pyrolysis of a Victorian brown coal at elevated pressures. Fuel 82, 1491–1497.

Sato, T., Imaeda, S., Sato, K., 1988. Thermal transformation of yttrium hydroxides to yttrium oxides. Thermochim. Acta 133, 79–85.

Sato, S., Hasegawa, M., Sodesawa, T., Nozaki, F., 1991. Silica-supported boron phosphate catalyst prepared by chemical vapor deposition. Bull. Chem. Soc. Jpn. 64, 516–522.

Satrio, J.A.B., Jagtap, S.B., Wheelock, T.D., 2002. Utilization of sulfur oxides for the production of sodium sulfate. Ind. Eng. Chem. Res. 41, 3540–3547.

Savel'ev, A.M., Starik, A.M., 2018. The formation of $(Al_2O_3)_n$ clusters as a probable mechanism of aluminum oxide nucleation during the combustion of aluminized fuels: numerical analysis. Combust. Flame 196, 223–236.

Savenko, V.S., 2001. Global hydrological cycle and geochemical balance of phosphorus in the ocean. Okeanologiya 41 (3), 379–385.

Saw, W.L., Hupa, M., Nathan, G.J., Ashman, P.J., 2010. Influence of stoichiometry on the release of atomic sodium from a burning black liquor droplet in a flat flame with and without boron. Fuel 89, 2608–2616.

Saw, W.L., Nathan, G.J., Ashman, P.J., Alwahabi, Z.T., Hupa, M., 2010a. Simultaneous measurement of the surface temperature and the release of atomic sodium from a burning black liquor droplet. Combust. Flame 157, 769–777.

Saw, W.L., Nathan, G.J., Ashman, P.J., Hupa, M., 2010b. Influence of droplet size on the release of atomic sodium from a burning black liquor droplet in a flat flame. Fuel 89, 1840–1848.

Scaccia, S., Mecozzi, R., 2012. Trace Cd, Co and Pb elements distribution during Sulcis coal pyrolysis: graphite furnace atomic absorption spectrometry determination with slurry sampling technique. Microchem. J. 100, 48–54.

Scala, F., Clack, H.L., 2008. Mercury emissions from coal combustion: modeling and comparison of Hg capture in a Fabric Filter versus an Electrostatic Precipitator. J. Hazard. Mater. 152, 616–623.

Schaefer, S.C., Gokcen, N.A., 1980. Electrochemical determination of thermodynamic properties of molybdenite, MoS_2. High Temp. Sci. 12, 267–276.

Schamps, J., Lefebvre-Brion, H., 1972. SCF calculations of the electronic states of magnesium monoxide. J. Chem. Phys. 56 (1), 573–585.

Schaub, G., 1996. Report No. EHS 182. Environmental Health Criteria 182: Thallium. World Health Organization, International Program on Chemical Safety. Available from: www.inchem.org.

Scheehle, E. (Ed.), 2006. Global Anthropogenic Non-CO_2 Greenhouse Gas Emissions: 1990–2020. Office of Atmospheric Programs, U.S. Environmental Protection Agency, Washington, DC, 274 pp.

Schimmelpfennig, B., Wahlgren, U., Gropen, O., Haaland, A., 2001. The gas phase structures of tungsten chlorides: density functional theory calculations on WCl_6,WCl_5, WCl_4, WCl_3 and W_2Cl_6. J. Chem. Soc. Dalton Trans. 10, 1616–1620.

Schissel, P.O., Trulson, O.C., 1965. Mass spectrometric study of the oxidation of tungsten. J. Chem. Phys. 43 (2), 737–743.

Schlesinger, W.H., Klein, E.M., Vengosh, A., 2017. Global biogeochemical cycle of Vanadium. Proc. Natl. Acad. Sci. USA 114 (52), E11092–E11100.

Schmahl, N.G., Schwitzgebel, G., Kling, H., Speck, E., 1979. Thermodynamic investigations of the solid solution of lead zirconate/lead titanate. Mater. Res. Bull. 14, 1213–1218.

Schofield, K., 1967. The bond dissociation energies of Group IIA diatomic oxides. Chem. Rev. 67 (6), 707–715.

Schofield, K., 1978. Rate constants for the gaseous interactions of O(2 $^1D_2$) and O(2 $^1S_0$): a critical evaluation. J. Photochem. 9, 55–68.

Schofield, K., 1979. Critically evaluated rate constants for gaseous reactions of several electronically excited species. J. Phys. Chem. Ref. Data 8, 723–798.

Schofield, K., 1992. The flame chemistry of alkali and alkaline earth metals. In: Fontijn, A. (Ed.), Gas Phase Metal Reactions. Elsevier Science Publishers, Amsterdam, pp. 529–571.

Schofield, K., 2003. A new method to minimize high-temperature corrosion resulting from alkali sulfate and chloride deposition in combustion systems. I. Tungsten salts. Energy Fuels 17, 191–203.

Schofield, K., 2003a. A new method for the direct flame calibration of nebulized additive concentrations. Combust. Flame 133, 147–156.

Schofield, K., 2003b. A new method to control high temperature corrosion in gas turbines resulting from alkali sulfates. In: Proceedings of ASME Turbo Expo 2003: Power for Land, Sea and Air held in Atlanta Georgia, USA: American Society of Mechanical Engineers, Paper No. GT2003-38782, 7 pp.

Schofield, K., 2004. Let them eat fish: hold the mercury. Chem. Phys. Lett. 386, 65–69.

Schofield, K., 2005. A new method to minimize high-temperature corrosion resulting from alkali sulfate and chloride deposition in combustion systems. II. Molybdenum salts. Energy Fuels 19, 1898–1905.

Schofield, K., 2005a. Mercury emission chemistry: the similarities or are they generalities of mercury and alkali combustion deposition processes? Proc. Combust. Inst. 30, 1263–1271.

Schofield, K., 2006. An overlooked series of gas phase diatomic metal oxide ions that are long-lived. J. Phys. Chem. A 110, 6938–6947.

Schofield, K., 2008. Fuel-mercury combustion emissions: an important heterogeneous mechanism and an overall review of its implications. Environ. Sci. Technol. 42, 9014–9030. Suppl Inform S1–S16.

Schofield, K., 2009. Alkali control for gas turbines by a combined hot gas clean-up approach. Proceedings of ASME Turbo Expo 2009: Power for Land, Sea and Air held in Orlando FL. American Society of Mechanical Engineers, New York NY, Paper No. GT2009-60254, 8 pp.

Schofield, K., 2012. Mercury emission control from coal combustion systems: a modified air preheater solution. Combust. Flame 159, 1741–1747.

Schofield, K., 2012a. The chemical nature of combustion deposition and corrosion: the case of alkali chlorides. Combust. Flame 159, 1987–1996.

Schofield, K., 2016. Autism, chemicals, probable cause and mitigation: a new examination. Autism: Open access. 6: 184. 27pp. Available from: <https://doi.org/10.4172/2165-7890.1000184>. OMICS International.

Schofield, K., 2016a. Comment on Ernest et al. Programmable thermal dissociation of reactive gaseous mercury, a potential approach to chemical speciation: results from a field study. Atmosphere 2014, 5, 575–596. 7:119–121.

Schofield, K., 2017. Mercury emission control: phase II. Let's now go passive. Energy 122, 311–318.

Schofield, K., 2017a. The metal neurotoxins: an important role in current human neural epidemics? Int. J. Environ. Res. Public Health 14, 1511. 23 pp., Open access.

Schofield, K., Sugden, T.M., 1965. Some observations on the ionization of alkali and alkaline earth elements in hydrogen flames. Proc. Combust. Inst. 10, 589–604.

Schofield, K., Sugden, T.M., 1971. Interaction of halogens and alkaline earth elements in $H_2/O_2/N_2$ flames. Trans. Faraday Soc. 67, 1054–1068.

Schoonmaker, R.C., Porter, R.F., 1958. Mass spectrometric study of ferrous chloride vapor. J. Chem. Phys. 29, 116–120.

Schoonmaker, R.C., Friedman, A.H., Porter, R.F., 1959. Mass spectrometric and thermodynamic study of gaseous transition metal(II) halides. J. Chem. Phys. 31, 1586–1589.

Schoonmaker, R.C., Venkitaraman, A.R., Lee, P.K., 1967. The vaporization of zinc phosphide. J. Phys. Chem. 71, 2676–2683.

Schrauzer, G.N., 2009. Selenium and selenium-antagonistic elements in nutritional cancer prevention. Crit. Rev. Biotechnol. 29 (1), 10–17.

Schrauzer, G.N., Surai, P.F., 2009. Selenium in human and animal nutrition: resolved and unresolved issues. A partly historical treatise in commemoration of the 50th anniversary of the discovery of the biological essentiality of selenium, dedicated to the memory of Klaus Schwarz (1914–1978) on the occasion of the 30th anniversary of his death. Crit. Rev. Biotechnol. 29 (1), 2–9.

Schroder, D., 2008. Gaseous rust: thermochemistry of neutral and ionic iron oxides and hydroxides in the gas phase. J. Phys. Chem. A 112, 13215–13224.

Schroder, D., Schwarz, H., 2003. Thermochemistry of neutral and cationic iron hydroxides $Fe(OH)_{1,2}$ and Fe $(OH)^{+}{}_{1,2}$ in the gas phase. Int. J. Mass Spectrom. 227, 121–134.

Schulz, A., Smith, B.J., Radom, L., 1999. Heats of formation of alkali and alkaline earth oxides and hydroxides: some dramatic failures of the G2 method. J. Phys. Chem. A 103, 7522–7527.

Schurmann, H., Unterberger, S., Hein, K.R.G., Monkhouse, P.B., Gottwald, U., 2001. The influence of fuel additives on the behavior of gaseous alkali-metal compounds during pulverized coal combustion. Faraday Discuss. 119, 433–444. 458–459.

Schurmann, H., Monkhouse, P.B., Unterberger, S., Hein, K.R.G., 2007. In situ parametric study of alkali release in pulverized coal combustion: effects of operating conditions and gas composition. Proc. Combust. Inst. 31, 1913–1920.

Schwartz, C.P., Prendergast, D., 2015. On the difficulty of reproducibly measuring $PbCl_2$ X-ray absorption spectra. J. Chem. Phys. 143, 111102. 4 pp.

Schwartz, G.E., Rivera, N., Lee, S.-W., et al., 2016. Leaching potential and redox transformations of arsenic and selenium in sediment microcosms with fly ash. Appl. Geochem. 67, 177–185.

Seames, W.S., 2003. An initial study of the fine fragmentation fly ash particle mode generated during pulverized coal combustion. Fuel Process. Technol. 81, 109–125.

Seames, W.S., Wendt, J.O.L., 2000. Partitioning of arsenic, selenium and cadmium during the combustion of Pittsburgh and Illinois #6 coals in a self-sustained combustor. Fuel Process. Technol. 63, 179–196.

Seames, W.S., Wendt, J.O.L., 2007. Regimes of association of arsenic and selenium during pulverized coal combustion. Proc. Combust. Inst. 31, 2839–2846.

Sebald, P., Vennekate, H., Oswald, R., et al., 2010. A theoretical study of ZnH_2: a case of very strong Darling-Dennison resonance. Mol. Phys. 180 (3-4), 487–499.

Sebbar, N., Rutz, L., Bockhorn, H., 2012. Thermochemical properties for hydrogenated and oxy-hydrogenated aluminum species. Soft Mater. 10, 313–343.

Sedo, O., Alberti, M., Havel, J., 2005. Laser ablation synthesis of new binary chalcogen molecules from the selenium-sulfur system. Polyhedron 24 (5), 639–644.

Segadaes, A.M., Morelli, M.R., Kiminami, R.G.A., 1998. Combustion synthesis of aluminum titanate. J. Eur. Ceram. Soc. 18, 771–781.

Selcuk, N., Gogebakan, Y., Gogebakan, Z., 2005. Partitioning of trace elements during fluidized bed combustion of high ash content lignite. Proceedings of 18th International Conference Fluidized Bed Combustion, held in Toronto, Canada. American Society of Mechanical Engineers, New York, pp. 359–366.

Self, D.E., Plane, J.M.C., 2003. A kinetic study of the reactions of iron oxides and hydroxides relevant to the chemistry of iron in the upper mesosphere. PhysChemChemPhys 5, 1407–1418.

Semeniuk, K., Dastoor, A., 2017. Development of a global ocean mercury model with a methylation cycle: outstanding issues. Global Biogeochem. Cycles 31, 400–433.

Semenov, G.A., 1969. Evaporation of titanium dioxide. Inorg. Mater. USSR 5, 54–59.

Semenov, G.A., 1985. Mass spectrometric determination of the enthalpy of formation of gaseous antimony oxide. Russ. J. Phys. Chem. 59 (6), 895–896.

Sengelov, L.W., Hansen, T.B., Bartolome, C., et al., 2013. Sulfation of condensed potassium chloride by SO_2. Energy Fuels 27, 3283–3289.

Senior, C.L., 2000. Development of a mechanistic model for prediction of emission of trace elements from coal-fired power plants. Fuel Process. Technol. 63, 75–77.

Senior, C.L., Johnson, S.A., 2005. Impact of carbon-in-ash on mercury removal across particulate control devices in coal-fired power plants. Energy Fuels 19, 859–863.

Senior, C.L., Bool III, L.E., Morency, J.R., 2000. Laboratory study of trace element vaporization from combustion of pulverized coal. Fuel Process. Technol. 63, 109–124.

Senior, C.L., Bool III, L.E., Srinivasachar, S., Pease, B.R., Porle, K., 2000a. Pilot scale study of trace element vaporization and condensation during combustion of a pulverized sub-bituminous coal. Fuel Process. Technol. 63, 149–165.

Senior, C.L., Helble, J.J., Sarofim, A.F., 2000b. Emissions of mercury, trace elements and fine particles from stationary combustion sources. Fuel Process. Technol. 65/66, 263–288.

Senior, C.L., Sarofim, A.F., Zeng, T., Helble, J.J., Mamani-Paco, R., 2000c. Gas phase transformations of mercury in coal fired power plants. Fuel Process. Technol. 63, 197–213.

Senior, C.L., Lignell, D.O., Sarofim, A.F., Mehta, A., 2006. Modeling arsenic partitioning in coal-fired power plants. Combust. Flame 147, 209–221.

Senior, C., Montgomery, C.J., Sarofim, A., 2010. Transient model for behavior of mercury in Portland cement kilns. Ind. Eng. Chem. Res. 49, 1436–1443.

Senior, C., Van Otten, B., Wendt, J.O.L., Sarofim, A., 2010a. Modeling the behavior of selenium in pulverized coal combustion systems. Combust. Flame 157, 2095–2105.

Senior, C.L., Tyree, C.A., Meeks, N.D., et al., 2015. Selenium partitioning and removal across a wet FGD scrubber at a coal-fired power plant. Environ. Sci. Technol. 49, 14376–14382.

Senneca, O., Urciuolo, M., Bareschino, P., et al., 2016. Pyrolysis, combustion and fragmentation model of coal particles: preliminary results. Combust. Sci. Technol. 188 (4–5), 759–768.

Senneca, O., Heuer, S., Bareschino, P., Urciuolo, M., et al., 2019. Fragmentation of pulverized coal in a laminar drop-tube reactor: experiments and model. Proc. Combust. Inst. 37, 2849–2855.

Serebrennikov, L.V., Sekachev, Yu.N., Maltsev, A.A., 1983. The spectra and structure of diboron trioxide. High Temp. Sci. 16, 23–34.

Setzer, K.-D., Kalb, M., Fink, E.H., 2008. The $b^1\Sigma^+(b0^+)$-$X^3\Sigma^-(X_10^+, X_21)$ and $a^1\Delta(a_2)$-$X^3\Sigma^-(X_21)$ transitions of SeS. J. Mol. Spectrosc. 251, 16–21.

Setzer, K.-D., Fink, E.H., Hill, C., Brown, J.M., 2015. Electronic states and spectra of BiH. J. Mol. Spectrosc. 312, 97–109.

Sevy, A., Huffaker, R.F., Morse, M.D., 2017. Bond dissociation energies of tungsten molecules: WC, WSi, WS, WSe and WCl. J. Phys. Chem. A 121, 9446–9457.

Shah, P., Strezov, V., Stevanov, C., Nelson, P.F., 2007. Speciation of arsenic and selenium in coal combustion products. Energy Fuels 21, 506–512.

Shah, P., Strezov, V., Prince, K., Nelson, P.F., 2008. Speciation of As, Cr, Se and Hg under coal fired power station conditions. Fuel 87, 1859–1869.

Shah, P., Strezov, V., Nelson, P.F., 2009. X-ray absorption near edge structure spectrometry study of nickel and lead speciation in coals and coal combustion products. Energy Fuels 23, 1518–1525.

Shah, P., Strezov, V., Nelson, P.F., 2012. Speciation of chromium in Australian coals and combustion products. Fuel 102, 1–8.

Shaheen, W.M., Selim, M.M., 1998. Effect of thermal treatment on physicochemical properties of pure and mixed manganese carbonate and basic copper carbonate. Thermochim. Acta 322, 117–128.

Shaheen, S.M., Hooda, P.S., Tsadilas, C.D., 2014. Opportunities and challenges in the use of coal fly ash for soil improvements: a review. J. Environ. Manage. 145, 249–267.

Shahid, M., Shamshad, S., Rafiq, M., Khalid, S., et al., 2017. Chromium speciation, bioavailability, uptake, toxicity and detoxification in soil-plant system: a review. Chemosphere 178, 513–533.

Shannon, G.N., Matsuura, H., Rozelle, P., Fruehan, R.J., Pisupati, S., Sridhar, S., 2009. Effect of size and density on the thermodynamic predictions of coal particle phase formation during coal gasification. Fuel Process. Technol. 90, 1114–1121.

Shao, J., Yan, R., Chen, H., Yang, H., Lee, D.H., Liang, D.T., 2008. Emission characteristics of heavy metals and organic pollutants from the combustion of sewage sludge in a fluidized bed combustor. Energy Fuels 22, 2278–2283.

Sharipov, A.S., Titova, N.S., Starik, A.M., 2012. Evaluation of the reaction rate constants for the gas phase Al-CH_4-air combustion chemistry. Combust. Theory Model 16, 842–868.

Sharkas, K., Gagliardi, L., Truhlar, D.G., 2017. Multi-configuration pair-density functional theory and complete active space second-order perturbation theory: bond dissociation energies of FeC, NiC, FeS, NiS, FeSe and NiSe. J. Phys. Chem. A 121, 9392–9400.

Sharma, S.D., Dolan, M., Ilyushechkin, A.Y., McLennan, K.G., Nguyen, T., Chase, D., 2010. Recent developments in dry hot syngas cleaning processes. Fuel 89, 817–826.

Sharma, V.K., McDonald, T.J., Sohn, M., et al., 2015. Biogeochemistry of selenium: a review. Environ. Chem. Lett. 13, 49–58.

Sharonova, O.M., Anshits, N.N., Fedorchak, M.A., et al., 2015. Characterization of ferrospheres recovered from high-calcium fly ash. Energy Fuels 29, 5404–5414.

Shaulov, Yu.Kh, Mosin, A.M., 1973. Thermodynamic functions of arsenic chloride, arsenic dichloride and arsenic trichloride. Russ. J. Phys. Chem. 47 (5), 644–645.

Shay, E., De Gandiaga, E., Madl, A.K., 2013. Considerations for the development of health-based surface dust cleanup criteria for beryllium. Crit. Rev. Toxicol. 43, 220–243.

Shayesteh, A., Ghazizadeh, 2015. Dunham coefficients for the $X^1\Sigma^+$ ground state of BH and BD. J. Mol. Spectrosc. 312, 110–112.

Shayesteh, A., Appadoo, D.R.T., Gordon, I., Bernath, P.F., 2003. The vibration-rotation emission spectrum of MgH_2. J. Chem. Phys. 119, 7785–7788.

Shayesteh, A., Gordon, I.E., Appadoo, D.R.T., Bernath, P.F., 2005. Infrared emission spectra and equilibrium bond lengths of gaseous ZnH_2 and ZnD_2. PhysChemChemPhys 7, 3132–3142.

Shayesteh, A., Yu, S., Bernath, P.F., 2005a. Gaseous HgH_2, CdH_2 and ZnH_2. Chem. Eur. J. 11, 4709–4712.

Shayesteh, A., Yu, S., Bernath, P.F., 2005b. Infrared emission spectra and equilibrium structures of gaseous HgH_2 and HgD_2. J. Phys. Chem. A 109 (17), 10280–10286.

Shayesteh, A., Henderson, R.D.E., Le Roy, R.J., Bernath, P.F., 2007. Ground state potential energy curve and dissociation energy of MgH. J. Phys. Chem. 111, 12495–12505.

Shayesteh, A., Alavi, S.F., Rahman, M., Gharib-Nezhad, E., 2017. *Ab initio* transition dipole moments and potential energy curves for the low-lying electronic states of CaH. Chem. Phys. Lett. 667, 345–350.

Shefa, S.T., Heroux, P., 2017. Both physiology and epidemiology support zero tolerable blood lead levels. Toxicol. Lett. 280, 232–237.

Sheldon, R.I., Gilles, P.W., 1977. The high temperature vaporization and thermodynamics of the titanium oxides. XI. Stoichiometric titanium monoxide. J. Chem. Phys. 66, 3705–3711.

Shelton, R.A.J., 1961. Vapor pressures of the solid copper (I) halides. Trans. Faraday Soc. 57, 2113–2118.

Shen, F., Liu, J., Zhang, Z., Dai, J., 2015. On-line analysis and kinetic behavior of arsenic release during coal combustion and pyrolysis. Environ. Sci. Technol. 49, 13716–13723.

Shen, F.H., Liu, J., Zhang, Z., Yang, Y.J., 2016. Temporal measurements and kinetics of selenium release during coal combustion and gasification in a fluidized bed. J. Hazard. Mater. 310, 40–47.

Shen, F.H., Liu, J., Dong, Y.C., Gu, C.K., 2018. Insights into the effect of chlorine on arsenic release during MSW incineration: an online analysis and kinetic study. Waste Manage. 75, 327–332.

Shenyavskaya, E.A., Mal'tsev, A.A., Gurvich, L.V., 1967. Electronic-vibrational-rotational spectrum of scandium monochloride molecule and some additional data on the molecular constants of scandium, yttrium and lanthanum monofluorides. J. Mol. Spectrosc. 22, 104–105.

Shepler, B.C., Peterson, K.A., 2003. Mercury oxide: a systematic investigation of its ground electronic state. J. Phys. Chem. A 107, 1783–1787.

Shepler, B.C., Peterson, K.A., 2006. Chemically accurate thermochemistry of cadmium: an *ab initio* study of Cd + XY (X = H, O, Cl, Br; Y = Cl, Br). J. Phys. Chem. A 110, 12321–12329.

Sheppard, S.C., Sanipelli, B., 2012. Trace elements in feed, manure and manured soils. J. Environ. Qual. 41 (6), 1846–1856.

Shepler, B.C., Balabanov, N.B., Peterson, K.A., 2005. *Ab initio* thermochemistry involving heavy atoms: an investigation of the reactions Hg + IX (X = I, Br, Cl, O). J. Phys. Chem. A 109, 10363–10372.

Shestakov, O., Gielen, R., Fink, E.H., 1998. LIF studies of the A_1A' states of BiOH and BiOD. J. Mol. Spectrosc. 192, 111–118.

Shi, Q., Ran, Q., Tam, W.S., et al., 2001. Laser induced fluorescence spectroscopy of CrS. Chem. Phys. Lett. 339, 154–160.

Shi, D., Liu, H., Sun, J., et al., 2009. Spectroscopic investigations on AsH($X^3\Sigma^-$) radical using CCSD(T) theory in combination with correlation-consistent quintuple basis set. J. Mol. Struct. Theochem. 911, 8–12.

Shi, D.-H., Liu, H., Zhang, X.-N., et al., 2011. MRCI investigations on dissociation energy and molecular constants of BCl($X^1\Sigma^+$) radical. Int. J. Quantum Chem. 111, 2825–2834.

Shi, D., Song, Z., Niu, X., et al., 2016. Accurate predictions of spectroscopic and molecular properties of 27 Λ-S and 73 Ω states of AsS radical. Spectrochim. Acta A Mol. Biomol. Spectrosc 153, 30–44.

Shi, Y.-L., Chen, W.-Q., Wu, S.-L., Zhu, Y.-G., 2017. Anthropogenic cycles of arsenic in mainland China: 1990–2010. Environ. Sci. Technol. 51, 1670–1678.

Shigegaki, Y., Basu, S.K., Wakihara, M., Taniguchi, M., 1988. Thermal analysis and kinetics of oxidation of molybdenum sulfides. J. Thermal Anal. 34, 1427–1440.

Shim, I., Mandix, K., Gingerich, K.A., 1991. Theoretical and experimental investigations of the Ga_2 molecule. J. Phys. Chem. 95, 5435–5442.

Shim, Y.K., Lewin, M.D., Ruiz, P., Eichner, J.E., Mumtaz, M.M., 2017. Prevalence and associated demographic characteristics of exposure to multiple metals and their species in human populations: the United States NHANES 2007–2012. J. Toxicol. Environ. Health A 80 (9), 502–512.

Shimazaki, E., Matsumoto, N., Niwa, K., 1957. The vapor pressure of germanium dioxide. Bull. Chem. Soc. Jpn. 30, 969–971.

Shirley, R., Liu, Y., Totton, T.S., West, R.H., Kraft, M., 2009. First principles thermochemistry for the combustion of a $TiCl_4$ and $AlCl_3$ mixture. J. Phys. Chem. A 113, 13790–13796.

Shkodin, V.G., Abisheva, R.U., 1986. Thermolysis of normal silver sulfate. Zhur. Neorg. Khim. 31 (2), 301–304.

Shmakov, A.G., Korobeinichev, O.P., Shvartsberg, V.M., et al., 2008. Suppression of hydrocarbon flames by organophosphorus compounds and their based mixtures. Combust. Explos. Shock Waves 44 (3), 266–272.

Shmakov, A.G., Korobeinichev, O.P., Knyazkov, D.A., et al., 2013. Combustion chemistry of $Ti(OC_3H_7)_4$ in premixed flat burner-stabilized $H_2/O_2/Ar$ flame at 1 atm. Proc. Combust. Inst. 34, 1143–1149.

Shoji, T., Huggins, F.E., Huffman, G.P., Linak, W.P., Miller, C.A., 2002. XAFS spectroscopy analysis of selected elements in fine particulate matter derived from coal combustion. Energy Fuels 16, 325–329.

Shotyk, W., Cheburkin, A.K., Appleby, P.G., Fankhauser, A., Kramers, J.D., 1996. Two thousand years of atmospheric arsenic, antimony and lead deposition recorded in an ombrotrophic peat bog profile, Jura Mountains, Switzerland. Earth Planet Sci. Lett. 145, E1–E7.

Shugurov, S.M., Lopatin, S.I., 2006. The thermodynamic properties of gaseous salts formed by some 3d metal oxides. Russ. J. Phys. Chem. 80, 1749–1753.

Shugurov, S.M., Lopatin, S.I., 2011. Thermochemical study of gaseous salts of oxygen-containing acids: XXVII. Antimonites of alkali metals. Russ. J. Gen. Chem. 81, 1411–1416.

Shugurov, S.M., Panin, A.I., Lopatin, S.I., Emelyanova, K.A., 2015. Thermodynamic study of gaseous tin molybdates by high temperature mass spectrometry. Rapid Commun. Mass Spectrom. 29 (15), 1427–1436.

Shukla, A.R., Asthana, B.P., Kushawaha, V.S., 1975. On the potential energy curves and dissociation energy of BaO molecule. Spectrosc. Lett. 8, 731–743.

Shvartsberg, V.M., Bolshova, T.A., Korobeinichev, O.P., 2010. Numerical study of inhibition of hydrogen/air flames by atomic iron. Energy Fuels 24, 1552–1558.

Shvartsberg, V.M., Shmakov, A.G., Bolshova, T.A., Korobeinichev, O.P., 2012. Mechanism for inhibition of atmospheric pressure syngas/air flames by trimethylphosphate. Energy Fuels 26, 5528–5536.

Si, J.P., Liu, X.W., Xu, M.H., et al., 2014. Effect of kaolin additive on $PM_{2.5}$ reduction during pulverized coal combustion: importance of sodium and its occurrence in coal. Appl. Energy 114, 434–444.

Sidorov, L.N., Minayeva, I.I., Zasorin, E.Z., Sorokin, I.D., Borshchevskiy, A.Ya, 1980. Mass spectrometric investigation of gas phase equilibria over bismuth trioxide. High Temp. Sci. 12, 175–196.

Siegbahn, P.E.M., 1993. A comparative study of the bond strengths of the second row transition metal hydrides, fluorides and chlorides. Theor. Chim. Acta 86, 219–228.

Siegbahn, P.E.M., 1993a. A comparison of the bonding in the second-row transition-metal oxides and carbenes. Chem. Phys. Lett. 201, 15–23.

Siegbahn, P.E.M., 1993b. Binding in second-row transition metal dioxides, trioxides, tetraoxides, pentoxides and superoxides. J. Phys. Chem. 97, 9096–9102.

Siegbahn, P.E.M., 1994. The bonding in second row transition metal dihydrides, difluorides and dichlorides. Theor. Chim. Acta 87, 441–452.

Siegbahn, P.E.M., 1996. Electronic structure calculations for molecules containing transition metals. Adv. Chem. Phys. 93, 333–387.

Siegbahn, P.E.M., Blomberg, M.R.A., Bauschlicher Jr, C.W., 1984. Potential energy surfaces of MH_2 (M = Co, Fe and Cu). J. Chem. Phys. 81, 1373–1382.

Sievers, M.R., Chen, Y.-M., Armentrout, P.B., 1996. Metal oxide and carbide thermochemistry of Y^+, Zr^+, Nb^+ and Mo^+. J. Chem. Phys. 105, 6322–6333.

Sigel, A., Sigel, H., Sigel, R.K.O. (Eds.), 2013. Cadmium: From Toxicity to Essentiality. 16 Chapters, 589 pp. Springer, Dordrecht.

Sikorska, C., Skurski, P., 2009. Cationic and anionic daughters of AlOH and GaOH: an *ab initio* study. Chem. Phys. Lett. 477, 259–265.

Silk, J.E., Hansen, L.D., Eatough, D.J., Hill, M.W., Mangelson, N.F., Lytle, F.W., et al., 1989. Chemical characterization of vanadium, nickel and arsenic in oil flyash samples using EXAFS and XANES spectroscopy. Physica B 158, 247–248.

Sillerova, H., Chrastny, V., Cadkova, E., Komarek, M., 2014. Isotope fractionation and spectroscopic analysis as an evidence of Cr(VI) reduction during biosorption. Chemosphere 95, 402–407.

Silva, L.F.O., Oliveira, M.L.S., Sampaio, C.H., de Brum, I.A.S., Hower, J.C., 2013. Vanadium and nickel speciation in pulverized coal and petroleum coke co-combustion. Energy Fuels 27, 1194–1203.

Silvennoinen, J., Roppo, J., Nurminen, R.-V., Aho, M., Vainikka, P., Ferrer, E., 2005. Co-combustion of coal with refuse derived fuel and biomass: prevention of chlorine deposition by using coal ash alkali absorption ability. Paper No. ASME FBC 2005-78120. Proceedings of 18th Int Conf Fluidized Bed Combustion held in Toronto, Ontario, Canada. American Society of Mechanical Engineers, New York NY, pp. 525–534.

Silver, J.A., Zahniser, M.S., Stanton, A.C., Kolb, C.E., 1984. Temperature dependent termolecular reaction rate constants for potassium and sodium superoxide formation. Proc. Combust. Inst. 20, 605–612.

Simeonsson, J.B., Ezer, M., Pacquette, H.L., Preston, S.L., Swart, D.J., 1997. Laser-induced fluorescence of As, Se and Sb in the inductively coupled plasma. Spectrochim. Acta B 52, 1955–1963.

Simms, J.A., 2011. Dark side of manganese. Chem. Eng. News 89 (37), 4.

Simonsson, J., Olofsson, N.-E., Hosseinnia, A., Bengtsson, P.-E., 2018. Influence of potassium chloride and other metal salts on soot formation studied using imaging LII and ELS, and TEM techniques. Combust. Flame 190, 188–200.

Singh, H., Kolay, P.K., 2009. Analysis of coal ash for trace elements and their geo-environmental implications. Water. Air. Soil Pollut. 198, 87–94.

Singh, R., Singh, L., Singh, S.V., 2015. Beneficiation of iron and aluminum oxides from fly ash at lab scale. Int. J. Mineral Process. 145, 32–37.

Singhal, A., Skandan, G., Wang, A., Glumac, N., Kear, B.H., Hunt, R.D., 1999. On nanoparticle aggregation during vapor phase synthesis. Nanostruct. Mater. 11, 545–552.

Sinyoung, S., Songsiriritthigul, P., Asavapisit, S., Kajitvichyanukul, P., 2011. Chromium behavior during cement production processes: a clinkerization hydration and leaching study. J. Hazard. Mater 191, 296–305.

Sipp, E., Langlais, F., Naslain, R., 1992. CVD of Y_2O_3 from YCl_3-CO_2-H_2-Ar gas mixtures: an experimental study. J. Alloys Compds. 186, 77–87.

Sippula, O., Lind, T., Jokiniemi, J., 2008. Effects of chlorine and sulfur on particle formation in wood combustion performed in a laboratory scale reactor. Fuel 87, 2425–2436.

Sjostrom, S., Durham, M., Bustard, C.J., Martin, C., 2010. Activated carbon injection for mercury control: overview. Fuel 89, 1320–1322.

Skalny, A.V., Kaminskaya, G.A., Krekesheva, T.I., Abikenova, S.K., et al., 2018. Assessment of hair metal levels in aluminum plant workers using scalp hair ICP-DRC-MS analysis. J. Trace Elem. Med. Biol. 50, 658–663.

Skodras, G., Grammelis, P., Prokopidou, M., Kakaras, E., Sakellaropoulos, G., 2009. Chemical, leaching and toxicity characteristics of circulating fluidized bed combustion residues. Fuel 88, 1201–1209.

Skoglund, N., Bafver, L., Fahlstrom, J., et al., 2016. Fuel design in co-combustion of demolition wood chips and municipal sewage sludge. Fuel Process. Technol. 141 (2), 196–201.

Skulan, A.J., Nielsen, I.M.B., Melius, C.F., Allendorf, M.D., 2006. BAC-MP4 predictions of thermochemistry for gas phase antimony compounds in the Sb-H-C-O-Cl system. J. Phys. Chem. A 110, 5919–5928.

Slemr, F., Brunke, E.-G., Ebinghaus, R., Kuss, J., 2011. Worldwide trend of atmospheric mercury since 1995. Atmos. Chem. Phys. 11, 4779–4787.

Sliger, R.N., Kramlich, J.C., Marinov, N.M., 2000. Towards the development of a chemical kinetic model for the homogeneous oxidation of mercury by chlorine species. Fuel Process. Technol. 65-66, 423–438.

Sliznev, V.V., Vogt, N., Vogt, J., 2004. *Ab initio* study of the ground and lower-lying excited electronic states of CoF_2, $CoCl_2$, $CoBr_2$, CoI_2. Mol. Phys. 102, 1767–1776.

Slobodin, B.V., Fotiev, A.A., Miller, I.I., 1976. Thermal properties of iron orthovanadate. Zhur. Neorg. Khim. 21, 327–330.

Slupczynska, M., Jamroz, D., Orda, J., Wiliczkiewicz, A., Krol, B., 2018. Long-term supplementation of laying hen diets with various selenium sources as a method for the fortification of eggs with selenium. J. Chem. Article 7986591. Available from: https://doi.org/10.1155/2018/7986591.

Smith, T.C., Clouthier, D.J., 2018. Detection and characterization of the tin dihydride (SnH_2 and SnD_2) molecule in the gas phase. J. Chem. Phys. 148, 024302. 8 pp.

Smith, R.D., Campbell, J.A., Nielson, K.K., 1979. Concentration dependence upon particle size of volatilized elements in fly ash. Environ. Sci. Technol. 13, 553–558.

Smith, C.A., Krishnakumar, B., Helble, J.J., 2011. Homo- and heterogeneous mercury oxidation in a bench-scale flame-based flow reactor. Energy Fuels 25, 4367–4376.

Smith, A.L., Colle, J.-Y., Raison, P.E., et al., 2015. Thermodynamic investigation of $Na_2U_2O_7$ using Knudsen effusion mass spectrometry and high temperature X-ray diffraction. J. Chem. Thermodyn. 90, 1990208.

Smith, A.L., Raison, P.E., Martel, L., et al., 2015a. A new look at the structural properties of trisodium urinate, Na_3UO_4. Inorg. Chem. 54, 3552–3561.

Smoes, S., Drowart, J., 1977. Determination of the atomization energies of the molecules CSe(g) and CSe_2(g) by the mass spectrometric Knudsen cell method. J. Chem. Soc. Faraday Trans. II 73, 1746–1754.

Smoes, S., Drowart, J., 1984. Determination of the dissociation energies of gaseous iron monoxide and manganese monoxide by the mass spectrometric Knudsen cell method. High Temp. Sci. 17, 31–52.

Smoes, S., Drowart, J., 1984a. Determination of the dissociation energy of selenium monoxide by the mass spectrometric Knudsen cell method. J. Chem. Soc. Faraday Trans. II 80, 1171–1180.

Smoes, S., Coppens, P., Bergman, C., Drowart, J., 1969. Mass spectroscopic determination of the dissociation energies of the gaseous rare earth monosulfides. Trans. Faraday Soc. 65, 682–687.

Smoes, S., Mandy, F., Auwera-Mahieu, A.V., Drowart, J., 1972. Determination by the mass spectrometric Knudsen cell method of the dissociation energies of the Group IB chalcogenides. Bull. Soc. Chim. Belges 81, 45–56.

Smyly, D.S., Townsend, W.P., Zeegers, P.J.Th, Winefordner, J.D., 1971. Measurement of the free-atom fraction in hydrogen/oxygen/argon flames using the relative absorption technique. Spectrochim. Acta B 26, 531–546.

Sneddon, A.A., 2012. Selenium and vascular health. Pure Appl. Chem. 84 (2), 239–248.

Sobolev, O., Gutyj, B., Petryshak, R., Pivtorak, J., et al., 2018. Biological role of selenium in the organism of animals and humans. Ukrainian J. Ecol. 8 (1), 654–665.

Soldan, P., Lee, E.P.F., Gamblin, S.D., Wright, T.G., 2000. Na_2O and Na_2O^+: thermodynamics and low-lying electronic states. J. Phys. Chem. A 104, 3317–3325.

Solovyev, N.D., 2015. Importance of selenium and selenoprotein for brain function: from antioxidant protection to neuronal signaling. J. Inorg. Biochem. 153, 1–12.

Solovyev, N., Drobyshev, E., Bjorklund, G., Dubrovskii, Y., et al., 2018. Selenium, selenoprotein P, and Alzheimer's disease: is there a link? Free Radiol. Biol. Med. 127, 124–133.

Sommer, A., Walsh, P.N., White, D., 1960. Mass spectrometric and infrared emission investigation of the vapor species in the B-S system at elevated temperatures. J. Chem. Phys. 33, 296–297.

Sommersacher, P., Kienzl, N., Brunner, T., Obernberger, I., 2016. Simultaneous online determination of S, Cl, K, Na, Zn and Pb release from a single particle during biomass combustion. II. Results from test runs with spruce and straw pellets. Energy Fuels 30, 3428–3440.

Sonchik, S.M., Andrews, L., Carlson, K.D., 1983. Matrix reactions of molecular oxygen and ozone with aluminum atoms. J. Phys. Chem. 87, 2004–2011.

Song, J., Apra, E., Khait, Y.G., et al., 2006. High level *ab initio* calculations on the NiO_2 system. Chem. Phys. Lett. 428, 277–282.

Song, P., Guan, W., Yao, C., Su, Z.M., Wu, Z.J., Feng, J.D., et al., 2007. Electronic structures of 4d transition metal monoxides by density functional theory. Theor. Chem. Acc. 117, 407–415.

Song, C., Gao, T., Han, H., Wan, M., Yu, Y., 2008. *Ab initio* study of spin-orbit interaction in the ground electronic states of XH (X = C, Si, Ge and Sn) molecules. J. Mol. Struct. Theochem. 870, 65–71.

Song, W., Jiao, F., Yamada, N., et al., 2013. Condensation behavior of heavy metals during oxy-fuel combustion: deposition, species distribution and their particle characteristics. Energy Fuels 27, 5640–5652.

Song, S., Selin, N.E., Soerensen, A.L., et al., 2015. Top-down constraints on atmospheric mercury emissions and implications for global biogeochemical cycling. Atmos. Chem. Phys. 15 (12), 7103–7125.

Sorensen, T.E., England, W.B., 2000. Valence states of BeO Feynman's way. Int. J. Quantum Chem. 76, 259–279.

Soria, J., Gauthier, D., Falcoz, Q., et al., 2013. Local CFD kinetic model of cadmium vaporization during fluid bed incineration of municipal solid waste. J. Hazard. Mater. 248, 276–284.

Sorum, L., Frandsen, F.J., Hustad, J.E., 2004. On the fate of heavy metals in municipal solid waste combustion. II. From furnace to filter. Fuel 83, 1703–1710.

Sorvajarvi, T., DeMartini, N., Rossi, J., Toivonen, J., 2014. In situ measurement technique for simultaneous detection of K, KCl, and KOH vapors released during combustion of solid biomass fuel in a single particle reactor. Appl. Spectrosc. 68 (2), 179–184.

Sorvajarvi, T., Viljanen, J., Toivonen, J., Marshall, P., Glarborg, P., 2015. Rate constant and thermochemistry for $K + O_2 + N_2 = KO_2 + N_2$. J. Phys. Chem. A 119, 3329–3336.

Soton, E., de Lauzon, M., Baud, F., 2018. Contribution of cutlery and serving dishes to the human daily intake of silver: an *in vitro* assessment. Ann. Biol. Clin. 76 (2), 165–172.

Spears, D.A., Martinez-Tarrazona, M.R., 2004. Trace elements in combustion residues from a UK power station. Fuel 83, 2265–2270.

Spiridonov, V.P., Gershikov, A.G., Zasorin, E.Z., et al., 1983. Refinement of the molecular structure and determination of the force field and vibrational frequencies of diboron trioxide, B_2O_3. A high temperature electron diffraction study. High Temp. Sci. 16, 325–339.

Spiro, C.L., Chen, C.C., Kimura, S.G., Lavigne, R.G., Schields, P.W., 1990. Deposit remediation on coal-fired gas turbines through the use of additives. Prog. Energy Combust. Sci. 16, 213–220.

Spirko, V., 2016. Morphing *ab initio* potential energy curve of beryllium monohydride. J. Mol. Spectrosc. 330, 89–95.

Squitti, R., Ventriglia, M., Siotto, M., Salustri, 2017. Copper in Alzheimer's disease. Chapter 2, pp. 19–34. In: White, A.R., Aschner, M., Costa, L.G., Bush, A.I. (Eds.), Biometals in Neurodegenerative Diseases. Academic Press, Elsevier, London, 415 pp.

Srikesh, G., Nesaraj, A.S., 2015. Synthesis and characterization of phase pure NiO nanoparticles via the combustion route using different organic fuels for electrochemical capacitor applications. J. Electrochem. Sci. Technol. 6 (1), 16–25.

Srivastava, R.D., Farber, M., 1978. Thermodynamic properties of Group 3 oxides. Chem. Rev. 78, 627–638.

Srivastava, R.K., Miller, C.A., Erickson, C., Jambhekar, R., 2004. Emissions of sulfur trioxide from coal-fired power plants. J. Air Waste Manage. Assoc. 54, 750–762.

Srivastava, S., Desai, M., Merchan-Merchan, W., Saveliev, A.V., 2015. Volumetric flame synthesis of 1-D molybdenum oxide nanostructures. Proc. Combust. Inst. 35, 2307–2314.

Stagg, N.J., Richards, W.G., 1974. Spectroscopic dissociation energies of diatomic alkaline earth oxides. Mol. Phys. 27, 787–788.

Stam, A.F., Meij, R., te Winkel, H., van Eijk, R.J., Huggins, F.E., Brem, G., 2011. Chromium speciation in coal and biomass co-combustion products. Environ. Sci. Technol. 45, 2450–2456.

Stanger, R., Ting, T., Wall, T., 2014. High pressure conversion of NO_x and Hg and their capture as aqueous condensates in a piston-compressor simulating oxy-fuel CO_2 compression. Int. J. Greenhouse Gas Control 29, 209–220.

Starovoitov, E.M., Ryabova, V.G., Gurvich, L.V., Khitrov, A.N., Nazarenko, I.I., Belyaev, A.V., 1977. Spectrometric determination of the dissociation energies of molecules. V. Hydroxides of alkaline earth metals CaOH, Ca $(OH)_2$ and calcium oxide. High Temp. USSR 15, 770–772.

Stearns, C.A., Miller, R.A., Kohl, F.J., Fryburg, G.C., 1977. Investigation of the formation of gaseous sodium sulfate in a doped methane-oxygen flame. J. Electrochem. Soc. 124, 1145–1146.

Steblevskii, A.V., Alikhanyan, A.S., Sokolova, I.D., Gorgoraki, V.I., 1977. Thermal stability of alkali metal metaphosphates. Russ. J. Inorg. Chem. 22, 11–13.

Steblevskii, A.V., Alikhanyan, A.S., Sokolova, I.D., Gorgoraki, V.I., 1978. Thermodynamics of the volatilization of sodium pyrophosphate and sodium and lithium orthophosphates. Russ. J. Inorg. Chem. 23, 173–176.

Steenari, B.-M., Fedje, K.K., 2010. Addition of kaolin as potassium sorbent in the combustion of wood fuel: effects on fly ash properties. Fuel 89, 2026–2032.

Stefaniak, J., Dutta, A., Verbinnen, B., Shakya, M., Rene, E.R., 2018. Selenium removal from mining and process wastewater: a systematic review of available technologies. J. Water Supply Res. Technol.-Aqua 67 (8), 903–918.

Steiger, R.A., Cater, E.D., 1975. Vaporization, thermodynamics and dissociation energy of yttrium monosulfide. High Temp. Sci. 7, 204–214.

Steiger, R.P., Cater, E.D., 1975a. Mass spectrometric determination of dissociation energies of high temperature molecules: scandium sulfide, yttrium sulfide, lanthanum sulfide, zirconium sulfide, uranium(II) oxide. High Temp. Sci. 7, 288–300.

Steinberg, M., Schofield, K., 1987. The high temperature chemistry and thermodynamics of alkali metals (lithium, sodium and potassium) in oxygen-rich flames. Fall Joint Technical Meeting of the Western States and Japanese Sections of the Combustion Institute. Honolulu HI, pp. 135–137.

Steinberg, M., Schofield, K., 1991. A reevaluation of the vaporization behavior of sodium oxide and the bond strengths of NaO and Na_2O: implications for the mass spectrometric analyses of alkali/oxygen systems. J. Chem. Phys. 94 (5), 3901–3907.

Steinberg, M., Schofield, K., 1996. The controlling chemistry in flame generated surface deposition of Na_2SO_4 and the effects of chlorine. Proc. Combust. Inst. 26, 1835–1843.

Steinberg, M., Schofield, K., 2002. The controlling chemistry of surface deposition from sodium and potassium seeded flames free of sulfur or chlorine impurities. Combust Flame 129, 453–470.

Steinbrenner, H., Speckmann, B., Klotz, L.O., 2016. Selenoproteins: antioxidant selenoenzymes and beyond. Arch. Biochem. Biophys. 595 (Special Issue), 113–119.

Steinbrunner, B.S., Edwards, J.G., 1991. Vaporization studies of the CdS-ZnS solid-solid solution system. High Temp. Sci. 32, 167–192.

Sterckeman, T., Puschenreiter, M., 2018. Phytoextraction of cadmium: feasibility in field applications and potential use of harvested biomass. In: Van der Ent, A., Echevarria, G., Baker, A.J.M., Morel, J.L. (Eds.), Agromining: Farming for Metals: Extracting Unconventional Resources Using Plants. Mineral Resource Reviews Book Series Springer International Publishing. pp. 205–219.

Sterling, R.O., Helble, J.J., 2003. Reaction of arsenic vapor species with fly ash compounds: kinetics and speciation of the reaction with calcium silicates. Chemosphere 51, 1111–1119.

Stern, K.H., 1972. High temperature properties and decomposition of inorganic salts. III. Nitrates and nitrites. J. Phys. Chem. Ref. Data 1, 747–772.

Stern, K.H., 2001. High Temperature Properties and Thermal Decomposition of Inorganic Salts with Oxyanions. CRC Press, Boca Raton,FL.

Stern, K.H., Weise, E.L., 1966. High temperature properties and decomposition of inorganic salts: Part 1. Sulfates. National Standard Reference Data Series: National Bureau of Standards, vol. 7, 38 pp.

Stern, K.H., Weise, E.L., 1969. High temperature properties and decomposition of inorganic salts. II. Carbonates. Natl. Stand. Ref. Data Ser. Nat. Bur Stand (US) 30, 32 pp.

Stevens, A.E., Feigerle, C.S., Lineberger, W.C., 1983. Laser photoelectron spectroscopy of MnH^- and FeH^-: electronic structures of the metal hydrides, identification of a low-spin excited state of MnH, and evidence for a low-spin ground state of FeH. J. Chem. Phys. 78, 5420–5431.

Stiakaki, M.-A.D., Tsipis, A.C., Tsipis, C.A., Xanthopoulos, C.E., 1994. Molecular geometries and energetics of metal containing systems using an improved ASED-MO model: a systematic test to Group 1 and 2 metal containing systems in the gas phase. Chem. Phys. 189, 533–555.

Stoddard, J.L., Van Sickle, J., Herlihy, A.T., et al., 2016. Continental-scale increase in lake and stream phosphorus: are oligotrophic systems disappearing in the United States? Environ. Sci. Technol. 50, 3409–3415.

Stoffaneller, R., Morse, N.L., 2015. A review of dietary selenium intake and selenium status in Europe and the Middle East. Nutrients 7 (3), 1494–1537.

Stoll, H., Metz, B., Dolg, M., 2002. Relativistic energy-consistent pseudopotentials: recent developments. J. Comput. Chem. 23 (8), 767–778.

Stollberg, D.W., Carter, W.B., Hampikian, J.M., 1997. Combustion CVD of magnesium spinel and nickel spinel. Surf. Coat. Technol. 94–95, 137–143.

Stolyarova, V.L., 2016. Mass spectrometric thermodynamic studies of oxide systems and materials. Russ. Chem. Rev. 85 (1), 60–80.

Stolyarova, V.L., Lopatin, S.I., Shugurov, S.M., 2004. Gaseous manganese molybdates and tungstates. Dokl. Phys. Chem. 395 (1), 80–83.

Stolyarova, V.L., Lopatin, S.I., Shugurov, S.M., 2004a. Thermodynamic properties of gaseous iron(II) salts. Dokl. Phys. Chem. 398 (1), 208–210.

Stolyarova, V.L., Shilov, A.L., Lopatin, S.I., Shugurov, S.M., 2014. High temperature mass spectrometric study and modeling of thermodynamic properties of binary glass-forming systems containing Bi_2O_3. Rapid Commun. Mass Spectrom. 28 (7), 801–810.

Strain, J.J., Davidson, P.W., Bonham, M.P., Duffy, E.M., Stokes-Riner, A., Thurston, S.W., et al., 2008. Associations of maternal long-chain polyunsaturated fatty acids, methylmercury, and infant development in the Seychelles child development nutrition study. Neurotoxicology 29, 776–782.

Strand, M., Bohgard, M., Swietlicki, E., Gharibi, A., Sanati, M., 2004. Laboratory and field test of a sampling method for characterization of combustion aerosols at high temperatures. Aerosol Sci. Technol. 38, 757–765.

Streets, D.G., Devane, M.K., Lu, Z., et al., 2011. All-time releases of mercury to the atmosphere from human activities. Environ. Sci. Technol. 45, 10485–10491.

Streets, D.G., Horowitz, H.M., Jacob, D.J., Lu, Z., Levin, L., et al., 2017. Total mercury released to the environment by human activities. Environ. Sci. Technol. 51, 5969–5977.

Streets, D.G., Lu, Z., Levin, L., ter Schure, A.F.H., Sunderland, E.M., 2018. Historical releases of mercury to air, land and water from coal combustion. Sci. Total Environ. 615, 131–140.

Streets, D.G., Horowitz, H.M., Lu, Z.F., Levin, L., Thackray, C.P., Sunderland, E.M., 2019. Global and regional trends in mercury emissions and concentration, 2010–2015. Atmos. Environ. 201, 417–427.

Struck, C.W., Baglio, J.A., 1991. An estimation of rare earth monohalide enthalpies of formation using the Rittner energy expression. High Temp. Sci. 30, 113–135.

Stucki, S., Jakob, A., 1997. Thermal treatment of incinerator fly ash: factors influencing the evaporation of $ZnCl_2$. Waste Manage. 17, 231–236.

Stwalley, W.C., 1975. Mass reduced quantum numbers: application to the isotopic mercury hydrides. J. Chem. Phys. 63 (7), 3062–3080.

Stwalley, W.C., Zemke, W.T., Yang, S.C., 1991. Spectroscopy and structure of the alkali hydride diatomic molecules and their ions. J. Phys. Chem. Ref. Data 20 (1), 153–187.

Su, T., Wang, J., 2011. Modeling batch leaching behavior of arsenic and selenium from bituminous coal fly ashes. Chemosphere 85, 1368–1374.

Su, S., Liu, L., Wang, L., Syed-Hassan, S.S.A., Kong, F., et al., 2017. Mass flow analysis of mercury transformation and effect of seawater flue gas desulfurization on mercury removal in a full scale coal-fired powerplant. Energy Fuels 31, 11109–11116.

Suarez-Ruiz, I., Hower, J.C., Thomas, G.A., 2007. Hg and Se capture and fly ash carbons from combustion of complex pulverized feed blends mainly of anthracitic coal rank in Spanish power plants. Energy Fuels 21, 59–70.

Suetens, T., Guo, M., Van Acker, K., Blanpain, B., 2015. Gas-solid reaction kinetics of $ZnFe_2O_4$ formation from 907–1100°C. J. Phys. Chem. A 119, 4718–4722.

Sugden, T.M., 1962. Excited species in flames. Ann. Rev. Phys. Chem. 13, 369–390.

Sugden, T.M., Schofield, K., 1966. Heats of dissociation of gaseous alkali earth dihydroxides. Trans. Faraday Soc. 62, 566–575.

Sukandar, S., Yasuda, K., Tanaka, M., Aoyama, I., 2006. Metals leachability from medical waste incinerator fly ash: a case study on particle size comparison. Environ. Pollut. 144, 726–735.

Sullivan, M.B., Iron, M.A., Redfern, P.C., Martin, J.M.L., Curtiss, L.A., Radom, L., 2003. Heats of formation of alkali metal and alkaline earth metal oxides and hydroxides: surprisingly demanding targets for high-level *ab initio* procedures. J. Phys. Chem. A 107, 5617–5630.

Sullivan, P.A., Sumathi, R., Green, W.H., Tester, J.W., 2004. *Ab initio* modeling of organophosphorus combustion chemistry. PhysChemChemPhys 6, 4296–4309.

Sun, L., Bu, Y., Yan, S., 2004. Structures and bonding character of cyclic AsO_2 and AsS_2 systems in the doublet state. Chem. Phys. Lett. 387, 160–164.

Sun, L., Luo, K., Fan, J., Lu, H., 2017. Experimental study of extracting alumina from coal fly ash using fluidized beds at high temperature. Fuel 199, 22–27.

Sun, D., Hale, L., Kar, G., Soolanayakanahally, R., Adl, S., 2018. Phosphorus recovery and reuse by pyrolysis: applications for agriculture and environment. Chemosphere 194, 682–691.

Sun, X.J., Jiang, Y.Q., Xia, W., Jin, S.N., et al., 2018a. Association between prenatal nickel exposure and preterm low birth weight: possible effect of selenium. Environ. Sci. Pollut. Res. 25 (26), 25888–25895.

Sunahori, F.X., Zhang, X., Clouthier, D.J., 2006. The electronic spectrum of jet-cooled copper hydrosulfide (CuSH). J. Chem. Phys. 125, 084310. 9 pp.

Sunahori, F.X., Gharaibeh, M., Clouthier, D.J., Tarroni, R., 2015. BH_2 revisited: new extensive measurements of laser induced fluorescence transitions and *ab initio* calculations of near-spectroscopic accuracy. J. Chem. Phys. 142, 174302. 13 pp.

Sunderland, E.M., Driscoll Jr, C.T., Hammitt, J.K., et al., 2016. Benefits of regulating hazardous air pollutants from coal and oil-fired utilities in the United States. Environ. Sci. Technol. 50, 2117–2120.

Sunderlin, L.S., Armentrout, P.B., 1990. Reactions of Mn^+ with i-C_4H_{10}, neo-C_5H_{12}, $(CH_3)_2CO$, c-C_3H_6 and c-C_2H_4O: bond energies for $MnCH_2^+$, MnH and $MnCH_3$. J. Phys. Chem. 94, 3589–3597.

Suvarapu, L.N., Seo, Y.-K., Baek, S.-O., 2013. Speciation and determination of mercury by various analytical techniques. Rev. Anal. Chem. 32 (3), 225–245.

Suzumura, T., Nakajima, T., Hirao, K., 1999. Ground state properties of MH, MCl and M_2 (M = Cu, Ag and Au) calculated by a scalar relativistic density functional theory. Int. J. Quantum Chem. 75, 757–766.

Svane, M., Hagstrom, M., Davidsson, K.O., Boman, J., Pettersson, J.B.C., 2006. Cesium as a tracer for alkali processes in a circulating fluidized bed reactor. Energy Fuels 20, 979–985.

Swanson, S.M., Engle, M.A., Ruppert, L.F., Affolter, R.H., Jones, K.B., 2013. Partitioning of selected trace elements in coal combustion products from two coal-burning power plants in the United States. Int. J. Coal Geol. 113, 116–126.

Swietlik, R., Molik, A., Molenda, M., et al., 2011. Chromium (III/VI) speciation in urban aerosol. Atmos. Environ. 45, 1364–1368.

Swietlik, R., Trojanowska, M., Jozwiak, M.A., 2012. Evaluation of the distribution of heavy metals and their chemical forms in ESP fractions of fly ash. Fuel Process. Technol. 95, 109–118.

Swietlik, R., Trojanowska, M., Lozynska, M., Molik, A., 2014. Impact of solid fuel combustion technology on valence speciation of chromium in fly ash. Fuel 137, 306–312.

Swihart, M.T., Catoire, L., 2000. Thermochemistry of aluminum species for combustion modeling from *ab initio* molecular orbital calculations. Combust. Flame 121, 210–222.

Syc, M., Pohorely, M., Jeremias, M., Vosecky, M., Kamenikova, P., Skoblia, S., et al., 2011. Behavior of heavy metals in steam fluidized bed gasification of contaminated biomass. Energy Fuels 25, 2284–2291.

Syversen, T., Kaur, P., 2012. The toxicology of mercury and its compounds. J. Trace Elem. Med. Biol. 26, 215–226.

Szidarovszky, T., Csaszar, A.G., 2015. Toward accurate thermochemistry of the ^{24}MgH,^{25}MgH and ^{26}MgH molecules at elevated temperatures: corrections due to unbound states. J. Chem. Phys. 142, 014103. 8 pp.

Tackett, B.S., Clouthier, D.J., 2002. HPO does not follow Walsh's rules! Improved molecular structures from the spectroscopy of jet-cooled HPO and DPO. J. Chem. Phys. 117, 10604–10612.

Taerakul, P., Sun, P., Golightly, D.W., et al., 2006. Distribution of arsenic and mercury in lime spray dryer ash. Energy Fuels 20, 1521–1527.

Tagawa, H., 1984. Thermal decomposition temperatures of metal sulfates. Thermochim. Acta 80, 23–33.

Tagawa, H., Kawabe, K., 1987. Thermal decomposition of $CdSO_4$ and $Zr(SO_4)_2$. Proc. Electrochem. Soc. High Temp. Mater. Chem. IV 88–5, 388–396.

Tagawa, H., Saijo, H., 1985. Kinetics of the thermal decomposition of some transition metal sulfates. Thermochim. Acta. 91, 67–77.

Taggart, R.K., Hower, J.C., Dwyer, G.S., Hsu-Kim, H., 2016. Trends in the rare earth element content of US-based coal combustion fly ashes. Environ. Sci. Technol. 50, 5919–5926.

Taggart, R.K., Rivera, N.A., Levard, C., Ambrosi, J.P., Borschneck, D., et al., 2018. Differences in bulk and microscale yttrium speciation in coal combustion fly ash. Environ. Sci.-Process. Impacts 20 (10), 1390–1403.

Taher-Mansour, F., Allouche, A., Korek, M., 2008. Theoretical electronic structure of the lowest-lying states of ScCl molecule below 22,500 cm^{-1}. J. Mol. Spectrosc. 248, 61–65.

Takahashi, K., Kunz, A., Woiki, D., Roth, P., 2000. Thermal dissociation of tin tetrachloride base on Cl- and Sn-concentration measurements. J. Phys. Chem. A 104, 5246–5253.

Takahashi, N., Terada, K., Nakamura, T., 2000a. Atmospheric pressure chemical vapor deposition of tin nitride films using a halide source. J. Mater. Chem. 10, 2835–2837.

Takahashi, F., Kida, A., Shimaoka, T., 2010. Statistical estimate of mercury removal efficiencies for air pollution control devices of municipal solid waste incinerators. Sci. Total Environ. 408, 5472–5477.

Takahashi, F., Shimaoka, T., Kida, A., 2012. Atmospheric mercury emissions from waste combustions measured by continuous monitoring devices. J. Air Waste Manage. Assoc. 62 (6), 686–695.

Takahashi, F., Katta, V.R., Linteris, G.T., Babushok, V.I., 2019. Numerical study of gas phase interactions of phosphorus-containing compounds with co-flow diffusion flames. Proc. Combust. Inst. 37, 4145–4153.

Takaoka, M., Yamamoto, T., Takeda, N., Oshita, K., Tanaka, T., Uruga, T., 2007. Determination of chemical states of mercury on activated carbon using X-ray absorption near-edge spectroscopy. Am. Inst. Phys. Conf. Proc. 882, 283–285.

Takeshita, T.Y., Dunning Jr, T.H., 2015. Generalized valence bond description of chalcogen-nitrogen compounds. I. NS, F(NS) and H(NS). J. Phys. Chem. A 119, 1446–1455.

Takeuchi, T., Doteshita, I., Asami, S., 2004. Epitaxial growth of sub-nanometer thick tin oxide films on sapphire substrates by pulsed atomic layer chemical vapor deposition. Surf. Interface Anal. 36, 1133–1135.

Takiguchi, H., Tamura, T., 2018. Mercury emission control in Japan. Asian J. Atmos. Environ. 12 (1), 37–46.

Takuwa, T., Naruse, I., 2007. Emission control of sodium compounds and their formation mechanisms during coal combustion. Proc. Combust. Inst. 31, 2863–2870.

Takuwa, T., Mkilaha, I.S.N., Naruse, I., 2006. Mechanisms of fine particulates formation with alkali metal compounds during coal combustion. Fuel 85, 671–678.

Tamadon, F., Seppelt, K., 2013. The elusive halides VCl_5, $MoCl_6$ and $ReCl_6$. Angew. Chem. Int. Ed. 52, 767–769.

Tang, J., Steenari, B.-M., 2016. Leaching optimization of municipal solid waste incineration ash for resource recovery: a case study of Cu, Zn, Pb, and Cd. Waste Manage. 48, 315–322.

Tang, J., Xiao, T., Wang, S., Lei, J., Zhang, M., Gong, Y., et al., 2009. High cadmium concentrations in areas with endemic fluorosis: a serious hidden toxic? Chemosphere 76, 300–305.

Tang, T., Xu, J., Lu, R., Wo, J., Xu, X., 2010. Enhanced Hg^{2+} removal and Hg^0 re-emission control from wet fuel gas desulfurization liquors with additives. Fuel 89, 3613–3617.

Tang, Q., Liu, G., Yan, Z., Sun, R., 2012. Distribution and fate of environmentally sensitive elements (arsenic, mercury, stibium, selenium) in coal-fired power plants at Huainan, Anhui, China. Fuel 95, 334–339.

Tang, S., Wang, L., Feng, X.-B., et al., 2016. Actual mercury speciation and mercury discharges from coal-fired power plants in Inner Mongolia, Northern China. Fuel 180, 194–204.

Tanrikut, E., Karaer, A., Celik, O., et al., 2014. Role of endometrial concentrations of heavy metals (cadmium, lead, mercury and arsenic) in the aetiology of unexplained infertility. Eur. J. Obstetrics Gynecol. Reprod. Biol. 179, 187–190.

Tarczay, G., Miller, T.A., Czako, G., Csaszar, A.G., 2005. Accurate *ab initio* determination of spectroscopic and thermochemical properties of mono- and dichlorocarbenes. PhysChemChemPhys 7, 2881–2893.

Tariq, S.A., Hill, J.O., 1981. Thermal analysis of mercury(I) sulfate and mercury(II) sulfate. J. Thermal Anal. 21, 277–281.

Taylor, R.M., 1984. Influence of chloride on the formation of iron oxides from Fe(II) chloride. I. Effect of [Cl]/[Fe] on the formation of magnetite. Clays Clay Miner. 32, 167–174.

Taylor, C.M., Chaudhuri, R.K., Freed, K.F., 2005. Electronic structure of the calcium monohydroxide radical. J. Chem. Phys. 122, 044317. 13 pp.

Taylor, C.M., Golding, J., Emond, A.M., 2014. Lead, cadmium and mercury levels in pregnancy: the need for international consensus on levels of concern. J. Dev. Origins Health Dis. 5 (1), 16–30.

Taylor, C.M., Emmett, P.M., Emond, A.M., Golding, J., 2018. A review of guidance on fish consumption in pregnancy: is it fit for purpose? Public Health Nutr. 21 (11), 2149–2159.

Teixeira, F.B., de Oliveira, A.C.A., Leao, L.K.R., Fagundes, N.C.F., et al., 2018. Exposure to inorganic mercury causes oxidative stress, cell death, and functional deficits in the motor cortex. Front. Mol. Neurosci. 11, article 125.

Tellinghuisen, J., Ashmore, J.G., 1983. Mixed representations for diatomic spectroscopic data: applications to HgBr. Chem. Phys. Lett. 102 (1), 10–16.

Tellinghuisen, J., Tellinghuisen, P.C., Davies, S.A., Berwanger, P., Viswanathan, K.S., 1982. (B-X) transitions in HgCl and HgI. Appl. Phys. Lett. 41 (9), 789–791.

ten Brink, H.M., Eenkhoorn, S., Hamburg, G., 1996. A mechanistic study of the formation of slags from iron-rich coals. Fuel 75, 952–958.

Teng, Y.-L., Jiang, L., Han, S., Xu, Q., 2007. Matrix isolation infrared spectroscopic and density functional theory studies on reactions of laser ablated lead and tin atoms with water molecules. Bull. Chem. Soc. Jpn. 80 (11), 2149–2156.

Teramae, T., Takarada, T., 2009. Fine ash during pulverized coal combustion. Energy Fuels 23, 2018–2024.

Thayer, J.S., 2010. Roles of organometal(loid) compounds in environmental cycles. Chapter 1. In: Sigel, A., Sigel, H., Sigel, R.K.O. (Eds.), Metal Ions in Life Sciences: Volume 7. Organometallics in Environment and Toxicology. Royal Society of Chemistry, London, UK, pp. 1–32.

Theard, L.P., Hildenbrand, D.L., 1964. Heat of formation of Be_2O(g) by mass spectrometry. J. Chem. Phys. 41, 3416–3420.

Theodorakopoulos, G., Petsalakis, I.D., Hamilton, I.P., 1999. *Ab initio* calculations on the ground and excited states of BeOH and MgOH. J. Chem. Phys. 111, 10484–10490.

Thiebaud, J., Thomson, M.J., Mani, R., Morrow, W.H., Morris, E.A., Jia, C.Q., 2009. In situ optical absorption mercury continuous emission monitor. Environ. Sci. Technol. 43, 9294–9299.

Thompson, C.A., Andrews, L., 1994. Reactions of laser ablated Be atoms with O_2: infrared spectra of beryllium oxides in solid argon. J. Chem. Phys. 100, 8689–8699.

Thompson, C.A., Andrews, L., 1996. Reactions of laser ablated Be atoms with H_2O: infrared spectra and density functional calculations of HOBeOH, HBeOH and HBeOBeH. J. Phys. Chem. 100, 12214–12221.

Thompson, D., Argent, B.B., 1999. The mobilization of sodium and potassium during coal combustion and gasification. Fuel 78, 1679–1689.

Thorneloe, S.A., Kosson, D.S., Sanchez, F., Garrabrants, A.C., Helms, G., 2010. Evaluating the fate of metals in air pollution control residues from coal-fired power plants. Environ. Sci. Technol. 44, 7351–7356.

Thummel, H., Klotz, R., Peyerimhoff, S.D., 1989. Study of the $a^3\Pi$ and $X^1\Sigma^+$ states of the MgO molecule: spin-forbidden transitions and breakdown of the Born-Oppenheimer approximation. Chem. Phys. 135, 229–245.

Thurston, G.D., Ito, K., Lall, R., 2011. A source apportionment of U.S. fine particulate matter air pollution. Atmos. Environ. 45, 3924–3936.

Thy, P., Jenkins, B.M., Lesher, C.E., Grundvig, S., 2006. Compositional constraints on slag formation and potassium volatilization from rice straw blended wood fuel. Fuel Process. Technol. 87, 383–408.

Tian, H.Z., Wang, Y., Xue, Z.G., Cheng, K., Qu, Y.P., Chai, F.H., et al., 2010. Trend and characteristics of atmospheric emissions of Hg, As and Se from coal combustion in China, 1980–2007. Atmos. Chem. Phys. 10, 11905–11919.

Tian, H., Cheng, K., Wang, Y., Zhao, D., Chai, F., Xue, Z., et al., 2011. Quantitative assessment of variability and uncertainty of hazardous trace element (Cd, Cr and Pb) contents in Chinese coals by using bootstrap simulation. J. Air Waste Manage. Assoc. 61, 755–763.

Tian, H.Z., Zhao, D., He, M.C., Wang, Y., Cheng, K., 2011a. Temporal and spatial distribution of atmospheric antimony emission inventories from coal combustion in China. Environ. Pollut. 159 (6), 1613–1619.

Tian, H., Cheng, K., Wang, Y., Zhao, D., Lu, L., Jia, W., et al., 2012. Temporal and spatial variation characteristics of atmospheric emissions of Cd, Cr and Pb from coal in China. Atmos. Environ. 50, 157–163.

Tian, H.Z., Zhao, D., Cheng, K., et al., 2012a. Anthropogenic atmospheric emissions of antimony and its spatial distribution characteristics in China. Environ. Sci. Technol. 46, 3973–3980.

Tian, H.Z., Lu, L., Hao, J.M., Gao, J.J., Cheng, K., Liu, K.Y., et al., 2013. A review of key hazardous trace elements in Chinese coals: abundance, occurrence, behavior during coal combustion and their environmental impacts. Energy Fuels 27, 601–614.

Tian, H., Liu, K., Zhou, J., et al., 2014. Atmospheric emission inventory of hazardous trace elements from China's coal-fired power plants: temporal trends and spatial variation characteristics. Environ. Sci. Technol. 48, 3575–3582.

Tian, H.Z., Zhou, J.R., Zhu, C., et al., 2014a. A comprehensive global inventory of atmospheric antimony emissions from anthropogenic activities, 1995–2010. Environ. Sci. Technol. 48, 10235–10241.

Tian, H.Z., Zhu, C.Y., Gao, J.J., et al., 2015. Quantitative assessment of atmospheric emissions of toxic heavy metals from anthropogenic sources in China: historical trend, spatial distribution, uncertainties and control policies. Atmos. Chem. Phys. 15, 10127–10147.

Tian, C., Gupta, R., Zhao, Y., Zhang, J., 2016. Release behaviors of arsenic in fine particles generated from a typical high arsenic coal at a high temperature. Energy Fuels 30, 6201–6209.

Ticknor, J.L., Hsu-Kim, H., Deshusses, M.A., 2014. A robust framework to predict mercury speciation in combustion flue gases. J. Hazard. Mater. 264, 380–385.

Tientega, F., Harrison, J.F., 1994. Electronic and geometric structure of the diatomics ScN, ScP and ScAs. Chem. Phys. Lett. 223, 202–206.

Tillman, D.A., 1994. Trace Metals in Combustion Systems. Academic Press, Inc, San Diego, CA.

Tillman, D.A., Duong, D., Miller, B., 2009. Chlorine in solid fuels fired in pulverized fuel boilers - Sources, forms, reactions and consequences: a literature review. Energy Fuels 23, 3379–3391.

Tirez, K., Silversmit, G., Bleux, N., Adriaensens, E., Roekens, E., Servaes, K., et al., 2011. Determination of hexavalent chromium in ambient air: a story of method induced Cr(III) oxidation. Atmos. Environ. 45, 5332–5341.

Tischer, R.L., Scheller, K., 1970. The behavior of uranium oxide particles in reducing flames. Combust. Flame 15, 199–202.

Tissari, J., Sippula, O., Kouki, J., Vuorio, K., Jokiniemi, J., 2008. Fine particle and gas emissions from the combustion of agricultural fuels fired in a 20 kW burner. Energy Fuels 22, 2033–2042.

Titov, A.A., Klimenko, M.A., Goryacheva, E.G., Opolchenova, N.L., Stepareva, N.N., Sokolova, N.P., 2008. Preparation of ceria and yttria nanopowders via thermal decomposition of oxalates, carbonates and hydroxides. Inorg. Mater. 44, 1101–1104.

Tiwari, S.N., Prakash, S., 1998. Magnesium oxide as inhibitor of hot oil ash corrosion. Mater. Sci. Technol. 14, 467–472.

Tolbert, M.A., Beauchamp, J.L., 1986. Homolytic and heterolytic bond dissociation energies of the second row group 8, 9 and 10 diatomic transition-metal hydrides: correlation with electronic structure. J. Phys. Chem. 90, 5015–5022.

Tomljenovic, L., 2011. Aluminum and Alzheimer's disease: after a century of controversy, is there a plausible link. J. Alzheimer's Disease 23, 567–598.

Tomljenovic, L., Shaw, C.A., 2011. Do aluminum vaccine adjuvants contribute to the rising prevalence of autism? J. Inorg. Biochem. 105, 1489–1499.

Tomonari, M., Nagashima, U., Hirano, T., 2009. *Ab initio* calculation of the electronic structures of the $^7\Sigma^+$ ground and $A^7\Pi$ and $a^5\Sigma^+$ excited states of MnH. J. Chem. Phys. 130, 154105. 9 pp.

Tong, G.S.M., Jeung, G.H., Cheung, A.S.-C., 2003. Theoretical studies of the first-row transition metal phosphides. J. Chem. Phys. 118 (20), 9224–9232.

Torkmahalleh, M.A., Lin, L., Holsen, T.M., Rasmussen, D.H., Hopke, P.K., 2013. Chromium speciation changes in the presence of ozone and reactive oxygen species at low relative humidity. Atmos. Environ. 71, 92–94.

Torkmahalleh, M.A., Yu, C.-H., Lin, L., et al., 2013a. Improved atmospheric sampling of hexavalent chromium. J. Air Waste Manage. Assoc. 63 (11), 1313–1323.

Torres-Ordonez, R.J., Longwell, J.P., Sarofim, A.F., 1989. Intrinsic kinetics of calcium sulfide oxidation. Energy Fuels 3, 506–515.

Torres-Ordonez, R.J., Wall, T.F., Longwell, J.P., Sarofim, A.F., 1993. Sulfur retention as solid calcium sulfide during coal combustion: a modeling study to define mechanisms and possible technologies. Fuel 72, 633–643.

Tossell, J.A., 2003. Calculation of the energetics for oxidation of gas phase elemental Hg by Br and BrO. J. Phys. Chem. A 107, 7804–7808.

Tossell, J.A., 2006. Calculation of the energetics for the oligomerization of gas phase HgO and HgS and for the solvolysis of crystalline HgO and HgS. J. Phys. Chem. A 110, 2571–2578.

Totton, T.S., Shirley, R., Kraft, M., 2011. First-principles thermochemistry for the combustion of $TiCl_4$ in a methane flame. Proc. Combust. Inst. 33, 493–500.

Toxicological Profile for Aluminum, 2008. 336 pp. Agency for Toxic Substances and Disease Registry. U.S. Department of Health and Human Services.

Toxicological Profile for Antimony and Compounds, 2017. 328 pp., 1992 Report, 148 pp. Agency for Toxic Substances and Disease Registry. U.S. Department of Health and Human Services.

Toxicological Profile for Arsenic, 2007. 559 pp. Agency for Toxic Substances and Disease Registry. U.S. Department of Health and Human Services.

Toxicological Profile for Barium and Barium Compounds, 2007. 231 pp. Agency for Toxic Substances and Disease Registry. U.S. Department of Health and Human Services.

Toxicological Profile for Beryllium, 2002. 290 pp. Agency for Toxic Substances and Disease Registry. U.S. Department of Health and Human Services.

Toxicological Profile for Boron, 2010. 247 pp. Agency for Toxic Substances and Disease Registry. U.S. Department of Health and Human Services.

Toxicological Profile for Cadmium, 2012. 487 pp. Agency for Toxic Substances and Disease Registry. U.S. Department of Health and Human Services.

Toxicological Profile for Cesium, 2004. 306 pp. Agency for Toxic Substances and Disease Registry. U.S. Department of Health and Human Services.
Toxicological Profile for Chromium, 2012. 592 pp. Agency for Toxic Substances and Disease Registry. U.S. Department of Health and Human Services.
Toxicological profile for Cobalt, 2004. 486 pp. Agency for Toxic Substances and Disease Registry. U.S. Department of Health and Human Services.
Toxicological Profile for Copper, 2004. 314 pp. Agency for Toxic Substances and Disease Registry. U.S. Department of Health and Human Services.
Toxicological Profile for Lead, 2007. 582 pp. Agency for Toxic Substances and Disease Registry. U.S. Department of Health and Human Services.
Toxicological profile for Manganese, 2012. tox profiles tp151, 556 pp. Agency for Toxic Substances and Disease Registry. U.S. Department of Health and Human Services.
Toxicological Profile for Mercury, 1999. 676 pp. Addendum 2013. Addendum (Alkyl and dialkyl compounds) 168 pp. Agency for Toxic Substances and Disease Registry. U.S. Department of Health and Human Services.
Toxicological Profile for Nickel, 2005. 397 pp. U.S. Agency for Toxic Substances and Disease Registry. U.S. Department of Health and Human Services.
Toxicological Profile for Phosphate Ester Flame Retardants, 2012. 431 pp. Agency for Toxic Substances and Disease Registry. U.S. Department of Health and Human Services.
Toxicological Profile for Selenium, 2003. 457 pp. Agency for Toxic Substances and Disease Registry. U.S. Department of Health and Human Services.
Toxicological Profile for Silver, 1990. 157 pp. Agency for Toxic Substances and Disease Registry. U.S. Department of Health and Human Services.
Toxicological Profile for Strontium, 2004. 445 pp. Agency for Toxic Substances and Disease Registry. U.S. Department of Health and Human Services.
Toxicological Profile for Thallium, 1992. 114 pp. Agency for Toxic Substances and Disease Registry. U.S. Department of Health and Human Services.
Toxicological Profile for Tin and Tin Compounds, 2005. 426 pp. Agency for Toxic Substances and Disease Registry. U.S. Department of Health and Human Services.
Toxicological Profile for Titanium Tetrachloride, 1997. 125 pp. Agency for Toxic Substances and Disease Registry. U.S. Department of Health and Human Services.
Toxicological Profile for Tungsten, 2005. 203 pp. Agency for Toxic Substances and Disease Registry. U.S. Department of Health and Human Services.
Toxicological Profile for Uranium, 2013. 526 pp. Agency for Toxic Substances and Disease Registry. U.S. Department of Health and Human Services.
Toxicological Profile for Vanadium, 2012. 256 pp. Agency for Toxic Substances and Disease Registry. U.S. Department of Health and Human Services.
Toxicological Profile for Zinc, 2005. 352 pp. Agency for Toxic Substances and Disease Registry. U.S. Department of Health and Human Services.
Toxvaerd, S., 2015. Dynamics of homogeneous nucleation. J. Chem. Phys. 143, 154705. 6 pp.
Tran, Q.K., Steenari, B.-M., Iisa, K., Lindqvist, O., 2004. Capture of potassium and cadmium by kaolin in oxidizing and reducing atmospheres. Energy Fuels 18, 1870–1876.
Trautvetter, U., Ditscheid, B., Jahreis, G., Glei, M., 2018. Habitual intakes, food sources and excretions of phosphorus and calcium in three German study collectives. Nutrients 10 (2), article 171. Available from: https://doi.org/10.3390/nu10020171.
Tribelhorn, M.J., Brown, M.E., 1995. Thermal dissociation of barium and strontium peroxides. Thermochim. Acta. 255, 143–154.
Tridot, G., 1966. Application of thermal methods of analysis to the behavior of vanadyl, uranyl, uranium (IV) and titanium sulfates and their derivatives in various atmospheres. Pure Appl. Chem. 13 (4), 543–568.
Trkula, M., Harris, D.O., 1983. Laser spectroscopy of the $^1A''$-X^1A' system of CuOH and CuOD. J. Chem. Phys. 79 (3), 1138–1144.
Truex, T.J., Pierson, W.R., McKee, D.E., Shelef, M., Baker, R.E., 1980. Effects of barium fuel additive and fuel sulfur level on diesel particulate emissions. Environ. Sci. Technol. 14, 1121–1124.

Tsona, N.T., Li, J., Du, L., 2018. From O_2^- initiated SO_2 oxidation to sulfate formation in the gas phase. J. Phys. Chem. A 122, 5781–5788.

Tuenge, R.T., Laabs, F., Franzen, H.F., 1976. The high temperature vaporization of defect scandium monosulfide. J. Chem. Phys. 65, 2400–2406.

Tufano, G.L., Stein, O.T., Wang, B., Kronenburg, A., Rieth, M., Kempf, A.M., 2018. Coal particle volatile combustion and flame interaction. I. Characterization of transient and group effects. Fuel 229, 262–269.

Tufano, G.L., Stein, O.T., Wang, B., Kronenburg, A., Rieth, M., Kempf, A.M., 2018a. Coal particle volatile combustion and flame interaction. II. Effects of particle Reynolds number and turbulence. Fuel 234, 723–731.

Tumarev, A.S., Baimakov, A.Yu, 1972. Thermodynamic properties of stannous oxide. J. Appl. Chem. USSR 45, 2657–2658.

Turley, C.D., Brenchley, D.L., Landolt, R.R., 1973. Barium additives as diesel smoke suppressants. J. Air Pollut. Control Assoc. 23, 783–787.

Twarowski, A., 1995. Reduction of the phosphorus oxide and acid reaction set. Combust. Flame 102, 41–54.

Tyagi, R., Zhang, Z., Pitzer, R.M., 2014. Electronic spectrum of the UO and UO^+ molecules. J. Phys. Chem. A 118, 11758–11767.

Uberoi, M., Punjak, W.A., Shadman, F., 1990. The kinetics and mechanism of alkali removal from flue gases by solid sorbents. Prog. Energy Combust. Sci. 16, 205–211.

Uddin, M.A., Ozaki, M., Sasaoka, E., Wu, S., 2009. Temperature-programed decomposition desorption of mercury species over activated carbon sorbents for mercury removal from coal-derived flue gas. Energy Fuels 23, 4710–4716.

Uekawa, N., Mochizuki, N., Kajiwara, J., et al., 2003. Nonstoichiometric properties of zinc oxide nanoparticles prepared by decomposition of zinc peroxide. PhysChemChemPhys 5, 929–934.

Uglietti, C., Gabrielli, P., Cooke, C.A., Vallelonga, P., Thompson, L.G., 2015. Widespread pollution of the South American atmosphere predates the industrial revolution by 240 years. Proc. Natl. Acad. Sci. USA 112 (8), 2349–2354.

Umemoto, H., Kanemitsu, T., Tanaka, A., 2014. Production of B atoms and BH radicals from $B_2H_6/He/H_2$ mixtures activated on heated W wires. J. Phys. Chem. A 118, 5156–5163.

United Nations-Cd, 2010. Final review of scientific information on cadmium. United Nations Environment Programme: Chemicals Branch, 324 pp. <www.unep.org>.

United Nations Review-Pb, 2010. Final review of scientific information on lead. Environment Program: Chemicals Branch DTIE, 332 pp. <www.unep.org>.

United States Environmental Protection Agency, 1984. Health effects assessment for iron and compounds. Washington, DC, Report EPA-540/1-86-054, 34 pp.

Unkel, W., Freedman, A., 1983. Conductivity measurements of phosphorus-doped potassium-seeded MHD plasmas. AIAA J. 21, 1648–1651.

Unver, I.K., Terzi, M., 2018. Distribution of trace elements and coal and coal fly ash and their recovery with mineral processing practices: a review. J. Mining Environ. 9 (3), 641–655.

Urban, D.R., Wilcox, J., 2006. A theoretical study of properties and reactions involving arsenic and selenium compounds present in coal combustion flue gases. J. Phys. Chem. A 110, 5847–5852.

Urban, D.R., Wilcox, J., 2006a. Theoretical study of the kinetics of the reactions $Se + O_2 \rightarrow SeO + O$ and $As + HCl \rightarrow AsCl + H$. J. Phys. Chem. A 110, 8797–8801.

Uribe, E.A., Daza, M.C., Villaveces, J.L., 2010. CoH_n (n = 1-3): classical and non-classical cobalt polyhydride. Chem. Phys. Lett. 490, 143–147.

US Geological Survey, 2015. Mineral Commodity Summaries. US Department of the Interior, Restin VA.

Uy, O.M., Drowart, J., 1969. Mass spectrometric determination of the dissociation energies of the molecules BiO, BiS, BiSe and BiTe. Trans. Faraday Soc. 65, 3221–3230.

Uy, O.M., Drowart, J., 1971. Determination by the mass spectrometric Knudsen cell method of the atomization energies of the gaseous aluminum chalcogenides, Al_2, AlCu, AlCuS and $AlCuS_2$. Trans. Faraday Soc. 67, 1293–1301.

Uy, O.M., Srivastava, R.D., Farber, M., 1971. Mass spectrometric determination of the heats of formation of gaseous BO_2 and BOF_2. High Temp. Sci. 3, 462–468.

Uzunova, E.L., Nikolov, G.St, Mikosch, H., 2002. Electronic structure and vibrational modes of cobalt oxide clusters CoO_n(n = 1-4) and their monoanions. J. Phys. Chem. A 106, 4104–4114.

Uzunova, E.L., Mikosch, H., Nikolov, G.St, 2008. Electronic structure of oxide, peroxide and superoxide clusters of the 3d elements: a comparative density functional study. J. Chem. Phys. 128, 094307. 12 pp.

Vaccari, D.A., Strigul, N., 2011. Extrapolating phosphorus production to estimate resource reserves. Chemosphere 84, 792–797.

Vahidinia, A., Samiee, F., Faradmal, J., Rahmani, A., et al., 2019. Mercury, lead, cadmium, and barium levels in human breast milk and factors affecting their concentrations in Hamadan, Iran. Biol. Trace Elem. Res. 187 (1), 32–40.

Vainikka, P., Bankiewicz, D., Frantsi, A., Silvennoinen, J., Hannula, J., Yrjas, P., et al., 2011. High temperature corrosion of boiler waterwalls induced by chlorides and bromides. I. Occurrence of the corrosive ash forming elements in a fluidized bed boiler co-firing solid recovered fuel. Fuel 90, 2055–2063.

Vainio, E., Fleig, D., Brink, A., et al., 2013. Experimental evaluation and field application of a salt method for SO_3 measurement in flue gases. Energy Fuels 27 (5), 2767–2775.

Valcke, M., Bourgault, M.-H., Haddad, S., Bouchard, M., et al., 2018. Deriving a drinking water guideline for a non-carcinogenic contaminant: the case of manganese. Int. J. Environ. Res. Public Health 15 (6), article 1293.

Valmari, T., Lind, T.M., Kauppinen, E.I., Sfiris, G., Nilsson, K., Maenhaut, W., 1999. Field study on ash behavior during circulating fluidized bed combustion of biomass. I. Ash formation. Energy Fuels 13, 379–389.

Van Caneghem, J., Brems, A., Lievens, P., Block, C., Billen, P., Vermeulen, I., et al., 2012. Fluidized bed waste incinerators: design, operational and environmental issues. Prog. Energy Combust. Sci. 38, 551–582.

van der Hurk, J., Hollander, T.J., Alkemade, C.Th.J., 1974. Excitation energies of strontium mono-hydroxide bands measured in flames. J. Quant. Spectrosc. Rad. Transfer 14, 1167–1178.

van der Veen, I., de Boer, J., 2012. Phosphorus flame retardants: properties, production, environmental occurrence, toxicity and analysis. Chemosphere 88, 1119–1153.

Vander Wal, R.L., 2002. Fe-catalyzed single-walled carbon nanotube synthesis within a flame environment. Combust. Flame 130, 37–47.

Vander Wal, R.L., Hall, L.J., 2002. Ferrocene as a precursor reagent for metal-catalyzed carbon nanotubes: competing effects. Combust. Flame 130, 27–36.

Van de Velden, M., Dewil, R., Baeyens, J., Josson, L., Lanssens, P., 2008. The distribution of heavy metals during fluidized bed combustion of sludge. J. Hazard. Mater. 151, 96–102.

Van Dyk, J.C., Benson, S.A., Laumb, M.L., Waanders, B., 2009. Coal and coal ash characteristics to understand mineral transformations and slag formation. Fuel 88, 1057–1063.

Vanek, A., Grosslova, Z., Mihaljevic, M., et al., 2016. Isotopic tracing of thallium contamination in soils affected by emissions from coal-fired power plants. Environ. Sci. Technol. 50, 9864–9871.

van Eyk, P.J., Ashman, P.J., Alwahabi, Z.T., Nathan, G.J., 2009. Simultaneous measurements of the release of atomic sodium, particle diameter and particle temperature for a single burning coal particle. Proc. Combust. Inst. 32, 2099–2106.

van Eyk, P.J., Ashman, P.J., Alwahabi, Z.T., Nathan, G.J., 2011. The release of water-bound and organic sodium from Loy Yang coal during the combustion of single particles in a flat flame. Combust. Flame 158, 1181–1192.

van Eyk, P.J., Ashman, P.J., Nathan, G.J., 2011a. Mechanism and kinetics of sodium release from brown coal char particles during combustion. Combust. Flame 158, 2512–2523.

Vangelatos, I., Angelopoulos, G.N., Boufounos, D., 2009. Utilization of ferroalumina as raw material in the production of Ordinary Portland Cement. J. Hazard. Mater. 168, 473–478.

van Lith, S.C., Jensen, P.A., Frandsen, F.J., Glarborg, P., 2008. Release to the gas phase of inorganic elements during wood combustion. II. Influence of fuel composition. Energy Fuels 22, 1598–1609.

van Lith, S.C., Frandsen, F.J., Montgomery, M., Vilhelmsen, T., Jensen, S.A., 2009. Lab-scale investigation of deposit-induced chlorine corrosion of superheater materials under simulated biomass-firing conditions. I. Exposure at 560°C. Energy Fuels 23, 3457–3468.

Van Otten, B., Buitrago, P.A., Senior, C.L., Silcox, G.D., 2011. Gas phase oxidation of mercury by bromine and chlorine in flue gas. Energy Fuels 25, 3530–3536.

Vanpee, M., Shirodkar, P.P., 1979. A study of flame inhibition by metal compounds. Symp. (Int.) Combust. Proc. 17, 787–795.

van Velzen, D., Langenkamp, H., Herb, G., 2002. Review: mercury in waste incineration. Waste Manage. Res. 20, 556–568.

van Wullen, C., 1998. Molecular density functional calculations in the regular relativistic approximation: method, application to coinage metal diatomics, hydrides, fluorides and chlorides, and comparison with first-order relativistic calculations. J. Chem. Phys. 109 (2), 392–399.

VanZee, R.J., Khan, A.U., 1976. Infrared emission from ultraviolet and vacuum ultraviolet states of PO in the cool phosphorus chemiluminescence flame. Chem. Phys. Lett. 41 (1), 180–182.

VanZee, R.J., Khan, A.U., 1976a. Transient emitting species in phosphorus chemiluminescence. J. Chem. Phys. 65 (5), 1764–1772.

Van Zee, R.J., Hamrick, Y.M., Li, S., Weltner Jr, W., 1992. Cobalt, rhodium and iridium dioxide molecules and Walsh-type rules. J. Phys. Chem. 96, 7247–7251.

Varberg, T.D., Hill, E.J., Field, R.W., 1989. Laser spectroscopy of CoH: spin-orbit splitting of the ground state. J. Mol. Spectrosc. 138, 630–637.

Varrica, D., Bardelli, F., Dongarra, G., Tamburo, E., 2013. Speciation of Sb in airborne particulate matter, vehicle brake linings and brake pad wear residues. Atmos. Environ. 64, 18–24.

Varsi, K., Bolann, B., Torsvik, I., Eik, T.C.R., Hol, P.J., Bjorke-Monsen, A.L., 2017. Impact of maternal selenium status on infant outcome during the first six months of life. Nutrients 9 (5), article 486.

Vasil'ev, V.G., Ershova, Z.V., 1972. Thermal dissociation of anhydrous beryllium sulfate. Zhur. Neorgan. Khim. 17, 631–633.

Vasiliu, M., Feller, D., Gole, J.L., Dixon, D.A., 2010. Structures and heats of formation of simple alkaline earth metal compounds: fluorides, chlorides and hydroxides of Be, Mg and Ca. J. Phys. Chem. A 114, 9349–9358.

Vasiliu, M., Li, S., Peterson, K.A., et al., 2010a. Structures and heats of formation of simple alkali metal compounds: hydrides, chlorides, fluorides, hydroxides and oxides for Li, Na and K. J. Phys. Chem. A 114, 4272–4281.

Vasiliu, M., Grant, D.J., Feller, D., Dixon, D.A., 2012. Heats of formation of MH_xCl_y (M = Si, P, As, Sb) compounds and main group fluorides from high level electronic structure calculations. J. Phys. Chem. A 116, 3717–3727.

Vasiliu, M., Hill, J.G., Peterson, K.A., Dixon, D.A., 2018. Structures and heats of formation of simple alkaline earth metal compounds. II. Fluorides, chlorides, oxides and hydroxides for Ba, Sr and Ra. J. Phys. Chem. A 122, 316–327.

Vassilev, S.V., Vassileva, C.G., 2009. A new approach for the combined chemical and mineral classification of the inorganic matter in coal. 1. Chemical and mineral classification systems. Fuel 88, 235–245.

Vassilev, S.V., Vassileva, C.G., Baxter, D., Andersen, L.K., 2009. A new approach for the combined chemical and mineral classification of the inorganic matter in coal. 2. Potential applications of the classification systems. Fuel 88, 246–254.

Vassilev, S.V., Baxter, D., Andersen, L.K., Vassileva, C.G., 2013. An overview of the composition and application of biomass ash. I. Phase-mineral and chemical composition and classification. Fuel 105, 40–76.

Vassilev, S.V., Baxter, D., Andersen, L.K., Vassileva, C.G., 2013a. An overview of the composition and application of biomass ash. II. Potential utilization, technological and ecological advantages and challenges. Fuel 105, 19–39.

Vassilev, S.V., Baxter, D., Vassileva, C.G., 2014. An overview of the behavior of biomass during combustion. II. Ash fusion and ash formation mechanisms of biomass types. Fuel 117, 152–183.

Vassilev, S.V., Vassileva, C.G., Baxter, D., 2014a. Trace element concentrations and associations in some biomass ashes. Fuel 129, 292–313.

Veits, I.V., Gurvich, L.V., 1956. The dissociation energy for the oxides of magnesium, calcium, strontium and barium. Optik. Spektrosc. 1, 22–33.

Vejahati, F., Xu, Z., Gupta, R., 2010. Trace elements in coal: associations with coal and minerals and their behavior during coal utilization - a review. Fuel 89, 904–911.

Vengosh, A., Coyte, R., Karr, J., et al., 2016. Origin of hexavalent chromium in drinking water wells from the Piedmont aquifors of North Carolina. Environ. Sci. Technol. Lett. 3, 409–414.

Venkateswarlu, P., Khanna, B.N., 1960. Emission spectrum of bismuth monochloride. I. Vibrational analysis of the 617-422 nm system. Proc. Indian Acad. Sci. A 51, 14–27.

Vent-Schmidt, T., Andrews, L., Riedel, S., 2015. Reactions of laser-ablated U atoms with HF: infrared spectra and quantum calculations of HUF, UH and UF in noble gas solids. J. Phys. Chem. A 119, 2253–2261.

Verbinnen, B., Billen, P., Van Coninckxloo, M., Vandecasteele, C., 2013. Heating temperature dependence of Cr (III) oxidation in the presence of alkali and alkaline earth salts and subsequent Cr(VI) leaching behavior. Environ. Sci. Technol. 47, 5858–5863.

Vergnon, P., Bianchi, D., Chaoui, R.B., Coudurier, G., 1980. Preparation and characterization of finely divided molybdenum trioxide: infrared study of the water interaction. J. Chim. Phys. 77, 1043–1049.

Verhaegen, G., Colin, R., Exsteen, G., Drowart, J., 1965. Mass spectrometric determination of the stability of gaseous molybdites, tungstites, molybdates and tungstates of magnesium, calcium, strontium and tin. Trans. Faraday Soc. 61, 1372–1375.

Verhulst, D., Buekens, A., Spencer, P.J., Eriksson, G., 1996. Thermodynamic behavior of metal chlorides and sulfates under the conditions of incineration furnaces. Environ. Sci. Technol. 30, 50–56.

Viana, R.B., 2017. Unimolecular rearrangement of the simplest compound models with the selenium-oxygen, selenium-sulfur and selenium-selenium bond: SeXH and HSeXH (X = O, S, Se). Mol. Phys. 115 (8), 1004–1013.

Vinckier, C., Helaers, J., 1998. Kinetic investigation of the reaction $Sr(^1S) + O_2 + He$ in the temperature range from 303 to 968 K. J. Phys. Chem. A 102, 8333–8338.

Vinckier, C., Remeysen, J., 1994. A kinetic study of the association reaction $Ca(^1S) + O_2 + He \rightarrow CaO_2 + He$ in the temperature range from 303 to 986 K. J. Phys. Chem. 98, 10535–10538.

Vinckier, C., Corthouts, J., De Jaegere, S., 1988. A kinetic study of the reaction of copper atoms in the gas phase with molecular oxygen at 300 K in microwave-induced plasma afterglows. J. Chem. Soc. Faraday Trans. II 84 (12), 1951–1960.

Vinckier, C., Christiaens, P., Hendrickx, 1992. Kinetics of the molecular oxygen reactions with sodium, magnesium and copper atoms. In: Fontijn, A. (Ed.), Gas Phase Metal Reactions. Elsevier Science Publishers, Amsterdam, pp. 57–72.

Viswanadham, P., Edwards, J.M., 1975. High temperature thermodynamic properties of manganese(II) sulfide by the dynamic Knudsen-torsion effusion method. J. Chem. Phys. 62, 3875–3882.

Vogel, C., Adam, C., 2011. Heavy metal removal from sewage sludge ash by thermochemical treatment with gaseous hydrochloric acid. Environ. Sci. Technol. 45, 7445–7450.

Vogel, M., Wenzel, W., 2005. Multireference calculations of the electronic structure of VF_2 and VCl_2. J. Chem. Phys. 123, 194110. 5 pp.

Vogel, C., Exner, R.M., Adam, C., 2013. Heavy metal removal from sewage sludge ash by thermochemical treatment with polyvinylchloride. Environ. Sci. Technol. 47, 563–567.

Vogel, C., Adam, C., Kappen, P., et al., 2014. Chemical state of chromium in sewage sludge ash bases phosphorus-fertilizers. Chemosphere 103, 250–255.

Von Schnering, H.-G., Honle, W., 1988. Bridging chasms with polyphosphides. Chem. Rev. 88, 243–273.

von Szentpaly, L., 2008. Atom-based thermochemistry: predictions of the sublimation enthalpies of group 12 chalcogenides and the formation enthalpies of their polonides. J. Phys. Chem. 112, 12695–12701.

Wag, K.J., Reis, V.V., Frederick, W.J., Grace, T.M., 1997. Mathematical model for the release of inorganic emissions during black liquor char combustion. TAPPI J. 80 (5), 135–145.

Wagener, S., Dommershausen, N., Jungnickel, H., et al., 2016. Textile functionalization and its effects on the release of silver nanoparticles into artificial sweat. Environ. Sci. Technol. 50, 5927–5934.

Wagman, D.D., Evans, W.H., Parker, V.B., Schumm, R.H., Halow, I., Bailey, S.M., et al., 1982. The NBS tables of chemical thermodynamic properties: selected values for inorganic and C_1 and C_2 organic substances in SI units. J. Phys. Chem. Ref. Data 11 (Suppl 2), 392 pp.

Wahrenberger, C., Seward, T.M., Volker, D., 2002. Volatile trace-element transport in high temperature gases from Kudriavy volcano (Iturup, Kurile Islands, Russia), Geochem. Soc. Spec. Publ. Water-Rock Interact. Ore Deposits Environ. Geochem., 7. pp. 307–327.

Wai, K.-M., Wu, S., Li, X., et al., 2016. Global atmospheric transport and source-receptor relationships for arsenic. Environ. Sci. Technol. 50, 3714–3720.

Waindich, A., Muller, M., 2014. Alkali removal at 1400°C under gasification conditions. Fuel 116, 889–893.

Walch, S.P., Bauschlicher Jr, C.W., 1983. CASSCF/CI calculations for first row transition metal hydrides: the TiH $(^4\Phi)$, $VH(^5\Delta)$, $CrH(^6\Sigma^+)$, $MnH(^7\Sigma^+)$, $FeH(^{4,6}\Delta)$ and $NiH(^2\Delta)$ states. J. Chem. Phys. 78, 4597–4605.

Walji, S.-D., Sentjens, K.M., Le Roy, R.J., 2015. Dissociation energies and potential energy functions for the ground $X^1\Sigma^+$ and "avoided-crossing" $A^1\Sigma^+$ states of NaH. J. Chem. Phys. 142, 044305. 11 pp.

Wallschlager, D., Feldmann, J., 2010. Formation, occurrence, significance and analysis of organoselenium and organotellurium compounds in the environment. Chapter 10. In: Sigel, A., Sigel, H., Sigel, R.K.O. (Eds.), Metal Ions in Life Sciences: Volume 7. Organometallics in Environment and Toxicology. Royal Society of Chemistry, London, UK, pp. 319–364.

Walsh, P.N., Goldstein, H.W., White, D., 1960. Vaporization of rare earth oxides. J. Am. Ceram. Soc. 43, 229–233.

Wan, S.L., Wrixon, A.D., 1988. Radiation doses from coal fired plants in Oxfordshire and Berkshire. National Radiological Protection Board, Report NRPB-R203, 19 pp.

Wan, Y.N., Camara, A.Y., Huang, Q.Q., Yu, Y., et al., 2018. Arsenic uptake and accumulation in rice (*Oryza sativa* L.) with selenite fertilization and water management. Ecotoxicol. Environ. Saf. 156, 67–74.

Wang, B., Du, Y., 2013. Cadmium and its neurotoxic effects. Oxid. Med. Cell Longev. Available from: https://doi.org/10.1155/2013/898034. Art.No.2013:898034.

Wang, J., Anthony, E.J., 2005. Analysis of the reaction rate for mercury vapor and chlorine. Chem. Eng. Technol. 28, 569–573.

Wang, J., Tomita, A., 2003. A chemistry on the volatility of some trace elements during coal combustion and pyrolysis. Energy Fuels 17, 954–960.

Wang, S.G., Schwarz, W.H.E., 1998. Density functional study of first row transition metal dihalides. J. Chem. Phys. 109, 7252–7262.

Wang, X., Andrews, L., 2001. Precious metal-molecular oxygen complexes: neon matrix infrared spectra and density functional calculations for $M(O_2)$, $M(O_2)_2$ (M = Pd, Pt, Ag, Au). J. Phys. Chem. A 105, 5812–5822.

Wang, X., Andrews, L., 2002. Neon matrix infrared spectra and DFT calculations of tungsten hydrides WH_x (x = 1-4, 6). J. Phys. Chem. A 106, 6720–6729.

Wang, X., Andrews, L., 2003. Chromium hydrides and dihydrogen complexes in solid neon, argon and hydrogen: matrix infrared spectra and quantum chemical calculations. J. Phys. Chem. A 107, 570–578.

Wang, X., Andrews, L., 2003a. Matrix infrared spectra and density functional theory calculations of manganese and rhenium hydrides. J. Phys. Chem. A 107, 4081–4091.

Wang, X., Andrews, L., 2004. Infrared spectra of magnesium hydride molecules, complexes and solid magnesium dihydride. J. Phys. Chem. A 108, 11511–11520.

Wang, X., Andrews, L., 2004a. Infrared spectra of Zn and Cd hydride molecules and solids. J. Phys. Chem. A 108, 11006–11013.

Wang, X., Andrews, L., 2004b. Metal dihydride (MH_2) and dimer (M_2H_4) structures in solid argon, neon and hydrogen (M = Ca, Sr, Ba): infrared spectra and theoretical calculations. J. Phys. Chem. A 108, 11500–11510.

Wang, X., Andrews, L., 2005. Infrared spectra and electronic structure calculations for the Group II metal $M(OH)_2$ dihydroxide molecules. J. Phys. Chem. A 109, 2782–2792.

Wang, X., Andrews, L., 2005a. Infrared spectra and structures for group IV dihydroxide and tetrahydroxide molecules. J. Phys. Chem. 109, 10689–10701.

Wang, X., Andrews, L., 2005b. Infrared spectra and structures of the coinage metal dihydroxide molecules. Inorg. Chem. 44 (24), 9076–9083.

Wang, X., Andrews, L., 2005c. Infrared spectra of $M(OH)_{1,2,4}$ (M = Pb, Sn) in solid argon. J. Phys. Chem. A 109, 9013–9020.

Wang, X., Andrews, L., 2005d. Infrared spectrum of $Hg(OH)_2$ in solid neon and argon. Inorg. Chem. 44 (1), 108–113.

Wang, X., Andrews, L., 2005e. Matrix infrared spectra and density functional theory calculations of molybdenum hydrides. J. Phys. Chem. A 109, 9021–9027.

Wang, X., Andrews, L., 2005f. Mercury dihydride forms a covalent molecular solid. PhysChemChemPhys 7, 750–759.

Wang, X., Andrews, L., 2005g. Zinc and cadmium dihydroxide molecules: matrix infrared spectra and theoretical calculations. J. Phys. Chem. A 109, 3849–3857.

Wang, X., Andrews, L., 2005h. One dimensional BeH_2 polymers: infrared spectra and theoretical calculations. Inorg. Chem. 44, 610–614.

Wang, X., Andrews, L., 2006. Contrasting products in the reactions of Cr, Mo and W atoms with H_2O_2: argon matrix infrared spectra and theoretical calculations. J. Phys. Chem. A 110, 10409–10418.

Wang, X., Andrews, L., 2006a. Infrared spectra and density functional calculations for the $Sc(OH)_{2,3}$ and HOScO molecules and the $Sc(OH)_2^+$ cation in solid argon. J. Phys. Chem. A 110, 1850–1858.

Wang, X., Andrews, L., 2006b. Infrared spectra of $M(OH)_{1,2,3}$ (M = Mn, Fe, Co, Ni) molecules in solid argon and the character of first row transition metal hydroxide bonding. J. Phys. Chem. A 110, 10035–10045.

Wang, X., Andrews, L., 2007. Infrared spectroscopic observation of the group 13 metal hydroxides, $M(OH)_{1,2,3}$ (M = Al, Ga, In, Tl) and $HAl(OH)_2$. J. Phys. Chem. A 111, 1860–1868.

Wang, X., Andrews, L., 2009. Infrared spectra and theoretical calculations for Fe, Ru and Os metal hydrides and dihydrogen complexes. J. Phys. Chem. A 113, 551–563.

Wang, X., Wang, W.X., 2017. Selenium induces the demethylation of mercury in marine fish. Environ. Pollut. 231, 1543–1551. Part 2.

Wang, J.Z., Das, K.K., Balasubramanian, K., 1990. Potential energy surfaces for insertion of Zr into H_2. Mol. Phys. 69 (1), 147–159.

Wang, J., Tomlinson, M.J., Caruso, J.A., 1995. Extraction of trace elements in coal flyash and subsequent speciation by high-performance liquid chromatography with inductively coupled plasma mass spectrometry. J. Anal. Atom. Spectrom. 10, 601–607.

Wang, X., Chertihin, G.V., Andrews, L., 2002. Matrix infrared spectra and DFT calculations of the reactive MH_x (x = 1-3), $(H_2)MH_2$, MH_2^+ and MH_4^- (M = Sc, Y, La) species. J. Phys. Chem. A 106, 9213–9225.

Wang, X., Andrews, L., Manceron, L., Marsden, C., 2003. Infrared spectra and DFT calculations for the coinage metal hydrides MH, $(H_2)MH$, MH_2, M_2H, M_2H^- and $(H_2)CuHCu$ in solid argon, neon and hydrogen. J. Phys. Chem. A 107 (41), 8492–8505.

Wang, X., Souter, P.F., Andrews, L., 2003a. Infrared spectra of antimony and bismuth hydrides in solid matrixes. J. Phys. Chem. A 107, 4244–4249.

Wang, Y.-F., Huang, K.-L., Li, C.-T., Mi, H.-H., Luo, J.-H., Tsai, P.-J., 2003b. Emissions of fuel metals content from a diesel vehicle engine. Atmos. Environ. 37, 4637–4643.

Wang, H.L., Hao, Z.P., Zhuang, Y.H., Wang, W., Liu, X.Y., 2008. Characterization of inorganic components of size-segregated particles in the flue gas of a coal-fired power plant. Energy Fuels 22, 1636–1640.

Wang, J., Gupta, A., Klein, T.M., 2008a. Plasma enhanced chemical vapor deposition of Cr_2O_3 thin films using chromium hexacarbonyl, $Cr(CO)_6$, precursor. Thin Solid Films 516, 7366–7372.

Wang, J., Wang, T., Burken, J.G., et al., 2008b. Adsorption of arsenic (V) onto fly ash: a speciation-based approach. Chemosphere 72, 381–388.

Wang, J., Yamada, O., Nakazato, T., Zhang, Z.-G., Suzuki, Y., Sakanishi, K., 2008c. Statistical analysis of the concentrations of trace elements in a wide diversity of coals and its implications for understanding elemental modes of occurrence. Fuel 87, 2211–2222.

Wang, L.J., Stolyarova, V.L., Lopatin, S.I., Seetharaman, S., 2009. High temperature mass spectrometric study of the vaporization processes in the system CaO-MgO-Al_2O_3-Cr_2O_3-FeO-SiO_2. Rapid Commun. Mass Spectrom. 23, 2233–2239.

Wang, Q., Zhang, L., Sato, A., Ninomiya, Y., Yamashita, T., Dong, Z., 2009a. Mineral interactions and their impacts on the reduction of PM_{10} emissions during co-combustion of coal with sewage sludge. Proc. Combust. Inst. 32, 2701–2708.

Wang, X., Liang, B., Andrews, L., 2009b. Infrared spectra and density functional theory calculations of coinage metal sulfide molecules and complexes. Dalton Trans. 21, 4190–4198.

Wang, L., Ju, Y.W., Liu, G.J., Chou, C.L., Zheng, L.G., Qi, C.C., 2010. Selenium in Chinese coals: distribution, occurrence and health impact. Environ. Earth Sci. 60, 1641–1651.

Wang, Q., Yan, J.-H., Chi, Y., Li, X.-D., Lu, S.-Y., 2010a. Application of thermal plasma to vitrify fly ash from municipal solid waste incinerators. Chemosphere 78, 626–630.

Wang, S., Zhang, L., Wu, Y., Ancora, M.P., Zhao, Y., Hao, J., 2010b. Synergistic mercury removal by conventional pollutant control strategies for coal fired power plants in China. J. Air Waste Manage. Assoc. 60, 722–730.

Wang, S.X., Zhang, L., Li, G.H., Wu, Y., Hao, J.M., Pirrone, N., et al., 2010c. Mercury emission and speciation of coal-fired power plants in China. Atmos. Chem. Phys. 10, 1183–1192.

Wang, H., Qiu, P., Li, Y., Han, Z., Wu, S., Zhao, G., 2012. Effect of quenching temperature on melting characteristics of coal ash in a reducing atmosphere. Energy Fuels 26, 2204–2219.

Wang, H., Qiu, P., Wu, S., Zhu, Y., Li, Y., Zhao, G., 2012a. Melting behavior of typical ash particles in reducing atmosphere. Energy Fuels 26, 3527–3541.

Wang, C., Chen, M., Li, Z.H., Zhou, M., 2013. Formation and infrared spectroscopic characterization of three oxygen-rich BiO_4 isomers in solid argon. J. Phys. Chem. A 117 (44), 11217–11224.

Wang, J.-C., Zhang, Y.-P., Han, L.-N., Chang, L.-P., Bao, W.-R., 2013a. Simultaneous removal of hydrogen sulfide and mercury from simulated syngas by iron-based sorbents. Fuel 103 (special issue), 73–79. Available from: https://doi.org/10.1016/j.fuel.2011.10.056.

Wang, F., Wang, S., Zhang, L., et al., 2014. Mercury enrichment and its effects on atmospheric emissions in cement plants of China. Atmos. Environ. 92, 421–428.

Wang, H., Ho, K.T., Scheckel, K.G., et al., 2014a. Toxicity, bioaccumulation and biotransformation of silver nanoparticles in marine organisms. Environ. Sci. Technol. 48, 13711–13717.

Wang, R., Balkanski, Y., Boucher, O., et al., 2015. Significant contribution of combustion-related emissions to the atmospheric phosphorus budget. Nat. Geosci. 8, 48–54.

Wang, Z., Marin, G., Naterer, G.F., Gabriel, K.S., 2015a. Thermodynamics and kinetics of the thermal decomposition of cupric chloride in its hydrolysis reaction. J. Thermal Anal. Calorim. 119 (2), 815–823.

Wang, R., Balkanski, Y., Boucher, O., et al., 2015b. Sources, transport and deposition of iron in the global atmosphere. Atmos. Chem. Phys. 15 (11), 6247–6270.

Wang, B., Cao, Y., Li, J., et al., 2016. Migration and redistribution of sulfur species during chemical looping combustion of coal with $CuFe_2O_4$ combined oxygen carrier. Energy Fuels 30, 8499–8510.

Wang, B., Kurian, V., Mahapatra, N., et al., 2016a. Investigation of corrosion and fouling in syngas cooler tubes. Fuel Process. Technol. 141, 202–209.

Wang, F., Tian, G., Wang, X.-F., et al., 2016b. Application of genetic algorithm-back propagation for prediction of mercury speciation in combustion flue gas. Clean Technol. Environ. Policy 18 (4), 1211–1218.

Wang, F., Wang, S., Meng, Y., et al., 2016c. Mechanisms and roles of fly ash compositions on the adsorption and oxidation of mercury in flue gas from coal combustion. Fuel 163, 232–239.

Wang, F., Wang, S., Zhang, L., et al., 2016d. Mercury mass flow in iron and steel production process and its implications for mercury emission control. J. Environ. Sci. 43, 293–301.

Wang, H., Duan, Y., Li, Y.-N., Xue, Y., Liu, M., 2016e. Inner relationship between CO, NO and Hg in a 6 kW_{th} circulating fluidized bed combustor under an O_2/CO_2 atmosphere. Energy Fuels 30, 4221–4228.

Wang, X.Y., Huang, Y.J., Zhong, Z.P., Pan, Z.G., Liu, C.Q., 2016f. Theoretical investigation of cadmium vapor adsorption on kaolinite surfaces with DFT calculations. Fuel 166, 333–339.

Wang, Y., Chen, L., Gao, Y., et al., 2016g. Effects of prenatal exposure to cadmium on neurodevelopment of infants in Shandong, China. Environ. Pollut. 211, 67–73.

Wang, Y., Xiang, Y., Wang, D., et al., 2016h. Effect of sodium oxides in ash composition on ash fusibility. Energy Fuels 30, 1437–1444.

Wang, X., Lin, C.-J., Yuan, W., Sommar, J., et al., 2016j. Emission dominated gas exchange of elemental mercury vapor over natural surfaces in China. Atmos. Chem. Phys. 16, 11125–11143.

Wang, H., Duan, Y., Li, Y.-N., Xue, Y., Liu, M., 2016k. Investigation of mercury emission and its speciation from an oxy-fuel circulating fluidized bed combustor with recycled warm flue gas. Chem. Eng. J. 300, 230–235.

Wang, F., Shen, B., Yang, J., Singh, S., 2017. Review of mercury formation and capture from CO_2 enriched oxy-fuel combustion flue gas. Energy Fuels 31, 1053–1064.

Wang, C., Zhang, Y., Shi, Y., Liu, H., Zou, C., et al., 2017a. Research on collaborative control of Hg, As, Pb, and Cr by electrostatic-fabric-integrated precipitator and wet flue gas desulfurization in coal-fired power plants. Fuel 210, 527–534.

Wang, X., Ruan, R., Yang, T., Adeosun, A., et al., 2017b. Sulfate removal by kaolin addition to address fouling in a full-scale furnace burning high-alkaline Zhundong coal. Energy Fuels 31, 12823–12830.

Wang, J.W., Zhang, Y.S., Liu, Z., Norris, P., et al., 2017c. Effective coordinated air pollution control devices in coal-fired power plants on arsenic emissions. Energy Fuels 31 (7), 7309–7316.

Wang, B., Liu, J., Liu, B., Liu, X.Y., Yu, X.D., 2018. Prenatal exposure to arsenic and neurobehavioral development of newborns in China. Environ. Int. 121, 421–427. Part 1.

Wang, C.-A., Zhao, L., Han, T., Chen, W., Yan, Y., et al., 2018a. Release and transformation behaviors of sodium, calcium, and iron during oxy-fuel combustion of Zhundong coals. Energy Fuels 32, 1242–1254.

Wang, M., Binns, J., Donnelly, M.-E., Pena-Alvarez, M., et al., 2018b. High pressure synthesis and stability of cobalt hydrides. J. Chem. Phys. 148, 144310. 4 pp.

Wang, Q., Liu, Y., Wu, Z., 2018c. Laboratory study on mercury release of the gypsum from the mercury co-removal wet flue gas desulfurization system with additives. Energy Fuels 32, 1005–1011.

Wang, G., Jensen, P.A., Wu, H., Frandsen, F.J., Sander, B., Glarborg, P., 2018d. Potassium capture by kaolin. I. KOH. Energy Fuels 32, 1851–1862.

Wang, G., Jensen, P.A., Wu, H., Frandsen, F.J., Sander, B., Glarborg, P., 2018e. Potassium capture by kaolin. II. K_2CO_3, KCl, and K_2SO_4. Energy Fuels 32, 3566–3578.

Wang, C., Liu, H., Zhang, Y., Zu, C., Anthony, E.J., 2018f. Review of arsenic behavior during coal combustion: volatilization, transformation, emission and removal technologies. Prog. Energy Combust. Sci. 68, 1–28.

Wang, D., Xue, M.Y., Wang, Y.K., Zhou, D.Z., Tang, L., et al., 2019. Effects of straw amendment on selenium aging in soils: mechanism and influential factors. Sci. Total Environ. 657, 871–881.

Washburn, C.A., 1963. The vaporization of iron oxides. United States Atomic Energy Commission, Report UCRL-10991, 20 pp.

Washburn, E.B., Trivedi, J.N., Catoire, L., Beckstead, M.W., 2008. The simulation of the combustion of micrometer-sized aluminum particles with steam. Combust. Sci. Technol. 180, 1502–1517.

Wasson, S.J., Linak, W.P., Gullett, B.K., et al., 2005. Emissions of chromium, copper, arsenic and PCDDs/Fs from open burning of CCA-treated wood. Environ. Sci. Technol. 39, 8865–8876.

Waterland, L.R., Fournier Jr, D.J., Lee, J.W., Carroll, G.J., Thurnau, R.C., 1994. The fate of trace metals in a rotary kiln incinerator: tests with two different wet scrubber systems. In: Seeker, W.R., Koshland, C.R. (Eds.), Incineration of Hazardous Waste. II. Toxic Combustion By-Products. Gordon and Breach Science Publishers, Switzerland, pp. 557–574.

Watson, L.R., Thiem, T.L., Dressler, R.A., Salter, R.H., Murad, E., 1993. High temperature mass spectrometric studies of the bond energies of gas phase ZnO, NiO and CuO. J. Phys. Chem. 97, 5577–5580.

Watt, J.A.J., Burke, I.T., Edwards, R.A., Malcolm, H.M., Mayes, W.M., et al., 2018. Vanadium: a re-emerging environmental hazard. Environ. Sci. Technol. 52, 11973–11974.

Wayne, F.D., Colbourn, E.A., 1977. Effective spin-spin interactions and singlet-triplet transition intensities in a class of open-shell diatomic molecules. Mol. Phys. 34 (4), 1141–1155.

Wayne, F.D., Colbourn, E.A., 1984. Effective spin-spin interactions and forbidden transition intensities in a class of open-shell diatomic molecules: erratum. Mol. Phys. 51 (2), 531–532.

Wee, H.L., Wu, H., Zhang, D.-K., French, D., 2005. The effect of combustion conditions on mineral matter transformation and ash deposition in a utility boiler fired with a sub-bituminous coal. Proc. Combust. Inst. 30, 2981–2989.

Wei, X., Lopez, C., von Puttkamer, T., Schnell, U., Unterberger, S., Hein, K.R.G., 2002. Assessment of chlorine/alkali/mineral interactions during co-combustion of coal and straw. Energy Fuels 16, 1095–1108.

Wei, X., Schnell, U., Hein, K.R.G., 2005. Behavior of gaseous chlorine and alkali metals during biomass thermal utilization. Fuel 84, 841–848.

Wei, X., Wang, Y., Liu, D., et al., 2009. Release of sulfur and chlorine during co-firing RDF and coal in an internally circulating fluidized bed. Energy Fuels 23, 1390–1397.

Wei, B.G., Yu, J.P., Wang, J., Li, H.R., et al., 2018. Trace metals in the urine and hair of the population in an endemic arsenism area. Biol. Trace Elem. Res. 182 (2), 209–216.

Weigelt, A., Ebinghaus, R., Manning, A.J., et al., 2015. Analysis and interpretation of 18 years of mercury observations since 1996 at Mace Head, Ireland. Atmos. Environ. 100, 85–93.

Weir, A., Westerhoff, P., Fabricius, L., et al., 2012. Titanium dioxide nanoparticles in food and personal care products. Environ. Sci. Technol. 46, 2242–2250.

Wen, H., Carignan, J., 2007. Reviews on atmospheric selenium: emissions, speciation and fate. Atmos. Environ. 41, 7151–7165.

Wendlandt, W.W., 1974. The thermal properties of inorganic compounds. I. Some mercury(I) and (II) compounds. Thermochim. Acta. 10, 101–107.

Wendlandt, H.G., Glemser, O., 1963. Reaction of oxides with water at high pressures and temperatures. Angew. Chem. 75, 949–957.

Weng, W., Chen, S., Wu, H., Glarborg, P., Li, Z., 2018. Optical investigation of gas-phase KCl/KOH sulfation in post flame conditions. Fuel 224, 461–468.

Werner, J.H., Cool, T.A., 1999. Kinetic model for the decomposition of dimethyl-methylphosphonate in a hydrogen/oxygen flame. Combust. Flame 117, 78–98.

Werther, J., Ogada, T., 1999. Sewage sludge combustion. Prog. Energy Combust. Sci. 25, 55–116.

Westblom, U., Fernandez-Alonso, F., Mahon, C.R., Smith, G.P., Jeffries, J.B., Crosley, D.R., 1994. Laser induced fluorescence diagnostics of a propane/air flame with a manganese fuel additive. Combust. Flame 99, 261–268.

Westerhoff, P., Lee, S., Yang, Y., et al., 2015. Characterization, recovery opportunities and valuation of metals in municipal sludges from US wastewater treatment plants. Environ. Sci. Technol. 49, 9479–9488.

Wey, M.-Y., Su, J.-L., Chen, J.-C., 1999. Influence of operating conditions on the formation of heavy metal compounds during incineration. J. Air Waste Manage. Assoc. 49, 444–453.

White, A.R. (Ed.), 2017. Metals in the Brain: Measurement and Imaging. Humana Press, Springer Nature, New York, Neuromethods Book Series, 270 pp.

White, D., Mann, D.E., Walsh, P.N., Sommer, A., 1960. Infrared emission spectrum of gaseous HBO_2. J. Chem. Phys. 32, 488–492.

Whitham, C.J., Ozeki, H., Saito, S., 2000. Microwave spectra of CuOD and AgOD: molecular structure and harmonic force field of CuOH and AgOH. J. Chem. Phys. 112 (2), 641–646.

Wibberley, L.J., Wall, T.F., 1982. Alkali ash reactions and deposit formation in pulverized coal-fired boilers: experimental aspects of sodium silicate formation and the formation of deposits. Fuel 61, 93–99.

Widener, A., 2015. Tighter limit for beryllium proposed. Chem. Eng. News 93 (32), 5.

Wiedemeier, H., Csillag, F.J., 1979. Equilibrium sublimation and thermodynamic properties of SnS. Thermochim. Acta 34, 257–265.

Wiedemeier, H., Csillag, F.J., 1980. The decomposition and thermodynamic properties of SnS_2. High Temp. Sci. 12, 277–288.

Wiedemeier, H., Gilles, P.W., 1965. Mass spectrometric study of the sublimation of MnS(s) and the dissociation energy of MnS(g). J. Chem. Phys. 42, 2765–2769.

Wiedmann, I., Choy, K.-L., Derby, B., 1994. Flame-assisted deposition of lead titanate-based thin films: correlation of deposition process, microstructure and electrical properties. In: Sale, F.R. (Ed.), British Ceramic Proceedings No.53: Novel Synthesis and Processing of Ceramics. The Institute of Materials, London, pp. 133–141.

Wiinikka, H., Gebart, R., Boman, C., Bostrom, D., Nordin, A., Ohman, M., 2006. High temperature aerosol formation in wood pellets flames: spatially resolved measurements. Combust. Flame 147, 278–293.

Wiinikka, H., Gebart, R., Boman, C., Bostrom, D., Nordin, A., Ohman, M., 2007. Influence of fuel ash composition on high temperature aerosol formation in fixed bed combustion of woody biomass pellets. Fuel 86, 181–193.

Wiinikka, H., Gronberg, C., Ohrman, O., Bostrom, D., 2009. Influence of TiO_2 additive on vaporization of potassium during straw combustion. Energy Fuels 23, 5367–5374.

Wilcomb, B.E., Bernstein, R.B., 1976. Dissociation energies of ground state HgX molecules (X = I, Br, Cl) from analysis of vibrational level spacings. J. Mol. Spectrom. 62, 442–448.

Wilcox, B.P., Chrysostom, E.T.H., McIlroy, A., Daily, J.W., Kennedy, I.M., 2003. Measurement of CrO in flames by cavity ringdown spectroscopy. Appl. Phys. B 77, 535–540.

Wilcox, J., Sasmaz, E., Kirchofer, A., Lee, S.-S., 2011. Heterogeneous mercury reaction chemistry on activated carbon. J. Air Waste Manage. Assoc. 61, 418–426.

Wilcox, J., Rupp, E., Ying, S.C., Lim, D.H., Negreira, A.S., Kirchofer, A., et al., 2012. Mercury adsorption and oxidation in coal combustion and gasification processes. Int. J. Coal Geol. 90, 4–20.

Wilhelm, S.M., 2001. Estimate of mercury emissions to the atmosphere from petroleum. Environ. Sci. Technol. 35, 4704–4710.

Willenborg, W., Muller, M., Hilpert, K., 2006. Alkali removal at about 1400°C for the pressurized pulverized coal combustion combined cycle. I. Thermodynamics and concept. Energy Fuels 20, 2593–2598.

Williams, R.J.P., Frausto da Silva, J.J.R., 2002. The involvement of molybdenum in life. Biochem. Biophys. Res. Commun. 292 (2), 293–299.

Willis, J.B., 1970. Atomization problems in atomic absorption spectroscopy. II. Determination of degree of atomization in premixed flames. Spectrochim. Acta B 25, 487–512.

Willis, J.B., 1971. Atomization problems in atomic absorption spectroscopy. III. Absolute atomization efficiencies of sodium, copper, silver and gold in a Meker-type air-acetylene flame. Spectrochim. Acta B 26, 177–189.

Wilson, A.K., Dunning Jr, T.H., 2004. The HSO-SOH isomers revisited: the effect of tight *d* functions. J. Phys. Chem. A 108, 3129–3133.

Winter, R.M., Mallepalli, R.R., Hellem, K.P., Szydlo, S.W., 1994. Determination of As, Cd, Cr and Pb species formed in a combustion environment. Combust. Sci. Technol. 101, 45–58.

Wiseman, C.L.S., Parnia, A., Chakravartty, D., Archbol, J., Copes, R., Cole, D., 2019. Total, methyl and inorganic mercury concentrations in blood and environmental exposure sources in newcomer women in Toronto, Canada. Environ. Res. 169, 261–271.

Witek, H.A., Nakijima, T., Hirao, K., 2000. Relativistic and correlated all-electron calculations on the ground and excited states of AgH and AuH. J. Chem. Phys. 113 (18), 8015–8025.

Witek, H.A., Fedorov, D.G., Hirao, K., et al., 2002. Theoretical study of the unusual potential energy curve of the $A^1\Sigma^+$ state of AgH. J. Chem. Phys. 116 (19), 8396–8406.

Wlokas, I., Faccinetto, A., Tribalet, B., Schulz, C., Kempf, A., 2013. Mechanism of iron oxide formation from iron pentacarbonyl doped low-pressure hydrogen/oxygen flames. Int. J. Chem. Kinet. 45, 487–498.

Wolf, R.E., Morman, S.A., Hageman, P.L., Hoefen, T.M., Plumlee, G.S., 2011. Simultaneous speciation of arsenic, selenium and chromium: species stability, sample preservation and analysis of ash and soil leachates. Anal. Bioanal. Chem. 401, 2733–2745.

Wolfe, P.J., Giang, A., Ashok, A., et al., 2016. Costs of IQ loss from leaded aviation gasoline emissions. Environ. Sci. Technol. 50, 9026–9033.

Won, J.H., Park, J.Y., Lee, T.G., 2007. Mercury emissions from automobiles using gasoline, diesel and liquefied petroleum gas. Atmos. Environ. 41, 7547–7552.

Wong, A.L., Tam, W.S., Cheung, A.S.-C., 2003. Rotational and hyperfine analysis of the near infrared $^3\Phi_4$-$X^3\Phi_4$ transitions of CoCl and CoI. J. Chem. Phys. 119, 3234–3239.

Wooldridge, M.S., 1998. Gas phase combustion synthesis of particles. Prog. Energy Combust. Sci. 24, 63–87.

Woon, D.E., Dunning Jr, T.H., 2009. A comparison between polar covalent bonding and hypervalent recoupled pair bonding in diatomic chalcogen halide species {O, S, Se}x{F, Cl, Br}. Mol. Phys. 107 (8–12), 991–998.

World Health Organization, 1982. Titanium. Environmental Health Criteria 24: International Program on Chemical Safety, Geneva.

World Health Organization, 1987. Selenium. Environmental Health Criteria 58, Geneva, 195 pp.

World Health Organization, 1988. Phosphine and selected metal phosphides. Environmental Health Criteria 73, Geneva, 69 pp.

World Health Organization, 1989. Mercury: environmental aspects. Environmental Health Criteria 86, Geneva, 60 pp.

World Health Organization, 1989a. Lead: environmental aspects. Environmental Health Criteria 85, Geneva, 76 pp.

World Health Organization, 1990. Barium. Environmental Health Criteria 107: International Program on Chemical Safety, Geneva.

World Health Organization, 1991. Inorganic mercury. Environmental Health Criteria 118, Geneva, 105 pp.

World Health Organization, 1992. Cadmium. Environmental Health Criteria 134, Geneva, 141 pp.

World Health Organization, 1992a. Cadmium: environmental aspects. Environmental Health Criteria 135, Geneva, 66 pp.

World Health Organization, 1995. Inorganic lead. Environmental Health Criteria 165, Geneva, 150 pp.

World Health Organization, 1996. Trace elements in human nutrition and health, Geneva, 360 pp.

World Health Organization, 1997. Aluminum. Environmental Health Criteria 194, Geneva, 146 pp.

World Health Organization, 1998. Boron. Environmental Health Criteria 204, Geneva, 102 pp.

World Health Organization, 2000. Safety evaluation of certain food additives and contaminants. Food additive series 44. International program on chemical safety, Geneva, 41 pp.

World Health Organization, 2001. Arsenic and arsenic compounds. Environmental Health Criteria 224, Geneva, 280 pp.

World Health Organization, 2001a. Barium and barium compounds. Concise International Chemical Assessment Document 33, Geneva, 57 pp.

World Health Organization, 2001b. Vanadium pentoxide and other inorganic vanadium compounds. Concise International Chemical Assessment Document 29, Geneva, 514 pp.

World Health Organization, 2001c. Zinc. Environmental Health Criteria 221, Geneva, 232 pp.

World Health Organization, 2004. Copper in drinking water. WHO/SDE/WSH/03.04/88, 31 pp.

World Health Organization, 2004a. Manganese and its compounds: environmental aspects. Concise International Chemical Assessment Document 63, Geneva, 65 pp.

World Health Organization, 2005. Tin and inorganic tin compounds. Concise International Chemical Assessment Document 65, Geneva, 98 pp.

World Health Organization, 2006. Cobalt in hard metals and cobalt sulfate, gallium arsenide, indium phosphide and vanadium pentoxide. International Agency for Research on Cancer, Monograph 86, 353 pp.

World Health Organization, 2006a. Guidelines for drinking water quality. First addendum to third edition, vol. 1. Recommendations, 595 pp.

World Health Organization, 2009. Inorganic chromium (III) compounds. Concise International Chemical Assessment Document 76. International Program on Chemical Safety, Geneva, 93 pp.

World Health Organization, 2010. Strontium and strontium compounds, Concise International Chemical Assessment Document 77, Geneva, 77 pp.

World Health Organization, 2013. Inorganic chromium (VI) compounds. Concise International Chemical Assessment Document 78. International program on chemical safety. Geneva, 116 pp.

Wu, C.Y., Biswas, P., 1993. An equilibrium analysis to determine the speciation of metals in an incinerator. Combust. Flame 93, 31–40.

Wu, H., Li, Y., 2019. Ash cenosphere fragmentation during pulverized pyrite combustion: importance of cooling. Proc. Combust. Inst. 37, 2773–2780.

Wu, H., Wang, L.-S., 1997. A study of nickel oxide (NiO), nickel dioxide (ONiO), and $Ni(O_2)$ complex by anion photoelectron spectroscopy. J. Chem. Phys. 107 (1), 16–21.

Wu, H., Wang, L.-S., 1998. Photoelectron spectroscopy and electronic structure of ScO_n^- (n = 1-4) and YO_n^- (n = 1-5): strong electron correlation effects in ScO^- and YO^-, J. Phys. Chem. A, 102, 9129–9135.

Wu, H.Y., Wahlbeck, P.G., 1972. Vapor pressures of TiO(g) in equilibrium with Ti_2O_3(s) and Ti_3O_5(s, β): dissociation energy of TiO(g). J. Chem. Phys. 56, 4534–4540.

Wu, H., Desai, S.R., Wang, L.-S., 1997. Chemical bonding between Cu and Oxygen. Copper oxides vs O_2 complexes: a study of CuO_x (x = 0-6) species by anion photoelectron spectroscopy. J. Phys. Chem. A 101 (11), 2103–2111.

Wu, Z.J., Wang, M.Y., Su, Z.M., 2007. Electronic structures and chemical bonding in diatomic ScX to ZnX (X = S, Se, Te). J. Comput. Chem. 28, 703–714.

Wu, Y., Wu, S., Li, Y., Gao, J., 2009. Physico-chemical characteristics and mineral transformation behavior of ashes from crop straw. Energy Fuels 23, 5144–5150.

Wu, C.-L., Cao, Y., He, C.-C., Dong, Z.-B., Pan, W.-P., 2010. Study of elemental mercury re-emission through a lab-scale simulated scrubber. Fuel 89, 2072–2080.

Wu, H., Glarborg, P., Frandsen, F.J., Dam-Johansen, K., Jensen, P.A., 2011. Dust-firing of straw and additives: ash chemistry and deposition behavior. Energy Fuels 25, 2862–2873.

Wu, H., Pedersen, A.J., Glarborg, P., Frandsen, F.J., Dam-Johansen, K., Sander, B., 2011a. Formation of fine particles in co-combustion of coal and solid recovered fuel in a pulverized coal-fired power station. Proc. Combust. Inst. 33, 2845–2852.

Wu, S., Azhar Uddin, M., Nagano, S., Ozaki, M., Sasaoka, E., 2011b. Fundamental study on decomposition characteristics of mercury compounds over solid powder by temperature-programmed decomposition desorption mass spectrometry. Energy Fuels 25, 144–153.

Wu, H., Glarborg, P., Frandsen, F.J., Dam-Johansen, K., Jensen, P.A., Sander, B., 2011c. Co-combustion of pulverized coal and solid recovered fuel in an entrained flow reactor: general combustion and ash behavior. Fuel 90, 1980–1991.

Wu, H., Glarborg, P., Frandsen, F.J., Dam-Johansen, K., Jensen, P.A., Sander, B., 2013. Trace elements in co-combustion of solid recovered fuel and coal. Fuel Process. Technol. 105, 212–221.

Wu, H., Liu, H., Wang, Q., et al., 2013a. Experimental study of homogeneous mercury oxidation under O_2/CO_2 atmosphere. Proc. Combust. Inst. 34, 2847–2854.

Wu, H., Pedersen, M.N., Jespersen, J.B., et al., 2014. Modeling the use of sulfate additives for potassium chloride destruction in biomass combustion. Energy Fuels 28, 199–207.

Wu, M.-H., Lin, C.-L., Zeng, W.-Y., 2014a. Effect of waste incineration and gasification processes on heavy metal distribution. Fuel Process. Technol. 125, 67–72.

Wu, D.-Y., Wang, Y.-H., Wang, Y., et al., 2016. Release of alkali metals during co-firing biomass and coal. Renew. Energy 96 (A), 91–97.

Wu, H., Yrjas, P., Vainikka, P., et al., 2016a. Sulfation of alkali halides in a bench scale bubbling fluidized bed reactor. Fuel 177, 173–179.

Wu, H., Zhang, B., Qiu, Y., et al., 2016b. Experiment study on mercury migration across wet flue gas desulfurization slurry under oxy-coal combustion atmosphere. Fuel 181, 1184–1190.

Wu, Q.-R., Wang, S.-X., Li, G.-L., et al., 2016c. Temporal trend and spatial distribution of speciated atmospheric mercury emissions in China during 1978–2014. Environ. Sci. Technol. 50 (24), 13428–13435.

Wu, Q., Gao, W., Wang, S., Hao, J., 2017. Updated atmospheric speciated mercury emissions from iron and steel production in China during 2000–2015. Atmos. Chem. Phys. 17, 10423–10433.

Wu, Q., Wang, S., Liu, K., Li, G., Hao, J., 2018. Emission-limit-oriented strategy to control atmospheric mercury emissions in coal-fired power plants toward the implementation of the Minamata convention. Environ. Sci. Technol. 52, 11087–11093.

Wu, J., Liu, D.J., Shou, X.J., Zhang, J.S., Meng, F.C., et al., 2018a. Chinese children with autism: a multiple chemical elements profile in erythrocytes. Autism Res. 11 (6), 834–845.

Wu, H.Y., Qin, M.L., Cao, Z.Q., Li, X.L., Jia, B.R., et al., 2018b. Direct synthesis of vanadium oxide nano powders by the combustion approach. Chem. Phys. Lett. 706, 7–13.

Wu, B., Tian, H., Hao, Y., Liu, S., Liu, X., Liu, W., et al., 2018c. Effects of wet flue gas desulfurization and wet electrostatic precipitators on emission characteristics of particulate matter and its ionic compositions from four 300 MW level ultralow coal-fired power plants. Environ. Sci. Technol. 52, 14015–14026.

Wu, Q., Li, G., Wang, S., Liu, K., Hao, J., 2018d. Mitigation options of atmospheric mercury emissions in China. Environ. Sci. Technol. 52, 12368–12375.

Wyslouzil, B.E., Wolk, J., 2016. Overview: homogeneous nucleation from the vapor phase. The experimental science. J. Chem. Phys. 145, 211702. 26 pp.

Xantheas, S.S., Dunning Jr, T.H., Mavridis, A., 1997. The dissociation energies of NF($X^3\Sigma^-$) and NCl($X^3\Sigma^-$). J. Chem. Phys. 106 (8), 3280–3286.

Xian, H., Cao, Z.X., Xu, X., et al., 2000. Theoretical study of low-lying electronic states of CuO and CuO^-. Chem. Phys. Lett. 326, 485–493.

Xiao, Z., Shang, T., Zhuo, J., Yao, Q., 2016. Study on the mechanisms of ultrafine particle formation during high sodium coal combustion in a flat flame burner. Fuel 181, 1257–1264.

Xiao, G.X., Hu, Y.L., Li, N., Yang, D.J., 2018. Spatial autocorrelation analysis of monitoring data of heavy metals in rice in China. Food Control 89, 32–37.

Xin, J., Klynning, L., Royen, P., 1996. Laser spectroscopy of YCl: observation of the $d^3\Phi$-$a^3\Pi$ transition. J. Mol. Spectrosc. 176, 1–8.

Xin, M., Gustin, M., Johnson, D., 2007. Laboratory investigation of the potential for re-emission of atmospherically derived mercury from soils. Environ. Sci. Technol. 41, 4946–4951.

Xing, Y., Koylu, U.O., Rosner, D.E., 1996. Synthesis and restructuring of inorganic nano-particles in counterflow diffusion flames. Combust. Flame 107, 85–102.

Xiong, S., Burvall, J., Orberg, H., Kalen, G., Thyrel, M., Ohman, M., et al., 2008. Slagging characteristics during combustion of corn stovers with and without kaolin and calcite. Energy Fuels 22, 3465–3470.

Xu, L.T., Dunning Jr, T.H., 2015. Generalized valence bond description of the ground states ($X^1\Sigma_g^+$) of homonuclear pnictogen diatomic molecules: N_2, P_2 and As_2. J. Chem. Theory Comput. 11, 2496–2507.

Xu, X., Truhlar, D.G., 2011. Accuracy of effective core potentials and basis sets for density functional calculations, including relativistic effects, as illustrated by calculations on arsenic compounds. J. Chem. Theory Comput. 7, 2766–2779.

Xu, M., Yan, R., Zheng, C., Qiao, Y., Han, J., Sheng, C., 2003. Status of trace element emission in a coal combustion process: a review. Fuel Process. Technol. 85, 215–237.

Xu, C.H., Shi, S.Q., Surya, C., Woo, C.H., 2007. Synthesis of antimony oxide nano-particles by vapor transport and condensation. J. Mater. Sci. 42, 9855–9858.

Xu, F., Liu, X., Tse, S.D., et al., 2007a. Flame synthesis of zinc oxide nanowires. Chem. Phys. Lett. 449, 175–181.

Xu, M., Yu, D., Yao, H., Liu, X., Qiao, Y., 2011. Coal combustion generated aerosols: formation and properties. Proc. Combust. Inst. 33, 1681–1697.

Xu, Y., Cao, X., Zhang, Y., 2014. Silver nanoparticle modified MoO_3 with enhanced electrochemical properties for a lithium-ion battery. Can. J. Chem. 92, 16–18.

Xu, L.L., Kang, Y., Zhang, G.Q., et al., 2015. Study of alkali emission and control with firing a high alkali coal. Combust. Sci. Technol. 187 (12), 1959–1973.

Xu, X., Zhao, J., Li, Y., et al., 2016. Demethylation of methylmercury in growing rice plants: an evidence of self-detoxification. Environ. Pollut. 210, 113–120.

Xu, W., Shao, M., Yang, Y., et al., 2017. Mercury emission from sintering in the iron and steel industry. Fuel Process. Technol. 159, 340–344.

Xu, Q., Wang, M., Zhao, Y., Yang, C., Ma, X., 2017a. *Ab initio* studies on spectroscopic constants for the HAsO molecule. J. Phys. Chem. A 121, 7009–7015.

Xu, J., Bao, J., Tang, J., Du, M., Liu, H., et al., 2018. Characteristics and inhibition of Hg^{o} re-emission in a wet flue gas desulfurization system. Energy Fuels 32, 6111–6118.

Xu, X., Liao, W., Lin, Y., Dai, Y., Shi, Z., Huo, X., 2018a. Blood concentrations of lead, cadmium, mercury and their association with biomarkers of DNA oxidative damage in preschool children living in an e-waste recycling area. Environ. Geochem. Health 40, 1481–1494.

Xue, Z.Y., Zhong, Z.P., Zhang, B., 2019. Experimental studies on co-combustion of sludge and wheat straw. Catalysts 9 (2), article 182.

Yamawaki, M., Oka, T., Yasumoto, M., Sakurai, H., 1993. Thermodynamics of vaporization of cesium molybdate by means of mass spectrometry. J. Nucl. Mater. 201, 257–260.

Yamdagni, R., Pupp, C., Porter, R.F., 1970. Mass spectrometric study of the evaporation of lithium and sodium molybdates and tungstates. J. Inorg. Nucl. Chem. 32, 3509–3523.

Yan, R., Gauthier, D., Flamant, G., Badie, J.M., 1999. Thermodynamic study of the behavior of minor coal elements and their affinities to sulfur during coal combustion. Fuel 78, 1817–1829.

Yan, R., Gauthier, D., Flamant, G., 2000. Possible interactions between As, Se and Hg during coal combustion. Combust. Flame 120, 49–60.

Yan, R., Gauthier, D., Flamant, G., 2001. Partitioning of trace elements in the flue gas from coal combustion. Combust. Flame 125, 942–954.

Yan, R., Gauthier, D., Flamant, G., 2001a. Volatility and chemistry of trace elements in a coal combustor. Fuel 80, 2217–2226.

Yan, R., Gauthier, D., Flamant, G., Wang, Y., 2004. Behavior of selenium in the combustion of coal or coke spiked with Se. Combust. Flame 138, 20–29.

Yan, N.-Q., Liu, S.-H., Chang, S.-G., Miller, C., 2005. Method for the study of gaseous oxidants for the oxidation of mercury gas. Ind. Eng. Chem. Res. 44, 5567–5574.

Yan, G., Cho, H.-M., Lee, I., Kim, G., 2012. Significant emissions of ^{210}Po by coal burning into the urban atmosphere of Seoul, Korea. Atmos. Environ. 54, 80–85.

Yan, D., Peng, Z., Ding, Q., et al., 2015. Distribution of Hg, As and Se in material and flue gas streams from preheater-precalciner cement kilns and vertical shaft cement kilns in China. J. Air Waste Manage. Assoc. 65 (8), 1002–1010.

Yang, H., 2015. Lake sediments may not faithfully record decline of atmospheric pollutant deposition. Environ. Sci. Technol. 49, 12607–12608.

Yang, X., Boggs, J.E., 2005. Extensive *ab initio* study of the electronic states of SCl including spin-orbit couplings. J. Chem. Phys. 123, 184304. 7 pp.

Yang, W., Zhou, J., Zhou, Z., et al., 2008. Characterization of sodium compounds on NO reduction at high temperature in NO_x control technologies. Fuel Process. Technol. 89, 1317–1323.

Yang, Y., Weaver, M.N., Merz Jr, K.M., 2009. Assessment of the 6-31 + G** + LANL2DZ mixed basis set coupled with density functional theory methods and the effective core potential: predictions of heats of formation and ionization potentials for first-row-transition-metal complexes. J. Phys. Chem. A 113, 9843–9851.

Yang, Y.G., Li, F.-L., Bi, X.-Y., Sun, L., Liu, T.-Z., Jin, Z.-S., et al., 2011. Lead, zinc and cadmium in vegetable/crops in a zinc smelting region and its potential human toxicity. Bull. Environ. Contam. Toxicol. 87, 586–590.

Yang, G., Ma, L., Xu, D., et al., 2012. Levels and speciation of arsenic in the atmosphere in Beijing, China. Chemosphere 87, 845–850.

Yang, H., Hou, W., Zhang, H., Zhou, L., 2013. Kinetic interpretation on mercury oxidation and transformation in simulated flue gases. Int. J. Environ. Sci. Technol. 10, 689–696.

Yang, J., Liu, D., Wang, Y., et al., 2016. Release and the interaction mechanism of uranium and alkali/alkaline-earth metals during coal combustion. Fuel 186, 405–413.

Yang, Y., Liu, H., Zhang, P., 2016a. Structural and electronic properties of U_nO_m (n = 1-3, m = 1-3n) clusters: a theoretical study using screened hybrid density functional theory. J. Chem. Phys. 144, 184304. 11 pp.

Yang, Y., Liu, J., Shen, F., et al., 2016b. Kinetic study of heterogeneous mercury oxidation by HCl on fly ash surface in coal-fired flue gas. Combust. Flame 168, 1–9.

Yang, T.-H., Jia, K., Kai, X.-P., et al., 2016d. A study of the migration behavior of K, Na, and Cl during biomass gasification. Bioresources 11 (3), 7133–7144.

Yang, J., Ma, S., Zhao, Y., et al., 2017. Mercury emission and speciation in fly ash from a 35 MW_{th} large pilot boiler of oxyfuel combustion with different flue gas recycle. Fuel 195, 174–181.

Yang, Y., Liu, J., Wang, Z., Zhang, Z., 2017a. Homogeneous and heterogeneous reaction mechanisms and kinetics of mercury oxidation in coal fired flue gas with bromine addition. Proc. Combust. Inst. 36 (3), 4039–4049.

Yang, Y., Lin, X., Chen, X., Guo, W., Wang, Y., et al., 2018. Investigation on the effects of different forms of sodium, chlorine and sulfur and various pretreatment methods on the deposition characteristics of sodium species during pyrolysis of a sodium-rich coal. Fuel 234, 872–885.

Yang, M., Zhao, W.Y., Xu, W.H., He, Z.M., Feng, D.Y., 2018a. Absorption, bioaccumulation and transportation of selenium in three vegetables differing in selenium transport distances. Int. J. Agric. Biol. 20 (10), 2301–2306.

Yang, Z., Li, H., Liu, X., Li, P., Yang, J., Lee, P.-H., et al., 2018b. Promotional effect of CuO loading on the catalytic activity and SO_2 resistance of MnO_x/TiO_2 catalyst for simultaneous NO reduction and Hg^o oxidation. Fuel 227, 79–88.

Yang, Y., Hu, H., Xie, K., Huang, Y., Huang, Y., Liu, H., et al., 2019. Insight of arsenic transformation behavior during high-arsenic coal combustion. Proc. Combust. Inst. 37, 4443–4450.

Yao, H., Naruse, I., 2005. Control of trace metal emissions by sorbents during sewage sludge combustion. Proc. Combust. Inst. 30, 3009–3016.

Yao, Z.T., Xia, M.S., Sarker, P.K., Chen, T., 2014. A review of the alumina recovery from coal fly ash with a focus in China. Fuel 120, 74–85.

Yates, H.M., Brook, L.A., Sheel, D.W., et al., 2008. The growth of copper oxides on glass by flame assisted chemical vapor deposition. Thin Solid Films 517 (2), 517–521.

Ye, J., Pang, H.F., Cheung, A.S.-C., 2007. Optical-optical double resonance spectrometry of YBr and YCl. Chem. Phys. Lett. 442, 251–258.

Yetter, R.A., Rabitz, H., Dryer, F.L., Brown, R.C., Kolb, C.E., 1991. Kinetics of high temperature B/O/H/C chemistry. Combust. Flame 83, 43–62.

Yilmaz, A., Hindiyarti, L., Jensen, A.D., Glarborg, P., Marshall, P., 2006. Thermal dissociation of SO_3 at 1000–1400 K. keep together. J. Phys. Chem. A 110, 6654–6659.

Yim, S.H., Tobe, R., Turanov, A.A., Carlson, B.A., 2018. Radioactive Se-75 labeling and detection of selenoproteins. In: Selenoproteins: Methods and Protocols, Chavatte, L. (Ed.), Method in Molecular Biology Series. vol. 1661. Humana Press, New Jersey USA, pp. 177–192.

Yoneda, S., Suzuki, K.T., 1997. Detoxification of mercury by selenium by binding of equimolar Hg-Se complex to a specific plasma protein. Toxicol. Appl. Pharmacol. 143, 274–280.

Yoon, Y., Reyes, C., Jones, A.D., Kelly, P., Chang, D.P.Y., Kennedy, I.M., 1996. The behavior of chromium in laminar hydrogen and ethene diffusion flames. Proc. Symp. (Int.) Combust. 26, 1953–1959.

Youcai, Z., Stucki, S., Ludwig, C., Wochele, J., 2004. Impact of moisture on volatility of heavy metals in municipal solid waste incinerated in a laboratory scale simulated incinerator. Waste Manage. 24, 581–587.

Younes, C., Nguyen, L.D., Pattoret, A., 1981. Studies of the thermodynamic systems U-O, Ce-O, U-Ce-O, U-La-O, Ce-La-O, and Ce-La-Y-O by mass spectrometry: improved statistical assessment of uncertainties in derived thermodynamic data. High Temp. High Press. 13, 105–111.

Young, G., Sullivan, K., Zachariah, M.R., Yu, K., 2009. Combustion characteristics of boron nanoparticles. Combust. Flame 156, 322–333.

Young, G., Jacob, R., Zachariah, M.R., 2015. High pressure ignition and combustion of aluminum hydride. Combust. Sci. Technol. 187 (9), 1335–1350.

Young, G., Jian, G., Jacob, R., Zachariah, M.R., 2015a. Decomposition and ignition characteristics of titanium hydride at high heating rates. Combust. Sci. Technol. 187 (8), 1182–1194.

Yrjas, P., Skrifvars, B.-J., Hupa, M., Roppo, J., Nylund, M., Vainikka, P., 2005. Chlorine in deposits during co-firing of biomass, peat and coal in a full-scale CFBC boiler. In: Proceedings of 18th International Conference Fluidized Bed Combustion held in Toronto, Ontario, Canada. American Society of Mechanical Engineers; New York NY. Paper No. ASME FBC 2005-78097, pp. 679–687.

Yu, S., Kennedy, I.M., 2000. A preliminary kinetic model of chromium in a hydrogen/air flame. Combust. Sci. Technol. 160, 35–45.

Yu, S., Jones, A.D., Chang, D.P.Y., Kelly, P.B., Kennedy, I.M., 1998. The transformation of chromium in a laminar premixed hydrogen-air flame. Proc. Symp. (Int.) Combust. 27, 1639–1645.

Yu, S., Gordon, I.E., Sheridan, P.M., Bernath, P.F., 2006. Infrared emission spectroscopy of the $A^4\Phi_i$- $X^4\Delta_i$ and $B^4\Pi_i$-$X^4\Delta_i$ transitions of CoS. J. Mol. Spectrosc. 236, 255–259.

Yu, D., Xu, M., Yao, H., Liu, X., Zhou, K., 2008. A new method for identifying the modes of particulate matter from pulverized coal combustion. Powder Technol. 183, 105–114.

Yu, D.-X., Xu, M.-H., Yao, H., Liu, X.-W., 2009. Size distributions of major elements in residual ash particles from coal combustion. Chin. Sci. Bull. 54 (6), 958–964.

Yu, B., Fu, X., Yin, R., et al., 2016. Isotopic composition of atmospheric mercury in China: new evidence for sources and transformation processes in air and in vegetation. Environ. Sci. Technol. 50, 9262–9269.

Yu, S., Zhang, B., Wei, J., Zhang, T., et al., 2017. Effects of chlorine on the volatilization of heavy metals during the co-combustion of sewage sludge. Waste Manage. 62, 204–210.

Yu, W., Andrews, L., Wang, X., 2017a. Infrared spectroscopic and electronic structure investigations of beryllium halide molecules, cations, and anions in noble gas matrices. J. Phys. Chem. A 121, 8843–8855.

Yuan, C.-S., Lin, H.-Y., Wu, C.-H., Liu, M.-H., 2005. Partition and size distribution of heavy metals in the flue gas from municipal solid waste incinerators in Taiwan. Chemosphere 59, 135–145.

Yuan, H., Liang, Q., Gong, X., 2012. Crystallization of coal ash slags at high temperatures and effects on the viscosity. Energy Fuels 26, 3717–3722.

Yuan, Y., Li, S., Yao, Q., 2015. Dynamic behavior of sodium release from pulverized coal combustion by phase-selective laser-induced breakdown spectroscopy. Proc. Combust. Inst. 35, 2339–2346.

Yunusa, I.A.M., Loganathan, P., Nissanka, S.P., et al., 2012. Application of coal fly ash in agriculture: a strategic perspective. Crit. Rev. Environ. Sci. Technol. 42 (6), 559–600.

Zachara, B.A., 2018. Selenium in complicated pregnancy: a review. In: Makowski, G.S. (Ed.), Advances in Clinical Chemistry, vol. 86. Elsevier Academic Press, San Diego CA, pp. 157–178.

Zachariah, M.R., Huzarewicz, S., 1991. Aerosol processing of YBaCuO superconductors in a flame reactor. J. Mater. Res. 6 (2), 264–269.

Zachariah, M.R., Huzarewicz, S., 1991a. Flame synthesis of high T_c superconductors. Combust. Flame 87, 100–103.

Zack, L.N., Sun, M., Bucchino, M.P., Clouthier, D.J., Ziurys, L.M., 2012. Gas phase synthesis and structure of monomeric ZnOH: a model species for metalloenzymes and catalytic surfaces. J. Phys. Chem. 116, 1542–1550.

Zagorodnyuk, A.V., Magunov, R.L., Kovalevskaya, I.P., 1972. High temperature study of germanium monoxide. Izv. Akad. Nauk. SSSR. Neorg. Mater. 8, 1864–1865.

Zahra, N., Kalim, I., Mahmood, M., Naeem, N., 2017. Perilous effects of heavy metals contamination on human health. Pak. J. Anal. Environ. Chem. 18 (1), 1–17.

Zaidi, A., Lahmar, S., Lakhdar, Z.B., Rosmus, P., Chambaud, G., 2006. Theoretical study of HBeO. Chem. Phys. 321, 41–47.

Zaitsev, A.I., Dobrokhotova, Zh.V., Mogutnov, B.M., 1995. Thermodynamics of iron phosphides. High Temp. Mater. Sci. 34, 173–185.

Zamansky, V.M., Ho, L., Maly, P.M., Seeker, W.R., 1996. Re-burning promoted by nitrogen and sodium containing compounds. Symp. (Int.) Combust. Proc. 26 (Part 2), 2075–2082.

Zang, J., Zhang, Q., Zhang, D., et al., 2015. The laser-induced fluorescence spectroscopy of yttrium monosulfide. J. Mol. Spectrosc. 313, 49–53.

Zaplatynsky, I., 1977. Volatilization of oxides during oxidation of some superalloys at 1200°C. Oxid. Metals 11, 289–305.

Zayed, J., Hong, B., L'esperance, G., 1999. Characterization of manganese-containing particles collected from the exhaust emissions of automobiles running with MMT additive. Environ. Sci. Technol. 33 (19), 3341–3346.

Ze, Y., Zheng, L., Zhao, X., et al., 2013. Molecular mechanism of titanium dioxide nanoparticles-induced oxidative injury in the brain of mice. Chemosphere 92, 1183–1189.

Zeegers, P.J.Th, Townsend, W.P., Winefordner, J.D., 1969. Estimation of free-atom fraction and dissociation energies of compounds in flames by an atomic absorption method. Spectrochim. Acta B 24, 243–254.

Zemba, S., Ames, M., Green, L., Botelho, M.J., Gossman, D., Linkov, I., et al., 2011. Emissions of metals and polychlorinated dibenzo(p)dioxins and furans from Portland cement manufacturing plants: inter-kiln variability and dependence on fuel types. Sci. Total Environ. 499 (20), 4198–4205.

Zeng, T., Sarofim, A.F., Senior, C.L., 2001. Vaporization of arsenic, selenium and antimony during coal combustion. Combust. Flame 126, 1714–1724.

Zeng, T., Helble, J.J., Bool, L.E., Sarofim, A.F., 2009. Iron transformations during combustion of Pittsburgh no. 8 coal. Fuel 88, 566–572.

Zeng, H.-L., Li, H., Lu, J., Guan, Q., Cheng, L., 2019. Assessment of 12 metals and metalloids in blood of general populations living in Wuhan, China, by ICP-MS. Biol. Trace Elem. Res. 189, 344–353.

Zeuthen, J.H., Jensen, P.A., Jensen, J.P., Livbjerg, H., 2007. Aerosol formation during the combustion of straw with addition of sorbents. Energy Fuels 21, 699–709.

Zhang, J., Megaridis, C.M., 1996. Soot suppression by ferrocene in laminar ethylene/air non-premixed flames. Combust. Flame 105, 528–540.

Zhang, L., Ninomiya, Y., 2007. Transformation of phosphorus during combustion of coal and sewage sludge and its contributions to PM_{10}. Proc. Combust. Inst. 31, 2847–2854.

Zhang, Q., Kang, S., 2016. Comment on "Ice core perspective on mercury pollution during the past 600 years. Environ. Sci. Technol. 2015; 49: 7641–7647 by Beal et al.". Environ. Sci. Technol. 50, 1065–1067. Response 1068–1069.

Zhang, Z., Zhang, B., Long, J., Zhang, X., Chen, G., 1998. Thallium pollution associated with mining of thallium deposits. Sci. China D 41, 75–81.

Zhang, L., Dong, J., Zhou, M., 2000. Matrix-isolation Fourier transform infrared and theoretical studies of laser-ablated Sc atom reactions with water molecules. J. Phys. Chem. A 104, 8882–8886.

Zhang, L., Shao, L., Zhou, M., 2001. Reactions of laser ablated Y and La atoms with H_2O: infrared spectra and density functional calculations of the HMO, HMOH and $M(OH)_2$ molecules in solid argon. Chem. Phys. 272, 27–36.

Zhang, F.-S., Yamasaki, S., Kimura, K., 2002. Waste ashes for use in agricultural production. II. Contents of minor and trace metals. Sci. Total Environ. 286 (1–3), 111–118.

Zhang, L., Ito, M., Sato, A., Ninomiya, Y., 2003. Fate of alkali elements during pyrolysis and combustion of Chinese coals. J. Chem. Eng. Jpn. 36, 759–768.

Zhang, L., Masui, M., Mizukoshi, H., Ninomiya, Y., Kanaoka, C., 2007. Formation of submicron particulates (PM_1) from the oxygen-enriched combustion of dried sewage sludge and their properties. Energy Fuels 21, 88–98.

Zhang, Y., Chen, Y., Meng, A., Li, Q., Cheng, H., 2008. Experimental and thermodynamic investigation on transfer of cadmium influenced by sulfur and chlorine during municipal solid waste incineration. J. Hazard. Waste 153, 309–319.

Zhang, L., Jiao, F., Binner, E., et al., 2011. Evolution of organically bound metals during coal combustion in air and O_2/CO_2 mixtures: a case study of Victorian brown coal. Proc. Combust. Inst. 33, 2795–2802.

Zhang, Q., Liu, H., Li, W., Xu, J., Liang, Q., 2012. Behavior of phosphorus during co-gasification of sewage sludge and coal. Energy Fuels 26, 2830–2836.

Zhang, R., Khalizov, A., Wang, L., Hu, M., Xu, W., 2012a. Nucleation and growth of nanoparticles in the atmosphere. Chem. Rev. 112, 1957–2011.

Zhang, G., Hai, J., Ren, M., Zhang, S., Cheng, J., Yang, Z., 2013. Emission, mass balance and distribution characteristics of PCDD/Fs and heavy metals during co-combustion of sewage sludge and coal in power plants. Environ. Sci. Technol. 47, 2123–2130.

Zhang, L., Daukoru, M., Torkamani, S., et al., 2013a. Measurements of mercury speciation and fine particle size distribution on combustion of China coal seams. Fuel 104, 732–738.

Zhang, H., Feng, X., Chan, H.M., Larssen, T., 2014. New insights into traditional health risk assessments of mercury exposure: implications of selenium. Environ. Sci. Technol. 48, 1206–1212.

Zhang, Y.-X., Jaegle, L., Thompson, L., Streets, D.G., 2014a. Six centuries of changing oceanic mercury. Global Biogeochem. Cycles 28 (11), 1251–1261.

Zhang, B., Liu, J., Dai, G., Chang, M., Zheng, C., 2015. Insights into the mechanism of heterogeneous mercury oxidation by HCl over V_2O_5/TiO_2 catalyst: periodic density functional theory study. Proc. Combust. Inst. 35, 2855–2865.

Zhang, X.-N., Li, C.-T., Zhao, L.-K., et al., 2015a. Simultaneous removal of elemental mercury and NO from flue gas by V_2O_5-CeO_2/TiO_2 catalysts. Appl. Surf. Sci. 347, 392–400.

Zhang, Y., Li, S., Ren, Y., et al., 2015b. A new diagnostic for volume fraction measurement of metal-oxide nanoparticles in flames using phase-selective laser induced breakdown spectroscopy. Proc. Combust. Inst. 35, 3681–3688.

Zhang, Z.-H., Song, Q., Alwahabi, Z.T., et al., 2015c. Temporal release of potassium from pinewood particles during combustion. Combust. Flame 162, 496–505.

Zhang, K., Zhang, D., Zhang, K., Cao, Y., 2016. Capture of gas phase arsenic by ferrospheres separated from fly ashes. Energy Fuels 30, 8746–8752.

Zhang, L., Wang, S., Wu, Q., et al., 2016a. Mercury transformation and speciation in flue gases from anthropogenic emission sources: a critical review. Atmos. Chem. Phys. 16, 2417–2433.

Zhang, X., Yan, P., Li, R., et al., 2016b. Extensive spin-orbit multi-reference computations on the excited states of the phosphorus monochloride molecule. J. Quant. Spectrosc. Radiat. Trans. 180, 154–166.

Zhang, Y., Shi, M., Wang, J., et al., 2016c. Occurrence of uranium in Chinese coals and its emissions from coal-fired power plants. Fuel 166, 404–409.

Zhang, Y.-X., Jacob, D.J., Horowitz, H.M., et al., 2016d. Observed decrease in atmospheric mercury explained by global decline in anthropogenic emissions. Proc. Natl. Acad. Sci. USA 113 (3), 526–531.

Zhang, H., Holmes, C.D., Wu, S., 2016e. Impacts of changes in climate, land-use and land cover on atmospheric mercury. Atmos. Environ. 141, 230–244.

Zhang, Y., Yang, J., Yu, X., et al., 2017. Migration and emission characteristics of Hg in coal fired power plant of China with ultra low emission air pollution control devices. Fuel Process. Technol. 158, 272–280.

Zhang, Y., Shang, P., Wang, J., Norris, P., et al., 2017a. Trace element (Hg, As, Cr, Cd, Pb) distribution and speciation in coal-fired power plants. Fuel 208, 647–654.

Zhang, G., Zhang, L., Fan, H., Hu, E., 2017b. Concentration, enrichment, and partitioning behavior of heavy metals in ash from a down-fired furnace burning anthracite coal. Energy Fuels 31, 9381–9392.

Zhang, W., Zhen, G., Chen, L., Wang, H., Li, Y., et al., 2017c. Economic evaluation of health benefits of mercury emission controls for China and the neighboring countries in East Asia. Energy Policy 106, 579–587.

Zhang, W.B., Li, F.Q., Wu, J., Li, H.X., 2017d. Mercury removal technologies of the flue gas from power plants. Progress Chem. 29 (12), 1435–1445.

Zhang, B., Bogush, A., Wei, J.-X., Xu, W., et al., 2018. Influence of chlorine on the fate of lead and copper during clinkerization. Energy Fuels 32, 7718–7726.

Zhang, Y.-F., Zhang, S.-Y., Mao, Q., Li, H., et al., 2018c. Volatility and partitioning of Cd and Pb during sewage sludge thermal conversion. Waste Manage. 75, 333–339.

Zhang, S., Jiang, X., Lv, G., Nixiang, A., Jin, Y., et al., 2019. Effect of chlorine, sulfur, moisture and ash content on the partitioning of As, Cr, Cu, Mn, Ni, and Pb during bituminous coal and pickling sludge co-combustion. Fuel 239, 601–610.

Zhao, Y., Setser, D.W., 1995. Quenching of $PF(b^1\Sigma^+)$ by F_2, Cl_2, Br_2, NO_2, and N_2O. The energy-pooling reaction with $NF(b^1\Sigma^+)$ and the thermochemistry of PF, PCl and PBr. J. Phys. Chem. 99, 12179–12187.

Zhao, Z.-Q., Liu, C.-Q., 2010. Anthropogenic inputs of boron into urban atmosphere: evidence from boron isotopes of precipitations in Guiyang City, China. Atmos. Environ. 44, 4165–4171.

Zhao, B., Uchikawa, K., McCormick, J.R., Ni, C.Y., Chen, J.G., Wang, H., 2005. Ultrafine anatase TiO_2 nanoparticles produced in premixed ethylene stagnation flame at 1 atm. Proc. Combust. Inst. 30, 2569–2576.

Zhao, Y., Mann, M.D., Pavlish, J.H., Mibeck, B.A.F., Dunham, G.E., Olson, E.S., 2006. Application of gold catalyst for mercury oxidation by chlorine. Environ. Sci. Technol. 40, 1603–1608.

Zhao, Y., Zhang, J., Sun, J., et al., 2006a. Mineralogy, chemical composition and microstructure of ferrospheres in fly ashes from coal combustion. Energy Fuels 20 (4), 1490–1497.

Zhao, L., Zhang, F.-S., Wang, K., Zhu, J., 2008. Chemical properties of heavy metals in typical hospital waste incinerator ashes in China. Waste Manage. 29, 1114–1121.

Zhao, H., Liu, X., Tse, S.D., 2009. Effects of pressure and precursor loading in the flame synthesis of titania nanoparticles. Aerosol Sci. 40, 919–937.

Zhao, Y.-C., Zhang, J.-Y., Liu, J., et al., 2010. Study on mechanism of mercury oxidation by fly ash from coal combustion. Chin. Sci. Bull. 55 (2), 163–167.

Zhao, P.S., Dong, F., He, D., Zhao, X.J., Zhang, X.L., et al., 2013. Characteristics of concentrations and chemical compositions for $PM_{2.5}$ in the region of Beijing, Tianjin and Hebei, China. Atmos. Chem. Phys. 13, 4631–4644.

Zhao, Y., Zhang, J., Zheng, C., 2013a. Release and removal using sorbents of chromium from a high-Cr lignite in Shenbei coalfield, China. Fuel 109, 86–93.

Zhao, L.-K., Li, C.-T., Zhang, X.-A., et al., 2015. A review on oxidation of elemental mercury from coal-fired flue gas with selective catalytic reduction catalysts. Catal Sci Technol 5 (7), 3459–3472.

Zhao, S., Duan, Y., Tan, H., et al., 2016. Migration and emission characteristics of trace elements in a 660 MW coal-fired power plant of China. Energy Fuels 30, 5937–5944.

Zhao, S., Duan, Y., Yao, T., Liu, M., Lu, J., et al., 2017. Study on the mercury emission and transformation in an ultra-low emission coal-fired power plant. Fuel 199, 653–661.

Zhao, S.L., Duan, Y.F., Chen, L., Li, Y.N., Yao, T., et al., 2017a. Study on emission of hazardous trace elements in a 350MW coal-fired power plant. II. Arsenic, chromium, barium, manganese, lead. Environ. Pollut. 226, 404–411.

Zhao, Y.Z., Jia, H.J., Ren, Q.Q., 2017b. The characteristics of zinc and arsenic from co-firing of municipal sewage sludge with biomass in a fluidized bed. Energy Fuels 31 (1), 755–762.

Zhao, T., Wang, X.R., Yang, X.J., Yan, X.H., et al., 2017c. Thermogravimetric and XRD study of the effects of chloride salts on the thermal decomposition of mercury compounds. J. Mater. Cycles Waste Manage. 19 (2), 712–717.

Zhao, S., Li, R., Zhang, H., Li, H., 2017d. MRCI + Q calculations on spectroscopic properties of excited states of PbH including spin-orbit coupling. Chem. Phys. Lett. 671, 92–99.

Zhao, S., Duan, Y., Chen, C., Wu, H., Liu, D., Liu, M., et al., 2018. Distribution and speciation transformation of hazardous trace element arsenic in particulate matter of a coal-fired power plant. Energy Fuels 32, 6049–6055.

Zhao, Y., Ren, Q., Na, Y., 2018a. Effect of the addition of biomass on the fate and speciation of arsenic during sewage sludge combustion. Energy Fuels 32, 9513–9518.

Zhao, S., Duan, Y., Lu, J., Gupta, R., Pudasainee, D., et al., 2018b. Thermal stability, chemical speciation and leaching characteristics of hazardous trace elements in FGD gypsum from coal-fired power plants. Fuel 231, 94–100.

Zhao, S.L., Duan, Y.F., Lu, J.C., Gupta, R., et al., 2018d. Chemical speciation and leaching characteristics of hazardous trace elements in coal and fly ash from coal-fired power plants. Fuel 232, 463–469.

Zheng, Y., Jensen, P.A., Jensen, A.D., Sander, B., Junker, H., 2007. Ash transformation during co-firing coal and straw. Fuel 86, 1008–1020.

Zheng, Y., Jensen, P.A., Jensen, A.D., 2008. A kinetic study of gaseous potassium capture by coal minerals in a high temperature fixed-bed reactor. Fuel 87, 3304–3312.

Zheng, Y., Jensen, A.D., Windelin, C., Jensen, F., 2012. Review of technologies for mercury removal from flue gas from cement production processes. Prog. Energy Combust. Sci. 38, 599–629.

Zheng, C., Wang, L., Zhang, Y., Zhang, J., Zhao, H., et al., 2017. Partitioning of hazardous trace elements among air pollution control devices in ultra-low-emission coal-fired power plants. Energy Fuels 31, 6334–6344.

Zheng, C., Xu, L., Liu, S., Wang, L., Liang, C., et al., 2018. Speciation and thermal stability of mercury in solid products from ultralow emission air pollution control devices. Energy Fuels 32, 12655–12664.

Zhitkevich, V.F., Lyuty, A.I., Nesterko, N.A., Rossikhin, V.S., Tsikora, I.L., 1963. A spectroscopic study of dissociation and ionization in a flame. Opt. Spectrosc. 14, 17–19.

Zhong, L.P., Cao, Y., Li, W.Y., et al., 2011. Selenium speciation in flue desulfurization residues. J. Environ. Sci. China 23 (1), 171–176.

Zhong, M.-M., Kuang, X.-Y., Wang, Z.-H., et al., 2013. Probing the structural and electronic properties of aluminum-sulfur Al_nS_m ($2 \le n + m \le 6$) clusters and their oxides. J. Mol. Model. 19, 263–274.

Zhou, M., Andrews, L., Ismail, N., Marsden, C., 2000. Infrared spectra of UO_2, UO_2^+ and UO_2^- in solid neon. J. Phys. Chem. A 104, 5495–5502.

Zhou, M., Zhang, L., Dong, J., Qin, Q., 2000a. Reactions of group IV metal atoms with water molecules: matrix isolation FTIR and theoretical studies. J. Am. Chem. Soc. 122, 10680–10688.

Zhou, M., Dong, J., Zhang, L., Qin, Q., 2001. Reactions of group V metal atoms with water molecules: matrix isolation FTIR and quantum chemical studies. J. Am. Chem. Soc. 123 (1), 135–141.

Zhou, M., Zhang, L., Shao, L., Wang, W., Fan, K., Qin, Q., 2001a. Reactions of Mn with H_2O and MnO with H_2: matrix isolation FTIR and quantum chemical studies. J. Phys. Chem. A 105, 5801–5807.

Zhou, K., Xu, M., Yu, D., Liu, X., Wen, C., Zhan, Z., et al., 2010. Formation and control of fine potassium-enriched particulates during coal combustion. Energy Fuels 24, 6266–6274.

Zhou, K., Xu, M.-H., Yu, D.-X., Wen, C., Zhan, Z.-H., Yao, H., 2010a. The effects of coal blending on the formation and properties of particulate matter during combustion. Chin. Sci. Bull. 55, 3448–3455.

Zhou, J., Tian, H.Z., Zhu, C., et al., 2015. Future trends of global antimony emissions from anthropogenic activities until 2050. Atmos. Environ. 120, 385–392.

Zhou, J.-Z., Wu, S.-M., Pan, Y., et al., 2015a. Mercury in municipal solids waste incineration fly ash in China: chemical speciation and risk assessment. Fuel 158, 619–624.

Zhou, Q., Duan, Y.-F., Hong, Y.-G., et al., 2015b. Experimental and kinetic studies of gas phase mercury adsorption by raw and bromine activated carbon. Fuel Process. Technol. 134, 325–332.

Zhou, H., Zhou, C., Lynam, M.M., et al., 2017. Atmospheric mercury temporal trends in the Northeastern United States from 1992 to 2014: are measured concentrations responding to decreasing regional emissions? Environ. Sci. Technol. Lett. 4, 91–97.

Zhou, C.C., Liu, G.J., Xu, Z.Y., Sun, H., Lam, P.K.S., 2017a. Effect of ash composition on the partitioning of arsenic during fluidized bed combustion. Fuel 204, 91–97.

Zhou, Z., Liu, X., Hu, Y., Xu, J., Cao, X.E., Liao, Z., 2018. Investigation on synergistic oxidation behavior of NO and Hg^o during the newly designed fast SCR process. Fuel 225, 134–139.

Zhou, C., Liu, G., Xu, Z., Sun, H., Lam, P.K.S., 2018a. Retention mechanisms of ash compositions on toxic elements (Sb, Se and Pb) during fluidized bed combustion. Fuel 213, 98–105.

Zhou, H., Zhao, W.Q., Ye, L., Chen, Z.H., Cui, Y.X., 2018b. Postnatal low-concentration arsenic exposure induces autism-like behavior and affects frontal cortex neurogenesis in rats. Environ. Toxicol. Pharmacol. 62, 188–198.

Zhou, J., Du, B.Y., Hu, Y.M., Liang, J., Liu, H.L., et al., 2019. A new criterion for the health risk assessment of selenium and lead exposure to residents near a smelter. Environ. Pollut. 244, 218–227.

Zhu, W., Pratsinis, S.E., 1997. Synthesis of SiO_2 and SnO_2 particles in diffusion flame reactors. AIChE J. 43, 2657–2664.

Zhu, Z., Niu, X., Shi, D., Sun, J., 2013. Spin-orbit coupling effects in the ground state and four low-lying excited valence states $a^1\Delta$, $B^3\Sigma^-$, $1^3\Delta$ and $1^3\Sigma^+$ of SeS molecule. Comput. Theor. Chem. 1004, 38–46.

Zhu, Z., Zhuo, Y., Fan, Y., Wang, Z., 2016. Fate of mercury in flue gas desulfurization gypsum determined by temperature programmed decomposition and sequential chemical extraction. J. Environ. Sci. 43, 169–176.

Zhu, W., Lin, C.-J., Wang, X., Sommar, J., Fu, X., Feng, X., 2016a. Global observations and modeling of atmosphere-surface exchange of elemental mercury: a critical review. Atmos. Chem. Phys. 16, 4451–4480.

Zhu, Z.-W., Zhuo, Y.-Q., Fan, Y.-M., Wang, Z.-P., 2017. Mechanism of mercury vapor release from flue gas desulfurization gypsum. Fuel Process. Technol. 156, 485–490.

Zhu, Y., Wang, Y., Meng, F., Li, L., Wu, S., et al., 2018. Distribution of metal and metalloid elements in human scalp hair in Taiyuan, China. Ecotoxicol. Environ. Safety 148, 538–545.

Zhu, S., Zhang, M., Li, Z., Zhang, Y., Yang, H., et al., 2019. Influence of combustion temperature and coal types on alumina crystal phase formation of high-alumina coal ash. Proc. Combust. Inst. 37, 2919–2926.

Zhukovskii, V.M., Tkachenko, E.V., Gur'ev, A.V., 1972. Thermodynamic analysis and preparation conditions for iron molybdates in solid phase synthesis reactions. Zhurnal Fiz. Khim. 46, 1545–1546.

Zhuo, J., Li, S., Duan, L., Yao, Q., 2012. Effect of phosphorus transformation on the reduction of particulate matter formation during co-combustion of coal and sewage sludge. Energy Fuels 26, 3162–3166.

Zidek, A., Macey, K., MacKinnon, L., Patel, M., et al., 2017. A review of human biomonitoring data used in regulatory risk assessment under Canada's chemicals management program. Int. J. Hyg. Environ. Health A 220 (2), 167–178.

Ziegler, T., Li, J., 1995. Density functional study of C-H and O-H activation and methanol oxidation by chromyl chloride. Organometallics 14, 214–223.

Zimmermann, E., Konigs, S., Neuschutz, D., 1996. Determination by mass spectrometry of the partial pressures of SnO, Sn_2O_2, Sn_4O_4 and Sn_6O_6 in equilibrium with oxygen saturated tin melts. Zeit. Phys. Chem. 193, 195–206.

Zimmermann, E., Konigs, S., Neuschutz, D., 1999. Mass spectrometric determination of the partial pressures of SnO, Sn_2O_2 and O_2 in equilibrium with solid SnO_2. Zeit. Phys. Chem. 209, 271–280.

Zmbov, K.F., Miletic, M., 1978. Mass spectrometric determination of the dissociation energy of PbO_2 and ionization potentials of PbO and PbO_2 molecules. Adv. Mass Spectrom. 7A, 573–577.

Zollner, J.H., Glasoe, W.A., Panta, B., et al., 2012. Sulfuric acid nucleation: power dependences, variations with relative humidity and effects of bases. Atmos. Chem. Phys. 12, 4399–4411.

Zong, Y., Jacob, R.J., Li, S., Zachariah, M.R., 2015. Size resolved high temperature oxidation kinetics of nano-sized titanium and zirconium particles. J. Phys. Chem. A 119, 6171–6178.

Zou, W., Boggs, J.E., 2008. Theoretical study on low-lying electronic states of NiH_2. J. Phys. Chem. A 112 (17), 4100–4105.

Zou, W., Boggs, J.E., 2009. Theoretical study of the electronic states of $CuCl_2$. J. Chem. Phys. 130, 154313. 10 pp.

Zou, R., Zeng, X., Luo, G., et al., 2016. Mercury stability of biproducts from wet flue gas desulfurization. Fuel 186, 215–221.

Zou, C., Wang, C., Liu, H., Wang, H., Zhang, Y., 2017. Effect of volatile and ash contents in coal on the volatilization of arsenic during isothermal coal combustion. Energy Fuels 31, 12831–12838.

Zubero, M.B., Aurrekoetxea, J.J., Ibarluzea, J.M., Arenaza, M.J., Rodriguez, C., Saenz, J.R., 2010. Heavy metal levels (Pb,Cd,Cr and Hg) in the adult general population near an urban solid waste incinerator. Sci. Total Environ. 408, 4468–4474.

Index

Note: Page numbers followed by "*f*" and "*t*" refer to figures and tables, respectively.

A

ABS. *See* Ammonium bisulfate (ABS)
Additives, 17
ADHD. *See* Attention deficit hyperactivity disorder (ADHD)
Alkali chloride
  deposits, 40–41
  molecule, 41
Alkali hydrides, 30–31
Alkali zirconates, 462
Alkali(s), 7–8, 15–16, 23–24
  aluminosilicate formation, 46
  basic fundamental discoveries, 38–40
  bronzes, 305, 412
  combustion power plant data, 47–54
  gas phase stabilities, 28–31
  gaseous transport, 44–47
  health concerns with, 25
  initial implications, 40–43
  laboratory combustion studies, 31–38
  manipulations, 43–44
  metals, 7
  release
    in coal combustors, 48
    from fuels, 50–51
  sodium and potassium's basic chemistry, 25–28
AlO. *See* Aluminum oxide (AlO)
Aluminium trichloride ($AlCl_3$), 62–64
Aluminosilicates, 46, 52–53
Aluminum (Al), 54–55, 58–60
  basic chemistry, 56–58
  combustion power plant data, 65–66
  gas phase stabilities and transport, 58–63
  health concerns with, 55–56
  laboratory combustion related studies, 63–65
  nanoparticle/water mixtures, 61–62
  properties, 57*t*
  stability trends for gaseous molecules of B/Al/Ga/In/Tl group, 60*t*
Aluminum carbonate ($Al_2(CO_3)_3$), 37, 57–58, 201–203
Aluminum hydroxide, 55–58
Aluminum oxide (AlO), 61–62
Aluminum titanates, 395–397
Aluminum trihydride (AlH_3), 57–58
Alzheimer's diseases, 200–201, 246, 259–261
Amalgamation, 288–290, 292
Ammonium bisulfate (ABS), 20–21, 35–37
Antimony, 67
  basic chemistry, 69–72
  combustion power plant data, 74–77
  gas phase stabilities and transport, 72–74
  health concerns with, 68
  laboratory scale combustion studies, 74
  properties, 69*t*
  stability trends for gaseous molecules of N/P/As/Sb/Bi group, 70*t*
Antimony dioxide (SbO_2), 70–71
Antimony oxide (SbO), 77
Apatite, 322–323
Arsenates, 80–82, 87–88
Arsenic, 15, 77
  basic chemistry, 80–82
  combustion power plant data, 87–93
  gas phase stabilities and transport, 82–86
  health concerns with, 78–80
  laboratory scale combustion studies, 86–87
  properties, 81*t*
Arsenic oxide (AsO), 93
  AsO(g), 83–84, 86, 88
Arsenites, 80–82
Arsine (AsH_3), 85–86, 122
$AsCl_3$, 80–82
$AsCl_5$, 80–82
Ash, 19–20, 51, 447–448
AsH_3, 92–93
As_2O_3, 77, 80–82
As_2O_5(s), 83–84, 88
As_4O_6(g), 83–84, 90–91
As_2S_3, 80–82
As_2S_5, 80–82
Atmospheric/atmosphere, 19–20
  chemistry, 19–20
  deposition to soils, 273–275
  emissions problem, 7
  nuclei, 19–20
  phosphorus budget model, 322–323

Atmospheric/atmosphere (*Continued*)
pollution, 447–448
Atomic
alkali ions, 31–33
mercury, 283
nickel thermal emission, 316
Attention deficit hyperactivity disorder (ADHD), 68
Auger electron spectroscopy method, 90–91

B

BaCl(g), 103–104
$BaCO_3$, 105
$Ba(OH)_2$, 101–103
$BaO_2(s)$, 100–101, 106
BaOH(g), 101–103
Barite ($BaSO_4$), 94–97
Barium, 94
basic chemistry, 95–97
bond strength values and trends for Group IIA gaseous molecules, 102*t*
combustion power plant data, 105–107
dissociation energy of BaO, 97–100
gas phase stabilities and transport, 100–101
health concerns with, 94–95
laboratory combustion studies, 101–105
properties, 96*t*
Barium oxide (BaO)
BaO(g), 95–97, 103
BaO(s), 95–97, 100–101, 105–106
vaporization, 97
dissociation energy of, 97–100
Barium sulfate, 94
Barium sulfide (BaS), 95–97, 100–101
Barium sulphite ($BaSO_3$), 95–97
Barium thiosulfate (BaS_2O_3), 95–97
BCl_3, 126
BeO, 109–110, 113
dissociation energy, 111–113
$Be_2O(g)$, 109–110
$BeO_2(s)$, 110–111
$Be(OH)_2$, 110–111
BeOH(g), 109–110
BeS(g), 115
Beryl ($Be_3Al_2(SiO_3)_6$), 108
Beryllium, 108
basic chemistry, 109–111
combustion power plant data, 116–117
dissociation energy of BeO, 111–113
gas phase stabilities and transport, 113–115
health concerns with, 108–109
properties, 110*t*
Beryllium titanate, 395–397
B_2H_6, 126
BH_3CO, 126
Binary nanoparticles, 214–215
Biofuels, 47
Biological chemistry, 231
Biomass, 46, 48, 458–459
particle emission, 48–50
pyrolysis, 46
Bismuth (Bi), 117
basic chemistry, 118–120
combustion power plant data, 122–124
gas phase stabilities and transport, 120–122
health concerns with, 118
properties, 119*t*
Bismuth carbonate ($(BiO)_2CO_3$), 112–113
Bismuth sulfate ($Bi_2(SO_4)_3$), 118–119
Bismuthine (BiH_3), 120, 122
Black liquor
droplet combustion, 51–52
waste fuels, 47, 47*t*
Blood, 275–277
Bluefin Tuna in North Atlantic, 271–272
Body-organ effects, 259–261
Bond strengths, 34–35
for chlorides, 139–142
Bonding energy, 164–166
Borax, 124–125
Born–Oppenheimer breakdown in spectroscopy, 114–115
Boron, 2–4, 17, 124–125
basic chemistry, 126–127
combustion power plant data, 132–134
gas phase stabilities and transport, 128–132
health concerns with, 125–126
properties, 126*t*
Boron chlorides, 129–130
Brazil nuts, 344–347
Bromine, 17, 288–290
B_2S_3, 67, 71, 126–127

C

Cadmium (Cd), 134–135
basic chemistry, 137–139
chemical kinetic constraints, 142–143
combustion power plant data, 144–148
gas phase stabilities and transport, 139–142
health concerns with, 135–137
laboratory combustion studies, 143–144
properties, 138*t*
stability trends
for gaseous molecules, 140*t*
for gaseous Zn/Cd/Hg molecules, 141*t*
CaH(g), 155–156
Calcium (Ca), 71–72, 148–149

basic chemistry, 149–151
combustion power plant data, 158–160
gas phase stabilities and transport, 154–157
health concerns with, 149
laboratory combustion systems, 157–158
properties, 150*t*
Calcium borate, 124–125
Calcium carbonate ($CaCO_3$), 149–151
Calcium chloride ($CaCl_2$), 155
Calcium dihydrogen phosphate ($Ca(H_2PO_4)_2$), 324
Calcium dioxide (CaO_2), 154
CaO_2(s), 149–151
Calcium metasilicate ($CaSiO_3$), 149–151
Calcium oxide (CaO), 354–355
dissociation energy of, 152–153
Calcium sulfite ($CaSO_3$), 149–151
Calcium titanate ($CaTiO_3$), 149–151
Capture/retention, 54
Carbon nanotubes, 224–225
Carbonate deposits, 42–43
Carcinogenic, 1–2
Catalysis, 288–290
Catalyst, 288–290
CDC. *See* Centers for Disease Control (CDC)
CdH, 142
Cell metabolism, 187–188
CEMs. *See* Continuous emission monitors (CEMs)
Cenospheres, 64
Centers for Disease Control (CDC), 310
Cesium (Cs), 23–24
Chemical analysis of coal fly ash samples, 90–91
Chemical kinetic constraints, 23
of cadmium, 142–143
chromium, 178–180
of mercury, 280–282
Chemical kinetics, 12
Chemical thermodynamics, 12
modelling, 239
Chemideposition, 13, 20, 282–283
Chemiionization, 104–105, 375–376
Chemiluminescence, 61–62, 322–323
Chloride salts, 337–338
Chlorides, 452–453
Chlorine, 12, 14, 17, 42–43, 93, 146–147, 366
deposition and corrosion, 44
Chromates (Cr^{VI}), 161
Chromite (Cr^{III}), 161
Chromium (Cr), 161
basic chemistry, 164–166
chemical kinetic constraints, 178–180
combustion power plant, 180–186
first row transition elements, kJ mol^1, 168*t*
gas phase stabilities and transport, 167–171
health concerns with, 162–164
laboratory combustion studies and gas phase stabilities, 171–178
properties of, 165*t*
Cinnabar (HgS), 277–278
Clean Air Act (1990), 1–2
Coal, 7, 11–12, 48, 52–53, 285–286, 366
capture, 90
combustion, 16, 144
Cobalt (Co), 186–187
basic chemistry, 188–191
combustion power plant data, 197–199
gas phase stabilities and transport, 191–193
health concerns with, 187–188
laboratory combustion studies, 193–197
properties of, 190*t*
Cobaltite (CoAsS), 186–187
Cold water scrubbing method, 357–358
Colemanite, 124–125
Combustion, 1, 17, 19–20, 23, 447–448
mode, 15
molecular gas phase transport in, 334
sources, 322–323
Combustion power plant data
alkalis, 47–54
aluminum, 65–66
antimony, 74–77
arsenic, 87–93
barium, 105–107
beryllium, 116–117
bismuth, 122–124
boron, 132–134
cadmium, 144–148
calcium, 158–160
chromium, 180–186
cobalt, 197–199
copper, 209–213
iron, 225–229
lead, 238–246
magnesium, 255–258
manganese, 267–270
mercury, 285–292
molybdenum, 308–309
nickel, 318–322
phosphorus, 335–336
scandium, 342–343
selenium, 353–358
silver, 365–366
strontium, 376–377
thallium, 383–384
tin, 391–392
titanium, 401–403
tungsten, 415–416

Combustion power plant data (*Continued*)
uranium, 425–426
vanadium, 436–439
yttrium, 446
zinc, 456–460
zirconium, 466–467
Computational thermochemistry, 433–434
Condensation, 40
Condensation nuclei, 19–20
Condensed phase
chemistry, 12
species, 11–12
Condensed phase sulfide, 76–77
Constraints, 16
Continuous emission monitors (CEMs), 287–288
Copper (Cu), 199–200
basic chemistry, 201–203
combustion power plant data, 209–213
gas phase stabilities and transport, 203–207
health concerns with, 200–201
laboratory combustion studies, 208–209
properties of, 202*t*
Core ash, 20
CVD (chemical vapor deposition), 63–64, 400–401, 465–466
Cylindrical Inconel-600 stainless steel, 37–38
Cysteine, 135–137

D
Danish mass balance of barium, 107
Dew point, 13, 40, 46–47
Diatomic hydride, 453–454
Diatomic oxide, 251–252
Dichloride, 103–104, 194–195, 315–316
Dihalides, 8
Dissociation energy, 250–251
of BaO, 97–100
of BeO, 111–113
of CaO, 152–153
of SrO, 370–372
Dissociative methods, 31–33
DNA synthesis, 187–188
Domestic fire, 92

E
Early Knudsen cell-mass spectrometric measures, 462–464
Earth's crust, 343–344
elements, 393–394
Effusion mass spectrometric method, 30–31
Electron beam method, 438–439
Elemental phosphorus, 325–328
Environmental aquatic wild life, 20
EPA. *See* United States Environmental Protection Agency (EPA)
ESPs, 65, 145
EXAFS method, 438–439

F
Fabric filters (FF), 291–292
FF. *See* Fabric filters (FF)
FGD system. *See* Flue gas desulfurization system (FGD system)
Flame
chemistry, 7–8, 17
CVD, 64–65
gases free of nonmetallic content, 17
particle synthesis, 465–466
suppression, 105
synthesis, 63–64
Flue gas desulfurization system (FGD system), 65–66, 255–257, 267–270
Fluidized bed combustion of coal, 436–437
Fluidized bed methods, 15
Fluorophosphate esters, 324–325
Fly ash, 19–21, 132–133
Fossil-fueled flames, 41
Fourier-transform infrared spectroscopy (FTIR)
absorption, 44–45
spectrum of $MgH_2(g)$, 253
Fourier-transform Raman spectroscopy, 412
Fuel ash, 20
Fuel particle, 19–20
Fuel pyrolysis/gasification processes, 52

G
Gas phase kinetics, 11
Gas phase stabilities and transport
alkalis, 28–31
aluminum, 58–63
antimony, 72–74
arsenic, 82–86
barium, 100–101
beryllium, 113–115
bismuth, 120–122
boron, 128–132
cadmium, 139–142
calcium, 154–157
chromium, 167–178
cobalt, 191–193
copper, 203–207
iron, 219–221
lead, 235–238
magnesium, 251–255
manganese, 264–266
molybdenum, 298–302

nickel, 312–318
phosphorus, 327–331
scandium, 338–342
selenium, 349–353
silver, 362–364
strontium, 372–375
thallium, 380–383
tin, 388–390
titanium, 398–400
tungsten, 405–408
uranium, 420–425
yttrium, 442–445
zinc, 451–454
zirconium, 462–464
Gaseous
alkali cleanup methods, 52–53
alkali molecules, 7–8
BeO, 111–112
$HgCl_2$, 288–290
mercury, 8
molybdates and tungstates, 302–303
monosulfide, 130
NaCl, 31–33
oxides, 28–29
SeO_2, 358
SnO, 388–389
superoxide, 154
transport of alkalis, 44–47
transport of zinc, 454–456
ZnS, 453
Gaseous chromic acid ($CrO_2(OH)_2$), 171–172
Gasification, 52, 76–77
Geiger Counters, 416–418
Germanium (Ge), 392–393
Gibbs free energy of formation, 296–298
Global anthropogenic emissions, 231
Global atmosphere, 214
Global mining, 199–200
Governmental Safe Drinking Water Acts, 1–2
Guided ion-beam mass spectrometry, 317, 364

H

Halogens, 7–8
$HBO_2(g)$, 126–129
Heterogeneous chemistry, 7–8
Heterogeneous conditions of alkalis, 37–38
Heterogeneous Deakin Process, 286–287
HOM-OH/ClM-Cl ratio, 103
Homogeneous conditions of alkalis, 31–37
Homogeneous gas phase kinetics, 129
Homogeneous nucleation, 19–20, 35–37
Huntington's disease, 200–201, 259–261
Hydrides, 77
Hydrogen, 2–4, 61–62
flames, 13
studies, 154–155
Hydroxides, 42–43, 55
Hydroxyoxides, 170, 175–176, 178–180
Hypophosphorous acid (H_3PO_2), 325–326

I

ICP. *See* Inductively coupled plasma (ICP)
IGCC gasifier, 197–198
Ilmenite ($FeTiO_3$), 393–394
Incinerator disposal method, 335–336
Inductively coupled plasma (ICP), 37, 48
analysis, 342–343
ICP-MS method, 353–354
Inorganic salts, 105
Ionic sodium acetylide (NaCCH), 25–27
Iron (Fe), 214
basic chemistry, 216–219
combustion power plant data, 225–229
gas phase stabilities and transport, 219–221
health concerns with, 214–216
laboratory combustion studies, 222–225
nanoparticles, 215–216
properties of, 217*t*

J

Japanese pyrolysis study, 132–133

K

Kaolin clays, 52–54
Kernite, 124–125
Knudsen cell-mass
spectral analysis of vanadium, 429–431
spectrometric equilibrated experiment, 100–101
spectrometric method, 398
spectrometric study, 441–442
of mixed metal oxides, 35–37
Kronig's selection rules, 100, 115

L

Laboratory combustion studies
alkalis, 31–38
aluminum, 63–65
barium, 101–105
cadmium, 143–144
calcium, 157–158
chromium, 171–178
cobalt, 193–197
copper, 208–209
iron, 222–225
lead, 238

Laboratory combustion studies (*Continued*)
- magnesium, 255
- manganese, 267
- mercury, 282–285
- molybdenum, 303–308
- nickel, 318
- phosphorus, 331–333
- strontium, 375–376
- titanium, 400–401
- tungsten, 411–415
- vanadium, 435–436
- zirconium, 465–466

Laboratory scale combustion studies
- of antimony, 74
- of arsenic, 86–87

Landfill, 92, 184–185, 213
Laser-induced breakdown spectroscopy, 31–33
Laser-induced fluorescence (LIF), 31–33, 206–207
- of atomic As, 86

Lead (Pb), 216
- basic chemistry, 233–235
- bond strength trends for relevant gaseous molecules, 235*t*
- combustion power plant data, 238–246
- gas phase stabilities and transport, 235–238
- health concerns with, 231–233
- laboratory combustion studies, 238
- properties of, 234*t*

LIF. *See* Laser-induced fluorescence (LIF)
Lime/Limestone, 149–151, 318–320, 336–337, 357–358, 436–437
Lithium (Li), 7–8, 23–25

M

M-Cl/M-OH ratio, 103
Magnesium (Mg), 246
- basic chemistry, 247–248
- combustion power plant data, 255–258
- dissociation energy of, 248–251
- environmental concerns for, 246
- gas phase stabilities and transport, 251–255
- health concerns with, 246
- laboratory combustion studies, 255
- properties of, 247*t*

Magnetohydrodynamic (MHD), 331–332
Manganese (Mn), 258–259
- basic chemistry, 261–263
- combustion power plant data, 267–270
- gas phase stabilities and transport, 264–266
- health concerns with, 259–261
- laboratory combustion studies, 267
- properties of, 262*t*

Marine selenium cycle, 343–344
Mass balances, 14, 89–90
Mass spectrometric
- effusion
  - beam method, 315–316
  - cell equilibrium method, 317
- method, 406–407
- sampling, 375

Matrix isolation, 58–60, 340, 369, 434–435, 451
Mercaptan, 135–137
Mercury (Hg), 8, 13, 15, 42–43, 271–272
- basic chemistry, 277–278
- chemical kinetic constraints, 280–282
- combustion power plant studies, 285–292
- gas phase stabilities and transport, 278–280
- health concerns with, 275–277
- laboratory combustion studies, 282–285
- properties, 274*t*

Metal borates, 126–127, 130–132
Metal oxides, 196–197, 228
Metallic arsenic, 80–82
Metallic nickel, 311–312
Metallic titanium, 395–397
Metallic tungsten, 406–407
Metalloid, 67
Metallurgical slags, 19–20
Methyl cyclopentadienyl manganese tricarbonyl (MMT), 258–259
Methyl mercury, 275–277
MgO(s), 248, 252–253
MHD. *See* Magnetohydrodynamic (MHD)
Mineral lead chromate, 161
Minimum risk level (MRL), 78–80, 367
Mixed metal compounds, 15–16, 420
MMT. *See* Methyl cyclopentadienyl manganese tricarbonyl (MMT)
Molecular gas phase transport in combustion, 334
Molybdates, 296–298, 302–303
Molybdenum (Mo), 293, 403
- basic chemistry, 294–298
- combustion power plant data, 308–309
- gas phase stabilities and transport, 298–302
- health concerns with, 294
- laboratory combustion studies, 303–308
- potential role for molybdates, 302–303
- properties, 295*t*
- thermochemistry of gaseous molybdates, 297*t*

Molybdic acid, 296–298
MRL. *See* Minimum risk level (MRL)
Municipal solid waste (MSW), 86–87
Municipal wastes, 47–48
- incineration, 46

N

"Nascent" chlorine transport, 54
Natural biomethylation, 231
Natural gas, 2–4, 285–286
Neurotoxins, 275–277
$(NH_4)HSO_4$. *See* Ammonium bisulfate (ABS)
Nickel (Ni), 309–310
  basic chemistry, 311–312
  combustion power plant data, 318–322
  gas phase stabilities and transport, 312–318
  health concerns with, 310–311
  laboratory combustion studies, 318
Nicolet Fourier transform Raman spectrometer, 37
Nitrates, 17
No observable adverse effect level (NOAEL), 200–201

O

Optical spectroscopy, 464–465
Organic-coupled salts, 25–27
Organomercurials, 275–277
Orion Model 94-17B Combination Chloride Electrode, 37
Orthostannates, 386–387
Oxy-burn, 290–292
Oxygen-lean modes, 76–77
Oxygen-rich flame, 8

P

Partial equilibrium balance, 33
Particles, 19–20
  combustion, 11–12
Pentavalent oxide (As_2O_5), 80–82
Pentoxide, 80–82
Petcoke, 336–337
Petroleum, 2–4
Phosphate esters, 324–325
Phosphine (PH_3), 324, 330
Phosphorus (P), 17, 322–323
  basic chemistry, 325–327
  combustion power plant data, 335–336
  gas phase stabilities, 327–331
  health concerns with, 324–325
  laboratory combustion studies, 331–333
  molecular gas phase transport in combustion, 334
Photoionization study, 279–280
Polymolybdates, 296–298
Potassium (K), 7–8, 23–25, 33–34
  basic chemistry, 25–28
  condensed phase molecules, 26*t*
  potassium-seeded MHD system, 331–332
Potassium antimonate, 71–72
Potassium antimony tartrate hydrate, 68
Potential gas phase molecules, 251–252
Pyrite (As^{III}), 87–88
Pyrolysis, 52, 76–77, 270

Q

Quantitative flame photometric techniques, 2

R

Radioactive ^{137}Cs, 23–25
Raman scattering, 63–64
Rare earth elements, 439–440
Realgar (As_4S_4), 80–82
Red Mud, 54–55
Reference Dosage (RfD), 294
River runoff, 322–323
Rubidium (Rb), 23–24

S

Scandia, 336–337
Scandium (Sc), 336–337
  basic chemistry, 337–338
  combustion power plant data, 342–343
  gas phase stabilities and transport, 338–342
  health concerns with, 337
Schuetteite, 283
SCRs. *See* Selective catalytic converters (SCRs)
Selective catalytic converters (SCRs), 35–37, 286–287
Selenites, 348–349
Selenium (Se), 15, 135–137, 343–344
  basic chemistry, 348–349
  combustion power plant data, 353–358
  gas phase stabilities and transport, 349–353
  health aspects of, 344–347
  Se-cysteine detoxicant products, 275–277
Selenium chlorides, 352
Sensitive elements, 2–4, 3*t*
Sesquioxide, vaporization of, 338–340
Sewage sludge, 2–4, 47, 87–88, 147
–SH group, 344–347
Shale oil, 447–448
Silica (SiO_2), 52–53
Silicon, 17
Silver (Ag), 358
  basic chemistry, 360–362
  combustion power plant data, 365–366
  gas phase stabilities and transport, 362–364
  health concerns with, 359–360
Sintering process, 271–272
Smaltite ($CoAs_2$), 186–187
Sodium (Na), 7–8, 23–25, 33–34
  basic chemistry, 25–28
  condensed phase molecules, 26*t*

Sodium (Na) (*Continued*)
  equilibrium molecular or dissociative vapor pressures, 27*t*
  gas phase bond strengths, 27*t*
  melting and boiling points, 27*t*
Sodium acetate, 25–27
Sodium antimonate, 71–72
Sodium borates, 124–125
Sodium carbonate (Na_2CO_3), 38–39, 43
  rates of deposition for, 38*f*
Sodium chloride (NaCl), 38–39
  deposit, 44
  rates of deposition for, 38*f*
Sodium hydroxide (NaOH), 33
Sodium sulfate (Na_2SO_4), 38–39, 43
  rates of deposition for, 38*f*
Solid fuels, 436–437
Solid particle fragmentation models, 19–20
Soot formation, 34–35, 255
Spectroscopic identifications, 2
Spinels, 64–65
Spin–orbit interactions, 112–113
"Steam-iron" chemical looping mechanism, 149–151
Stibnite (Sb_2S_3), 67, 71, 73–74
Stove burning, 92, 184–185, 213
Strontium (Sr), 367
  basic chemistry, 368–370
  combustion power plant data, 376–377
  dissociation energy of SrO and meaning, 370–372
  gas phase stabilities and transport, 372–375
  health concerns with, 367–368
  laboratory combustion studies, 375–376
Strontium oxide (SrO), 369
  dissociation energy of, 370–372
Sulfate, 20–21, 57–58, 94, 164–166
  deposits, 52
Sulfide, 73–74, 395–397
Sulfur, 12, 14, 17, 27–28, 43, 146–147
Sulfur dioxide (SO_2), 20–21, 44
Sulfur trioxide (SO_3), 20–21
Surface
  loss rates, 40
  reactive surface, 19–20
  thermodynamically driven chemical deposition on, 20
Synthetic biogases, 87–88, 144
Syphilis, 78–80

T

Tarter emetic, 68
Temperature, importance of, 15–16
Tetramethyl/ethyl additive compounds, 238–239
TGA. *See* Thermogravimetric analysis (TGA)
Thallium (Tl), 378–379
  basic chemistry, 379–380
  combustion power plant data, 383–384
  gas phase stabilities and transport, 380–383
  health concerns with, 379
  thallium-rich soils, 378–379
Thermal ionization, 34–35, 152, 158, 443–444
Thermochemistry
  of BO, 129
  of titanium's chlorides, 399
Thermogravimetric analysis (TGA), 27–28, 441–442
Time constraints, 15–16, 186
Tin (Sn), 384–385
  basic chemistry, 386–387
  combustion power plant data, 391–392
  gas phase stabilities and transport, 388–390
  health concerns with, 385–386
Tincal, 124–125
Tire-derived fuels, 447–448
Titanic acid, 395–397
Titanium (Ti), 393–394
  basic chemistry, 395–397
  combustion power plant data, 401–403
  gas phase stabilities and transport, 398–400
  health concerns with, 394–395
  laboratory combustion studies, 400–401
Titanium phosphate, 395–397
Titanium tetrachloride ($TiCl_4$), 394–395
Torsion-mass loss method, 363
Toxic metals, 1–2, 344–347
Toxicity assessments of Se, 344–347
Trace metals, 2–5, 13, 17, 309–310
  in combustion, 7, 13–14
Transition metal hydrides, 364, 445
Transpiration study, 315
Trichloride, 85
  vaporization, 340–341
Trihydroxide, 61–62, 120, 441–442
Trimethyl borate ($B(OCH_3)_3$), 130–131
Trivalent meta-arsenites, 82
Troe formalism, 33–34
Tungstates
  potential role for, 408–410
  thermochemistry of gaseous, 409*t*
Tungsten (W), 403
  basic chemistry, 404
  combustion power plant data, 415–416
  gas phase stabilities and transport, 405–408
  health concerns with, 403–404
  laboratory combustion studies, 411–415
  potential role for tungstates, 408–410
Two-photon ionization spectroscopy method, 316–317, 408

U

Ulexite, 124–125
Ultrafine coal fly ash particles, 48–50
Ultraviolet absorption, 44–45
United States Environmental Protection Agency (EPA), 1–2, 294
Uranium (U), 416–418
  basic chemistry, 419–420
  combustion power plant data, 425–426
  gas phase stabilities and transport, 420–425
  health concerns with, 418–419
Uranyl nitrate ($UO_2(NO_3)_2$), 420, 425

V

Vanadates, 432, 436–439
Vanadium (V), 426–427
  basic chemistry, 429–435
  combustion power plant data, 436–439
  health concerns with, 428–429
  laboratory combustion studies, 435–436
Vaporization
  of chromium, 175–176
  of Co, 191–193
  of metal tungsten, 406–407
  of MgO, 248
  of polymolybdates, 302–303
  of sesquioxide, 338–340
  of tungstates, 408–410
  of yttrium oxide, 441–442

W

Waste heat recovery, 15–16
Water
  insoluble trihydroxide, 57–58
  vapor, 19–20
Witherite ($BaCO_3$), 94
World Health Organization (WHO), 68, 231–233, 259–261, 418–419

X

X-ray absorption fine structure method (XAFS method), 90–91
X-ray absorption method, 240–241, 356–357, 438–439
X-ray absorption spectroscopy (XANES), 212–213, 438–439
  analysis of ash, 354–355
X-ray diffraction (XRD), 92–93
X-ray photoelectron method, 90–91

Y

YAG lasers. *See* Yttrium aluminum garnet lasers (YAG lasers)
Yttria (Y_2O_3), 441–442
Yttrium (Y), 336–337, 439–440
  basic chemistry, 440–442
    properties of yttrium's condensed phase molecules, 442*t*
  combustion power plant data, 446
  gas phase stabilities and transport, 442–445
  health concerns with, 440
Yttrium aluminum garnet lasers (YAG lasers), 440–441

Z

Zhundong coal reserves in China, 48
Zinc (Zn), 447–448
  basic chemistry, 449–451
  chemical kinetic constraints, 454
  combustion power plant data, 456–460
  deficiency, 448–449
  gas phase stabilities, 451–454
  gaseous transport of zinc, 454–456
  health concerns with zinc, 448–449
Zirconium sulfide spectroscopy (ZrS spectroscopy), 464–465
Zirconium (Zr), 460–461
  basic chemistry, 461–462
  combustion power plant data, 466–467
  gas phase stabilities and transport, 462–464
  health concerns with, 461
  laboratory combustion studies, 465–466
  molecular spectroscopy, 464–465
$ZnCl_4(s)$, 462–464

9780128191262